Probability and Statistical Theory for Applied Researchers

Probability and
Statistical Theory for
Applied Researchers

T. W. Epps
University of Virginia, USA

World Scientific

NEW JERSEY · LONDON · SINGAPORE · BEIJING · SHANGHAI · HONG KONG · TAIPEI · CHENNAI

Published by

World Scientific Publishing Co. Pte. Ltd.

5 Toh Tuck Link, Singapore 596224

USA office: 27 Warren Street, Suite 401-402, Hackensack, NJ 07601

UK office: 57 Shelton Street, Covent Garden, London WC2H 9HE

Library of Congress Cataloging-in-Publication Data
Epps, T. W.
    Probability and statistical theory for applied researchers / T. W. Epps, University of Virginia, USA.
      pages cm
    Includes bibliographical references and index.
    ISBN 978-9814513159 (hard cover : alk. paper)
    1. Social sciences--Research--Statistical methods. 2. Probabilities. I. Titl
    HA29.E67 2013
    519.5--dc23
                                        2013027332

British Library Cataloguing-in-Publication Data
A catalogue record for this book is available from the British Library.

In-house Editor: Ms. Sandhya Venkatesh

Printed in Singapore

Contents

For Misses L., M., S., and T.—and Mr. Z.

Preface

This book is designed for graduate courses in applied disciplines that require a rigorous background in probability and mathematical statistics, such as actuarial science, biostatistics, economics, finance, mathematical psychology, and systems engineering. It is also intended for the active researcher who desires a deeper understanding of the foundations of statistical practice and seeks better ways to model stochastic behavior. Multivariate calculus and an elementary course in statistics are minimal requirements for using this book. Matrix algebra is helpful but is not used extensively until the final chapter. The more advanced mathematical tools that are needed—basic measure and integration theory, a little complex analysis—are taught along the way.

Besides the math tutoring that accompanies the principal thematic material, what distinguishes this book from the many fine texts on probability and statistics now on the market? For both the student and the practicing researcher there is no better way to learn than to solve problems. In fact, there is really no *other* way to master mathematical concepts. With that in mind, I have sprinkled the text liberally with examples and with exercises that allow one to apply new concepts as they are introduced. The problems range in difficulty from simple questions of fact or interpretation, to proofs of theoretical results, to extensive analytical and numerical problem solving. To be sure, the exercises themselves scarcely make the present work distinctive. What does distinguish it is that complete, detailed solutions of *all* exercises appear at the end of each chapter. Working out the exercises and then consulting the solutions provide immediate feedback that either confirms one's efforts or shows the need for correction. And the detailed solutions do far more than just providing the "answers". By showing (one way) to approach problems and to apply the tools at hand, they aid the reader in developing critical skills of analysis and logical thinking.

Notes for Students: "Facts" alone are of little use to the applied researcher. One needs the judgment of when to apply them and the analytical technique for doing so. In writing this book I have tried to help you develop the skills needed to deduce your *own* facts, some of which (one hopes) will be new to the world, and maybe even important. Toward this goal I have tried to make explanations of new concepts abundantly clear and have shown how to apply them through the many examples, exercises, and solutions. However, as you progress in the book, more and more will be expected of you. Explanations at any point will be "abundantly clear" only if you have digested the material that has gone before. In order to do this, you must work the exercises—not all, necessarily, as some of them are, quite frankly, very hard—but as many as you have the patience and time to do. For those that you just cannot solve, at least study the solutions and try to see how you might be led to find them on your own. In some cases—perhaps in many—you will find better ways than I have shown. If that happens, my goal will have been achieved.

Here are a warning and some advice. Statisticians and the books they produce are famous (infamous?) for their inability to express two or more consecutive sentences without relying on some arcane abbreviation. You will not escape them here. Eventually, strings of letters and dots such as a.s., b.l.u.e., c.d.f., i.i.d., m.l.e., and u.m.p. will become familiar. (When that happens, you can start to worry about yourself.) Of course, these arcana are defined when first introduced, and there are occasional reminders later. However, in case you forget as you plod through the pages or as you jump around in using the book as a reference, you might like to have a way to look them up. That is what Appendix A is for. It is ordered alphabetically by symbol to make it easy to find things.

Likewise, as in any lengthy treatment of mathematics, you will encounter here an absolutely bewildering array of symbols. Insofar as possible I have given each of these a consistent meaning throughout the book. The benefit is that one does not have continually to ask "What does that mean *here*?" The cost is that it required the creative use of fonts, faces, cases, superscripts, subscripts, primes, stars, bars, carats, tildes, circumflexes, and—*of course!*—Greek characters. On the off chance that you might forget the meaning of one or two of these artistic creations, I have supplied a comprehensive list in Appendix B. Having no way to arrange these alphabetically, I placed them in the order in which they were introduced. Each symbol is accompanied by a short description and the number of the page on which it was defined. Thus, when you run into symbol Ψ' on page 511, you can look through earlier page numbers in the list and see that it was defined on page 497.

Glancing at these two appendices before starting Chapter 1—and perhaps bookmarking or dog-earing the pages—will improve the ambience for those who are within earshot as you make your way through this book.

Notes for Instructors: Providing solutions to *all* the problems rather than, say, to just the odd-numbered ones, increases the book's instructional value, but it does require you to develop original problems for exams and assignments. Does this just add to the instructional burden? On the contrary, it actually *reduces* it by encouraging a far more efficient means of assessment than through graded "homework". Replacing homework with frequent short quizzes, conducted either in class or in a separate problem session, has two significant benefits: (i) It allows students (honorably) to learn from and coach each other, and (ii) it greatly lightens the awful, tedious chore of grading. An approach that rewards both diligence and creative thinking is to give weekly quizzes of just two or three problems, some newly concocted and some selected from the exercises.

Naturally, deciding whether and how to use this book in teaching depends on your students' preparation in mathematics. The first three chapters of Part I explain and apply the basic concepts of measure theory and integration that are essential for probability, but students without prior exposure to real analysis will need to take this material at a slower pace. For them it would be appropriate to use Parts I and II for separate courses in probability and statistics. For those with more advanced math skills a one-semester or two-quarter course comprising Part I and Chapters 5-8 of Part II is well within reach. Indeed, for many years earlier versions of these chapters served as the main text for an intensive, one-semester doctoral course in economics at University of Virginia. From this course students moved on to specialized offerings in econometrics, mathematical economics, and mathematical finance. For the last field especially the study of stochastic processes is essential, and Part I provides a foundation and a brief introduction via conditional expectation, martingales, and Brownian motion.

General Caveats: All forms of art—painting, sculpture, music, theater, film, etc.—are widely recognized to reflect the artist's own personal experience and world view. It is less obvious, but nonetheless true, that scientific endeavor, and scholarly work generally, carries the same personal imprint. My own background is that of an applied researcher whose study and modeling of stochastic phenomena has generated questions and ideas that led, from time to time, to contributions to the theoretical literature. The astute reader will soon see in these pages indications of my special applied interest: financial economics. This is expressed largely through examples, which (it is hoped) will still be found instructive—perhaps even entertaining—by those in other fields. A more pressing concern for some readers is that my own applied and theoretical interests have influenced the *scope* of this work. Although "experimental" economics is now an important subfield, it is still true that most economists who do applied research take their data from the hands of patient government statisticians or off the financial network, never delving into issues of experimental design. Accordingly, no treatment of that topic, or of sampling design generally, is to be found in this work. Likewise, ANOVA (analysis of variance), which is the premier tool of many experimental researchers, is treated here just as a narrow subtopic of regression with normal errors. Finally, some may question the need for the extensive treatment of goodness-of-fit tests in Chapter 9, but I justify the emphasis on this special research interest as a partial offset to its neglect by other writers.

Acknowledgements: Over the years hundreds of students have contributed to this enterprise through their questions, suggestions, constructive criticisms, and their sometimes-baffled, sometimes-excited, sometimes-bored facial expressions. (Yes, there has been some griping as well, but I abjure to dwell on that.) Among the multitude of helpful critics I cannot fail to single out and offer special thanks to (now Professor) Sangwon Suh, who corrected many errors in the first circulated version of the lecture notes from which the book developed. A more recent—and enormous—debt is owed to my colleague Ron Michener, who used later versions of the notes for several years in his own teaching. Much of what is good here is due to his sometimes severe (but always apt) critiques and his knack for explaining complex ideas in simple ways.

Part I

The Theory of Probability

Chance is merely the name for a cause unperceived by the human intellect.—
Anaxagoras

*Given the initial state of a system, nature determines its future state
through a process that is fundamentally uncertain.*—Stephen Hawking

These two quotations seem to represent totally incompatible views
about the origin of uncertainty, and yet the mathematical theory of probability reveals that they are actually consistent within the proper spheres
of application. Although the laws of quantum physics do make uncertainty
"fundamental" at scales far below our direct perception, it is only imperfect
knowledge of the initial state that prevents us from knowing *almost* perfectly the future states of processes encountered in everyday life. How can
this impenetrable uncertainty "in the small" be consistent with potentially
unerring predictability "in the large"? What resolves this paradox is the
"law of large numbers"—an intellectual triumph of the theory of probability that we here begin to study, and the very foundation of the *statistical*
theory taken up in Part II.

Chapter 1

Probability on Abstract Sets

We all speak often of "probabilities", but what do they actually represent? There are two common intuitive interpretations, each associated with its own school of statistical reasoning—"Bayesian" *vs.* "frequentist". The *subjective* view, which underlies the Bayesian approach to statistics, is that probabilities exist solely in the mind of the individual as "degrees of belief". The *objective* view, maintained by frequentist statisticians, is that probabilities represent frequencies of occurrence in repeated *chance experiments*. Remarkably, the mathematical theory of probability that we here undertake to study is consistent with both interpretations and takes no stand whatever on the intuitive meaning. Nevertheless, in seeking to learn the theory it helps to have in mind some concrete interpretation that relates to personal experience. Although we do discuss Bayesian methods in Part II, our emphasis there is on the frequentist approach that still forms the mainstream of statistical practice. For that reason, and because the graphic sense of it is clearer, let us adopt for now the objective—but still vague— conception of probabilities as frequencies of results of chance experiments, putting off until later (page 24) a precise statement of what this means.

We shall never define "chance experiment" precisely, but everyday experience provides many examples, such as flipping a coin one or more times, dealing a card from a well shuffled deck, and picking at random from some collection of individuals or objects. The common feature of such actions is that they can be repeated indefinitely under initial conditions that appear identical within the limits of our perception, yet they yield results that are nevertheless in some degree *uncertain*. There are also what can be called "experiments of nature". These are natural, technological, or socioeconomic processes that respond to conditions that are influential but are not fully controlled, and which impart to the results an aspect of *randomness*. For

3

these, replication under fixed initial conditions must be regarded as merely conceptual. The terms "chance", "uncertain", and "random", as they are applied to either type of experiment, indicate that the results are unpredictable, in the sense that not all relevant questions about them can be answered in advance.

Although they are inherently unpredictable, the results of chance experiments often *matter* to us. We may need to make decisions whose consequences depend on what happens, or we may just want to improve our understanding of a stochastic (random) process. At first, the prospect of managing or learning under conditions of pure randomness may seem hopeless. Fortunately, despite the intrinsic uncertainty, an emerging sense of order—a predictable regularity—can often be detected upon repeating chance experiments many times or through prolonged observation of stable systems. By characterizing the predictable regularity in a consistent, logical way it is in fact possible to make productive decisions and to gain useful knowledge. This logical consistency is attained within the mathematical framework by assigning to the various uncertain results—or to classes or *sets* of them—numbers with certain properties. Although our intuition is served by thinking of these numbers loosely as prospective frequencies of occurrence, a satisfactory mathematical theory requires a precise definition. Below we shall define probabilities formally as *measures* with certain special properties, measures in turn being special types of *set functions* that map collections of sets into the real numbers. To prepare for the formal definitions, it is necessary first to become familiar with sets and set operations and to master the specific terms that *probabilists*—specialists in the theory of probability—use to describe them.

1.1 Spaces, Sets, and Collections of Sets

To begin we need some special notation for conciseness and clarity. Sets, which can be represented by various symbols, are defined symbolically using braces together with enclosed and, sometimes, appended characters that indicate the sets' contents. For example, $S := \{a, b, c\}$ *defines* S as the set comprising the entities represented by the three letters—the colon between "S" and "$=$" indicating that S is here *being* defined. How the elements are arranged in the display does not matter, so $\{a, b, c\}$, $\{c, b, a\}$, $\{b, a, c\}$, ... all denote precisely the same set. The symbols $x \in S$ indicate that x is an element of set S. The set $\{x\}$ that contains just a single element is called a "singleton". Sets with many (or arbitrarily many) elements are described more compactly. Thus,

- $\{a_1, a_2, ..., a_n\}$ or $\{a_j\}_{j=1}^n$ denotes a set of n objects;
- $\{1, 2, ...\}$ or $\{j\}_{j=1}^\infty$ denotes the set of positive integers; and
- $\{x : a \le x \le b\}$ for $a < b$ denotes the closed interval of real numbers a through b.

For intervals there is a still more compact notation:

- (a, b) for the open interval $\{x : a < x < b\}$ that excludes the points a, b (to be distinguished from the set $\{a, b\}$, which includes *only* those two points);
- $[a, b)$ and $(a, b]$ for closed–open and open–closed intervals, respectively;
- $[a, b]$ for closed interval $\{x : a \le x \le b\}$.

Here are other short ways of designating certain sets that are commonly encountered:

- $\Re := (-\infty, \infty)$ for the entire set of (finite) real numbers;
- $\Re_+ := [0, \infty)$ for the set of nonnegative reals;
- $\aleph := \{1, 2, ...\}$ for the positive integers—the "natural" numbers;
- $\aleph_0 := \{0, 1, 2, ...\}$ for $\aleph$ with $\{0\}$ added;
- $\mathbb{Q} := \{\pm m/n : m \in \aleph_0, n \in \aleph\}$ for the rational numbers; and
- $\mathcal{C} := \{x + iy : x \in \Re, y \in \Re, i = \sqrt{-1}\}$ for the complex numbers.

The colons inside the braces in these definitions can be read as "such that" or "in which". Thus, in words, $\mathcal{C}$ is the set of numbers $x + iy$ *such that* x and y are real and i is the "imaginary" unit. Some other essential notation:

- $S_1 \times S_2 \times \cdots \times S_k$ and $\times_{j=1}^k S_j$ represent the Cartesian product of sets $S_1, S_2, ..., S_k$, which is the set of k-tuples of objects $\{(x_1, x_2, ..., x_k) : x_1 \in S_1, x_2 \in S_2, ..., x_k \in S_k\}$.
- $S^k := \times_{j=1}^k S$ represents the k-fold Cartesian product of S with itself. In particular,
- $\Re^k := \times_{j=1}^k \Re$ represents the k-dimensional space of real numbers.

The distinction between *countable* and *uncountable* sets is basic. Literally, a set is countable if its elements can be counted—that is, put into one-to-one correspondence with the natural numbers $\aleph$ or a subset thereof. For a set to be considered countable there must be some system for removing its elements one by one in such a way that no element would ultimately be left behind if the process were continued indefinitely. While finite sets are obviously countable, not all countable sets are finite. For example, $\mathbb{Q}$ is a countable set, since there does exist a system—namely, Cantor's diagonalization principle[1]—for counting them out one at a time. A set is uncountable if its elements cannot be put into correspondence with the natural numbers. The set of real numbers belonging to any open interval is one example of an uncountable set. Another is the set of irrational numbers in any such interval. Another is the set of transcendental numbers in an interval.

We will occasionally refer to the fact that $\mathbb{Q}$ is a *dense* subset of $\Re$, meaning that any open interval $I \subset \Re$ contains some rational number. Equivalently, each real $x \in \Re$ is either a rational number or else can be expressed as the limit of a sequence of rationals $\{q_j\}_{j=1}^{\infty}$. Thus, "completing" $\mathbb{Q}$ by adding all such limit points produces $\Re$ itself.

1.1.1   *Outcomes and Sample Spaces*

The totality of all possible results from a chance experiment that are considered relevant is called the *sample space* or *outcome set*. Specifying exactly what this space comprises—i.e., contemplating all the possible *relevant* results—is the first step in obtaining a probabilistic description of the experiment. "Outcome" is the probabilist's term for some such relevant result. Thus, if we flip a coin a single time and are interested just in which side turns up—heads or tails—then we can take the sample space to be, symbolically, the set $\{H, T\}$. We use the symbol Ω to represent the sample space and the symbol ω to represent some generic outcome; hence, $\Omega = \{H, T\}$ for this experiment, and we would write $\omega = H$ to indicate that the coin came up heads.

While this description of Ω for the coin flip experiment seems natural, there are in fact many other possibilities. The representation that is adopted depends ultimately on what aspects of the experiment are of interest and, therefore, relevant. For example, if one cares just about how many times the coin flips over as it is tossed, then $\Omega = \aleph_0$. If what matters is the orientation of the settled coin with respect to some specified coordi-

---

[1]For a nice discussion of countability and the diagonalization principle in particular, see Simmons (2003, pp. 31-35).

nate system, then the relevant outcome is an angle (in radians, say), and $\Omega = [0, 2\pi)$. Alternatively, if angles can be measured just to m decimal places, then $\Omega = \{\langle 2\pi j \rangle / 10^m\}_{j=0}^{10^m - 1}$, where "$\langle \cdot \rangle$" signifies nearest integer. Note that the number of flips and the orientation of the coin would be irrelevant if only the face were of interest, and adding these extra descriptions by taking $\Omega = \{(f, n, a) : f \in \{H, T\}, n \in \aleph_0, a \in [0, 2\pi)\}$ would be an unnecessary complication.

Here are other examples of chance experiments, relevant interests, and appropriate descriptions of sample spaces. A common feature is that one and *only* one outcome—one and only one ω—occurs each time the experiment is performed. This is an absolute requirement.

(i) If we flip a coin three times and are interested in the possible head–tail sequences, then the sample space could be defined as the set of eight such possible sequences:

$$\Omega_1 = \{HHH, HHT, HTH, THH, TTH, THT, HTT, TTT\}.$$

(ii) If we flip a coin three times and care only about how many heads turn up, we could just take $\Omega_2 = \{0, 1, 2, 3\}$. Notice that this is simply a coarser description of the experiment than Ω_1, equivalent to grouping outcomes in Ω_1 by numbers of heads, as

$$\{(TTT), (TTH, THT, HTT), (HHT, HTH, THH), (HHH)\}.$$

Learning that outcome $\omega = THT$ occurred in Ω_1 would tell us also that outcome $\omega = 1$ occurred in Ω_2, but not conversely. In this sense Ω_1 is said to be a "refinement" of Ω_2—a more detailed description of the result.

(iii) If we flip a coin until the first occurrence of heads and care just about how many flips there will be, then we could take $\Omega_3 = \aleph$. Alternatively, listing all the possible sequences and describing Ω_3 as $\{H, TH, TTH, ...\}$ would amount to the same thing.

(iv) If we choose a single person at random from the students in a class, and if we care just about the identity of that person, then Ω_4 could be any list of unique identifiers, such as names or student IDs.

(v) If we draw at random with replacement a sample of size n from a collection of N students, and if our interest is in the precise composition and order of such a sample, then Ω_5 could comprise the set of all ordered n-tuples of the identifying features.

1.1.2   *Subsets and Events*

Given an experiment and a definition of Ω, we will think of any subset of Ω as an *event*. This term facilitates reference to some collection of outcomes ω that have a common property of interest. For instance, we might speak of the event of getting precisely one head in three flips of a coin. This designates the subset $\{TTH, THT, HTT\}$ of Ω_1 and the subset $\{1\}$ of Ω_2. In the latter case the event is a singleton set, but that does not matter, for both the Ω_1 description and the Ω_2 description work perfectly well. However, the result "there are at least two heads in succession" is an event in Ω_1 (the set $\{HHH, HHT, THH\}$), but it is *not* an event in Ω_2. This underscores the fact that the description of Ω must be detailed enough so that we know whether each thing of interest has occurred once we learn the "outcome". In other words *each relevant event A must be representable as a set of outcomes* ω. The entire Ω of any experiment is, of course, an event in itself—the event that *one* of the possible things occurs. It is also useful to regard as an event a thing or set of things that *cannot* occur in an experiment; for example, getting four tails in three flips of a coin. Such an event contains no elements of Ω and is therefore an *empty* set, designated $\varnothing$.

Note that an event A is said to *occur* if the experiment's outcome, ω, is an element of A.

1.1.3   *Operations and Relations among Events*

Most readers will already be familiar with the basic set operations and relations, but here is a quick review of concepts and notation.

(i) $\cup$ or *union* of sets. If A and B are sets (events), then $A \cup B$ contains all elements of either or both. That is, we just combine or *unite* the two sets. Thus, $\{a, b, c\} \cup \{c, d, e\} = \{a, b, c, d, e\}$. A probabilist describes $A \cup B$ as the event that either A *or* B occurs, meaning that an outcome occurs that is in either or both of these sets.

(ii) $\cap$ or *intersection* of sets. If A and B are sets (events), then $A \cap B$ contains all elements that are in common to A and B. Thus, $\{a, b, c\} \cap \{c, d, e\} = \{c\}$. The probabilist describes $A \cap B$ as the event that both A *and* B occur, meaning that an outcome occurs that is in both these sets.

(iii) $\setminus$ or *difference* of sets. $A \setminus B$ is the set of elements of A that are *not* in B. Thus, $\{a, b, c\} \setminus \{c, d, e\} = \{a, b\}$. To the probabilist this is the event that A occurs but B does not. (The symbol "$-$" is sometimes used in the literature.)

(iv) ^{c} or *complement* of a set. A^c is the set of outcomes *not* in A or, to the probabilist, the event that A does *not* occur. Thus, $A^c = \Omega \backslash A$, $A \backslash B = A \cap B^c$, and $\Omega^c = \Omega \backslash \Omega = \varnothing$. (Other notations in the literature: $\bar{A}, A', A^*, ...$)

(v) $\subset$ or *subset* relation. $A \subset B$ indicates that each element of A is also an element of B. Thus, $A \subset B$ if and only if $A \cap B = A$. In probabilistic terms, A *implies* B since the occurrence of A (of an outcome belonging to A) implies also the occurrence of B (of an outcome also in B). $A \supset B$ means the same as $B \subset A$. Clearly, $(A \cap B) \subset A$, $(A \cap B) \subset B$, and both A and B are subsets of $A \cup B$. Note that since $\varnothing \cap A = \varnothing$ for all A, $\varnothing$ is considered to be a subset of every set.

(vi) $=$ or *equality* relation. $A = B$ means that $A \subset B$ *and* $B \subset A$. Conversely, $A \neq B$ indicates that equality does not hold. (We do not take "$\subset$" to signify "proper" subset, as is sometimes done; thus, $A \subset B$ allows $A = B$, and the notation $A \subseteq B$ is not used here.)

The operations of union and intersection are commutative and associative. Thus, $A \cup B$ is the same as $B \cup A$, and $A \cup B \cup C$ signifies either $(A \cup B) \cup C$ or $A \cup (B \cup C)$; likewise for intersections. There are also the distributive properties $A \cap (B \cup C) = (A \cap B) \cup (A \cap C)$ and $A \cup (B \cap C) = (A \cup B) \cap (A \cup C)$. Drawing a picture with sets as closed curves (a "Venn diagram") helps to see that these are true. One more special term: If $A \cap B = \varnothing$ then the sets are said to be "disjoint". In probability disjoint sets are called *exclusive* events, since if A occurs (an outcome in A occurs) then it is not also possible for B to occur.

1.1.3.1 *Sequences of Events*

Often we deal with collections of events that have some natural ordering. Such collections are called *sequences*.[2] For example, an experiment may itself comprise a sequence of steps or "trials", like drawing repeatedly from an urn containing various numbers of red and white balls. Thus, we may start out with one of each color and draw at random, then replace what was drawn, add one more red, and draw again; and so forth. We may then designate the events of getting red on draws one, two, ... as the sequence

---

[2] "Collection" and "set" have exactly the same meaning. We speak of *collections* of sets or *classes* of sets merely to avoid repetition.

$R_1, R_2, \ldots$ More generally, finite and countably infinite sequences of events $A_1, A_2, \ldots$ are represented as $\{A_j\}_{j=1}^{n}$ and $\{A_j\}_{j=1}^{\infty}$, respectively. Of course, the choice of symbol j as index is completely arbitrary.[3] An alternative representation is $\{A_u\}_{u \in \mathbb{U}}$, where the index set $\mathbb{U}$ could be finite, countably infinite, or even uncountable. Mostly, we shall deal with countable sequences. Unions and intersections of such sequences are denoted $\cup_{j=1}^{\infty} A_j$ and $\cap_{j=1}^{\infty} A_j$. The first is the event that *at least* one of the events occurs (an outcome occurs that belongs to at least one); the second is the event that *all* occur (an outcome occurs that belongs to all). When the range of the index is understood from the context, we sometimes omit it and write $\cup_j A_j$ or simply $\cup A_j$.

Monotone sequences are sets whose members have subset/superset relations to adjacent members. By $\{A_j\} \uparrow$ we designate an ordered collection each of whose members is contained in the *next* member of the sequence; i.e., $A_j \subset A_{j+1}$. Such sets are said to be *monotone increasing*. Conversely, $\{A_j\} \downarrow$ signifies sets that are *monotone decreasing*, which means that $A_j \supset A_{j+1}$ for each j in the list. Note that the terms "increasing" and "decreasing" are in the weak sense, in that the subset relations are not necessarily proper.[4]

Example 1.1. If A_j represents the set of individuals (in any population, real or conceptual) whose age to the nearest whole year is at most j, then the countable collection $\{A_j\}_{j=0}^{\infty}$ is monotone increasing. If A_u represents the set of intervals of real numbers with length greater than u, then the uncountable collection $\{A_u\}_{u \in (0,\infty)}$ is monotone decreasing.

1.1.3.2 *Limits of Sequences of Sets*

For infinite monotone sequences there is a very intuitive notion of *limit*. Thus, by $\lim_{j \to \infty} A_j = A$ we mean, in the case $\{A_j\} \uparrow$, that $\cup_{j=1}^{\infty} A_j = A$. Conversely, if $\{A_j\} \downarrow$ then $\lim_{j \to \infty} A_j = A$ signifies that $\cap_{j=1}^{\infty} A_j = A$. To provide some examples, we can take $\Omega = [0, 1]$ (the set of real numbers $\{x : 0 \leq x \leq 1\}$). Then if $A_n := [0, 1 - 1/n]$ we have $\lim_{n \to \infty} A_n = \cup_{n=1}^{\infty} A_n = [0, 1)$. The limiting set is open on the right because unity is never

---

[3]Strictly, one should think of a sequence of subsets of Ω as a *function*—a mapping from natural numbers $\mathbb{N}$ (or a subset thereof) *into* the collection of all subsets of Ω or *onto* the specific collection of subsets whose members appear in the sequence. Thus, sequence $\{A_j\}_{j=1}^{\infty}$ assigns to the integer 1 the set A_1, to the integer 2 the set A_2, and so forth.

[4]Yes, this does mean that a sequence such as $A, A, A, \ldots$ is *both* increasing and decreasing, but we rarely have occasion to consider such trivialities.

included in *any* member of $\{A_n\}_{n=1}^{\infty}$, yet for any $\varepsilon > 0$ the set $[0, 1 - \varepsilon]$ ultimately *is* included when n is sufficiently large. The same limit obtains if $A_n = [0, 1 - 1/n)$. Likewise, for either $B_n = [0, 1/n]$ or $B_n = [0, 1/n)$ we have $\lim_{n \to \infty} B_n = \cap_{n=1}^{\infty} B_n = \{0\}$, which is the singleton containing the number zero only (*not* the empty set).

Exercise 1.1. *What is* $\lim_{n \to \infty} B_n$ *if* $B_n = (0, 1/n].$? *What is* $\lim_{n \to \infty} A_n$ *if* $A_n = (0, 1 + 1/n)$?

While the concept of limit is clear intuitively for monotone sequences, it is less obvious how the idea could apply more generally. However, an example suggests that the concept should have broader application if only we could get the right definition. With $\Omega = \Re$ (the real numbers), consider the sequence $A_1 = [-1, 0], A_2 = [0, 1/2], A_3 = [-1/3, 0], ...$ and so on, with

$$A_n = \begin{cases} [-1/n, 0], & n \text{ odd} \\ [0, 1/n], & n \text{ even} \end{cases} \tag{1.1}$$

for arbitrary $n \in \aleph$. The sequence is clearly not monotone, yet it seems intuitively clear that the concept of limit here is meaningful and, indeed, that $\lim_{n \to \infty} A_n = \{0\}$ is the right answer. In fact, we can define limits in a broader way that includes such sequences using the concepts *limit inferior* and *limit superior* of sets. For countable sequences $\{A_n\}$ these are denoted $\liminf_n A_n$ and $\limsup_n A_n$, respectively.

Definition 1.1.

(i) $\liminf_n A_n := \cup_{n=1}^{\infty} \cap_{m=n}^{\infty} A_m$.
(ii) $\limsup_n A_n := \cap_{n=1}^{\infty} \cup_{m=n}^{\infty} A_m$.

At first encounter these expressions can seem daunting, but it is actually not hard to get a working understanding. To get a sense of what (i) means, let $B_n := \cap_{m=n}^{\infty} A_m$. This is the set of outcomes that are contained in *all* events of the sequence from n onward. Then $\liminf_n A_n = \cup_{n=1}^{\infty} B_n$ is the set of outcomes that are *either* in all events from 1 forward *or* in all events from 2 forward *or* from 3 forward, and so on. In short, $\liminf_n A_n$ is the set of outcomes that are in all events from *some*, possibly indefinite, point onward in the sequence. To say that $\liminf_n A_n$ "occurs" is then to say that *all but finitely many of the events occur.*[5]

---

[5]While this is the usual, *literal* interpretation of lim inf, one must not think that a *specific* n can necessarily be found beyond which all events occur (or that a *specific* finite number of events can be found that do *not* occur). To see why not, consider the increasing sequence $\{A_n = [0, 1 - 1/n]\}_{n=1}^{\infty}$, for which $\cap_{m=n}^{\infty} A_m = A_n$ and $\liminf_n A_n = \cup_{n=1}^{\infty} A_n = [0, 1)$.

To see what (ii) means, let $C_n := \cup_{m=n}^{\infty} A_m$. This is the set of outcomes that are contained in *some* event from n forward—that is, in *one or more* of those events. $\limsup_n A_n = \cap_{n=1}^{\infty} C_n$ is thus the set of outcomes that are in some event from 1 forward *and* in some event from 2 forward *and* from 3 forward, and so on. In short, $\limsup_n A_n$ is the set of outcomes that are in one or more events of the remaining infinite sequence no matter how far along in the sequence we are. To state it negatively, if $\limsup_n A_n$ occurs we will never reach a point in the sequence beyond which *none* of the remaining events will occur. But if such be the case the number of events that do occur cannot be finite. Thus, $\limsup_n A_n$ represents the set of outcomes that are in *infinitely many* of the events, and so the occurrence of $\limsup_n A_n$ means that infinitely many of events $\{A_n\}$ occur. For this reason $\limsup_n A_n$ is often represented by the phrase "A_n i.o.", where "i.o." stands for "infinitely often".

Exercise 1.2. *Explain why both of the sequences $\{B_n\}$ and $\{C_n\}$ in the definitions are monotone. Determine whether each sequence is increasing or decreasing.*

Now if all but finitely many events of infinite sequence $\{A_n\}$ occur, then the number that do occur is clearly infinite. Thus, the occurrence of $\liminf_n A_n$ *implies* the occurrence of $\limsup_n A_n$, so that $\liminf_n A_n \subset \limsup_n A_n$. On the other hand, it is not necessarily true that $\limsup_n A_n \subset \liminf_n A_n$, since the occurrence of infinitely many of $\{A_n\}$ does not rule out the occurrence of infinitely many of $\{A_n^c\}$. Therefore, it is not *generally* the case that the limit inferior and limit superior are equal (i.e., the same sets). When this *does* happen, we have our broader concept of limit that encompasses sequences of sets that are not necessarily monotone.

Definition 1.2. If there exists a set A such that $\liminf_n A_n = \limsup_n A_n = A$, then we regard A as the limit of the sequence $\{A_n\}$ and write $\lim_{n\to\infty} A_n = A$.

Exercise 1.3.

(i) If $\{A_n\} \uparrow$ show that

$$\liminf_n A_n = \limsup_n A_n = \bigcup_{n=1}^{\infty} A_n.$$

(ii) If $\{A_n\} \downarrow$ show that

$$\liminf_n A_n = \limsup_n A_n = \bigcap_{n=1}^{\infty} A_n.$$

(iii) For the sequence $\{A_n\}$ defined in (1.1) show that

$$\liminf_n A_n = \limsup_n A_n = \{0\}.$$

(iv) If $A_n = [-1, 0]$ for n odd and $A_n = [0, 1]$ for n even, what are $\liminf_n A_n$ and $\limsup_n A_n$? Does $\lim_{n \to \infty} A_n$ exist?

(v) If $A_n = [0, 1 - 1/n]$ for n odd and $A_n = [0, 1 + 1/n]$ for n even, what are $\liminf_n A_n$ and $\limsup_n A_n$? Does $\lim_{n \to \infty} A_n$ exist?

1.1.3.3 *De Morgan's Laws*

Let A and B be two subsets of Ω. The following relations are known as De Morgan's Laws:

- $(A \cup B)^c = A^c \cap B^c$
- $(A \cap B)^c = A^c \cup B^c.$

Expressed in words, these become obvious and easy to remember. (i) If $(A \cup B)^c$ occurs then $A \cup B$ does *not* occur, meaning that neither A nor B occurs, meaning that A does not occur *and* B does not occur; hence, $A^c \cap B^c$ occurs. (ii) If $(A \cap B)^c$ occurs then A and B do not *both* occur, meaning that either A does not occur or B does not occur; hence, $A^c \cup B^c$. The laws extend to arbitrary collections; thus, (for countable collections) $\left(\cup_{j=1}^{\infty} A_j\right)^c = \cap_{j=1}^{\infty} A_j^c$ and $\left(\cap_{j=1}^{\infty} A_j\right)^c = \cup_{j=1}^{\infty} A_j^c$.

Exercise 1.4. *Apply De Morgan's laws to show that (i) $(\liminf_n A_n)^c = \limsup_n A_n^c$ and (ii) $(\limsup_n A_n)^c = \liminf_n A_n^c$. Then explain these relations intuitively.*

1.1.4 *Fields and σ Fields of Sets*

We have described sequences of sets (sequences of *events* in the context of chance experiments) as collections of subsets of Ω that have a natural ordering. In general, there is no implication that these exhaust the entire sample space, in the sense that their union is the entire Ω; nor is it necessarily true that uniting or intersecting sets in the sequence produces other sets in the sequence. However, collections of sets that do have these properties of exhaustion and closure under standard operations are of paramount importance in defining probabilities (and measures generally). In the discussion below we take for granted that all the sets represented are subsets of some nonempty space Ω, and we freely apply the same $\{A_j\}$ notation used for sequences to collections of sets that are not necessarily ordered and may even be uncountable.

Definition 1.3. A collection $\mathcal{F}$ of sets is called a *field* if (i) $(A \cup B) \in \mathcal{F}$ whenever $A \in \mathcal{F}$ and $B \in \mathcal{F}$ and (ii) $A^c \in \mathcal{F}$ whenever $A \in \mathcal{F}$.

Thus, a field of sets is defined as a collection that contains unions of all pairs of its members and complements of its members; but these minimal requirements actually imply much more. First, if all of A, B, and C are in $\mathcal{F}$ then $(A \cup B) \cup C = A \cup B \cup C$ is also in $\mathcal{F}$, and so on (by induction) for the union of any *finite* number of sets. Thus, if $\{A_j\}_{j=1}^n$ is a collection of $\mathcal{F}$ sets—again, not necessarily having a natural ordering—then $\cup_{j=1}^n A_j$ is in the collection. Second, since $A^c \in \mathcal{F}$ if A is in $\mathcal{F}$, then so is $A \cup A^c = \Omega$, so the collection contains the entire space. It also contains the empty set, since $\Omega^c = \varnothing$. Next, an application of De Morgan shows that finite intersections of sets in $\mathcal{F}$ remain in the collection as well. Finally, there is also closure under differencing, since $A \backslash B = A \cap B^c$. Thus, a field is closed under *finitely* many of the standard set operations.

Each collection $\{A_j\}$ of sets has associated with it a *special* field that will be of importance in the sequel.

Definition 1.4. Suppose that $\mathcal{F}$ is a field containing sets $\{A_j\}$ and that $\mathcal{F} \subset \mathcal{G}$ for any other field $\mathcal{G}$ that contains $\{A_j\}$. Then $\mathcal{F}$ is said to be the field *generated* by $\{A_j\}$.

Note that the generated field $\mathcal{F}$ is the *smallest* field associated with any collection $\{A_j\}$ and so is unique to that collection; however, the *same* $\mathcal{F}$ corresponds to indefinitely many other collections, since to any $\{A_j\}$ can be added arbitrarily many copies of its members.

Exercise 1.5. *If sets $\{A_j\}_{j=1}^n$ are in field F, show that $\cap_{j=1}^n A_j \in \mathcal{F}$.*

Exercise 1.6. *$\{A_j\}_{j=1}^n$ (for some integer $n \geq 1$) is a* monotone *sequence of sets, among which are included $\varnothing$ and Ω. Can such a collection constitute a field?*

Example 1.2. Take $\Omega = [0, 1)$ and, with the objective of constructing a field $\mathcal{F}$, begin with the (uncountable) collection of all closed–open intervals contained in Ω; i.e., sets of the form $[a, b)$ where $0 \leq a < b \leq 1$. We shall construct the field that is generated by these fundamental constituent sets. Finite unions of such closed–open intervals are not necessarily of this form— e.g., $[0, .2) \cup [.3, 1)$ is not an interval at all—so we must add all such finite unions to the collection. Doing this automatically takes care of adding the complements of its members as well—e.g., $[.2, .3)^c = [0, .2) \cup [.3, 1)$—except that we must specifically add the empty set. Adding finite unions and $\varnothing$ to our fundamental class therefore produces a field of subsets of $[0, 1)$.

Example 1.3. The smallest field of subsets of any nonempty space Ω is the collection $\{\varnothing, \Omega\}$. This is often called the "trivial" field. The largest field is the collection comprising *all* subsets of Ω.

Exercise 1.7.

(i) *If a coin is flipped once and $\Omega = \{H, T\}$, what is the smallest field containing each of the individual events H and T?*

(ii) *If a coin is flipped twice and $\Omega = \{HH, TH, HT, TT\}$, what is the smallest field of subsets?*

(iii) *If Ω comprises n distinct elements, how many sets will be contained in the field generated by these elements? (Hint: The composition of a set can be determined by asking, for each of the n elements, "Is this element in the set?")*

It will turn out that fields are the smallest collections of sets over which it is useful to define probability measures, since one always wants to be able to work out probabilities of events constructed from the basic constituent class by finite set operations. For instance, knowing the probability of event A and the probability of event B is not enough; we would want to know the probability that *either* occurs $(A \cup B)$, that *both* occur $(A \cap B)$, that *neither* occurs $((A \cup B)^c = A^c \cap B^c)$, and that one occurs but not the other (e.g., $A \backslash B$). This would not always be possible if probabilities were defined on a collection smaller than a field. However, in the case that a sample space contains uncountably many outcomes—for example, if Ω is an interval of real numbers—then it happens that fields may still not be extensive enough to cover all cases of interest. In such cases a more comprehensive collection is needed, known as a *sigma field* or *sigma algebra* of sets.

Definition 1.5. A collection of sets $\mathcal{F}$ is called a *sigma field* (abbreviated "σ field") if it is closed under *countable* unions and under complementation.

Notice that this extension of the concept of field merely ensures that we remain in the collection if we pool together any countably infinite number of sets in the collection. In other words, the σ field is "closed" under countable unions. De Morgan's laws and closure under complementation further imply closure under countably many intersections and differences. To appreciate the distinction between field and σ field, we must see an example of a set that can be built up only from a countably infinite number of operations.

Example 1.4. As in Example 1.2 take $\Omega = [0, 1)$ and begin with a fundamental class of closed–open intervals $[a, b)$ on Ω. Neither this fundamental class nor the field generated by it contains individual points in Ω. For example, the singleton $\{.5\}$ is clearly not in the fundamental class, nor can it be constructed by finitely many operations on those closed–open sets. However, all members of the monotone sequence $\{[.5, .5 + .5/n)\}_{n=1}^{\infty}$ are in the fundamental class and therefore in the field that it generates. The limit of this sequence, $\{.5\}$, is attainable by countably many operations as $\cap_{n=1}^{\infty}[.5, .5 + .5/n)$. Thus, if we knew how to assign probabilities to the closed–open intervals and how to find probabilities of events constructed through countably many set operations (as we will ultimately see how to do), then we could indeed deduce the probabilities associated with singleton sets—and many others also.

Exercise 1.8.

(i) $\mathcal{F}$ is the field comprising all subsets of a space Ω that contains finitely many elements. Show that $\mathcal{F}$ is also a σ field. (Hint: $A \cup A = A$, etc.)

(ii) If $\mathcal{G}$ and $\mathcal{H}$ are two σ fields of subsets of an arbitrary space Ω, show that $\mathcal{G} \cap \mathcal{H}$ (the collection comprising sets that belong to both $\mathcal{G}$ and $\mathcal{H}$) is also a σ field.

(iii) In the experiment of flipping two coins, take $\Omega := \{HH, HT, TH, TT\}$, put $H_1 := \{HH, HT\}$, $T_1 := \{TH, TT\}$, $S_1 := \{HH, TT\}$, $S_2 := \{HT, TH\}$, and let $\mathcal{G} := \sigma(H_1, T_1)$ (the field generated by H_1 and T_1) and $\mathcal{H} := \sigma(S_1, S_2)$. Is $\mathcal{G} \cup \mathcal{H}$ a field?

As the result of the first part of the exercise implies, we can work in perfect generality with fields alone whenever sample spaces are finite. The broader concept of σ field is needed only when the nature of the experiment and our interests require the specification of infinitely many outcomes. This extra complexity is, however, often required in statistical applications and in applied probability models. In such cases, so long as the number of outcomes is *countably* infinite, we can still work with the σ field of *all* subsets of Ω. On the other hand, when the sample space is *uncountable* it is not generally possible to associate probabilities with all events. There may be subsets of such an Ω that cannot be constructed by *countably* many operations on a fundamental class whose probabilities are readily determinable, and (as we shall see) the prevailing mathematical theory affords no sure way of assigning to such events numbers with the properties we want probabilities to have. Fortunately, the existence of these "nonmeasurable" sets is rarely of any consequence in applied work.

Just as we refer to *fields* that are generated by some fundamental class of sets, we refer also to generated σ fields. When $\Omega = \Re$ (or some interval contained in $\Re$), it is customary to regard the open–closed intervals $(a, b]$ as the fundamental class, but in fact the open intervals, the closed intervals, and the closed–open intervals—indeed, the classes of open and closed *sets*—all generate precisely the same σ field, which has a special name and symbol.[6]

Definition 1.6. The σ field $\mathcal{B}$ generated by the intervals $(a, b]$ (or $[a, b)$, or (a, b), or $[a, b]$) with $-\infty \leq a < b \leq \infty$ is called the *Borel sets* of the real line.

1.2 Set Functions and Measures

In developing a rigorous theory of probability one must start with the more general concepts of *set functions* and *measures*. Throughout the discussion we take for granted that we are endowed with a nonempty space Ω (at this point not necessarily comprising outcomes of a chance experiment) and a σ field of subsets $\mathcal{F}$. In the definitions we pair these together as $(\Omega, \mathcal{F})$, since $\mathcal{F}$ has no meaning except in relation to a particular space and since different σ fields can be associated with the same space. (For example, with $\Omega = \Re$ we could have $\mathcal{F} = \mathcal{B}$ or $\mathcal{F} = \{\varnothing, \Re\}$.) While set functions and measures can be defined on smaller classes of sets than σ fields, this is the relevant class for our subsequent interests.

Definition 1.7. A *set function* φ defined on $(\Omega, \mathcal{F})$ is a mapping from $\mathcal{F}$ to $\Re$ (the real numbers); i.e., $\varphi : \mathcal{F} \to \Re$.

---

[6]A more informative notation would be $\mathcal{B}_\Re$, but we stick with $\mathcal{B}$ for brevity. Such detail will be added to denote Borel subsets of other spaces, as $\mathcal{B}_{[0,1)}$ for the subsets of $[0, 1)$ and $\mathcal{B}^k$ for the Borel sets of $\Re^k$.

In other words, φ assigns to each set of $\mathcal{F}$ (the domain) a single real number. There are all sorts of possibilities.

Example 1.5.

(i) Take Ω to be a group of individuals, such as the students in a particular class. Since the space is finite, we can work with the field $\mathcal{F}$ that comprises all subsets of Ω. Define set function φ_1 as follows, where A is any such subset:

$$\varphi_1(A) = \begin{cases} -1, & \text{if all members of } A \text{ are male} \\ +1, & \text{if all members of } A \text{ are female} \\ 0, & \text{otherwise.} \end{cases}$$

(ii) Take $\Omega = \mathbb{Q}$ (the rational numbers) and $\mathcal{F}$ as the σ field comprising all the subsets of $\mathbb{Q}$. Define set function φ_2 as

$$\varphi_2(A) = \begin{cases} 0, & \text{if } A \text{ has finitely many elements or is empty} \\ +\infty, & \text{if } A \text{ has infinitely many elements} \end{cases}.$$

Definition 1.8.

(i) A set function φ is *additive* if $\varphi(\varnothing) = 0$ and $\varphi(A \cup B) = \varphi(A) + \varphi(B)$ whenever $A \cap B = \varnothing$.
(ii) Set function φ is *countably additive* if $\varphi(\varnothing) = 0$ and $\varphi(\cup_{n=1}^{\infty} A_n) = \sum_{n=1}^{\infty} \varphi(A_n)$ whenever $\{A_n\}_{n=1}^{\infty}$ are disjoint.

Thus, an additive set function assigns to the union of two disjoint sets the sum of the numbers it assigns to the components. Induction shows that additive set functions thus defined are also *finitely* additive; i.e., $\varphi(\cup_{j=1}^{n} A_j) = \sum_{j=1}^{n} \varphi(A_j)$ for any finite n whenever $\{A_j\}_{j=1}^{n}$ are disjoint sets. Countable additivity just extends the property to $n = \infty$, but it is in fact an additional requirement that does not follow from finite additivity.

Exercise 1.9.
   (i) Is φ_1 in Example 1.5 an additive set function?
   (ii) Is φ_2 countably additive?

We are now ready to focus on the particular class of set functions that are most relevant in probability theory.

Definition 1.9.

(i) A *measure* $\mathcal{M}$ defined on $(\Omega, \mathcal{F})$ is a nonnegative, countably additive set function. In symbols, $\mathcal{M} : \mathcal{F} \to [0, \infty]$.

(ii) Measure $\mathcal{M}$ is *finite* if $\mathcal{M} : \mathcal{F} \to [0, \infty) =: \Re_+$.

Together, the three objects Ω, $\mathcal{F}$, and $\mathcal{M}$ define what is called a *measure space*. Ω specifies the basic elements of all sets that are to be measured; $\mathcal{F}$ specifies which sets are measurable; and $\mathcal{M}$ specifies the specific measure. Infinitely many measures could in principle be defined on any given *measurable space* $(\Omega, \mathcal{F})$. The nonnegativity restriction on $\mathcal{M}$ is not crucial and is sometimes relaxed, but all measures that we shall encounter do have this property.

Although the definition of measure seems fairly abstract, there are some very familiar examples. Aside from probability measures, the two most important ones for our purposes are *counting measure* and *Lebesgue measure*. Later we shall use these to construct some probabilities.

1. Take Ω to be a countable set and $\mathcal{F}$ the σ field comprising all its subsets. If a given set $A \in \mathcal{F}$ is *finite*, set $N(A)$ equal to the number of elements of the set; otherwise, put $N(A) = +\infty$. Thus, the measure $N : \mathcal{F} \to \aleph_0 \cup \{+\infty\}$ merely counts or enumerates the elements of sets. Singletons $\{0\}$ and $\{+\infty\}$ are included in the range of N to extend it to empty and infinite sets.

Exercise 1.10. *Verify that N satisfies the requirements for a measure.*

2. Take $\Omega = \Re$ and $\mathcal{F} = \mathcal{B}$ (the Borel sets of $\Re$), and set $\lambda((a, b]) := b - a$ when $+\infty \geq b \geq a \geq -\infty$. This is *Lebesgue* or *length measure*. Since $a = -\infty$ and $b = +\infty$ are allowed, λ maps $\mathcal{F}$ onto $\Re_+ \cup \{+\infty\} \equiv [0, +\infty]$. The definition shows how to measure just the open–closed intervals—a "fundamental" class that generates $\mathcal{B}$. However, basic theorems in measure theory show that λ has a unique extension to $\mathcal{B}$, some members of which are extremely complicated.[7] From countable additivity one can see at once how λ measures some other types of sets besides open–closed intervals.

---

[7]For the theorems consult any basic text on real analysis/measure theory; e.g., Ash (1972), Billingsley (1986,1995), Royden (1968), Taylor (1966). An example of a complicated Borel set is the *Cantor set*, which is constructed progressively as follows. From the closed interval $[0, 1]$ remove the open interval $(1/3, 2/3)$. leaving the closed intervals $[0, 1/3]$ and $[2/3, 1]$ Then remove the center third from each of these; i.e., remove $(1/9, 2/9)$ from $[0, 1/3]$ and $(7/9, 8/9)$ from $[2/3, 1]$. Continue in this way *ad infinitum*, removing the open center third of each closed interval that remains from the preceding step. The resulting set is measurable, being constructed by countably many differencing operations, and indeed has Lebesgue measure zero. Nevertheless, the set contains uncountably many points.

a. For open intervals (a, b) use the fact that

$$(a, b) = \bigcup_{n=1}^{\infty} \left(a + \frac{n-1}{n}(b-a), a + \frac{n}{n+1}(b-a) \right]$$

(a union of *disjoint* sets) plus countable additivity:

$$\lambda\left((a, b)\right) = \lim_{N \to \infty} \sum_{n=1}^{N} \lambda\left(\left(a + \frac{n-1}{n}(b-a), a + \frac{n}{n+1}(b-a)\right]\right)$$

$$= (b-a) \lim_{N \to \infty} \sum_{n=1}^{N} \left(\frac{n}{n+1} - \frac{n-1}{n} \right)$$

$$= (b-a) \lim_{N \to \infty} \sum_{n=1}^{N} \left(\frac{1}{n} - \frac{1}{n+1} \right)$$

$$= (b-a) \lim_{N \to \infty} \left[\left(1 - \frac{1}{2}\right) + \left(\frac{1}{2} - \frac{1}{3}\right) + \cdots + \left(\frac{1}{N} - \frac{1}{N+1}\right) \right]$$

$$= (b-a) \lim_{N \to \infty} \left(1 - \frac{1}{N+1} \right)$$

$$= b - a.$$

Thus, $\lambda\left((a, b)\right) = \lambda\left((a, b]\right)$.

b. For a single point $\{b\}$ deduce that $\lambda(\{b\}) = 0$ from the previous result and the fact that

$$\lambda\left((a, b]\right) = \lambda\left((a, b) \cup \{b\}\right) = \lambda\left((a, b)\right) + \lambda(\{b\}).$$

c. For a *countable* set of single points, such as the rational numbers, deduce from countable additivity that the Lebesgue measure is zero:

$$\lambda(\mathbb{Q}) = \lambda\left(\bigcup_{q \in \mathbb{Q}} \{q\}\right) = \sum_{q \in \mathbb{Q}} \lambda(\{q\}) = \sum_{q \in \mathbb{Q}} 0 = 0.$$

d. For the entire real line, we have for any finite number a that

$$\lambda(\Re) = \lambda\left((-\infty, a]\right) + \lambda\left(\bigcup_{n=1}^{\infty} (a + n - 1, a + n]\right).$$

Thus, $\lambda(\Re) = \infty + \sum_{n=1}^{\infty} 1 = \infty$.

e. For the irrational numbers $\Re\backslash\mathbb{Q}$ we also have $\lambda(\Re\backslash\mathbb{Q}) = \infty$, since

$$\lambda(\Re\backslash\mathbb{Q}) = \lambda(\Re) - \lambda(\mathbb{Q}). \tag{1.2}$$

Exercise 1.11. *Verify (1.2) by showing that if $\mathcal{M}$ is a measure on $(\Omega, \mathcal{F})$ and if $A \subset B$ are $\mathcal{M}$ measurable with $\mathcal{M}(A) < \infty$, then $\mathcal{M}(B\backslash A) = \mathcal{M}(B) - \mathcal{M}(A)$.*

Notice that trying to extend beyond *countably* many set operations could lead to some absurdities, such as

$$1 = \lambda\left((0,1]\right) = \lambda\left(\bigcup_{x\in(0,1]}\{x\}\right) = \sum_{x\in(0,1]}\lambda\left(\{x\}\right) = 0.$$

This illustrates why measures have to be defined on collections of sets generated only by *countably many* operations on some "fundamentally" measurable class. Naturally, the same restriction applies to the specific class of measures that represent probabilities.

Although counting measure applies only to countable spaces, an extension to $(\Re, \mathcal{B})$ is often useful. If $\mathbb{X} \in \mathcal{B}$ is a specified, countable set of real numbers, such as the integers or the rationals, we simply take $N(B) := N(B \cap \mathbb{X})$ for each Borel set B.

1.3 Probability Measures

Stated most succinctly, a probability measure—or probability *function*—is just a measure $\mathcal{M}$ on some measurable space $(\Omega, \mathcal{F})$ with the special property that $\mathcal{M}(\Omega) = 1$. However, we shall use the special symbol $\mathbb{P}$ for probability measures, and—abiding some redundancy—will state the definition in terms of three essential properties, which are known as *Kolmogorov's axioms*.[8] As usual, we take for granted that there are a nonempty space Ω and a σ field of subsets $\mathcal{F}$, and that all events (subsets of Ω) that we attempt to measure belong to $\mathcal{F}$.[9]

Definition 1.10. A probability measure $\mathbb{P}$ on $(\Omega, \mathcal{F})$ is a mapping from $\mathcal{F}$ into the reals (i.e., a set function) with the following properties:

1. $\mathbb{P}(\Omega) = 1$
2. $\mathbb{P}(A) \geq 0$ for all events A

---

[8] Renowned Russian probabilist A. N. Kolmogorov (1903-1987) was the first to adopt the modern set-theoretic framework that gives the theory of probability its proper place in mathematics. We shall see more of his profound influence in the sequel in connection with the concept of conditional expectation, the law of large numbers, and the law of the "iterated" logarithm.

[9] Notice that we refer to probabilities in two senses—as *functions* and as the *numbers* those functions assign to specific events; thus, probability function $\mathbb{P}$ assigns to event A the probability *number* $\mathbb{P}(A)$. The same convenient looseness of terminology is involved in referring to a real-valued function $f : \Re \to \Re$ and to its value $f(x)$ at a particular $x \in \Re$.

3. $\mathbb{P}(\cup_{j=1}^{\infty} A_j) = \sum_{j=1}^{\infty} \mathbb{P}(A_j)$ for all collections $\{A_j\}_{j=1}^{\infty}$ of exclusive events.

Of course, the second and third properties, nonnegativity and countable additivity, apply to more general measures, except that we now refer to sets as events and to disjoint sets as exclusive events. The three objects $\Omega, \mathcal{F}$, and $\mathbb{P}$ define a *probability space*, denoted by the triple $(\Omega, \mathcal{F}, \mathbb{P})$. We will be able to determine many other vital properties of $\mathbb{P}$ from Kolmogorov's axioms, but let us look at some examples first. (These also illustrate how other measures on certain spaces can be used to create probabilities.)

Example 1.6. Flipping a coin three times, take

$$\Omega = \{TTT, TTH, THT, HTT, HTH, HHT, THH, HHH\},$$

and $\mathcal{F}$ as the collection of all 2^8 subsets. For any such subset A set

$$\mathbb{P}(A) := \frac{N(A)}{N(\Omega)} = \frac{N(A)}{8},$$

where N is counting measure on $(\Omega, \mathcal{F})$. Thus, for

$$H_1 := \{HTT, HTH, HHT, HHH\},$$

the event "heads on the first flip", we have $\mathbb{P}(H_1) = \mathbb{P}(H_1^c) \equiv \mathbb{P}(T_1) = \frac{1}{2}$. Since the same probability applies to events "heads on flip two" and "heads on flip three", this model for $\mathbb{P}$ corresponds to our notion that the coin is "fair".

Example 1.7. Taking $\Omega = [0, 1)$, $\mathcal{F} = \mathcal{B}_{[0,1)}$ (the Borel sets of the unit interval), and $\mathbb{P} = \lambda$ (Lebesgue measure) gives the probability space $([0, 1), \mathcal{B}_{[0,1)}, \lambda)$. This serves to model a chance experiment that will be used in many examples. Figure 1.1 depicts a circular scale of unit circumference with a pointer attached to the center by one end, allowing it to rotate. The chance experiment consists of spinning the pointer and allowing it to come to rest. Regard an outcome of the experiment as the clockwise distance of the tip of the settled pointer from some origin ω_0, as measured around the perimeter of the circle. With this convention the sample space comprises the real numbers on $[0, 1)$. (How many times the spinning pointer might pass ω_0 is irrelevant; only the final position matters.) Since the circumference has unit length, it is plausible to set $\mathbb{P}(A) = \lambda(A)$ for any $A \in \mathcal{B}_{[0,1)}$. Indeed, this would describe in the most rigorous way our notion of a "fair" pointer. Notice, however, that an implication of this eminently reasonable model is that $\mathbb{P}(\{\omega\}) = 0$ for each $\omega \in [0, 1)$. In other words, although we are certain *ex ante* that the pointer will stop *somewhere*, the probability is zero that it will stop at any *specific* point. Indeed, it also follows that $\mathbb{P}(D) = 0$, where D is any countable set. In particular, even though there are infinitely many rational numbers on $[0, 1)$, the properties of Lebesgue measure imply that $\mathbb{P}(\mathbb{Q} \cap [0, 1)) = 0$.

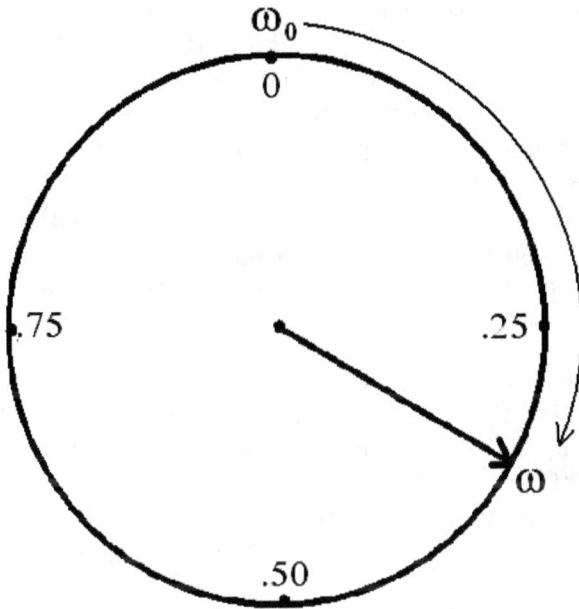

Fig. 1.1   The unit circular scale.

Definition 1.11. An event A on a probability space $(\Omega, \mathcal{F}, \mathbb{P})$ is said to be $\mathbb{P}$-*null* (or simply *null*) if (and only if) $\mathbb{P}(A) = 0$. If A is null, then $\mathbb{P}(A^c) = 1$, in which case we say that A^c occurs "almost surely" (a.s.) or "almost everywhere (a.e.) with respect to measure $\mathbb{P}$".

As Example 1.7 suggests, having a consistent theory of probability requires swallowing the unpalatable fact that events of probability zero are not *necessarily* impossible. In other words, $\mathbb{P}(A) = 0$ does not imply that

$A = \varnothing$; likewise, $\mathbb{P}(A^c) = 1$ does not imply that $A^c = \Omega$.[10] On the other hand, the first axiom reassures us that $A^c = \Omega$ does imply $\mathbb{P}(A^c) = 1$, and we are about to be further comforted that empty sets are also $\mathbb{P}$-null.

1.3.1 *Fundamental Properties of* $\mathbb{P}$ *Measures*

From Kolmogorov's three axioms follow many other properties of $\mathbb{P}$ that are essential to applications of probability theory. Although no intuitive interpretation is needed to deduce these properties, they may seem more compelling if one thinks of probabilities in the frequentist sense referred to in the introduction. To make that loose concept now precise, let us imagine performing a chance experiment n times, keeping the initial conditions as nearly constant as possible, and counting the number of times $n(A)$ that event A occurs. Then in the frequentist view $\mathbb{P}(A)$ represents $\lim_{n \to \infty} n(A)/n$. Thus, rendered roughly in words, $\mathbb{P}(A)$ is the relative frequency of occurrence in *infinitely many* trials.

Here now are some of the most basic properties of $\mathbb{P}$, along with their proofs.

4. The complement rule: $\mathbb{P}(A^c) = 1 - \mathbb{P}(A)$.
   Proof: $1 = \mathbb{P}(\Omega) = \mathbb{P}(A \cup A^c) = \mathbb{P}(A) + \mathbb{P}(A^c)$.
5. Probability of the empty set (an impossible event): $\mathbb{P}(\varnothing) = 0$.
   Proof: $\mathbb{P}(\Omega^c) = 1 - \mathbb{P}(\Omega)$.
6. Subset inequality: $\mathbb{P}(A) \leq \mathbb{P}(B)$ whenever $A \subset B$.
   Proof: $B = A \cup (B \backslash A)$ and $A \cap (B \backslash A) = \varnothing$ imply $\mathbb{P}(B) = \mathbb{P}(A) + \mathbb{P}(B \backslash A) \geq \mathbb{P}(A)$.
7. Bounds on probabilities: $\mathbb{P}(A) \leq 1$.
   Proof: $A \subset \Omega \Longrightarrow \mathbb{P}(A) \leq \mathbb{P}(\Omega)$.
8. Subtraction rule: $\mathbb{P}(B \backslash A) = \mathbb{P}(B) - \mathbb{P}(A \cap B)$.
   Proof: $\mathbb{P}(B) = \mathbb{P}((B \backslash A) \cup (A \cap B))$.
9. Addition rule: $\mathbb{P}(A \cup B) = \mathbb{P}(A) + \mathbb{P}(B) - \mathbb{P}(A \cap B)$.
   Proof: $\mathbb{P}(A \cup B) = \mathbb{P}(A \cup (B \backslash A)) = \mathbb{P}(A) + \mathbb{P}(B \backslash A)$.
10. Monotone property: If $\{A_n\}$ is a monotone sequence of events approaching limit A, then $\lim_{n \to \infty} \mathbb{P}(A_n) = \mathbb{P}(A)$. Another way to put this is that $\lim_{n \to \infty} \mathbb{P}(A_n) = \mathbb{P}(\lim_{n \to \infty} A_n)$ if $\{A_n\}$ is monotone.
    Proof: (i) If $\{A_n\} \uparrow A$ then $\mathbb{P}(A) = \mathbb{P}(\cup_{n=1}^{\infty} A_n) = \mathbb{P}(\cup_{n=1}^{\infty}(A_n \backslash A_{n-1}))$,

---

[10] Null events may well contain nonmeasurable subsets, whose probabilities are therefore not defined. However, it is common practice in modeling to "complete" the probability space by adding to the σ field $\mathcal{F}$ all subsets of null sets and extending the domain of $\mathbb{P}$ accordingly, putting $\mathbb{P}(B) = 0$ whenever $B \subset A$, $A \in \mathcal{F}$, and $\mathbb{P}(A) = 0$.

where we take $A_0 = \varnothing$. $\{A_n \setminus A_{n-1}\}_{n=1}^{\infty}$ being exclusive,

$$\mathbb{P}\left(\bigcup_{n=1}^{\infty} A_n\right) = \sum_{n=1}^{\infty} \mathbb{P}(A_n \setminus A_{n-1}) = \lim_{n\to\infty} \sum_{j=1}^{n} \mathbb{P}(A_j \setminus A_{j-1})$$

$$= \lim_{n\to\infty} \mathbb{P}\left(\bigcup_{j=1}^{n}(A_j \setminus A_{j-1})\right) = \lim_{n\to\infty} \mathbb{P}(A_n).$$

(ii) If $\{A_n\} \downarrow A$ then $\{A_n^c\} \uparrow A^c$, so (i) and the complement rule imply

$$\lim_{n\to\infty} \mathbb{P}(A_n) = 1 - \lim_{n\to\infty} \mathbb{P}(A_n^c) = 1 - \mathbb{P}(A^c) = \mathbb{P}(A).$$

11. Limits generally: If $A_n \to A$ then $\lim_{n\to\infty} \mathbb{P}(A_n) = \mathbb{P}(A)$. In other words

$$\lim_{n\to\infty} \mathbb{P}(A_n) = \mathbb{P}\left(\lim_{n\to\infty} A_n\right)$$

applies also to sequences that are not monotone, so long as they have limits.

Proof: $A_n \to A$ implies $\cap_{n=1}^{\infty} B_n = \cup_{n=1}^{\infty} C_n = A$, where $\{B_n := \cup_{m=n}^{\infty} A_m\} \downarrow$ and $\{C_n := \cap_{m=n}^{\infty} A_m\} \uparrow$. But $\mathbb{P}(C_n) \le \mathbb{P}(A_n) \le \mathbb{P}(B_n)$ by the subset inequality, and $\mathbb{P}(C_n) \to \mathbb{P}(A)$ and $\mathbb{P}(B_n) \to \mathbb{P}(A)$ by monotonicity.

12. Boole's inequality: $\mathbb{P}(\cup_{j=1}^{\infty} A_j) \le \sum_{j=1}^{\infty} \mathbb{P}(A_j)$.
Proof: $\mathbb{P}(A_1 \cup A_2) \le \mathbb{P}(A_1) + \mathbb{P}(A_2)$ by the addition rule, whence by induction $\mathbb{P}(\cup_{j=1}^{n} A_j) \le \sum_{j=1}^{n} \mathbb{P}(A_j)$ for any finite n. For the infinite case, let $\{B_n\} = \{\cup_{j=1}^{n} A_j\}$. This is monotone, with $\cup_{j=1}^{\infty} A_j = \lim_{n\to\infty} B_n$, so

$$\mathbb{P}\left(\bigcup_{j=1}^{\infty} A_j\right) = \lim_{n\to\infty} \mathbb{P}(B_n) = \lim_{n\to\infty} \mathbb{P}\left(\bigcup_{j=1}^{n} A_j\right)$$

$$\le \lim_{n\to\infty} \sum_{j=1}^{n} \mathbb{P}(A_j) = \sum_{j=1}^{\infty} \mathbb{P}(A_j).$$

13. Bonferroni's inequality: $\mathbb{P}(\cap_{j=1}^{n} A_j) \ge 1 - \sum_{j=1}^{n} \mathbb{P}(A_j^c) = \sum_{j=1}^{n} \mathbb{P}(A_j) - (n-1)$. In particular, $\mathbb{P}(A_1 \cap A_2) \ge \mathbb{P}(A_1) + \mathbb{P}(A_2) - 1$.
Proof:

$$\mathbb{P}\left(\bigcap_{j=1}^{n} A_j\right) = 1 - \mathbb{P}\left[\left(\bigcap_{j=1}^{n} A_j\right)^c\right] = 1 - \mathbb{P}\left(\bigcup_{j=1}^{n} A_j^c\right)$$

$$\ge 1 - \sum_{j=1}^{n} \mathbb{P}(A_j^c) = 1 - \sum_{j=1}^{n}[1 - \mathbb{P}(A_j)].$$

14. Law of total probability (part 1): Suppose $A_1, A_2, \ldots$ are exclusive and that $\cup_{j=1}^{\infty} A_j = \Omega$. (Events with this property are said to *partition* the sample space.) Then $\mathbb{P}(B) = \sum_{j=1}^{\infty} \mathbb{P}(B \cap A_j)$ for any event B.[11]
    Proof: $B = B \cap \Omega = B \cap \left(\cup_{j=1}^{\infty} A_j \right) = \cup_{j=1}^{\infty} \left(B \cap A_j \right)$.

15. Borel–Cantelli lemma (convergence part): If $\sum_{n=1}^{\infty} \mathbb{P}(A_n) < \infty$ then $\mathbb{P}(A_n \text{ i.o.}) = 0$.
    Proof:

$$\mathbb{P}(A_n \text{ i.o.}) = \mathbb{P}(\limsup_n A_n) = \mathbb{P}\left(\bigcap_{n=1}^{\infty} B_n \right) = \mathbb{P}(\lim_{n \to \infty} B_n),$$

where $B_n := \cup_{m=n}^{\infty} A_m$. Since $\{B_n\} \downarrow$ the monotone property implies

$$\mathbb{P}(A_n \text{ i.o.}) = \lim_{n \to \infty} \mathbb{P}(B_n) = \lim_{n \to \infty} \mathbb{P}\left(\bigcup_{m=n}^{\infty} A_m \right) \leq \lim_{n \to \infty} \sum_{m=n}^{\infty} \mathbb{P}(A_m),$$

and the last limit is necessarily zero if $\sum_{n=1}^{\infty} \mathbb{P}(A_n)$ converges to a finite number.

Most of these properties should be familiar from a first course in probability or statistics. Here is an application of the Borel–Cantelli lemma.

Example 1.8. A random draw is made from each of a sequence of urns, each urn containing one red ball and various numbers of white ones. The first urn contains one red and one white; the second, one red and two white; the third, one red and $2^{3-1} = 4$ white; and so on, with the nth urn containing one red and 2^{n-1} white balls. Assume that the probability of drawing red from the nth urn is $\mathbb{P}(R_n) = \left(2^{n-1} + 1 \right)^{-1}$, which is the inverse of the total number of balls in the urn. If we start with the first urn, proceed to the next, and continue indefinitely, what is the probability of obtaining infinitely many red balls? We are interested, therefore, in $\mathbb{P}(R_n \text{ i.o.})$, which is to say in $\mathbb{P}(\limsup_n R_n)$. Since $\sum_{n=1}^{\infty} \mathbb{P}(R_n) = \sum_{n=1}^{\infty} \left(2^{n-1} + 1 \right)^{-1} < \sum_{n=1}^{\infty} 2^{-(n-1)} = 2 < \infty$, Borel–Cantelli implies that $\mathbb{P}(R_n \text{ i.o.}) = 0$. Recalling De Morgan's laws, this implies that $\mathbb{P}\left((\limsup_n R_n)^c \right) = \mathbb{P}(\liminf_n R_n^c) = 1$, which means that we can be "almost" sure that only white balls will be drawn beyond some point.

---

[11]If the sense of this is not immediately obvious, imagine a bucket of paint being dropped inside a house and flowing from room to room. The collection of rooms partitions the house, and by measuring the soiled areas in all the rooms and adding them up we get the total area that is affected.

1.3.2 *Modeling Certain Finite Experiments*

Before proceeding with properties and definitions, we should consider some specific examples of probability models. A particularly simple situation, yet one that often arises, is that the chance experiment can be assumed to have a finite number of *equally likely* outcomes. Specifically, suppose our interests make it possible to define outcomes in such a way that (i) there are only finitely many of them and (ii) all $N(\Omega) < \infty$ of them have the same probability. One might refer to (ii) as the "principle of complete ignorance". Of course, that the probabilities are equal in any particular setting can never be precisely verified, although one could *test* such an hypothesis by applying statistical techniques to be described in later chapters. In the absence of prior experience, however, the symmetry that characterizes games of chance and statistical sampling experiments often makes us comfortable in *assuming* that outcomes are equally likely—once they are appropriately defined. For example, it is customary to assume that the two faces of a coin have the same chance of turning up on each flip, that each card has the same chance of being drawn from a well shuffled deck of 52, that each side of a 6-sided die has the same chance of being shown, that each of the possible samples of size n without replacement from a population of size N has the same chance of being drawn, and so forth.

Finding the probabilities of events in the equally-likely setting is just a matter of counting. If there are $N(\Omega)$ possible outcomes $\{\omega_j\}_{j=1}^{N(\Omega)}$, and if $\mathbb{P}(\{\omega_1\}) = \mathbb{P}(\{\omega_2\}) = \cdots = \mathbb{P}(\{\omega_{N(\Omega)}\})$, then the additivity axiom implies that each outcome has probability $\mathbb{P}(\{\omega_1\}) = 1/N(\Omega)$; i.e.,

$$1 = \mathbb{P}(\Omega) = \mathbb{P}\left(\bigcup_{j=1}^{N(\Omega)} \omega_j\right) = \sum_{j=1}^{N(\Omega)} \mathbb{P}(\{\omega_j\}) = N(\Omega)\mathbb{P}(\{\omega_1\}).$$

Taking the domain of $\mathbb{P}$ to be the field $\mathcal{F}$ comprising all subsets of the finite Ω, we complete the definition of the measure by putting

$$\mathbb{P}(A) = \mathbb{P}\left(\bigcup_{\omega \in A} \omega\right) = \sum_{j:\omega_j \in A} \mathbb{P}(\{\omega_j\}) = \sum_{j=1}^{N(A)} \mathbb{P}(\{\omega_1\}) = \frac{N(A)}{N(\Omega)}$$

for any $A \in \mathcal{F}$.

Exercise 1.12. Show that $\mathbb{P}$ as thus defined satisfies Kolmogorov's three axioms for probability measures.

Example 1.9. The experiment involves flipping two fair coins. We care only about the number of heads. While $\{0, 1, 2\}$ would suffice to describe the sample space, considerations of symmetry warn us that these outcomes are not equally likely, since $\omega = 0$ and $\omega = 2$ both require the same face on both coins, whereas $\omega = 1$ corresponds to either HT or TH. We choose instead the more refined description $\Omega = \{TT, TH, HT, HH\}$, whereupon the probability of zero heads is $\mathbb{P}(\{TT\}) = \frac{1}{4}$, of one head is $\mathbb{P}(\{TH, HT\}) = \frac{2}{4}$, and of two heads is $\mathbb{P}(\{HH\}) = \frac{1}{4}$.

Example 1.10. The experiment involves flipping 100 coins, and our interest is just in the number of heads. Again, it is not reasonable to treat each of $\{0, 1, 2, ..., 100\}$ as equally likely. Instead, take Ω to be the set of all possible strings of 100 H's and T's. In principle, one could enumerate all the possibilities and simply count how many strings contain each number of heads of interest.

Example 1.11. The experiment involves choosing at random four of 100 people to receive first, second, third, and fourth prizes. We want to know the chance that Lillie, Madeline, Susie, and Tasha will win the prizes in that particular order, and also the probability that they would win in *any* order. For this let Ω comprise all the ordered subsets of four taken without replacement from a set of 100. The probability that Lillie, Madeline, Susie, and Tasha win in that order would then be the inverse of $N(\Omega)$, while the chance that the four win in any order whatever would $N(\Omega)^{-1}$ multiplied by the number of arrangements of the four individuals.

Example 1.12. The experiment involves dealing (drawing without replacement) five cards from a deck of 52. Our interest is in various poker hands—that is, in the composition of the five cards by suit and denomination but without regard to the order in which the cards were dealt. To know the chance of getting precisely three aces (in any order) we would need to be able to count the number of subsets of five taken from 52 that contained just that many such cards.

As the examples illustrate, the task of determining probabilities even in finite experiments can easily get out of hand, since the relevant sample spaces are often very large. Some shortcuts are needed to assist in counting, and here they are.

1.3.2.1 *Counting Rules*

Much can be done with just three rules. All three flow from the following elementary principle of enumeration: If operations A and B can be performed independently in a ways and b ways, respectively, then the two together can be performed in $a \cdot b$ ways.

(i) **Permutations with replacement.** We draw k objects from n objects, replacing each one after it is drawn so that it can be drawn again. In how many ways can this be done if we distinguish different arrangements of the k objects? Consider that there are k empty bins or slots, as

The first slot can be filled in n ways. For each of these ways, the second can be filled in n ways also, giving $n \cdot n = n^2$ ways of filling the first two bins. Continuing, there are n^k ways of filling all k bins. **Examples:** There are 2^{100} ordered sequences of H's and T's in 100 flips of a coin. There are 2^n possible subsets of n objects, since a subset is determined by answering the question "Included, or not?" for each item. There are 52^5 possible sequences of five cards drawn from a deck *with* replacement.

(ii) **Permutations without replacement.** Suppose we draw k objects from n objects but do *not* replace each one after it is drawn. In how many ways can this be done if we continue to distinguish different arrangements of the k objects? Obviously, we must now have $k \le n$. Again consider the k bins. The first can be filled in n ways, as before, but the second can be filled in only $n-1$ ways, the third in $n-2$ ways, and so forth, there being $n-k+1$ ways for the last bin. Thus, the number of permutations is

$$P_k^n = n(n-1) \cdot \ \cdots \ \cdot (n-k+1).$$

In particular, there are $P_n^n = n!$ ways of *arranging* n objects. For $k \in \{0, 1, 2, ..., n-1\}$ there is the more compact expression

$$P_k^n = \frac{n!}{(n-k)!}.$$

This extends to include $k = n$ upon defining $0! := 1$. **Examples:** There are $100 \cdot 99 \cdot 98$ ways to assign prizes to three of 100 people in a particular order, and there are $P_{100}^{100} = 100!$ ways that 100 people could line up.

(iii) **Combinations.** We draw k objects from n without replacement but do not wish to distinguish different arrangements of a given set of k. The number of such combinations, represented as C_k^n or $\binom{n}{k}$ (read "n

choose k"), is given by the number of permutations without replacement *divided by* the number of ways of rearranging the k objects:

$$\binom{n}{k} = \frac{P_k^n}{P_k^k} = \frac{P_k^n}{k!} = \frac{n!}{k!(n-k)!}.$$

Note that $\binom{n}{k} \equiv \binom{n}{n-k}$ since $n - k$ objects are left behind when k are selected; also, the definition $0! := 1$ gives $\binom{n}{0} = \binom{n}{n} = 1$. Thus, there are $\binom{52}{5}$ possible poker hands, and there are $\binom{100}{3}$ ways of assigning prizes to three of 100 people if we do not keep track of who wins what.

These rules make it easy to estimate the probabilities of some very complicated events.

Example 1.13. Flipping 100 coins and considering the 2^{100} head/tail sequences to be equally likely, the probability of obtaining $j \in \{0, 1, ..., 100\}$ heads is the number of arrangements of j heads and $100 - j$ tails in a sequence of 100 H's and T's, divided by 2^{100}. Since an arrangement is determined by choosing j of the 100 bins in which to place H's, the probability is $\binom{100}{j}/2^{100}$.

Example 1.14. The probability of getting *at least* $m \in \{0, 1, ..., 100\}$ heads in 100 coins is $\sum_{j=m}^{100} \binom{100}{j}/2^{100}$. When $m < 50$, this is easier to find via the complement rule, as $1 - \sum_{j=0}^{m-1} \binom{100}{j}/2^{100}$.

Example 1.15. The event of getting precisely three aces in a five-card poker hand involves getting the three aces and then getting two other cards that are not aces. There are $\binom{4}{3}$ ways of getting three aces from among the four in the deck, $\binom{48}{2}$ ways of getting the other two cards, and $\binom{4}{3}\binom{48}{2}$ ways that both things can happen. Since there are $\binom{52}{5}$ possible 5-card hands, the probability is $\binom{4}{3}\binom{48}{2}/\binom{52}{5}$. The probability of getting *at least* three aces is $\binom{4}{3}\binom{48}{2}/\binom{52}{5} + \binom{4}{4}\binom{48}{1}/\binom{52}{5}$.

Example 1.16. The probability that any particular individual will be included in a committee of 10 drawn at random from 30 individuals is $\binom{1}{1}\binom{29}{9}/\binom{30}{10}$.

Exercise 1.13. *Two-person teams are to be formed by choosing a person at random from one of two groups and pairing with someone chosen at random from the second group. If each group consists of five men and five women, what is the probability that all ten teams will comprise individuals of the same sex—that is, that all are of the form MM or WW?*

1.4 Conditional Probability and Independence

1.4.1 *Probabilities with Partial Information*

Logically, one's assessment of the probability of an event depends on what is already known about the outcome of the chance experiment. If the experiment is yet to be performed, then nothing at all is known about what will happen apart from what is possible and what is not.[12] In this circumstance one's information consists of the *trivial* field $\mathcal{F}_0 := \{\Omega, \varnothing\}$. One has this "information" in the sense that one can give definite "yes" or "no" answers to the questions "Will ω turn out to be in Ω?" and "Will ω be in $\varnothing$?" In some cases, however, one has (or can envision having) more information about the outcome of the experiment, and when that happens the information can sometimes be used to advantage in sharpening estimates of probabilities of certain events. These sharper probabilities are called *conditional* probabilities.

To look at a concrete example, consider drawing one person at random from a classroom of students, the interest being in whether the person drawn has a particular characteristic, such as occupying a desk in the first row, having brown hair, having a name containing a particular letter, etc. In other words, the interest is in the *event* that the person drawn will be from a set of students having that particular characteristic. Ω is now the group of students, and the probability space is $(\Omega, \mathcal{F}, \mathbb{P})$, where $\mathcal{F}$ comprises all subsets of Ω and $\mathbb{P}(\{\omega\}) = 1/N(\Omega)$ for each outcome ω. *Ex ante* (before the experiment has been performed), we know only that some member of the class will be drawn and that no one *not* in the class *can* be drawn— the trivial information just described. Knowing only this, an event $A \in \mathcal{F}$ would be assigned probability $\mathbb{P}(A) = N(A)/N(\Omega)$. Although not usually done except for emphasis, this *could* also be represented as $\mathbb{P}(A \mid \Omega)$ or as $\mathbb{P}(A \mid \mathcal{F}_0)$ where the mark "|" indicates "conditioning". These symbols would be rendered in words as "the probability of A *given* that the outcome is in Ω" or as "the probability of A given trivial information". At the other extreme, suppose the draw had already been made and that the identity of the person drawn was known. In that case for *each* of the sets in $\mathcal{F}$ we could give a definite "yes" or "no" answer to the question "Does ω lie in this set?" This is the case of *full* information about the experiment. Clearly, given

---

[12]This, by the way, illustrates one of the pitfalls of applying the axiomatic theory of probability to "experiments of nature": Things sometimes happen whose possibility was entirely unforeseen.

full information, we would assign to a set A a probability of either 0 or 1, depending on whether or not the known ω was in A. That is, $\mathbb{P}(A \mid \omega) = 0$ if $\omega \in A^c$ and $\mathbb{P}(A \mid \omega) = 1$ if $\omega \in A$. To represent probabilities based on full information without having to specify a particular ω, we would write just $\mathbb{P}(A \mid \mathcal{F})$. Think of this as representing the *collection* of probabilities of A conditional on each of the sets that *generate* $\mathcal{F}$.

Now, taking an intermediate case, suppose that there is *partial* information about the outcome. For example, suppose we know whether or not the last name of the person drawn contains the letter b. Let B and B^c represent the events "contains b" and "does not contain b". The information set can now be identified with the sub-σ field $\mathcal{F}_B := \{\Omega, B, B^c, \varnothing\}$ of $\mathcal{F}$ that is *generated* by B. That is, we can say for sure whether ω lies in each of these four sets but cannot answer such a question for any other set. Given this partial information, how would one determine the probability that ω lies in an arbitrary set $A \in \mathcal{F}$?

Consider the probability of A given the occurrence of B, $\mathbb{P}(A \mid B)$. Thinking of the equally-likely principle and regarding B now as the relevant sample space (since ω has come from B), we would judge $\mathbb{P}(A \mid B)$ to be the number of outcomes in B that are also in A, divided by the total number of outcomes in B. For example, if A is the set of individuals whose last name contains the letter a and if there are 11 individuals whose name contains both a and b out of 20 individuals whose name contains b, then $\mathbb{P}(A \mid B) = N(A \cap B)/N(B) = 11/20$. Thus, taking $\mathbb{P}(A \mid B) = N(A \cap B)/N(B)$ as the intuitive answer, let us verify that the set function $\mathbb{P}(\cdot \mid B) : \mathcal{F} \to [0,1]$ is a probability measure on $(\Omega, \mathcal{F})$. This just requires checking that $\mathbb{P}(\cdot \mid B)$ satisfies Kolmogorov's axioms. First, $\mathbb{P}(\Omega \mid B) = N(\Omega \cap B)/N(B) = N(B)/N(B) = 1$. Next, $\mathbb{P}(A \mid B) \geq 0$ for each $A \in \mathcal{F}$, since $N(A \cap B) \geq 0$ and $N(B) > 0$. Finally, for any collection $\{A_j\}$ of exclusive events the additivity of counting measure assures that

$$\mathbb{P}\left(\bigcup_j A_j \mid B\right) = \frac{N\left[\bigcup_j (A_j \cap B)\right]}{N(B)} = \sum_j \frac{N(A_j \cap B)}{N(B)} = \sum_j \mathbb{P}(A_j \mid B).$$

(1.3)

Of course, none of this makes sense unless B is an event with at least one outcome. Likewise, if $N(B^c) > 0$ we would take $\mathbb{P}(A \mid B^c) = N(A \cap B^c)/N(B^c)$. The more general symbol $\mathbb{P}(A \mid \mathcal{F}_B)$ represents the probability of A given the information $\mathcal{F}_B$—that is, the collection of probabilities of A conditional on the events in the field generated by B.

So far so good, but how can this idea be extended to cases in which Ω

is infinite or is finite but with outcomes having different probabilities? We desire a definition that covers the general case yet still delivers the result intuition demands when Ω is finite and outcomes are equally likely.

Definition 1.12.

(i) Given a probability space $(\Omega, \mathcal{F}, \mathbb{P})$ and an event B with $\mathbb{P}(B) > 0$, the conditional probability of $A \in \mathcal{F}$ given the event B is

$$\mathbb{P}(A \mid B) = \frac{\mathbb{P}(A \cap B)}{\mathbb{P}(B)}.$$

(ii) The triple $(\Omega, \mathcal{F}, \mathbb{P}(\cdot \mid B))$ is the conditional probability space given the event B.

(iii) Given any sub-σ field $\mathcal{G}$ of $\mathcal{F}$, $\mathbb{P}(\cdot \mid \mathcal{G})$ represents the collection of probability measures that result from conditioning on the non-null events that generate $\mathcal{G}$.

There are three points to note about this definition. First, verifying that $\mathbb{P}(\cdot \mid B)$ satisfies Kolmogorov's axioms shows that this is indeed a legitimate probability measure for any $B \in \mathcal{F}$ such that $\mathbb{P}(B) > 0$. Second, since $\mathbb{P}(A \cap B) = N(A \cap B)/N(\Omega)$ and $\mathbb{P}(B) = N(B)/N(\Omega)$ in the equally-likely case, it does deliver the intuitive result (1.3) when that case applies. Finally, although we have not defined $\mathbb{P}(A \mid B)$ when $\mathbb{P}(B) = 0$, we do sometimes want to extend the definition to cover such cases. This can be done by setting

$$\mathbb{P}(A \mid B) = \lim_{n \to \infty} \frac{\mathbb{P}(A \cap B_n)}{\mathbb{P}(B_n)},$$

provided such limit exists (and is the same) for any sequence of non-null events $\{B_n\}_{n=1}^{\infty}$ converging down to B. The relevance of this situation will become clear in Section 2.4.4 when we discuss conditional distributions of continuous random variables.

1.4.2 *Applications of Conditional Probability*

Probabilities of complicated events can often be deduced more easily by conditioning on other events. One way this can be done is to make use of the *multiplication rule* for probabilities, which we adjoin to the list of properties in Section 1.3.1:

16. Multiplication rule: $\mathbb{P}(A \cap B) = \mathbb{P}(A \mid B)\mathbb{P}(B)$ if $\mathbb{P}(B) > 0$; otherwise $\mathbb{P}(A \cap B) = 0$.

Proof: Immediate from definition of $\mathbb{P}(A \mid B)$ when $\mathbb{P}(B) > 0$; otherwise, follows from $0 \leq \mathbb{P}(A \cap B) \leq \mathbb{P}(B)$.

Example 1.17. Two cards are dealt from a deck of 52. The probability of getting two aces can be expressed directly as $\binom{4}{2}/\binom{52}{2} = \frac{3}{51} \cdot \frac{4}{52}$. Alternatively, we can think of the experiment as having two steps: Draw one card from the 52, then draw another from the remaining 51. Writing $A_1 \cap A_2$ as the event of ace on first draw and ace on the second, there are two ways of expressing the probability via the multiplication rule: (i) $\mathbb{P}(A_1 \cap A_2) = \mathbb{P}(A_1 \mid A_2)\mathbb{P}(A_2)$ and (ii) $\mathbb{P}(A_1 \cap A_2) = \mathbb{P}(A_2 \mid A_1)\mathbb{P}(A_1)$. Way (i) is not very helpful, since it is not clear without further analysis how to determine either of the factors. But (ii) provides an immediate answer that exploits the simplicity of the sample space in each step of the experiment and thereby makes the counting trivial: $\mathbb{P}(A_1 \cap A_2) = \frac{3}{51} \cdot \frac{4}{52}$.

Example 1.18. Dealing three cards, the event of three aces is found directly as $\binom{4}{3}/\binom{52}{3}$ or from the multiplication rule as

$$\mathbb{P}(A_1 \cap A_2 \cap A_3) = \mathbb{P}((A_1 \cap A_2) \cap A_3)$$
$$= \mathbb{P}(A_3 \mid A_1 \cap A_2)\,\mathbb{P}(A_1 \cap A_2)$$
$$= \mathbb{P}(A_3 \mid A_1 \cap A_2)\,\mathbb{P}(A_2 \mid A_1)\mathbb{P}(A_1)$$
$$= \frac{2}{50} \cdot \frac{3}{51} \cdot \frac{4}{52}.$$

Another application of conditioning comes with what we shall call "part 2" of the law of total probability. This is useful when it is hard to find the probability of some event directly but when it is easy to find the probabilities conditional on events that partition Ω. (Recall that events $\{A_j\}$ form a partition if they are exclusive and if $\cup A_j = \Omega$.)

17. Law of total probability (part 2): If non-null events $\{A_j\}$ form a partition, then for any event B

$$\mathbb{P}(B) = \sum_j \mathbb{P}(B \cap A_j) = \sum_j \mathbb{P}(B \mid A_j)\mathbb{P}(A_j).$$

Proof: Express each $\mathbb{P}(B \cap A_j)$ via the multiplication rule.

Example 1.19. Two cards are dealt from a normal deck. What is the probability that the *second* card is an ace? With A_1 and A_2 as the events of ace on the first and second draws, we want $\mathbb{P}(A_2)$. Applying the law of total probability, partition the sample space (all the 52^2 ordered pairs of cards) as A_1 and A_1^c. Then $\mathbb{P}(A_2) = \mathbb{P}(A_2 \mid A_1)\mathbb{P}(A_1) + \mathbb{P}(A_2 \mid A_1^c)\mathbb{P}(A_1^c) =$

$\frac{3}{51} \cdot \frac{4}{52} + \frac{4}{51} \cdot \frac{48}{52} = \frac{4}{52}$. (Having now arrived at the conclusion $\mathbb{P}(A_2) = \mathbb{P}(A_1)$, we realize that we might have seen this intuitively. Why should the order of the two cards matter? If they were dealt and laid face down, we could just switch them without changing the chance that either was an ace.)

The following identity, which provides a way of conditioning in reverse, is really trivial; yet it is fundamental to the Bayesian approach to statistics and is often useful in its own right. (Bayesian methods are surveyed in Chapter 10.)

18. Bayes' rule: If A is a "nontrivial" event (one such that $0 < \mathbb{P}(A) < 1$) and if B is not $\mathbb{P}$-null, then

$$\mathbb{P}(A \mid B) = \frac{\mathbb{P}(B \mid A)\,\mathbb{P}(A)}{\mathbb{P}(B)} = \frac{\mathbb{P}(B \mid A)\,\mathbb{P}(A)}{\mathbb{P}(B \mid A)\,\mathbb{P}(A) + \mathbb{P}(B \mid A^c)\,\mathbb{P}(A^c)}.$$

Proof: The numerator equals $P(A \cap B)$ by the multiplication rule, and the denominator in the last expression equals $\mathbb{P}(B)$ by the law of total probability.

Example 1.20. Continuing the example of two cards dealt from a deck of 52, the probability of ace on the first draw given ace on the second is

$$\mathbb{P}(A_1 \mid A_2) = \frac{\mathbb{P}(A_2 \mid A_1)\,\mathbb{P}(A_1)}{\mathbb{P}(A_2 \mid A_1)\mathbb{P}(A_1) + \mathbb{P}(A_2 \mid A_1^c)\mathbb{P}(A_1^c)}$$

$$= \frac{3}{51}.$$

Exercise 1.14. *According to one estimate, left-handed people account for about 10% of the population and about 20% of schizophrenics. Thus, ignoring the tiny proportion of people who are even handed (ambidextrous), the probability of selecting a left-hander in a random draw of schizophrenics is about one-fourth the probability of selecting a right-hander—i.e.,* $\mathbb{P}(L \mid S)/\mathbb{P}(R \mid S) = .2/.8$.[13]
(i) Now if we choose at random from among left-handers and again at random among right-handers, there will be some probability of choosing a person with schizophrenia in each experiment. What is the ratio of these probabilities—i.e., $\mathbb{P}(S \mid L)/\mathbb{P}(S \mid R)$*?*
(ii) If $\mathbb{P}(S) = .04$ *what are the individual probabilities,* $\mathbb{P}(S \mid L)$ *and* $\mathbb{P}(S \mid R)$*?*

---

[13]Sources for the rough figures are Wang (2011) and Saha *et. al* (2005).

While conditioning is a powerful tool, applying it correctly is sometimes rather tricky, as the following example shows.[14]

Example 1.21. A contestant in a TV game show is told that a valuable prize lies behind one of three closed doors marked A, B, C; that a "booby prize" in the form of a goat is behind another door; and that the third door has nothing at all. The contestant chooses one door. The game master then walks to one of the other doors not chosen and opens it to reveal the goat (placed there by an assistant behind the scenes after the contestant's choice was announced). The contestant, able now to eliminate one door as the location of the prize, is given the option to switch from the first choice to the remaining unopened door. For example, if A were chosen first and the goat revealed at C, the contestant could either stick with A or switch to B. The question is, "Does making the switch increase the probability of winning?"

Our first thought may be that the new information is irrelevant to the choice. Thinking that it amounts to C^c (prize *not* behind door C), we might reason (correctly) that $\mathbb{P}(A \mid C^c) = \mathbb{P}(B \mid C^c) = \frac{1}{2}$ and conclude that there is no advantage to switching. The fallacy is that the opening of door C actually conveys *sharper* information than C^c; that is, it limits the outcomes to a set, C^R, say, where $C^R \subset C^c$ and $C^c \backslash C^R \neq \varnothing$. This can be understood intuitively as follows. If the prize were really behind door A, the contestant's first choice, then the goat could be taken either to B or to C, indifferently; whereas if the prize were not behind A then the goat would *have* to be taken to the only remaining door. Thus, in all replications of the experiment in which the contestant does not *initially* choose the right door, the game master is implicitly revealing the location of the prize by opening the only remaining door that does not contain it. We can see how this affects the contestant's chances as follows. In $\frac{1}{3}$ of repeated games the first guess will have been right. A contestant who sticks with the first choice will therefore win $\frac{1}{3}$ of the time. But in $\frac{2}{3}$ of such plays the prize will be behind one of the two other doors and therefore behind the one *not* opened. Thus, in $\frac{2}{3}$ of all games switching makes one *certain* of gaining the prize, so switching each time raises the chance of winning from $\frac{1}{3}$ to $\frac{2}{3}$.

---

[14]People's off-hand answers to this problem divide them up in an interesting way. Those who are exceptionally bright tend to get it right, whether or not they know anything about probability theory. Among the rest, those who are entirely ignorant of theory often seem to have an intuitive sense of the right answer, which often elicits some arrogant ridicule from those who know a little.

To show this formally, let us calculate and compare $\mathbb{P}(A \mid C^R)$ and $\mathbb{P}(B \mid C^R)$, assuming that the first choice was A. (Clearly, $\mathbb{P}(C \mid C^R) = \mathbb{P}(C^R \mid C) = 0$, since the prize is never revealed.) First, Bayes' rule gives

$$\mathbb{P}(A \mid C^R) = \frac{\mathbb{P}(C^R \mid A)\mathbb{P}(A)}{\mathbb{P}(C^R)}.$$

The values in the numerator are $\mathbb{P}(A) = \frac{1}{3}$ and $\mathbb{P}(C^R \mid A) = \frac{1}{2}$, since if the prize is behind A the game master can choose at random which of B or C to open. The denominator comes via the law of total probability:

$$\mathbb{P}(C^R) = \mathbb{P}(C^R \mid A)\mathbb{P}(A) + \mathbb{P}(C^R \mid B)\mathbb{P}(B) + \mathbb{P}(C^R \mid C)\mathbb{P}(C)$$
$$= \frac{1}{2} \cdot \frac{1}{3} + 1 \cdot \frac{1}{3} + 0 \cdot \frac{1}{3} = \frac{1}{2}.$$

Thus, $\mathbb{P}(A \mid C^R) = \left(\frac{1}{2} \cdot \frac{1}{3} \right) \div \frac{1}{2} = \frac{1}{3}$ as deduced intuitively, whereas

$$\mathbb{P}(B \mid C^R) = \frac{\mathbb{P}(C^R \mid B)\mathbb{P}(B)}{\mathbb{P}(C^R)} = \left(1 \cdot \frac{1}{3} \right) \div \frac{1}{2} = \frac{2}{3}.$$

Notice that $\mathbb{P}(C^R) < \mathbb{P}(C^c) = \frac{2}{3}$ together with $C^R \cap C = \varnothing$ show that C^R is in fact a proper subset of C^c. The revelation does indeed narrow the set of possible outcomes relative to knowing just that the prize is not behind C.[15]

Exercise 1.15. *Among individuals who confront the problem in Example 1.21 about 80% of those in the top decile by intelligence get the right answer, whether or not they have had any formal instruction in probability theory. Of those in the bottom nine deciles by intelligence, about 90% have had no formal instruction in probability, yet still about 80% of this group get the right answer. Remarkably, those in the bottom nine deciles who do know a little probability almost never get it right. What is the probability that someone who does get the right answer ranks in the top decile by intelligence?*

1.4.3 *Stochastic Independence*

While partial information *can* sharpen estimates of probabilities, it does not *necessarily* do so; that is, sometimes conditioning information does turn out to be irrelevant. Suppose two cards are drawn from a deck of 52 *with* replacement. The probability of two aces can be calculated directly

---

[15]Another way to see this: Although C^R does imply (is a subset of) C^c, event C^c does *not* imply C^R since door B could instead have been opened if the prize were at A.

using counting rules for permutations with replacement. Thus, there are 4^2 ways to draw two aces (to fill two bins with aces) and 52^2 ways to draw two cards without regard to denomination. Therefore, $\mathbb{P}(A_1 \cap A_2) = \frac{4 \cdot 4}{52 \cdot 52} = \mathbb{P}(A_1) \cdot \mathbb{P}(A_2)$. In this case $\mathbb{P}(A_2 \mid A_1) = \frac{4^2}{52^2} / \frac{4}{52} = \frac{4}{52} = \mathbb{P}(A_2)$ and likewise $\mathbb{P}(A_1 \mid A_2) = \mathbb{P}(A_1)$. Intuitively, if the draws are with replacement, what happens on the first draw has no bearing on what happens on the second, and *vice versa*. Events with this property are said to be *stochastically independent* or simply *independent*, a term defined formally as follows:

Definition 1.13. Events A, B on a probability space $(\Omega, \mathcal{F}, \mathbb{P})$ are independent if and only if $0 < \mathbb{P}(A \cap B) = \mathbb{P}(A)\mathbb{P}(B) < \min\{\mathbb{P}(A), \mathbb{P}(B)\}$.

The first inequality restricts the definition to events of positive probability, which together with the equality implies that $\mathbb{P}(B \mid A) = \mathbb{P}(B)$ and that $\mathbb{P}(A \mid B) = \mathbb{P}(A)$ when A and B are independent. The second inequality restricts to events of probability less than unity, which implies that their *complements* are not $\mathbb{P}$-null. In other words, the definition applies to *nontrivial* events.[16] The intuition here is at least as important as the formalism. If events are independent, then the occurrence of one does not change the assessment of the probability of the other. Thus, what turns up on one draw with replacement from a deck of cards should not affect what happens subsequently; getting heads on one flip of a coin should not change the odds of getting heads on the next flip; and so forth. This intuitive sense often makes us comfortable in *assuming* independence of events in certain experiments. Doing so usually simplifies our calculations, since

---

[16]If either $\mathbb{P}(A) = 0$ or $\mathbb{P}(B) = 0$, it is clear (subset inequality!) that $\mathbb{P}(A \cap B) = 0$. Thus, to include zero-probability (null) events in the definition would require us to accept that any zero-probability event is independent of all other events in $\mathcal{F}$. But this would often conflict with our intuitive notion of the concept. For example, reflect on the experiment depicted in Fig. 1.1, where a pointer on a circular scale of unit circumference is to be spun and $\mathbb{P} = \lambda$ (Lebesgue measure). Let $B = \{.1\}$ (the single point at a clockwise distance 0.1 from the origin), and take $A = [0, .2]$. We would certainly not want to think of these events as independent, since knowing that B has occurred changes our assessment of the probability of A from 0.2 to

$$\lim_{n \to \infty} \frac{\mathbb{P}(A \cap (.1 - 1/n, .1])}{\mathbb{P}((.1 - 1/n, .1])} = \lim_{n \to \infty} \frac{\mathbb{P}((.1 - 1/n, .1])}{\mathbb{P}((.1 - 1/n, .1])} = 1,$$

yet $\mathbb{P}(A \cap B) = \mathbb{P}(B) = 0$. Excluding zero-probability (null) events from the definition is costless, because the importance of independence is to facilitate calculation of probabilities of the joint occurrence of two or more events, and the answer is obvious when one or more of these is null.

Likewise, the restriction $\mathbb{P}(A) < 1, \mathbb{P}(B) < 1$ guarantees that the complements are not null. The significance of this is made apparent by Exercise 1.16.

when nontrivial events A, B are independent, we have an immediate way to find the *joint* probability: $\mathbb{P}(A \cap B) = \mathbb{P}(A)\mathbb{P}(B)$. (Of course, if either of them *is* trivial, then the answer is the same.)

Exercise 1.16. *If A, B are independent, show that A^c, B are independent and that A^c, B^c are independent. Explain the intuition.*

Exercise 1.17. *What is the connection between independence and exclusivity? Are independent events always exclusive? Are they ever exclusive? What is the sense of it?*

Exercise 1.18. *Events A and B are independent, and event C is exclusive of both. If $\mathbb{P}(A) = .3$, $\mathbb{P}(B) = .4$, and $\mathbb{P}(C) = .2$ find (i) $\mathbb{P}(A^c \cap B^c)$; (ii) $\mathbb{P}(A^c \cup B^c)$; (iii) $\mathbb{P}(A \backslash B)$; (iv) $\mathbb{P}((A \backslash B) \cup (B \backslash A))$; (v) $\mathbb{P}((A \backslash B) \cup (A^c \backslash B))$; (vi) $\mathbb{P}(A \backslash B^c)$; (vii) $\mathbb{P}(A \cup B \cup C)$; (viii) $\mathbb{P}(C^c \backslash (A \cup B))$; (ix) $\mathbb{P}(A \cup B \cup C^c)$; (x) $\mathbb{P}(A^c \cup B^c \cup C^c)$.*

The concept of independence extends to more than two events, but the formal condition is a little more complicated.

Definition 1.14. Nontrivial events $A_1, A_2, \dots$ are *mutually* independent if and only if for *each* finite subcollection $A_{i_1}, A_{i_2}, \dots, A_{i_k}$ of $k \geq 2$ distinct events

$$\mathbb{P}(A_{i_1} \cap A_{i_2} \cap \cdots \cap A_{i_k}) = \mathbb{P}(A_{i_1})\mathbb{P}(A_{i_2}) \cdot \cdots \cdot \mathbb{P}(A_{i_k}).$$

In words, nontrivial events are independent when the probability of the joint occurrence of *any* subcollection of two or more equals the product of their probabilities. Mutual independence clearly implies, but is stronger than, independence between all pairs. Again, the utility of this concept relies on having the right intuition about conditions for independence. If we are *sure* that a 6-sided die is "fair"—meaning that the probability of turning up any one side is $1/6$ each time the die is tossed, then we can assert that the probability of 10 successive rolls of 6's is $1/6^{10}$ and that the *conditional* probability of a 6 on the 11th roll given 10 successive 6's is still

$$
\begin{aligned}
\mathbb{P}(6_{11} \mid 6_1 \cap \cdots \cap 6_{10}) &= \frac{\mathbb{P}(6_1 \cap \cdots \cap 6_{10} \cap 6_{11})}{\mathbb{P}(6_1 \cap \cdots \cap 6_{10})} \\
&= \frac{\mathbb{P}(6_1) \cdot \cdots \cdot \mathbb{P}(6_{10})\mathbb{P}(6_{11})}{\mathbb{P}(6_1) \cdot \cdots \cdot \mathbb{P}(6_{10})} \\
&= \mathbb{P}(6_{11}) = \frac{1}{6}.
\end{aligned}
$$

Recall that the convergence part of the Borel–Cantelli lemma gives a sufficient condition that infinitely many of a sequence of events occur with probability zero; namely, that $\sum_{n=1}^{\infty} \mathbb{P}(A_n) < \infty$ implies $\mathbb{P}(\limsup_n A_n) =: \mathbb{P}(A_n \text{ i.o.}) = 0$. This is true whether or not the events in the sequence are mutually independent. In the independence case, however, a special result applies, known as the *divergence* part of Borel–Cantelli. We add this to our list of fundamental properties of $\mathbb{P}$ measure.

19. If events $\{A_n\}_{n=1}^{\infty}$ are mutually independent and $\sum_{n=1}^{\infty} \mathbb{P}(A_n) = \infty$, then $\mathbb{P}(\limsup_n A_n) = 1$.

    Proof: Applying in succession the complement rule, De Morgan's law, and Boole's inequality gives for $\mathbb{P}(\limsup_n A_n)$

$$1 - \mathbb{P}\left[\left(\limsup_n A_n\right)^c\right] = 1 - \mathbb{P}(\liminf_n A_n^c)$$

$$= 1 - \mathbb{P}\left(\bigcup_{n=1}^{\infty} \bigcap_{m=n}^{\infty} A_m^c\right)$$

$$\geq 1 - \sum_{n=1}^{\infty} \mathbb{P}\left(\bigcap_{m=n}^{\infty} A_m^c\right).$$

Since sequence $\left\{\bigcap_{m=n}^{N} A_m^c\right\}_{N=n}^{\infty}$ is monotone decreasing, the monotone property of probability implies

$$\mathbb{P}\left(\bigcap_{m=n}^{\infty} A_m^c\right) = \mathbb{P}\left(\lim_{N\to\infty} \bigcap_{m=n}^{N} A_m^c\right) = \lim_{N\to\infty} \mathbb{P}\left(\bigcap_{m=n}^{N} A_m^c\right).$$

The independence of the $\{A_n\}$ (and hence of the complements) now gives

$$\mathbb{P}(\limsup_n A_n) \geq 1 - \sum_{n=1}^{\infty} \lim_{N\to\infty} \prod_{m=n}^{N} \mathbb{P}(A_m^c)$$

$$= 1 - \sum_{n=1}^{\infty} \lim_{N\to\infty} \prod_{m=n}^{N} [1 - \mathbb{P}(A_m)]$$

$$\geq 1 - \sum_{n=1}^{\infty} \lim_{N\to\infty} \prod_{m=n}^{N} \exp[-\mathbb{P}(A_m)]$$

$$= 1 - \sum_{n=1}^{\infty} \exp\left[-\sum_{m=n}^{\infty} \mathbb{P}(A_n)\right]$$

$$= 1,$$

where in the third line we have used the fact that $e^{-x} \geq 1 - x$ for any real x, and where the last inequality follows from the divergence of $\sum_{n=1}^{\infty} \mathbb{P}(A_n)$. Since no event has probability exceeding unity, the conclusion follows.

Example 1.22. An experiment consists of draws from a sequence of urns. Urn n, for $n = 1, 2, ...$, contains one red ball and $n - 1$ white ones. Let R_n be the event of obtaining a red ball from the nth urn. Assume that the events $\{R_n\}_{n=1}^{\infty}$ are independent and that $\mathbb{P}(R_n) = \frac{1}{n}$. Since the sum $\sum_{n=1}^{\infty} \frac{1}{n}$ does not converge, one is "almost" sure to draw infinitely many red balls if the experiment is carried on indefinitely. Stated differently, there will never be a point beyond which all remaining draws are almost certain to yield white.

Exercise 1.19. *Find the flaw in the following argument. Define a sequence of events $\{A_n\}$ such that $A_n = A$ for each n, where $0 < \mathbb{P}(A) < 1$. Then $\sum_{n=1}^{\infty} \mathbb{P}(A_n) = +\infty$, yet $\mathbb{P}(A_n \; i.o.) = \mathbb{P}(A) < 1$, disproving Borel–Cantelli.*

1.5 Solutions to Exercises

1. $\lim_{n\to\infty} B_n = \varnothing$. $\lim_{n\to\infty} A_n = (0, 1]$.
2. Since the intersection of any two sets is contained in both of them, $B_n = \cap_{m=n}^{\infty} A_m = A_n \cap_{m=n+1}^{\infty} A_m \subset \cap_{m=n+1}^{\infty} A_m = B_{n+1}$, so that $\{B_n\} \uparrow$. Since any set is a subset of its union with any other set, $C_{n+1} = \cup_{m=n+1}^{\infty} A_m \subset A_n \cup (\cup_{m=n+1}^{\infty} A_m) = C_n$, so that $\{C_n\} \downarrow$.
3. (i) If $\{A_n\} \uparrow A$ then $\cap_{m=n}^{\infty} A_m = A_n$, so $\liminf_n A_n = \cup_{n=1}^{\infty} A_n = A$ by the definition of limits for monotone sequences. For the same reason $\cup_{m=n}^{\infty} A_m = A$, so $\limsup_n A_n = \cap_{n=1}^{\infty} A = A$.
   (ii) If $\{A_n\} \downarrow A$ then $\cap_{m=n}^{\infty} A_m = A$ by the definition of limits for monotone sequences, so $\liminf_n A_n = \cup_{n=1}^{\infty} A = A$. Likewise, if $\{A_n\} \downarrow$ then $\cup_{m=n}^{\infty} A_m = A_n$, so $\limsup_n A_n = \cap_{n=1}^{\infty} A_n = A$.
   (iii) $\cap_{m=n}^{\infty} A_m = \{0\}$ for each n, and so $\liminf_n A_n = \cup_{n=1}^{\infty} \{0\} = \{0\}$. But $\{0\} \subset \cup_{m=n}^{\infty} A_m \subset \left[-\frac{1}{n}, \frac{1}{n}\right]$ for each n and $\cap_{n=1}^{\infty} \left[-\frac{1}{n}, \frac{1}{n}\right] = \{0\}$, so $\limsup_n A_n = \cap_{n=1}^{\infty} (\cup_{m=n}^{\infty} A_m) = \{0\} = \liminf_n A_n = \lim_{n\to\infty} A_n$.
   (iv) $\cap_{m=n}^{\infty} A_m = \{0\}$ for each n and so $\liminf_n A_n = \{0\}$. But $\cup_{m=n}^{\infty} A_n = [-1, 1]$ for each n, and so $\limsup_n A_n = [-1, 1] \neq \liminf A_n$. Sequence $\{A_n\}$ has no limit.
   (v) $[0, 1) \supset \cap_{m=n}^{\infty} A_m \supset [0, 1 - 1/n]$ for each n, and $\cup_{n=1}^{\infty} [0, 1 - 1/n] = [0, 1)$, so $\liminf_n A_n = \cup_{n=1}^{\infty} (\cap_{m=n}^{\infty} A_m) = [0, 1)$. However, $[0, 1] \subset$

$\cup_{m=n}^{\infty} A_m \subset [0, 1 + 1/n]$ for each n, and $\cap_{n=1}^{\infty} [0, 1 + 1/n] = [0, 1]$, so $\limsup_n A_n = [0, 1] \neq \liminf A_n$. Sequence $\{A_n\}$ has no limit.

4. Formally,

$$\left(\liminf_n A_n\right)^c = \left(\bigcup_{n=1}^{\infty} \bigcap_{m=n}^{\infty} A_m\right)^c = \bigcap_{n=1}^{\infty} \left(\bigcap_{m=n}^{\infty} A_m\right)^c$$

$$= \bigcap_{n=1}^{\infty} \bigcup_{m=n}^{\infty} A_m^c = \limsup_n A_n^c.$$

$$\left(\limsup_n A_n\right)^c = \left(\bigcap_{n=1}^{\infty} \bigcup_{m=n}^{\infty} A_m\right)^c = \bigcup_{n=1}^{\infty} \left(\bigcup_{m=n}^{\infty} A_m\right)^c$$

$$= \bigcup_{n=1}^{\infty} \bigcap_{m=n}^{\infty} A_m^c = \liminf_n A_n^c.$$

Intuitively,

(i) If $\liminf_n A_n$ does not occur, then there will be infinitely many *non*occurrences; hence, A_n^c i.o.

(ii) If $\limsup_n A_n$ does not occur, then only finitely many of the $\{A_n\}$ do occur, which means that all but finitely many of $\{A_n^c\}$ occur.

5. $\{A_j\}_{j=1}^n \in \mathcal{F} \implies \{A_j^c\}_{j=1}^n \in \mathcal{F} \implies \cup_{j=1}^n A_j^c \in \mathcal{F} \implies \left(\cup_{j=1}^n A_j^c\right)^c \in \mathcal{F} \implies \cap_{j=1}^n A_j \in \mathcal{F}$.

6. The monotone sequences $\{\varnothing, \Omega\}$ and $\{\Omega, \varnothing\}$ (each "sequence" comprising just one element) are the only ones that are fields, for only they contain the complements of their members.

7. (i) $\mathcal{F} = \{\Omega, H, T, \varnothing\}$

(ii) $\mathcal{F}$ contains the following 16 sets:

$\varnothing, \Omega,$
$HH, TH, HT, TT,$
$(HH, TH), (HH, HT), (HH, TT), (TH, HT), (TH, TT), (HT, TT),$
$(HH, TH, HT), (HH, TH, TT), (HH, HT, TT), (TH, HT, TT).$

(iii) There are two answers ("yes" or "no") to each of the n questions, "Is element $\cdot$ contained in $\mathcal{F}$?" There are 2^n possible strings of answers.

8. (i) The collection of *all* subsets of Ω contains complements of all its members and unions of all its members. By repeating sets, the union operation can be carried out arbitrarily many times; e.g., $A \cup A \cup \cdots$.

(ii) If $A \in \mathcal{G}$ and $A \in \mathcal{H}$ then $A^c \in \mathcal{G}$ and $A^c \in \mathcal{H}$, so $A \in \mathcal{G} \cap \mathcal{H}$ and $A^c \in \mathcal{G} \cap \mathcal{H}$. If $\cup_{n=1}^{\infty} A_n \in \mathcal{G}$ and $\cup_{n=1}^{\infty} A_n \in \mathcal{H}$ then $\cup_{n=1}^{\infty} A_n \in \mathcal{G} \cap \mathcal{H}$.

(iii) $\mathcal{G} = \{\Omega, H_1, T_1, \varnothing\}$, $\mathcal{H} = \{\Omega, S_1, S_2, \varnothing\}$, and neither $H_1 \cup S_1 = \{HH, HT, TT\}$ nor $T_1 \cup S_2 = \{TH, HT, TT\}$ is included in $\mathcal{G} \cup \mathcal{H} = \{\Omega, H_1, T_1, S_1, S_2, \varnothing\}$, so unions of fields are not necessarily fields. (Likewise for σ fields.)

9. (i) Set function φ_1 is not additive. For example, if disjoint sets A and B both contain only males, then $A \cup B$ contains only males, so $\varphi_1(A \cup B) = -1 \neq \varphi_1(A) + \varphi_1(B) = -2$.

   (ii) If disjoint sets A, B are finite (i.e., have finitely many elements), then $A \cup B$ is finite, so $\varphi_2(A \cup B) = 0 = \varphi_2(A) + \varphi_2(B)$. If A, B are infinite, then so is $A \cup B$, so $\varphi_2(A \cup B) = +\infty = \varphi_2(A) + \varphi_2(B)$. The same is true if one is finite and the other infinite, since $A \cup B$ is still infinite. Thus, φ_2 is *finitely* additive. However, φ_2 is not *countably* additive, since if each of disjoint sets $\{A_n\}_{n=1}^{\infty}$ is finite and nonempty, then $\cup_{n=1}^{\infty} A_n$ is infinite, so $\varphi_2(\cup_{n=1}^{\infty} A_n) = +\infty \neq \sum_{n=1}^{\infty} \varphi_2(A_n) = 0$.

10. Clearly N is nonnegative. If any of the disjoint sets $\{A_n\}_{n=1}^{\infty}$ is infinite, then so is the union, and $N(\cup_{n=1}^{\infty} A_n) = +\infty = \sum_{n=1}^{\infty} N(A_n)$. The same result holds if all of the sets are finite and infinitely many are nonempty. If all the sets are finite and finitely many are nonempty, then $N(\cup_{n=1}^{\infty} A_n) = \sum_{n=1}^{\infty} N(A_n) < \infty$. Thus, set function N is nonnegative and countably additive; hence, a measure.

11. This follows trivially from finite additivity, since $B = (A \cap B) \cup (B \backslash A) = A \cup (B \backslash A)$ when $A \subset B$.

12. $\mathbb{P}$ is nonnegative since N is. If Ω has finitely many elements, then there can be at most finitely many disjoint sets $\{A_j\}_{j=1}^{n}$, and

$$\mathbb{P}\left(\bigcup_{j=1}^{n} A_j\right) = \frac{N\left(\cup_{j=1}^{n} A_j\right)}{N(\Omega)} = \sum_{j=1}^{n} \frac{N(A_j)}{N(\Omega)} = \sum_{j=1}^{n} \mathbb{P}(A_j).$$

Extension to countable additivity is made by including in the union infinitely many copies of the empty set, which does not change the result. Finally, $\mathbb{P}(\Omega) = N(\Omega)/N(\Omega) = 1$.

13. The ten selections from the first group can be arrayed as a sequence of M's and W's. The sequence from the second group must match, and $\binom{10}{5}$ such sequences are possible, so the probability is $1/\binom{10}{5} = 1/252$.

14. We are given that $\mathbb{P}(L) = .1$, that $\mathbb{P}(L \mid S) = .2$, and (by inference, ignoring the few who are even handed) that $\mathbb{P}(R) = .9$ and $\mathbb{P}(R \mid S) = .8$.

    (i) By Bayes' rule $\mathbb{P}(S \mid L) = \mathbb{P}(L \mid S)\mathbb{P}(S)/\mathbb{P}(L)$ and $\mathbb{P}(S \mid R) =$

$\mathbb{P}(R \mid S)\,\mathbb{P}(S)\,/\mathbb{P}(R)$, so that

$$\frac{\mathbb{P}(S \mid L)}{\mathbb{P}(S \mid R)} = \frac{\mathbb{P}(L \mid S)}{\mathbb{P}(R \mid S)} \cdot \frac{\mathbb{P}(R)}{\mathbb{P}(L)} = \frac{.2}{.8} \cdot \frac{.9}{.1} = 2.25.$$

(ii) $\mathbb{P}(S \mid L) = .2\,(.04/.10) = .08.$ $\mathbb{P}(S \mid R) = .8\,(.04/.90) \doteq .0356.$

15. Adopting some symbols, let I, C, and P^c represent, respectively, the events that a person is in the top decile by intelligence, gets the correct answer, and has had no instruction in probability theory. We are given that $\mathbb{P}(I) = .1$, $\mathbb{P}(P^c \mid I^c) = .9$, and $\mathbb{P}(C \mid I) = \mathbb{P}(C \mid I^c \cap P^c) = .8$; and we can take $\mathbb{P}(C \mid I^c \cap P) = 0$. We thus have

$$
\begin{aligned}
\mathbb{P}(I \mid C) &= \frac{\mathbb{P}(I \cap C)}{\mathbb{P}(C)} = \frac{\mathbb{P}(C \mid I)\,\mathbb{P}(I)}{\mathbb{P}(C \cap I) + \mathbb{P}(C \cap I^c)} \\
&= \frac{\mathbb{P}(C \mid I)\,\mathbb{P}(I)}{\mathbb{P}(C \cap I) + \mathbb{P}(C \cap I^c \cap P^c) + \mathbb{P}(C \cap I^c \cap P)} \\
&= \frac{\mathbb{P}(C \mid I)\,\mathbb{P}(I)}{\mathbb{P}(C \mid I)\,\mathbb{P}(I) + \mathbb{P}(C \mid I^c \cap P^c)\,\mathbb{P}(I^c \cap P^c) + 0} \\
&= \frac{\mathbb{P}(C \mid I)\,\mathbb{P}(I)}{\mathbb{P}(C \mid I)\,\mathbb{P}(I) + \mathbb{P}(C \mid I^c \cap P^c)\,\mathbb{P}(P^c \mid I^c)\,\mathbb{P}(I^c)} \\
&= \frac{(.8)\,(.1)}{(.8)\,(.1) + (.8)\,(.9)\,(.9)} \doteq .1099.
\end{aligned}
$$

16. (i) $B = (A \cap B) \cup (A^c \cap B)$. This being a union of exclusive events, we have

$$\mathbb{P}(B) = \mathbb{P}(A \cap B) + \mathbb{P}(A^c \cap B) = \mathbb{P}(A)\,\mathbb{P}(B) + \mathbb{P}(A^c \cap B),$$

so $\mathbb{P}(A^c \cap B) = [1 - \mathbb{P}(A)]\,\mathbb{P}(B) = \mathbb{P}(A^c)\mathbb{P}(B).$
(ii) We have

$$
\begin{aligned}
\mathbb{P}(A^c \cap B^c) &= 1 - \mathbb{P}(A \cup B) \\
&= 1 - [\mathbb{P}(A) + \mathbb{P}(B) - \mathbb{P}(A \cap B)] \\
&= [1 - \mathbb{P}(A)]\,[1 - \mathbb{P}(B)] \\
&= \mathbb{P}(A^c)\mathbb{P}(B^c).
\end{aligned}
$$

17. "A, B exclusive" implies $\mathbb{P}(A \cap B) = \mathbb{P}(\varnothing) = 0$, while "$A, B$ independent" implies $\mathbb{P}(A \cap B) = \mathbb{P}(A)\mathbb{P}(B) > 0$. These two conditions cannot simultaneously hold. Thus, independence and exclusivity are inconsistent, either implying the converse of the other. When non-null events A, B are exclusive, then $\mathbb{P}(A \mid B) = \mathbb{P}(B \mid A) = 0$. When events A, B are independent, $\mathbb{P}(A \mid B) = \mathbb{P}(A) > 0$ and $\mathbb{P}(B \mid A) = \mathbb{P}(B) > 0$.

18. (i) $\mathbb{P}(A^c \cap B^c) = \mathbb{P}(A^c)\mathbb{P}(B^c) = (.7)(.6) = .42$

    (ii) $\mathbb{P}(A^c \cup B^c) = \mathbb{P}((A \cap B)^c) = 1 - (.3)(.4) = .88$

    (iii) $\mathbb{P}(A \backslash B) = \mathbb{P}(A \cap B^c) = \mathbb{P}(A)\mathbb{P}(B^c) = (.3)(.6) = .18$

    (iv) $\mathbb{P}((A \backslash B) \cup (B \backslash A)) = \mathbb{P}(A)\mathbb{P}(B^c) + \mathbb{P}(A^c)\mathbb{P}(B) = .18 + .28 = .46$

    (v) $\mathbb{P}((A \backslash B) \cup (A^c \backslash B)) = \mathbb{P}(A \cap B^c) + \mathbb{P}(A^c \cap B^c) = \mathbb{P}(B^c) = .6$

    (vi) $\mathbb{P}(A \backslash B^c) = \mathbb{P}(A \cap B) = .12$

    (vii) $\mathbb{P}(A \cup B \cup C) = \mathbb{P}(A \cup B) + \mathbb{P}(C) = .58 + .2 = .78$

    (viii) $\mathbb{P}(C^c \backslash (A \cup B)) = \mathbb{P}(C^c \cap A^c \cap B^c) = \mathbb{P}((A \cup B \cup C)^c) = 1 - .78 = .22$

    (ix) $\mathbb{P}(A \cup B \cup C^c) = \mathbb{P}(C^c) = .8$

    (x) $\mathbb{P}(A^c \cup B^c \cup C^c) = \mathbb{P}((A \cap B \cap C)^c) = 1 - \mathbb{P}(A \cap B \cap C) = 1 - 0.$

19. Since events $\{A_n\}_{n=1}^{\infty}$ are all the *same*, they are not *independent*, a condition without which the conclusion of Borel–Cantelli does not always hold.

Chapter 2

Probability on Sets of Real Numbers

In the previous chapter we encountered a variety of examples of chance experiments. For some of these—rolling a die, counting flips of a coin, spinning a pointer on a $[0, 1)$ scale—the outcomes were represented most naturally by ordinary real numbers. In other examples certain "abstract" forms seemed more appropriate—strings of H's and T's, names of individuals, suits and denominations of playing cards. Although such abstract forms could always be mapped into the real numbers just by enumerating the outcomes, it would usually be pointless to apply such arbitrary labels. In statistical applications, however, the numerical labeling usually comes about automatically, because it is some quantitative feature of the outcomes that typically concerns us; e.g., the incomes of households selected at random, the longevities of a sample of machine parts, the changes in blood pressure in a clinical trial. In such cases we naturally characterize events in terms of the quantitative features; e.g., the event that a randomly chosen household will be below some "poverty level", that a machine part will function for less than a specified time, that a patient's blood pressure will decline.

2.1 Random Variables and Induced Measures

To attach probabilities to such inherently quantifiable events—indeed, even to assure that they *are* events in the formal sense—requires establishing a one-to-one correspondence between sets of numbers and $\mathbb{P}$-measurable sets of outcomes in underlying probability space $(\Omega, \mathcal{F}, \mathbb{P})$. This correspondence is established formally by regarding the numerical values as being assigned via a single-valued transformation from outcome set Ω to some set of numbers. Mathematicians refer to any such single-valued mapping from one set to another as a *function*. In probability theory the special functions that assign numbers to outcomes of chance experiments are called *random variables (r.v.s)*.

From this point r.v.s will be central to everything we do, and it is in this chapter that we begin to get an understanding of them. We consider initially just mappings to $\Re$, the *real* numbers. *Vector-valued* r.v.s—also called *random vectors*—that map Ω into $\Re^k$ and *complex-valued* r.v.s that map to $\mathcal{C}$ (the complex numbers) will be treated in Sections 2.4 and 3.4.2.

2.1.1   *Random Variables as Mappings*

Definition 2.1. A (real, scalar-valued, "proper") random variable on a probability space $(\Omega, \mathcal{F}, \mathbb{P})$ is an $\mathcal{F}$-measurable mapping from Ω to $\Re$.

We will see shortly what the measurability requirement entails and will explain the meaning of "proper" in the next section. For the moment, just think of a r.v. as a function that assigns numbers to outcomes, and let us set up notation and look at examples. Typically, capital letters such as $X, Y, Z, \ldots$ or $X_1, X_2, \ldots$ are used to represent r.v.s. The functional relation is expressed symbolically in the usual way as $X : \Omega \to \Re$, and the expression $X(\omega) = x$ indicates that X assigns to generic outcome ω the generic real number x. The image set $X(\Omega) := \{X(\omega) : \omega \in \Omega\}$ represents a r.v.'s *range*. Function X thus maps Ω to $\Re$ and *onto* $X(\Omega)$.

Example 2.1. In the experiment of flipping three coins represent the sample space as

$$\Omega = \{TTT, TTH, THT, THH, HTT, HTH, HHT, HHH\}, \qquad (2.1)$$

and let $X(\omega)$ assign to outcome ω the total number of H's it contains. Then X maps Ω onto $X(\Omega) = \{0, 1, 2, 3\} \subset \Re$. Specifically,

$$X(\omega) := \begin{cases} 0, \omega = TTT \\ 1, \omega \in \{TTH, THT, HTT\} \\ 2, \omega \in \{THH, HTH, HHT\} \\ 3, \omega = HHH \end{cases}.$$

Notice that while the functional relation $X : \Omega \to \Re$ requires that each outcome be assigned one and only one real number, the function need not be one to one.

Example 2.2. Suppose that a coin is flipped repeatedly until the first occurrence of heads and that our interest is in the total number of flips required. If outcomes are represented as strings of zero or more T's terminating in H's, as $H, TH, TTH, TTTH, ...$, r.v. X could be defined by putting $X(\omega)$ equal to the total number of letters in string ω. Alternatively, we could identify outcomes directly with the number of flips and represent the sample space as $\Omega = \aleph$, in which case $X(\omega) := \omega$. In either case $X : \Omega \rightarrow X(\Omega) = \aleph$.

Example 2.3. For the pointer experiment in Example 1.7 outcomes are represented naturally as clockwise distances from the origin, and so $\Omega = [0, 1)$ is already a subset of $\Re$. With $X(\omega) := \omega$ we have $X(\Omega) = [0, 1)$.

Let us now see what it means that a random variable is an "$\mathcal{F}$-measurable" mapping. Thinking about the coin flip experiment of Example 2.1, we would want to be able to assign probabilities to statements such as "X takes the value 3" and "X takes a value in $[1, 2)$" and "X takes a value in the set $[1, 2) \cup \{3\}$". Indeed, for an arbitrary experiment and arbitrary r.v. X we would like to be able to attach a probability to the statement $X \in S$ ("X takes a value in set S") for *any* $S \subset \Re$. Unfortunately, in complicated sample spaces this is not always possible, given our fundamental notions of how probabilities should behave, as represented in Kolmogorov's axioms. In most applications we have to settle for attaching probabilities to the Borel sets, $\mathcal{B}$. The question is, "What sorts of functions $X : \Omega \rightarrow \Re$ make this possible?" Let us write $X^{-1}(B)$ to represent the inverse image of a Borel set B. This is the set of outcomes—that is, the subset of Ω—to which X assigns a value contained in the set of real numbers B. In symbols, $X^{-1}(B) = \{\omega : X(\omega) \in B\}$. Note that even when $B = \{x\}$ is just a singleton, $X^{-1}(B)$ does not *necessarily* represent a single outcome ω, since function X is not necessarily one-to-one. For example, in the coin experiment, we have $X^{-1}([1, 2)) = X^{-1}(\{1\}) = \{TTH, THT, HTT\}$. A probability can certainly be assigned to this set—the value $\frac{3}{8}$ if the eight outcomes are considered equally likely, and therefore the same value can be assigned to the *events* that $X \in [1, 2)$ and that $X = 1$. Thus, we could write $\mathbb{P}\left[X^{-1}([1, 2))\right] = \mathbb{P}\left(X^{-1}\{1\}\right) = \frac{3}{8}$ or, in the more concise and intuitive shorthand used continually hereafter, $\Pr(X \in [1, 2)) = \Pr(X = 1) = \frac{3}{8}$.

What, then, does the $\mathcal{F}$-measurability of a r.v. X mean, in general? It means formally that to *each* Borel set B of real numbers there corresponds an inverse image $X^{-1}(B)$ that belongs to our σ field $\mathcal{F}$. Belonging to $\mathcal{F}$ makes the set $\mathbb{P}$-measurable so that a probability number can be assigned to it. Thus, measurability guarantees that $X^{-1}(B)$ is a set of outcomes whose probability can be determined. In short, when X is measurable we can attach a probability number to the *event* $X \in B$ for any Borel set B.

Given the importance of measurability, it would be nice to know whether a given $X : \Omega \to \Re$ has this special property. Fortunately, to verify measurability one does not have to check that $X^{-1}(B) \in \mathcal{F}$ for *every* Borel set, because it suffices that this hold for all sets in any collection that *generates* $\mathcal{B}$. Of course, this condition is *necessary* also. Thus, X is measurable if and only if $X^{-1}(I) \in \mathcal{F}$ for all open intervals I (*or* for all closed intervals *or* for all half-open intervals). Note, too, that measurability *implies* that $X^{-1}(\{x\}) \in \mathcal{F}$ for each real number x. (Why?) Should this necessary condition *fail* for some x, we would know that X is *not* a r.v. on $(\Omega, \mathcal{F})$.

One rarely has to worry that a function that arises in applied work will fail this measurability test, but two (rather contrived) examples will illustrate the possibility.

Example 2.4. In the example of flipping three coins our field $\mathcal{F}$ would ordinarily comprise all 2^8 subsets of the outcomes in Expression (2.1), but consider instead the *sub*-field $\mathcal{H} = (\Omega, T_1, H_1, \varnothing)$ generated by events

$$T_1 := \{TTT, TTH, THT, THH\}, \; H_1 := \{HTT, HTH, HHT, HHH\}$$

and the sub-field $\mathcal{G}$ generated by

$$\{TTT, TTH\}, \{THT, THH\}, \{HTT, HTH\}, \{HHT, HHH\}.$$

The r.v. Z, say, that assigns to each of the eight outcomes in Ω the total number of heads on the *first* flip is $\mathcal{H}$-measurable, since for any interval I we have

$$Z^{-1}(I) = \begin{cases} T_1, & \{0\} \in I, \{1\} \notin I \\ H_1, & \{0\} \notin I, \{1\} \in I \\ \Omega, & \{0,1\} \in I \\ \varnothing, & \{0\} \notin I, \{1\} \notin I \end{cases}.$$

Likewise, the r.v. Y that counts the number of heads on the first *two* flips is $\mathcal{G}$ measurable since $Y^{-1}(\{0\})$, $Y^{-1}(\{1\})$, and $Y^{-1}(\{2\})$ are all $\mathcal{G}$ sets. Of course, since $\mathcal{H} \subset \mathcal{G} \subset \mathcal{F}$ the r.v. Z is $\mathcal{G}$-measurable also, and both Z and Y are also $\mathcal{F}$-measurable. But X (the number of heads after three flips) is not $\mathcal{G}$-measurable, and neither Y nor X is $\mathcal{H}$-measurable. Thus, X is not a r.v. on $(\Omega, \mathcal{G})$, and neither Y nor X is a r.v. on $(\Omega, \mathcal{H})$. For example, no set in $\mathcal{H}$ corresponds to $Y(\omega) = 2$, and none in either $\mathcal{H}$ or $\mathcal{G}$ corresponds to $X(\omega) = 3$.

Example 2.5. Consider the probability space $([0,1), \mathcal{B}_{[0,1)}, \lambda)$ comprising outcome set $\Omega = [0,1)$, σ field $\mathcal{F} = \mathcal{B}_{[0,1)}$, and $\mathbb{P} = \lambda$ (Lebesgue measure).

While all the sets in $\mathcal{B}_{[0,1)}$ are Lebesgue measurable (and some outside as well), there are sets outside of $\mathcal{B}_{[0,1)}$ that are *not* measurable.[1] Let B^* be a *particular* one of the many such nonmeasurable sets, and define the mapping $X : \Omega \to \Re$ as $X(\omega) = 1$ if $\omega \in B^*$ and $X(\omega) = 0$ otherwise. Since $B^* = X^{-1}(\{1\}) \notin \mathcal{B}_{[0,1)}$, this function is not a r.v. on measurable space $([0,1), \mathcal{B}_{[0,1)})$.

Fortunately, measurability is a robust property, since r.v.s remain r.v.s when subjected to standard algebraic operations.[2] For example, if $\{X_j\}_{j=1}^{\infty}$ is any countable collection of r.v.s on some $(\Omega, \mathcal{F})$, then so are (i) $aX_1 + b$ and $aX_1 + bX_2$ for any real a, b; (ii) $X_1 X_2$; (iii) $|X_1|$; (iv) $\max\{X_1, X_2\} =: X_1 \vee X_2$ and $\min\{X_1, X_2\} =: X_1 \wedge X_2$; (v) $\sup_n \{X_n\}$ and $\inf_n \{X_n\}$; and (vi) $g(X_1)$ for any Borel-measurable g (one such that $g^{-1}(B) \in \mathcal{B}$ $\forall$ $B \in \mathcal{B}$). However, the r.v.s $\sup_n \{X_n\}$ and $\inf_n \{X_n\}$ may not be "proper", in the sense to be described, nor is

$$g(X_1) := \begin{cases} 1/X_1, & X_1 \neq 0 \\ +\infty, & X_1 = 0 \end{cases}$$

proper unless $\Pr(X_1 = 0) = 0$. Also, $aX_1 + b$ would be considered a *degenerate* r.v. if $a = 0$, since it is then measurable with respect to the "trivial" field $(\Omega, \varnothing)$. [3]

2.1.2 *Induced Probability Measures*

We now know that a random variable takes us from sets of possibly "abstract" outcomes to comfortable sets of real numbers. Once a r.v. has been defined, we can often jettison the original probability space and work with a new one that the variable *induces*. Indeed, in applied statistical work we typically live entirely in the induced space and give little or no thought

---

[1]For examples of nonmeasurable sets of $\Re$ see Billingsley (1986, pp. 41-42), Royden (1968, pp. 63-64), or Taylor (1966, pp. 93-94).

[2]For proofs of the following see any standard text on measure theory; e.g., Taylor (1966).

[3]In some advanced applications—particularly with continuous-time stochastic processes—the σ field $\mathcal{F}$ is sometimes "completed" to ensure that all relevant functions are measurable. Thus, suppose there were a $B^* \in \mathcal{B}$ such that $X^{-1}(B^*) \notin \mathcal{F}$, but that $X^{-1}(B^*) \subset A^* \in \mathcal{F}$ and $\mathbb{P}(A^*) = 0$. In words, the inverse image of some Borel set might not be $\mathbb{P}$-measurable but might be contained *within* an $\mathcal{F}$ set of measure zero. In such a case the idea is just to ignore the "quasi-event" $X \in B^*$. "Completing" $\mathcal{F}$—that is, proclaiming that all such nonmeasurable subsets of null sets have been added to it, makes X measurable and qualifies it as a r.v. on $(\Omega, \mathcal{F}, \mathbb{P})$.

to the precursory $(\Omega, \mathcal{F}, \mathbb{P})$. It is therefore a good idea to understand just what induced spaces are.

Definition 2.2. To any proper r.v. X on a probability space $(\Omega, \mathcal{F}, \mathbb{P})$ there corresponds an *induced* probability space $(\Re, \mathcal{B}, \mathbb{P}_X)$, where $\mathbb{P}_X(B) := \mathbb{P}(X^{-1}(B))$ for each $B \in \mathcal{B}$.

Just as $\mathbb{P}(A)$ in the original space represents the probability of getting an outcome in some $A \in \mathcal{F}$, $\mathbb{P}_X(B)$ represents the probability of getting a value of X in some $B \in \mathcal{B}$; that is, it represents $\Pr(X \in B)$ in our short-hand. With X and $\mathbb{P}_X$ in hand one can work just with spaces and sets of real numbers. In applied work $\mathbb{P}_X$ is usually modeled directly or through other functions described in the next section (c.d.f.s, p.m.f.s, and p.d.f.s) instead of being deduced from the original space, but understanding its properties requires connecting it to $(\Omega, \mathcal{F}, \mathbb{P})$. An essential step in understanding is to verify that $\mathbb{P}_X$ is a valid probability on measurable space $(\Re, \mathcal{B})$. As always, to verify this requires checking that $\mathbb{P}_X$ satisfies Kolmogorov's axioms. In doing this one sees immediately that the measurability of X is an essential requirement:

(i) $\mathbb{P}_X(\Re) = 1$, since $\mathbb{P}_X(\Re) := \mathbb{P}[X^{-1}(\Re)] = \mathbb{P}(\Omega)$.
(ii) $\mathbb{P}_X(B) := \mathbb{P}[X^{-1}(B)] \geq 0$.
(iii) $\mathbb{P}_X$ is countably additive: If $\{B_j\}_{j=1}^{\infty}$ are disjoint Borel sets, then

$$\mathbb{P}_X\left(\bigcup_{j=1}^{\infty} B_j\right) = \mathbb{P}\left[X^{-1}\left(\bigcup_{j=1}^{\infty} B_j\right)\right] = \mathbb{P}\left(\bigcup_{j=1}^{\infty} X^{-1}(B_j)\right),$$

since the set of outcomes that map to B_1 or B_2 or ... is the union of the outcomes that map to B_1 together with the outcomes that map to B_2, etc. The inverse-image sets are disjoint since function X is single-valued, so the last probability is $\sum_{j=1}^{\infty} \mathbb{P}\left[X^{-1}(B_j)\right] = \sum_{j=1}^{\infty} \mathbb{P}_X(B_j)$.

At last we can explain the meaning of "proper". The fact that a proper r.v. X maps Ω to $\Re$ implies that its realizations are a.s. *finite*, meaning that $\lim_{n \to \infty} \Pr(|X| \geq n) = 0$. One does occasionally encounter "improper" r.v.s that have positive probability of surpassing any finite bound. Thus, an improper r.v. Y would be represented as $Y : \Omega \to [-\infty, \infty]$ and have $\mathbb{P}_Y(\Re) < 1$. For example, with X as the number of heads in the 3-coin experiment and Y given by $1/X$ for $X > 0$ and $+\infty$ otherwise, we would have $\mathbb{P}_Y(\Re) = 7/8$. In what follows, all r.v.s should be regarded as proper unless otherwise specified.

2.2 Representing Distributions of Random Variables

Working from the initial probability space $(\Omega, \mathcal{F}, \mathbb{P})$, evaluating $\mathbb{P}(A)$ for each $A \in \mathcal{F}$ (or just for members of a partition $\{A_j\}$) would show how the unit probability "mass" associated with Ω is apportioned or *distributed* over the abstract space. Similarly, working from the space $(\Re, \mathcal{B}, \mathbb{P}_X)$ induced by a r.v. X, the probabilities $\{\mathbb{P}_X(B)\}_{B \in \mathcal{B}}$ show how the unit mass is distributed over the real numbers. Once we know the *distribution* of a r.v., we know as much as our prior "trivial" information $(\Omega, \varnothing)$ can tell us about the values it will take. Indeed, as may be inferred from the sample of topics in Part II, the principal objective of the field of scientific endeavor known as "statistical inference" is to learn about distributions of r.v.s that arise in real chance experiments. However, in statistics—and in applied probability modeling generally—induced measures are usually not the most convenient representations for these distributions, and we generally put them aside and work entirely with *cumulative distribution functions (c.d.f.s)* and other functions derived from them.

2.2.1 *Cumulative Distribution Functions*

Recall that the σ field $\mathcal{B}$ of subsets of $\Re$ can be generated by intervals such as $(a, b]$ for real numbers $a < b$. Since $(a, b] = (-\infty, b] \backslash (-\infty, a] = (-\infty, b] \cap (-\infty, a]^c$, infinite intervals like $(-\infty, b]$ also generate $\mathcal{B}$. If the measure of each such infinite interval is known, then the corresponding measures of finite intervals can be deduced via the subtraction rule. Thus, $(a, b] = (-\infty, b] \backslash (-\infty, a]$ implies that $\mathbb{P}_X((a, b]) = \mathbb{P}_X((-\infty, b]) - \mathbb{P}_X((-\infty, a])$. Indeed, if $\mathbb{P}_X((-\infty, b])$ be known for every $b \in \Re$, it is possible to find $\mathbb{P}_X(B)$ for any $B \in \mathcal{B}$ by adding, subtracting, and/or taking limits. As examples, the probability attached to *open* interval (a, b) is found as

$$\mathbb{P}_X((a, b)) = \mathbb{P}_X((-\infty, b)) - \mathbb{P}_X((-\infty, a])$$
$$= \mathbb{P}_X(\lim_{n \to \infty}(-\infty, b - \frac{1}{n}]) - \mathbb{P}_X((-\infty, a])$$
$$= \lim_{n \to \infty}\mathbb{P}_X((-\infty, b - \frac{1}{n}]) - \mathbb{P}_X((-\infty, a]),$$

while the probability residing at a single point b is found as

$$\mathbb{P}_X\left(\{b\}\right) = \mathbb{P}_X\left(\bigcap_{n=1}^{\infty}(b - \frac{1}{n}, b]\right)$$

$$= \lim_{n\to\infty}\mathbb{P}_X\left((b - \frac{1}{n}, b]\right)$$

$$= \mathbb{P}_X\left((-\infty, b]\right) - \lim_{n\to\infty}\mathbb{P}_X\left(\left(-\infty, b - \frac{1}{n}\right]\right),$$

or simply as $\mathbb{P}_X\left(\{b\}\right) = \mathbb{P}_X((a, b]) - \mathbb{P}_X((a, b))$.

Thus, everything we would want to know about the induced distribution of probability over $\Re$ can be determined by knowing just the value of $\mathbb{P}_X((-\infty, x])$ for each $x \in \Re$. And here is the payoff for this realization: Since the set of probabilities $\{\mathbb{P}_X((-\infty, x]) : x \in \Re\}$ is merely a mapping from $\Re$ to $[0, 1]$, we can put measures aside and characterize the behavior of a r.v. using just an ordinary (point) function of a real variable. Now all we need is a simpler notation for such a function.

Definition 2.3. The function $F : \Re \to [0, 1]$ defined by $F(x) := \mathbb{P}_X((-\infty, x]), x \in \Re$, is called the *cumulative distribution function (c.d.f.)* of the r.v. X.

The value of the c.d.f. at some x thus represents the probability that X takes on a value up to and *including* x; that is, using our convenient shorthand, $F(x) = \Pr(X \le x)$. Once the function F has been determined, we can use it alone (or others functions derived from it) to attach probabilities to Borel sets. To do this correctly requires further understanding of the properties of c.d.f.s., but we shall look at some examples before proceeding with the theory. The examples will also illustrate the use of "indicator" functions to express functions that have different forms on disjoint subsets of their domains.[4] Here are a definition and some properties of indicators.

Definition 2.4. Let $\mathfrak{G}$ be some space and $\mathcal{S}$ a collection of subsets of $\mathfrak{G}$. For $A \in \mathcal{S}$ and $s \in \mathfrak{G}$ the indicator function $\mathbf{1}_A(s)$ takes the value unity when $s \in A$ and zero otherwise.

For example, with $\mathfrak{G} = \Re$, $\mathcal{S} = \mathcal{B}$, and $A = [0, 1)$,

$$\mathbf{1}_{[0,1)}(x) = \begin{cases} 1, & 0 \le x < 1 \\ 0, & x \in [0, 1)^c \end{cases}.$$

---

[4]These are sometimes called "characteristic functions" in other fields of mathematics, but in probability and statistics that term is reserved for a far more important concept, introduced in Section 3.4.2.

Properties: (i) $\mathbf{1}_{\mathfrak{S}} = 1$; (ii) $\mathbf{1}_A \geq 0$; (iii) $\mathbf{1}_{\cup_{j=1}^{\infty} A_j} = \sum_{j=1}^{\infty} \mathbf{1}_{A_j}$ if $\{A_j\}$ are disjoint; (iv) $\mathbf{1}_{A^c} = 1 - \mathbf{1}_A$; (v) $\mathbf{1}_{\varnothing} = 0$; (vi) $\mathbf{1}_A \leq \mathbf{1}_B$ whenever $A \subset B$; (vii) $\mathbf{1}_A \leq 1$; (viii) $\mathbf{1}_{B \setminus A} = \mathbf{1}_B - \mathbf{1}_{A \cap B}$; (ix) $\mathbf{1}_{A \cup B} = \mathbf{1}_A + \mathbf{1}_B - \mathbf{1}_{A \cap B}$; (x) $\mathbf{1}_{A \cap B} = \mathbf{1}_A \mathbf{1}_B$.[5]

Example 2.6. Working from the abstract probability space as an illustration, the c.d.f. of X, the number of heads in three flips of a coin, is found as follows. For any fixed $x < 0$

$$
\begin{aligned}
F(x) &= \mathbb{P}_X((-\infty, x]) \\
&= \mathbb{P}(X^{-1}((-\infty, x])) \\
&= \mathbb{P}(\{\omega : X(\omega) \leq x\}) \\
&= \mathbb{P}(\varnothing) = 0.
\end{aligned}
$$

Similarly, for any fixed $x \in [0, 1)$, $F(x) = \mathbb{P}(\{TTT\}) = \frac{1}{8}$; for any fixed $x \in [1, 2)$, $F(x) = \mathbb{P}(\{TTT, TTH, THT, HTT\}) = 4/8$; and so on. Thus,

$$
F(x) = \begin{cases}
0, & x < 0 \\
\frac{1}{8}, & 0 \leq x < 1 \\
\frac{4}{8}, & 1 \leq x < 2 \\
\frac{7}{8}, & 2 \leq x < 3 \\
1, & x \geq 3
\end{cases}
\tag{2.2}
$$

$$
= \frac{1}{8} \mathbf{1}_{[0,1)}(x) + \frac{4}{8} \mathbf{1}_{[1,2)}(x) + \frac{7}{8} \mathbf{1}_{[2,3)}(x) + \mathbf{1}_{[3,\infty)}(x)
$$

$$
= \frac{1}{8} \mathbf{1}_{[0,\infty)}(x) + \frac{3}{8} \mathbf{1}_{[1,\infty)}(x) + \frac{3}{8} \mathbf{1}_{[2,\infty)}(x) + \frac{1}{8} \mathbf{1}_{[3,\infty)}(x).
$$

This is a *step function* that is constant between positive jumps, as shown in Fig. 2.1.[6]

Example 2.7. Flipping a coin until the first occurrence of heads, express Ω as the countably infinite set $\{H, TH, TTH, TTTH, ...\}$. Outcome ω_j that represents the string of length j is the intersection of j events $T_1 \cap T_2 \cap \cdots \cap T_{j-1} \cap H_j$; that is, it is the event of getting T on the first flip and

---

[5]If we take $\mathcal{S}$ to be a σ field of subsets of $\mathfrak{S}$ and $\mathbb{P}$ to be a measure on $(\mathfrak{S}, \mathcal{S})$, then $\mathbb{P}(A) = \int \mathbf{1}_A \mathbb{P}(d\omega)$ for $A \in \mathcal{S}$, where this is an "abstract" integral, to be defined in Section 3.2. The first nine properties of probability measures in Section 1.3.1 then follow from relations (i)-(ix) by integration.

[6]One should take care not to misinterpret the expressions in (2.2). For example, the statement "$F(x) = \frac{4}{8}$ for $1 \leq x < 2$" does *not* mean that $\Pr(X \in [1, 2)) = \frac{4}{8}$ (even though this happens to be true in the example). It *does* mean that $\Pr(X \leq x) = \frac{4}{8} = .5$ for *each* fixed number $x \in [1, 2)$. Thus, $F(1.0) = .5$, $F(1.5) = .5$, $F(1.9) = .5$, $F(1.999) = .5$, and so on.

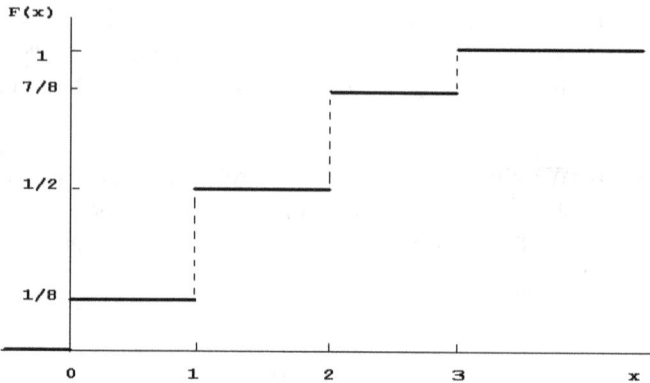

Fig. 2.1   C.d.f. for X, the number of heads in three flips of a fair coin.

on the second flip and so on up to flip $j-1$, then getting H. Assuming these events to be mutually independent and each to have probability $\frac{1}{2}$, we deduce that $\mathbb{P}(\omega_j) = 2^{-j}$ for $j \in \aleph$. Defining $X(\omega_j) := j$ (the number of flips until the first heads), we have $F(x) = 0$ for $x < 1$, $F(x) = \frac{1}{2}$ for $x \in [1,2)$, $F(x) = \frac{1}{2}+\frac{1}{4}$ for $x \in [2,3)$, and so on, with $F(x) = \frac{1}{2}+\frac{1}{4}+\cdots+\frac{1}{2^j}$ for $x \in [j, j+1)$ and $j \in \aleph$. Two more compact representations are

$$F(x) = \mathbf{1}_{[1,\infty)}(x) \sum_{j=1}^{[x]} 2^{-j} = \mathbf{1}_{[0,\infty)}(x)\left(1 - 2^{-[x]}\right), \qquad (2.3)$$

where $[x]$ signifies the greatest integer contained in x. This is also a step function, but the set of discontinuity points is now the countably infinite set $\aleph$.

Example 2.8. Model the results of the pointer experiment in Example 1.7 by the probability space $([0,1), \mathcal{B}_{[0,1)}, \lambda)$, where λ is Lebesgue measure, and put $X(\omega) = \omega$. Although this probability space is based on a *subset* of the reals, an induced measure can be extended to all of $(\Re, \mathcal{B})$ by taking $\mathbb{P}_X(B) = \lambda(B \cap [0,1))$ for each $B \in \mathcal{B}$; that is, the probability mass associated with any Borel set is taken to be the length of the part of the set that lies in $[0,1)$. Expressing the c.d.f., we have for $x < 0$

$$F(x) = \mathbb{P}_X((-\infty, x]) = \lambda((-\infty, x] \cap [0,1)) = \lambda(\varnothing) = 0;$$

for $x \in [0,1)$

$$F(x) = \lambda((-\infty, x] \cap [0,1)) = \lambda([0,x]) = x;$$

$b\mathbf{x} \geq 1$

$$) = \lambda((-\infty, x] \cap [0, 1)) = \lambda([0, 1)) = 1.$$

essions are

er is

$$F(x) = \begin{cases} 0, & x < 0 \\ x, & 0 \leq x < 1 \\ 1, & x \geq 1 \end{cases} \tag{2.4}$$

$$= x\mathbf{1}_{[0,1)}(x) + \mathbf{1}_{[1,\infty)}(x)$$

$$= \max\{0, \min\{x, 1\}\}.$$

This is sketched in Fig. 2.2. Unlike those in the previous examples this c.d.f. has no discontinuities.

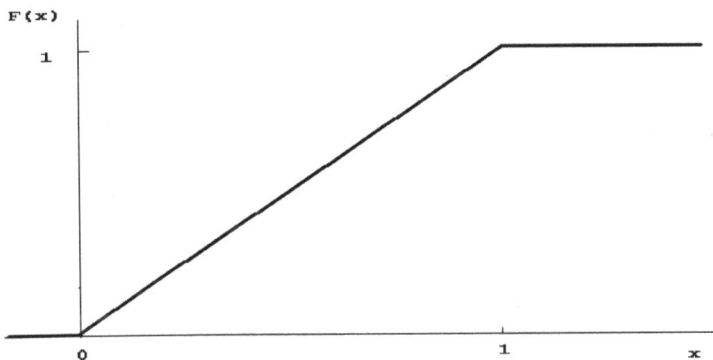

Fig. 2.2   A continuous c.d.f. for the pointer experiment.

Example 2.9. Keep the probability space the same as in the previous example, but define X as

$$X(\omega) := \begin{cases} 0, & 0 \leq \omega \leq .5 \\ \omega, & .5 < \omega < 1 \end{cases}.$$

Thus, X takes the value zero whenever the pointer stops in the first half of the scale; otherwise, it takes a value equal to the clockwise distance from the origin. As before, $F(x) = \lambda(\varnothing) = 0$ for $x < 0$ and $F(x) = 1$ for $x > 1$. However, for any $x \in [0, .5]$ we have $F(x) = \lambda([0, .5]) = .5$, and for $x \in (.5, 1)$

$$F(x) = \lambda([0, x) \cap [0, 1)) = \lambda([0, x)) = x.$$

A representation with indicator functions is

$$F(x) = .5\mathbf{1}_{[0,.5)}(x) + x\mathbf{1}_{[.5,1)}(x) + \mathbf{1}_{[1,\infty)}(x).$$

This is depicted in Fig. 2.3. Clearly, F is not a step function, but neither is it everywhere continuous. There is probability "mass" .5 at $x = 0$ because X takes the value zero whenever the pointer stops anywhere in the first half of the scale. Since $\Pr(X \in (0,.5)) = 0$, there is no additional mass until we reach .5, but from there it accumulates continuously up to $x = 1$.

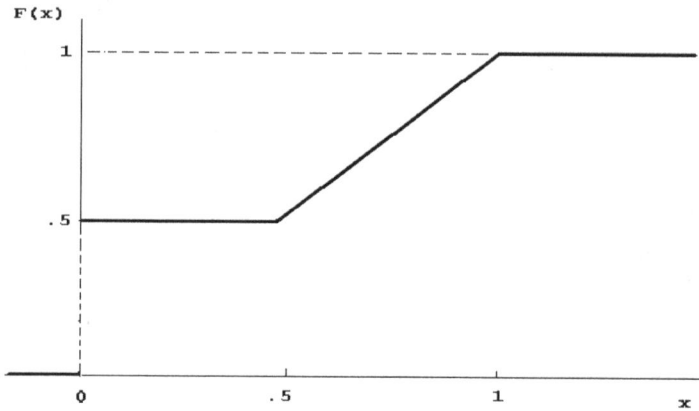

Fig. 2.3   A c.d.f. that is neither continuous nor a step function.

Let us now look at the properties a function F must have to be the c.d.f. of a "proper" r.v.—and which we must know in order to deduce from it the $\mathbb{P}_X$ measures of various Borel sets.

(i) F is nondecreasing. This is clear intuitively since $F(x)$ registers the *cumulative* probability up to each x. To show it, we have for $a < b$,

$$
\begin{aligned}
F(b) &= \mathbb{P}_X\left((-\infty, a] \cup (a, b]\right) \\
&= \mathbb{P}_X\left((-\infty, a]\right) + \mathbb{P}_X\left((a, b]\right) \\
&= F(a) + \mathbb{P}_X\left((a, b]\right) \geq F(a).
\end{aligned}
$$

This shows as well that $F(b) - F(a) = \mathbb{P}_X\left((a, b]\right) = \Pr(a < X \leq b)$.

(ii) $F(x)$ approaches unity as $x \uparrow \infty$; in symbols, $\lim_{x\to\infty} F(x) = 1$. Thus, for a proper r.v., as x increases $F(x)$ will ultimately pick up all the probability mass that resides in each bounded-above (measurable) subset of $\Re$. To show it,

$$
\begin{aligned}
F(+\infty) &:= \lim_{n\to\infty} F(n) \\
&= \lim_{n\to\infty} \mathbb{P}_X\left((-\infty, n]\right) \\
&= \mathbb{P}_X\left(\lim_{n\to\infty} (-\infty, n]\right) \\
&= \mathbb{P}_X\left(\bigcup_{n=1}^{\infty}(-\infty, n]\right) = \mathbb{P}_X(\Re) = 1.
\end{aligned}
$$

Notice that even for a proper r.v. F may not actually *attain* the value unity at any finite x, as in Example 2.7 above.

(iii) $F(x)$ approaches zero as $x \downarrow -\infty$. Again, the meaning is that, for proper r.v.s, the probability mass on any bounded-below (measurable) subset of $\Re$ is ultimately left behind as x decreases. To show it,

$$
\begin{aligned}
F(-\infty) &:= \lim_{n\to\infty} F(-n) \\
&= \lim_{n\to\infty} \mathbb{P}_X\left((-\infty, -n]\right) \\
&= \mathbb{P}_X\left(\lim_{n\to\infty} (-\infty, -n]\right) \\
&= \mathbb{P}_X\left(\bigcap_{n=1}^{\infty}(-\infty, -n]\right) = \mathbb{P}_X(\varnothing) = 0.
\end{aligned}
$$

Although this limiting value of zero *is* attained in all the above examples, we shall see many examples later in which it is not.

(iv) F is *right continuous* at each point: $F(x+) := \lim_{n\to\infty} F(x+1/n) = F(x)$. This means that $F(x+\varepsilon) - F(x) = \Pr(x < X \le x + \varepsilon)$ can be made arbitrarily close to zero by taking ε sufficiently small.[7] This too follows from the monotone property of $\mathbb{P}_X$:

$$
\begin{aligned}
F(x+) - F(x) &= \lim_{n\to\infty}\left[F(x+\frac{1}{n}) - F(x)\right] = \lim_{n\to\infty} \mathbb{P}_X\left((x, x+\frac{1}{n}]\right) \\
&= \mathbb{P}_X\left(\lim_{n\to\infty}(x, x+\frac{1}{n}]\right) = \mathbb{P}_X\left(\bigcap_{n=1}^{\infty}(x, x+\frac{1}{n}]\right) \\
&= \mathbb{P}_X(\varnothing) = 0.
\end{aligned}
$$

---

[7] For a visual interpretation, choose *any* number x, and in any of Figs. 2.1-2.3 follow along the path $\{x+\varepsilon, F(x+\varepsilon) : \varepsilon \downarrow 0\}$. There is always an ε small enough that the subsequent part of the path is continuous.

(v) *F may be* discontinuous from the left. This means that

$$F(x) - F(x-) = \lim_{n \to \infty} F(x) - F(x - 1/n) = \lim_{n \to \infty} \Pr(x - 1/n < X \le x)$$

may be positive. Here the interval $(x - \frac{1}{n}, x]$ contracts not to the empty set as did $(x, x + \frac{1}{n}]$ but to the single point $\{x\}$, where there *may* be positive probability mass. Thus, in Example 2.6 there is positive mass at each $x \in \{0, 1, 2, 3\}$; in Example 2.7, at each $x \in \mathbb{N}$; and in Example 2.9, at $x = 0$. To show it,

$$F(x) - F(x-) = \lim_{n \to \infty} \left[F(x) - F\left(x - \frac{1}{n}\right) \right] = \lim_{n \to \infty} \mathbb{P}_X \left(\left(x - \frac{1}{n}, x\right] \right)$$

$$= \mathbb{P}_X \left(\lim_{n \to \infty} \left(x - \frac{1}{n}, x\right] \right) = \mathbb{P}_X \left(\bigcap_{n=1}^{\infty} \left(x - \frac{1}{n}, x\right] \right)$$

$$= \mathbb{P}_X (\{x\}) = \mathbb{P}(X^{-1}(x)) = \Pr(X = x) \ge 0.$$

A point at which occurs a jump or "saltus" of the c.d.f. is referred to as an "atom". The number of such atoms may be infinite, as in Example 2.7, but it is always *countable*, since for arbitrarily large $n \in \mathbb{N}$ the number of atoms of mass n^{-1} or greater cannot exceed n.

Notice that $F(x) = \mathbb{P}_X((-\infty, x]) = \Pr(X \le x)$ implies that

$$1 - F(x) = \mathbb{P}_X((-\infty, x]^c) = \mathbb{P}_X((x, \infty)) = \Pr(X > x).$$

Also, to find the probability on the *open* interval $(-\infty, x)$, one subtracts from $F(x) = \mathbb{P}_X((-\infty, x])$ the probability mass at $\{x\}$, as given by the saltus at that point:

$$\mathbb{P}_X((-\infty, x)) = \Pr(X < x) = \Pr(X \le x) - \Pr(X = x)$$
$$= F(x) - [F(x) - F(x-)] = F(x-).$$

While left discontinuities are possible, Example 2.8 shows that some c.d.f.s are continuous everywhere. In such cases we have $\mathbb{P}_X(\{x\}) = \Pr(X = x) = 0$ for each x, even when $X^{-1}(\{x\}) = \{\omega : X(\omega) = x\} \neq \emptyset$. This illustrates once again the fact that *possible* events may have zero probability—be "$\mathbb{P}$-null". Thus, while $A = \emptyset$ always implies that $\mathbb{P}(A) = 0$, and, for induced measures, $B = \emptyset$ implies $\mathbb{P}_X(B) = 0$, the converses are not universally true. The distinction between "$\mathbb{P}$-null" and "impossible" makes precise definitions of certain concepts a bit tricky, as the following discussion indicates.

Definition 2.5. A Borel set $\mathbb{S}$ is called a *support* of the random variable X if $\mathbb{P}_X(\mathbb{S}) = 1$.

Thus, a support is any set of real numbers on which X lies with unit probability. For example, in the pointer experiment any (measurable) set $\mathbb{S}$ that contains the range $X(\Omega) = [0,1)$ is a support. It may seem that $[0,1)$ itself is in some sense a *minimal* support in this experiment, since there is no probability mass elsewhere. However, since $\mathbb{P}_X(\{x\}) = 0$ for each x and $\mathbb{P}_X$ is countably additive, it is also true that "smaller" sets than $[0,1)$, such as $[0,1)\setminus\{.5\}$, $[0,1)\setminus\{\frac{1}{n}\}_{n=1}^{\infty}$, and even $[0,1)\setminus\mathbb{Q}$ are all supports as well. Nevertheless, for any r.v. it is convenient to be able to refer to some specific set $\mathbb{X}$ as its "canonical" support. While the range would be a reasonable choice in most cases, $X(\Omega)$ *need* not be a measurable set when Ω is uncountable, even though $X^{-1}(B) \in \mathcal{F}$ for all *Borel* sets B. We shall adopt the following convention.

- When X is supported on a *countable* set, we shall take $\mathbb{X}$ to be the "smallest" of these—i.e., the intersection of all countable sets $\mathbb{S}$ for which $\mathbb{P}_X(\mathbb{S}) = 1$.
- Otherwise, we shall regard $\mathbb{X}$ as the set of the r.v.'s support *points*; that is, points $\{s\} \in \Re$ such that $\mathbb{P}_X((s - \varepsilon, s + \varepsilon)) > 0$ for each $\varepsilon > 0$. (Note that $\{1\}$ is such a support point in the pointer experiment.)

Thus, in the four examples above the canonical supports are (i) $\mathbb{X} = \{0, 1, 2, 3\}$, (ii) $\mathbb{X} = \aleph$, (iii) $\mathbb{X} = [0, 1]$, (iv) $\mathbb{X} = \{0\} \cup [.5, 1]$.

Example 2.10. Let $\mathcal{P}$ be the set of real polynomials with rational coefficients, not all of which are equal to zero. That is, $\mathcal{P}$ comprises functions of the form $P_n(t) = a_0 + a_1 t + \cdots + a_n t^n$ for some integer n and some set of coefficients $\{a_j\}_{j=0}^{n} \in \mathbb{Q}$, not all zero. In algebra we learn that a complex number t is called an *algebraic* number[8] if it solves the equation $P_n(t) = 0$ for *some* $P_n \in \mathcal{P}$; otherwise, t is said to be a *transcendental* number. Clearly, all rational numbers are algebraic, since $t = m/n$ for some $m \in \aleph_0, n \in \aleph$ implies that $nt - m = 0$; however, an irrational number (an element of $\Re\setminus\mathbb{Q}$) can be either algebraic (e.g., $\sqrt{2}$) or transcendental (e.g., π). Thus, if $\mathbb{A} \subset \Re$ represents the set of algebraic real numbers, we have $\mathbb{Q} \subset \mathbb{A}$ and $\mathbb{A}\setminus\mathbb{Q} \neq \varnothing$. Nevertheless, it is known that the set $\mathbb{A}$ is countable. Thus, for any *continuous* r.v. X with canonical support $\mathbb{X} \subset \Re$ we have $\mathbb{P}_X(\mathbb{X}\setminus\mathbb{A}) = 1$, so that the transcendentals of $\mathbb{X}$ are also a support.

Example 2.11. Let $\{a_j\}_{j=1}^{\infty}$ be some sequential representation of the

---

[8]Specifically, it is an algebraic number over the *rational* field—i.e., over the field of rational numbers $\mathbb{Q}$.

rational numbers and $\{b_j\}_{j=1}^{\infty}$ be positive constants such that $\sum_{j=1}^{\infty} b_j = 1$. Then the r.v. $X : \Omega \to \mathbb{Q}$ with $\{\mathbb{P}_X(\{a_j\}) = b_j\}_{j=1}^{\infty}$ has canonical support $\mathbb{X} = \mathbb{Q}$; but since $\mathbb{Q}$ is dense in $\Re$, the set of its support *points* is the entire real line. (The example is due to Chung (1974).)

Exercise 2.1. *A fair coin is flipped. If it comes up tails, r.v. X takes the value zero. If it comes up heads, the pointer on the unit circular scale is spun and X is assigned a value equal to the clockwise distance from origin to pointer. Find the c.d.f. of X.*

Exercise 2.2. *A faucet supplies water at the rate of 5 gallons per minute. A pointer on a circular scale running from 0 to 60 is spun, and the faucet is allowed to run for the time (in minutes) indicated by the pointer when it comes to rest. Letting X be the volume of water supplied, find the c.d.f. of X.*

Exercise 2.3. *Let F be a c.d.f. Determine which of the following functions is also a c.d.f.:*

(i) $G_1(x) := F(x)^{\alpha}, \alpha > 0$
(ii) $G_2(x) := F(x)^{\beta}, \beta \leq 0$
(iii) $G_3(x) := \left[e^{F(x)} - 1\right] / (e - 1)$
(iv) $G_4(x) := \log\left[1 + F(x)\right] / \log 2$
(v) $G_5(x) := F(x)\mathbf{1}_{[.5,\infty)}(x).$

Note on notation: Here and throughout "log" represents the natural logarithm function, base e. Arguments appear in parentheses only when necessary to avoid ambiguity; e.g., $\log(x + 1)$ *vs.* $\log x + 1$ and $\log(2\pi)$ *vs.* $(\log 2)\pi$.

We end this general discussion of distributions of r.v.s by defining an important category of these.

Definition 2.6. Random variable X is said to be distributed *symmetrically* about $a \in \Re$ if and only if $\Pr(X - a > x) = \Pr(X - a < -x)$ for each real x. In terms of c.d.f.s this requires $1 - F(x + a) = F((-x + a)-)$ for each x. The distribution is symmetric about the *origin* if and only if $F(x) + F(-x-) = 1$ for each $x \in \Re$.

2.2.2 A Taxonomy of Random Variables

The c.d.f. of any r.v. can be expressed uniquely as a convex combination of three special types of c.d.f.s, as $F = \alpha F_D + \alpha' F_{AC} + \alpha'' F_{CS}$. Here the α's are nonnegative and sum to unity, F_D increases only by jumps, and F_{AC} and F_{CS} are continuous (abbreviations explained below). A r.v. is said to be of a "pure" type if its c.d.f. has just one of these components. Classifying r.v.s in this way is important because for the two most important pure types there are other useful ways to represent their distributions besides their c.d.f.s.

Definition 2.7. X is said to be a *discrete* r.v. if its c.d.f. increases only by jumps. It is said to be *continuous* if its c.d.f. is a continuous function.

As the definition implies, a discrete r.v. is supported on the atoms of the countable set $\mathbb{X}$ at which the c.d.f. $F = F_D$ is discontinuous.[9] By contrast, for a continuous r.v. with support $\mathbb{X}$ the set $\mathbb{X}\backslash\mathcal{D}$ remains a support for any countable $\mathcal{D}$. That is, $\mathbb{P}_X(\mathbb{X}\backslash\mathcal{D}) = 1$ and $\mathbb{P}_X(\mathcal{D}) = 0$ whenever X is continuous and $\mathcal{D}$ is countable. Example 2.9 illustrates a r.v. that is neither purely discrete nor purely continuous. The c.d.f. of such a "mixed" r.v. has one or more discontinuities but also increases continuously on some interval. Thus, X is of the mixed type if $0 < \mathbb{P}_X(\mathcal{D})$ for some countable $\mathcal{D}$ but $\mathbb{P}_X(\mathcal{D}) < 1$ for all such $\mathcal{D}$, or, equivalently, if $0 < \alpha < 1$ in the decomposition of its c.d.f.

Continuous r.v.s are further classified as *absolutely continuous (a.c.)*, *continuous singular (c.s.)*, or mixtures thereof. All have c.d.f.s that are everywhere continuous and that are differentiable almost everywhere (a.e.) with respect to Lebesgue measure, but absolutely continuous c.d.f.s have an additional smoothness requirement beyond what is needed for simple continuity. What is relevant for our purposes is the fact that to an a.c. function $F = F_{AC}$ there corresponds a "density" function f such that $f(t) = F'(t)$ a.e. and $F(x) = \int_{-\infty}^{x} f(t) \cdot dt$, $x \in \Re$. In other words, a.c. functions can be represented as integrals with respect to Lebesgue measure.

No such f exists for members of the continuous-singular subclass of continuous r.v.s. Indeed, although $F = F_{CS}$ may increase continuously on an interval $[a, b]$, its derivative equals zero a.e., and so $F(b) - F(a) \neq \int_a^b F'(x) \cdot dx = 0$. Integrals of such functions exist in the Lebesgue sense

---

[9]The c.d.f.s of Examples 2.6 and 2.7 are "step" functions—i.e., constant on the open intervals between jumps. All such c.d.f.s correspond to discrete r.v.s, but not all discrete r.v.s have c.d.f.s of this form. The discrete r.v. in Example 2.11 has c.d.f. $F(x) = \sum_{j=1}^{\infty} b_j \mathbf{1}_{[a_j,\infty)}(x)$, and there is no open interval on which F is constant.

(to be described in Section 3.2) but not in the Riemann sense, because F' fails to exist at a countably infinite number of points of $[a, b]$. This is why the conclusion of the "fundamental theorem of integral calculus" does not apply here.

Note that the discontinuous c.d.f.s of discrete r.v.s also satisfy $F'(x) = 0$ a.e. and $\int_a^b F'(x) \cdot dx = 0$. The measures induced by both discrete and continuous-singular r.v.s are said to be *singular* with respect to Lebesgue measure, meaning that there is a set $\mathbb{X}$ with $\mathbb{P}_X(\mathbb{X}) = 1$ but $\lambda(\mathbb{X}) = 0$. Generally, if $\mathcal{M}$ and $\mathcal{N}$ are measures on the same measurable space $(\Omega, \mathcal{F})$, then $\mathcal{M}$ is singular with respect to $\mathcal{N}$ if there is an $S \in \mathcal{F}$ such that $\mathcal{N}(S) = 0$ and $\mathcal{M}(A) = \mathcal{M}(A \cap S)$ for all $A \in \mathcal{F}$. Thus, the value of $\mathcal{M}$ on any set is not increased by adding points outside of "support" S. The polar case to singularity is *absolute continuity*. $\mathcal{M}$ is absolutely continuous with respect to $\mathcal{N}$ if $\mathcal{M}(A) = 0$ whenever $\mathcal{N}(A) = 0$. In this case measure $\mathcal{N}$ is said to *dominate* measure $\mathcal{M}$. One of course wants to know how the absolute continuity of *measures* relates to the absolute continuity of *c.d.f.s*. Corresponding to the c.d.f. F of any r.v. X is an induced probability measure $\mathbb{P}_X$ on $(\Re, \mathcal{B})$ such that $\mathbb{P}_X((a, b]) = F(b) - F(a)$. If r.v. X is either discrete or continuous singular, then $\mathbb{P}_X$ is supported on a countable set and is thus singular with respect to (w.r.t.) λ. By contrast, if $\mathbb{P}_X$ is absolutely continuous w.r.t. λ, then the corresponding F must be an absolutely continuous function (relative to λ). In the purely discrete case $\mathbb{P}_X$ is dominated by counting measure, and so we would say that F is absolutely continuous w.r.t. N.

Note: Examples of continuous-singular distributions can be found in Billingsley (1995, pp. 407-409), Chung (pp. 12-13), and Feller (1971, pp. 141-142). Although they are of theoretical interest, these pathological cases are rarely explicitly encountered in applied work,[10] and we shall have no further dealings with them. Henceforth, any r.v. described without qualification as "continuous" should be considered "absolutely continuous" w.r.t. Lebesgue measure.

2.2.3   *Probability Mass and Density Functions*

Other representations of the distributions of purely discrete and continuous r.v.s can be developed from their c.d.f.s, and these turn out to be the most useful ones of all.

---

[10]But see Feller (1971, p. 142) for an instance in which the encounter is *inexplicit*.

Definition 2.8. The *probability mass function (p.m.f.)* of a discrete r.v. X is the function $f(x) := F(x) - F(x-), x \in \Re$.

Thus, $f(x) = \mathbb{P}_X(\{x\}) = \Pr(X = x)$ registers the probability mass at x. For any p.m.f. f there is a *countable* set $\mathbb{X}$ such that (i) $f(x) = F(x) - F(x-) > 0$ for $x \in \mathbb{X}$, (ii) $\sum_{x \in \mathbb{X}} f(x) = 1$, and (iii) $f(x) = 0$ for $x \in \mathbb{X}^c$. These imply that $0 \leq f(x) \leq 1$ for each $x \in \Re$. Any function with these properties can represent the probability distribution of a discrete r.v.

Example 2.12. For the number X of heads in three flips of a fair coin,

$$f(x) = \left\{ \begin{array}{l} \frac{1}{8}, x = 0 \\ \frac{3}{8}, x = 1 \\ \frac{3}{8}, x = 2 \\ \frac{1}{8}, x = 3 \\ 0, \text{ elsewhere} \end{array} \right\} = 1_{\{0,1,2,3\}}(x) \frac{\binom{3}{x}}{8}.$$

Example 2.13. For the number X of flips of a fair coin until the first occurrence of heads, $f(x) = 2^{-x} 1_{\mathbb{N}}(x)$.

Since $f(x)$ equals the $\mathbb{P}_X$ measure of the singleton $\{x\}$, the probability mass on any Borel set B is just the sum of values of f over the countably many points $x \in \mathbb{X} \cap B$:

$$\mathbb{P}_X(B) = \Pr(X \in B) = \sum_{x \in \mathbb{X} \cap B} f(x). \tag{2.5}$$

Thus, in Example 2.12 $\mathbb{P}_X((0,2]) = \Pr(0 < X \leq 2) = f(1) + f(2) = \frac{6}{8}$. For brevity, we often omit reference to the support and write $\Pr(X \in B) = \sum_{x \in B} f(x)$.

Now consider a continuous r.v. X with c.d.f. F, for which F' is positive on a set of unit $\mathbb{P}_X$ measure and exists except possibly on some countable set $\mathcal{C} \subset \Re$. For example, $\mathcal{C} = \{0, 1\}$ for the c.d.f. in Fig. 2.2. Integral representations such as $\mathbb{P}_X((-\infty, x]) = F(x) = \int_{-\infty}^{x} f(t) \cdot dt$ or, more generally, $\mathbb{P}_X(B) = \int_{x \in B} f(x) \cdot dx$ for any Borel set B, implicitly *define* the probability density function of X w.r.t. Lebesgue measure; however such representations do not specify function f uniquely. Since the value of such an integral remains unchanged if f is altered on any countable set, the p.d.f. should be considered an *equivalence class* of functions, any two members of which coincide a.e. with respect to Lebesgue measure.[11] The

---

[11] A modified f with infinitely many discontinuities on a finite interval might still be integrable, although not in the *Riemann* sense. Again, we anticipate the discussion of integration in Section 3.2, which extends the class of integrable functions.

standard convention, however, is to adopt a working definition based on a particular subclass, as follows:

Definition 2.9. If r.v. X has (absolutely) continuous c.d.f. F, differentiable except on some countable set $\mathcal{C}$, its *probability density function (p.d.f.)* with respect to Lebesgue measure is the function

$$f(x) = \begin{cases} F'(x), & x \in \mathcal{C}^c \\ q(x), & x \in \mathcal{C} \end{cases},$$

where $q(x)$ can be assigned any nonnegative value.

In words, we assign to f the value of F' where the derivative exists but are free to choose any nonnegative value elsewhere. Restricting q to nonnegative values is customary simply because F' is never negative on $\mathcal{C}^c$. It is convenient and natural to pick values that yield a concise and pleasing expression for f.

Example 2.14. In the pointer experiment of Example 2.8, $F'(x) = f(x) = 1$ for $x \in (0,1)$ and $F'(x) = f(x) = 0$ for $x \in (-\infty, 0) \cup (1, \infty)$, while F' does not exist on $\mathcal{C} := \{0, 1\}$. Setting $f(0) = f(1) = 1$ gives the concise expression

$$f(x) = \left\{ \begin{array}{ll} 1, & 0 \le x \le 1 \\ 0, & \text{elsewhere} \end{array} \right\} = \mathbf{1}_{[0,1]}(x) \tag{2.6}$$

and makes the density positive on the canonical support. Since the density is constant there, this is called the *uniform* distribution on $[0, 1]$.

To see why the term "density" is appropriate for the p.d.f., recall that $F(x) - F(x - h) = \mathbb{P}_X((x - h, x]) = \Pr(x - h < X \le x)$ is the probability mass on the interval $(x - h, x]$. The ratio of the interval's mass to its Lebesgue measure (its length) is

$$\frac{\mathbb{P}_X((x - h, x])}{\lambda((x - h, x])} = \frac{F(x) - F(x - h)}{h}.$$

This ratio thus represents the average density of probability on $(x - h, x]$ relative to Lebesgue measure—*average* because the probability need not be the same for every equal subinterval of $(x - h, x]$. Sending $h \to 0$ gives, if the limit exists,

$$\lim_{h \downarrow 0} \frac{F(x) - F(x - h)}{h} = F'(x-),$$

which is the left-hand derivative of F at x. Likewise, the average density in $(x, x + h]$ is $[F(x + h) - F(x)]/h$, and the limit of this, if it exists, is the right-hand derivative $F'(x+)$. The two derivatives agree if F is differentiable at x, and so $F'(x-) = F'(x+) = F'(x) = f(x)$.

As is customary in the statistics literature, we use the same symbol f for p.m.f. and p.d.f. These functions have very different properties, but the context and the specific definition will always make the meaning clear. Indeed, although distinct terms are often used for clarity, we frequently refer to p.m.f.s as "densities" also, for the value of $f(x) = \mathbb{P}_X(\{x\})$ is the limit of the average density of probability w.r.t. *counting* measure. Specifically, for any x in countable support $\mathbb{X}$

$$\mathbb{P}_X(\{x\}) = \lim_{h \downarrow 0} \frac{\mathbb{P}_X([x - h, x + h])}{N([x - h, x + h])},$$

where $N(B) \equiv N(B \cap \mathbb{X})$ counts the number of "atoms" on set B—that is, the number of points in B at which c.d.f. F is discontinuous. Thus, by specifying the applicable measure—Lebesgue or counting, one can appropriately refer to f in either role as a "p.d.f." And just as $\mathbb{P}_X(B)$ has the integral representation $\int_{x \in B} f(x) \cdot dx$ when r.v. X is continuous, we will see in Section 3.2 that probabilities for discrete r.v.s can be expressed as integrals with respect to counting measure, as $\int_{x \in B} f(x) \, N(dx)$.[12]

The fundamental properties of p.d.f.s of continuous r.v.s—and how to derive probabilities from them—are obvious from the definition. Thus, $\Pr(X \leq x) = F(x) = \int_{-\infty}^{x} f(t) \cdot dt = \Pr(X < x)$ since $\mathbb{P}_X(\{x\}) = 0$, and for any interval $I \in \{(a, b), (a, b], [a, b), [a, b]\}$

$$\Pr(X \in I) = F(b) - F(a) = \int_a^b f(t) \cdot dt.$$

Likewise, for any Borel set B, whether open or closed,

$$\Pr(X \in B) = \int_{x \in B} f(x) \cdot dx := \int f(x) \mathbf{1}_B(x) \cdot dx. \tag{2.7}$$

Any function f can serve as model for a p.d.f. if (i) $f(x) \geq 0, x \in \Re$, and (ii) $\int_{-\infty}^{\infty} f(x) \cdot dx = 1$. Although probability *mass* functions must satisfy $f(x) = \Pr(X = x) \leq 1$, p.d.f.s of continuous r.v.s are not necessarily bounded above. After all, their values represent *densities* of probability with respect to length measure, and any amount of probability mass up to unity can be packed into an arbitrarily short interval.

---

[12]In fact, a common integral expression $\mathbb{P}_X(B) = \int_{x \in B} f(x) \mathcal{M}(dx)$ applies when X is purely continuous, purely discrete, and even *mixed*—that is, when c.d.f. F has discontinuities on a countable set $\mathcal{D}$ with $0 < \mathbb{P}_X(\mathcal{D}) < 1$. In this representation f is a density with respect to $\mathcal{M}(\cdot) := N(\cdot \cap \mathcal{D}) + \lambda(\cdot)$, the sum of the two mutually singular measures, counting and Lebesgue. Such a general representation will be found useful in Part II.

Example 2.15. The nonnegative function $f(x) := e^{-x}\mathbf{1}_{[0,\infty)}(x)$ is a p.d.f., since

$$\int_{-\infty}^{\infty} f(x)\cdot dx = \int_{-\infty}^{0} f(x)\cdot dx + \int_{0}^{\infty} f(x)\cdot dx = \int_{-\infty}^{0} 0\cdot dx + \int_{0}^{\infty} e^{-x}\cdot dx = 1.$$

The corresponding c.d.f. is

$$F(x) = \int_{-\infty}^{x} f(t)\cdot dt$$

$$= \begin{cases} \int_{-\infty}^{x} 0\cdot dt = 0, & x < 0 \\ \int_{0}^{x} e^{-t}\cdot dt = 1 - e^{-x}, & x \ge 0 \end{cases}$$

$$= \left(1 - e^{-x}\right)\mathbf{1}_{[0,\infty)}(x)$$

and

$$\Pr(a \le X \le b) = \Pr(a < X < b) = \begin{cases} 0, & b \le 0 \\ 1 - e^{-b}, & a \le 0 < b \\ e^{-a} - e^{-b}, & 0 \le a < b \end{cases}.$$

Example 2.16. $x^{-1/2}e^{-x} > 0$ on $(0, \infty)$ and (as shown in Section 4.3.2) $\int_{0}^{\infty} x^{-1/2}e^{-x}\cdot dx = \sqrt{\pi}$, so

$$f(x) := \frac{1}{\sqrt{\pi}}x^{-1/2}e^{-x}\mathbf{1}_{(0,\infty)}(x)$$

is a p.d.f. Notice that $\lim_{x\downarrow 0} f(x) = +\infty$.

Exercise 2.4. *Five cards are drawn with replacement from a deck of 52 cards containing 4 aces. Find the p.m.f. and the c.d.f. of X, the number of aces drawn.*

Exercise 2.5. *Five cards are dealt without replacement from a deck of 52 cards containing 4 aces. Find the p.m.f. and the c.d.f. of X, the number of aces drawn.*

Exercise 2.6. *X has p.d.f. $f(x) = \mathbf{1}_{[1,e)}(x)k\log x$ for some k (where $\log e = 1$).*

(i) *Find the value of k.*
(ii) *Find the c.d.f.*
(iii) *Calculate (a) $\Pr(X > 1)$; (b) $\Pr(X < e/2)$; (c) $\Pr(X \in \mathbb{Q} \cap [1, e))$, where $\mathbb{Q}$ is the set of rational numbers.*

2.2.4   *Quantiles*

In statistics we often need to find specific numbers that bound realizations of r.v.s with known probabilities. Such numbers are called *quantiles*. In the definition below X is a r.v. with c.d.f. F, and $p \in (0,1)$.

Definition 2.10. The p *quantile* of a r.v. X (or of its distribution) is a number x_p such that $\Pr(X \leq x_p) = F(x_p) \geq p$ and $\Pr(X \geq x_p) = 1 - F(x_p-) \geq 1 - p$.

Definition 2.11. The *median* of a r.v. X (or of its distribution) is the .5 quantile, $x_{.5}$.

Example 2.17.

(i) If X is the number of heads in two flips of a fair coin, then $x_{.5} = 1$ since $\Pr(X \leq 1) = \Pr(X \geq 1) = .75 > .5$.

(ii) If X is a discrete r.v. with p.m.f. $f(-1) = .2$, $f(0) = .5$, and $f(1) = .3$, then $x_{.5} = 0$ since $\Pr(X \leq 0) = .7$ and $\Pr(X \geq 0) = .8$.

Medians of r.v.s, and quantiles generally, are not necessarily unique points. For example, if X is the number of heads in *three* flips of a coin, then (2.2) in Example 2.6 gives $F(x) \geq .5$ and $1 - F(x-) \geq .5$ for each $x \in [1,2]$. Uniqueness need not hold even for continuous r.v.s. For example, if X has p.d.f. $f(x) = 3/2$ for $x \in [0, 1/3] \cup [2/3, 1]$ and $f(x) = 0$ elsewhere, then any point on $[1/3, 2/3]$ is a median. In such cases a unique median is sometimes specified arbitrarily as $x_{.5} = (\sup \mathfrak{M} + \inf \mathfrak{M})/2$, where $\mathfrak{M}$ represents the set of points that qualify as medians by Definition 2.11. In the two cases just described we have (i) $\mathfrak{M} = [1,2]$ and $x_{.5} = (1+2)/2 = 1.5$ and (ii) $\mathfrak{M} = [1/3, 2/3]$ and $x_{.5} = (1/3 + 2/3)/2 = .5$, respectively. However, all quantiles of a continuous r.v. *are* uniquely determined if its c.d.f. F is strictly increasing (and therefore one to one) on $\mathbb{X}$, in which case $x_\alpha = F^{-1}(\alpha)$. This is the situation that commonly applies in statistical applications.

The following exercises demonstrate properties of the median that will be referred to in Chapter 3. The result of the first exercise implies that

$$\{x \in \mathfrak{M} : \Pr(X \leq x) - \Pr(X \geq x) \neq 0\}$$

is a set of Lebesgue measure zero; the second exercise implies that $\Pr(X \leq x) - \Pr(X \geq x) \neq 0$ for $x \notin \mathfrak{M}$; and the third shows that $\Pr(X > x^*) = \Pr(X < x^*)$ is sufficient but not necessary for $x^* \in \mathfrak{M}$.

Exercise 2.7. *Show that there can be at most one x^* such that (i)* $\Pr\left(X \le x^*\right) > .5$ *and* $\Pr\left(X \ge x^*\right) \ge .5$ *and at most one x^* such that (ii)* $\Pr\left(X \le x^*\right) \ge .5$ *and* $\Pr\left(X \ge x^*\right) > .5$.

Exercise 2.8. *Putting $x_{.5}^- := \inf \mathfrak{M}$ and $x_{.5}^+ := \sup \mathfrak{M}$, show that* $\Pr\left(X \le x\right) - \Pr\left(X \ge x\right) > 0$ *for each $x > x_{.5}^+$ and that* $\Pr\left(X \le x\right) - \Pr\left(X \ge x\right) < 0$ *for each $x < x_{.5}^-$.*

Exercise 2.9. *Show (i) that* $\Pr\left(X > x^*\right) = \Pr\left(X < x^*\right)$ *for some x^* implies that $x^* = x_{.5}$ (i.e., that $x^* \in \mathfrak{M}$); (ii) that $x^* = x_{.5}$ does not imply that* $\Pr\left(X > x^*\right) = \Pr\left(X < x^*\right)$*; but (iii) that* $\Pr\left(X > x_{.5}\right) = \Pr\left(X < x_{.5}\right)$ *does hold for continuous X.*

In statistics we often work with *upper* quantiles, denoted x^p and defined as follows.

Definition 2.12. A $1 - p$ quantile of a r.v. X is also an upper-p quantile of X (or of its distribution); i.e., $x^p = x_{1-p}$.

Thus, $p = 1 - F\left(x^p\right)$ for a continuous r.v. whose c.d.f. strictly increases on $\mathbb{X}$.

Example 2.18. If $f\left(x\right) = e^{-x} \mathbf{1}_{[0,\infty)}\left(x\right)$, the open interval $(0,\infty)$ is a support and $F\left(x\right) = \left(1 - e^{-x}\right) \mathbf{1}_{[0,\infty)}\left(x\right)$ is strictly increasing there, with $F^{-1}\left(p\right) = -\log\left(1-p\right)$. Therefore,
   (i) $x_{.25} = x^{.75} \doteq 0.288$,
   (ii) $x_{.5} = -\log .5 \doteq 0.693$,
   (iii) $x_{.75} = x^{.25} \doteq 1.386$.

2.2.5 *Modality*

Definition 2.13. A *mode* of a discrete or continuous r.v. X is a point at which the probability mass or density function f has a local maximum on support $\mathbb{X}$.

In words, a modal value is a point in $\mathbb{X}$ having *locally* high frequency or density. As we shall see in Chapter 4, most of the standard models for p.d.f.s and p.m.f.s are in fact *unimodal*, there being a single point m such that $f(m) > f(x)$ for all $x \in \mathbb{X}$ with $x \ne m$. The models have this feature because most data that we describe do exhibit a single "hump" of highest frequency. Figure 2.4 illustrates a discrete distribution with this property: $f\left(x\right) = \theta^x e^{-\theta} / x!$ for $x \in \mathbb{X} = \aleph_0$ and $\theta = 5.5$.

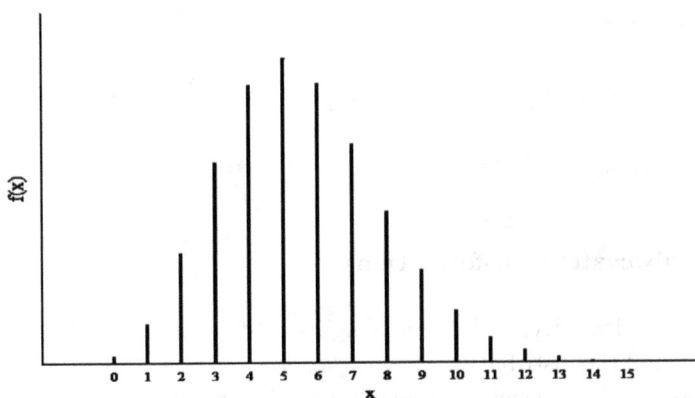

Fig. 2.4   A unimodal discrete distribution.

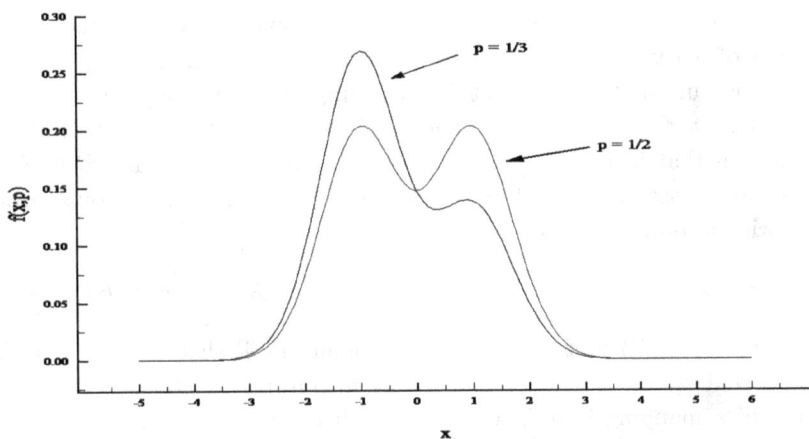

Fig. 2.5   Two bimodal density functions.

The presence of bi- or multimodality often indicates that the data are a mixture of disparate types and that further analysis is needed to describe them. For example, the distribution of heights of a mixed group of varsity wrestlers and basketball players is apt to be bimodal. Whether it is depends on the difference between the modes of the distinct groups, the dispersion within each group, and the relative numbers in each. Figure 2.5 depicts bimodal densities

$$f(x;p) = p\frac{e^{-(x+1)^2/2}}{\sqrt{2\pi}} + (1-p)\frac{e^{-(x-1)^2/2}}{\sqrt{2\pi}}$$

with $p = 1/3$ and $p = 1/2$.

2.3   Univariate Transformations

Having modeled a r.v. X by specifying its c.d.f., p.m.f., or p.d.f., one often wants to find the distribution of some *function* of X. For example, suppose X represents the clockwise distance from origin to pointer in Example 1.7, with p.d.f. modeled as $f(x) = \mathbf{1}_{[0,1]}(x)$. Adopting this device for a casino game, suppose the player wins $Y := -\log X$ units of currency at each play. Obviously, the player is most vitally interested not in the distribution of X itself but in that of Y. How can this be found?

To see how, let us consider the general problem of finding the distribution of a r.v. Y that depends on X as $Y = g(X)$, where g is some Borel-measurable function. Let X have support $\mathbb{X}$ and suppose g maps $\mathbb{X}$ onto the set $\mathbb{Y}$; i.e., $X : \Omega \to \mathbb{X}$ and $g : \mathbb{X} \to \mathbb{Y} \subset \Re$. The measurability of g requires that inverse image $g^{-1}(B) := \{x : g(x) \in B\}$ of any $B \in \mathcal{B}$ be also a Borel set, and therefore one whose $\mathbb{P}_X$ measure can be found from the original probability space, as

$$\mathbb{P}_X\left[g^{-1}(B)\right] = \mathbb{P}\left(X^{-1}[g^{-1}(B)]\right) = \mathbb{P}\left(\{\omega : X(\omega) \in g^{-1}(B)\}\right).$$

In words, $g^{-1}(B)$ is simply the set of points in $\Re$ that g maps into B. (Note that "g^{-1}" here does *not* denote an inverse *function* in the usual sense of a mapping from $\mathbb{Y}$ to $\mathbb{X}$, for both the argument of g^{-1} and the value of g^{-1} itself are *sets* in the present context.) Were g not measurable, $Y = g(X)$ would not be a r.v., but fortunately all the ordinary functions one encounters in mathematical analysis satisfy this measurability property. Indeed, in most applied problems finding the set $g^{-1}(B)$ is straightforward, and from there it is an easy step to find the measure $\mathbb{P}_Y$ that is induced by Y.

Here is how to do it. Since events $Y \in B$ and $X \in g^{-1}(B)$ are precisely the same, so are their probabilities. That is,

$$\begin{aligned}\mathbb{P}_Y(B) &= \mathbb{P}\left(\{\omega : Y(\omega) \in B\}\right) = \mathbb{P}\left(\{\omega : g[X(\omega)] \in B\}\right) \quad\quad (2.8)\\ &= \mathbb{P}\left(\{\omega : X(\omega) \in g^{-1}(B)\}\right) = \mathbb{P}_X\left[g^{-1}(B)\right].\end{aligned}$$

Thus, the $\mathbb{P}_Y$ measure of B is just the $\mathbb{P}_X$ measure of the set of real numbers to which g assigns a value in B. Upon taking sets of the specific form $B = (-\infty, y]$ for arbitrary real y, (2.8) delivers the c.d.f. of Y:

$$F_Y(y) = \mathbb{P}_Y\left((-\infty, y]\right) = \mathbb{P}_X\left[g^{-1}\left((-\infty, y]\right)\right], y \in \Re. \qquad (2.9)$$

Notice that just as the original r.v. $X : \Omega \to \Re$ took us from probability space $(\Omega, \mathcal{F}, \mathbb{P})$ to induced space $(\Re, \mathcal{B}, \mathbb{P}_X)$, the additional transformation $g : \mathbb{X} \to \mathbb{Y} \subset \Re$ takes us on to $(\Re, \mathcal{B}, \mathbb{P}_Y)$; and just as we found $\mathbb{P}_X(B)$ as $\mathbb{P}\left[X^{-1}(B)\right]$, we find $\mathbb{P}_Y(B)$ as $\mathbb{P}_X\left[g^{-1}(B)\right]$. In both cases we move from one probability space to another by working with inverse images, and this is made possible by the measurability of the functions X and g. Notice, too, that none of the arguments leading to (2.8) or (2.9) depends on whether X is discrete, continuous, or mixed. Thus, these expressions suffice to determine distributions of measurable functions of all r.v.s; however, some very useful shortcuts do apply when X is either purely discrete or purely continuous.[13]

2.3.1   Functions of a Discrete Random Variable

When X is supported on a countable set $\mathbb{X}$, then $\mathbb{Y}$, the image of $\mathbb{X}$ under g, is necessarily countable as well, since *function* g assigns to each $x \in \mathbb{X}$ only a single value. Therefore, Y is also a discrete r.v., and its p.m.f. can be found directly from (2.8) without first deriving the c.d.f. Thus, suppose X has p.m.f. f_X, where $f_X(x) := \mathbb{P}_X(\{x\}) > 0$ for each $x \in \mathbb{X}$, and let $g : \mathbb{X} \to \mathbb{Y}$. In (2.8) take $B = \{y\}$, the singleton set containing some (any) element y of $\mathbb{Y}$. Then $g^{-1}(\{y\}) := \{x : g(x) = y\}$ is the set of all x values—there can be arbitrarily many—to which g assigns the specific value y, and

$$\mathbb{P}_Y(\{y\}) = \mathbb{P}_X\left[g^{-1}(\{y\})\right] = \sum_{x \in g^{-1}(\{y\})} \mathbb{P}_X(\{x\}) = \sum_{x \in g^{-1}(\{y\})} f_X(x).$$

Thus, the p.m.f. of Y at each $y \in \mathbb{Y}$ is found by summing the values of f_X at all points x to which g assigns the value y, as $f_Y(y) = \sum_{x \in g^{-1}(\{y\})} f_X(x)$. (Remember that the summation here actually runs over the set $g^{-1}(\{y\}) \cap \mathbb{X}$—the *countably* many points in $g^{-1}(\{y\})$ at which $f_X > 0$.)

We have emphasized that the set $g^{-1}(\{y\})$ may well contain more than one value of x. However, when function g is one to one on $\mathbb{X}$, then a

---

[13]We shall return to the introductory gambling example in Example 2.23, which will use one of the shortcuts to show that $f_Y(y) = e^{-y}1_{(0,\infty)}(y)$ when $Y = -\log X$ and $f_X(x) = 1_{[0,1)}(x)$.

single $x \in \mathbb{X}$ does correspond to each $y \in \mathbb{Y}$, and there exists a real-valued $h : \mathbb{Y} \to \mathbb{X}$. There is then a simple formula for the p.m.f. of Y:

$$f_Y(y) = f_X[h(y)].$$

Example 2.19. With $f_X(x) = 2^{-x}1_{\aleph}(x)$ take $g(X) = X^2$, so $g : \mathbb{X} = \aleph \to \mathbb{Y} = \{1, 4, 9, ...\}$. While g is not one to one on all of $\Re$, it *is* one to one on $\mathbb{X}$, with $h(y) = +\sqrt{y}$. Thus,

$$f_Y(y) = f_X(\sqrt{y}) = 2^{-\sqrt{y}}1_{\mathbb{Y}}(y).$$

Example 2.20. With $f_X(x) = \frac{1}{4}1_{\{-1,0,1,2\}}(x)$ take $g(X) = X^2$. Now $g : \mathbb{X} = \{-1, 0, 1, 2\} \to \mathbb{Y} = \{0, 1, 4\}$ is not one to one on the support of X. Since $g^{-1}(\{0\}) = \{0\}$, $g^{-1}(\{1\}) = \{-1, 1\}$, and $g^{-1}(\{4\}) = \{2\}$, the p.m.f. of Y is

$$f_Y(y) = \begin{cases} f_X(0) & = \frac{1}{4}, y = 0 \\ f_X(-1) + f_X(1) & = \frac{1}{2}, y = 1 \\ f_X(2) & = \frac{1}{4}, y = 4 \\ 0, \text{ else.} \end{cases} = \frac{1}{4}1_{\{0,4\}}(y) + \frac{1}{2}1_{\{1\}}(y).$$

Exercise 2.10. *If X is the number of heads in three flips of a fair coin and $Y := X^2 - 2X$, find the p.m.f. of Y.*

Exercise 2.11. *Whenever the support of a discrete r.v. is unbounded, its c.d.f. has a countable infinity of discontinuities; for example, the r.v. X with $\mathbb{X} = \aleph$ and p.m.f. $f_X(x) = 2^{-x}1_{\aleph}(x)$. Construct a r.v. $Y = g(X)$ (for this particular X) whose c.d.f. F_Y has infinitely many discontinuities on an interval of finite length.*

2.3.2　*Functions of a Continuous Random Variable*

When X is continuous, the new variable $Y = g(X)$ may be continuous, discrete, or mixed, depending on g. In any case the distribution of Y can be represented through the c.d.f., which is found by expressing $\mathbb{P}_X\left[g^{-1}((-\infty, y])\right]$ in (2.9) for each real y as an integral of the p.d.f.:

$$F_Y(y) = \mathbb{P}_Y((-\infty, y]) = \mathbb{P}_X\left[g^{-1}((-\infty, y])\right] = \int_{g^{-1}((-\infty,y])} f_X(x) \cdot dx.$$

$$(2.10)$$

Whether there is another representation than the c.d.f. depends on what type of r.v. Y turns out to be. Here is an example in which Y has a p.m.f.

Example 2.21. With $f_X(x) = 1_{[0,1]}(x)$ let

$$g(x) = \begin{cases} 0, & x < \frac{1}{3} \\ 1, & \frac{1}{3} \le x < \frac{2}{3} \\ 2, & \frac{2}{3} \le x \end{cases} = 1_{[1/3,\infty)}(x) + 1_{[2/3,\infty)}(x).$$

Thus, $g : \mathbb{X} = [0,1] \to \mathbb{Y} = \{0,1,2\}$, so Y is a discrete r.v. The p.m.f. can be worked out directly. Corresponding to the elements of $\mathbb{Y}$ are inverse images $g^{-1}(\{0\}) = (-\infty, \frac{1}{3})$, $g^{-1}(\{1\}) = [\frac{1}{3}, \frac{2}{3})$, $g^{-1}(\{2\}) = [\frac{2}{3}, \infty)$, while $g^{-1}(\{y\}) = \varnothing$ for $y \in \mathbb{Y}^c$. Using

$$f_Y(y) = \mathbb{P}_Y(\{y\}) = \mathbb{P}_X\left[g^{-1}(\{y\})\right] = \int_{g^{-1}(\{y\})} f_X(x) \cdot dx,$$

the p.m.f. is

$$f_Y(y) = \begin{cases} \int_{-\infty}^{\frac{1}{3}} f_X(x) \cdot dx = \int_{-\infty}^0 0 \cdot dx + \int_0^{\frac{1}{3}} 1 \cdot dx = \frac{1}{3}, & y = 0 \\ \int_{\frac{1}{3}}^{\frac{2}{3}} f_X(x) \cdot dx = \int_{\frac{1}{3}}^{\frac{2}{3}} 1 \cdot dx = \frac{1}{3}, & y = 1 \\ \int_{\frac{2}{3}}^{\infty} f_X(x) \cdot dx = \int_{\frac{2}{3}}^1 1 \cdot dx + \int_1^\infty 0 \cdot dx = \frac{1}{3}, & y = 2 \\ 0, & \text{else.} \end{cases}$$

$$= \frac{1}{3} 1_{\{0,1,2\}}(y).$$

Alternatively, one can work out c.d.f. F_Y and find f_Y from that. The inverse images under g of sets $g^{-1}((-\infty, y])$ are

$$g^{-1}((-\infty, y]) = \begin{cases} \varnothing, & y < 0 \\ (-\infty, \frac{1}{3}), & 0 \le y < 1 \\ (-\infty, \frac{2}{3}), & 1 \le y < 2 \\ (-\infty, \infty), & 2 \le y \end{cases}.$$

Using $F_y(y) = \mathbb{P}_Y((-\infty, y]) = \mathbb{P}_X\left[g^{-1}((-\infty, y])\right] = \int_{x \in g^{-1}((-\infty, y])} f_X(x) \cdot dx$, the c.d.f. is

$$F_Y(y) = \begin{cases} 0, & y < 0 \\ \int_{-\infty}^0 0 \cdot dx + \int_0^{1/3} 1 \cdot dx = \frac{1}{3}, & 0 \le y < 1 \\ \int_{-\infty}^0 0 \cdot dx + \int_0^{2/3} 1 \cdot dx = \frac{2}{3}, & 1 \le y < 2 \\ \int_{-\infty}^0 0 \cdot dx + \int_0^1 1 \cdot dx + \int_1^\infty 0 \cdot dx = 1, & 2 \le y \end{cases}.$$

Exercise 2.12. *Taking $Y = g(X) = X \cdot 1_{[.5,\infty)}(X)$ and $f_X(x) = 1_{[0,1]}(x)$, find a representation of the distribution of Y.*

In the example and the exercise $\mathbb{Y}$ contains points with positive probability mass because g is constant on one or more sets of positive $\mathbb{P}_X$ measure.

When there are no such sets on which g is constant, the new variable Y will also be continuous. One can then find the p.d.f. by first finding the c.d.f., then differentiating.

Example 2.22. With $Y := X^2$ and $f_X(x) = \frac{1}{2}\mathbf{1}_{[-1,1]}(x)$ we find f_Y by first applying (2.9) to find F_Y. Since $g^{-1}((-\infty, y]) = [-\sqrt{y}, \sqrt{y}]$ for $y > 0$ and $g^{-1}((-\infty, y]) = \varnothing$ elsewhere, we have

$$F_Y(y) = \mathbb{P}_X\left[g^{-1}((-\infty, y])\right] = \begin{cases} \int_{-\sqrt{y}}^{\sqrt{y}} f_X(x) \cdot dx = \min\left\{\sqrt{y}, 1\right\}, & y > 0 \\ 0, & y \le 0 \end{cases}$$

$$= \sqrt{y}\mathbf{1}_{(0,1)}(y) + \mathbf{1}_{[1,\infty)}(y).$$

Setting $f_Y(0) = 0$ and $f_Y(y) = F'_y(y)$ elsewhere gives

$$f_Y(y) = \frac{1}{2\sqrt{y}}\mathbf{1}_{(0,1)}(y).$$

Notice that the density is unbounded above.

Exercise 2.13. *Taking $Y := X^2$ and $f_X(x) = (2\pi)^{-1/2}e^{-x^2/2}$ for $x \in \Re$, find c.d.f. F_Y and, from it, p.d.f. f_Y. (Here X has the standard normal distribution, to be described at length in Section 4.3.6.)*

In the last example and exercise g is not constant on any set of positive measure, but neither is it one to one on the support of X. When g *is* one to one and differentiable almost surely (that is, except perhaps on $\mathbb{P}_X$-null sets), the p.d.f. can be obtained in a single step from the *change-of-variable (c.o.v.) formula*, which we now derive. First, however, a word of caution. The c.o.v. formula is an extremely useful shortcut *when it applies*, but working out the c.d.f. first, as done in the example and exercise, is in most cases the surest and most insightful way to find the p.d.f. of a function of a continuous r.v. The c.d.f. approach applies for *any* measurable g, whether one to one or not, and it conveys a sense of understanding that cannot be gained by just plugging into a formula.

Proceeding to the derivation, begin with the case that $g : \mathbb{X} \to \mathbb{Y}$ is almost surely differentiable and strictly *increasing* on $\mathbb{X}$. Thus, g' exists and is positive except possibly on a null set $\mathbb{X}_0 \subset \mathbb{X}$. Then (i) $g : \mathbb{X}\backslash\mathbb{X}_0 \to \mathbb{Y}\backslash\mathbb{Y}_0$ has unique inverse $h : \mathbb{Y}\backslash\mathbb{Y}_0 \to \mathbb{X}\backslash\mathbb{X}_0$ ($\mathbb{Y}_0$ being the $\mathbb{P}_Y$-null image of $\mathbb{X}_0$ under g); (ii) $dh(y)/dy$ also exists and is positive on $\mathbb{Y}\backslash\mathbb{Y}_0$; and (iii) $g^{-1}((-\infty, y]) = \{x : x \le h(y)\}$ on $\mathbb{Y}\backslash\mathbb{Y}_0$. Point (iii) follows since small values of y are associated with small values of x when g is strictly increasing. Then for any $y \in \mathbb{Y}\backslash\mathbb{Y}_0$

$$F_Y(y) = \mathbb{P}_X\left[g^{-1}((-\infty, y])\right] = \mathbb{P}_X((-\infty, h(y)]) = F_X[h(y)] \qquad (2.11)$$

and

$$f_Y(y) = F_Y'(y) = \frac{d}{dy} F_X[h(y)] = f_X[h(y)] \frac{dh(y)}{dy}.$$

This is easier to recall as $f_Y(y) = f_X(x) \cdot dx/dy$, just keeping in mind that x has to be expressed as a function of y. As usual, f_Y can be assigned any nonnegative value on null set $\mathbb{Y}_0$.

If g is almost surely strictly *decreasing* and differentiable, then $g^{-1}((-\infty, y]) = [h(y), \infty)$, since small values of y are now associated with large values of x. Moreover, the continuity of r.v. X implies that $\mathbb{P}_X([h(y), \infty)) = \mathbb{P}_X((h(y), \infty))$, so that

$$F_Y(y) = \mathbb{P}_X[(h(y), \infty)] = 1 - F_X[h(y)]$$

and

$$f_Y(y) = -f_X[h(y)] \frac{dh(y)}{dy} = -f_X(x) \cdot \frac{dx}{dy}.$$

Finally, merging the increasing and decreasing cases delivers the iconic *change-of-variable formula*:

$$f_Y(y) = f_X(x) \cdot \left| \frac{dx}{dy} \right|. \tag{2.12}$$

It may be that one-to-one function g is sufficiently complicated that an analytical expression for the inverse h cannot be found; for example, $g(x) = x^2 \log(1+x)$ for $x \geq 0$. The change-of-variable procedure still applies in this case, but f_Y must be evaluated computationally at each $y \in \mathbb{Y}$. For this one would find the unique root x_y such that $y - g(x_y) = 0$ and express (2.12) as

$$f_Y(y) = f_X(x_y) \left| \frac{1}{g'(x_y)} \right|.$$

The c.o.v. formula can seem a bit mysterious and be hard to recall—is it $|dx/dy|$ or $|dy/dx|$?—unless one goes through the derivation each time it is needed. Fortunately, it loses its mystery and becomes easily remembered once one understands why the factor $|dx/dy|$ appears. One should think of $|dx/dy| \equiv |dh(y)/dy| = |1/g'(x)|$ in (2.12) as a scale factor that accounts for any concentration or attenuation of probability mass as it is shifted from $\mathbb{X}$ to $\mathbb{Y}$ (or, more precisely, as it is shifted from $\mathbb{X}\backslash\mathbb{X}_0$ to $\mathbb{Y}\backslash\mathbb{Y}_0$). For example, if $g'(x) < 1$, the mass associated with interval $[x, x + \Delta x)$ for some small Δx is pushed into the shorter interval of length $\Delta y := g(x + \Delta x) - g(x) \doteq g'(x) \Delta x$. Accordingly, the average *density* over $[y, y + \Delta y)$ is higher than

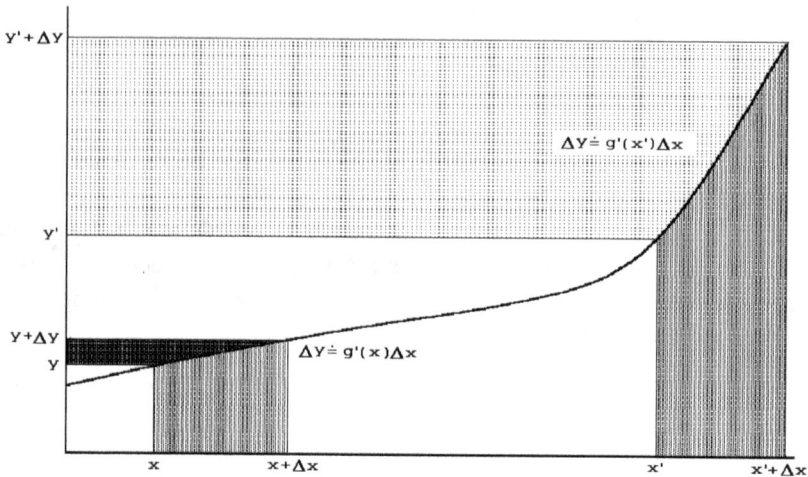

Fig. 2.6   Concentration and attenuation of probability mass via transformation.

that over $[x, x + \Delta x)$. Conversely, if at some x' we have $g'(x') > 1$, then the density around $y' = g(x')$ is reduced, since the mass on $[x', x' + \Delta x)$ is stretched over the longer interval $[g(x'), g(x') + g'(x') \Delta x)$. See Fig. 2.6.

Example 2.23. Let $f_X(x) = \mathbf{1}_{[0,1]}(x)$ and $g(X) = -\log X$. Although g is not defined at $x = 0$, the (singleton) set $\{0\}$ has zero $\mathbb{P}_X$ measure and can be ignored. In other words, $\mathbb{X} \backslash \{0\} = (0,1]$ is a support of X and g is strictly decreasing there, with inverse $X = h(Y) = e^{-Y}$ that maps $\mathbb{Y} = [0, \infty)$ onto $(0,1]$. Formula (2.12) gives

$$f_Y(y) = f_X(e^{-y}) \left| \frac{de^{-y}}{dy} \right| = e^{-y} \mathbf{1}_{[0,\infty)}(y).$$

Exercise 2.14. *Find f_Y if $Y = e^X$ and $f_X(x) = (2\pi)^{-1/2} e^{-x^2/2}, x \in \Re$. (Y is said to have a* lognormal *distribution, which is described in Section 4.3.8.)*

In order to determine whether a function $g : \mathbb{X} \to \mathbb{Y}$ is one to one on a domain $\mathbb{X}$, it is necessary to ascertain whether there is a unique inverse— that is, whether to each $y \in \mathbb{Y}$ there corresponds a single x. If it is difficult to judge this directly by inspection, graphing g over $\mathbb{X}$ (or some extensive subset thereof) can help. Of course, when g is *affine*, as $g(x) := a + bx$, then clearly g is one to one if and only if $g' = b \neq 0$. In general, however, not much can be inferred from g' alone. Example 2.22 makes clear that $g'(X) \neq 0$ a.s. does not imply that g is one to one, and neither does the fact that $g' = 0$ *somewhere* on $\mathbb{X}$ imply the converse. For example, $g(x) = x^3$ is one to one on $\Re$ even though $g'(0) = 0$. It is not enough even to know whether g' has uniform *sign* on all sets of positive measure. For example, g' is strictly positive on $(-1, 1] \backslash \{0\}$ when $g(x) := (1 + x)\mathbf{1}_{(-1,0]}(x) + x\mathbf{1}_{(0,1]}(x)$, yet g is not one to one; whereas g' changes sign for the one-to-one function $g(x) := (1 + x)\mathbf{1}_{(-1,0]}(x) + (2 - x)\mathbf{1}_{(0,1]}(x)$. Beyond the affine case, there are two situations in which correct inferences can be drawn from g'. First, g *is* one to one if g' exists, is continuous, and does not vanish throughout an interval containing $\mathbb{X}$. Second, g is *not* one to one if $g' = 0$ on a set of positive $\mathbb{P}_X$ measure.

When X is continuous and $g : \mathbb{X} \to \mathbb{Y}$ is not one to one, we have seen how to obtain the density from the c.d.f., as in Example 2.22, and we have emphasized that this is usually the surest method. Nevertheless, with due care it is possible to make use of the c.o.v. formula even when g is *not* one to one, and sometimes (as in Example 2.24 below) this is actually the most feasible approach. This is done by summing the contributions $f_X(x) |dx/dy|$ to $f_Y(y)$ from *all* points in $\mathbb{X}$ such that $g(x) = y$. However, this works *only if*, for the given y, the set $\{x \in \mathbb{X} : g(x) = y\}$ has zero $\mathbb{P}_X$ measure and $g' \neq 0$ there. Since X is continuous, this inverse image set will have zero $\mathbb{P}_X$ measure so long as it is *countable*. For example, suppose for some $y \in \mathbb{Y}$ there are points $\{x_j\}_{j=1}^{\infty} \in \mathbb{X}$ such that $g(x_j) = y$ and $g'(x_j) \neq 0$. Then to each x_j there correspond a distinct inverse function $h^j : y \to x_j$ and a contribution $f_X[h^j(y)] |dh^j(y)/dy|$ to the density at y. Aggregating the contributions,

$$f_Y(y) = \sum_{j=1}^{\infty} f_X[h^j(y)] \left| \frac{dh^j(y)}{dy} \right| = \sum_{j=1}^{\infty} f_X(x_j) \left| \frac{dx_j}{dy} \right|. \tag{2.13}$$

As in Example 2.22 take $Y = X^2$ and $f_X(x) = \frac{1}{2}\mathbf{1}_{[-1,1]}(x)$. Corresponding to each $y \in (0, 1]$ are $x_1 = +\sqrt{y} =: h^1(y)$ and $x_2 = -\sqrt{y} =: h^2(y)$, and so

$$f_Y(y) = f_X(\sqrt{y}) \left| \frac{d\sqrt{y}}{dy} \right| + f_X(-\sqrt{y}) \left| -\frac{d\sqrt{y}}{dy} \right|$$

$$= \left(\frac{1}{2} \left| \frac{1}{2\sqrt{y}} \right| + \frac{1}{2} \left| -\frac{1}{2\sqrt{y}} \right| \right) \mathbf{1}_{(0,1]}(y)$$

$$= \frac{1}{2\sqrt{y}} \mathbf{1}_{(0,1]}(y).$$

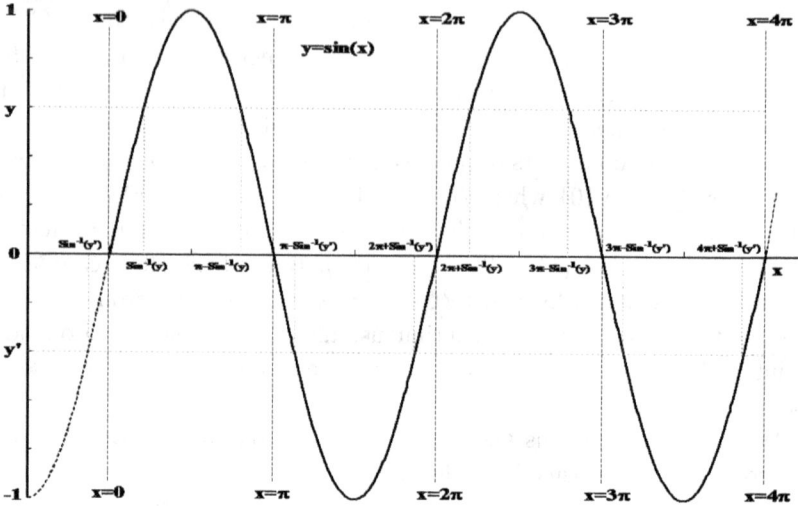

Fig. 2.7   Finding the p.d.f. of $Y = \sin X$ when $\mathbb{X} = [0, \infty)$.

Example 2.24. Let X have p.d.f. $f_X(x)$ and support $\mathbb{X} = [0, \infty)$, and take $Y = g(X) = \sin X$. Then $g : \mathbb{X} \to \mathbb{Y} = [-1, 1]$, as depicted in Fig. 2.7. The set of points $\mathbb{X}_0 := g^{-1}(\{-1, 1\}) \cap \mathbb{X} = \{j\pi/2\}_{j=1}^{\infty}$ at which $g' = 0$ is countable and therefore $\mathbb{P}_X$-null, so we may consider the restricted support $\mathbb{Y} \backslash \mathbb{Y}_0 = (-1, 1)$. For each $y \in \mathbb{Y} \backslash \mathbb{Y}_0$ there are also countably many points $x \in \mathbb{X}$ such that $y = \sin(x)$. Let $\sin^{-1}(y) \in (-\pi/2, \pi/2)$ represent the principal value of $\sin^{-1}(y)$. Referring to the figure, one can see that for any $y \in [0, 1)$ the set $\sin^{-1}(y) \cap \mathbb{X}$ comprises the points $\{2\pi j + \sin^{-1}(y), 2\pi j + \pi - \sin^{-1}(y)\}_{j=0}^{\infty}$, while for $y' \in (-1, 0)$ we have

$$\sin^{-1}(y') \cap \mathbb{X} = \{2\pi j + \pi - \sin^{-1}(y'), 2\pi(j+1) + \sin^{-1}(y')\}_{j=0}^{\infty}.$$

Since $|d\sin^{-1}(y)/dy| = |d\sin^{-1}(y)/dy| = (1 - y^2)^{-1/2}$ at each such point, Expression (2.13) gives for $f_Y(y)$

$$\sum_{j=0}^{\infty} \frac{f_X[2\pi j + \sin^{-1}(y)] + f_X[2\pi j + \pi - \sin^{-1}(y)]}{\sqrt{1 - y^2}}, y \in [0, 1)$$

$$\sum_{j=0}^{\infty} \frac{f_X[2\pi j + \pi - \sin^{-1}(y)] + f_X[2\pi(j+1) + \sin^{-1}(y)]}{\sqrt{1 - y^2}}, y \in (-1, 0)..$$

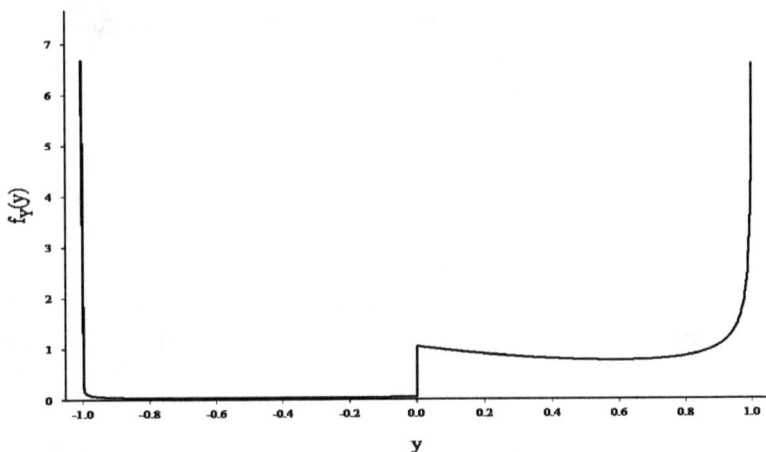

Fig. 2.8   P.d.f. of $Y = \sin(X)$ when X is distributed as exponential.

For example, with $f_X(x) = e^{-x}\mathbf{1}_{[0,\infty)}(x)$ this reduces to

$$f_Y(y) = \begin{cases} \frac{1}{1-e^{-2\pi}} \left[e^{-\sin^{-1}(y)} + e^{-\pi+\sin^{-1}(y)} \right] \frac{1}{\sqrt{1-y^2}}, & y \in [0,1) \\ \frac{e^{-\pi}}{1-e^{-2\pi}} \left[e^{\sin^{-1}(y)} + e^{-\pi-\sin^{-1}(y)} \right] \frac{1}{\sqrt{1-y^2}}, & y \in (-1,0) \end{cases}.$$

The rather strange looking result is depicted in Fig. 2.8. Notice that the p.d.f. is unbounded toward each end of the support.

2.4   Vector-valued Random Variables

We now extend the concept of random variable to the multivariate case. While there are indeed some new concepts, this section also serves as a review of the first two sections of the chapter.

To begin, here are a few words about terminology and notation. Throughout, the term "vector" is restricted to mean a k-tuple of objects— such as real or complex numbers or real- or complex-valued functions—all in the same class (i.e., all numbers or all functions, etc.). A vector with just one element is referred to simply as a "scalar". "Matrix" refers to a rectangular array comprising two or more vectors all having the same number of elements, all in the same class. Thus, we can consider arrays of numbers, or of real-valued functions of real variables, or of random variables on the same probability space, and so on. Symbols representing vectors and matrices are in bold face, and vectors are in column form unless transposition is indicated by a prime, as

$$\mathbf{v} = \begin{pmatrix} v_1 \\ v_2 \\ \vdots \\ v_k \end{pmatrix} \equiv (v_1, v_2, ..., v_k)'.$$

If the k elements of $\mathbf{v}$ are real numbers, then we write $\mathbf{v} \in \Re^k$. Thus, in this context the expression (v_1, v_2) represents a 2-vector—a point in $\Re^2$, rather than an open interval in $\Re$. The "inner" or "dot" product of vectors, $\mathbf{v}'\mathbf{v} : = \mathbf{v} \cdot \mathbf{v} := \sum_{j=1}^{k} v_j^2$, is a scalar, while $\mathbf{vv}' := \{v_i v_j\}_{i,j=1}^{k}$ is a $k \times k$ matrix of real numbers. Symbol $|\mathbf{v}| := \sqrt{\mathbf{v}'\mathbf{v}}$ represents the "length" of $\mathbf{v}$—the Euclidean distance of $\mathbf{v} \equiv (v_1, v_2, ..., v_k)'$ from $\mathbf{0} \equiv (0, 0, ..., 0)'$ (the origin); likewise, $|\mathbf{v}_1 - \mathbf{v}_2|$ represents the distance between $\mathbf{v}_1$ and $\mathbf{v}_2$. Finally, if $\mathbf{M}$ is a square *matrix*, then $|\mathbf{M}|$ represents its determinant and $\|\mathbf{M}\|$ represents the absolute value of $|\mathbf{M}|$.

Now given the model $(\Omega, \mathcal{F}, \mathbb{P})$ for a chance experiment, let us associate with each outcome ω the vector $\mathbf{X}(\omega) = [X_1(\omega), X_2(\omega), ..., X_k(\omega)]' \in \Re^k, k > 1$. That is, $\mathbf{X}$ assigns to ω the value $\mathbf{X}(\omega) = (x_1, x_2, ..., x_k)'$ comprising k real numbers. As before, $\mathbb{X}$ represents the support of $\mathbf{X}$, but now $\mathbb{X} \subset \Re^k$. Attaching superscript k, let $\mathcal{B}^k$ represent the Borel sets of $\Re^k$. This is the σ field generated by the "rectangles"—sets of the form $(a_1, b_1] \times (a_2, b_2] \times \cdots \times (a_k, b_k] =: \times_{j=1}^{k}(a_j, b_j]$, where "$\times$" represents Cartesian product. To all these sets of $\mathcal{B}^k$ it is possible to assign consistent measures of area (for $k = 2$), or volume (for $k = 3$), or their higher-dimensional analogs.

Definition 2.14. A (proper) vector-valued random variable on a probability space $(\Omega, \mathcal{F}, \mathbb{P})$ is an $\mathcal{F}$-measurable mapping from Ω to $\Re^k$.

As in the univariate case, the definition requires the inverse image $\mathbf{X}^{-1}(B_k) = \{\omega : \mathbf{X}(\omega) \in B_k\}$ of any $B_k \in \mathcal{B}^k$ to be a set in $\mathcal{F}$, and thus one whose probability can be determined. Also corresponding to the scalar case, it is enough that $\mathbf{X}^{-1}(B_k) \in \mathcal{F}$ for all sets in a collection that *generates* $\mathcal{B}^k$, such as the open or closed rectangles (product sets). Moreover, for $\mathbf{X}$ to be measurable it is necessary and sufficient that each scalar component X_j be measurable.

Example 2.25. One person is to be drawn at random from the students in a class. With Ω as the (finite) set of students, let $\mathbf{X} = (X_1, X_2)'$ assign to individual ω the ordered *pair* of numbers representing ω's height in inches and weight in pounds. With $B_2 := (64, 70] \times (130, 180]$, we interpret $\mathbf{X}^{-1}(B_2)$ as the set of individuals whose heights are greater than 64 but no greater than 70 inches *and* whose weights range from 130 through 180 pounds. With $B_2' := (64, 70] \times \Re$ we interpret $\mathbf{X}^{-1}(B_2')$ as the set of individuals whose heights are between 64 and 70 inches, without regard to weight. To $B_2'' := \Re \times (130, 180]$ we associate the set of individuals of any height whose weights are in the stated interval.

As in the scalar case the introduction of a vector r.v. on $(\Omega, \mathcal{F}, \mathbb{P})$ induces the new space $(\Re^k, \mathcal{B}^k, \mathbb{P}_\mathbf{X})$, where "joint" induced measure $\mathbb{P}_\mathbf{X}$ is defined as

$$\mathbb{P}_\mathbf{X}(B_k) := \mathbb{P}\left[\mathbf{X}^{-1}(B_k)\right] = \mathbb{P}(\{\omega : \mathbf{X}(\omega) \in B_k\}).$$

Any ordered collection of r.v.s $\{X_1, X_2, ..., X_k\}$ that are defined on *the same probability space* can be regarded as a vector r.v. $\mathbf{X}$ whose behavior is described by a joint measure $\mathbb{P}_\mathbf{X}$ on $(\Re^k, \mathcal{B}^k)$. Such component r.v.s are then said to be *jointly distributed*. Thus, if each of $X_1, X_2, ..., X_k$ are r.v.s—$\mathcal{F}$-measurable functions—on $(\Omega, \mathcal{F}, \mathbb{P})$, then vector $\mathbf{X} : \Omega \to \Re^k$ is also $\mathcal{F}$-measurable, meaning that $\{\omega : \mathbf{X}(\omega) \in B_k\} \in \mathcal{F}$ for each $B_k \in \mathcal{B}^k$.[14] If vector-valued r.v. $\mathbf{X}$ is "proper"—as we henceforth always assume—then $\lim_{n\to\infty} \mathbb{P}_\mathbf{X}(B_n) = 1$ for any increasing sequence $\{B_n\}$ of $\mathcal{B}^k$ sets such that $\cup_{n=1}^\infty B_n = \Re^k$.

Exercise 2.15. *If X and Y are r.v.s on $(\Omega, \mathcal{F}, \mathbb{P})$, show that*

$$\Pr(X + Y > r) \leq \Pr\left(X > \frac{r}{2}\right) + \Pr\left(Y > \frac{r}{2}\right), r \in \Re.$$

Although $\mathbb{P}_\mathbf{X}$ does afford a representation of the distribution of probability over $\Re^k$, there are other, more convenient representations, just as in the scalar case.

---

[14]To elaborate a bit, if $\{X_j\}_{j=1}^k$ are r.v.s on *different* probability spaces $\{(\Omega_j, \mathcal{F}_j, \mathbb{P}_j)\}$, then there is no way to link their behaviors probabilistically. For example, although we can determine separately for *each* X_j the probability that $X_j \in B_j$ (some Borel set of $\Re$), we cannot say with what chance events $\{X_j \in B_j\}_{j=1}^k$ *simultaneously* occur. Such a collection of r.v.s on different spaces is thus *not* a vector-valued r.v. They could be made such by embedding the separate spaces $\{(\Omega_j, \mathcal{F}_j, \mathbb{P}_j)\}$ into one, as, for example, by taking $\Omega = \times_{j=1}^k \Omega_j$, $\mathcal{F}$ the σ field generated by events of the form $\times_{j=1}^k S_j$ for $S_j \in \mathcal{F}_j$, and $\mathbb{P}\left(\times_{j=1}^k S_j\right) = \prod_{j=1}^k \mathbb{P}_j(S_j)$. (This *particular* way of doing it would make r.v.s $\{X_j\}$ both jointly distributed and *independent*, as explained below.)

2.4.1 *Representing Bivariate Distributions*

The essential features of multivariate distributions can be understood by looking just at the bivariate case, $k = 2$; and subscripts can be avoided by replacing $\mathbf{X} := (X_1, X_2)'$ with $\mathbf{Z} := (X, Y)'$. The general k-dimensional case is treated briefly in Section 2.4.5. Restricting now to $k = 2$, let $\mathbf{Z}$ have support $\mathbb{Z} \subset \Re^2$. The induced measure is now $\mathbb{P}_{\mathbf{Z}}$, where $\mathbb{P}_{\mathbf{Z}}(B_2) = \mathbb{P}\left[\mathbf{Z}^{-1}(B_2)\right] = \mathbb{P}\left(\{\omega : (X, Y)'(\omega) \in B_2\}\right)$. The *joint c.d.f.* of $(X, Y)'$ is defined analogously to the scalar case as the $\mathbb{P}_{\mathbf{Z}}$ measure of right-closed infinite rectangles:

$$F_{\mathbf{Z}}(x, y) := \mathbb{P}_{\mathbf{Z}}\left((-\infty, x] \times (-\infty, y]\right) = \Pr\left(X \leq x, Y \leq y\right).$$

Thus, $F_{\mathbf{Z}}(x, y)$ represents the probability that X takes a value no greater

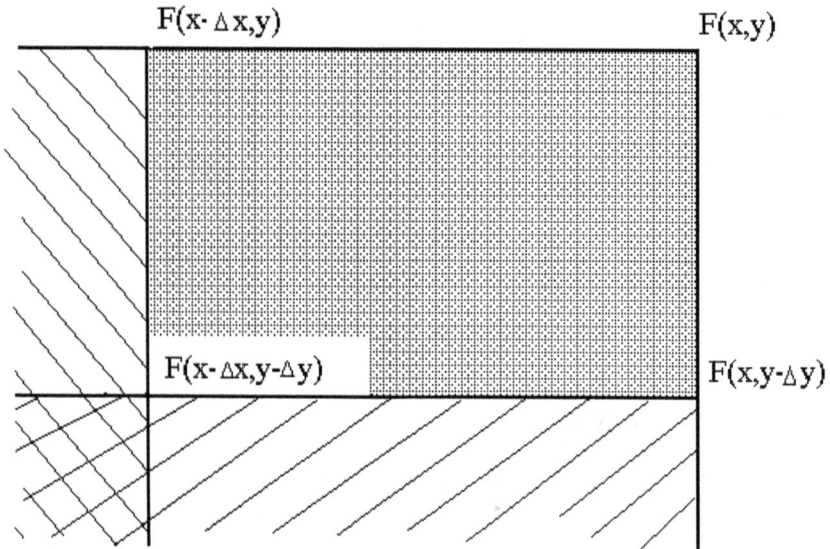

Fig. 2.9   Probability mass on a rectangle from the c.d.f.

than x and, simultaneously, Y takes a value no greater than y. Again, the Borel sets of $\Re^2$ can be generated from these infinite rectangles just as they can from the finite ones. Figure 2.9 illustrates how to find the measure of $(x - \Delta x, x] \times (y - \Delta y, y]$ from the joint c.d.f.:

$$\mathbb{P}_{\mathbf{Z}}\left((x - \Delta x, x] \times (y - \Delta y, y]\right) = F_{\mathbf{Z}}(x, y) - F_{\mathbf{Z}}(x, y - \Delta y)$$
$$-F_{\mathbf{Z}}(x - \Delta x, y) + F_{\mathbf{Z}}(x - \Delta x, y - \Delta y). \tag{2.14}$$

In the figure the probability of the stippled rectangle is found by the following steps: (i) subtract from the value of the c.d.f. at the upper right corner (which is the probability of the infinite rectangle extending to the left and below) the values of the c.d.f. at the upper left and lower right corners (the probabilities associated with the two regions marked $\backslash\backslash\backslash$ and $///$); then (ii) add back the value of the c.d.f. at the lower left corner (the probability in the cross-hatched region $\times\times\times$), which otherwise would have been subtracted twice from $F_{\mathbf{Z}}(x, y)$.

Exercise 2.16. *A coin is flipped and then a pointer on a unit circular scale is spun. X represents the number of heads on the coin, with $\mathbb{X} = \{0, 1\}$, and Y represents the clockwise distance of pointer from origin, with $\mathbb{Y} = [0, 1]$. Assuming that events $X \in B$ and $Y \in B'$ are independent, where B and B' are any Borel sets, show that the joint c.d.f. of $\mathbf{Z} = (X, Y)'$ is*

$$F_{\mathbf{Z}}(x, y) = \begin{cases} 0, & x < 0 \ or \ y < 0 \\ \max\{0, .5 \min\{y, 1\}\}, & 0 \leq x < 1 \\ \max\{0, \min\{y, 1\}\}, & x \geq 1 \end{cases}.$$

The exercise concerns a vector r.v. that is neither discrete nor continuous, since it combines two different types. When vector valued r.v.s *are* either discrete or continuous, the joint distribution can be represented by joint p.m.f.s and p.d.f.s, respectively.

2.4.1.1 *Joint P.M.F.s of Discrete R.V.s*

Letting Δx and Δy approach zero in (2.14), the sequence of rectangles converges to the point (x, y), the only point in $\Re^2$ that is contained in all members of the sequence. Thus, one may determine $\Pr(X = x, Y = y) = \mathbb{P}_{\mathbf{Z}}(\{(x, y)\})$ as

$$\Pr(X = x, Y = y) = F_{\mathbf{Z}}(x, y) - F_{\mathbf{Z}}(x, y-) - F_{\mathbf{Z}}(x-, y) + F_{\mathbf{Z}}(x-, y-),$$

where arguments "$x-$" and "$y-$" denote left-hand limits. In the discrete case the support of $(X, Y)'$ is a countable set $\mathbb{Z}$, at each point of which the probability mass, as computed above, is positive. The distribution of $(X, Y)'$ in the discrete case can therefore be represented by a *joint* p.m.f. on $\mathbb{Z}$, as

$$f_{\mathbf{Z}}(x, y) := \mathbb{P}_{\mathbf{Z}}(\{(x, y)\}) = \Pr(X = x, Y = y).$$

$f_{\mathbf{Z}}$ has the properties $0 < f_{\mathbf{Z}}(x, y) \leq 1$ for $(x, y) \in \mathbb{Z}$ and $\sum_{(x,y) \in \mathbb{Z}} f_{\mathbf{Z}}(x, y) = 1$. In the bivariate case these joint p.m.f.s can often be expressed compactly as a table. It is usually easier to deduce the joint p.m.f. directly, rather than from the c.d.f.

Example 2.26. A fair coin is flipped three times. Letting Ω comprise the resulting 3-character strings of H's and T's, let $X(\omega)$ be the number of heads in outcome ω and $Y(\omega)$ be the number of changes of sequence from H to T or from T to H. The table below shows how (X, Y) pairs are assigned to the eight outcomes, and display (2.15) gives a representation of $f_{\mathbf{Z}}$.

$\omega:$	TTT	THT	TTH	HTT	HHT	HTH	THH	HHH
$X(\omega):$	0	1	1	1	2	2	2	3
$Y(\omega):$	0	2	1	1	1	2	1	0

$$
\begin{array}{c|cccc}
y\backslash x & 0 & 1 & 2 & 3 \\
\hline
0 & \frac{1}{8} & 0 & 0 & \frac{1}{8} \\
1 & 0 & \frac{2}{8} & \frac{2}{8} & 0 \\
2 & 0 & \frac{1}{8} & \frac{1}{8} & 0
\end{array}
\tag{2.15}
$$

Thus, $f_{\mathbf{Z}}(0,0) = \frac{1}{8}$ because the single outcome TTT has both zero heads and zero changes of sequence, while $f_{\mathbf{Z}}(1,1) = \frac{2}{8}$ since the two outcomes TTH and HTT have one of each.

Given a Borel set $B_2 \subset \Re^2$, summing f_Z over the countably many elements belonging to both the support and B_2 gives $\mathbb{P}_{\mathbf{Z}}(B_2) = \Pr(\mathbf{Z} \in B_2) = \sum_{(x,y) \in B_2 \cap Z} f_{\mathbf{Z}}(x,y)$. As in the scalar case, we abbreviate the notation as $\mathbb{P}_{\mathbf{Z}}(B_2) = \sum_{(x,y) \in B_2} f_{\mathbf{Z}}(x,y)$ or simply as $\sum_{B_2} f_{\mathbf{Z}}(x,y)$, its being understood that the sum extends just over the appropriate *countable* set. Taking $B_2 = (-\infty, x] \times (-\infty, y]$ gives the c.d.f. at (x, y):

$$
F_{\mathbf{Z}}(x,y) = \sum_{(-\infty,x] \times (-\infty,y]} f_{\mathbf{Z}}(s,t) =: \sum_{s \leq x} \sum_{t \leq y} f_{\mathbf{Z}}(s,t).
$$

As in the scalar case joint p.m.f. $f_{\mathbf{Z}}$ can be regarded as a density with respect to *counting measure*, $N(\cdot)$, which assigns to Borel set B_2 the number of points in $B_2 \cap Z$. We then have the integral representation $\mathbb{P}_{\mathbf{Z}}(B_2) = \int f_Z(s,t) N(ds \times dt)$.

Example 2.27. In the 3-coin example with $B_2 = (.5, 2] \times [1, 2.5)$
$$
\begin{aligned}
\mathbb{P}_{\mathbf{Z}}(B_2) &= \Pr(.5 < X \leq 2, 1 \leq Y < 2.5) \\
&= f_{\mathbf{Z}}(1,1) + f_{\mathbf{Z}}(1,2) + f_{\mathbf{Z}}(2,1) + f_{\mathbf{Z}}(2,2) \\
&= \frac{6}{8}.
\end{aligned}
$$
With B_2' as the set of all ordered pairs of positive integers such that the first is even and the second is odd we have
$$
\mathbb{P}_{\mathbf{Z}}(B_2') = f_{\mathbf{Z}}(2,1) = \frac{2}{8}.
$$

2.4.1.2   Joint P.D.F.s of Continuous R.V.s

If the joint c.d.f. of $\mathbf{Z} = (X, Y)'$ is (absolutely) continuous, then $\mathbf{Z}$ is a *continuous* vector-valued r.v. There then exists a density $f_{\mathbf{Z}}$ with respect to Lebesgue measure such that

$$F_{\mathbf{Z}}(x, y) = \int_{-\infty}^{x} \int_{-\infty}^{y} f_{\mathbf{Z}}(s, t) \cdot dt\, ds$$

for each $(x, y) \in \Re^2$. Density $f_{\mathbf{Z}}$ is called the *joint p.d.f.* of $\mathbf{Z} = (X, Y)'$. To interpret it, look again at Fig. 2.9 and Expression (2.14). The latter can be written as

$$\mathbb{P}_{\mathbf{Z}}\left((x - \Delta x, x] \times (y - \Delta y, y]\right) = \Delta_y F_{\mathbf{Z}}(x, y) - \Delta_y F_{\mathbf{Z}}(x - \Delta x, y),$$

where $\Delta_y F_{\mathbf{Z}}(\cdot, y) := F_{\mathbf{Z}}(\cdot, y) - F_{\mathbf{Z}}(\cdot, y - \Delta y)$. With like convention for changes in the x direction the above probability is just $\Delta_x \Delta_y F_{\mathbf{Z}}(x, y)$. The ratio of this to the area of the rectangle, $\Delta x \cdot \Delta y$, represents the average *density* of probability over this region. The limit as $\Delta x \to 0$ and $\Delta y \to 0$ (if it exists) equals the partial derivative $\partial^2 F_{\mathbf{Z}}(x, y) / \partial x \partial y$, which is interpreted as the density of probability at (x, y). Absolute continuity implies that the derivative does exist a.e. $\mathbb{P}_{\mathbf{Z}}$; that is, everywhere except perhaps on a set of zero $\mathbb{P}_{\mathbf{Z}}$ measure. For other than such exceptional points we can take $f_{\mathbf{Z}}(x, y) = \partial^2 F_{\mathbf{Z}}(x, y) / \partial x \partial y$, while $f_{\mathbf{Z}}(x, y)$ can be given any nonnegative value on the exceptional sets themselves. The resulting *version* of the joint p.d.f. has the properties (i) $f_{\mathbf{Z}}(x, y) \geq 0$; (ii) $\int_{\Re^2} f_{\mathbf{Z}}(x, y) \cdot dx\, dy = 1$; and (iii) $\Pr\left((X, Y)' \in B_2\right) = \int_{B_2} f_{\mathbf{Z}}(x, y) \cdot dx\, dy$.

When B_2 is a rectangle—that is, a product set such as $(a, b] \times (c, d]$, Fubini's theorem from real analysis authorizes the integral in (iii) to be evaluated as a repeated integral, and in either order, as

$$\Pr\left((X, Y) \in (a, b] \times (c, d]\right) = \Pr(a < X \leq b, c < Y \leq d)$$
$$= \int_{c}^{d} \int_{a}^{b} f_{\mathbf{Z}}(x, y) \cdot dx\, dy$$
$$= \int_{a}^{b} \int_{c}^{d} f_{\mathbf{Z}}(x, y) \cdot dy\, dx.$$

When the intersection of B_2 and the support $\mathbb{Z}$ is not rectangular, or if the density function is defined piecewise in nonrectangular regions, the limits in the inner integral have to be expressed as functions of the variable in the outer integral. This is illustrated in Example 2.29 and Exercise 2.21 below, but here first are some simpler cases.

Example 2.28. The pointer on the unit circular scale is spun twice, and $(X, Y)'$ is the vector of clockwise distances from origin to pointer on the first and second spins, respectively. Take the probability space to be $([0,1]^2, \mathcal{B}_{[0,1]}^2, \lambda)$, where $[0,1]^2 := [0,1] \times [0,1]$, $\mathcal{B}_{[0,1]}^2$ are the Borel sets of the unit rectangle, and λ is two-dimensional Lebesgue measure—i.e., area. Outcomes ω being themselves vectors of distances (real numbers), define $\mathbf{Z}(\omega) := (X, Y)'(\omega) = \omega$ and extend the probability space to $(\Re^2, \mathcal{B}^2)$ by setting $\mathbb{P}_{\mathbf{Z}}(B_2) = \lambda\left(B_2 \cap [0,1)^2\right)$. The c.d.f. of $\mathbf{Z}$ is

$$
F_{\mathbf{Z}}(x,y) = \begin{cases} 0, & x < 0 \text{ or } y < 0 \\ xy, & x \in [0,1], y \in [0,1] \\ x, & x \in [0,1], y \geq 1 \\ y, & y \in [0,1], x \geq 1 \\ 1, & x > 1, y > 1 \end{cases}.
$$

Exercise 2.17. *Show that the above c.d.f. can be represented compactly as*

$$
F_{\mathbf{Z}}(x,y) = \min\left\{\max\left\{x, 0\right\}, 1\right\} \cdot \min\left\{\max\left\{y, 0\right\}, 1\right\}, (x,y) \in \Re^2.
$$

Exercise 2.18. *Find the joint p.d.f. corresponding to $F_{\mathbf{Z}}$.*

Exercise 2.19. *Show that $f_{\mathbf{Z}}(x,y) = e^{-x-y}\mathbf{1}_{[0,\infty)^2}(x,y)$ has the properties of a joint p.d.f.*

Exercise 2.20. *Find the joint c.d.f. corresponding to the p.d.f. in the previous exercise.*

Example 2.29. With $f_{\mathbf{Z}}(x,y) = e^{-x-y}\mathbf{1}_{(0,\infty)^2}(x,y)$ we find $\Pr(Y > X) = \mathbb{P}_{\mathbf{Z}}\left(\{(x,y)\} : y > x\right)$. This can be evaluated as a repeated integral of the p.d.f., in either order. Integrating first over y and restricting y to be greater than x give

$$
\begin{aligned}
\Pr(Y > X) &= \int_{x=-\infty}^{\infty} \int_{y=x}^{\infty} f_{\mathbf{Z}}(x,y) \cdot dy\, dx \\
&= \int_{x=-\infty}^{0} \int_{y=x}^{\infty} 0 \cdot dy\, dx + \int_{x=0}^{\infty} \int_{y=x}^{\infty} e^{-x-y} \cdot dy\, dx \\
&= \int_{x=0}^{\infty} e^{-x} \left[\int_{y=x}^{\infty} e^{-y} \cdot dy\right] \cdot dx \\
&= \int_{x=0}^{\infty} e^{-x} \left[e^{-x}\right] \cdot dx = \frac{1}{2}.
\end{aligned}
$$

Alternatively, integrating first over x and constraining it to be less than y give

$$\Pr(Y > X) = \int_{y=-\infty}^{\infty} \int_{x=-\infty}^{y} f_{\mathbf{Z}}(x,y) \cdot dx \, dy$$

$$= \int_{y=0}^{\infty} \int_{x=0}^{y} e^{-x-y} \cdot dx \, dy$$

$$= \int_{y=0}^{\infty} e^{-y} \left(\int_{x=0}^{y} e^{-x} \cdot dx \right) \cdot dy$$

$$= \int_{y=0}^{\infty} e^{-y} \left(1 - e^{-y} \right) \cdot dy = \frac{1}{2}.$$

Exercise 2.21. *(i) Show that $f_{\mathbf{Z}}(x,y) = 2e^{-x-y} 1_{\{(x,y):0<x<y\}}(x,y)$ is a p.d.f., and (ii) find $\Pr(X > 1)$.*

2.4.2 Marginal Distributions

Given the vector-valued $\mathbf{Z} = (X,Y)'$ and bivariate measure $\mathbb{P}_{\mathbf{Z}}$, consider the event that X takes a value in some *one-dimensional* Borel set B, without restriction on Y. Since this is the event that $\mathbf{Z}$ takes a value in the subset $B \times \Re$ of $\Re^2$, its probability is simply $\mathbb{P}_{\mathbf{Z}}(B \times \Re)$. Likewise the event that Y takes a value in a one-dimensional set B', without restriction on X, has probability $\mathbb{P}_{\mathbf{Z}}(\Re \times B')$. Since B and B' are arbitrary members of the one-dimensional Borel sets $\mathcal{B}$, it follows that both $\mathbb{P}_{\mathbf{Z}}(B \times \Re)$ and $\mathbb{P}_{\mathbf{Z}}(\Re \times B')$ are nonnegative, countably-additive set functions—that is, *measures*—on the one-dimensional measure space $(\Re, \mathcal{B})$. Moreover, they are probability measures since $\mathbb{P}_{\mathbf{Z}}(\Re \times \Re) = 1$.

Definition 2.15. The measures $\mathbb{P}_X(\cdot) := \mathbb{P}_{\mathbf{Z}}(\cdot \times \Re)$ on $(\Re, \mathcal{B})$ and $\mathbb{P}_Y(\cdot) := \mathbb{P}_{\mathbf{Z}}(\Re \times \cdot)$ on $(\Re, \mathcal{B})$ are the *marginal* probability measures induced by the r.v.s X and Y, respectively.

The marginal measures are the easiest ones to use to find probabilities of events involving either X or Y alone. As usual, however, there are more convenient functions from which such probabilities can be deduced.

Definition 2.16. Given $\mathbf{Z} = (X,Y)'$ and the associated joint and marginal measures, the marginal c.d.f.s of X and Y are

$$F_X(x) = \mathbb{P}_X((-\infty, x]) = \mathbb{P}_{\mathbf{Z}}((-\infty, x] \times \Re), \ x \in \Re$$
$$F_Y(y) = \mathbb{P}_Y((-\infty, y]) = \mathbb{P}_{\mathbf{Z}}(\Re \times (-\infty, y]), \ y \in \Re.$$

Exercise 2.22. *Let $F_\mathbf{Z}$ be the joint c.d.f. of $\mathbf{Z} = (X, Y)'$. Show that the marginal c.d.f.s are*

$$F_X(x) = F_\mathbf{Z}(x, +\infty), x \in \Re$$
$$F_Y(y) = F_\mathbf{Z}(+\infty, y), y \in \Re.$$

Exercise 2.23. *Find the marginal c.d.f.s corresponding to*

$$F_\mathbf{Z}(x, y) = \left(1 - e^{-x} - e^{-y} + e^{-x-y}\right) \mathbf{1}_{(0,\infty)^2}(x, y).$$

As stated earlier, a *discrete* vector-valued r.v. $\mathbf{Z} = (X, Y)'$ is supported on a countable set $\mathbb{Z}$. It follows that the set of points $\{x_j\}$ such that $(x_j, y)' \in \mathbb{Z}$ is countable for each real y; likewise, the set of points $\{y_j\}$ such that $(x, y_j)' \in \mathbb{Z}$ is countable for each real x. The marginal measures $\mathbb{P}_X$ and $\mathbb{P}_Y$ thus having countable supports $\mathbb{X}$ and $\mathbb{Y}$, it follows that (i) both X and Y are discrete r.v.s.; (ii) marginal c.d.f.s F_X and F_Y increase only by jumps, of which there are countably many; and (iii) the variables have marginal p.m.f.s f_X and f_Y. The marginal p.m.f.s can be found either from the marginal c.d.f.s, as $f_X(x) = F_X(x) - F_X(x-)$, or directly from the joint p.m.f. of $\mathbf{Z}$, as

$$f_X(x) = \mathbb{P}_X(\{x\}) = \sum_{y:(x,y)\in\mathbb{Z}} \mathbb{P}_\mathbf{Z}(\{(x,y)\}) = \sum_{y:(x,y)\in\mathbb{Z}} f_\mathbf{Z}(x, y),$$

where summation is over all values of y such that $(x, y) \in \mathbb{Z}$.

Example 2.30. The joint p.m.f. of X and Y, the numbers of heads and sequence changes in three flips of a fair coin, is shown below. The extra column in the right margin and the extra row in the bottom display the marginal p.m.f.s. (That they appear in the margins of such tables explains the origin of the term "marginal".) $f_Y(y)$ is found by summing $f_\mathbf{Z}(\cdot, y)$ over all cells of row y, and $f_X(x)$ is found by summing $f_\mathbf{Z}(x, \cdot)$ over all cells of column x. The supports of X and Y are $\mathbb{X} = \{0, 1, 2, 3\}$ and $\mathbb{Y} = \{0, 1, 2\}$. $\mathbb{Z}$, the support of $\mathbf{Z}$, comprises the points (x, y) corresponding to cells with positive entries. Thus, $\mathbb{Z}$ is here a proper subset of $\mathbb{X} \times \mathbb{Y}$.

$y \setminus x$	0	1	2	3	$f_Y(y)$
0	$\frac{1}{8}$	0	0	$\frac{1}{8}$	$\frac{2}{8}$
1	0	$\frac{2}{8}$	$\frac{2}{8}$	0	$\frac{4}{8}$
2	0	$\frac{1}{8}$	$\frac{1}{8}$	0	$\frac{2}{8}$
$f_X(x)$	$\frac{1}{8}$	$\frac{3}{8}$	$\frac{3}{8}$	$\frac{1}{8}$	

(2.16)

Since marginal p.m.f.s can be deduced from the joint p.m.f., the latter clearly contains all the information that they contain about the behavior of each variable alone. However, the joint p.m.f. also contains additional information about the interaction between X and Y. Thus, the marginal distributions alone would not tell us that event $(X, Y) = (0, 1)$ in Example 2.30 has probability zero.

As stated earlier, the vector-valued r.v. $\mathbf{Z} = (X, Y)'$ is (absolutely) continuous if the joint c.d.f. is (absolutely) continuous. In this case there exists a joint p.d.f. $f_{\mathbf{Z}}$ such that $\mathbb{P}_{\mathbf{Z}}(B_2) = \int_{B_2} f_{\mathbf{Z}}(x, y) \cdot dx\, dy$ for $B_2 \in \mathcal{B}^2$. Since the joint c.d.f. can be constructed from the p.d.f. as $F_{\mathbf{Z}}(x, y) = \int_{-\infty}^{y} \int_{-\infty}^{x} f_{\mathbf{Z}}(s, t) \cdot ds\, dt$, it follows that the marginal c.d.f.s are given by the repeated integrals

$$F_X(x) = F_{\mathbf{Z}}(x, +\infty) = \int_{s=-\infty}^{x} \int_{y=-\infty}^{\infty} f_{\mathbf{Z}}(s, y) \cdot dy\, ds$$

$$F_Y(y) = F_{\mathbf{Z}}(+\infty, y) = \int_{t=-\infty}^{y} \int_{x=-\infty}^{\infty} f_{\mathbf{Z}}(x, t) \cdot dx\, dt.$$

Differentiating with respect to the upper limits of the outer integrals shows that the marginal p.d.f.s can be found from the joint p.d.f. as

$$f_X(x) = F'_X(x) = \int_{y=-\infty}^{\infty} f_{\mathbf{Z}}(x, y) \cdot dy$$

$$f_Y(y) = F'_Y(y) = \int_{x=-\infty}^{\infty} f_{\mathbf{Z}}(x, y) \cdot dx.$$

As thus defined, f_X and f_Y are *versions* of the marginal p.d.f.s corresponding to the version of the joint density represented by $f_{\mathbf{Z}}$. Note that "integrating out" y and x, respectively, in these expressions is akin to summing across columns and rows of the tabular presentations of joint p.m.f.s.

Example 2.31. With $f_{\mathbf{Z}}(x, y) = 2e^{-x-y} \mathbf{1}_{\{(x,y):0<x<y\}}(x, y)$ the marginal p.d.f. of X is found as follows. For $x < 0$, $f_X(x) = \int_{-\infty}^{\infty} 0 \cdot dy = 0$. For $x > 0$

$$f_X(x) = \int_{-\infty}^{\infty} f_{\mathbf{Z}}(x, y) \cdot dy = \int_{-\infty}^{x} 0 \cdot dy + \int_{x}^{\infty} 2e^{-x-y} \cdot dy$$

$$= 2e^{-2x}.$$

Exercise 2.24.

(i) Verify that $f_X(x) = 2e^{-2x} \mathbf{1}_{(0,\infty)}(x)$ is a p.d.f.

(ii) Find $\Pr(X > 1)$.

(iii) Find the marginal p.d.f. of Y in Example 2.31.

Exercise 2.25. *Let f_X be a continuous p.d.f. with single mode m and such that f_X is strictly increasing for $x < m$ and strictly decreasing for $x > m$. Let $\mathbf{Z} = (X, Y)'$ be uniformly distributed over the region between f_X and the horizontal axis, in the sense that $\Pr(\mathbf{Z} \in R_1) = \Pr(\mathbf{Z} \in R_2)$ for all sub-regions R_1, R_2 having the same Lebesgue measure (area). Show that*

(i) The marginal p.d.f. of X is f_X.
(ii) The marginal p.d.f. of Y is given by
$$f_Y(y) = \left[f_{X-}^{-1}(y) - f_{X+}^{-1}(y) \right] \mathbf{1}_{(0, f_X(m))}(y),$$
where f_{X+}^{-1} and f_{X-}^{-1} are, respectively, the inverse functions on the ascending and descending segments of f_X.
(iii) $f_Y(y) = 2\left[(\pi y)^{-1} - 1 \right]^{1/2} \mathbf{1}_{(0, \pi^{-1})}(y)$ if $f_X(x) = \left[\pi (1 + x^2) \right]^{-1}, x \in \Re$.

2.4.3 *Independence of Random Variables*

Recall that events A, A' on a probability space $(\Omega, \mathcal{F}, \mathbb{P})$ are independent if and only if (i) they are "nontrivial", meaning that they and their complements have positive $\mathbb{P}$ measure, and (ii) $\mathbb{P}(A \cap A') = \mathbb{P}(A)\mathbb{P}(A')$. Introducing the bivariate r.v. $\mathbf{Z} = (X, Y)'$ and the induced probability space $(\Re^2, \mathcal{B}^2, \mathbb{P}_{\mathbf{Z}})$, consider the two-dimensional sets $B \times \Re$ and $\Re \times B'$, where B, B' are one-dimensional Borel sets of $\Re$. Figure 2.10 depicts such sets as intersecting strips in the case that B and B' are simple intervals. As the figure illustrates, the intersection of $B \times \Re$ with $\Re \times B'$ is itself a product set $B \times B'$—in this case a single rectangle. Of course, $B \times B'$ is still an element of $\mathcal{B}^2$ and therefore $\mathbb{P}_{\mathbf{Z}}$-measurable. Now $\mathbb{P}_{\mathbf{Z}}(B \times \Re)$ is the same as the *marginal* measure $\mathbb{P}_X(B)$, since the event $\mathbf{Z} := (X, Y)' \in B \times \Re$ occurs if and only if $X \in B$. Likewise $\mathbb{P}_{\mathbf{Z}}(\Re \times B') = \mathbb{P}_Y(B')$ since event $\mathbf{Z} \in \Re \times B' \iff Y \in B'$. Consider sets B, B' such that the events $X \in B$ and $Y \in B'$ are nontrivial. Then by our standard definition the events $\mathbf{Z} \in B \times \Re$ and $\mathbf{Z} \in \Re \times B'$ are independent if and only if $\mathbb{P}_{\mathbf{Z}}\left[(B \times \Re) \cap (\Re \times B') \right] = \mathbb{P}_{\mathbf{Z}}(B \times \Re)\mathbb{P}_{\mathbf{Z}}(\Re \times B')$ or, equivalently,

$$\mathbb{P}_{\mathbf{Z}}(B \times B') = \mathbb{P}_X(B)\mathbb{P}_Y(B'). \tag{2.17}$$

Expressed in the usual shorthand, the above is

$$\Pr\left[(X, Y) \in B \times B' \right] =: \Pr(X \in B, Y \in B') = \Pr(X \in B)\Pr(Y \in B').$$

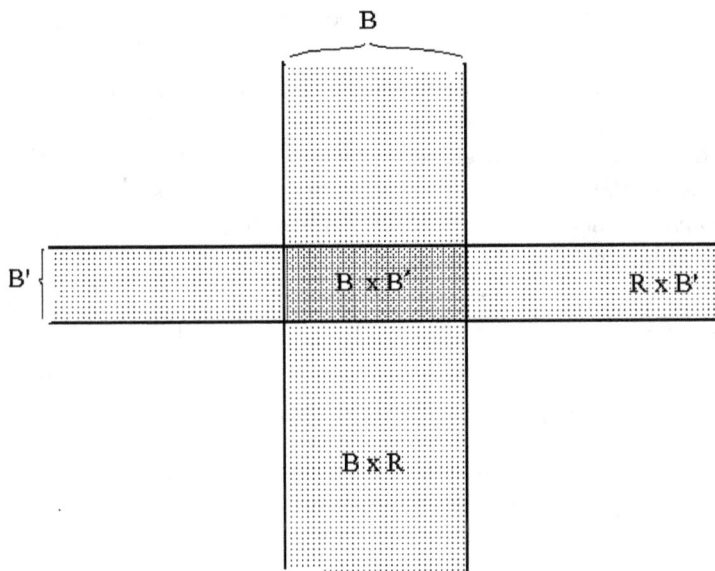

Fig. 2.10   Intersection of two infinite product spaces in $\Re_2$.

Note that (2.17) continues to hold even for events we have styled by inference as "trivial", even though we decline to consider them "independent". For example, if $\mathbb{P}_X(B) = \mathbb{P}_Z(B \times \Re) = 1$, then the set $(\Re \times \Re) \setminus (B \times \Re) = (\Re \setminus B \times \Re)$ is $\mathbb{P}_Z$-null, and so $\mathbb{P}_Z(B \times B') = \mathbb{P}_Z(\Re \times B') = \mathbb{P}_Y(B')$. Likewise, if $\mathbb{P}_X(B) = \mathbb{P}_Z(B \times \Re) = 0$, then $B \times B' \subset B \times \Re$ implies that $\mathbb{P}_Z(B \times B') = 0$ also. For a "degenerate" r.v. X—one such that $X(\omega) = c$ (a fixed constant) for almost all ω—all events are trivial. The following definition excludes degenerate r.v.s, which we simply do not classify as either dependent or independent.

Definition 2.17. Nondegenerate r.v.s X, Y are statistically (or stochastically) *independent* if and only $\mathbb{P}_\mathbf{Z}(B \times B') = \mathbb{P}_X(B)\mathbb{P}_Y(B')$ for each pair of Borel sets B, B'.

The above implies by the definition of conditional probability that $\Pr(X \in B \mid Y \in B') = \Pr(X \in B)$ and that $\Pr(Y \in B' \mid X \in B) = \Pr(Y \in B')$ for each pair B, B' such that $\Pr(X \in B) \in (0, 1)$ and $\Pr(Y \in B') \in (0, 1)$. Here, "$\Pr(\cdot \mid \cdot\cdot)$" stands for "probability of $\cdot$ *given* $\cdot\cdot$". The condition "for each" is crucial: Knowledge that *any* nontrivial event concerning X has occurred must not require us to reassess the probability of a Y event, and *vice versa.*

Independence of r.v.s. can be deduced from the relations between their joint and marginal c.d.f.s or, if both are of the same type, from relations between their joint and marginal p.m.f.s or p.d.f.s. More importantly, when the nature of the experiment that generates realizations of X and Y justifies *assuming* independence, then we can use these relations to deduce the joint distributions from the marginals. Let us see what these relations are. Taking $B = (-\infty, x]$ and $B' = (-\infty, y]$, we have $\mathbb{P}_{\mathbf{Z}}(B \times B') = F_{\mathbf{Z}}(x, y)$, $\mathbb{P}_X(B) = F_X(x)$, and $\mathbb{P}_Y(B') = F_Y(y)$. If relation (2.17) holds for all measurable B, B', it follows that

$$F_{\mathbf{Z}}(x, y) = F_X(x)F_Y(y) \text{ for each } (x, y) \in \Re^2. \tag{2.18}$$

Conversely, since the right-closed sets $(-\infty, x]$ generate $\mathcal{B}$ and the product sets $(-\infty, x] \times (-\infty, y]$ generate $\mathcal{B}^2$, the converse also holds, so (2.18) is also sufficient for independence of X and Y. Finally, if bivariate r.v. $\mathbf{Z} = (X, Y)'$ is either discrete or continuous, the same product relation for the p.m.f.s or p.d.f.s is necessary and sufficient for independence; that is,

$$f_{\mathbf{Z}}(x, y) = f_X(x)f_Y(y) \text{ for each } (x, y) \in \Re^2. \tag{2.19}$$

In the case of continuous r.v.s, (2.19) simply has to hold for some *version* of the respective p.d.f.s.

Exercise 2.26. *Show that (2.17) implies (2.19) in the discrete case. (Hint: Take $B = \{x\}$ and $B' = \{y\}$.)*

Exercise 2.27. *Show that (2.18) implies (2.19) in the continuous case. (Hint: Differentiate.)*

Exercise 2.28. *Show that r.v.s X and Y are independent if and only if the joint p.d.f. can be written as*

$$f_{\mathbf{Z}}(x, y) = g(x)h(y), \tag{2.20}$$

where the right side is positive on a product set (rectangle) $\mathbb{X} \times \mathbb{Y}$.

Exercise 2.29. *Determine whether the following r.v.s are independent:*

(i) Those whose joint p.m.f. is shown in (2.16).
(ii) Those with joint p.d.f. $f_{\mathbf{Z}}(x, y) = e^{-x-y}\mathbf{1}_{[0,\infty)^2}(x, y)$.
(iii) Those with joint p.d.f. $f_{\mathbf{Z}}(x, y) = 2e^{-x-y}\mathbf{1}_{\{0<x<y\}}(x, y)$.

Exercise 2.30. *Use your judgment to determine whether the following pairs of r.v.s should be modeled as independent:*

(i) *R.v.s whose values correspond to the numbers that turn up on two successive rolls of a six-sided die.*

(ii) *R.v.s that measure the heights of two successive persons drawn at random, with replacement, from the class.*

(iii) *R.v.s that measure the heights of two successive persons drawn at random, without replacement, from the class.*

(iv) *R.v.s that measure the height and weight of a person drawn at random.*

Random variables X and Y are said to be *functionally* related if there is some measurable g such that $g[X(\omega), Y(\omega)] = 0$ for almost all ω. Clearly, *statistical* dependence does not imply the existence of a functional relation. (Consider the dependent X and Y in Example 2.30.) It is less obvious—but still true—that functionally related variables are not *necessarily* statistically dependent. Thus, functional and statistical dependence are distinct conditions.

Example 2.32. Let Y be such that $\Pr(Y < 0) > 0$ and $\Pr(Y > 0) > 0$; let Z be statistically independent of Y and distributed symmetrically about the origin; and put $X := ZY/|Y|$ if $Y \neq 0$ and $X = 0$ otherwise. Then X and Y are functionally related, yet they are statistically independent if (and only if) $\Pr(Y = 0) = 0$.

2.4.4 *Conditional Distributions*

When (nontrivial) events in an abstract probability space are not independent, the probability associated with one event has to be reassessed upon learning that the other has occurred. In the same way, if r.v.s X, Y are not independent, the probability of event $Y \in B'$ may depend on the realization of X. For example, suppose the experiment is to draw at random one member of a class of students and that $\mathbf{Z} = (X, Y)'$ assigns to the outcome the person's height (X) and weight (Y). Before the experiment, having just the trivial information $\{\Omega, \varnothing\}$, the marginal distribution of Y would determine the assessment of $\Pr(Y > 160)$, the probability that the person would weigh more than 160 pounds. However, this assessment might well change upon acquiring the partial information that $X = 72$ inches. Thus, $\Pr(Y > 160)$ and $\Pr(Y > 160 \mid X = 72)$ might well differ. *Conditional distributions* are used to make such probability statements about one r.v.

when the realization of the other is known.

To set up notation, let $(\Omega, \mathcal{F}, \mathbb{P})$ be the probability space and $\mathbf{Z} = (X, Y)' : \Omega \to \Re^2$ induce space $(\Re^2, \mathcal{B}^2, \mathbb{P}_{\mathbf{Z}})$. If B_2, B_2' are sets in $\mathcal{B}^2$ with $\mathbb{P}_{\mathbf{Z}}(B_2) > 0$, then the conditional probability that $\mathbf{Z} \in B_2'$ given $\mathbf{Z} \in B_2$ is determined in the usual way as $\Pr(\mathbf{Z} \in B_2' \mid \mathbf{Z} \in B_2) := \mathbb{P}_{\mathbf{Z}}(B_2' \mid B_2) = \mathbb{P}_{\mathbf{Z}}(B_2 \cap B_2')/\mathbb{P}_{\mathbf{Z}}(B_2)$. Now consider events such that B_2 involves X alone and B_2' restricts Y alone, with B_2 still having positive $\mathbb{P}_{\mathbf{Z}}$ measure. Thus, with B, B' some (one-dimensional) Borel sets of $\Re$, take $B_2 = B \times \Re$ and $B_2' = \Re \times B'$. Now

$$\mathbb{P}_{\mathbf{Z}}(B_2' \mid B_2) = \frac{\mathbb{P}_{\mathbf{Z}}((B \times \Re) \cap (\Re \times B'))}{\mathbb{P}_{\mathbf{Z}}(B \times \Re)} = \frac{\mathbb{P}_{\mathbf{Z}}(B \times B')}{\mathbb{P}_X(B)}, \qquad (2.21)$$

where the denominator is now the (positive) marginal probability that $X \in B$. The numerator in the last expression is the probability that X takes a value in B *and* that Y takes a value in B', so each expression in (2.21) represents the conditional probability that $Y \in B'$ given $X \in B$, or $\Pr(Y \in B' \mid X \in B)$ in our notational shorthand. We want to interpret such expressions in the case that $B = \{x\}$, a singleton, as in the weight/height example with $x = 72$, so as to deduce the conditional distribution of Y given the event $X = x$. The concept is more straightforward in the discrete case.

2.4.4.1   *The Discrete Case*

Expression (2.21) can be used directly to obtain the conditional p.m.f. of Y given $X = x$ when $\mathbf{Z} = (X, Y)'$ is a discrete bivariate r.v. Take $B' = \{y\}$ and $B = \{x\}$. Now $\mathbb{P}_{\mathbf{Z}}(B \times \Re) = \mathbb{P}_X(B) = f_X(x)$ (the marginal probability that $X = x$) and $\mathbb{P}_{\mathbf{Z}}(B \times B') = \mathbb{P}_{\mathbf{Z}}(\{(x, y)\}) = f_{\mathbf{Z}}(x, y)$ (the joint probability that $X = x$ and $Y = y$). Let $\mathbb{Z}$ be the support of $\mathbf{Z}$, the countable set of points $(x, y) \in \Re^2$ at which $f_{\mathbf{Z}}(x, y) > 0$, and choose x such that $f_X(x) > 0$.

Definition 2.18. When $f_X(x) > 0$ the function $f_{Y \mid X}(\cdot \mid x) : \Re \to [0, 1]$ with

$$f_{Y \mid X}(y \mid x) := \frac{f_{\mathbf{Z}}(x, y)}{f_X(x)} \mathbf{1}_{\mathbb{Z}}(x, y)$$

is the conditional p.m.f. of Y given the event $X = x$.

Of course, if $f_Y(y) > 0$, reversing symbols x and y gives the conditional p.m.f. of X given the event $Y = y$. Verifying that this is a *good* definition requires showing that $f_{Y|X}$ has the properties of a p.m.f. First, $f_{Y|X}(y \mid x) \geq 0$, since it is positive on $\mathbb{Z}$ and zero elsewhere; and, second, the conditional probabilities sum to unity because

$$\sum_{y:(x,y)\in\mathbb{Z}} f_{Y|X}(y \mid x) = \sum_{y:(x,y)\in\mathbb{Z}} \frac{f_{\mathbf{Z}}(x,y)}{f_X(x)} = \frac{\sum_{y:(x,y)\in\mathbb{Z}} f_{\mathbf{Z}}(x,y)}{f_X(x)} = \frac{f_X(x)}{f_X(x)}.$$

Example 2.33. From the table in (2.16) that shows the joint and marginal distributions of the numbers of heads and sequence changes in three flips of a fair coin, the conditional distribution of Y given $X = 2$ is found to be

$$f_{Y|X}(y \mid 2) = \begin{cases} \frac{f_{\mathbf{Z}}(2,1)}{f_X(2)} = \frac{2}{8} \div \frac{3}{8} = \frac{2}{3}, \ y = 1 \\ \frac{f_{\mathbf{Z}}(2,2)}{f_X(2)} = \frac{1}{8} \div \frac{3}{8} = \frac{1}{3}, \ y = 2 \end{cases}$$

and zero elsewhere. The conditional distribution of X given $Y = 0$ is

$$f_{X|Y}(x \mid 0) = \begin{cases} \frac{f_{\mathbf{Z}}(0,0)}{f_Y(0)} = \frac{1}{8} \div \frac{2}{8} = \frac{1}{2}, \ x = 0 \\ \frac{f_{\mathbf{Z}}(3,0)}{f_Y(0)} = \frac{1}{8} \div \frac{2}{8} = \frac{1}{2}, \ x = 3 \end{cases}$$

and zero elsewhere.

Exercise 2.31. *A fair, 6-sided die is rolled twice. Take an outcome ω to be an ordered pair of faces, one of the set $\Omega = \left\{ (\cdot,\cdot), ..., (\vdots\kern-2pt\cdot\kern2pt, \vdots\kern-2pt\cdot\kern2pt) \right\}$, and let $\mathbf{Z} = (X,Y)'$ assign to ω the ordered pair of integers corresponding to the number of dots on the faces. Find joint p.m.f. $f_{\mathbf{Z}}$ and marginal p.m.f.s f_X and f_Y. Without making any calculations, what would the conditional distributions $f_{Y|X}(\cdot \mid x)$ and $f_{X|Y}(\cdot \mid y)$ be for arbitrary $x \in \{1,2,...,6\}$ and $y \in \{1,2,...,6\}$? What is special about this case?*

Letting B be any one-dimensional Borel set, one finds the probability that $Y \in B$ conditional on $X = x$ by summing the conditional probabilities over those values of y in B and the support, as $\Pr(Y \in B \mid X = x) = \sum_{y\in B:(x,y)\in\mathbb{Z}} f_{Y|X}(y \mid x)$. To simplify, this is usually shortened to $\sum_{y\in B} f_{Y|X}(y \mid x)$. As in the univariate case, we shall see that it is possible to represent such sums as integrals with respect to counting measure, as $\int_{y\in B} f_{Y|X}(y \mid x) N(dy)$.

Exercise 2.32. *Find the conditional p.m.f. of Y given $X = 2$ in the 3-coin example.*

2.4.4.2   *The Continuous Case*

In the continuous case $\Pr(Y \in B \mid X = x)$ is found by integrating over B a certain function $f_{Y|X}(\cdot \mid x)$ called the conditional p.d.f. However, the situation is now rather delicate, since the event $X = x$ is an event of zero probability; i.e., $\Pr(X = x) = \mathbb{P}_{\mathbf{Z}}(\{x\} \times \Re) = \mathbb{P}_X(\{x\}) = 0$. On the one hand it might seem meaningless—or, at best, irrelevant—to condition on null events. Yet the facts that zero probability events do occur in the continuous case—indeed, that *some* zero probability event *must* occur—and that such events can convey meaningful information lead us to search for a sensible definition. Here is an example of the kinds of situations that arise.

Example 2.34. A pointer on a circular scale of unit circumference is spun twice. X represents the clockwise distance of pointer from origin after the first spin, and Y is the sum of the clockwise distances on the two spins (assumed to be independent). Below in Example 2.38 we shall see how to deduce the marginal distribution of Y. One would use that distribution to determine $\Pr(Y \in B)$ given the trivial information one had before the experiment began. Now, spinning the pointer once, suppose realization $X(\omega) = .32$ is observed. Although, this was an event of zero probability *ex ante*, it did in fact occur, and knowing that it did should alter the assessment of $\Pr(Y \in B)$. In particular, it is clear without any calculation that $\Pr(Y \geq .32 \mid X = .32) = 1 > \Pr(Y \geq .32)$.

To develop a useful interpretation of conditional density $f_{Y|X}(\cdot \mid x)$, begin as in the discrete case with the expression $\mathbb{P}_{\mathbf{Z}}(B_2' \mid B_2) = \mathbb{P}_{\mathbf{Z}}(B \times B')/\mathbb{P}_X(B)$ in (2.21), which is the conditional probability that $Y \in B'$ given $X \in B$. Now take $B' = (-\infty, y]$ and $B = (x - h, x]$ for some $h > 0$ and for (x, y) such that derivatives $F_X'(x)$ and $\partial F_{\mathbf{Z}}(x, y)/\partial x$ exist (as they do a.e. $\mathbb{P}_{\mathbf{Z}}$) and such that $F_X'(x) = f_X(x) > 0$. $\mathbb{P}_{\mathbf{Z}}(B_2' \mid B_2)$ is now the conditional probability that $Y \leq y$ given that X takes a value in $(x - h, x]$; i.e.,

$$\Pr(Y \leq y \mid x - h < X \leq x) = \frac{\mathbb{P}_{\mathbf{Z}}((x - h, x] \times (-\infty, y])}{\mathbb{P}_{\mathbf{Z}}((x - h, x] \times \Re)}.$$

The ratio on the right has a simple geometrical interpretation. With x values on the horizontal axis and y values on the vertical, the denominator represents the probability mass in the infinitely long vertical strip of width h, while the numerator represents the probability mass in that portion of the strip that lies at or below y. We want to consider the limit of this expression as $h \to 0$ and the vertical strip degenerates to the line $\{x\} \times \Re$. Although both numerator and denominator approach zero as $h \to 0$, one can see that the *ratio* might very well approach a nonnegative, finite value that depends, in general, on x and y. Indeed, under the stated conditions this is the case, and it is not hard to determine what this limit is. We have

$$\frac{\mathbb{P}_{\mathbf{Z}}\left((x-h,x] \times (-\infty,y]\right)}{\mathbb{P}_{\mathbf{Z}}\left((x-h,x] \times \Re\right)} = \frac{F_{\mathbf{Z}}\left(x,y\right) - F_{\mathbf{Z}}\left(x-h,y\right)}{F_X\left(x\right) - F_X\left(x-h\right)}.$$

Taking limits, and recalling that $F_{\mathbf{Z}}\left(\cdot, y\right)$ and $F_X\left(\cdot\right)$ are differentiable at x, we have

$$\lim_{h \to 0} \Pr\left(Y \leq y \mid x-h < X \leq x\right) = \lim_{h \to 0} \frac{\left[F_{\mathbf{Z}}\left(x,y\right) - F_{\mathbf{Z}}\left(x-h,y\right)\right]/h}{\left[F_X\left(x\right) - F_X\left(x-h\right)\right]/h}$$

$$= \frac{\partial F_{\mathbf{Z}}\left(x,y\right)/\partial x}{dF_X\left(x\right)/dx}$$

$$= \frac{\frac{\partial}{\partial x} \int_{-\infty}^{x} \int_{-\infty}^{y} f_{\mathbf{Z}}\left(s,t\right) \cdot dt\, ds}{f_X\left(x\right)}$$

$$= \frac{\int_{-\infty}^{y} f_{\mathbf{Z}}\left(x,t\right) \cdot dt}{f_X\left(x\right)}$$

$$=: F_{Y|X}\left(y \mid x\right).$$

The last expression is the conditional c.d.f. of Y at y given that $X = x$. Differentiating this with respect to y gives our useful definition of the conditional *density*:

Definition 2.19. For x such that $F'_X\left(x\right) = f_X(x) > 0$ the function $f_{Y|X}\left(\cdot \mid x\right) : \Re \to \Re_+$ given by

$$f_{Y|X}(y \mid x) = \frac{d}{dy} F_{Y|X}\left(y \mid x\right) \mathbf{1}_Z(x,y) = \frac{f_{\mathbf{Z}}(x,y)}{f_X(x)} \mathbf{1}_Z(x,y)$$

is the conditional p.d.f. of Y given the event $X = x$.

Note that $f_{Y|X}$ as thus defined is indeed a legitimate p.d.f. since $f_{Y|X}(y \mid x) \geq 0$ and

$$\int_{-\infty}^{\infty} f_{Y|X}(y \mid x) \cdot dy = F_{Y|X}\left(+\infty \mid x\right) = \frac{\int_{-\infty}^{\infty} f_{\mathbf{Z}}(x,y) \cdot dy}{f_X(x)} = \frac{f_X(x)}{f_X(x)}.$$

There is, however, a technical issue concerning the interpretation of $f_{Y|X}$. Just as densities $f_{\mathbf{Z}}$ and f_X may be altered on sets of zero Lebesgue measure without altering the probabilities of events determined from them, the same is true of the conditional density as well. Thus, in the continuous case one must regard any given $f_{Y|X}$ just as a *version* of the conditional density.

If B is any (one-dimensional) Borel set, $\Pr(Y \in B \mid X = x)$ is found by integrating (any version of) the conditional density over this set, as

$$\Pr(Y \in B \mid X = x) = \int_B f_{Y|X}(y \mid x) \cdot dy.$$

However, just as there are other versions of conditional densities in the continuous case, a conditional probability $\Pr(Y \in B \mid X = x)$ must itself be regarded merely as an equivalence class, each member of which produces the same value of

$$\mathbb{P}_Y(B) = \int \Pr(Y \in B \mid X = x) f_X(x) \cdot dx = \int_B \int f_{Y|X}(y \mid x) f_X(x) \cdot dx\, dy$$

for each $B \in \mathcal{B}$.

Example 2.35. If $f_{\mathbf{Z}}(x, y) = 2e^{-x-y} \mathbf{1}_{\{(x,y):0<x<y\}}(x, y)$, the marginal p.d.f. of X is

$$f_X(x) = 2e^{-2x} \mathbf{1}_{(0,\infty)}(x),$$

and the conditional p.d.f. of Y given $X = x$ is

$$f_{Y|X}(y \mid x) = \frac{2e^{-x-y}}{2e^{-2x}} \mathbf{1}_{\{(x,y):0<x<y\}}(x, y)$$

$$= e^{x-y} \mathbf{1}_{\{(x,y):0<x<y\}}(x, y).$$

Exercise 2.33. *Find the conditional p.d.f. of X given $Y = y$ in Example 2.35.*

Exercise 2.34. *Find the conditional c.d.f.s of Y given $X = x$ and of X given $Y = y$ in Example 2.35.*

Exercise 2.35. *Random variables X and Y have joint p.d.f. $f(x, y) = k\left(x^2 + y^2\right) \mathbf{1}_{[-1,1]^2}(x, y)$ for some $k > 0$. Find*

(i) *The value of k.*
(ii) *The marginal p.d.f. of X.*
(iii) *The conditional p.d.f. of Y given $X = x$ (for appropriate values of x).*
(iv) *The conditional probability that $Y < 1/2$ given that $X = 1/2$.*
(v) *The conditional probability that $Y < 1/2$ given that $X < 1/2$.*

Exercise 2.36. *Random variables X and Y have joint p.d.f. $f(x, y) = \pi^{-1} \mathbf{1}_{\{(x,y):x^2+y^2\leq 1\}}(x, y)$ (i.e., constant density on a circle of unit radius centered at the origin). Find*

(i) $f_X(x)$

(ii) $F_X(x)$

(iii) $f_{Y|X}(y \mid x)$ *(for appropriate values of x)*

(iv) The conditional probability that $Y < 1/2$ given that $X = 1/2$.

(v) The conditional probability that $Y < 1/2$ given that $X < 1/2$.

2.4.4.3 *Interpreting Conditional Distributions and Probabilities Ex Ante*

While it is easy to see the meaning of conditional probabilities and conditional distributions given partial knowledge of events that have already occurred, we sometimes need to think of conditioning on events that are merely prospective. This requires thinking of conditional probabilities and conditional densities as random variables, represented as $\Pr(Y \in B \mid X)$, $f_{Y|X}(y \mid X)$, and $f_{X|Y}(x \mid Y)$. The introductory example of choosing a person at random from the class can help us to see what such expressions mean. If it is possible to calculate $\Pr(Y > 160 \mid X = x)$ for $x = 72$ (the conditional probability that the person weighs more than 160 pounds given a height of 72 inches), then this can also be done for each value in the support of X. That is, $\Pr(Y > 160 \mid X = x)$ can be found for each $x \in \mathbb{X}$. *Ex ante*—that is, before the person is selected and the height determined—this conditional probability would be regarded as a function of the *as-yet-to-be-realized* value of X. It is therefore a r.v. whose numerical value would be determined by the outcome of the experiment. That is, just as X itself takes outcomes $\omega \in \Omega$ into the reals, the function $\Pr(Y > 160 \mid X)$ maps Ω into $[0, 1]$. The same interpretation applies to values of the conditional p.m.f. and to (versions of) the conditional p.d.f., except that the latter can take values on $[0, \infty)$.

2.4.5 *Extension to the k-variate Case*

Having considered bivariate r.v.s in detail, it is easy to extend the concepts of joint, marginal, and conditional distributions to the multivariate case. Adopting subscript notation now, think of $\mathbf{X} = (X_1, X_2, ..., X_k)'$ as a mapping from Ω to the k-dimensional reals $\Re^k$. With $B_k \in \mathcal{B}^k$ (the k-dimensional Borel sets) the induced probability $\mathbb{P}_{\mathbf{X}}(B_k)$ is the probability assigned by $\mathbb{P}$ to the set of outcomes $\mathbf{X}^{-1}(B_k)$ that $\mathbf{X}$ maps into B_k. From the induced measure the joint c.d.f. is developed as $F_{\mathbf{X}}(\mathbf{x}) = \mathbb{P}_{\mathbf{X}}\left(\times_{j=1}^k (-\infty, x_j]\right)$. When there is a countable set $\mathbb{X}$ of points $\mathbf{x} = (x_1, x_2, ..., x_k)'$ such that $\mathbb{P}_{\mathbf{X}}(\{\mathbf{x}\}) > 0$ for $\mathbf{x} \in \mathbb{X}$ and $\mathbb{P}_{\mathbf{X}}(\mathbb{X}) = 1$, then $\mathbf{X}$ is a *discrete* multivariate r.v., and the joint p.m.f. is given by $f_\mathbf{X}(\mathbf{x}) = \mathbb{P}_\mathbf{X}(\{\mathbf{x}\})$. When c.d.f. $F_\mathbf{X}(\mathbf{x})$ is absolutely continuous, then $\mathbf{X}$ is a *continuous* multivariate r.v. and $F_\mathbf{X}$ is differentiable almost everywhere with respect to Lebesgue measure. $\mathbf{X}$ then has joint p.d.f. given by $f_\mathbf{X}(\mathbf{x}) = \partial^k F_\mathbf{X}(\mathbf{x})/\partial x_1 \cdots \partial x_k$ for "almost all" $\mathbf{x}$. If $B_k \in \mathcal{B}^k$, then $\Pr(\mathbf{X} \in B_k) = \mathbb{P}_\mathbf{X}(B_k)$ is determined in the discrete case by summing $f_\mathbf{X}(\mathbf{x})$ over the countable set of points in $B_k \cap \mathbb{X}$, and in the continuous case it is determined by integrating $f_\mathbf{X}(\mathbf{x})$ over B_k.

The joint *marginal* c.d.f. corresponding to any subset of the variables in $\mathbf{X}$ is obtained by evaluating $F_\mathbf{X}$ at $+\infty$ for each of the excluded variables. For example, if $\mathbf{Y} := (X_1, X_2, ..., X_m)'$ contains the first $m < k$ of the variables in $\mathbf{X}$, then

$$F_\mathbf{Y}(x_1, ..., x_m) = F_\mathbf{X}(x_1, ..., x_m, +\infty, ..., +\infty).$$

Taking $m = 1$ gives the marginal c.d.f. of one particular member of $\mathbf{X}$. $f_\mathbf{Y}(\mathbf{y})$, the joint marginal p.m.f./p.d.f. corresponding to $F_\mathbf{Y}(\mathbf{y})$, can be obtained from the joint p.m.f./p.d.f. of all k variables by summing out or integrating out the remaining $k - m$ of them. Thus, in the discrete case

$$f_\mathbf{Y}(\mathbf{y}) = \sum_{x_{m+1}} \cdots \sum_{x_k} f_\mathbf{X}(x_1, ..., x_m, x_{m+1}, ..., x_k),$$

where the multiple sum is over countably many points $(x_{m+1}, ..., x_k)'$ such that

$$(x_1, ..., x_m, x_{m+1}, ..., x_k)' \in \mathbb{X};$$

and in the continuous case

$$f_\mathbf{Y}(\mathbf{y}) = \int_{-\infty}^{\infty} \cdots \int_{-\infty}^{\infty} f_\mathbf{X}(x_1, ..., x_m, x_{m+1}, ..., x_k) \cdot dx_{m+1} \cdots dx_k.$$

Of course, in the continuous case $f_\mathbf{Y}$ can also be found by differentiating the marginal c.d.f.; i.e., $f_\mathbf{Y}(\mathbf{y}) = \partial^m F_\mathbf{Y}(\mathbf{y})/\partial y_1 \cdots \partial y_m$. For brevity marginal c.d.f.s, p.m.f.s, and p.d.f.s are often denoted just with numerical subscripts that identify the pertinent variables; e.g., as F_j or f_j (rather than F_{X_j} or f_{X_j}) for the marginal distribution of X_j, and as F_{ijk} or f_{ijk} for the joint marginal distribution of X_i, X_j, X_k.

Generalizing the concept of independence from the bivariate to the k-variate case is not entirely trivial, since there is a distinction between *pairwise, three-wise*, etc. independence and *mutual* independence. R.v.s $\{X_j\}_{j=1}^k$ are said to be *mutually* independent if and only if for all sequences of one-dimensional Borel sets $\left\{B^{(j)}\right\}_{j=1}^{k}$ the $\mathbb{P}_{\mathbf{X}}$ measure of the product set equals the product of the marginal measures; i.e.,

$$\mathbb{P}_{\mathbf{X}}\left(\times_{j=1}^{k} B^{(j)}\right) = \prod_{j=1}^{k} \mathbb{P}_{X_j}(B^{(j)}). \tag{2.22}$$

Taking $B^{(j)} = \Re$ for all j belonging to any proper subset of $\{1, 2, ..., k\}$ (in which case $\mathbb{P}_{X_j}(B^{(j)}) = 1$) shows that mutual independence of all k r.v.s. implies the mutual independence of all subsets of the variables—all pairs, all triples, and so forth. Mutual independence is thus a stronger condition than pairwise independence, three-wise independence, etc. (Note: In the sequel, "independent" by itself should be understood to mean *mutually* independent.) In practical terms the implication of mutual independence is that learning the realization of any subset of the variables does not change our assessment of the probabilities of events pertaining to any or all of the remaining ones.

Extending to *infinite* collections, r.v.s $\{X_j\}_{j=1}^{\infty}$ are said to be mutually independent if and only if those belonging to *each* finite subcollection are independent. (Why restrict to *finite* subcollections? With $k = \infty$ in (2.22) both sides might well be zero—and hence equal—even though the relation failed to hold for finite subsets.) A necessary and sufficient condition for mutual independence of any k r.v.s is that the joint c.d.f. of $\mathbf{X} \in \Re^k$ factor as the product of the marginal c.d.f.s. That is, $\{X_j\}_{j=1}^{k}$ are independent if and only if $F_{\mathbf{X}}(\mathbf{x}) = F_1(x_1)F_2(x_2) \cdots F_k(x_k)$ for *each* $\mathbf{x} = (x_1, x_2, ..., x_k)' \in \Re^k$. If $\mathbf{X}$ is either purely discrete or purely continuous, then it is also necessary and sufficient that $f_{\mathbf{X}}(\mathbf{x}) = f_1(x_1)f_2(x_2) \cdots f_k(x_k)$ for each $\mathbf{x} = (x_1, x_2, ..., x_k)'$. As in the bivariate case (Exercise 2.28), it is necessary and sufficient for mutual independence of $\{X_j\}_{j=1}^{k}$ with supports $\{\mathbb{X}_j\}_{j=1}^{k}$ that the p.d.f./p.m.f. be expressible as $f_{\mathbf{X}}(x) = \prod_{j=1}^{k} g_j(x_j)$ on the product space, $\times_{j=1}^{k} \mathbb{X}_j$, where g_j is some (measurable) function of x_j alone but is not necessarily a p.d.f. or p.m.f. As usual, in applied work one typically just (i) *assumes* mutual independence when the nature of the experiment makes this plausible, (ii) models the individual marginal distributions, then (iii) deduces the joint c.d.f./p.m.f./p.d.f. by taking the product of the respective marginals.

Generalizing the concept of independence to random vectors whose component variables are not necessarily independent, let $\mathbf{X} = (\mathbf{X}_1', \mathbf{X}_2', ..., \mathbf{X}_m')' \in \Re^M$ be a collection of m vector-valued r.v.s, with $\mathbf{X}_j : \Omega \to \Re^{k_j}$ for each j and $M = \sum_{j=1}^{m} k_j$. Let $\mathbb{P}_{\mathbf{X}}$ be the joint measure on $\mathcal{B}^M$

(the Borel sets of $\Re^M$) induced by $\mathbf{X}$ and $\left\{\mathbb{P}_{\mathbf{X}_j}\right\}_{j=1}^m$ the marginal measures induced by the m subvectors. Then random vectors $\{\mathbf{X}_j\}_{j=1}^m$ are mutually stochastically independent if and only if $\mathbb{P}_{\mathbf{X}}\left(\times_{j=1}^m B_{k_j}\right) = \prod_{j=1}^m \mathbb{P}_{\mathbf{X}_j}\left(B_{k_j}\right)$ for each collection $\left\{B_{k_j} \in \mathcal{B}^{k_j}\right\}_{j=1}^m$. This is equivalent to the condition for c.d.f.s that $F_{\mathbf{X}}(\mathbf{x}) = \prod_{j=1}^m F_{\mathbf{X}_j}(\mathbf{x}_j)$ for each $\mathbf{x} = (\mathbf{x}_1', \mathbf{x}_2', ..., \mathbf{x}_m')' \in \Re^M$, and there are analogous conditions for joint p.m.f.s/p.d.f.s when the random vectors are all either discrete or continuous.

Finally, we extend the definition of conditional distributions to the multivariate case. Let $\mathbf{Y} \in \Re^\ell$ and $\mathbf{W} \in \Re^m$ ($\ell + m \leq k$) be subvectors of $\mathbf{X} = (X_1, X_2, ..., X_k)'$ having no common elements (e.g., $\mathbf{Y} = (X_1, X_2, X_3)'$ and $\mathbf{W} = (X_4, X_5, X_6, X_7)'$), and let $f_{\mathbf{Y},\mathbf{W}}(\mathbf{y}, \mathbf{w})$, $f_{\mathbf{Y}}(\mathbf{y})$, $f_{\mathbf{W}}(\mathbf{w})$ represent the joint and marginal p.d.f.s/p.m.f.s. Then the conditional p.d.f./p.m.f. of $\mathbf{W}$ given the realization $\mathbf{Y} = \mathbf{y}$ is defined as $f_{\mathbf{W}|\mathbf{Y}}(\mathbf{w} \mid \mathbf{y}) = f_{\mathbf{Y},\mathbf{W}}(\mathbf{y}, \mathbf{w}) / f_{\mathbf{Y}}(\mathbf{y})$ for $\mathbf{y}$ such that $f_{\mathbf{Y}}(\mathbf{y}) > 0$. If $\{X_j\}_{j=1}^k$ happen to be mutually independent, then $f_{\mathbf{W}|\mathbf{Y}}(\mathbf{w} \mid \mathbf{y}) = f_{\mathbf{W}}(\mathbf{w})$ for *all* such subvectors.

Exercise 2.37. *For* $\mathbf{X} \in \Re^3$ *let* $f_{\mathbf{X}}(\mathbf{x}) = ce^{-x_1-x_2-x_3}1_{\{\mathbf{x}:0<x_1<x_2<x_3\}}(\mathbf{x})$

   (i) Find the constant c such that $f_{\mathbf{X}}(\mathbf{x})$ is a p.d.f.
   (ii) Find f_{12}, the joint marginal p.d.f. of X_1 and X_2.
   (iii) Find F_1, the marginal c.d.f. of X_1.
   (iv) Determine whether X_1, X_2, X_3 are mutually independent.
   (v) Find $\Pr(X_3 > 3 \mid X_2 = 2, X_1 = 1)$.
   (vi) Find $\Pr(X_3 > 3 \mid X_2 > 2, X_1 > 1)$.

2.5   Multivariate Transformations

Given the k-vector-valued r.v. $\mathbf{X} = (X_1, X_2, ..., X_k)'$, suppose we create k new r.v.s $\mathbf{Y} = (Y_1, Y_2, ..., Y_k)'$ as $\{Y_j = g_j(\mathbf{X})\}_{j=1}^k$. In vector notation the transformation is

$$\mathbf{Y} = \mathbf{g}(\mathbf{X}) = (g_1(\mathbf{X}), g_2(\mathbf{X}), ..., g_k(\mathbf{X}))'.$$

For example, with $k = 2$ we might have $\mathbf{Y} = (Y_1, Y_2)' = (X_1^2 - X_2^2, X_1 + X_2)'$. With $\mathbb{X} \subset \Re^k$ as the support of $\mathbf{X}$, let $\mathbf{g}$ map $\mathbb{X}$ onto the (image) set $\mathbb{Y} \subset \Re^k$. As usual, we require that $\mathbf{g}$ be a measurable function, in the sense that the inverse image of any Borel set B_k of $\Re^k$ is itself a Borel set: $\mathbf{g}^{-1}(B_k) = \{\mathbf{x} : \mathbf{g}(\mathbf{x}) \in B_k\} \in \mathcal{B}^k$. At the most general level, representing the measure induced by $\mathbf{Y}$ in terms of that induced by $\mathbf{X}$

works just as in the univariate case. Thus, corresponding to (2.8), measure $\mathbb{P}_{\mathbf{Y}}$ is given by

$$\mathbb{P}_{\mathbf{Y}}(B_k) = \mathbb{P}\left(\{\omega : \mathbf{Y}(\omega) \in B_k\}\right) = \mathbb{P}_{\mathbf{X}}\left[\mathbf{g}^{-1}\left(B_k\right)\right].$$

In particular, putting $B_k := \times_{j=1}^{k} (-\infty, y_j]$ gives the joint c.d.f. of $\mathbf{Y}$ at the point $\mathbf{y} = (y_1, y_2, ..., y_k)'$:

$$F_{\mathbf{Y}}(\mathbf{y}) = \mathbb{P}_{\mathbf{Y}}\left(\times_{j=1}^{k} (-\infty, y_j]\right) = \mathbb{P}_{\mathbf{X}}\left[\mathbf{g}^{-1}\left(\times_{j=1}^{k} (-\infty, y_j]\right)\right]. \qquad (2.23)$$

Although the concept is easy enough, it is often not easy to describe the set $\mathbf{g}^{-1}\left(\times_{j=1}^{k} (-\infty, y_j]\right)$ precisely, much less to work out its $\mathbb{P}_{\mathbf{X}}$ measure. How to do it depends on the specific features of the problem. Here we consider just the common cases that the elements of $\mathbf{X}$ are either all discrete or all continuous, with joint p.m.f. or joint p.d.f. $f_{\mathbf{X}}$. As for univariate transformations, we deal separately with the discrete and continuous cases. However, the general setting of induced measures will be revisited at the end of the section to establish the very useful result that (measurable) functions of independent r.v.s are themselves independent.

2.5.1 *Functions of Discrete Random Variables*

$\mathbf{X}$ is now a discrete vector-valued r.v. with joint p.m.f. $f_{\mathbf{X}}$, where $f_{\mathbf{X}}(\mathbf{x}) > 0$ for each of the countably many points $\mathbf{x} \in \mathbb{X}$, and $\mathbf{g}$ maps $\mathbb{X}$ onto the k-dimensional set $\mathbb{Y}$; i.e., $\mathbf{g} : \mathbb{X} \to \mathbb{Y} \subset \Re^k$. Of course, $\mathbb{Y}$ is necessarily countable as well, so that the new r.v. $\mathbf{Y} = \mathbf{g}\left(\mathbf{X}\right)$ is also discrete. Following the general approach just described, for any particular $\mathbf{y} \in \mathbb{Y}$ the set of $\mathbf{x}$ values that $\mathbf{g}$ maps to the point $\mathbf{y} \in \Re^k$ (i.e., the inverse image of $\mathbf{y}$) is $\mathbf{g}^{-1}\left(\{\mathbf{y}\}\right) := \{\mathbf{x} : \mathbf{g}(\mathbf{x}) = \mathbf{y}\}$. Then it is clear that $\mathbf{Y}$ takes on the value $\mathbf{y}$ with the same probability that $\mathbf{X}$ takes on a value in the set $\mathbf{g}^{-1}\left(\{\mathbf{y}\}\right)$. That is,

$$f_{\mathbf{Y}}\left(\mathbf{y}\right) = \mathbb{P}_{\mathbf{Y}}(\{\mathbf{y}\}) = \sum_{\mathbf{x} \in \mathbf{g}^{-1}(\{\mathbf{y}\})} \mathbb{P}_{\mathbf{X}}(\{\mathbf{x}\}) = \sum_{\mathbf{x} \in \mathbf{g}^{-1}(\{\mathbf{y}\})} f_{\mathbf{X}}(\mathbf{x}).$$

Describing the set $\mathbf{g}^{-1}\left(\{\mathbf{y}\}\right)$ precisely is usually easier when $\mathbf{g}$ is one to one on $\mathbb{X}$ and thus has a unique inverse, $\mathbf{h} : \mathbb{Y} \to \mathbb{X}$. The set $\mathbf{g}^{-1}\left(\{\mathbf{y}\}\right)$ then contains the single point $\{\mathbf{h}(\mathbf{y})\}$, and

$$f_{\mathbf{Y}}(\mathbf{y}) = f_{\mathbf{X}}\left[\mathbf{h}(\mathbf{y})\right].$$

Example 2.36. Flipping three fair coins, take X_1, X_2 to be the numbers of heads and sequence changes, respectively. The joint p.m.f., $f_{\mathbf{X}}$, is given in tabular form as

$x_2 \backslash x_1$	0	1	2	3
0	$\frac{1}{8}$	0	0	$\frac{1}{8}$
1	0	$\frac{2}{8}$	$\frac{2}{8}$	0
2	0	$\frac{1}{8}$	$\frac{1}{8}$	0

The support $\mathbb{X}$ comprises the ordered pairs of integers corresponding to cells with positive entries. Define the new vector r.v. $\mathbf{Y}$ as

$$\mathbf{Y} = \begin{pmatrix} Y_1 \\ Y_2 \end{pmatrix} = \mathbf{g}(\mathbf{X}) = \begin{pmatrix} g_1(X_1, X_2) \\ g_2(X_1, X_2) \end{pmatrix} = \begin{pmatrix} X_1^2 - X_2^2 \\ X_1 + X_2 \end{pmatrix}.$$

To determine $\mathbb{Y}$, the support of $\mathbf{Y} = \mathbf{g}(\mathbf{X})$, evaluate (Y_1, Y_2) at each cell of the $f_\mathbf{X}$ table that contains a positive entry. The result is a table with (y_1, y_2) pairs in the (x_1, x_2) cells:

$x_2 \backslash x_1$	0	1	2	3
0	$(0,0)$			$(9,3)$
1		$(0,2)$	$(3,3)$	
2		$(-3,3)$	$(0,4)$	

This transformation is one to one, since each nonempty cell has a unique ordered pair (y_1, y_2), so we attach to each of these (y_1, y_2) pairs the probability assigned by $f_\mathbf{X}$ to the cell that contains it. Thus, $f_\mathbf{Y}(0,0) = f_\mathbf{X}(0,0) = \frac{1}{8}$, $f_\mathbf{Y}(0,2) = f_\mathbf{X}(1,1) = \frac{2}{8}$, and so on, leading to the following tabular display for $f_\mathbf{Y}$:

$y_2 \backslash y_1$	-3	0	3	9
0		$\frac{1}{8}$		
2		$\frac{2}{8}$		
3	$\frac{1}{8}$		$\frac{2}{8}$	$\frac{1}{8}$
4		$\frac{1}{8}$		

2.5.2   *Functions of Continuous Random Variables*

Finding distributions of transformations $\mathbf{g}$ of continuous random vectors is usually easiest by working from the joint p.d.f.s. Treating first the case that $\mathbf{g}$ is one to one on support $\mathbb{X}$, we extend change-of-variable formula (2.12) to the vector case. Let $\mathbf{h}$ be the unique inverse transformation that maps $\mathbb{Y}$ onto $\mathbb{X}$; that is, $\mathbf{h}(\mathbf{y}) = (h_1(\mathbf{y}), h_2(\mathbf{y}), ..., h_k(\mathbf{y}))'$ gives the solutions for $\mathbf{x}$ in terms of $\mathbf{y}$. The *Jacobian* of the inverse transformation, denoted $J_\mathbf{h}$, is defined as the determinant of the $k \times k$ matrix of partial derivatives; i.e.,

$$J_{\mathbf{h}} := \left| \frac{d\mathbf{x}}{d\mathbf{y}} \right| = \begin{vmatrix} \frac{\partial h_1(\mathbf{y})}{\partial y_1} & \frac{\partial h_1(\mathbf{y})}{\partial y_2} & \cdots & \frac{\partial h_1(\mathbf{y})}{\partial y_k} \\ \frac{\partial h_2(\mathbf{y})}{\partial y_1} & \frac{\partial h_2(\mathbf{y})}{\partial y_2} & \cdots & \frac{\partial h_2(\mathbf{y})}{\partial y_k} \\ \cdots & \cdots & \ddots & \cdots \\ \frac{\partial h_k(\mathbf{y})}{\partial y_1} & \frac{\partial h_k(\mathbf{y})}{\partial y_2} & \cdots & \frac{\partial h_k(\mathbf{y})}{\partial y_k} \end{vmatrix}.$$

Being a determinant, $J_{\mathbf{h}}$ is a scalar, and it is nonzero and finite a.e. because $\mathbf{g}$ is one to one; however, like a derivative in the univariate case, $J_{\mathbf{h}}$ may well be negative. The multivariate change-of-variable formula gives as the joint p.d.f. of $\mathbf{Y}$ the joint p.d.f. of $\mathbf{X}$ at $\mathbf{h}(\mathbf{y})$ rescaled by the *absolute value* of $J_{\mathbf{h}}$. That is,

$$f_{\mathbf{Y}}(\mathbf{y}) = f_{\mathbf{X}}[\mathbf{h}(\mathbf{y})] \, |J_{\mathbf{h}}| \, . \tag{2.24}$$

As in the univariate case the factor $|J_{\mathbf{h}}|$ accounts for the concentration or attenuation of probability mass caused by the transformation. Of course, $|J_{\mathbf{h}}| \equiv |J_{\mathbf{h}}(\mathbf{y})|$ may vary from one point in $\mathbb{Y}$ to another, according as $\mathbf{g}$ thins or thickens the mass in the vicinity of $\mathbf{x} = \mathbf{h}(\mathbf{y})$.

Here, then, are the steps one can follow to find the joint p.d.f. of $\mathbf{Y}$ when $\mathbf{g} : \mathbb{X} \to \mathbb{Y}$ is one to one:

(i) If it is possible to do so, express the inverse transformation $\mathbf{h} : \mathbb{Y} \to \mathbb{X}$ analytically by solving for $\mathbf{X}$ in terms of $\mathbf{Y}$; that is, express X_j explicitly as $h_j(Y_1, Y_2, ..., Y_k)$ for each $j \in \{1, 2, ..., k\}$. In the affine case, $\mathbf{Y} = \mathbf{g}(\mathbf{X}) = \mathbf{a} + \mathbf{B}\mathbf{X}$, the solution is simply $\mathbf{X} = \mathbf{h}(\mathbf{Y}) = \mathbf{B}^{-1}(\mathbf{Y} - \mathbf{a})$ so long as $|\mathbf{B}| = |d\mathbf{y}/d\mathbf{x}| = J_{\mathbf{g}}$ is not zero. If $\mathbf{h}$ cannot be expressed analytically, one will have to settle for an implicit expression to be evaluated computationally.

(ii) Determine the set $\mathbb{Y}$ that is the image of $\mathbb{X}$ under $\mathbf{g}$; that is, the set of values onto which $\mathbf{g}$ maps the support of $\mathbf{X}$.

(iii) Calculate the Jacobian of the inverse transformation as the determinant $J_{\mathbf{h}}$ of the $k \times k$ matrix of partial derivatives. If an analytical expression for $\mathbf{h}$ is available, this would be found as $J_{\mathbf{h}} = |d\mathbf{x}/d\mathbf{y}|$; otherwise, as $J_{\mathbf{h}} = 1/|d\mathbf{y}/d\mathbf{x}| = 1/J_{\mathbf{g}}$. In the affine case $J_{\mathbf{h}} = |\mathbf{B}^{-1}| = 1/|\mathbf{B}|$.

(iv) Evaluate joint p.d.f. $f_{\mathbf{X}}$ at the point $\mathbf{h}(\mathbf{y}) = (h_1(\mathbf{y}), h_2(\mathbf{y}), ..., h_k(\mathbf{y}))'$ and multiply by $|J_{\mathbf{h}}|$.

Sometimes step 2 can be tricky, but it helps to have already determined the inverse transformation in step 1, as the following example shows.

Example 2.37. Spinning a fair pointer twice, let X_1 and X_2 measure the clockwise distances from the origin on the two spins. Model the joint distribution of $\mathbf{X} = (X_1, X_2)'$ as uniform over the unit square, with p.d.f. $f_{\mathbf{X}}(x_1, x_2) = 1_{[0,1]^2}(x_1, x_2)$. Defining $Y_1 := X_1 + X_2$ and $Y_2 := X_2$, the joint p.d.f. of

$$\mathbf{Y} = \begin{pmatrix} Y_1 \\ Y_2 \end{pmatrix} = \begin{pmatrix} 1 & 1 \\ 0 & 1 \end{pmatrix} \begin{pmatrix} X_1 \\ X_2 \end{pmatrix} = \begin{pmatrix} X_1 + X_2 \\ X_2 \end{pmatrix}$$

is found by following the four steps listed above.

(i) Express the X's in terms of the Y's, as

$$\begin{pmatrix} X_1 \\ X_2 \end{pmatrix} = \begin{pmatrix} 1 & 1 \\ 0 & 1 \end{pmatrix}^{-1} \begin{pmatrix} Y_1 \\ Y_2 \end{pmatrix} = \begin{pmatrix} 1 & -1 \\ 0 & 1 \end{pmatrix} \begin{pmatrix} Y_1 \\ Y_2 \end{pmatrix} = \begin{pmatrix} Y_1 - Y_2 \\ Y_2 \end{pmatrix}.$$

(ii) Find $\mathbb{Y}$. A first thought might be that $\mathbb{Y}$ is the product space $[0, 2] \times [0, 1]$, since Y_1 *can* take any value on $[0, 2]$ and Y_2 lies on $[0, 1]$; however, considering the inverse transformation shows that not all values in $[0, 2] \times [0, 1]$ can be attained from $\mathbb{X} = [0, 1]^2$. The fact that $0 \le Y_1 - Y_2 = X_1 \le 1$ implies that $Y_1 - 1 \le Y_2 \le Y_1$. Adding the further constraint $Y_2 = X_2 \in [0, 1]$ gives $\max\{0, Y_1 - 1\} \le Y_2 \le \min\{1, Y_1\}$. Thus,

$$\mathbb{Y} = \{\mathbf{y} : y_1 \in [0, 2], y_2 \in [\max\{0, y_1 - 1\}, \min\{1, y_1\}]\}$$
$$\equiv \{\mathbf{y} : y_1 \in [y_2, 1 + y_2], y_2 \in [0, 1]\},$$

which is depicted by the shaded parallelogram in Fig. 2.11.

(iii) Find the Jacobian of inverse transformation, as

$$J_{\mathbf{h}} = \begin{vmatrix} \frac{\partial(y_1 - y_2)}{\partial y_1} & \frac{\partial(y_1 - y_2)}{\partial y_2} \\ \frac{\partial y_2}{\partial y_1} & \frac{\partial y_2}{\partial y_2} \end{vmatrix} = \begin{vmatrix} 1 & -1 \\ 0 & 1 \end{vmatrix} = 1 = 1/J_{\mathbf{g}}.$$

In this case $\mathbf{g}$ merely transfers mass uniformly from one region of the plane to another with the same area, neither concentrating nor attenuating it. (Had one wrongly concluded that $\mathbb{Y} = [0, 2] \times [0, 1]$ in step 2, finding that $J_{\mathbf{h}} = 1$ would immediately reveal the error.)

(iv) Evaluate $f_{\mathbf{X}}$ at $\mathbf{x} = \mathbf{h}(\mathbf{y})$ and rescale. Since $f_{\mathbf{X}}(x_1, x_2) = 1$ for $(x_1, x_2) \in \mathbb{X}$ and $|J_{\mathbf{h}}| = 1$, we also have $f_{\mathbf{Y}}(y_1, y_2) = 1$ for $(y_1, y_2) \in \mathbb{Y}$. (Note that Y_1, Y_2 are not independent since $\mathbb{Y}$ is not a product space.)

Exercise 2.38. *Find the joint p.d.f. of $Y_1 = X_1 + 2X_2$ and $Y_2 = X_2$ when X_1, X_2 have p.d.f. $f_{\mathbf{X}}(x_1, x_2) = 1_{[0,1]^2}(x_1, x_2)$.*

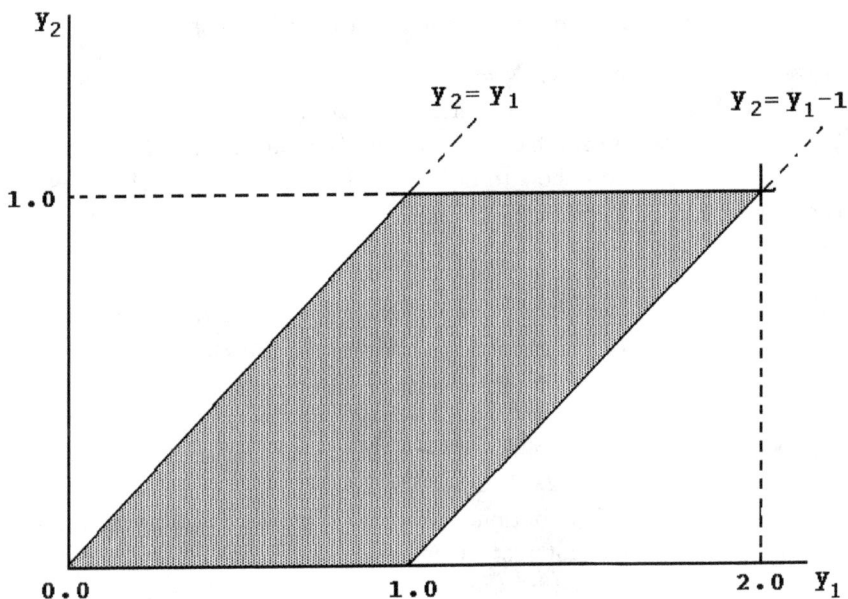

Fig. 2.11   Range $\mathbb{Y}$ for $\mathbf{Y} = (Y_1, Y_2)'$.

The discussion at the end of Section 2.3.2 and Example 2.3.2 showed how to extend the univariate change-of-variable formula to certain functions g : $\mathbb{X} \to \mathbb{Y}$ that are not one to one. The multivariate procedure can be adapted in the same way. Thus, suppose that for some $\mathbf{y} \in \mathbb{Y}$ there are countably many points $\{\mathbf{x}_j\}_{j=1}^{\infty} \in \mathbb{X}$ such that $\mathbf{g}(\mathbf{x}_j) = \mathbf{y}$ and such that $|J_{\mathbf{g}}| \neq 0$. Then to each such $\mathbf{x}_j$ there correspond a unique inverse transformation $\mathbf{h}^j(\mathbf{y}) = \mathbf{x}_j$ and a contribution $f_{\mathbf{X}}\left[\mathbf{h}^j(\mathbf{y})\right]|J_{\mathbf{h}^j}|$ to the density at $\mathbf{y}$. Aggregating the contributions gives

$$f_{\mathbf{Y}}(\mathbf{y}) = \sum_{j=1}^{\infty} f_{\mathbf{X}}\left[\mathbf{h}^j(\mathbf{y})\right]|J_{\mathbf{h}^j}|.$$

Note, however, that the above does not apply if $\mathbf{g}(\mathbf{x})$ equals some fixed $\mathbf{y}$ for all $\mathbf{x}$ in a set $\mathbb{X}_{\mathbf{y}}$ of positive $\mathbb{P}_{\mathbf{X}}$ measure, for then we have a situation in which $\mathbf{Y}$ is not absolutely continuous. Example 2.21 shows an analogous situation in the one-dimensional case.

2.5.3 *Distributions of* $m < k$ *Functions of* k *Variables*

Given a k-dimensional r.v. $\mathbf{X} = (X_1, X_2, ..., X_k)'$, one often wants to find the joint distribution of *fewer* than k functions of $\mathbf{X}$. For instance, in Example 2.37 one might care only about the (marginal) distribution of $Y_1 = X_1 + X_2$. Here is how to derive the joint distribution of discrete or continuous r.v.s. $\mathbf{Y} = (Y_1, Y_2, ..., Y_m)' = (g_1(\mathbf{X}), g_2(\mathbf{X}), ..., g_m(\mathbf{X}))'$ when $m < k$.

(i) *Invent* $k - m$ new functions $\{g_j(\mathbf{X})\}_{j=m+1}^{k}$, and define $k - m$ new variables as $Y_{m+1} = g_{m+1}(\mathbf{X}), ..., Y_k = g_k(\mathbf{X})$ in such a way that the augmented collection of k variables is either purely discrete or purely continuous.

(ii) Determine the joint p.m.f. or p.d.f. of $(Y_1, Y_2, ..., Y_m, Y_{m+1}, ..., Y_k)'$ as explained in Section 2.5.1 or Section 2.5.2.

(iii) Derive from the distribution in $\Re^k$ the (joint) *marginal* distribution of the m variables of interest by summing or integrating out the $k - m$ new variables created in step 1.

While the procedure seems straightforward, some care is needed in choosing the $k - m$ new variables. Ideally, $g_{m+1}, ..., g_k$ should make the augmented transformation one to one and be such that the final integration/summation step is easy to carry out. This may require some experimentation.

Example 2.38. Finding the marginal distribution of Y_1 in Example 2.37 requires integrating out Y_2, as $f_1(y_1) = \int_{-\infty}^{\infty} f_{\mathbf{Y}}(y_1, y_2) \cdot dy_2$. One has to be careful, because support set $\mathbb{Y}$ is not a product space. (See Fig. 2.11.) The constraint

$$y_2 \in [\max\{0, y_1 - 1\}, \min\{1, y_1\}]$$

implies that $y_2 \in [0, y_1]$ for $y_1 \in [0, 1]$, while $y_2 \in [y_1 - 1, 1]$ for $y_1 \in [1, 2]$. Thus,

$$f_1(y_1) = \begin{cases} \int_{-\infty}^{0} 0 \cdot dy_2 + \int_{0}^{y_1} 1 \cdot dy_2 & = y_1, \; 0 \le y_1 \le 1 \\ \int_{-\infty}^{y_1-1} 0 \cdot dy_2 + \int_{y_1-1}^{1} 1 \cdot dy_2 & = 2 - y_1, \; 1 \le y_1 \le 2 \end{cases}$$
$$= y_1 \mathbf{1}_{[0,1)}(y_1) + (2 - y_1)\mathbf{1}_{[1,2]}(y_1).$$

The sum of two independent r.v.s distributed uniformly on $[0, 1]$ thus has a *triangular* p.d.f., symmetric about 1.0, as depicted in Fig. 2.12.

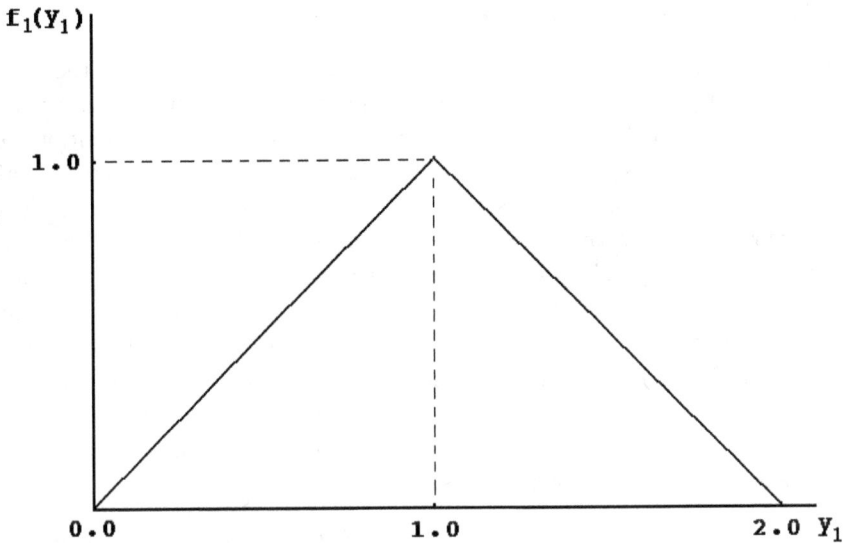

Fig. 2.12   P.d.f. of $Y_1 = X_1 + X_2$.

Example 2.39. To find the p.d.f. of $Y_1 = g_1(\mathbf{X}) = X_1 + X_2$ when $f_{\mathbf{X}}(x_1, x_2) = 1_{[0,1]^2}(x)$, let us choose $Y_2 = g_2(\mathbf{X}) = X_1 - X_2$ instead of $Y_2 = X_2$. Transformation $\mathbf{g}$ now rotates and extends $\mathbb{X} = [0,1]^2$ onto a square $\mathbb{Y}$ with vertices at $(0,0), (1,1), (2,0),$ and $(1,-1)$ and area 2. The inverse transformation is $X_1 = (Y_1 + Y_2)/2, X_2 = (Y_1 - Y_2)/2$ with $|J_{\mathbf{h}}| = \frac{1}{2}$, and the joint p.d.f. of $\mathbf{Y}$ is $f_{\mathbf{Y}}(\mathbf{y}) = \frac{1}{2}$ for $y_1 \in [0,1), y_2 \in (-y_1, y_1)$ and for $y_1 \in [1,2), y_2 \in (y_1 - 2, 2 - y_1)$. Integrating out y_2 gives as before

$$f_1(y_1) = y_1 1_{[0,1)}(y_1) + (2 - y_1) 1_{[1,2]}(y_1).$$

Exercise 2.39. *Find the marginal p.d.f. of $Y_1 = X_1 + 2X_2$ from the joint p.d.f. of Y_1 and $Y_2 = X_2$ in Exercise 2.38.*

Exercise 2.40. *When Z_1 and Z_2 are independent, each with (standard normal) p.d.f. $f_Z(z) = (2\pi)^{-1/2} e^{-z^2/2}$, show that the p.d.f. of $Y := Z_1/Z_2$ is $f_Y(y) = [\pi(1 + y^2)]^{-1}$. (This is the standard Cauchy p.d.f., which we shall encounter often in the sequel.)*

2.5.4   *Convolution Sums/Integrals*

There are some special shortcuts for finding the distribution of the *sum* of two jointly distributed r.v.s. To explain them, we again dispense with subscripts, taking $\mathbf{Z} = (X, Y)'$ as a bivariate r.v. with joint distribution $f_{\mathbf{Z}}$ and $S := X + Y$ as the sum. To find the distribution of S with the general procedure for transformations in Section 2.5.3 involves (i) introducing a new variable T, say, that makes the transformation one to one; (ii) finding the joint p.m.f./p.d.f. $f_{S,T}(s, t)$; then (iii) summing or integrating out t. Typically, the best choice of T for this purpose is either $T = X$ or $T = Y$, since each makes the transformation one to one with Jacobian of unity. With $T = Y$ the inverse transformation is $X = S - T$ and $Y = T$, the joint p.m.f./p.d.f. is $f_{\mathbf{Z}}(s - t, t)$, and the marginal p.m.f./p.d.f. is $f_S(s) = \sum_t f_{\mathbf{Z}}(s - t, t)$ or $f_S(s) = \int_{-\infty}^{\infty} f_{\mathbf{Z}}(s - t, t) \cdot dt$, depending on whether $\mathbf{Z}$ is discrete or continuous.

Exercise 2.41. *R.v.s X_1, X_2 are independent with marginal distributions*

$$f_j(x) = \left(\frac{\lambda_j^x e^{-\lambda_j}}{x!} \right) 1_{\aleph_0}(x), j \in \{1, 2\},$$

where $\lambda_1 > 0, \lambda_2 > 0$ and $\aleph_0 := \{0, 1, 2, ...\}$. Find the p.m.f. of $S = X_1 + X_2$. (Distributions $\{f_j\}$ are members of the Poisson *family, which are discussed in Section 4.2.6.)*

Exercise 2.42. *Find the p.d.f. of $S := X + Y$ if $\mathbf{Z} = (X, Y)'$ has joint p.d.f. $f_{\mathbf{Z}}(x, y) = 2e^{-x-y} 1_{\{(x,y):0<x<y\}}(x, y)$.*

Exercise 2.43. *R.v.s X_1, X_2 are independent and identically distributed (i.i.d.) with (standard normal) p.d.f. $f(x) = (2\pi)^{-1/2} e^{-x^2/2}$. Find the p.d.f. of $S = X_1 + X_2$. (Hint: Manipulate the convolution integral until it has the form $g(s) \int_{-\infty}^{\infty} (2\pi\phi^2)^{-1/2} e^{-(x-\theta)^2/(2\phi^2)} \cdot dx$ for some function $g(s)$, some quantity θ, and some positive ϕ, and use the fact—to be proved in Chapter 4—that integrals of the form appearing here equal unity.)*

When the original X, Y are independent, the joint p.m.f./p.d.f. factors as $f_{\mathbf{Z}}(x, y) = f_X(x)f_Y(y)$, yielding the special formulas

$$f_S(s) = \sum_y f_X(s - y)f_Y(y)$$

$$f_S(s) = \int_{-\infty}^{\infty} f_X(s - y)f_Y(y) \cdot dy. \qquad (2.25)$$

Note that we have just retained "y" as the dummy operand instead of introducing a new symbol. These *convolution* formulas offer a direct means of finding distributions of sums of independent r.v.s. Section 3.4.2 introduces some indirect methods for this that are often easier to use, and in Section 3.5.3 we will see that there is a much more intuitive way of deriving (and remembering) expression (2.25).

Exercise 2.44. *If* $\mathbf{Z} = (X, Y)'$ *has p.d.f.* $f_{\mathbf{Z}}(x, y) = e^{-x-y}1_{(0,\infty)^2}(x, y)$, *find the p.d.f. of* $S = X + Y$.

2.5.5 *Independence of Functions of Independent R.V.s*

We return to the general setting of induced measures to establish a fact that will be used repeatedly in the sequel. With $\mathbf{X} = (X_1, X_2, ..., X_k)'$ as a vector-valued r.v. with support $\mathbb{X}$ and induced measure $\mathbb{P}_{\mathbf{X}}$, suppose now that $\{X_j\}_{j=1}^{k}$ are mutually independent. Thus, for all sequences $\{B^{(j)}\}_{j=1}^{k}$ of (one-dimensional) Borel sets we have $\mathbb{P}_{\mathbf{X}}\left(\times_{j=1}^{k}B^{(j)}\right) = \prod_{j=1}^{k}\mathbb{P}_{X_j}(B^{(j)})$. Now introduce transformation $\mathbf{Y} = \mathbf{g}(\mathbf{X})$ of specific form $\{Y_j = g_j(X_j)\}_{j=1}^{k}$, such that each distinct element Y_j of $\mathbf{Y}$ is a (Borel) measurable function of a single, distinct element X_j of $\mathbf{X}$, and such that each Y_j is nondegenerate (*not* such that $\Pr(Y_j = c) = 1$ for some constant c). As always, measurability guarantees that the inverse image under g_j of any Borel set B be itself a Borel set. Let $g_j^{-1}(B) = \{x : g_j(x) \in B\}$ represent this inverse image. Then the inverse image under $\mathbf{g}$ of product set $\times_{j=1}^{k}B^{(j)}$ is the product set $\times_{j=1}^{k}g_j^{-1}(B^{(j)})$. For example, when $k = 2$

$$\mathbf{g}^{-1}\left(B^{(1)} \times B^{(2)}\right) = \left\{\mathbf{x} : (g_1(x_1), g_2(x_2)) \in B^{(1)} \times B^{(2)}\right\}$$

$$= \left\{x_1 : g_1(x_1) \in B^{(1)}\right\} \times \left\{x_2 : g_2(x_2) \in B^{(2)}\right\}$$

$$= g_1^{-1}\left(B^{(1)}\right) \times g_2^{-1}\left(B^{(2)}\right),$$

and in general $\mathbf{g}^{-1}\left(\times_{j=1}^{k}B^{(j)}\right) = \times_{j=1}^{k}g_j^{-1}(B^{(j)})$. It then follows that the $\mathbb{P}_{\mathbf{Y}}$ measure of the product of any sequence of Borel sets $\{B^{(j)}\}_{j=1}^{k}$ is

$$\mathbb{P}_{\mathbf{Y}}\left(\times_{j=1}^{k}B^{(j)}\right) = \mathbb{P}_{\mathbf{X}}\left(\times_{j=1}^{k}g_j^{-1}\left(B^{(j)}\right)\right)$$

$$= \prod_{j=1}^{k}\mathbb{P}_{X_j}\left(g_j^{-1}\left(B^{(j)}\right)\right) = \prod_{j=1}^{k}\mathbb{P}_{Y_j}(B^{(j)}),$$

where the second equality follows from the mutual independence of $\{X_j\}_{j=1}^k$. This shows that nondegenerate functions $\{Y_j = g_j\,(X_j)\}_{j=1}^k$ of individual independent r.v.s are themselves mutually independent. In particular, if X, Y are independent and $g : \Re \to \Re$ and $h : \Re \to \Re$ are measurable and nondegenerate, then r.v.s $G = g(X), H = h(Y)$ are independent also, with joint c.d.f. $F_{G,H}(g,h) = F_G(g)F_H(h)$ and joint p.d.f./p.m.f. $f_{G,H}(g,h) = f_G(g)f_H(h)$ for each $(g,h) \in \Re^2$.

Generalizing, let $\mathbf{X} = (\mathbf{X}_1', \mathbf{X}_2', ..., \mathbf{X}_m')' \in \Re^M$ be a collection of m mutually independent vector-valued r.v.s $\left\{\mathbf{X}_j \in \Re^{k_j}\right\}_{j=1}^m$, with $M = \sum_{j=1}^m k_j$; and let $\left\{g_j : \Re^{k_j} \to \Re\right\}_{j=1}^m$ be nondegenerate, measurable functions. Then r.v.s $\{Y_j = g_j\,(\mathbf{X}_j)\}_{j=1}^m$ are mutually independent.

2.6   Solutions to Exercises

1. The support of X is $\mathbb{X} = [0,1]$. Half the probability mass will be at $\{0\}$ since the probability of tails is $\frac{1}{2}$. The remaining mass of $\frac{1}{2}$ is spread uniformly on $(0,1]$. The c.d.f. is

$$F(x) = \left\{ \begin{array}{l} 0,\, x < 0 \\ \frac{1+x}{2},\, 0 \leq x < 1 \\ 1,\, x \geq 1 \end{array} \right\} = \min\left\{\frac{1+x}{2}, 1\right\} \mathbf{1}_{[0,\infty)}\,(x).$$

2. The support of X is $\mathbb{X} = [0,300]$ and the unit probability mass is distributed uniformly: $F(x) = \max\left\{\min\left\{\frac{x}{300}, 1\right\}, 0\right\}, x \in \Re$.

3. Let g be a nondecreasing, right-continuous map from $[0,1]$ onto $[0,1]$, and set $G\,(x) = g\,[F\,(x)]$. Then
   (i) $G\,(+\infty) = g\,(1) = 1$
   (ii) $G\,(-\infty) = g\,(0) = 0$
   (iii) $G\,(x+h) - G\,(x) = g\,[F\,(x+h)] - g\,[F\,(x)] \geq 0$ for $h > 0$, and (by right continuity)
   (iv) $\lim_{h \to 0} G\,(x+h) - G\,(x) = g\,[\lim_{h \to 0} F\,(x+h)] - g\,[F\,(x)] = 0$.
   Thus, $G\,(x) = g\,[F\,(x)]$ is a c.d.f. All of $\{G_j\}_{j=1}^5$ except G_2 is a function $g(F)$ having the required properties.

4. The support of X is $\mathbb{X} = \{0, 1, ..., 5\}$. For $x \in \mathbb{X}$ there are $4^x \cdot 48^{5-x}$ ways to draw x aces and $5 - x$ other cards in a *particular* order, and $\binom{5}{x}$ such arrangements, out of 52^5 ways of drawing five cards. Thus, $f(x) = \binom{5}{x}\left(\frac{4}{52}\right)^x\left(\frac{48}{52}\right)^{5-x} = \binom{5}{x}\left(\frac{1}{13}\right)^x\left(\frac{12}{13}\right)^{5-x}$ for $x \in \mathbb{X}$. The c.d.f. can be represented as $F(x) = \min\left\{\sum_{j=0}^{[x]} f(j), 1\right\}$ where $[x]$ is the greatest

integer contained in x and where the sum is interpreted as being equal to zero when $[x] < 0$.

5. The support of X is $\mathbb{X} = \{0, 1, 2, 3, 4\}$. For $x \in \mathbb{X}$ there are $\binom{4}{x}$ ways of drawing x aces and $\binom{48}{5-x}$ ways of drawing the remaining cards, out of $\binom{52}{5}$ possible 5-card hands. Thus, $f(x) = \binom{4}{x}\binom{48}{5-x}/\binom{52}{5}$ for $x \in \mathbb{X}$. The c.d.f. can be represented as $F(x) = \min\left\{\sum_{j=0}^{[x]} f(j), 1\right\}$ where $[x]$ is the greatest integer contained in x and where the sum is interpreted as being equal to zero when $[x] < 0$.

6. (i) $1 = \int_1^e k \log x \cdot dx = k \left[x \log x - x\right]_1^e = k$.

(ii)

$$F(x) = \begin{cases} 0, & x < 1 \\ \int_1^x \log t \cdot dt = x \log x - x + 1, & 1 \le x < e \\ 1, & x \ge e \end{cases}$$

$$= (x \log x - x + 1)\, \mathbf{1}_{[1,e)}(x) + \mathbf{1}_{[e,\infty)}(x).$$

(iii) (a) $\Pr(X > 1) = 1 - F(1) = 1$; (b) $\Pr(X < e/2) = F(e/2) \doteq .058$; (c) $\Pr(X \in \mathbb{Q} \cap [1, e)) = 0$ since X is continuous and $\mathbb{Q} \cap [1, e)$ is countable.

7. Under condition (i) there is some $\varepsilon > 0$ such that $F(x^*) = .5 + \varepsilon$ and $F(x^* -) = 1 - \Pr(X \ge x^*) \le .5$. Thus, at x^* c.d.f. F takes a jump from .5 or less to some point above .5. Since F is nondecreasing, no such jump can occur elsewhere. For example, if some $x^{**} > x^*$ were asserted to satisfy the same condition, then we would have $F(x^{**} -) \ge F(x^*) > .5$ and hence $F(x^{**}) = 1 - F(x^{**} -) < .5$, a contradiction. A similar argument applies under condition (ii).

8. For $x > x_{.5}^+$ we have $\Pr(X \le x) \ge \Pr(X \le x_{.5}^+) \ge .5$. But x is not a median, so $\Pr(X \ge x) < .5$ and therefore $\Pr(X \le x) - \Pr(X \ge x) > 0$. For $x < x_{.5}^-$ we have $\Pr(X \ge x) \ge \Pr(X \ge x_{.5}^-) \ge .5$. But x is not a median, so $\Pr(X \le x) < .5$ and therefore $\Pr(X \le x) - \Pr(X \ge x) < 0$.

9. (i) If $\Pr(X > x^*) = \Pr(X < x^*)$ then $2\Pr(X > x^*) + \Pr(X = x^*) = 1 = 2\Pr(X < x^*) + \Pr(X = x^*)$. The first equality implies $\Pr(X > x^*) = \frac{1}{2} - \frac{1}{2}\Pr(X = x^*) \ge \frac{1}{2} - \Pr(X = x^*)$, so that $\Pr(X \ge x^*) \ge \frac{1}{2}$; likewise, $\Pr(X \le x^*) \ge \frac{1}{2}$ follows from the second equality.

(ii) Consider Example 2.17(ii).

(iii) Suppose $\Pr(X < x^*) > \Pr(X > x^*)$ for continuous X and some x^*. Then x^* is not a median, since $1 = \Pr(X \le x^*) + \Pr(X \ge x^*) > 2\Pr(X \ge x^*)$ implies $\Pr(X \ge x^*) < \frac{1}{2}$.

10. $g(x) = x^2 - 2x$ maps $\mathbb{X} = \{0, 1, 2, 3\}$ onto $\mathbb{Y} = \{-1, 0, 3\}$. The inverse images $g^{-1}(\{y\})$ for each $y \in \mathbb{Y}$ are $g^{-1}(\{-1\}) = \{1\}$, $g^{-1}(\{0\}) =$

$\{0, 2\}$, $g^{-1}(\{3\}) = 3$. Therefore, $f_Y(y) = \mathbb{P}_Y(\{y\})$ is given by

$$f_Y(y) = \begin{cases} f_X(1) = \frac{3}{8}, \, y = -1 \\ f_X(0) + f_X(2) = \frac{4}{8}, \, y = 0 \\ f_X(3) = \frac{1}{8}, \, y = 3 \end{cases}$$

$$= \frac{3}{8}1_{\{-1\}}(y) + \frac{4}{8}1_{\{0\}}(y) + \frac{1}{8}1_{\{3\}}(y).$$

11. As one example, $Y = 1 - 2^{-X}$ maps $\aleph$ onto $\mathbb{Y} = \{\frac{1}{2}, \frac{3}{4}, \frac{7}{8}...\} \subset [.5, 1)$.
12. The support is $\mathbb{Y} = \{0\} \cup [.5, 1]$. The inverse images, $g^{-1}((-\infty, y])$, are $\varnothing$ for $y < 0$, $[0, .5)$ for $y \in [0, .5)$, $[0, y)$ for $y \in [.5, 1)$, and $[0, 1]$ for $y \geq 1$. Thus,

$$F_Y(y) = \begin{cases} \mathbb{P}_X(\varnothing) = 0, \, y < 0 \\ \mathbb{P}_X([0, .5)) = .5, \, 0 \leq y < .5 \\ \mathbb{P}_X([0, y)) = y, \, .5 \leq y < 1 \\ \mathbb{P}_X([0, 1]) = 1, \, y \geq 1 \end{cases}.$$

13. The support of Y is $\mathbb{Y} = [0, \infty)$, so $F_Y(y) = 0$ for $y < 0$. For each $y \geq 0$ we have $g^{-1}((-\infty, y]) = [-\sqrt{y}, \sqrt{y}]$, so $F_Y(y) = \mathbb{P}_X([-\sqrt{y}, \sqrt{y}])$ $= F_X(\sqrt{y}) - F_X(-\sqrt{y}-) = F_X(\sqrt{y}) - F_X(-\sqrt{y})$ (since F_X is continuous). Differentiating,

$$f_Y(y) = F'_X(\sqrt{y})\frac{d\sqrt{y}}{dy} - F'_X(-\sqrt{y})\frac{d(-\sqrt{y})}{dy}$$

$$= 2f_X(\sqrt{y})(2\sqrt{y})^{-1} = \frac{1}{\sqrt{2\pi}}y^{-1/2}e^{-y/2}1_{[0,\infty)}(y).$$

14. The support of Y is $\mathbb{Y} = [0, \infty)$ and transformation $g(X) = e^X$ is one to one. With $h(y) = \log y$ and $dh(y)/dy = y^{-1}$, the change–of–variable formula gives $f_Y(y) = \frac{1}{y\sqrt{2\pi}}e^{-(\log y)^2/2}1_{[0,\infty)}(y)$.
15. $X + Y \leq 2\max\{X, Y\}$ implies

$$\Pr(X + Y > r) \leq \Pr(\max\{X, Y\} > r/2)$$
$$= \Pr[(X > r/2) \cup (Y > r/2)]$$
$$\leq \Pr(X > r/2) + \Pr(Y > r/2).$$

16. Since the support of Z is $\mathbb{Z} = \{0, 1\} \times [0, 1]$, it is clear that $F_Z(x, y) = 0$ if either $x < 0$ or $y < 0$. Independence implies $F_Z(x, y) = \mathbb{P}_Z((-\infty, x] \times (-\infty, y]) = \mathbb{P}_X((-\infty, x])\mathbb{P}_Y((-\infty, y]) = F_X(x)F_Y(y)$ for all (x, y). For $(x, y) \in [0, 1) \times [0, 1]$ we have $F_Z(x, y) = .5y$, and for $(x, y) \in [0, 1) \times (1, \infty)$ we have $F_Z(x, y) = .5 \cdot 1$. The two cases are handled jointly as $F_Z(x, y) = .5\min\{y, 1\}$, and extending to

$F_Z(x, y) = \max\{0, .5\min\{y, 1\}\}$ makes the expression valid even for $y < 0$. Similarly, for $(x, y) \in [1, \infty) \times [0, 1]$ we have $F_Z(x, y) = y$ and writing as $F_Z(x, y) = \max\{0, \min\{y, 1\}\}$ makes the expression valid for any y.

17. We have
$$\min\{\max\{x, 0\}, 1\} = \begin{cases} 0, & x < 0 \\ x, & x \in [0, 1] \\ 1, & x > 1 \end{cases} .$$
Likewise for y.

18. Differentiating partially with respect to x and y gives $f_Z(x, y) = \mathbf{1}_{[0,1]^2}(x, y)$.

19. $f_Z(x, y)$ is nonnegative, and
$$\int_0^\infty \int_0^\infty e^{-x-y} \cdot dy\, dx = \int_0^\infty e^{-x} \left[\int_0^\infty e^{-y} \cdot dy \right] \cdot dx$$
$$= \int_0^\infty e^{-x} \cdot 1 \cdot dx = 1.$$

20. $F_Z(x, y) = 0$ for $x \le 0$ or $y \le 0$. When both are positive, $F_Z(x, y) = \int_0^x \int_0^y e^{-s-t} \cdot dt\, ds = \int_0^x e^{-s} [-e^{-t}]_0^y \cdot ds = (1 - e^{-y}) \int_0^x e^{-s} \cdot ds = (1 - e^{-y})(1 - e^{-x})$.

21. (i) $f_Z(x, y)$ is nonnegative. It integrates to unity since
$$\int_0^\infty \int_0^y 2e^{-x-y} \cdot dx\, dy = \int_0^\infty 2e^{-y}(1 - e^{-y}) \cdot dy$$
$$= 2 \int_0^\infty e^{-y} \cdot dy - \int_0^\infty 2e^{-2y} \cdot dy = 1.$$
(ii) $\Pr(X > 1)$ can be found by integrating in either order:
$$\int_1^\infty \int_1^y 2e^{-x-y} \cdot dx\, dy = \int_1^\infty 2e^{-y}(e^{-1} - e^{-y}) \cdot dy = e^{-2}$$
$$\int_1^\infty \int_x^\infty 2e^{-x-y} \cdot dy\, dx = \int_1^\infty 2e^{-x}(e^{-x}) \cdot dx = e^{-2}.$$
Drawing a sketch of the relevant region of $\Re^2$ makes this easy to see.

22. $F_X(x) = \mathbb{P}_X((-\infty, x]) = \mathbb{P}_Z((-\infty, x] \times \Re) = F_Z(x, +\infty)$.
$F_Y(y) = \mathbb{P}_Y((-\infty, y]) = \mathbb{P}_Z(\Re \times (-\infty, y]) = F_Z(+\infty, y)$.

23. $F_X(x) = F_Z(x, +\infty) = (1 - e^{-x})\mathbf{1}_{(0,\infty)}(x)$.
$F_Y(y) = F_Z(+\infty, y) = (1 - e^{-y})\mathbf{1}_{(0,\infty)}(y)$.

24. (i) $f_X(x) \ge 0$ and $\int_0^\infty 2e^{-2x} \cdot dx = 1$.
(ii) $\Pr(X > 1) = \int_1^\infty 2e^{-2x} \cdot dx = e^{-2}$. Note the connection to Exercise 2.21 (ii), where evaluating the double integral first with respect to y simply produced the marginal p.d.f. of X.
(iii) $f_Y(y) = \int_0^y 2e^{-x-y} \cdot dx = 2e^{-y}(1 - e^{-y})\mathbf{1}_{(0,\infty)}(y)$.

25. (i) Random vector (X, Y) is uniformly distributed over the region between f_X and the horizontal axis; and since $\int f_X(x) \cdot dx = 1$, the joint density in that region is unity. Thus, the marginal p.d.f. of X at each $x \in \Re$ is $\int_0^{f_X(x)} 1 \cdot dy = f_X(x)$.

(ii) Clearly, $f_Y(y) = 0$ for $y < 0$ and $y > f_X(m)$, while for $0 < y < f_X(m)$ we have

$$f_Y(y) = \int_{f_{X+}^{-1}(y)}^{f_{X-}^{-1}(y)} 1 \cdot dx = f_{X-}^{-1}(y) - f_{X+}^{-1}(y).$$

Since Y is continuous (and since the inverses need not be defined at the extremities), we can set $f_Y(0) = f_Y(m) = 0$ without loss of generality.

(iii) P.d.f. $f_X(x) = \left[\pi \left(1 + x^2\right)\right]^{-1}$ is continuous, with $f_X(m) = \pi^{-1}$ at single mode $m = 0$. Setting $y = f_X(x)$ and solving for x give $f_{X\pm}^{-1}(y) = \pm\sqrt{(\pi y)^{-1} - 1}$ for $y \in \left(0, \pi^{-1}\right)$, so the result follows from (ii).

26. Following the hint,

$$f_{\mathbf{Z}}(x, y) = \mathbb{P}_{\mathbf{Z}}(\{(x, y)\}) := \mathbb{P}_{\mathbf{Z}}(\{x\} \times \{y\})$$
$$= \mathbb{P}_X(\{x\}) \, \mathbb{P}_Y(\{y\})$$
$$= f_X(x) f_Y(y).$$

27. Following the hint,

$$f_{\mathbf{Z}}(x, y) = \frac{\partial^2 F_{\mathbf{Z}}(x, y)}{\partial y \partial x} = \frac{\partial^2 F_X(x) F_Y(y)}{\partial y \partial x}$$
$$= \frac{\partial F_X(x)}{\partial x} \cdot \frac{\partial F_Y(y)}{\partial y} = f_X(x) f_Y(y).$$

28. If X, Y are continuous r.v.s, it is necessary and sufficient for independence that the joint p.d.f. factors as the product of the marginals: $f_{\mathbf{Z}}(x, y) = f_X(x) f_Y(y)$, where $f_X(x) > 0$ for $x \in \mathbb{X}$ (the support of X) and $f_Y(y) > 0$ for $y \in \mathbb{Y}$ (the support of Y). Taking $g = f_X$ and $h = f_Y$ shows that independence does imply (2.20). Conversely, if (2.20) holds, then the marginal p.d.f.s must be proportional to g and h, since $f_X(x) = \int_{y \in \mathbb{Y}} g(x) h(y) \cdot dy = g(x) \int_{y \in \mathbb{Y}} h(y) \cdot dy =: g(x) k_X$ for some constant k_X that does not depend on x, and $f_Y(y) = \int_{x \in \mathbb{X}} g(x) h(y) \cdot dx = h(y) \int_{x \in \mathbb{X}} g(x) \cdot dx = h(y) k_Y$ for some constant k_Y that does not depend on y. But the fact that the marginals integrate to unity implies that the product of the two normalizing constants is unity; i.e., $1 = \int_{x \in \mathbb{X}} f_X(x) \cdot dx = k_X \int_{x \in \mathbb{X}} g(x) \cdot dx = k_X k_Y$ and $1 = \int_{y \in \mathbb{Y}} f_Y(y) \cdot dy = k_Y \int_{y \in \mathbb{Y}} h(y) \cdot dy = k_Y k_X$. Therefore, for each $(x, y) \in \mathbb{X} \times \mathbb{Y}$, $f_{\mathbf{Z}}(x, y) = g(x) h(y) = [f_X(x)/k_X] \cdot [f_Y(y)/k_Y] = f_X(x) f_Y(y)$.

29. (i) Not independent: For example $f_Z(0,1) = 0 \neq \frac{1}{8} \cdot \frac{4}{8} = f_X(0)f_Y(1)$.

(ii) Independent: $f_Z(x,y)$ is of the form $g(x)h(y)$, is positive on a product set, and vanishes elsewhere.

(iii) Not independent: $f_Z(x,y)$ is of the form $g(x)h(y)$ but is positive on other than a product set.

30. (i) Yes—so long as the die is not so "loaded" as to turn up the same face every time, in which case the r.v.s would be degenerate and all events trivial.

(ii) Yes—so long as not everyone in the class is of precisely the same height.

(iii) No: The proportions of individuals of various heights would typically change after the first draw. (If all were of the same height, we would be back to the trivial case.)

(iv) No: Weights and heights tend to be related.

31. We are justified in *assuming* that X and Y are independent and have the same marginal distributions, with $f_X(t) = f_Y(t) = \frac{1}{6}\mathbf{1}_{\{1,2,...,6\}}(t)$. From independence the joint p.m.f. is just the product of marginals, so that $f_Z(x,y) = \frac{1}{36}$ for $(x,y) \in \{1,2,...,6\}^2 := \{1,2,...,6\} \times \{1,2,...,6\}$; and all conditional distributions are the same as the marginals.

32. From the table in (2.16) we have $f_{Y|X}(y|2) = f_Z(2,y)/f_X(2)$ for $y \in \{0,1,2\}$. Thus, $f_{Y|X}(0|2) = 0$, $f_{Y|X}(1|2) = \frac{2}{8} \div \frac{3}{8}$, $f_{Y|X}(2|2) = \frac{1}{8} \div \frac{3}{8}$.

33. The marginal p.d.f. of Y is $f_Y(y) = \int_0^y 2e^{-x-y} \cdot dx = 2e^{-y}(1 - e^{-y})\mathbf{1}_{(0,\infty)}(y)$, so $f_{X|Y}(x|y) = f_Z(x,y)/f_Y(y) = e^{-x}/(1 - e^{-y})\mathbf{1}_{\{(x,y):0<x<y\}}(x,y)$.

34. Working from the conditional p.d.f.s, $F_{Y|X}(y|x) = \int_{-\infty}^y f_{Y|X}(t|x) \cdot dt = \int_x^y e^{x-t} \cdot dt = 1 - e^{x-y}$ for $y > x > 0$ and $F_{Y|X}(y|x) = 0$ elsewhere; and

$$F_{X|Y}(x|y) = \int_{-\infty}^x f_{X|Y}(s|y) \cdot ds = \int_0^{\min\{x,y\}} e^{-s}/\left(1 - e^{-y}\right) \cdot ds$$
$$= \frac{1 - e^{-\min\{x,y\}}}{1 - e^{-y}}$$

for $x > 0$ and $F_{X|Y}(x|y) = 0$ elsewhere.

35. (i) Integrating, $\int_{-1}^1 \int_{-1}^1 k\left(x^2 + y^2\right) \cdot dy\, dx = 8k/3$, so $k = 3/8$.

(ii) For $x \in [-1,1]$

$$f_X(x) = \int_{-1}^1 \frac{3}{8}\left(x^2 + y^2\right) \cdot dy = \frac{1}{4}\left(3x^2 + 1\right).$$

(iii) For $(x,y) \in [-1,1]^2$

$$f_{Y|X}(y \mid x) = \frac{3\left(x^2 + y^2\right)/8}{(3x^2 + 1)/4} = \frac{1}{2}\left(\frac{x^2 + y^2}{x^2 + 1/3}\right).$$

(iv) $\Pr\left(Y < 1/2 \mid X = 1/2\right) = \int_{-1}^{1/2} f_{Y|X}\left(y \mid 1/2\right) \cdot dy = 9/14.$
(v)

$$\Pr\left(Y < 1/2 \mid X < 1/2\right) = \frac{\Pr\left(X < 1/2, Y < 1/2\right)}{\Pr\left(X < 1/2\right)}$$

$$= \frac{\int_{-1}^{1/2}\int_{-1}^{1/2}\frac{3}{8}\left(x^2 + y^2\right) \cdot dx\, dy}{\int_{-1}^{1/2}\frac{1}{4}\left(3x^2 + 1\right) \cdot dx}$$

$$= \frac{27/64}{21/32} = \frac{9}{14}.$$

36. (i) For $x \in [-1, 1]$

$$f_X\left(x\right) = \int_{-\sqrt{1-x^2}}^{+\sqrt{1-x^2}} \pi^{-1} \cdot dy = \frac{2}{\pi}\sqrt{1 - x^2}.$$

(ii) Clearly, $F_X\left(x\right) = 0$ for $x < -1$ and $F_X = 1$ for $x > 1$. For $x \in [-1, 1]$ we have

$$F_X\left(x\right) = \int_{-1}^{x} \frac{2}{\pi}\sqrt{1 - t^2} \cdot dt$$

$$= \frac{1}{\pi}\left[t\sqrt{1 - t^2} + \sin^{-1}\left(t\right)\right]_{-1}^{x}$$

$$= \frac{1}{\pi}\left[x\sqrt{1 - x^2} + \sin^{-1}\left(x\right) + \frac{\pi}{2}\right].$$

(iii) For (x, y) on the unit circle we have

$$f_{Y|X}\left(y \mid x\right) = \frac{1}{2\sqrt{1 - x^2}}.$$

(iv)

$$\Pr\left(Y < 1/2 \mid X = 1/2\right) = \int_{-\sqrt{1-(1/2)^2}}^{1/2} \frac{1}{2\sqrt{1 - (1/2)^2}} \cdot dy$$

$$= \frac{1}{2}\left(\frac{1}{\sqrt{3}} + 1\right)$$

(v) We have

$$\Pr\left(Y < 1/2 \mid X < 1/2\right) = \frac{\Pr\left(X < 1/2, Y < 1/2\right)}{\Pr\left(X < 1/2\right)}$$

$$= \frac{\Pr\left(Y < 1/2\right) - \Pr\left(X \geq 1/2, Y < 1/2\right)}{\Pr\left(X < 1/2\right)}$$

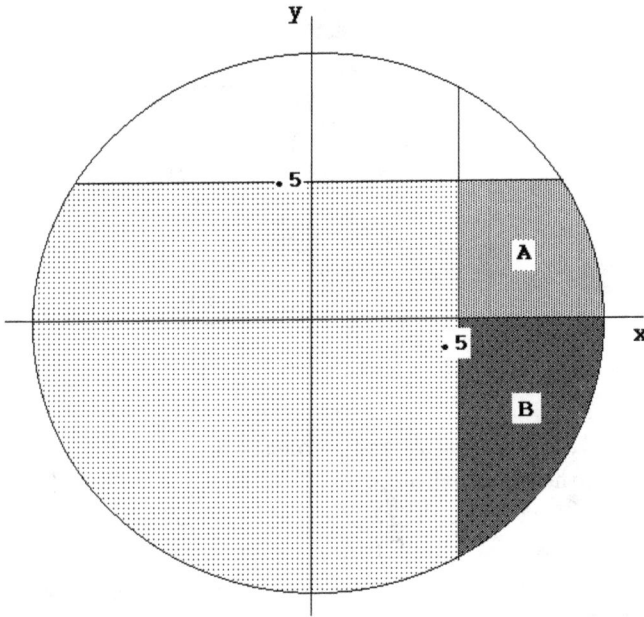

Fig. 2.13

By symmetry, $\Pr\left(Y < 1/2\right) = \Pr\left(X < 1/2\right) = F_X\left(1/2\right)$, so

$$\Pr\left(Y < 1/2 \mid X < 1/2\right) = 1 - \frac{\Pr\left(X \geq 1/2, Y < 1/2\right)}{F_X\left(1/2\right)}.$$

Referring to the figure, we see that $\Pr\left(X \geq 1/2, Y < 1/2\right)$ is the probability mass associated with areas A and B, which are, respectively,

$$\Pr\left(X \geq 1/2, 0 < Y < 1/2\right) = \int_0^{1/2} \int_{1/2}^{\sqrt{1-y^2}} \pi^{-1} \cdot dx\, dy$$

$$\Pr\left(X \geq 1/2, Y \leq 0\right) = \int_{1/2}^{1} \int_{-\sqrt{1-x^2}}^{0} \pi^{-1} \cdot dy\, dx.$$

Working out the integrals and expressing $F_X\left(1/2\right)$ give as the answer

$$\Pr\left(Y < 1/2 \mid X < 1/2\right) = 1 - \frac{(1 - \pi^{-1})/4}{.5 + \pi^{-1}\left[\frac{\sqrt{3}}{4} + \sin^{-1}\left(1/2\right)\right]}$$

$$\doteq .788.$$

37. (i) $c = 6$ since

$$\int_{x_3=0}^{\infty}\int_{x_2=0}^{x_3}\int_{x_1=0}^{x_2} e^{-x_1-x_2-x_3}\, dx_1\, dx_2\, dx_3$$

$$= \int_{x_3=0}^{\infty}\int_{x_2=0}^{x_3} e^{-x_2-x_3}(1 - e^{-x_2})\, dx_2\, dx_3$$

$$= \int_{x_3=0}^{\infty} e^{-x_3}\left[\int_{x_2=0}^{x_3}(e^{-x_2} - e^{-2x_2})\, dx_2\right] dx_3$$

$$= \int_0^{\infty}\left(\frac{e^{-x_3}}{2} - e^{-2x_3} + \frac{e^{-3x_3}}{2}\right) dx_3 = \frac{1}{6}.$$

(ii) $f_{12}(x_1, x_2) = \int_{-\infty}^{\infty} f_{\mathbf{X}}(x_1, x_2, x_3)\, dx_3 = \int_{x_2}^{\infty} 6e^{-x_1-x_2-x_3}\, dx_3 = 6e^{-x_1-2x_2}$ for $0 < x_1 < x_2$ and zero elsewhere.

(iii) First, the marginal p.d.f. of X_1 is $f_1(x_1) = \int_{-\infty}^{\infty} f_{12}(x_1, x_2)\, dx_2 = \int_{x_1}^{\infty} 6e^{-x_1-2x_2}\, dx_2 = 3e^{-3x_1}\mathbf{1}_{(0,\infty)}(x_1)$. Then $F_1(x_1) = \int_0^{x_1} 3e^{-3s}\, ds = 1 - e^{-3x_1}$ for $x_1 > 0$ and $F_1(x_1) = 0$ else.

(iv) They are not mutually independent because $f_{\mathbf{X}}(x_1, x_2, x_3)$ cannot be represented as the product of functions that are positive on a *product set*.

(v) We must first find the conditional p.d.f. of X_3 given $X_1 = 1$ and $X_2 = 2$:

$$f_{3|1,2}(x_3 \mid 1, 2) = f_{\mathbf{X}}(1, 2, x_3)/f_{12}(1, 2)$$
$$= 6e^{-1-2-x_3}\mathbf{1}_{(2,\infty)}(x_3) / \left(6e^{-1-4}\right)$$
$$= e^{2-x_3}\mathbf{1}_{(2,\infty)}(x_3).$$

Then $\Pr(X_3 > 3 \mid X_2 = 2, X_1 = 1) = e^2 \int_3^{\infty} e^{-x_3}\, dx_3 = e^{-1}$.

(vi) Now we do not need a conditional p.d.f. but must find $\Pr(X_3 > 3 \mid X_2 > 2, X_1 > 1)$ as

$$\mathbb{P}_{\mathbf{X}}((1,\infty) \times (2,\infty) \times (3,\infty)) / \mathbb{P}_{X_1 X_2}((1,\infty) \times (2,\infty)).$$

The denominator is found from the joint marginal p.d.f. of X_1, X_2 as

$$\int_{x_2=2}^{\infty}\int_{x_1=1}^{x_2} 6e^{-x_1-2x_2}\, dx_1\, dx_2 = \int_{x_2=2}^{\infty} 6e^{-2x_2}(e^{-1} - e^{-x_2})\, dx_2$$
$$= 3e^{-5} - 2e^{-6}.$$

The numerator is found from the joint p.d.f. of X_1, X_2, X_3 as

$$\int_{x_3=3}^{\infty} \int_{x_2=2}^{x_3} \int_{x_1=1}^{x_2} 6e^{-x_1-x_2-x_3} \, dx_1 \, dx_2 \, dx_3$$

$$= \int_{x_3=3}^{\infty} 6e^{-x_3} \left[\int_{x_2=2}^{x_3} e^{-x_2} \left(\int_{x_1=1}^{x_2} e^{-x_1} \, dx_1 \right) dx_2 \right] dx_3$$

$$= \int_{x_3=3}^{\infty} 6e^{-x_3} \left[\int_{x_2=2}^{x_3} e^{-x_2} \left(e^{-1} - e^{-x_2} \right) dx_2 \right] dx_3$$

$$= \int_{x_3=3}^{\infty} 6e^{-x_3} \left[e^{-1}(e^{-2} - e^{-x_3}) - \frac{1}{2} \left(e^{-4} - e^{-2x_3} \right) \right] dx_3$$

$$= 6e^{-6} - 6e^{-7} + e^{-9}.$$

Thus,

$$\Pr(X_3 > 3 \mid X_2 > 2, X_1 > 1) = \frac{6e^{-6} - 6e^{-7} + e^{-9}}{3e^{-5} - 2e^{-6}}$$

$$= \frac{6e^{-1} - 6e^{-2} + e^{-4}}{3 - 2e^{-1}}.$$

38. Transformation $Y_1 = X_1 + 2X_2, Y_2 = X_2$ is one to one, with inverse $X_1 = Y_1 - 2Y_2, X_2 = Y_2$ and $|J_h| = 1$. Now $X_1 \in [0,1]$ requires $(Y_1 - 1)/2 \le Y_2 \le Y_1/2$, while $X_2 \in [0,1]$ requires $0 \le Y_2 \le 1$. Thus,

$$\mathbb{Y} = \left\{ \mathbf{y} : 0 \le y_1 \le 3, \max\left\{0, \frac{y_1-1}{2}\right\} \le y_2 \le \min\left\{\frac{y_1}{2}, 1\right\} \right\},$$

and $f_{\mathbf{Y}}(y_1, y_2) = 1_{\mathbb{Y}}(y_1, y_2)$.

39. $\max\left\{0, \frac{y_1-1}{2}\right\} = 0$ for $0 \le y_1 \le 1$ and $\max\left\{0, \frac{y_1-1}{2}\right\} = \frac{y_1-1}{2}$ for $1 \le y_1 \le 3$, while $\min\left\{\frac{y_1}{2}, 1\right\} = \frac{y_1}{2}$ for $0 \le y_1 \le 2$ and $\min\left\{\frac{y_1}{2}, 1\right\} = 1$ for $2 \le y_1 \le 3$. Thus, the marginal p.d.f. of Y_1 is

$$f_1(y_1) = \left(\int_0^{y_1/2} 1 \cdot dy_2 \right) 1_{[0,1]}(y_1) + \left(\int_{(y_1-1)/2}^{y_1/2} 1 \cdot dy_2 \right) 1_{[1,2]}(y_1)$$

$$+ \left(\int_{(y_1-1)/2}^{1} 1 \cdot dy_2 \right) 1_{[2,3]}(y_1)$$

$$= \frac{y_1}{2} 1_{[0,1]}(y_1) + \frac{1}{2} 1_{[1,2]}(y_1) + \frac{3-y_1}{2} 1_{[2,3]}(y_1).$$

40. Set $Y = g_1(Z_1, Z_1) = Z_1/Z_2$ and $W = g_2(Z_1, Z_2) = Z_2$. One can see by inspection that $\mathbf{g}$ is a one-to-one mapping of $\Re^2$ onto itself. The

inverse functions are $Z_1 = h_1(Y, W) = YW$ and $Z_2 = h_2(Y, W) = W$ with $J_h = W$, so the joint p.d.f. is

$$f_{Y,W}(y, w) = f_Z(yw, w)|w| = (2\pi)^{-1}|w|e^{-(1+y^2)w^2/2}.$$

The marginal p.d.f. of Y is $f_Y(y) = \int_{-\infty}^{\infty} f_{Y,W}(y, w) \cdot dw$. The integrand being an even function of w, we have

$$f_Y(y) = 2\int_0^\infty f_{Y,W}(y, w) \cdot dw$$

$$= \pi^{-1}\int_0^\infty we^{-(1+y^2)w^2/2} \cdot dw$$

$$= \frac{1}{\pi(1+y^2)}\int_0^\infty d\left[-e^{-(1+y^2)w^2/2}\right]$$

$$= \frac{1}{\pi(1+y^2)}\left[-e^{-(1+y^2)w^2/2}\right]_0^\infty$$

$$= \frac{1}{\pi(1+y^2)}.$$

41. Applying the convolution formula, we have, for $s \in \aleph_0$,

$$f_S(s) = \sum_{x=0}^{\infty} f_1(s-x)f_2(x) = \sum_{x=0}^{s} \frac{\lambda_1^{s-x}e^{-\lambda_1}}{(s-x)!} \cdot \frac{\lambda_2^x e^{-\lambda_2}}{x!}$$

$$= \frac{\lambda_1^s e^{-(\lambda_1+\lambda_2)}}{s!}\sum_{x=0}^{s}\binom{s}{x}\left(\frac{\lambda_2}{\lambda_1}\right)^x = \frac{\lambda_1^s e^{-(\lambda_1+\lambda_2)}}{s!}\left(1+\frac{\lambda_2}{\lambda_1}\right)^s$$

$$= \frac{(\lambda_1+\lambda_2)^s e^{-(\lambda_1+\lambda_2)}}{s!},$$

showing that the sum of independent Poisson variates is distributed as Poisson.

42. Setting $S = X + Y$ and $T = Y$ the inverse transformation is $Y = T$, $X = S - T$, and the Jacobian of the inverse transformation is

$$J_h = \begin{vmatrix} 1 & -1 \\ 0 & 1 \end{vmatrix} = 1.$$ Since $0 < X$ requires $T < S$ and $X < Y$ requires $T > S/2$, the transformation maps $\mathbb{Z} = \{(x, y) : 0 < x < y\}$ onto $\mathbb{U} = \{(s, t) : 0 < s/2 < t < s\}$. Thus, the joint p.d.f. of $\mathbb{U} = (S, T)'$ is $f_{\mathbb{U}}(s, t) = f_Z(s-t, t)|J_h| = 2e^{-(s-t)-t} = 2e^{-s}1_{\mathbb{U}}(s)$. Integrating out t, the marginal p.d.f. of S is $f_S(s) = \int_{-\infty}^{\infty} f_{\mathbb{U}}(s, t) \cdot dt = \int_{s/2}^{s} 2e^{-s} \cdot dt = se^{-s}1_{(0,\infty)}(s)$.

43. The convolution formula gives

$$f_S(s) = \int_{-\infty}^{\infty} \frac{1}{\sqrt{2\pi}} e^{-(s-x)^2/2} \frac{1}{\sqrt{2\pi}} e^{-x^2/2} \cdot dx$$

$$= \int_{-\infty}^{\infty} \frac{1}{2\pi} e^{(s^2-2sx+2x^2)/2} \cdot dx$$

$$= \frac{1}{\sqrt{2\pi}} e^{-s^2/4} \int_{-\infty}^{\infty} \frac{1}{\sqrt{2\pi}} e^{-(s^2/2-2sx+2x^2)/2} \cdot dx$$

$$= \frac{\sqrt{\frac{1}{2}}}{\sqrt{2\pi}} e^{-s^2/4} \int_{-\infty}^{\infty} \frac{1}{\sqrt{2\pi\left(\frac{1}{2}\right)}} e^{-(x-s/2)^2/\left(2\cdot\frac{1}{2}\right)} \cdot dx.$$

The integral is of the form $\int_{-\infty}^{\infty} \left(2\pi\phi^2\right)^{-1/2} e^{-(x-\theta)^2/(2\phi^2)} \cdot dx$ with $\theta = s/2$ and $\phi^2 = 1/2$. Following the hint, we conclude that $f_S(s) = \frac{1}{\sqrt{4\pi}} e^{-s^2/4}$.

44. X, Y are independent, with $f_X(x) = e^{-x}1_{(0,\infty)}(x)$ and $f_Y(y) = e^{-y}1_{(0,\infty)}(y)$. Applying the convolution formula $f_S(s) = \int_{-\infty}^{\infty} f_X(s - y)f_y(y)\,dy = \int_0^s e^{-(s-y)}e^{-y}\,dy = se^{-s}$ for $s > 0$ and zero elsewhere. (The upper limit of the integral is s because $f_X(s-y) = 0$ for $s-y \leq 0$.)

Chapter 3

Mathematical Expectation

3.1 Introduction

We have defined probability functions as measures that map σ fields of sets of outcomes—events—into $[0, 1]$. In the classical view one thinks of $\mathbb{P}(A)$ as the *relative frequency* with which event A would occur if an experiment were to be performed infinitely many times while holding constant the conditions under one's control. Equivalently, $\mathbb{P}(A)$ is the "long-run average" value taken on by an *indicator function* $\mathbf{1}_A(\omega)$ that has the value unity when outcome ω is in A and is zero otherwise. (Think of the indicator function just as registering "hits" of A.) Here, "long-run" refers to the hypothetical, infinitely replicated experiment. Since such an indicator function maps the outcomes in Ω into $\Re$—specifically, $\mathbf{1}_A : \Omega \to \{0, 1\}$, it is just a special type of random variable whose long-run average value happens to have a natural interpretation. Long-run average values of other types of r.v.s besides indicator functions are also very much of interest. As a practical example, if the real number $X(\omega)$ registers the amount of money won when outcome ω occurs in a game of chance, then the hypothetical long-run average value of X serves as an estimate of the average winning on some *finite* number of plays. Indeed, even if the game were to be played only once, one might plausibly regard this long-run average as a "best guess" of the winning on that one play. This long-run average value of X is called the *mathematical expectation* or *expected value* of X and is represented by the symbol "EX".[1]

---

[1] The more common notation is $E(X)$, but the parentheses are really superfluous unless there is some ambiguity about what is being averaged. Thus, in the convention followed here the expression $EX + Y$ represents a new r.v. with realizations $EX + Y(\omega)$, whereas $E(X + Y)$ represents the long-run average of the sum—a scalar constant. However, EXY always represents $E(XY)$.

Example 3.1. In the experiment of flipping three coins assume that the eight possible strings of three H's and/or T's are equally likely, and take $X(\omega)$ as the number of H's in outcome ω. In infinitely many trials of the experiment any one of the eight outcomes would occur with relative frequency $\mathbb{P}(\{\omega\}) = N(\{\omega\})/N(\Omega) = 1/8$. Since each of the eight realizations $X(TTT) = 0$, $X(HTT) = 1$, $X(THT) = 1$, $X(TTH) = 1$, $X(HHT) = 2$, $X(HTH) = 2$, $X(THH) = 2$, $X(HHH) = 3$ would occur in 1/8th of these trials, the long-run average value of X would be

$$EX = (0 + 1 + 1 + 1 + 2 + 2 + 2 + 3) \cdot \frac{1}{8} = \frac{12}{8}.$$

In symbols, this is

$$EX = \sum_{\omega \in \Omega} X(\omega)\mathbb{P}(\{\omega\}). \tag{3.1}$$

Grouping by values of X, the calculation appears as

$$EX = 0 \cdot \frac{1}{8} + 1 \cdot \left(\frac{1}{8} + \frac{1}{8} + \frac{1}{8}\right) + 2 \cdot \left(\frac{1}{8} + \frac{1}{8} + \frac{1}{8}\right) + 3 \cdot \frac{1}{8}$$
$$= 0 \cdot \frac{1}{8} + 1 \cdot \frac{3}{8} + 2 \cdot \frac{3}{8} + 3 \cdot \frac{1}{8},$$

and recognizing $\mathbb{X} = \{0, 1, 2, 3\}$ as the support of X and the fractions as the probabilities induced by X give another representation, as

$$EX = \sum_{x \in \mathbb{X}} x\mathbb{P}_X(\{x\}). \tag{3.2}$$

Finally, noting that X is a discrete r.v. whose p.m.f. takes the value $f(x) = F(x) - F(x-) = \mathbb{P}_X(\{x\})$ at each $x \in \mathbb{X}$, we have two more ways of expressing EX:

$$EX = \sum_{x \in \mathbb{X}} x\left[F(x) - F(x-)\right],$$

$$EX = \sum_{x \in \mathbb{X}} xf(x). \tag{3.3}$$

Of the four representations of EX in the example, (3.1) conveys most naturally the concept of averaging over outcomes. Of course, this expression applies even when outcomes are not equally likely, $\mathbb{P}(\{\omega\})$ then taking different values for different elements of Ω. On the other hand, formulas (3.2) or (3.3) are clearly easier to use for actual calculations.

This chapter explains the concept and properties of mathematical expectation in detail and develops many tools that are essential in probability modeling and in statistical applications. For this it is necessary to work in a general setting that accommodates continuous and even mixed (discrete and continuous) r.v.s, and at this level of generality the sums that appear in the expressions above must be replaced by integrals. This in turn requires a concept of integration that is more general than the familiar Riemann form.

3.2 Integration

3.2.1 *Integrals on Abstract Measure Spaces*

We will now construct an integral that will enable us to represent mathematical expectation as an average over outcomes, as in Expression (3.1), but which allows for averaging over other than countable spaces. While the ultimate interest is in probability spaces and integrals of functions of outcomes of chance experiments, it is useful to start in a more general setting. For this let $(\Omega, \mathcal{F}, \mathcal{M})$ be a measure space, where Ω is some arbitrary, non-empty space of elements ω, $\mathcal{F}$ is a σ field of subsets of Ω, and $\mathcal{M}$ is a measure on $(\Omega, \mathcal{F})$. Here Ω might comprise abstract outcomes of a chance experiment, but it could also be the real numbers, or k-dimensional space, or even a space of functions. The measure $\mathcal{M}$ can be a probability measure on abstract sets, such as $\mathbb{P}$; or a probability measure on the Borel sets of $\Re$ or $\Re^k$, such as $\mathbb{P}_X$ or $\mathbb{P}_{\mathbf{X}}$; or it can be something other than a probability, such as Lebesgue measure λ on $(\Re, \mathcal{B})$. In this general setting, when g is an $\mathcal{F}$-measurable function that maps Ω into $\Re$, an integral $I := \int_\Omega g(\omega) \cdot d\mathcal{M}(\omega) \equiv \int_\Omega g(\omega) \, \mathcal{M}(d\omega)$ (both notations being used) can be regarded roughly as a weighted sum of the values g over Ω, the weights being governed by measure $\mathcal{M}$. If $\mathcal{M}(\Omega) = 1$, as is true when $\mathcal{M}$ is a probability measure, then I can be thought of intuitively as a weighted *average* of the values of g. However, when the space Ω is not countable, the precise interpretation will be that I is a *limit* of a sequence of weighted sums or averages.

3.2.1.1 *Construction of the Integral*

The abstract integral is defined in stages that allow integrand g to become progressively more general. Step one is to define I for nonnegative *simple* functions, which are functions that take a constant, nonnegative, finite value on each of finitely many sets that partition Ω. Thus, g is nonnegative and simple if it takes the value $g(\omega) = \gamma_j \geq 0$ for each $\omega \in A_j$, and where sets $\{A_j\}_{j=1}^m \in \mathcal{F}$ are a finite partition of Ω. (That is, $\cup_{j=1}^m A_j = \Omega$ and $A_i \cap A_j = \varnothing$ for $i \neq j$.) Simple function g can be written compactly with indicator functions as $g(\omega) = \sum_{j=1}^m \gamma_j \mathbf{1}_{A_j}(\omega)$. The integral with respect to $\mathcal{M}$ of such a nonnegative simple function is defined as

$$I \equiv \int_\Omega g(\omega) \mathcal{M}(d\omega) := \sum_{j=1}^m \gamma_j \mathcal{M}(A_j).$$

This is indeed just a weighted sum of the m possible values of g, using as weights the measures of the sets on which g takes the respective values.

Step two is to extend beyond simple functions to functions g that are just nonnegative, finite, and measurable; that is, to functions $g : \Omega \to \Re_+$ such that for all $B \in \mathcal{B}$ the inverse image $g^{-1}(B) = \{\omega : g(\omega) \in B\}$ is an $\mathcal{F}$ set—a set that $\mathcal{M}$ can measure. While simple functions are of course measurable, measurability is a much less restrictive requirement. Still, it can be shown that any such nonnegative measurable function is the limit of a sequence of nonnegative simple functions. A standard (but by no means unique) construction of such a sequence of functions starts by defining a sequence of partitions of Ω based on the values of g. To form the partition at stage $n \in \aleph$ one begins with the sets

$$A_{j,n} := \left\{\omega : \frac{j-1}{2^n} \leq g(\omega) < \frac{j}{2^n}\right\}, j \in \left\{1, 2, ..., 2^{2n}\right\}.$$

Since for each j the interval of real numbers $[\frac{j-1}{2^n}, \frac{j}{2^n})$ is a Borel set, and since g is measurable, the sets $\{A_{j,n}\}$ do belong to $\mathcal{F}$, and they are clearly disjoint. (If it is *not* clear, remember that function $g : \Omega \to \Re_+$ assigns a *single* real number to a given ω.) On each of these sets g varies by no more than 2^{-n} and remains strictly less than $2^{2n}/2^n = 2^n$. To this collection at stage n is added one more set, $A_{0,n} := \Omega \backslash \cup_{j=1}^{2^{2n}} A_{j,n}$, which comprises all the remaining ω's for which $g(\omega) \geq 2^n$. For each $n \in \aleph$ the augmented collection $\{A_{j,n}\}_{j=0}^{2^{2n}}$ is then a finite partition of Ω. As $n \to \infty$ the allowed variation in g in the first 2^{2n} sets declines, and the residual sets $\{A_{0,n}\}_{n=1}^\infty$ on which g becomes arbitrarily large decrease monotonically. It should be clear that the simple function

$$g_n(\omega) := \sum_{j=1}^{2^{2n}} \frac{j-1}{2^n} \mathbf{1}_{A_{j,n}}(\omega) + 2^n \mathbf{1}_{A_{0,n}}(\omega)$$

is less than or equal to g at each ω, and it is easy to believe (and true!) that it converges up to any nonnegative, measurable g as $n \to \infty$. For these nonnegative measurable functions the integral is defined as the limit of the sequence of integrals of the approximating simple functions; i.e.,

$$\int_\Omega g(\omega)\mathcal{M}(d\omega) \; : \; = \lim_{n\to\infty} \int g_n(\omega)\mathcal{M}(d\omega)$$

$$= \lim_{n\to\infty} \sum_{j=1}^{2^{2n}} \left[\frac{j-1}{2^n} \mathcal{M}(A_{j,n}) \right] + 2^n \mathcal{M}(A_{0,n}).$$

While there are many other possible constructions of simple functions that converge up to g, it can be shown that the limit is the same for any such sequence—something that clearly must happen in order for the definition to make sense. Note that nothing in the definition prevents $\int_\Omega g(\omega)\,\mathcal{M}(d\omega)$ from having the value $+\infty$. This happens if either (i) there is some $b > 0$ such that $\mathcal{M}(\{\omega : g(\omega) \geq b\}) = +\infty$ or if (ii) $\mathcal{M}(\{\omega : g(\omega) \geq b\}) < \infty$ for all $b > 0$ but $\lim_{n\to\infty} \int g_n(\omega)\,\mathcal{M}(d\omega) = +\infty$ for all simple functions converging up to g.[2] Whether $\int_\Omega g(\omega)\,\mathcal{M}(d\omega)$ is finite or not, there is at least no ambiguity about its value, and so the integral of a nonnegative function with respect to any measure $\mathcal{M}$ always *exists*; however, the convention is to say that a nonnegative function g is "$\mathcal{M}$-integrable"—that is, integrable with respect to $\mathcal{M}$—if and only if $\int_\Omega g(\omega)\mathcal{M}(d\omega) < +\infty$.

The last step in the construction of the abstract integral is to extend to measurable functions that are not necessarily nonnegative. This is done by treating the nonnegative and negative parts separately. Thus, suppose there is a set $\Omega^+ \subset \Omega$ such that $g(\omega) \geq 0$ for all $\omega \in \Omega^+$ and $g(\omega) < 0$ for all $\omega \in \Omega^- := \Omega \backslash \Omega^+$. Then both $g^+ := g \cdot 1_{\Omega^+}$ and $g^- := -g \cdot 1_{\Omega^-}$ are nonnegative, measurable functions, and $g^+ - g^- = g(1_{\Omega^+} + 1_{\Omega^-}) \equiv g$. The integral of g is then defined piecewise as

$$\int_\Omega g(\omega)\mathcal{M}(d\omega) := \int_\Omega g^+(\omega)\mathcal{M}(d\omega) - \int_\Omega g^-(\omega)\mathcal{M}(d\omega), \qquad (3.4)$$

provided at least one of g^+ and g^- is integrable. Here each piece is evaluated as the limit of integrals of nonnegative simple functions, as in step two. A measurable function g is said to be *integrable* if *both* g^+ and g^- are integrable; that is, if integrals of both the positive and the negative components are finite. Since $g^+(\omega) + g^-(\omega) = |g(\omega)|$, this is equivalent to specifying that g be *absolutely integrable*; that is, that $\int_\Omega |g(\omega)|\,\mathcal{M}(d\omega) < \infty$. Henceforth, we shall regard this condition as the criterion for integrability of

---

[2]Examples of the two cases are (i) $(\Omega, \mathcal{F}, \mathcal{M}) = (\Re, \mathcal{B}, \lambda)$ and $g(x) = 1$ and (ii) $(\Omega, \mathcal{F}, \mathcal{M}) = ((0,1], \mathcal{B}_{(0,1]}, \lambda)$ and $g(x) = x^{-1}$.

any function g with respect to any measure $\mathcal{M}$. Although this is the criterion for integrability, the integral *exists*—has an unambiguous value—under broader conditions; specifically,

- $\int_\Omega g(\omega)\mathcal{M}(d\omega) = +\infty$ if $\int_\Omega g^+(\omega)\mathcal{M}(d\omega) = +\infty$ and $\int_\Omega g^-(\omega)\mathcal{M}(d\omega) < \infty$
- $\int_\Omega g(\omega)\mathcal{M}(d\omega) = -\infty$ if $\int_\Omega g^+(\omega)\mathcal{M}(d\omega) < \infty$ and $\int_\Omega g^-(\omega)\mathcal{M}(d\omega) = +\infty$.

The integral *does not exist* when integrals of both the positive and negative parts are infinite, because $+\infty - (+\infty)$ has no definite value.[3]

A crucial feature of the construction of the abstract integral is that the space Ω on which g is defined need have no natural order. That is because the partition of Ω that supports the approximating sums is based not on the values of its elements—the ω's—but on values of integrand g. These values, being real numbers, do have a natural order, whereas the ω's could be outcomes of a chance experiment or other abstract entities that have no natural arrangement. Below we emphasize the contrast with the familiar Riemann construction of the integral of functions $g : \Re \to \Re$, which would make no sense at all for a space Ω that was not totally ordered.

Notation: The abbreviated notation $\int g \cdot d\mathcal{M}$ is often used when integration is over the entire space. The range of integration can be restricted to measurable sets $A \subset \Omega$ by using indicator functions, as $\int g 1_A \cdot d\mathcal{M}$. This is implied by the simpler expression $\int_A g \cdot d\mathcal{M}$.

Exercise 3.1. *(i) In the three-coin experiment of Example 3.1 show that r.v. X is a simple function and identify the partition $\{A_j\}$ in its construction.*

(ii) Taking Ω as the set of outcomes in the coin experiment, $\mathcal{F}$ as the collection of all subsets of Ω, and $\mathcal{M} = \mathbb{P}$, use the definition of the integral to show that EX in Expression (3.1) can be written as $EX = \int X(\omega)\mathbb{P}(d\omega)$.

---

[3]One who thinks that $+\infty - (+\infty)$ does have a definite value (e.g., 0?) should answer the following questions: (i) How many positive integers are there? (ii) How many *even* positive integers are there? (iii) How many positive integers are left when one subtracts the answer to (ii) from the answer to (i)? German mathematician Georg Cantor (1845-1918) was the first to investigate systematically the *cardinality* (countability) of infinite sets. Simmons (2003, pp. 21-42) gives a fascinating (and readable) account of these ideas and thereby leads us to some seemingly bizarre but logically inescapable conclusions.

3.2.1.2 *Properties of the Integral*

Here are some important properties that follow from the definition of the abstract integral. In each case the relevant functions $g, h, \{g_j\}$ are assumed to be $\mathcal{M}$-integrable, and we express integrals in the compact form $\int g \cdot d\mathcal{M}$.

(i) Integrals of constants: If $g : \Omega \to \{a\} \in \Re$ (i.e., $g(\omega) = a$ for all ω) then $\int g \cdot d\mathcal{M} = \int a \cdot d\mathcal{M} = a\mathcal{M}(\Omega)$. More generally, the same result holds if $g(\omega) = a$ for all ω in a measurable set $\Omega_a \subset \Omega$ such that $\mathcal{M}(\Omega \backslash \Omega_a) = 0$ (that is, if $g = a$ a.e. with respect to $\mathcal{M}$).

(ii) Integrals of products of constants and integrable functions: $\int ag \cdot d\mathcal{M} = a \int g \cdot d\mathcal{M}$.

(iii) Linearity:

$$\int (g + h) \cdot d\mathcal{M} = \int g \cdot d\mathcal{M} + \int h \cdot d\mathcal{M},$$

assuming (as is implied by the integrability of g and h) that the right side is not of a form such as $(+\infty) + (-\infty)$. Combined with points 1 and 2, this implies that, for finite constants a, b, c,

$$\int (a + bg + ch) \cdot d\mathcal{M} = a\mathcal{M}(\Omega) + b \int g \cdot d\mathcal{M} + c \int h \cdot d\mathcal{M}. \quad (3.5)$$

Thus, the operation of integration "passes through" affine functions.[4] More generally, if $\{a_j\}_{j=0}^m$ are finite constants and $\{g_j\}_{j=1}^m$ is a finite collection of integrable functions, then

$$\int \left(a_0 + \sum_{j=1}^m a_j g_j \right) \cdot d\mathcal{M} = a_0 \mathcal{M}(\Omega) + \sum_{j=1}^m a_j \int g_j \cdot d\mathcal{M}.$$

(iv) Integration over unions of disjoint, measurable subsets of Ω: If $A \cap A' = \varnothing$, then

$$\int_{A \cup A'} g \cdot d\mathcal{M} = \int g 1_{A \cup A'} \cdot d\mathcal{M} = \int g (1_A + 1_{A'}) \cdot d\mathcal{M}$$
$$= \int_A g \cdot d\mathcal{M} + \int_{A'} g \cdot d\mathcal{M}.$$

More generally, if $\{A_j\}_{j=1}^\infty$ is a collection of disjoint, measurable sets, then

$$\int_{\cup_{j=1}^\infty A_j} g \cdot d\mathcal{M} = \sum_{j=1}^\infty \int_{A_j} g \cdot d\mathcal{M}.$$

---

[4]If $\mathcal{M}$ is not a *finite* measure, the first term on the right-hand side of (3.5) would equal $+\infty$ or $-\infty$, depending on the sign of a. In that case, given that g and h are integrable, the integral on the left would *exist* and equal the right-hand side, even though $a + bg + ch$ would not be $\mathcal{M}$-integrable. Of course, this situation does not arise for probability measures.

(v) Absolute value inequality:

$$\int |g| \cdot d\mathcal{M} \geq \left| \int g \cdot d\mathcal{M} \right|. \tag{3.6}$$

(vi) Dominance of integral of dominant function: If $h(\omega) \geq g(\omega)$ for all $\omega \in \Omega$, then

$$\int h \cdot d\mathcal{M} \geq \int g \cdot d\mathcal{M}.$$

Of course, the same inequality holds under the weaker condition that $h \geq g$ a.e. $\mathcal{M}$ (almost everywhere with respect to $\mathcal{M}$).

Two other important properties are the conclusions of the *monotone convergence* and *dominated convergence theorems*, which are proved in standard texts in real analysis.

Theorem 3.1. *(Monotone convergence) If $\{g_n\}$ is an (a.e.-$\mathcal{M}$) monotone increasing sequence of nonnegative measurable functions converging (a.e. $\mathcal{M}$) up to a function g, then*

$$\lim_{n \to \infty} \int g_n \cdot d\mathcal{M} = \int \lim_{n \to \infty} g_n \cdot d\mathcal{M} = \int g \cdot d\mathcal{M}. \tag{3.7}$$

Theorem 3.2. *(Dominated convergence) If $\{g_n\}_{n=1}^{\infty}$ is a sequence of integrable functions converging pointwise (a.e. $\mathcal{M}$) to a function g, and if there is an integrable function h such that $|g_n| \leq h$ for each n, then g itself is integrable and (3.7) holds.*

These two theorems give conditions under which the operations of taking limits and integrating can be interchanged. In the monotone case, where the restriction is that $0 \leq g_n(\omega) \leq g_{n+1}(\omega)$ for almost all ω, there is no guarantee that the limit function g is $\mathcal{M}$-integrable. If it is not, then both sides of (3.7) are equal to $+\infty$. This would happen if either (i) $g_{n^*}, g_{n^*+1}, \ldots$ fail to be integrable for some n^*, or (ii) all the $\{g_n\}$ are integrable but sequence $\{\int g_n \cdot d\mathcal{M}\}$ diverges.

Example 3.2. Take $\Omega = \Re_+$ and $g_n(x) := \mathbf{1}_{[0,n]}(x)$. Then $g_n(x) \uparrow g(x) \equiv 1$ as $n \to \infty$ for each $x \in \Re_+$. If $\mathcal{M} = \lambda$ (Lebesgue measure) then $\int g_n(x) \lambda(dx) = 1 \cdot n = n$, so that $\int g_n(x) \lambda(dx) \to \int g(x) \lambda(dx) = +\infty$.

Example 3.3. To see why domination is essential in Theorem 3.2, again take $\Omega = \Re_+$ and $\mathcal{M} = \lambda$, but now set $g_n(x) = \mathbf{1}_{(n,\infty)}(x)$. Then $\int g_n \cdot d\lambda = 0 \cdot \lambda([0,n]) + 1 \cdot \lambda((n,+\infty)) = 0 + (+\infty)$ for each n, and so $\lim_{n \to \infty} \int g_n \cdot d\lambda = +\infty$. On the other hand $\{g_n(x)\} \downarrow g(x) \equiv 0$ for each x, and therefore $\int \lim_{n \to \infty} g_n \cdot d\lambda = 0 \neq \lim_{n \to \infty} \int g_n \cdot d\lambda$.

Example 3.4. Divide $[0,1]$ into $n \geq 2$ equal intervals of length $1/n$, and consider a piecewise linear function F_n (a c.d.f.) that increases monotonically and continuously from $(0,0)$ to $(1,1)$ as follows. Each interval $j \in \{1, 2, ..., n\}$ is further divided into two subintervals, $I_{j,n} := \left[\frac{j-1}{n}, \frac{j}{n} - \frac{1}{n^2}\right)$ and $I'_{j,n} := \left[\frac{j}{n} - \frac{1}{n^2}, \frac{j}{n}\right)$. On interval j function F_n connects linearly from $\left(\frac{j-1}{n}, \frac{j-1}{n}\right)$ to $\left(\frac{j}{n} - \frac{1}{n^2}, \frac{j-1}{n} + \frac{1}{n^2}\right)$ and thence linearly to $\left(\frac{j}{n}, \frac{j}{n}\right)$, as depicted in Fig. 3.1. The derivatives on the two subintervals are $f_n(x) := F'_n(x) = (n-1)^{-1} \mathbf{1}_{I_{j,n}}(x)$ and $f_n(x) = (n-1) \mathbf{1}_{I'_{j,n}}(x)$, the same for all j. Taken together, the flatter segments on the n intervals take up $n\left(1/n - 1/n^2\right) = 1 - 1/n$ of the $[0,1]$ domain, so $F_\infty(x) := \lim_{n \to \infty} F_n(x)$ has derivative $f_\infty(x) = 0$ for almost all x. Accordingly, $F_\infty(1) - F_\infty(0) = 1 \neq \int f_\infty(x)\lambda(dx) = 0$, even though $F_n(1) - F_n(0) = \int f_n(x)\lambda(dx) = 1$ for each n. Here sequence $\{f_n\}_{n=2}^\infty$ is not monotone, nor are the members dominated by an integrable function, since $\left\{(n-1)^{-1} \mathbf{1}_{I_{j,n}}(x)\right\} \downarrow 0$ and $\left\{(n-1) \mathbf{1}_{I'_{j,n}}(x)\right\} \uparrow \infty$ for each j. Figure 3.2 shows the progression of shapes for $n \in \{5, 10, 50\}$.

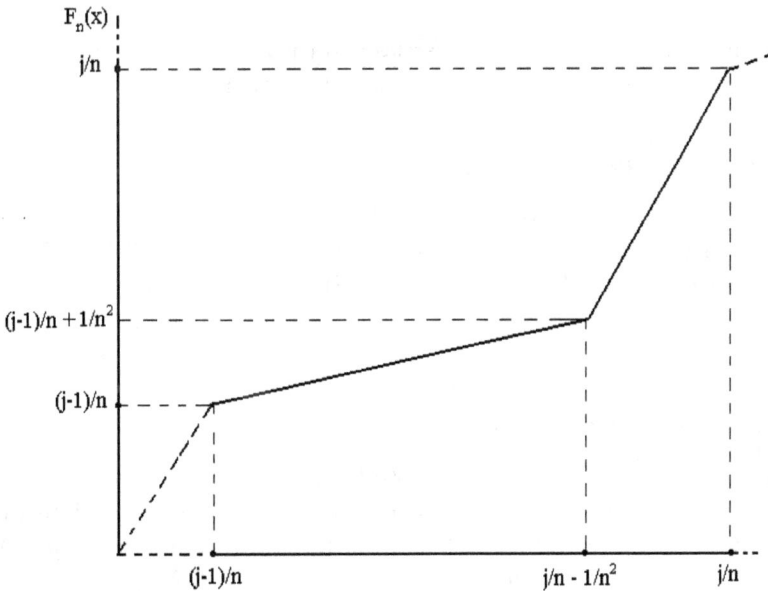

Fig. 3.1   The two segments of $F_n(x)$ for $x \in \left[\frac{j-1}{n}, \frac{j}{n}\right]$.

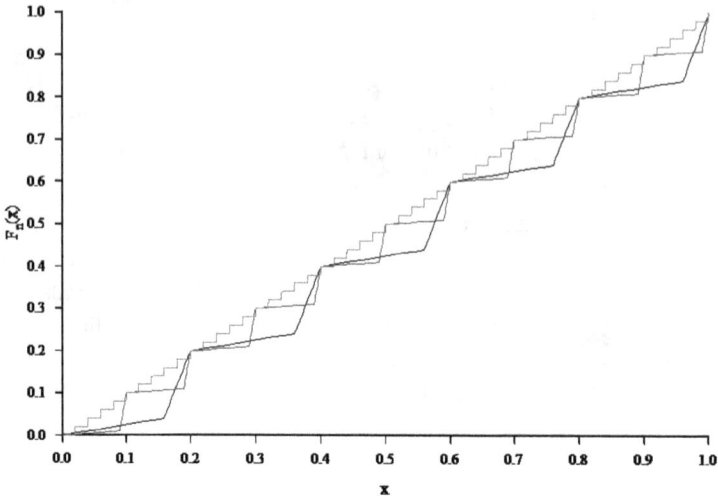

Fig. 3.2   Functions F_n in Example 3.4 for $n \in \{5, 10, 50\}$.

Other examples of these properties and theorems will be given later as they apply specifically to mathematical expectation.

3.2.2   *Integrals on Sets of Real Numbers*

The definition of the abstract integral provides new constructions for integrals of real-valued functions. For these we take $\Omega = \Re$ and $\mathcal{F} = \mathcal{B}$, introduce a measure $\mathcal{M}$ on the space $(\Re, \mathcal{B})$, and consider measurable functions $g : \Re \to \Re$. Various special cases arise depending on the choice of $\mathcal{M}$.

3.2.2.1   *The Lebesgue Integral*

When $\mathcal{M} = \lambda$ (Lebesgue measure) we have the *Lebesgue integral*, written variously as $\int_\Re g(x)\lambda(dx)$, as $\int g \cdot d\lambda$, or—just as for the Riemann integral— as $\int g(x) \cdot dx$. Recalling the definition of the abstract integral, $\int g \cdot d\lambda$ is constructed as the difference between integrals of the nonnegative, measurable functions $g^+ := g\mathbf{1}_{[0,\infty)}$ and $g^- := -g\mathbf{1}_{(-\infty,0)}$, provided that at least one of these integrals is finite. In turn, each of $\int g^+ \cdot d\lambda$ and $\int g^- \cdot d\lambda$ is the limit of a sequence of integrals of simple functions that converges up to the integrand. In the Lebesgue context integrals of simple functions are just sums of areas of approximating rectangles, $\sum_j \gamma_j \lambda(A_j)$. The crucial feature is that the rectangles are formed by subdividing the space that contains the values of g^+ and g^- rather than the set over which g is to be integrated.

To see clearly the distinction between the Lebesgue and Riemann concepts, recall from elementary calculus the Riemann construction of definite integral $\int_a^b g(x) \cdot dx$. This is also defined as a limit of sums of areas of approximating rectangles, but these are formed by subdividing $[a, b]$. Thus, the Riemann construction partitions the x axis—the *domain* of g—directly, whereas in the Lebesgue construction the domain is partitioned based on the values of g itself. Specifically, the Riemann integral is the limit of the approximating sums $\left\{ \sum_{j=1}^n g(x_j^*)(x_j - x_{j-1}) \right\}_{n=1}^\infty$, where $x_j^* \in [x_{j-1}, x_j]$ and $\{x_j\}_{j=0}^n$ satisfy $a =: x_0 < x_1 < \cdots < x_n := b$ and $\max_{1 \le j \le n}\{x_j - x_{j-1}\} \to 0$ as $n \to \infty$. For g to have a Riemann integral, the limit of the sum must exist and be the same for all points x_j^* contained in each interval $[x_{j-1}, x_j]$. In particular, the limits must be the same when each x_j^* is such that g attains either its supremum or its infimum in $[x_{j-1}, x_j]$. The conditions are met if g is continuous on $[a, b]$ or if it is bounded and has only finitely many points of discontinuity on this finite interval. Extending to the real line, $\int_{-\infty}^\infty g(x) \cdot dx$ is defined as the limit of $\int_a^b g(x) \cdot dx$ as $a \to -\infty$ and $b \to \infty$, provided that the limits are the same taken in either order.

We have seen that the construction of the abstract integral extends integration to functions defined on *unordered* spaces—clearly a radical improvement on the Riemann concept. In practice, what is the advantage of the corresponding Lebesgue construction for integrals of functions of real numbers—which, of course, *are* ordered? The answer, in two words, is "greater generality". There are badly behaved functions on $\Re$ that are Lebesgue integrable but not Riemann integrable, and the conclusions of Theorems 3.1 and 3.2 do not hold for the Riemann construction without further conditions on the integrands.

Example 3.5. Take $g(x) = 1_{\mathbb{Q}}(x)$ (that is, g takes the value unity on the rationals and the value zero elsewhere). Then the Riemann integral $\int_0^1 g(x) \cdot dx$ does not exist, since on any subinterval $[x_{j-1}, x_j]$ of $[0, 1]$ the supremum (also the maximum) of g is unity and the infimum (also the minimum) is zero. That is, if each intermediate $x_j^* \in \mathbb{Q}^c$ (the irrationals), then $g(x_j^*) = 0$ and $\lim_{n \to \infty} \sum_{j=1}^n g(x_j^*)(x_j - x_{j-1}) = 0$, whereas $\lim_{n \to \infty} \sum_{j=1}^n g(x_j^*)(x_j - x_{j-1}) = 1$ when each $x_j^* \in \mathbb{Q}$. By contrast, the Lebesgue construction gives an unambiguous and sensible answer. Since $g = 1_\mathbb{Q}$ is a simple function (equal to zero on $\mathbb{Q}^c$ and unity on $\mathbb{Q}$), we have

$$\int_0^1 g \cdot d\lambda = 0 \cdot \lambda\left([0,1] \cap \mathbb{Q}^c\right) + 1 \cdot \lambda\left([0,1] \cap \mathbb{Q}\right)$$

$$= 0 \cdot 1 + 1 \cdot 0 = 0.$$

Exercise 3.2. *Putting* $\mathbb{F}_k := \left\{0, \frac{1}{k}, \frac{2}{k}, ..., \frac{k-1}{k}, 1\right\}$ *for* $k \in \aleph$ *and* $\mathbb{X}_n :=$ $\cup_{k=1}^n \mathbb{F}_k$, *consider the sequence of functions* $\{g_n : [0,1] \to \Re\}_{n=1}^\infty$ *given by*

$$g_n(x) = \begin{cases} 1, \ x \in \mathbb{X}_n \\ 0, \ x \in [0,1] \setminus \mathbb{X}_n \end{cases}.$$

Clearly, g_n *is dominated on* $[0,1]$ *by* $h \equiv 1$.

(i) *Show that* $\int_0^1 g_n(x) \cdot dx = 0$ *for each* n *in both the Riemann and Lebesgue constructions.*

(ii) *Determine the function* g *such that* $\lim_{n \to \infty} g_n(x) = g(x)$ *for each* $x \in [0,1]$.

(iii) *Show that* $\int_0^1 g(x) \cdot dx = 0$ *in the Lebesgue construction but that* g *is not Riemann integrable.*

(iv) *Explain what this implies about the applicability to Riemann integrals of the conclusion of Theorem 3.2.*

Happily, when $g : \Re \to \Re$ is Riemann integrable on some finite interval $[a, b]$, the Lebesgue and Riemann constructions give the same answer, and the standard integration formulas from elementary calculus still apply. This is the usual situation in applied work. However, for analytical purposes we ordinarily assume the Lebesgue construction, largely because Theorems 3.1 and 3.2 facilitate crucial limiting arguments.

The Riemann/Lebesgue distinction has another implication that is important in statistical applications. When integrating over the entire real line, the Riemann integral must be defined as the limit of the sequence of integrals over expanding but *finite* sets (a_n, b_n), with $a_n \downarrow -\infty$ and $b_n \uparrow +\infty$. This is so because the Riemann construction requires partitioning the interval of integration into finite pieces. No such additional limiting operation is required in the Lebesgue case, since the approximating sums are based on values of the integrand. The extra limiting operation needed for the Riemann construction can make the value of a Riemann integral ambiguous, because different ways of taking limits can produce different answers. For example, evaluating

$$\int_{-\infty}^{\infty} \frac{x}{1+x^2} \cdot dx \qquad (3.8)$$

as $\lim_{a_n \to -\infty} \left(\lim_{b_n \to \infty} \int_{a_n}^{b_n} \frac{x}{1+x^2} \cdot dx \right)$ gives the answer $+\infty$ (the value of the inner limit for each a_n), while $\lim_{b_n \to \infty} \left(\lim_{a_n \to -\infty} \int_{a_n}^{b_n} \frac{x}{1+x^2} \cdot dx \right) = -\infty$ and $\lim_{b_n \to \infty} \int_{-b_n}^{b_n} \frac{x}{1+x^2} \cdot dx = 0$. The different ways of taking limits thus give different, but, in each case, *definite* answers. On the other hand, since it requires the integral to be evaluated separately over regions on which the integrand has opposite signs, the Lebesgue construction shows that the integral simply has *no* definite value:

$$\int_{-\infty}^{\infty} \frac{x}{1+x^2} \cdot dx = \left[\int_0^{\infty} \frac{x}{1+x^2} \cdot dx \right] - \left[\int_{-\infty}^0 \frac{-x}{1+x^2} \cdot dx \right]$$

$$= +\infty - (+\infty) = ?.$$

As explained in defining the abstract form $\int g \cdot d\mathcal{M}$, integrability of g requires that $\int |g| \cdot d\mathcal{M}$ be *finite*, a condition that is violated when $g(x) = x/(1+x^2)$ and $\mathcal{M} = \lambda$. Indeed, not only is this g not integrable (over $\Re$) with respect to λ, the integral actually fails to exist.[5]

Exercise 3.3. *Show that $\int_{-\infty}^{\infty} \frac{|x|}{1+x^2} \cdot dx = +\infty$ in both the Lebesgue and Riemann constructions, regardless of the order in which the Riemann limits are evaluated.*

Exercise 3.4. *Determine whether integrals (i) $\int_0^{\infty} x^{-1} e^{-x} \cdot dx$ and (ii) $\int_{-\infty}^{\infty} x^{-1} e^{-x^2} \cdot dx$ exist in the Lebesgue sense.*

3.2.2.2 *Cauchy Principal Value*

Integrals such as (3.8), in which one or both limits is infinite, are often referred to as "improper". The same term is used for integrals over finite or infinite intervals that contain isolated points at which the integrand is infinite, such as $\int_{-1}^2 x^{-1} \cdot dx$. As we have seen, improper integrals may or may not converge to finite values in the Lebesgue sense. An alternative construction of the integral is sometimes useful, especially when integrating *complex-valued* functions. (We shall see why in Section 3.4.2 in connection

---

[5]Later (p. 163) we will give an intuitive interpretation of an integral $\int x f(x) \cdot dx$ as being (proportional to) the *center of mass* of function f. One will then have a tangible sense of why $\int x/(1+x^2) \cdot dx$ simply *does not exist*.

with inverting characteristic functions.) In particular, if $I_n := \int_{-n}^{n} g(x) \cdot dx$ exists for each n (in the Lebesgue sense), and if the sequence $\{I_n\}$ approaches a finite limit as $n \to \infty$, then we refer to the limiting value as the *Cauchy principal value* of the integral. Likewise, if $\lim_{x \to c} |g(x)| = +\infty$ for $c \in (a, b)$, we define the Cauchy principal value of $\int_a^b g(x) \cdot dx$ as $\lim_{n \to \infty} \left[\int_{c+n^{-1}}^{b} g(x) \cdot dx + \int_a^{c-n^{-1}} g(x) \cdot dx \right]$. For example, we have already seen that the Cauchy principal value of (3.8) is zero. Indeed, if g be any *odd* function (one such that $g(x) = -g(-x)$), then the Cauchy principal value of $\int_{-\infty}^{\infty} g(x) \cdot dx$ is zero whenever $\int_{-n}^{n} g(x) \cdot dx$ exists for each n in the Lebesgue sense, because

$$
\int_{-n}^{n} g(x) \cdot dx = \int_{-n}^{0} g(x) \cdot dx + \int_{0}^{n} g(x) \cdot dx
$$
$$
= \int_{n}^{0} g(-x) \cdot d(-x) + \int_{0}^{n} g(x) \cdot dx
$$
$$
= - \int_{0}^{n} g(x) \cdot dx + \int_{0}^{n} g(x) \cdot dx.
$$

However, it is true in general that if $\int |g(x)| \cdot dx$ exists in the Cauchy sense, then $\int g(x) \cdot dx$ exists and has the same value in both the Lebesgue and Cauchy constructions.[6]

Example 3.6. While $\int_{-1}^{2} x^{-1} \cdot dx$ does not exist as a Lebesgue integral, its Cauchy principal value is

$$
\int_{-1}^{2} x^{-1} \cdot dx = \lim_{n \to \infty} \left[\int_{-1}^{-n^{-1}} x^{-1} \cdot dx + \int_{n^{-1}}^{2} x^{-1} \cdot dx \right]
$$
$$
= \lim_{n \to \infty} \left[\int_{1}^{n^{-1}} (-x)^{-1} \cdot d(-x) + \int_{n^{-1}}^{2} x^{-1} \cdot dx \right]
$$
$$
= \lim_{n \to \infty} \left[-\int_{n^{-1}}^{1} x^{-1} \cdot dx + \int_{n^{-1}}^{2} x^{-1} \cdot dx \right]
$$
$$
= \log 2.
$$

Note, however, that $\int_{-1}^{2} |x^{-1}| \cdot dx$ does not exist in the Cauchy sense.

3.2.2.3 *Lebesgue–Stieltjes Integrals*

We have seen that the Lebesgue construction of $\int g(x) \cdot dx \equiv \int g \cdot d\lambda \equiv \int_{\Re} g(x) \lambda(dx)$ involves weighting values of g by the Lebesgue measures of cer-

---

[6]See Taylor (1973, pp. 129-132) for further details on the relation among Lebesgue, Riemann, and Cauchy forms of the integral.

tain sets. That is, in the sums $\{\sum_{j=0}^{2^{2n}} \gamma_{j,n} \lambda (A_{j,n})\}_{n=1}^{\infty}$ of which $\int g(x)\lambda(dx)$ is the limit (for nonnegative g) each value $\gamma_{j,n}$ of the approximating simple function at stage n is weighted by the *length* measure of the set $A_{j,n}$. As in the earlier construction of the abstract integral on an arbitrary space Ω, using other measures than Lebesgue measure leads to a still more general concept of integration over the reals. The idea is very simple: Integrating with respect to some measure $\mathcal{M}$ on $(\Re, \mathcal{B})$ just requires attaching to $\gamma_{j,n}$ the weight $\mathcal{M}(A_{j,n})$. Integrals with respect to measures other than length are called *Lebesgue–Stieltjes (L-S)* integrals, and we shall see that these are of fundamental importance in probability and statistics.

Example 3.7. Take $\mathbb{X}$ to be a countable set of distinct real numbers, and for arbitrary $B \in \mathcal{B}$ take $N(B) := N(B \cap \mathbb{X})$, thus extending counting measure (for this particular $\mathbb{X}$) to measurable space $(\Re, \mathcal{B})$. Then for $g : \Re \to \Re_+$ the Lebesgue-Stieltjes construction gives

$$\int g(x) N(dx) \equiv \int g \cdot dN = \sum_{x \in \mathbb{X}} g(x) N(\{x\}) = \sum_{x \in \mathbb{X}} g(x).$$

Thus, if f is the p.m.f. of a r.v. supported on $\mathbb{X}$, $\Pr(X \in B) = \sum_{x \in B \cap \mathbb{X}} f(x)$ can be expressed in integral form as

$$\sum_{x \in B \cap \mathbb{X}} f(x) N(\{x\}) = \int_B f(x) N(dx) \equiv \int_B f \cdot dN.$$

Likewise, for some integrable g,

$$\sum_{x \in \mathbb{X}} g(x) f(x) = \int gf \cdot dN.$$

In this context p.m.f. f is regarded a density with respect to counting measure. Alternatively, since $f(x) \equiv \mathbb{P}_X(\{x\})$ we can write

$$\Pr(X \in B) \equiv \mathbb{P}_X(B) = \int_B \mathbb{P}_X(dx)$$

and

$$\sum_{x \in \mathbb{X}} g(x) f(x) = \int g(x) \mathbb{P}_X(dx) = \int g \cdot d\mathbb{P}_X.$$

Our principal interest is Lebesgue-Stieltjes integrals with respect to probability measures. When $\mathcal{M} : \mathcal{B} \to [0,1]$ is a probability measure on $(\Re, \mathcal{B})$, then $\int g \cdot d\mathcal{M}$ becomes, in effect, a weighted *average* of values of g, and hence a mathematical expectation.

Exercise 3.5. *Taking $\mathcal{M}$ as $\mathbb{P}_X$, use the definition of the integral to show that EX in Expression (3.2) can be written as $\int x\mathbb{P}_X(dx)$.*

In the probability context when the c.d.f. F corresponds to an induced measure $\mathbb{P}_X$, integrals $\int g(x)\mathbb{P}_X(dx) \equiv \int g \cdot d\mathbb{P}_X$ are usually written as $\int g(x) \cdot dF(x)$ or simply as $\int g \cdot dF$. Either way, the *intuitive* interpretation is that $g(x)$ is being weighted not by $dx = \lambda\left((x - dx, x]\right)$ but by the *probability* associated with $(x - dx, x]$. Thus, one can think of symbols $\mathbb{P}_X(dx)$ and $dF(x)$ as representing $\mathbb{P}_X\left((x - dx, x]\right)$ and $F(x) - F(x - dx)$, respectively.

Given a suitably well behaved function g and a c.d.f. F one can actually see how to calculate a definite integral $\int_{(a,b]} g(x) \cdot dF(x)$ over some interval $(a, b]$ by following the Riemann–Stieltjes construction. For this, partition $(a, b]$ into n intervals $(x_{j-1}, x_j]$ such that $\max_{1 \leq j \leq n} \{|x_j - x_{j-1}|\} \to 0$ as $n \to \infty$. With $x_j^* \in (x_{j-1}, x_j]$, the integral is defined as

$$\lim_{n \to \infty} \sum_{j=1}^{n} g(x_j^*) \left[F(x_j) - F(x_{j-1}) \right] \tag{3.9}$$

provided the limit exists and is the same for each choice of x_j^*.

To see what such a limiting value would be in a common situation, let us suppose (i) that g is continuous; (ii) that c.d.f. F has finitely many (left) discontinuities $N(\mathcal{D})$ on $(a, b]$; and (iii) that F is absolutely continuous elsewhere, with density $f = F' \geq 0$. Thus, F could be the c.d.f. of a discrete, continuous, or mixed r.v., depending on the value of $N(\mathcal{D})$ and the specification of f. Now at each discontinuity point $x \in \mathcal{D}$ there is positive probability mass $F(x) - F(x-)$. By taking n sufficiently large, the partition of intervals can be arranged so that each point of discontinuity coincides with the right end point of one of the intervals. In this way $N(\mathcal{D})$ of the intervals $(x_{j-1}, x_j]$ will have right end point x_j as its sole point of discontinuity, while function F will be continuous on each of the remaining $n - N(\mathcal{D})$ intervals. (Figure 3.3 depicts a partition with $n = 7$ and $N(\mathcal{D}) \equiv N\left(\{x_1, x_3, x_5, x_7\}\right) = 4$.) On each continuity interval the mean value theorem guarantees that there exists $x_j^{**} \in (x_{j-1}, x_j)$ such that $F(x_j) - F(x_{j-1}) = f(x_j^{**})(x_j - x_{j-1})$. Letting $n \to \infty$ and $\max_{1 \leq j \leq n} \{|x_j - x_{j-1}|\} \to 0$, the sum of the $N(\mathcal{D})$ terms of (3.9) corresponding to the intervals with discontinuities will approach $\sum_{x \in \mathcal{D}} g(x) \left[F(x) - F(x-) \right]$, while the limit of the remaining sum of $n - N(\mathcal{D})$ terms, $\sum_j g(x_j^*) f(x_j^{**})(x_j - x_{j-1})$, approaches the Riemann integral $\int_a^b g(x) f(x) \cdot dx$. Notice that the sum over the elements of $\mathcal{D}$ merely weights the value of g at each discontinuity point by the probability mass that resides there. Note, too, that with Stieltjes integrals the notation must make clear whether intervals of integration are open or closed, as

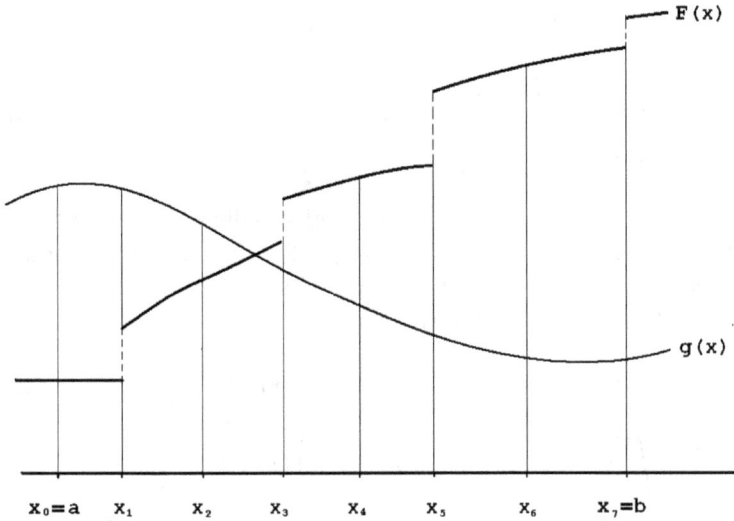

Fig. 3.3   Stieltjes construction of $\int_a^b g \cdot dF$.

$\int_{(a,b]} g(x) \cdot dF(x)$, $\int_{[a,b)} g(x) \cdot dF(x)$, $\int_{(a,b)} g(x) \cdot dF(x)$, etc., since the singletons $\{a\}$, $\{b\}$ can themselves have positive measure. (Thus, if $g(b) > 0$ in Fig. 3.3, we would have $\int_{(a,b]} g(x) \cdot dF(x) > \int_{(a,b)} g(x) \cdot dF(x)$.)

Here, then, is the simple recipe for evaluating the Stieltjes integral w.r.t. F of a continuous function g over a finite interval:

- At each point $x \in \mathcal{D}$ where there is positive probability mass $F(x) - F(x-)$ multiply $g(x)$ by this probability (which is the density of $\mathbb{P}_X$ with respect to counting measure), and accumulate the results; then
- Add to this sum the integral of g times $f = F'$ (the density of $\mathbb{P}_X$ with respect to Lebesgue measure). Thus, for some Borel set A,

$$\int_A g \cdot dF = \sum_{x \in A \cap \mathcal{D}} g(x) \left[F(x) - F(x-) \right] + \int_A g(x) f(x) \cdot dx. \quad (3.10)$$

In the second integral the gaps in A at the finitely many points of $\mathcal{D}$ where $f = F'$ fails to exist do not affect the value, for one simply evaluates the integral piecewise between the gaps.

Example 3.8. With

$$F(x) = \begin{cases} 0, \, x < 0 \\ \frac{1}{2}, \, 0 \le x < .5 \\ x, \, .5 \le x < 1 \\ 1, \, x \ge 1 \end{cases} = \frac{1}{2} 1_{[0,.5)}(x) + x 1_{[.5,1)}(x) + 1_{[1,\infty)}(x),$$

the value of $\int_{[0,1)} (x^2 + 1) \cdot dF(x)$ is found as follows. F has a single discontinuity at 0, with $F(0) - F(0-) = \frac{1}{2}$. Elsewhere it is continuous, with $F'(x) = f(x) = 0$ for $x \in (0, .5)$ and $F'(x) = f(x) = 1$ for $x \in (.5, 1)$. Following the steps above, the value is

$$\left(0^2 + 1\right) \cdot [F(0) - F(0-)] + \int_0^{.5} (x^2 + 1) f(x) \cdot dx + \int_{.5}^1 (x^2 + 1) f(x) \cdot dx$$

$$= 1 \cdot \frac{1}{2} + \int_0^{.5} (x^2 + 1) \cdot 0 \cdot dx + \int_{.5}^1 (x^2 + 1) \cdot 1 \cdot dx = \frac{31}{24}.$$

Exercise 3.6. *Find $\int_{(0,1)} (x^2 + 1) \cdot dF(x)$ for the c.d.f. in Example 3.8.*

Riemann–Stieltjes integrals over finite intervals exist for functions g that are continuous and for functions g that have finitely many discontinuities, provided no discontinuity points of g and F coincide. As for the ordinary Riemann integral, the Riemann–Stieltjes integral over an infinite interval is evaluated as a limit of integrals over expanding but finite intervals. Likewise, to prevent ambiguity, g is said to be integrable with respect to F if and only if $\int_{\Re} |g(x)| \cdot dF(x) < \infty$. The annotation "$\Re$" is usually omitted when integration is over the entire real line, as are the arguments of g and F, as $\int g \cdot dF$.

Here is what can happen when g and F have discontinuities at one or more points in common.

Example 3.9. Let $g(x) = F(x) = 1_{[b,\infty)}(x)$, both being left discontinuous at b. For arbitrary $a < b < c$

$$\int_{(a,c)} g \cdot dF = \lim_{n \to \infty} \sum_{j=1}^n g(x_j^*) [F(x_j) - F(x_{j-1})],$$

where $a = x_0 < x_1 < \cdots < x_n = c$. Following the usual construction, we will fix the discontinuity point b at the right end point of some interval $(x_{j-1}, x_j]$ for each n. As more intervals are added below b, the value of j such that $x_j = b$ also increases, so let us write it as $x_{j_n} = b$. Since $g(x) = 1_{[b,\infty)}(x) = 0$ for $x < b$, the terms $g(x_j^*) [F(x_j) - F(x_{j-1})]$ all vanish for $j < j_n$. They also vanish for $j > j_n + 1$ because $F(x) = 1$ for

$x \geq b$. Therefore, for each n the approximating sum collapses to a single term,

$$g(x_{j_n}^*) [F(x_{j_n}) - F(x_{j_n-1})] = g(x_{j_n}^*) [F(b) - F(x_{j_n-1})] = g(x_{j_n}^*).$$

Now g is Riemann–Stieltjes integrable with respect to F if and only if $\lim_{n\to\infty} g(x_{j_n}^*)$ exists and has a common value for each $x_{j_n}^* \in (x_{j_n-1}, x_{j_n}]$. But this condition fails, since $g(x_{j_n}) = 1$ for each n while $g(x) = 0$ for all $x < x_{j_n}$.

The Lebesgue–Stieltjes construction immediately overcomes the difficulty encountered in this example. Since $g = \mathbf{1}_{[b,\infty)}$ is a simple function, its (constant) value of unity on $[b, c)$ gets weighted by the probability mass that resides there. Thus,

$$\int_{(a,c)} \mathbf{1}_{[b,\infty)} \cdot dF = \int_{(a,b)} 0 \cdot \mathbb{P}_X\left((a,b)\right) + \int_{\{b\}} 1 \cdot \mathbb{P}_X\left(\{b\}\right) + \int_{(b,c)} 1 \cdot \mathbb{P}_X\left((b,c)\right)$$
$$= 0 + 1 + 0 = 1.$$

Besides being more widely applicable, the Lebesgue–Stieltjes construction lets us invoke Theorems 3.1 and 3.2 to determine limits of certain sequences of integrals. A particularly important and common such application is that of differentiating "under the integral sign". Thus, for some $g : \Re^2 \to \Re$, suppose we know that the integral $\int g(x,t) \cdot dF(x)$ exists (in the Lebesgue sense) for all t in some interval $[a, b]$. Letting $I : [a, b] \to \Re$ represent the integral as a function of t, we would like to know whether its derivative can be evaluated as

$$I'(t) := \frac{d}{dt} \int g(x,t) \cdot dF(x) = \int \frac{\partial g(x,t)}{\partial t} \cdot dF(x); \qquad (3.11)$$

that is, whether it is valid to interchange the operations of differentiation and integration. Writing $I'(t) = \lim_{\eta\downarrow 0} [I(t+\eta) - I(t-\eta)]/(2\eta)$, we have (from the linearity property of the integral) that

$$I'(t) = \lim_{\eta\downarrow 0} \int \left[\frac{g(x,t+\eta) - g(x,t-\eta)}{2\eta} \right] \cdot dF(x).$$

Thus, the interchange of differentiation and integration amounts to an interchange of operations of taking limits. Obviously, a necessary condition for this to be valid is that $g(x,t)$ be differentiable with respect to t for "almost all" x, but this in itself is not enough. According to Theorem 3.2 what *does* suffice is for there to exist an integrable function $D : \Re^2 \to \Re$ such that for all sufficiently small η

$$\left| \frac{g(x,t+\eta) - g(x,t-\eta)}{2\eta} \right| \leq D(x,t)$$

on all sets of positive $\mathbb{P}_X$ measure. Thus, if we can find such a dominating function we will know that the operation in (3.11) is indeed valid.

Example 3.10. Anticipating an application in Section 3.4.2, let us suppose that the function $\int e^{tx} \cdot dF(x) =: M(t)$ is known to have a finite value for all t in some neighborhood of the origin; i.e., for all $t \in [-2\delta, 2\delta]$ and some $\delta > 0$. We would like to know whether $M'(t)$ can be expressed as

$$\int \frac{\partial e^{tx}}{\partial t} \cdot dF(x) = \int x e^{tx} \cdot dF(x)$$

for sufficiently small $|t|$. Now

$$M'(t) = \lim_{\eta \downarrow 0} \frac{\int e^{(t+\eta)x} \cdot dF(x) - \int e^{(t-\eta)x} \cdot dF(x)}{2\eta}$$

$$= \lim_{\eta \downarrow 0} \int \frac{e^{(t+\eta)x} - e^{(t-\eta)x}}{2\eta} \cdot dF(x)$$

$$= \lim_{\eta \downarrow 0} \int \frac{e^{\eta x} - e^{-\eta x}}{2\eta} e^{tx} \cdot dF(x)$$

$$= \lim_{\eta \downarrow 0} \int \eta^{-1} \sinh(\eta x) e^{tx} \cdot dF(x),$$

so to apply dominated convergence we must find a dominating function D such that $D(x,t) \geq \eta^{-1} |\sinh(\eta x)| e^{tx}$ for all sufficiently small η and such that $\int D(x,t) \cdot dF(x) < \infty$ in some neighborhood of $t = 0$. Recognizing that $\sinh(\eta x) = -\sinh(-\eta x)$ and expanding $\sinh(\eta x)$ in Taylor series[7] about $\eta = 0$, we find that

$$\eta^{-1} |\sinh(\eta x)| = |x| \cdot \sinh(\eta x)/(\eta x)$$

$$= |x| \left[1 + \frac{(\eta x)^2}{3!} + \frac{(\eta x)^4}{5!} + \cdots + \frac{(\eta x)^{2j}}{(2j+1)!} + \cdots \right].$$

Thus, $\eta^{-1} |\sinh(\eta x)|$ strictly decreases as $\eta \downarrow 0$ and $\eta^{-1} \sinh(\eta x) \downarrow x$, so for $\eta \in (0, \delta]$

$$\eta^{-1} |\sinh(\eta x)| \leq \delta^{-1} |\sinh(\delta x)| \leq \frac{e^{\delta x} + e^{-\delta x}}{2\delta} = \delta^{-1} \cosh(\delta x).$$

Putting $D(x,t) := \delta^{-1} \cosh(\delta x) e^{tx}$, we see that

$$\int D(x,t) \cdot dF(x) = \frac{M(t+\delta) + M(t-\delta)}{2\delta} < \infty$$

for $|t| \leq \delta$ (since $M(t)$ exists for $|t| \leq 2\delta$). Thus, it follows that $M'(t) = \int x e^{tx} \cdot dF(x)$ and, in particular, that $M'(0) \equiv M'(t)|_{t=0} = \int x \cdot dF(x) = EX$.

---

[7] For some alternative statements of Taylor's theorem, see Section 6.2.2.

Exercise 3.7. X *is a continuous r.v. with c.d.f.* F. *Find*

(i) $EF(X) = \int_{\Re} F(x) \cdot dF(x)$.

(ii) $EF(X)^k = \int_{\Re} F(x)^k \cdot dF(x)$ *for any* $k \in \aleph_0$.

Exercise 3.8. $\{X_n\}_{n=1}^{\infty}$ *are discrete r.v.s with p.m.f.* $f_n(x) = n^{-1}1_{\{1,2,\dots,n\}}(x)$ *and c.d.f.* $F_n(x) = \min\{n^{-1}[x], 1\} 1_{[0,\infty)}(x)$ *for each* $n \in \aleph$. *("$[x]$" represents "greatest integer in x".)*

(i) *Find* $EF_n(X) = \int_{\Re} F_n(x) \cdot dF_n(x)$ *and determine the limiting value as* $n \to \infty$.

(ii) *Express* $EF_n(X)^k$ *for any* $k \in \aleph_0$ *and use (or prove) the fact that* $\lim_{n\to\infty} n^{-(k+1)} \sum_{j=1}^{n} j^k = (k+1)^{-1}$ *to find* $\lim_{n\to\infty} EF_n(X)^k$.

Exercise 3.9. X *is a continuous r.v. distributed symmetrically about the origin. Show that* $\Pr(|X| \leq a) = 2\int_{-a}^{a} F(x) \cdot dF(x)$. *(Hint: Recall Definition 2.6.)*

3.3 Properties of $E(\cdot)$

While $\int X(\omega)\mathbb{P}(d\omega)$ is the most insightful representation of EX, the Stieltjes construction $\int x \cdot dF(x)$ offers a more practical means of calculation. Moreover, as does the abstract form, it gives a single expression for EX that applies to discrete, continuous, and mixed r.v.s alike. When X is purely discrete with support $\mathbb{X}$ and p.m.f. f, the construction (3.10) with $\mathcal{D} = \mathbb{X}$ gives

$$\int x \cdot dF(x) = \sum_{x \in \mathbb{X}} x\,[F(x) - F(x-)] = \sum_{x \in \mathbb{X}} xf(x).$$

Thus, in the discrete case EX is just the sum of the countably many real numbers in the support weighted by the probability masses that reside there. Since the probabilities add to unity, this is simply a weighted average of the values in $\mathbb{X}$. In the other common case that X is purely (absolutely) continuous with density f the Stieltjes construction gives $\int x \cdot dF(x) = \int xf(x) \cdot dx$, which is evaluated as an ordinary Lebesgue (or, in most applications, Riemann) integral. Since $\int f(x) \cdot dx = 1$ this still can be thought of as a weighted average. When X is of the mixed type and has no p.m.f. or p.d.f., then $\int x \cdot dF(x)$ must be evaluated by the recipe for Stieltjes integrals—i.e., as a weighted sum over $\mathcal{D}$, the discontinuity points of F, plus the Lebesgue integral of $xF'(x)$ where F' exists (as it does a.e. λ). Still, the interpretation as weighted average applies.

It is also possible to represent EX in terms of a *density* in all three cases—discrete, continuous, and mixed, as

$$EX = \int xf(x)\,\mathcal{M}(dx). \tag{3.12}$$

Here $\mathcal{M}$ represents a measure with respect to which f *is* a density—i.e., a measure that *dominates* the relevant probability measure $\mathbb{P}_X$, in the sense that $\mathbb{P}_X(B) = 0$ whenever $\mathcal{M}(B) = 0$. In the discrete case counting measure N dominates; in the continuous case, Lebesgue measure λ. Since these two measures are mutually singular, taking $\mathcal{M} = N + \lambda$ gives a representation that covers both these cases and the mixed case as well. (Here, $N + \lambda$ assigns to Borel set B the value $N(B \cap \mathcal{D}) + \lambda(B)$, where $\mathcal{D}$ is the set of discontinuities of F.) While the expression $\int x \cdot dF(x)$ affords the same generality as (3.12), having a uniform representation in terms of densities will often be helpful in developing the statistical theory of Part II.

Here are some examples and exercises to illustrate the actual computation of EX.

Example 3.11. If $f(x) = 1_{[0,1]}(x)$, $EX = \int_{-\infty}^{\infty} xf(x) \cdot dx = \int_0^1 x \cdot dx = \frac{1}{2}$.

Example 3.12. If $f(x) = e^{-x}1_{[0,\infty)}(x)$, $EX = \int_{-\infty}^{\infty} xf(x) \cdot dx = \int_0^{\infty} xe^{-x} \cdot dx$. Integrating by parts with $u = x$, $dv = e^{-x}$, $du = dx$, and $v = -e^{-x}$ gives $EX = -xe^{-x}\big|_0^{\infty} + \int_0^{\infty} e^{-x} \cdot dx = 0 + 1$.

Example 3.13. If $f(x) = 2^{-x}1_{\aleph}(x)$, then $EX = \sum_{x=1}^{\infty} x2^{-x}$. To evaluate the sum, consider $\sum_{x=1}^{\infty} xz^x$ for $z \in (0,1)$. Written as $z\sum_{x=1}^{\infty} xz^{x-1}$, each term of the sum equals $d(z^x)/dz$, so[8] $z\sum_{x=1}^{\infty} xz^{x-1} = z\frac{d}{dz}\sum_{x=1}^{\infty} z^x = z\frac{d}{dz}\left(\frac{1}{1-z} - 1\right) = \frac{z}{(1-z)^2}$. Putting $z = \frac{1}{2}$ gives $EX = 2$.

---

[8]The interchange of summation and differentiation in the first step is justified as follows. Since $d\left(\sum_{x=1}^{\infty} z^x\right)/dz = \lim_{\eta \to 0}\sum_{x=1}^{\infty} \eta^{-1}\left[(z+\eta)^x - z^x\right]$ and $(z+\eta)^x = z^x + \eta x(z+\eta^*)^{x-1}$ (mean value theorem, $|\eta^*| < |\eta|$), we have

$$\frac{d}{dz}\sum_{x=1}^{\infty} z^x = \lim_{\eta \to 0}\sum_{x=1}^{\infty} x(z+\eta^*)^{x-1}.$$

For each $z \in (0,1)$ there is an $\varepsilon \in (0,1)$ such that $z \in (0, 1-2\varepsilon)$ and an η such that $|\eta^*| < \varepsilon$. Thus, for each $z \in (0,1)$ we have

$$\sum_{x=1}^{\infty} x(z+\eta^*)^{x-1} < \sum_{x=1}^{\infty} x(1-\varepsilon)^{x-1}.$$

The last sum converges by the ratio test, since

$$\lim_{x \to \infty} \frac{(x+1)(1-\varepsilon)^x}{x(1-\varepsilon)^{x-1}} = 1 - \varepsilon < 1. \quad \text{(continued)}$$

Exercise 3.10.

(i) X is the number of heads in three flips of a fair coin. Find EX.

(ii) Y is the number of sequence changes in three flips of a fair coin. Find EY.

Exercise 3.11. *If $f(x) = (2\pi)^{-1/2} \exp(-x^2/2)$, find EX.*

Exercise 3.12. *If $f(x) = \left(\lambda^x e^{-\lambda}/x!\right) 1_{\aleph_0}(x)$ for some $\lambda > 0$, find EX.*

Representing EX in abstract form $\int X(\omega)\mathbb{P}(d\omega)$ shows at once how to interpret the expectation of a measurable function of X. Thus, if $g : \Re \to \Re$ is measurable and $Y = g(X)$,

$$EY = \int Y(\omega)\mathbb{P}(d\omega) = \int g\left[X(\omega)\right]\mathbb{P}(d\omega) = Eg(X).$$

Calculation of EY as $\int y\mathbb{P}_Y(dy) = \int y \cdot dF_Y$ requires deducing F_Y by the techniques of Section 2.3. However, there is usually an easier way. Since $\mathbb{P}_Y(B) = \mathbb{P}_X[g^{-1}(B)]$, where $g^{-1}(B) = \{x : g(x) \in B\}$ for any Borel set B, we also have $EY = \int g(x)\mathbb{P}_X(dx) = \int g(x) \cdot dF_X(x)$. When X is discrete this is $\sum_X g(x)f_X(x)$; when continuous, $\int g(x)f_X(x) \cdot dx$.

Example 3.14. Find Ee^X if X has c.d.f.

$$F(x) = \begin{cases} 0, & x < 0 \\ \frac{1}{2}, & 0 \le x < \frac{1}{2} \\ x, & \frac{1}{2} \le x < 1 \\ 1, & x \ge 1 \end{cases}.$$

As in Example 3.8 , we add (i) the sum of values times probability masses at discontinuity point(s) to (ii) the integral of value times density over the continuous region. Thus,

$$Ee^X = e^0\left[F(0) - F(0-)\right] + \int_{.5}^1 e^x \cdot dx = \frac{1}{2} + e^1 - e^{1/2}.$$

Exercise 3.13. *X, Y are the numbers of heads and sequence changes, respectively, in three flips of a fair coin. Find EX^2 and EY^3.*

Exercise 3.14. *Find $E(-\log X)$ if $f(x) = 1_{[0,1]}(x)$.*

Thus, by dominated convergence,

$$\frac{d}{dz}\sum_{x=1}^{\infty} z^x = \sum_{x=1}^{\infty} \lim_{\eta \to 0} x\left(z + \eta^*\right)^{x-1} = \sum_{x=1}^{\infty} xz^{x-1}.$$

The mathematical expectation of some $g(X)$ can fail to *exist* in two general ways. First, g itself may be undefined at some point or points on which $\mathbb{P}_X$ places positive probability mass. In this case $g(X)$ is simply not a measurable *function* on any support of X. Second, if g is a measurable function on $\mathbb{X}$ but if neither $g^+ := \max\{g,0\}$ nor $g^- := -\min\{g,0\}$ is integrable—i.e., if $\int g^+ \cdot dF = +\infty$ and $\int g^- \cdot dF = +\infty$—then the Lebesgue construction gives no definite value to $\int g \cdot dF$. On the other hand, $Eg(X)$ can exist but be infinite. Specifically, $Eg(X) = +\infty$ if $\int g^+ \cdot dF = +\infty$ and $\int g^- \cdot dF < \infty$, while $Eg(X) = -\infty$ if $\int g^+ \cdot dF < \infty$ and $\int g^- \cdot dF = +\infty$. Whenever $\int |g| \cdot dF$ fails to converge—i.e., if either $\int g^+ \cdot dF = +\infty$ or $\int g^- \cdot dF = +\infty$—then we say that g itself is not integrable with respect to $\mathbb{P}_X$. Note that a function $g : \mathbb{X} \to \Re$ that is bounded on $\mathbb{X}$ always has a mathematical expectation and it is always finite, a consequence of the dominance property of the integral (Section 3.2.1.2) and the finiteness of probability measure. In short, all bounded functions of any r.v. are $\mathbb{P}$-integrable.

Example 3.15. (i) If $f(x) = \binom{3}{x}\frac{1}{8}\mathbf{1}_{\{0,1,2,3\}}(x)$, then $E(X-a)^{-1}$ fails to exist when $a \in \{0,1,2,3\}$ because $(x-a)^{-1}$ is undefined at some point on the support of X.

(ii) If $f(x) = 2^{-x}\mathbf{1}_{\aleph}(x)$, then $E2^X = +\infty$, since $\sum_{x\in\aleph} 2^x \cdot 2^{-x} = \lim_{n\to\infty}\sum_{x=1}^n 1 = +\infty$.

(iii) If $f(x) = \mathbf{1}_{[0,1]}(x)$, then EX^{-1} exists and equals $+\infty$. The fact that $\{0\} \subset [0,1]$ and that division by zero is undefined is not the issue, since subset $(0,1]$ is a support of the continuous r.v. X. The problem is that x^{-1} is unbounded on $(0,1]$ (and on any other support) and grows so quickly as $x \to 0$ that $\int_\varepsilon^1 x^{-1} \cdot dx$ does not converge as $\varepsilon \downarrow 0$.[9]

(iv) If $f(x) = \mathbf{1}_{[0,1]}(x)$, then $E\log X$ exists and is finite, even though $\log x$ is undefined at $x = 0$ (a null set, as before) and $|\log x|$ is unbounded on $(0,1]$. Here $\log x \to -\infty$ slowly enough as $x \downarrow 0$ that $\int_\varepsilon^1 \log x \cdot dx = x\log x - x|_\varepsilon^1 = -1 - \varepsilon\log\varepsilon + \varepsilon$ does converge (to -1) as $\varepsilon \to 0$.

Exercise 3.15. X has p.d.f. $f(x) = e^{-x}\mathbf{1}_{(0,\infty)}(x)$. Find (i) EX^2 and (ii) Ee^X.

Exercise 3.16. X is the number of flips of a fair coin until the first occurrence of heads. For what values of z is Ez^X finite?

---

[9]The Cauchy interpretation of $\int_0^1 x^{-1} \cdot dx$ as $\lim_{\varepsilon\to 0}\int_\varepsilon^1 x^{-1} \cdot dx$ gives the same answer as the Lebesgue construction. Since the former has an infinite limit, so does any sequence of approximating simple functions in the Lebesgue construction.

Exercise 3.17. X has p.d.f. $f(x) = (1+x)^{-2} 1_{(0,\infty)}(x)$. Find EX.

Exercise 3.18. X has p.d.f. $f(x) = \left[\pi(1+x^2)\right]^{-1}$ *(Recall from Exercise 2.40 that this is the p.d.f. of the standard Cauchy distribution.) Determine whether EX exists. Why does evaluating EX as $\lim_{c \to \infty} \int_{-c}^{c} \frac{x}{\pi(1+x^2)} \cdot dx$. give the wrong answer?*

How can our intuitive notion of expectation as "long-run average" accommodate the mathematically demonstrable fact that $E|g(X)| = +\infty$ for certain functions g and r.v.s X? Is it that an infinite value of $|g|$ will eventually turn up if we keep performing the chance experiment? But this cannot happen when $Y \equiv g(X)$ is a "proper" r.v., for all proper r.v.s are *finite* with probability one. (Since $\mathbb{P}_Y(\Re) = \lim_{n \to \infty} \mathbb{P}_Y((-n, n)) = 1$ if Y is proper, there can be no positive residue of probability mass *at* infinity.) So what does $E|Y| = +\infty$ really imply about our long-run averages? What it really means is that the average over realizations $|Y(\omega)|$ will *eventually* surpass any number $b > 0$—no matter how large b is. In other words, the long-run average of realizations simply cannot be bounded, and such unbounded quantities are precisely the ones that we call "infinite".

Example 3.16. A random variable Y with c.d.f. $F_Y(y) = \left(1 - y^{-1/2}\right) 1_{[1,\infty)}(y)$ has expected value $EY = +\infty$. Twenty independent realizations are shown in the table, along with the running averages. Note how the occasional large values spike up the average, which would (were it not for computational limitations) indeed ultimately exceed any given bound.

n	y_n	$\frac{1}{n}\sum_{j=1}^{n} y_j$	n	y_n	$\frac{1}{n}\sum_{j=1}^{n} y_j$
1	15.23	15.23	11	5.83	279.15
2	1.44	8.34	12	8.51	256.59
3	200.48	72.38	13	3.53	237.13
4	2680.96	724.53	14	3.00	220.40
5	2.81	580.18	15	4.21	205.99
6	124.66	504.26	16	269.39	209.95
7	10.89	433.78	17	120410.62	7280.58
8	1.54	379.75	18	5.32	6876.40
9	1.63	337.74	19	2.54	6514.62
10	25.15	306.48	20	1.19	6188.95

The concept of mathematical expectation extends formally to the multivariate case simply by interpreting r.v.s as k-vectors and regarding induced probability measures as defined on sets of $\mathcal{B}^k$. Thus, if $\mathbf{X} = (X_1, X_2, ..., X_k)' : \Omega \to \Re^k$ and $g : \Re^k \to \Re$ is measurable, then[10]

$$Eg(\mathbf{X}) = \int_\Omega g\left[\mathbf{X}(\omega)\right] \mathbb{P}(d\omega) = \int_{\Re^k} g(\mathbf{x}) \mathbb{P}_{\mathbf{X}}(d\mathbf{x}) = \int_{\Re^k} g(\mathbf{x}) \cdot dF_{\mathbf{X}}(\mathbf{x}).$$

The above extends to allow g itself to be vector valued, the notation signifying that the operation of expectation is to be carried out element by element. That is, if $\mathbf{g} : \Re^k \to \Re^m$ is measurable, then

$$E\mathbf{g}(\mathbf{X}) = \int_\Omega \mathbf{g}\left[\mathbf{X}(\omega)\right] \mathbb{P}(d\omega) = \int_{\Re^k} \mathbf{g}(\mathbf{x}) \mathbb{P}_{\mathbf{X}}(d\mathbf{x}) = \int_{\Re^k} \mathbf{g}(\mathbf{x}) \cdot dF_{\mathbf{X}}(\mathbf{x})$$

$$\equiv \left(\int g_1 \cdot dF_{\mathbf{X}}, \int g_2 \cdot dF_{\mathbf{X}}, ..., \int g_k \cdot dF_{\mathbf{X}} \right)'.$$

Fubini's theorem from analysis implies that multiple integrals such as $\int g(\mathbf{x}) \cdot dF_{\mathbf{X}}(\mathbf{x})$ can evaluated as *repeated* integrals in any order, provided that either (i) g is nonnegative on X or (ii) any such repeated integral is finite when g is replaced by $|g|$. In case (i) the value of the integral might be $+\infty$, but it would be the same regardless of the order of integration. In any case, the expectation of an *integrable* function—one for which $E\left|g(\mathbf{X})\right|$ is finite—can always be found as a repeated integral. To check that $g(\mathbf{X})$ is integrable, simply determine whether the repeated integral of $|g|$ is finite when evaluated in any one order. Thus, for a vector $\mathbf{X}$ of discrete r.v.s

$$E\mathbf{g}(\mathbf{X}) = \sum_{x_1} \cdots \sum_{x_k} \mathbf{g}(x_1, x_2, ..., x_k) f_{\mathbf{X}}(x_1, x_2, ..., x_k),$$

and the jth member is

$$Eg_j(\mathbf{X}) = \sum_{x_1} \cdots \sum_{x_k} g_j(x_1, x_2, ..., x_k) f_{\mathbf{X}}(x_1, x_2, ..., x_k),$$

where the sums run over the countably many values of $\mathbf{x} := (x_1, x_2, ..., x_k)'$ such that $f_{\mathbf{X}}(\mathbf{x}) > 0$. In the continuous case the jth member is an ordinary repeated integral with respect to Lebesgue measure:

$$Eg_j(\mathbf{X}) = \int_{-\infty}^\infty \cdots \int_{-\infty}^\infty g_j(x_1, x_2, ..., x_k) f_{\mathbf{X}}(x_1, x_2, ..., x_k) \cdot dx_1 \cdots dx_k.$$

---

[10]Although standard, the last expression below is a slight abuse of notation. Interpreting $\mathbb{P}_X(d\mathbf{x})$ loosely as the probability mass in the hypercube $\times_{j=1}^k (x_j - dx_j, x_j]$, the representation $\int_{\Re_k} g(\mathbf{x}) \cdot d^k F_X(\mathbf{x})$ would be more appropriate. In any case it is usually best to regard "$\int g \cdot dF$" as just a shorthand for an expression in terms of the induced measure, which exists more generally than does the Riemann–Stieltjes construction.

The special case with $m = k$ and $\mathbf{g}(\mathbf{X}) = \mathbf{X}$ is

$$EX = \int_\Omega \mathbf{X}(\omega)\mathbb{P}(d\omega) = \int_{\Re^k} \mathbf{x}\mathbb{P}_\mathbf{X}(d\mathbf{x}) = \int_{\Re^k} \mathbf{x} \cdot dF_\mathbf{X}(\mathbf{x}).$$

The result is a vector of constants, $(EX_1, EX_2, ..., EX_k)'$, corresponding to the "long-run averages" of the elements of $\mathbf{X}$.

When conditions for Fubini's theorem are not met, evaluating repeated sums or integrals in different orders may lead to different answers.

Example 3.17. Let X, Y be independent with $f_X(x) = \frac{1}{2}1_{\{-1,1\}}(x)$ and $f_Y(y) = 2^{-y}1_\aleph(y)$. The joint p.m.f. is $f_\mathbf{Z}(x,y) = 2^{-(y+1)}1_{\{-1,1\}\times\aleph}(x,y)$. With $g(X,Y) = X \cdot 2^Y$, evaluating $\int_{\Re^2} g(x,y) \cdot dF(x,y)$ by summing first over $\{-1,1\}$ gives

$$\sum_{y\in\aleph}\left[\sum_{x\in\{-1,1\}} x \cdot 2^y \cdot 2^{-(y+1)}\right] = \sum_{y\in\aleph}\left(\sum_{x\in\{-1,1\}}\frac{x}{2}\right) = \sum_{y\in\aleph}0 = 0,$$

while summing first over $\aleph$ gives

$$\sum_{x\in\{-1,1\}}\left[\sum_{y\in\aleph}x \cdot 2^y \cdot 2^{-(y+1)}\right] = \sum_{x\in\{-1,1\}}\left(x\sum_{y\in\aleph}2^{-1}\right)$$

$$= \sum_{x\in\{-1,1\}} x(+\infty) = (-\infty) + (+\infty),$$

which has no definite value. The condition for Fubini fails because $\sum\sum|x|\,2^y f_\mathbf{Z}(x,y) = +\infty$ for either order of summation.

Exercise 3.19. *X, Y are the numbers of heads and sequence changes in three flips of a fair coin. Find (i) $E(X+Y)$, (ii) EXY, and (iii) $E\left(\frac{X}{1+Y}\right)$.*

From the properties of the integral in Section 3.2.1.2 follow many properties of mathematical expectation that are of vital importance in applied probability and statistics. In each case we take for granted that the expectations involved in these relations do exist, and we express $\int g[X(\omega)]\mathbb{P}(d\omega)$, $\int g(x)\mathbb{P}_X(dx)$, and $\int g(x)\cdot dF(x)$ more compactly as $\int g(X)\cdot d\mathbb{P}$, $\int g\cdot d\mathbb{P}_X$, and $\int g \cdot dF$ unless arguments are essential to clarity. Since all three forms are valid expressions for $Eg(X)$, they are all used at some point below.

(i) Expectations of constants: If $X : \Omega \to a$ for some real constant a (X then being a *degenerate* r.v.), or more generally if $X = a$ almost surely, then

$$EX = \int a \cdot d\mathbb{P} = a\mathbb{P}(\Omega) = \int a \cdot d\mathbb{P}_X = a\mathbb{P}_X(\Re) = a.$$

(ii) Expectations of constants times r.v.s: If b is a constant, then

$$E\left(bX\right) = \int bX \cdot d\mathbb{P} = b \int X \cdot d\mathbb{P} = bEX.$$

(iii) Linearity of E: If (X, Y) is a bivariate r.v., then

$$
\begin{aligned}
E\left(X + Y\right) &= \int \left(X + Y\right) \cdot d\mathbb{P} \\
&= \int X \cdot d\mathbb{P} + \int Y \cdot d\mathbb{P} \\
&= EX + EY,
\end{aligned}
$$

(provided that these are not infinite with opposite sign). With points 1 and 2 this has extensive and vitally important consequences, which are usually referred to collectively as the "linearity property" of the E operator:

(a) If a, b are constants, then

$$E(a + bX) = \int (a + bX) \cdot d\mathbb{P} = a + bEX.$$

Thus, the expectation of an affine function of a r.v. equals the function of its expectation; i.e., *if $g(X)$ is affine in X, then $Eg(X) = g(EX)$.*

(b) If a_0 and $\mathbf{a} := (a_1, a_2, ..., a_k)'$ are constants and $\mathbf{X}$ is a k-vector-valued r.v., then

$$E(a_0 + \mathbf{a}'\mathbf{X}) = E\left(a_0 + \sum_{j=1}^{k} a_j X_j\right) = a_0 + \sum_{j=1}^{k} a_j EX_j = a_0 + \mathbf{a}'E\mathbf{X}.$$

Thus, *if $g\left(\mathbf{X}\right)$ is affine in $\mathbf{X}$, then $Eg(\mathbf{X}) = g(E\mathbf{X})$*

(c) If $\mathbf{h} : \Re^k \to \Re^m$, a_0 and $\mathbf{a} := (a_1, a_2, ..., a_m)'$ are constants, and $\mathbf{X}$ is a k-vector-valued r.v., then

$$E\left[a_0 + \mathbf{a}'\mathbf{h}(\mathbf{X})\right] = a_0 + \sum_{j=1}^{m} a_j Eh_j(\mathbf{X}) = a_0 + \mathbf{a}'E\mathbf{h}(\mathbf{X}).$$

Thus, *if $g\left[\mathbf{h}(\mathbf{X})\right] = g\left[h_1(\mathbf{X}), h_2(\mathbf{X}), ..., h_k(\mathbf{X})\right]$ is an affine composition of functions* (here, the h's), *then $Eg\left[\mathbf{h}(\mathbf{X})\right] = g\left[E\mathbf{h}(\mathbf{X})\right]$.*

Caveat: If in any of these cases g is not *affine* in its arguments— i.e., not of the form $a + bX$ or $a_0 + \sum_j a_j X_j$ or $a_0 + \sum_j a_j h_j(\mathbf{X})$, where $a, b, a_0, a_1, a_2, ...$ are constants—then the $E(\cdot)$ operator *does not pass through the function* g, in general. That is, to take the simplest case of a scalar function of one r.v., one cannot expect the two operations $Eg(X) = \int g\left(X\right) \cdot d\mathbb{P} = \int g \cdot dF$ and $g(EX) = g\left(\int X \cdot d\mathbb{P}\right) = g\left(\int x \cdot dF\right)$ always to yield the same result. The key points are

- *If g is an affine function of the arguments then $E(\cdot)$ does pass through, in which case Eg can be evaluated directly if the expectations of its arguments are already known.*
- If g is *not* affine, then $Eg(X)$, $Eg(\mathbf{X})$, $Eg[\mathbf{h}(\mathbf{X})]$,... must be worked out explicitly as $\int_{\Re} g(x) \cdot dF(x)$, $\int_{\Re^k} g(\mathbf{x}) \cdot dF(\mathbf{x})$, $\int_{\Re^k} g[\mathbf{h}(\mathbf{x})] \cdot dF(\mathbf{x})$ ... . (See point (x) below for the one exception to this rule.)

Example 3.18. X is the number of heads in three flips of a fair coin, and Y is the number of sequence changes. Having already calculated $EX = \frac{3}{2}$ and $EY = 1$, no more work is required to find $E(1 + 2X + 3Y) = 1 + 2EX + 3EY = 1 + 2(\frac{3}{2}) + 3(1) = 7$.

Example 3.19. X is the number of heads in three flips of a fair coin, and Y is the number of sequence changes. Having already calculated $EX^2 = 3$ and $EY^3 = \frac{5}{2}$, no more work is required to find $E(1 + 2X^2 + 3Y^3) = 1 + 2EX^2 + 3EY^2 = 1 + 2(3) + 3(\frac{5}{2}) = 14\frac{1}{2}$.

Example 3.20. X is the number of heads in three flips of a fair coin. $EX^2 = 3 \neq (EX)^2 = \frac{9}{4}$. Here $g(X) = X^2$ is not affine in X!

Example 3.21. With $f(x) = \frac{1}{2}\mathbf{1}_{\{-1,1\}}(x)$ and $g(X) = \mathbf{1}_{(-\infty,0)}(X) - \mathbf{1}_{(0,\infty)}(X)$, we have $Eg(X) = 1 \cdot \frac{1}{2} - 1 \cdot \frac{1}{2} = 0 = EX = g(EX)$, despite the fact that g is not affine. That g be affine is *sufficient* but not necessary for $Eg(X) = g(EX)$. Still, the expectation of a function that is not affine must **always** be worked out from scratch.

(iv) Integration over disjoint subsets: If $\mathcal{F}$-sets $\{A_j\}_{j=1}^{\infty}$ on probability space $(\Omega, \mathcal{F}, \mathbb{P})$ partition Ω, then

$$Eg(X) = \int_{\Omega} g(X) \cdot d\mathbb{P} = \sum_{j=1}^{\infty} \int_{A_j} g(X) \cdot d\mathbb{P}.$$

If Borel sets $\{B_j\}_{j=1}^{\infty}$ partition $\Re$, then

$$Eg(X) = \int_{\Re} g \cdot dF = \sum_{j=1}^{\infty} \int_{B_j} g \cdot dF = \sum_{j=1}^{\infty} \int_{B_j} g \cdot d\mathbf{P}_X.$$

Of course, integrability of g requires that $E|g(X)| < \infty$, which in turn requires that $\int_{A_j} |g(X)| \cdot d\mathbb{P} < \infty$ and $\int_{B_j} |g| \cdot dF < \infty$ for each $\mathcal{F}$-set A_j and each Borel set B_j.

Example 3.22. If $f(x) = \left[\pi(1 + x^2)\right]^{-1}$ (the standard Cauchy p.d.f.), then, recalling Exercise 3.3,

$$
E\,|X| = \int_{\Re} \frac{|x|}{\pi(1 + x^2)} \cdot dx
$$

$$
= \int_{(-\infty, 0)} \frac{-x}{\pi(1 + x^2)} \cdot dx + \int_{\{0\}} \frac{0}{\pi} \cdot dx + \int_{(0, \infty)} \frac{x}{\pi(1 + x^2)} \cdot dx
$$

$$
= (+\infty) + 0 + (+\infty) = \infty.
$$

The thick, quadratically vanishing tails in this distribution correspond to high probabilities of extreme values or "outliers".

(v) Absolute-value inequality: If EX exists, then

$$
E\,|X| = \int |x| \cdot dF \geq \left| \int x \cdot dF \right| = |EX|.
$$

Example 3.23. If $f(x) = \frac{1}{\sqrt{2\pi}} e^{-x^2/2}$ (the standard normal p.d.f.), then

$$
E\,|X| = \int \frac{|x|}{\sqrt{2\pi}} e^{-x^2/2} \cdot dx
$$

$$
= \int_{-\infty}^{0} \frac{-x}{\sqrt{2\pi}} e^{-x^2/2} \cdot dx + \int_{0}^{\infty} \frac{x}{\sqrt{2\pi}} e^{-x^2/2} \cdot dx
$$

$$
= \int_{0}^{\infty} \frac{y}{\sqrt{2\pi}} e^{-y^2/2} \cdot dy + \int_{0}^{\infty} \frac{x}{\sqrt{2\pi}} e^{-x^2/2} \cdot dx \quad \text{(putting } y := -x\text{)}
$$

$$
= -\frac{2}{\sqrt{2\pi}} e^{-x^2/2} \Big|_{0}^{\infty} = \sqrt{\frac{2}{\pi}}
$$

$$
> \left| \int \frac{x}{\sqrt{2\pi}} e^{-x^2/2} \cdot dx \right| = 0 = |EX|.
$$

(vi) Domination of integrals of dominated functions: If X and Y are integrable and $X(\omega) \geq Y(\omega)$ for almost all $\omega \in \Omega$ (all except those in a $\mathbb{P}$-null set), then

$$
EX = \int X \cdot d\mathbb{P} \geq \int Y \cdot d\mathbb{P} = EY.
$$

Likewise, if $g(x) \geq h(x)$ for all $x \in \mathbb{X}$ where $\mathbb{P}_X(\mathbb{X}) = 1$, and if g and h are integrable,

$$
Eg\,(X) = \int g \cdot dF \geq \int h \cdot dF = Eh(X).
$$

Example 3.24. Properties 2 and 4 imply a result that is important in connection with "moments" of r.v.s., which are treated in Section 3.4.1:

$$E\,|X|^{k} < \infty \Longrightarrow E\,|X|^{t} < \infty \text{ for } t \in [0, k].\qquad (3.13)$$

The proof is as follows:

$$
\begin{aligned}
E\,|X|^{k} &= \int_{[-1,1]} |x|^{k} \cdot dF + \int_{\Re\setminus[-1,1]} |x|^{k} \cdot dF \\
&\geq \int_{[-1,1]} 0 \cdot dF + \int_{\Re\setminus[-1,1]} |x|^{k} \cdot dF \\
&\geq \int_{\Re\setminus[-1,1]} |x|^{t} \cdot dF, t \in [0, k] \\
&= \int_{\Re} |x|^{t} \cdot dF - \int_{[-1,1]} |x|^{t} \cdot dF \\
&\geq \int_{\Re} |x|^{t} \cdot dF - 1 = E\,|X|^{t} - 1,
\end{aligned}
$$

and so $E\,|X|^{t} \leq 1 + E\,|X|^{k} < \infty$.

Example 3.25. Probabilistic arguments often simplify proofs in other areas of analysis. The following theorem justifies a particular form of the remainder term in Taylor's theorem (see Exercise 6.1 in Chapter 6): If functions g and h are continuous on $[a, b]$, and if h does not change sign on $[a, b]$, then there exists $c \in [a, b]$ such that

$$\int_{a}^{b} g(x)\,h(x) \cdot dx = g(c) \int_{a}^{b} h(x) \cdot dx.$$

A probabilistic proof runs as follows. Since h is either strictly positive or strictly negative, $f(x) := h(x) / \int_{a}^{b} h(t) \cdot dt$ is defined on $[a, b]$, is strictly positive, and integrates to unity. f is therefore the density of some r.v. X, and $\int_{a}^{b} g(x) f(x) \cdot dx = Eg(X)$. Since g is continuous (therefore bounded), there are numbers m, M such that $m \leq g(x) \leq M$ for $x \in [a, b]$. The domination property of E then implies that $m \leq Eg(X) \leq M$, so that, by continuity, there exists at least one point $c \in [a, b]$ such that

$$g(c) = Eg(X) = \frac{\int_{a}^{b} g(x)\,h(x) \cdot dx}{\int_{a}^{b} h(x) \cdot dx}.$$

(vii) Monotone convergence: Let $\{X_n\}_{n=1}^{\infty}$ be a sequence of *nonnegative* r.v.s converging a.s. *up to* the r.v. X, meaning that $X_n(\omega) \le X_{n+1}(\omega)$ for each n and $\lim_{n\to\infty} X_n(\omega) = X(\omega)$ for almost all ω. Then $\lim_{n\to\infty} EX_n = E\left(\lim_{n\to\infty} X_n\right) = EX$. Thus, if EX is finite then EX_n is finite for each n and the sequence $\{EX_n\}_{n=1}^{\infty}$ converges to it, while if either of EX or $\lim_{n\to\infty} EX_n$ is infinite so is the other.

This has two sorts of applications. Expressing EX_n for each n and taking limits give the value of EX when this is hard to find directly. Conversely, if EX is easily determined, there is no need to work out the limit.

Example 3.26. X has p.d.f. $f(x) = (1+x)^{-2}\mathbf{1}_{(0,\infty)}(x)$, and $X_n = X\mathbf{1}_{(0,n^{\varepsilon})}(X)$ for $n \in \aleph$ and some $\varepsilon > 0$. Thus, $X_n(\omega) = X(\omega)$ when $X(\omega) \in (0, n^{\varepsilon})$ and $X_n(\omega) = 0$ otherwise. Since $\mathbf{1}_{(0,n^{\varepsilon})}(x) \uparrow \mathbf{1}_{\Re}(x) \equiv 1$ for each x, it follows that $X_n(\omega) \uparrow X(\omega)$ for each ω and therefore, from the result of Exercise 3.17, that $\lim_{n\to\infty} EX_n = \infty$.

(viii) Dominated convergence: Let $\{X_n\}_{n=1}^{\infty}$ be a sequence of r.v.s converging pointwise a.s. to a r.v. X—meaning that $X_n(\omega) \to X(\omega)$ for almost all ω, not necessarily monotonically. Now let Y be a r.v. on the same probability space such that $|X_n(\omega)| \le |Y(\omega)|$ for each n and almost all ω. Then $E|Y| < \infty$ implies that $E|X| < \infty$ and that $\lim_{n\to\infty} EX_n = EX$.

Example 3.27. X has p.d.f. $f(x) = \frac{1}{\sqrt{2\pi}}e^{-x^2/2}$ and $X_n = X\mathbf{1}_{(-\sqrt{n},n^2)}(X)$ for $n \in \aleph$. Example 3.23 showed that $E|X| = \sqrt{2/\pi} < \infty$. Since $X_n \to X$ and $|X_n| \le |X|$, it follows that $EX_n \to EX = 0$.

Example 3.28. X has p.d.f. $f(x) = \frac{1}{\pi(1+x^2)}$ and $X_n = |X|/n$ for $n \in \aleph$. Then $\lim_{n\to\infty} X_n(\omega) = 0$ for almost all ω, so X_n converges pointwise to the *degenerate* r.v. Y for which $\mathbb{P}_Y(\{0\}) = 1$ and $EY = 0$. However, $E|X| = +\infty$ (Example 3.22) implies that $EX_n = n^{-1}E|X| = +\infty$ for each finite n, so $\lim_{n\to\infty} EX_n \ne EY = 0$. Here, members of sequence $\{X_n\}$ are *not* dominated by an integrable r.v. (i.e., one with finite expectation). This example also shows that the conclusion of the monotone convergence theorem need not hold for monotone *decreasing* sequences.

Example 3.29. X is a r.v. whose c.d.f. F is continuous and strictly increasing. Hence, there is a unique inverse function F^{-1} such that

$F\left[F^{-1}(y)\right] = y$ for each $y \in \Re$. Then

$$\Pr\left[F(X) \leq y\right] = \begin{cases} 0, \, y < 0 \\ F\left[F^{-1}(y)\right] = y, \, 0 \leq y < 1 \,, \\ 1, \, y \geq 1 \end{cases}$$

so that $Y := F(X)$ has a uniform distribution on $[0,1]$.[11] Thus, if

$$A_\varepsilon := \{\omega : 0 \leq F[X(\omega)] \leq 1 - \varepsilon\}$$

then $\mathbb{P}(A_\varepsilon) = 1 - \varepsilon$ for any $\varepsilon \in [0,1]$. It follows that $\mathbb{P}(\{\omega : \lim_{n\to\infty} F[X(\omega)]^n = 0\}) > 1 - \varepsilon$ for any $\varepsilon > 0$ and hence that $F(X)^n \to 0$ almost surely. Since $|F(X)^n| = F(X)^n \leq 1 = \int 1 \cdot dF$, dominated convergence implies that $EF(X)^n \to 0$ as $n \to \infty$.

Exercise 3.20. *Verify the conclusion of Example 3.29 by direct calculation of $EF(X)^n$.*

(ix) Expectations of indicator functions: When B is any Borel set, the function $1_B(X)$ is a measurable (indeed, a *simple*) function, so the Lebesgue–Stieltjes construction gives

$$\int 1_B(x) \cdot dF(x) = \int 1_B(x) \cdot d\mathbb{P}_X = 1 \cdot \mathbb{P}_X(B) = \mathbb{P}_X(B).$$

Thus, $\Pr(X \in B) = E1_B(X)$ when B is any Borel set.[12] In particular, $E1_\Re(X) = 1$, so that if a is any constant and X any r.v., then $Ea = E(a1_\Re(X)) = aE1_\Re(X) = a$, as in point 1 above.

(x) Expectations of products of independent r.v.s: When X, Y are independent r.v.s. and $g : \Re \to \Re$ and $h : \Re \to \Re$ are such that $Eg(X)$ and $Eh(Y)$ exist, then

$$E[g(X)h(Y)] = Eg(X) \cdot Eh(Y). \tag{3.14}$$

This product rule for independent r.v.s. is an exception to the usual injunction against passing E through functions that are not affine. It is valid because the joint c.d.f. of $\mathbf{Z} = (X,Y)'$ is itself the product

---

[11] The stated restrictions on F make the uniformity of $F(X)$ easier to establish, but they are not necessary. See the discussion of the probability integral transform on page 258.

[12] With X defined on space $(\Omega, \mathcal{F}, \mathbb{P})$ and $A := \{\omega : X(\omega) \in B\}$, the statement $\mathbb{P}_X(B) = E1_B(X)$ corresponds to $\mathbb{P}(A) = E1_A$, which was the motivating example in the chapter's introductory paragraph.

of functions of x and y alone when X, Y are independent; that is, $F_{\mathbf{Z}}(x, y) = F_X(x) F_Y(y)$, and so

$$
\int_{\Re^2} g(x) h(y) \cdot dF_{\mathbf{Z}}(x, y) = \int_{\Re} g(x) \left[\int_{\Re} h(y) \cdot dF_Y(y) \right] \cdot dF_X(x)
$$
$$
= \left[\int h(y) \cdot dF_Y(y) \right] \left[\int g(x) \cdot dF_X(x) \right].
$$

Extending (3.14), if $\{X_j\}_{j=1}^k$ are mutually independent and $\{g_j\}_{j=1}^k$ are such that $\{E g_j(X_j)\}_{j=1}^k$ exist, then

$$
E \prod_{j=1}^k g_j(X_j) = \prod_{j=1}^k E g_j(X_j). \tag{3.15}
$$

Extending, if $\{\mathbf{X}_j\}_{j=1}^m$ are mutually independent random vectors with $\mathbf{X}_j \in \Re^{k_j}$, and if $g_j : \Re^{k_j} \to \Re$, then

$$
E \prod_{j=1}^k g_j(\mathbf{X}_j) = \prod_{j=1}^k E g_j(\mathbf{X}_j)
$$

if the expectations on the right exist.

Exercise 3.21. *A fair pointer on a circular scale of unit circumference is to be spun n times, and $X_1, X_2, ..., X_n$ are the clockwise origin-to-pointer distances to be recorded. Find the smallest value of n such that $E \prod_{j=1}^n X_j < .01$.*

Exercise 3.22. *A fair pointer on a circular scale of unit circumference is to be spun twice, and X_1, X_2 are the clockwise origin-to-pointer distances to be recorded. Find each of the following and explain why Expressions (i) and (ii) have the same value but differ from that for (iii).*

(i) $E\left(\frac{1+X_1}{1+X_2}\right) = \int_0^1 \int_0^1 \frac{1+x_1}{1+x_2} \cdot dx_1 dx_2$

(ii) $E(1 + X_1) E\left(\frac{1}{1+X_2}\right) = \left[\int_0^1 (1 + x_1) \cdot dx_1\right] \left[\int_0^1 \frac{1}{1+x_2} \cdot dx_2\right]$

(iii) $\frac{E(1+X_1)}{E(1+X_2)} = \frac{\int_0^1 (1+x_1) \cdot dx_1}{\int_0^1 (1+x_2) \cdot dx_2}.$

3.4   Expectations as Informative Functionals

Given a r.v. X and an integrable function $g : \Re \to \Re$, one can think of the operation $E g(X) = \int g \cdot d\mathbb{P}_X = \int g \cdot dF$ in two ways. First, since it assigns a real number to g, one can think of it as a mapping from some space $\mathfrak{G}$ of (integrable) functions to the real numbers. From this viewpoint the probability law of X itself (as represented by $\mathbb{P}_X$ or F) is regarded as fixed and g could be any element of $\mathfrak{G}$. Thus, we would regard $Eg(X)$ as conveying information about g itself—namely, its long-run average value under $\mathbb{P}_X$. Alternatively, one can fix some particular g and regard the operation as a mapping from a space $\mathfrak{F}$ of *distributions*. Let us now view expectation from this latter perspective and think of the real number $Eg(X)$ as conveying some particular sort of information about the probability law of X—the nature of that information depending, of course, on g. More generally, with other classes of functions $g : \Re \times \Re \to \Re$ or $g : \Re \times \Re \to C$ (the space of complex numbers) the transformation $Eg(X, t) = \int g(x; t) \cdot dF(x) =: \gamma(t)$ becomes not just a real number but a real- or complex-valued function of a real variable. In such cases the expectation operation becomes a mapping from the space of distributions into the space of real or complex functions. We shall see that such real or complex functions can potentially convey a great deal of information about the underlying probability law.

To bring this rather abstract view of things into clearer focus, reflect that r.v. X could have one of infinitely many probability distributions over $\Re$. For instance, X could be discrete with p.m.f. $\binom{3}{x}\frac{1}{8}1_{\{0,1,2,3\}}(x)$, as for the number of heads among three coins, or X could be continuous with p.d.f. $f(x) = 1_{[0,1]}(x)$, as in the pointer experiment. Let us then think of there being a space $\mathfrak{F}$ of distributions, characterized for definiteness as a space of c.d.f.s. The pointer and coin c.d.f.s are now viewed just as particular elements of $\mathfrak{F}$. From this perspective $Eg : \mathfrak{F} \to \Re$ can indeed be considered a mapping from $\mathfrak{F}$ to the reals or from $\mathfrak{F}$ to a space of real- or complex-valued functions, depending on the nature of g. Since in any case the domain of Eg is a space of functions, the transformation Eg is termed a *functional*.

Learning the value of Eg—the point in $\Re$ or in some function space to which a particular F is mapped—may indeed communicate some information about F. Precisely what it conveys, if anything, depends on g. Clearly, trivial functions such as $g : \Re \to \{c\}$, where c is a constant, provide no information about F at all, since $Eg = \int c \cdot dF = c$ maps *all* members of $\mathfrak{F}$ to the same point of $\Re$. However, certain choices of $g : \Re \to \Re$ do provide specific information that can restrict F to a certain subspace of $\mathfrak{F}$. Indeed, appropriate choices of $g : \Re \times \Re \to \Re$ or $g : \Re \times \Re \to C$ can even pick out the single element of $\mathfrak{F}$ to which F corresponds, the mapping Eg from $\mathfrak{F}$ to the real or complex functions then being one to one. In the latter case, in which Eg *characterizes* F, it is often easier to deduce certain properties of F itself by working with Eg instead.

We shall learn about these powerful characterizations in Section 3.4.2—and apply them many, many times in later chapters; but there is much to be said first about simpler functionals $Eg : \mathfrak{F} \to \mathfrak{R}$ that convey more modest information. As we proceed through the lengthy but vitally important discussion of "moments", keep in mind the big picture that we are simply using mathematical expectation to learn specific things about distributions.

3.4.1   *Moments of Random Variables*

3.4.1.1   *Moments about the Origin*

Definition 3.1. Taking $g(X) = X^k$, the functional $\mu'_k := EX^k$, when it exists, is called the kth *moment* of X about the origin.

Thus, μ'_k is defined as $\int x^k \cdot dF$ when the integral exists. The finiteness of the kth *absolute* moment, $\nu'_k := E|X|^k = \int |x|^k \cdot dF$, guarantees that μ'_k exists and is finite. Of course, $\mu'_k = \pm\infty$ if one of $\int_{[0,\infty)} x^k \cdot dF$ and $-\int_{(-\infty,0)} x^k \cdot dF$ is finite and the other not. Note that μ'_k always exists when k is an even, positive integer, although it may be infinite. The primes distinguish these moments about the origin from the *central* moments, μ_k and ν_k, which are discussed below. Only positive, integer-valued moments are relevant unless $\Pr(X > 0) = 1$. Fractional moments cannot necessarily be represented as real numbers when F places positive mass on the negative axis, and negative integer moments do not exist if there is positive mass at the origin. Of course, $\mu'_0 \equiv 1$ always. Expression (3.13) in Example 3.24 shows that if $\nu'_k < \infty$ for some positive integer k, then all lower-order moments $\mu'_{k-1}, \mu'_{k-2}, ..., \mu'_1$ exist and are finite as well. If they do exist, odd moments of r.v.s whose distributions are symmetric about the origin are equal to zero,[13] since when k is odd the Lebesgue construction gives

$$EX^k = \int_{(0,\infty)} x^k \mathbb{P}_X(dx) - \int_{(-\infty,0)} (-x)^k \mathbb{P}_X(dx)$$

$$= \int_{(0,\infty)} x^k \mathbb{P}_X(dx) - \int_{(0,\infty)} x^k \mathbb{P}_X(dx).$$

---

[13]Recall that a distribution is symmetric about the origin if and only if $\mathbb{P}_X((-\infty,-x)) = \mathbb{P}_X((x,\infty))$ or, equivalently, $F(-x-) = 1 - F(x)$, for each real x. If X is purely discrete this implies $f(x) = f(-x)$ for each $x \in \mathbb{X}$, and if X is purely continuous it implies that there is a version of the density such that $f(x) = f(-x)$ a.e. with respect to Lebesgue measure.

Because it appears so often in quantitative expressions, the *first* moment about the origin gets a special symbol, $\mu := EX$, and a special name, the *mean*. The name applies because μ is in fact the arithmetic mean or average of the values X would take on in indefinitely many replications of the chance experiment. It is common also, although somewhat imprecise, to refer to the mean or other moment of a *distribution* rather than to the mean or other moment of the r.v. that has that distribution. In the purely discrete and continuous cases $\mu = \sum_{x \in \mathbb{X}} x f(x)$ and $\mu = \int_{-\infty}^{\infty} x f(x) \cdot dx$, respectively, and there is also the general integral representation $\mu = \int x f(x) \mathcal{M}(dx)$ as in (3.12).

Example 3.30. Earlier examples have shown that

(i) The mean number of heads in three flips of a fair coin is $\mu = \sum_{x=0}^{3} x \binom{3}{x} \frac{1}{8} = \frac{3}{2}$.
(ii) The mean clockwise distance of pointer from origin on the unit circular scale is $\mu = \int_0^1 x \cdot dx = \frac{1}{2}$.
(iii) The mean number of flips of a fair coin until the first occurrence of heads is $\mu = \sum_{x=1}^{\infty} x 2^{-x} = 2$.
(iv) $E \max \{0, X\} = +\infty$ and $E \min \{0, X\} = -\infty$ if X has (Cauchy) p.d.f. $f(x) = \pi^{-1}(1 + x^2)^{-1}$, so that $\mu = EX$ does not exist in this case.

When it does exist and is finite, the mean has the geometric interpretation as the *center of mass* of the probability distribution.[14] Plotting p.m.f. or p.d.f. $f(x)$ against x and imagining a uniform physical structure with the corresponding profile, the mean would be located at the apparent point of balance. Such a balance point for an hypothetical distribution is depicted in Fig. 3.4, along with the *modal* value m and the median $x_{.5}$ (discussed below). Recalling the introductory discussion, knowing the value of Eg with $g(X) = X$ just identifies the (infinite) subfamily of distributions in $\mathfrak{F}$ that have the same center of mass.

Exercise 3.23. *For each of Examples 3.30 (i)–(iii) give an example of another r.v. with the same mean, μ.*

---

[14] If n symmetrical physical objects of uniform density and masses $\{m_j\}_{j=1}^n$ are arranged with their centers at vector positions $\{\mathbf{x}_j\}_{j=1}^n$ in some coordinate system, then the *center of mass* of the collection is located at $\mu_n := \sum_{j=1}^n f_j \mathbf{x}_j$, where $f_i := m_i / \sum_{j=1}^n m_j$ is object i's share of the total mass. By progressively subdividing a single physical structure into vanishingly small units, one finds that $\mu_n \to \int \mathbf{x} f(\mathbf{x}) \cdot d\mathbf{x}$, where $f(\mathbf{x})$ is the limiting ratio of mass to volume at $\mathbf{x}$—i.e., the *density*. The probabilist's concept of center of mass of a distribution thus corresponds precisely to that of the physicist. For a wonderfully clear explanation of the latter see Feynman *et al.* (1963, Ch. 19).

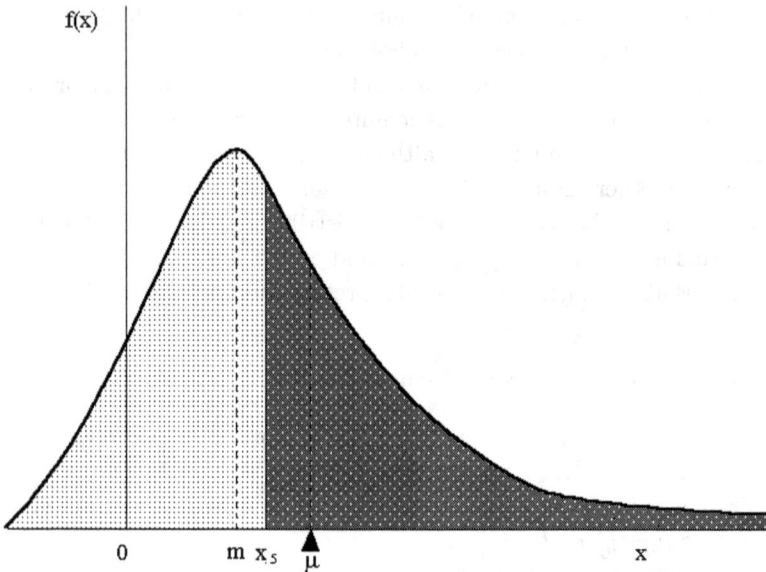

Fig. 3.4   A density function with mode m, median $x_{.5}$, and mean μ.

Exercise 3.24. *Present another continuous distribution like that in Example 3.30 (iv) for which $E\left|X\right| = +\infty$.*

The mean of a r.v. with finite second moment has the important property of being the "best" predictor of the realization of X, in the particular sense that it minimizes the mathematical expectation of the squared prediction error. That is, if η is any constant used to "predict" realization $X\left(\omega\right)$, the *mean squared error* $E\left(X - \eta\right)^2$ is at least as large as $E\left(X - \mu\right)^2$. To see this, suppose $EX^2 < \infty$, and let η, a known constant, be some putative predictor. Then

$$
\begin{aligned}
E0\left(X - \eta\right)^2 &= E\left[\left(X - \mu\right) + \left(\mu - \eta\right)\right]^2 \\
&= E(X - \mu)^2 + 2\left(\mu - \eta\right)E\left(X - \mu\right) + \left(\mu - \eta\right)^2 \\
&= E(X - \mu)^2 + \left(\mu - \eta\right)^2,
\end{aligned}
$$

and this attains its minimum value when $\eta = \mu$. Clearly, what makes this true is that μ is the unique constant such that $E\left(X - \mu\right) = 0$. While a useful result, one must bear in mind that μ's status as "best" predictor depends crucially on the arbitrary—but widely applied—mean-squared-error criterion. We shall see directly that μ is not necessarily the minimizer of $E|X - \eta|$; nor is it generally the minimizer of $E|X - \eta|^k$ for any value of k other than 2.

Exercise 3.25. *If X has p.m.f.* $f(x) = \frac{1}{3}1_{\{-2\}}(x) + \frac{2}{3}1_{\{1\}}(x)$, *show that the mean* quartic *error, $E(X - \eta)^4$, is not* minimized *at $\eta = \mu = 0$. (Hint: Express $E(X - \eta)^4$ as a polynomial in η and show that*
$$dE(X - \eta)^4 / d\eta \Big|_{\eta=0} < 0.)$$

The relation between the mean and the median is fundamental. The median of a r.v. X (or of its distribution) was defined (Definition 2.11) as the .5 quantile—a number $x_{.5}$ such that $\Pr(X \geq x_{.5}) \geq .5$ and $\Pr(X \leq x_{.5}) \geq .5$. We have seen that the median is not necessarily unique unless the c.d.f. is strictly increasing, but the common convention in such cases is to designate as "the" median the average of $x_{.5}^+ := \sup \mathfrak{M}$ and $x_{.5}^- := \inf \mathfrak{M}$, where $\mathfrak{M} := \{x : \Pr(X \geq x) \geq .5, \Pr(X \leq x) \geq .5\}$. For example, if $f(x) = \frac{1}{2}1_{\{-1,1\}}(x)$, then $\Pr(X \geq x) \geq .5$ and $\Pr(X \leq x) \geq .5$ for each $x \in \mathfrak{M} = [-1, 1]$, but a specific value can be assigned as $(x_{.5}^+ + x_{.5}^-)/2 = (1 - 1)/2 = 0$.

The median is truly unique for the continuous distribution depicted in Fig. 3.4, since there is no central interval without probability mass. In general, shifting mass among locations above $x_{.5}^+$ or among locations below $x_{.5}^-$ does not alter $x_{.5}$, but such shifts do change the balance point μ. Thus, "right-skewed" unimodal distributions—those with longer right tails, as in the figure—have $\mu > x_{.5}$, while "left-skewed" unimodal distributions have $\mu < x_{.5}$. From cases (i)–(iii) of Example 3.30 it may appear that μ always coincides with $x_{.5}$ whenever the probability mass is distributed symmetrically about $x_{.5}$. This is true provided that μ exists, but case (iv) shows the need for this condition. There the distribution as a whole has no unique center of mass because the center of mass to the right of any $x^* \in \Re$ is $+\infty$ while that to the left of x^* is $-\infty$; i.e.,

$$[1 - F(x^*)]^{-1} \int_{x^*}^{\infty} \frac{x}{\pi(1 + x^2)} \cdot dx = +\infty,$$

$$F(x^*)^{-1} \int_{-\infty}^{x^*} \frac{x}{\pi(1 + x^2)} \cdot dx = -\infty.$$

Of course, no definite value can be assigned to $+\infty + (-\infty)$.

Exercise 3.26. *Find the set of numbers that qualify as medians of the distribution in case (iii) of Example 3.30 and show that $\left(x_{.5}^{+} + x_{.5}^{-}\right)/2 < \mu$.*

Exercise 3.27. *If a distribution places positive probability mass on an interval $(x^{*}, x^{**}]$, where $-\infty \leq x^{*} < x^{**} \leq \infty$, show that its center of mass on this interval is*

$$[F(x^{**}) - F(x^{*})]^{-1} \int_{(x^{*}, x^{**}]} x \cdot dF.$$

While $EX = \int x \cdot dF$ represents μ as the Stieltjes integral of x with respect to F, it is possible also to calculate the mean of any r.v.—discrete, continuous, or mixed—by integrating F and/or $1 - F$ with respect to Lebesgue measure. We demonstrate this for nonnegative r.v.s and leave the general case as an exercise. Assuming now that $F(0-) = \Pr(X < 0) = 0$, we have, for $n > 0$,

$$
\begin{aligned}
\int_{0}^{n} [1 - F(x)] \cdot dx &= n - \int_{0}^{n} F(x) \cdot dx \\
&= n - \left[xF(x)\big|_{0}^{n} - \int_{0}^{n} x \cdot dF(x) \right] \\
&= n - \left[nF(n) - \int_{0}^{n} x \cdot dF(x) \right] \\
&= n - \int_{0}^{n} (n - x) \cdot dF(x) \\
&= n - E \max\{n - X, 0\} \\
&= E \min\{n, X\}.
\end{aligned}
$$

Since $\min\{n, X\}$ converges up to X as $n \to \infty$, the monotone convergence theorem gives

$$EX = \lim_{n \to \infty} E \min\{n, X\} = \int_{0}^{\infty} [1 - F(x)] \cdot dx = \int_{0}^{\infty} \Pr(X > x) \cdot dx \tag{3.16}$$

when r.v. X is almost surely nonnegative. Note that this not a Stieltjes integral but merely an integral with respect to Lebesgue measure. Since the set $\{x : 0 < F(x) - F(x-)\}$ is countable, it is also true that

$$EX = \int_{0}^{\infty} [1 - F(x-)] \cdot dx = \int_{0}^{\infty} \Pr(X \geq x) \cdot dx. \tag{3.17}$$

In geometric terms these results indicate that the mean of a nonnegative r.v. can be visualized as the area between the c.d.f. and the line

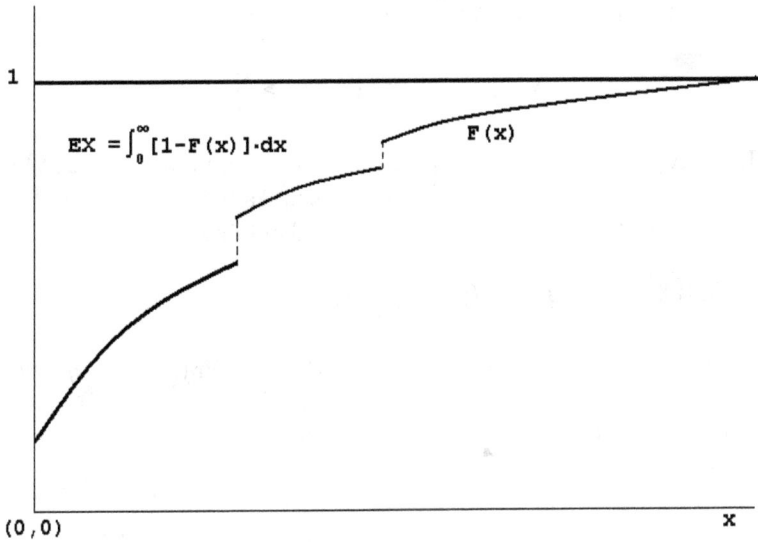

$$EX = \int_0^\infty [1 - F(x)] \cdot dx$$

$F(x)$

(0,0)   x

Fig. 3.5   Mean of nonnegative r.v. X from its c.d.f.

$g(x) = 1$, as in Fig. 3.5. If $F(x)$ approaches unity slowly enough that $\lim_{t \to \infty} \int_0^t [1 - F(x)] \cdot dx = +\infty$, then $EX = +\infty$ also.

Two other useful results follow from relation (3.17). The first provides a pair of bounds for the expected value of a nonnegative r.v., and the second establishes an important property of a median of *any* r.v.

(i) If $\Pr(X \geq 0) = 1$,

$$\sum_{n=1}^{\infty} \Pr(X > n) \leq EX \leq 1 + \sum_{n=1}^{\infty} \Pr(X > n). \qquad (3.18)$$

The proof follows from (3.17) and the identities

$$\int_0^\infty \Pr(X > x) \cdot dx = \sum_{n=0}^{\infty} \int_n^{n+1} [1 - F(x)] \cdot dx = \sum_{n=1}^{\infty} \int_{n-1}^n [1 - F(x)] \cdot dx.$$

For $x \in [n, n+1]$ we have $1 - F(x) \leq 1 - F(n)$, and so

$$EX \leq \sum_{n=0}^{\infty} \int_n^{n+1} [1 - F(n)] \cdot dx = \sum_{n=0}^{\infty} [1 - F(n)] \leq 1 + \sum_{n=1}^{\infty} \Pr(X > n).$$

Similarly, since $1 - F(x) \geq 1 - F(n)$ for $x \in [n-1, n]$, we have

$$EX \geq \sum_{n=1}^{\infty} \int_{n-1}^{n} [1 - F(x)] \cdot dx \geq \sum_{n=1}^{\infty} [1 - F(n)].$$

(ii) If $E|X| < \infty$, $E|X - \eta|$ attains its minimum value at $\eta = x_{.5}$, where $x_{.5}$ is any median; i.e., any element of $\mathfrak{M} = \{x : \Pr(X \leq x) \geq .5, \Pr(X \geq x) \geq .5\}$. Applying (3.17),

$$E|X - \eta| = \int_0^{\infty} \Pr(|X - \eta| \geq t) \cdot dt$$
$$= \int_0^{\infty} \Pr(X \leq t - \eta) \cdot dt + \int_0^{\infty} \Pr(X \geq t + \eta) \cdot dt.$$

Changing variables ($x \to t - \eta$ in the first integral, $x \to t + \eta$ in the second) gives

$$E|X - \eta| = \int_{-\infty}^{\eta} \Pr(X \leq x) \cdot dx + \int_{\eta}^{\infty} \Pr(X \geq x) \cdot dx$$
$$= \int_{-\infty}^{x_{.5}} \Pr(X \leq x) \cdot dx + \int_{x_{.5}}^{\infty} \Pr(X \geq x) \cdot dx + \Delta(\eta),$$

where

$$\Delta(\eta) := \int_{x_{.5}}^{\eta} [\Pr(X \leq x) - \Pr(X \geq x)] \cdot dx \qquad (3.19)$$

and $x_{.5}$ is any element of $\mathfrak{M}$. With $x_{.5}^- = \inf \mathfrak{M}$ and $x_{.5}^+ = \sup \mathfrak{M}$, the results of Exercises 2.7 and 2.8 imply $\int_{\mathfrak{M}} [\Pr(X \leq x) - \Pr(X \geq x)] \cdot dx = 0$, while $\int_{x_{.5}}^{\eta} [\Pr(X \leq x) - \Pr(X \geq x)] \cdot dx > 0$ if either $\eta < x_{.5}^-$ or $\eta > x_{.5}^+$. Accordingly, $E|X - \eta|$ attains its minimum at any $\eta \in \mathfrak{M}$.

Exercise 3.28. *In the special case that r.v. X is absolutely continuous show that the minimizer of $E|X - \eta|$ is a number $x_{.5}$ such that $\Pr(X \leq x_{.5}) = \Pr(X \geq x_{.5}) = .5$. (Hint: Differentiate $\Delta(\eta)$ in (3.19) and equate to zero.)*

Exercise 3.29. *Let $\mathfrak{F}$ be the space of all possible c.d.f.s of a r.v. X, and let $\mathfrak{F}_\mu \subset \mathfrak{F}$ be the collection of all c.d.f.s of X for which EX exists and is finite. For $F \in \mathfrak{F}$ put $p_F(x_{.5}) := F(x_{.5}-) = \Pr(X < x_{.5})$, where $x_{.5}$ is a median. For $\mathfrak{F} \in \mathfrak{F}_\mu$ put $p_F(\mu) := F(\mu-)$. Determine the values of (i) $\sup_{F \in \mathfrak{F}} \{p_F(x_{.5})\}$ and (ii) $\sup_{F \in \mathfrak{F}_\mu} \{p_F(\mu)\}$. In each case prove whether or not there exists a distribution in the class for which the supremum is actually attained.*

Exercise 3.30. *Extending (3.16), show that* $EX^k = \int_0^\infty kx^{k-1}\left[1 - F(x)\right] \cdot$ *dx for $k \geq 1$ if* $\Pr\left(X \geq 0\right) = 1$.

Exercise 3.31. *Extending (3.16), show that if X is any r.v. (not necessarily nonnegative) such that $E\left|X\right| < \infty$ then*

$$EX = -\int_{-\infty}^0 F(x) \cdot dx + \int_0^\infty \left[1 - F(x)\right] \cdot dx. \qquad (3.20)$$

Exercise 3.32. *Find EX for the mixed r.v. in Example 3.14 by*

(i) Evaluating $EX = \int x \cdot dF$ as a Stieltjes integral.
(ii) Applying (3.16).

Exercise 3.33. *Verify that the function $F(x) = (1 - x^{-1})\mathbf{1}_{[1,\infty)}(x)$ is a c.d.f., and determine whether a r.v. with this distribution possesses a finite mean.*

The kth moment of $Y = a + bX$, an affine transformation of X, can be expressed in terms of moments of X up through order k, assuming that these are finite.[15] This is true because the binomial formula,

$$(c + d)^k = \sum_{j=0}^k \binom{k}{j} c^j d^{k-j} \text{ for } k = 0, 1, 2, ...,$$

and the linearity property of $E\left(\cdot\right)$ imply

$$E\left(a + bX\right)^k = E\left[\sum_{j=0}^k \binom{k}{j} a^j b^{k-j} X^{k-j}\right] = \sum_{j=0}^k \binom{k}{j} a^j b^{k-j} EX^{k-j}$$

$$= \sum_{j=0}^k \binom{k}{j} a^j b^{k-j} \mu'_{k-j}, \qquad (3.21)$$

where $\mu'_0 := EX^0 = 1$. For example, for $k = 1$

$$E(a + bX) = a + bEX$$

$$= \sum_{j=0}^1 \binom{1}{j} a^j b^{1-j} \mu'_{1-j}$$

$$= a + b\mu,$$

---

[15]If they were *not* all finite, some of the terms in expression (3.21) could be infinite but of opposite sign.

and for $k = 2$

$$E(a + bX)^2 = E(b^2 X^2 + 2abX + a^2)$$

$$= \sum_{j=0}^{2} \binom{2}{j} a^j b^{k-j} \mu'_{k-j}$$

$$= b^2 \mu'_2 + 2ab\mu + a^2.$$

3.4.1.2 Central Moments

Definition 3.2. If r.v. X has finite mean μ, the functional $\mu_k :=$ $E\left(X - \mu\right)^k$, when it exists, is called the kth *central moment* of X or the kth moment *about the mean*.

Thus, μ_k is defined as the Lebesgue integral $\int (x - \mu)^k \cdot dF$ *provided* (i) that μ exists and is finite and (ii) that the integral exists in the Lebesgue sense. The integral does exist if corresponding central *absolute* moment $\nu_k := \int |x - \mu|^k \cdot dF$ takes a finite value. Clearly, only integer-valued central moments with $k \geq 2$ are relevant. (Why?) Applying (3.21) with $a = -\mu$ and $b = 1$ gives μ_k in terms of moments about the origin of order up to k when these are finite:

$$\mu_k := E\left(X - \mu\right)^k = \sum_{j=0}^{k} \binom{k}{j} (-\mu)^j \mu'_{k-j}. \tag{3.22}$$

Since all of $\left\{\mu'_j\right\}_{j=1}^{k}$ exist and are finite if absolute moment $\nu'_k < \infty$, this suffices also for the finiteness of μ_k.

The importance of *second* central moment $\mu_2 = E(X - \mu)^2$ warrants giving it a special symbol, σ^2, and a special name, the *variance* of X. With our usual interpretation of $E\left(\cdot\right)$, σ^2 represents the "long-run average" of the squared deviations of X from its mean; hence, the variance of X conveys a sense of the *dispersion*, or spread, of its distribution. The notations VX and $Vg(X)$ are often used for the variance of X and of some function of X. Of course, VX can be calculated as $\sum_{x \in \mathbb{X}} (x-\mu)^2 f(x)$ and as $\int_{-\infty}^{\infty} (x-\mu)^2 f(x) \cdot dx$ in the purely discrete and continuous cases, respectively. Applying (3.22) with $k = 2$ gives the alternative formula

$$\sigma^2 \equiv VX = EX^2 - \mu^2 = \mu'_2 - \mu^2. \tag{3.23}$$

This is usually easier to use in numerical computations. It leads trivially to another useful formula:

$$EX^2 = \sigma^2 + \mu^2 = VX + (EX)^2. \tag{3.24}$$

In words, if r.v. X has a mean, then the expected value of its square (the second moment about the origin) equals its variance plus the square of its mean. Note that this implies (i) that $EX^2 \geq \max\{\sigma^2, \mu^2\}$ and (ii) that $EX^2 \geq (EX)^2$, with equality if and only if $\mathbb{P}_X(\{c\}) = 1$ for some real constant c.

Example 3.31.

(i) The variance of the number of heads in three flips of a fair coin is
$\sigma^2 = \sum_{x=0}^3 \left(x - \frac{3}{2}\right)^2 \binom{3}{x}\frac{1}{8} = \sum_{x=0}^3 x^2 \binom{3}{x}\frac{1}{8} - \left(\frac{3}{2}\right)^2 = \frac{3}{4}$.

(ii) The variance of the clockwise distance of pointer from origin on the unit circular scale is $\sigma^2 = \int_0^1 \left(x - \frac{1}{2}\right)^2 \cdot dx = \int_0^1 x^2 \cdot dx - \left(\frac{1}{2}\right)^2 = \frac{1}{12}$.

(iii) The variance of the number of flips of a fair coin until the first occurrence of heads is $\sigma^2 = \sum_{x=1}^\infty (x-2)^2 2^{-x} = \sum_{x=1}^\infty x^2 2^{-x} - 2^2 = 2$.

(iv) If X has p.d.f. $f(x) = 2x^{-3} 1_{[1,\infty)}(x)$ then $\mu = 2$ and $\mu_2' = +\infty$, and so $VX = +\infty$ also.

(v) If X has p.d.f. $f(x) = \pi^{-1}(1+x)^{-2}$ then $\mu_2' = +\infty$ but VX is undefined (does not exist) since $\mu \equiv \int_{-\infty}^\infty xf(x) \cdot dx$ does not exist in the Lebesgue sense. (Although the second moment about the *origin* is well defined, one cannot specify moments about a *mean* if there is no mean.)

Exercise 3.34. *Extend the trick used in Example 3.13 to verify that* $VX = 2$ *when* X *is the number of flips of a fair coin until the first occurrence of heads. (Hint: Express* $\sum_x x^2 z^x$ *as* $\sum_x x(x-1)z^x + \sum_x xz^x$.)

While the mean of a r.v. is measured in the same units as the variable itself, the variance is measured in squared units. A related quantity that is commensurate with μ is the *standard deviation*, σ, which is the positive square root of the variance: $\sigma := +\sqrt{\sigma^2} = +\sqrt{E(X - \mu)^2}$. Clearly, σ exists and is finite if and only if the same is true of σ^2. When it does exist, σ is the quantity most often used to indicate the dispersion of a probability distribution. Visually, if one plots $f(x)$ against x and imagines a physical structure with this profile, the standard deviation (if finite) would increase (decrease) as the distribution was spread out (concentrated) about its mean. Thus, recalling the introductory discussion on page 160, knowing the value of Eg with $g(X) = (X - \mu)^2$ identifies the (infinite) subfamily of distributions in $\mathfrak{F}$ that have the same degree of dispersion, as measured in this customary way.

Exercise 3.35. Gini's mean difference *is an alternative measure of dispersion that is commensurate with the mean and that can be finite even when* *the variance is not. This is defined as*

$$\mathfrak{g} := E\,|X - Y|, \tag{3.25}$$

where X and Y are i.i.d. r.v.s. (For example, realizations of X and Y might be generated by independent draws with replacement from some population.) If X and Y have mean μ and continuous c.d.f. F with $F(a) = 0$ for some finite a, show that

$$\frac{\mathfrak{g}}{2} = \mu - a - \int_a^\infty [1 - F(x)]^2 \cdot dx.$$

(Hint: Show that $E\,|X - Y|$ can be expressed as $2 \int_a^\infty \left[\int_a^x (x - y) \cdot dF(y) \right] \cdot dF(x)$, then do some integrating by parts.)

If a new r.v. Y is constructed as $a + bX$, we have seen that $\mu_Y = a + b\mu_X$. We can also express σ_Y^2 in terms of σ_X^2:

$$\begin{aligned}
\sigma_Y^2 &= E\,(Y - \mu_Y)^2 = E\,[(a + bX) - E(a + b\mu_X)]^2 \tag{3.26} \\
&= b^2 E(X - \mu_X)^2 \\
&= b^2 \sigma_X^2.
\end{aligned}$$

The standard deviations are then related as $\sigma_Y = +\sqrt{b^2 \sigma_X^2} = |b|\,\sigma_X$. Thus, if $|b| > 1$ the distribution of Y is more dispersed than that of X, while if $|b| < 1$ it is more concentrated about its mean. Notice that σ_Y does not depend on a, since this merely shifts the distribution of X but does not change its dispersion. In general, such affine transformations change the kth-order central moment by the factor b^k.

Here is an important special case. Taking $a = -\mu/\sigma$ and $b = 1/\sigma$ gives $Y = (X - \mu)/\sigma$, a r.v. with zero mean and unit variance. Such a transformation is said to put X into *standard form* or to "standardize" X. We will see why standardization is often useful when we take up the normal distribution in Section 4.3.6.

Two of the higher central moments are also of some value in describing distributions. The sign of third central moment $\mu_3 = E(X - \mu)^3$ can indicate the direction of skewness. Typically, unimodal distributions with longer right tails, such as that depicted in Fig. 3.4, have $\mu_3 > 0$, while left-skewed distributions have $\mu_3 < 0$. Of course, $\mu_3 = 0$ for distributions that are symmetric about the mean and for which $\nu_3 = E\,|X - \mu|^3 < \infty$. The dimensionless, mean/variance-invariant quantity

$$\alpha_3 := \frac{\mu_3}{\sigma^3}, \tag{3.27}$$

known as the *coefficient of skewness*, is a simple units-free indicator of asymmetry, whose sign matches that of μ_3. Notice that $\alpha_3 = EY^3$, where Y is X in standard form. Of course, α_3 is defined only if $0 < \sigma^2 < \infty$.

Under the same condition on σ^2 one can construct the standardized *fourth* central moment,

$$\alpha_4 := \frac{\mu_4}{\sigma^4} = EY^4, \tag{3.28}$$

which is called the *coefficient of kurtosis*.[16] Among distributions having the same variance, those with higher kurtosis typically have more probability mass in the tails and therefore a higher relative frequency of *outliers*— observations far from the mean in units of σ. In particular, symmetric distributions with bounded support necessarily have $\alpha_4 < 3$, while symmetric densities whose tails vanish more slowly than in proportion to $e^{-x^2/2}$ as $|x| \to \infty$ have $\alpha_4 > 3$. Empirically, many economic variables appear to have such "leptokurtic" distributions; e.g., logarithms of annual incomes of a population of individuals and daily changes in market prices (or logs of prices) of common stocks, foreign currencies, and futures contracts. As shown in Section 4.3.6, r.v.s. in the normal family all have $\alpha_4 = 3$, and so the "excess" kurtosis, $\alpha_4 - 3$, is often considered to indicate the frequency of outliers relative to what is "normal". However, one must be cautious in interpreting α_4 for multimodal or asymmetric distributions. There is, after all, no way to convey all aspects of a distribution's shape with one or two numbers. Figure 3.6 depicts the classic form of a symmetric, unimodal, leptokurtic distribution. This *logistic* distribution, which is described further in Section 9.2.5, has $\alpha_4 = 4.2$. The version plotted here as the solid curve is parameterized to have $\mu = 0$ and $\sigma^2 = 1$ for comparison with the standard normal, shown as the dashed line. The relative "peakedness" of the curve—the higher density around the mean—is characteristic of such distributions, as is the lower density between one and two standard deviations from the mean. (Clearlly, since areas under the two curves are the same, higher density in some regions must be offset by lower density elsewhere.) Separate plots of the logistic and normal upper tails (scaled to the right axis) show more clearly that the logistic really does have more outliers.

Exercise 3.36. *Show that α_4, when it exists, is strictly greater than unity.*

---

[16]Note that $\mu_4 = E(X - \mu)^4$ always *exists* so long as the mean is finite, although it may equal $+\infty$. Accordingly, α_4 exists for any r.v. with positive, finite variance.

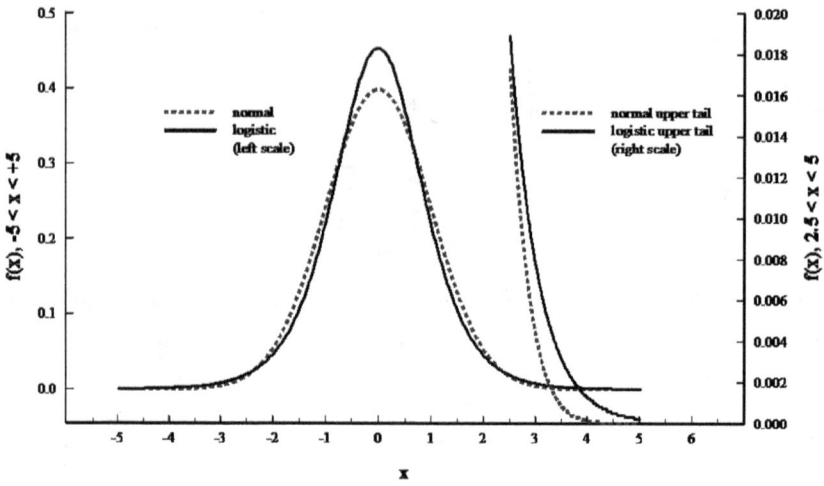

Fig. 3.6   P.d.f.s of normal and logistic r.v.s, both in standard form.

3.4.1.3   *Three Inequalities Involving Moments*

Several important analytic inequalities involve moments. We shall see that these are useful both for the development of theory and for applications in modeling real-world phenomena. In what follows the existence of the various moments and other expectations is implicitly assumed.

Chebychev's Inequality   This inequality, of which a generalized version is given here, has several alternative forms. It will be used in Chapter 6 to prove certain laws of large numbers.

Proposition 3.1. *If (i) X is a r.v. with support $\mathbb{X}$, (ii) $g : \mathbb{X} \to \Re_{+}$ is a nonnegative function, and (iii) c is any positive number, then*

$$\Pr\left[g(X) \geq c\right] \leq \frac{Eg(X)}{c}. \tag{3.29}$$

Proof. Let $\mathbb{X}_{[c,\infty)} = \{x : g(x) \geq c\} \cap \mathbb{X}$, and $\mathbb{X}_{[0,c)} = \{x : g(x) < c\} \cap \mathbb{X}$. These are, respectively, the (disjoint) subsets of $\mathbb{X}$ on which g is greater than or equal to c and on which g is less than c. Writing $Eg(X)$ as

$$\int_{\mathbb{X}} g(x)\,\mathbb{P}_X(dx) = \int_{\mathbb{X}_{[c,\infty)}} g(x)\,\mathbb{P}_X(dx) + \int_{\mathbb{X}_{[0,c)}} g(x)\,\mathbb{P}_X(dx)$$

and recalling that g is nonnegative, we have

$$Eg(X) \geq \int_{\mathbb{X}_{[c,\infty)}} g(x) \, \mathbb{P}_X(dx) \geq c \int_{\mathbb{X}_{[c,\infty)}} \mathbb{P}_X(dx)$$
$$= c\mathbb{P}_X\left(\mathbb{X}_{[c,\infty)}\right) = c\Pr\left[g(X) \geq c\right]. \qquad \square$$

Here are two useful special cases. The first is the usual form in which Chebychev's inequality is presented; the second is called *Markov's inequality*.

Corollary 3.1. *If* $k > 0$

$$\Pr\left[|X - \mu| \geq k\right] \leq \frac{\sigma^2}{k^2} \qquad (3.30)$$

$$\Pr\left[|X| \geq k\right] \leq \frac{E|X|^m}{k^m}, m \geq 0. \qquad (3.31)$$

Proof. For (3.30) put $g(X) = (X - \mu)^2$ and $c = k^2$. (If $(X - \mu)^2$ is not integrable with respect to $\mathbb{P}_X$, take the right side to be $+\infty$.) For (3.31) put $g(X) = |X|^m$ and $c = k^m$. $\qquad \square$

Inequality (3.30) is obviously not useful if $k \leq \sigma$, and even when $k \gg \sigma$ the bound is often highly conservative; nevertheless, it is the best bound that applies simultaneously to *all* distributions and *all* positive values of k.

Exercise 3.37. *Show that (3.30) holds as an equality when* $k = 1$ *and* $f(x) = \frac{1}{2}1_{\{-1,1\}}(x)$.

Jensen's Inequality This inequality, important both as a theoretical tool and in applied disciplines, pertains to expectations of *convex functions* of r.v.s.

Definition 3.3. A real-valued function g is convex on an interval $\mathbb{I}$ if, for each $p \in [0,1]$ and each $x', x'' \in \mathbb{I}$,

$$g\left[(1-p)x' + px''\right] \leq (1-p)g(x') + pg(x''). \qquad (3.32)$$

g is said to be *strictly* convex if this holds as a strict inequality, and g is *concave* if the inequality is reversed.

A convex (concave) function is strictly convex (concave) unless it is linear over some interval. The defining condition for convexity has several implications:[17] (i) Convex functions are continuous and have left and right derivatives $g'(x-)$ and $g'(x+)$ at each point; (ii) if g is convex and differentiable

---

[17] See Royden (1968, pp. 108-110) for proofs.

everywhere (left and right derivatives being equal) then $g'(x') \leq g'(x'')$ for $x' \leq x''$; and (iii) if g is convex and twice differentiable, then $g'' \geq 0$. Condition (3.32) has a simple geometric interpretation, which is easier to see by putting $p := (x - x') / (x'' - x')$ and rewriting as

$$g(x) \leq g(x') + \frac{g(x'') - g(x')}{x'' - x'}(x - x') \text{ for } x \text{ between } x' \text{ and } x''. \quad (3.33)$$

The interpretation is that a straight line connecting any two points on the graph of a convex function lies on or above the function at all points in between. On the other hand, the inequality is reversed if we extend the line and consider points x in $\mathbb{I}$ that are *not* between x' and x''; i.e.,[18]

$$g(x) \geq g(x') + \frac{g(x'') - g(x')}{x'' - x'}(x - x') \text{ elsewhere in } \mathbb{I}. \quad (3.34)$$

Now let x'' approach x' so as to bring the points together. Setting $x'' = x' + h > x'$ in (3.34) and letting $h \downarrow 0$ give $g(x) \geq g(x') + g'(x'+)(x - x')$, while setting $x'' = x' + h < x'$ and letting $h \uparrow 0$ give $g(x) \geq g(x') + g'(x'-)(x - x')$. Thus, for each $x' \in \mathbb{I}$ there are numbers $a(x') \equiv a, b(x') \equiv b$—not necessarily unique—such that $g(x') = a + bx'$ and $g(x) \geq a + bx$ for all $x \in \mathbb{I}$. This means that at any point on a convex function a straight line can be drawn that just touches the function there and lies on or below the function elsewhere in $\mathbb{I}$. The tangent line is the unique such linear "support" at points where g is differentiable, but there are infinitely many support lines where left and right derivatives are unequal. See Fig. 3.7. We are now ready to state and prove Jensen's inequality.

Proposition 3.2. *If r.v. X has support $\mathbb{X}$ and mean μ, and if g is a convex function on an interval $\mathbb{I}$ containing $\mathbb{X}$, then*

$$Eg(X) \geq g(\mu). \quad (3.35)$$

Proof. Constructing a support line at $x = \mu$, choose a, b such that $g(\mu) = a + b\mu$ and $g(x) \geq a + bx$ for all $x \in \mathbb{I}$ (and therefore all $x \in \mathbb{X}$). Then by the dominance and linearity properties of the integral

$$Eg(X) \geq \int (a + bx) \, \mathbb{P}_X(dx) = a + b\mu = g(\mu). \qquad \square$$

---

[18]Suppose, to the contrary, that (3.34) failed to hold for some x^* not between x' and x''. Pick an x between x' and x'' and connect $(x, g(x))$ and $(x^*, g(x^*))$ with a straight line. Such a line must lie below the function somewhere in between, contradicting the hypothesis that g is convex on $\mathbb{I}$.

g(x)

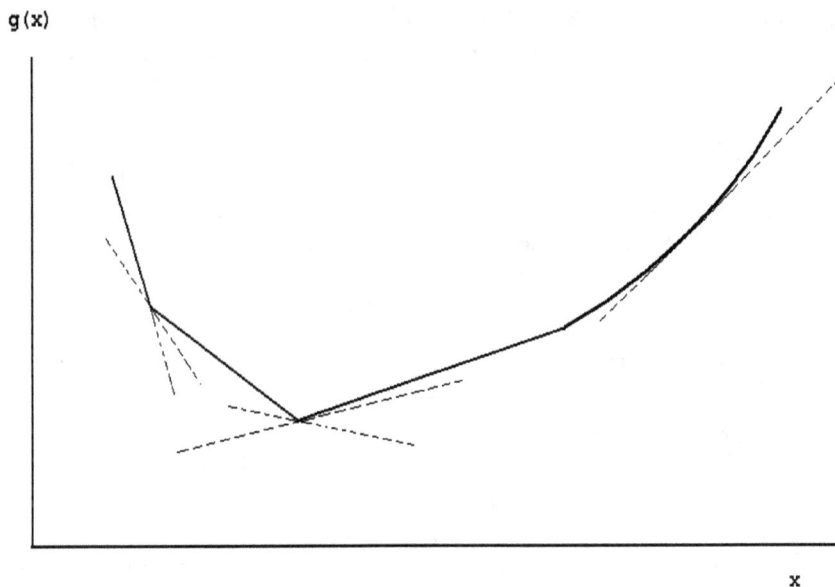

Fig. 3.7 A convex function with support lines.

This result shows qualitatively just how the "pass-through" feature of $E(\cdot)$ fails for a large class of functions. That is, instead of $Eg(X) = g(EX)$, as is always true when g is *affine* in its argument and EX exists, there is a definite direction of *inequality* for convex functions. The inequality in (3.35) is strict when r.v. X is nondegenerate and g is *strictly* convex (each support line touching at just one point).

Example 3.32. If r.v. X is nondegenerate, then $EX^2 = \mu^2 + \sigma^2 > \mu^2$ and $Ee^{tX} > e^{t\mu}, t \in \Re$.

Example 3.33. Example 3.24 used certain properties of the integral to prove that $\nu'_k := E\,|X|^k < \infty$ implies $\nu'_t := E\,|X|^t < \infty$ for $t \in [0, k]$. Of course, for $t = 0$ this is trivial, but for $t \in (0, k]$ Jensen's inequality delivers a sharper conclusion:

$$E\,|X|^k = E\left[\left(|X|^t\right)^{k/t}\right] \geq \left(E\,|X|^t\right)^{k/t}.$$

This gives a definite upper bound for the tth absolute moment: $\nu'_t \leq (\nu'_k)^{t/k}$. Notice that $\nu'_k \equiv \mu'_k$ when k is an even integer.

Exercise 3.38. *If $h : \mathbb{X} \to \Re$ is a concave integrable function of r.v. X with mean μ, show that $Eh(X) \leq h(\mu)$. (Hint: g is concave if and only if $-g$ is convex!)*

Example 3.34.

(i) The *harmonic mean* η of a nonnegative r.v. X is the inverse of the first *negative-integer* moment, or

$$\eta := \left(\mu'_{-1} \right)^{-1} = \left(EX^{-1} \right)^{-1}.$$

X^{-1} being a strictly convex function of X, Jensen implies that $\mu'_{-1} > \mu^{-1}$ if X is nondegenerate, and so $\eta < \mu$.

(ii) The *geometric mean* of a nonnegative r.v. X is $\gamma := \exp\left(E \log X \right)$. Since $\log X$ is strictly concave and the exponential is strictly increasing, Jensen implies that $\gamma < \exp(\log EX) = \mu$ if X is nondegenerate.

Jensen's inequality finds important application in modeling rational decisions under uncertainty; for example, in describing an individual's willingness to take gambles, to choose among alternative investment projects, or to decide whether to insure property against damage. To take a particular case, suppose an individual has to consider different insurance plans to compensate losses to a home through storm or fire. Assume that the probability distribution of monetary loss from damage in the coming year—the random variable L—is known and represented by c.d.f. $F_L (\ell) = \Pr (L \leq \ell)$. Abstracting from other sources of uncertainty, assume also that end-of-year wealth in the absence of loss is a known value w_0, so that wealth at end of year if uninsured would be the r.v. $w_0 - L$. Different insurance plans, indexed by $j \in \{1, 2, ..., n\}$, would cover different proportions of loss $\{p_j L\}$, have different annual premiums $\{c_j\}$, and therefore would lead to different distributions of uncertain end-of-year wealth $\{W_j = w_0 - c_j - (1 - p_j) L\}$; namely,

$$F_j (w) = \Pr [w_0 - c_j - (1 - p_j) L \leq w]$$
$$= 1 - F_L \left(\frac{w_0 - c_j - w}{1 - p_j} - \right), \, j \in \{1, 2, ..., n\}.$$

A choice among the n plans would simply be a choice among the n distributions. It is reasonable to suppose that a rational individual could order these according to preference, from most preferred to least preferred, with ties being permitted. Given this and other such rationality postulates for choice under uncertainty, it can be shown[19] that there exists a function u of wealth such that the preference ordering is indicated by the ranking of expected values, $\{Eu(W_j)\}$. Thus, if the individual strictly prefers plan j to plan $j+1$, then $\int u(w) \cdot dF_j(w) > \int u(w) \cdot dF_{j+1}(w)$. Function $u(\cdot)$ is known as the individual's *utility function* for wealth.

Adopting this expected-utility paradigm, a researcher could predict a "rational" person's choice among any set of uncertain prospects merely by discovering the function u that characterizes the person's preferences. In principle, u could be so determined to any desired degree of accuracy by observing choices among sufficiently many proposed gambles, but predictions of a general nature could be based just on certain qualitative features of u. For example, u would have to be strictly increasing if the individual always prefers more wealth to less. Suppose, in addition, that the individual is risk averse, preferring a sure prospect of wealth w' to any risky prospect F_R having expected value w'. This is where Jensen's inequality applies. Letting $F_S(w) = \mathbf{1}_{[w',\infty)}(w)$ be the c.d.f. of the safe prospect, we want

$$\int u(w) \cdot dF_S(w) = u(w') \geq \int u(w) \cdot dF_R(w) \qquad (3.36)$$

for any distribution F_R such that $w' = \int w \cdot dF_R(w)$. By Jensen's inequality such risk-averse preferences are consistent with *concave* utility functions.

Exercise 3.39. *If utility function u correctly ranks uncertain prospects $\{F_j\}_{j=1}^n$ according to preference, show that any function of the form $u' = a + bu$ with $b > 0$ also correctly ranks the prospects.*

Exercise 3.40. *Let F_1, F_2 be distribution functions for future wealth under two alternative risky prospects. F_1 is said to* stochastically dominate *F_2 if all individuals who prefer more wealth to less would choose F_1. Assuming that individuals satisfy the postulates of expected utility theory, this requires that*

$$\int u(w) \cdot dF_1(w) - \int u(w) \cdot dF_2(w) > 0 \qquad (3.37)$$

for all strictly increasing utility functions u whose expected values under F_1 and F_2 are finite. Restricting to continuous utilities, finite on $[0, \infty)$,

---

[19]The original axiomatic development of expected-utility theory is due to von Neuman and Morgenstern (1944). Luce and Raiffa (1957) give a particularly lucid description. Among the many extensions of the theory are those of Savage (1954) and Fishburn (1969), who axiomatize subjective probability along with utility. Applications to financial decisions in particular are given by Borch (1968), Mossin (1973), Ingersoll (1987), and Epps (2009).

and to distributions with $F_1(0) = F_2(0) = 0$, show that condition (3.37) holds if and only if $\int_0^\infty [F_2(w) - F_1(w)] \cdot du(w) > 0$. Conclude from this that $F_1(w) \le F_2(w)$ for all $w > 0$ and $F_1(w) < F_2(w)$ for some $w > 0$ is a sufficient condition for F_1 to dominate F_2.

Exercise 3.41. *A person with utility function $u(w) = 1 - e^{-w/100}$ and current wealth \$100 is exposed to a risk that makes future wealth W a r.v. with c.d.f. $F(w) = \left(1 - e^{-w/80}\right) \mathbf{1}_{[0,\infty)}$. Find the largest insurance premium ι that the individual would willingly pay to insure against this risk; i.e., the value that makes the individual indifferent between having $\$(100 - \iota)$ for sure and having \$100 subject to risk F.*

Schwarz' Inequality The Schwarz inequality (or "Cauchy–Schwarz" or "Cauchy–Bienaymé–Schwarz" inequality) places an upper bound on the expected value of the product of two r.v.s. Our first application comes immediately following, in connection with the concepts of covariance and correlation.

Proposition 3.3. *If (X, Y) are jointly distributed r.v.s, and if $g : \Re \to \Re$ and $h : \Re \to \Re$ are measurable functions, then*

$$|Eg(X)h(Y)| \le E|g(X)h(Y)| \le \sqrt{Eg(X)^2 Eh(Y)^2}. \qquad (3.38)$$

Proof. The first inequality follows from (3.6). The second clearly holds (as an equality) if either $g(X) = 0$ or $h(Y) = 0$ almost surely, so let us consider the case that $Eg(X)^2 > 0$ and $Eh(Y)^2 > 0$. Now pick any $n \in \aleph$ and set $G_n := \min\{|g(X)|, n\}$ and $H_n := \min\{|h(X)|, n\}$. These are bounded (by 0 and n), and they converge up to $|g(X)|$ and $|h(X)|$, respectively, as $n \to \infty$. Then for any finite t

$$0 \le E(G_n - tH_n)^2 = EG_n^2 - 2tEG_n H_n + t^2 EH_n^2.$$

Putting $t = \sqrt{EG_n^2/EH_n^2}$ gives $EG_n H_n \le \sqrt{EG_n^2 EH_n^2} < \infty$, where finiteness follows from the boundedness of G_n and H_n. Taking limits as $n \to \infty$, monotone convergence implies that $EG_n^2 \to Eg^2, EH_n^2 \to Eh^2, EG_n H_n \to E|gh|$ and therefore that $(E|gh|)^2 \le Eg^2 Eh^2$. $\qquad \square$

Note that the right side of (3.38) would be $+\infty$ unless both g^2 and h^2 were integrable. Besides guaranteeing that $E|gh| < \infty$ (and that Egh exists) when g and h are square integrable, the Schwarz inequality places a definite bound on the expectation of the product. Both the fact of boundedness and the precise knowledge of the bound are useful.

Exercise 3.42. *Use the Schwarz inequality to prove that $E\,|X| \le \sqrt{EX^2}$ (a special case of the result in Example 3.33).*

3.4.1.4 *Covariance and Correlation*

Expected values of products of r.v.s are called *product moments*. Only one specific product moment merits special consideration. The *covariance* between jointly distributed r.v.s X and Y, denoted σ_{XY}, is defined as the expectation of the product of deviations from the respective means, provided that the means exist and are finite. Thus, if $E\,|X| \cdot E\,|Y| < \infty$ then $\sigma_{XY} := E(X - \mu_X)(Y - \mu_Y)$; otherwise, σ_{XY} is undefined (does not exist for X and Y).[20] Two other symbols are used: (i) σ_{jk} when the r.v.s are designated with subscripts, as X_j, X_k; and (ii) $Cov(X, Y)$ or $Cov(X_j, X_k)$, corresponding to the operator notations EX and VX. From the definition it is obvious that

$$\sigma_{jk} = Cov(X_j, X_k) = Cov(X_k, X_j) = \sigma_{kj}. \qquad (3.39)$$

Notice that $VX \equiv Cov(X, X)$, so we sometimes write $VX_j = \sigma_j^2$ instead as σ_{jj}. Two computational shortcuts are helpful: (i) $\sigma_{XY} = EXY - \mu_X\mu_Y$, and (ii) $\sigma_{XY} = EXY$ if *either* $\mu_X = 0$ or $\mu_Y = 0$. Point (i) comes by expanding $(X - \mu_X)(Y - \mu_Y)$ and exploiting the linearity of $E(\cdot)$, and (ii) follows from (i). Point (i) also yields the trivial but often useful relation $EXY = \mu_X\mu_Y + \sigma_{XY}$, corresponding to $EX^2 = \mu_X^2 + \sigma_X^2$ in the univariate case.

Taking $g(X) = X - \mu_X$ and $h(Y) = Y - \mu_Y$, Schwarz' inequality shows that $|Cov(X, Y)| < \infty$ if both VX and VY are finite; however, σ_{XY} can be finite even without this condition. In particular, for independent r.v.s with finite means we have

$$E(X - \mu_X)(Y - \mu_Y) = E(X - \mu_X) \cdot E(Y - \mu_Y) = 0 \cdot 0$$

whether second moments are finite or not.[21] Partly because of this the covariance is used as a qualitative indicator of the nature of dependence or "stochastic relatedness" between two r.v.s, its sign indicating whether the variables tend to move together or at odds. The intuition

---

[20] σ_{XY} may fail to exist even though $E\,|XY| < \infty$. Mukhopadhyay (2010) gives this example: If $X \sim N(0, 1)$ and $Y = X^{-1}$, then $E\,|XY| = 1$ but EY does not exist, since (Exercise 3.4) $EX^{-1}\mathbf{1}_{(0,\infty)}(X) = +\infty$ and $EX^{-1}\mathbf{1}_{(-\infty,0)}(X) = -\infty$.

[21] Dependent r.v.s with infinite variances can also have finite covariance. Let W, T_1, T_2 be mutually independent with zero means, and put $EW^2 = 1$ and $ET_1^2 = +\infty$, $ET_2^2 = +\infty$. (Student's t variates with 2 degrees of freedom are examples of such $\{T_j\}$; see Section 5.3.3.) Then $V(W + T_1)$ and $V(W + T_2)$ are infinite but $Cov(W + T_1, W + T_2) = 1$.

is that when there is a positive association between X and Y the signs of realizations $X(\omega) - \mu_X$ and $Y(\omega) - \mu_Y$ usually agree, so one expects $\sigma_{XY} = \int [X(\omega) - \mu_X][Y(\omega) - \mu_Y]\,\mathbb{P}(d\omega)$ to be positive. Conversely, when $\sigma_{XY} < 0$ large values of X tend to be associated with small values of Y, and *vice versa*. While this is merely heuristic, we shall soon see more precisely how to interpret σ_{XY} as a measure of association.

Exercise 3.43. *X and Y are independent r.v.s having the same distribution, with mean μ and variance σ^2. (For example, realizations of X and Y might be generated by independent draws with replacement from some population.) Show that $\sigma^2 = E(X - Y)^2/2$.*

Just as (3.20) represents EX in terms of c.d.f. F_X, there is a way to express $Cov(X, Y)$ in terms of joint and marginal c.d.f.s. The result, attributed to Hoeffding (1940), is as follows:[22]

$$Cov(X, Y) = \int_{-\infty}^{\infty} \int_{-\infty}^{\infty} [F_{XY}(x, y) - F_X(x) F_Y(y)] \cdot dx\, dy. \qquad (3.40)$$

Proof. When (X_1, Y_1) and (X_2, Y_2) are i.i.d. as (X, Y), and when $E|XY|$, $E|X|$, and $E|Y|$ are all finite, we have

$$E|(X_1 - X_2)(Y_1 - Y_2)| \leq E|X_1 Y_1| + E|X_2| E|Y_1| + E|X_1| E|Y_2| + E|X_2 Y_2|$$
$$= 2E|XY| + 2E|X| E|Y| < \infty$$

and

$$E(X_1 - X_2)(Y_1 - Y_2) = 2EXY - 2\mu_X \mu_Y = 2Cov(X, Y).$$

Since $\int_{-\infty}^{\infty} [\mathbf{1}_{(-\infty,b]}(x) - \mathbf{1}_{(-\infty,a]}(x)] \cdot dx = b - a$ for each $a, b \in \Re$, $E(X_1 - X_2)(Y_1 - Y_2)$ can be expressed in terms of indicator functions as

$$E\{\int [\mathbf{1}_{(-\infty,X_1]}(x) - \mathbf{1}_{(-\infty,X_2]}(x)]\, dx \cdot \int [\mathbf{1}_{(-\infty,Y_1]}(y) - \mathbf{1}_{(-\infty,Y_2]}(y)]\, dy\}$$
$$= E\{\int \int [\mathbf{1}_{(-\infty,X_1]}(x)\,\mathbf{1}_{(-\infty,Y_1]}(y) - \mathbf{1}_{(-\infty,X_2]}(x)\,\mathbf{1}_{(-\infty,Y_1]}(y)$$
$$- \mathbf{1}_{(-\infty,X_1]}(x)\,\mathbf{1}_{(-\infty,Y_2]}(y) + \mathbf{1}_{(-\infty,X_2]}(x)\,\mathbf{1}_{(-\infty,Y_2]}(y)] \cdot dx\, dy\}.$$

The quantity in $\{\}$ being absolutely integrable, Fubini's theorem authorizes the interchange of expectation and integration. Thus follow

$$E(X_1 - X_2)(Y_1 - Y_2) = 2 \int \int [F_{XY}(x, y) - F_X(x) F_Y(y)] \cdot dx\, dy$$

and (3.40).   □

----

[22]The proof below is from Shea (1983), who gives further details and applications of Hoeffding's lemma.

The *coefficient of correlation* of r.v.s X and Y, denoted ρ_{XY} (or ρ_{jk} for r.v.s X_j, X_k), is a dimensionless quantity that is expressed in terms of the covariance and hence is defined only when σ_{XY} exists. When σ_{XY} does exist, the definition is

$$\rho_{XY} := \begin{cases} 0, & \sigma_{XY} = 0 \\ \frac{\sigma_{XY}}{\sigma_X \sigma_Y} = \frac{Cov(X,Y)}{\sqrt{VX \cdot VY}}, & 0 < \sigma_X < \infty, 0 < \sigma_Y < \infty \end{cases} \tag{3.41}$$

Whenever the standard deviations are finite and positive, ρ_{XY} is thus the covariance between the standardized versions of X and Y:

$$\rho_{XY} = \frac{E\left(X - \mu_X\right)\left(Y - \mu_Y\right)}{\sigma_X \sigma_Y} = Cov\left(\frac{X - \mu_X}{\sigma_X}, \frac{Y - \mu_Y}{\sigma_Y}\right).$$

An immediate consequence of Schwarz' inequality is that $|\rho_{XY}| \leq 1$, and it is clear that the signs of ρ_{XY} and σ_{XY} always agree. In particular, $\rho_{XY} = 0$ if X, Y are independent with finite means. Whenever $\rho_{XY} = 0$ r.v.s X and Y are said to be *uncorrelated*; the term *orthogonal* is also sometimes used.

Treating the case $\sigma_{XY} = 0$ separately in (3.41) is not the usual convention in the literature, but it extends the definition to allow X and Y to be independent with finite means but without finite second moments and also to allow either or both to be *degenerate*, in the sense of having zero variance. In either case we consider X and Y to be uncorrelated. Note that ρ_{XY} is left undefined for variables with infinite variances but with *nonzero* covariance. The reason is that dividing σ_{XY} by $\sigma_X \sigma_Y$ is intended just to create an index of "relatedness" that is units-free and bounded, and one could hardly consider r.v.s with nonzero covariance to be stochastically unrelated, regardless of their variances. Still, there are ways of extending the concept of correlation to r.v.s without either first or second moments.

Exercise 3.44. *Let X and Y be jointly distributed r.v.s with unique lower and upper (marginal) p quantiles (x_p, x^p), (y_p, y^p), $p \in (0, 1/2)$. For functions $g : \Re \to \Re$ and $h : \Re^2 \to \Re$ that are bounded on sets of finite Lebesgue measure define the p-truncated expectations*

$$E_p g\left(X\right) := \frac{Eg\left(X\right) \mathbf{1}_{(x_p, x^p)}\left(X\right)}{E \mathbf{1}_{(x_p, x^p)}\left(X\right)} = \frac{Eg\left(X\right) \mathbf{1}_{(x_p, x^p)}\left(X\right)}{1 - 2p}$$

$$E_p h\left(X, Y\right) := \frac{E\left[h\left(X, Y\right) \mathbf{1}_{(x_p, x^p)}\left(X\right) \mathbf{1}_{(y_p, y^p)}\left(Y\right)\right]}{E\left[\mathbf{1}_{(x_p, x^p)}\left(X\right) \mathbf{1}_{(y_p, y^p)}\left(Y\right)\right]}.$$

In particular, $E_p X$ is the p-truncated mean, $V_p X = E_p\left(X - E_p X\right)^2$ is the truncated variance, and $Cov_p\left(X, Y\right) = E_p\left(X - E_p X\right)\left(Y - E_p Y\right)$ is the truncated covariance. Finally, let $\rho_p\left(X, Y\right) := Cov_p\left(X, Y\right) / \sqrt{V_p X \cdot V_p Y}$

be the p-truncated correlation, and put $\rho_0(X,Y) := \lim_{p\to 0}\rho_p(X,Y)$*, provided that the limit exists and lies on* $[-1,1]$*.*

Now considering a specific case, let X *and* Y *be i.i.d. as Cauchy with median* δ*, scale parameter* $\gamma > 0$*, and marginal p.d.f.*

$$f(x) = \frac{1}{\pi\left[1+(x-\delta)^2/\gamma^2\right]};$$

and for $\alpha \in [0,1]$ *put* $W_\alpha := \alpha X + (1-\alpha)Y$*. Find* $\rho_p(X,W_\alpha)$ *and* $\rho_0(X,W_\alpha)$*. (Hint: Use the fact—established in Example 3.49 below—that the marginal distribution of* W_α *is the same as that of* X *and* Y *for each* $\alpha \in [0,1]$*.)*

The proper sense in which covariance and correlation can be considered to measure stochastic relatedness is actually quite limited. While independent variables with finite means are necessarily uncorrelated, variables that are uncorrelated are not *necessarily* independent, as the following shows.

Exercise 3.45. X, Y *are discrete r.v.s with joint p.m.f.*

$y\backslash x$	-1	0	$+1$
-1	0	$\frac{1}{4}$	0
0	$\frac{1}{4}$	0	$\frac{1}{4}$
$+1$	0	$\frac{1}{4}$	0

Show that $\rho_{XY} = 0$ *but that* X *and* Y *are not independent.*

Although $\rho_{XY} = 0$ does not imply independence, it *is* true that there is a very special type of relation between X and Y when they are "perfectly" correlated—i.e., when $|\rho_{XY}| = 1$. In particular, $|\rho_{XY}| = 1$ if and only if there are a constant a and a constant $b \neq 0$ such that $Y(\omega) = a + bX(\omega)$ for almost all ω (i.e., all except for those in a null set). The only-if part ($|\rho_{XY}| = 1$ implies existence of a, b) is shown as follows. Letting $s := \rho_{XY}/|\rho_{XY}| = \pm 1$ be the sign of ρ_{XY}, we have $\sigma_{XY} = s\sigma_X\sigma_Y$ when $|\rho_{XY}| = 1$ and

$$E\left[s\frac{\sigma_Y}{\sigma_X}(X-\mu_X)-(Y-\mu_Y)\right]^2 = s^2\frac{\sigma_Y^2}{\sigma_X^2}\sigma_X^2 - 2s\frac{\sigma_Y}{\sigma_X}\sigma_{XY} + \sigma_Y^2$$
$$= s^2\sigma_Y^2 - 2s\frac{\sigma_Y}{\sigma_X}(s\sigma_X\sigma_Y) + \sigma_Y^2$$
$$= \sigma_Y^2(1-s^2) = 0.$$

But since $\left|s\frac{\sigma_Y}{\sigma_X}(X-\mu_X)-(Y-\mu_Y)\right| \geq 0$ for all ω, the above implies that $s\frac{\sigma_Y}{\sigma_X}(X-\mu_X)-(Y-\mu_Y) = 0$ for almost all ω, since the expectation of a nonnegative quantity can be zero only if the quantity is zero on all sets of positive measure. Thus, the conclusion follows with $a = \mu_Y - s\frac{\sigma_Y}{\sigma_X}\mu_X$ and $b = s\frac{\sigma_Y}{\sigma_X}$.

Exercise 3.46. *Show that $|\rho_{XY}| = 1$ if X has finite variance and $Y = a + bX$ a.s. for any $b \neq 0$.*

Exercise 3.47. *If X, Y have correlation ρ_{XY} and if $W = a + bX$ and $U = c + dY$, find ρ_{WU}.*

Since $|\rho_{XY}| = 1$ if and only if (almost all) realizations $\{X(\omega), Y(\omega)\}$ plot on a straight line with nonzero slope, it is proper to interpret covariance and correlation as indicating the extent of *linear* (or affine) association. With this interpretation it is logically clear that "X, Y independent" (*no* systematic relation between X and Y) implies $\rho_{XY} = 0$ (no degree of *linear* association) but that $\rho_{XY} = 0$ does not imply independence. Indeed, X and Y can be *functionally* related—realizations of one being perfectly predictable from those of the other—and yet have $|\rho_{XY}| < 1$ or even $\rho_{XY} = 0$.

Exercise 3.48. *Calculate ρ_{XY} if (i) X has p.d.f. $f_X(x) = 1_{[1,2]}(x)$, and $Y = X^{-1}$; and (ii) X has p.d.f. $f_X(x) = .51_{[-1,1]}(x)$, and $Y = X^2$.*

3.4.1.5 *Covariance and Correlation Matrices*

While covariance and correlation are inherently *bivariate* concepts, one often needs to work with covariances and correlations among all pairs of some collection of more than two r.v.s. These are usually displayed and manipulated in matrix form. As we shall see shortly, these matrices arise naturally when determining variances of linear or affine functions of r.v.s.

Let $\mathbf{X} := (X_1, X_2, ..., X_m)'$ be a vector of jointly distributed r.v.s with means $\boldsymbol{\mu_X} := (\mu_1, \mu_2, ..., \mu_m)'$ and finite variances $\{\sigma_j^2\}_{j=1}^m$. The m-vector of deviations from means is then $\mathbf{X} - \boldsymbol{\mu_X}$, and outer product $(\mathbf{X} - \boldsymbol{\mu_X})(\mathbf{X} - \boldsymbol{\mu_X})'$ is an $m \times m$ matrix with entry $(X_j - \mu_j)(X_k - \mu_k)$ in the jth row and kth column. The matrix $\boldsymbol{\Sigma_X} := E(\mathbf{X} - \boldsymbol{\mu_X})(\mathbf{X} - \boldsymbol{\mu_X})'$ of expected values of these products of deviations is called the *variance-covariance matrix* or simply the *covariance matrix* of $\mathbf{X}$:

$$\boldsymbol{\Sigma_X} = \begin{pmatrix} \sigma_{11} & \sigma_{12} & \cdots & \sigma_{1m} \\ \sigma_{21} & \sigma_{22} & \cdots & \sigma_{2m} \\ \vdots & \vdots & \ddots & \vdots \\ \sigma_{m1} & \sigma_{m2} & \cdots & \sigma_{mm} \end{pmatrix} = \begin{pmatrix} \sigma_{11} & \sigma_{12} & \cdots & \sigma_{1m} \\ \sigma_{12} & \sigma_{22} & \cdots & \sigma_{2m} \\ \vdots & \vdots & \ddots & \vdots \\ \sigma_{1m} & \sigma_{2m} & \cdots & \sigma_{mm} \end{pmatrix}.$$

We sometimes just write $V\mathbf{X}$ for the covariance matrix, corresponding to the "VX" notation for the variance of a scalar r.v. Note that elements $\{\sigma_{jj} = \sigma_j^2\}_{j=1}^m$ on the principal diagonal of $\mathbf{\Sigma_X}$ are variances while off-diagonal elements are covariances; also, that the symmetry of $\mathbf{\Sigma_X}$ (apparent in the second expression) follows from relation (3.39). A related entity, the *correlation matrix*, replaces the covariances with correlations and has units on the main diagonal—correlations of the variables with themselves:

$$\mathbf{R_X} := \begin{pmatrix} 1 & \rho_{12} & \cdots & \rho_{1m} \\ \rho_{12} & 1 & \cdots & \rho_{2m} \\ \vdots & \vdots & \ddots & \vdots \\ \rho_{1m} & \rho_{2m} & \cdots & 1 \end{pmatrix}.$$

Clearly, $\mathbf{R_X}$ is nothing more than the covariance matrix of the standardized variables $\{(X_j - \mu_j)/\sigma_j\}_{j=1}^m$.

3.4.1.6 *Moments of Linear Functions of R.V.s*

Letting $\mathbf{a} := (a_1, a_2, ..., a_m)'$ be a vector of finite constants, not all zero, the linear functional (inner product) $\sum_{j=1}^m a_j X_j = \mathbf{a'X} =: Y$ defines a new r.v. with mean and variance

$$EY = E\mathbf{a'X} = \mathbf{a'}E\mathbf{X} = \mathbf{a'}\boldsymbol{\mu_X}$$
$$VY = E(Y - EY)^2 = E[\mathbf{a'}(\mathbf{X} - \boldsymbol{\mu_X})]^2.$$

As a scalar, $\mathbf{a'}(\mathbf{X} - \boldsymbol{\mu_X})$ equals its transpose, so VY can be expressed also in the following ways:

$$\begin{aligned} VY &= E\left[\mathbf{a'}(\mathbf{X} - \boldsymbol{\mu_X})(\mathbf{X} - \boldsymbol{\mu_X})'\mathbf{a}\right] \\ &= \mathbf{a'}E(\mathbf{X} - \boldsymbol{\mu_X})(\mathbf{X} - \boldsymbol{\mu_X})'\mathbf{a} \\ &= \mathbf{a'}\mathbf{\Sigma_X}\mathbf{a} \\ &= \sum_{j=1}^m \sum_{k=1}^m a_j a_k \sigma_{jk} \\ &= \sum_{j=1}^m a_j^2 \sigma_j^2 + \sum_{j=1}^m \sum_{k=1, k\neq j}^m a_j a_k \sigma_{jk} \\ &= \sum_{j=1}^m a_j^2 \sigma_j^2 + 2\sum_{j=2}^m \sum_{k=1}^{j-1} a_j a_k \sigma_{jk}. \end{aligned} \tag{3.42}$$

The double sums in the last three expressions can be written more compactly as $\sum_{j,k=1}^m$, $\sum_{j\neq k=1}^n$, and $\sum_{k<j=2}^m$, respectively, or even more simply as $\sum_{j,k}$, $\sum_{j\neq k}$, and $\sum_{k<j}$.

The fact that $V\mathbf{a}'\mathbf{X} \geq 0$ for all $\mathbf{a}$ implies that covariance matrix $\mathbf{\Sigma_X}$ is *positive semidefinite*, so that $|\mathbf{\Sigma_X}|$ and all the principal *minors* are nonnegative. (The principal minors of a square matrix $\mathbf{M}$ are determinants of the square *sub*matrices formed by deleting corresponding rows and columns of $\mathbf{M}$.) Thus, covariance and correlation matrices are symmetric and have nonnegative determinants. If $|\mathbf{\Sigma_X}| = 0$ (equivalently, $|\mathbf{R_X}| = 0$) then the argument by which $|\rho_{XY}| = 1$ was shown to imply that $Y = a + bX$ extends to show that the elements of $\mathbf{X}$ are almost surely linearly dependent; that is, for any j there are constants $a_j, \{b_{ij}\}_{i \neq j}$ such that $X_j(\omega) = a_j + \sum_{i \neq j} b_{ij} X_i(\omega)$ for almost all ω. If there is no such linear dependence, then $|\mathbf{\Sigma_X}| > 0$ and covariance matrix $\mathbf{\Sigma_X}$ is *positive definite*.

In the case $m = 2$ the expressions for VY reduce to

$$V(a_1 X_1 + a_2 X_2) = E\left[(a_1 X_1 + a_2 X_2) - (a_1 \mu_1 + a_2 \mu_2)\right]^2$$
$$= E\left[a_1(X_1 - \mu_1) + a_2(X_2 - \mu_2)\right]^2$$
$$= a_1^2 \sigma_1^2 + 2a_1 a_2 \sigma_{12} + a_2^2 \sigma_2^2$$
$$= a_1^2 \sigma_1^2 + 2a_1 a_2 \rho_{12} \sigma_1 \sigma_2 + a_2^2 \sigma_2^2.$$

In particular, the variance of the sum of two r.v.s is

$$V(X_1 + X_2) = \sigma_1^2 + 2\rho_{12}\sigma_1\sigma_2 + \sigma_2^2.$$

This increases monotonically with the correlation and, since $-1 \leq \rho_{12} \leq 1$, is bounded as

$$(\sigma_1 - \sigma_2)^2 \leq V(X_1 + X_2) \leq (\sigma_1 + \sigma_2)^2.$$

To aid the intuition, think of the condition $\rho_{12} > 0$ as implying that realizations of X_1 and X_2 are "usually" either both above average or both below average, in which case the sum deviates more from its mean than if the variables were uncorrelated. On the other hand, when $\rho_{12} < 0$ the sum tends to be closer to its mean, since large values of one variable tend to be offset by below-average values of the other.

Exercise 3.49. *We desire to estimate an unobservable quantity γ, which is known to be the sum of two other unobservable, positive quantities, α and β. Realizations of random variables A, B, and C are observable and are known to have the properties $EA = \alpha$, $EB = \beta$, $EC = \gamma$, $VA = \alpha^2$, $VB = \beta^2$, $VC = \gamma^2$, and $Cov(A, B) = \rho\alpha\beta$ for some unknown $\rho \in [-1, 1]$. We can use either the realization C itself as an estimate of γ, or else the realization of the sum $A + B$. Let C^* represent whichever of these is chosen. If the goal is to minimize the mean squared error, $E(C^* - \gamma)^2$, should we take $C^* = C$ or $C^* = A + B$? In other words, is it better to estimate γ directly or as the sum of noisy estimates of its components?*

Exercise 3.50. *Suppose that the* $m \geq 2$ *jointly distributed r.v.s* $\{X_j\}_{j=1}^m$ *all have the same variance and that the correlation between any pair is the same constant* ρ. *Show that* $\rho \geq -(m-1)^{-1}$. *(In particular, not all pairs of three or more r.v.s can be perfectly negatively correlated.)*

Exercise 3.51. *R.v.s* X, Y, Z *have* $\rho_{XY} = +1$ *and* $\rho_{XZ} = -1$. *Find* ρ_{YZ}.

Exercise 3.52. *If* X *and* Y *have standard deviations* σ_X *and* σ_Y, *show that* $\sigma_{X-Y} \geq |\sigma_X - \sigma_Y|$.

When the elements of $\mathbf{X}$ are uncorrelated, matrix $\mathbf{\Sigma_X}$ is diagonal and

$$VY = \sum_{j=1}^m a_j^2 \sigma_{jj} \equiv \sum_{j=1}^m a_j^2 \sigma_j^2$$

when $Y = \mathbf{a'X}$. In words, *the variance of a weighted sum of uncorrelated variables is the sum of squared weights times variances.* When $a_j = 1$ for all j, this specializes further as $VY = \sum_{j=1}^m \sigma_j^2$; that is, *the variance of a sum of uncorrelated variables equals the sum of their variances.* Since mutual independence implies uncorrelatedness (assuming that the means are finite), these statements obviously apply when $\{X_j\}_{j=1}^m$ are independent and have finite variances, a situation that is often encountered in statistical applications.

Extending further, let $\mathbf{A} = \{a_{jk}\}_{j=1,2,\ldots,p;k=1,2,\ldots,m}$ be a $p \times m$ matrix of finite constants, and define a new p-vector r.v. by the linear transformation $\mathbf{Y} = \mathbf{AX}$. Thus, the jth element is $Y_j = \sum_{k=1}^m a_{jk} X_k$. The $p \times 1$ mean vector $\boldsymbol{\mu_Y}$ and $p \times p$ covariance matrix $\mathbf{\Sigma_Y}$ are $\boldsymbol{\mu_Y} = E\mathbf{AX} = \mathbf{A}E\mathbf{X} = \mathbf{A}\boldsymbol{\mu_X}$ and

$$\begin{aligned}
\mathbf{\Sigma_Y} &= E(\mathbf{Y} - \boldsymbol{\mu_Y})(\mathbf{Y} - \boldsymbol{\mu_Y})' \\
&= \mathbf{A}E\left[(\mathbf{X} - \boldsymbol{\mu_X})(\mathbf{X} - \boldsymbol{\mu_X})'\right]\mathbf{A}' \\
&= \mathbf{A}\mathbf{\Sigma_X}\mathbf{A}'.
\end{aligned}$$

In particular, the jkth element of $\mathbf{\Sigma_Y}$ is

$$Cov(Y_j, Y_k) = Cov\left(\mathbf{a}_j'\mathbf{X}, \mathbf{a}_k'\mathbf{X}\right) = \mathbf{a}_j'\mathbf{\Sigma_X}\mathbf{a}_k = \mathbf{a}_k'\mathbf{\Sigma_X}\mathbf{a}_j,$$

where $\mathbf{a}_j', \mathbf{a}_k'$ are the jth and kth *rows* of $\mathbf{A}$. Since the rank of $\mathbf{A}$ can be no greater than $\min\{m, p\}$, it follows that $\mathbf{\Sigma_Y}$ is necessarily singular if $p > m$. In that case, provided that no row of $\mathbf{A}$ is a zero vector (and thus that no element of $\mathbf{Y}$ is identically zero), one or more of the $\{Y_j\}$ can be expressed as a linear combination of the remaining elements—and is therefore perfectly correlated with such a combination. For example, in the case $p = 2$ and $m = 1$ the r.v.s $Y_1 = a_{11}X_1$ and $Y_2 = a_{21}X_1$ are perfectly correlated so long as $a_{11} \neq 0$ and $a_{21} \neq 0$.

3.4.2 *Generating Functions*

Moments of r.v.s are one type of informative functional of the c.d.f. That is, they are mappings from the space of distributions $\mathfrak{F}$ to the reals whose values tell us something about the distributions. For r.v.s whose moments exist and are finite, we have seen that these can reveal certain properties of the distributions—center of mass, dispersion, symmetry, and tail thickness. However, even knowing *all* the positive integer moments does not necessarily identify a distribution uniquely.[23] Here we consider some other types of mappings that, under certain conditions, are one to one and therefore do identify the element of $\mathfrak{F}$ precisely. These mappings, known as *integral transforms*, are expected values of certain real- or complex-valued functions of the form $g(X, t)$, where X is a r.v. and t is either real ($t \in \Re$) or complex ($t \in \mathcal{C}$). When $Eg(X, t) = \int g(x, t) \cdot dF(x)$ exists and is finite for all t in some set, the transformation maps the space of c.d.f.s onto a subspace of functions of t. Some of the many transforms of this sort are

 (i) Mellin transforms, with $g(x, t) = x^t$ and $t \in \mathcal{C}$;
 (ii) Probability generating functions (p.g.f.s), with $g(x, t) = t^x$ and $t \in \Re$;
 (iii) Moment generating functions (m.g.f.s), with $g(x, t) = e^{tx}$ and $t \in \Re$; and
 (iv) Characteristic functions (c.f.s), with $g(x, t) = e^{itx}, i = \sqrt{-1}, t \in \Re$.

The p.g.f. is used with models for "count" (i.e., integer-valued) data, and is discussed in Chapter 4. For all other purposes our needs are met entirely by c.f.s, but we describe first the m.g.f., which is closely related but a little easier to understand.

3.4.2.1 *Moment Generating Functions*

Definition 3.4. The function $M(t) := Ee^{tX} = \int e^{tx} \cdot dF(x)$, if it is finite for all t in some neighborhood of the origin, is called the *moment generating*

---

[23]The lognormal distribution, discussed in Chapter 4, has p.d.f.

$$f(x) = (2\pi)^{-1/2} x^{-1} e^{-(\log x)^2/2} \mathbf{1}_{(0,\infty)}(x).$$

All moments are finite, with $EX^q = e^{q^2/2}$ for all real q. The function

$$f^*(x) := f(x) [1 + \sin(2\pi \log x)]$$

is also a p.d.f. Although it is obviously distinct from f, it is known to have precisely the same positive integer moments. On the other hand, the positive integer moments do uniquely identify distributions having bounded support.

function (m.g.f.) of the (scalar) r.v. X.

Of course, $M(0) = Ee^{0 \cdot X} = E1 = 1$ is finite for any X, and the integral always *exists* for any t since $e^y > 0$ for all real y. The issue is whether the integral is finite in some open interval containing $\{0\}$; that is, there must be some $\delta > 0$ such that the integral converges for all $t \in [-\delta, \delta]$. The problem is that when $t > 0$ the function e^{tx} blows up quickly as $x \uparrow \infty$ and likewise for $t < 0$ as $x \downarrow -\infty$. Unless the support of X is bounded, this explosive behavior *may* make $\int e^{tx} \cdot dF = +\infty$ for all $t > 0$ or for all $t < 0$ or for both. Whether this happens depends on how much probability mass resides in the tails of the distribution; and when it does happen X simply does not have an m.g.f.

Example 3.35.

(i) For $f(x) = \mathbf{1}_{[0,1]}(x)$ (the uniform distribution on $[0,1]$)

$$Ee^{tX} = \int_0^1 e^{tx} \cdot dx = \frac{e^t - 1}{t}, t \neq 0.$$

Since $Ee^{tX} = 1$ when $t = 0$, we see that $M(t)$ is finite for *all* real t, and so the uniform distribution does possess an m.g.f. Expanding e^t in Taylor series about $t = 0$ shows that $\lim_{t \to 0} \frac{e^t - 1}{t} = 1$ and hence that $M(\cdot)$ is continuous at the origin.

(ii) For $f(x) = \theta e^{-\theta x} \mathbf{1}_{[0,\infty)}(x)$, where $\theta > 0$ (the exponential distribution),

$$Ee^{tX} = \int_0^\infty \theta e^{-(\theta - t)x} \cdot dx = \frac{\theta}{\theta - t}, t < \theta.$$

Since $\theta > 0$, the integral does converge in a neighborhood of the origin, and so the exponential distribution does possess an m.g.f.

(iii) For $f(x) = \frac{1}{2} e^{-|x|}$ (the Laplace or double-exponential distribution)

$$Ee^{tX} = \frac{1}{2} \left[\int_{-\infty}^0 e^{tx+x} \cdot dx + \int_0^\infty e^{tx-x} \cdot dx \right].$$

When $t > -1$ the first integral is $(1+t)^{-1}$, and when $t < 1$ the second integral is $(1-t)^{-1}$. Thus, $M(t) = \frac{1}{2} \left[(1+t)^{-1} + (1-t)^{-1} \right] = (1-t^2)^{-1}$ for $t \in (-1, 1)$, so the Laplace distribution does possess an m.g.f.

(iv) For $f(x) = \theta^{x-1}(1-\theta)\mathbf{1}_{\aleph}(x), \theta \in (0,1)$ (the geometric distribution)

$$Ee^{tX} = \sum_{x=1}^{\infty} e^{tx}\theta^{x-1}(1-\theta)$$

$$= e^t(1-\theta)\sum_{x=1}^{\infty}\left(\theta e^t\right)^{x-1}$$

$$= \frac{e^t(1-\theta)}{1-\theta e^t}, t < -\log\theta.$$

Since $-\log\theta > 0$, the geometric distribution does possess an m.g.f.

(v) For $f(x) = \left[\pi(1+x^2)\right]^{-1}$ (the standard Cauchy distribution)

$$Ee^{tX} = \int_{-\infty}^{\infty}\frac{e^{tx}}{\pi(1+x^2)}\cdot dx > \int_{0}^{\infty}\frac{e^{tx}}{\pi(1+x^2)}\cdot dx.$$

For any $t > 0$ the last integral is greater than

$$t\int_{0}^{\infty}\frac{x}{\pi(1+x^2)}\cdot dx = +\infty,$$

and a similar argument gives the same result for all $t < 0$. Thus, the Cauchy distribution has no m.g.f.

The last example shows that functional Ee^{tX} does not map all distributions $\mathfrak{F}$ into the space of functions whose domains include an interval about the origin. However, it does map into such a space all distributions whose support is bounded (such as the uniform) as well as distributions with unbounded support whose tails are sufficiently thin (such as the exponential, Laplace, and geometric).

When the m.g.f. does exist, it has two principal uses. First, it does characterize the distribution, in that the mapping from the subfamily of $\mathfrak{F}$ is one to one. This is known as the *uniqueness property* of the m.g.f. In fact, there are inversion formulas—inverse transforms—that express F in terms of M. While these formulas are not easy to use, it is easy just to learn the forms of M that corresponds to various models for F. Thus, from Example 3.35 one learns that $M(t) = \theta/(\theta - t)$ and $M(t) = e^t(1-\theta)/(1-\theta e^t)$ correspond, respectively, to the exponential and geometric distributions with parameter θ—and to those distributions alone. The importance of being able to recognize such correspondences will become apparent in later chapters. The second principal use of m.g.f.s is that they do "generate" moments; that is, it is often easier to deduce a distribution's positive integer moments from the m.g.f.—once this is known—than to evaluate integrals such as $\int x^k \cdot dF(x)$. We will see first how this works with characteristic functions, related transforms that have all the desirable properties of m.g.f.s and yet exist for *all* distributions $\mathfrak{F}$.

3.4.2.2 *Characteristic Functions*

Vector-valued r.v.s have been defined as mappings from an outcome set Ω into $\Re^k$, $k > 1$. Because of the intimate connection between $\Re^2$ and the space $\mathcal{C}$ (the complex plane) *complex-valued r.v.s* can be defined in terms of the bivariate r.v.s that map Ω into $\Re^2$. Likewise, just as we can have random *functions* such as e^{tX} that take us from $\Omega \times \Re$ to $\Re$, we can visualize complex-valued random functions from $\Omega \times \Re$ to $\mathcal{C}$ as mappings to $\Re^2$. Characteristic functions are mathematical expectations of such complex-valued random functions. They are among the most useful concepts in the theory of probability.

Just what sort of connection is there between $\Re^2$ and $\mathcal{C}$? It is clear that to any vector $\boldsymbol{v} = (v_1, v_2)' \in \Re^2$ there corresponds one and only one point in $\mathcal{C}$; namely, $\phi = v_1 + iv_2$, where $i := \sqrt{-1}$ is the "imaginary" unit. Likewise, the vector $(v_1, -v_2)'$ corresponds to $\bar{\phi} = v_1 - iv_2$, which is the *conjugate* of ϕ. Moreover, the one-to-one correspondence is such that both linear relations and distances are preserved. Linearity is preserved since $\alpha \boldsymbol{v} \leftrightarrow \alpha v_1 + i\alpha v_2 = \alpha \phi$ for $\alpha \in \Re$ and $v + v^* \leftrightarrow (v_1 + v_1^*) + i(v_2 + v_2^*) = \phi + \phi^*$. Such a one-to-one correspondence between $\Re^2$ and $\mathcal{C}$ is called an *isomorphism*. Further, if we define the distance between complex numbers ϕ, ϕ^* as the modulus of the difference, or

$$|\phi - \phi^*| := \sqrt{(\phi - \phi^*) \cdot (\bar{\phi} - \bar{\phi}^*)} = \sqrt{(v_1 - v_1^*)^2 + (v_2 - v_2^*)^2},$$

then this is precisely the Euclidean distance between the corresponding points $\boldsymbol{v}, \boldsymbol{v}^*$ in $\Re^2$. Likewise, the space of bivariate functions $v(t) = (v_1(t), v_2(t))'$ that map $\Re$ to $\Re^2$ is isomorphic to the space $\mathfrak{C}$ of complex-valued functions, $\phi(t) = v_1(t) + iv_2(t)$. The isomorphic relation and the preservation of distance qualify the correspondence between $\Re_2$ and $\mathcal{C}$ as an *isometry*. However, despite the strong geometric connections, the spaces have very different *algebraic* properties. For example, multiplication works very differently for vectors and complex numbers.

Figure 3.8 illustrates the geometry of $\mathcal{C}$. Note that for $v_2 \neq 0$ the angle θ formed by $v_1 + iv_2$ and the real axis is the inverse tangent of v_2/v_1, so that v_1 and v_2 correspond to $r\cos\theta$ and $r\sin\theta$, respectively, where $r := |\phi| = \sqrt{v_1^2 + v_2^2}$. Taylor series expansions of $\cos\theta$ and $i\sin\theta$ about $\theta = 0$ are

$$\cos\theta = 1 - \frac{\theta^2}{2} + \frac{\theta^4}{4!} - + ... = 1 + \frac{(i\theta)^2}{2} + \frac{(i\theta)^4}{4!} + ...$$

$$i\sin\theta = i\left(\theta - \frac{\theta^3}{3!} + \frac{\theta^5}{5!} - + ...\right) = i\theta + \frac{(i\theta)^3}{3!} + \frac{(i\theta)^5}{5!} + ...$$

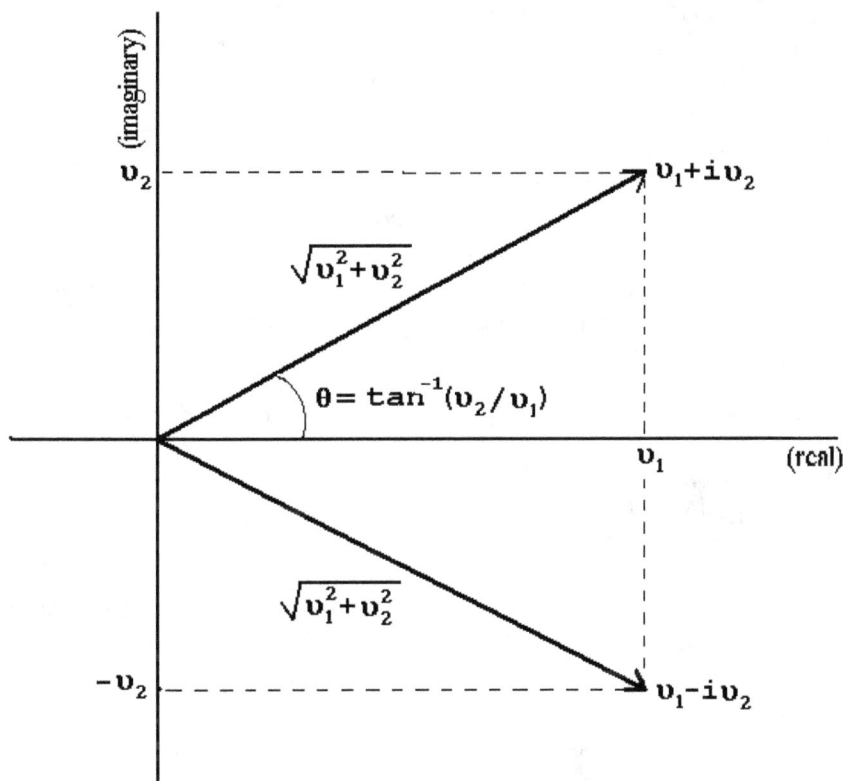

Fig. 3.8 Complex number as vector in (r,i) plane.

Summing and arranging in order of powers of $i\theta$ give $\cos\theta + i\sin\theta = \sum_{j=0}^{\infty} (i\theta)^j / j!$, which is the corresponding expansion of $e^{i\theta}$. All three series being convergent for all real θ, we have what is known as *Euler's identity*:

$$e^{i\theta} = \cos\theta + i\sin\theta, \theta \in \Re. \qquad (3.43)$$

Thus, any complex number $v_1 + iv_2$ can be expressed as $re^{i\theta}$ with $r = |\phi|$.

Now we can introduce some random variables. If $\mathbf{Y} = (Y_1, Y_2)'$ is a bivariate r.v. on probability space $(\Omega, \mathcal{F}, \mathbb{P})$, a complex-valued r.v. C can be defined on the same space via the correspondence that exists between the realizations, outcome by outcome: $\mathbf{Y}(\omega) \leftrightarrow Y_1(\omega) + iY_2(\omega) =: C(\omega)$. Likewise, the complex-valued random function $C(t) := Y_1(t) + iY_2(t)$ corresponds to the vector-valued random function $\mathbf{Y}(t) = (Y_1(t), Y_2(t))'$. Morever, if there is some $\mathcal{T} \subset \Re$ such that expectations $v_1(t) = EY_1(t)$ and

$v_2(t) = E Y_2(t)$ exist for $t \in \mathcal{T}$, then for such t the complex-valued function $v_1(t) + i v_2(t)$ corresponds to $EC(t)$. Characteristic functions are the expected values of a particular *class* of such complex-valued random functions.

Specifically, let X be a r.v., and put $Y_1(t) := \cos(tX)$, $Y_2(t) := \sin(tX)$, and $C(t) := Y_1(t) + i Y_2(t)$ for $t \in \Re$. The sine and cosine functions are bounded and therefore integrable, so that the expectations $v_1(t) := E \cos(tX)$, $v_2(t) := E \sin(tX)$, and $\phi(t) := EC(t) = E \cos(tX) + i E \sin(tX)$ are necessarily finite for each real t. Finally, by Euler's identity, each realization of $C(t)$ can be represented as $e^{itX(\omega)}$, and so we can write

$$\phi(t) = \int e^{itX(\omega)} \, \mathbb{P}(d\omega) = \int e^{itx} \, \mathbb{P}_X(dx) = \int e^{itx} \cdot dF(x) = E e^{itX}.$$

Definition 3.5. If X is a (scalar-valued) r.v. with c.d.f. F, the function $\phi(t) := E e^{itX} = \int e^{itx} \cdot dF(x), t \in \Re$, is called the *characteristic function (c.f.)* of X.

While it is clear enough from the boundedness of the sine and cosine that $\phi(t)$ is finite at each $t \in \Re$, there is another way to see it. Since the cosine is an even function ($\cos \theta = \cos(-\theta)$) and the sine is odd ($\sin \theta = -\sin(-\theta)$), $e^{i\theta}$ and $e^{-i\theta}$ are complex conjugates, and so $|e^{i\theta}| = \sqrt{e^{i\theta} e^{-i\theta}} = \sqrt{\cos^2 \theta + \sin^2 \theta} = 1$ for all real θ. Thus, the absolute-value inequality and the dominance property of the integral imply that

$$|\phi(t)| \le E |e^{itX}| = \int |e^{itx}| \cdot dF(x) = \int 1 \cdot dF(x) = 1.$$

Besides confirming the finiteness of $|\int e^{itx} \cdot dF(x)|$ for *every* distribution F at *each* $t \in \Re$, this shows that all values of $\phi(\cdot)$ lie on or within a circle of unit radius about the origin in the complex plane. Thus, the integral transform $\phi(\cdot)$ sends each element of $\mathfrak{F}$ (the space of distributions) onto a subspace of the space $\mathfrak{C}$ of complex-valued *functions* (those that take values in the space C of complex *numbers*). The c.f. thus avoids the convergence problem that is associated with the m.g.f.

Finding the c.f. is easy—indeed, trivial—if the m.g.f. does exist and has a known form, for in that case $\phi(t) = M(it)$. In effect, i is simply treated as a constant with the special property $i^2 = -1$ in working out the integral $\int e^{itx} \cdot dF$. When the m.g.f. does not exist, some special integration tricks (or a good table of integrals) may be required. The c.f.s of many distributions encountered in applied work do not have simple "closed" forms and can be expressed only as convergent series. On the other hand, there are also distributions whose c.f.s have simple forms but whose densities do not.

Example 3.36. The c.f.s corresponding to the first four m.g.f.s in Example 3.35 are (i) $\phi(t) = \frac{e^{it}-1}{it}$, (ii) $\phi(t) = \frac{\theta}{\theta-it}$, (iii) $\phi(t) = (1+t^2)^{-1}$, and (iv) $\phi(t) = e^{it}(1-\theta)/(1-\theta e^{it})$. Exercise 3.59 below establishes that the Cauchy c.f. is $\phi(t) = e^{-|t|}$.

Example 3.37. The m.g.f. corresponding to standard normal p.d.f. $f(x) = (2\pi)^{-1/2}\exp(-x^2/2)$ is

$$M(t) = \int_{-\infty}^{\infty} \frac{1}{\sqrt{2\pi}}e^{tx-x^2/2}\cdot dx$$

$$= e^{t^2/2}\int_{-\infty}^{\infty} \frac{1}{\sqrt{2\pi}}e^{-(x^2-2tx+t^2)/2}\cdot dx$$

$$= e^{t^2/2}\int_{-\infty}^{\infty} \frac{1}{\sqrt{2\pi}}e^{-(x-t)^2/2}\cdot dx.$$

Changing variables as $z = x - t$ gives

$$M(t) = e^{t^2/2}\int_{-\infty}^{\infty} \frac{1}{\sqrt{2\pi}}e^{-z^2/2}\cdot dz = e^{t^2/2}. \tag{3.44}$$

Hence,

$$\phi(t) = M(it) = e^{(it)^2/2} = e^{-t^2/2}. \tag{3.45}$$

Example 3.38. Here is another way to establish (3.45). Since $\int e^{itx}\frac{1}{\sqrt{2\pi}}e^{-x^2/2}\cdot dx$ is finite for all $t \in \mathfrak{R}$, the dominated convergence theorem implies (recall Example 3.10) that

$$\phi'(t) = \int \frac{\partial e^{itx}}{\partial t}\frac{1}{\sqrt{2\pi}}e^{-x^2/2}\cdot dx$$

$$= \int e^{itx}\frac{ix}{\sqrt{2\pi}}e^{-x^2/2}\cdot dx, t \in \mathfrak{R}.$$

Integrating by parts with $u = e^{itx}$, $du = ite^{itx}\cdot dx$, $dv = \frac{ix}{\sqrt{2\pi}}e^{-x^2/2}\cdot dx$, and $v = -\frac{i}{\sqrt{2\pi}}e^{-x^2/2}$, we have

$$\phi'(t) = -ie^{itx}\frac{e^{-x^2/2}}{\sqrt{2\pi}}\Bigg|_{-\infty}^{\infty} + i^2t\int_{-\infty}^{\infty} e^{itx}\frac{1}{\sqrt{2\pi}}e^{-x^2/2}\cdot dx$$

$$= 0 - t\phi(t).$$

Solving the differential equation $\phi'(t) + t\phi(t) = 0$ subject to $\phi(0) = 1$ gives (3.45).

Example 3.39. The symmetric "stable" laws have served importantly as models for various noisy random processes, such as day-to-day changes in prices of financial assets; e.g., Mandelbrot(1963), McCulloch (1996). These have c.f.

$$\phi(t) = \exp\left(i\delta t - \gamma|t|^{\alpha}\right), \qquad (3.46)$$

where $\delta \in \Re$ is the median, $\gamma > 0$ is a scale parameter, and $\alpha \in (0, 2]$ governs the thickness of the tails. The normal distributions ($\alpha = 2$) and the Cauchy distributions ($\alpha = 1$) are the only ones whose densities have closed forms. Distributions with $\alpha > 1$ possess means (equal to δ), but the normals are the only members with finite variances.

Exercise 3.53. *Random variable X has the standard normal distribution, as in Example 3.37. Use (3.45) to determine the values of (i) $E \sin X$ and (ii) $E \cos X$ without doing any calculations.*

Here are a few important features of c.f.s. First, the conjugate relation between e^{itX} and e^{-itX} implies that $\overline{\phi(t)}$ (the conjugate of $\phi(t)$) is simply $\phi(-t)$, which is the c.f. of $-X$ at t. The conjugate of any c.f. is therefore also a c.f. Next, if r.v. X is distributed symmetrically about the origin, then X and $-X$ have the same distribution, in which case $\phi(t) = Ee^{itX} = Ee^{it(-X)} = \overline{\phi(t)}$. Thus, c.f.s of r.v.s distributed symmetrically about the origin are real valued. The converse—that X is symmetric about the origin if Ee^{itX} is real-valued—is also true and follows from the uniqueness property of c.f.s, described below. Of course, $\phi(0) = Ee^{0 \cdot X} = 1$ for any r.v. X. Since tX and X have the same distribution for all t only if $X = 0$ a.s., the uniqueness property also implies that such degenerate r.v.s are the only ones for which $\phi(t) \equiv 1$. Finally, the continuity of e^{itx} as a function of t confers the same property on the c.f. Indeed, it is *uniformly* continuous, since

$$|\phi(t+h) - \phi(t)| = \left|Ee^{itX}\left(e^{ihX} - 1\right)\right|$$
$$\leq E\left[\left|e^{itX}\right|\left|e^{ihX} - 1\right|\right] = E\left[\left|e^{ihX} - 1\right|\right],$$

and the last expression converges to zero as $h \to 0$ at a rate that does not depend on t.

Exercise 3.54. *X has p.d.f. $f(x) = \frac{1}{2\pi}\mathbf{1}_{(-\pi,\pi)}(x)$. Show that the c.f. is $\phi(t) = \frac{\sin(\pi t)}{\pi t}$ for $t \neq 0$, and verify its continuity at $t = 0$.*

The value of the c.f. at an arbitrary $t_0 \neq 0$ has a simple geometrical interpretation. To illustrate, consider an absolutely continuous X with

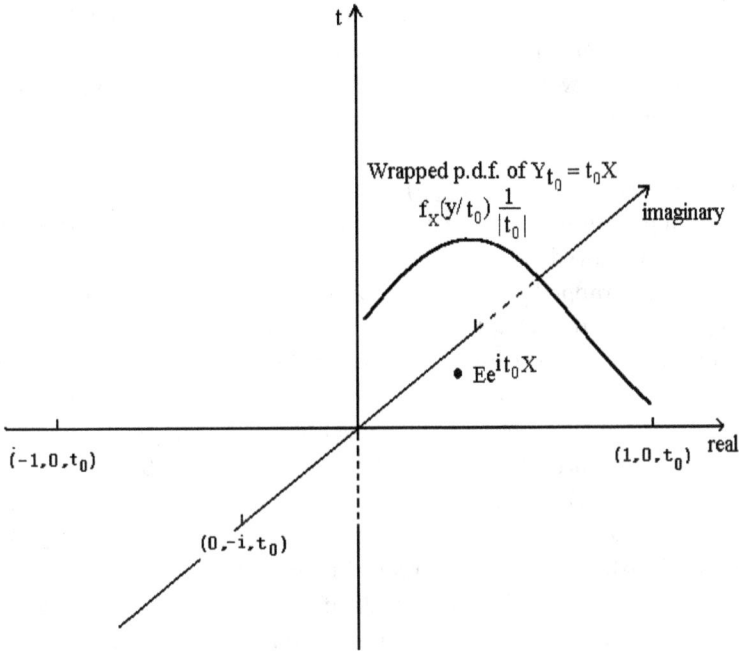

Fig. 3.9 A "wrapped-around" p.d.f. and its center of mass on the complex unit circle.

p.d.f. f_X, and put $Y_{t_0} := t_0 X$. Change-of-variable formula (2.12) shows the p.d.f. of Y_{t_0} to be $f_{t_0}(y) := f_X(y/t_0)\frac{1}{|t_0|}$. Now visualize the unit circle in the complex plane—a circle of unit radius, centered at $(0,0)$, that crosses the real axis at $(1,0)$ and $(-1,0)$ and crosses the imaginary axis at $(0,i)$ and $(0,-i)$. This is depicted in Fig. 3.9 from an oblique perspective as the lightly shaded disk. The t axis is represented in the figure as a line perpendicular to the plane of the disk and passing through the center of the disk at the point t_0 on that axis. Visualizing p.d.f. $f_{t_0}(\cdot)$ as a two-dimensional figure standing up with its base in the complex plane, imagine wrapping $f_{t_0}(\cdot)$ around the edge of the unit disk. The result is depicted by the darkly shaded, bell-shaped curve in the figure. The process of wrapping simply places the density $f_X(x)$ associated with any real number x at the point $e^{it_0 x} = \cos(t_0 x) + i\sin(t_0 x)$ in the complex plane—that is, on the edge of the disk at an angle $t_0 x$ from the positive real axis.

Now just as EX can be regarded as the center of mass of the two-dimensional f_X, the value of the c.f. at t_0, $\phi(t_0) = Ee^{it_0 X}$, can be visualized as the center of mass of this three-dimensional wrapped-around distribution f_{t_0}. This is the single point on the unit disk—the black dot in the figure—at which the disk would be balanced if supported there from below. To visualize how ϕ changes as t increases from some positive t_0, think of sliding the disk upward along the vertical axis. This increases the scale of $Y_t = tX$ and *stretches* the distribution $f_t(\cdot)$; that is, the density $f_X(x)$ at any $x > 0$ is shifted in a counterclockwise direction about the circle, while the density at any $x < 0$ is moved clockwise. Of course, such a change moves the center of mass of the wrapped-around distribution. As t continues to increase, the tails of $f_t(\cdot)$ progressively lap around and around the circle, making the distribution more nearly uniform and pushing the center of mass toward the circle's center at $(0,0)$.[24] Going the other way and moving the disk down the vertical axis toward $t = 0$ concentrates the wrapped-around p.d.f. more and more tightly. Ultimately, at $t = 0$ *all* the probability mass of e^{itX} is at $(1,0)$, and so $\phi(0) = 1 + i \cdot 0 = 1$. Of course, as the disk is slid further down and t becomes negative the mass again spreads out and $\phi(t)$ drifts away from $(1,0)$. Tracking the center of mass as t varies continuously from $-\infty$ to ∞, one can visualize the c.f. itself as a continuous curve contained within an infinitely long cylinder of unit radius. See Fig. 3.10.

Generating Moments with C.F.s and M.G.F.s.   Finding a moment of a r.v. from the definition requires working out a definite integral, as when one evaluates $\int_{-\infty}^{\infty} x^k \cdot dF$ to find $\mu'_k = EX^k$. Finding simple expressions for such definite integrals is sometimes hard, and it has to be done all over again for each value of k. Of course, finding the c.f. also requires working out an integral; however, once the function ϕ has been found, any finite moment μ'_k can be recovered just by *differentiation*. Specifically, μ'_k is found by differentiating the c.f. k times and evaluating the derivative at $t = 0$.

   To get a sense of why this is so, consider a discrete r.v. X whose support $\mathbb{X}$ is a finite set of real numbers. For such a r.v. all positive integer moments are finite; moreover, they and the c.f. can be calculated as finite

---

[24]C.f.s of absolutely continuous distributions, and of *some* continuous-singular ones, do in fact approach $(0,0)$ as $|t| \to \infty$, while c.f.s of discrete r.v.s satisfy $\lim_{|t| \to \infty} \sup |\phi(t)| = 1$. Thus, any r.v. whose c.f. vanishes in the limit is continuous, although not necessarily *absolutely* so. Discrete r.v.s with "lattice" distributions (those supported on equally spaced points) have strictly periodic c.f.s. For example, c.f.s of models for *count* data, such as the Poisson, binomial, and negative binomial, repeat at intervals of 2π, with $\phi(t) = 1$ for $t \in \{\pm 2\pi j\}_{j \in \mathbb{N}_0}$. For other properties of c.f.s that can be visualized geometrically see Epps (1993).

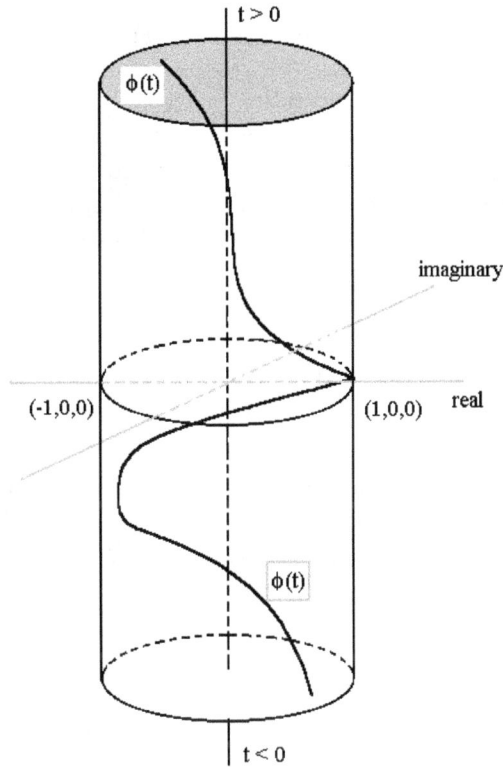

Fig. 3.10 $\phi(t)$ as a curve contained in the unit complex cylinder.

sums, as $\mu'_k = \sum_{x \in \mathbb{X}} x^k f(x)$ and $\phi(t) = \sum_{x \in \mathbb{X}} e^{itx} f(x)$. Differentiating ϕ with respect to t gives

$$\phi'(t) = \sum_{x \in \mathbb{X}} \frac{d\left(e^{itx}\right)}{dt} f(x) = \sum_{x \in \mathbb{X}} ixe^{itx} f(x),$$

so that

$$\phi'(0) := \phi'(t)|_{t=0} = i \sum_{x \in \mathbb{X}} xf(x) = i\mu,$$

or i times the first moment. (Because the sum has finitely many terms, there is no concern about reversing the order of differentiation and summation.)

Differentiating $k - 1$ more times and evaluating at $t = 0$ give

$$\phi^{(k)}(0) = i^k \sum_{x \in X} x^k f(x) = i^k \mu_k'.$$

In fact, the moment-generating property of the c.f. is not limited to discrete r.v.s with finite support but applies whenever moments are finite. As we know, the finiteness of moments is tied to the tail behavior of the distribution, in that $f(x)$—the density or mass at x—must drop off rapidly enough as $|x| \to \infty$ so that integrals $\int |x|^k f(x) \cdot dx$ or sums $\sum |x|^k f(x)$ converge. Corresponding to thinness of tails in the distribution is the *smoothness* of ϕ at the origin.[25] The more finite moments there are of X, the more derivatives $\phi(t)$ has at $t = 0$. The following gives the precise connection between differentiability and the existence of finite moments:[26]

Theorem 3.3.

(i) *If r.v. X has k finite moments, then its c.f. has at least k uniformly continuous derivatives.*

(ii) *If the c.f. of X has k derivatives at the origin, then all positive integer moments to order k are finite if k is even. If k is odd, only moments to order $k - 1$ are necessarily finite.*

Example 3.40. The first moment of the standard Cauchy variate fails to exist. Its c.f., $\phi(t) = e^{-|t|}$, is not differentiable at the origin since (by the mean-value theorem, where $|t^*| < |t|$) $\left(e^{-|t|} - e^0\right)/t = -|t|/t + e^{-|t^*|}|t|/2$ tends to $+1$ as $t \uparrow 0$ and to -1 as $t \downarrow 0$. Thus, the left-hand and right-hand derivatives at $t = 0$ are not the same. Symmetric stable c.f.s (3.46) with $\alpha \in (1, 2)$ do have finite means, although they are only once differentiable at $t = 0$. Normal r.v.s (stable with $\alpha = 2$) have infinitely differentiable c.f.s and possess all positive integer moments. (For examples of r.v.s with differentiable c.f.s but without first moments see Bisgaard and Sasvari (2000, p. 51) or Lukacs (1970, p. 22).)

Exercise 3.55. *If the c.f. of X is twice differentiable at the origin, show that $VX = -d^2 \log \phi(t)/dt^2\big|_{t=0}$.*

---

[25] Generally, the behavior of ϕ near the origin is influenced most strongly by the behavior of f in the extreme tails, while the behavior of ϕ for large $|t|$ responds to that of f at small values of $|x|$. One can see this intuitively as follows. When $|t| \doteq 0$ it is only when $|x|$ is large that variations in t much alter the values of e^{itx} and, therefore, of $|\phi(t)| = |\int e^{itx} f(x) \cdot dx|$. Conversely, it is when $|t|$ is large that $|\phi(t)|$ is sensitive to variations of f near the origin.

[26] For a proof see any of Billingsley (1986,1995), Chung (1974), Laha and Rohatgi (1979), or Lukacs (1970).

If a random variable has a moment generating function, then *all* its moments are finite, because for $t \neq 0$ the exponential $e^{|tx|}$ dominates any polynomial of finite degree in tx as $|x| \to \infty$. Therefore, only r.v.s whose c.f.s have infinitely many derivatives at the origin possess m.g.f.s. When the m.g.f. does exist, moments about the origin can be found as with c.f.s by differentiating $M(t)$ and evaluating at $t = 0$ as in Example 3.10, and there are no imaginary units to deal with in this case. Thus, $M^{(k)}(t)\big|_{t=0} = \mu'_k$. Sometimes useful for generating central moments directly is the *central m.g.f.*,

$$m(t) := e^{-t\mu} M(t) = E e^{t(X-\mu)}, \tag{3.47}$$

from which $m^{(k)}(t)\big|_{t=0} = \mu_k$.

Example 3.41. The m.g.f. corresponding to p.m.f. $f(x) = \left(\theta^x e^{-\theta}/x!\right) 1_{\aleph_0}(x)$, $\theta > 0$ (the Poisson distribution) is

$$M(t) = e^{-\theta} \sum_{x=0}^{\infty} \frac{(\theta e^t)^x}{x!} = e^{-\theta} e^{\theta e^t} = \exp\left[\theta\left(e^t - 1\right)\right]. \tag{3.48}$$

Then $\mu = M'(t)\big|_{t=0} = \theta e^0 M(0) = \theta$, $\mu'_2 = M''(t)\big|_{t=0} = \left(\theta^2 e^{2 \cdot 0} + \theta e^0\right) M(0) = \theta^2 + \theta$, and $\sigma^2 = \mu'_2 - \mu^2 = \theta$. The variance comes directly by differentiating twice $m(t) = \exp\left[\theta\left(e^t - 1 - t\right)\right]$ and evaluating $m''(t)$ at $t = 0$.

A Taylor expansion of $\log M(t)$ about $t = 0$ gives

$$\log M(t) = \sum_{\ell=1}^{k} \kappa_\ell \frac{t^\ell}{\ell!} + R_{k+1}.$$

Here R_{k+1} is a remainder term, and $\kappa_\ell := d^\ell \log M(t)/dt^\ell\big|_{t=0}$ is the ℓth *cumulant* of X. For this reason the log of the m.g.f. is referred to as the *cumulant generating function* (c.g.f.). We shall see later that cumulants are a particularly useful way of describing distributions of sums of independent r.v.s. When the r.v. has k moments, the first k cumulants can also be obtained from a kth-order expansion of the c.f., as[27]

$$\log \phi(t) = \sum_{\ell=1}^{k} \kappa_\ell \frac{(it)^\ell}{\ell!} + R_{k+1}. \tag{3.49}$$

---

[27] Recall that the existence of the kth derivative of ϕ confirms the existence of k moments only if k is *even*.

Table 3.1   Central moments in terms of cumulants.

$$\mu_2 = \kappa_2$$
$$\mu_3 = \kappa_3$$
$$\mu_4 = \kappa_4 + 3\kappa_2^2$$
$$\mu_5 = \kappa_5 + 10\kappa_2\kappa_3$$
$$\mu_6 = \kappa_6 + 15\kappa_2\kappa_4 + 10\kappa_3^2 + 15\kappa_2^3$$
$$\mu_7 = \kappa_7 + 21\kappa_2\kappa_5 + 35\kappa_3\kappa_4 + 105\kappa_2^2\kappa_3$$
$$\mu_8 = \kappa_8 + 28\kappa_2\kappa_6 + 56\kappa_3\kappa_5 + 35\kappa_4^2 + 210\kappa_2^2\kappa_4 + 280\kappa_2\kappa_3^2 + 105\kappa_2^4$$
$$\mu_9 = \kappa_9 + 36\kappa_2\kappa_7 + 84\kappa_3\kappa_6 + 126\kappa_4\kappa_5 + 378\kappa_2^2\kappa_5 + 1260\kappa_2\kappa_3\kappa_4$$
$$\qquad + 280\kappa_3^3 + 1260\kappa_2^3\kappa_3$$
$$\mu_{10} = \kappa_{10} + 45\kappa_2\kappa_8 + 120\kappa_3\kappa_7 + 210\kappa_4\kappa_6 + 630\kappa_2^2\kappa_6 + 126\kappa_5^2 + 2520\kappa_2\kappa_3\kappa_5$$
$$\qquad + 1575\kappa_2\kappa_4^2 + 2100\kappa_3^2\kappa_4 + 3150\kappa_2^3\kappa_4 + 6300\kappa_2^2\kappa_3^2 + 945\kappa_2^5$$

It is easy to see that $\kappa_1 = \mu$ and (Exercise 3.55) that $\kappa_2 = \sigma^2$. Likewise, from

$$\frac{d^3 \log M}{dt^3} = \frac{M'''}{M} - 3\frac{M''(M')^2}{M^2} + 2\frac{(M')^3}{M^3}$$

one sees that $\kappa_3 = \mu_3' - 3\mu_2'\mu^2 + 2\mu^3 = E(X - \mu)^3 = \mu_3$. At this point the exact correspondence between moments and cumulants ends, for differentiating once again shows that $\kappa_4 = \mu_4 - 3\mu_2^2 = \mu_4 - 3\sigma^4$. Thus, $\mu_4 = \kappa_4 + 3\kappa_2^2$. In general, a moment of any order k can be expressed in terms of cumulants up to the same order. Table 3.1 shows the correspondence for central moments μ_k up to $k = 10$. Note that if X is in standard form ($EX = 0$ and $VX = 1$), then $\kappa_1 = 0$, $\kappa_2 = 1$, $\kappa_3 = \alpha_3$ (the coefficient of skewness), and $\kappa_4 = \alpha_4 - 3$ (the *excess* kurtosis).

If a r.v.'s m.g.f. exists and has a tractable form, differentiating c.g.f. $\log M(t)$ to find the cumulants is often the fastest way to obtain the moments. For example, by expressing $\log M(t)$ in Example 3.41 one can see at once that the cumulants of *all* orders equal θ.

Exercise 3.56. *If X is normally distributed, show that all its cumulants of order three and higher are identically zero.*

Exercise 3.57. *Express the cumulants $\{\kappa_\ell(Y)\}$ of $Y = aX$ in terms of the cumulants of X.*

The Uniqueness Property of C.F.s.   We have seen that the integral transform $Ee^{itX} = \int e^{itx} \cdot dF$ exists for all c.d.f.s F and therefore maps the entire space $\mathfrak{F}$ of distributions into the space $\mathfrak{C}$ of complex-valued functions on $\mathfrak{R}$. Let us denote by $\mathbb{C}$ the subset of $\mathfrak{C}$ that corresponds to c.f.s. Obviously, not just any complex-valued function of a real variable can be a c.f.—for example, all c.f.s satisfy $\phi(0) = 1$.[28] However, the onto map from $\mathfrak{F}$ to $\mathbb{C}$ is in fact one to one. Thus, knowing the value of $\phi(t)$ for *each* real t tells us precisely the distribution that corresponds.

As it does for m.g.f.s, this uniqueness property of the c.f. follows from the fact that there is a specific recipe for recovering F from the generating function. However, this inversion process is often easier for c.f.s than for m.g.f.s. Among several known inversion formulas here are two that are particularly useful.[29] Notice that they give values of the c.d.f. only at continuity points, where there is no probability mass. Of course, the value at any x where $F(x) - F(x-) > 0$ can be found by *right* continuity, as $F(x) = \lim_{n \to \infty} F(x + n^{-1})$.

Theorem 3.4.

(i) If $\mathbb{P}_X(\{a\}) = \mathbb{P}_X(\{b\}) = 0$,

$$F(b) - F(a) = \lim_{n \to \infty} \int_{-n}^{n} \frac{e^{-ita} - e^{-itb}}{2\pi it} \phi(t) \cdot dt. \tag{3.50}$$

(ii) If $\mathbb{P}_X(\{x\}) = 0$,

$$F(x) = \frac{1}{2} - \lim_{n \to \infty} \int_{-n}^{n} \frac{e^{-itx}}{2\pi it} \phi(t) \cdot dt. \tag{3.51}$$

Note that the integrals here represent Cauchy principal values of improper integrals from $-\infty$ to $+\infty$, which sometimes do not exist in the Lebesgue construction.

Example 3.42. For an increasing sequence of constants $\{x_j\}_{j=1}^{\infty}$ let $\mathbb{P}_X(x_j) = p_j > 0$, where $\sum_{j=1}^{\infty} p_j = 1$, so that $\phi(t) = Ee^{itX} = \sum_{j=1}^{\infty} p_j e^{itx_j}$. Then (3.51) gives

$$F(x) = \frac{1}{2} - \frac{1}{2\pi} \sum_{j=1}^{\infty} p_j I_j(x),$$

where

$$I_j(x) := \lim_{n \to \infty} \int_{-n}^{n} \frac{e^{it(x_j - x)}}{it} \cdot dt$$

$$= \frac{1}{i} \lim_{n \to \infty} \int_{-n}^{n} \frac{\cos t(x_j - x)}{t} \cdot dt + \lim_{n \to \infty} \int_{-n}^{n} \frac{\sin t(x_j - x)}{t} \cdot dt.$$

---

[28] Bochner's theorem (Lukacs, 1970) gives the precise necessary and sufficient conditions for a function to be a c.f.

[29] For proofs see Bisgaard and Sasvári (2000).

It is easy to see that the Cauchy principal value of the first integral equals zero for each n.[30] Putting $u := t\,(x_j - x)$ in the second integral, we have

$$I_j\,(x) = \lim_{n\to\infty} \int_{-n(x_j-x)}^{n(x_j-x)} \frac{\sin u}{u} \cdot du.$$

Now one can discover (e.g., from a table of definite integrals or as an exercise in complex integration) that $\int_{-\infty}^{\infty} \frac{\sin u}{u} \cdot du = \pi$; therefore, we have $I_j\,(x) = \pi \cdot sign\,(x_j - x)$. If we pick $m \in \mathbb{N}$ and $x \in (x_m, x_{m+1})$ we thus have

$$F\,(x) = \frac{1}{2} - \frac{1}{2\pi}\left(-\pi \sum_{j=1}^{m} p_j + \pi \sum_{j=m+1}^{\infty} p_j\right)$$

$$= \frac{1}{2} + \frac{1}{2}\left[\sum_{j=1}^{m} p_j - \left(1 - \sum_{j=1}^{m} p_j\right)\right]$$

$$= \sum_{j=1}^{m} p_j,$$

while for $x < x_1$ we have $F\,(x) = \frac{1}{2} - \frac{1}{2}\sum_{j=1}^{\infty} p_j = 0$ and for $x > x_\infty$ we have

$$F\,(x) = \frac{1}{2} + \frac{1}{2}\sum_{j=1}^{\infty} p_j = 1.$$

Example 3.43. From part (i) of Example 3.36 the c.f. corresponding to $f\,(x) = \mathbf{1}_{[0,1]}\,(x)$ is $\phi(t) = \frac{e^{it}-1}{it}$. Expression (3.50) gives

$$F\,(1) - F\,(0) = \lim_{n\to\infty} \int_{-n}^{n} \left(\frac{1 - e^{-it}}{2\pi it}\right) \frac{e^{it} - 1}{it} \cdot dt$$

$$= \frac{1}{\pi} \lim_{n\to\infty} \int_{-n}^{n} \frac{1 - \left(e^{it} + e^{-it}\right)/2}{t^2} \cdot dt$$

$$= \frac{1}{\pi} \lim_{n\to\infty} \int_{-n}^{n} \frac{1 - \cos t}{t^2} \cdot dt$$

$$= \frac{1}{\pi} \int_{-\infty}^{\infty} \frac{\sin^2\,(t/2)}{(t/2)^2} \cdot d\,(t/2) = 1.$$

---

[30]Putting $\delta := x_j - x$, the Cauchy principal value is

$$\lim_{\varepsilon\to\infty} \left[\int_{-n}^{-\varepsilon} t^{-1}\cos t\delta \cdot dt + \int_{\varepsilon}^{n} t^{-1}\cos t\delta \cdot dt\right] = 0.$$

A good table of integrals, such as Gradshteyn *et al.* (2000), can sometimes deliver simple forms for the integrals in (3.50) and (3.51), but they must often be evaluated numerically. The following special case applies to many continuous distributions and is often easier to apply.

Theorem 3.5. *If the c.f. of a r.v. X is absolutely integrable, meaning that $\int_{-\infty}^{\infty} |\phi(t)| \cdot dt = \int \sqrt{\phi(t)\,\phi(-t)} \cdot dt < \infty$, then X is a continuous r.v. with continuous, bounded p.d.f.*

$$f(x) = \frac{1}{2\pi} \int_{-\infty}^{\infty} e^{-itx} \phi(t) \cdot dt. \tag{3.52}$$

Example 3.44. The standard normal c.f., $\phi(t) = e^{-t^2/2}$, is absolutely integrable, since $\int \left| e^{-t^2/2} \right| \cdot dt = \int e^{-t^2/2} \cdot dt = \sqrt{2\pi}$. Thus,

$$f(x) = \frac{1}{2\pi} \int_{-\infty}^{\infty} e^{-itx} e^{-t^2/2} \cdot dt$$

$$= \frac{1}{2\pi} e^{-x^2/2} \int_{-\infty}^{\infty} e^{-(t^2+2itx+i^2x^2)/2} \cdot dt$$

$$= \frac{1}{\sqrt{2\pi}} e^{-x^2/2} \int_{-\infty}^{\infty} \frac{1}{\sqrt{2\pi}} e^{-(t+ix)^2/2} \cdot dt.$$

Here we are integrating over the line $(-\infty, ix)$ to $(+\infty, ix)$ in the complex plane, parallel to and x units from the real axis. The area under the bell curve is the same as if it had been placed along real axis itself, or $\int_{-\infty}^{\infty} (2\pi)^{-1/2} e^{-z^2/2} \cdot dz = 1$. Thus, the formal substitution $z = t + ix$ yields

$$f(x) = \frac{1}{\sqrt{2\pi}} e^{-x^2/2}, \; x \in \Re.$$

Note that not all continuous random variables have absolutely integrable c.f.s. For example, the function $(1 - it)^{-\alpha}$ (the c.f. of a *gamma* r.v.) is not absolutely integrable when $\alpha \in (0,1]$, since

$$\int_{-\infty}^{\infty} \left|(1-it)^{-\alpha}\right| \cdot dt = \int_{-\infty}^{\infty} \sqrt{(1-it)^{-\alpha}(1+it)^{-\alpha}} \cdot dt$$

$$= \int_{-\infty}^{\infty} \left(1+t^2\right)^{-\alpha/2} \cdot dt$$

$$= 2\int_0^{\infty} \left(1+t^2\right)^{-\alpha/2} \cdot dt$$

$$= 2\int_0^1 \left(1+t^2\right)^{-\alpha/2} \cdot dt + 2\int_1^{\infty} \left(1+t^2\right)^{-\alpha/2} \cdot dt$$

$$> 2^{1-\alpha/2} + 2\int_1^{\infty} \left(2t^2\right)^{-\alpha/2} \cdot dt$$

$$= 2^{1-\alpha/2}\left(1 + \int_1^{\infty} t^{-\alpha} \cdot dt\right) = +\infty$$

when $\alpha \leq 1$.[31]

Exercise 3.58. *Verify that the (standard) Cauchy c.f., $\phi(t) = e^{-|t|}$, is absolutely integrable, and use (3.52) to show that the p.d.f. is $\left[\pi\left(1+x^2\right)\right]^{-1}$ for $x \in \Re$.*

Exercise 3.59. *Recognizing from part (iii) of Example 3.36 that $\left(1+t^2\right)^{-1}$ is the c.f. of the Laplace distribution, apply Theorem 3.5 to show that the Cauchy c.f. is $\phi\left(t\right) = e^{-|t|}$.*

C.F.s of Vector-Valued R.V.s. Just as there are joint c.d.f.s, p.d.f.s, and p.m.f.s to describe the distribution over $\Re^k$ of vector-valued r.v.s, there is also a joint c.f. The joint c.f. of $\mathbf{X} = (X_1, X_2, ..., X_k)'$ at any $\mathbf{t} \in \Re^k$ is given by

$$\phi(\mathbf{t}) := Ee^{i\mathbf{t}'\mathbf{X}} = \int_{\Re^k} \exp\left(i\sum_{j=1}^k t_j x_j\right) \cdot dF(\mathbf{x}).$$

If we know the form of the joint c.f., we can do two useful things. First, we can find the (joint) *marginal* c.f. of any subset of $\ell < k$ variables merely

---

[31] However, while absolute integrability is sufficient for (3.52), it is not *necessary*; for example, it is possible to show that

$$\frac{1}{2\pi} \int_{-\infty}^{\infty} e^{itx}\left(1-it\right)^{-1} \cdot dt = e^{-x}\mathbf{1}_{[0,\infty)}\left(x\right),$$

which is the p.d.f. of a member of the exponential subfamily of gammas.

by setting to zero the $k - \ell$ elements of $\mathbf{t}$ corresponding to the excluded variables. For example, the marginal c.f. of X_1 is

$$\phi_1(t) = \phi(t, 0, ..., 0) = E \exp\left[i(tX_1 + 0 \cdot X_2 + \cdots + 0 \cdot X_k)\right] = Ee^{itX_1}.$$

Second, whenever $\phi(\mathbf{t})$ is differentiable, the product moment $E(X_1^{m_1} X_2^{m_2} \cdot \cdots \cdot X_k^{m_k})$ can be found by differentiating $\phi(\mathbf{t})$ partially m_1 times with respect to t_1, m_2 times with respect to t_2, and so on, evaluating at $\mathbf{t} = \mathbf{0}$, and dividing by $i^{m_1 + m_2 + \cdots + m_k}$.

Example

3.45. X_1, X_2 have p.d.f. $f(x_1, x_2) = 2e^{-x_1-x_2} 1_{\{(x_1,x_2):0\leq x_1\leq x_2\}}(x_1, x_2)$. The joint c.f. is

$$\phi(t_1, t_2) = 2 \int_0^\infty \int_0^{x_2} e^{it_1 x_1 + it_2 x_2} e^{-x_1-x_2} \cdot dx_1\, dx_2$$

$$= \frac{2}{(1 - it_2)(2 - it_1 - it_2)}.$$

The marginal c.f.s are

$$\phi_1(t) = \phi(t, 0) = \frac{2}{2 - it}$$

$$\phi_2(t) = \phi(0, t) = \frac{2}{(1 - it)(2 - it)}.$$

The first moments are

$$EX_1 = \left.\frac{\partial \phi(t_1, t_2)}{i \partial t_1}\right|_{t=0} = \frac{1}{2}$$

$$EX_2 = \left.\frac{\partial \phi(t_1, t_2)}{i \partial t_2}\right|_{t=0} = \frac{3}{2},$$

and the first product moment is

$$EX_1 X_2 = \left.\frac{\partial^2 \phi(t_1, t_2)}{i^2 \partial t_1 \partial t_2}\right|_{t=0} = 1.$$

Recall that the joint p.d.f./p.m.f./c.d.f. of mutually independent r.v.s equals the product of their respective marginal p.d.f.s/p.m.f.s/c.d.f.s. The same product rule applies for c.f.s, a consequence of the fact that expectations of products of functions of independent r.v.s equal products of expectations. Thus, when $\{X_j\}_{j=1}^k$ are independent,

$$\phi(\mathbf{t}) = Ee^{i(t_1 X_1 + t_2 X_2 + \cdots + t_k X_k)} = \prod_{j=1}^k Ee^{it_j X_j} = \prod_{j=1}^k \phi_j(t_j). \tag{3.53}$$

Naturally, joint m.g.f.s, when they exist, have properties that correspond to those of c.f.s.

Example 3.46. (i) If $\{X_j\}_{j=1}^k$ are independent r.v.s each distributed as standard normal, then the joint c.f. is $\phi(\mathbf{t}) = \prod_{j=1}^k e^{-t_j^2/2}$.

(ii) If X_1 is distributed as standard normal, X_2 is distributed as standard Cauchy, and X_1, X_2 are independent, then $\phi(\mathbf{t}) = e^{-t_1^2/2 - |t_2|}$.

C.F.s of Affine Functions of R.V.s. One of the most fruitful applications of c.f.s and other generating functions is in determining distributions of affine functions of r.v.s—particularly of mutually independent r.v.s. This is often much easier than working out convolution sums or integrals (Section 2.5.4). The most basic result pertains to affine functions of a single r.v. If X has c.f. ϕ_X and $Y := a + bX$ for arbitrary real constants a, b, then

$$\phi_Y(t) = Ee^{itY} = Ee^{ita+itbX} = e^{ita}Ee^{itbX} = e^{ita}\phi_X(tb). \qquad (3.54)$$

Example 3.47. If X is distributed as standard normal and $Y = a + bX$, then $\phi_Y(t) = e^{ita - t^2b^2/2}$. If $b \neq 0$, this is the c.f. of a normal variate with mean $\mu = a$, variance $\sigma^2 = b^2$, and standard deviation $\sigma = |b|$.

Exercise 3.60. *Random variables X_1 and X_2 are independent. Use Euler's identity (3.43) to verify that $Ee^{i(X_1+X_2)} = Ee^{iX_1} \cdot Ee^{iX_2}$.*

Extending to the vector case, let $\mathbf{X} = (X_1, X_2, ..., X_n)'$ be mutually independent with marginal c.f.s $\{\phi_j\}_{j=1}^n$. For constants a_0 and $\mathbf{a} = (a_1, a_2, ..., a_n)'$ the c.f. of $Y = a_0 + \mathbf{a}'\mathbf{X}$ is

$$\phi_Y(t) = E \exp\left(ita_0 + it\sum_{j=1}^n a_j X_j\right) = e^{ita_0}\prod_{j=1}^n \phi_j(ta_j). \qquad (3.55)$$

As an important special case, the c.f. of a sum of mutually independent and identically distributed (i.i.d.) r.v.s with c.f. ϕ is $\phi_Y = \phi^n$. Knowing ϕ_Y, the c.d.f. F_Y can always be found using an appropriate inversion formula, although sometimes this can be done only numerically. Fortunately, there are many cases in which no inversion is needed, for one can just recognize ϕ_Y as the c.f. of a particular distribution.

Example 3.48. If $\{X_j\}_{j=1}^n$ are independent standard normals, the c.f. of $Y = n^{-1/2}\sum_{j=1}^n X_j$ is

$$\phi_Y(t) = \prod_{j=1}^n \phi_j\left(t/\sqrt{n}\right) = \left[\exp\left(-\frac{t^2}{2n}\right)\right]^n = e^{-t^2/2},$$

showing that Y itself is distributed as standard normal.

Example 3.49. If $\{X_j\}_{j=1}^n$ are independent standard Cauchys, the c.f. of $Y = n^{-1} \sum_{j=1}^n X_j$ is

$$\phi_Y(t) = \prod_{j=1}^n \phi(t/n) = [\exp(-|t|/n)]^n = e^{-|t|}.$$

More generally, suppose $\{X_j\}_{j=1}^n$ are i.i.d. as Cauchy with median δ, scale γ, and c.f. $\phi(t) = \exp(it\delta - \gamma|t|)$, and put $Y := \sum_{j=1}^n p_j X_j$, where $\{p_j\}_{j=1}^n$ are nonnegative constants with $\sum_{j=1}^n p_j = 1$. Then

$$\phi_Y(t) = \prod_{j=1}^n \phi(p_j t) = \exp\left[\sum_{j=1}^n p_j(it\delta - \gamma|t|)\right] = \exp(it\delta - \gamma|t|).$$

This demonstrates that the average or weighted average of any number of i.i.d. Cauchy variates has the same distribution as each component.

Exercise 3.61. *X has the standard normal distribution.*

(i) Show (either directly or by differentiating the c.f.) that $VX = 1$.
(ii) If b is a positive constant, apply (3.26) to show that $V(bX) = b^2$.
(iii) Use the change-of-variable technique to show that $Y = bX$ has p.d.f.

$$f_Y(y) = \frac{1}{\sqrt{2\pi b^2}} \exp\left[-y^2/(2b^2)\right].$$

(iv) Apply (3.54) to show that Y has c.f. $\phi_Y(t) = e^{-b^2 t^2/2}$.

Exercise 3.62. *X_1, X_2 are independent standard normals. Show by calculating the marginal c.f.s that the marginal distribution of each of $Y_1 = X_1 + X_2$ and $Y_2 = X_1 - X_2$ is normal with mean zero and variance $\sigma^2 = 2$.*

A note of caution: Although $\phi_{X_1+X_2} = \phi_1 \phi_2$ when X_1, X_2 are independent, it is not always true that a c.f. that is the product of c.f.s corresponds to the sum of independent r.v.s.[32]

Exercise 3.63. *X_1 is distributed as standard Cauchy, and $X_2 = X_1$ almost surely (i.e., $X_2(\omega) = X_1(\omega)$ except possibly on null sets). Show that $\phi_{X_1+X_2} = \phi_1 \phi_2$ even though X_1, X_2 are not independent. Is it also true that $\phi_{X_1-X_2} = \phi_1 \overline{\phi_2}$?*

---

[32]Since independence would also imply that $\phi_{X_1-X_2} = \phi_1 \overline{\phi_2}$, this additional condition would be necessary for independence; yet not even the joint conditions $\phi_{X_1+X_2} = \phi_1 \phi_2$ and $\phi_{X_1-X_2} = \phi_1 \overline{\phi_2}$ are sufficient. For a counterexample, see Bisgaard and Sasvari (2000, p. 15).

The connection between c.f.s. of independent r.v.s and c.f.s of their sums implies a connection between their cumulants. Specifically, when $\{X_j\}_{j=1}^n$ are independent and $Y_n := \sum_{j=1}^n X_j$ then $\kappa_\ell (Y_n) = \sum_{j=1}^n \kappa_\ell (X_j)$ when all the cumulants exist. This follows from (3.49), since

$$\log \phi_{Y_n} (t) = \sum_{j=1}^n \log \phi_j (t)$$

$$= \sum_{j=1}^n \left[\sum_{\ell=1}^k \kappa_\ell (X_j) \frac{(it)^\ell}{\ell!} + R_{k+1} (X_j) \right]$$

$$= \sum_{\ell=1}^k \left[\sum_{j=1}^n \kappa_\ell (X_j) \right] \frac{(it)^\ell}{\ell!} + R_{k+1} (Y).$$

In particular, when $\{X_j\}_{j=1}^n$ are i.i.d. with ℓth cumulant $\kappa_\ell (X)$, we have $\kappa_\ell (Y_n) = n\kappa_\ell (X)$.[33] This pleasing relation has an important consequence for the *asymptotic* behavior of standardized sums as $n \to \infty$. Considering the sum Y_n of independent r.v.s with $EX = \kappa_1 (X) = \mu$, $VX = \kappa_2 (X) = \sigma^2$, and finite fourth moments, we have

$$Z_n := \frac{Y_n - EY_n}{\sqrt{VY_n}} = n^{-1/2} \sum_{j=1}^n \frac{X_j - \mu}{\sigma}.$$

Now the standardized summands $\{X_j' := (X_j - \mu)/\sigma\}$ have $\kappa_1 (X') = 0$, $\kappa_2 (X') = 1$, $\kappa_3 (X') = \alpha_3$, and $\kappa_4 (X') = \alpha_4 - 3$, where α_3 and α_4 are the coefficients of skewness and kurtosis of the $\{X_j\}$. Thus, for $\ell \in \{1, 2, 3, 4\}$ we have $\kappa_\ell (Z_n) = n\kappa_\ell (X'/\sqrt{n}) = n^{1-\ell/2}\kappa_\ell (X')$, and so

$$\log \phi_{Z_n} (t) = \frac{(it)^2}{2} + \frac{\alpha_3}{\sqrt{n}} \frac{(it)^3}{3!} + \frac{(\alpha_4 - 3)}{n} \frac{(it)^4}{4!} + R(n), \qquad (3.56)$$

where $R(n) \to 0$ faster than n^{-1} as $n \to \infty$. We shall prove in Section 6.7 that the distribution of a standardized sum of n i.i.d. r.v.s with finite variance approaches the standard normal as $n \to \infty$. The result $\lim_{n\to\infty} \log \phi_{Z_n} (t) = -t^2/2$ from (3.56) is clearly consistent with this. Moreover, (3.56) suggests, besides, that the rate of convergence to the normal depends on the skewness and kurtosis of the underlying distribution. Section 6.7.3 will show how to make practical use of this.

---

[33]Note that this *cumulative* property of the $\{\kappa_\ell\}$ does not generally hold for moments, except for those moments that always correspond to specific cumulants ($\mu = \kappa_1, \sigma^2 = \kappa_2, \mu_3 = \kappa_3$). This property accounts for both cumulants' name and their usefulness in analysis.

3.5 Conditional Expectation

3.5.1 *Moments of Conditional Distributions*

Section 2.4.4 introduced conditional distributions, as represented for jointly distributed r.v.s X, Y by the conditional c.d.f., $F_{Y|X}(\cdot \mid x)$, or conditional p.m.f. or p.d.f., $f_{Y|X}(\cdot \mid x)$. These functions allow us to adjust estimates of $\Pr(Y \in B)$ (B some Borel set) upon learning the specific value taken on by X. For example, $\Pr(a < Y \le b \mid X = x)$ would be calculated as $F_{Y|X}(b \mid x) - F_{Y|X}(a \mid x)$, or as $\sum_{y \in (a,b]} f_{Y|X}(y \mid x)$ or $\int_a^b f_{Y|X}(y \mid x) \cdot dy$ in the discrete and continuous cases. At the simplest level, the *conditional expectation* of Y given $X = x$ can be thought of just as the center of mass, or *mean*, of the corresponding conditional distribution, $E(Y \mid x) = \int y \cdot dF_{Y|X}(y \mid x)$. In the same way one can calculate expectations of various measurable functions $g(Y)$ conditional on $X = x$, as $\int g(y) \cdot dF_{Y|X}(y \mid x)$. Specifically, with $g(Y) = Y^k$ one gets conditional moments about the origin; with $g(Y) = [Y - E(Y \mid x)]^k$, conditional *central* moments; and with $g(Y,t) = e^{tY}$ or e^{itY}, conditional m.g.f.s or c.f.s. Since these conditional expectations are just expectations with respect to particular distributions, they have all the usual properties, including linearity, dominance, monotone and dominated convergence.

Example 3.50. Two fair coins are flipped (independently), and X_1, X_2 represent the total number of heads after the first and second flips, respectively. The joint and marginal p.m.f.s are

$x_1 \backslash x_2$	0	1	2	$f_1(\cdot)$
0	$\frac{1}{4}$	$\frac{1}{4}$	0	$\frac{1}{2}$
1	0	$\frac{1}{4}$	$\frac{1}{4}$	$\frac{1}{2}$
$f_2(\cdot)$	$\frac{1}{4}$	$\frac{1}{2}$	$\frac{1}{4}$	

,

and the conditional p.m.f.s of X_2 given $X_1 = 0$ and $X_1 = 1$ are

x_2	$f_{2|1}(\cdot|0)$	$f_{2|1}(\cdot|1)$
0	$\frac{1}{2}$	0
1	$\frac{1}{2}$	$\frac{1}{2}$
2	0	$\frac{1}{2}$

.

The conditional means are $E(X_2 \mid 0) = \frac{1}{2}$ and $E(X_2 \mid 1) = \frac{3}{2}$; the conditional variances are $V(X_2 \mid 0) = V(X_2 \mid 1) = \frac{1}{4}$; and the conditional m.g.f.s are $M_{2|1}(t \mid 0) = \frac{1}{2}(1 + e^t)$ and $M_{2|1}(t \mid 1) = \frac{1}{2}e^t(1 + e^t)$.

3.5.2 *Expectations Conditional on σ Fields*

For many purposes it is necessary to interpret conditional expectations more broadly than just as integrals with respect to conditional c.d.f.s. For example, we need to attach meaning to symbols like $E(Y \mid X)$, in which the value assumed by X is not yet known, and, more generally, to interpret expressions such as $E(Y \mid \mathcal{G})$, where $\mathcal{G}$ is a σ field.

For this let us return to the two-coin example and approach it from first principles. The probability space is $(\Omega, \mathcal{F}, \mathbb{P})$, where $\Omega = \{TT, HT, TH, HH\}$, $\mathcal{F}$ is the collection of all 2^4 subsets, and $\mathbb{P}(A) = N(A)/N(\Omega) = N(A)/4$ for any $A \subset \Omega$. X_1 and X_2, the total number of heads after one and two flips, are $\mathcal{F}$-measurable functions on Ω—i.e., random variables, the value of each being known as soon as the outcome is revealed. Equivalently, since we know the outcome if and only if we know whether each event in $\mathcal{F}$ occurred, we will know the values of both variables as soon as we have "information" $\mathcal{F}$. Now consider the *sub-σ* field $\mathcal{G} := \{\Omega, (TT, TH), (HT, HH), \varnothing\}$ of four events generated by the sets (TT, TH) and (HT, HH). This is a "sub" σ field because $\mathcal{G} \subset \mathcal{F}$, the four events of $\mathcal{G}$ being among the 2^4 events of $\mathcal{F}$. Now X_1 is $\mathcal{G}$-measurable, since "knowing" $\mathcal{G}$ (knowing which of (TT, TH) and (HT, HH) occurred) tells whether the first flip produced head or tail. However, X_2 is *not* $\mathcal{G}$-measurable, since neither of (TT, TH) or (HT, HH) pins down the value of X_2. Thus, $\mathcal{G}$ represents the information about what happened on the first flip, while $\mathcal{F}$ is the *full* information about the experiment—"full" in the sense that it conveys all the details of interest. Specifically, "knowing" $\mathcal{F}$ means that we know the outcome of the experiment at the level of detail consistent with our original description of Ω.

With this understanding, the meaning of $E(X_2 \mid \mathcal{G})$ is evident: It is the expected number of heads after two flips *given* the knowledge of what happened on the first. Now $\mathcal{G}$ is in fact the *smallest* field with respect to which X_1 is measurable, so we say that it is the field *generated* by X_1 and express this in symbols as $\mathcal{G} = \sigma(X_1)$. Thus, we can write $E(X_2 \mid \mathcal{G}) \equiv E(X_2 \mid \sigma(X_1))$, or simply as $E(X_2 \mid X_1)$, to represent the expected value of X_2 given knowledge of the value of X_1. Before the experiment takes place, when we "know" only the trivial information $\{\Omega, \varnothing\}$, the expressions $E(X_2 \mid \mathcal{G}) \equiv E(X_2 \mid \sigma(X_1)) \equiv E(X_2 \mid X_1)$ represent a $\mathcal{G}$-measurable *random variable*, since at that point information $\mathcal{G}$ (and hence knowledge of the value of X_1) is not yet in hand. For example, in the coin experiment r.v. $E(X_2 \mid X_1)$ can take one of the two values $E(X_2 \mid 0) = \frac{1}{2}$

(when $X_1(\omega) = 0$) or $E(X_2 \mid 1) = \frac{3}{2}$, each with probability $\frac{1}{2}$. By contrast, the ordinary or "unconditional" expectation EX_2 can be regarded as being conditional on the trivial field $\{\Omega, \varnothing\}$, which contains no information beyond what is possible and what is not.

This view of conditioning on a body of information—usually represented as a σ field of sets—is especially useful in applied modeling, particularly of stochastic processes that evolve through time or space. Examples include economic time series (stock prices, the rate of aggregate output, the money supply), physical time series (total rainfall from some time origin, temperature of reagents in a chemical reaction), or spatial variation (temperature at different points in a heated metal rod, electric charge on the surface of a capacitor). In each case, given information about the state of the system up to some point in time or space, one might formulate an expectation of its value farther out. In probabilistic modeling, this information about the state would be represented as a σ field with respect to which all relevant quantifiable features of the system were measurable. Thus, we might consider the expected rate of interest offered by 3-month Treasury bills at time $t + 1$ given the state of the economy at t, $E(R_{t+1} \mid \mathcal{F}_t)$. Here $\mathcal{F}_t$ would be the σ field generated by the histories of whatever variables are thought to be useful in predicting the T-bill rate.

Let us now look at an especially important property of conditional expectation. With $(\Omega, \mathcal{F}, \mathbb{P})$ as the probability space and Y an $\mathcal{F}$-measurable r.v., let $\mathcal{G}$ be a sub-σ field of $\mathcal{F}$ with respect to which Y is *not* measurable. $\mathcal{G}$ is thus a proper subset of $\mathcal{F}$, so that if we know $\mathcal{F}$ we know $\mathcal{G}$, but not conversely. In particular, knowing $\mathcal{G}$ does not pin down the value of Y. Now we have seen that the "unconditional" mean $EY \equiv E(Y \mid \{\Omega, \varnothing\})$ is the best predictor of the realization of Y, in the narrow sense that it minimizes the mean squared error given *trivial* information $\{\Omega, \varnothing\}$. In the same way $E(Y \mid \mathcal{G})$ minimizes the expected squared error of Y conditional on the possibly nontrivial information $\mathcal{G}$. With this interpretation of $E(Y \mid \mathcal{G})$, consider the identity

$$EY = E[Y - E(Y \mid \mathcal{G}) + E(Y \mid \mathcal{G})]$$
$$= E[Y - E(Y \mid \mathcal{G})] + E[E(Y \mid \mathcal{G})], \qquad (3.57)$$

where the second line just follows from the linearity property of $E(\cdot)$. Both the outer expectations here are unconditional, whereas $E(Y \mid \mathcal{G})$ is based on the partial knowledge about whether outcome ω either was or was not in each of the sets of $\mathcal{G}$. Although the value of $E(Y \mid \mathcal{G})$ is known once $\mathcal{G}$ is known, it is not *necessarily* known just from the trivial information that *some* outcome in Ω must occur—and it is *definitely* not known if $\mathcal{G} \backslash \{\Omega, \varnothing\} \neq \varnothing$. Thus, with respect to information $\{\Omega, \varnothing\}$, $E(Y \mid \mathcal{G})$ is a $\mathcal{G}$-measurable (and therefore an $\mathcal{F}$-measurable) r.v. that will take the value $E(Y \mid \mathcal{G})(\omega)$ when outcome ω occurs. Now $Y(\omega) - E(Y \mid \mathcal{G})(\omega)$ is the prediction error based on information $\mathcal{G}$. Since there is no reason based on information $\{\Omega, \varnothing\}$ to expect a prediction with (possibly) *better* information $\mathcal{G} \supset \{\Omega, \varnothing\}$ to be wrong, it is plausible that the expected prediction error given $\{\Omega, \varnothing\}$ would be zero; i.e., $E[Y - E(Y \mid \mathcal{G})] = 0$. That is, since trivial information $\{\Omega, \varnothing\}$ is also in $\mathcal{G}$, it cannot improve the prediction based on $\mathcal{G}$. It then follows from (3.57) that

$$EY = E[E(Y \mid \mathcal{G})]. \tag{3.58}$$

In words, the expectation of Y (given trivial information) equals the expected value (given trivial information) of the *conditional* expectation given partial information.

As an example, suppose Y is the price of a share of stock two days hence, that $\mathcal{F}$ represents what we will know in two days, that $\mathcal{G}$ represents what we shall know tomorrow, and that $E(Y \mid \mathcal{G})$ is tomorrow's best prediction of the price. As of today $E(Y \mid \mathcal{G})$ is a random variable because we do not yet have access to tomorrow's information. However, we do know that $E(Y \mid \mathcal{G})$ will be our best prediction of Y once that new information becomes available, so the current expectations of Y and of $E(Y \mid \mathcal{G})$ are just the same. In other words, today's best prediction is the same as the prediction of tomorrow's best prediction.

Relation (3.58) is referred to variously as the *law of iterated expectations* and the *tower* property of conditional expectation. It is actually a special case of the more general condition

$$EY\mathbf{1}_A = E[E(Y \mid \mathcal{G})\mathbf{1}_A], \tag{3.59}$$

where A is any $\mathcal{G}$-measurable set and $\mathbf{1}_A = \mathbf{1}_A(\omega)$ is an indicator function. Thus, one knows whether or not A has occurred when one "knows" $\mathcal{G}$, so that $\mathbf{1}_A$ is a $\mathcal{G}$-measurable r.v. Decomposing as in (3.57) gives

$$EY\mathbf{1}_A = E\{[Y - E(Y \mid \mathcal{G})]\mathbf{1}_A\} + E[E(Y \mid \mathcal{G})\mathbf{1}_A].$$

The first term can be interpreted as the *covariance* between the prediction error based on $\mathcal{G}$ information and the indicator that signals whether event $A \in \mathcal{G}$ has occurred. Were this covariance different from zero, it would be possible to use the knowledge that A occurred to reduce the mean-squared prediction error of Y given $\mathcal{G}$. For example, were the covariance positive, then the average prediction error given the event $\mathbf{1}_A = 1$ would be positive, in which case $E(Y \mid \mathcal{G})$ would, on average, be too small given this information. However, since A is $\mathcal{G}$-measurable, $E(Y \mid \mathcal{G})$ already takes into account the information about A. Thus, the first term is zero and (3.59) holds.

Exercise 3.64. *Explain why (3.58) is just a special case of (3.59).*

Of course, the arguments we have used to justify (3.58) and (3.59) are merely heuristic, depending as they do on the very special squared-error loss function for prediction errors. The rigorous approach, following the path laid down by Kolmogorov (1933), is simply to *define* $E(Y \mid \mathcal{G})$ as a $\mathcal{G}$-measurable r.v. that satisfies condition (3.59). However, when σ field $\mathcal{G}$ is generated by another r.v., the tower property can be demonstrated formally by working from conditional distributions. Suppose then that $\mathcal{G} := \sigma(X)$ (the σ field generated by r.v. X). The two expressions $E(Y \mid \mathcal{G}) = E(Y \mid X)$ now have the same meaning. Since $E(Y \mid x)$ can be calculated as $E(Y \mid x) = \int y \cdot dF_{Y\mid X}(y \mid x)$ for some particular x in the support of X, the expectation of $E(Y \mid X)$ is

$$E[E(Y \mid X)] = \int E(Y \mid x) \cdot dF_X(x) \tag{3.60}$$

$$= \int \left[\int y \cdot dF_{Y\mid X}(y \mid x) \right] \cdot dF_X(x)$$

$$= \int \int y \cdot dF_{\mathbf{Z}}(x, y),$$

where $F_{\mathbf{Z}}(x, y) = F_{Y\mid X}(y \mid x) F_X(x)$ is the joint c.d.f. of $\mathbf{Z} = (X, Y)'$. We conclude, then, that

$$EY = E[E(Y \mid X)]. \tag{3.61}$$

Note that when r.v. X is continuous the result of calculation (3.60) would remain the same if function $E(Y \mid \cdot)$ were altered on any set of zero Lebesgue measure. For this reason, just as we speak of *versions* and *equivalence classes* of conditional densities and probabilities in the continuous case—i.e., $f_{Y\mid X}(y \mid x)$ and $\Pr(Y \in B \mid X = x)$—a particular $E(Y \mid \cdot):$ $\mathbb{X} \to \Re$ merely represents a class of functions for which $E[E(Y \mid X)] = EY$ for any integrable Y. Fortunately, in applied work such technical distinctions are rarely important.

Exercise 3.65. *Write out the above steps in each of the special cases that $\mathbf{Z} = (X, Y)'$ is (i) discrete and (ii) continuous.*

Exercise 3.66. *If X_j is the total number of heads after j flips of a fair coin, show that $EX_2 = E[E(X_2 \mid X_1)] = 1$.*

3.5.3 *Applications*

Conditional expectation has a multitude of applications in applied probability modeling and statistical theory and practice. Here are some examples.

- Suppose Y is some r.v. whose value will be known at future date T given information $\mathcal{F}_T$. We insist also that Y be integrable, in that $E\,|Y| = \int |Y\,(\omega)| \;\mathbb{P}\,(d\omega) < \infty$. At $t < T$ we can form an expectation of Y given the incomplete information $\mathcal{F}_t \subset \mathcal{F}_T$, as $E\,(Y \mid \mathcal{F}_t)$. As we progress from initial time $t = 0$, we acquire more information (and forget nothing, let us assume), so that $\{\mathcal{F}_t\}_{0 \le t \le T}$ represents a non-decreasing sequence of σ fields—a "filtration".[34] Letting $X_t := E\,(Y \mid \mathcal{F}_t)$, the conditional expectations *process* $\{X_t\}_{0 \le t \le T}$ shows how our "best" prediction of Y evolves through time as we gain information. Naturally, $X_T = E\,(Y \mid \mathcal{F}_T) = Y$, since we have full information at T; and if information at $t = 0$ is trivial, then $X_0 = E\,(Y \mid \mathcal{F}_0) = EY$. The tower property of conditional expectation then implies that for any $s \in [0, t]$

$$E\,(X_t \mid \mathcal{F}_s) = E\,[E\,(Y \mid \mathcal{F}_t) \mid \mathcal{F}_s]$$
$$= E\,(Y \mid \mathcal{F}_s)$$
$$= X_s,$$

which (along with integrability) is the condition that defines a *martingale process* adapted to the filtration $\{\mathcal{F}_t\}$. From the properties of martingales follow many useful results in probability theory and a wide range of applications. In particular, martingales figure prominently in mathematical finance, especially in the analysis of "derivative" securities, such as stock options and futures contracts.[35]

- Relation (3.61) obviously extends to iterated expectations of integrable functions of Y, as $Eg(Y) = E\,\{E\,[g(Y) \mid X]\}$. This last expression leads to a useful relation between unconditional and conditional variances. Corresponding to the expression $VY = EY^2 - (EY)^2$ for the variance, the conditional variance of Y given (the as yet unobserved value of) X is

$$V(Y \mid X) = E(Y^2 \mid X) - E(Y \mid X)^2.$$

---

[34]So called because $\{\mathcal{F}_t\}_{0 \le t \le T}$ acts like a sequence of filters that catch progressively finer detail as time advances, thus affording sharper predictions of Y.

[35]Williams (1991) gives a splendid introduction to martingale theory. For financial applications see Baxter and Rennie (1998), Musiela and Rutkowski (2004), and Epps (2007, 2009).

Now since $V(Y \mid X)$ is a function of X, it is a r.v. with respect to trivial information. Its expected value is

$$E\left[V(Y \mid X)\right] = E\left[E(Y^2 \mid X)\right] - E\left[E(Y \mid X)^2\right]$$
$$= EY^2 - E\left[E(Y \mid X)^2\right]. \tag{3.62}$$

Of course, conditional mean $E\left(Y \mid X\right)$ is also a r.v., and its (unconditional) variance is

$$V\left[E\left(Y \mid X\right)\right] = E\left[E(Y \mid X)^2\right] - \left\{E\left[E(Y \mid X)\right]\right\}^2$$
$$= E\left[E(Y \mid X)^2\right] - (EY)^2 \tag{3.63}$$

Combining (3.62) and (3.63) gives

$$E\left[V(Y \mid X)\right] + V\left[E\left(Y \mid X\right)\right] = EY^2 - (EY)^2$$

and hence

$$VY = E\left[V(Y \mid X)\right] + V\left[E(Y \mid X)\right]. \tag{3.64}$$

Thus, the unconditional variance equals the expected value of the conditional variance plus the variance of the conditional mean. Since the latter variance is nonnegative, it follows that the mean squared prediction error given trivial information is at least as large as the *expected* prediction error given knowledge of X; that is, $VY \geq E\left[V(Y \mid X)\right]$. More generally, if Y is an $\mathcal{F}_T$-measurable r.v. and $\mathcal{F}_s$ and $\mathcal{F}_t$ are sub-σ fields with $\mathcal{F}_s \subset \mathcal{F}_t \subset \mathcal{F}_T$, then $V\left(Y \mid \mathcal{F}_s\right) \geq E\left[V\left(Y \mid \mathcal{F}_t\right) \mid \mathcal{F}_s\right]$. Thus, one *expects* the precision of predictions to increase as more information is acquired.

Exercise 3.67. *Find VX_2 in the coin flip example directly from the marginal distribution of X_2, then check that (3.64) replicates it.*

- Just as the law of total probability sometimes makes it easier to calculate probabilities by conditioning on events that partition Ω, it is often easier to find mathematical expectations of r.v.s in complicated settings by first conditioning on some other variable, then "iterating" the expectation. Sometimes, in working out an expected value, one has to be a little creative in order to exploit the power of conditioning. The next two examples illustrate these possibilities. (For another, look at Exercise 5.15 on page 354 to see how conditional expectations can be used to work out the p.d.f. of Student's t distribution.)

Example 3.51. A coin is flipped. If it turns up tails, a pointer on a unit circular scale is spun and the clockwise distance of pointer from origin is recorded after the pointer comes to rest. The process is repeated until the coin first turns up heads, at which point the experiment ends. Letting T represent the sum of the pointer distances obtained, let us find ET. Let N be the number of flips until first heads and $X_1, X_2, ..., X_j$ be the clockwise distances on the first $j \in \aleph$ spins. Setting $X_0 := 0$ in case $N = 1$, we have $T = \sum_{j=0}^{N-1} X_j$. The difficulty in finding ET is that N itself is a random variable. However, under the reasonable assumption that N is independent of $\{X_j\}_{j=1}^{\infty}$ a simple conditioning argument overcomes the problem. Conditioning on N, the expectation operator passes through the summation, as

$$E(T \mid N) = \sum_{j=0}^{N-1} E(X_j \mid N) = \sum_{j=0}^{N-1} EX_j = \frac{N-1}{2}.$$

Here, the second equality follows from independence; the last, from modeling $X_1, X_2, ...$ as uniformly distributed on $[0, 1]$. The tower property and the result of Example 3.30 now give

$$ET = E[E(T \mid N)] = \frac{EN - 1}{2} = \frac{1}{2}.$$

Exercise 3.68. *Find the m.g.f. of T in Example 3.51 and use it to verify that $ET = 1/2$.*

Example 3.52. Showing in the direct way that the c.f. of a standard Cauchy r.v. has the form $\phi(t) = e^{-|t|}$ requires working out the expectation $Ee^{itY} = \int_{-\infty}^{\infty} e^{ity} \left[\pi(1 + y^2)\right]^{-1} \cdot dy$. This can be done by applying Cauchy's residue theorem from complex analysis, integrating $e^{itz} \left[\pi(1 + z^2)\right]^{-1}$ with respect to the complex variable z around semicircular regions in the complex plane and finding the residues at $\pm i$. We side-stepped the need for complex integration in Exercise 3.59 by applying Theorem 3.5 and exploiting the conjugate relation between the Cauchy and Laplace distributions, but such devious methods rarely satisfy. Here is a clever alternative derivation (due to Datta and Ghosh (2007)) using conditional expectations and the form of the normal c.f. (from Examples 3.37 and 3.38).

Exercise 2.40 established that Z_1/Z_2 has a standard Cauchy distribution when Z_1, Z_2 are independent standard normals, and the distribution of $Y := Z_1/|Z_2|$ is precisely the same because of the symmetry about the origin of the standard normal distribution, the independence of Z_1, Z_2, and the fact that $\Pr(Z_2 = 0) = 0$. (This is verified in Exercise 3.69 below.) To show that Y has c.f. $e^{-|t|}$ we make use of the fact that $Y \sim N\left(0, 1/Z_2^2\right)$ conditional on Z_2, with conditional c.f. $\phi_{Y|Z_2}(t) = e^{-t^2/(2Z_2^2)}$. From the tower property we have

$$\phi(t) = E\left[E\left(e^{itZ_1/|Z_2|} \mid Z_2\right)\right]$$

$$= Ee^{-t^2/(2Z_2^2)}$$

$$= \int_{-\infty}^{\infty} e^{-t^2/(2z^2)} \frac{1}{\sqrt{2\pi}} e^{-z^2/2} \cdot dz$$

$$= \sqrt{2/\pi} \int_0^{\infty} e^{-(t^2/z^2 + z^2)/2} \cdot dz.$$

Letting I represent the integral, the change of variables $z = |t|/x$ gives

$$I = -\int_{\infty}^0 |t| x^{-2} e^{-(t^2/x^2 + x^2)/2} \cdot dx = \int_0^{\infty} |t| x^{-2} e^{-(t^2/x^2 + x^2)/2} \cdot dx.$$

Adding this to the original expression for I, we have

$$I + I = \int_0^{\infty} (1 + |t|/x^2) e^{-(t^2/x^2 + x^2)/2} \cdot dx$$

$$= e^{-|t|} \int_0^{\infty} (1 + |t|/x^2) e^{-(x - |t|/x)^2/2} \cdot dx.$$

Another change of variables as $x - |t|/x \to w$ gives $dw = 1 + |t|/x^2$ and

$$2I = e^{-|t|} \int_{-\infty}^{\infty} e^{-w^2/2} \cdot dw = \sqrt{2\pi} e^{-|t|},$$

from which follow $I = \sqrt{\pi/2} e^{-|t|}$ and $\phi(t) = e^{-|t|}$.

Exercise 3.69. *If (i) the marginal distribution of r.v. Z_1 is symmetric about the origin, (ii) $\Pr(Z_2 = 0) = 0$, and (iii) Z_1, Z_2 are independent, show that Z_1/Z_2 and $Z_1/|Z_2|$ have the same distribution. (Hint: Show that $\Pr(Z_1/Z_2 \le x) = \Pr(Z_1/|Z_2| \le x)$ for any $x \in \Re$.)*

- Recall from Section 2.5.4 the convolution formula for the density of the sum $S = X + Y$ of two independent, continuous r.v.s:

$$f_S(s) = \int_{-\infty}^{\infty} f_X(s - y) f_Y(y) \cdot dy. \tag{3.65}$$

This was deduced as a special case of the multivariate change-of-variable procedure, but we can come at it now in a much simpler way via conditional expectation. Conditioning on the σ field generated by Y, and recognizing the independence of X and Y, we have

$$\Pr\left(S \leq s \mid Y\right) = E\left[\mathbf{1}_{(-\infty,s]}\left(X+Y\right) \mid Y\right]$$
$$= E\left[\mathbf{1}_{(-\infty,s-Y]}\left(X\right) \mid Y\right]$$
$$= F_X\left(s-Y\right).$$

The c.d.f. of S then comes courtesy of the law of total probability, as $F_S\left(s\right) = \int F_X\left(s-y\right) \cdot dF_Y\left(y\right)$. Finally, F_X being dominated by a $\mathbb{P}_Y$-integrable function (e.g., unity), we can differentiate "under the integral sign" and obtain $f_S\left(s\right) = \int f_X\left(s-y\right) \cdot dF_Y\left(y\right)$. This specializes to (3.65) when Y is absolutely continuous but becomes $f_S\left(s\right) = \sum_{y \in \mathbb{Y}} f_X\left(s-y\right) f_Y\left(y\right)$ when Y is supported on a countable $\mathbb{Y}$. In either case S itself is absolutely continuous. It follows that the sum of finitely many independent r.v.s is absolutely continuous so long as at least one summand has this property.[36]

Example 3.53. X has p.d.f. $f_X\left(x\right) = \left(2\pi\right)^{-1/2} \exp\left(-x^2/2\right)$ (standard normal) and Y has p.m.f. $f_Y\left(y\right) = \theta^y e^{-\theta}/y!$ for $x \in \aleph_0$ (Poisson). The sum $S = X + Y$ then has p.d.f.

$$f_S\left(s; \theta, \sigma\right) = \sum_{y=0}^{\infty} \frac{1}{\sqrt{2\pi\sigma^2}} e^{-(s-y)^2/(2\sigma^2)} \frac{\theta^y e^{-\theta}}{y!}.$$

In essence, the probability mass at each atom $y \in \mathbb{Y}$ is smeared into a bell curve centered on y and with area equal to $f_Y\left(y\right)$. Summing the contributions from all the atoms has the effect of smoothing the discrete Poisson. How smooth is the resulting p.d.f. depends on the normal scale parameter (standard deviation) σ. Figure 3.11 depicts densities with $\theta = 3, \sigma = .25$ and $\theta = 5, \sigma = 1.0$.

Exercise 3.70. *X and Y are absolutely continuous. Give an example in which $S = X + Y$ is discrete.*

3.6 Solutions to Exercises

1. (i) X is a simple function because its support is the finite set $\mathbb{X} = \{0,1,2,3\}$. The sets $A_0 = TTT$, $A_1 = (HTT, THT, TTH)$, $A_2 =$

---

[36]This is also apparent from the characteristic function of S: If $S = \sum_{j=1}^{n} X_j$ and the $\{X_j\}$ are independent, then $|\phi_S\left(t\right)| = \prod_{j=1}^{n} |\phi_j(t)| \leq \min\left\{|\phi_j\left(t\right)|\right\}_{j=1}^{n}$, which approaches zero as $|t| \to \infty$ when at least one of the $\{X_j\}$ is absolutely continuous.

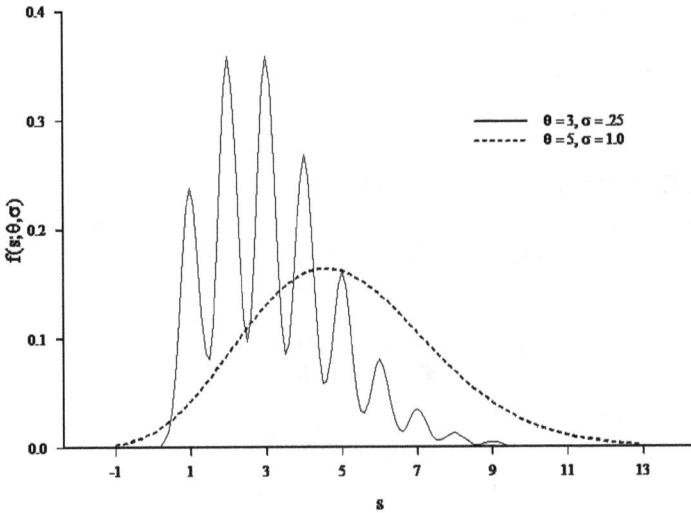

Fig. 3.11 Densities from convolution of Poisson variate with parameter θ and normal with variance σ^2.

(HHT, HTH, THH), $A_3 = HHH$ partition Ω, and X takes the same value for all ω's in a given set. However, any *finer* partition, including the eight elementary outcomes themselves, also gives a characterization of X as a simple function.

(ii) Given any partition of Ω into sets on which X is constant, the definition of the integral for simple functions shows that $\int X(\omega)\mathbb{P}(d\omega)$ can be calculated as the sum of the X values times the $\mathbb{P}$ measures of the sets. Thus, $\sum_{j=0}^{3} j\mathbb{P}(A_j) = 0\frac{1}{8} + 1\frac{3}{8} + 2\frac{3}{8} + 3\frac{1}{8}$ would be one such calculation. Another would be based on the partition of Ω into the elementary outcomes, as $\int X(\omega)\mathbb{P}(d\omega) = \sum_{\omega \in \Omega} X(\omega)\mathbb{P}(\{\omega\}) = 0\frac{1}{8} + 1\frac{1}{8} + 1\frac{1}{8} + 1\frac{1}{8} + 2\frac{1}{8} + 2\frac{1}{8} + 2\frac{1}{8} + 3\frac{1}{8}$.

2. (i) For each $n \in \mathbb{N}$ the set $\mathbb{X}_n := \cup_{k=1}^{n}\mathbb{F}_k = \{0,1\} \cup \{0,\frac{1}{2},1\} \cup \{0,\frac{1}{3},\frac{2}{3},1\}\cup\cdots\cup\{0,\frac{1}{n},\frac{2}{n},...,\frac{n-1}{n},1\}$ comprises an increasing sequence of fractions (rational numbers) $\{q_j\}_{j=0}^{N(n)}$, with $q_0 = 0$ and $q_{N(n)} = 1$. In the Riemann construction evaluate $\int_0^1 g_n(x) \cdot dx$ as the sum of $N(n)$ integrals $\left\{\int_{q_j}^{q_{j+1}} g_n(x) \cdot dx\right\}_{j=0}^{N(n)-1}$ over the open intervals $\{(q_j, q_{j+1})\}_{j=0}^{N(n)-1}$. Since $g_n(x) = 0$ on each such interval, we have $\int_0^1 g_n(x) \cdot dx = 0$ as well. In the Lebesgue construction, since function

g_n is simple for each n, we have

$$\int_0^1 g_n(x) \cdot dx = 1 \cdot \lambda(\mathbb{X}_n) + 0 \cdot \lambda([0,1] \setminus \mathbb{X}_n)$$
$$= 1 \cdot 0 + 0 \cdot 1 = 0.$$

(ii) Each rational number q on $[0,1]$—that is, each element of $\mathbb{Q} \cap [0,1]$—is of the form m/n for some $n \in \mathbb{N}$ and some $m \in \{0, 1, ..., n\}$. Thus, each such q is included in $\mathbb{X}_n$ for some n. Consequently, $\mathbb{X}_n \to \mathbb{Q} \cap [0,1]$ as $n \to \infty$, and

$$g_n(x) \to g(x) = \begin{cases} 1, & x \in \mathbb{Q} \cap [0,1] \\ 0, & x \in [0,1] \setminus \mathbb{Q} \end{cases}.$$

(iii) See Example 3.5.

(iv) The message is that the conclusion of the dominated convergence theorem does not always apply to Riemann integrals.

3. In each construction the integral can be decomposed as

$$\int_{-\infty}^{\infty} \frac{|x|}{1+x^2} \cdot dx = \int_{-\infty}^{0} \frac{-x}{1+x^2} \cdot dx + \int_0^{\infty} \frac{x}{1+x^2} \cdot dx$$

and evaluated in the Riemann sense as

$$\int_{-\infty}^{\infty} \frac{|x|}{1+x^2} \cdot dx = \lim_{a_n \to -\infty} \int_{a_n}^{0} \frac{-x}{1+x^2} \cdot dx + \lim_{b_n \to \infty} \int_0^{b_n} \frac{x}{1+x^2} \cdot dx$$

$$= \lim_{a_n \to -\infty} -\frac{\log(1+x^2)}{2} \Big|_{a_n}^{0} + \lim_{b_n \to \infty} \frac{\log(1+x^2)}{2} \Big|_0^{b_n}$$

$$= \lim_{a_n \to -\infty} \frac{\log(1+a_n^2)}{2} + \lim_{b_n \to \infty} \frac{\log(1+b_n^2)}{2}$$

$$= +\infty + \infty = +\infty.$$

4. (i) $\int_0^{\infty} x^{-1} e^{-x} \cdot dx > \int_0^1 x^{-1} e^{-x} \cdot dx > e^{-1} \int_0^1 x^{-1} \cdot dx = e^{-1} \log x \big|_0^1 = +\infty$. The integral does exist, but $x^{-1} e^{-x}$ is not Lebesgue integrable over $\mathfrak{R}_+$.

(ii) $\int_{-\infty}^{\infty} x^{-1} e^{-x^2} \cdot dx = \int_{-\infty}^{0} x^{-1} e^{-x^2} \cdot dx + \int_0^{\infty} x^{-1} e^{-x^2} \cdot dx$. The method of part (i) shows the value of the second integral to be $+\infty$. Changing variables as $y = -x$ in the first integral gives $\int_{-\infty}^{0} x^{-1} e^{-x^2} \cdot dx = -\int_0^{\infty} y^{-1} e^{-y^2} \cdot dy = -\infty$ and shows that $\int_{-\infty}^{\infty} x^{-1} e^{-x^2} \cdot dx$ does not exist in the Lebesgue sense.

5. Function $g(x) = x$ for $x \in \mathbb{X} = \{0, 1, 2, 3\}$ is *simple*. Taking $\mathbb{P}_X(\{x\}) = \binom{3}{x} \frac{1}{8}$ for $x \in \mathbb{X}$, the definition of integrals of simple functions gives $\int x \mathbb{P}_X(dx) = \sum_{x \in \mathbb{X}} x \mathbb{P}_X(\{x\})$.

6. Both $x^2 + 1$ and F are continuous on $(0, 1)$. F is not differentiable at $.5$, but $\{.5\}$ is a null set since F is continuous there. Therefore $\int_{(0,1)} (x^2 + 1) \cdot dF(x)$ can be evaluated as the Riemann integral $\int_0^1 (x^2 + 1) F'(x) \cdot dx = \int_0^{.5} (x^2 + 1) \cdot 0 \cdot dx + \int_{.5}^1 (x^2 + 1) \, 1 \cdot dx = \frac{19}{24}$.

7. (i) Integrating by parts,
$$\int_{\Re} F \cdot dF = F(x)^2 \Big|_{-\infty}^{\infty} - \int_{\Re} F \cdot dF = 1 - \int_{\Re} F \cdot dF,$$
from which $\int_{\Re} F \cdot dF = 1/2$. Similarly,
(ii) $\int_{\Re} F^k \cdot dF = 1 - k \int_{\Re} F^k \cdot dF$, and $\int_{\Re} F^k \cdot dF = (k+1)^{-1}$ for $k \geq 0$.

8. (i) $\int_{\Re} F_n \cdot dF_n = \sum_{j=1}^n \frac{j}{n} \cdot \frac{1}{n} = n^{-2} \sum_{j=1}^n j = (n+1)/(2n)$ and $\lim_{n \to \infty} EF_n(X) = 1/2$.
(ii) $\int_{\Re} F_n^k \cdot dF_n = n^{-(k+1)} \sum_{j=1}^n j^k = \sum_{j=1}^n (j/n)^k \, [j/n - (j-1)/n]$ (a Riemann sum), so that $\lim_{n \to \infty} EF_n(X)^k = \int_0^1 x^k \cdot dx = (k+1)^{-1}$.

9. Integrating by parts gives $\int_{-a}^a F \cdot dF = F^2(a) - F^2(-a) - \int_{-a}^a F \cdot dF$, so
$$\int_{-a}^a F \cdot dF = \frac{1}{2} [F(a) + F(-a)][F(a) - F(-a)]$$
$$= \frac{1}{2} [F(a) + 1 - F(a)] [\Pr(-a < X \leq a)]$$
$$= \frac{1}{2} \Pr(|X| \leq a).$$
Symmetry is used in the second step; absolute continuity, in steps 2 and 3.

10. (i) $EX = \sum_{x=0}^3 x f_X(x) = \sum_{x=0}^3 x \binom{3}{x} \frac{1}{8} = \frac{3}{2}$.
(ii) $EY = \sum_{y=0}^2 y f_Y(y) = \sum_{y=0}^2 y \binom{2}{y} \frac{1}{4} = 1$.

11. $EX = \int_{-\infty}^{\infty} \frac{x}{\sqrt{2\pi}} e^{-x^2/2} \cdot dx = \frac{1}{\sqrt{2\pi}} \left(-e^{-x^2/2} \right)_{-\infty}^{\infty} = 0$.

12. $EX = \sum_{x=0}^{\infty} x \cdot \lambda^x e^{-\lambda} / x! = \sum_{x=1}^{\infty} \lambda^x e^{-\lambda} / (x-1)!$. Putting $y := x - 1$ gives $EX = \lambda \sum_{y=0}^{\infty} \lambda^y e^{-y}/y! = \lambda$.

13. (i) $EX^2 = \sum_{x=0}^3 x^2 \binom{3}{x} \frac{1}{8} = 3$.
(ii) $EY^3 = \sum_{y=0}^2 y^3 \binom{2}{y} \frac{1}{4} = \frac{10}{4}$.

14. It does not matter that $\log x$ is undefined at $x = 0$, because $\{0\}$ is a null set. Thus,
$$E(-\log X) = -\int_0^1 \log x \cdot dx = -(x \log x - x)_0^1$$
$$= -(0 - 1) + \lim_{x \to 0} (x \log x - x)$$
$$= 1 + \lim_{x \to 0} x \log x = 1 - \lim_{n \to \infty} \frac{1}{n} \log n.$$

To evaluate the limit, use the definition of the natural log, as $\log n = \int_1^n \frac{1}{t} \cdot dt$, and the inequality $\frac{1}{t} \le \frac{1}{\sqrt{t}}$ for $t \ge 1$, whereby $0 \le \lim_{n\to\infty} \frac{1}{n} \log n = \lim_{n\to\infty} \frac{1}{n} \int_1^n \frac{1}{t} \cdot dt \le \lim_{n\to\infty} \frac{1}{n} \int_1^n \frac{1}{\sqrt{t}} \cdot dt = \lim_{n\to\infty} \frac{2(\sqrt{n}-1)}{n} = 0$. Alternatively, just apply l'Hospital's rule, whereby $\lim_{n\to\infty} \frac{1}{n} \log n = \lim_{n\to\infty} \left(\frac{d\log n}{dn} \div \frac{dn}{dn} \right) = \lim_{n\to\infty} \left(n^{-1} \div 1 \right) = 0$. Thus, $E(-\log X) = 1$. Still another approach is to use the transformation procedure to show that $Y = -\log X$ has p.d.f. $f_Y(y) = e^{-y} 1_{(0,\infty)}(y)$, whereby $EY = \int_0^\infty ye^{-y} \cdot dy = 1$.

15. (i) $EX^2 = \int_0^\infty x^2 e^{-x} \cdot dx$. Integrating by parts with $u = x^2$, $dv = e^{-x} \cdot dx$, $du = 2x \cdot dx$, $v = -e^{-x}$ gives $EX^2 = \left(-x^2 e^{-x} \right)_0^\infty + 2\int_0^\infty xe^{-x} \cdot dx = 2\int_0^\infty xe^{-x} \cdot dx$. That $\lim_{x\to\infty} xe^{-x} = 0$ follows from $e^x > 1+x+x^2$ for all $x > 0$. Integrating by parts again with $u = x$ and v as before gives $EX^2 = 2$.

(ii) $Ee^X = \int_0^\infty e^x e^{-x} \cdot dx = \int_0^\infty 1 \cdot dx = +\infty$.

16. Modeling the p.m.f. as $f(x) = 2^{-x} 1_\mathbb{N}(x)$, we have $Ez^X = \sum_{x=1}^\infty z^x 2^{-x} = \sum_{x=1}^\infty \left(\frac{z}{2} \right)^x$. This geometric series converges for $|z| < 2$.

17. $EX = \int_0^\infty \frac{x}{(1+x)^2} \cdot dx = \int_0^1 \frac{x}{(1+x)^2} \cdot dx + \int_1^\infty \frac{x}{(1+x)^2} \cdot dx > 0 + \int_1^\infty \frac{x}{(2x)^2} \cdot dx = \frac{1}{4} \int_1^\infty \frac{1}{x} \cdot dx = \frac{1}{4} \lim_{x\to\infty} \log x = +\infty$.

18. Evaluating $\int_{-\infty}^\infty \frac{x}{\pi(1+x^2)} \cdot dx$ via the Lebesgue construction as

$$\int_0^\infty \frac{x}{\pi(1+x^2)} \cdot dx - \int_{-\infty}^0 \frac{-x}{\pi(1+x^2)} \cdot dx = \left[\frac{\log(1+x^2)}{2\pi} \Big|_0^\infty \right] - \left[-\frac{\log(1+x^2)}{2\pi} \Big|_{-\infty}^0 \right]$$

shows that neither the integral of the nonnegative part, $g^+(x) = x1_{[0,\infty)}(x)$, nor that of the negative part, $g^-(x) = -x1_{(-\infty,0)}(x)$, converges. Therefore, the integral does not exist. Intuitively, the distribution $f(x)$ has no definite center of mass, because we cannot "balance" infinities. Since the distribution restricted to $|x| \le c$ does have definite center of mass (at 0), the construction $\lim_{c\to\infty} \int_{-c}^c \frac{x}{\pi(1+x^2)} \cdot dx$ gives the wrong answer.

19. From the joint p.m.f. in (2.15) we have
(i)

$$E(X+Y) = \sum_{x=0}^3 \sum_{y=0}^2 (x+y) f_\mathbf{Z}(x,y)$$

$$= 0 \cdot \frac{1}{8} + (1+1) \cdot \frac{2}{8} + (1+2) \cdot \frac{1}{8} + (2+1) \cdot \frac{2}{8}$$

$$+ (2+2) \cdot \frac{1}{8} + (3+0) \cdot \frac{1}{8}$$

$$= \frac{20}{8};$$

(ii)

$$EXY = \sum_{x=0}^{3}\sum_{y=0}^{2} xy f_{\mathbf{Z}}(x,y)$$

$$= (1 \cdot 1)\frac{2}{8} + (1 \cdot 2)\frac{1}{8} + (2 \cdot 1)\frac{2}{8} + (2 \cdot 2)\frac{1}{8}$$

$$= \frac{12}{8};$$

(iii)

$$E\left(\frac{X}{1+Y}\right) = \sum_{x=0}^{3}\sum_{y=0}^{2} \frac{x}{1+y} f_{\mathbf{Z}}(x,y)$$

$$= \left(\frac{0}{1+0}\right)\frac{1}{8} + \left(\frac{1}{1+1}\right)\frac{2}{8} + \left(\frac{1}{1+2}\right)\frac{1}{8}$$

$$+ \left(\frac{2}{1+1}\right)\frac{2}{8} + \left(\frac{2}{1+2}\right)\frac{1}{8} + \left(\frac{3}{1+0}\right)\frac{1}{8}$$

$$= \frac{21}{24}.$$

20. To solve this directly, integrate by parts to get

$$EF(X)^n = \int_{-\infty}^{\infty} F(x)^n \cdot dF(x)$$

$$= F(x)^{n+1}\Big|_{-\infty}^{\infty} - \int nF(x)^n \cdot dF(x)$$

$$= 1 - nEF(X)^n,$$

and thus $EF(X)^n = (n+1)^{-1}$. Alternatively, use the fact that $Y = F(X)$ is uniformly distributed on $[0,1]$ and evaluate EY^n as $\int_0^1 y^n \cdot dy$.

21. $E\prod_{j=1}^{n} X_j = 2^{-n} < \frac{1}{100}$ if and only if $n > \frac{\log 100}{\log 2} \doteq 6.644$. The smallest integer value consistent with the condition is $n = 7$.

22. (i) & (ii)

$$E\left(\frac{1+X_1}{1+X_2}\right) = \int_0^1 (1+x_1) \cdot dx_1 \int_0^1 \frac{1}{1+x_2} \cdot dx_2$$

$$= E(1+X_1)E\left(\frac{1}{1+X_2}\right) = \frac{3}{2}\log 2.$$

(iii) $E(1+X_1)/E(1+X_2) = \frac{3}{2}/\frac{3}{2} = 1$. (ii) is equivalent to (i) because expectations of products of functions of independent r.v.s equal products of expectations. (iii) differs from (ii) because $(1+X_2)^{-1}$ is not affine in $(1+X_2)$, so $E(1+X_2)^{-1}$ does not always equal $[E(1+X_2)]^{-1}$.

(Indeed, inequality (3.35) and the subsequent discussion imply that $E(1+X_2)^{-1} > [E(1+X_2)]^{-1}$ unless $\Pr(X = c) = 1$ for some constant c.)

23. (i) $f(x) = 1_{\{3/2\}}(x)$

(ii) $f(x) = 1_{\{1/2\}}(x)$

(iii) $f(x) = 1_{\{2\}}(x)$

24. $f(x) = \frac{1}{(1+x)^2} 1_{(0,\infty)}(x)$

25. From $\mu'_k = \frac{1}{3}(-2)^k + \frac{2}{3}$ we have $\mu = 0$, $\mu_2 = 2$, $\mu_3 = -2$, $\mu_4 = 6$ and

$$E(X - \eta)^4 = \mu_4 + 4\mu_3\eta + 6\mu_2\eta^2 + \eta^4$$
$$= 6 - 8\eta + 12\eta^2 + \eta^4.$$

The derivative evaluated at $\eta = \mu = 0$ is -8, so μ is not an extremal point. (The only real root of $dE(X - \eta)^4/d\eta^4 = -8 + 24\eta + 4\eta^3$ is at $\sqrt[3]{4} - \sqrt[3]{2} \doteq 0.32748$, at which the mean quartic error is easily seen to have its minimum.)

26. The p.m.f. is $f(x) = 2^{-x} 1_{\aleph}(x)$, so $\Pr(X \leq 1) = \frac{1}{2} = \Pr(X \geq 2)$. Thus, for any $x \in [1, 2]$ it is true that $\Pr(X \geq x) = \Pr(X \leq x)$. Then $x^+_{.5} = 2$, $x^-_{.5} = 1$ and $x_{.5} = (x^+_{.5} + x^-_{.5})/2 = 3/2$, whereas $\mu = EX = 2$.

27. Given c.d.f. F with $F(x^{**}) - F(x^*) > 0$, define a new c.d.f. F^* as

$$F^*(x) = \begin{cases} 0, & x < x^* \\ \frac{F(x)}{F(x^{**}) - F(x^*)}, & x^* \leq x < x^{**} \\ 1 & x \geq x^{**} \end{cases}.$$

The center of mass of distribution F^* is the same as that of distribution F on $(x^*, x^{**}]$; namely, $\int_{\Re} x \cdot dF^*(x) = \int_{(x^*, x^{**}]} x \cdot dF^* = [F(x^{**}) - F(x^*)]^{-1} \int_{(x^*, x^{**}]} x \cdot dF$.

28. Differentiating $\Delta(\eta)$ in (3.19) and equating $\Delta'(\eta)$ to zero give

$$0 = \Pr(X \leq \eta) - \Pr(X \geq \eta)$$
$$= \Pr(X \leq \eta) - [1 - \Pr(X \leq \eta)],$$

where the second equality follows from the continuity of X. Thus, any $x_{.5}$ such that $\Pr(X \leq x_{.5}) = \Pr(X \geq x_{.5}) = .5$ is a stationary value of $E|X - \eta|$. Moreover, since

$$\frac{[\Pr(X \leq x_{.5} + \delta) - \Pr(X \geq x_{.5} + \delta)]}{\delta} \geq 0$$

for any $\delta \neq 0$, any such stationary point is a minimum. (Note that there may be infinitely many such stationary values. For example, any $x_{.5} \in \mathfrak{M} = [.25, .75]$ is a median of a r.v. with p.d.f. $f(x) = 2[1_{[0,.25]}(x) + 1_{[.75,1]}(x)]$.)

29. (i) By definition, $x_{.5}$ satisfies $\Pr(X \geq x_{.5}) \geq .5$ and $\Pr(X \leq x_{.5}) \geq .5$, so $p_F(x_{.5}) = 1 - \Pr(X \leq x_{.5}) \leq .5$. The supremum is attained whenever there is no probability mass at $x_{.5}$; for in that case the equality $1 = \Pr(X > x_{.5}) + \Pr(X < x_{.5})$ and the definition of $x_{.5}$ imply that $\Pr(X < x_{.5}) = .5$.

(ii) Consider the sequence of r.v.s $\{X_n\}_{n=1}^{\infty}$, where X_n has c.d.f. $F_n(x) := (1 - n^{-1})\mathbf{1}_{[0,n)}(x) + \mathbf{1}_{[n,\infty)}(x)$. Since $\mu_n := EX_n = 1$ and $p_{F_n}(\mu_n) = 1 - n^{-1}$ for all n, we have $\sup_n \{p_{F_n}(\mu_n)\} = 1$. Since there is no way for $p_F(\mu)$ to exceed unity, we have $\sup_{F \in \mathfrak{F}_\mu} \{p_F(\mu)\} = 1$. However, there is no F for which the supremum is attained. For suppose $\Pr(X < \mu) = 1$. Then

$$\mu = \int x \cdot dF = \int_{(-\infty,\mu)} x \cdot dF + \int_{[\mu,\infty)} x \cdot dF$$

$$= \int_{(-\infty,\mu)} x \cdot dF < \mu \int_{(-\infty,\mu)} dF = \mu,$$

a contradiction.

30. Set $Y = X^k$, note that $\Pr(Y > x^k) = \Pr(X > x)$, and use (3.16).

31. $EX = \int_{\Re} x \cdot dF = \int_{(-\infty,0]} x \cdot dF + \int_{(0,\infty)} x \cdot dF = \int_{(-\infty,0]} x \cdot dF - \int_{(0,\infty)} x \cdot d(1 - F)$. Integrating by parts with $u = x, v = F$ then $u = x$, $v = 1 - F$ gives

$$EX = xF(x)\big|_{-\infty}^{0} - \int_{-\infty}^{0} F(x) \cdot dx - x[1 - F(x)]\big|_{0}^{\infty} + \int_{0}^{\infty} [1 - F(x)] \cdot dx.$$

The text demonstrates that $x[1 - F(x)]\big|_{0}^{\infty} = 0$ if $E|X| < \infty$. It is clear that $\lim_{x \to 0} xF(x) = 0$, but we must show that $\lim_{x \to -\infty} xF(x) = 0$ also. Now $0 \leq n \int_{(-\infty,-n]} dF = nF(-n) \leq \int_{(-\infty,-n]} |x| \cdot dF = \int_{(-\infty,0]} |x| \cdot dF - \int_{(-n,0]} |x| \cdot dF$. Letting $n \to \infty$ gives $0 \leq \lim_{n \to \infty} nF(-n) \leq \int_{(-\infty,0]} |x| \cdot dF - \int_{(-\infty,0]} |x| \cdot dF = 0$, where the last equality holds because $\int_{(-\infty,0]} |x| \cdot dF \leq \int_{\Re} |x| \cdot dF < \infty$. Therefore, $\lim_{n \to \infty} nF(-n) = 0 = \lim_{x \to -\infty} xF(x)$.

32. (i) Evaluating $\int x \cdot dF$ as a Stieltjes integral gives $EX = 0\left(\frac{1}{2}\right) + \int_{.5}^{1} x \cdot dx = \frac{3}{8}$, since $F(0) - F(0-) = \frac{1}{2}$ and $F'(x) = dx$ for $x \in (.5, 1)$.

(ii) Alternatively, $EX = \int_{0}^{\infty} [1 - F(x)] \cdot dx = \int_{0}^{.5} \frac{1}{2} \cdot dx + \int_{.5}^{1} (1 - x) \cdot dx + \int_{1}^{\infty} 0 \cdot dx = \frac{1}{4} + \frac{1}{8} = \frac{3}{8}$.

33. F is a c.d.f. since (i) $F(1) = 0$, (ii) F is nondecreasing and right-continuous, and (iii) $\lim_{x \to \infty} F(x) = 1$. Applying (3.20), $-\int_{-\infty}^{0} F(x) \cdot dx + \int_{0}^{\infty} [1 - F(x)] \cdot dx = 0 + \int_{1}^{\infty} x^{-1} \cdot dx = +\infty$.

34. Following the hint, write (for $|z| < 1$) $\sum_{x=1}^{\infty} x(x-1)z^x =$
$z^2 \sum_{x=0}^{\infty} d^2 z^x / dz^2 = z^2 \frac{d^2}{dz^2} \sum_{x=0}^{\infty} z^x = z^2 \frac{d^2}{dz^2}(1-z)^{-1} = 2z^2(1-z)^{-3}$.
Putting $z = \frac{1}{2}$ gives $EX(X-1) = 4$ and $VX = EX^2 - (EX)^2$
$= EX(X-1) + EX - (EX)^2 = 4+2-4 = 2$.

35. We have for $E|X-Y|$

$$\int_a^\infty \left[\int_a^x (x-y) \cdot dF(y)\right] \cdot dF(x) + \int_a^\infty \left[\int_a^y (y-x) \cdot dF(x)\right] \cdot dF(y),$$

so

$$\frac{g}{2} = \int_a^\infty \left[\int_a^x (x-y) \cdot dF(y)\right] \cdot dF(x)$$

$$= \int_a^\infty xF(x) \cdot dF(x) - \int_a^\infty \left[\int_a^x y \cdot dF(y)\right] \cdot dF(x).$$

In the second term integrate by parts, with $u := \int_a^x y \cdot dF(y)$, $v := F(x)$, and $du = x \cdot dF(x)$, to get

$$\int_a^\infty \left[\int_a^x y \cdot dF(y)\right] \cdot dF(x) = \left[\int_a^x y \cdot dF(y)\right] F(x) \Big|_a^\infty$$

$$- \int_a^\infty xF(x) \cdot dF(x)$$

$$= \mu - \int_a^\infty xF(x) \cdot dF(x)$$

and hence

$$\frac{g}{2} = 2 \int_a^\infty xF \cdot dF - \mu$$

$$= 2\left[\int_a^\infty x(1-F) \cdot d(1-F) - \int_a^\infty x \cdot d(1-F)\right] - \mu.$$

In the first bracketed term integrate by parts with $u := x$ and $v := (1-F)^2/2$, then write the second bracketed term as $+\int_a^\infty x \cdot dF$, obtaining

$$\frac{g}{2} = 2\left[\frac{x(1-F)^2}{2}\Big|_a^\infty - \frac{1}{2}\int_a^\infty (1-F)^2 \cdot dx + \int_a^\infty x \cdot dF\right] - \mu$$

$$= 2\left[-\frac{a}{2} - \frac{1}{2}\int_a^\infty (1-F)^2 \cdot dx + \mu\right] - \mu$$

$$= \mu - a - \int_a^\infty (1-F)^2 \cdot dx.$$

36. With Y as the standard form of any r.v. having positive, finite variance, we have

$$\alpha_4 = E\left[\left(Y^2\right)^2\right] = VY^2 + \left[E\left(Y^2\right)\right]^2 = VY^2 + 1 > 1.$$

The inequality is strict, since $VY^2 = 0$ only if $VY = 0$, in which case α_4 is undefined.

37. Distribution $f(x) = \frac{1}{2}\mathbf{1}_{\{-1,1\}}(x)$ has $\mu = 0$ and $\sigma^2 = 1$, so $\Pr\left[|X - \mu| \geq 1\right] = \Pr\left[|X| \geq 1\right] = \Pr\left[|X| = 1\right] = 1 = \sigma^2/k$ when $k = 1$.

38. That h is concave implies that $-h$ is convex. Jensen gives $E\left[-h(X)\right] \geq -h(\mu)$ and therefore $Eh(X) \leq h(\mu)$.

39. Suppose F_1 is preferred to F_2, so that $\int u \cdot dF_1 - \int u \cdot dF_2 > 0$. Then

$$\int u' \cdot dF_1 - \int u' \cdot dF_2 = \left(a + b\int u \cdot dF_1\right) - \left(a + b\int u \cdot dF_2\right)$$

$$= b\left(\int u \cdot dF_1 - \int u \cdot dF_2\right) > 0.$$

40. Writing (3.37) as $\int_{[0,\infty)} u(w) \cdot d[F_1(w) - F_2(w)]$ and integrating by parts give

$$u(w)\left[F_1(w) - F_2(w)\right]\big|_0^\infty - \int_0^\infty \left[F_1(w) - F_2(w)\right] \cdot du(w).$$

The conclusion follows if the first term vanishes. The limit as $w \downarrow 0$ is zero. For the upper limit (taking $u(w) > 0$ for sufficiently large w with no loss of generality),

$$\lim_{w \to \infty} u(w)\left[F_1(w) - F_2(w)\right] = \lim_{w \to \infty} u(w)\left\{\left[1 - F_2(w)\right] - \left[1 - F_1(w)\right]\right\}$$

$$\leq \lim_{w \to \infty} u(w)\left|\int_w^\infty dF_2(w) - \int_w^\infty dF_1(w)\right|$$

$$< \lim_{w \to \infty} \left(\int_w^\infty u \cdot dF_1 + \int_w^\infty u \cdot dF_2\right)$$

$$= 0,$$

where the third step uses the fact that u is increasing and the last equality follows from the finiteness of expected utility under both distributions.

41. The value ι such that $u(100 - \iota) = Eu(W) = \int \left(1 - e^{-w/100}\right) \cdot dF(w)$ solves $1 - e^{-(100-i)/100} = 1 - e^{-1+i/100} = \int_0^\infty \left(1 - e^{-w/100}\right)\frac{1}{80}e^{-w/80} \cdot$ $dw = 1 - \frac{1}{80}\int_0^\infty e^{-w\left(\frac{1}{80}+\frac{1}{100}\right)} \cdot dw = 1 - \frac{1}{1.8}$. Hence, $\iota = 100\left(1 - \log 1.8\right)$ $\doteq \$41.22$.

42. $E|X| = E(|X| \cdot 1) \leq \sqrt{E|X|^2 \cdot E1^2} = \sqrt{EX^2}$. More generally, $E|X|^t \leq \sqrt{E|X|^{2t}}$ for all t.

43. $E(X-Y)^2 = E[(X-\mu)-(Y-\mu)]^2 = VX - 2Cov(X,Y) + VY = \sigma^2 + 0 + \sigma^2$.

44. The marginal distributions of X, Y, and W_α are the same for each $\alpha \in [0,1]$, so the symmetry of the Cauchy distribution implies that $E_pX = E_pY = E_pW_\alpha = \delta$. Thus,

$$Cov_p(X, W_\alpha) = E_p(X-\delta)(W_\alpha - \delta)$$
$$= E_p(X-\delta)[\alpha(X-\delta) + (1-\alpha)(Y-\delta)]$$
$$= \alpha V_pX$$

and

$$\rho_p(X, W_\alpha) = \frac{Cov_p(X, W_\alpha)}{\sqrt{V_pX \cdot V_pW_\alpha}}$$
$$= \frac{\alpha V_pX}{V_pX}$$
$$= \rho_0(X, W_\alpha).$$

45. X and Y have the same marginal distributions, $f(x) = \binom{2}{x+1}\frac{1}{4}\mathbf{1}_{\{-1,0,1\}}$, and $EX = EY = 0$, so $\sigma_{XY} = EXY = (-1 \cdot 0)\frac{1}{4} + (0 \cdot -1)\frac{1}{4} + (0 \cdot 1)\frac{1}{4} + (1 \cdot 0)\frac{1}{4} = 0$. But X and Y are not independent since, for example, $f_{\mathbf{Z}}(-1,-1) = 0 \neq f_X(-1)f_Y(-1) = \left(\frac{1}{4}\right)^2$.

46. We have $EY = a + b\mu_X$, $VY = \sigma_Y^2 = b^2\sigma_X$, $\sigma_Y = |b|\sigma_X$, and $\sigma_{XY} = E(X-\mu_X)(Y-\mu_Y) = bE(X-\mu_X)^2 = b\sigma_X^2$. Therefore, $|\rho_{XY}| = \frac{|\sigma_{XY}|}{\sigma_X\sigma_Y} = \frac{|b|\sigma_X^2}{|b|\sigma_X^2} = 1$.

47. Clearly, $\rho_{WU} = 0$ if $b = 0$ or $d = 0$. Otherwise, $Cov(W,U) = bdCov(X,Y)$, $VW = b^2\sigma_X^2$, $VU = d^2\sigma_Y^2$, and $\rho_{WU} = \frac{bd}{|bd|}\rho_{XY} = \rho_{XY} \cdot sign(bd)$.

48. (i) $EX = \int_1^2 x \cdot dx = \frac{3}{2}$, $EY = \int_1^2 x^{-1} \cdot dx = \log 2$, $VX = EX^2 - (EX)^2 = \int_1^2 x^2 \cdot dx - \frac{9}{4} = \frac{1}{12}$, $VY = EY^2 - (EY)^2 = \int_1^2 x^{-2} \cdot dx - (\log 2)^2 = \frac{1}{2} - (\log 2)^2$. $\sigma_{XY} = EXY - EX \cdot EY = 1 - \frac{3}{2}\log 2$. $\rho_{XY} = \sigma_{XY}/(\sigma_X\sigma_Y) = (1 - \frac{3}{2}\log 2)/\left[\sqrt{\frac{1}{12}}\sqrt{\frac{1}{2} - (\log 2)^2}\right] \doteq -.984$.

    (ii) $EX = 0$, so $Cov(X,Y) = EXY = EX^3 = 0$.

49. Since $EC^* = \gamma$ for both choices of C^*, we have $E(C^* - \gamma)^2 = VC^*$ in both cases. Now

$$VC - V(A+B) = \gamma^2 - (\alpha^2 + 2\rho\alpha\beta + \beta^2)$$
$$= (\alpha^2 + 2\alpha\beta + \beta^2) - (\alpha^2 + 2\rho\alpha\beta + \beta^2)$$
$$= 2\alpha\beta(1-\rho).$$

This is nonnegative for all feasible values of ρ and strictly positive for $\rho < 1$, so that $A + B$ is the better estimator.

50. $0 \leq V\left(\sum_{j=1}^{m} X_j\right) = \sum_{j=1}^{m} VX_j + \sum_{i \neq j} Cov(X_i, X_j) = m\sigma^2 + (m^2 - m)\rho\sigma^2$, implies $\rho \geq -m/(m^2 - m) = -(m-1)^{-1}$.

51. $\rho_{XY} = +1$ implies that $Y = a + bX$ a.s. for some a and some $b > 0$, while $\rho_{XZ} = -1$ implies that $X = c - dZ$ a.s. for some c and some $d > 0$. Thus, $Y = (a + bc) - bdZ$ a.s., with $bd > 0$, so $\rho_{YZ} = -1$.

52. $\sigma^2_{X-Y} = \sigma^2_X - 2\sigma_{XY} + \sigma^2_Y \geq \sigma^2_X - 2\sigma_X\sigma_Y + \sigma^2_Y = (\sigma_X - \sigma_Y)^2$.

53. (i) Function $\sin x$ is *odd*, whereas p.d.f. $f(x)$ is *even*, and so $\sin x \cdot f(x)$ is also odd. Since $E|\sin X|$ is clearly finite, it follows that $E\sin X = \int_{-\infty}^{\infty} \sin x\, f(x) \cdot dx = 0$.
(ii) From (i) and the result of Example 3.37 one sees that $E\cos X = E\cos X + iE\sin X = \phi(1) = e^{-1/2}$.

54. If $t = 0$ then of course $\phi(t) = Ee^0 = 1$. For $t \neq 0$ we have $\phi(t) = \frac{1}{2\pi}\int_{-\pi}^{\pi} e^{itx} \cdot dx = \frac{1}{2\pi}\int_{-\pi}^{\pi}[\cos(xt) + i\sin(xt)] \cdot dx = \left[\frac{\sin(xt) - i\cos(xt)}{2\pi t}\right]_{-\pi}^{\pi} = \frac{\sin(\pi t)}{\pi t}$. Using Taylor's theorem with remainder after the linear term, $\sin(\pi t) = \sin(0) + \cos(0)\pi t - \frac{\sin(\pi t^*)}{2}\pi^2 t^2$, where $|t^*| < t$. Thus, $\frac{\sin(\pi t)}{\pi t} = 1 + \frac{\sin(\pi t^*)\pi t}{2}$, which approaches unity as $t \to 0$.

55. $-d^2 \log\phi(t)/dt^2 = -\frac{d}{dt}\left[\frac{\phi'(t)}{\phi(t)}\right] = -\frac{\phi''(t)}{\phi(t)} + \left[\frac{\phi'(t)}{\phi(t)}\right]^2$. Evaluating at $t = 0$ gives $-d^2 \log\phi(t)/dt^2\big|_{t=0} = -\phi''(0) + \phi'(0)^2 = -i^2\mu_2' + (i\mu)^2 = \mu_2' - \mu^2 = EX^2 - (EX)^2 = VX$.

56. If $X \sim N(\mu, \sigma^2)$ then $\log M(t) = \mu t + \sigma^2 t^2/2 = \sum_{\ell=1}^{\infty} \kappa_\ell (it)^\ell/\ell!$ for all $t \in \Re$, so that $\kappa_1 = \mu$, $\kappa_2 = \sigma^2$, and $\kappa_j = 0$ for all $j \geq 3$.

57. When X has cumulants through order k we have $\log\phi_Y(t) = \log\phi_X(at) = \sum_{\ell=1}^{k} \kappa_\ell (iat)^\ell/\ell! + R_{k+1} = \sum_{\ell=1}^{k}(\kappa_\ell a^\ell)(it)^\ell/\ell! + R_{k+1}$, so that $\kappa_\ell(aX) = a^\ell\kappa_\ell(X)$.

58. $\int_{\Re}|e^{-|t|}| \cdot dt = \int_{-\infty}^{\infty} e^{-|t|} \cdot dt = 2\int_0^{\infty} e^{-t} \cdot dt = 2 < \infty$. The inversion formula for absolutely integrable c.f.s gives

$$f(x) = \frac{1}{2\pi}\int e^{-itx}\phi(t) \cdot dt = \frac{1}{2\pi}\int_{-\infty}^{\infty} e^{-itx - |t|} \cdot dt$$

$$= \frac{1}{2\pi}\left[\int_{-\infty}^{0} e^{t(1-ix)} \cdot dt + \int_{0}^{\infty} e^{-t(1+ix)} \cdot dt\right]$$

$$= \frac{1}{2\pi}\left[\int_{0}^{\infty} e^{-t(1-ix)} \cdot dt + \int_{0}^{\infty} e^{-t(1+ix)} \cdot dt\right]$$

$$= \frac{1}{2\pi}\left(\frac{1}{1-ix} + \frac{1}{1+ix}\right) = \frac{1}{\pi(1+x^2)}.$$

59. The inversion theorem implies

$$\frac{1}{2\pi} \int_{-\infty}^{\infty} \frac{e^{-itx}}{1+t^2} \cdot dt = \frac{1}{2} e^{-|x|}.$$

Reversing the roles of x and t gives

$$e^{-|t|} = \int_{-\infty}^{\infty} \frac{e^{-itx}}{\pi(1+x^2)} \cdot dx = \int_{-\infty}^{\infty} \frac{e^{ity}}{\pi(1+y^2)} \cdot dy,$$

where the second equality follows on setting $y = -x$.

60. Put $e^{i(X_1+X_2)} = \cos(X_1+X_2) + i\sin(X_1+X_2)$, use the formulas for cosines and sines of sums, take expectations, and simplify to get

$$E e^{i(X_1+X_2)} = E(\cos X_1 + i\sin X_1) \cdot E(\cos X_2 + i\sin X_2).$$

61. (i) Working directly, $VX = EX^2 = \frac{1}{\sqrt{2\pi}} \int_{-\infty}^{\infty} x^2 e^{-x^2/2} \cdot dx$. Integrating by parts with $u = x$, $dv = xe^{-x^2/2}$, $du = dx$, $v = -e^{-x^2/2}$ gives $EX^2 = \frac{1}{\sqrt{2\pi}} -xe^{-x^2/2}\Big|_{-\infty}^{\infty} + \int_{-\infty}^{\infty} \frac{1}{\sqrt{2\pi}} e^{-x^2/2} \cdot dx = 1$, since $\lim_{|x|\to\infty} xe^{-x^2/2} = 0$ and the integral of the p.d.f. is unity. Working from the c.f., $EX^2 = i^{-2}\phi''(0) = i^{-2} d^2 e^{-t^2/2}/dt^2\Big|_{t=0} = i^{-2} d\left(-te^{-t^2/2}\right)/dt\Big|_{t=0} = -\left(-e^{-t^2/2} + t^2 e^{-t^2/2}\right)_{t=0} = 1$.

(ii) $V(bX) = b^2 VX = b^2$.

(iii) Transformation $Y = bX$ is one-to-one, with $|dx/dy| = 1/|b|$, so

$$f_Y(y) = \frac{f_X(y/b)}{|b|} = \frac{1}{\sqrt{2\pi b^2}} \exp\left(-\frac{y^2}{2b^2}\right).$$

(iv) $\phi_Y(t) = \phi_X(bt) = e^{-b^2 t^2/2}$.

62. $\phi_{\mathbf{Y}}(\mathbf{t})$, the joint c.f. of $\mathbf{Y} := (Y_1, Y_2)'$, is

$$
\begin{aligned}
E e^{i\mathbf{t}'\mathbf{Y}} &= E \exp\left[i(t_1 Y_1 + t_2 Y_2)\right] \\
&= E \exp\left[i(t_1 X_1 + t_1 X_2) + i(t_2 X_1 - t_2 X_2)\right] \\
&= E \exp\left[i(t_1 + t_2)X_1 + i(t_1 - t_2)X_2\right] \\
&= E e^{i(t_1+t_2)X_1} E e^{i(t_1-t_2)X_2} \\
&= e^{-(t_1+t_2)^2/2 - (t_1-t_2)^2/2} = e^{-t_1^2 - t_2^2}.
\end{aligned}
$$

The marginal c.f.s are $\phi_1(t_1) = \phi_{\mathbf{Y}}(t_1, 0) = e^{-t_1^2}$ and $\phi_2(t_2) = \phi_{\mathbf{Y}}(0, t_2) = e^{-t_2^2}$. If $X \sim N(0,1)$, we have seen that the variance of bX is b^2 and that the c.f. of bX is $e^{-b^2 t^2/2}$. Thus, both Y_1 and Y_2 have variance $b^2 = 2$.

63. $\phi_{X_1+X_2}(t) = \phi_{2X_1}(t) = \phi_1(2t) = e^{-2|t|} = \phi_1(t)\phi_2(t)$; but $\phi_{X_1-X_2}(t) = 1 \neq \phi_1(t)\bar{\phi}_2(t)$.

64. Take $A = \Omega$.

65. (i) In the discrete case

$$E\left[E\left(Y|X\right)\right] = \sum_{x \in X} E\left(Y|x\right) f_X(x)$$

$$= \sum_{x \in X} \left[\sum_{y:(x,y) \in Z} y f_{Y|X}(y|x) \right] f_X(x)$$

$$= \sum_{x \in X} \sum_{y:(x,y) \in Z} y f_{\mathbf{Z}}(x,y)$$

$$= \sum_{(x,y) \in Z} y f_{\mathbf{Z}}(x,y) = EY.$$

(ii) In the continuous case

$$E\left[E\left(Y|X\right)\right] = \int_{-\infty}^{\infty} E(Y|x) f_X(x) \cdot dx$$

$$= \int_{-\infty}^{\infty} \left[\int_{-\infty}^{\infty} y f_{Y|X}(y|x) \cdot dy \right] f_X(x) \cdot dx$$

$$= \int_{-\infty}^{\infty} \int_{-\infty}^{\infty} y f_{\mathbf{Z}}(x,y) \cdot dy\, dx = EY.$$

66. The text shows that $E\left(X_2|0\right) = \frac{1}{2}$ and $E\left(X_2|1\right) = \frac{3}{2}$, so $E(X_2|x_1) = \frac{1+2x_1}{2}$ for $x_1 \in \{0,1\}$. Marginal p.m.f.s of X_1 and X_2 are $f_1(x_1) = \frac{1}{2}\mathbf{1}_{\{0,1\}}(x)$ and $f_2(x_2) = \binom{2}{x}\frac{1}{4}\mathbf{1}_{\{0,1,2\}}(x)$. From these, $EX_2 = (0)\frac{1}{4} + (1)\frac{1}{2} + (2)\frac{1}{4} = 1$ and $E\left[E\left(X_2|X_1\right)\right] = \left(\frac{1}{2}\right)\frac{1}{2} + \left(\frac{3}{2}\right)\frac{1}{2} = 1$.

67. $VX_2 = EX_2^2 - (EX_2)^2 = (0^2)\frac{1}{4} + (1^2)\frac{1}{2} + (2^2)\frac{1}{4} - 1 = \frac{3}{2} - 1 = \frac{1}{2}$ and $VX_1 = EX_1^2 - (EX_1)^2 = (0^2)\frac{1}{2} + (1^2)\frac{1}{2} - \left(\frac{1}{2}\right)^2 = \frac{1}{4}$. The conditional variances are $V(X_2|0) = V(X_2|1) = \frac{1}{4}$, and we can represent $E\left(X_2|X_1\right)$ as $(1+2X_1)/2$. Thus, $E\left[V\left(X_2|X_1\right)\right] + V\left[E\left(X_2|X_1\right)\right] = E\left(\frac{1}{4}\right) + V\left(\frac{1+2X_1}{2}\right) = \frac{1}{4} + VX_1 = \frac{1}{2}$.

68. $M_T(t) = E\left[E\left(e^{tT} \mid N\right)\right] = EM_X(t)^{N-1}$, where $M_X(t) = \int_0^1 e^{tx} \cdot dx = (e^t - 1)t^{-1}\mathbf{1}_{\Re\backslash\{0\}}(t) + \mathbf{1}_{\{0\}}(t)$ (Example 3.35). Then $M_T(t) = \sum_{n=1}^{\infty} M_X(t)^{n-1}2^{-n} = 2^{-1}M_X(t)/\left[2 - M_X(t)\right]$ for t such that $M_X(t) < 2$ (t less than about 1.2564). Differentiating, and simplifying give $M_T'(t) = M_X'(t)/\left[2 - M_X(t)\right]^2$ and $ET = M_T'(0) = M_X'(0) = 1/2$.

69. Since $\Pr\left(Z_2 = 0\right) = 0$, both Z_1/Z_2 and $Z_1/|Z_2|$ are proper r.v.s with

$$\Pr\left(Z_1/Z_2 \in \Re\right) = \Pr\left(Z_1/|Z_2| \in \Re\right) = 1.$$

For any $x \in \Re$ we have

$$\Pr\left(\frac{Z_1}{Z_2} \le x\right) = \Pr\left(Z_1 \le Z_2 x, Z_2 > 0\right) + \Pr\left(-Z_1 \le -Z_2 x, Z_2 < 0\right)$$

$$= \Pr\left(Z_1 \le |Z_2|x, Z_2 > 0\right) + \Pr\left(-Z_1 \le |Z_2|x, Z_2 < 0\right)$$
$$= \Pr\left(Z_1 \le |Z_2|x, Z_2 > 0\right) + \Pr\left(Z_1 \le |Z_2|x, Z_2 < 0\right)$$
$$= \Pr\left(Z_1 \le |Z_2|x\right)$$
$$= \Pr\left(\frac{Z_1}{|Z_2|} \le x\right).$$

The equality in the third line follows from the symmetry of Z_1 and the independence of Z_1, Z_2. The first and fourth lines follow from the law of total probability.

70. Take $Y = -X$. Obviously, what is lacking here is independence.

Chapter 4

Models for Distributions

Examples given in Chapters 2-3 have introduced many specific models for distributions. This chapter gives a systematic account of those most often encountered in statistics and in applied probability generally. We group these in Sections 4.2 and 4.3 according to whether they model discrete or continuous r.v.s. Each model that is described constitutes a *family* of distributions indexed by $k \geq 1$ *parameters* and whose representations (in terms of densities, c.d.f.s, or c.f.s) have the same functional form, as $\{f(x; \boldsymbol{\theta}) : \boldsymbol{\theta} \in \boldsymbol{\Theta}\}$, $\{F(x; \boldsymbol{\theta}) : \boldsymbol{\theta} \in \boldsymbol{\Theta}\}$, or $\{\phi(t; \boldsymbol{\theta}) : \boldsymbol{\theta} \in \boldsymbol{\Theta}\}$ for all real x and t. Here $\boldsymbol{\Theta} \in \Re^k$ is the space of parameter values on which the respective functions of x or t have the required properties. As we shall see in Part II, a common goal of statistical inference is to determine which specific member of such a family best describes observed data, a process that involves using the data to obtain a *point estimate* of $\boldsymbol{\theta}$. Another common goal is to test various *a priori* hypotheses about $\boldsymbol{\theta}$. A special class of distributions will be of importance in devising such techniques of inference. Although most of the applications come later, we introduce it here because much can be inferred about a specific family just from knowing that it is in the class.

4.1   The Exponential Class of Distributions

Definition 4.1. A family of distributions is in the *exponential class* if their densities
$$\{f(x; \boldsymbol{\theta}) : \boldsymbol{\theta} \in \boldsymbol{\Theta} \subset \Re^k\}$$
(all with respect to the same dominating measure $\mathcal{M}$) can be expressed in the form
$$f(x; \boldsymbol{\theta}) = b(\boldsymbol{\theta}) c(x) e^{\boldsymbol{\eta}(\boldsymbol{\theta})' \mathbf{T}(x)}, \tag{4.1}$$

235

where $b \in \Re$ and $\boldsymbol{\eta}(\boldsymbol{\theta}) \in \Re^m$ do not involve x, and $c \in \Re$ and $\mathbf{T} \in \Re^m$ do not involve $\boldsymbol{\theta}$.

The exponential class includes discrete, (absolutely) continuous, and mixed families, but the common models are all of the first two types. The elements of $\boldsymbol{\eta}(\boldsymbol{\theta})$ are commonly referred to as the family's "natural" or "canonical" parameters. A representation such as (4.1) is clearly not unique—for example, $b(\boldsymbol{\theta})$ and $c(x)$ could be replaced by $ab(\boldsymbol{\theta})$ and $a^{-1}c(x)$ for $a \neq 0$, and there could be offsetting linear transformations of $\boldsymbol{\eta}(\boldsymbol{\theta})$ and $\mathbf{T}$—but there is always some representation with a *minimal* number of natural parameters, m. In the minimal representation m and k (the dimension of $\boldsymbol{\theta}$) are typically, but not necessarily, the same.

Example 4.1. (i) The family of distributions

$$\left\{ f(x;\boldsymbol{\theta}) = (2\pi\theta_2)^{-1/2} \exp\left[-\frac{(x-\theta_1)^2}{2\theta_2} \right] : \theta_1 \in \Re, \theta_2 > 0 \right\} \qquad (4.2)$$

has a minimal representation (4.1) with $m = k = 2$ and

$$b(\boldsymbol{\theta}) = \theta_2^{-1/2} \exp\left(-\frac{\theta_1^2}{2\theta_2} \right)$$

$$c(x) = (2\pi)^{-1/2}$$

$$\boldsymbol{\eta}(\boldsymbol{\theta})' = \left(-\frac{1}{2\theta_2}, \frac{\theta_1}{\theta_2} \right)$$

$$\mathbf{T}(x)' = (x^2, x).$$

(ii) The minimal representation for the subfamily with $\theta := \theta_1 = \sqrt{\theta_2}$ has $m = 2 > k = 1$ and $\boldsymbol{\eta}(\theta)' = (-1/(2\theta), \theta^{-1/2})$.

(iii) The minimal representation for the subfamily with $\theta := \theta_1 = \theta_2$ has $m = k = 1$, $\eta(\theta) = -1/(2\theta)$, $T(x) = x^2$, and $c(x) = (2\pi)^{-1/2} e^x$.

Example 4.2. The family

$$\left\{ f(x;\boldsymbol{\theta}) = \frac{1}{\theta_2} e^{-(x-\theta_1)/\theta_2} \mathbf{1}_{[\theta_1,\infty)}(x) : \theta_1 \in \Re, \theta_2 > 0 \right\}$$

is *not* in the exponential class: The p.d.f. cannot be written as (4.1) because the indicator function of x and θ_1 cannot be appropriately decomposed. On the other hand, the subfamily

$$\left\{ f(x;\theta) = \theta^{-1} e^{-x/\theta} \mathbf{1}_{[0,\infty)}(x) : \theta > 0 \right\} \qquad (4.3)$$

does belong to the class, with $b(\theta) = \theta^{-1}$, $c(x) = \mathbf{1}_{[0,\infty)}(x)$, $\eta(\theta) = -\theta^{-1}$, and $T = x$.

As Example 4.2 illustrates, a distribution whose support depends on a parameter cannot be in the exponential class.

Note: In much of the literature, distributions with the representation (4.1) are referred to as *exponential families*; however, we reserve the term "exponential family" for the specific subset (4.3) of the *gamma* family (discussed in Section 4.3.2). Thus, we state that *the* exponential family (4.3) is in the *exponential class* rather than that it is *an* exponential family.

It is sometimes convenient to write densities of members of the exponential class in *canonical form* as

$$f(x; \boldsymbol{\eta}) = a(\boldsymbol{\eta}) c(x) e^{\boldsymbol{\eta}' \mathbf{T}}, \qquad (4.4)$$

where

$$a(\boldsymbol{\eta}) := b[\boldsymbol{\theta}(\boldsymbol{\eta})] = \left[\int c(x) e^{\boldsymbol{\eta}' \mathbf{T}} \mathcal{M}(dx) \right]^{-1}$$

and $\mathcal{M}$ is the measure with respect to which f is a density. $\mathbf{H} := \left\{ \boldsymbol{\eta} : \int c(x) e^{\boldsymbol{\eta}' \mathbf{T}(x)} \mathcal{M}(dx) < \infty \right\}$ is called the family's *natural parameter space*. A simple representation of the joint distribution of "natural statistics" $\mathbf{T}$ can often be found from the canonical form. In particular, if $\boldsymbol{\theta}$ is such that $\boldsymbol{\eta}(\boldsymbol{\theta})$ is not on a boundary of $\mathbf{H}$, then $Ec(X) e^{(\boldsymbol{\eta}+\mathbf{t})' \mathbf{T}(X)} < \infty$ for $\mathbf{t}$ in some open set containing $\{\mathbf{0}\}$. In that case the m.g.f. of $\mathbf{T}$ exists and can be expressed very simply:

$$M_{\mathbf{T}}(\mathbf{t}) := \int e^{\mathbf{t}' \mathbf{T}} f(x; \boldsymbol{\eta}) \mathcal{M}(dx) = \frac{a(\boldsymbol{\eta})}{a(\boldsymbol{\eta} + \mathbf{t})}.$$

The cumulants of $\mathbf{T}$ can then be derived by differentiating cumulant generating function (c.g.f.)

$$\log M_{\mathbf{T}}(\mathbf{t}) = \log a(\boldsymbol{\eta}) - \log a(\boldsymbol{\eta} + \mathbf{t}). \qquad (4.5)$$

Thus, for natural statistics T_j and $T_{j'}$

$$ET_j = - \left. \frac{\partial \log a(\boldsymbol{\eta} + \mathbf{t})}{\partial t_j} \right|_{\mathbf{t}=0} \qquad (4.6)$$

$$Cov(T_j, T_{j'}) = - \left. \frac{\partial^2 \log a(\boldsymbol{\eta} + \mathbf{t})}{\partial t_j \, \partial t_{j'}} \right|_{\mathbf{t}=0}. \qquad (4.7)$$

For example, in model (4.3) with $m = 1$ we have $M_T(t) = \eta/(\eta+t)$, $EX = -\eta^{-1} = \theta$, and $VX = \eta^{-2} = \theta^2$.

Finally, since the *joint* distribution of i.i.d. r.v.s $\{X_i\}_{i=1}^n$ still has the exponential form, as

$$f(x_1, x_2, ..., x_n; \boldsymbol{\eta}) = a(\boldsymbol{\eta})^n \prod_{i=1}^n c(x_i) e^{\boldsymbol{\eta}' \sum_{i=1}^n \mathbf{T}(x_i)},$$

the m.g.f. of the *sum* of natural statistics $\sum_{i=1}^n \mathbf{T}(X_i)$ is simply $M_{\mathbf{T}}(\mathbf{t})^n$.

4.2 Discrete Distributions

A discrete r.v. is one whose distribution has a countable support $\mathbb{X}$, with $\mathbb{P}_X(\{x\}) = F(x) - F(x-) > 0$ for $x \in \mathbb{X}$ and $\mathbb{P}_X(\{x\}) = 0$ elsewhere. Distributions of discrete r.v.s are usually represented by probability mass functions (p.m.f.s) $f(\cdot)$—that is, densities with respect to counting measure N—with the properties $f(x) := \mathbb{P}_X(\{x\}) > 0$ for $x \in \mathbb{X}$ and $\int f(x) \mathcal{M}(dx) = \int f(x) N(dx) = \sum_{x \in \mathbb{X}} f(x) = 1$.

In fact, all the common discrete models pertain just to *count* data, for which support $\mathbb{X}$ is a subset of the nonnegative integers $\aleph_0$. We shall see that many of these are in the exponential class. In studying such count models it is helpful to use another transform with generating and uniqueness properties similar to those of the m.g.f. and c.f.

Definition 4.2. When X is a r.v. with support $\mathbb{X} \subset \aleph_0$ the functional $p(t) := Et^X, t \in \Re$, is called the *probability generating function (p.g.f.)* of X.

The relation $E|t|^X = \sum_{x \in \mathbb{X}} |t|^x f(x) \le \sum_{x \in \mathbb{X}} f(x) = 1$ for $|t| \le 1$ shows that the p.g.f. is always finite in that range. Whether it is so for other values of t depends on the model for $f(\cdot)$.

Differentiating the p.g.f. repeatedly and evaluating at $t = 0$ generates probabilities $\{f(x)\}_{x \in \mathbb{X}}$, a fact that explains the name. Thus, differentiating $\sum_{x \in \mathbb{X}} t^x f(x)$ zero or more times term by term (as is warranted by dominated convergence), we have

$$p(0) = \sum_{x=0}^{\infty} t^x f(x) \bigg|_{t=0} = \left[f(0) + t f(1) + t^2 f(2) + t^3 f(3) + \cdots \right]\big|_{t=0} = f(0)$$

$$p'(0) = \sum_{x=0}^{\infty} x t^{x-1} f(x) \bigg|_{t=0} = \left[f(1) + 2t f(2) + 3t^2 f(3) + \cdots \right]\big|_{t=0} = f(1)$$

$$p''(0) = \sum_{x=0}^{\infty} x(x-1) t^{x-2} f(x) \bigg|_{t=0} = \left[2f(2) + 6t f(3) + \cdots \right]\big|_{t=0} = 2f(2)$$

and so on, with

$$p^{(k)}(0) = k! f(k).$$

Since it directly generates values of the p.m.f., it is clear that the p.g.f. does characterize the distribution.

When $p(t)$ is finite in an open interval containing $t = 1$, all moments of X are finite as well. The connection with m.g.f.s is then obvious, since

$Et^X = Ee^{(\log t)X} = M(\log t)$. Still, even when the m.g.f. exists the different construction of the p.g.f. offers advantages in working with models for count data. Evaluating derivatives at $t = 1$ produces the *descending factorial moments*, defined as

$$\mu'_{[k]} := E\left[X(X-1) \cdots \cdot (X - k + 1)\right].$$

Thus,

$$p'(1) = \sum_{x \in X} \frac{dt^x}{dt} f(x) \bigg|_{t=1}$$

$$= \sum_{x \in X} x t^{x-1} f(x) \bigg|_{t=1}$$

$$= \sum_{x \in X} x f(x)$$

$$= EX = \mu'_{[1]} = \mu$$

$$p''(1) = \sum_{x \in X} \frac{d^2 t^x}{dt^2} f(x) \bigg|_{t=1}$$

$$= \sum_{x \in X} x(x-1) t^{x-2} f(x) \bigg|_{t=1}$$

$$= \sum_{x \in X} x(x-1) f(x)$$

$$= E\left[X(X-1)\right] = \mu'_{[2]}$$

and so on, with

$$p^{(k)}(1) = \sum_{x \in X} x(x-1) \cdots \cdot (x - k + 1) f(x)$$

$$= E\left[X(X-1) \cdots \cdot (X - k + 1)\right]$$

$$= \mu'_{[k]}.$$

Moments about the origin can then be recovered from the descending factorials in the obvious way. For example,

$$\mu = EX = \mu'_{[1]}$$
$$\mu'_2 = E\left[X(X-1)\right] + EX = \mu'_{[2]} + \mu$$
$$\mu'_3 = E\left[X(X-1)(X-2)\right] + 3EX^2 - 2EX = \mu'_{[3]} + 3\mu'_2 - 2\mu.$$

As is true for the m.g.f. and c.f., the p.g.f. of a sum of independent r.v.s is the product of their p.g.f.s. That is, if $Y = X_1 + X_2 + \cdots + X_k$ and $\{X_j\}_{j=1}^{k}$ are mutually independent with p.g.f.s $\{p_j(t)\}_{j=1}^{k}$, then

$$p_Y(t) = Et^{X_1 + X_2 + \cdots + X_k} = Et^{X_1} Et^{X_2} \cdots \cdot Et^{X_k} = \prod_{j=1}^{k} p_j(t). \qquad (4.8)$$

Of course, this specializes to $p_Y(t) = p_1(t)^k$ when the $\{X_j\}$ are identically distributed as well as independent.

Our survey of count distributions begins with an elementary model from which many others can be derived. In all cases there is some fixed probability space $(\Omega, \mathcal{F}, \mathbb{P})$ that is the basis for all the probability calculations, and $X : \Omega \to \aleph_0$ is $\mathcal{F}$-measurable.

4.2.1   *The Bernoulli Distribution*

For some $S \in \mathcal{F}$ let $\mathbb{P}(S) =: \theta \in (0, 1)$ and take $X(\omega) := 1_S(\omega)$. Then $\mathbb{P}_X(\{0\}) = \mathbb{P}(S^c) = 1 - \theta$ and $\mathbb{P}_X(\{1\}) = \mathbb{P}(S) = \theta$. As an indicator function, X just registers whether the outcome of some experiment was a "success" or a "failure", success being identified with event S. The p.m.f. of X is

$$f(x) = \left\{ \begin{array}{ll} 1 - \theta, & x = 0 \\ \theta, & x = 1 \\ 0, & \text{elsewhere} \end{array} \right\} = (1 - \theta)^{1-x} \theta^x 1_{\{0,1\}}(x),$$

and the m.g.f., c.f., and p.g.f. are, respectively,

$$M(t) = (1 - \theta) + \theta e^t = 1 + \theta(e^t - 1)$$
$$\phi(t) = 1 + \theta(e^{it} - 1)$$
$$p(t) = 1 + \theta(t - 1).$$

Either from these functions or directly from the definition one finds that $EX = \theta$ and $VX = \theta(1 - \theta)$. We write $X \sim B(1, \theta)$ to indicate concisely that X has the Bernoulli distribution with "success" probability θ, the symbol "$\sim$" standing for "is distributed as". Why "1" is included as a separate parameter will be seen directly.

Many real-world chance experiments can be regarded as sequences of independent Bernoulli experiments—called in this context Bernoulli "trials".

4.2.2 The Binomial Distribution

We conduct a fixed number n of independent Bernoulli trials, each having probability θ of success. Letting X_j represent the indicator for the jth trial, the r.v.

$$Y := X_1 + X_2 + \cdots + X_n \qquad (4.9)$$

represents the total number of successes. Expression (4.8) shows the p.g.f. of Y to be

$$p_Y(t) = [1 + \theta(t-1)]^n .$$

Thus, $f_Y(0) = p_Y(0) = (1-\theta)^n$, and differentiating $y \in \aleph$ times and evaluating at $t = 0$ yield

$$p_Y^{(y)}(0) = n(n-1) \cdot \cdots \cdot (n-y+1)\, \theta^y (1-\theta)^{n-y} \mathbf{1}_{\{0,1,\ldots,n\}}(y),$$

so that the p.m.f. of Y is

$$\frac{p_Y^{(y)}(0)}{y!} = f_Y(y) = \binom{n}{y} \theta^y (1-\theta)^{n-y} \mathbf{1}_{\{0,1,\ldots,n\}}(y). \qquad (4.10)$$

A r.v. with this p.m.f. is said to have a *binomial* distribution with parameters n and θ; for short, we write $Y \sim B(n,\theta)$. The binomial *family* comprises the distributions $\left\{ \binom{n}{y}\theta^y(1-\theta)^{n-y}\mathbf{1}_{\{0,1,\ldots,n\}}(y) : n \in \aleph, \theta \in (0,1) \right\}$. The Bernoulli distributions are the subfamily with $n = 1$, which explains the notation $B(1,\theta)$.

To approximate the binomial coefficients for large n (and for many other applications later on) it is helpful to use the bounds[1]

$$1 < \frac{n!}{\sqrt{2\pi n}\,(n/e)^n} < 1 + \frac{1}{12n-1} \qquad (4.11)$$

and the resulting *Stirling's approximation*:

$$n! \doteq \sqrt{2\pi n} \left(\frac{n}{e}\right)^n . \qquad (4.12)$$

Members of the binomial family serve as useful models for many common chance experiments. The key requirements for applying the model are (i) that the data be counts of occurrences of some event—our "success"—in a fixed number of repeated *independent* trials and (ii) that the probability of success not change from one trial to the next.

---

[1] For other forms, discussion, and references to proofs see "Stirling's Formula" in volume 8 of Kotz *et al.* (1988, pp. 779-780).

Example 4.3.

(i) A fair coin is flipped three times and the number of heads Y is recorded. Each toss of the coin is a Bernoulli trial with two classes of outcomes, H and T. It is usually reasonable to assume that the trials are independent and that the probability of heads is always $1/2$. In that case $Y \sim B(3, 1/2)$, with $f(y) = \binom{3}{y} 2^{-3} 1_{\{0,1,2,3\}}(y)$.

(ii) The number of aces obtained in five draws *with replacement* from an ordinary deck (*c.f.* Exercise 2.4) is distributed as $B(5, 1/13)$.

(iii) Twenty names are drawn *with replacement* from a class of 50 students, of whom 23 are male and 27 are female. Y is the number of female students drawn. Since the names are drawn with replacement, the trials are independent, each with probability $27/50$ of success. Therefore, $Y \sim B(20, 27/50)$.

(iv) Twenty names are drawn *without* replacement from a class of 50 students, of whom 23 are male and 27 are female. Y is the number of female students drawn. The Bernoulli trials are *not* independent, because the chance of success on each trial after the first depends on what was drawn previously. The appropriate model is *hypergeometric* (discussed below) rather than binomial.

Because the binomial support $\mathbb{Y} = \{0, 1, ..., n\}$ depends on parameter n, the full binomial family is not in the exponential class. However, in modeling real experiments, such as the first three in Example 4.3, the number of "trials" is typically known. If success probability θ is not also known, then we have a problem of statistical inference, which (as we shall see in Part II) would be approached by conducting some $n_0 \in \aleph$ independent trials, observing the outcomes, and using the data to estimate or restrict the possible values of θ. Such problems of inference amount to choosing the appropriate model from the subfamily $\{f_Y(y; n_0, \theta) : \theta \in (0, 1)\}$, which *is* a member of the exponential class. For example, for the representation (4.1) we can take $b(\theta) = (1 - \theta)^{n_0}$, $c(x) = \binom{n_0}{x} 1_{\{0, 1, ..., n_0\}}(x)$, $\eta(\theta) = \log \theta - \log(1 - \theta)$, and $T(x) = x$.

Here is a more insightful, "constructive" derivation of (4.10). Letting S_j be the event of success on the jth trial, the event of getting precisely y successes and $n - y$ failures in a *particular* order—say, the successes first and then the failures—can be written as

$$S_1 \cap S_2 \cap \cdots \cap S_y \cap S_{y+1}^c \cap \cdots \cap S_n^c.$$

By independence, the probability of this event is

$$\mathbb{P}(S_1)\mathbb{P}(S_2)\cdots\cdot\mathbb{P}(S_y)\mathbb{P}(S_{y+1}^c)\cdots\cdot\mathbb{P}(S_n^c) = \theta^y(1-\theta)^{n-y}.$$

Since there are $\binom{n}{y}$ possible rearrangements of the y successes and $n-y$ failures, the probability of y successes without regard to order is as given by (4.10). The familiar binomial formula, $(a+b)^n = \sum_{j=0}^{n} \binom{n}{j}a^{n-j}b^j$, shows that

$$\sum_{y=0}^{n} \binom{n}{y}\theta^y(1-\theta)^{n-y} = [(1-\theta)+\theta]^n = 1,$$

confirming that (4.10) is a legitimate p.m.f. and explaining the model's name.

The moments of $Y \sim B(n,\theta)$ can be deduced directly as

$$\mu_k' = \sum_{y=0}^{n} y^k \binom{n}{y}\theta^y(1-\theta)^{n-y},$$

but some manipulation is needed to express this in simple form. It is easier to differentiate the p.g.f. and find the factorial moments, or else to work from the canonical form (4.4) and differentiate the c.g.f. (4.5) of "natural statistic" $T(Y) = Y$. However, the most insightful approach for finding the mean and variance is to exploit the representation of Y as the sum of independent Bernoullis. Then

$$EY = EX_1 + EX_2 + \cdots + EX_n = \theta + \theta + \cdots + \theta = n\theta$$

and

$$VY = VX_1 + VX_2 + \cdots + VX_n = n\theta(1-\theta).$$

Exercise 4.1. *Show directly that $\sum_{y=0}^{n} y\binom{n}{y}\theta^y(1-\theta)^{n-y} = n\theta$.*

Exercise 4.2. *Derive the m.g.f. of $Y \sim B(n,\theta)$ and use it to express the coefficient of skewness in terms of n and θ.*

Exercise 4.3. *A pair of fair 6-sided dice are rolled 20 times. Find the probability of getting "snake eyes"—outcome $(\cdot,\cdot)$—at least once.*

Exercise 4.4. *Eight heads were obtained when a coin was flipped 10 times. Find the probability of getting such an extreme result (eight or more heads or two or fewer heads) with a fair coin.*

Exercise 4.5. *You are one of N voters who will be asked to vote "yes" or "no" on some issue. You are trying to decide whether it is worth the effort to cast a vote, given that the other $N-1$ voters will actually do so. What is the chance that your vote will be decisive (i.e., break a tie) if*

    (i) $N = 2n$ is an even number?

    (ii) $N = 2n - 1$ is odd, votes are decided independently, and you think each other voter has the same probability p of voting "yes"?

    (iii) For what value of p is the probability in part (ii) greatest?

The binomial subfamily whose members have the same success probability θ is closed under convolution; that is, the sum of two or more independent r.v.s in the subfamily remains in the subfamily. While this is obvious from the characterization of binomials as sums of Bernoulli r.v.s, it is instructive also to demonstrate the fact by means of generating functions.

Exercise 4.6. *If $Y_1 \sim B(n_1, \theta)$ and $Y_2 \sim B(n_2, \theta)$ are independent binomial variates, show that $S := Y_1 + Y_2 \sim B(n_1 + n_2, \theta)$. (Hint: Use the p.g.f. or m.g.f.)*

4.2.3 *The Multinomial Distribution*

The binomial is the right model for the number of successes in a fixed number of independent trials when counting just a *single* class of outcomes, but how can we extend to a multivariate model of counts of two or more classes? In the binomial setup the number of failures in n trials is just n minus the number of successes, but if there are $m+1 > 2$ categories one may want to keep track of the number of outcomes in m of them; for example, we may want to count both the number of snake eyes $(\cdot, \cdot)$ and the number of "box cars"—event $(\vdots, \vdots)$—in throws of a pair of dice. When the trials are independent and the probabilities associated with the various categories remain the same from one trial to the next, the *multinomial* distribution is the right model for the resulting vector-valued r.v. $\mathbf{Y} = (Y_1, Y_2, ..., Y_m)'$.

The p.m.f. is best derived constructively. To simplify the explanation, assume that there are $m + 1 = 3$ categories, with $\mathbf{Y} := (Y_1, Y_2)'$ and $\boldsymbol{\theta} := (\theta_1, \theta_2)'$ as the vectors of counts and probabilities. In n trials the probability of getting y_1 outcomes of the first type, y_2 of the second, and $n - y_1 - y_2$ of the third in that *particular* order is $\theta_1^{y_1} \theta_2^{y_2} (1 - \theta_1 - \theta_2)^{n - y_1 - y_2}$, where (y_1, y_2) can be any pair of nonnegative integers whose sum is at most n. In how many ways can these outcomes be rearranged? There are $\binom{y_1+y_2}{y_1} \equiv \binom{y_1+y_2}{y_2}$ ways of rearranging the outcomes of the first two types in any $y_1 + y_2$ positions, and for each such arrangement there are $\binom{n}{n-y_1-y_2} = \binom{n}{y_1+y_2}$ ways of choosing among all n positions where to place the outcomes in category three. The total number of possible arrangements is therefore

$$\binom{n}{y_1+y_2}\binom{y_1+y_2}{y_1} = \frac{n!}{(y_1+y_2)!\,(n-y_1-y_2)!}\frac{(y_1+y_2)!}{y_1!y_2!}$$

$$= \frac{n!}{y_1!y_2!\,(n-y_1-y_2)!}$$

$$=: \binom{n}{y_1,y_2}.$$

Thus,

$$\Pr\left(Y_1 = y_1, Y_2 = y_2\right) = \binom{n}{y_1,y_2}\theta_1^{y_1}\theta_2^{y_2}\left(1 - \theta_1 - \theta_2\right)^{n-y_1-y_2}$$

for $y_1 \in \{0, 1, 2, ..., n\}$, $y_2 \in \{0, 1, 2, ..., n - y_1\}$. Extending to any $m+1 \geq 3$ categories, a simple induction gives as the general form of the multinomial p.m.f.

$$f_{\mathbf{Y}}(\mathbf{y}; n, \boldsymbol{\theta}) = \binom{n}{y_1, y_2, ..., y_m}\theta_1^{y_1}\theta_2^{y_2}\cdots\cdot\theta_m^{y_m}\left(1 - \sum_{j=1}^{m}\theta_j\right)^{n-\sum_{j=1}^{m}y_j}, \mathbf{y} \in \mathbb{Y}_n.$$

(4.13)

Here $\mathbb{Y}_n$ is the set of m-vectors containing nonnegative integers with $\sum_{j=1}^{m} y_j \leq n$, and *vector* combinatoric

$$\binom{n}{y_1, y_2, ..., y_m} := \frac{n!}{y_1!y_2!\,\cdots\,\cdots\,y_m!(n - y_1 - y_2 - \cdots - y_m)!} =: \binom{n}{\mathbf{y}}$$

is the number of ways of distributing n objects among $m + 1$ categories. The p.m.f. can be represented more compactly as

$$f_{\mathbf{Y}}(\mathbf{y}) = \binom{n}{\mathbf{y}}\boldsymbol{\theta}^{\mathbf{y}}\left(1 - \boldsymbol{\theta}'\mathbf{1}\right)^{n-\mathbf{y}'\mathbf{1}}, \mathbf{y} \in \mathbb{Y}_n,$$

(4.14)

where $\mathbf{1}$ is a vector of units and $\boldsymbol{\theta}^{\mathbf{y}} := \theta_1^{y_1}\theta_2^{y_2} \cdots\cdot \theta_m^{y_m}$. Of course, when $m = 1$ (4.14) reduces to (4.10), so the expression holds for any positive integer m.[2] The name "multinomial" applies because the expression for $f_{\mathbf{Y}}(\mathbf{y})$ represents a generic term of the multinomial expansion

$$\left(\theta_1 + \theta_2 + \cdots + \theta_m + 1 - \sum_{j=1}^{m}\theta_j\right)^n = \left(\boldsymbol{\theta}'\mathbf{1} + 1 - \boldsymbol{\theta}'\mathbf{1}\right)^n,$$

---

[2]The multinomial distribution is sometimes defined with $Y_{m+1} = n - \mathbf{Y}'\mathbf{1}$ included as a separate (redundant) variable. The treatment here makes the binomial a special case of the multinomial and simplifies both notation and description of properties.

which shows that the probabilities $f_{\mathbf{Y}}(\mathbf{y})$ do indeed sum to unity.

As was true for the binomial family, only the subfamily of multi-nomials with fixed $n = n_0 \in \aleph$ is in the exponential class. Putting $b(\boldsymbol{\theta}) := (1 - \boldsymbol{\theta}'\mathbf{1})^{n_0}$, $c(\mathbf{y}) := \binom{n_0}{\mathbf{y}}$, $\eta_j(\boldsymbol{\theta}) := \log \theta_j - \log(1 - \boldsymbol{\theta}'\mathbf{1})$, and $T_j(\mathbf{y}) := y_j$ for $j \in \{1, 2, ..., m\}$ gives a representation corresponding to (4.1).

To learn more about the distributions' properties, start with the p.g.f. For $\mathbf{t} := (t_1, t_2, ..., t_m)'$

$$
\begin{aligned}
p_{\mathbf{Y}}(\mathbf{t}) = E\mathbf{t}^{\mathbf{Y}} &:= Et_1^{Y_1} t_2^{Y_2} \cdots \cdot t_m^{Y_m} \\
&= \sum_{\mathbf{y} \in \mathbb{Y}_n} \binom{n}{\mathbf{y}} (t_1 \theta_1)^{y_1} (t_2 \theta_2)^{y_2} \cdots \cdot (t_m \theta_m)^{y_m} (1 - \boldsymbol{\theta}'\mathbf{1})^{n - \mathbf{y}'\mathbf{1}} \\
&= [1 + \theta_1 (t_1 - 1) + \theta_2 (t_2 - 1) + \cdots + \theta_m (t_m - 1)]^n \\
&= [1 + \boldsymbol{\theta}' (\mathbf{t} - \mathbf{1})]^n.
\end{aligned}
$$

Setting the t's for any subset of the Y's to unity shows that the marginal distribution of the remaining variables is also multinomial. In particular, the p.g.f. of Y_j alone is

$$
p_{\mathbf{Y}}(1, ..., 1, t_j, 1, ..., 1) = [1 + \theta_j (t_j - 1)]^n,
$$

so that the marginal distribution of any single component is binomial. This amounts to counting outcomes in category j as successes and lumping all others together as failures.

To get the factorial moments of any Y_j, differentiate $p_{\mathbf{Y}}(\mathbf{t})$ partially with respect to t_j and evaluate all the t's at unity. Thus,

$$
\begin{aligned}
EY_j &= \left. \frac{\partial p_{\mathbf{Y}}(\mathbf{t})}{\partial t_j} \right|_{\mathbf{t}=1} \\
&= \left. n\theta_j \left[1 + \boldsymbol{\theta}' (\mathbf{t} - \mathbf{1}) \right]^{n-1} \right|_{\mathbf{t}=1} \\
&= n\theta_j \\
EY_j (Y_j - 1) &= \left. \frac{\partial^2 p_{\mathbf{Y}}(\mathbf{t})}{\partial t_j^2} \right|_{\mathbf{t}=1} \\
&= \left. n(n-1)\theta_j^2 \left[1 + \boldsymbol{\theta}' (\mathbf{t} - \mathbf{1}) \right]^{n-2} \right|_{\mathbf{t}=1} \\
&= n(n-1)\theta_j^2,
\end{aligned}
$$

and so on. The product moments are developed as

$$
\begin{aligned}
EY_i Y_j &= \left. \frac{\partial^2 p_{\mathbf{Y}}(\mathbf{t})}{\partial t_i \partial t_j} \right|_{\mathbf{t}=1} \\
&= \left. n(n-1)\theta_i \theta_j \left[1 + \boldsymbol{\theta}' (\mathbf{t} - \mathbf{1}) \right]^{n-2} \right|_{\mathbf{t}=1} = n(n-1)\theta_i \theta_j,
\end{aligned}
$$

and so on, with

$$EY_1 Y_2 \cdots Y_m = n(n-1) \cdots (n-m+1)\theta_1\theta_2 \cdots \theta_m \qquad (4.15)$$

for $n \geq m$. In particular, the covariance between Y_i and Y_j is

$$Cov(Y_i, Y_j) = EY_i Y_j - EY_i EY_j = \begin{cases} n\theta_j(1-\theta_j), & i = j \\ -n\theta_i\theta_j, & i \neq j \end{cases}. \qquad (4.16)$$

Exercise 4.7. *Why does formula (4.15) require $n \geq m$? What is the value of $EY_1 Y_2 \cdots Y_m$ if $n < m$?*

Exercise 4.8. *Show that the covariance matrix of multinomials is given by*

$$V\mathbf{Y} = n\left(\mathbf{\Theta} - \boldsymbol{\theta}\boldsymbol{\theta}'\right), \qquad (4.17)$$

where $\mathbf{\Theta}$ is an $m \times m$ diagonal matrix with $\boldsymbol{\theta}$ as principal diagonal.

4.2.4 The Negative Binomial Distribution

Let us now undertake another sequence of independent Bernoulli trials with success probability $\theta \in (0,1)$, but instead of stopping after a fixed number of trials we will keep going until the nth success is obtained, and we will keep track of the number of *failures*. The random variable Y that represents the number of failures before the nth success then has support $\mathbb{Y} = \aleph_0 :=$ $\{0, 1, 2, ...\}$, and its p.m.f. $f_Y(\cdot)$ can be determined constructively as follows. Any experiment that yielded some $y \in \mathbb{Y}$ must have required a total of $n+y$ trials, ending in a success. Thus, the first $n+y-1$ trials must have contained y failures and $n-1$ successes. Any particular arrangement of these could occur with probability $(1-\theta)^y\theta^{n-1}$, and the probability of any *particular* such arrangement followed by a success on trial $n+y$ is $(1-\theta)^y\theta^n$. Since there are $\binom{n+y-1}{n-1} = \binom{n+y-1}{y}$ ways of arranging the y failures and the first $n-1$ successes, it follows that

$$f_Y(y) = \binom{n+y-1}{n-1}(1-\theta)^y\theta^n \mathbf{1}_{\aleph_0}(y). \qquad (4.18)$$

A r.v. with this p.m.f. is said to have a *negative binomial* distribution with parameters n and θ—in symbols, $Y \sim NB(n, \theta)$. In the special case $Y \sim NB(1, \theta)$ the r.v. $X = Y+1$ is said to have the *geometric* distribution, with p.m.f.

$$f_X(x) = (1-\theta)^{x-1}\theta\mathbf{1}_\aleph(x).$$

Thus, as defined here (and most commonly elsewhere), a geometric r.v. represents the number of *trials* before the first success.

Note: One encounters in the literature a confusing variety of definitions and parameterizations of the general negative binomial distribution. The one given here is easy to remember and to generate constructively, and it offers the strongest parallels with the binomial model.

Example 4.4. A fair coin is flipped repeatedly until the first occurrence of heads. If Y is the number of *tails* that occur, then

$$f_Y(y) = \binom{y}{0} \left(\frac{1}{2}\right)^y \frac{1}{2} = \frac{1}{2^{y+1}}$$

for $y \in \{0, 1, 2, ...\}$. If X represents the total number of flips of the coin, then $X = Y+1$ and $f_X(x) = 2^{-x} 1_{\aleph}(x)$—a model that we have encountered in many previous examples.

To verify that (4.18) is a p.m.f. requires showing that $\sum_{y=0}^{\infty} f_Y(y) = 1$. For this, consider the function $h(t) = (1-t)^{-n}$ for $t < 1$. Fixing $y \in \aleph$ and differentiating y times produces

$$h^{(y)}(t) = n(n+1)\cdots\cdots(n+y-1)(1-t)^{-(n+y)} = \frac{(n+y-1)!}{(n-1)!}(1-t)^{-(n+y)}.$$

A Taylor expansion of $h(t)$ about $t = 0$ yields

$$h(t) = \sum_{y=0}^{\infty} \frac{h^{(y)}(0)}{y!} t^y = \sum_{y=0}^{\infty} \binom{n+y-1}{n-1} t^y,$$

convergent for $|t| < 1$. This representation of $(1-t)^{-n}$ is a *negative*-binomial expansion, corresponding to binomial expansion $(1+t)^n = \sum_{y=0}^{n} \binom{n}{y} t^y$. Taking $t := 1 - \theta$ gives $h(1-\theta) = [1-(1-\theta)]^{-n} = \theta^{-n}$ and

$$1 = \theta^n h(1-\theta) = \sum_{y=0}^{\infty} \binom{n+y-1}{n-1}(1-\theta)^y \theta^n = \sum_{y=0}^{\infty} f_Y(y).$$

Although the constructive derivation from Bernoulli trials requires n to be a positive integer, formula (4.18) does not, for we can put

$$\binom{\varphi+y-1}{\varphi-1} := \frac{(\varphi+y-1)(\varphi+y-2)\cdots\cdots\varphi}{y!}$$

for any nonnegative integer y and any real φ. Restricting to *positive* φ, this can be further expressed in terms of the *gamma function*, as

$$\binom{\varphi+y-1}{\varphi-1} = \frac{\Gamma(\varphi+y)}{\Gamma(\varphi)y!}.$$

The function $\Gamma\left(\cdot\right)$ is defined at any $\alpha > 0$ by the integral expression

$$\Gamma(\alpha) := \int_0^\infty t^{\alpha-1}e^{-t} \cdot dt.$$

It can be shown (Exercise 4.29) to satisfy $\Gamma\left(1+\alpha\right) = \alpha\Gamma\left(\alpha\right)$ and $\Gamma\left(1\right) = 1$, which imply that $\alpha! = \Gamma\left(1+\alpha\right)$ when $\alpha \in \aleph_0$. Replacing n in (4.18) by a positive real φ extends the negative binomial family as

$$\left\{ f_Y\left(y\right) = \frac{\Gamma\left(\varphi+y\right)}{\Gamma\left(\varphi\right)y!}\left(1-\theta\right)^y\theta^\varphi \mathbf{1}_{\aleph_0}\left(y\right) : \theta \in \left(0,1\right), \varphi > 0\right\} \qquad (4.19)$$

and greatly increases its flexibility in modeling count data.

The p.g.f. of Y can be worked out as follows:

$$\begin{aligned}
p_Y(t) = Et^Y &= \sum_{y=0}^\infty t^y \frac{\Gamma\left(\varphi+y\right)}{\Gamma\left(\varphi\right)y!}\left(1-\theta\right)^y\theta^\varphi \\
&= \sum_{y=0}^\infty \frac{\Gamma\left(\varphi+y\right)}{\Gamma\left(\varphi\right)y!}\left[(1-\theta)t\right]^y\theta^\varphi \\
&= \left[\frac{\theta}{1-(1-\theta)t}\right]^\varphi \sum_{y=0}^\infty \frac{\Gamma\left(\varphi+y\right)}{\Gamma\left(\varphi\right)y!}\left[(1-\theta)t\right]^y\left[1-(1-\theta)t\right]^\varphi.
\end{aligned}$$

Now if $0 < t < \frac{1}{1-\theta}$ the final bracketed factors are both between zero and unity, just like the factors $1-\theta$ and θ in the basic form of the p.m.f. Since the terms of the sum correspond to values of the p.m.f. of $NB[\varphi, 1-(1-\theta)t]$, the sum converges to unity, leaving

$$p_Y(t) = \left[\frac{\theta}{1-(1-\theta)t}\right]^\varphi.$$

Exercise 4.9. *By differentiating the p.g.f. show that the mean and variance of $Y \sim NB(\varphi,\theta)$ are $\mu_Y = \varphi(1-\theta)/\theta, \sigma_Y^2 = \varphi(1-\theta)/\theta^2$.*

Exercise 4.10. *Adopting the constructive interpretation of Y as the number of failures before the nth success, explain why it makes sense that μ_Y increases with n and $1-\theta$.*

Exercise 4.11. *X is the number of times a pair of fair, six-sided dice are rolled until the first occurrence of "snake eyes" (outcome $(\cdot,\cdot)$). Find EX and VX.*

The subfamily of (4.19) with fixed $\varphi = \varphi_0 > 0$ is in the exponential class, although the broader two-parameter family is not. The following exercise shows that the subfamily corresponding to fixed θ is closed under convolution; that is, the sum of two or more independent members is also a member.

Exercise 4.12. *If $Y_1 \sim NB(\varphi_1, \theta)$ and $Y_2 \sim NB(\varphi_2, \theta)$ are independent negative binomial variates, show that $S := Y_1 + Y_2 \sim NB(\varphi_1 + \varphi_2, \theta)$.*

4.2.5   The Hypergeometric Distribution

A common experiment that cannot be represented as a sequence of independent Bernoulli trials is one in which draws are made without replacement from a finite collection of objects. Part (iv) of Example 4.3 described one such scenario. Suppose n objects are to be drawn from $N \geq n$ objects without replacement and that on each draw the outcome is scored as either success or failure. Because sampling is without replacement, the events "success on draw j" and "success on draw k" are no longer independent. However, the distribution of X, the number of successes in n draws without replacement from $N \geq n$ objects, can be found as follows.

Of the N objects in the collection suppose S have the characteristic identified with success and the remaining $F = N - S$ do not. If n draws are made, there must be at least $n_S := \max\{0, n - F\}$ successes, and there can be no more than $n^S := \min\{n, S\}$. For example, if $n = 7$ balls are drawn without replacement from an urn containing $S = 6$ red and $F = 4$ white, then at least $n - F = 3$ red *must* be drawn and no more than $\min\{n, S\} = 6$ *could* be drawn. Therefore, X, the number of successes in n draws, has support $\mathbb{X} := \{n_S, n_S + 1, ..., n^S\}$. For $x \in \mathbb{X}$ there are $\binom{S}{x}$ ways of making x successful draws and $\binom{F}{n-x}$ ways that the remaining draws could be unsuccessful, for a total of $\binom{S}{x}\binom{F}{n-x}$ ways to get x successes out of $\binom{N}{n}$ total ways to draw n objects from among $N = S + F$. Thus,

$$f_X(x) = \frac{\binom{S}{x}\binom{F}{n-x}}{\binom{S+F}{n}} \mathbf{1}_{\{n_S, n_S+1, ..., n^S\}}(x). \qquad (4.20)$$

Note that if we put $\binom{m}{M} := 0$ when $M > m$ then (4.20) holds for $x \in \{0, 1, ..., n\}$. A r.v. with this p.m.f. is said to have the *hypergeometric distribution* with parameters S, F, and n; i.e., $X \sim HG(S, F, n)$.

Example 4.5.

(i) If X is the number of red balls obtained in 7 draws without replacement from an urn containing 6 red and 4 white, then $\mathbb{X} = \{3, 4, 5, 6\}$, and the p.m.f. (in tabular form) is

x	$f_X(x)$
3	$\binom{6}{3}\binom{4}{4}/\binom{10}{7} = \frac{1}{6}$
4	$\binom{6}{4}\binom{4}{3}/\binom{10}{7} = \frac{1}{2}$
5	$\binom{6}{5}\binom{4}{2}/\binom{10}{7} = \frac{3}{10}$
6	$\binom{6}{6}\binom{4}{1}/\binom{10}{7} = \frac{1}{30}$

(ii) The number of aces obtained when five cards are *dealt* from an ordinary deck of 52 is distributed as $HG(4, 48, 5)$ (c.f. Exercise 2.5).

There is no simple representation for the hypergeometric p.g.f., but kth descending factorial moment $\mu'_{[k]} := EX(X-1) \cdots (X-k+1)$ can be found by reducing the expression $\sum_{x=0}^{n} x(x-1) \cdots (x-k+1) f_X(x)$ to a constant times the sum of another hypergeometric with parameters $S - k, F$, and $n - k$. For example, to find $\mu'_{[1]} = \mu$ start with

$$EX = \sum_{x=0}^{n} x \left\{ \frac{\frac{S!}{x!(S-x)!} \frac{F!}{(n-x)!(F-n+x)!}}{\frac{(S+F)!}{n!(S+F-n)!}} \right\}, \tag{4.21}$$

begin the sum at unity, express $x \cdot 1/x!$ in the first factor as $1/(x-1)!$, and rearrange to get

$$EX = \frac{nS}{S+F} \sum_{x=1}^{n} \left\{ \frac{\frac{(S-1)!}{(x-1)![S-1-(x-1)]!} \frac{F!}{[n-1-(x-1)]![F-(n-1)+(x-1)]!}}{\frac{(S-1+F)!}{(n-1)![S-1+F-(n-1)]!}} \right\}. \tag{4.22}$$

Finally, setting $y = x - 1$ reduces the sum to

$$\sum_{y=0}^{n-1} \frac{\binom{S-1}{y}\binom{F}{n-1-y}}{\binom{S-1+F}{n-1}},$$

which equals unity because it is the sum over a support of $HG(S-1, F, n-1)$. Thus,

$$EX = \frac{nS}{S+F} = \frac{nS}{N}. \tag{4.23}$$

Exercise 4.13. *Verify that the right-hand sides of (4.21) and (4.22) are equal.*

Exercise 4.14. *Following the same approach as that to (4.23), show that the second descending factorial moment, $\mu'_{[2]} = EX(X-1)$, is*

$$\mu'_{[2]} = \frac{n(n-1)S(S-1)}{N(N-1)}.$$

Exercise 4.15. *Use (4.23) and the result of Exercise 4.14 to show that* $VX = nSF(N-n)/\left[N^2(N-1)\right]$. *Verify that the formulas for* EX *and* VX *hold in the case* $S = 6$, $F = 4$, $n = 7$ *by calculating* EX *and* VX *directly from the p.m.f. in part (i) of Example 4.5.*

Exercise 4.16. *(i) Show that* $\sum_{j=0}^{n} \binom{n}{j}^2 = \binom{2n}{n}$. *(Hint: Hypergeometric probabilities sum to unity.) (ii) Using this result, express the probability that two individuals who each flip a fair coin* n *times will obtain the same number of heads.*

Exercise 4.17. *A discriminating tea drinker claims that she can taste the difference between two cups of tea to which cream and tea were added in different orders. To test her claim, eight cups of tea are set before her. She is told that cream was inserted first in four of them, tea first in the others, and that the cups have been set before her in random order. Let* X *be the number of cream-first cups that she identifies correctly. If her claim is false and her choices purely random, what would be the distribution of* X?[3]

Exercise 4.18. *An urn contains* n *white balls and* n *black balls. Balls are drawn at random, one at a time, without replacement, until all* n *balls of one type or the other have been removed, leaving only those of the other type. For some fixed* $n \in \aleph$ *let* $X_{n,n}$ *represent the number of balls that remain.*

(i) What is the support, $\mathbb{X}_{n,n}$, *of* $X_{n,n}$?

(ii) For arbitrary $n \in \aleph$ *and* $x \in \mathbb{X}_{n,n}$ *develop an expression for* $f_{n,n}(x) = \Pr(X_{n,n} = x)$. *(Hint: Think of a given outcome of the experiment as a particular arrangement of the* $2n$ *balls. How many distinguishable outcomes are there? Consider what an arrangement corresponding to* $X_{n,n} = x \in \mathbb{X}_{n,n}$ *would look like. How many such arrangements are there?)*

(iii) Show that, for each fixed $x \in \{1, 2, ..., n-1\}$, $\lim_{n\to\infty} f_{n,n}(1) = 1/2$ *and* $\lim_{n\to\infty} f_{n,n}(x+1)/f_{n,n}(x) = 1/2$. *Conclude that the distribution of* $X_{n,n}$ *converges as* $n \to \infty$ *to the geometric distribution with* $\theta = 1/2$; *i.e.,* $f_{\infty,\infty}(x) = 2^{-x}1_{\aleph}(x)$. *Figure 4.1 compares the distributions for* $n = 5$ *and* $n = 10$ *with the limiting form.*

Exercise 4.19. *Extending Exercise 4.18, let* $X_{m,n}$ *be the number of remaining balls if the urn contains* m *white and* n *black. Find the distribution of* $X_{m,n}$.

---

[3]For the full story of the actual event—and the science behind the creaming of tea—see the entertaining article by Stephen Sen (2012).

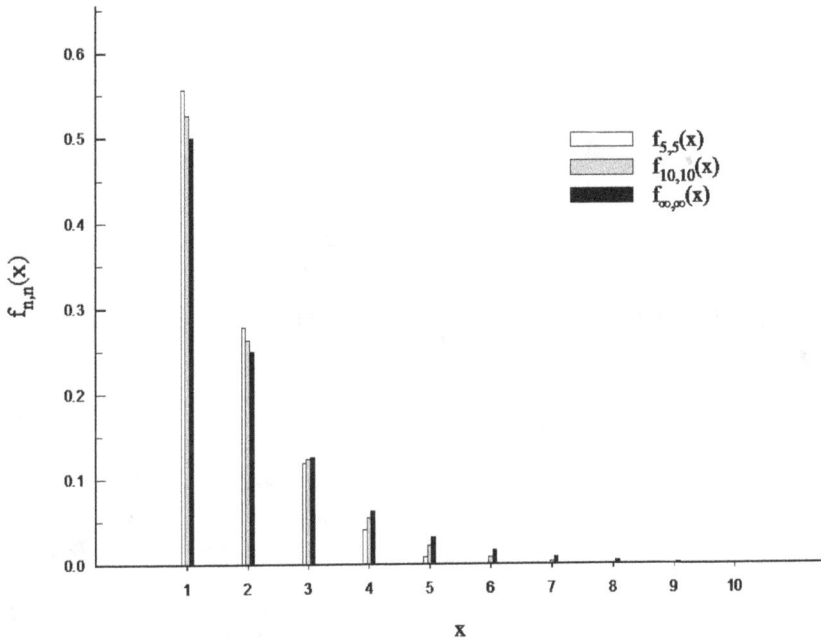

Fig. 4.1   Distributions of number of residual balls, for $n = 5, 10, \infty$.

4.2.6   *The Poisson Distribution*

We come now to a model for count data that cannot be derived just from simple counting rules. As motivation, consider the following example. The proprietor of a retail store wishes to model the number of customers X who enter during a fixed period of each market day. He reasons that by dividing the fixed period into some large number n of subperiods the distribution of X could be approximated as binomial. That is,

$$f(x) \doteq f_n(x) := \binom{n}{x} \theta_n^x (1 - \theta_n)^{n-x} \mathbf{1}_{\{0,1,\ldots,n\}}(x),$$

where θ_n is the probability that a single customer enters during any one of the n subperiods. Among several drawbacks of this idea the most obvious is that it ignores the possibility that more than one customer could enter during the same subperiod. However, since this becomes less unrealistic as the length of a subperiod diminishes, one is led to consider how the distribution of X_n would behave as $n \to \infty$ with θ_n declining proportionately as, say, $\theta_n = \lambda/n$ for some $\lambda > 0$. Using Stirling's formula (4.12) for the large factorials in $\binom{n}{x}$, we get

$$
\begin{aligned}
f_n(x) &\doteq \frac{\sqrt{2\pi n}\,n^n e^{-n}}{x!\sqrt{2\pi(n-x)}(n-x)^{n-x}e^{-n+x}}\left(\frac{\lambda}{n}\right)^x\left(1-\frac{\lambda}{n}\right)^{n-x} \\
&= \frac{(n-x)^x}{x!\sqrt{1-x/n}(1-x/n)^n e^x}n^{-x}\left(\frac{\lambda}{1-\lambda/n}\right)^x\left(1-\frac{\lambda}{n}\right)^n \\
&= \frac{(1-x/n)^x}{x!\sqrt{1-x/n}(1-x/n)^n e^x}\frac{\lambda^x}{(1-\lambda/n)^x}\left(1-\frac{\lambda}{n}\right)^n.
\end{aligned}
$$

Taking limits and using $(1+a/n)^n \to e^a$ for any $a \in \Re$ (see page 373 for the proof) shows that

$$
f_n(x) \to f(x) := \frac{\lambda^x e^{-\lambda}}{x!}1_{\aleph_0}(x). \tag{4.24}
$$

Note that since $n \to \infty$ the support has expanded to $\mathbb{X} = \aleph_0 := \{0,1,2,...\}$. The limiting function is indeed a p.m.f. on $\aleph_0$, since $\sum_{x=0}^{\infty}\lambda^x/x! = e^\lambda$.

A discrete r.v. X with p.m.f. (4.24) is said to have the *Poisson* distribution with "intensity" parameter λ; i.e., $X \sim P(\lambda)$.[4] The family $\{f(x;\lambda) : \lambda > 0\}$ clearly belongs to the exponential class, with $b(\lambda) = e^{-\lambda}$, $c(x) = 1_{\aleph_0}(x)/x!$, $\eta(\lambda) = \log\lambda$, and $T(x) = x$ in a representation corresponding to (4.1) and with $a(\eta) = \exp(-e^\eta)$ in canonical form (4.4).

The Poisson distribution can be a useful model for the number of events occurring in a span of time or space. For example, potential temporal applications include arrivals of customers, landings of aircraft, passages by vehicles, emissions of radioactive particles, and transactions on a stock exchange. Spatial applications include the number of visible stars in a small randomly selected region of the sky, individuals in a queue, craters on the lunar surface, bacteria on a culture medium, coins in the pocket of a person chosen at random. Critical conditions are (i) that there can be at most one occurrence at a given point in time or space and (ii) that the number of occurrences in different intervals of time or space be independent and identically distributed. While these are rarely met precisely in any real-world setting, the Poisson model is still often a useful approximation. When it is not, it can still be a good starting point for further analysis. We will see a common way to extend it at the end of the section.

---

[4]Those with some knowledge of the French language will need to be assured that there is nothing *fishy* about the distribution. It is attributed to Simeon Poisson, who in 1837 published a derivation along the lines given here. The distribution had appeared in earlier work by de Moivre (1711), whose name is associated with another famous "limit theorem" encountered in Chapter 6.

Exercise 4.20. *On average there are 4.3 arrivals of customers in a retail store during the hour from 1200–1300 on weekdays. Using the Poisson model, find the probability that the number of customers arriving during this period on a particular day will be below the average. (Hint: Exercise 3.12 has already established that $EX = \lambda$ when $X \sim P(\lambda)$.)*

The Poisson p.g.f., m.g.f., and central m.g.f. are

$$p_X(t) = \sum_{x=0}^{\infty} t^x \frac{\lambda^x e^{-\lambda}}{x!} = e^{\lambda(t-1)} \sum_{x=0}^{\infty} \frac{(\lambda t)^x e^{-\lambda t}}{x!} = e^{\lambda(t-1)}$$

$$M_X(t) = \exp\left[\lambda\left(e^t - 1\right)\right]$$

$$m_X(t) = e^{-t\lambda} M_X(t) = \exp\left[\lambda\left(e^t - 1 - t\right)\right].$$

Moments of Poisson r.v.s are most easily deduced from the cumulants, whose generating function is $\log M_X(t) = \lambda(e^t - 1)$. Since

$$\left.\frac{d^j \lambda(e^t - 1)}{dt^j}\right|_{t=0} = \lambda$$

for all $j \in \mathbb{N}$, all cumulants are simply equal to λ. In particular, $EX = \kappa_1 = \lambda$, $VX = \kappa_2 = \lambda$, $\mu_3 = \kappa_3 = \lambda$, and $\mu_4 = \kappa_4 + 3\kappa_2^2 = \lambda + 3\lambda^2$. Accordingly, the coefficients of skewness are

$$\alpha_3 = \lambda^{-1/2}, \alpha_4 = 3 + \lambda^{-1/2}.$$

Exercise 4.21. *Apply the results for canonical representations of families in the exponential class to show that the Poisson's natural statistic $T(X) = X$ has c.g.f. $\lambda(e^t - 1)$.*

The following recurrence relation aids in calculating Poisson probabilities, particularly on digital computers or calculators, and it has some theoretical value as well:

$$f(x) = \lambda \frac{f(x-1)}{x}, \quad x \in \mathbb{N}. \tag{4.25}$$

Thus, with $f(0) = e^{-\lambda}$ we have $f(1) = \lambda e^{-\lambda}/1$, $f(2) = \lambda^2 e^{-\lambda}/(1 \cdot 2)$, $f(3) = \lambda^3 e^{-\lambda}/(1 \cdot 2 \cdot 3)$, and so forth.

Exercise 4.22. *Use (4.25) to show that $EX = \lambda$.*

Exercise 4.23. *Use (4.25) to show that $\mu'_{[k]} := EX(X-1)\cdots\cdots(X-k+1) = \lambda^k$ for $k \in \mathbb{N}$.*

The next exercise shows that family $\{P(\lambda) : \lambda > 0\}$ is closed under convolution; that is, the sum of independent Poissons is also Poisson.

Exercise 4.24. *If $X_1 \sim P(\lambda_1)$ and $X_2 \sim P(\lambda_2)$ are independent, show that $S = X_1 + X_2 \sim P(\lambda_1 + \lambda_2)$.*

The Poisson model for arrivals of customers in Exercise 4.20 would be deficient if the *intensity* of arrivals (the average number per hour) varied from day to day. The number of arrivals X on any arbitrary day would then be a *mixture* of Poissons with different intensities. A way to capture such variation is to model intensity as a r.v. L with $\Pr(L > 0) = 1$ and to take $X \sim P(L)$ conditional on L. Replacing parameters with r.v.s in this way produces what is called an *hierarchical* model.[5] Clearly, there are as many possibilities as there are ways to model L.

Example 4.6. Let $X \sim P(L)$ conditional on L, a continuous r.v. with p.d.f. $f(\ell) = \Gamma(\beta)^{-1} \ell^{\lambda/\beta-1} e^{-\ell/\beta} \mathbf{1}_{(0,\infty)}(\ell)$ for positive λ, β. We will see in Section 4.3.2 that this is a member of the *gamma* family of distributions, with mean $EL = \lambda$ and m.g.f. $M_L(\zeta) := Ee^{\zeta L} = (1 - \zeta\beta)^{-\lambda/\beta}$ for $\zeta < \beta^{-1}$. The distribution of X itself can be deduced from the conditional p.g.f. We have $p_{X|L}(t) \equiv E(t^X \mid L) = \exp[L(t-1)]$ and hence

$$p_X(t) = E p_{X|L}(t) = M_L(t-1) = [1 - (t-1)\beta]^{-\lambda/\beta}, t < 1 + \beta^{-1}.$$

Setting $\beta =: (1 - \theta)/\theta$ for some $\theta \in (0,1)$ and $\lambda/\beta =: \varphi > 0$ puts this in the form

$$p_X(t) = \left[\frac{\theta}{1 - (1 - \theta)t}\right]^\varphi,$$

showing that $X \sim NB[\varphi = \lambda/\beta, \theta = 1/(\beta+1)]$.

Exercise 4.25. *Let $L \sim P(\lambda)$ for some $\lambda > 0$ and $X \sim P(L)$ conditional on L. (In the event $L = 0$ put $\Pr(X = 0 \mid L = 0) = 1$.)*

(i) Work out the p.g.f. of X and express derivatives p'_X and p''_X.

(ii) Find EX and VX using conditional expectations, then from the p.g.f.

(iii) Show that the derivatives of p_X satisfy the recurrence relation

$$d^{n+1} p_X(t) = \lambda e^{t-1}(d+1)^n p_X(t), \tag{4.26}$$

where differential operator d is such that $d^k p_X(t) := p_X^{(k)}(t)$.

(iv) Use the recursion to calculate $\{f_X(x)\}_{x=0}^5$ when $\lambda = 5.0$.

---

[5]The distribution of the convolution of independent Poisson and normal r.v.s in Example 3.53 can also be regarded as an hierarchical model. There the normal distribution has a Poisson-distributed mean.

Table 4.1 Common distributions for count data.

Name/ Abbrev.	P.m.f. $f(x)$	Support $\mathbb{X}$	Parameter Space	Mean, Variance	C.f. $\phi(t)$
Bernoulli* B$(1,\theta)$	$\theta^x \tau^{1-x}$ $\tau = 1-\theta$	$\{0,1\}$	$\theta \in (0,1)$	$\theta,$ $\theta\tau$	$\tau + \theta e^{it}$
Binomial* B(n,θ)	$\binom{n}{x}\theta^x \tau^{n-x}$ $\tau = 1-\theta$	$\aleph_0^n$	$\theta \in (0,1)$ $n \in \aleph$	$n\theta,$ $n\theta\tau$	$\left(\tau + \theta e^{it}\right)^n$
Multinomial* M$(n,\boldsymbol{\theta})$	$\binom{n}{\mathbf{x}}\boldsymbol{\theta}^{\mathbf{x}}\tau^{n-\mathbf{x}'\mathbf{1}}$ $\tau = 1-\mathbf{1}'\boldsymbol{\theta}$	$x_j \in \aleph_0^n$ $\mathbf{x}'\mathbf{1} \leq n$	$\boldsymbol{\theta} \in (0,1)^m$ $\boldsymbol{\theta}'\mathbf{1} \leq 1$	$n\boldsymbol{\theta},$ $n\left(\Theta - \boldsymbol{\theta}\boldsymbol{\theta}'\right)$	$\left(\tau + \boldsymbol{\theta}'e^{it}\right)^n$
Neg. Binomial* NB(φ,θ)	$\frac{\Gamma(\varphi+x)}{x!\Gamma(\varphi)}\theta^\varphi \tau^x$ $\tau = 1-\theta$	$\aleph_0$	$\theta \in (0,1)$ $\varphi > 0$	$\varphi\tau/\theta,$ $\varphi\tau/\theta^2$	$\frac{\theta^\varphi}{\left(1-\tau e^{it}\right)^\varphi}$
Hypergeom. HG(S,F,n)	$\frac{\binom{S}{x}\binom{F}{n-x}}{\binom{N}{n}}$ $N = S+F$	$\aleph_{n\wedge S}^{(n-F)^+}$	$S,F \in \aleph$ $n \in \aleph$	$\frac{nS}{N}$ $\frac{nSF(N-n)}{N^2(N-1)}$	n.s.f.
Poisson* P(θ)	$\frac{\theta^x e^{-\theta}}{x!}$	$\aleph_0$	$\theta > 0$	$\theta,$ θ	$e^{\theta(e^{it}-1)}$

4.2.7 Summary of Discrete Models

Table 4.1 summarizes the features of the six models for count data that we have just presented.[6] Several new symbols are introduced to make the table compact. Greek letter τ stands for $1-\theta$ in entries for Bernoulli, binomial, and negative binomial distributions and for $1-\mathbf{1}'\boldsymbol{\theta}$ in the multinomial. $\aleph_m^n$ represents the integers $\{m, m+1, ..., n\}$, but $\aleph_0$ includes all nonnegative integers as usual. For the multinomial entries (i) $\boldsymbol{\theta}^{\mathbf{x}} := \theta_1^{x_1}, \theta_2^{x_2}, ..., \theta_n^{x_n}$; (ii) Θ denotes a diagonal matrix with vector $\boldsymbol{\theta}$ on the principal diagonal; and (iii) $e^{it} := \left(e^{it_1}, e^{it_2}, ..., e^{it_m}\right)'$. The notations $n \wedge S := \min\{n, S\}$ and $(n-F)^+ := (n-F) \vee 0 := \max\{n-F, 0\}$ for the hypergeometric model are standard. The hypergeometric c.f. has no simple form. An asterisk indicates that the family is in the exponential class. For the binomial, multinomial, and negative binomial models this is true only for subfamilies with common values of n and φ.

4.3 Continuous Distributions

As we know, continuous r.v.s are those having absolutely continuous c.d.f.s F, densities f with respect to Lebesgue measure, and supports $\mathbb{X}$ comprising uncountably many points. We now examine some of the more important

---

[6]For a far more extensive survey of discrete models see Johnson *et al.* (1992).

models of this type. Not included here are some specific models such as the Cauchy that are less often used in applied work and have already appeared in examples. One other applied model, the logistic, was introduced in Section 3.4.1 and is described more fully in Chapter 9. Other continuous families that were developed to describe the behavior of sample *statistics* (as opposed to *data*) are presented in Chapter 5. We begin the present survey with a fundamental building block, a special case of which will at once be recognized from many examples.

4.3.1   *The Uniform Distribution*

A r.v. with p.d.f. $f(x) = \frac{1}{\beta-\alpha}\mathbf{1}_{[\alpha,\beta]}(x)$ is said to be *uniformly distributed* on interval $[\alpha, \beta]$. In symbols, $X \sim U(\alpha, \beta)$. The corresponding c.d.f. is

$$F(x) = \left\{ \begin{array}{ll} 0, & x < \alpha \\ \frac{x-\alpha}{\beta-\alpha}, & \alpha \le x \le \beta \\ 1, & x > \beta \end{array} \right\} = \frac{x-\alpha}{\beta-\alpha}\mathbf{1}_{[\alpha,\beta]}(x) + \mathbf{1}_{(\beta,\infty)}(x).$$

Of course, the uniform distribution on $[0, 1]$ has been our standard model for the clockwise distance from the origin in the pointer experiment (Example 1.7 on page 22).

The uniform distribution is a useful "seed" for generating other distributions. Letting Y be a r.v. with continuous, strictly increasing c.d.f. F_Y and setting $X := F_Y(Y)$, it is clear that $[0, 1]$ is a support of X and that

$$\Pr(X \le x) = \Pr[F_Y(Y) \le x] = \Pr\left[Y \le F_Y^{-1}(x)\right]$$
$$= F_Y\left[F_Y^{-1}(x)\right] = x$$

for $x \in [0,1]$. Note that F_Y^{-1} has the usual meaning here as an inverse (point) function, which exists because the strictly increasing F_Y is one to one. The c.d.f. of X is then

$$F_X(x) = \left\{ \begin{array}{ll} 0, & x < 0 \\ x, & 0 \le x \le 1 \\ 1, & x > 1 \end{array} \right\} = x\mathbf{1}_{[0,1]}(x) + \mathbf{1}_{(1,\infty)}(x),$$

so $X \sim U(0,1)$.[7] If instead r.v. Y is discrete, F_Y then being a step function, or if Y is continuous but has c.d.f. that is not strictly increasing,[8] then the same result applies upon defining $F_Y^{-1}(x)$ as $\sup\{y : F_Y(y) < x\} \equiv$

---

[7]At this point, take another look at Exercise 3.7 and see if you can think of another way to approach it.

[8]This would happen if the support of Y comprised intervals separated by one or more gaps of positive Lebesgue measure; for example, if Y had p.d.f. $f(y) = \mathbf{1}_{(0,.5)\cup(1,1.5)}(y)$.

inf $\{y : F_Y(y) \geq x\}$. Thus, the *probability integral transform* $F_Y(Y)$ of any r.v. Y is distributed as $U(0,1)$. To generate independent realizations of any r.v. Y, one applies to realizations of $X \sim U(0,1)$ the *inverse* probability integral transform: $Y = F_Y^{-1}(X)$. That is, if $X \sim U(0,1)$ then for $y \in \Re$

$$\Pr(Y \leq y) = \Pr\left[F_Y^{-1}(X) \leq y\right] = \Pr\left[X \leq F_Y(y)\right] = F_Y(y).$$

To get a visual impression of how it works, imagine that we have set up some apparatus for generating realizations of $X \sim U(0,1)$, such as our pointer on the $[0,1)$ scale or the appropriate computer software, and have generated realizations $\{x_j\}_{j=1}^{3}$. Plotting the c.d.f. F_Y as in Fig. 4.2 and identifying the points $\{y_j\}$ such that $x_j = F_Y(y_j)$, it is clear that $\Pr(Y \leq y_j) = \Pr(X \leq x_j) = x_j = F_Y(y_j)$. Thus, the $\{y_j\}$ are in fact realizations of a r.v. with c.d.f. F_Y.

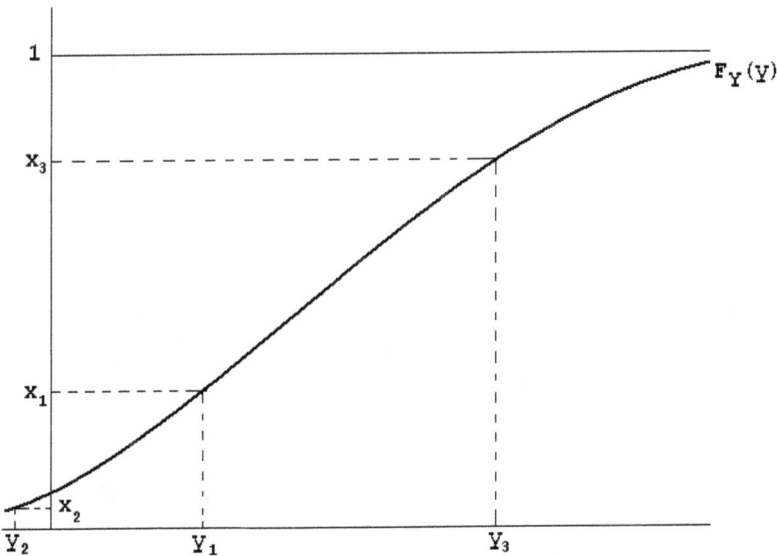

Fig. 4.2   Generating realizations of Y with c.d.f F_Y via inverse probability integral transform.

Example 4.7. Five independent realizations of X, distributed as uniform on $(0,1)$, are (to two decimal places) .34, .87, .05, .21, .63.

(i) If $F_Y(y) = (1 - e^{-y})\, \mathbf{1}_{(0,\infty)}(y)$, the inverse transform is $Y = F_Y^{-1}(X) = -\log(1 - X)$, and the realizations of Y are $0.42, 2.04, 0.05, 0.24, 0.99$.

(ii) Y has a Bernoulli distribution with success probability $1/2$, so that

$$F_Y(y) = \begin{cases} 0, & y < 0 \\ \frac{1}{2}, & 0 \le y < 1, \\ 1, & y \ge 1 \end{cases} = \frac{1}{2}\mathbf{1}_{[0,1)}(y) + \mathbf{1}_{[1,\infty)}(y).$$

The inverse, $F_Y^{-1}(X) = \sup\{y : F_Y(y) < X\}$, is

$$F_Y^{-1}(X) = \begin{cases} 0, & X \le \frac{1}{2} \\ 1, & \frac{1}{2} < X \end{cases},$$

and the realizations of Y are $0, 1, 0, 0, 1$. Figure 4.3 depicts F_Y and F_Y^{-1}.

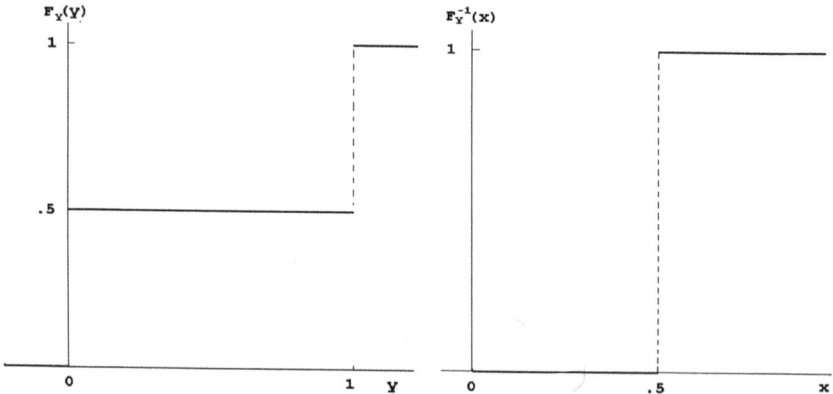

Fig. 4.3 Inverse c.d.f. of a discrete r.v.

Example 4.8. Y has c.d.f. $F_Y(y) = (1 - y^{-1/2})\, \mathbf{1}_{[0,\infty)}(y)$. The table shows 20 independent realizations of $X \sim U(0,1)$ and of the corresponding $Y = F_Y^{-1}(X) = (1 - X)^{-2}$ that appeared previously in Example 3.16. (The y's are calculated from x's in higher precision than those displayed.)

x	$y = (1-x)^{-2}$
0.7437	15.2288
0.1675	1.4430
0.9294	200.4767
0.9807	2680.9630
0.4033	2.8089
0.9104	124.6615
0.6969	10.8867
0.1929	1.5352
0.2167	1.6300
0.8006	25.1492
0.5859	5.8324
0.6573	8.5125
0.4678	3.5307
0.4222	2.9957
0.5126	4.2102
0.9391	269.3875
0.9971	120410.6200
0.5663	5.3173
0.3727	2.5413
0.0847	1.1936

Example 4.9. It is interesting that there may be *tighter* bounds than ± 1 for the correlation coefficient of two jointly distributed r.v.s X and Y if they have *specified* marginal distributions. Suppose X and Y are in standard form (zero means, unit variances), so that $\rho_{XY} = EXY$; and let F_X, F_Y be their marginal c.d.f.s. It is known that all joint distributions F_{XY} with these marginals satisfy the bounds

$$F_{XY}(x,y)^- := \max\{F_X(x) + F_Y(y) - 1, 0\} \le F_{XY}(x,y)$$
$$\le \min\{F_X(x), F_Y(y)\} =: F_{XY}(x,y)^+$$

for $(x,y) \in \Re^2$.[9] Moreover, the correlations corresponding to these boundary c.d.f.s also bound the possible correlations between X and Y; that is,

$$\rho_{XY}^- := \int_{\Re^2} xy \cdot dF_{XY}(x,y)^- \le \rho_{XY} \le \int_{\Re^2} xy \cdot dF_{XY}(x,y)^+ =: \rho_{XY}^+.$$

Simulation affords a simple way to estimate ρ_{XY}^- and ρ_{XY}^+ without evaluating these integrals, as follows. Let $U \sim U(0,1)$, and put $X := F_X^{-1}(U)$,

---

[9] See Demirtas and Hedeker (2011) and the references cited.

$Y^+ := F_Y^{-1}(U)$, $Y^- := F_Y^{-1}(1 - U)$. Then

$$\begin{aligned}
\Pr\left(X \le x, Y^+ \le y\right) &= \Pr\left(F_X^{-1}(U) \le x, F_Y^{-1}(U) \le y\right) \\
&= \Pr\left(U \le F_X(x), U \le F_Y(y)\right) \\
&= \Pr\left(U \le \min\{F_X(x), F_Y(y)\}\right) \\
&= \min\{F_X(x), F_Y(y)\} \\
&= F_{XY}(x, y)^+,
\end{aligned}$$

and

$$\begin{aligned}
\Pr\left(X \le x, Y^- \le y\right) &= \Pr\left(F_X^{-1}(U) \le x, F_Y^{-1}(1 - U) \le y\right) \\
&= \Pr\left(U \le F_X(x), 1 - U \le F_Y(y)\right) \\
&= \Pr\left(1 - F_Y(y) \le U \le F_X(x)\right) \\
&= \max\{F_X(x) + F_Y(y) - 1, 0\} \\
&= F_{XY}(x, y)^-.
\end{aligned}$$

Thus, $\rho_{XY}^- = EXY^-$ and $\rho_{XY}^+ = EXY^+$, and so from realizations $\{u_j\}_{j=1}^n$ of U and the corresponding $\{x_j, y_j^-, y_j^+\}_{j=1}^n$ (for some large n) one can estimate the boundary correlations as

$$\tilde{\rho}_{XY}^- = n^{-1} \sum_{j=1}^n x_j y_j^-, \quad \tilde{\rho}_{XY}^+ = n^{-1} \sum_{j=1}^n x_j y_j^+.$$

In this way correlations between two r.v.s, one distributed as *exponential* and the other as *normal* (see below) can be shown to lie between about ± 0.90.

The m.g.f. of the uniform distribution on (α, β) is

$$M(t) = \int_\alpha^\beta \frac{e^{tx}}{\beta - \alpha} \cdot dx = \mathbf{1}_{\{0\}}(t) + \frac{e^{t\beta} - e^{t\alpha}}{t(\beta - \alpha)} \mathbf{1}_{\Re \setminus \{0\}}(t),$$

and the c.f. is

$$\phi(t) = \mathbf{1}_{\{0\}}(t) + \frac{e^{it\beta} - e^{it\alpha}}{it(\beta - \alpha)} \mathbf{1}_{\Re \setminus \{0\}}(t).$$

When $\alpha = -\beta$ the distribution is symmetric about the origin and $\phi(t) = \sin(t\beta)/(t\beta)$ for $t \ne 0$, as in Exercise 3.54.

Exercise 4.26. *If $X \sim U(\alpha, \beta)$ show that $\mu = (\alpha + \beta)/2$ and $\sigma^2 = (\beta - \alpha)^2/12$.*

Exercise 4.27. *A continuously revolving beacon located one mile from a shoreline flashes its light at random intervals. If the flash occurs when the beacon points at an angle $\theta \in (-\pi/2, \pi/2)$ from a line drawn perpendicular to the shore, the ray of light then strikes the shore at a point $X = \tan \theta$, where $X = 0$ corresponds to the point on the shore nearest the beacon. (Refer to Fig. 4.4.) Assuming that $\theta \sim U(-\pi/2, \pi/2)$, find the p.d.f. of X.*

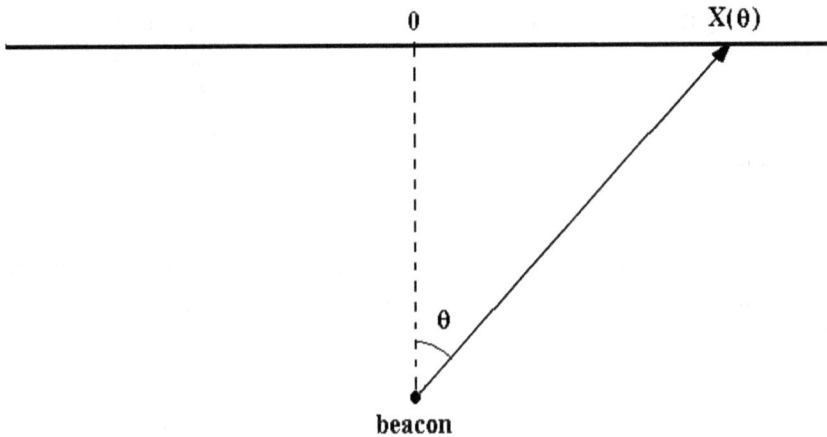

Fig. 4.4   Point illumined by beacon *vs.* its angle with respect to shore.

Exercise 4.28. *In Exercise 4.5 it was found that the chance that any one of $2n + 1$ voters will break a tie on a "yes-no" vote is $\binom{2n}{n} p^n (1-p)^n$ when each votes "yes" with probability p. Suppose now that the $2n$ voters have possibly different probabilities $\{P_i\}_{i=1}^{2n}$ of voting positive. Letting B be the event that any one person's vote will be decisive (i.e., break a tie), develop an expression for $\mathbb{P}(B)$ and determine its value under each of the following conditions.*

(i) $P_i = 1$ for n of the voters and $P_i = 0$ for the others.
(ii) $P_i = 1$ for more than n of the voters and $P_i = 0$ for the others.
(iii) $P_i = p$ for all $2n$ voters.
(iv) $\{P_i\}_{i=1}^{2n}$ are distributed independently as $U(0,1)$.

4.3.2 The Gamma Distribution

Section 4.2.4 introduced gamma function, defined for $\alpha > 0$ by the integral expression

$$\Gamma(\alpha) := \int_0^\infty y^{\alpha-1} e^{-y} \cdot dy.$$

This is easily seen to have the recurrence property

$$\Gamma(\alpha+1) = \alpha\Gamma(\alpha), \qquad (4.27)$$

and to satisfy $\Gamma(1) = 1$. Γ is defined only for positive α since the integral otherwise fails to converge. (Recall Exercise 3.4.) The function has a closed form when (and only when) α is an integer multiple of $1/2$; i.e., when $\alpha = m/2$ for $m \in \aleph$. When m is even, so that $m/2 = n$ is an integer, formula (4.27) implies that $\Gamma(n+1) = n!$ Notice that this is consistent with the definition $0! = 1$.

Exercise 4.29. *Prove the recurrence relation (4.27).*

The case $\alpha = m/2$ with m odd is also handled by recurrence beginning with a value for $\Gamma(1/2)$. Finding this is worth the effort, as we will make good use of it later. Writing out the integral for $\Gamma(1/2)$ and squaring give the double integral

$$\Gamma\left(\frac{1}{2}\right)^2 = \int_0^\infty \int_0^\infty x^{-1/2} y^{-1/2} e^{-(x+y)} \cdot dy\, dx.$$

This can be evaluated by two changes of variables.[10] First, set $x = w^2$ and $y = z^2$, a one-to-one mapping from $(w, z) \in [0, \infty)^2$ to $(x, y) \in [0, \infty)^2$. The Jacobian of the transformation from (x, y) to (w, z) is

$$J_{(x,y)\to(w,z)} = \begin{vmatrix} \frac{\partial x}{\partial w} & \frac{\partial y}{\partial w} \\ \frac{\partial x}{\partial z} & \frac{\partial y}{\partial z} \end{vmatrix} = \begin{vmatrix} 2w & 0 \\ 0 & 2z \end{vmatrix} = 4zw.$$

Making the substitution in the double integral and multiplying by J (which is nonnegative on $[0, \infty)^2$) reduces the integral to

$$\Gamma\left(\frac{1}{2}\right)^2 = 4 \int_0^\infty \int_0^\infty e^{-(w^2+z^2)} \cdot dw\, dz.$$

Now, transform to polar coordinates with $w = r\cos\theta$ and $z = r\sin\theta$. This is another one-to-one mapping, but now from $(r, \theta) \in [0, \infty) \times [0, \pi/2)$ to $(w, z) \in [0, \infty)^2$. The Jacobian of this transformation is

$$J_{(w,z)\to(r,\theta)} = \begin{vmatrix} \frac{\partial w}{\partial r} & \frac{\partial z}{\partial r} \\ \frac{\partial w}{\partial \theta} & \frac{\partial z}{\partial \theta} \end{vmatrix} = \begin{vmatrix} \cos\theta & \sin\theta \\ -r\sin\theta & r\cos\theta \end{vmatrix} = r,$$

---

[10]The two steps could be combined into one by setting $x = r^2 \cos^2\theta$ and $y = r^2 \sin^2\theta$ at the outset, but the logic is easier to follow this way.

so that

$$\Gamma\left(\frac{1}{2}\right)^2 = 4 \int_0^\infty \int_0^{\pi/2} re^{-r^2} \cdot d\theta\, dr = \pi.$$

The range of the gamma function being strictly positive, we take the positive root and declare $\Gamma(1/2) = +\sqrt{\pi}$. Finally, recurrence implies $\Gamma(3/2) = \frac{1}{2}\sqrt{\pi}$, $\Gamma(5/2) = \frac{3}{4}\sqrt{\pi}$, and so on.

The integral $\int_0^t y^{\alpha-1} e^{-y} \cdot dy$ with finite upper limit is called the *incomplete* gamma function. This increases from zero to $\Gamma(\alpha)$ as t increases from zero to infinity. From this it is clear that continuous function

$$F_T(t) := \left[\frac{1}{\Gamma(\alpha)} \int_0^t y^{\alpha-1} e^{-y} \cdot dy\right] \mathbf{1}_{(0,\infty)}(t)$$

is a c.d.f., corresponding to p.d.f.

$$f_T(t) = \frac{1}{\Gamma(\alpha)} t^{\alpha-1} e^{-t} \mathbf{1}_{(0,\infty)}(t). \tag{4.28}$$

Exercise 4.30. *Confirm (i) that F_T satisfies the conditions for being a c.d.f. and (ii) that the density is as given by (4.28).*

Taking r.v. T to have p.d.f. (4.28) and setting $X = \beta T$ for $\beta > 0$, the change-of-variable formula gives as the p.d.f. of X

$$f_X(x) = \frac{1}{\Gamma(\alpha)\beta^\alpha} x^{\alpha-1} e^{-x/\beta} \mathbf{1}_{(0,\infty)}(x). \tag{4.29}$$

A r.v. with this p.d.f. is said to have a *gamma distribution* with parameters α and β. For short, we write $X \sim \Gamma(\alpha, \beta)$, relying on the second argument to distinguish this from the gamma *function*. The gamma family comprises the collection $\{f_X(x; \alpha, \beta) : \alpha > 0, \beta > 0\}$. A member of the exponential class, an expression of its p.d.f. in canonical form

$$f_X(x; \boldsymbol{\eta}) = a(\boldsymbol{\eta}) c(x) e^{\boldsymbol{\eta}' \mathbf{T}(x)} \tag{4.30}$$

is obtained with

$$\boldsymbol{\eta}' = (\alpha, -\beta^{-1})$$

$$a(\boldsymbol{\eta}) = \frac{(-\eta_2)^{\eta_1}}{\Gamma(\eta_1)}$$

$$c(x) = e^{-\log x} \mathbf{1}_{(0,\infty)}(x)$$

$$\mathbf{T}(x)' = (\log x, x).$$

A *three-parameter* version of the gamma is sometimes encountered that adds an arbitrary location parameter, as

$$f_X(x; \alpha, \beta, \gamma) = \frac{1}{\Gamma(\alpha)\beta^\alpha} (x-\gamma)^{\alpha-1} e^{-(x-\gamma)/\beta} \mathbf{1}_{(\gamma,\infty)}(x). \tag{4.31}$$

Exercise 4.31. *Verify that $X = \beta T$ has p.d.f. (4.29).*

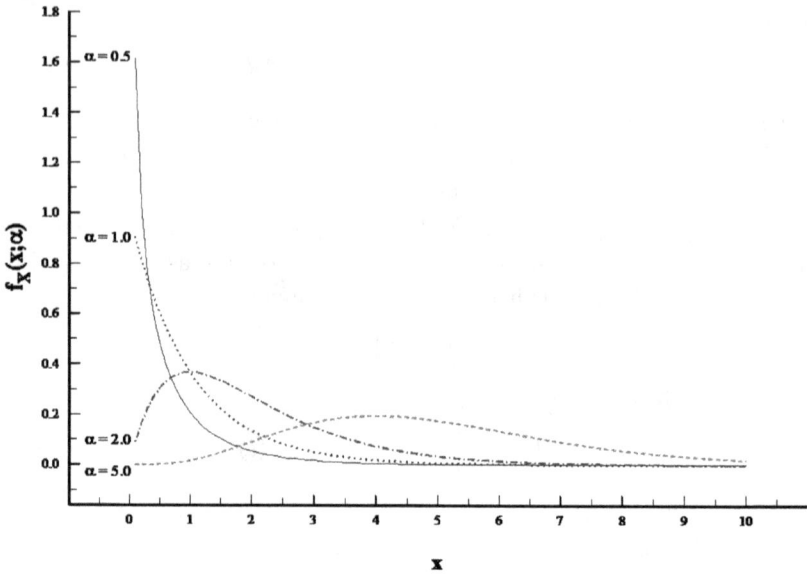

Fig. 4.5   Shapes of $\Gamma(\alpha, 1)$ p.d.f.s for $\alpha \in \{.5, 1., 2., 5\}$.

The gamma family is a flexible set of models for positive, continuous data. The *chi-squared* subfamily, $\{f_X(x; n/2, 2) : n \in \aleph\}$, is important in statistical applications and is discussed in Chapter 5. The *exponential* subfamily, $\{f_X(x; 1, \beta) : \beta > 0\}$, arises as a model of interarrival times for Poisson events, as explained below.[11]

Figure 4.5 depicts the shapes for several values of α when scale parameter $\beta = 1$. Notice that

$$\lim_{x \to 0} f_X(x) = \begin{cases} +\infty, \alpha < 1 \\ \beta^{-1}, \alpha = 1 \\ 0, \alpha > 1 \end{cases}.$$

To see one way that an exponential distribution can arise in applied modeling, put $N_t = 0$ for $t \leq 0$, and for $t > 0$ let N_t count the number of events of some sort that occur during the time span $(0, t]$. For example, these might be landings of aircraft at an airport, transactions on a stock exchange, or detections of radioactive particles. Thus, we would have $N_t = 2$ if there were events at some (unspecified) times $0 < \tau_1 < \tau_2 \leq t$. Given

[11]Recall that the exponential subfamily of gammas is to be distinguished from the exponential *class* of distributions.

some mean *rate* of events $\lambda > 0$, let us (i) model the number of events $N_{t_2} - N_{t_1}$ during any interval $(t_1, t_2]$ as Poisson with parameter $\lambda(t_2 - t_1)$ and (ii) require that increments $N_{t_2} - N_{t_1}$, $N_{t_4} - N_{t_3}$ over nonoverlapping intervals be independent. We would then have $N_t \sim P(\lambda t)$ for $t > 0$. The resulting integer-valued, nondecreasing stochastic process $\{N_t\}_{t \geq 0}$ is called a *Poisson process*. Now, counting from $t = 0$, let continuous r.v. T represent the time that elapses before the *first* Poisson event, and let us find its c.d.f. and p.d.f. Note that for any $t > 0$ the event $T \leq t$ occurs if and only if at least one Poisson event occurs during $(0, t]$. Thus, $\Pr(T \leq t) = \Pr(N_t > 0) = 1 - \Pr(N_t \leq 0)$. Since $N_t \sim P(\lambda t)$ and $f_{N_t}(n) = (\lambda t)^n e^{-\lambda t}/n!$ for $n \in \aleph_0$, we have $\Pr(N_t \leq 0) = \Pr(N_t = 0) = e^{-\lambda t}$ and

$$F_T(t) = \left\{ \begin{array}{c} 0, t \leq 0 \\ 1 - e^{-\lambda t}, t > 0 \end{array} \right\} = \left(1 - e^{-\lambda t}\right) \mathbf{1}_{(0,\infty)}(t).$$

Differentiating gives $f_T(t) = \lambda e^{-\lambda t} \mathbf{1}_{(0,\infty)}(t)$, the exponential p.d.f. with scale parameter $\beta = \lambda^{-1}$.

The exponential model for interarrival times (the times between successive events) has a characteristic "memoryless" property; that is, the chance of an occurrence during a given span of length t does not depend on the elapsed time since the previous event. Thus, if T is the time until the next occurrence and t' any positive number, $\Pr(T > t + t' \mid T > t') = \Pr(T > t) = e^{-\lambda t}$. This very special feature limits the applicability of the exponential model. For example, it would characterize the time between breakdowns of a machine only if breakdowns were just as likely to occur right after repairs as after prolonged periods of operation. It would characterize the time between rains only if the chance of rain tomorrow or next week were the same regardless of the length of the current dry spell.

Exercise 4.32. *Show that the exponential distribution has the "memoryless" property.*

Another way to express the memoryless feature is through the *hazard function*. The hazard function associated with a nonnegative, continuous r.v. T is the ratio of the density at t to the probability that $T > t$; i.e.,

$$h_T(t) := \frac{f_T(t)}{1 - F_T(t)}.$$

Since the density at t is

$$f_T(t) = \lim_{\delta \to 0} \frac{F(t + \delta) - F(t)}{\delta} = \lim_{\delta \to 0} \frac{\Pr(t < T \leq t + \delta)}{\delta},$$

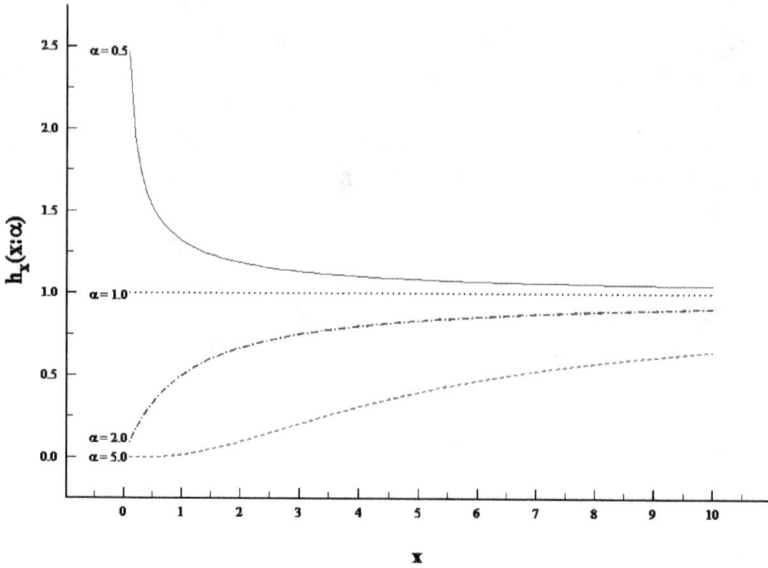

Fig. 4.6   Hazard functions for $\Gamma(\alpha, 1)$, $\alpha \in \{.5, 1., 2., 5\}$.

the product of $h_T(t)$ and some $\delta > 0$ approximates the conditional probability that T falls between t and $t + \delta$ *given* that T takes a value greater than t; i.e.,

$$h_T(t)\delta \doteq \frac{\Pr(t < T \leq t + \delta)}{\Pr(T > t)}.$$

The hazard function is thus approximately proportional to the chance that an event will occur in a short interval after t, given that it has not yet occurred by t. The memoryless property corresponds to a *constant* hazard rate—one that does not vary with t. This clearly applies to the exponential, $f_T(t) = \lambda e^{-\lambda t} \mathbf{1}_{(0,\infty)}(t)$, for which $h_T(t) = \lambda$. By contrast, gamma distributions with $\alpha < 1$ have decreasing hazard (the longer the dry spell, the less chance of rain), while those with $\alpha > 1$ have increasing hazard (a machine is more apt to break down the longer it has been in operation). The broader gamma family thus allows more flexibility in fitting data for occurrence times. Figure 4.6 depicts hazard functions corresponding to the distributions shown in Fig. 4.5.

The gamma family has another nice property. Letting $X \sim \Gamma(\alpha, 1)$ and $r \in (-\alpha, \infty)$, the rth moment about the origin is

$$\mu_r' = \int_0^\infty x^r \cdot \frac{x^{\alpha-1}e^{-x}}{\Gamma(\alpha)} \cdot dx = \frac{\Gamma(r+\alpha)}{\Gamma(\alpha)} \int_0^\infty \frac{x^{r+\alpha-1}e^{-x}}{\Gamma(r+\alpha)} \cdot dx = \frac{\Gamma(r+\alpha)}{\Gamma(\alpha)}.$$

Thus, gamma distributions possess moments of all real orders greater than the negative of their shape parameters. When $r = k \in \aleph$, the recurrence formula for the gamma function gives a simple expression for positive-integer moments: $\mu_k' = \alpha(\alpha+1) \cdots \cdots (\alpha+k-1)$.

Exercise 4.33. *When* $-\alpha < -k \leq 0$ *show that*

$$\mu_{-k}' = \frac{1}{(\alpha-1)(\alpha-2)\cdots\cdots(\alpha-k)}.$$

When scale parameter β is not unity, the rth moment is simply scaled by the factor β^r. In particular, the mean and variance of $X \sim \Gamma(\alpha, \beta)$ are $\mu = \alpha\beta$ and $\sigma^2 = \alpha\beta^2$, respectively, and the rth moment about the origin is

$$\mu_r' = \beta^r \frac{\Gamma(r+\alpha)}{\Gamma(\alpha)}, r \in (-\alpha, \infty). \tag{4.32}$$

The m.g.f. and c.f. are

$$M_X(t) = (1-t\beta)^{-\alpha}, t < 1/\beta \tag{4.33}$$
$$\phi_X(t) = (1-it\beta)^{-\alpha}, t \in \Re.$$

The cumulants of $\log X$ can be found from the expression for the moments of X, since $M_{\log X}(t) = E e^{t \log X} = EX^t$. Alternatively, they can be found by recognizing that $\log X$ appears as a "natural" statistic in canonical form (4.30).

Exercise 4.34. *If* $X \sim \Gamma(\alpha, \beta)$ *show that the m.g.f. is given by (4.33).*

Exercise 4.35. *Taking* $\mathbf{T}(X) = (\log X, X)'$ *as the gamma family's natural statistics, express joint cumulant generating function (c.g.f.)* $\log M_{\mathbf{T}}(\mathbf{t})$, *and use the c.g.f. to (i) verify that* $EX = \alpha\beta$ *and (ii) show that* $E \log X = \log \beta + \Psi(\alpha)$, *where* $\Psi(\alpha) := d \log \Gamma(\alpha)/d\alpha$ *is the digamma function. (iii) Confirm result (ii) by recognizing* μ_t' *as the c.g.f. of* $\log X$. *(iv) Use the results of (i) and (ii) to show that* $\Psi(\alpha) < \log \alpha$ *for* $\alpha > 0$.

Using the c.f. or m.g.f. it is easily verified that the subfamily of gamma variates with the same scale parameter is closed under finite convolution.

Thus, if $\{X_j\}_{j=1}^k$ are independently distributed with $X_j \sim \Gamma\left(\alpha_j, \beta_0\right)$, then

$$\sum_{j=1}^k X_j \sim \Gamma\left(\sum_{j=1}^k \alpha_j, \beta_0\right). \tag{4.34}$$

Exercise 4.36. *Prove that the subfamily $\{\Gamma(\alpha, \beta) : \alpha > 0, \beta = \beta_0\}$ is closed under (finite) convolution. Infer from this that any gamma distribution with $\alpha \in \aleph$ can be represented as the sum of independent exponential r.v.s.*

Exercise 4.37. *Let $\{T_1, T_2, ...\}$ be i.i.d. r.v.s with $\Pr\left(T_j > 0\right) = 1$ for each j, and put $N_t = 0$ for $t \le 0$ and $N_t = \max\left\{n : \sum_{j=1}^n T_j \le t\right\}$ for $t > 0$. The integer-valued stochastic process $\{N_t\}_{t \ge 0}$ is called a* renewal process. *It gets its name as follows. Think of the $\{T_j\}_{j=1}^\infty$ as lifetimes of physical entities, such as light bulbs. A first bulb is installed. When it burns out, it is replaced by another bulb having the same distribution of lifetimes, which in turn is replaced by another such bulb, and so on ad infinitum. Then N_t equals the number of bulbs that have been replaced (renewed) as of time t.*

(i) For each $n \in \aleph$ show that

$$\Pr\left(N_t = n\right) = \Pr\left(\sum_{j=1}^n T_j \le t\right) - \Pr\left(\sum_{j=1}^{n+1} T_j \le t\right).$$

(ii) Find the p.m.f. of N_t for $t > 0$ if $\{T_j\}_{j=1}^\infty$ are i.i.d. as exponential with $ET_j = \lambda^{-1}$ for some $\lambda > 0$.

Exercise 4.38. *Let $X_1 \sim \Gamma\left(\alpha, 1\right)$ and $X_2 \sim \Gamma\left(\beta, 1\right)$ be independent gamma variates, and put $Y_1 := X_1/\left(X_1 + X_2\right), Y_2 := X_1 + X_2$.*

(i) Show that Y_1, Y_2 are independent.
(ii) Show that the marginal p.d.f. of Y_1 is

$$f(y) = \frac{1}{\mathfrak{B}(\alpha, \beta)} y^{\alpha-1} (1-y)^{\beta-1} \mathbf{1}_{[0,1]}(y), \tag{4.35}$$

where

$$\mathfrak{B}(\alpha, \beta) := \int_0^1 y^{\alpha-1} (1-y)^{\beta-1} \cdot dy = \frac{\Gamma(\alpha)\Gamma(\beta)}{\Gamma(\alpha+\beta)} \tag{4.36}$$

is the beta function.
(iii) Show that the results in (i) and (ii) remain the same if $X_1 \sim \Gamma\left(\alpha, \gamma\right)$ and $X_2 \sim \Gamma\left(\beta, \gamma\right)$ for any positive scale parameter γ.

4.3.3 The Beta Distribution

The p.d.f.

$$f_Y\left(y; \alpha, \beta\right) = \frac{1}{\mathfrak{B}\left(\alpha, \beta\right)} y^{\alpha-1}\left(1-y\right)^{\beta-1} \mathbf{1}_{[0,1]}\left(y\right)$$

represents the standard form of the *beta distribution* with parameters $\alpha > 0, \beta > 0$. We write $Y \sim \mathfrak{Be}\left(\alpha, \beta\right)$ to indicate that r.v. Y has this distribution. Since $\mathfrak{B}\left(1,1\right) = 1$, we see that $f\left(y; 1, 1\right) = \mathbf{1}_{[0,1]}\left(y\right)$, so that the uniform distribution on $[0,1]$ is a special case. Putting $X := a + \left(b - a\right)Y$ with $b > a$ gives a four-parameter version,

$$f_X\left(x; \alpha, \beta, a, b\right) = \frac{1}{\mathfrak{B}\left(\alpha, \beta\right)\left(b-a\right)^{\alpha+\beta-1}}\left(x-a\right)^{\alpha-1}\left(b-x\right)^{\beta-1}\mathbf{1}_{[a,b]}\left(x\right),$$

with $X \sim U(a,b)$ when $\alpha = \beta = 1$. The beta is one of the few common models that have bounded support.[12]

The standard form of the beta distribution turns up often in applications of Bayesian methods of statistical inference. We will see some of these in Chapter 10. Here are some of the properties of the standard form ($a = 0, b = 1$). Moments about the origin are easily seen to be

$$\mu_r' = \frac{\mathfrak{B}\left(\alpha + r, \beta\right)}{\mathfrak{B}\left(\alpha, \beta\right)}, r > -\alpha,$$

with

$$EY = \alpha\left(\alpha + \beta\right)^{-1}$$

$$VY = \alpha\beta\left(\alpha + \beta + 1\right)^{-1}\left(\alpha + \beta\right)^{-2}.$$

The m.g.f. clearly exists since the support is *bounded*, but it has no closed form. The density function is symmetric about $y = .5$ when $\alpha = \beta$, being concave (shaped like an inverted U) when $\alpha = \beta > 1$ and convex (U-shaped) when $\alpha = \beta < 1$. Figures 4.7 and 4.8 illustrate some of the possible shapes. Note that interchanging α and β has the effect of interchanging y and $1-y$ and flipping the image about $y = .5$.

---

[12]Of course, there is no *inherent* scarcity of such models, for if X is any r.v. such that $\mathbb{P}_X\left([a,b]\right) \equiv \int_{[a,b]} dF\left(x\right) > 0$, then

$$g\left(x; a, b\right) := \frac{f\left(x\right)\mathbf{1}_{[a,b]}\left(x\right)}{\int_{[a,b]} dF\left(x\right)}$$

is supported on that bounded interval. For another explicit family with a range of shapes much like that of the beta see McCullagh (1989).

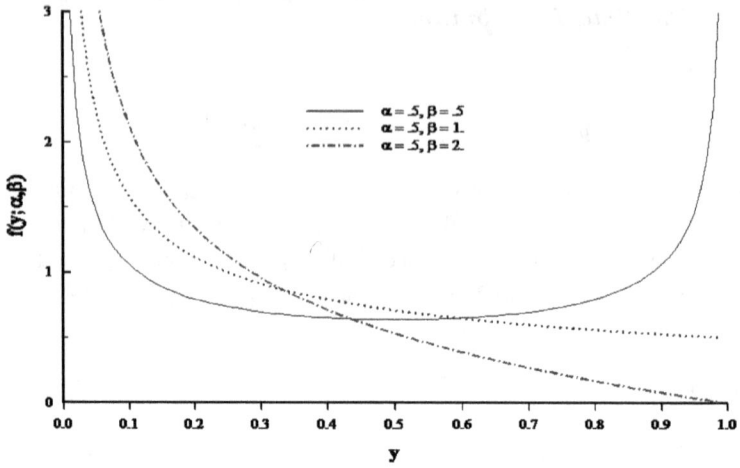

Fig. 4.7   Beta p.d.f.s for $\alpha = .5$ and various β.

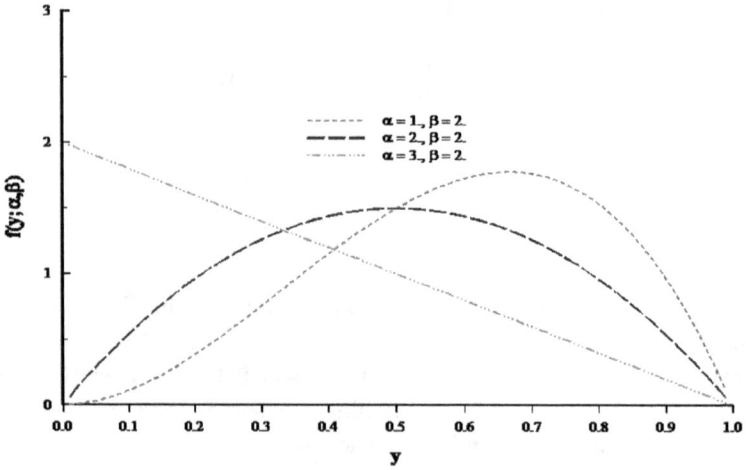

Fig. 4.8   Beta p.d.f.s. for $\beta = 2.$ and various α.

4.3.4 *The Weibull Distribution*

Weibull distributions are another generalization of the exponential distributions. Letting $Y \sim \Gamma(1, \beta)$ (exponential with scale parameter β), set $X^\alpha = Y$ for some $\alpha > 0$. Then

$$F_X(x) = \Pr(Y \leq x^\alpha) = \left[1 - \exp\left(-\frac{x^\alpha}{\beta}\right)\right] \mathbf{1}_{(0,\infty)}(x)$$

and

$$f_X(x) = \alpha \beta^{-1} x^{\alpha-1} e^{-x^\alpha/\beta} \mathbf{1}_{(0,\infty)}(x).$$

The collection $\{f_X(x; \alpha, \beta) : \alpha > 0, \beta > 0\}$ constitutes the *Weibull* family, another flexible class of models for positive-valued, continuous data. For short we write $X \sim W(\alpha, \beta)$ if X has a Weibull distribution with these parameters. Of course, with $\alpha = 1$ we are back to the exponential subfamily. Only subfamilies of the form $\{f_X(x; \alpha_0, \beta\} : \beta > 0\}$ with $\alpha_0 > 0$ are in the exponential class. One sometimes encounters a three-parameter version with arbitrary location parameter γ, corresponding to the extended version (4.31) of the gamma family. Our discussion will be limited to the standard two-parameter form.

Moments of Weibull variates are most easily deduced from their relation to exponentials. For any $r > -\alpha^{-1}$ (4.32) implies

$$\mu'_r = EX^r = EY^{r/\alpha} = \beta^{r/\alpha} \Gamma\left(1 + \frac{r}{\alpha}\right).$$

In particular,

$$\mu = \beta^{1/\alpha} \Gamma\left(1 + \frac{1}{\alpha}\right)$$

$$\sigma^2 = \beta^{2/\alpha} \left[\Gamma\left(1 + \frac{2}{\alpha}\right) - \Gamma\left(1 + \frac{1}{\alpha}\right)^2\right].$$

The m.g.f. does exist, but there is no simple expression for it.

Like the gamma family, the Weibull class is often a good model for waiting times. The hazard function,

$$h_T(t) := \frac{f_T(t)}{1 - F_T(t)} = \frac{\alpha}{\beta} t^{\alpha-1}, t > 0,$$

increases with t when $\alpha > 1$, is constant when $\alpha = 1$, and decreases for $0 < \alpha < 1$. Figures 4.9 and 4.10 show Weibull p.d.f.s and hazard functions for several values of α.

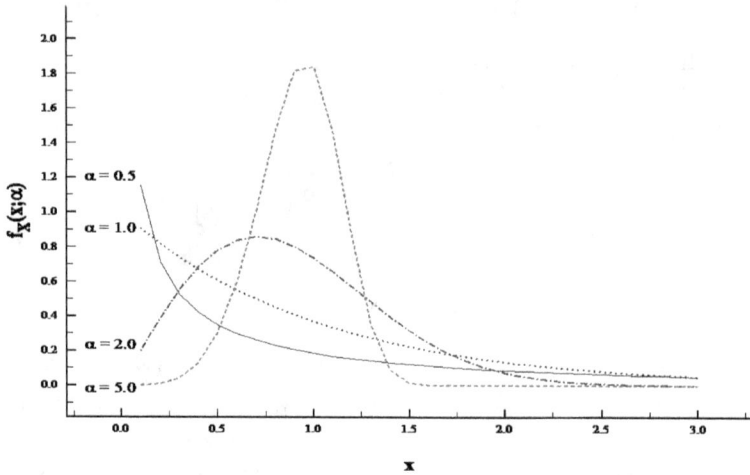

Fig. 4.9   Shapes of $W(\alpha, 1)$ p.d.f.s for $\alpha \in \{.5, 1., 2., 5\}$.

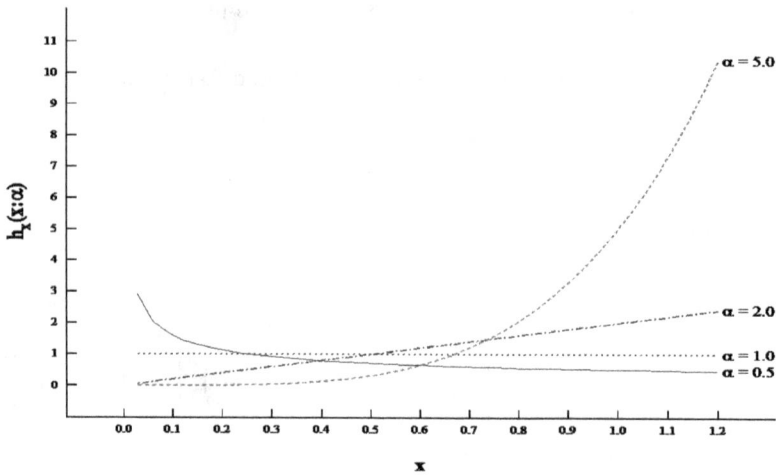

Fig. 4.10   Hazard functions for $W(\alpha, 1)$, $\alpha \in \{.5, 1., 2., 5\}$.

4.3.5   *The Laplace Distribution*

The Laplace or "double exponential" distribution is just a symmetrized version of the exponential. The p.d.f. of the standard form, $f(y) = \frac{1}{2}e^{-|y|}$, is symmetric about the origin. Setting $X := \alpha + \beta Y$ with $\beta > 0$ gives as

the general version

$$f(x; \alpha, \beta) = \frac{1}{2\beta} e^{-|x-\alpha|/\beta}.$$

If X has this p.d.f., we write $X \sim L(\alpha, \beta)$ for short. Expressing the exponential component of $f(x; \alpha, \beta)$ as

$$e^{-|x-\alpha|/\beta} = e^{(\alpha-x)/\beta} 1_{(-\infty, \alpha)}(x) + e^{(x-\alpha)/\beta} 1_{[\alpha, \infty)}(x)$$

shows that the Laplace family is not in the exponential class. The distribution is unimodal and symmetric about median α. Owing to the presence of a cusp at this point, the left- and right-hand derivatives $f'(\alpha-; a, \beta)$ and $f'(\alpha+; \alpha, \beta)$ are of different sign. Figure 4.11 shows the standard form of the p.d.f.

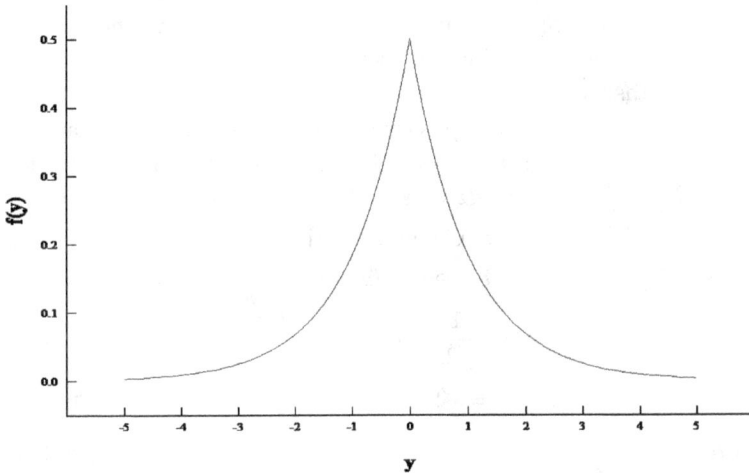

Fig. 4.11   The standard Laplace density.

The Laplace c.f. is $\phi_X(t) = e^{it\alpha}(1 + t^2\beta^2)^{-1}$, and the m.g.f. is $M_X(t) = e^{t\alpha}(1 - t^2\beta^2)^{-1}$ for $|t| < \beta^{-1}$. Thus, all moments are finite, so symmetry implies that the mean is α and that all odd central moments are zero. Central moments of even order are

$$\mu_{2k} = \beta^{2k}(2k)!, k \in \aleph. \tag{4.37}$$

Thus, the coefficient of kurtosis is $\alpha_4 = \mu_4/\mu_2^2 = 4!/4 = 6$, reflecting the relatively thick exponential tails compared with the normal distribution.

Exercise 4.39. *If $X \sim L(\alpha, \beta)$) show that even central moments are as given by (4.37).*

Exercise 4.40. X_1, X_2 are i.i.d. as exponential with p.d.f. $f(x; \beta) = \beta^{-1}e^{-x/\beta}1_{(0,\infty)}(x), \beta > 0$. Find the distribution of $Y := X_1 - X_2$. (Hint: Characteristic functions!)

4.3.6   *The Univariate Normal Distribution*

The normal or "Gaussian" distribution is the continuous model most often encountered in statistical applications and in applied work generally.[13] There are two principal reasons: (i) The distribution's properties make it very easy to work with; and (ii) the *central limit theorem*, discussed in Chapter 6, justifies its use to approximate empirical distributions of many data sets and the *sampling* distributions of many statistics. The normal model has already appeared in several examples, but we now fill in the details and establish some facts not yet verified.

The first task is to prove that the function $(2\pi)^{-1/2}\exp(-z^2/2)$ is a legitimate p.d.f. It is clearly positive for each $z \in \Re$, so all that remains is to show that it integrates to unity. There are several ways to do this. First, set $I := \int_{-\infty}^{\infty} e^{-z^2/2} \cdot dz$ and exploit the symmetry about the origin to write $I = 2\int_0^{\infty} e^{-z^2/2} \cdot dz$. Changing variables as $y = z^2/2$ (a one-to-one map from $[0, \infty)$ to itself), gives $dz/dy = (2y)^{-1/2}$ and

$$I = 2\int_0^{\infty} e^{-y}\left|\frac{1}{\sqrt{2y}}\right| \cdot dy = \sqrt{2}\int_0^{\infty} y^{-1/2}e^{-y} \cdot dy$$
$$= \sqrt{2}\Gamma(1/2) = \sqrt{2\pi}.$$

A more direct approach is to show that the double integral $I^2 = \int_{-\infty}^{\infty}\int_{-\infty}^{\infty} e^{-(x^2+y^2)/2} \cdot dx\,dy$ equals 2π. This was already done in Section 4.3.2 in the process of showing that $\Gamma(1/2) = \sqrt{\pi}$, but here is a more intuitive, geometrical argument that does not involve a change to polar coordinates. The double integral represents the volume under the three-dimensional surface depicted as contour plot and in profile in the two parts of Fig. 4.12.   It can be approximated as the sum of volumes of hollow cylinders of various radii, $r = \sqrt{x^2 + y^2}$. For example, referring to the figure, it is approximated by the volume of a cylinder with

---

[13]The term "normal" arose from the model's broad success in representing continuous or near-continuous data. Although the common appellation "Gaussian" suggests attribution to Carl Friedrich Gauss (1777-1855), the normal distribution appeared much earlier in the correspondence and published work of Abraham de Moivre (1667-1754), who used it to approximate the binomial distribution. The famous de Moivre–Laplace theorem, proved in Chapter 6, will illuminate the connection. For a detailed history of the normal distribution see Patel and Read (1982).

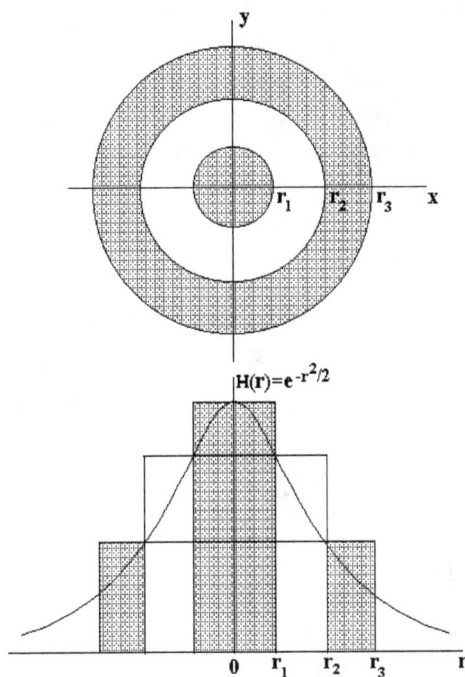

Fig. 4.12   Volume under $\exp\left[-\left(x^2+y^2\right)/2\right]$ as sum of volumes of annular cylinders.

height $H\left(0\right) = e^{-0^2/2} = 1$ and radius r_1 plus the volume of a hollow cylinder with height $H\left(r_1\right) = e^{-r_1^2/2}$ and thickness $r_2 - r_1$, and so on. Now
a cylinder with thickness dr and height $H\left(r\right) = \exp\left(-r^2/2\right)$ has cross-
sectional area $dA\left(r\right) = d\left(\pi r^2\right) = 2\pi r \cdot dr$. Its contribution to total volume
is $dV = H\left(r\right) \cdot dA\left(r\right) = 2\pi r e^{-r^2/2}$. Integrating from $r = 0$ to $r = \infty$ gives
$I^2 = 2\pi \int_0^\infty r e^{-r^2/2} \cdot dr = 2\pi$.

Exercise 4.41. *As an alternative geometrical argument, I^2 can be approximated as the sum of volumes of circular disks cut parallel to the x, y plane.
Show that this leads to*

$$I^2 = -\int_0^\infty A\left(r\right) \cdot dH\left(r\right) = \pi \int_0^\infty r^3 e^{-r^2/2} \cdot dr = 2\pi.$$

A r.v. Z with p.d.f. $f_Z(z) = (2\pi)^{-1/2} \exp(-z^2/2)$ is said to have a
standard normal distribution. In symbols we write $Z \sim N(0,1)$, where, as
we shall see shortly, the two numerals are values of the mean and variance.

Note: Henceforth the symbol "Z" will represent a standard normal r.v. unless it is otherwise specifically defined.

The density function of Z has the familiar symmetric bell shape, centered at the origin, concave on the interval $(-1, 1)$, and with convex tails tending rapidly toward zero as $|z| \to \infty$. The c.d.f., represented by the special symbol $\Phi(\cdot)$, is

$$\Phi(z) := \int_{-\infty}^{z} \frac{1}{\sqrt{2\pi}} e^{-y^2/2} \cdot dy.$$

Since $\Phi(0) = 0.5$ by symmetry, this can be written for $z > 0$ as

$$\Phi(z) = \frac{1}{2} + \int_{0}^{z} \frac{1}{\sqrt{2\pi}} e^{-y^2/2} \cdot dy.$$

Although the integral has no simple closed form, its value can be approximated by finite sums. Expanding $e^{-y^2/2}$ in Taylor series gives

$$\Phi(z) = \frac{1}{2} + \frac{1}{\sqrt{2\pi}} \int_0^z \left(1 - \frac{y^2}{2} + \frac{y^4}{2^2 \cdot 2!} - \frac{y^6}{2^3 \cdot 3!} + \cdots + (-1)^j \frac{y^{2j}}{2^j j!} + \cdots \right) \cdot dy$$

$$= \frac{1}{2} + \frac{1}{\sqrt{2\pi}} \sum_{j=0}^{\infty} (-1)^j \frac{z^{2j+1}}{(2j+1) \, 2^j j!}$$

$$= \frac{1}{2} + \frac{1}{\sqrt{2\pi}} \left(z - \frac{z^3}{6} + \frac{z^5}{40} - \frac{z^7}{336} + - \cdots \right).$$

Approximations of this sort are used to construct tables of Φ, such as that in Appendix C.1, and for calculations in statistical software. From any such source one finds that $\Phi(1) \doteq .8413$, $\Phi(2) \doteq .9772$, $\Phi(3) \doteq .9987$, and so on. Working in reverse from the table, one can approximate the values of quantiles $z_{.90} = \Phi^{-1}(.90) \doteq 1.282 = z^{.10}$, $z_{.95} = \Phi^{-1}(.95) \doteq 1.645 = z^{.05}$, $z_{.99} = \Phi^{-1}(.99) \doteq 2.326 = z^{.01}$, and so on. These can be found directly in Appendix C.2, which gives various upper quantiles $z^p = \Phi^{-1}(1 - p) = z_{1-p}$ to five decimal places. Note that the symmetry of f_Z about the origin implies that

$$\Phi(-z) = \mathbb{P}_Z((-\infty, -z]) = \mathbb{P}_Z([z, +\infty))$$
$$= 1 - \mathbb{P}_Z((-\infty, z)) = 1 - \mathbb{P}_Z((-\infty, z])$$
$$= 1 - \Phi(z);$$

hence, $\Phi(z)$ need be tabulated only for positive values of z. Symmetry also implies that lower and upper quantiles are related as $z_p = -z^p$.

The following calculations illustrate how to find $\mathbb{P}_Z\left((a,b]\right) = \Pr(a <$ $Z \le b)$ for various a, b. Of course, probabilities associated with open and closed intervals with the same end points are precisely the same. Drawing a sketch of the p.d.f. and representing probabilities as shaded areas, as in Fig. 4.13, helps to see the relations.

$$\mathbb{P}_Z\left((1,\infty)\right) = 1 - \mathbb{P}_Z\left((-\infty,1)\right) = 1 - \Phi(1) \doteq .1587 \doteq \mathbb{P}_Z\left((-\infty,-1)\right)$$

$$\mathbb{P}_Z\left((-1,\infty)\right) = \mathbb{P}_Z\left((-\infty,1)\right) = \Phi(1) \doteq .8413$$

$$\mathbb{P}_Z\left((1,2)\right) = \Phi(2) - \Phi(1) \doteq .1359 \doteq \mathbb{P}_Z\left((-2,-1)\right)$$

$$\mathbb{P}_Z\left((-2,1)\right) = \Phi(1) - \Phi(-2) = \Phi(1) - [1 - \Phi(2)] \doteq .8185 \doteq \mathbb{P}_Z\left((-1,2)\right)$$

$$\mathbb{P}_Z\left((-1,1)\right) = 2\left[\Phi(1) - \Phi(0)\right] = 2\Phi(1) - 1 \doteq .6826.$$

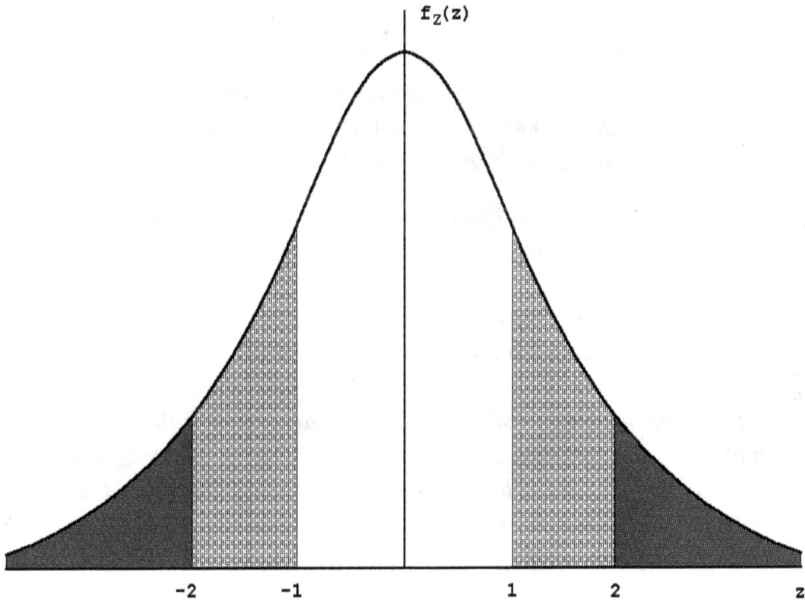

Fig. 4.13 Normal probabilities as areas under f_Z.

There are several formulas that provide useful bounds on $\mathbb{P}_Z\left((z,\infty)\right) = 1 - \Phi\left(z\right)$ for positive z. Here are two that are particularly simple. As shown in the table below, they are reasonably close when $z \geq 2$:

$$\left(z + z^{-1}\right)^{-1} f_Z\left(z\right) < 1 - \Phi\left(z\right) < z^{-1} f_Z\left(z\right).$$

To verify the upper bound for $z > 0$ write[14]

$$f_Z\left(z\right) = -\int_z^\infty df_Z\left(y\right) = \int_z^\infty y f_Z\left(y\right) \cdot dy$$

$$> z\int_z^\infty f_Z\left(y\right) \cdot dy$$

$$= z\left[1 - \Phi\left(z\right)\right].$$

z	$\frac{f_Z(z)}{z+z^{-1}}$	$1 - \Phi\left(z\right)$	$\frac{f_Z(z)}{z}$
1.0	.12098536	.15865525	.24197073
1.5	.05977735	.06680720	.08634506
2.0	.02159639	.02275013	.02699548
2.5	.00604424	.00620967	.00701132
3.0	.00132955	.00134990	.00147728
3.5	.00023052	.00023263	.00024934
4.0	.00003149	.00003167	.00003346
4.5	.00000338	.00000340	.00000355
5.0	.00000029	.00000029	.00000030

Exercise 4.42. *Use the relation* $z^{-1}f_Z\left(z\right) = -\int_z^\infty d\left[y^{-1}f_Z\left(y\right)\right]$ *to verify the lower bound for* $1 - \Phi\left(z\right)$ *when* $z > 0$.

Exercise 4.43. *Let* $\Pi_z^2 := \Pr\left(|Z| < z\right)^2 = \frac{1}{2\pi}\int_{-z}^z\int_{-z}^z e^{-(x^2+y^2)/2} \cdot dx\, dy$. *Following the argument with Fig. 4.12 one might try to evaluate* Π_z^2 *as* $\widehat{\Pi}_z^2 := \int_0^z re^{-r^2/2} \cdot dr = 1 - e^{-z^2/2}$. *Were this valid, there would be a simple closed form for* Φ *at any real* z. *However, comparing with the tables, one sees that* $\widehat{\Pi}_z < \Pi_z$ *for each* $z > 0$; *for example* $\widehat{\Pi}_1 \doteq .6273 < .6826 \doteq \Pi_1$. *Explain why the argument based on Fig. 4.12 gives the correct answer in the limit as* $z \to \infty$ *but underestimates* Π_z *for finite, positive* z. *Show that a better approximation is* $\widetilde{\Pi}_z = \sqrt{1 - e^{-2z^2/\pi}}$.

Exercise 4.44. *If* $Z \sim N(0,1)$, *show that* $\int_{-\infty}^x \Phi\left(t\right) \cdot dt = x\Phi\left(x\right) + f_Z\left(x\right)$ *and conclude that* $\int_{-\infty}^0 \Phi\left(t\right) \cdot dt = \left(2\pi\right)^{-1/2}$.

---

[14] Just differentiate $f_Z\left(y\right)$ to verify the second equality.

Example 3.37 showed that the standard normal m.g.f. and c.f. are $M_Z(t) = e^{t^2/2}$ and $\phi_Z(t) = e^{-t^2/2}, t \in \Re$. Since M_Z exists, all positive integer moments of Z are finite. That $\mu = EZ = 0$ then follows from the symmetry of the p.d.f. about the origin, or by working out $\int_\Re z (2\pi)^{-1/2} e^{-z^2/2} \cdot dz$, or by evaluating M_Z' at $t = 0$. Indeed, all the higher odd moments ($\mu_3' = EZ^3$, $\mu_5' = EZ^5$, etc.) are zero as well, since $Eg(Z) = 0$ when g is any *odd* function (one such that $g(-z) = -g(z)$) for which $E|g(Z)| < \infty$. The even moments can be obtained from the following recurrence relation (to be proved in Exercise 4.46 below as a consequence of Stein's lemma):

$$\mu_k' = (k-1)\mu_{k-2}', \ k \in \{2, 3, 4, ...\}. \tag{4.38}$$

Starting from $\mu_0' = EZ^0 \equiv 1$, this gives $\mu_2' = 1 \cdot \mu_0' = 1$, $\mu_4' = 3\mu_2' = 3$, $\mu_6' = 5 \cdot \mu_4' = 15$, $\mu_8' = 7 \cdot \mu_6' = 105$, and so on, with

$$\begin{aligned}
\mu_{2k}' &= (2k-1)(2k-3) \cdots \cdot 1 \\
&= \frac{(2k)(2k-1)(2k-2)(2k-3) \cdots \cdot 1}{(2k)(2k-2) \cdots \cdot 2} \\
&= \frac{(2k)!}{2^k k!}.
\end{aligned} \tag{4.39}$$

Of course, since $\mu = 0$ the central moments $\{\mu_k\}$ are the same as those about the origin. In particular, the variance is $\sigma^2 = 1$ and the coefficients of skewness and kurtosis are $\alpha_3 = 0$ and $\alpha_4 = 3$.

Exercise 4.45. *Taking* $g(Z) = e^{tZ}$ *and* $c = e^{tz}$ *for some* $z > 0$ *and* $t > 0$ *show that the Chebychev inequality (3.29) implies the upper bound* $\Pr(Z \geq z) \leq e^{-z^2/2}$.

From the normal c.f. and a little algebra one can get expressions for expectations of various trigonometric functions of standard normals.

Example 4.10.

(i) $e^{-2} = \phi(2) = Ee^{2iZ}$ implies

$$\begin{aligned}
e^{-2} &= E(\cos Z + i \sin Z)^2 \\
&= E \cos^2 Z + 2iE \sin Z \cos Z - E \sin^2 Z \\
&= E \cos^2 Z - E \sin^2 Z \ (\sin z \cos z \text{ is odd}) \\
&= E \cos^2 Z - (1 - E \cos^2 Z),
\end{aligned}$$

whence $E \cos^2 Z = (1 + e^{-2})/2$ and $E \sin^2 Z = (1 - e^{-2})/2$.

(ii) $e^{-8} = \phi(4) = Ee^{4iZ}$ implies

$$e^{-8} = E\left(\cos Z + i\sin Z\right)^4$$
$$= E\cos^4 Z - 6E\sin^2 Z \cos^2 Z + E\sin^4 Z;$$

and subtracting $1 = E(\cos^2 Z + \sin^2 Z)^2 = E\cos^4 Z + 2E\sin^2 Z\cos^2 Z + E\sin^4 Z$ leads to $E\sin^2 Z\cos^2 Z = \left(1 - e^{-8}\right)/8$.

(iii) Identities

$$E\sin^2 Z = E\sin^2 Z\left(\cos^2 Z + \sin^2 Z\right)$$
$$E\cos^2 Z = E\cos^2 Z\left(\cos^2 Z + \sin^2 Z\right)$$

and the results of (i) and (ii) imply

$$E\cos^4 Z = \frac{1 + e^{-2}}{2} - \frac{1 - e^{-8}}{8}$$
$$E\sin^4 Z = \frac{1 - e^{-2}}{2} - \frac{1 - e^{-8}}{8}.$$

In Exercise 3.53, as in the example, we used the form of the normal c.f. to work out expectations of trigonometric functions of standard normals. Relation (4.38) makes it possible to go the other way.

Example 4.11. Since $\sin tz$ is an odd function of z for each fixed $t \in \Re$, we know that $\phi(t) = E\cos tZ + iE\sin tZ = E\cos tZ$. Expressing $\cos tZ$ as the uniformly convergent Taylor series $\sum_{k=0}^{\infty}(-1)^k(tZ)^{2k}/(2k)!$ and using (4.39), we have

$$\phi(t) = \sum_{k=0}^{\infty}(-1)^k\frac{t^{2k}}{(2k)!}EZ^{2k}$$
$$= \sum_{k=0}^{\infty}\frac{(-t^2)^k}{(2k)!}\frac{(2k)!}{2^k k!}$$
$$= \sum_{k=0}^{\infty}\frac{1}{k!}\left(\frac{-t^2}{2}\right)^k$$
$$= e^{-t^2/2}.$$

Relation (4.39) is a consequence of a broader result known as Stein's lemma, which can be used to work out expectations of many other functions of standard normals.

Lemma 4.1. *If* $Z \sim N(0,1)$ *and if* $g : \Re \to \Re$ *is differentiable a.e. with respect to Lebesgue measure and such that* $\int_{-\infty}^{\infty}|g(z)|\cdot d\Phi(z) < \infty$ *and* $\int_{-\infty}^{\infty}|g'(z)|\cdot d\Phi(z) < \infty$, *then* $EZg(Z) = Eg'(Z)$.

Example 4.12.

(i) With $g(Z) = e^Z = g'(Z)$ we have $EZe^Z = Ee^Z = M_Z(1) = e^{1/2}$.

(ii) With $g(Z) = \sin Z, g'(Z) = \cos Z$ we have $EZ \sin Z = E \cos Z = E \cos Z + iE \sin Z = Ee^{iZ} = e^{-1/2}$.

(iii) With $g(Z) = Z|Z|, g'(Z) = -2Z1_{(-\infty,0)}(Z) + 2Z1_{(0,\infty)}(Z)$ we have

$$EZ^2|Z| = -2 \int_{-\infty}^0 z \frac{1}{\sqrt{2\pi}} e^{-z^2/2} \cdot dz + 2 \int_0^\infty z \frac{1}{\sqrt{2\pi}} e^{-z^2/2} \cdot dz$$

$$= 2\sqrt{2/\pi}.$$

Exercise 4.46. *Prove Stein's lemma and show that it implies (4.38).*

Exercise 4.47. *If g is k times differentiable and $\int_{-\infty}^\infty |d^k g(z)/dz^k| \cdot d\Phi(z) < \infty$ for $k \in \aleph_0$, show that $EZ^k g(Z) = (k-1)EZ^{k-2}g(Z) + EZ^{k-1}g'(Z)$ for $k \in \{2, 3, ...\}$.*

Exercise 4.48. *Use Stein's lemma to evaluate $EZ^2 \cos Z$.*

Exercise 4.49. *Prove that $\int_{-\infty}^\infty z\Phi(z) \cdot d\Phi(z) = (2\sqrt{\pi})^{-1}$.*

Exercise 4.50. *Show that $EZ1_{(a,b)}(Z) = f_Z(a) - f_Z(b)$ for $-\infty \le a < b \le \infty$.*

Generalizing the standard normal distribution to allow more general centering and scale, let $X := \delta + \gamma Z$ where $Z \sim N(0,1)$. Then $\phi_X(t) = e^{it\delta}\phi_Z(t\gamma) = e^{it\delta - t^2\gamma^2/2}$. Since $|\phi_X(t)| = e^{-t^2\gamma^2/2}$ is integrable, the inversion formula (3.52) gives as the p.d.f. of X

$$f_X(x) = \frac{1}{2\pi} \int_{-\infty}^\infty e^{-itx} e^{it\delta - t^2\gamma^2/2} \cdot dt$$

$$= \frac{1}{\sqrt{2\pi\gamma^2}} e^{-(x-\delta)^2/2\gamma^2}, x \in \Re.$$

The simple formulas for means and variances of affine functions of r.v.s imply

$$\mu = EX = \delta + \gamma EZ = \delta$$

$$\sigma^2 = VX = \gamma^2 VZ = \gamma^2$$

$$\sigma = |\gamma|.$$

Given these interpretations of δ and γ, it is customary to write the p.d.f. and c.f. of X as

$$f_X(x; \mu, \sigma^2) = \frac{1}{\sqrt{2\pi\sigma^2}} e^{-(x-\mu)^2/2\sigma^2}$$

$$\phi_X(t) = e^{it\mu - t^2\sigma^2/2}.$$

A r.v. with this p.d.f. and this c.f. is said to have a *normal distribution* with mean μ and variance σ^2. In symbols we write $X \sim N(\mu, \sigma^2)$. The normal *family* comprises the collection $\{ f_X(x; \mu, \sigma^2) : \mu \in \Re, \sigma^2 > 0 \}$.

Example 4.1 (p. 236) showed that the normal family belongs to the exponential class. A convenient version of canonical form $f(x; \boldsymbol{\eta}) = a(\boldsymbol{\eta}) c(x) e^{\boldsymbol{\eta}' \mathbf{T}(x)}$ has

$$\boldsymbol{\eta} = \left(-\frac{1}{2\sigma^2}, \frac{\mu}{\sigma^2} \right)$$

$$\mathbf{T}(x) = (x^2, x)$$

$$c(x) = (2\pi)^{-1/2}$$

$$\alpha(\boldsymbol{\eta}) = (-2\eta_1)^{1/2} \exp\left(\frac{\eta_2^2}{4\eta_1} \right).$$

The particular affine transformation $-\mu/\sigma + X/\sigma = (X - \mu)/\sigma$ puts X in standard form and takes us back to $N(0, 1)$. Thus, defining a r.v. $Z := (X - \mu)/\sigma$, it follows that $Z \sim N(0, 1)$ if $X \sim N(\mu, \sigma^2)$. A consequence that is profoundly obvious—but too useful to risk being overlooked—is that any r.v. X distributed as $N(\mu, \sigma^2)$ can be expressed as $X = \mu + \sigma Z$, where $Z \sim N(0, 1)$ and where we interpret "=" as meaning "equal in distribution".

Exercise 4.51. *Extend Stein's lemma to show that $EXh(X) = \mu Eh(X) + \sigma^2 Eh'(X)$ when $X \sim N(\mu, \sigma^2)$ and h is differentiable with $E|h(X)| < \infty$ and $E|h'(X)| < \infty$.*

Exercise 4.52. *X and Y are i.i.d. as $N(\mu, \sigma^2)$. Show that Gini's mean difference, $\mathfrak{g} := E|X - Y|$, is given by*

$$\mathfrak{g} = \frac{2\sigma}{\sqrt{\pi}}.$$

Exercise 4.53. *If $X \sim N(\mu, \sigma^2)$ express (i) $E\sin X$ (ii) $E\cos X$ (iii) $E\sin^2 X$ (iv) $E\cos^2 X$ (v) $E\sin X \cos X$ (vi) $EX\cos X$ (vii) $EX\sin X$.*

Exercise 4.54. *Show that the coefficients of skewness and kurtosis for any member of the normal family are $\alpha_3 = 0$ and $\alpha_4 = 3$, respectively.*

The fact that $(X - \mu)/\sigma \sim N(0, 1)$ when $X \sim N(\mu, \sigma^2)$ makes it possible to use tabulated values of standard normal c.d.f. Φ to determine probabilities for X. Thus,

$$\Pr(X \leq x) = \Pr\left(\frac{X - \mu}{\sigma} \leq \frac{x - \mu}{\sigma} \right) = \Pr\left(Z \leq \frac{x - \mu}{\sigma} \right) = \Phi\left(\frac{x - \mu}{\sigma} \right).$$

Example 4.13. If $X \sim N(5, 100)$ then $\Pr(15 < X \le 25) = \Phi(2) - \Phi(1) \doteq .1359$.

Here is a useful relation, to be proved in Section 4.3.7:

$$E\Phi(a \pm bZ) = \int_{-\infty}^{-\infty} \Phi(a \pm bz) \cdot d\Phi(z) = \Phi\left(\frac{a}{\sqrt{1+b^2}}\right). \qquad (4.40)$$

Hence, $E\Phi(X) = \Phi\left(\mu/\sqrt{1+\sigma^2}\right)$ when $X \sim N\left(\mu, \sigma^2\right)$.

Exercise 4.55. Let $X \sim N\left(\mu, \sigma^2\right)$ and, for some finite $a < b$, set $Y(\omega) := X(\omega)$ if $X(\omega) \in [a, b]$; otherwise, Y is not observed. Express EY. (Data are said to be truncated if realizations are reported only when they lie within a certain range. Truncation is common in statistical applications. For example, a survey of welfare recipients might record incomes only of those below some "poverty" threshold.)

Exercise 4.56. With X, a, and b as above, set $W := a\mathbf{1}_{(-\infty,a]}(X) + X\mathbf{1}_{(a,b)}(X) + b\mathbf{1}_{[b,\infty)}(X)$. Express EW. (Data are said to be censored if all realizations beyond a certain level are reported at that level. For example, a machine's operating lifetime might be assigned the fixed value $\tau > 0$ if it was still functioning after that much time on test.)

The normal family is closed under finite convolution as well as under affine transformation. Indeed, any (nondegenerate) affine transformation of independent normals remains in the normal family; and, as we shall see shortly, the statement is true even for an important class of jointly dependent normals.

Exercise 4.57. $\{X_j\}_{j=1}^k$ are mutually independent with $X_j \sim N(\mu_j, \sigma_j^2)$, and $Y := a_0 + \sum_{j=1}^k a_j X_j$. Show that $\mu_Y = a_0 + \sum_{j=1}^k a_j \mu_j$, $\sigma_Y^2 = \sum_{j=1}^k a_j^2 \sigma_j^2$, and $Y \sim N(\mu_Y, \sigma_Y^2)$.

Exercise 4.58. Z_1, Z_2 are independent standard normal variates, and $S := Z_1 + Z_2$. Find $E(Z_1 \mid S)$ and confirm that $E[E(Z_1 \mid S)] = EZ_1 = 0$.

4.3.7 *The Multivariate Normal Distribution*

The multivariate extension of the Gaussian model retains much of the elegance and simplicity of the univariate version. This fact and the model's wide applicability to real data have made it the preeminent framework for multivariate analysis in statistics. As for the univariate case the development begins with a "standard" multinormal and proceeds to the general form. The usefulness of vector/matrix notation in this context becomes apparent right away.

4.3.7.1   The "Spherical" and General Forms

The joint p.d.f. of $\mathbf{Z} := (Z_1, Z_2, ..., Z_k)'$, a vector of mutually independent standard normals, is

$$f_{\mathbf{Z}}(\mathbf{z}) = \prod_{j=1}^{k} \left(\frac{1}{\sqrt{2\pi}} e^{-z_j^2/2} \right)$$

$$= (2\pi)^{-k/2} \exp\left(-\frac{1}{2} \sum_{j=1}^{k} z_j^2 \right)$$

$$= (2\pi)^{-k/2} e^{-\mathbf{z}'\mathbf{z}/2}, \mathbf{z} \in \Re^k.$$

This represents the "spherical normal" distribution, so called because the *level curves*—the loci of points $\mathbf{z}'\mathbf{z} = r^2$ for positive constants r—describe "hyperspheres" of radius r in $\Re^k$. In particular, $\mathbf{z}'\mathbf{z} = r^2$ defines a circle of radius r when $k = 2$ and a sphere when $k = 3$. The mean vector and covariance matrix are $E\mathbf{Z} = \mathbf{0}$ and $V\mathbf{Z} := E\mathbf{Z}\mathbf{Z}' = \mathbf{I}_k$ (the $k \times k$ identity matrix).

The joint m.g.f. of the spherical normal distribution can be found either by multiplying marginal m.g.f.s $\left\{ M_{Z_j}(t_j) = e^{t_j^2/2} \right\}_{j=1}^{k}$, as

$$M_{\mathbf{Z}}(\mathbf{t}) = \prod_{j=1}^{k} e^{t_j^2/2} = e^{\mathbf{t}'\mathbf{t}/2}, \mathbf{t} \in \Re^k,$$

or by direct integration, as

$$M_{\mathbf{Z}}(\mathbf{t}) = Ee^{\mathbf{t}'\mathbf{Z}} = \int_{\Re^k} (2\pi)^{-k/2} e^{\mathbf{t}'\mathbf{z} - \mathbf{z}'\mathbf{z}/2} \cdot d\mathbf{z}$$

$$= e^{\mathbf{t}'\mathbf{t}/2} \int_{\Re^k} (2\pi)^{-k/2} e^{-(\mathbf{z}'\mathbf{z} - 2\mathbf{t}'\mathbf{z} + \mathbf{t}'\mathbf{t})/2} \cdot d\mathbf{z}$$

$$= e^{\mathbf{t}'\mathbf{t}/2} \int_{\Re^k} (2\pi)^{-k/2} e^{-(\mathbf{z}-\mathbf{t})'(\mathbf{z}-\mathbf{t})/2} \cdot d\mathbf{z}$$

$$= e^{\mathbf{t}'\mathbf{t}/2}, \mathbf{t} \in \Re^k. \tag{4.41}$$

The last step follows by changing variables as $\mathbf{y} = \mathbf{z} - \mathbf{t}$ in the integral and recognizing the integrand of $(2\pi)^{-k/2} e^{-\mathbf{y}'\mathbf{y}/2}$ as the spherical-normal p.d.f. The c.f. of the spherical normal is $\phi(\mathbf{t}) = M(i\mathbf{t}) = e^{-\mathbf{t}'\mathbf{t}/2}$.

Generalizing to allow nonzero means and a covariance matrix other than the identity matrix, let $\mathbf{A}$ be a $k \times k$ nonsingular matrix, $\boldsymbol{\mu}$ a k-vector, and set $\mathbf{X} = \mathbf{g}(\mathbf{Z}) := \mathbf{AZ} + \boldsymbol{\mu}$. The nonsingularity of $\mathbf{A}$ makes the transformation one to one, with $\mathbf{Z} = \mathbf{h}(\mathbf{X}) := \mathbf{A}^{-1}(\mathbf{X} - \boldsymbol{\mu})$ as the inverse. Change-of-variable formula (2.24) yields

$$f_{\mathbf{X}}(\mathbf{x}) = f_{\mathbf{Z}}\left[\mathbf{A}^{-1}(\mathbf{x} - \boldsymbol{\mu})\right]|J_{\mathbf{h}}|,$$

where

$$J_{\mathbf{h}} := \left|\frac{\partial \mathbf{z}}{\partial \mathbf{x}}\right| = \begin{vmatrix} \partial z_1/\partial x_1 & \partial z_2/\partial x_1 & \cdots & \partial z_k/\partial x_1 \\ \partial z_1/\partial x_2 & \partial z_2/\partial x_2 & \cdots & \partial z_k/\partial x_2 \\ \vdots & \vdots & \ddots & \vdots \\ \partial z_1/\partial x_k & \partial z_2/\partial x_k & \cdots & \partial z_k/\partial x_k \end{vmatrix} = \left|\mathbf{A}^{-1}\right|.$$

With the identity $\left(\mathbf{A}^{-1}\right)'\mathbf{A}^{-1} = (\mathbf{A}')^{-1}\mathbf{A}^{-1} = \left(\mathbf{AA}'\right)^{-1}$ this becomes

$$f_{\mathbf{X}}(\mathbf{x}) = (2\pi)^{-k/2}\left\|\mathbf{A}^{-1}\right\|e^{-(\mathbf{x}-\boldsymbol{\mu})'\left(\mathbf{AA}'\right)^{-1}(\mathbf{x}-\boldsymbol{\mu})/2}, \mathbf{x} \in \Re^k, \qquad (4.42)$$

where $\left\|\mathbf{A}^{-1}\right\|$ is the absolute value of determinant $\left|\mathbf{A}^{-1}\right|$. Note that $\boldsymbol{\mu}$ is the expected value of $\mathbf{X}$ and that $\boldsymbol{\Sigma} := \mathbf{AA}'$ is the covariance matrix, since

$$V\mathbf{X} = E(\mathbf{X} - \boldsymbol{\mu})(\mathbf{X} - \boldsymbol{\mu})' = E\left(\mathbf{AZZ}'\mathbf{A}'\right)$$
$$= \mathbf{A}E\left(\mathbf{ZZ}'\right)\mathbf{A}' = \mathbf{AI}_k\mathbf{A}' = \mathbf{AA}'.$$

$\boldsymbol{\Sigma}$ is nonsingular since

$$|\boldsymbol{\Sigma}| = \left|\mathbf{AA}'\right| = |\mathbf{A}|\left|\mathbf{A}'\right| = |\mathbf{A}|^2 > 0,$$

and it is therefore positive definite. Noting that $\|\mathbf{A}\| = |\boldsymbol{\Sigma}|^{1/2}$ and $\left\|\mathbf{A}^{-1}\right\| = 1/\|\mathbf{A}\| = |\boldsymbol{\Sigma}|^{-1/2}$, (4.42) can be expressed as

$$f_{\mathbf{X}}(\mathbf{x}) = (2\pi)^{-k/2}|\boldsymbol{\Sigma}|^{-1/2}e^{-(\mathbf{x}-\boldsymbol{\mu})'\boldsymbol{\Sigma}^{-1}(\mathbf{x}-\boldsymbol{\mu})/2}, \mathbf{x} \in \Re^k. \qquad (4.43)$$

With $\boldsymbol{\mu}$ as a k-vector of finite constants and $\boldsymbol{\Sigma}$ as a $k \times k$ symmetric, positive-definite matrix, this is the usual representation of the *multivariate normal* distribution. We write $\mathbf{X} \sim N(\boldsymbol{\mu}, \boldsymbol{\Sigma})$ to indicate that $\mathbf{X}$ has this distribution.

The m.g.f. and c.f. follow at once from (4.41):

$$M_{\mathbf{X}}(\mathbf{t}) = Ee^{\mathbf{t}'(\boldsymbol{\mu}+\mathbf{AZ})} = e^{\mathbf{t}'\boldsymbol{\mu}}M_{\mathbf{Z}}(\mathbf{A}'\mathbf{t}) = e^{\mathbf{t}'\boldsymbol{\mu}}e^{\mathbf{t}'\mathbf{AA}'\mathbf{t}/2} = e^{\mathbf{t}'\boldsymbol{\mu}+\mathbf{t}'\boldsymbol{\Sigma}\mathbf{t}/2}$$
$$\phi_{\mathbf{X}}(\mathbf{t}) = e^{i\mathbf{t}'\boldsymbol{\mu}-\mathbf{t}'\boldsymbol{\Sigma}\mathbf{t}/2}. \qquad (4.44)$$

Like its univariate counterpart, the multivariate normal family is in the exponential class.

4.3.7.2   *Special Cases*

When $k = 1$ Expression (4.43) reduces to the univariate normal p.d.f.,

$$f_1(x_1) = \left(2\pi\sigma_1^2\right)^{-1/2} e^{-\frac{1}{2}(x_1 - \mu_1)^2/\sigma_1^2}, \quad x_1 \in \Re.$$

When $k = 2$ the distribution is the *bivariate* normal. In scalar notation the p.d.f. is

$$f_{1,2}(x_1, x_2) = \frac{1}{2\pi\sigma_1\sigma_2\sqrt{1 - \rho^2}} \exp\left[-\frac{z_1^2 - 2\rho z_1 z_2 + z_2^2}{2(1 - \rho^2)}\right], \qquad (4.45)$$

where $z_j := (x_j - \mu_j)/\sigma_j$ and $\rho := \sigma_{12}/(\sigma_1\sigma_2)$ is the coefficient of correlation. The level curves of $f_{1,2}(x_1, x_2)$ are *ellipses*, the slopes of whose principal axes have the same sign as does ρ. When $\rho = 0$ and $\sigma_1 = \sigma_2$ the ellipses degenerate to circles, as in the spherical-normal case. In higher dimensions the level curves are *ellipsoids*.

Exercise 4.59. *Show that formula (4.43) reduces to (4.45) when $k = 2$.*

We can now verify the relation (4.40) from the previous section. We first show that

$$\int_{-\infty}^c \Phi\left(a - bz\right) \cdot d\Phi\left(z\right) = \Phi\left(c, \frac{a}{\sqrt{1 + b^2}}; \frac{b}{\sqrt{1 + b^2}}\right), \qquad (4.46)$$

where $\Phi\left(z_1, z_2; \rho\right) := \Pr\left(Z_1 \le z_1, Z_2 \le z_1\right)$ is the bivariate c.d.f. of standard normals having correlation $\rho \in (-1, 1)$. Write the left side of (4.46) as

$$\int_{-\infty}^c \left(\int_{-\infty}^{a - bz} \frac{1}{\sqrt{2\pi}} e^{-x^2/2} \cdot dx\right) \frac{1}{\sqrt{2\pi}} e^{-z^2/2} \cdot dz$$

$$= \int_{-\infty}^c \int_{-\infty}^{a - bz} \frac{1}{2\pi} e^{-(x^2 + z^2)/2} \cdot dx \, dz$$

and change variables as

$$\begin{pmatrix} x \\ z \end{pmatrix} = \begin{pmatrix} \sqrt{1 + b^2} & -b \\ 0 & 1 \end{pmatrix} \begin{pmatrix} y \\ w \end{pmatrix} = \begin{pmatrix} \sqrt{1 + b^2}\, y - bw \\ w \end{pmatrix},$$

so that

$$J_{(x,z)\to(y,w)} = \begin{vmatrix} \partial x/\partial y & \partial x/\partial w \\ \partial z/\partial y & \partial z/\partial w \end{vmatrix} = \sqrt{1 + b^2}.$$

Putting $b =: \rho/\sqrt{1 - \rho^2}$ (so that $\sqrt{1 + b^2} = 1/\sqrt{1 - \rho^2}$ and $b/\sqrt{1 + b^2} = \rho$) now gives

$$\int_{-\infty}^c \Phi(a - bz) \cdot d\Phi(z) = \int_{-\infty}^c \int_{-\infty}^{a/\sqrt{1 + b^2}} \frac{1}{2\pi\sqrt{1 - \rho^2}} \exp\left[-\frac{y^2 - 2\rho yz + z^2}{2(1 - \rho^2)}\right] \cdot dy \, dz,$$

which corresponds to (4.46). Now letting $c \to +\infty$, the right side of (4.46) becomes the marginal c.d.f. of the second variate, and since the sign of b is then irrelevant, we have

$$\int_{-\infty}^{\infty} \Phi\left(a \pm bz\right) \cdot d\Phi\left(z\right) = \Phi\left(\frac{a}{\sqrt{1+b^2}}\right).$$

Exercise 4.60. *Let* $\{Z_j\}_{j=1}^{\infty}$ *be i.i.d. as* $N\left(0,1\right)$ *and put* $S_n := \sum_{j=1}^{n} Z_j$ *for* $n \in \aleph$. *Show that for any real* k

$$\Pr\left(S_n \leq k, S_{n+1} > k\right) = \Phi\left(\frac{k}{\sqrt{n}}\right) - \Phi\left(\frac{k}{\sqrt{n}}, \frac{k}{\sqrt{n+1}}; \sqrt{\frac{n}{n+1}}\right),$$

and express the limiting values as $k \to +\infty$ *for fixed* n, *as* $n \to \infty$ *for fixed* k, *and as* $n \to \infty$ *when* $k = n$.

4.3.7.3 *Important Properties*

Here are some important features of the multivariate normal distribution.

(i) Nondegenerate affine functions of multinormals are multinormal.
   If (a) k-vector $\mathbf{X}$ is distributed as $N\left(\boldsymbol{\mu}, \boldsymbol{\Sigma}\right)$; (b) $\mathbf{A}$ is an $m \times k$ matrix with rank $m \leq k$; and (c) $\mathbf{a}_0 \in \Re^m$, then m-vector $\mathbf{Y} := \mathbf{a}_0 + \mathbf{A}\mathbf{X}$ has mean $\mathbf{a}_0 + \mathbf{A}\boldsymbol{\mu}$ and nonsingular—hence positive-definite—covariance matrix $\mathbf{A}\boldsymbol{\Sigma}\mathbf{A}'$. That $\mathbf{Y}$ is *multinormal* with these parameters is easily seen in terms of characteristic functions. For $\mathbf{t} \in \Re^m$ we have

$$\phi_{\mathbf{Y}}\left(\mathbf{t}\right) = E e^{i\mathbf{t}'\left(\mathbf{a}_0 + \mathbf{A}\mathbf{X}\right)} = e^{i\mathbf{t}'\mathbf{a}_0} \phi_{\mathbf{X}}\left(\mathbf{A}'\mathbf{t}\right)$$
$$= e^{i\mathbf{t}'\left(\mathbf{a}_0 + \mathbf{A}\boldsymbol{\mu}\right) - \mathbf{t}'\mathbf{A}\boldsymbol{\Sigma}\mathbf{A}'\mathbf{t}/2},$$

   which is of the form (4.44).
   Recall that Exercise 4.57 showed $a_0 + \mathbf{a}'\mathbf{X}$ to be normally distributed when elements of $\mathbf{X}$ are *independent* normals (and $\mathbf{a} \neq \mathbf{0}$). Such an $\mathbf{X}$ is multivariate normal with a *diagonal* covariance matrix; but we now recognize that the independence condition is not necessary, for $Y := a_0 + \mathbf{a}'\mathbf{X}$ is (univariate) normal whenever $\mathbf{X} \sim N(\boldsymbol{\mu}, \boldsymbol{\Sigma})$—whether $\boldsymbol{\Sigma}$ is diagonal or not.

(ii) Marginal distributions of components of a multinormal vector $\mathbf{X}$ are also multinormal.
   Thus, if $\mathbf{X}_1 \in \Re^{k_1}$ is a subvector of $\mathbf{X}$, then it is also normally distributed. To see this, arrange $\mathbf{X}$ so that $\mathbf{X}_1$ comprises the *first* k_1 of its $k = k_1 + k_2$ elements, with $\mathbf{X}_2 \in \Re^{k_2}$ as those that remain. Then, with $\mathbf{0}$ a $k_1 \times k_2$ matrix of zeroes and $\mathbf{A}$ as the $k_1 \times k$ matrix $(\mathbf{I}_{k_1}, \mathbf{0})$,

it follows from property (i) that $\mathbf{AX} = \mathbf{X}_1$ is multinormal with mean and covariance matrix

$$(\mathbf{I}_{k_1}, \mathbf{0}) \begin{pmatrix} \boldsymbol{\mu}_1 \\ \boldsymbol{\mu}_2 \end{pmatrix} = \boldsymbol{\mu}_1$$

$$(\mathbf{I}_{k_1}, \mathbf{0}) \begin{pmatrix} \boldsymbol{\Sigma}_{11} & \boldsymbol{\Sigma}_{12} \\ \boldsymbol{\Sigma}_{21} & \boldsymbol{\Sigma}_{22} \end{pmatrix} \begin{pmatrix} \mathbf{I}_{k_1} \\ \mathbf{0}' \end{pmatrix} = \boldsymbol{\Sigma}_{11}.$$

In particular, any single element X_j of $\mathbf{X}$ is distributed as $N\left(\mu_j, \sigma_j^2\right)$.

(iii) Uncorrelated multivariate normal r.v.s are *independent*.

Although uncorrelated r.v.s are not independent *in general*, they *are* independent when they are distributed as multivariate normal. In other words, the absence of stochastic *linear* association between multinormals implies the absence of any form of stochastic dependence. To see this, observe that $\boldsymbol{\Sigma}$ is diagonal when all covariances are zero, and when $\boldsymbol{\Sigma}$ is diagonal we have

$$|\boldsymbol{\Sigma}|^{1/2} = \sigma_1 \sigma_2 \cdots \cdots \sigma_k,$$

$$\boldsymbol{\Sigma}^{-1} = \begin{pmatrix} \frac{1}{\sigma_1^2} & 0 & \cdots & 0 \\ 0 & \frac{1}{\sigma_2^2} & \cdots & 0 \\ \vdots & \vdots & \ddots & \vdots \\ 0 & 0 & \cdots & \frac{1}{\sigma_k^2} \end{pmatrix},$$

$$(\mathbf{x} - \boldsymbol{\mu})' \boldsymbol{\Sigma}^{-1} (\mathbf{x} - \boldsymbol{\mu}) = \sum_{j=1}^{k} \left(\frac{x_j - \mu_j}{\sigma_j} \right)^2,$$

and

$$f_{\mathbf{X}}(\mathbf{x}; \boldsymbol{\mu}, \boldsymbol{\Sigma}) = \frac{1}{(2\pi)^{k/2} \sigma_1 \sigma_2 \cdots \cdots \sigma_k} \exp\left[-\frac{1}{2} \sum_{j=1}^{k} \left(\frac{X_j - \mu_j}{\sigma_j} \right)^2 \right]$$

$$= \prod_{j=1}^{k} \left\{ \frac{1}{\sqrt{2\pi\sigma_j^2}} \exp\left[-\frac{1}{2} \left(\frac{X_j - \mu_j}{\sigma_j} \right)^2 \right] \right\}$$

$$= \prod_{j=1}^{k} f_j(x_j; \mu_j, \sigma_j^2).$$

Likewise, if multinormal k-vector $\mathbf{X}$ is decomposed into subvectors $\left\{ \mathbf{X}_j \in \Re^{k_j} \right\}_{j=1}^{\ell}$ with covariance matrices $\left\{ \boldsymbol{\Sigma}_{jj} \right\}_{j=1}^{\ell}$, and if covariance matrix $\boldsymbol{\Sigma}$ is block diagonal, as

$$\boldsymbol{\Sigma} = \begin{pmatrix} \boldsymbol{\Sigma}_{11} & \mathbf{0} & \cdots & \mathbf{0} \\ \mathbf{0}' & \boldsymbol{\Sigma}_{22} & \cdots & \mathbf{0} \\ \mathbf{0}' & \cdots & \ddots & \vdots \\ \mathbf{0}' & \cdots & \mathbf{0}' & \boldsymbol{\Sigma}_{\ell\ell} \end{pmatrix},$$

then multinormal vectors $\{\mathbf{X}_j\}_{j=1}^{\ell}$ are also independent. Example 2.32 showed that two r.v.s can be functionally related, yet statistically independent. The next exercise provides another example of this fact.

Exercise 4.61. X_1, X_2 *are independent normal variates with zero means and equal variances. Show that $S := X_1 + X_2$ and $D := X_1 - X_2$ are independent.*[15]

(iv) That jointly distributed r.v.s $\{X_j\}_{j=1}^{k}$ have normal marginals does not *by itself* imply that $\mathbf{X} = (X_1, X_2, ..., X_k)'$ is multinormal.

If the $\{X_j\}$ are normal *and* independent, then indeed $\mathbf{X}$ *is* multinormal, since

$$\prod_{j=1}^{k} \frac{1}{\sqrt{2\pi\sigma_j^2}} e^{-\frac{1}{2}(x_j - \mu_j)^2/\sigma_j^2} = (2\pi)^{-k/2} |\boldsymbol{\Sigma}|^{-1/2} e^{-(\mathbf{x}-\boldsymbol{\mu})' \boldsymbol{\Sigma}^{-1}(\mathbf{x}-\boldsymbol{\mu})/2}$$

$$= f(\mathbf{x}; \boldsymbol{\mu}, \boldsymbol{\Sigma})$$

with

$$\boldsymbol{\mu} = (\mu_1, \mu_2, ..., \mu_k)'$$

$$\boldsymbol{\Sigma} = \begin{pmatrix} \sigma_1^2 & 0 & \cdots & 0 \\ 0 & \sigma_2^2 & \cdots & 0 \\ \vdots & \vdots & \ddots & \vdots \\ 0 & 0 & \cdots & \sigma_k^2 \end{pmatrix}.$$

However, just having normal marginals is not enough, as the following exercise shows.

Exercise 4.62. $\mathbf{X} := (X_1, X_2)'$ *has p.d.f.*

$$f_{\mathbf{X}}(x_1, x_2) = \pi^{-1} e^{-(x_1^2 + x_2^2)/2} \mathbf{1}_{\{(x_1, x_2) : x_1 x_2 > 0\}}(x_1, x_2)$$

---

[15]The converse statement is also true: If X_1, X_2 are independent and identically distributed r.v.s with zero means and equal variances, and if $X_1 + X_2$ and $X_1 - X_2$ are independent, then X_1 and X_2 are distributed as normal. For the proof see Chung (1974, pp. 171-2). There are many such *characterizations* of the normal distribution. Patel and Read (1982, Ch. 4) give an extensive survey.

(i.e., $f_\mathbf{X}$ is positive when x_1 and x_2 have the same sign and is zero otherwise). Show that $\mathbf{X}$ is not bivariate normal but that the marginal distribution of each component is univariate normal.

(v) The distribution of a subvector $\mathbf{X}_2$ of $\mathbf{X} = (\mathbf{X}_1', \mathbf{X}_2')'$ *conditional* on values of $\mathbf{X}_1$ is multinormal, and the conditional mean of $\mathbf{X}_2$ has an *affine* structure.

With $\mathbf{X}_1 \in \Re^{k_1}$, $\mathbf{X}_2 \in \Re^{k_2}$, and $k_1 = k - k_2 \in \{1, 2, ..., k-1\}$ we have

$$\mathbf{X} = \begin{pmatrix} \mathbf{X}_1 \\ \mathbf{X}_2 \end{pmatrix} \sim N\left[\begin{pmatrix} \boldsymbol{\mu}_1 \\ \boldsymbol{\mu}_2 \end{pmatrix}, \begin{pmatrix} \Sigma_{11} & \Sigma_{12} \\ \Sigma_{21} & \Sigma_{22} \end{pmatrix} \right].$$

If we put $\mathbf{U} := (\mathbf{X}_2 - \boldsymbol{\mu}_2) - \mathbf{B}(\mathbf{X}_1 - \boldsymbol{\mu}_1)$ for some $k_2 \times k_1$ matrix of constants $\mathbf{B}$, then $E\mathbf{U} = \mathbf{0}$, and the $k_2 \times k_1$ matrix of covariances of the elements of $\mathbf{U}$ with those of $\mathbf{X}_1$ is

$$E\mathbf{U}(\mathbf{X}_1 - \boldsymbol{\mu}_1)' = E(\mathbf{X}_2 - \boldsymbol{\mu}_2)(\mathbf{X}_1 - \boldsymbol{\mu}_1)' - \mathbf{B}E(\mathbf{X}_1 - \boldsymbol{\mu}_1)(\mathbf{X}_1 - \boldsymbol{\mu}_1)'$$
$$= \Sigma_{21} - \mathbf{B}\Sigma_{11}.$$

Choosing $\mathbf{B} := \Sigma_{21}\Sigma_{11}^{-1}$ makes elements of $\mathbf{U}$ and $\mathbf{X}_1$ uncorrelated, with

$$\Sigma_\mathbf{U} := E\mathbf{U}\mathbf{U}' = \Sigma_{22} - \Sigma_{21}\Sigma_{11}^{-1}\Sigma_{12}.$$

$\Sigma_\mathbf{U}$ is positive definite, since the positive definiteness of Σ rules out exact linear relations between $\mathbf{X}_2$ and $\mathbf{X}_1$ and thus makes $V(\mathbf{a}'\mathbf{U}) = \mathbf{a}'\Sigma_\mathbf{U}\mathbf{a} > 0$ for all $\mathbf{a} \neq \mathbf{0}$. Now the k vector comprising $\mathbf{X}_1 - \boldsymbol{\mu}_1$ and $\mathbf{U}$ is an affine function of $\mathbf{X} - \boldsymbol{\mu}$, since

$$\begin{pmatrix} \mathbf{X}_1 - \boldsymbol{\mu}_1 \\ \mathbf{U} \end{pmatrix} = \begin{pmatrix} \mathbf{I}_{k_1} & \mathbf{0} \\ -\mathbf{B}\mathbf{I}_{k_1} & \mathbf{I}_{k_2} \end{pmatrix} \begin{pmatrix} \mathbf{X}_1 - \boldsymbol{\mu}_1 \\ \mathbf{X}_2 - \boldsymbol{\mu}_2 \end{pmatrix};$$

and the covariance matrix of this random vector has the positive-definite form

$$E\begin{pmatrix} \mathbf{X}_1 - \boldsymbol{\mu}_1 \\ \mathbf{U} \end{pmatrix}(\mathbf{X}_1' - \boldsymbol{\mu}_1', \mathbf{U}') = \begin{pmatrix} \Sigma_{11} & \mathbf{0} \\ \mathbf{0} & \Sigma_\mathbf{U} \end{pmatrix}.$$

Thus, the uncorrelated $\mathbf{U}$ and $\mathbf{X}_1$ are jointly normal and, therefore, independent. Now with $\mathbf{t} \in \Re^{k_2}$ the conditional c.f. of $\mathbf{X}_2$ given $\mathbf{X}_1$ is

$$\phi_{\mathbf{X}_2|\mathbf{X}_1}(\mathbf{t}) = E\left(e^{i\mathbf{t}'\mathbf{X}_2} \mid \mathbf{X}_1\right)$$
$$= e^{i\mathbf{t}'\left[\boldsymbol{\mu}_2 + \Sigma_{21}\Sigma_{11}^{-1}(\mathbf{X}_1 - \boldsymbol{\mu}_1)\right]} E\left(e^{i\mathbf{t}'\mathbf{U}} \mid \mathbf{X}_1\right)$$
$$= e^{i\mathbf{t}'\left[\boldsymbol{\mu}_2 + \Sigma_{21}\Sigma_{11}^{-1}(\mathbf{X}_1 - \boldsymbol{\mu}_1)\right]} Ee^{i\mathbf{t}'\mathbf{U}}$$
$$= e^{i\mathbf{t}'\left[\boldsymbol{\mu}_2 + \Sigma_{21}\Sigma_{11}^{-1}(\mathbf{X}_1 - \boldsymbol{\mu}_1)\right]} e^{-\mathbf{t}'\Sigma_\mathbf{U}\mathbf{t}/2},$$

which is of the form (4.44). The conditional distribution of $\mathbf{X}_2$ given $\mathbf{X}_1$ is therefore normal with mean vector and covariance matrix

$$E\left(\mathbf{X}_2 \mid \mathbf{X}_1 = \mathbf{x}_1\right) = \boldsymbol{\mu}_2 + \boldsymbol{\Sigma}_{21}\boldsymbol{\Sigma}_{11}^{-1}\left(\mathbf{x}_1 - \boldsymbol{\mu}_1\right)$$
$$V\left(\mathbf{X}_2 \mid \mathbf{X}_1 = \mathbf{x}_1\right) = \boldsymbol{\Sigma}_{\mathbf{U}} = \boldsymbol{\Sigma}_{22} - \boldsymbol{\Sigma}_{21}\boldsymbol{\Sigma}_{11}^{-1}\boldsymbol{\Sigma}_{12}.$$

Note that the conditional mean has the *affine* structure $\mathbf{a} + \mathbf{B}\mathbf{x}_1$ and that the conditional variance does not depend on $\mathbf{x}_1$.

In the bivariate case $\mathbf{X} := (X, Y)'$, it is easy to work out the conditional distribution of Y given $X = x$ directly from the p.d.f.s. Setting $x_1 = x$, $x_2 = y$, $\mu_1 = \mu_X$, $\mu_2 = \mu_Y$, $\sigma_1 = \sigma_X$, and $\sigma_2 = \sigma_Y$ in (4.45), algebra reduces the exponent of e from

$$-\frac{1}{2(1-\rho^2)}\left[\left(\frac{x-\mu_X}{\sigma_X}\right)^2 - 2\rho\left(\frac{y-\mu_Y}{\sigma_Y}\right)\left(\frac{x-\mu_X}{\sigma_X}\right) + \left(\frac{y-\mu_Y}{\sigma_Y}\right)^2\right]$$

to

$$-\frac{1}{2}\left[\left(\frac{y-\alpha-\beta x}{\sigma_Y\sqrt{1-\rho^2}}\right)^2 + \left(\frac{x-\mu_X}{\sigma_X}\right)^2\right], \qquad (4.47)$$

where $\beta := \rho\sigma_Y/\sigma_X$ and $\alpha := \mu_Y - \beta\mu_X$. The conditional p.d.f. is then

$$f_{Y|X}(y \mid x) = \frac{\frac{1}{2\pi\sigma_Y\sigma_X\sqrt{1-\rho^2}}\exp\left\{-\left[\frac{1}{2}\left(\frac{y-\alpha-\beta x}{\sigma_Y\sqrt{1-\rho^2}}\right)^2 + \frac{1}{2}\left(\frac{x-\mu_X}{\sigma_X}\right)^2\right]\right\}}{\frac{1}{\sqrt{2\pi}\sigma_X}\exp\left[-\frac{1}{2}\left(\frac{x-\mu_X}{\sigma_X}\right)^2\right]}$$

$$= \frac{1}{\sqrt{2\pi}\sigma_Y\sqrt{1-\rho^2}}\exp\left[-\frac{1}{2}\left(\frac{y-\alpha-\beta x}{\sigma_Y\sqrt{1-\rho^2}}\right)^2\right],$$

showing that $Y \sim N\left[\alpha + \beta x, \sigma_Y^2\left(1 - \rho^2\right)\right]$ conditional on $X = x$.

Exercise 4.63. *Work out the algebra to show that the equality holds in (4.47).*

Notice again that the conditional mean is affine in x and that the conditional variance is the same for all x—a condition sometimes referred to as "homoskedasticity". The conditional mean is called the *regression function* of Y on X. The relation between Y and X can be expressed as

$$Y = \alpha + \beta X + U, \qquad (4.48)$$

where $U := Y - (\alpha + \beta X) \sim N\left[0, \sigma_Y^2 (1 - \rho^2)\right]$ is independent of X. In the general k-variate case with $Y := X_1$ and $\mathbf{W} := (X_2, X_3, ..., X_k)'$ there is the comparable expression $Y = \alpha + \boldsymbol{\beta}'\mathbf{W} + U$ for some $(k-1)$-vector of constants $\boldsymbol{\beta}$. Here $E(Y \mid \mathbf{W} = \mathbf{w}) = \alpha + \boldsymbol{\beta}'\mathbf{w} \equiv \alpha + \mathbf{w}'\boldsymbol{\beta}$ is the *multivariate* linear regression function, and U is again univariate normal with zero mean and variance that does not depend on $\mathbf{w}$.

4.3.7.4 *A Characterization of the Multinormal Distribution*

We know that r.v.s $\{X_j\}_{j=1}^k$ on a probability space $(\Omega, \mathcal{F}, \mathbb{P})$ are distributed as multivariate normal if they have normal marginal distributions *and* are mutually independent. On the other hand, variables distributed as $N(\boldsymbol{\mu}, \boldsymbol{\Sigma})$ are not *necessarily* independent—indeed, they are not unless $\boldsymbol{\Sigma}$ is diagonal. How can one determine whether a vector-valued r.v. $\mathbf{X}$ is multivariate normal when its components are *not* independent? The following theorem provides the answer.

Theorem 4.1. *(Cramér–Wold) The r.v.* $\mathbf{X} \in \Re^k$ *with mean* $\boldsymbol{\mu}$ *and non-singular covariance matrix* $\boldsymbol{\Sigma}$ *is distributed as* $N(\boldsymbol{\mu}, \boldsymbol{\Sigma})$ *if and only if* $\mathbf{a}'\mathbf{X}$ *is distributed as* $N(\mathbf{a}'\boldsymbol{\mu}, \mathbf{a}'\boldsymbol{\Sigma}\mathbf{a})$ *for all constant vectors* $\mathbf{a} \in \Re^k$ *such that* $\mathbf{a}'\mathbf{a} > 0$.

Here the restriction $\mathbf{a}'\mathbf{a} > 0$ just rules out the degenerate case that $\mathbf{a} = \mathbf{0}$. The theorem tells us that a collection of jointly distributed r.v.s. is multivariate normal if and only if *all* (nondegenerate) linear functions of them are univariate normal.

Proof. That $\mathbf{X} \sim N(\boldsymbol{\mu}, \boldsymbol{\Sigma})$ implies $\mathbf{a}'\mathbf{X} \sim N(\mathbf{a}'\boldsymbol{\mu}, \mathbf{a}'\boldsymbol{\Sigma}\mathbf{a})$ follows from the normality of affine functions of multinormals. For the converse, note that $\mathbf{a}'\mathbf{X} \sim N(\mathbf{a}'\boldsymbol{\mu}, \mathbf{a}'\boldsymbol{\Sigma}\mathbf{a})$ for *all* non-null $\mathbf{a} \in \Re^k$ implies that $Ee^{it\mathbf{a}'\mathbf{X}} = e^{it\mathbf{a}'\boldsymbol{\mu} - t^2\mathbf{a}'\boldsymbol{\Sigma}\mathbf{a}/2}$ for each $t \in \Re$. But, taking $\mathbf{t} := t\mathbf{a}$, this means that $Ee^{it'\mathbf{X}} = e^{it'\boldsymbol{\mu} - t'\boldsymbol{\Sigma}t/2}$ for all $\mathbf{t} \in \Re^k$. The conclusion then follows from the uniqueness theorem for c.f.s. $\qquad\square$

Example 4.14. Exercise 4.62 showed that r.v.s X_1, X_2 with p.d.f., $f_{\mathbf{X}}(x_1, x_2) = \pi^{-1}e^{-x_1^2/2 - x_2^2/2}$ for $x_1 x_2 \geq 0$ are each standard normal even though the pair is not bivariate normal. In this case $\mathbf{a}'\mathbf{X} = a_1 X_1 + a_2 X_2$ is (standard) normal in the two cases $\mathbf{a} = (1, 0)'$ and $\mathbf{a} = (0, 1)'$, but the condition for Cramér-Wold fails since normality does not hold for *all* $\mathbf{a}$. In particular, it does not hold for $\mathbf{a} = (1, 1)'$, as can be seen by working out the m.g.f., $M_{X_1+X_2}(t) = Ee^{t(X_1+X_2)}$. This is

$$M_{X_1+X_2}(t) = \int_0^\infty \int_0^\infty e^{tx_1+tx_2} \frac{e^{-(x_1^2+x_2^2)/2}}{\pi} dx_2\, dx_1$$

$$+ \int_{-\infty}^0 \int_{-\infty}^0 e^{tx_1+tx_2} \frac{e^{-(x_1^2+x_2^2)/2}}{\pi} dx_2\, dx_1$$

$$= \left[\int_0^\infty \frac{e^{tx-x^2/2}}{\sqrt{\pi}} dx \right]^2 + \left[\int_{-\infty}^0 \frac{e^{tx-x^2/2}}{\sqrt{\pi}} dx \right]^2$$

$$= 2\left[e^{t^2/2} \int_0^\infty \frac{e^{-(x-t)^2/2}}{\sqrt{2\pi}} dx \right]^2 + 2\left[e^{t^2/2} \int_{-\infty}^0 \frac{e^{-(x-t)^2/2}}{\sqrt{2\pi}} dx \right]^2$$

$$= 2e^{t^2} \left[\left(\int_{-t}^\infty \frac{1}{\sqrt{2\pi}} e^{-z^2/2} dz \right)^2 + \left(\int_{-\infty}^{-t} \frac{1}{\sqrt{2\pi}} e^{-z^2/2} dz \right)^2 \right]$$

$$= 2e^{t^2} \left\{ [1 - \Phi(-t)]^2 + \Phi(-t)^2 \right\}$$

$$= 2e^{t^2} \left[\Phi(t)^2 + \Phi(-t)^2 \right],$$

which is not the m.g.f. of a normally distributed r.v.

Exercise 4.64. X_1, X_2 *have joint p.d.f.* $f_{\mathbf{X}}(x_1, x_2) = \pi^{-1} e^{-x_1^2/2 - x_2^2/2}$ *for* $x_1 x_2 \geq 0$.

(i) *Use the results of Example 4.14 to show that X_1 and X_2 have correlation $\rho = 1/\pi$.*

(ii) *Show that $Y_1 := X_1 - X_2$ and $Y_2 := X_1 + X_2$ are uncorrelated.*

(iii) *Show that Y_1, Y_2 are not independent by working out the joint p.d.f.*

4.3.7.5 *Brownian Motion*

Imagine a game that involves a sequence of plays, on each of which there is some uncertain monetary gain or loss. Specifically, starting with known value $W_0 = 0$, let the total winning after $t \in \mathbb{N}$ plays be $W_t = Z_1 + Z_2 + \dots + Z_t$, where the $\{Z_j\}$ are i.i.d. as $N(0,1)$. Clearly, the values of all of $W_0, W_1, \dots, W_t$ would be known at t, so $\{W_t\}_{t=0}^\infty$ as thus constructed is a discrete-time stochastic process adapted to an information structure (filtration) $\{\mathcal{F}_t\}_{t=0}^\infty$. Moreover, it has the following properties:

(i) $EW_t = 0$, $VW_t = t$, and $W_t \sim N(0, t)$ for $t \in \mathbb{N}$; thus, W_t has the same distribution as $\sqrt{t}Z$ where $Z \sim N(0,1)$.

(ii) If $\mathcal{F}_s$ represents the information available at $s < t$, then

$$E\left(W_t \mid \mathcal{F}_s\right) = E\left(W_s + Z_{s+1} + ... + Z_t \mid \mathcal{F}_s\right) = W_s.$$

Moreover, since $E\left|W_T\right| = \sqrt{T}E\left|Z\right| < \infty$ for any $T \in \aleph$, the process $\{W_t\}_{t=0}^{T}$ is a *martingale* adapted to $\{\mathcal{F}_t\}_{t=0}^{T}$.

(iii) For $s \leq t$ the covariance between W_s and W_t (given trivial information $\mathcal{F}_0$) is

$$E\left(W_s W_t\right) = E\left[E\left(W_s W_t \mid \mathcal{F}_s\right)\right] = EW_s^2 = s.$$

More generally, $Cov\left(W_t, W_u\right) = \min\left\{t, u\right\}$.

(iv) The joint distribution of $\mathbf{W}_{\{t_j\}} := \left(W_{t_1}, W_{t_2}, ... W_{t_n}\right)'$—the sequence of total winnings at $t_1 < t_2 < ... < t_n$—is multivariate normal with mean $\mathbf{0}$ and covariance matrix

$$E\mathbf{W}_{\{t_j\}}\mathbf{W}'_{\{t_j\}} = \begin{pmatrix} t_1 & t_1 & \cdots & t_1 \\ t_1 & t_2 & \cdots & t_2 \\ \vdots & \vdots & \ddots & \vdots \\ t_1 & t_2 & \cdots & t_n \end{pmatrix}.$$

Extending, let us consider now a process $\{W_t\}_{t \geq 0}$ that evolves in *continuous* time in such a way that

- $W_0 = 0$
- $W_u - W_t \sim N\left(0, u - t\right)$ for $0 \leq t < u$
- $W_u - W_t$ is independent of $W_s - W_r$ for $0 \leq r < s \leq t < u$.

These properties define a continuous-time process adapted to a filtration $\{\mathcal{F}_t\}_{t \geq 0}$. It is known variously as the *Wiener* process or as *Brownian motion*.[16] It is easy to see that properties (i)-(iv) of the discrete-time process $\{W_t\}_{t=0}^{\infty}$ carry over to continuous version $\{W_t\}_{t \geq 0}$, but there are many new (and more subtle) features also; for example,

(v) Sample paths $\{W_t\left(\omega\right)\}_{t \geq 0}$ of the process are almost surely *continuous*. This is a deep property that applies for all $t \geq 0$, but we can get a sense of its truth from the fact that for any *particular* t (and any $\varepsilon > 0$)

$$\lim_{\Delta t \to 0} \Pr\left(\left|W_{t+\Delta t} - W_t\right| < \varepsilon\right) = \lim_{\Delta t \to 0} \Pr\left(\left|Z\right|\sqrt{\Delta t} < \varepsilon\right)$$

$$= \lim_{\Delta t \to 0} \Pr\left(\left|Z\right| < \frac{\varepsilon}{\sqrt{\Delta t}}\right) = 1.$$

---

[16]Norbert Wiener (1923) presented the first systematic theoretical account of the process. Einstein, in one of his three celebrated 1905 papers, used it to model molecular vibration, to which he attributed the incessant, random movement of microscopic particles that botanist Robert Brown had reported early in the 19th century.

(vi) Sample paths are nowhere *differentiable*. Again, one gets a sense of this strong result by noting that

$$\lim_{\Delta t \to 0} \Pr\left(\frac{|W_{t+\Delta t} - W_t|}{\Delta t} > B \right) = \Pr(|Z| > B\sqrt{\Delta t}) \to 1$$

for any particular t and any $B < \infty$.

Figure 4.14 plots a computer-generated track that approximates one realization of a Brownian motion on $[0, t]$. Although the track has no jumps, it is so spiky and rough that it is impossible to construct a unique tangent line at any point. Despite these and other odd features (two of which will be described in Chapter 6), the Wiener process has found extensive use in applied modeling.[17] The applications usually involve extended versions of the "standard" process that allow for arbitrary initial value, nonzero average change over time, and more general volatility, as $\{Y_t = Y_0 + \alpha t + \beta W_t\}_{t \geq 0}$. A classic example of such an application in mathematical finance appears in Exercise 4.65 (p. 301). Here are two, somewhat more involved examples that show how the continuity and symmetry of the process can be exploited to derive distributions of maximal values and first-passage times. We will revisit and extend the second of these in Section 4.3.9.

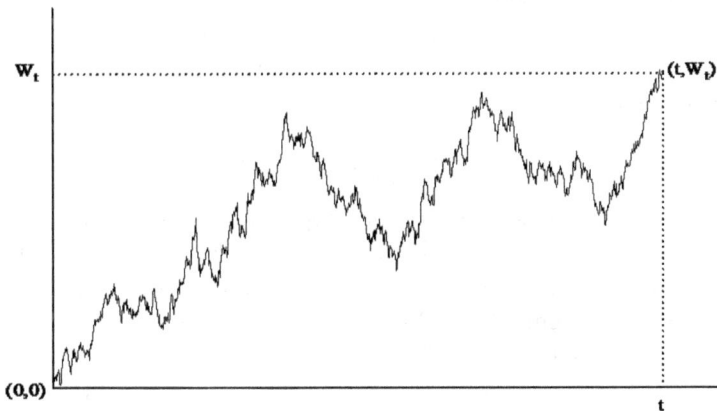

Fig. 4.14   A simulated realization of a standard Brownian motion over $[0, t]$.

---

[17]Karlin and Taylor (1975) give an accessible, non-measure-theoretic account. Among many advanced treatments of merit are Durrett (1984) and Karatzas and Shreve (1991).

Example 4.15. Let $\mathfrak{S}_T := \sup \{W_t : 0 \le t \le T\}$ be the maximum value attained by a standard Brownian motion that starts at the origin and evolves over the finite time span $[0, T]$. We will work out the distribution function $F_{\mathfrak{S}_T}(\mathfrak{s}) := \Pr(\mathfrak{S}_T \le \mathfrak{s})$. Fixing some $\mathfrak{s} \ge 0$, consider a sample path $\{W_t(\omega)\}_{0 \le t \le T}$ for which $\mathfrak{S}_T(\omega) > \mathfrak{s}$. The path being continuous, there is a $t_{\mathfrak{s}} \in (0, T)$ such that $W_{t_{\mathfrak{s}}}(\omega) = \mathfrak{s}$, and given the information $\mathcal{F}_{t_{\mathfrak{s}}}$ the random variable $W_T - W_{t_{\mathfrak{s}}} \equiv W_T - \mathfrak{s}$ is distributed symmetrically about 0. Thus,

$$\Pr(W_T - \mathfrak{s} \ge \mathfrak{s} - w \mid \mathcal{F}_{t_{\mathfrak{s}}}) = \Pr(W_T - \mathfrak{s} \le w - \mathfrak{s} \mid \mathcal{F}_{t_{\mathfrak{s}}}), w \in \Re, \mathfrak{s} \ge 0,$$

or, equivalently, $\Pr(W_T \ge 2\mathfrak{s} - w \mid \mathcal{F}_{t_{\mathfrak{s}}}) = \Pr(W_T \le w \mid \mathcal{F}_{t_{\mathfrak{s}}})$. But there is always *some* such $t_{\mathfrak{s}}$ whenever $\mathfrak{S}_T > \mathfrak{s}$, and so

$$\Pr(\mathfrak{S}_T > \mathfrak{s}, W_T \le w) = \Pr(\mathfrak{S}_T > \mathfrak{s}, W_T \ge 2\mathfrak{s} - w).$$

Taking $w = \mathfrak{s}$, this implies that

$$\Pr(\mathfrak{S}_T > \mathfrak{s}, W_T \le \mathfrak{s}) = \Pr(\mathfrak{S}_T > \mathfrak{s}, W_T \ge \mathfrak{s}) = \Pr(\mathfrak{S}_T > \mathfrak{s}, W_T > \mathfrak{s})$$

(since event $\{\omega : W_T(\omega) = \mathfrak{s}\}$ is $\mathbb{P}$-null) and hence, by the law of total probability, that

$$\Pr(\mathfrak{S}_T > \mathfrak{s}) = \Pr(\mathfrak{S}_T > \mathfrak{s}, W_T \le \mathfrak{s}) + \Pr(\mathfrak{S}_T > \mathfrak{s}, W_T > \mathfrak{s})$$
$$= 2\Pr(\mathfrak{S}_T > \mathfrak{s}, W_T > \mathfrak{s}).$$

Finally, since $\{\omega : W_T(\omega) > \mathfrak{s}\} \subset \{\omega : \mathfrak{S}_T(\omega) > \mathfrak{s}\}$, we conclude that

$$\Pr(\mathfrak{S}_T > \mathfrak{s}) = 2\Pr(W_T > \mathfrak{s}) = 2\Phi\left(-\frac{\mathfrak{s}}{\sqrt{T}}\right)\mathbf{1}_{[0,\infty)}(\mathfrak{s})$$

and hence that

$$F_{\mathfrak{S}_T}(\mathfrak{s}) = \left[2\Phi\left(\frac{\mathfrak{s}}{\sqrt{T}}\right) - 1\right]\mathbf{1}_{[0,\infty)}(\mathfrak{s}).$$

Example 4.16. It is now but a short step to work out the distribution of the *first-passage time* $T_{\mathfrak{s}} := \inf\{t : W_t \ge \mathfrak{s}\}$; that is, the first time at which the Brownian motion attains the fixed value $\mathfrak{s} > 0$. Clearly, $T_{\mathfrak{s}}$ is less than or equal to t if and only if the Brownian motion's maximum value on $[0, t]$ is greater than or equal to $\mathfrak{s}$. Thus,

$$F_{T_{\mathfrak{s}}}(t) := \Pr(T_{\mathfrak{s}} \le t) = \Pr(\mathfrak{S}_t \ge \mathfrak{s}) = 2\Phi\left(-\frac{\mathfrak{s}}{\sqrt{t}}\right)\mathbf{1}_{(0,\infty)}(t),$$

and so

$$f_{T_{\mathfrak{s}}}(t) = 2 \cdot \frac{d}{dt}\Phi\left(-\frac{\mathfrak{s}}{\sqrt{t}}\right)\mathbf{1}_{(0,\infty)}(t)$$
$$= \frac{\mathfrak{s}}{\sqrt{2\pi t^3}}e^{-\mathfrak{s}^2/(2t)}\mathbf{1}_{(0,\infty)}(t).$$

4.3.8   *The Lognormal Distribution*

We now introduce a flexible, continuous family of distributions derived from the normal that is supported on the positive reals. The *lognormal* model has been applied to many observable phenomena, including heights of male draftees, incomes of a population of households, values of business firms' assets, and ratios of future to current prices of financial assets. Under standard measurement conventions and institutional arrangements all of these have the property of being nonnegative. The defining characteristic of a lognormal r.v. is that its *logarithm* is normally distributed.

The standard form of the model is represented by the distribution of $Y := e^Z$ where $Z \sim N(0,1)$. Exercise 2.14 showed the p.d.f. of Y to be

$$f_Y(y) = f_Z(\log y) \left| \frac{d \log y}{dy} \right|$$

$$= \frac{1}{y\sqrt{2\pi}} e^{-(\log y)^2/2} \mathbf{1}_{(0,\infty)}(y).$$

Since $\log Y = Z$ is standard normal, Y itself is called a standard *lognormal* variate. ("Exponential-normal" would be a more apt term, but the "log" terminology is firmly established in the literature.)

Generalizing from the standard form, put $X := e^\mu Y^\sigma = e^{\mu + \sigma Z}$ with $\mu \in \Re$ and $\sigma \neq 0$, so that $\log X \sim N(\mu, \sigma^2)$. The change-of-variable formula now leads to

$$f_X(x) = \frac{1}{x\sqrt{2\pi\sigma^2}} e^{-(\log x - \mu)^2/(2\sigma^2)} \mathbf{1}_{[0,\infty)}(x).$$

The collection $\{f_X(x; \mu, \sigma^2) : \mu \in \Re, \sigma^2 > 0\}$ constitutes the (two-parameter) lognormal family. We write either $\log X \sim N(\mu, \sigma^2)$ or $X \sim LN(\mu, \sigma^2)$ to indicate that X has such a distribution. As for the gamma and Weibull distributions, there is also a three-parameter, location-shifted version.

Many properties of the lognormal are obvious from its connection with normals. The family is closed under transformations of the form $e^a X^b$ with $b \neq 0$, because the normal family is closed under affine transformations; i.e., $\log(e^a X^b) = a + b \log X \sim N(a + b\mu, b^2\sigma^2)$. Taking $a = 0, b = -1$ shows that the *inverse* of a lognormal variate is still lognormal. Moments of lognormals about the origin correspond to values of m.g.f.s of normals. Thus, if $\log X \sim N(\mu, \sigma^2)$,

$$\mu'_r := EX^r = Ee^{r \log X} = M_{\log X}(r) = e^{r\mu + r^2\sigma^2/2},$$

a result that holds for *all* $r \in \Re$. Taking $r = 1$ gives $EX = e^{\mu + \sigma^2/2}$ for the mean, while the variance is

$$VX = \mu_2' - (EX)^2 = e^{2\mu + 2\sigma^2} - \left(e^{\mu + \sigma^2/2}\right)^2 = e^{2\mu + \sigma^2}(e^{\sigma^2} - 1).$$

Although all moments of the lognormal are finite, the m.g.f. does not exist. (If this seems paradoxical, consider that, for any finite r, $\lim_{x \to \infty} x^{-r} e^{tx} = +\infty$ when $t > 0$.) Of course, the c.f. does exist, but its form is too complicated to be useful in applied work.

Extension to the multivariate case parallels that for the normal distribution. When $\log \mathbf{X} := (\log X_1, \log X_2, ..., \log X_k)' \sim N(\boldsymbol{\mu}, \boldsymbol{\Sigma})$, the vector-valued r.v. $\mathbf{X} := (X_1, X_2, ..., X_k)'$ has the multivariate lognormal distribution with parameters $\boldsymbol{\mu}, \boldsymbol{\Sigma}$. The joint p.d.f. is

$$f_{\mathbf{X}}(\mathbf{x}) = (2\pi)^{-k/2} |\boldsymbol{\Sigma}|^{-1/2} \left(\prod_{j=1}^{k} x_j^{-1}\right) \exp\left[-\frac{1}{2}(\log \mathbf{x} - \boldsymbol{\mu})' \boldsymbol{\Sigma}^{-1}(\log \mathbf{x} - \boldsymbol{\mu})\right]$$

for $\mathbf{x} \in (0, \infty)^k$. Abiding some abuse of notation, we write either $\log \mathbf{X} \sim N(\boldsymbol{\mu}, \boldsymbol{\Sigma})$ or $\mathbf{X} \sim LN(\boldsymbol{\mu}, \boldsymbol{\Sigma})$ to signify that $\mathbf{X}$ is multivariate lognormal.

Now suppose that $\log \mathbf{X} \sim N(\boldsymbol{\mu}, \boldsymbol{\Sigma})$ and that a_0 and $\mathbf{a} = (a_1, a_2, ..., a_k)'$ are constants with $\mathbf{a}'\mathbf{a} > 0$, and define $Q := e^{a_0} X_1^{a_1} X_2^{a_2} \cdots \cdots X_k^{a_k}$. Then

$$\log Q = a_0 + \sum_{j=1}^{k} a_j \log X_j = a_0 + \mathbf{a}' \log \mathbf{X} \sim N(a_0 + \mathbf{a}'\boldsymbol{\mu}, \mathbf{a}'\boldsymbol{\Sigma}\mathbf{a}),$$

so Q itself is (univariate) lognormal with parameters $a_0 + \mathbf{a}'\boldsymbol{\mu}, \mathbf{a}'\boldsymbol{\Sigma}\mathbf{a}$. Thus, the fact that affine transformations of multivariate normals are normal implies that products of arbitrary *powers* of joint lognormals are distributed as lognormal. This closure under multiplicative transformations makes the lognormal distribution particularly attractive in financial modeling.

Example 4.17. For $t \in (0, 1, 2, ...)$ let P_t be the price at time t of a "primary" financial asset (e.g., a share of stock, a bond, or a commodity held primarily for investment), and let $R_t := P_t/P_{t-1}$ be the one-period price relative. R_t represents the total return per currency unit from buying the asset at $t - 1$ and holding until t; and $\log R_t$ represents the average continuously compounded *rate* of return over the unit time interval, in the sense that $P_t \equiv P_{t-1} e^{\log R_t}$. Taking $t = 0$ as the current time and P_0 as the known current price, let us model future one-period returns $\{R_t\}_{t=1}^{\infty}$ conditional on $\mathcal{F}_0$ (the information available at $t = 0$) as independent and identically distributed lognormals: $\log R_t \sim N\left(\mu, \sigma^2\right)$. Thus, μ and σ^2 are the mean and variance of the continuously compounded rate of return in each period. Fixing some future time $T \in \aleph$, let us see what the model implies about the distribution of the time-T price. Writing

$$P_T \equiv P_0 \cdot \frac{P_1}{P_0} \cdots \cdots \frac{P_T}{P_{T-1}} = P_0 \prod_{t=1}^{T} R_t$$

shows that P_T is proportional to the product of independent lognormals and hence that $\log P_T = \log P_0 + \sum_{t=1}^{T} \log R_t$ is an affine transformation of multivariate normals. It follows that $P_T \sim LN(\log P_0 + T\mu, T\sigma^2)$.

If $\mu = .01$ and $\sigma^2 = .04$ for the one-period return, the probability that price would at least double in 12 periods is estimated to be

$$\Pr\left(\frac{P_{12}}{P_0} > 2\right) = \Pr\left(\log P_{12} - \log P_0 > \log 2\right)$$

$$= \Pr\left(\sum_{t=1}^{12} \log R_t > \log 2\right)$$

$$= \Pr\left(\frac{\sum_{t=1}^{12} \log R_t - 12(.01)}{\sqrt{12(.04)}} > \frac{\log 2 - 12(.01)}{\sqrt{12(.04)}}\right)$$

$$\doteq \Pr\left(Z > .827\right) = 1 - \Phi\left(.827\right)$$

$$\doteq .204.$$

The probability that an investor would *lose* at least half the initial investment is

$$\Pr\left(\frac{P_{12}}{P_0} < \frac{1}{2}\right) = \Pr\left(\sum_{t=1}^{12} \log R_t < -\log 2\right)$$

$$\doteq \Pr\left(Z < -1.174\right) = \Phi\left(-1.174\right)$$

$$\doteq .120.$$

Exercise 4.65. *The price of a share of stock evolves continuously through time as* $\{S_t = S_0 R_t\}_{t \geq 0}$, *where* $S_0 > 0$ *is the initial price and* R_t *is the stock's total return per dollar over* $[0, t]$. *We model the total return as* $R_t = \exp\left(\mu t + \sigma W_t\right)$, *where* $\{W_t\}_{t \geq 0}$ *is a standard Brownian motion; hence,* $R_0 = 1$ *and* $\log R_t \sim N\left(\mu t, \sigma^2 t\right)$ *for* $t > 0$. *The price of a default-free* *discount bond that matures and returns one currency unit at T evolves as $\left\{B_t = e^{r(t-T)}\right\}_{0 \leq t \leq T}$, where $r > 0$ is the continuously compounded riskless rate of interest—assumed to be the same for all t.*

(i) *Suppose investors in a financial market all understand the dynamics of $\{S_t\}$ and are "risk-neutral"—indifferent to risk. The stock's initial price should then be such that its* expected *total return per dollar equals the* sure *return per dollar of the riskless bond. In other words, S_0 should satisfy $ES_t/S_0 = B_t/B_0 = e^{rt}$. What value would μ have in such a market of risk-neutral, well-informed investors?*

(ii) *A T-expiring European-style put option on the stock gives its owner the right—but not the obligation—to sell one share of the stock at a predetermined "strike" price $K > 0$ at a fixed future date T. The option would be "exercised" at T if $S_T < K$, in which case it would be worth $K - S_T$; otherwise it would expire and have no value. Thus, the value of the put at T would be $P_T := (K - S_T)^+ \equiv \max\{K - S_T, 0\}$. Find an expression for P_0, the put's initial value in a risk-neutral market. (The solution is the famous formula of Black and Scholes (1973) for the price of a European put on a stock that pays no dividend. The validity of the formula does not actually require investors to be risk neutral, since in a market free from opportunities for* arbitrage—*trades that yield sure, costless gains—such "derivative" assets would be priced as though investors were indifferent to risk.)*

(iii) *A T-expiring European-style call option on the stock gives its owner the right to buy one share of the stock at strike price K at future date T. Its value at that date would thus be $C_T = (S_T - K)^+$. Find C_0, the call's initial value in a risk-neutral market. (Hint: Show that $C_T - P_T \equiv S_T - K$ with probability one.)*

(iv) *Still reckoning in terms of a risk-neutral market, find the initial value of a T-expiring straddle that can be sold at T for $S_T := |S_T - K|$.*

(v) *One who has a "long" position in a T-expiring futures contract has an obligation to purchase an underlying asset at time T for a predetermined price, regardless of the asset's market price at that time. Still reckoning in terms of a risk-neutral market, find the initial value F_0 of a position that commits one to pay K currency units for one share of the stock at T. (Again, the resulting expression for the value of a futures contract on a financial asset with zero "cost of carry" does not actually require risk neutrality.)*

4.3.9 The Inverse Gaussian Distribution

Recalling Example 4.16, the p.d.f. of the first time $T_{\mathfrak{s}}$ that a standard Brownian motion $\{W_t\}_{t \geq 0}$ attains a value $\mathfrak{s} > 0$ is

$$f_{T_{\mathfrak{s}}}(t) = \frac{\mathfrak{s}}{\sqrt{2\pi t^3}} \exp\left(-\frac{\mathfrak{s}^2}{2t}\right) \mathbf{1}_{(0,\infty)}(t).$$

This was found by showing that the distribution of the maximal value on $[0, t]$, $\mathfrak{S}_t := \sup\{W_s : 0 \leq s \leq t\}$, satisfies

$$\Pr(\mathfrak{S}_t \geq \mathfrak{s}) = 2\Pr(W_t \geq \mathfrak{s}) = 2\Phi\left(-\frac{\mathfrak{s}}{\sqrt{t}}\right) \mathbf{1}_{(0,\infty)}(\mathfrak{s})$$

and hence that $F_{T_{\mathfrak{s}}}(t) := \Pr(T_{\mathfrak{s}} \leq t) = \Pr(\mathfrak{S}_t \geq \mathfrak{s})$ for $t > 0$. Now consider the corresponding situation for a *nonstandard* Brownian motion $\{Y_t = \delta t + \gamma W_t\}_{t \geq 0}$ that has nonzero drift and arbitrary volatility $\gamma > 0$, but whose initial value still equals 0. The effect of the volatility parameter is trivial; but when $\delta \neq 0$ the distribution of increments $Y_t - Y_s$ is no longer symmetric about the origin, and this alters the result substantially. Specifically, we now have

$$F_{T_{\mathfrak{s}}}(t) = \left[e^{2\delta\mathfrak{s}/\gamma^2}\Phi\left(-\frac{\delta t + \mathfrak{s}}{\gamma\sqrt{t}}\right) + \Phi\left(\frac{\delta t - \mathfrak{s}}{\gamma\sqrt{t}}\right)\right] \mathbf{1}_{(0,\infty)}(t) \qquad (4.49)$$

and

$$f_{T_{\mathfrak{s}}}(t) = F'_{T_{\mathfrak{s}}}(t) = \frac{\mathfrak{s}}{\sqrt{2\pi\gamma^2 t^3}} \exp\left[-\frac{(\mathfrak{s} - \delta t)^2}{2\gamma^2 t}\right] \mathbf{1}_{(0,\infty)}(t),$$

a result first derived by Schrödinger (1915).[18]

Now putting $\delta = \mathfrak{s}/\mu$ with $\mu > 0$ and $\gamma = \mathfrak{s}/\sqrt{\lambda}$ with $\lambda > 0$, and letting x take the place of t, we have the usual forms of the c.d.f. and p.d.f. of the *inverse Gaussian distribution*:

$$F_X(x; \mu, \lambda) = \left\{e^{2\lambda/\mu}\Phi\left[-\sqrt{\frac{\lambda}{x}}\left(\frac{x}{\mu}+1\right)\right] + \Phi\left[\sqrt{\frac{\lambda}{x}}\left(\frac{x}{\mu}-1\right)\right]\right\} \mathbf{1}_{(0,\infty)}(x)$$

$$f_X(x; \mu, \lambda) = \sqrt{\frac{\lambda}{2\pi x^3}} \exp\left[-\frac{\lambda(x-\mu)^2}{2\mu^2 x}\right] \mathbf{1}_{(0,\infty)}(x). \qquad (4.50)$$

For short, we write $X \sim IG(\mu, \lambda)$ if r.v. X has density (4.50). The family $\{f_X(x; \mu, \lambda) : \mu > 0, \lambda > 0\}$ is another member of the exponential

---

[18]A more accessible source for the derivation is Epps (2007, pp. 342-343).

class. The relation to the Gaussian (normal) distribution is "inverse" in the sense that (4.50) applies to the time for a Brownian motion to cover a given distance, whereas the distance covered within a given time is Gaussian. Since the distribution was derived in a different context by Wald (1944), it is also known as the *Wald distribution*.

To work out the m.g.f., put $\phi := \lambda/\mu$ and write (4.50) in the form

$$f_X(x; \mu, \phi) = \sqrt{\frac{\mu\phi}{2\pi x^3}} \exp\left[-\phi\left(\frac{x}{2\mu} - 1 + \frac{\mu}{2x}\right)\right] \mathbf{1}_{(0,\infty)}(x).$$

Then

$$M_X(t) = \int_0^\infty e^{tx} f_X(x; \mu, \phi) \cdot dx$$

$$= \int_0^\infty \sqrt{\frac{\mu\phi}{2\pi x^3}} \exp\left\{-\phi\left[\frac{x(1 - 2\mu t/\phi)}{2\mu} - 1 + \frac{\mu}{2x}\right]\right\} \cdot dx$$

$$= e^{-\phi\left(1 - \sqrt{1 - 2\mu t/\phi}\right)} \int_0^\infty \sqrt{\frac{\mu'\phi'}{2\pi x^3}} \exp\left[-\phi'\left(\frac{x}{2\mu'} - 1 + \frac{\mu'}{2x}\right)\right] \cdot dx,$$

where $\mu' := \mu(1 - 2\mu t/\phi)^{-1/2}$ and $\phi' := \phi(1 - 2\mu t/\phi)^{1/2}$ for $t < \phi/(2\mu)$. The integrand corresponds to $f_X(x; \mu', \phi')$ and thus integrates to unity, and with λ/μ again in place of ϕ we have

$$M_X(t) = \exp\left[\frac{\lambda}{\mu}\left(1 - \sqrt{1 - 2\mu^2 t/\lambda}\right)\right] \tag{4.51}$$

for $t < \lambda/(2\mu^2)$. Thus, all moments exist, and this statement applies even to negative moments. One can determine from c.g.f. $\log M_X(t)$ that $EX = \kappa_1 = \mu$ and $VX = \kappa_2 = \mu^3/\lambda$, that the coefficient of skewness is $\alpha_3 = 3\sqrt{\mu/\lambda}$—always positive, and that the coefficient of kurtosis is $\alpha_4 = 3 + 15\mu/\lambda$. Parameter λ affects both the scale and the overall shape, as shown by the plots of $f_X(x; 1, \lambda)$ in Fig. 4.15.

Exercise 4.66. *Putting* $Y := \mu^2 X^{-1}$, *show that* $f_Y(y) = \mu^{-1} y f_X(y; \mu, \lambda)$ *and conclude that* $EX^{-k} = \mu^{-2k-1} EX^{k+1}$. *Using this result, show that* $EX^{-1} = \mu^{-1} + \lambda^{-1}$ *and* $EX^{-2} = \mu^{-2} + 3\mu^{-1}\lambda^{-1} + 3\lambda^{-2}$.

Exercise 4.67. *(i) Show that the m.g.f. of* X^{-1} *is*

$$M_{X^{-1}}(t) = \left(1 - \frac{2t}{\lambda}\right)^{-1/2} \exp\left[\frac{\lambda}{\mu}\left(1 - \sqrt{1 - 2t/\lambda}\right)\right] \tag{4.52}$$

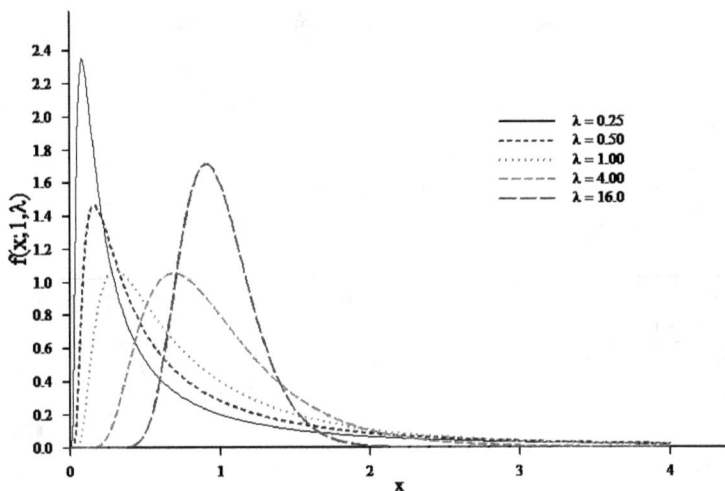

Fig. 4.15 Inverse Gaussian p.d.f.s with $\mu = 1$ and $\lambda \in \{.25, .5, 1.0, 4.0, 16.0\}$.

for $t < \lambda/2$. Then (ii) use this result to verify that $EX^{-1} = \mu^{-1} + \lambda^{-1}$ and (iii) show that the distribution of X^{-1} converges to $\Gamma(1/2, 2/\lambda)$ as $\mu \to \infty$.

Exercise 4.68. *In canonical form for the exponential class the inverse Gaussian p.d.f. is $f(x; \eta_1, \eta_2) = a(\boldsymbol{\eta}) c(x) e^{-\eta_1 x - \eta_2 x^{-1}}$, with $c(x) \propto x^{-3/2} \mathbf{1}_{(0,\infty)}(x)$ and $\eta_1 = \lambda/(2\mu^2), \eta_2 = \lambda/2$. Show that the natural parameter space $\mathbf{H} := \left\{ \boldsymbol{\eta} : \int c(x) e^{-\eta_1 x - \eta_2 x^{-1}} \cdot dx < \infty \right\}$ is $[0, \infty) \times (0, \infty)$, a set that contains some of its boundary points.*

4.3.10 Summary of Continuous Models

Table 4.2 summarizes the features of the continuous models described in this chapter. The annotation "n.s.f." stands for "no simple form", and "$\mathbf{B} >> 0$" means that $\mathbf{B}$ is positive definite. An asterisk after the name indicates that the family, in the form given here, is in the exponential class. Note that $\mathbb{X}$ represents the "canonical" support—the set of support *points*. It is implicit that $f(x) = 0$ at any $x \in \mathbb{X}$ at which the form in column 2 is invalid—e.g., at $x = 0$ when $X \sim \Gamma(\alpha, \beta)$ with $\alpha < 1$.

Table 4.2 Common continuous distributions.

Name/ Abbrev.	P.d.f. $f(x)$	Support X	Parameter Space	Mean, μ Variance, σ^2	C.f. $\phi(t)$
Uniform $U(\alpha,\beta)$	$(\beta-\alpha)^{-1}$	$[\alpha,\beta]$	$\alpha<\beta$	$\frac{\alpha+\beta}{2}$ $\frac{(\beta-\alpha)^2}{12}$	$\frac{e^{it\beta}-e^{it\alpha}}{it(\beta-\alpha)}$
Gamma* $\Gamma(\alpha,\beta)$	$\frac{x^{\alpha-1}e^{-x/\beta}}{\Gamma(\alpha)\beta^\alpha}$	$\Re_+$	$\alpha>0$ $\beta>0$	$\alpha\beta$ $\alpha\beta^2$	$\frac{1}{(1-it\beta)^\alpha}$
Exponential* $\Gamma(1,\beta)$	$\beta^{-1}e^{-x/\beta}$	$\Re_+$	$\beta>0$	β β^2	$\frac{1}{1-it\beta}$
Beta* $\mathcal{B}e(\alpha,\beta)$	$\frac{x^{\alpha-1}(1-x)^{\beta-1}}{\mathcal{B}(\alpha,\beta)}$	$[0,1]$	$\alpha>0$ $\beta>0$	$\frac{\alpha}{\alpha+\beta}$ $\frac{\alpha\beta}{(\alpha+\beta+1)(\alpha+\beta)^2}$	n.s.f.
Weibull $W(\alpha,\beta)$	$\frac{\alpha}{\beta}x^{\alpha-1}e^{-x^\alpha/\beta}$	$\Re_+$	$\alpha>0$ $\beta>0$	$\beta^{\frac{1}{\alpha}}\Gamma\left(\frac{\alpha+1}{\alpha}\right)$ $\beta^{\frac{2}{\alpha}}\Gamma\left(\frac{\alpha+2}{\alpha}\right)-\mu^2$	n.s.f.
Laplace $L(\alpha,\beta)$	$\frac{e^{-|x-\alpha|/\beta}}{2\beta}$	$\Re$	$\alpha\in\Re$ $\beta>0$	α $2\beta^2$	$\frac{e^{it\alpha}}{1+t^2\beta^2}$
Normal* $N(\alpha,\beta^2)$	$\frac{e^{-(x-\alpha)^2/(2\beta^2)}}{\sqrt{2\pi\beta^2}}$	$\Re$	$\alpha\in\Re$ $\beta>0$	α β^2	$e^{it\alpha-t^2\beta^2/2}$
Multinormal* $N(\boldsymbol{\alpha},\mathbf{B})$	$\frac{e^{-(\mathbf{x}-\boldsymbol{\alpha})'\mathbf{B}^{-1}(\mathbf{x}-\boldsymbol{\alpha})/2}}{\sqrt{(2\pi)^k|\mathbf{B}|}}$	$\Re^k$	$\boldsymbol{\alpha}\in\Re^k$ $\mathbf{B}>>0$	$\boldsymbol{\alpha}$ $\mathbf{B}$	$e^{it'\boldsymbol{\alpha}-t'\mathbf{B}t/2}$
Lognormal* $LN(\alpha,\beta^2)$	$\frac{e^{-(\log x-\alpha)^2/\left(2\beta^2\right)}}{x\sqrt{2\pi\beta^2}}$	$\Re_+$	$\alpha\in\Re$ $\beta>0$	$e^{\alpha+\beta^2/2}$ $\mu^2(e^{\beta^2}-1)$	n.s.f.
Inv. Gaussian* $IG(\alpha,\beta)$	$\frac{e^{-\beta(x-\alpha)^2/\left(2\alpha^2 x\right)}}{\sqrt{2\pi x^3/\beta}}$	$\Re_+$	$\alpha>0$ $\beta>0$	α $\frac{\alpha^3}{\beta}$	$e^{\frac{\beta}{\alpha}\left(1-\sqrt{1-2it\frac{\alpha^2}{\beta}}\right)}$

4.4 A Concluding Example

Part I of this work has dealt entirely with the theory of probability, albeit with a focus on those parts of the theory that constitute the essential foundation for statistics. In the next part we begin the study of statistical theory *per se*, but we conclude the present part with a capstone example that applies many of the tools that have been learned.

Let $\{X_j\}_{j=1}^n$, $n\geq 2$, be independent and identically distributed (i.i.d.) *continuous* r.v.s with c.d.f. F and p.d.f. f, and let $\{X_{(j:n)}\}_{j=1}^n$ denote the same variables when arranged in increasing order of their realizations. That is, $X_{(1:n)}(\omega)$ will represent the minimum of $\{X_j(\omega)\}_{j=1}^n$ once the experiment has concluded with outcome ω, and $X_{(n:n)}(\omega)$ will represent the maximum. (In the sampling context taken up in Chapter 5 $\{X_{(j:n)}\}_{j=1}^n$ are regarded as the *order statistics* of a random sample.) For our concluding exercise, let us work out a general expression for the coefficient of correlation between $X_{(1:n)}$ and $X_{(n:n)}$ and apply it in two special cases.

To begin, we will need to express the joint distribution of $X_{(1:n)}$ and $X_{(n:n)}$, $F_{1,n}(x, y) := \Pr\left(X_{(1:n)} \leq x, X_{(n:n)} \leq y\right)$, and the two marginals, $F_1(x) := \Pr\left(X_{(1:n)} \leq x\right)$ and $F_n(y) := \Pr\left(X_{(n:n)} \leq y\right)$. The marginals are easy. First, recognize that the largest of the $\{X_j\}$ is no larger than some $y \in \Re$ if and only if *all* the $\{X_j\}$ are no larger than y. Thus,

$$\begin{aligned} F_n(y) &= \Pr\left(X_1 \leq y, X_2 \leq y, ..., X_n \leq y\right) \\ &= \Pr\left(X_1 \leq y\right)\Pr\left(X_2 \leq y\right) \cdot \cdots \cdot \Pr\left(X_n \leq y\right) \\ &= F(y)^n. \end{aligned}$$

Differentiating, we find the p.d.f. of $X_{(n:n)}$ at $y \in \Re$ to be

$$f_n(y) = nF(y)^{n-1} f(y).$$

Similarly, the event that smallest value $X_{(1:n)}$ is no larger than some $x \in \Re$ is the complement of the event that all the $\{X_j\}$ are strictly greater than x, so

$$\begin{aligned} F_1(x) &= 1 - \Pr\left(X_1 > x, X_2 > x, ..., X_n > x\right) \\ &= 1 - \Pr\left(X_1 > x\right)\Pr\left(X_2 > x\right) \cdot \cdots \cdot \Pr\left(X_n > x\right) \\ &= 1 - [1 - F(x)]^n \end{aligned}$$

and

$$f_1(x) = n[1 - F(x)]^{n-1} f(x).$$

For the joint distribution we apply the law of total probability to get

$$\Pr\left(X_{(n:n)} \leq y\right) = \Pr\left(X_{(1:n)} \leq x, X_{(n:n)} \leq y\right) + \Pr\left(X_{(1:n)} > x, X_{(n:n)} \leq y\right),$$

or

$$\begin{aligned} F_n(y) &= F_{1,n}(x, y) + \Pr\left(x < X_1 \leq y, x < X_2 \leq y, ..., x < X_n \leq y\right) \\ &= F_{1,n}(x, y) + [F(y) - F(x)]^n \end{aligned}$$

for $x < y$. Thus,

$$F_{1,n}(x, y) = \{F_n(y) - [F(y) - F(x)]^n\}\, \mathbf{1}_{\{(x,y):x<y\}}(x, y).$$

Differentiating with respect to x and y, we have for the joint p.d.f.

$$f_{1,n}(x, y) = n(n-1)[F(y) - F(x)]^{n-2} f(x) f(y)\, \mathbf{1}_{\{(x,y):x<y\}}(x, y).$$

Now the coefficient of correlation is

$$\begin{aligned} \rho_{1,n} :&= \frac{Cov\left(X_{(1:n)}, X_{(n:n)}\right)}{\sqrt{VX_{(1:n)}VX_{(n:n)}}} \\ &= \frac{EX_{(1:n)}X_{(n:n)} - EX_{(1:n)}EX_{(n:n)}}{\sqrt{VX_{(1:n)}VX_{(n:n)}}}. \end{aligned} \tag{4.53}$$

The means and variances and the product moment are found as follows:

$$EX_{(1:n)} = n \int_{-\infty}^{\infty} x \left[1 - F(x)\right]^{n-1} f(x) \cdot dx$$

$$EX_{(n:n)} = n \int_{-\infty}^{\infty} y F(y)^{n-1} f(y) \cdot dy$$

$$VX_{(1:n)} = n \int_{-\infty}^{\infty} x^2 \left[1 - F(x)\right]^{n-1} f(x) \cdot dx - \left(EX_{(1:n)}\right)^2$$

$$VX_{(n:n)} = n \int_{-\infty}^{\infty} y^2 F(y)^{n-1} f(y) \cdot dy - \left(EX_{(n:n)}\right)^2$$

$$EX_{(1:n)} X_{(n:n)} = n(n-1) \int_{-\infty}^{\infty} \int_{-\infty}^{y} xy \left[F(y) - F(x)\right]^{n-2} f(x) f(y) \cdot dx\, dy.$$

Working out the integrals with the appropriate model for F and plugging into (4.53) complete the problem.

As one specific example, take $\{X_j\}_{j=1}^{n}$ to be i.i.d. as uniform on $[0,1]$, so that $F(x) = x \mathbf{1}_{[0,1]}(x) + \mathbf{1}_{(1,\infty)}(x)$ and $f(x) = \mathbf{1}_{[0,1]}(x)$. Then the means can be worked out as

$$EX_{(n:n)} = n \int_0^1 y^n \cdot dy = n(n+1)^{-1}$$

and

$$
\begin{aligned}
EX_{(1:n)} &= n \int_0^1 x (1-x)^{n-1} \cdot dx \\
&= n \left[-\frac{1}{n} x (1-x)^n \right]_0^1 + n \int_0^1 \frac{1}{n} (1-x)^n \cdot dx \\
&= (n+1)^{-1}.
\end{aligned}
$$

Alternatively, we could have just written down the answers without any work at all by recognizing that the realizations of $\{X_j\}_{j=1}^{n}$ divide $[0,1]$ into $n+1$ segments and that, by symmetry, the segments must be of the same length on average. The variances, however, do require a little work:

$$
\begin{aligned}
VX_{(n:n)} &= n \int_0^1 y^{n+1} \cdot dy - n^2 (n+1)^{-2} \\
&= n(n+1)^{-2} (n+2)^{-1} \\
VX_{(1:n)} &= n \int_0^1 x^2 (1-x)^{n-1} \cdot dx - (n+1)^{-2}.
\end{aligned}
$$

The integral in the last line can be handled with two integrations by parts, but there is an easier way. Note that if $X \sim U(0,1)$ then $Y := 1 - X$

has precisely the same distribution. Since the minimum of n realizations of X corresponds to the maximum of n realizations of Y, it follows that $EX_{(1:n)} = 1 - EX_{(n:n)}$ (as already verified directly) and that $VX_{(1:n)} = VX_{(n:n)}$. Finally, the product moment is

$$EX_{(1:n)}X_{(n:n)} = n(n-1) \int_0^1 \int_0^y xy \, (y-x)^{n-2} \cdot dx \, dy$$

$$= n \int_0^1 y \left[-x \, (y-x)^{n-1} \Big|_0^1 + \int_0^y (y-x)^{n-1} \cdot dx \right] \cdot dy$$

$$= \int_0^1 y^{n+1} \cdot dy$$

$$= (n+2)^{-1},$$

and the covariance is

$$Cov\left(X_{(1:n)}, X_{(n:n)}\right) = (n+2)^{-1} - n \, (n+1)^{-2} = (n+1)^{-2} \, (n+2)^{-1}.$$

Putting the pieces together gives the pleasingly simple result

$$\rho_{1,n} = \frac{(n+1)^{-2} \, (n+2)^{-1}}{n(n+1)^{-2} \, (n+2)^{-1}} = \frac{1}{n}.$$

Since for any $\alpha < \beta$ the r.v. $Y := \alpha + (\beta - \alpha) X$ is distributed as $U(\alpha, \beta)$ when $X \sim U(0,1)$, and since the correlation coefficient is invariant under changes in location and scale, the last result applies to all members of the uniform family.

With a little reflection, it does make sense that the correlation is positive, for if the smallest of the $\{X_j\}$ happens to be above its average, then the largest is more apt to be as well, and *vice versa*. Indeed, it is easy to see that the conditional expectations are

$$E\left(X_{(n:n)} \mid X_{(1:n)}\right) = 1 - n^{-1} \left(1 - X_{(1:n)}\right)$$

$$E\left(X_{(1:n)} \mid X_{(n:n)}\right) = n^{-1} X_{(n:n)},$$

both of which are linear and increasing in values of the other variable.

Finding the correlation of the minimum and maximum of *two* independent uniforms—the case $n = 2$—is made easier by noting that $g(X_{(1:2)}, X_{(2:2)}) \equiv g(X_1, X_2)$ for any function g that is permutation invariant; i.e., for g such that $g(x, y) = g(y, x)$. Thus,

$$EX_{(1:2)}X_{(2:2)} = EX_1X_2 = EX_1 \cdot EX_2 = \frac{1}{4}$$

$$E\left(X_{(1:2)} + X_{(2:2)}\right) = E(X_1 + X_2) = 1$$

$$EX_{(1:2)} = \frac{1}{3}$$

$$EX_{(2:2)} = \frac{2}{3}$$

$$Cov\left(X_{(1:2)}, X_{(2:2)}\right) = \frac{1}{4} - \frac{1}{3} \cdot \frac{2}{3} = \frac{1}{36}$$

$$VX_{(1:2)} = VX_{(2:2)} = 2\int_0^1 y^3 \cdot dy - \frac{4}{9} = \frac{2}{36}$$

$$\rho_{1,2} = \frac{1}{36} \div \frac{2}{36} = \frac{1}{2}.$$

Using permutation invariance, we can also find the correlation of the minimum and maximum of two independent normals. If Z_1 and Z_2 are i.i.d. as $N(0, 1)$ and $Z_{(1:2)}$ and $Z_{(2:2)}$ are the minimum and maximum, then $E\left(Z_{(1:2)} + Z_{(2:2)}\right)^2 = E(Z_1 + Z_2)^2 = 2$ and $EZ_{(1:2)}Z_{(2:2)} = EZ_1Z_2 = 0$. Since $Y := -Z$ has the same distribution as Z, it follows that $Z_{(1:2)} = -Z_{(2:2)}$, so $EZ_{(1:2)}^2 = EZ_{(2:2)}^2 = 1$. To find the expected values, we have $f_2(z) = 2\Phi(z) f_Z(z)$ as the p.d.f. of $Z_{(2:2)}$, where Φ and f_Z are the standard normal c.d.f. and p.d.f. Thus,

$$EZ_{(2:2)} = 2\int_{-\infty}^{\infty} \Phi(z) z f_Z(z) \cdot dz$$

$$= 2\int_{-\infty}^{\infty} \Phi(z) z \frac{1}{\sqrt{2\pi}} e^{-z^2/2} \cdot dz$$

$$= -2\int_{-\infty}^{\infty} \Phi(z) \cdot df_Z(z)$$

$$= -2\Phi(z) f_Z(z)\big|_{-\infty}^{\infty} + 2\int_{-\infty}^{\infty} f_Z(z)^2 \cdot dz$$

$$= 0 + \frac{2}{\sqrt{2\pi}} \int_{-\infty}^{\infty} \frac{1}{\sqrt{2\pi}} e^{-z^2} \cdot dz$$

$$= \frac{1}{\sqrt{\pi}}$$

and $EZ_{(1:2)} = -1/\sqrt{\pi}$. The variances are $VZ_{(1:1)} = VZ_{(2:2)} = EZ^2_{(2:2)} - (EZ_{(2:2)})^2 = 1 - 1/\pi$ and the covariance is $Cov\left(Z_{(1:2)}, Z_{(2:2)}\right) = -EZ_{(1:2)}EZ_{(2:2)} = 1/\pi$, so

$$\rho_{1,2} = \frac{1/\pi}{1 - 1/\pi} = \frac{1}{\pi - 1},$$

a (striking!) result that applies to all members of the normal family.

4.5 Solutions to Exercises

1. Writing $y\binom{n}{y}$ as $n!/\left[(n-y)!(y-1)!\right]$, the sum is

$$\sum_{y=1}^{n} \frac{n!}{(n-y)!(y-1)!} \theta^y (1-\theta)^{n-y}$$

$$= n\theta \sum_{y=1}^{n} \frac{(n-1)!}{[(n-1)-(y-1)]!(y-1)!} \theta^{y-1} (1-\theta)^{(n-1)-(y-1)}$$

$$= n\theta.$$

2. The m.g.f. is

$$M(t) = \sum_{y=0}^{n} e^{ty} \binom{n}{y} \theta^y (1-\theta)^{n-y}$$

$$= \sum_{y=0}^{n} \binom{n}{y} \left(\theta e^t\right)^y (1-\theta)^{n-y}$$

$$= \left[1 + \theta(e^t - 1)\right]^n,$$

using the binomial formula to reduce the sum. Differentiating gives

$$M'(t) = n\theta \left[1 + \theta(e^t - 1)\right]^{n-1} e^t$$

$$M''(t) = M'(t) + n(n-1)\theta^2 \left[1 + \theta(e^t - 1)\right]^{n-2} e^{2t}$$

$$M'''(t) = M''(t) + 2n(n-1)\theta^2 \left[1 + \theta(e^t - 1)\right]^{n-2} e^{2t}$$
$$+ n(n-1)(n-2)\theta^3 \left[1 + \theta(e^t - 1)\right]^{n-3} e^{3t}.$$

Evaluating at $t = 0$,

$$M'(0) = \mu = n\theta$$

$$M''(0) = \mu'_2 = n\theta + n(n-1)\theta^2$$

$$M'''(0) = \mu'_3 = n\theta + 3n(n-1)\theta^2 + n(n-1)(n-2)\theta^3.$$

The variance is thus $\sigma^2 = \mu_2' - \mu^2 = n\theta(1 - \theta)$, and the third central moment is $\mu_3 = \mu_3' - 3\mu_2'\mu + 2\mu^3 = n\theta(1 - \theta)(1 - 2\theta)$. The coefficient of skewness, $\alpha_3 = \mu_3/\sigma^3$, is therefore

$$\alpha_3 = \frac{n\theta(1 - \theta)(1 - 2\theta)}{[n\theta(1 - \theta)]^{3/2}}$$

$$= \frac{1 - 2\theta}{\sqrt{n\theta(1 - \theta)}}.$$

3. Modeling Y as $B(20, \frac{1}{36})$, we have $\Pr(Y \geq 1) = 1 - \Pr(Y = 0) = 1 - \binom{20}{0} \left(\frac{1}{36}\right)^0 \left(\frac{35}{36}\right)^{20} \doteq 0.431$.

4. Taking the number of heads in 10 flips of a coin to be distributed as $B(10, \frac{1}{2})$,

$$\Pr(Y \leq 2) + \Pr(Y \geq 8)$$
$$= \frac{1}{2^{10}} \left[\binom{10}{0} + \binom{10}{1} + \binom{10}{2} + \binom{10}{8} + \binom{10}{9} + \binom{10}{10} \right]$$
$$= \frac{112}{2^{10}}.$$

5. (i) There is no chance that you will break a tie if the number of other voters is odd.

(ii) The probability that there will be n "yes" votes and n "no" votes in independent trials is $P(p) := \binom{2n}{n} p^n (1 - p)^n$.

(iii) $P(p)$ has a unique maximum at $p = .5$; i.e., when other voters appear to be most ambivalent.

6. $p_S(t) = Et^{Y_1 + Y_2} = Et^{Y_1} Et^{Y_2} = p_1(t)p_2(t)$ by independence, so

$$p_S(t) = [1 + \theta(t - 1)]^{n_1} [1 + \theta(t - 1)]^{n_2}$$
$$= [1 + \theta(t - 1)]^{n_1 + n_2},$$

which is the p.g.f. of $B(n_1 + n_2, \theta)$. By induction, the sum of any finite number of independent binomials with the same parameter θ is binomially distributed.

7. If $n < m$ at least one of the Y_j is zero with probability one.

8. It is apparent from expression (4.16) that the covariance matrix has elements $n\theta_j - n\theta_j^2$ on the diagonal and $-n\theta_i\theta_j$ in the ith row and jth column.

9. With $p_Y(t) = \theta^\varphi [1 - (1 - \theta)t]^{-\varphi}$

$$p_Y'(t) = \varphi\theta^\varphi (1 - \theta) [1 - (1 - \theta)t]^{-\varphi - 1}$$

and

$$p_Y''(t) = (\varphi + 1)\varphi\theta^\varphi(1 - \theta)^2 \left[1 - (1 - \theta) t\right]^{-\varphi - 2}.$$

Evaluating at $t = 1$ gives $\mu_Y = p_Y'(1) = \varphi(1 - \theta)/\theta$ and

$$\sigma_Y^2 = EY(Y - 1) + EY - (EY)^2$$
$$= p_Y''(1) + \varphi\frac{1 - \theta}{\theta} - \varphi^2\frac{(1 - \theta)^2}{\theta^2}$$
$$= \varphi\frac{(1 - \theta)}{\theta^2}.$$

10. More trials would be needed, on average, to obtain more successes, and more failures would typically be experienced. There would be more failures, on average, the greater the probability of failure, $1 - \theta$.

11. With $X - 1 \sim NB(1, \frac{1}{36})$ we have $EX = 1\left(1 - \frac{1}{36}\right)/\frac{1}{36} + 1 = 36$ and $VX = 35 \cdot 36 = 1260$.

12. Working from the p.g.f.,

$$p_S(t) = p_1(t)p_2(t)$$
$$= \left[\frac{\theta}{1 - (1 - \theta)t}\right]^{\varphi_1} \left[\frac{\theta}{1 - (1 - \theta)t}\right]^{\varphi_2}$$
$$= \left[\frac{\theta}{1 - (1 - \theta)t}\right]^{\varphi_1 + \varphi_2},$$

which is the p.g.f. of $NB(\varphi_1 + \varphi_2, \theta)$. By induction, the sum of any finite number of independent negative binomials with the same parameter θ is distributed as negative binomial.

13. Factoring S from the first factor in the numerator of (4.21) and rearranging the denominator yields the first factor in the numerator of (4.22). The second factor in the numerator of (4.22) is just a rearrangement of that in (4.21). Factoring $(S + F)/n$ from the denominator of (4.21) yields the denominator of (4.22). Thus, the sum in (4.21) equals that in (4.22) scaled by the factor $S \div (S + F)/n$.

14. Factoring and rearranging as before give for $EX(X - 1)$

$$\frac{n(n - 1)S(S - 1)}{N(N - 1)} \sum_{x=2}^{n} \left\{ \frac{\frac{(S-2)!}{(x-2)![S-2-(x-2)]!} \frac{F!}{[n-2-(x-2)]![F-(n-2)+(x-2)]!}}{\frac{(S-2+F)!}{(n-2)![S-2+F-(n-2)]!}} \right\}$$
$$= \frac{n(n - 1)S(S - 1)}{N(N - 1)}.$$

15. We have

$$VX = EX^2 - (EX)^2 = EX(X-1) + EX - (EX)^2$$
$$= \frac{n(n-1)S(S-1)}{N(N-1)} + \frac{nS}{N} - \frac{n^2S^2}{N^2}$$
$$= \frac{nSF(N-n)}{N^2(N-1)}.$$

From the example, $EX = \sum_{x=3}^{6} x f_X(x) = 21/5$, which equals $nS/N = 7 \cdot 6/10$; and $VX = \sum_{x=3}^{6} x^2 f_X(x) - (EX)^2 = \frac{273}{15} - \frac{441}{25} = .56$, which equals $nSF(N-n)/[N^2(N-1)] = (7 \cdot 6 \cdot 4 \cdot 3)/(100 \cdot 9)$.

16. (i) This follows from (4.20) with $S = F = n$.

(ii) Both obtain precisely $j \in \{0, 1, ..., n\}$ heads with probability $\left[\binom{n}{j} 2^{-n}\right]^2$, so they obtain the same number with probability $\sum_{j=0}^{n} \binom{n}{j}^2 2^{-2n} = \binom{2n}{n} 2^{-2n}$. For example, with $n = 5$ this is $\binom{10}{5} 2^{-10} \doteq .246$.

17. The problem amounts to choosing four of the eight possible positions occupied by the cups to which cream was added first. There are $\binom{8}{4} = 70$ possible arrangements. Identifying (correctly) precisely $x \in \{0, 1, ..., 4\}$ with cream first amounts to choosing (incorrectly) $4 - x$ of those with cream last. This would be done at random in $\binom{4}{x}\binom{4}{4-x}$ ways. Thus, $X \sim HG(4, 4, 4)$ and

$$f(x) = \frac{\binom{4}{x}\binom{4}{4-x}}{\binom{8}{4}} 1_{\{0,1,...,4\}}(x).$$

18. (i) At least one ball of the other type *must* remain, else all of that type would have been the first to be removed; and no more than n *can* be left. Thus, $\mathbb{X}_{n,n}$ comprises the first n positive integers.

(ii) There are altogether $\binom{2n}{n}$ distinguishable arrangements of the n white balls and the n black balls, so we can regard the experiment has having $\binom{2n}{n}$ outcomes. Event $X_{n,n} = x$ occurs when (a) x balls of one type are in the last x positions (draws $2n - x + 1, ..., 2n$), (b) one ball of the other type is in position $2n - x$, and (c) $n - 1$ balls of the other type are among the first $2n - x - 1$ positions. There are $\binom{2n-x-1}{n-1}\binom{1}{1}\binom{x}{x} = \binom{2n-x-1}{n-1}$ such arrangements for each of the $\binom{2}{1} = 2$ assignments of the type to be left as the residual. Thus,

$$f_{n,n}(x) = \frac{2\binom{2n-x-1}{n-1}}{\binom{2n}{n}} 1_{\{1,2,...,n\}}(x).$$

(iii) $f_{n,n}(1) = 2\binom{2n-2}{n-1}/\binom{2n}{n} = n^2/(2n^2 - n) \to 1/2$, and

$$\frac{f_{n,n}(x+1)}{f_{n,n}(x)} = \frac{\binom{2n-x-2}{n-1}}{\binom{2n-x-1}{n-1}}$$

$$= \frac{n-x}{2n-x-1} \to \frac{1}{2}$$

for each fixed x. Thus, $f_{n,n}(x) \to f_{\infty,\infty}(x) := 2^{-x}\mathbf{1}_{\mathbb{N}}(x)$ as $n \to \infty$. To illustrate, the distributions for $n \in \{2, 4, 6, 8, 10, \infty\}$ are

x	$f_{2,2}(x)$	$f_{4,4}(x)$	$f_{6,6}(x)$	$f_{8,8}(x)$	$f_{10,10}(x)$	$f_{\infty,\infty}(x)$
1	.6667	.5714	.5455	.5333	.52632	.50000
2	.3333	.2857	.2727	.2667	.26316	.25000
3		.1143	.1212	.1231	.12384	.12500
4		.0286	.0455	.0513	.05418	.06250
5			.0130	.0186	.02167	.03125
6			.0022	.0056	.00774	.01562
7				.0012	.00238	.00781
8				.0002	.00060	.00391
9					.00011	.00195
10					.00001	.00098

19. For $x \in \{1, 2, ..., n\}$ there are $\binom{m+n-x-1}{m-1}$ arrangements in which x black balls come last, and for $x \in \{1, 2, ..., m\}$ there are $\binom{m+n-x-1}{n-1}$ arrangements in which x white balls are last. Thus,

$$f_{m,n}(x) = \frac{\binom{m+n-x-1}{m-1}\mathbf{1}_{\{1,2,...,n\}}(x) + \binom{m+n-x-1}{n-1}\mathbf{1}_{\{1,2,...,m\}}(x)}{\binom{m+n}{m}}.$$

20. Taking $X \sim P(4.3)$, $\Pr(X < 4.3) = \Pr(X \le 4) = \sum_{x=0}^{4} 4.3^x e^{-4.3}/x! \doteq .570$.

21. Canonical representation $f(x) = \alpha(\eta)c(x)e^{\eta T}$ has $\eta = \log \lambda$ and $\alpha(\eta) = \exp(-e^\eta)$, and natural statistic $T = X$ has c.g.f.

$$\log \alpha(\eta) - \log \alpha(\eta + t) = e^{\eta+t} - e^\eta = \lambda(e^t - 1).$$

22. $EX = \sum_{x=0}^{\infty} xf(x) = \sum_{x=1}^{\infty} xf(x) = \lambda \sum_{x=1}^{\infty} f(x-1) = \lambda$.

23. Applying the recurrence formula k times gives

$$f(x) = \frac{\lambda^k f(x-k)}{x(x-1)\cdots\cdots(x-k+1)},$$

from which it follows that

$$\mu'_{[k]} = \sum_{x=k}^{\infty} x(x-1)\cdots(x-k+1)f(x)$$

$$= \lambda^k \sum_{x=k}^{\infty} f(x-k) = \lambda^k.$$

24. Working from the p.g.f.,

$$p_S(t) = p_1(t)p_2(t)$$
$$= e^{\lambda_1(t-1)}e^{\lambda_2(t-1)}$$
$$= e^{(\lambda_1+\lambda_2)(t-1)},$$

which is the p.g.f. of $P(\lambda_1+\lambda_2)$. By induction, $\sum_{j=1}^{k} X_j \sim P(\sum_{j=1}^{k} \lambda_j)$ when $\{X_j\}_{j=1}^{k}$ are independent Poisson variates with parameters $\{\lambda_j\}_{j=1}^{k}$.

25. (i) L has p.g.f. $p_L(t) = e^{\lambda(t-1)}$ and X has conditional p.g.f. $p_{X|L}(t) = e^{L(t-1)}$, so $p_X(t) = Ee^{L(t-1)} = p_L(e^{t-1}) = \exp[\lambda(e^{t-1}-1)]$. The first two derivatives are $p'_X(t) = \lambda e^{t-1}p_X(t)$ and $p''_X(t) = \lambda e^{t-1}[p'_X(t) + p_X(t)]$.

(ii) $EX = E[E(X \mid L)] = EL = \lambda = p'_X(1)$ and $VX = E[V(X \mid L)] + V[E(X \mid L)] = EL + VL = \lambda + \lambda = 2\lambda$. Alternatively, $VX = EX(X-1) + EX - (EX)^2 = p''_X(1) + \lambda - \lambda^2 = (\lambda^2 + \lambda) + \lambda - \lambda^2 = 2\lambda$.

(iii) Expressed with the d operator, the first two derivatives are $dp_X(t) = \lambda e^{t-1}(d+1)^0 p_X(t)$ and $d^2 p_X(t) = \lambda e^{t-1}(d+1)^1 p_X(t)$, so (4.26) does hold for $n = 0$ and $n = 1$. Proceeding by induction, assume that $d^n p_X(t) = \lambda e^{t-1}(d+1)^{n-1}p_X(t)$ for $n \in \{1, 2, ...\}$. Then

$$d^{n+1}p_X(t) = d[d^n p_X(t)]$$
$$= \lambda e^{t-1}(d+1)^{n-1} \cdot dp_X(t) + (d+1)^{n-1}p_X(t) \cdot d[\lambda e^{t-1}]$$
$$= \lambda e^{t-1}[(d+1)^{n-1}dp_X(t) + (d+1)^{n-1}p_X(t)]$$
$$= \lambda e^{t-1}(d+1)^n p_X(t),$$

as was to be proved.

(iv) Applying the binomial formula,

$$d^{n+1}p_X(0) \equiv p_X^{(n+1)}(t)\Big|_{t=0} = \lambda e^{-1}\sum_{j=0}^{n}\binom{n}{j}p_X^{(n-j)}(0)$$

for $n \in \aleph_0$. The table and Fig. 4.16 show $f_X(x) = p_X^{(x)}(0)/x!$ and $f_L(x) = \lambda^x e^{-\lambda}/x!$ for $x \in \{0, 1, ..., 15\}$ and $\lambda = 5$.

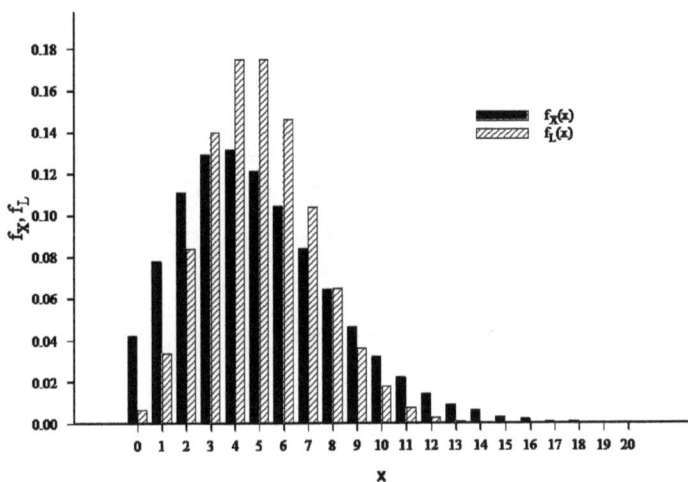

Fig. 4.16 Poisson *vs.* Poisson–Poisson mixture probabilities.

x	$f_X(x)$	$f_L(x)$
0	0.042	0.007
1	0.078	0.034
2	0.111	0.084
3	0.129	0.140
4	0.131	0.175
5	0.121	0.175
6	0.104	0.146
7	0.084	0.104
8	0.064	0.065
9	0.046	0.036
10	0.032	0.018
11	0.022	0.008
12	0.014	0.003
13	0.009	0.001
14	0.006	0.000
15	0.003	0.000

26. We have

$$\mu = \int_a^b \frac{x}{b-a} \cdot dx = \frac{a+b}{2}$$

$$\mu_2' = \int_a^b \frac{x^2}{b-a} \cdot dx = \frac{b^2 + ab + a^2}{3}$$

$$\sigma^2 = \mu_2' - \mu^2 = \frac{(b-a)^2}{12}.$$

27. Since $\theta \sim U(-\pi/2, \pi/2)$ we have

$$F_\theta(\theta) = \pi^{-1}\left(\theta + \frac{\pi}{2}\right) \mathbf{1}_{(-\pi/2, \pi/2)}(\theta) + \mathbf{1}_{[\pi/2, \infty)}(\theta).$$

Then

$$F_X(x) = \Pr(X \le x) = \Pr[\tan(\theta) \le x]$$

$$= \Pr[\theta \le \tan^{-1}(x)] = \pi^{-1}\left[\tan^{-1}(x) + \frac{\pi}{2}\right]$$

for $x \in \Re$; and $f_X(x) = F_X'(x) = [\pi(1 + x^2)]^{-1}$ (the Cauchy distribution).

28. Let $\mathbf{s} = (s_1, \dots s_n, s_{n+1}, \dots, s_{2n})$ be any of the $\binom{2n}{n}$ arrangements of voters of whom n vote "yes" and the remainder vote "no". The probability of a tie among the $2n$ voters is then

$$\mathbb{P}(B) = \sum_{\mathbf{s}} P_{s_1} \cdot \dots \cdot P_{s_n}(1 - P_{s_{n+1}}) \cdot \dots \cdot (1 - P_{s_{2n}}). \qquad (4.54)$$

For example, if $n = 1$, $\mathbb{P}(B) = P_1(1 - P_2) + P_2(1 - P_1)$, and if $n = 2$

$$\mathbb{P}(B) = P_1 P_2 (1 - P_3)(1 - P_4) + P_1 P_3 (1 - P_2)(1 - P_4)$$

$$+ P_1 P_4 (1 - P_2)(1 - P_3) + P_2 P_3 (1 - P_1)(1 - P_4)$$

$$+ P_2 P_4 (1 - P_1)(1 - P_3) + P_3 P_4 (1 - P_1)(1 - P_2).$$

(i) If n voters vote "yes" with probability one and the others vote "yes" with probability zero, then precisely one term in the summation is unity and all others are zero, so that $\mathbb{P}(B) = 1$.

(ii) If more than n voters are certain to vote "yes" and the remainder (if any) "no", then *all* terms will be zero and $\mathbb{P}(B) = 0$.

(iii) If all voters have the same probability p of voting yes, $\mathbb{P}(B) = \binom{2n}{n} p^n (1 - p)^n$.

(iv) Put $\mathbb{P}(B) = \int_B \mathbb{P}(d\omega) = E\mathbf{1}_B$ and $\mathbf{P} := (P_1, ..., P_{2n})$. Then $\mathbb{P}(B \mid \mathbf{P}) = E(\mathbf{1}_B \mid \mathbf{P})$ is given by the right side of (4.54), and

$$\mathbb{P}(B) = E[E(\mathbf{1}_B \mid \mathbf{P})]$$

$$= \sum_s EP_{s_1} \cdot ... \cdot EP_{s_n} (1 - EP_{s_{n+1}}) \cdot ... \cdot (1 - EP_{s_{2n}})$$

$$= \binom{2n}{n} 2^{-2n}.$$

29. Set $\Gamma(\alpha + 1) = \int_0^\infty y^\alpha e^{-y} \cdot dy$ and integrate by parts with $u = y^\alpha$, $dv = e^{-y}$, $du = \alpha y^{\alpha-1}$, $v = -e^{-y}$ to get $\Gamma(\alpha + 1) = -y^\alpha e^{-y}\big|_0^\infty + \alpha \int_0^\infty y^{\alpha-1} e^{-y} \cdot dy = 0 + \alpha \Gamma(\alpha)$.

30. (i) F_T is continuous (and therefore right continuous) since definite integrals are continuous functions of their limits, and it is nondecreasing since the integrand is nonnegative. Finally, $F_T(0)$ is clearly zero, while $\lim_{t \to \infty} F(t) = \frac{1}{\Gamma(\alpha)} \int_0^\infty y^{\alpha-1} e^{-y} \cdot dy = \Gamma(\alpha)/\Gamma(\alpha) = 1$.

(ii) Differentiating and applying Leibnitz' formula shows that $F'(t) = \frac{1}{\Gamma(\alpha)} t^{\alpha-1} e^{-t}$. This is defined and positive for all $\alpha > 0$ when $t > 0$.

31. The transformation being one to one from $(0, \infty)$ to itself, the change-of-variable formula gives, for $x > 0$,

$$f_X(x) = f_Z(x/\beta) \left| \frac{dx}{dz} \right|$$

$$= \frac{1}{\Gamma(\alpha)} \left(\frac{x}{\beta} \right)^{\alpha-1} e^{-x/\beta} \cdot \frac{1}{\beta}$$

$$= \frac{1}{\Gamma(\alpha)\beta^\alpha} x^{\alpha-1} e^{-x/\beta}.$$

32. Taking $F_T(t) = 1 - e^{-\lambda t}$, we have for $t \geq 0$ and $t' \geq 0$

$$\Pr(T > t' + t \mid T > t') = \frac{\mathbb{P}_T[(t' + t, \infty) \cap (t', \infty)]}{\mathbb{P}_T[(t', \infty)]}$$

$$= \frac{\mathbb{P}_T[(t' + t, \infty)]}{\mathbb{P}_T[(t', \infty)]} = \frac{1 - F_T(t' + t)}{1 - F_T(t')}$$

$$= e^{-\lambda t} = \Pr(T > t).$$

33. Using $\mu'_r = \Gamma(r + \alpha)/\Gamma(\alpha)$ for $r = -k$ and $-k + \alpha > 0$, plus the recurrence relation for the gamma function,

$$\mu'_{-k} = \frac{\Gamma(\alpha - k)}{\Gamma(\alpha)}$$

$$= \frac{\Gamma(\alpha - k)}{(\alpha - 1)(\alpha - 2) \cdots \cdots (\alpha - k)\Gamma(\alpha - k)}.$$

34. When $Z \sim \Gamma(\alpha, 1)$

$$M_Z(t) = \int_0^\infty e^{tz} \cdot \frac{z^{\alpha-1}e^{-z}}{\Gamma(\alpha)} \cdot dz = \int_0^\infty \frac{z^{\alpha-1}e^{-(1-t)z}}{\Gamma(\alpha)} \cdot dz.$$

Changing variables as $y = (1 - t)z$, which is positive when $t < 1$, gives

$$M_Z(t) = \int_0^\infty \frac{1}{\Gamma(\alpha)} \left(\frac{y}{1-t}\right)^{\alpha-1} e^{-y} \left|\frac{1}{1-t}\right| \cdot dy$$

$$= (1-t)^{-\alpha} \int_0^\infty \frac{y^{\alpha-1}e^{-y}}{\Gamma(\alpha)} \cdot dy$$

$$= (1-t)^{-\alpha}, \ t < 1.$$

Since $M_X(t) = M_Z(\beta t)$ when $X = \beta Z$, we have $M_X(t) = (1 - \beta t)^{-\alpha}, \ t < \beta^{-1}$.

35. From (4.30) we have $\eta_1 = \alpha - 1$, $\eta_2 = -\beta^{-1}$ and, for $j \in \{1, 2\}$,

$$\frac{\partial \log M_{\mathbf{T}}(\mathbf{t})}{\partial t_j} = -\frac{\partial \log \alpha(\boldsymbol{\eta} + \mathbf{t})}{\partial t_j}$$

$$= -\frac{\partial}{\partial t_j} \left[(1 + \eta_1 + t_1)\log(-\eta_2 - t_2) - \log\Gamma(1 + \eta_1 + t_1)\right].$$

Hence,
(i)

$$EX = \frac{\partial \log M_{\mathbf{T}}(\mathbf{t})}{\partial t_2}\bigg|_{\mathbf{t}=0} = \frac{1 + \eta_1}{-\eta_2} = \alpha\beta$$

(ii)

$$E\log X = \frac{\partial \log M_{\mathbf{T}}(\mathbf{t})}{\partial t_1}\bigg|_{\mathbf{t}=0} = -\log(-\eta_2) + \frac{\Gamma'(1 + \eta_1)}{\Gamma(1 + \eta_1)} = \log\beta + \Psi(\alpha).$$

(iii) This is replicated by evaluating at $t = 0$

$$\frac{d\log\mu'_t}{dt} = \frac{d}{dt}\left[t\log\beta + \log\Gamma(t + \alpha) - \log\Gamma(\alpha)\right].$$

(iv) $\alpha\beta = EX \equiv Ee^{\log X} > e^{E\log X} = \beta e^{\Psi(\alpha)} \implies \log\alpha > \Psi(\alpha)$.

36. Setting $S_k := \sum_{j=1}^{k} X_j$,

$$\phi_{S_k}(t) = \prod_{j=1}^{k} \phi_j(t) = \prod_{j=1}^{k} (1 - i\beta_0 t)^{-\alpha_j} = (1 - i\beta_0 t)^{-\sum_{j=1}^{k} \alpha_j}.$$

Recognize the last expression as the c.f. of $\Gamma(\sum_{j=1}^{k} \alpha_j, \beta_0)$ and appeal to the uniqueness theorem to conclude that $S_k \sim \Gamma(\sum_{j=1}^{k} \alpha_j, \beta_0)$. It follows that if $Y \sim \Gamma(k, \beta)$, then Y can be represented as $Y = \sum_{j=1}^{k} X_j$, where the X_j are i.i.d. as $\Gamma(1, \beta)$.

37. (i) With $S_n := \sum_{j=1}^{n} T_j$, we have $\Pr(N_t = n) = \Pr(S_n \le t, S_{n+1} > t)$, since n bulbs must have failed by t and $n+1$ must *not* have failed. The result then follows from the law of total probability:

$$\Pr(S_n \le t) = \Pr(S_n \le t, S_{n+1} > t) + \Pr(S_n \le t, S_{n+1} \le t)$$
$$= \Pr(N_t = n) + \Pr(S_{n+1} \le t).$$

(ii) If $\{T_j\}_{j=1}^{\infty}$ are i.i.d. as exponential with mean λ^{-1}—hence, as $\Gamma(1, \lambda^{-1})$, then $S_n \sim \Gamma(n, \lambda^{-1})$ by (4.34). Therefore, for $n \in \mathbb{N}$ we have

$$\Pr(N_t = n) = \int_0^t \frac{s^{n-1} e^{-\lambda s}}{\Gamma(n) \lambda^{-n}} \cdot ds - \int_0^t \frac{s^n e^{-\lambda s}}{\Gamma(n+1) \lambda^{-n-1}} \cdot ds.$$

Changing variables as $x = \lambda s$ gives

$$\Pr(N_t = n) = \int_0^{\lambda t} \frac{x^{n-1} e^{-x}}{(n-1)!} \cdot dx - \int_0^{\lambda t} \frac{x^n e^{-x}}{n!} \cdot dx$$

$$= \frac{1}{n!} \left(\int_0^{\lambda t} n x^{n-1} e^{-x} \cdot dx - \int_0^{\lambda t} x^n e^{-x} \cdot dx \right)$$

$$= \frac{1}{n!} \left(\int_0^{\lambda t} n x^{n-1} e^{-x} \cdot dx + \left[x^n e^{-x} \right]_0^{\lambda t} - \int_0^{\lambda t} n x^{n-1} e^{-x} \cdot dx \right)$$

$$= \frac{(\lambda t)^n e^{-\lambda t}}{n!}.$$

Thus, N_t for $t > 0$ is distributed as $P(\lambda t)$. (The independence of $\{T_j\}_{j=1}^{\infty}$ implies that increments $N_{t_2} - N_{t_1}$ and $N_{t_4} - N_{t_3}$ over nonoverlapping intervals are independent also, and so $\{N_t\}_{t \ge 0}$ is a Poisson process.)

38. (1) Transformation $\mathbf{y} = \mathbf{g}(\mathbf{x}) = [x_1/(x_1+x_2),(x_1+x_2)]'$ is a one-to-one mapping of $(0,\infty)^2$ onto $(0,1)\times(0,\infty)$, with inverse $\mathbf{x} = \mathbf{h}(\mathbf{y}) = (y_1 y_2, y_2 - y_1 y_2)'$ and Jacobian $J_{\mathbf{h}} = y_2$. Change-of-variable formula $f_{\mathbf{Y}}(\mathbf{y}) = f_{\mathbf{X}}(\mathbf{x})|J_{\mathbf{h}}|$ gives

$$f_{\mathbf{Y}}(y_1,y_2) = \frac{1}{\Gamma(\alpha)\Gamma(\beta)} y_1^{\alpha-1}(1-y_1)^{\beta-1} y_2^{\alpha+\beta-1} e^{-y_2} 1_{(0,1)\times(0,\infty)}(y_1,y_2)$$

$$= \frac{1}{\mathcal{B}(\alpha,\beta)} y_1^{\alpha-1}(1-y_1)^{\beta-1} 1_{(0,1)}(y_1)$$

$$\cdot \frac{1}{\Gamma(\alpha+\beta)} y_2^{\alpha+\beta-1} e^{-y_2} 1_{(0,\infty)}(y_2),$$

a product of nonnegative factors that take positive values on a product space. Independence of Y_1, Y_2 follows by the result of Exercise 2.28.

(ii) Integrating out y_2 gives (4.35), and the support can be extended to $[0,1]$.

(iii) If $X_1 \sim \Gamma(\alpha,1)$ and $X_2 \sim \Gamma(\beta,1)$, then $\gamma X_1 \sim \Gamma(\alpha,\gamma)$ and $X_2 \sim \Gamma(\beta,\gamma)$, and $Y_1 = \gamma X_1/(\gamma X_1 + \gamma X_2) \equiv X_1/(X_1+X_2)$. In particular, taking $\gamma = 2$ shows that the result holds when X_1, X_2 are distributed as chi-squared.

39. We have $\mu_{2k} = E(X-\alpha)^{2k} = \int_{-\infty}^{\infty} \frac{1}{2\beta}(x-\alpha)^{2k} e^{-|x-\alpha|/\beta} \cdot dx$. Changing variables as $Y = (X-\alpha)/\beta$ gives

$$\mu_{2k} = \beta^{2k} \int_{-\infty}^{\infty} \frac{1}{2} y^{2k} e^{-|y|} \cdot dy$$

$$= \beta^{2k} \int_{0}^{\infty} y^{2k} e^{-y} \cdot dt$$

$$= \beta^{2k} \Gamma(2k+1).$$

40. Y has c.f.

$$\phi_Y(t) = E e^{it(X_1-X_2)} = E e^{itX_1} E e^{-itX_2}$$
$$= (1-it\beta)^{-1}(1+it\beta)^{-1} = (1+t^2\beta^2)^{-1},$$

which is the c.f. of Laplace with scale β and median 0.

41. The area of a circle cut at height $H(r) = e^{-r^2/2}$ above the x,y plane is $A(r) = \pi r^2$. A disk with thickness $-dH(r) = re^{-r^2/2}$ has volume $-A(r) \cdot dH(r) = \pi \int_0^\infty r^3 e^{-r^2/2} \cdot dr$. Integrating by parts with $u = r^2$ and $dv = re^{-r^2/2} \cdot dr$ gives $I^2 = 2\pi$.

42. We have

$$z^{-1} f_Z(z) = - \int_z^\infty d\left[y^{-1} f_Z(y) \right]$$

$$= \int_z^\infty \left(y^{-2} + 1 \right) f_Z(y) \cdot dy$$

$$< \left(z^{-2} + 1 \right) \left[1 - \Phi(z) \right].$$

43. Π_z^2 represents the volume above the square $(-z, z) \times (-z, z)$ in the x, y plane, whereas $\widehat{\Pi}_z^2$ is the volume above the circle with radius z. Therefore, $\widehat{\Pi}_z^2 < \Pi_z^2$ for each positive, finite z. However, the limiting values as $z \to \infty$ are the same, since the plane is covered in the limit by either the circle or the square. A better approximation—but always an overestimate—is the volume under a circle with radius $z' = 2z/\sqrt{\pi}$, which has the same area as the square:

$$\widetilde{\Pi}_z^2 = \int_0^{z'} r e^{-r^2/2} \cdot dr = \left(1 - e^{-2z^2/\pi} \right).$$

For example, $\widetilde{\Pi}_1 \doteq .6862$. This latter approximation is due to Pólya (1949).

44. Integrating by parts,

$$\int_{-\infty}^x \Phi(t) \cdot dt = \left[t\Phi(t) \right]_{-\infty}^x - \int_{-\infty}^x t f_Z(t) \cdot dt$$

$$= x\Phi(x) - \lim_{t \to -\infty} t\Phi(t) - \int_{-\infty}^x \frac{1}{\sqrt{2\pi}} t e^{-t^2/2} \cdot dt$$

$$= x\Phi(x) - \left[-\frac{1}{\sqrt{2\pi}} e^{-t^2/2} \right]_{-\infty}^x$$

$$= x\Phi(x) + f_Z(x).$$

Exercise 3.31 showed that $\lim_{t \to -\infty} t\Phi(t) = 0$.

45. $\Pr(Z \geq z) = \Pr\left(e^{tZ} \geq e^{tz} \right)$ for $t > 0$. Applying (3.29) with $g(Z) = e^{tZ}$ and $c = e^{tz}$ gives $\Pr(Z \geq z) \leq e^{-tz} E e^{tZ} = e^{-tz + t^2/2}$ for $t > 0, z > 0$. The right side is minimized at $t = z$. The bound is not very useful for small z but becomes increasingly tight as z increases; for example, $\Pr(Z \geq 4) \doteq .00032 < e^{-8} \doteq .00034$.

46. Take $u = g(z)$, $du = g'(z) \cdot dz$, $dv = ze^{-z^2/2} \cdot dz$, $v = -e^{-z^2/2}$ and integrate by parts:

$$E[Zg(Z)] = \int_{-\infty}^{\infty} zg(z) \frac{1}{\sqrt{2\pi}} e^{-z^2/2} \cdot dz = \frac{1}{\sqrt{2\pi}} \int_{-\infty}^{\infty} g(z) ze^{-z^2/2} \cdot dz$$

$$= -g(z) \frac{e^{-z^2/2}}{\sqrt{2\pi}} \Big|_{-\infty}^{\infty} + \int_{-\infty}^{\infty} g'(z) \frac{1}{\sqrt{2\pi}} e^{-z^2/2} \cdot dz$$

$$= 0 + Eg'(Z).$$

The last equality follows from the fact that $E|g(Z)| < \infty$ implies $\lim_{z \to \pm\infty} g(z) e^{-z^2/2} = 0$. To prove the recurrence relation for moments, take $g(Z) = Z^{k-1}$ for $k \geq 2$ and apply Stein's lemma to get $EZ^k = (k-1) EZ^{k-2}$.

47. Stein's lemma implies that $EZg(Z)$ is finite so long as g and g' are themselves $\mathbb{P}_Z$ integrable (i.e., integrable with respect to standard normal measure). Repeated application, as

$$EZ^2 g = EZ(Zg) = E(g + Zg') = Eg + Eg''$$
$$EZ^3 g = EZ(Z^2 g) = \cdots = 3Eg' + Eg^{(3)}$$
$$EZ^4 g = EZ(Z^3 g) = \cdots = 3Eg + 6Eg'' + Eg^{(4)}$$

and so forth, shows that $Z^k g$ is integrable so long as $\{g^{(j)}\}_{j=0}^{k}$ are. The relation

$$EZ^k g(Z) = (k-1) EZ^{k-2} g(Z) + EZ^{k-1} g'(Z) \qquad (4.55)$$

for $k \in \{2, 3, ...\}$ follows on applying Stein's lemma to $h(Z) = Z^{k-1} g(Z)$.

48. With $g(Z) = Z \cos Z$ Stein's lemma gives

$$EZ^2 \cos Z = E \frac{d(Z \cos Z)}{dZ}$$

$$= E \cos Z - EZ \sin Z = 0$$

by part (ii) of Example 4.12.

49. Apply Stein's lemma:

$$\int_{-\infty}^{\infty} z\Phi(z) \cdot d\Phi(z) = \int_{-\infty}^{\infty} \Phi'(z) \cdot d\Phi(z)$$

$$= E f_Z(Z)$$

$$= \frac{1}{\sqrt{2\pi}} \int_{-\infty}^{\infty} \frac{1}{\sqrt{2\pi}} e^{-z^2} \cdot dz$$

$$= \frac{1}{2\sqrt{\pi}}.$$

50. We have

$$E Z \mathbf{1}_{(a,b)}(Z) = \int_a^b \frac{z}{\sqrt{2\pi}} e^{-z^2/2} \cdot dz = -\frac{1}{\sqrt{2\pi}} e^{-z^2/2} \Big|_a^b = f_Z(a) - f_Z(b).$$

51. Put $h(X) = h(\mu + \sigma Z) = g(Z)$ so that $g'(Z) = \sigma h'(X)$. Stein's lemma then gives $E\left(\frac{X-\mu}{\sigma}\right) h(X) = \sigma E h'(X)$ and thus $E X h(X) = \mu E h(X) + \sigma^2 E h'(X)$.

52. Putting $X =: \mu + \sigma Z$ and $Y =: \mu + \sigma W$, where W and Z are i.i.d. as $N(0,1)$, we have

$$\mathfrak{g} = \sigma E |Z - W|$$

$$= \sigma \int_{-\infty}^{\infty} \int_{-\infty}^{\infty} |z - w| \cdot d\Phi(w) \, d\Phi(z)$$

$$= 2\sigma \int_{-\infty}^{\infty} \left[\int_{-\infty}^z (z - w) \cdot d\Phi(w) \right] \cdot d\Phi(z)$$

$$= 2\sigma \int_{-\infty}^{\infty} \left[z\Phi(z) - \int_{-\infty}^z w \cdot d\Phi(w) \right] \cdot d\Phi(z).$$

But Stein's lemma gives $\int_{-\infty}^{\infty} z\Phi(z) \cdot d\Phi(z) = E f_Z(Z)$, and

$$\int_{-\infty}^z w \cdot d\Phi(w) = \int_{-\infty}^z \frac{1}{\sqrt{2\pi}} w e^{-w^2/2} \cdot dw$$

$$= -\int_{-\infty}^z d f_Z(w)$$

$$= -f_Z(z).$$

Thus,

$$\mathfrak{g} = 4\sigma E f_Z(Z) = \frac{2\sigma}{\sqrt{\pi}}.$$

53. (i) From $\sin X = \left(e^{iX} - e^{-iX}\right)/(2i)$ we have

$$E \sin X = \frac{\phi_X(1) - \phi_X(-1)}{2i} = \frac{e^{i\mu - \sigma^2/2} - e^{-i\mu - \sigma^2/2}}{2i}$$

$$= e^{-\sigma^2/2} \sin \mu.$$

(ii) $E \cos X = e^{-\sigma^2/2} \cos \mu$

(iii) $E \sin^2 X = \left[1 - e^{-2\sigma^2} \cos(2\mu)\right]/2$

(iv) $E \cos^2 X = 1 - E \sin^2 X$

(v) $E \sin X \cos X = \frac{1}{2} E \sin(2X) = \frac{1}{2} e^{-2\sigma^2} \sin(2\mu)$

(vi) From the extension of Stein's lemma,

$$EX \cos X = \mu E \cos X - \sigma^2 E \sin X$$
$$= e^{-\sigma^2/2} \left(\mu \cos \mu - \sigma^2 \sin \mu\right)$$

(vii) $EX \sin X = e^{-\sigma^2/2} \left(\mu \sin \mu + \sigma^2 \cos \mu\right)$.

54. Since $\alpha_k = \mu_k/\sigma^k = E\left(\frac{X-\mu}{\sigma}\right)^k = EZ^k, k \in \{3, 4\}$, the coefficients of skewness and kurtosis are the same for all members of the normal family, including the standard normal. Thus, $\alpha_3 = EZ^3 = 0$ and $\alpha_4 = EZ^4 = 3$.

55. X has the same distribution as $\mu + \sigma Z$, where $Z \sim N(0, 1)$; and $X \in [a, b]$ if and only if $Z \in [(a - \mu)/\sigma =: \alpha, (b - \mu)/\sigma =: \beta]$. Hence,

$$EY = E\left(\mu + \sigma Z \mid \alpha \leq Z \leq \beta\right)$$

$$= \mu + \sigma \int_\alpha^\beta \frac{z f_Z(z)}{\Phi(\beta) - \Phi(\alpha)} \cdot dz$$

$$= \mu + \sigma \left[\frac{f_Z(\alpha) - f_Z(\beta)}{\Phi(\beta) - \Phi(\alpha)}\right].$$

56. With α, β as above we have

$$EW = aE\mathbf{1}_{(-\infty, \alpha]}(Z) + E\left(\mu + \sigma Z\right) \mathbf{1}_{(\alpha, \beta)}(Z) + bE\mathbf{1}_{[\beta, +\infty)}(Z)$$
$$= a\Phi(\alpha) + E\left(\mu + \sigma Z\right) \mathbf{1}_{(\alpha, \beta)}(Z) + b\left[1 - \Phi(\beta)\right]$$
$$= a\Phi(\alpha) + \mu\left[\Phi(\beta) - \Phi(\alpha)\right] + \sigma\left[f_Z(\alpha) - f_Z(\beta)\right] + b\left[1 - \Phi(\beta)\right].$$

57. By linearity of $E(\cdot)$, $\mu_Y = a_0 + \sum_{j=1}^k a_j EX_j = a_0 + \sum_{j=1}^k a_j \mu_j$. By the general formula for variances of affine functions of uncorrelated

variables, $\sigma_Y^2 = \sum_{j=1}^k a_j^2 V X_j = \sum_{j=1}^k a_j^2 \sigma_j^2$. Working with the c.f.,

$$\phi_Y(t) = E \exp\left(ita_0 + \sum_{j=1}^k ita_j X_j\right)$$

$$= e^{ita_0} \prod_{j=1}^k \phi_j(ta_j)$$

$$= e^{ita_0} \prod_{j=1}^k e^{ita_j \mu_j - t^2 a_j^2 \sigma_j^2/2}$$

$$= \exp\left[it\left(a_0 + \sum_{j=1}^k a_j \mu_j\right) - \frac{t^2}{2}\sum_{j=1}^k a_j^2 \sigma_j^2\right]$$

$$= e^{it\mu_Y - t^2 \sigma_Y^2/2},$$

which is the specific member $N(\mu_Y, \sigma_Y^2)$ of the normal family.

58. Given that Z_1 and Z_2 are independent and identically distributed, it seems clear that $E(Z_1 \mid S) = E(Z_2 \mid S)$; and from $S = E(S \mid S) = E(Z_1 + Z_2 \mid S) = E(Z_1 \mid S) + E(Z_2 \mid S)$ one can infer that the common value of the two expectations is $S/2$. However, to develop our skills we can work out $E(Z_1 \mid S)$ as follows. The conditional p.d.f. of Z_1 given S is the ratio of joint p.d.f. $f_{Z_1,S}(z_1, s)$ to marginal p.d.f. $f_S(s)$. Applying the change-of-variable formula with $S = Z_1 + Z_2$ and Z_1 equal to itself, the Jacobian of the transformation from (z_1, z_2) to (z_1, s) is unity, so

$$f_{Z_1,S}(z_1, s) = f_{Z_1,Z_2}(z_1, s - z_1) \cdot 1$$

$$= \frac{1}{\sqrt{2\pi}} e^{-z_1^2/2} \cdot \frac{1}{\sqrt{2\pi}} e^{-(s-z_1)^2/2}$$

$$= \frac{1}{\sqrt{2\pi}} e^{-s^2/4} \frac{1}{\sqrt{2\pi}} e^{-(z_1 - s/2)^2}.$$

Rather than integrate out z_1 to get $f_S(s)$, just note that the closure of the normal family under affine transformations implies that

$S \sim N(ES, VS) = N(0, 2)$, so that $f_S(s) = \frac{1}{2\sqrt{\pi}} e^{-s^2/4}$ and

$$f_{Z_1|S}(z_1 \mid s) = \frac{\frac{1}{\sqrt{2\pi}} e^{-s^2/4} \frac{1}{\sqrt{2\pi}} e^{-(z_1 - s/2)^2}}{\frac{1}{2\sqrt{\pi}} e^{-s^2/4}}$$

$$= \frac{1}{\sqrt{2\pi}\sqrt{1/2}} \exp\left[-\frac{(z_1 - s/2)^2}{2(1/2)} \right].$$

This being the p.d.f. of $N(s/2, 1/2)$, it follows that $E(Z_1|S = s) = s/2$ and $E[E(Z_1|S)] = E(S/2) = E(Z_1 + Z_2)/2 = 0$.

59. When $k = 2$

$$\Sigma = \begin{pmatrix} \sigma_1^2 & \sigma_{12} \\ \sigma_{12} & \sigma_2^2 \end{pmatrix} = \begin{pmatrix} \sigma_1^2 & \rho_{12}\sigma_1\sigma_2 \\ \rho_{12}\sigma_1\sigma_2 & \sigma_2^2 \end{pmatrix},$$

$|\Sigma| = \sigma_1^2 \sigma_2^2 (1 - \rho_{12}^2)$, and

$$\Sigma^{-1} = \frac{1}{|\Sigma|} \begin{pmatrix} \sigma_2^2 & -\rho_{12}\sigma_1\sigma_2 \\ -\rho_{12}\sigma_1\sigma_2 & \sigma_1^2 \end{pmatrix}$$

$$= \begin{pmatrix} \frac{1}{\sigma_1^2(1-\rho_{12}^2)} & -\frac{\rho_{12}}{\sigma_1\sigma_2(1-\rho_{12}^2)} \\ -\frac{\rho_{12}}{\sigma_1\sigma_2(1-\rho_{12}^2)} & \frac{1}{\sigma_2^2(1-\rho_{12}^2)} \end{pmatrix}.$$

With $(\mathbf{x} - \boldsymbol{\mu})' = (x_1 - \mu_1, \ x_2 - \mu_2)$, multiplying out $(\mathbf{x} - \boldsymbol{\mu})' \Sigma^{-1} (\mathbf{x} - \boldsymbol{\mu})$ in the exponent and writing norming factor $(2\pi)^{-k/2} |\Sigma|^{-1/2}$ as $\left(2\pi\sigma_1\sigma_2\sqrt{1 - \rho_{12}^2}\right)^{-1}$ yields the bivariate formula.

60. $S_n \sim \sqrt{n}Z$, where $Z \sim N(0, 1)$, so

$$\Pr(S_n < k, S_{n+1} \geq k) = \Pr\left(Z \leq \frac{k}{\sqrt{n}}, Z_{n+1} > k - \sqrt{n}Z \right)$$

$$= \int_{-\infty}^{k/\sqrt{n}} \left[1 - \Phi\left(k - \sqrt{n}z\right) \right] \cdot d\Phi(z)$$

$$= \Phi\left(\frac{k}{\sqrt{n}}\right) - \int_{-\infty}^{k/\sqrt{n}} \Phi\left(k - \sqrt{n}z\right) \cdot d\Phi(z)$$

$$= \Phi\left(\frac{k}{\sqrt{n}}\right) - \Phi\left(\frac{k}{\sqrt{n}}, \frac{k}{\sqrt{n+1}}; \sqrt{\frac{n}{n+1}}\right),$$

where the last step follows from (4.46). The limiting values are zero in all three cases.

61. S and D are bivariate normal since $S = (1, 1)\mathbf{X}$ and $D = (1, -1)\mathbf{X}$, and they are uncorrelated, since $ESD = EX_1^2 - EX_2^2 = 0$.

62. $f_{\mathbf{X}}(x_1, x_2)$ is positive only on $[0, \infty)^2$ and on $(-\infty, 0]^2$ rather than on all of $\mathcal{R}^2$, and where it is positive the density is twice that of the bivariate normal. Integrating out x_2, the marginal p.d.f. of X_1 is

$$
\begin{aligned}
f_1(x_1) &= \int_{-\infty}^{\infty} f_{\mathbf{X}}(x_1, x_2) \cdot dx_2 \\
&= \begin{cases} \int_0^{\infty} \frac{1}{\pi} e^{-(x_1^2 + x_2^2)/2} \cdot dx_2, & x_1 \ge 0 \\ \int_{-\infty}^0 \frac{1}{\pi} e^{-(x_1^2 + x_2^2)/2} \cdot dx_2, & x_1 < 0 \end{cases} \\
&= \begin{cases} \frac{1}{\sqrt{2\pi}} e^{-x_1^2/2} \cdot 2 \int_0^{\infty} \frac{1}{\sqrt{2\pi}} e^{-x_2^2/2} \cdot dx_2, & x_1 \ge 0 \\ \frac{1}{\sqrt{2\pi}} e^{-x_1^2/2} \cdot 2 \int_{-\infty}^0 \frac{1}{\sqrt{2\pi}} e^{-x_2^2/2} \cdot dx_2, & x_1 < 0 \end{cases} \\
&= \frac{1}{\sqrt{2\pi}} e^{-x_1^2/2}, \quad x_1 \in \mathcal{R}.
\end{aligned}
$$

63. Add and subtract $\rho^2 \left(\frac{x - \mu_X}{\sigma_X} \right)^2$ in the expression in brackets to get

$$
\begin{aligned}
& \frac{\left(\frac{y - \mu_Y}{\sigma_Y} \right)^2 - 2\rho \left(\frac{y - \mu_Y}{\sigma_Y} \right) \left(\frac{x - \mu_X}{\sigma_X} \right) + \rho^2 \left(\frac{x - \mu_X}{\sigma_X} \right)^2}{1 - \rho^2} + \left(\frac{x - \mu_X}{\sigma_X} \right)^2 \\
&= \frac{\left[\left(\frac{y - \mu_Y}{\sigma_Y} \right) - \rho \left(\frac{x - \mu_X}{\sigma_X} \right) \right]^2}{1 - \rho^2} + \left(\frac{x - \mu_X}{\sigma_X} \right)^2 \\
&= \frac{\left(y - \mu_Y + \rho \frac{\sigma_Y}{\sigma_X} \mu_X - \rho \frac{\sigma_Y}{\sigma_X} x \right)^2}{\sigma_Y^2 (1 - \rho^2)} + \left(\frac{x - \mu_X}{\sigma_X} \right)^2 \\
&= \frac{(y - \alpha - \beta x)^2}{\sigma_Y^2 (1 - \rho^2)} + \left(\frac{x - \mu_X}{\sigma_X} \right)^2.
\end{aligned}
$$

64. (i) X_1 and X_2 have standard normal marginals, so $E(X_1 + X_2)^2 = 2 + 2\rho$. Differentiating $M_{X_1 + X_2}(t)$ twice and evaluating at $t = 0$ give $M''_{X_1 + X_2}(0) = E(X_1 + X_2)^2 = 2 + 2/\pi$, so that $\rho = 1/\pi$.
(ii) $EY_1 Y_2 = EX_1^2 - EX_2^2 = 0$.
(iii) $\mathbf{X} := (X_1, X_2)'$ is supported on $\mathbb{X} := \{(x_1, x_2) : x_1 x_2 > 0\}$, which comprises the first and third quadrants of the plane. Linear transformation $Y_1 = X_1 - X_2, Y_2 = X_1 + X_2$ rotates $\mathbb{X}$ counterclockwise by $\pi/4$, giving $\mathbb{Y} := \{(y_1, y_2) : |y_2| > |y_1|\}$. Although joint p.d.f. $f_{\mathbf{Y}}(y_1, y_2) = \frac{1}{2\pi} e^{-y_1^2/4} e^{-y_2^2/4} 1_{\mathbb{Y}}(y_1, y_2)$ is the product of functions of y_1 and y_2 alone, the support is not a product space, so Y_1, Y_2 are dependent. Thus, Y_1 and Y_2 are uncorrelated but nevertheless *dependent* linear functions of normals. This is possible because X_1, X_2 are not distributed as *bivariate* normal.

65. (i) Since $R_t \sim \exp\left(\mu t + \sigma\sqrt{t}Z\right)$, we have $ES_t/S_0 = Ee^{\mu t + \sigma\sqrt{t}Z} = e^{\mu t + \sigma^2 t/2} = B_t/B_0 = e^{rt}$, so $\mu = r - \sigma^2/2$ in a risk-neutral market.

(ii) Under risk neutrality we would have $P_0 = e^{-rT} E\left(K - S_T\right)^+$, with $\log S_T/S_0 \sim \left(r - \sigma^2/2\right) T + \sigma\sqrt{T}Z$. Now $S_T < K$ if and only if

$$Z < \frac{\log\left(K/S_0\right) - \left(r - \sigma^2/2\right) T}{\sigma\sqrt{T}} =: q_T,$$

so

$$
\begin{aligned}
P_0 &= e^{-rT} \int_{-\infty}^{q_T} \left[K - S_0 e^{(r-\sigma^2/2)T + \sigma\sqrt{T}z}\right] \frac{1}{\sqrt{2\pi}} e^{-z^2/2} \cdot dz \\
&= e^{-rT} K \Phi\left(q_T\right) - S_0 \int_{-\infty}^{q_T} e^{-\left(z - \sigma\sqrt{T}\right)^2/2} \cdot dz \\
&= B_0 K \Phi\left(q_T\right) - S_0 \Phi\left(q_T - \sigma\sqrt{T}\right).
\end{aligned}
$$

Notice that P_0 is the market value at $t = 0$ of a portfolio comprising $K \Phi\left(q_T\right) > 0$ T-maturing unit bonds and $-\Phi\left(q_T - \sigma\sqrt{T}\right) < 0$ shares of stock—i.e., a *long* position in bonds and a *short* position in stock. Assuming that $\{B_t\}$ and $\{S_t\}$ follow the specified dynamics and that bonds and stocks can be bought and sold free of brokerage fees or other costs, Black and Scholes (1973) showed that the portfolio can be continuously and costlessly rebalanced over time (purchases of bonds being financed by sales of stock, and *vice versa*) so as to be worth $(K - S_T)^+$ at time T *with probability one*. Thus, anyone who offers to sell the put for less than P_0 at $t = 0$ or who offers to buy it for more than P_0 affords the counterparty an opportunity for *arbitrage*—a riskless profit. Since astute traders quickly exploit such opportunities, the selling price of the put is essentially determined as a function of S_0 and B_0.

(iii) $(S_T - K)^+ - (K - S_T)^+ = S_T - K$ (draw a picture!), so

$$
\begin{aligned}
C_0 &= e^{-rT} E\left(P_T + S_T - K\right) \\
&= P_0 + S_0 - e^{-rT} K \\
&= e^{-rT} K \left[\Phi\left(q_T\right) - 1\right] + S_0 \left[1 - \Phi\left(q_T - \sigma\sqrt{T}\right)\right] \\
&= S_0 \Phi\left(-q_T + \sigma\sqrt{T}\right) - B_0 K \Phi\left(-q_T\right).
\end{aligned}
$$

(iv) $|S_T - K| = (S_T - K)^+ + (K - S_T)^+$, so the straddle's initial price would be $S_0 = C_0 + P_0$.

(v) The value of the "long" side of the contract at T is $\mathsf{F}_T := S_T - K$, so its initial value is $\mathsf{F}_0 = e^{-rT} E\left(S_T - K\right) = S_0 - B_0 K = C_0 - P_0$.

(The relation $C_0 - P_0 = F_0$ is known as "European put-call parity". Any departure from parity offers an opportunity for arbitrage—a result that does *not* depend on the dynamics assumed by Black and Scholes.)

66. We have

$$f_Y(y) = \left| \frac{d\left(\mu^2 y^{-1}\right)}{dy} \right| f_X\left(\mu^2 y^{-1}; \mu, \lambda\right)$$

$$= \left(\frac{\mu^2}{y^2} \right) \sqrt{\frac{\lambda y^3}{2\pi\mu^6}} \exp\left[-\frac{\lambda}{2\mu^4 y} \left(\frac{\mu^2}{y} - \mu \right)^2 \right] \mathbf{1}_{(0,\infty)}(y)$$

$$= \sqrt{\frac{\lambda}{2\pi y^3}} \frac{y}{\mu} \exp\left[-\frac{\lambda(y-\mu)^2}{2\mu^2 y} \right] \mathbf{1}_{(0,\infty)}(y)$$

$$= \mu^{-1} y f_X(y; \mu, \lambda).$$

Thus,

$$EX^{-k} = \mu^{-2k} EY^k = \int_0^\infty \mu^{-2k-1} y^{k+1} f_X(y; \mu, \lambda) \cdot dy = \mu^{-2k-1} EX^{k+1}.$$

From the expressions for EX, $VX := \mu_2$, and $\alpha_3 := \mu_3/\mu_2^{3/2}$ one obtains $\mu_2' = \mu^2\left(\mu\lambda^{-1} + 1\right)$ and $\mu_3' = \mu^3\left(3\mu^2\lambda^{-2} + 3\mu\lambda^{-1} + 1\right)$. Hence, $\mu_{-1}' = \mu^{-3}\mu_2' = \lambda^{-1} + \mu^{-1}$ and $\mu_{-2}' = \mu^{-5}\mu_3' = 3\lambda^{-2} + 3\mu^{-1}\lambda^{-1} + \mu^{-2}$.

67. (i) With $\phi := \lambda/\mu$ we have

$$M_{X^{-1}}(t) = \int_0^\infty \sqrt{\frac{\mu\phi}{2\pi x^3}} \exp\left\{ -\phi\left[\frac{x}{2\mu} - 1 + \frac{\mu}{2x}\left(1 - \frac{2t}{\mu\phi} \right) \right] \right\} \cdot dx.$$

Putting $\mu' := \mu\sqrt{1 - 2t/(\mu\phi)}$ and $\phi' := \phi\sqrt{1 - 2t/(\mu\phi)}$, the above can be expressed as

$$\left(1 - \frac{2t}{\mu\phi} \right)^{-1/2} e^{\phi-\phi'} \int_0^\infty \sqrt{\frac{\mu'\phi'}{2\pi x^3}} \exp\left[-\phi'\left(\frac{x}{2\mu'} - 1 + \frac{\mu'}{2x} \right) \right] \cdot dx.$$

The definite integral equals unity, and the rest simplifies to (4.52).

(ii) The derivative can be put in the form

$$M_{X^{-1}}'(t) = M_{X^{-1}}(t)\left[\mu^{-1}\left(1 - \frac{2t}{\lambda} \right)^{-1/2} + \lambda^{-1}\left(1 - \frac{2t}{\lambda} \right)^{-1} \right],$$

from which it is clear that $M_{X^{-1}}'(0) = EX^{-1} = \mu^{-1} + \lambda^{-1}$.

(iii) $\lim_{\mu\to\infty} M_{X^{-1}}(t) = (1 - 2t/\lambda)^{-1/2}$, which is the m.g.f. of $\Gamma(1/2, 2/\lambda)$.

68. $\eta_2 > 0$ is necessary for convergence since $\int_0^\infty x^{-3/2} e^{-\eta_1 x} \cdot dx = +\infty$ for $\eta_1 \in \Re$, and since $x^{-3/2} e^{-\eta_1 x - \eta_2 x^{-1}} \mathbf{1}_{(0,\infty)}(x) > x^{-3/2} e^{-\eta_1 x} \mathbf{1}_{(0,\infty)}(x)$ when $\eta_2 < 0$. If $\eta_2 = b > 0$ and $\eta_1 = -a < 0$, $\int_0^\infty x^{-3/2} e^{ax} \cdot dx > \int_b^\infty x^{-3/2} e^{ax-1} \cdot dx = +\infty$, so $\eta_1 \geq 0$ is necessary. Taking $\eta_2 = 1$ with no loss of generality, we have, for $\eta_1 \geq 0$, $\int_0^\infty x^{-3/2} e^{-\eta_1 x - x^{-1}} \cdot dx \leq \int_0^\infty x^{-3/2} e^{-x^{-1}} \cdot dx = \int_0^\infty y^{-1/2} e^{-y} \cdot dy = \Gamma(1/2) = \sqrt{\pi} < \infty$. Thus, $\mathbf{H} = [0, \infty) \times (0, \infty)$.

Part II

The Theory of Statistics

The magic of statistics embraces all phases of life...—William Feller

I don't want any of your statistics; I took your whole batch and lit my pipe with it. —Mark Twain

We use the mathematical theory of probability to infer the unknown results of a known random process. We use the theory of *statistics* to infer the unknown process from known results. As indicated by the divergent views expressed in the above quotations, the practice of statistical inference can be perceived in very different ways, depending on whether it is performed objectively and ethically or is used just to persuade in furtherance of selfish aims. The ethical researcher with a firm grasp of statistical theory is equipped both to extend human knowledge and to refute the statistical "evidence" advanced by those who are unethical or unqualified.

Chapter 5

Sampling Distributions

Statistical inference involves using observed data to produce an explicit model for the mechanism that generated them and that will generate future observations. We refer to this mechanism as the *data-generating process* (*d.g.p.*). The data on which the inference is based, called a *sample*, can be obtained in two general ways. First, it may be drawn from some existing larger collection, called a *population*. For example, one may select a sample from among currently unemployed individuals in the United States in order to learn about current durations of unemployment. In such a case the characteristics of the data will depend both on the method of sampling— that is, on the process by which the individuals are chosen—and on the population itself. Second, the sample may be collected over time as some stochastic process evolves. For example, one may observe at discrete points in time the national unemployment rate, the price of a share of stock, the number of business failures during an interval, or the number of traffic accidents. From such observations one can hope to model the underlying and evolving stochastic processes that govern these quantities.

Using a sample to learn about a population at a moment in time is called *cross-sectional* inference, while learning about an evolving stochastic process from observations over time is inference from *time series*. Our focus here is on cross-sectional inference. While most of the concepts and methods are applicable to time-series analysis, we will not treat the special issues that arise in characterizing the stochastic dependence of observations taken through time. For the same reason we confine the study of cross-sectional inference to that based on *simple random sampling*. By this we mean that, when sampling from a finite population, the observations are drawn at random and *with* replacement. There are in fact many other sampling schemes—sampling without replacement, stratified sampling, systematic sampling, cluster sampling, and so forth. While these other designs are usually more practical to carry out and often produce sharper inferences, random sampling is a benchmark to which the others are compared and is where an entire course on sampling theory would begin.

5.1 Random Sampling

We need to define carefully the concept of a random sample in order both to build a rigorous foundation and to develop the right intuition. This requires considering in more detail than heretofore the structure of models for experiments of repeated trials. As a first step, consider drawing just one element at random from a finite population. For this chance experiment Ω, the sample space, is the population itself, since its members are the potential outcomes. Drawing "at random" means specifically that each of the $N(\Omega)$ entities has the same chance of being drawn; i.e., $\mathbb{P}(\{\omega\}) = N(\Omega)^{-1}$. The equally-likely rule for assigning probabilities then gives $\mathbb{P}(A) = N(A)/N(\Omega)$ for a subset A of Ω. This measure being defined on the collection $\mathcal{F}$ of all subsets, we take the probability space to be $(\Omega, \mathcal{F}, \mathbb{P})$. Now introduce r.v. $X : \Omega \to \Re$. Given the finiteness of Ω, X is necessarily a discrete r.v. Its probability distribution can be represented in any of the usual ways: through its induced measure $\mathbb{P}_X$, its c.d.f. F_X, or its p.m.f. f_X.

Now think of a "meta-experiment" in which the above procedure is replicated some number of times n. That is, in each of n "trials" a single entity is drawn at random from the population and then replaced before the next draw. In this experiment the sample space becomes $\Omega^n = \Omega \times \Omega \times \cdots \times \Omega$, which is the set of all ordered n-tuples $\omega := (\omega_1, \omega_2, ..., \omega_n)$ of the elements of Ω; and the relevant σ field becomes $\mathcal{F}^n$, which represents the collection of all subsets of Ω^n. The sample space now contains $N(\Omega^n) = N(\Omega)^n$ outcomes, since any of $N(\Omega)$ elements could be selected on each of the n draws. To complete the description of the probability space requires a probability measure $\mathbb{P}^n$ on $(\Omega^n, \mathcal{F}^n)$. Still following the equally-likely principle, take the measure of any given ω to be $\mathbb{P}^n(\{\omega\}) = N(\Omega)^{-n}$, so that $\mathbb{P}^n(A^n) = N(A^n)/N(\Omega)^n$ for any subset A^n of Ω^n.

With this background we can begin to learn some of the properties of $\mathbb{P}^n$. We shall see that they are surprisingly benign. Letting A be any subset of Ω, the finite population, denote by $A_j \in \mathcal{F}^n$ the event that the outcome of the jth draw is an element of A. Thus, A_j is the particular subset of Ω^n

(and the particular element of $\mathcal{F}^n$) corresponding to all the ordered n-tuples $\omega = (\omega_1, ..., \omega_j, ..., \omega_n)$ in which $\omega_j \in A$ and in which ω_i is unrestricted for $i \neq j$. When $j = 1$ this is represented as $A_1 = A \times \Omega \times \cdots \times \Omega$; when $j = 2$, as $A_2 = \Omega \times A \times \cdots \times \Omega$; and so on up to $A_n = \Omega \times \cdots \times \Omega \times A$. Since for each j the number of outcomes in A_j is $N(A)N(\Omega)^{n-1}$, the probability measure of event A_j is

$$\mathbb{P}^n(A_j) = \frac{N(A)N(\Omega)^{n-1}}{N(\Omega)^n} = \frac{N(A)}{N(\Omega)} = \mathbb{P}(A),$$

just as for the original probability space for a single draw! If A' is another subset of Ω and $A'_k \in \mathcal{F}^n$ is the event that A' occurs on the kth draw, then likewise $\mathbb{P}^n(A'_k) = N(A')/N(\Omega) = \mathbb{P}(A')$.

Now consider the meaning of the event $A_1 \cap A'_2$. Represented as $A \times A' \times \Omega \times \cdots \times \Omega$, this is the event that an element in A is chosen first and that an element in A' is chosen second. The probability of $A_1 \cap A'_2$ is

$$\mathbb{P}^n(A_1 \cap A'_2) = \frac{N(A)N(A')N(\Omega)^{n-2}}{N(\Omega)^n}$$
$$= \mathbb{P}(A)\mathbb{P}(A')$$
$$= \mathbb{P}^n(A_1)\mathbb{P}^n(A'_2).$$

Generalizing, it should now be clear how to interpret an event of the form $A_1 \cap A'_2 \cap \cdots \cap A_n^{(n-1)}$ for any subsets $A, A', ..., A^{(n-1)}$ of Ω, and one should be able to see that

$$\mathbb{P}^n\left(A_1 \cap A'_2 \cap \cdots \cap A_n^{(n-1)}\right) = \mathbb{P}(A)\mathbb{P}(A') \cdots \cdots \mathbb{P}\left(A^{(n-1)}\right)$$
$$= \mathbb{P}^n(A_1)\mathbb{P}^n(A'_2) \cdots \cdots \mathbb{P}^n(A_n^{(n-1)}).$$

Thus, the probability measure that is appropriate for our scheme of random sampling is such that nontrivial events corresponding to different draws in the meta-experiment are *independent*. (Recall that "nontrivial" in this context refers to events with probabilities on the open interval $(0, 1)$.) What we gain by considering samples with replacement is the analytical simplicity that results from this independence. However, when sampling without replacement from a population that is very large relative to the sample size, not much is lost by ignoring the slight dependence among events and basing inferences on the simpler scheme.

Let us now define some random variables on the space Ω^n. First, let X be a r.v. defined on the original space Ω—i.e., on the "population". As usual, X induces from the original probability space $(\Omega, \mathcal{F}, \mathbb{P})$ the new space $(\Re, \mathcal{B}, \mathbb{P}_X)$, with $\mathbb{P}_X(B) := \mathbb{P}\left(X^{-1}(B)\right) := \mathbb{P}(\{\omega : X(\omega) \in B\})$ for any $B \in \mathcal{B}$ (the Borel sets of the line). Now for $\omega := (\omega_1, ..., \omega_j, ..., \omega_n)$ and any $j \in \{1, 2, ..., n\}$ let $X_j(\omega) := X(\omega_j)$. Thus, $X_j : \Omega^n \to \Re$ assigns to outcome $(\omega_1, \omega_2, ..., \omega_j, ..., \omega_n)$ of the sampling experiment the value that $X : \Omega \to \Re$ assigns to the entity selected on the jth draw. Then, beginning with $j = 1$, for any $B \in \mathcal{B}$ let $A_1 := X_1^{-1}(B) = \{\omega : X_1(\omega) \in B\} \in \mathcal{F}^n$ be the set of outcomes that X_1 maps into B. The measure induced by X_1 is then

$$\mathbb{P}_{X_1}(B) = \mathbb{P}^n(A_1) = \mathbb{P}^n\left(X_1^{-1}(B)\right) = \mathbb{P}(\omega_1 : X(\omega_1) \in B) = \mathbb{P}_X(B).$$

This can be expressed in the usual shorthand as $\Pr(X_1 \in B) = \Pr(X \in B)$. One must realize, however, that $\mathbb{P}_{X_1}(B)$ and $\mathbb{P}_X(B)$ are calculated in different probability spaces—the first pertaining to the meta-experiment of n trials and the second to the experiment of a single draw. Now introduce additional Borel sets $B', B'', ..., B^{(n-1)}$ and the associated inverse images

$$A_2 = X_2^{-1}(B'), A_3 = X_3^{-1}(B''), ..., A_n = X_n^{-1}\left(B^{(n-1)}\right).$$

Since events $\{A_j\}_{j=1}^n$ are independent for all collections $B, B', B'', ... \in \mathcal{B}$, it follows that r.v.s $\{X_j\}_{j=1}^n$ are themselves mutually independent. As well, they have the same marginal distributions, since $\mathbb{P}_{X_j}(B) = \mathbb{P}_X(B)$ for any Borel set B.

To summarize, the process of (i) sampling with replacement from a finite population and (ii) associating with each element a real number yields realizations of mutually *independent and identically distributed (i.i.d.)* random variables. In short, r.v.s $\{X_j\}_{j=1}^n$ created by sampling with replacement from a finite population are i.i.d. as $\mathbb{P}_X$ (or F_X or f_X), where $\mathbb{P}_X$ (or F_X or f_X) represents the distribution of a generic r.v. X that pertains to a single draw from the population. A collection of such i.i.d. r.v.s is called a *simple random sample*. Generalizing, if $\mathbf{X} : \Omega \to \Re^k$ assigns k attributes to each element of the population, the simple random sample $\{\mathbf{X}_j\}_{j=1}^n$ then comprises i.i.d. random *vectors*.

While this model covers all real experiments of sampling from fixed populations, it does not cover all the cases to which the methods of cross-sectional inference can be applied. For example, one might use a sample of items produced by a manufacturing process—40-watt incandescent light bulbs, say—to draw inferences about the indefinitely many items that *could* be produced in this way. R.v.s $\{X_j\}_{j=1}^n$ generated by the sampling process would measure one or (if vector-valued) more quantitative characteristics, such as the number of hours the bulbs last in continuous usage or the number of lumens produced from given current. The sample itself might be taken over time at one particular site or it might be a mix of bulbs selected over time from a several manufacturing sites that use the same process. In either case one should think of the sample as coming from an infinite population of all the potential items that could be produced. To regard the time-series sample as i.i.d., one would have to assume that the process remained the same during the sample period and that random fluctuations in longevity or brightness of bulbs produced at different times were stochastically independent. If sampling from different sites, the processes would have to be invariant over time and space, and chance variations in the attributes at different sites would have to be independent. Under these conditions the sample could be assumed to be simple random, even though its elements were not drawn at random with replacement from a finite population. To handle also this conceptual process of sampling from "infinite" populations, we adopt the following nonconstructive definition.

Definition 5.1. A *simple random sample* of size n is a collection $\{X_j\}_{j=1}^{n}$ of i.i.d. r.v.s or, in the multivariate case, a collection $\{\mathbf{X}_j\}_{j=1}^{n}$ of i.i.d. random vectors.

Some remarks about terminology and notation are in order before we proceed.

(i) The term "random sample" by itself should be understood henceforth to mean "simple random sample".

(ii) Random samples are often *represented* as vectors, $\mathbf{X} := (X_1, X_2, ..., X_n)'$, $\mathbf{Y} := (Y_1, Y_2, ..., Y_n)'$, etc., with generic, *ex post* realizations in lower case, as $\mathbf{X}(\omega) = \mathbf{x}$. Depending on the context, each element of such a vector might itself be vector valued.

(iii) As above, "n" usually denotes a sample size, subscripts being added when a discussion involves two or more samples, as n_X, n_Y, n_1, n_2, etc.

(iv) Recalling the generic r.v. $X : \Omega \to \Re$ that pertains to a single draw from population Ω, it is convenient to describe properties of the population itself in terms of corresponding properties of X. Thus, when we speak of the "distribution" of the population and the "moments" of the population we really have in mind the distribution and moments of X. In particular, "population mean" and "population variance" correspond to EX and VX.

Exercise 5.1. *A sample of size n is drawn with replacement from a population of size $N \geq n$. Find the probability that there are no repetitions—*

i.e., that no member of the population appears in the sample more than once. Using Stirling's formula, find the approximate numerical value when $N = 10,000$ and $n = 10$.

Exercise 5.2. *Continuous r.v. X is defined over some (infinite) population Ω. A random sample of size $n + 1$ is obtained sequentially, starting with an initial value X_0 and followed by $n \in \aleph$ additional independent replicates $X_1, X_2, ..., X_n$. The jth "draw" is said to produce a "record" if $X_j > \max\{X_0, X_1, ..., X_{j-1}\}$. Let R_n be the number of records attained. Find $P(n, 0) := \Pr(R_n = 0)$ and $P(n, n) := \Pr(R_n = n)$.*

Exercise 5.3. *Extending the previous exercise, develop a recursive formula for $P(n, j) := \Pr(R_n = j)$ that applies for each $j \in \{1, 2, ..., n\}$—that is, a formula that expresses $P(n, j)$ in terms of the distributions of R_m for $m < n$ and is computable, ultimately, from the values $\{P(m, 0)\}$ that can be found directly. (Hint: In order for there to be some specific number of records j, in which of the $n + 1$ positions can the largest observation be? If the largest observation is in some such feasible position i, in how many ways can there be j records?)*

5.2   Sample Statistics and Their Distributions

As a singular noun "statistics" refers to the realm of intellectual endeavor that deals with (i) summarizing and revealing salient properties of sample data and (ii) inferring from samples the properties of the data-generating processes. The two aspects of the discipline are styled *descriptive* statistics and *inferential* statistics. Thus, in the singular the term "statistics" encompasses a broad range of techniques and subtopics. As a plural noun, on the other hand, the word has a narrow, technical meaning.

Definition 5.2. Sample *statistics* are (measurable) functions of elements of samples.

Thus, for a random sample $\mathbf{X} := (X_1, X_2, ..., X_n)'$ and a function $g : \Re^n \to \Re$, the quantity $\mathfrak{G} := g(\mathbf{X})$ is a sample statistic if $g^{-1}(B) \in \mathcal{B}^n$ for each Borel set B. Similarly, if $\mathbf{g} : \Re^n \to \Re^k$, then $\mathfrak{G} = \mathbf{g}(\mathbf{X})$ is a vector-valued sample statistic if $\mathbf{g}^{-1}(B_k) \in \mathcal{B}^n$ for each $B_k \in \mathcal{B}^k$.[1] The measurability

---

[1] We do sometimes encounter complex-valued statistics $g : \Re^n \to \mathcal{C}$, such as the "empirical" characteristic function introduced in Chapter 9. For these functions measurability is defined via the isomorphic relation between $\mathcal{C}$ and $\Re^2$.

requirement in the definition is not onerous. Any continuous or piecewise continuous function satisfies this requirement, as do "order statistics" such as $X_{(1:n)} := \min\{X_1, X_2, ..., X_n\}$ and $X_{(n:n)} := \max\{X_1, X_2, ..., X_n\}$. The requirement just ensures that $\mathfrak{G}$ is a random variable on $(\Omega^n, \mathcal{F}^n, \mathbb{P}^n)$, and thus makes it possible to determine $\Pr(\mathfrak{G} \in B)$ for any Borel set B.

Note on notation: We represent sample statistics by capital letters because they are random variables, reserving lower case for generic values of the realizations, as $\mathfrak{G}(\omega) = \mathfrak{g}$. Thus, if $\mathfrak{G} = g(\mathbf{X})$ and $\mathbf{X}(\omega) = \mathbf{x}$, then $\mathfrak{g} = g(\mathbf{x})$.

5.2.1 *Sample Moments*

The sample statistics encountered most often are the sample *moments*. The kth sample moment about the origin is $M_k' := n^{-1} \sum_{j=1}^{n} X_j^k$, with $m_k' := n^{-1} \sum_{j=1}^{n} x_j^k$ as the sample realization. Just as the first moment of a r.v. X (e.g., of the population itself) gets the special symbol $EX = \mu$ and a special name, the "mean", so does the first sample moment. Thus, if the sample is represented as $\mathbf{X} := (X_1, X_2, ..., X_n)'$ we write $\bar{X} := n^{-1} \sum_{j=1}^{n} X_j$ for the *sample mean*, with $\bar{x} := n^{-1} \sum_{j=1}^{n} x_j$ as the realization; and if the sample is represented as $\mathbf{Y} = (Y_1, Y_2, ..., Y_n)'$ we write $\bar{Y} := n^{-1} \sum_{j=1}^{n} Y_j$ and $\bar{y} := n^{-1} \sum_{j=1}^{n} y_j$. Corresponding to population central moment μ_k, the kth sample moment about the (sample) mean is $M_k := n^{-1} \sum_{j=1}^{n} (X_j - \bar{X})^k$. A very basic fact—corresponding to $\mu_1 := E(X - \mu) = 0$ whenever $\mu = \mu_1' = EX$ is finite—is that

$$M_1 = n^{-1} \sum_{j=1}^{n} (X_j - \bar{X}) = n^{-1} \sum_{j=1}^{n} X_j - n^{-1} \sum_{j=1}^{n} \bar{X} = \bar{X} - n^{-1}(n\bar{X}) = 0.$$

In fact, $\bar{X}$ is the unique statistic A that solves $n^{-1} \sum_{j=1}^{n} (X_j - A) = 0$. It is also the unique A such that the sum (and, hence, the average) of squared deviations from A is a minimum.

Exercise 5.4. *Show that $\bar{X}$ is the unique solution to* $\min_{A \in \Re} \left\{ \sum_{j=1}^{n} (X_j - A)^2 \right\}$.

For reasons to be given later, the *sample variance* is defined a little differently from $M_2 := n^{-1} \sum_{j=1}^{n} (X_j - \bar{X})^2$, and it is also given a special symbol:

$$S^2 := \frac{1}{n-1} \sum_{j=1}^{n} (X_j - \bar{X})^2 = \frac{n}{n-1} M_2.$$

$S = +\sqrt{S^2}$ is called the *sample standard deviation*. The identities

$$M_2 = M_2' - \bar{X}^2 \tag{5.1}$$

$$M_2 = \frac{1}{2n^2} \sum_{i=1}^{n} \sum_{j=1}^{n} (X_i - X_j)^2 \tag{5.2}$$

$$S^2 = \frac{1}{\binom{n}{2}} \sum_{i<j} \frac{(X_i - X_j)^2}{2} \tag{5.3}$$

will be found useful later in understanding the properties of S^2 and M_2.

Exercise 5.5. *Verify the equalities in (5.1)-(5.3).*

5.2.2   *Moments of Sample Moments*

Not surprisingly, realizations of sample moments can be used to draw inferences about the moments and other features of populations; however, drawing *intelligent* inferences requires recognizing that these sample statistics are themselves random variables. As such, their realizations would ordinarily vary from one sample to the next in replications of the sampling experiment, just as would the numbers of heads on repeated flips of three coins, the clockwise distances from origin to pointer on repeated spins, and the number of dots on the upturned faces of a pair of dice. To use sample moments—and sample statistics generally—for inference requires an understanding of their probability distributions. Of course, not much can be said about these without knowing at least the family of models that governs the population itself. Nevertheless, it is possible to express the mean, variance, and higher moments of the sample moments in terms of population moments, assuming that these exist. While those population moments are typically unknown, we can at least determine how they influence and relate to the distributions of sample moments. In what follows we take for granted that all the relevant population moments are finite.

Expressing the first few moments of sample moments about the origin is relatively easy, since any M_k' is just a linear function of powers of the i.i.d. elements of the sample. Thus, the first moment is

$$EM_k' = n^{-1} \sum_{j=1}^{n} EX_j^k = n^{-1} \sum_{j=1}^{n} \mu_k' = \mu_k'.$$

In particular, $E\bar{X} = \mu$, the population mean. Thinking of expectation as "long-run average", the fact that $E\bar{X} = \mu$ implies that the average realization of $\bar{X}$ in infinitely many replications of the sampling experiment would equal the mean of the population. In other words, although no single realization would be apt to equal μ, the differences from μ are random and would eventually average out. Notice that this particular fact does not depend on the mutual independence of elements of the sample, nor does it depend on the sample size.

Now that M'_k is known to equal μ'_k on average, the next concern is about how *close* its realizations are apt to be. For example, one might want to know the long-run average absolute deviation, $E\,|M'_k - \mu'_k|$. Unfortunately, evaluating this requires a parametric model for the population, but it is at least possible to express $E\,(M'_k - \mu'_k)^2 = VM'_k$—and from it, the standard deviation—in terms of population moments alone. For this the independence of the sample elements *does* matter, and the result *does* depend on the sample size. Independence of $\{X_j\}_{j=1}^n$ implies the independence, and therefore the uncorrelatedness, of $\{X_j^k\}_{j=1}^n$; and since the variance of a sum of uncorrelated r.v.s equals the sum of their variances, we have

$$VM'_k = n^{-2} \sum_{j=1}^n VX_j^k = n^{-2}\left[nE\left(X^k - \mu'_k\right)^2\right] = n^{-1}\left[\mu'_{2k} - (\mu'_k)^2\right].$$

With $k = 1$ this gives the familiar result that the variance of the sample mean equals the population variance divided by the sample size:

$$V\bar{X} = n^{-1}\left(\mu'_2 - \mu^2\right) = \frac{\sigma^2}{n}.$$

Thus, while the exact distribution of any M'_k may be unknown, we do know that its dispersion (as measured by standard deviation) is inversely proportional to $\sqrt{n}$. Accordingly, if realizations of M'_k are to be used as *estimates* of μ'_k, a quadrupling of the sample size would be needed to increase their *precision* by a factor of two.

Finding moments of the *central* sample moments $\{M_k\}$ is more difficult, in part because the terms $\{(X_j - \bar{X})^k\}_{j=1}^n$ are correlated. Working through the calculations is easier if the population mean is zero; and no generality is lost by just *assuming* this, because $E(X_j - \bar{X}) = \mu - \mu = 0$ regardless of the value of μ. For M_2 either of relations (5.1) or (5.2) can be used to find the expected value. Using (5.1) and setting $\mu = 0$ (so that $E\bar{X}^2 = V\bar{X}$ and $\mu'_2 = \mu_2 = \sigma^2$) give

$$EM_2 = E\left(M'_2 - \bar{X}^2\right) = \mu'_2 - V\bar{X}$$

$$= \mu'_2 - \frac{\sigma^2}{n} = \frac{n-1}{n}\sigma^2.$$

Similarly, from (5.2) with $\mu = 0$,

$$EM_2 = \frac{1}{2n^2} \sum_{i,j=1}^{n} E\left(X_i - X_j\right)^2 = \frac{1}{2n^2} \sum_{i=1, j\neq i}^{n} \left(EX_i^2 - 2EX_iX_j + EX_j^2\right)$$

$$= \frac{2(n^2 - n)\sigma^2}{2n^2} = \frac{n-1}{n}\sigma^2.$$

Thus, on average in infinitely many replications of the sampling experiment, the second sample moment about the mean falls short of the population variance by the factor $(n-1)/n$. On the other hand, the statistic $S^2 = n/(n-1)M_2$ corrects this "bias", which accounts for its wide usage in statistical inference and the name "sample variance".

Working out VM_2 from (5.1) with $\mu = 0$, begin with

$$VM_2 = EM_2^2 - (EM_2)^2$$

$$= E\left(M_2' - \bar{X}^2\right)^2 - \left(\frac{n-1}{n}\sigma^2\right)^2.$$

Next, express M_2' and $\bar{X}$ and square to get

$$\left(M_2' - \bar{X}^2\right)^2 = \left(\frac{1}{n}\sum_{j=1}^{n} X_j^2 - \frac{1}{n^2}\sum_{j,k=1}^{n} X_jX_k\right)^2$$

$$= \frac{1}{n^2}\sum_{i,j=1}^{n} X_i^2X_j^2 - \frac{2}{n^3}\sum_{i,j,k=1}^{n} X_i^2X_jX_k + \frac{1}{n^4}\sum_{i,j,k,l=1}^{n} X_iX_jX_kX_l.$$

Now take expectations term by term. In the first sum $EX_i^2X_j^2 = \mu_4$ for the n terms with $i = j$, and $EX_i^2X_j^2 = \sigma^2 \cdot \sigma^2 = \sigma^4$ for the $n^2 - n$ remaining terms. In the second sum $EX_i^2X_jX_k = \mu_4$ for the n terms with $i = j = k$; $EX_i^2X_jX_k = \sigma^4$ for the $n^2 - n$ terms with $j = k \neq i$; and the remaining terms with $j \neq k$ are zero because the variables are independent and (as assumed) have zero mean. Finally, in the last sum $EX_iX_jX_kX_l = \mu_4$ for the n terms having all subscripts the same, and $EX_iX_jX_kX_l = \sigma^4$ for all terms whose subscripts are equal in pairs. There are $3n(n-1)$ such terms, since i can be paired with any one of j, k, l, and then the remaining two must match. Putting it all together and simplifying give

$$VM_2 = \left(\frac{n-1}{n}\right)^2 \frac{\mu_4 - \sigma^4}{n} + \frac{2(n-1)\sigma^4}{n^3}.$$

Since $VS^2 = \left(\frac{n}{n-1}\right)^2 VM_2$, the variance of the sample variance has the simpler expression

$$VS^2 = \frac{\mu_4 - \sigma^4}{n} + \frac{2\sigma^4}{n(n-1)}. \tag{5.4}$$

Notice that the dominant term for large n is proportional to n^{-1}, as was true for $V\bar{X}$. Such is the case generally for variances of sample moments.

Exercise 5.6. *For random samples of size n from populations with finite third moment show that (i) $Cov\left(\bar{X}, M_2\right) = \mu_3\left(n-1\right)/n^2$ and (ii) $Cov\left(\bar{X}, S^2\right) = \mu_3/n$.*

Exercise 5.7. *Express the coefficient of correlation between $\bar{X}$ and S^2 for random samples of size n from populations with finite fourth moment. Applying the bounds for correlations, show that the coefficients of skewness (α_3) and kurtosis (α_4) of any such population satisfy $\alpha_4 \geq \alpha_3^2 + 1$.[2]*

5.3 Sampling from Normal Populations

Traditional, "parametric" statistical inference begins with a sample and a family of models, one of which is assumed to have generated the data under study. For example, one might have adopted the Poisson family for data that represent the number of events in time or space or the lognormal family for incomes in a population of households.[3] The next step is to use the sample at hand to draw inferences about the values of the parameters that apply to these particular data. Inferences about parameters are of several types. One may seek to find sample statistics called "estimators" that give reliable point estimates. Alternatively, one may want to produce "confidence intervals", realizations of pairs of statistics that bracket the true but unknown values of parameters with known probability. Finally, one may want to test the hypothesis that a parameter lies in a specific region of the allowable space. Choosing estimators, building confidence intervals, and testing hypotheses (the subjects of Chapters 7 and 8) all require knowing the distributions of various sample statistics. Often, depending on the model and the statistics themselves, working out exact distributions analytically can be extremely difficult, which is why we often try to find approximations

---

[2]This derivation of the inequality is due to Sen (2012).

[3]The topic of *model selection* deals with statistical techniques for choosing among specific parametric families the one that best characterizes the data. The Akaike and Schwartz criteria are the best known procedures. Although these are beyond our present scope, we do describe ways of restricting models to specific *subfamiles* (Section 8.6) and of testing the "goodness of fit" of parametric families against unspecified alternatives (Section 9.2)

Also beyond our scope are topics of *experimental design, sampling design,* and *sequential analysis* that affect the composition and sizes of the samples with which we work.

based on large-sample asymptotics (the subject of Chapter 6). However, things are often easier when sampling from *normal* populations, for two reasons. First, the properties of the model itself facilitate mathematical analysis; for example, the independence of uncorrelated normals and the closure under affine transformations. Second, the two parameters of the univariate normal correspond to mean and variance, so knowing just the first two moments identifies the particular member of the family without further analysis. In this section we work out exact distributions of several statistics that will be found useful in drawing inferences about normal populations. The development of these results for the normal distribution will also illustrate techniques that can be applied more broadly.

Throughout the section the elements of sample $\mathbf{X} := (X_1, X_2, ..., X_n)'$ are assumed to be independently and identically distributed (i.i.d.) as $N\left(\mu, \sigma^2\right)$. $\mathbf{X}$ itself is therefore multivariate normal with mean $\boldsymbol{\mu} = (\mu, \mu, ..., \mu)' = \mu\mathbf{1}_n$ and covariance matrix $\boldsymbol{\Sigma} = \sigma^2\mathbf{I}_n$, where $\mathbf{1}_n$ is an n-vector of units and $\mathbf{I}_n$ is the $n \times n$ identity matrix. Thus, $\boldsymbol{\Sigma}$ is a diagonal matrix with σ^2 repeated along the principal diagonal.

5.3.1   *The Sample Mean and Sample Variance*

We first work out the marginal distribution of $\bar{X}$, then show that $\bar{X}$ and S^2 are stochastically independent, and then deduce the marginal distribution of S^2. We will often apply the following fact (proved in Section 4.3.7.1): If n-vector $\mathbf{X}$ is distributed as $N\left(\boldsymbol{\mu}, \boldsymbol{\Sigma}\right)$, and if $\mathbf{A}$ is an $m \times n$ matrix of rank $m \leq n$, then m-vector $\mathbf{AX}$ is distributed as $N\left(\mathbf{A}\boldsymbol{\mu}, \mathbf{A}\boldsymbol{\Sigma}\mathbf{A}'\right)$.

5.3.1.1   *The Marginal Distribution of $\bar{X}$*

Writing $\bar{X} = n^{-1}\sum_{j=1}^n X_j$ as the inner product $n^{-1}\mathbf{1}_n'\mathbf{X}$ puts $\bar{X}$ in the form $\mathbf{AX}$ for $\mathbf{A} = n^{-1}\mathbf{1}_n' \in \Re^n$ and shows that $\bar{X} \sim N\left(n^{-1}\mathbf{1}_n'\boldsymbol{\mu}, n^{-2}\mathbf{1}_n'\boldsymbol{\Sigma}\mathbf{1}_n\right)$. This simplifies to $\bar{X} \sim N\left(\mu, \sigma^2/n\right)$ since

$$\frac{1}{n}\mathbf{1}_n'\boldsymbol{\mu} = \frac{1}{n}\mathbf{1}_n'\mathbf{1}_n\mu = \frac{n\mu}{n} = \mu$$

and

$$\frac{1}{n^2}\mathbf{1}_n'\boldsymbol{\Sigma}\mathbf{1}_n = \frac{1}{n^2}\mathbf{1}_n'\left(\sigma^2\mathbf{I}_n\right)\mathbf{1}_n = \frac{\sigma^2}{n^2}\mathbf{1}_n'\mathbf{1}_n = \frac{\sigma^2}{n}.$$

Of course, it has already been shown that $E\bar{X} = \mu$ and $V\bar{X} = \sigma^2/n$ for random samples from any population with finite variance. What is new here is that $\bar{X}$ *is normally distributed when the population is normal.*

Example 5.1. Taking a random sample of size n from a normal population, we have for any $\varepsilon > 0$

$$\Pr\left(\left|\bar{X} - \mu\right| \geq \varepsilon\right) = \Pr\left(\left|\frac{\bar{X} - \mu}{\sigma/\sqrt{n}}\right| \geq \frac{\varepsilon\sqrt{n}}{\sigma}\right)$$

$$= \Pr\left(|Z| \geq \frac{\varepsilon\sqrt{n}}{\sigma}\right) \text{ for } Z \sim N\left(0, 1\right)$$

$$= 2\left[1 - \Phi\left(\frac{\varepsilon\sqrt{n}}{\sigma}\right)\right].$$

For $\sigma = .5$ and $n = 100$ this gives $\Pr\left(\left|\bar{X} - \mu\right| \geq 0.1\right) \doteq 0.0456$, which substantially improves the Chebychev bound of 0.25. Letting $n \to \infty$ and applying the monotonicity property of probability shows that $\Pr\left(\left|\bar{X} - \mu\right| \geq \varepsilon\right) \to 2\left[1 - \Phi(+\infty)\right] = 0$. By taking a large enough sample we can therefore be as confident as we like that $\bar{X}$ will take a value arbitrarily close to μ. This is a special case of the weak law of large numbers, which is discussed in Section 6.6. We shall see there that the result applies much more generally than just to samples from normal populations.

Exercise 5.8. *If $\sigma = 0.5$ find the smallest n such that $\Pr\left(\left|\bar{X} - \mu\right| \geq 0.1\right) \leq 0.05$.*

Exercise 5.9. *When $\{X_j\}_{j=1}^{n}$ are i.i.d. as $N\left(\mu, \sigma^2\right)$, we have seen that $\bar{X} \sim N\left(\mu, \sigma^2/n\right)$. Show that an analogous result holds for the inverse Gaussian distribution; specifically, that $\bar{X} \sim IG\left(\mu, \lambda n\right)$ if $\{X_j\}_{j=1}^{n}$ are i.i.d. as $IG\left(\mu, \lambda\right)$. (Hint: Apply expression (4.51).)*

5.3.1.2 The Independence of $\bar{X}$ and S^2

Setting $D_j := X_j - \bar{X}$, sample variance S^2 can be expressed compactly as $(n-1)^{-1}\sum_{j=1}^{n} D_j^2$. Since the deviations from $\bar{X}$ sum to zero, any one of $\{D_j\}_{j=1}^{n}$ equals the negative of the sum of the remaining ones. Putting $D_1 = -\sum_{j=2}^{n} D_j$ gives

$$(n-1)S^2 = \left(-\sum_{j=2}^{n} D_j\right)^2 + \sum_{j=2}^{n} D_j^2,$$

a function of $\{D_j\}_{j=2}^{n}$ alone. Adopting vector notation, write $D_j = \boldsymbol{\delta}_j'\mathbf{X}$, where $\boldsymbol{\delta}_j := \left(-\frac{1}{n}, -\frac{1}{n}, ..., 1 - \frac{1}{n}, ..., -\frac{1}{n}\right)'$ (an n-vector with $1 - \frac{1}{n}$ in the jth position and $-\frac{1}{n}$ elsewhere), and note that $\mathbf{1}_n'\boldsymbol{\delta}_j = \boldsymbol{\delta}_j'\mathbf{1}_n = \sum_{j=1}^{n} D_j = 0$. Writing the vector $(\bar{X}, D_2, ..., D_n)'$ as $\mathbf{DX}$, where $\mathbf{D} := \left(n^{-1}\mathbf{1}_n', \boldsymbol{\delta}_2', ..., \boldsymbol{\delta}_n'\right)'$, shows it to be multinormal with mean vector $\mathbf{D}\boldsymbol{\mu}$ and covariance matrix $\mathbf{D}\boldsymbol{\Sigma}\mathbf{D}'$. The off-diagonal elements of the first column and first row of $\mathbf{D}\boldsymbol{\Sigma}\mathbf{D}'$ are covariances between $\bar{X}$ and the D_j. Since

$$
\begin{aligned}
Cov\left(\bar{X}, D_j\right) &= E\left[n^{-1}\mathbf{1}_n'\left(\mathbf{X} - \boldsymbol{\mu}\right)\left(\mathbf{X} - \boldsymbol{\mu}\right)'\boldsymbol{\delta}_j\right] \\
&= n^{-1}\mathbf{1}_n'\left(\sigma^2\mathbf{I}_n\right)\boldsymbol{\delta}_j \\
&= n^{-1}\sigma^2\mathbf{1}_n'\boldsymbol{\delta}_j \\
&= 0,
\end{aligned}
$$

we conclude that $\bar{X}$ is uncorrelated with and—since they are jointly normal—*independent of* each of $\{D_j\}_{j=2}^n$. Finally, recalling that S^2 is a function of $\{D_j\}_{j=2}^n$ alone and that (non-constant) functions of individual independent r.v.s are themselves independent, we conclude that $\bar{X}$ and S^2 are stochastically independent when the sample is drawn from a normal population.[4] We have seen already (Example 2.32 and Exercise 4.61) that it is possible for *functionally* related r.v.s to be statistically independent. Here we have another example—and one that turns out to be of considerable importance in statistical inference.

Exercise 5.10. *Show directly, without vector notation, that* $Cov\left(\bar{X}, D_j\right) = 0$. *Is normality necessary for this result? Does it hold without any restriction whatsoever? (Hint: Read the footnote.)*

5.3.1.3 *The Marginal Distribution of* S^2

We will work out the distribution of a scale transformation of S^2, namely $C := (n-1)S^2/\sigma^2$, by establishing the following facts:

(i) If $\{Z_j\}_{j=1}^\infty$ are i.i.d. as $N(0,1)$, then for any positive integer ν the r.v. $C_\nu := \sum_{j=1}^\nu Z_j^2$, has a gamma distribution with shape parameter $\alpha = \nu/2$ and scale parameter $\beta = 2$.

(ii) When the population is normal and $n \geq 2$, $(n-1)S^2/\sigma^2$ has the same distribution as C_{n-1}.

Members of the gamma subfamily $\{\Gamma(\nu/2, 2) : \nu \in \aleph\}$ are known as the "chi-squared" distributions. They are important in both theoreti-

---

[4]Exercise 5.6 shows that $\bar{X}$ and S^2 have zero covariance whenever $\mu_3 = 0$. Thus, means and variances of samples from symmetric populations (with finite third moment) are always uncorrelated. However, it is known [Lukacs (1942)] that the *independence* of $\bar{X}$ and S^2 from random samples holds *only* for normal populations—and thus characterizes the Gaussian law. A much simpler proof of independence of $\bar{X}$ and S^2 will be given as an application of Basu's theorem (p. 470).

cal and applied statistics. Development of the chi-squared model begins by finding the distribution of $C_1 := Z^2$ when $Z \sim N(0,1)$. Since $F_{C_1}(c) = \mathbb{P}_Z([-\sqrt{c}, \sqrt{c}]) = \Pr(-\sqrt{c} \leq Z \leq \sqrt{c})$ for $c \geq 0$, the density is

$$
\begin{aligned}
f_{C_1}(c) &= \frac{d}{dc} \int_{-\sqrt{c}}^{\sqrt{c}} \frac{1}{\sqrt{2\pi}} e^{-z^2/2} \cdot dz \\
&= \sqrt{\frac{2}{\pi}} \frac{d}{dc} \int_0^{\sqrt{c}} e^{-z^2/2} \cdot dz \\
&= \sqrt{\frac{2}{\pi}} e^{-c/2} \frac{d\sqrt{c}}{dc} \\
&= \frac{1}{\Gamma\left(\frac{1}{2}\right) 2^{1/2}} c^{-1/2} e^{-c/2}.
\end{aligned}
$$

(Recall that $\Gamma\left(\frac{1}{2}\right) = \sqrt{\pi}$.) This is a gamma p.d.f. with shape parameter $\alpha = 1/2$ and scale parameter $\beta = 2$. Recalling from Expression (4.34) that any subfamily of gammas with the same scale parameter is closed under convolution, we see that

$$
C_\nu := \sum_{j=1}^{\nu} Z_j^2 \sim \Gamma\left(\sum_{j=1}^{\nu} \frac{1}{2}, 2\right) = \Gamma\left(\frac{\nu}{2}, 2\right), \nu \in \aleph.
$$

Definition 5.3. A r.v. C_ν distributed as $\Gamma\left(\frac{\nu}{2}, 2\right)$ for $\nu \in \aleph$ is said to have a *chi-squared* distribution with ν "degrees of freedom". In symbols, $C_\nu \sim \chi^2(\nu)$.

The rationale for the term "degrees of freedom" (d.f.) is given below. Properties of chi-squared variates follow from those of the larger gamma family. In particular, the m.g.f. is $M_{C_\nu}(t) = (1-2t)^{-\nu/2}$ for $t < 1/2$, the mean is $EC_\nu = (\nu/2) \cdot 2 = \nu$, and the variance is $VC_\nu = (\nu/2) \cdot 2^2 = 2\nu$. Appendix C.3 gives a table of upper-p quantiles of $\chi^2(\nu)$ for various ν and p—that is, values c_ν^p such that $\Pr(C_\nu \geq c_\nu^p) = p$. Software to compute both the quantiles and values of the c.d.f. is also widely available.

We can now develop the distribution of $C := (n-1)S^2/\sigma^2 = nM_2/\sigma^2$ when elements of sample $\{X_j\}_{j=1}^n$ are i.i.d. as $N(\mu, \sigma^2)$. The deviations $\{X_j - \bar{X}\}_{j=1}^n$ on which S^2 depends being mean-invariant, we can take $\mu = 0$ without loss of generality. The identity

$$
C = \frac{nM_2}{\sigma^2} = \sum_{j=1}^n \left(\frac{X_j}{\sigma}\right)^2 - n \left(\frac{\bar{X}}{\sigma}\right)^2,
$$

implies that

$$C + \left(\frac{\bar{X}}{\sigma/\sqrt{n}}\right)^2 = \sum_{j=1}^{n} \left(\frac{X_j}{\sigma}\right)^2.$$

Since $\mu = 0$, the summands $\{X_j/\sigma\}_{j=1}^{n}$ are i.i.d. as $N(0,1)$ and so the sum of squares is distributed as $\chi^2(n)$. Likewise, since $\bar{X} \sim N\left(0, \sigma^2/n\right)$ when $\mu = 0$, the squared standardized sample mean on the left side is distributed as $\chi^2(1)$. The two terms on the left are statistically independent because S^2 is independent of $\bar{X}$. Writing C_n for the sum on the right and $\left(\frac{\bar{X}}{\sigma/\sqrt{n}}\right)^2 =: C_1$, independence of C and C_1 implies that $Ee^{t(C+C_1)} = M_C(t)M_{C_1}(t)$, and the equality $C + C_1 = C_n$ implies $M_C(t)M_{C_1}(t) = M_{C_n}(t)$. M.g.f.s being strictly positive, we have

$$M_C(t) = \frac{M_{C_n}(t)}{M_{C_1}(t)} = \frac{(1-2t)^{-n/2}}{(1-2t)^{-1/2}} = (1-2t)^{-(n-1)/2}, t < \frac{1}{2}.$$

Accordingly, $C := (n-1)S^2/\sigma^2$ and $C_{n-1} := \sum_{j=1}^{n-1} Z_j^2$ have the same $\chi^2(n-1)$ distribution. The marginal distribution of S^2 itself is not needed: In applications it is easy to work with this rescaled, parameter-free version, whose quantiles it is practical to tabulate.

Although it now has many uses in statistics, the chi-squared model was developed specifically to describe the behavior of sample variances from normal populations. Since the deviations $\{X_j - \bar{X}\}_{j=1}^{n}$ that enter S^2 sum to zero, these n pieces of data do have only $n-1$ "degrees of freedom". That is, having extracted from sample $\{X_j\}_{j=1}^{n}$ an estimate of the population mean, we are left with just $n-1$ pieces of information about the population's dispersion. (Clearly, *one* observation gives no information at all about dispersion unless the mean is known.) This both explains the name attached to the chi-squared parameter and provides some intuition for why dividing $\sum_{j=1}^{n} (X_j - \bar{X})^2$ by $n-1$ instead of n makes $ES^2 = \sigma^2$.

Example 5.2. In a sample of size $n = 25$ from a normal population the probability that the sample variance exceeds the population variance by more than a factor of 2.0 is

$$\Pr\left(S^2 > 2\sigma^2\right) = \Pr\left(24 \cdot \frac{S^2}{\sigma^2} > 24 \cdot 2\right)$$

$$= \Pr\left(C_{24} > 48\right),$$

where $C_{24} \sim \chi^2(24)$. A table of χ^2 quantiles shows that this probability is less than .005.

Exercise 5.11. *A sample of size $n = 25$ is taken from a normal population. From a χ^2 table find numbers $s_{.05}$ and $s_{.95}$ such that $\Pr\left(S^2 \leq s_{.05} \cdot \sigma^2\right) = .05$ and $\Pr\left(S^2 \geq s_{.95} \cdot \sigma^2\right) = .05$.*

Exercise 5.12.

(i) *Show that $(n-1)\, S^2/\sigma^2 \sim \chi^2(n-1)$ leads directly to the result $ES^2 = \sigma^2$ when sampling from normal populations and that it also implies $VS^2 = 2\sigma^4/(n-1)$.*

(ii) *Show that under normality this expression for VS^2 is consistent with that given for the general case in (5.4).*

Exercise 5.13. *When $\{X_j\}_{j=1}^{n}$ are i.i.d. as $N\left(\mu, \sigma^2\right)$, we have seen that S^2 and $\bar{X}$ are independent and that $(n-1)\, S^2/\sigma^2 \sim \chi^2(n-1)$. Analogous results hold for the inverse Gaussian distribution; specifically, if $\{X_j\}_{j=1}^{n}$ are i.i.d. as $IG\left(\mu, \lambda\right)$ it is known (Tweedie (1957)) (i) that $\bar{X}$ and $\bar{Y} - \bar{X}^{-1} := n^{-1}\sum_{j=1}^{n}\left(X_j^{-1} - \bar{X}^{-1}\right)$ are independent and (ii) that $R := n\lambda\left(\bar{Y} - \bar{X}^{-1}\right) \sim \chi^2(n-1)$. Apply the first of these facts to prove the second. (Hint: Recall the results of Exercises 4.67 and 5.9.)*

5.3.2 *Quadratic Forms in Multinormals*

In statistical analysis one often works with quadratic forms in multivariate normal r.v.s. Here are two cases in which such quadratic forms have chi-squared distributions.[5]

Theorem 5.1. *If $\mathbf{X} : \Omega \to \Re^k$ is distributed as $N\left(\boldsymbol{\mu}, \boldsymbol{\Sigma}\right)$, where $\boldsymbol{\Sigma}$ has rank k, then*

$$\left(\mathbf{X} - \boldsymbol{\mu}\right)' \boldsymbol{\Sigma}^{-1}\left(\mathbf{X} - \boldsymbol{\mu}\right) \sim \chi^2\left(k\right).$$

Proof. Covariance matrix $\boldsymbol{\Sigma}$ is symmetric and positive definite, so there exits a nonsingular square-root matrix $\mathbf{R}$ such that $\mathbf{RR}' = \boldsymbol{\Sigma}$. Random vector $\mathbf{Z} := \mathbf{R}^{-1}\left(\mathbf{X} - \boldsymbol{\mu}\right)$, an affine function of multivariate normals, is itself multivariate normal with mean vector

$$E\mathbf{Z} = \mathbf{R}^{-1}\left(E\mathbf{X} - \boldsymbol{\mu}\right) = \mathbf{R}^{-1}\left(\boldsymbol{\mu} - \boldsymbol{\mu}\right) = \mathbf{0}$$

and covariance matrix

$$V\mathbf{Z} = E\mathbf{Z}\mathbf{Z}' = E\mathbf{R}^{-1}\left(\mathbf{X} - \boldsymbol{\mu}\right)\left(\mathbf{X} - \boldsymbol{\mu}\right)' \mathbf{R}'^{-1}$$
$$= \mathbf{R}^{-1}\boldsymbol{\Sigma}\mathbf{R}'^{-1} = \mathbf{R}^{-1}\left(\mathbf{RR}'\right)\mathbf{R}'^{-1} = \mathbf{I}_k.$$

---

[5]Some background in matrix algebra is needed to follow the proofs. Schott (1997) is a particularly good reference, since it focuses on statistical applications.

But $\mathbf{Z} \sim N\left(\mathbf{0}, \mathbf{I}_k\right)$ implies that its components $\{Z_j\}_{j=1}^k$ are i.i.d. as $N\left(0, 1\right)$ and therefore that

$$
\begin{aligned}
\left(\mathbf{X} - \boldsymbol{\mu}\right)' \boldsymbol{\Sigma}^{-1}\left(\mathbf{X} - \boldsymbol{\mu}\right) &= \left(\mathbf{X} - \boldsymbol{\mu}\right)' \left(\mathbf{RR}'\right)^{-1}\left(\mathbf{X} - \boldsymbol{\mu}\right) \\
&= \left(\mathbf{X} - \boldsymbol{\mu}\right)' \mathbf{R}'^{-1}\mathbf{R}^{-1}\left(\mathbf{X} - \boldsymbol{\mu}\right) \\
&= \left[\mathbf{R}^{-1}\left(\mathbf{X} - \boldsymbol{\mu}\right)\right]' \left[\mathbf{R}^{-1}\left(\mathbf{X} - \boldsymbol{\mu}\right)\right] \\
&= \mathbf{Z}'\mathbf{Z} \sim \chi^2\left(k\right).
\end{aligned}
$$
□

Theorem 5.2. *If* $\mathbf{Z} \sim N\left(\mathbf{0}, \mathbf{I}_k\right)$ *and* $\mathbf{D}$ *is a* $k \times k$ *symmetric* idempotent *matrix of rank* $m \leq k$, *then* $\mathbf{Z}'\mathbf{DZ} \sim \chi^2\left(m\right)$.

Proof. Being idempotent, $\mathbf{D}$ is its own square-root matrix; and, being symmetric and of rank m, it has m characteristic roots equal to unity and $k - m$ roots equal to zero. Let $\mathbf{Q}$ be the $k \times k$ matrix of eigenvectors. $\mathbf{Q}$ is an orthogonal matrix (meaning that $\mathbf{QQ}' = \mathbf{Q}'\mathbf{Q} = \mathbf{I}_k$) such that $\mathbf{Q}'\mathbf{DQ} = \boldsymbol{\Lambda}$, where $\boldsymbol{\Lambda}$ is a diagonal matrix with the k characteristic roots of $\mathbf{D}$ along the diagonal. Now $\mathbf{Z}'\mathbf{DZ} = \mathbf{Z}'\mathbf{QQ}'\mathbf{DQQ}'\mathbf{Z} = \mathbf{Z}'\mathbf{Q}\boldsymbol{\Lambda}\mathbf{Q}'\mathbf{Z}$. But linear transformation $\mathbf{Q}'\mathbf{Z}$ is also multivariate normal with mean $E\mathbf{Q}'\mathbf{Z} = \mathbf{0}$ and covariance matrix

$$
E(\mathbf{Q}'\mathbf{ZZ}'\mathbf{Q}) = \mathbf{Q}'E(\mathbf{ZZ}')\mathbf{Q} = \mathbf{Q}'\mathbf{I}_k\mathbf{Q} = \mathbf{Q}'\mathbf{Q} = \mathbf{I}_k.
$$

Since $\mathbf{Q}'\mathbf{Z}$ and $\mathbf{Z}$ have the same distributions, it follows that

$$
\mathbf{Z}'\mathbf{Q}\boldsymbol{\Lambda}\mathbf{Q}'\mathbf{Z} \sim \mathbf{Z}'\boldsymbol{\Lambda}\mathbf{Z} = \sum_{j=1}^m Z_j^2 \sim \chi^2\left(m\right).
$$
□

Theorem 5.1 will be applied in Chapters 7 and 8 in building confidence intervals and testing parametric hypotheses. Theorem 5.2 leads to a simple proof that $(n - 1) S^2 / \sigma^2 \sim \chi^2(n - 1)$ when sampling from $N\left(\mu, \sigma^2\right)$.

Example 5.3. $\sum_{j=1}^n \left(X_j - \bar{X}\right)^2 / \sigma^2$ and $C := \sum_{j=1}^n \left(Z_j - \bar{Z}\right)^2$ have the same distribution, where $\{Z_j\}_{j=1}^n$ are i.i.d. as $N\left(0, 1\right)$ and $\bar{Z} := n^{-1}\sum_{j=1}^n Z_j$. Writing $\bar{Z}$ as $n^{-1}\mathbf{1}'_n\mathbf{Z}$, where $\mathbf{Z} := \left(Z_1, Z_2, ..., Z_n\right)'$ and $\mathbf{1}_n$ is an n-vector of units, C can be expressed as the quadratic form

$$
C = \mathbf{Z}'\left(\mathbf{I}_n - n^{-1}\mathbf{1}_n\mathbf{1}'_n\right)\mathbf{Z},
$$

where $\mathbf{I}_n$ is the identity matrix. Since $\mathbf{I}_n - n^{-1}\mathbf{1}_n\mathbf{1}'_n$ is idempotent with rank equal to its trace, $n - 1$ (the sum of its diagonal elements), Theorem 5.2 implies that $C \sim \chi^2\left(n - 1\right)$.

Exercise 5.14. *Show that* $C = \mathbf{Z}'\left(\mathbf{I}_n - n^{-1}\mathbf{1}_n\mathbf{1}'_n\right)\mathbf{Z}$ *and verify that* $\mathbf{I}_n - n^{-1}\mathbf{1}_n\mathbf{1}'_n$ *is idempotent.*

5.3.3   *Certain Functions of Sample Means and Variances*

We now work out two more important results for samples from normal populations: the distributions of (i) the "studentized" sample mean, $\sqrt{n}\left(\bar{X} - \mu\right)/S$, and (ii) the ratio $\left(S_1^2/\sigma_1^2\right) \div \left(S_2^2/\sigma_2^2\right)$ of sample/population variances from independent samples from two different normal populations. The new models, "Student's" t and Snedecor's F, will be used in Chapters 7 and 8 to draw inferences about the mean of a single normal population and about the relation between variances of different populations.

5.3.3.1   *Distribution of "Studentized" Mean*

Although $\sqrt{n}\left(\bar{X} - \mu\right)/\sigma \sim N\left(0, 1\right)$, the function

$$T := \frac{\sqrt{n}\left(\bar{X} - \mu\right)}{S} = \frac{\sqrt{n}\left(\bar{X} - \mu\right)}{\sigma} \div \left(\frac{S}{\sigma}\right)$$

clearly has a different distribution, since sample standard deviation $S := +\sqrt{S^2}$ is itself a r.v. One expects the distribution of T still to be centered at the origin but to have more probability mass in the tails because of the randomness in S/σ; and, in fact, this is so. The p.d.f. was worked out in 1908 by W.S. Gosset, who published the result under the pen name "Student", and so we say that T is distributed as "Student's t".[6]

Since $C := \left(n - 1\right)S^2/\sigma^2 \sim \chi^2(n-1)$ is independent of $\bar{X}$, the quantity

$$T = \frac{\sqrt{n}\left(\bar{X} - \mu\right)/\sigma}{S/\sigma} = \frac{\sqrt{n}\left(\bar{X} - \mu\right)/\sigma}{\sqrt{C/(n-1)}} \tag{5.5}$$

is distributed as the ratio of a standard normal to the square root of an independent chi-squared variate divided by its degrees of freedom. This will be referred to as the "constructive" definition of Student's t. With this characterization, the distribution of T comes by applying the multivariate change-of-variable (c.o.v.) formula.

To simplify notation we set $\nu := n - 1$ and work out the distribution of $T = Z \div \sqrt{C/\nu}$, where $Z \sim N\left(0, 1\right)$ and $C \sim \chi^2(\nu) \equiv \Gamma\left(\nu/2, 2\right)$. Since Z and C are independent, their joint p.d.f. is the product of their marginals,

$$f_{ZC}(z, c) = \frac{1}{\sqrt{2\pi}}e^{-z^2/2} \cdot \frac{1}{\Gamma\left(\nu/2\right)2^{\nu/2}}c^{\nu/2-1}e^{-c/2}\mathbf{1}_{\Re_+}\left(c\right).$$

Introducing $W = C$ as a second variable, transformation $(z, c) \to (t, w)$ takes $\Re \times (0, \infty)$ back to itself, and inverse transformation $(t, w) \to (z =$

---

[6]The verb "studentize" has been coined to describe the process of dividing the deviation of a statistic from its mean by an estimate of its standard deviation; hence, we call T a "studentized" sample mean.

$t\sqrt{w/\nu}, c = w)$ is one to one with Jacobian $J = \sqrt{w/\nu} > 0$. The joint p.d.f. of (T, W) is therefore

$$
\begin{aligned}
f_{TW}(t, w) &= f_{ZC}(t\sqrt{w/\nu}, w)\,|J| \\
&= \frac{1}{\sqrt{2\pi}}e^{-t^2 w/(2\nu)} \cdot \frac{w^{\nu/2-1}e^{-w/2}}{\Gamma(\nu/2)\,2^{\nu/2}}1_{\Re_+}(w)\sqrt{w/\nu} \\
&= \frac{w^{(\nu+1)/2-1}e^{-w(1+t^2/\nu)/2}}{\Gamma\left(\frac{1}{2}\right)\Gamma\left(\frac{\nu}{2}\right)\sqrt{\nu}2^{(\nu+1)/2}}1_{\Re_+}(w). \quad\quad (5.6)
\end{aligned}
$$

Now we must integrate out w (treating t as constant) in order to find the marginal p.d.f. of T. The numerator in (5.6) has the form of a gamma p.d.f. with shape $\alpha := (\nu + 1)/2$ and scale $\beta := 2/\left(1 + t^2/\nu\right)$, apart from a normalizing constant $[\Gamma(\alpha)\beta^{\alpha}]^{-1}$. Its integral over $\Re_+$ therefore equals the inverse of the normalizing constant; i.e.,

$$
\Gamma(\alpha)\beta^{\alpha} = \Gamma\left(\frac{\nu+1}{2}\right)\left(\frac{2}{1+t^2/\nu}\right)^{(\nu+1)/2}.
$$

This leaves

$$
\begin{aligned}
f_T(t) &= \frac{\Gamma\left(\frac{\nu+1}{2}\right)}{\Gamma\left(\frac{1}{2}\right)\Gamma\left(\frac{\nu}{2}\right)\sqrt{\nu}}\left(1 + \frac{t^2}{\nu}\right)^{-(\nu+1)/2} \quad\quad (5.7) \\
&= \frac{1}{\mathfrak{B}\left(\frac{1}{2}, \frac{\nu}{2}\right)\sqrt{\nu}}\left(1 + \frac{t^2}{\nu}\right)^{-(\nu+1)/2}.
\end{aligned}
$$

A r.v. T with this p.d.f. is said to be distributed as Student's t "with ν d.f.", written $T \sim t(\nu)$ for short. Parameter ν takes its name and value from the parameter of the chi-squared variate used in the construction. (Recall that $\mathfrak{B}(\alpha, \beta)$ for $\alpha > 0, \beta > 0$ is the beta function, with $\mathfrak{B}(\alpha, \beta) = \Gamma(\alpha)\Gamma(\beta)/\Gamma(\alpha+\beta)$.)

Exercise 5.15. *Obtain another derivation of (5.7) using conditional expectations, as follows:*

(i) *Show that $F_T(t) = \Pr(T \le t) = E1_{(-\infty, t\sqrt{C/\nu}]}(Z)$, where both the argument and the set on which the indicator is unity are stochastic.*

(ii) *Use the tower property of conditional expectation (with independence of C and Z) to write*

$$
F_T(t) = E\left\{E\left[1_{(-\infty, t\sqrt{C/\nu}]}(Z)\mid C\right]\right\}
$$

as $E\Phi\left(t\sqrt{C/\nu}\right)$, where $\Phi(\cdot)$ is the standard normal c.d.f.

(iii) Differentiate with respect to t to express the p.d.f., $f_T(t)$, applying the dominated convergence theorem to justify the equality

$$\frac{dE\Phi\left(t\sqrt{C/v}\right)}{dt} = E\left[\frac{d\Phi\left(t\sqrt{C/v}\right)}{dt}\right].$$

(iv) Work out the expected value of $d\Phi\left(t\sqrt{C/v}\right)/dt$ to obtain (5.7).

The construction (5.5) implies that $\sqrt{n}\left(\bar{X}-\mu\right)/S$ is distributed as Student's t with $n-1$ d.f. when random sample $\{X_j\}_{j=1}^{n}$ is from a population distributed as $N\left(\mu,\sigma^2\right)$. We will use this fact frequently as we proceed in the study of statistical inference. Appendix C.4 presents upper quantiles of Student's t—numbers t_ν^p such that $\Pr\left(T \geq t_\nu^p\right) = p$ when $T \sim t(\nu)$. Also, software to calculate these directly is widely available. Note that the symmetry of $f_T(t)$ about $t=0$ implies that lower and upper quantiles are related as $t_{\nu,p} = -t_\nu^p$.

Example 5.4. A particular normal population is claimed to have mean $\mu = 10$. A sample of size $n = 25$ produces realizations $\bar{x} = 13, s^2 = 53$. With $\mu = 10$ the implied realization of the studentized mean is $(13 - 10) \div \sqrt{53/25} \doteq 2.06$. The probability of obtaining such a large deviation from zero (in either direction) is

$$\Pr\left(\left|\frac{\bar{X}-10}{S/\sqrt{25}}\right| \geq 2.06\right) = \Pr\left(|T| \geq 2.06\right)$$

$$= \Pr\left(T \geq 2.06\right) + \Pr\left(T \leq -2.06\right)$$

$$= 2\Pr\left(T \geq 2.06\right),$$

where $T \sim t(24)$. The upper-.025 quantile of $t(24)$ being about 2.064, we conclude that the probability of such an extreme sample realization would be only about 0.05.

Besides its symmetry about the origin, several properties of the t distribution are apparent from the construction and the form of the p.d.f. Since $f_T(t)$ is approximately proportional to $|t|^{-(v+1)}$ for large $|t|$, the tails vanish according to a power law as $|t| \to \infty$, rather than exponentially. This signals trouble with respect to the finiteness of moments. The fact that $e^{\varepsilon|t|}/|t|^{\nu+1} \to \infty$ for any $\varepsilon > 0$ shows that the Student distribution has no m.g.f. Indeed, since

$$\int_{-\infty}^{\infty} \frac{|t|^k}{|t|^{\nu+1}} \cdot dt = \int_{-\infty}^{\infty} |t|^{-(\nu+1-k)} \cdot dt$$

fails to converge for $k \geq \nu$, at most the first $\nu - 1$ moments are finite if ν is an integer. However, since the $\chi^2 (\nu)$ variate in the construction is the same as $\Gamma (\nu/2, 2)$, the d.f. parameter is not necessarily restricted to integer values. Thus, the Student family can be considered to comprise the densities $\{f_T(\cdot; \nu) : \nu > 0\}$, and when v is not an integer $E |T|^k < \infty$ for all nonnegative integers $k < \nu$. By symmetry, the odd moments less than ν are all zero. Finally, when $\nu = 1$ the p.d.f. specializes to $f_T(t) = \pi^{-1} (1 + t^2)^{-1}$, which is the standard Cauchy.

Exercise 5.16. *If $T \sim t(\nu)$ with $\nu > 1$ show that $E \log \left[1 + p \left(e^T - 1\right)\right] < \infty$ for all $p \in [0, 1]$.*

Two other worthwhile results are not so obvious. First, we can get an explicit expression for even moments ET^{2k} for $2k < \nu$ as

$$\mu_{2k} = \nu^k \frac{\mathcal{B} (\nu/2 - k, k + 1/2)}{\mathcal{B}(1/2, \nu/2)} = \frac{\nu^k \Gamma (\nu/2 - k) \Gamma (k + 1/2)}{\Gamma (1/2) \Gamma (\nu/2)}. \tag{5.8}$$

This comes from expressing ET^{2k} as

$$\frac{2}{\mathcal{B} (1/2, \nu/2) \sqrt{\nu}} \int_0^\infty \frac{t^{2k}}{(1 + t^2/\nu)^{(\nu+1)/2}} \cdot dt,$$

changing variables as $y = \left(1 + t^2/\nu\right)^{-1}$, and using the definition of the beta function. It gives as the variance the particular value

$$VT = \frac{\nu}{\nu - 2}, \nu > 2. \tag{5.9}$$

Second, letting $\nu \to \infty$ and using Stirling's approximation, $\Gamma (x + 1) \doteq (x/e)^x \sqrt{2\pi x}$, it is not hard to show that

$$\lim_{\nu \to \infty} f_T(t; \nu) = \frac{1}{\sqrt{2\pi}} e^{-t^2/2}. \tag{5.10}$$

Thus, as the d.f. parameter approaches infinity, Student's t approaches the standard normal. The approximation $T \dot\sim N (0, 1)$ being quite good for $\nu \geq 30$, many published tables present quantiles of $t(\nu)$ only for $\nu \in \{1, 2, ..., 30\}$.

Exercise 5.17. *Verify the formula (5.8) for the even moments of the t distribution.*

Exercise 5.18. *Verify that (5.8) specializes to (5.9) when $k = 1$.*

Exercise 5.19. *Verify that $\lim_{\nu \to \infty} f_T(t; \nu) = \frac{1}{\sqrt{2\pi}} e^{-t^2/2}$.*

We usually think of Student's t as a model for functions of *statistics*, but it is also a useful model for noisy or inhomogeneous *data*. To see how it could apply in such cases, consider first a simpler setup in which a population comprises two disparate groups of entities. Sampling from the first group only would produce realizations of a r.v. $X_1 \sim N\left(\mu, \sigma_1^2\right)$, while realizations of $X_2 \sim N\left(\mu, \sigma_2^2\right)$ would come from the other. However, if the two groups could not be distinguished *a priori*, a sample from the mixed population would yield realizations of a r.v. Y that followed a *mixture* of normals with different variances. This would have p.d.f.

$$f\left(y; \mu, \sigma_1^2, \sigma_2^2, p\right) = \frac{p}{\sqrt{2\pi\sigma_1^2}} e^{-(x-\mu)^2/(2\sigma_1^2)} + \frac{(1-p)}{\sqrt{2\pi\sigma_2^2}} e^{-(x-\mu)^2/(2\sigma_2^2)},$$

where p is the proportion of the total population that is of the first type. Such a model amounts to regarding Y as having the distribution $N\left(\mu, Q\right)$ *conditional on* the discrete r.v. Q with $\Pr\left(Q = \sigma_1^2\right) = p = 1 - \Pr\left(Q = \sigma_2^2\right)$. Now such *finite* mixtures can be generalized to continuous ones that afford more flexibility and, at the same time, make it easier to do statistical inference. In particular, if we put $Y := Z/\sqrt{Q}$ with $Z \sim N\left(0, 1\right)$ and $Q \sim \Gamma\left(\alpha, \beta\right)$, then $Y \sim N\left(0, Q^{-1}\right)$ conditional on Q, and the *unconditional* distribution is that of a rescaled Student variate with d.f. parameter 2α (not necessarily an integer). Y, in turn, can be rescaled and centered in any way one desires in order to characterize noisy data with lots of outliers.[7]

Exercise 5.20. *An alternative and more direct way of mixing normals is to take $Y \sim N\left(0, Q\right)$ conditional on $Q \sim \Gamma\left(\alpha, \beta\right)$. The p.d.f. of the resulting mixture has no closed form, but $f_Y\left(y\right)$ can be evaluated numerically at any $y \in \Re$ by inverting the c.f., which does have a simple expression. Find the c.f. and verify that, unlike the Student form, this particular mixture has all its moments.*

5.3.3.2 *Distribution of the Variance Ratio*

Suppose now that we have independent random samples of sizes n_1 and n_2 from two normal populations with variances σ_1^2 and σ_2^2 and arbitrary means. Setting $C_j := (n_j - 1) S_j^2/\sigma_j^2$ for $j \in \{1, 2\}$ shows that the variance ratio,

$$R := \frac{S_1^2/\sigma_1^2}{S_2^2/\sigma_2^2} = \frac{C_1/(n_1 - 1)}{C_2/(n_2 - 1)},$$

---

[7]Praetz (1972) and Blattberg and Gonedes (1974) proposed such modified Student forms to model daily rates of returns of common stocks, whose distributions are known to be *roughly* symmetrical and highly *leptokurtic*—i.e., having $\alpha_4 >> 3$.

is the ratio of two independent chi-squared variates, each divided by its d.f. Putting $\nu_j := n_j - 1$, the distribution of $R = (C_1/\nu_1) \div (C_2/\nu_2)$ can be worked out from the change-of-variable formula. The resulting p.d.f.,

$$f_R(r) = \frac{1}{\mathcal{B}(v_1/2, \nu_2/2)} \left(\frac{\nu_1}{\nu_2}\right)^{\nu_1/2} \frac{r^{v_1/2-1}}{(1 + r\nu_1/\nu_2)^{(\nu_1+\nu_2)/2}} \mathbf{1}_{(0,\infty)}(r), \quad (5.11)$$

represents what is called the "F distribution with ν_1 and ν_2 d.f."[8] For short, we write $R \sim F(\nu_1, \nu_2)$ if R has p.d.f. (5.11). ν_1 and ν_2 are referred to as the "d.f. of the numerator" and the "d.f. of the denominator", respectively.

Exercise 5.21. *Verify that (5.11) is the p.d.f. of the ratio of independent chi-squareds divided by their d.f.s (the constructive definition of the F distribution).*

Exercise 5.22. *Using the constructive definitions of the F and t distributions, show that $T^2 \sim F(1, \nu)$ if $T \sim t(\nu)$.*

Exercise 5.23. *Using the constructive definition of F, show that $R^{-1} \sim F(\nu_2, \nu_1)$ if $R \sim F(\nu_1, \nu_2)$.*

Appendix C.5 presents upper-p quantiles of R for various v_1, ν_2 and for $p \in \{.10, .05, .01\}$. Since there are two parameters, displaying such tables takes up a lot of space, which is conserved by making use of the following relation. If $r^p_{\nu_1, \nu_2}$ is such that $\Pr\left(R \geq r^p_{\nu_1, \nu_2}\right) = p$ when $R \sim F(\nu_1, \nu_2)$, then

$$p = \Pr\left(R^{-1} \leq 1/r^p_{\nu_1, \nu_2}\right) = 1 - \Pr\left(R^{-1} \geq 1/r^p_{\nu_1, \nu_2}\right),$$

so that $\Pr\left(R^{-1} \geq 1/r^p_{\nu_1, \nu_2}\right) = 1 - p$. Thus, if $r^p_{\nu_1, \nu_2}$ is the upper-p quantile of $F(\nu_1, \nu_2)$, the conclusion of Exercise 5.23 shows that $1/r^p_{\nu_1, \nu_2}$ is the p quantile (upper-$(1 - p)$ quantile) of $F(\nu_2, \nu_1)$. For this reason it suffices to tabulate upper quantiles for $p < .5$ only. Even so, the tables are unavoidably cumbersome. Fortunately, software to calculate the c.d.f. and quantiles of F distributions is now widely available.

Example 5.5. The upper-.05 quantile of $F(5, 10)$ is approximately 3.33. Therefore, the .05 quantile of $F(10, 5)$ is approximately $1/3.33 \doteq .300$.

5.4 Solutions to Exercises

1. There are $N(N-1)(N-2)\cdots\cdots(N-n+1)$ ways of drawing n objects from N without repetition and N^n ways of drawing the samples with

---

[8]The distribution is often referred to as Snedecor's F, after G.W. Snedecor (1881-1974).

replacement. Thus, the probability is

$$p_{n,N} = \frac{N(N-1)(N-2)\cdots\cdots(N-n+1)}{N^n} = \frac{N!}{(N-n)!N^n}.$$

Taking logs, using $\log m! \doteq \log\sqrt{2\pi m} + m(\log m - 1)$ (from Stirling's formula), and then exponentiating give $p_{10,10^5} \doteq .9955$.

2. Since X is a continuous r.v., the probability of ties is zero; and since the $\{X_j\}$ are i.i.d., each of the $(n+1)!$ arrangements of the $n+1$ observations has the same probability. Of these, there are $1 \cdot n!$ arrangements in which the largest value comes first and just one arrangement in which they appear in ascending order. Thus, $P(n,0) = n!/(n+1)! = 1/(n+1)$ and $P(n,n) = 1/(n+1)!$

3. For $j \in \{1, 2, ..., n\}$ let $N(n,j)$ be the number of the $(n+1)!$ possible arrangements of $\{X_0, X_1, ..., X_n\}$ that produce j records, so that $P(n,j) = N(n,j)/(n+1)!$. To obtain a particular number of records j, two conditions must hold: First, the largest observation, $X^* := \max\{X_0, X_1, ..., X_n\}$, must occupy some position $i \in \{j, j+1, ..., n\}$; and, second, the observations in positions $0, 1, ..., i-1$ must have been so arranged as to produce $j-1$ records. Once X^* has been installed in some such position i, the Xs that will fill positions $0, 1, ..., i-1$ can be chosen from the n that remain in $\binom{n}{i}$ ways. For each of these $\binom{n}{i}$ choices there are $N(i-1, j-1)$ ways of obtaining $j-1$ records and then $(n-i)!$ ways of filling positions $i+1, i+2, ..., n$. Thus, conditional on X^* being in position $i \in \{j, j+1, ..., n\}$, there are

$$\binom{n}{i}N(i-1, j-1)(n-i)! = \frac{n!}{i!}N(i-1, j-1)$$

ways of getting j records. Summing over i gives

$$N(n,j) = n! \sum_{i=j}^{n} \frac{1}{i!}N(i-1, j-1)$$

and

$$P(n,j) = \frac{1}{n+1}\sum_{i=j}^{n}\frac{1}{i!}N(i-1, j-1) = \frac{1}{n+1}\sum_{i=j}^{n}P(i-1, j-1)$$

$$(5.12)$$

for $j \in \{1, 2, ..., n\}$ and $n \in \aleph$, with $P(n,0) = 1/(n+1)$ for $n \in \aleph_0$. (Although $P(n,n) = 1/(n+1)!$ can be found directly, as in the previous solution, the recursive formula (5.12) leads to the same result.) The distributions $\{P(n,j)\}_{j=0}^{n}$ up to $n = 10$ are displayed in the matrix below. Notice that the calculation in (5.12) for $P(n, j)$ at each $n \geq 1$ and $j \geq 1$ involves summing the first $n-1$ entries in column $j-1$ of the matrix (then dividing by $n+1$).

$n \backslash j$	0	1	2	3	4	5	6	7	8	9	10
0	1.000										
1	.5000	.5000									
2	.3333	.5000	.1667								
3	.2500	.4583	.2500	.0417							
4	.2000	.4167	.2917	.0833	.0083						
5	.1667	.3806	.3125	.1181	.0208	.0014					
6	.1429	.3500	.3222	.1458	.0347	.0042	.0002				
7	.1250	.3241	.3257	.1679	.0486	.0080	.0007	.0000			
8	.1111	.3020	.3255	.1854	.0619	.0125	.0015	.0001	.0000		
9	.1000	.2829	.3232	.1994	.0742	.0174	.0026	.0002	.0000	.0000	
10	.0909	.2663	.3195	.2107	.0856	.0226	.0040	.0005	.0000	.0000	.0000

4. Expand the sum of squares as $\sum_{j=1}^{n}(X_j - A)^2 = \sum_{j=1}^{n} X_j^2 - 2A \sum_{j=1}^{n} X_j + nA^2$, differentiate with respect to A, equate to zero, and solve to get $A = \bar{X}$. The second derivative being a positive constant, $\bar{X}$ is the unique minimizer. Alternatively, write

$$\sum_{j=1}^{n}(X_j - A)^2 = \sum_{j=1}^{n}\left[(X_j - \bar{X}) + (\bar{X} - A)\right]^2$$

$$= \sum_{j=1}^{n}(X_j - \bar{X})^2 + 2(\bar{X} - A)\sum_{j=1}^{n}(X_j - \bar{X}) + n(\bar{X} - A)^2$$

$$= \sum_{j=1}^{n}(X_j - \bar{X})^2 + n(\bar{X} - A)^2$$

and see directly that this is minimized at $A = \bar{X}$.

5. (i) We have

$$M_2 = n^{-1}\sum_{j=1}^{n}(X_j - \bar{X})^2$$

$$= n^{-1}\left(\sum_{j=1}^{n} X_j^2 - 2\bar{X}\sum_{j=1}^{n} X_j + n\bar{X}^2\right)$$

$$= M_2' - 2\bar{X}^2 + \bar{X}^2 = M_2' - \bar{X}^2.$$

(ii)

$$\frac{1}{2n^2} \sum_{i,j=1}^{n} (X_i - X_j)^2 = \frac{1}{2n^2} \left(\sum_{i,j=1}^{n} X_i^2 - 2 \sum_{i,j=1}^{n} X_i X_j + \sum_{i,j=1}^{n} X_j^2 \right)$$

$$= \frac{1}{2n^2} \left(n \sum_{i=1}^{n} X_i^2 - 2 \sum_{i=1}^{n} X_i \sum_{j=1}^{n} X_j + n \sum_{j=1}^{n} X_j^2 \right)$$

$$= \frac{1}{2n^2} \left(n^2 M_2' - 2n^2 \bar{X}^2 + n^2 M_2' \right)$$

$$= M_2' - \bar{X}^2 = M_2.$$

(iii)

$$\frac{1}{\binom{n}{2}} \sum_{i<j} \frac{(X_i - X_j)^2}{2} = \frac{1}{2n(n-1)} \sum_{i,j=1}^{n} (X_i - X_j)^2$$

$$= \frac{n}{n-1} M_2 = S^2.$$

6. M_2 being invariant with respect to μ take $EX = 0$ to simplify the algebra. Then (i)

$$Cov\left(\bar{X}, M_2\right) = E\bar{X}M_2 = E\bar{X}\left(M_2' - \bar{X}^2\right)$$

$$= E\left[\frac{1}{n^2} \sum_{i,j=1}^{n} X_i X_j^2 - \frac{1}{n^3} \sum_{i,j,k=1}^{n} X_i X_j X_k \right].$$

Taking expectations, in the first sum $EX_i X_j^2 = \mu_3$ when $i = j$ and $EX_i X_j^2 = 0$ otherwise. In the second sum $EX_i X_j X_k = \mu_3$ in the n terms with equal subscripts, while remaining terms vanish. Collecting terms and simplifying yield the desired result.

(ii) Use $S^2 = \frac{n}{n-1} M_2$ to obtain $Cov\left(\bar{X}, S^2\right) = \frac{n}{n-1} E\bar{X}M_2 = \mu_3/n$.

7. The expressions for $Cov\left(\bar{X}, S^2\right)$, $V\bar{X}$, and VS^2 yield, after simplification,

$$\rho_{\bar{X},S^2}^2 = \frac{Cov\left(\bar{X}, S^2\right)^2}{V\bar{X} \cdot VS^2} = \frac{\mu_3^2/\sigma^6}{\mu_4/\sigma^4 - 1 + 2/(n-1)} = \frac{\alpha_3^2}{\alpha_4 - 1 + 2/(n-1)},$$

and the inequality follows since $\rho_{\bar{X},S^2}^2 \le 1$ holds for any integer $n > 1$.

8. $\Pr\left(|\bar{X} - \mu| \ge .1\right) = \Pr\left(\left|\frac{\bar{X}-\mu}{\sigma/\sqrt{n}}\right| \ge .1\sqrt{n}/.5\right) = \Pr\left(|Z| \ge .2\sqrt{n}\right)$. A table of the standard normal c.d.f. shows that $\Pr\left(|Z| \ge 1.96\right) \doteq .05$, so $n \ge (1.96/.2)^2 \doteq 96$.

9. $M_{\bar{X}}(t) = M_X(t/n)^n = \exp\left[\frac{\lambda n}{\mu}\left(1 - \sqrt{1 - 2\mu^2 t/\lambda n}\right)\right]$, which is the m.g.f. of $IG(\mu, \lambda n)$.

10. Since $E(X_j - \bar{X})$ is invariant with respect to μ, take $\mu = 0$. Assuming that $VX = \sigma^2 < \infty$,

$$Cov(\bar{X}, X_j - \bar{X}) = E\bar{X}X_j - E\bar{X}^2 = \frac{1}{n}E\sum_{i=1}^{n}X_iX_j - V\bar{X}$$

$$= \frac{1}{n}EX_j^2 - V\bar{X} = \frac{\sigma^2}{n} - \frac{\sigma^2}{n} = 0,$$

where the first equality in the second line follows from the independence of X_i and X_j when $i \neq j$ and the assumption that $\mu = 0$. Finiteness of VX is the only restriction required.

11. $\Pr\left(S^2 < s_{.05}\sigma^2\right) = \Pr\left(24S^2/\sigma^2 < 24s_{.05}\right) = \Pr\left(C_{24} < 24s_{.05}\right)$. A table of χ^2 quantiles shows that $\Pr\left(C_{24} < 13.848\right) \doteq .05$, so $s_{.05} \doteq .577$. Also, $\Pr\left(S^2 > s_{.95}\sigma^2\right) = \Pr\left(C_{24} > 24s_{.95}\right)$. The upper-.05 quantile of $\chi^2(24)$ is approximately 36.415, so $s_{.95} \doteq 2.094$.

12. (i) $C = (n-1)S^2/\sigma^2 \sim \chi^2(n-1)$, so $EC = n-1$ and $VC = 2(n-1)$ imply $ES^2 = EC \cdot \sigma^2/(n-1) = \sigma^2$ and $VS^2 = VC \cdot \sigma^4/(n-1)^2 = 2\sigma^4/(n-1)$.

(ii) Since $\mu_4 = 3\sigma^4$ for normal r.v.s, the general expression for VS^2 reduces to $2\sigma^4/(n-1)$ when the population is normal.

13. $R := \lambda n\left(\bar{Y} - \bar{X}^{-1}\right)$ and $\bar{X}$ being independent, the m.g.f.s satisfy

$$M_{R+\lambda n\bar{X}^{-1}}(t) = M_R(t)M_{\lambda n\bar{X}^{-1}}(t),$$

and hence (m.g.f.s having no zeroes) $M_R(t) = M_{\lambda n\bar{Y}}(t)/M_{\lambda n\bar{X}^{-1}}(t)$. Exercise 5.9 establishes that $\bar{X} \sim IG(\mu, \lambda n)$. This and Exercise 4.67 imply that

$$M_{\bar{X}^{-1}}(t) = \left(1 - \frac{2t}{\lambda n}\right)^{-1/2}\exp\left[\frac{\lambda n}{\mu}\left(1 - \sqrt{1 - \frac{2t}{\lambda n}}\right)\right], t > \lambda n/2,$$

and hence that

$$M_{\lambda n\bar{X}^{-1}}(t) = M_{\bar{X}^{-1}}(t\lambda n)$$

$$= (1 - 2t)^{-1/2}\exp\left[\frac{\lambda n}{\mu}\left(1 - \sqrt{1 - 2t}\right)\right], t > 1/2.$$

Similarly,

$$M_{\lambda n\bar{Y}}(t) = (1 - 2t)^{-n/2}\exp\left[\frac{\lambda n}{\mu}\left(1 - \sqrt{1 - 2t}\right)\right], t > 1/2,$$

so that $M_R(t) = (1 - 2t)^{-(n-1)/2}, t > 1/2$, which is the m.g.f. of $\chi^2(n-1)$.

14. $C = \sum_{j=1}^{n} Z_j^2 - n\bar{Z}^2$, or in matrix form,

$$\mathbf{Z'Z} - n^{-1}\left(\mathbf{Z'1}_n\right)\left(\mathbf{1'_n Z}\right) = \mathbf{Z'}\left(\mathbf{I}_n - n^{-1}\mathbf{1}_n\mathbf{1'_n}\right)\mathbf{Z}.$$

The matrix is idempotent since

$$
\begin{aligned}
\left(\mathbf{I}_n - n^{-1}\mathbf{1}_n\mathbf{1'_n}\right)\left(\mathbf{I}_n - n^{-1}\mathbf{1}_n\mathbf{1'_n}\right) &= \mathbf{I}_n - 2n^{-1}\mathbf{1}_n\mathbf{1'_n} + n^{-2}\mathbf{1}_n\mathbf{1'_n}\mathbf{1}_n\mathbf{1'_n} \\
&= \mathbf{I}_n - 2n^{-1}\mathbf{1}_n\mathbf{1'_n} + n^{-2}\mathbf{1}_n\left(n\right)\mathbf{1'_n} \\
&= \mathbf{I}_n - n^{-1}\mathbf{1}_n\mathbf{1'_n}.
\end{aligned}
$$

15. (i) We have

$$F_T\left(t\right) = \Pr\left(\frac{Z}{\sqrt{C/\nu}} \leq t\right) = \Pr\left(Z < t\sqrt{\frac{C}{\nu}}\right) = E1_{\left(-\infty, t\sqrt{C/\nu}\right]}\left(Z\right).$$

(ii) C and Z being independent, conditioning on C gives

$$
\begin{aligned}
F_T\left(t\right) &= E\left\{E\left[1_{\left(-\infty, t\sqrt{C/\nu}\right]}\left(Z\right) \mid C\right]\right\} \\
&= E\Phi\left(t\sqrt{\frac{C}{\nu}}\right).
\end{aligned}
$$

(iii) Since $0 \leq \Phi\left(\cdot\right) \leq 1$ the integrand in

$$E\Phi\left(t\sqrt{\frac{C}{\nu}}\right) = \int \Phi\left(t\sqrt{\frac{c}{\nu}}\right) f_C\left(c\right) \cdot dc$$

is dominated by an integrable function—namely, $1 \cdot f_C(c)$, where $f_C\left(c\right)$ is the p.d.f. of C. Since differentiation is just a limiting operation, we can therefore invoke dominated convergence and write

$$
\begin{aligned}
f_T\left(t\right) &= \frac{d}{dt}E\Phi\left(t\sqrt{\frac{C}{\nu}}\right) = \int \frac{d\Phi\left(t\sqrt{c/\nu}\right)}{dt} f_C\left(c\right) \cdot dc \\
&= \int \frac{d}{dt}\left[\int_{-\infty}^{t\sqrt{c/\nu}} \frac{1}{\sqrt{2\pi}}e^{-z^2/2} \cdot dz\right] f_C\left(c\right) \cdot dc.
\end{aligned}
$$

Applying Leibnitz' formula gives

$$f_T\left(t\right) = \int \frac{1}{\sqrt{2\pi\nu}}c^{1/2}e^{-c/(2\nu)} f_C(c) \cdot dc.$$

(iv) Since $C \sim \chi^2\left(\nu\right)$, we have

$$f_C(c) = \frac{1}{2^{\nu/2}\Gamma\left(\nu/2\right)}c^{\nu/2-1}e^{-c/2}1_{(0,\infty)}\left(c\right),$$

so that, using $\sqrt{\pi} = \Gamma(1/2)$,

$$f_T(t) = \frac{1}{2^{(\nu+1)/2}\Gamma(1/2)\,\Gamma(\nu/2)\,\sqrt{\nu}} \int_0^\infty c^{\frac{\nu-1}{2}} e^{-\frac{c}{2}\left(1+t^2/\nu\right)} \cdot dc.$$

Dividing the integrand by $\Gamma\left(\frac{\nu+1}{2}\right)\left(\frac{2}{1+t^2/\nu}\right)^{(\nu+1)/2}$ and multiplying by the same constant outside the integral turns the integrand into the p.d.f. of $\chi^2(\nu+1)$, which integrates to unity. We are left with

$$f_T(t) = \frac{\Gamma\left(\frac{\nu+1}{2}\right)}{2^{(\nu+1)/2}\Gamma\left(\frac{1}{2}\right)\Gamma\left(\frac{\nu}{2}\right)\sqrt{\nu}}\left(\frac{2}{1+t^2/\nu}\right)^{(\nu+1)/2},$$

which simplifies to the expression in the text for the Student p.d.f.

16. When $T > 0$, $\log\left[1 + p\left(e^T - 1\right)\right]$ increases from 0 at $p = 0$ to T at $p = 1$; when $T < 0$, it decreases from 0 at $p = 0$ to T at $p = 1$; and $\log\left[1 + p\left(e^T - 1\right)\right] = 0$ for all p when $T = 0$. Thus, $\left|\log\left[1 + p\left(e^T - 1\right)\right]\right| \leq |T|$ for all p, and since $E\,|T| < \infty$ when $\nu > 1$ the same is true for $E\left|\log\left[1 + p\left(e^T - 1\right)\right]\right|$.

17. Starting with $ET^{2k} = \frac{2}{\mathcal{B}(1/2,\nu/2)\sqrt{\nu}}\int_0^\infty \frac{t^{2k}}{(1+t^2/\nu)^{(\nu+1)/2}}\cdot dt$, change variables as $y = \left(1 + t^2/\nu\right)^{-1}$. Noting that this maps $(0,\infty)$ onto $(1,0)$ and that

$$\frac{dt}{dy} = -\frac{1}{2}\sqrt{\nu}\left(\frac{1}{y} - 1\right)^{-1/2} y^{-2},$$

we have

$$\begin{aligned}
ET^{2k} &= \frac{\nu^k}{\mathcal{B}(1/2,\nu/2)}\int_0^1 y^{(\nu+1)/2-2}\left(\frac{1}{y} - 1\right)^{k-1/2}\cdot dy \\
&= \frac{\nu^k}{\mathcal{B}(1/2,\nu/2)}\int_0^1 y^{(\nu/2-k)-1}(1-y)^{(k+1/2)-1}\cdot dy \\
&= \nu^k\frac{\mathcal{B}(\nu/2 - k, k + 1/2)}{\mathcal{B}(1/2,\nu/2)}, k < \nu/2.
\end{aligned}$$

18. Write the beta functions in terms of gamma functions and use the gamma recurrence relation:

$$\begin{aligned}
ET^2 &= \nu\cdot\frac{\Gamma(\nu/2 - 1)\,\Gamma(3/2)}{\Gamma(1/2)\,\Gamma(\nu/2)} \\
&= \nu\cdot\frac{\Gamma(\nu/2 - 1)\,\frac{1}{2}\Gamma(1/2)}{\Gamma(1/2)\,(\nu/2 - 1)\,\Gamma(\nu/2 - 1)} \\
&= \frac{\nu}{\nu - 2}.
\end{aligned}$$

19. Using Stirling, the normalizing factor of $f_T(t; \nu)$ is

$$\frac{\Gamma\left(\frac{\nu+1}{2}\right)}{\sqrt{\pi}\Gamma\left(\frac{\nu}{2}\right)\sqrt{\nu}} \doteq \frac{\left(\frac{\nu-1}{2e}\right)^{(\nu-1)/2}\sqrt{\pi(\nu-1)}}{\sqrt{\pi}\left(\frac{\nu-2}{2e}\right)^{(\nu-2)/2}\sqrt{\pi(\nu-2)}\sqrt{\nu}}$$

$$= \frac{1}{\sqrt{2\pi}e^{1/2}}\frac{1}{\left(1-\frac{1}{\nu-1}\right)^{(\nu-1)/2}}\sqrt{\frac{\nu-1}{\nu}}.$$

Since $\lim_{n\to\infty}(1+a/n)^{n/2} = \sqrt{\lim_{n\to\infty}(1+a/n)^n} = e^{a/2}$ for $a \in \Re$, we have $\lim_{\nu\to\infty}\left(1-\frac{1}{\nu-1}\right)^{\frac{\nu-1}{2}} = e^{-1/2}$, so the normalizing factor approaches

$$\lim_{\nu\to\infty}\frac{\Gamma\left(\frac{\nu+1}{2}\right)}{\sqrt{\pi}\Gamma\left(\frac{\nu}{2}\right)\sqrt{\nu}} = \frac{1}{\sqrt{2\pi}}.$$

Combining this with the fact that

$$\lim_{\nu\to\infty}\frac{1}{(1+t^2/\nu)^{(\nu+1)/2}} = \lim_{\nu\to\infty}\frac{1}{\sqrt{1+t^2/\nu}}\cdot\lim_{\nu\to\infty}\frac{1}{(1+t^2/\nu)^{\nu/2}}$$

$$= 1\cdot e^{-t^2/2}$$

shows that $f_T(t;\nu) \to \frac{1}{\sqrt{2\pi}}e^{-t^2/2}$ as $\nu \to \infty$.

20. We have

$$\phi_Y(t) = E\left[E\left(e^{itY}\mid Q\right)\right] = Ee^{-t^2Q/2} = \left(1+\beta\frac{t^2}{2}\right)^{-\alpha}$$

for the c.f. and $M_Y(t) = \left(1-\beta t^2/2\right)^{-\alpha}$, $(|t| \le \sqrt{2/\beta})$ for the m.g.f.

21. Introducing as second variable $S = C_2$, transformation $(c_1, c_2) \to (r, s)$ maps $(0,\infty)^2$ to $(0,\infty)^2$, with inverse $c_1 = rs\nu_1/\nu_2, c_2 = s$. This is one to one with Jacobian $J = s\nu_1/\nu_2$. Substituting for c_1, c_2 in the joint p.d.f. of C_1, C_2, multiplying by $|J|$, and simplifying give

$$f_{RS}(r,s) = \frac{\left(\frac{\nu_1}{\nu_2}\right)^{\frac{\nu_1}{2}}r^{\frac{\nu_1}{2}-1}}{\Gamma\left(\frac{\nu_1}{2}\right)\Gamma\left(\frac{\nu_2}{2}\right)2^{\frac{\nu_1+\nu_2}{2}}}\left[s^{\frac{\nu_1+\nu_2}{2}-1}e^{-s\left(1+\frac{\nu_1}{\nu_2}r\right)/2}\right].$$

The bracketed factor is the p.d.f. of $\Gamma\left(\frac{\nu_1+\nu_2}{2}, 2/\left(1+\frac{\nu_1}{\nu_2}r\right)\right)$ apart from normalizing factor $\left(1+\frac{\nu_1}{\nu_2}r\right)^{(\nu_1+\nu_2)/2}/\left[\Gamma\left(\frac{\nu_1+\nu_2}{2}\right)2^{(\nu_1+\nu_2)/2}\right]$, so integrating out s from 0 to ∞ and using the definition of $\mathfrak{B}(\nu_1/2, \nu_2/2)$ in terms of gamma functions yields (5.11).

22. T can be constructed as $Z/\sqrt{C/\nu}$, where $Z \sim N(0,1)$ and $C \sim \chi^2(\nu)$ are independent. Since $Z^2 \sim \chi^2(1)$, $T^2 = Z^2/(C/\nu)$ is the ratio of independent chi-squareds relative to their d.f., and so is distributed as $F(1,\nu)$.

23. $R \sim F(\nu_1, \nu_2)$ implies that $R = (C_1/\nu_1) \div (C_2/\nu_2)$, where C_1, C_2 are independent chi-squareds. Thus $R^{-1} = (C_2/\nu_2) \div (C_1/\nu_1) \sim F(\nu_2, \nu_1)$.

Chapter 6

Asymptotic Distribution Theory

6.1 Preliminaries

In Chapter 5 we saw how distributions of certain sample statistics can be found with change-of-variable formulas and techniques based on m.g.f.s or c.f.s. While the focus there was on samples from normal populations, the same methods can be applied more generally whenever *some* specific family of models is believed to describe the population. However, one often has little information on which to base a choice of parametric families and is content to draw inferences about just population moments or values of the c.d.f. at one or more points. Even when a parametric family has been assumed, it is often infeasible to work out the distribution of a statistic that is a complicated function of the data. In such cases one hopes at least to find some way of approximating the sampling distribution. There are three principal ways to do this: using Monte Carlo simulation, "bootstrapping", and developing analytical approximations that apply when samples are large.

If a family of models for the population is known, Monte Carlo methods can be used to estimate the distribution of a complicated statistic. This involves generating repeated, simulated random samples from a specific model by computer and calculating from each the realization of the statistic in question. For example, one could simulate a large number N of independent samples of arbitrary size n from a normal population with specific μ and σ^2 and calculate the N sample variances. The realizations $\{s_i^2\}_{i=1}^N$ would afford experimental evidence about the theoretical properties of S^2 that were worked out in Chapter 5; e.g., that $ES^2 = \sigma^2$ and that $C := (n-1) S^2/\sigma^2 \sim \chi^2(n-1)$. As an illustration, Fig. 6.1 summarizes a Monte Carlo experiment in which $N = 10,000$ pseudorandom samples

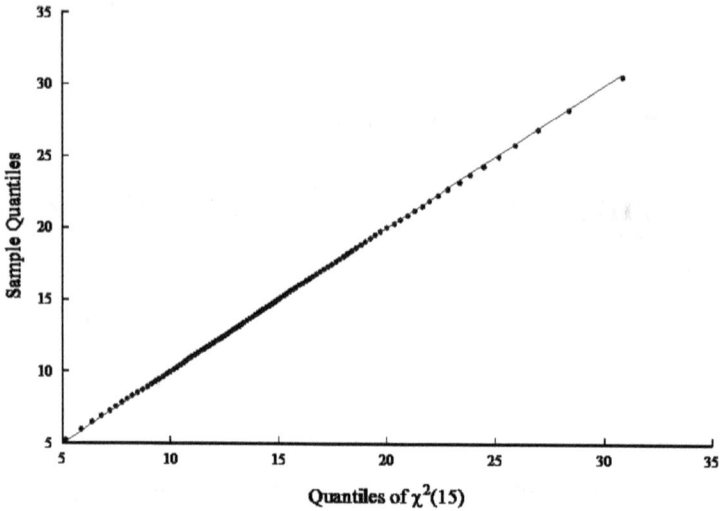

Fig. 6.1   Q-Q plot of quantiles of 10,000 replicates of $15S^2$ *vs.* $\chi^2\,(15)$.

of $n = 16$ were taken from $N\,(0,1)$. Calculating the realization of $15S^2$ from each set of 16 standard normals produced a sample of 10,000 values of the r.v. C. Arranging these in ascending order, the values at intervals of 100 are sample estimates of the quantiles of C at intervals of .01; i.e., $\{c_{j/100}\}_{j=1}^{99}$. For example, the 500th element can be regarded as the $500 \div 10,000 = .05$ *sample quantile*, $C_{.05,N}$. The figure plots the 99 sample quantiles $\{C_{.01j,N}\}_{j=1}^{99}$ against the corresponding theoretical quantiles of the $\chi^2\,(15)$ distribution. The latter are found using software that calculates the inverse of the cumulative distribution function, as $c_{j/100} = F_C^{-1}\,(j/100)$. Observing that the dots in this quantile–quantile (Q-Q) plot almost coincide with the line $c = c$, we surmise that there is no marked inconsistency between theory and empirical evidence. Also reassuring is that the mean of the 10,000 estimates of S^2 is 0.998834, close to the theoretical value $ES^2 = 1.0$.

In the same way one can simulate the distribution of some more complicated function of the data for which analytical results are unavailable or difficult to work out. However, while Monte Carlo methods are of great value, there are some disadvantages. Conclusions drawn from simulations are specific to the proposed family of models for the data, (often) to the particular values of the parameters, and also to the sample sizes used in the simulation experiment. One is never completely confident of their generality, regardless of how many parameter sets and sample sizes are tried. The reliability of even the specific findings depends on the skill of the programmer, the computing platform, and the quality of the pseudorandom number generators that were used. One who has had some programming experience is disinclined to accept at face value the simulation evidence presented by other researchers. Of course, as with any experimental findings, the real test is whether they can be replicated.

When generality is not the goal, but just inference from a particular data set, one can simulate using parameters estimated from the sample itself—a method known as *parametric bootstrapping*. For example, pseudorandom draws could be taken from $N(\bar{x}, s^2)$, using realizations $\bar{x}, s^2$ of the mean and variance of the primary sample. Another computer-intensive method, *nonparametric* bootstrapping, does not even require specifying a family of distributions. For this one takes repeated draws with replacement from the sample itself, calculating the statistic each time from the "resampled" data. There is now a large theoretical literature that guides the application and supports the usefulness of such methods.[1]

This chapter deals with the third principal way of approximating sampling distributions—"asymptotic" methods. These enable us to determine analytically the *limiting* form that a statistic's distribution may approach as the sample size n increases. If such a limiting distribution exists, it can serve as a useful approximation when the sample is sufficiently large. Whether it does so depends on (i) the rate of convergence to the limiting form and (ii) what information is to be extracted from the approximating distribution. Regarding the first point, for some problems there are theoretical guides to rates of convergence, and Monte Carlo can always be used to validate large-sample approximations worked out from theory. The second point refers to the difficulty of generalizing about how large n must be even when theoretical rates of convergence are known or Monte Carlo evidence is available. For example, estimates of probabilities for intervals in the tails of a distribution are usually less accurate in *proportional* terms than are those associated with intervals nearer the mode. Thus, anticipating applications to hypothesis testing in Chapter 8, when $n = 100$ asymptotic theory may be a reasonable guide to the true "size" of a test at nominal significance level $\alpha = .10$ but not to a test at level .01. Backing up the theory with Monte Carlo is always a good idea. This chapter presents the theory.

---

[1] Efron and Tibshirani (1993) give an accessible introduction to the bootstrap. The book by Fishman (1996) is a comprehensive guide to Monte Carlo methods.

6.2   Basic Concepts and Tools

We begin by setting out for reference and review some basic mathematical concepts that are of particular relevance for asymptotic theory. We then collect a few essential facts about characteristic functions, which turn out to be our most useful tools.

6.2.1   *Definitions*

- A **sequence of numbers** $\{a_n\}_{n=1}^{\infty}$ is a mapping from natural numbers $\aleph := \{1, 2, ...\}$ into $\Re$ or (if vector-valued) into $\Re^k$ or (if complex valued) into $\mathcal{C}$. For example, $1 - 1/n$ could represent the nth member of a sequence of real numbers. The sequence converges to a limit a when a_n can be made arbitrarily close to a by taking n sufficiently large. This means that for each $\varepsilon > 0$ there is a number $N(\varepsilon)$ such that $|a_n - a| < \varepsilon$ for *all* $n > N(\varepsilon)$. In this case we write $\lim_{n \to \infty} a_n = a$ or just $a_n \to a$.

- A **sequence of functions** $\{g_n(\cdot)\}_{n=1}^{\infty}$ on some domain $\mathcal{G}$ is a map from $\aleph \times \mathcal{G}$ into $\Re$ (in the case of real-valued functions) or into $\Re^k$ (in the case of vector-valued functions) or into $\mathcal{C}$ (if complex-valued); for example, $g_n(x) = x - x^2/n$ represents the nth member of a particular sequence of real-valued functions evaluated at $x \in \Re$. There are two senses in which g_n may converge to some limiting function g. Convergence is *pointwise* if for each x in the domain of g the sequence of numbers $\{g_n(x)\}$ converges to $g(x)$, just as $a_n \to a$ in the previous definition. Given $\varepsilon > 0$, then, there must be a number $N(\varepsilon, x)$ depending on ε and possibly on x such that $|g_n(x) - g(x)| < \varepsilon$ for all $n > N(\varepsilon, x)$. Convergence is *uniform* if g_n and g can be made close everywhere at once on the domain of g. Specifically, if $g : \mathcal{G} \to \Re$, then $g_n \to g$ uniformly if and only if for each $\varepsilon > 0$ there exists $N(\varepsilon)$ (*not* depending on x, obviously) such that $\sup_{x \in \mathcal{G}} |g_n(x) - g(x)| < \varepsilon$ whenever $n > N(\varepsilon)$. The condition is often written as $\lim_n \sup_{\mathcal{G}} |g_n - g| = 0$. For example, $\{g_n(x) = x - 1/(nx)\}_{n=1}^{\infty}$ for $x \in \mathcal{G} = (0, \infty)$ converges pointwise to $g(x) = x$ as $n \to \infty$, but it does not converge uniformly, since $(nx)^{-1}$ exceeds any $\varepsilon > 0$ for any n whenever $0 < x < (n\varepsilon)^{-1}$.

- **Order notation** is used in limiting arguments to represent quantities that are unspecified but whose minimum rate of convergence to zero is known as some independent variable approaches zero. One can justify ignoring these quantities when they are vanishing faster than others and so are ultimately dominated by them. A quantity that is $O(t)$—

read "of order t"—as $t \to 0$ is one whose dominant part approaches zero at the same rate as t. Specifically, the expression "$f(t) = O(t)$" means that $\lim_{t\to 0} f(t)/t = c$, where c is a *nonzero* constant. The expression "$f(t) = o(t)$" ("of order *less than* t") means that $f(t)$ vanishes faster than t, in which case $\lim_{t\to 0} f(t)/t = 0$. For example, if a, b, c, d are nonzero constants independent of t, then $at^{3/2} + bt^2 = o(t)$ as $t \to 0$, while $ct + dt^{3/2} = O(t)$. More generally $f(t) = O(t^p)$ and $f(t) = o(t^p)$ mean, respectively, that $\lim_{t\to 0} f(t)/t^p = c \neq 0$ and $\lim_{t\to 0} f(t)/t^p = 0$. An expression that is $O(t^0) \equiv O(1)$ does not vanish as $t \to 0$, whereas anything that is $o(1)$ *does* vanish. For example, $3 + 2t = O(1)$ while $\log(1+t) = o(1)$.

Here are some rules of arithmetic for order notation. If $f(t) = O(t^p)$, $g(t) = O(t^q)$, and $p < q$, then (i) $f(t) \pm g(t) = O(t^p)$, (ii) $f(t)g(t) = O(t^{p+q})$, and (iii) $g(t)/f(t) = O(t^{q-p})$. If $f(t) = O(t^p)$, $g(t) = o(t^q)$, and $p \leq q$, then (iv) $f(t) \pm g(t) = O(t^p)$, (v) $f(t)g(t) = o(t^{p+q})$, and (vi) $g(t)/f(t) = o(t^{q-p})$. If $f(t) = o(t^p)$, $g(t) = o(t^q)$, then (vii) $f(t)g(t) = o(t^{p+q})$.

In asymptotic statistical theory order notation is used mainly to approximate expressions involving the sample size n as it approaches *infinity*. In this case $1/n$ takes the place of t, and expressions such as $f(n) = O(n^{-p})$, $g(n) = o(n^{-p})$, and $h(n) = o(1)$ indicate the minimum rates at which these quantities vanish as $n \to \infty$. Specifically, $f(n) = O(n^{-p})$ and $g(n) = o(n^{-p})$ if and only if $\lim_{n\to\infty} n^p f(n) = c \neq 0$ and $\lim_{n\to\infty} n^p g(n) = 0$.[2]

6.2.2 *Important Theorems and Relations*

- **Taylor's theorem**, a basic result from elementary calculus that we have applied often heretofore, allows differentiable functions to be approximated (in some domain) by polynomials. If f and its first $m+1$ derivatives $\left\{ f^{(j)}(x) \right\}_{j=1}^{m+1}$ are continuous on a closed interval $[a, c]$, then for each x and b in $[a, c]$

$$f(x) = f(b) + \sum_{j=1}^{m} \frac{f^{(j)}(b)}{j!}(x-b)^j + R_{m+1}. \qquad (6.1)$$

---

[2]Notations $O_P(n^{-q})$ and $o_P(n^{-q})$ (orders *in probability*) are also encountered in the statistical literature. These are defined in Section 6.5.

Here, $f^{(j)}(b)$ is the jth derivative evaluated at b, and remainder term R_{m+1} is expressed as

$$R_{m+1} = \int_b^x \frac{(t-b)^m}{m!} f^{(m+1)}(t) \cdot dt. \qquad (6.2)$$

An alternative and often more useful form of the remainder is

$$R_{m+1} = \frac{f^{(m+1)}(b^*)}{(m+1)!}(x-b)^{m+1}, \qquad (6.3)$$

where b^* is some (unspecified) point between x and b. When $b = 0$ this can be written as

$$R_{m+1} = f^{(m+1)}(\theta x)\frac{x^{m+1}}{m+1)!},$$

for some $\theta \in (0,1)$. Defining the differential operator $(d)^j f := f^{(j)}$ gives another expression for (6.1) that generalizes to functions of more than one variable:

$$f(x) = f(b) + \sum_{j=1}^m \frac{[(x-b)\,d]^j}{j!} f(b) + R_{m+1}.$$

A function $f : \Re^k \to \Re$ that has continuous partial derivatives through order $m+1$ in a neighborhood of some $\mathbf{b} \in \Re^k$ can be expressed as

$$f(\mathbf{x}) = f(\mathbf{b}) + \sum_{j=1}^m \frac{[(\mathbf{x}-\mathbf{b})'\,\nabla]^j}{j!} f(\mathbf{b}) + \frac{[(\mathbf{x}-\mathbf{b})'\,\nabla]^{m+1}}{(m+1)!} f(\mathbf{b}^*),$$

where gradient operator $\nabla := (\partial/\partial x_1, \partial/\partial x_2, ..., \partial/\partial x_k)'$, and $\mathbf{b}^*$ is a point on the line segment connecting $\mathbf{x}$ and $\mathbf{b}$. When $\mathbf{b} = \mathbf{0}$ this is

$$f(\mathbf{x}) = f(\mathbf{0}) + \sum_{j=1}^m \frac{(\mathbf{x}'\nabla)^j}{j!} f(\mathbf{0}) + \frac{(\mathbf{x}'\nabla)^{m+1}}{(m+1)!} f(\theta\mathbf{x}), \qquad (6.4)$$

where $\theta \in (0,1)$. For example, with $k = 2$ this is

$$f(x_1, x_2) = f(0,0) + [x_1 \partial/\partial x_1 + x_2 \partial/\partial x_2] f(0,0)$$
$$+ \frac{1}{2!} (x_1 \partial/\partial x_1 + x_2 \partial/\partial x_2)^2 f(0,0)$$
$$+ \cdots + \frac{1}{m!} (x_1 \partial/\partial x_1 + x_2 \partial/\partial x_2)^m f(0,0)$$
$$+ \frac{1}{(m+1)!} (x_1 \partial/\partial x_1 + x_2 \partial/\partial x_2)^{m+1} f(\theta x_1, \theta x_2)$$
$$= f(0,0) + x_1 \frac{\partial f(0,0)}{\partial x_1} + x_2 \frac{\partial f(0,0)}{\partial x_2}$$
$$+ \frac{1}{2!} \left(x_1^2 \frac{\partial^2 f(0,0)}{\partial x_1^2} + 2x_1 x_2 \frac{\partial^2 f(0,0)}{\partial x_1 \partial x_2} + x_2^2 \frac{\partial^2 f(0,0)}{\partial x_2^2} \right)$$
$$+ \cdots + \frac{1}{m!} \sum_{j=0}^{m} \binom{m}{j} x_1^{m-j} x_2^j \frac{\partial^m f(0,0)}{\partial x_1^{m-j} \partial x_2^j}$$
$$+ \frac{1}{(m+1)!} \sum_{j=0}^{m+1} \binom{m+1}{j} x_1^{m+1-j} x_2^j \frac{\partial^{m+1} f(\theta x_1, \theta x_2)}{\partial x_1^{m+1-j} \partial x_2^j}.$$

- If $a_n \to a$ and if $f : \Re \to \Re$ is continuous at a, then $f(a_n) \to f(a)$. For example, if the $\{a_n\}$ and a are scalars, then $a_n^2 \to a^2$, $e^{a_n} \to e^a$, $\log(1 + a_n^2) \to \log(1 + a^2)$, etc.; and if $\mathbf{a}_n := (a_{1n}, a_{2n}, ..., a_{kn})'$ and $\mathbf{a} := (a_1, a_2, ..., a_k)'$, then $\sum_{j=1}^{k} a_{jn} \to \sum_{j=1}^{k} a_j$, $\prod_{j=1}^{k} a_{jn}^2 \to \prod_{j=1}^{k} a_j^2$, $\exp\left(\sum_{j=1}^{k} a_{jn}\right) \to \exp\left(\sum_{j=1}^{k} a_j\right)$, etc. The conclusion may fail to hold if f is not continuous at a; for example, if $a_n = 1 - 1/n$ and $f(x) = \mathbf{1}_{[1,\infty)}(x)$ then $\lim_{n\to\infty} f(a_n) = 0$ whereas $f(a) = f(1) = 1$.
- For any complex number z

$$\lim_{n\to\infty} \left[1 + \frac{z}{n} + o(n^{-1}) \right]^n = e^z. \tag{6.5}$$

More generally, if $\{o_j(n^{-1})\}_{j=1}^{n}$ represent possibly different quantities each of order less than n^{-1} as $n \to \infty$, then $\lim_{n\to\infty} \prod_{j=1}^{n} [1 + z/n + o_j(n^{-1})] = e^z$.

Proof: By the previous fact about limits of continuous functions, it suffices to show that

$$z = \lim_{n\to\infty} \log \prod_{j=1}^{n} \left[1 + \frac{z}{n} + o_j(n^{-1}) \right] = \lim_{n\to\infty} \sum_{j=1}^{n} \log \left[1 + \frac{z}{n} + o_j(n^{-1}) \right].$$

Expanding the logarithm about unity gives

$$
\lim_{n\to\infty} \left| \sum_{j=1}^{n} \log\left[1 + \frac{z}{n} + o_j\left(n^{-1}\right)\right] - z \right| = \lim_{n\to\infty} \left| \sum_{j=1}^{n} \left[\frac{z}{n} + o_j\left(n^{-1}\right)\right] - z \right|
$$

$$
= \lim_{n\to\infty} \left| \sum_{j=1}^{n} o_j\left(n^{-1}\right) \right|
$$

$$
\leq \lim_{n\to\infty} n \cdot \max_{1\leq j\leq n} \left\{ o_j\left(n^{-1}\right) \right\} = 0.
$$

- The infinite sum $\sum_{n=1}^{\infty} n^{-\alpha} =: \zeta(\alpha)$ (the "Riemann zeta" function) converges to a finite value if and only if $\alpha > 1$. Values of $\zeta(\alpha)$ are tabulated in various sources, such as Abramowitz and Stegun (1970), and explicit expressions exist for certain α; for example, $\zeta(2) = \pi^2/6$.

Exercise 6.1. *Use the theorem proved in Example 3.25 in Chapter 3 to show that Expression (6.3) for the remainder term in Taylor's theorem follows from Expression (6.2).*

6.2.3 *Essential Properties of Characteristic Functions*

Here are some of the specific facts about c.f.s that we will need to keep in mind.

- $\phi(t) := Ee^{itX} = e^{ita}$ for some real constant a if and only if X has a degenerate distribution, with $\mathbb{P}_X(\{a\}) = 1$. Thus, a r.v. with this c.f. is just a constant, in the sense that $X(\omega) = a$ for "almost all" ω.
- If $X \sim N\left(\mu, \sigma^2\right)$ then $\phi(t) = e^{i\mu t - t^2\sigma^2/2}$.
- If r.v. X has at least m moments, then (Theorem 3.3) its c.f. $\phi(t)$ has at least m (uniformly) continuous derivatives, and

$$
\phi(t) = 1 + \sum_{j=1}^{m} \phi^{(j)}(0) \frac{t^j}{j!} + o(t^m), \tag{6.6}
$$

where $\phi^{(j)}(0)$ (the jth derivative of $\phi(t)$ evaluated at $t = 0$) equals $i^j \mu_j'$. The important feature here is that the remainder can be thus controlled even if moments and derivatives beyond the mth do not exist.

Proof: Expand $\phi(t)$ as

$$
\phi(t) = 1 + \sum_{j=1}^{m-1} \phi^{(j)}(0) \frac{t^j}{j!} + \phi^{(m)}(\theta t) \frac{t^m}{m!}
$$

for $\theta \in (0,1)$, then add and subtract $\phi^{(m)}(0) t^m/m!$ to get

$$\phi(t) = 1 + \sum_{j=1}^{m} \phi^{(j)}(0) \frac{t^j}{j!} + \left[\phi^{(m)}(\theta t) - \phi^{(m)}(0)\right] \frac{t^m}{m!}.$$

The last term is $o(t^m)$ since $\phi^{(m)}(\theta t) \to \phi^{(m)}(0)$ as $t \to 0$.

- If each of the k elements of random vector $\mathbf{X}$ has m moments, the joint c.f. $\phi(\mathbf{t}) := Ee^{i\mathbf{t}'\mathbf{X}}$ for $\mathbf{t} \in \Re^k$ has at least m continuous partial derivatives, and

$$\phi(\mathbf{t}) = 1 + \sum_{j=1}^{m} (\mathbf{t}'\nabla)^j \frac{\phi(\mathbf{0})}{j!} + o(|\mathbf{t}|^m),$$

$|\mathbf{t}| := \sqrt{\mathbf{t}'\mathbf{t}}$. Partial derivatives $\partial^{j_1} \partial^{j_2} \cdots \partial^{j_p} \phi(\mathbf{0}) / \partial t_1^{j_1} \partial t_2^{j_2} \cdots \partial t_p^{j_p}$ in the expansions of gradient operator ∇ (that is, derivatives evaluated at $\mathbf{t} = \mathbf{0}$) are powers of i times the corresponding product moments: $i^{j_1+j_2+\cdots+j_p} E X_1^{j_1} X_2^{j_2} \cdots \cdots X_p^{j_p}$. Thus, for $k = 2$

$$\phi(t_1, t_2) = 1 + t_1 \frac{\partial \phi(0,0)}{\partial t_1} + t_2 \frac{\partial \phi(0,0)}{\partial t_2}$$

$$+ \frac{1}{2!} \left(t_1^2 \frac{\partial^2 \phi(0,0)}{\partial t_1^2} + 2 t_1 t_2 \frac{\partial^2 \phi(0,0)}{\partial t_1 \partial t_2} + t_2^2 \frac{\partial^2 \phi(0,0)}{\partial t_2^2} \right) + \cdots.$$

$$+ \frac{1}{m!} \sum_{j=0}^{m} \binom{m}{j} t_1^{m-j} t_2^j \frac{\partial^m \phi(0,0)}{\partial t_1^{m-j} \partial t_2^j} + o(|\mathbf{t}|^m)$$

$$= 1 + (i t_1 E X_1 + i t_2 E X_2)$$

$$- \frac{1}{2!} \left(t_1^2 E X_1^2 + 2 t_1 t_2 E X_1 X_2 + t_2^2 E X_2^2 \right)$$

$$+ \cdots + \frac{i^m}{m!} \sum_{j=0}^{m} \binom{m}{j} t_1^{m-j} t_2^j E X_1^{m-j} X_2^j + o(|\mathbf{t}|^m).$$

6.3 Convergence in Distribution

Let us consider a situation in which the marginal distributions of random variables in a sequence $\{X_n\}_{n=1}^{\infty}$ change progressively and approach some limiting form as n increases. Two such convergent sequences of distributions have already been encountered.

(i) Section 4.2.6: If $X_n \sim B(n, \lambda/n)$ for each $n \in \aleph$ (binomial with "success" probability λ/n), then the sequence of p.m.f.s $\{f_n(x)\}$ converges as $n \to \infty$ to $f(x) = \lambda^x e^{-\lambda}/x!$ for each $x \in \aleph_0$ (Poisson with intensity λ).

(ii) Section 5.3.3: If $X_n \sim t(n)$ (Student's t with n d.f.) for each n, then the sequence of densities $\{f_n(x)\}$ converges as $n \to \infty$ to $f(x) = (2\pi)^{-1/2} e^{-x^2/2}$ for each $x \in \Re$ (standard normal).

While these examples do illustrate the concept of convergence in distribution (and are useful in their own right), our interests in statistical applications are usually in sequences of *sample statistics* or of functions of statistics and parameters, as indexed by sample size n. For such sequences the pertinent question is to what the distributions would tend in the hypothetical situation that n was progressively increased—the point of the exercise being to see how to *approximate* the distribution that applies to a sample of the particular, fixed size that is in hand. In such applications we typically do not know the distribution for any finite n, which is why we desire approximations. For example, although the distributions of the sample mean, sample variance, and "studentized" sample mean are all known exactly when the population is normal, they are *not* known—not even *knowable*—unless some such parametric model is assumed. Nevertheless, we shall see that their distributions all have proper and explicit limiting forms so long as the population counterparts of the sample moments are finite.

Limiting distributions can *sometimes* be determined from the behavior of p.m.f.s or p.d.f.s, as in the binomial and Student's t examples above; however, since sequences of continuous r.v.s sometimes have limiting distributions of discrete form, and *vice versa*, it is necessary to *define* convergence in distribution in terms of limits of sequences of c.d.f.s or, equivalently, in terms of sequences of induced measures. Note that the following definitions are indeed equivalent, since each of the two necessary and sufficient conditions implies the other.

Definition 6.1.

(i) Let F be a c.d.f. and $\{X_n\}_{n=1}^{\infty}$ a sequence of r.v.s with c.d.f.s $\{F_n\}_{n=1}^{\infty}$. Sequence $\{X_n\}_{n=1}^{\infty}$ **converges in distribution** to F if and only if $F_n(x) \to F(x)$ for each $x \in \Re$ that is a *continuity point* of F.

(ii) Let $\mathbb{P}_X$ be a probability measure on $(\Re, \mathcal{B})$ and $\{X_n\}_{n=1}^{\infty}$ a sequence of r.v.s with induced measures $\{\mathbb{P}_{X_n}\}_{n=1}^{\infty}$. Sequence $\{X_n\}_{n=1}^{\infty}$ **converges in distribution** to $\mathbb{P}_X$ if and only if $\mathbb{P}_{X_n}(B) \to \mathbb{P}_X(B)$ for each Borel set B whose boundary has $\mathbb{P}_X$ measure zero.

Before elaborating on these definitions, let us introduce some alternative terminology and notation. Other terms for convergence in distribution are "weak convergence" and "convergence in law". (The sense in which the convergence is "weak" will be explained in Section 6.5.) Various concise notations for this are seen in the literature; among them, "$X_n \to F$ in distribution", "$X_n \to^d F$", "$X_n \to^{law} F$", "$X_n \Longrightarrow F$", and "$X_n \rightsquigarrow F$". Often, an abbreviation for the distribution will be used instead of the c.d.f. itself, as "$X_n \rightsquigarrow N(0,1)$" if $X_n \rightsquigarrow \Phi$ (the standard normal c.d.f.), or "$X_n \rightsquigarrow \chi^2(k)$" if the limiting F is the c.d.f. of chi-squared with k d.f. Both of these versions of the "$\rightsquigarrow$" notation will be used in what follows. When $X_n \rightsquigarrow F$, where F is some specified c.d.f., then F is called either the "limiting distribution" or the "asymptotic distribution" of sequence $\{X_n\}_{n=1}^{\infty}$. The symbolic expression "$X_n \rightsquigarrow \chi^2(k)$" could be read as "$\{X_n\}$ converges in distribution to chi-squared with k d.f." or as "$\{X_n\}$ is distributed asymptotically as $\chi^2(k)$" or as "the limiting distribution of $\{X_n\}$ is $\chi^2(k)$".

Returning to the definitions, note that version (i) requires only pointwise convergence of the c.d.f.s, and then only for points at which F is continuous. In terms of the measures, convergence is required only for Borel sets whose boundaries are $\mathbb{P}_X$-null. (The *boundary* of a set B comprises those points in B that are limits of sequences of points in B^c. For example, the endpoints of a closed interval are its boundaries.) An obvious question is why the limiting behaviors at the discontinuity points of F and on the boundary of B are excluded from consideration. We know that c.d.f.s can have left discontinuities and that these occur at points where there is positive probability mass. Thus, the definition requires convergence of F_n to F except at the countably many points ("atoms") where the limiting distribution has positive probability mass. In terms of Borel sets of $\Re$, this is the same as excluding any point x at which

$$\mathbb{P}_X(\{x\}) = \mathbb{P}_X((-\infty, x] \setminus (-\infty, x))$$
$$= \mathbb{P}_X((-\infty, x]) - \mathbb{P}_X((-\infty, x))$$
$$= F(x) - F(x-) > 0.$$

If there is no probability mass at x, then the measures $\mathbb{P}_X((-\infty, x))$ and $\mathbb{P}_X((-\infty, x])$ are the same. Otherwise, the boundary set $\{x\}$ must be excluded. The reason for the exclusion is that sequences $\{F_n(x)\}_{n=1}^{\infty}$ of perfectly valid c.d.f.s sometimes converge pointwise to functions $F_\infty(x)$ that are right discontinuous at one or more isolated points. Since such an F_∞ is not a valid c.d.f., it cannot be used to represent a limiting distribution.

The idea is simply to modify F_∞ at such exceptional points so as to make it right-continuous, designate the modified function as "F", and regard F or the corresponding $\mathbb{P}_X$ as the limiting distribution. These anomalies arise most often when each X_n in the sequence is continuous, yet the appropriate asymptotic approximation is discrete. Since there can be no smooth transition between these two types, limits of c.d.f.s of continuous r.v.s often behave strangely at points about which the probability mass concentrates. Of course, at any such point the limit of the sequence of *densities* would be infinite, and this would tell us nothing about how much probability mass resides there—which is why we work with c.d.f.s.

Such patchwork "fixes" at isolated points cannot rescue all such limits of c.d.f.s, however; and when it cannot the sequence $\{X_n\}_{n=1}^\infty$ simply has no limiting distribution. Determining from the definition whether there is a limiting distribution requires either finding a legitimate c.d.f. F that agrees with F_∞ except at points of discontinuity or showing that one does not exist. This requires checking that each candidate F is nondecreasing, right continuous, and satisfies $\lim_{x\to\infty} F(-x) = 0$ and $\lim_{x\to\infty} F(x) = 1$.[3] Here are some examples.

Example 6.1. Section 4.4 showed that if X has c.d.f. F, the nth order statistic $X_{(n:n)} := \max\{X_1, ..., X_n\}$ of a random sample has c.d.f. $F_{X_{(n:n)}}(x) = F(x)^n$.

(i) If $\{X_j\}_{j=1}^n$ are i.i.d. as $U(0,\theta)$, then $F_{X_{(n:n)}}(x) = (x/\theta)^n \mathbf{1}_{[0,\theta]}(x) + \mathbf{1}_{(\theta,\infty)}(x)$, so that $\lim_{n\to\infty} F_{X_{(n:n)}}(x) = \mathbf{1}_{[\theta,\infty)}(x)$—a legitimate (albeit *degenerate*) c.d.f. Here, as n increases the probability mass piles up closer and closer to θ.

(ii) The r.v. $Y_n := nX_{(n:n)}$ has c.d.f.

$$F_{Y_n}(y) = F_{X_{(n:n)}}\left(\frac{y}{n}\right) = \left(\frac{y}{n\theta}\right)^n \mathbf{1}_{[0,n\theta]}(x) + \mathbf{1}_{(n\theta,\infty)}(x).$$

This converges to $F_\infty(y) = 0$ for all $y \in \Re$, so that $nX_{(n:n)}$ simply has no limiting distribution. Here, all the probability mass progressively exits stage right.

(iii) R.v. $Z_n := n\left(\theta - X_{(n:n)}\right)$ exhibits a third type of behavior. Noting that continuity of $F_{X_{(n:n)}}$ implies $\Pr\left(X_{(n:n)} \geq x\right) = 1 - F_{X_{(n:n)}}(x)$, the

---

[3]If a weak limit F does exist, there remains the question of *uniqueness*. That is, would the same modification F apply to some G_∞ such that $G_\infty = F_\infty = F$ at continuity points but $G_\infty \neq F_\infty$ on the set $\mathcal{D}$ of discontinuities? Now $\mathcal{D}$ is necessarily countable since c.d.f. F can have only countably many jumps, so that $\Re\backslash\mathcal{D}$ is dense in $\Re$. Then for any $x \in \Re$ and any sequence $\{x_n\} \in \Re\backslash\mathcal{D}$ such that $\{x_n\}_{n=1}^\infty \downarrow x$ we have $0 = \lim_{n\to\infty}[G_\infty(x_n) - F(x_n)] = \lim_{n\to\infty} G_\infty(x_n) - F(x)$, showing that F is also the unique right-continuous modification of G_∞.

c.d.f. of Z_n is

$$
\begin{aligned}
F_{Z_n}(z) &= 1 - F_{X_{(n:n)}}\left(\theta - \frac{z}{n}\right) \\
&= \left[\mathbf{1}_{(-\infty,0)} + \mathbf{1}_{[0,n\theta]} + \mathbf{1}_{(n\theta,\infty)}\right](z) \\
&\quad - \left[\mathbf{1}_{(-\infty,0)} + \left(1 - \frac{z}{n\theta}\right)^n \mathbf{1}_{[0,n\theta]}\right](z) \\
&= \left[1 - \left(1 - \frac{z}{n\theta}\right)^n\right]\mathbf{1}_{[0,n\theta]}(z) + \mathbf{1}_{(n\theta,\infty)}(z).
\end{aligned}
$$

This converges to $F_{Z_\infty}(z) = \left(1 - e^{-z/\theta}\right)\mathbf{1}_{[0,\infty)}(z)$, the legitimate and (obviously) nondegenerate c.d.f. of the exponential distribution with scale θ.

Although one can often find limiting c.d.f.s directly in such ways, the following "continuity" theorem for characteristic functions (c.f.s) usually affords an easier way to determine whether and to what a sequence of r.v.s. converges in distribution.[4]

Theorem 6.1. *R.v.s* $\{X_n\}_{n=1}^{\infty}$ *with c.f.s* $\{\phi_n\}_{n=1}^{\infty}$ *converge in distribution if and only if sequence* $\{\phi_n(t)\}_{n=1}^{\infty}$ *converges (pointwise) for each* $t \in \Re$ *to a function* ϕ_∞ *that is continuous at* $t = 0$. *In this case the function* ϕ_∞ *is a c.f., and the corresponding c.d.f. is the (unique) F such that* $\phi_\infty(t) = \int e^{itx} \cdot dF(x)$.

The theorem tells us that if the c.f. of X_n can be expressed or *approximated* for each n in such a way that the limiting value ϕ_∞ can be found, then one can judge whether there is a limiting distribution merely by checking whether ϕ_∞ is continuous at the origin. Remarkably, no other feature besides continuity at the origin is relevant. If ϕ_∞ is continuous at $t = 0$, then (i) it *is* a c.f.; (ii) there *is* a limiting distribution; and (iii) the corresponding c.d.f. can be determined either by recognizing the form of ϕ_∞ or by inversion.

Example 6.2. Consider the sequence of p.d.f.s

$$
\left\{ f_n(x) = \sum_{j=1}^{n} \left[(n-1)^{-1}\mathbf{1}_{I_{j,n}}(x) + (n-1)\mathbf{1}_{I'_{j,n}}(x) \right] \right\}_{n=2}^{\infty},
$$

where $I_{j,n} := \left[\frac{j-1}{n}, \frac{j}{n} - \frac{1}{n^2}\right)$ and $I'_{j,n} := \left[\frac{j}{n} - \frac{1}{n^2}, \frac{j}{n}\right)$. Example 3.4 showed that these had the property that $\int f_\infty(x) \cdot dx := \int \lim_{n\to\infty} f_n(x) \cdot dx = 0$, even though $\int f_n(x) \cdot dx = 1$ for each $n \geq 2$. Letting X_n represent a r.v. with p.d.f. f_n, we shall find the limiting distribution of $\{X_n\}_{n=2}^{\infty}$

---

[4]For proof see e.g, Chung (1974, pp. 160-164) or Laha and Rohatgi (1979, pp. 153-160).

by evaluating $\phi_\infty(t) := \lim_{n\to\infty} \phi_n(t) = \lim_{n\to\infty} \int e^{itx} f_n(x) \cdot dx$ and confirming its continuity at the origin. For each n, r.v. X_n is supported on $\bigcup_{j=1}^{n} (I_{j,n} \cup I'_{j,n}) = [0,1)$. The measure associated with each of the $I_{j,n}$ components is $(n-1)^{-1} \left[\frac{j}{n} - \frac{1}{n^2} - \frac{j-1}{n} \right] = n^{-2}$, so these n subintervals of $[0,1)$ contribute nothing to the value of the limiting c.f. Accordingly,

$$\phi_\infty(t) = \lim_{n\to\infty} (n-1) \sum_{j=1}^{n} \int_{I'_{j,n}} e^{itx} \cdot dx$$

$$= \lim_{n\to\infty} \frac{n-1}{it} \left[\exp\left(\frac{itj}{n}\right) - \exp\left(\frac{itj}{n} - \frac{it}{n^2}\right) \right]$$

$$= \lim_{n\to\infty} \frac{n-1}{it} \left(1 - e^{-it/n^2}\right) \sum_{j=1}^{n} e^{itj/n}.$$

Expanding the exponentials about $t = 0$ gives

$$\phi_\infty(t) = \lim_{n\to\infty} \frac{n-1}{it} \left[\frac{it}{n^2} + o\left(n^{-2}\right) \right] \sum_{j=1}^{n} \sum_{k=0}^{\infty} \frac{1}{k!} \left(\frac{itj}{n}\right)^k$$

$$= \sum_{k=0}^{\infty} \frac{(it)^k}{k!} \lim_{n\to\infty} \left[\frac{1}{n} \sum_{j=1}^{n} \left(\frac{j}{n}\right)^k + o(1) \right]$$

$$= \sum_{k=0}^{\infty} \frac{(it)^k}{k!} \int_0^1 x^k \cdot dx$$

$$= \sum_{k=0}^{\infty} \frac{(it)^k}{(k+1)!}$$

$$= \frac{e^{it} - 1}{it}, t \neq 0.$$

This is a c.f. since $\lim_{t\to 0} \left(e^{it} - 1\right) / (it) = 1 = \phi(0)$. In fact, it is the c.f. of $X \sim U(0,1)$, which is, therefore, the limiting distribution.

Note: Since $\phi_n(t) := \int e^{itx} \cdot dF_n(x) = \int \cos(tx) \cdot dF_n(x) + i \int \sin(tx) \cdot dF_n(x)$, the continuity theorem for c.f.s implies that $\{X_n\}$ converges weakly to F if and only if the sequences of integrals of the *particular* bounded, continuous functions $\cos(tx)$ and $\sin(tx)$ converge to $\int \cos(tx) \cdot dF(x)$ and $\int \sin(tx) \cdot dF(x)$. In fact, it is true that $X_n \rightsquigarrow F$ if and only if $\int g(x) \cdot dF_n(x) \to \int g(x) \cdot dF(x)$ for *all* bounded, continuous, real functions g—the conclusion of what is known as the Helly–Bray theorem. Note, however, that $X_n \rightsquigarrow F$ does *not* imply that the sequence of moments $\{EX_n^k\}$ converges to $EX^k = \int x^k \cdot dF(x)$, because $g(X_n) = X_n^k$ is not bounded when $k \neq 0$. We will see later what extra conditions are needed for sequences of moments to converge to moments of limiting distributions.

The six examples below further illustrate the concept of convergence in distribution (and in some cases, *nonconvergence*) and show how to apply the definition and the continuity theorem. For now we stick with examples in which exact distributions of the $\{X_n\}$ are known, just to show more clearly how they approach the various limiting forms. As mentioned before, finite-sample distributions are typically *not* known in the relevant statistical applications. We will come to those relevant applications shortly and will see that it is still often possible to determine the limits indirectly. In fact, Examples 6.7 and 6.8 below give hints of how this will be done.

In the first three examples $\{Y_j\}_{j=1}^{\infty}$ represents a sequence of i.i.d. standard normals and X_n is in each case a linear function of the first n of these.

Example 6.3. If $X_n := \sum_{j=1}^{n} Y_j$ we have $EX_n = 0$ and $VX_n = n$, so $X_n \sim N(0, n)$ and $X_n / \sqrt{n} \sim N(0, 1)$. One sees intuitively that $\{X_n\}_{n=1}^{\infty}$ could have no limiting distribution, since the unit probability mass under the bell curve becomes more and more dispersed as n increases. To prove this, write the c.d.f. of X_n as

$$F_n(x) = \Pr(X_n \leq x) = \Pr\left(\frac{X_n}{\sqrt{n}} \leq \frac{x}{\sqrt{n}}\right) = \Phi\left(\frac{x}{\sqrt{n}}\right) = \mathbb{P}_Z\left((-\infty, \frac{x}{\sqrt{n}}]\right),$$

where Φ is the standard normal c.d.f. and $\mathbb{P}_Z$ is the corresponding induced measure. Sets $\{(-\infty, x/\sqrt{n}]\}_{n=1}^{\infty}$ converge monotonically to $(-\infty, 0]$ for each $x \in \Re$ ($\uparrow$ if $x < 0$; $\downarrow$ if $x > 0$), so by the monotone property of measures $F_n(x)$ converges pointwise to $F_{\infty}(x) = \mathbb{P}_Z((-\infty, 0]) = \Phi(0) = .5$ for each $x \in \Re$. F_{∞} is continuous, but it is clearly not a c.d.f., since $\lim_{x \to \infty} F(-x) = 1 - \lim_{x \to \infty} F(x) = .5$.

As in part (ii) of Example 6.1, the c.d.f.s $\{F_n\}$ here (or, more properly, the corresponding *measures*) lack a necessary property for convergence known as "tightness", which confines most of the probability mass associated with *all* the $\{X_n\}$ to some single finite interval. More precisely, for each $\varepsilon > 0$ there must be some finite $a > 0$ such that $\inf_{n \in \aleph} \mathbb{P}_{X_n}((-a, a]) \geq 1 - \varepsilon$. The lack of tightness is seen also in the c.f.s, $\{\phi_n(t) = \exp(-t^2 n/2)\}_{n=1}^{\infty}$, whose pointwise limit, $\phi_{\infty}(t) = \mathbf{1}_{\{0\}}(t)$, is manifestly discontinuous at $t = 0$.

Example 6.4. If $X_n := n^{-1/2} \sum_{j=1}^{n} Y_j$, then $EX_n = 0$ and $VX_n = 1$ for each n. Thus, each X_n is standard normal and so (trivially) has a standard normal limiting form. Formally, $F_n(x) = F_\infty(x) = \Phi(x)$ and $\phi_n(t) = \phi_\infty(t) = \exp(-t^2/2)$ for each n.

Example 6.5. With $X_n := \bar{Y} = n^{-1}\sum_{i=1}^n Y_i$ we have $EX_n = 0$, $VX_n = n^{-1}$, and $X_n \sim N(0, 1/n)$ for each n. Thus,

$$F_n(x) = \Pr\left(X_n\sqrt{n} \le x\sqrt{n}\right) = \Phi\left(\sqrt{n}x\right) \to F_\infty(x) := .5\mathbf{1}_{\{0\}}(x) + \mathbf{1}_{(0,\infty)}(x).$$

$F_\infty(x)$ is not right-continuous at zero and is therefore not a c.d.f.; but it does agree with c.d.f. $F(x) = \mathbf{1}_{[0,\infty)}(x)$ except at $x = 0$ where F is discontinuous. Thus, by the definition $X_n \rightsquigarrow F$. Working from the c.f.s instead of the c.d.f.s,

$$\phi_n(t) = e^{-t^2/2n} \to \phi_\infty(t) \equiv 1, \ t \in \Re.$$

This is clearly continuous at $t = 0$, and is in fact the c.f. that corresponds to $F(x) = \mathbf{1}_{[0,\infty)}(x)$:

$$Ee^{itX} = \int e^{itx} \cdot dF(x) = \int e^{itx} \cdot d\mathbf{1}_{[0,\infty)}(x) = e^{it \cdot 0} \cdot 1 = 1, t \in \Re.$$

The examples have shown that sequences of continuous r.v.s can behave very differently, either having no limiting distribution at all or approaching limiting forms that are discrete or continuous. Although each X_n in Example 6.5 is normally distributed, the probability mass in any arbitrarily short interval about 0 approaches unity as n increases, making X_n more and more like a degenerate r.v. that takes the value zero for sure. This is an instance of the sometimes awkward transition from continuous to discrete types.

Sequences of discrete distributions can also behave in different ways. In each of the next three examples $\{Y_n\}_{n=1}^\infty$ is a sequence of binomials, with $Y_n \sim B(n, \theta)$ for some $\theta \in (0, 1)$. Here it is much easier to work with c.f.s rather than with c.d.f.s.

Example 6.6. With $X_n := Y_n$ we have $\phi_n(t) = \left[1 + \theta\left(e^{it} - 1\right)\right]^n = \left(\tau + \theta e^{it}\right)^n$, where $\tau := 1 - \theta$. To determine the limiting behavior of $\{X_n\}$, consider the squared modulus:

$$\begin{aligned}
|\phi_n(t)|^2 &= \phi_n(t)\phi_n(-t) \\
&= \left[\tau^2 + \tau\theta\left(e^{it} + e^{-it}\right) + \theta^2\right]^n \\
&= \left(\tau^2 + 2\tau\theta\cos t + \theta^2\right)^n \\
&=: \psi(t;\theta)^n.
\end{aligned}$$

Of course, $0 \le |\phi_n(t)|^2 \le 1$ for all t and all θ, and so the same bounds apply to $\psi(t;\theta)$, with $\psi(t;\theta) = (\tau + \theta)^2 = 1$ for $t \in \mathbb{T} := \{\pm 2j\pi\}_{j=0}^\infty$ and

$0 \le \psi(t; \theta) < 1$ elsewhere. Thus, $|\phi_n(t)| \to |\phi_\infty(t)| = 1_{\mathbb{T}}(t)$. This being discontinuous at $t = 0$, it follows that $\{X_n\}$ has no limiting distribution. As in Example (6.3), the measures $\{\mathbb{P}_{X_n}\}$ lack tightness.

Example 6.7. With $X_n := Y_n/n$ we have

$$\phi_n(t) = \left[1 + \theta\left(e^{it/n} - 1\right)\right]^n$$

$$= \left\{1 + \theta\left[1 + \frac{it}{n} + o\left(n^{-1}\right) - 1\right]\right\}^n$$

$$= \left\{1 + \theta\left[\frac{it}{n} + o\left(n^{-1}\right)\right]\right\}^n$$

and hence by (6.5) $\phi_n(t) \to \phi_\infty(t) = e^{it\theta}$. The limiting distribution of X_n is now degenerate, with unit mass at θ.

Example 6.8. With $X_n := (Y_n - n\theta)/\sqrt{n}$ we have

$$\phi_n(t) = e^{-it\sqrt{n}\theta} \left[1 + \theta\left(e^{it/\sqrt{n}} - 1\right)\right]^n$$

$$= e^{-it\sqrt{n}\theta} \left\{1 + \theta\left[\frac{it}{\sqrt{n}} - \frac{t^2}{2n} + o\left(n^{-1}\right)\right]\right\}^n.$$

Taking logs and expanding $\log(1+x)$ as $x - x^2/2 + o(x^2)$ as $|x| \to 0$ give

$$\log \phi_n(t) = -it\sqrt{n}\theta + n\log\left\{1 + \theta\left[\frac{it}{\sqrt{n}} - \frac{t^2}{2n} + o\left(n^{-1}\right)\right]\right\}$$

$$= -it\sqrt{n}\theta + n\left[\left(\frac{it\theta}{\sqrt{n}} - \frac{t^2\theta}{2n}\right) + \left(\frac{t^2\theta^2}{2n}\right) + o\left(n^{-1}\right)\right]$$

$$= -\frac{t^2\theta(1-\theta)}{2} + o(1).$$

Thus, $\phi_n(t) \to e^{-t^2\theta(1-\theta)/2}$, which shows that $X_n \rightsquigarrow N[0, \theta(1-\theta)]$, or equivalently that $(Y_n - n\theta)/\sqrt{n\theta(1-\theta)} \rightsquigarrow N(0,1)$.

Example 6.8 proves the *De Moivre–Laplace theorem*, one of the earliest and most famous limit theorems in probability. It shows that for large n the distributions $B(n, \theta)$ and $N(n\theta, n\theta(1-\theta))$ place approximately the same probability mass on any open interval (a, b), despite the fact that the binomial distribution is discrete (unit mass distributed over a finite number of points) and the normal is continuous (no mass at all on *any* countable set of points). For given n the normal approximation improves as θ approaches

1/2. The Poisson approximation, $B(n, \theta) \doteq P(n\theta)$, is better when θ is very small or very close to unity. In the latter case the Poisson approximates the number of *failures*, $n - Y_n \sim B(n, 1 - \theta)$.

Exercise 6.2. *Use the conclusion of De Moivre–Laplace to show that the sequence of binomial distributions $\{B(n, \theta)\}_{n=1}^{\infty}$ lacks tightness. Specifically, show that if $Y_n \sim B(n, \theta)$ then $\Pr(a < Y_n < b) \to 0$ as $n \to \infty$ for each finite interval $(a, b]$. Hint:*

$$a < Y_n < b \iff \frac{a - n\theta}{\sqrt{n\theta(1 - \theta)}} < \frac{Y_n - n\theta}{\sqrt{n\theta(1 - \theta)}} < \frac{b - n\theta}{\sqrt{n\theta(1 - \theta)}}.$$

Exercise 6.3. *$\{Y_j\}_{j=1}^{\infty}$ are i.i.d. as $\Gamma(\alpha, \beta)$ (gamma with shape parameter α and scale parameter β). Find the limiting distribution of $X_n := n^{-1}\sum_{j=1}^{n} Y_j$.*

Exercise 6.4. *$Y_n \sim P(n)$ (Poisson with parameter n) for $n \in \aleph$ and $X_n := Y_n/n$. Find the limiting distribution of X_n.*

Exercise 6.5. *$Y_n \sim P(n)$ for $n \in \aleph$ and $X_n := (Y_n - n)/\sqrt{n}$. Find the limiting distribution of X_n.*

Exercise 6.6. *For each $n \in \aleph$ X_n is a discrete r.v. with p.m.f.*

$$f_n(x) = n^{-1} 1_{\{1/n, 2/n, \ldots, (n-1)/n, 1\}}(x).$$

Find the limiting distribution as $n \to \infty$. (Hint: Find the limiting c.d.f.)

6.4   Stochastic Convergence to a Constant

Example 6.5 involved a sequence of r.v.s whose distributions become increasingly concentrated about zero. In such a case can one just say that the r.v.s themselves converge to zero? Likewise, can one say that the r.v.s in Example 6.7 and Exercises 6.3 and 6.4 converge to nonzero constants? In fact, there are several senses in which sequences of r.v.s can be considered to converge to constants. The three most prominent ones are (i) convergence *in probability,* (ii) convergence *in mean square,* and (iii) convergence *almost surely* or *with probability one.* We define these first and then explain how they are related. Example 6.12 below illustrates some other forms of convergence.

6.4.1 *Convergence in Probability*

Definition 6.2. The sequence of r.v.s $\{X_n\}_{n=1}^{\infty}$ converges **in probability** to the constant c if and only if $\{X_n\}$ converges in distribution to $F(x) = \mathbf{1}_{[c,\infty)}(x)$.

Recall that convergence in distribution requires the sequence of c.d.f.s $\{F_n\}_{n=1}^{\infty}$ to converge pointwise to F at each continuity point—in this case, everywhere other than at c. One would therefore say of $X_n := \bar{Y}$ in Example 6.5 that $\{X_n\}$ converges in probability to zero, and of X_n in Example 6.7 that $\{X_n\}$ converges in probability to θ.

As for convergence in distribution generally, several notations are used in statistics and econometrics to indicate convergence in probability. Econometricians usually write "$P\lim_{n\to\infty} X_n = c$", while statisticians use "$X_n \to c$ in probability" or "$X_n \to c$ i.p." or "$X_n \to^p c$". Each would say that c is the *probability limit* of X_n. Both the $P\lim_{n\to\infty}$ and $\to^p$ notations will be used in what follows.

There are several ways to set about to prove convergence in probability. One can use the definition and see whether $F_n(x) \to \mathbf{1}_{[c,\infty)}(x)$ for $x \neq c$, as in Example 6.5. Alternatively, one can try to show that $\phi_n(t) \to e^{itc}$ and apply the continuity theorem for c.f.s, as was done in Examples 6.5 and 6.7. The following theorem gives another approach that is sometimes useful. It just formalizes the notion that when n is large enough X_n will "probably" be close to c or, equivalently, that X_n is "unlikely" to be far from c.

Theorem 6.2. *For $X_n \to^p c$ it is necessary and sufficient that, for each $\varepsilon > 0$,*

$$\Pr(|X_n - c| < \varepsilon) \to 1, \tag{6.7}$$

or equivalently that

$$\Pr(|X_n - c| \geq \varepsilon) \to 0. \tag{6.8}$$

Note that sequences of probabilities such as $\{\Pr(|X_n - c| \geq \varepsilon)\}_{n=1}^{\infty}$ are merely sequences of numbers, like the sequence $\{a_n\}_{n=1}^{\infty}$ in the introduction to the chapter. Applying the theorem just requires finding the limit of the sequence. Remarkably, this can often be done even when the probabilities themselves cannot be specifically determined for finite n.

Example 6.9. $\{Y_j\}_{j=1}^{\infty}$ are i.i.d. with mean μ and finite, positive variance σ^2, and $X_n := n^{-1}\sum_{j=1}^{n} Y_j$. Since the common distribution of the $\{Y_j\}$

is not specified, there is no way to determine $\Pr\left(|X_n - \mu| \geq \varepsilon\right)$ for any n. However, this probability can be *bounded* using Chebychev's inequality,

$$0 \leq \Pr\left(|X_n - \mu| \geq \varepsilon\right) \leq \frac{E\left(X_n - \mu\right)^2}{\varepsilon^2} = \frac{\sigma^2}{n\varepsilon^2}.$$

Since the upper bound approaches zero for any positive ε, it follows that $X_n \to^p \mu$.

Exercise 6.7. *Prove that (6.8) is necessary and sufficient for* $X_n \to^p c$.

Exercise 6.8. *Y has c.d.f.* $F_Y(y) = \left(1 - y^{-1}\right)\mathbf{1}_{[1,\infty)}(y)$ *and* $X_n = Y/n$. *Use Theorem 6.2 to prove that* $X_n \to^p 0$.

6.4.2 *Convergence in Mean Square*

Definition 6.3. The sequence of r.v.s $\{X_n\}_{n=1}^{\infty}$ converges **in mean square** to the constant c if $E(X_n - c)^2 \to 0$.

To express this concisely, one writes "$X_n \to c$ in mean square" or "$X_n \to^{m.s.} c$". Again, we are dealing here just with a sequence of numbers—in this case the expected values of the squared differences between X_n and c. When it does take place, convergence in mean square is usually easy to establish, since it involves just the moments of X_n. Sometimes, as in Example 6.9, one can see how the moments behave even when the c.d.f. is not specified. Here is a widely applicable theorem that illustrates how easy it is to prove mean-square convergence under appropriate conditions.

Theorem 6.3. *If* $\{Y_j\}_{j=1}^{\infty}$ *are i.i.d. r.v.s with finite 2kth moments* μ'_{2k}, *and if* $X_n := n^{-1}\sum_{j=1}^{n} Y_j^k = M'_k$, *then* $X_n \to^{m.s.} \mu'_k$.

Proof. $EX_n = \mu'_k$ and $VX_n = n^{-1}\left[\mu'_{2k} - (\mu'_k)^2\right]$, so $E\left(X_n - \mu'_k\right)^2 = VX_n \to 0$, and therefore $X_n \to^{m.s.} \mu'_k$. □

An obvious corollary is that $\bar{Y} \to^{m.s.} \mu$ when $\{Y_j\}_{j=1}^{n}$ are i.i.d. with mean μ and finite variance. We will see shortly that $X_n \to^{m.s.} c$ implies $X_n \to^p c$ (as Example 6.9 has already illustrated), so establishing mean-square convergence is yet another way to prove convergence in probability.

Exercise 6.9. *Suppose* $EX_n = \mu_n$ *and* $VX_n = \sigma_n^2$. *Prove that* $X_n \to^{m.s.} c$ *if and only if* $(\mu_n - c)^2 \to 0$ *and* $\sigma_n^2 \to 0$ *as* $n \to \infty$.

6.4.3 Almost-Sure Convergence

As we have seen, a sequence $\{X_n\}_{n=1}^{\infty}$ converges in probability to a constant c if and only if the chance of any threshold deviation of X_n from c approaches zero as n increases. In other words, $X_n \to^p c$ implies that, for any fixed but arbitrarily small tolerance ε, $\Pr(|X_n - c| \geq \varepsilon)$ can itself be made arbitrarily small by taking n sufficiently large. To put this in formal terms, if $X_n \to^p c$ then to any $\varepsilon > 0$ and $\delta > 0$ there corresponds $N(\varepsilon, \delta)$ such that $\Pr(|X_n - c| \geq \varepsilon) \leq \delta$ for each $n \geq N(\varepsilon, \delta)$. While this is a powerful result, it is not all that we might desire, because making the probability small at *each* individual stage beyond N does not by itself limit the chance that a significant deviation will *ever* subsequently occur. That is, bounding $\Pr(|X_n - c| \geq \varepsilon)$ for each individual $n \geq N$ does not bound $\Pr(\cup_{n=N}^{\infty} \{|X_n - c| \geq \varepsilon\})$—the probability that one or more large deviations will at *some* point occur. After all, there are infinitely many such subsequent stages and therefore infinitely many opportunities for one or more of the individual low-probability events to occur. As a trite example, by loading an urn with enough balls of another color, we could make the probability of getting a red ball on any single draw arbitrarily small. Still, if we kept drawing indefinitely, even with replacement, it is *certain* that we would eventually come up with the red. *Almost-sure* convergence is a stronger condition that does limit the probability of an ε deviation at *any* stage from some point forward.

Definition 6.4. The sequence of r.v.s $\{X_n\}_{n=1}^{\infty}$ converges **almost surely** to the constant c if and only if, for each $\varepsilon > 0$,

$$\Pr(|X_m - c| \geq \varepsilon \text{ for any } m \geq n) \to 0 \text{ as } n \to \infty, \tag{6.9}$$

or equivalently that

$$\Pr(|X_m - c| < \varepsilon \text{ for all } m \geq n) \to 1 \text{ as } n \to \infty. \tag{6.10}$$

The common notations are "$X_n \to c$ almost surely" or "$X_n \to^{a.s.} c$". Sometimes this is referred to as "strong" convergence or as convergence with probability one.

In case the concept of a.s. convergence is still a bit elusive, let us come at it in another way. Think of $B_n(\varepsilon) := \{\omega : |X_n(\omega) - c| \geq \varepsilon\}$ and $S_n(\varepsilon) := B_n(\varepsilon)^c$ as the events "big deviation at stage n" and "small deviation at stage n". Then (6.9) and (6.10) can be restated as

$$\lim_{n \to \infty} \mathbb{P}[\cup_{m=n}^{\infty} B_m(\varepsilon)] = 0$$

and

$$\lim_{n \to \infty} \mathbb{P}\left[\cap_{m=n}^{\infty} S_m\left(\varepsilon\right)\right] = 1, \tag{6.11}$$

respectively. Now the sequence $\{\cup_{m=n}^{\infty} B_m\left(\varepsilon\right)\}$ decreases monotonically as n increases, and the definition of limit for monotone sequences (p. 10) implies

$$\lim_{n \to \infty} \bigcup_{m=n}^{\infty} B_m\left(\varepsilon\right) = \bigcap_{n=1}^{\infty} \bigcup_{m=n}^{\infty} B_m\left(\varepsilon\right) = \limsup_n B_n\left(\varepsilon\right).$$

It then follows by the monotone property of probability that

$$\lim_{n \to \infty} \mathbb{P}\left[\bigcup_{m=n}^{\infty} B_m\left(\varepsilon\right)\right] = \mathbb{P}\left[\lim_{n \to \infty} \bigcup_{m=n}^{\infty} B_m\left(\varepsilon\right)\right] = \mathbb{P}\left[\limsup_n B_n\left(\varepsilon\right)\right].$$

Thus, condition (6.9) is equivalent to

$$\mathbb{P}\left[\limsup_n B_n\left(\varepsilon\right)\right] =: \mathbb{P}\left[B_n\left(\varepsilon\right) \text{ i.o.}\right] = 0$$

for each $\varepsilon > 0$. In other words, $X_n \to^{a.s.} c$ if and only if it is almost certain that only finitely many members of the sequence $\{X_n\}_{n=1}^{\infty}$ will differ from c by more than an arbitrary tolerance level. Equivalently, from some (unspecified) point forward it is almost certain that all members of the sequence will take values "close" to c; i.e., $\mathbb{P}\left[\liminf_n S_n\left(\varepsilon\right)\right] = 1$. (As usual, "almost certain" means that the probability is unity.)

Exercise 6.10. *Present an argument like that for events $\{B_n\left(\varepsilon\right)\}$ to show that (6.11) holds if and only if $\mathbb{P}\left[\liminf_n S_n\left(\varepsilon\right)\right] = 1$.*

Here is yet another way to look at almost-sure convergence—and one that really helps the intuition. Think of an outcome of a chance experiment as generating the entire sequence of realizations of X_n from $n = 1$ forward. If we fix a particular outcome ω, then the infinite sequence of realizations $\{X_n\left(\omega\right)\}_{n=1}^{\infty}$ that is determined by this outcome is simply a sequence of numbers, while another such sequence $\{X_n(\omega')\}_{n=1}^{\infty}$ would correspond to some other outcome ω'. Such sequences of numbers may converge to c for some outcomes; for others, not. If $\mathfrak{C}$ is the set of outcomes for which convergence does occur, then a.s. convergence just means that $\mathbb{P}\left(\mathfrak{C}\right) = 1$.[5] This is why we also speak of almost-sure convergence as convergence with probability one.

---

[5] Set $\mathfrak{C}$ is measurable with respect to the σ field generated by $\{X_n\}_{n=1}^{\infty}$, since knowing the values of these r.v.s tells us whether they are, from some point on, all within ε of c.

The following gives a sufficient condition for a.s. convergence that is often easy to verify.

Theorem 6.4. $X_n \to^{a.s.} c$ *if for each* $\varepsilon > 0$

$$\sum_{n=1}^{\infty} \Pr\left(|X_n - c| \geq \varepsilon\right) < \infty. \tag{6.12}$$

Proof. Apply the convergence part of the Borel–Cantelli lemma. Letting $B_n(\varepsilon) = \{\omega : |X_n(\omega) - c| \geq \varepsilon\}$ as above, (6.12) implies $\mathbb{P}\left[\limsup_n B_n(\varepsilon)\right] = 0$. $\square$

Note that (6.12) is sufficient for a.s. convergence but may be stronger than necessary. When it holds, $\{X_n\}_{n=1}^{\infty}$ is said to converge "completely" to c. However, the divergence part of Borel–Cantelli (page 40) shows that (6.12) is also necessary when the $\{X_n\}_{n=1}^{\infty}$ are independent.

Example 6.10. $\{X_n\}_{n=1}^{\infty}$ are independent r.v.s with c.d.f.s

$$F_n(x) = \left(1 - \frac{1}{nx}\right) \mathbf{1}_{[n^{-1}, \infty)}(x).$$

Then $\Pr\left(|X_n| \geq \varepsilon\right) = 1$ when $n < \varepsilon^{-1}$ and $\Pr\left(|X_n| \geq \varepsilon\right) = (n\varepsilon)^{-1}$ when $n \geq \varepsilon^{-1}$. But since $\sum_{n=1}^{\infty} n^{-1} = \infty$, Borel–Cantelli implies that infinitely many ε deviations are certain to occur. Thus, convergence is *not* almost sure even though $X_n \to^p 0$.

Exercise 6.11. $\{X_n\}_{n=1}^{\infty}$ *have c.d.f.s* $F_n(x) = \left(1 - n^{-2}/x\right) \mathbf{1}_{[n^{-2}, \infty)}(x)$. *Show that* $X_n \to^{a.s.} 0$. *How does* EX_n^2 *behave as* $n \to \infty$?

6.4.4 *Relations among the Modes of Convergence*

The connection between convergence in probability, convergence in mean square, and almost-sure convergence is very simple:

- Convergence in mean square implies convergence in probability ($m.s. \implies p$);
- Almost-sure convergence implies convergence in probability ($a.s. \implies p$); and
- There are no other connections.

Thus, convergence in probability does *not* imply either convergence in mean square or convergence a.s., nor does either of these imply the other. Statements 1 and 2 are easily verified by formal argument, while 3 is proved by counterexamples.

- m.s. $\implies p$: This follows from Chebychev along the lines of Example 6.9. Given that $X_n \to^{m.s.} c$ we have

$$0 \le \lim_{n\to\infty} \Pr(|X_n - c| \ge \varepsilon) \le \lim_{n\to\infty} \frac{E\left[(X_n - c)^2\right]}{\varepsilon^2} = 0$$

for any $\varepsilon > 0$.

- $p \nRightarrow$ m.s.: M.s. convergence depends on the behavior of moments, and these can behave wildly even as probability mass becomes concentrated about c. Exercise 6.8 provides an example. With $F_Y(y) = \left(1 - y^{-1}\right) \mathbf{1}_{[1,\infty)}(y)$ and $X_n = Y/n$ it is true that $X_n \to^p 0$, yet $EX_n^2 = n^{-2}EY^2 = n^{-2}\int_1^\infty y^2 \cdot dF_Y(y) = \infty$ for each n, so X_n does not converge in m.s. The following situation is even more revealing.

Example 6.11. For each $n \in \aleph$ X_n is a discrete r.v. with p.m.f.

$$f_n(x; \theta) = \left(1 - n^{-1}\right) \mathbf{1}_{\{0\}}(x) + n^{-1}\mathbf{1}_{\{n^\theta\}}(x)$$

for some real θ. (Thus, there is probability mass $1 - 1/n$ at the origin and mass $1/n$ at the point n^θ.) That $X_n \to^p 0$ can be shown in several ways. First, expressing the c.d.f. as $F_n(x) = \left(1 - n^{-1}\right) \mathbf{1}_{[0,n^\theta)}(x) + \mathbf{1}_{[n^\theta,\infty)}(x)$, it is apparent that $F_n(x) \to F_\infty(x) = \mathbf{1}_{[0,\infty)}(x)$ regardless of the value of θ. The same conclusion follows from Theorem 6.2, since $\Pr(|X_n - 0| \ge \varepsilon) \le n^{-1} \to 0$ for any $\varepsilon > 0$. Finally, since $\phi_n(t) = (1 - n^{-1})e^{it0} + n^{-1}e^{itn^\theta} \to e^{it0} = 1$ for all θ, the result also follows from the continuity theorem for c.f.s. On the other hand, whether $X_n \to^{m.s.} 0$ depends on the value of θ, since

$$\lim_{n\to\infty} E(X_n^2) = \lim_{n\to\infty} n^{2\theta-1} = \begin{cases} 0, & \theta < \frac{1}{2} \\ 1, & \theta = \frac{1}{2} \\ +\infty, & \theta > \frac{1}{2} \end{cases}.$$

- a.s. $\implies p$: This is really obvious, since $\Pr\left(|X_n - c| \ge \varepsilon\right)$ is necessarily small if the probability is small that $|X_m - c| \ge \varepsilon$ for *all* $m \ge n$. However, to show it take $B_n(\varepsilon) := \{\omega : |X_n(\omega) - c| \ge \varepsilon\} \subset \cup_{m=n}^\infty B_m(\varepsilon)$ and observe that

$$0 \le \lim_{n\to\infty} \mathbb{P}\left[B_n(\varepsilon)\right] \le \lim_{n\to\infty} \mathbb{P}\left[\bigcup_{m=n}^\infty B_m(\varepsilon)\right]$$

$$= \mathbb{P}\left[\bigcap_{n=1}^\infty \bigcup_{m=n}^\infty B_m(\varepsilon)\right]$$

$$= \mathbb{P}\left[\limsup_n B_n(\varepsilon)\right] = 0.$$

- $p \not\Rightarrow$ a.s.: There are in fact cases in which $\Pr\left(|X_n - c| \geq \varepsilon\right)$ at stage n can be controlled even though the event of a subsequent ε deviation at some point is likely or even "almost" certain. Example 6.10 showed one such. Here are two more illustrations of the fact.

Exercise 6.12. *Random variables $\{X_n\}_{n=1}^{\infty}$ are independent with $f_n(x) = \left(1 - n^{-1}\right) \mathbf{1}_{\{0\}}(x) + n^{-1} \mathbf{1}_{\{1\}}(x)$. Show that $X_n \to^p 0$ but that when $\varepsilon < 1$ the event $|X_n| \geq \varepsilon$ occurs for infinitely many n.*

Exercise 6.13. *Let $\{Y_n\}_{n=1}^{\infty}$ be i.i.d. r.v.s, and for each n put $X_n := Y_n/n$. Show that $X_n \to^p 0$ without any further condition but that $X_n \to^{a.s.} 0$ if and only if $E|Y_1| < \infty$. (Hint: Use relation (3.18) on page 167.)*

- m.s. $\not\Leftrightarrow$ a.s.: Neither of these modes of convergence implies the other. A.s. convergence puts no restriction on the moments of X_n. For example, X_n in Exercise 6.11 converges a.s. to zero, yet $EX_n^2 = \infty$ for each n. Conversely, convergence in m.s. controls $E(X_n - c)^2$ at stage n and thereby does limit $\Pr\left(|X_n - c| \geq \varepsilon\right)$, but this does not necessarily make small the chance of an eventual ε deviation.

Exercise 6.14. *$\{X_n\}_{n=1}^{\infty}$ are independent r.v.s with c.d.f.s*

$$F_n(x) = \left(1 - \frac{1}{nx^3}\right) \mathbf{1}_{[n^{-1/3}, \infty)}(x).$$

Show that $X_n \to^{m.s.} 0$ but that convergence is not a.s.

Exercise 6.15. *Show that $\sum_{n=1}^{\infty} E(X_n - c)^2 < \infty$ implies $X_n \to^{a.s.} c$.*

The following further elucidates the relations among the three principal modes of convergence, provides examples of other senses in which sequences of r.v.s might be said to converge to constants, and illustrates some possibilities that might surprise.

Example 6.12. Let $\{X_n\}_{n=1}^{\infty}$ be a sequence of r.v.s on a probability space $(\aleph, \mathcal{F}, \mathbb{P})$, with $\mathbb{P}\left(\{j\}\right) > 0$ for $j \in \aleph := \{1, 2, ...\}$. Note that $\mathbb{P}(\aleph) = \sum_{j=1}^{\infty} \mathbb{P}\left(\{j\}\right) = 1 < \infty$ implies that $Q_n := \sum_{j=n+1}^{\infty} \mathbb{P}\left(\{j\}\right) \downarrow 0$ as $n \to \infty$. With $X_n(j)$ as the realization of X_n at outcome j, put $X_n(j) = \mathbf{1}_{\{n+1, n+2, ...\}}(j)$. (Thus, in particular, $X_n(j) = 0$ for $j \in \{1, 2, ..., n\}$ and $X_n(j) = 1$ for integers $j > n$.) We can now draw the following conclusions about the sequence $\{X_n\}_{n=1}^{\infty}$.

(i) $X_n \to^p 0$ since for any $\varepsilon > 0$ we have $0 \leq \Pr\left(|X_n| \geq \varepsilon\right) \leq Q_n$.

(ii) $X_n \to^{m.s.} 0$ since $EX_n^2 = Q_n$. Indeed, $E|X_n|^p = Q_n \to 0$ for each $p > 0$.

(iii) For each integer $m \geq n$ the event $X_m \geq \varepsilon$ implies (is a subset of) $X_n \geq \varepsilon$, both sets being empty when $\varepsilon > 1$. Thus, letting $A_m(\varepsilon) := \{j : |X_m(j)| \geq \varepsilon\}$,

$$\Pr(|X_m| \geq \varepsilon \text{ for any } m \geq n) = \mathbb{P}\left[\bigcup_{m=n}^{\infty} A_m(\varepsilon)\right] = \mathbb{P}[A_n(\varepsilon)] \leq Q_n,$$

and so $X_n \to^{a.s.} 0$. Note that this would be true even if $\mathbb{P}$ were such that $\sum_{n=1}^{\infty} Q_n = \infty$, in which case convergence would not be "complete".[6]

(iv) $X_n(j) \to 0$ as $n \to \infty$ for each fixed integer j. Thus, for each fixed $m \in \aleph$ the set $\mathfrak{C}^c$ of outcomes $\{j\}$ for which $X_n(j)$ does *not* converge to zero is a subset of $\aleph \backslash \{1, 2, ..., m\}$. But $\{1, 2, ..., m\} \uparrow \aleph$ as $m \to \infty$ implies $\mathfrak{C}^c = \varnothing$, so we can say that $\{X_n\}$ converges to zero not just *almost* surely but *surely*—a super-strong mode of convergence that is rarely verifiable in statistical applications. However, even this strong result has its limitations, since

(v) $\sup_{m>n}\{|X_m|\} = 1$ for all n and thus fails to converge to zero as $n \to \infty$. Indeed, $\lim_{n \to \infty} X_n(n+1) = 1$.

The result $E|X_n|^p \to 0$ in part (ii) of Example 6.12 is referred to as "convergence in $\mathcal{L}^p$", where (in the probability context) the $\mathcal{L}^p$ spaces (usually considered just for $p \geq 1$) comprise the r.v.s X for which $E|X|^p < \infty$. Convergence of $\sup\{|X_n|\}$ is referred to as "convergence in $\mathcal{L}^\infty$", where $\mathcal{L}^\infty$ is the space of r.v.s that are almost surely bounded. Thus, in our example $\{X_n\}$ converges to zero in $\mathcal{L}^p$ for any $p > 0$ but does not converge to zero in $\mathcal{L}^\infty$.

---

[6]As an example of a $\mathbb{P}$ for which $\sum_{n=1}^{\infty} Q_n = \infty$, set $\mathbb{P}(\{j\}) = \frac{4}{3}\left[(1+j)^2 - 1\right]^{-1}$ for $j \in \aleph$. Induction shows that

$$\sum_{j=1}^{n} \mathbb{P}(\{j\}) = 1 - \frac{2(2n+3)}{3(n+1)(n+2)},$$

whence $\sum_{j=1}^{\infty} \mathbb{P}(\{j\}) = 1$. However, for each $N \in \aleph$ we have

$$\sum_{n=1}^{N} Q_n = \sum_{n=1}^{N} \frac{2(2n+3)}{3(n+1)(n+2)}$$

$$> \sum_{n=1}^{N} \frac{1}{n+2}$$

and $\lim_{N \to \infty} \sum_{n=1}^{N} (n+2)^{-1} = +\infty$.

Note that if $\{X_n\}_{n=1}^{\infty}$ is a sequence of *degenerate* r.v.s—i.e., constants with $\Pr(X_n = x_n) = 1$, and if sequence $\{x_n\}_{n=1}^{\infty}$ converges to the finite constant x, then we could say that $\{X_n\} \to x$ in *all* the above senses: in distribution, in probability, in mean square, almost surely, completely, surely, and in $\mathcal{L}^{\infty}$.

6.5 Convergence to a Random Variable

If a sequence $\{X_n\}_{n=1}^{\infty}$ of r.v.s converges to a constant c, then we could say equivalently that it converges to a degenerate r.v. X for which $\mathbb{P}_X(\{c\}) = 1$. This may happen even if the $\{X_n\}$ are defined on different probability spaces $\{(\Omega_n, \mathcal{F}_n, \mathbb{P}_n)\}$. A sequence of r.v.s can also converge to a r.v. that is not of this degenerate form, but for this to make sense all of $\{X_n\}_{n=1}^{\infty}$ and X must be defined on a common $(\Omega, \mathcal{F}, \mathbb{P})$.

Definition 6.5. The sequence $\{X_n\}_{n=1}^{\infty}$ converges (in some sense) to the r.v. X if the sequence $\{X_n - X\}_{n=1}^{\infty}$ converges to zero (in the same sense).

Thus, we can speak of $\{X_n\}_{n=1}^{\infty}$ converging almost surely, in probability, or in mean square to the r.v. X. Specifically,

- $X_n \to^p X$ if and only if $\lim_{n\to\infty} \Pr(|X_n - X| > \varepsilon) = 0$ for any $\varepsilon > 0$.
- $X_n \to^{m.s.} X$ if and only if $\lim_{n\to\infty} E(X_n - X)^2 = 0$.
- $X_n \to^{a.s.} X$ if and only if $\Pr(|X_n - X| > \varepsilon \text{ i.o.}) = 0$ for any $\varepsilon > 0$.

Another version of order notation is sometimes used in connection with convergent sequences of r.v.s.

Definition 6.6. The sequence of r.v.s $\{X_n\}_{n=1}^{\infty}$ is said to be $O_P(n^q)$, read "of order n^q *in probability*", if there is a r.v. X (possibly degenerate, but not identically zero) such that $n^{-q} X_n \to^p X$. Sequence $\{X_n\}_{n=1}^{\infty}$ is said to be $o_P(n^q)$, read "of order *less than* n^q in probability", if $n^{-q} X_n \to^p 0$.

Thus, X_n is (i) $O_P(n)$ in Example 6.6; (ii) $O_P(n^{1/2})$ in Example 6.3; (iii) $O_P(n^0) \equiv O_P(1)$ in Examples 6.4, 6.7, and 6.8; and (iv) $O_P(n^{-1/2})$ and $o_P(1)$ in Example 6.5.

Noting that both a.s. convergence and m.s. convergence imply convergence in probability, it is clear that if $X_n \to X$ in any of the three senses, then $\Pr(|X_n - X| > \varepsilon) \to 0$ for any $\varepsilon > 0$. If F is the c.d.f. of X, it thus seems plausible that c.d.f.s $\{F_n\}$ should converge to F, at least at continuity points. Indeed, this is so, the formal statement being as follows.[7]

Theorem 6.5. *Suppose $X_n \to X$ in any of the three senses, and let F be the c.d.f. of X. Then $X_n \rightsquigarrow F$.*

On the other hand, the facts that X_n converges in distribution to F and that a particular r.v. X has the distribution F do *not* imply that X_n converges to X. This is why $X_n \rightsquigarrow F$ is styled "weak" convergence. That is, nothing requires the realizations of X_n and X to be close with high probability just because the distributions are close. (Think of two i.i.d. random variables, such as the numbers that turn up on two successive rolls of a die.) Indeed, the joint condition that $X_n \rightsquigarrow F$ and $F(x) = \Pr(X \le x)$ for all x would allow all of $\{X_n\}_{n=1}^\infty$ and X to be defined on different probability spaces. However, if $X_n \rightsquigarrow F$ one can construct r.v.s $\{X_n\}_{n=1}^\infty$ and X on the *same* probability space to which $\{X_n\}$ does converge a.s.—a fact whose importance will be seen below in connection with the Mann–Wald theorem.[8]

6.5.1 *Continuity Theorem for Stochastic Convergence*

We already know that if a sequence of constants $\{a_n\}_{n=1}^\infty$ converges to a constant a and if $g : \Re \to \Re$ is continuous at a, then $g(a_n) \to g(a)$. There is a corresponding result for *stochastic* convergence, although the concept and proof are a little more involved. A joint property of a function g and r.v. X that we shall call "X continuity" will be found helpful for this.

Definition 6.7. Let $\mathbb{P}_X$ be the probability measure induced by a r.v. X having support $\mathbb{X} \subset \Re$, and let $g : \mathbb{X} \to \Re$ be a measurable function. If $\mathcal{D}_g \subset \Re$ is the (measurable, possibly empty) set of points at which g is discontinuous, then we say that g is *X-continuous* if $\mathbb{P}_X(\mathcal{D}_g) = 0$. Equivalently, but less concisely, we could say that g is *continuous almost surely* with respect to $\mathbb{P}_X$.

For example, if X is a continuous r.v. and g is either continuous on $\mathbb{X}$ or has at most countably many discontinuities, then g is X-continuous. The

---

[7]For a proof see Billingsley (1995, p. 330).

[8]Briefly, let $Y \sim U(0,1)$ be defined on some space $(\Omega, \mathcal{F}, \mathbb{P})$; and put $X_n(\omega) := F_n^{-1}(Y(\omega))$ and $X(\omega) := F^{-1}(Y(\omega))$, with inverse functions defined as on page 258. Then $F_n \to F$ a.e. implies $F_n^{-1} \to F^{-1}$ a.e. and $|X_n - X| \to 0$. For details, see Theorem 25.6 in Billingsley (1995).

following result, known as the *continuity theorem for stochastic convergence*, is extremely useful. Henceforth we refer to this just as the "continuity theorem", which is to be distinguished from the continuity theorem for c.f.s (Theorem 6.1).

Theorem 6.6. *If* $g : \Re \to \Re$ *is* X-*continuous, then (i)* $X_n \to^{a.s.} X$ *implies* $g(X_n) \to^{a.s.} g(X)$ *and (ii)* $X_n \to^p X$ *implies* $g(X_n) \to^p g(X)$.

Proof. (i) Since $\mathbb{P}_X (\mathcal{D}_g) = 0$, we can consider only outcomes at which $g(X(\omega))$ is continuous. Then for any $\varepsilon > 0$ there is a $\delta > 0$ such that $|g(X_n(\omega)) - g(X(\omega))| < \varepsilon$ whenever n is large enough that $|X_n(\omega) - X(\omega)| < \delta$. The set $\mathcal{S}_n(\delta) := \{\omega : |X_n(\omega) - X(\omega)| < \delta\}$ (the outcomes for which X_n and X are close) is therefore a subset of $\mathcal{G}_n(\varepsilon) := \{\omega : |g(X_n(\omega)) - g(X(\omega))| < \varepsilon\}$ (outcomes for which $g(X_n)$ and $g(X)$ are close). But then $\liminf_n \mathcal{S}_n(\delta) \subset \liminf_n \mathcal{G}_n(\varepsilon)$ (since if all $\{\mathcal{S}_n\}$ occur from some point forward then all $\{\mathcal{G}_n\}$ do). Finally, since $X_n \to^{a.s.} X$ implies $\mathbb{P}(\liminf_n \mathcal{S}_n(\delta)) = 1$, it follows that $\mathbb{P}(\liminf_n \mathcal{G}_n(\varepsilon)) = 1$ as well, so that $g(X_n) \to^{a.s.} g(X)$. (ii) $\mathcal{S}_n(\delta) \subset \mathcal{G}_n(\varepsilon)$ and $\mathbb{P}(\mathcal{S}_n(\delta)) \to 1$ imply $\mathbb{P}(\mathcal{G}_n(\varepsilon)) \to 1$. $\qquad\square$

Note that since X is allowed to be degenerate with $\mathbb{P}_X(\{c\}) = 1$ for some c, the theorem implies that $g(X_n) \to^p g(c)$ so long as g is continuous at c. On the other hand, if g has a discontinuity at c it should be clear that making $X_n(\omega)$ close to c does not necessarily make $g(X_n)$ close to $g(c)$.

The continuity theorem leads to another powerful result, on which we shall rely frequently in later chapters.

Theorem 6.7. *(Mann–Wald) Suppose the sequence* $\{X_n\}_{n=1}^{\infty}$ *converges weakly to* F_X *(that is, the induced measures* $\{\mathbb{P}_{X_n}\}$ *converge weakly to the* $\mathbb{P}_X$ *corresponding to* F_X*) and that* $g : \Re \to \Re$ *is* X-*continuous. Then* $g(X_n) \rightsquigarrow F_Y$*, where* $F_Y(y) := \Pr(g(X) \le y)$ *for* $y \in \Re$.

Proof. If $X_n \rightsquigarrow F_X$ there exist r.v.s $\{X_n\}, X$ on the same probability space such that X has c.d.f. F_X and $X_n \to^{a.s.} X$. If g is X continuous then $g(X_n) \to^{a.s.} g(X)$ by the continuity theorem. The conclusion follows since convergence of $\{g\{X_n\}\}$ to $g(X)$ implies convergence in distribution. $\qquad\square$

Stated less precisely but more simply, the Mann–Wald theorem tells us that if $\{X_n\}$ converges in distribution to F_X and if r.v. X has c.d.f. F_X, then the distribution of $g(X_n)$ approaches the distribution of $g(X)$ for an a.s.-continuous g. Some examples show how useful this can be.

Example 6.13. If $X_n \rightsquigarrow N(0,1)$ then $X_n^2 \rightsquigarrow \chi^2(1)$ and $e^{X_n} \rightsquigarrow LN(0,1)$ (lognormal).

The continuity and Mann–Wald theorems extend to sequences of vector-valued functions $\mathbf{g} : \Re^m \to \Re^k$ of vector-valued r.v.s $\mathbf{X}_n \in \Re^m$. Thus, the continuity theorem assures us that $\mathbf{g}(\mathbf{X}_n) \to^{a.s./p} \mathbf{g}(\mathbf{X})$ whenever $\mathbf{X}_n \to^{a.s./p} \mathbf{X}$ and $\mathbf{g}$ is $\mathbf{X}$-continuous. However, one must be a little careful. A vector version of the statement $X_n \rightsquigarrow F$ specifies the limiting *joint* distribution of $\{\mathbf{X}_n\}$, not just the limiting marginal distributions of the components.

Example 6.14. Suppose $\{X_n\}_{n=1}^{\infty}$ and $\{Y_n\}_{n=1}^{\infty}$ are defined on the same probability space, and that $\{(X_n, Y_n)\}_{n=1}^{\infty}$ is known to converge in distribution to bivariate standard normal with correlation coefficient ρ equal to zero. Then Mann–Wald does imply that $g_1(X_n, Y_n) := X_n + Y_n \rightsquigarrow N(0,2)$, that $g_2(X_n, Y_n) := X_n^2 + Y_n^2 \rightsquigarrow \chi^2(2)$, and so forth. However, nothing can be said about the limiting distribution of $g(X_n, Y_n)$ if we know only that $X_n \rightsquigarrow N(0,1)$ and that $Y_n \rightsquigarrow N(0,1)$. For example, $Y_n \equiv -X_n$ would also be consistent with this information, and in that case $g_1(X_n, Y_n) \equiv 0$ for each n and $g_2(X_n, Y_n) \rightsquigarrow \Gamma(1/2, 4)$.

The following result is usually referred to as "Slutsky's Theorem". It is merely a corollary to Mann-Wald if $\{X_n\}$ and $\{Y_n\}$ are defined on the same probability space, but this is not necessary for the conclusion.

Theorem 6.8. *If $X_n \rightsquigarrow F_X$ and $Y_n \to^p c$, then (i) $X_n \pm Y_n$, (ii) $X_n Y_n$, and (iii) X_n/Y_n have the same limiting distributions, respectively, as (i') $X_n \pm c$, (ii') cX_n, and (iii') X_n/c $(c \neq 0)$.*

Proof. Take $\{X_n\}$ and $\{Y_n\}$ to be sequences of r.v.s on $(\Omega, \mathcal{F}, \mathbb{P})$ and $(\Omega', \mathcal{F}', \mathbb{P}')$, respectively. Define $U_n := F_n(X_n) \sim U(0,1)$ on $(\Omega, \mathcal{F}, \mathbb{P})$, where F_n is the c.d.f. of X_n, and put $Y_n'(\omega) = G_n^{-1}(U_n(\omega))$, where G_n is the c.d.f. of Y_n. The result now follows from Mann-Wald. $\square$

Note that $\{Y_n\}$ here may be *degenerate* r.v.s—i.e., constants. An extremely useful consequence of Slutsky's theorem is the following:

Corollary 6.1. *If $X_n \rightsquigarrow F_X$ and $Y_n \to^p c$, then $\Pr(X_n \leq Y_n) \to F_X(c)$.*

6.5.2 Limits of Moments vs. Moments of Limiting Distributions

Confusion often arises about the distinction between limits of sequences of moments and moments of limiting distributions. Suppose we know that $X_n \to^p X$ and that X has c.d.f. F, in which case it follows that $X_n \rightsquigarrow F$. Alternatively, suppose we know *just* that $X_n \rightsquigarrow F$, in which case we can exhibit *some* r.v. X with this c.d.f. Does it then necessarily follow that $\lim_{n\to\infty} EX_n^k \equiv \lim_{n\to\infty} \int x^k \cdot dF_n$ equals $\int x^k \cdot dF \equiv EX^k$ for positive integers k? In fact, we have already seen in Example 6.11 that it does *not* follow. There, the sequence of r.v.s $\{X_n\}$ with p.m.f.s

$$f_n(x; \theta) = \left(1 - n^{-1}\right) \mathbf{1}_{\{0\}}(x) + n^{-1} \mathbf{1}_{\{n^\theta\}}(x)$$

was seen to converge in probability to a degenerate r.v. with c.d.f. $F(x) = \mathbf{1}_{[0,\infty)}(x)$ that puts unit mass at the origin. Obviously, then, $EX^r = 0$ for any r.v. with this limiting distribution and any $r > 0$, but $EX_n^r \to 0$ if and only if $\theta < 1/r$.

Whether $EX_n^r = \int x^r \cdot dF_n$ does approach $\int x^r \cdot dF$ when $X_n \rightsquigarrow F$ turns out to depend on whether x^r is *uniformly* integrable (u.i.) with respect to $\{F_n\}_{n=1}^\infty$.

Definition 6.8. Function $g : \Re \to \Re$ is **uniformly integrable** with respect to the measures corresponding to distribution functions $\{F_n\}_{n=1}^\infty$ if and only if

$$\lim_{h\to\infty} \sup_{n\in\aleph} \int_{\{x:|g(x)|\geq h\}} |g(x)| \cdot dF_n(x) = 0.$$

A little thought is needed to see just what this means. Clearly, the usual integrability requirement $\int |g| \cdot dF_n < \infty$ implies, in itself, that $\lim_{h\to\infty} \int_{\{x:|g(x)|\geq h\}} |g| \cdot dF_n = 0$ for each n, since

$$\int_\Re |g| \cdot dF_n = \int_{|g|<h} g \cdot dF_n + \int_{|g|\geq h} g \cdot dF_n \text{ (each } h > 0)$$

$$= \lim_{h\to\infty} \left(\int_{|g|<h} g \cdot dF_n + \int_{|g|\geq h} g \cdot dF_n \right)$$

$$= \int_\Re |g| \cdot dF_n + \lim_{h\to\infty} \int_{|g|\geq h} g \cdot dF_n.$$

Thus, integrability alone implies that for each n the quantity $\int_{|g|\geq h} g \cdot dF_n$ can be made less than any $\varepsilon > 0$ by taking h sufficiently large—greater than some $H(n, \varepsilon)$, say. The additional requirement for uniform integrability is that *all* members of sequence $\left\{\int_{|g|\geq h} g \cdot dF_n\right\}_{n=1}^{\infty}$ can be made small simultaneously when h is beyond some common threshold; that is, that there be a single $H(\varepsilon)$ such that $\sup_{n\in\aleph}\int_{|g|\geq h}|g|\cdot dF_n < \varepsilon$ when $h > H(\varepsilon)$.

Example 6.15. For each $n \in \{2,3,...\}$ let X_n have c.d.f.

$$F_n(x) = \frac{nx-1}{n-1}1_{[n^{-1},1)}(x) + 1_{[1,\infty)}(x)$$

and p.d.f.

$$f_n(x) = \frac{n}{n-1}1_{[n^{-1},1)}(x).$$

Then $g(x) = x^{-1}$ is integrable with respect to each F_n, with

$$EX_n^{-1} = \frac{n}{n-1}\int_{n^{-1}}^{1}\frac{1}{x}\cdot dx = \frac{n}{n-1}\log n.$$

However, x^{-1} is not u.i., since

$$\int_{\{x:|x^{-1}|\geq h\}} x^{-1}f_n(x)\cdot dx = \frac{n}{n-1}\int_0^{h^{-1}} x^{-1}1_{[n^{-1},1)}(x)\cdot dx$$

$$= \begin{cases} 0, & n \leq h \\ \frac{n}{n-1}\int_{n-1}^{h^{-1}} x^{-1}\cdot dx, & n > h \end{cases}$$

$$= \frac{n}{n-1}\log\left(\frac{n}{h}\right)1_{(h,\infty)}(n),$$

which clearly cannot be made small for all n.

Although obvious, it is worth pointing out that g is u.i. with respect to $\{F_n\}$ if g is a.s. *uniformly* bounded—meaning that there is some $H > 0$ such that $\Pr(|g(X_n)| \leq H) = 1$ for all n; for in that case $\sup_n \int_{|g|\geq h}|g|\cdot dF_n = 0$ for all $h > H$. Clearly, $g = x^{-1}$ is *not* uniformly bounded in Example 6.15, nor are the (individually) bounded functions $\{g_n(x) = n1_{[0,1]}(x)\}_{n=1}^{\infty}$ bounded *uniformly*.

The following theorem illuminates the connection between uniform integrability and limits of moments.[9]

Theorem 6.9. *Suppose* $\{X_n\}_{n=1}^{\infty}$ *converges in distribution to* F, *and let* X *have c.d.f.* F. *Then for finite* $r > 0$

$$E|X_n^r| = \int |x^r|\cdot dF_n \to \int |x^r|\cdot dF = E|X^r| < \infty$$

as $n \to \infty$ *if and only if* $|x^r|$ *is uniformly integrable with respect to* $\{F_n\}_{n=1}^{\infty}$.

---

[9]For a proof see Chung (1974, pp. 97-98).

Notes: (i) The finiteness of $E\,|X^r|$ when the $\{|X_n^r|\}$ are u.i. is part of the conclusion. (ii) The conclusion holds for positive integers r without the absolute value notation. In particular, $EX_n \to EX$ if $\{|X_n|\}$ are u.i.

Example 6.16. If X_n is a discrete r.v. with p.m.f.

$$f_n(x) = n^{-1}\mathbf{1}_{\{1/n,2/n,\ldots,(n-1)/n,1\}}(x)$$

for each $n \in \aleph$, then we have seen (Exercise 6.6) that $\{X_n\}$ converges in distribution to uniform on $[0,1]$; i.e., that $F_n(x) \to \mathbf{1}_{[0,1)}(x) + \mathbf{1}_{[1,\infty)}(x) =:$ $F(x)$. Since $\Pr\left(|X_n^r| \le 1\right) = 1$ for each $r \ge 0$, $g(x) = x^r$ is a.s. uniformly bounded and therefore u.i. Thus, Theorem 6.9 implies

$$\lim_{n\to\infty} EX_n^r = \lim_{n\to\infty} \int_{[0,1]} x^r \cdot dF_n = \int_{[0,1]} x^r \cdot dF$$

$$= \int_0^1 x^r \cdot dx = (r+1)^{-1}.$$

Since $EX_n^r = n^{-1}\sum_{j=1}^n (j/n)^r$ for each n, we have obtained a probabilistic proof that

$$\lim_{n\to\infty} \frac{1}{n^{r+1}} \sum_{j=1}^n j^r = \frac{1}{r+1}$$

for $r \ge 0$.[10] For example, $\lim_{n\to\infty} n^{-2}\sum_{j=1}^n j = 1/2$, $\lim_{n\to\infty} n^{-3}\sum_{j=1}^n j^2 = 1/3$, and $\lim_{n\to\infty} n^{-9.3}\sum_{j=1}^n j^{10.3} \doteq .088495575$.

Interestingly, Theorem 6.9 has a useful implication even for weakly convergent sequences $\{X_n\}$ that are *not* uniformly integrable; namely, that limits of absolute moments are always *at least as great* as the corresponding moments of the limiting distribution (when these are finite); in particular, $\lim_{n\to\infty} VX_n \ge VX$.

Corollary 6.2. *Suppose* $\{X_n\}_{n=1}^\infty$ *converges in distribution to* F *and that* $E\,|X^r| := \int |x^r| \cdot dF < \infty$, $r > 0$. *Then*

$$\lim_{n\to\infty} E\,|X_n^r| \ge E\,|X^r|. \tag{6.13}$$

---

[10]One should not be too impressed: A more general result could have been obtained more directly, since $n^{-1}\sum_{j=1}^n (j/n)^r$ is simply the Riemann sum that converges for all $r > -1$ to $\int_0^1 x^r \cdot dx = (1+r)^{-1}$. (Recall the solution to Exercise 3.8.) In the same way $n^{-1}\sum_{j=1}^n g(j/n) \to \int_0^1 g(x) \cdot dx$ for any Riemann-integrable g.

Proof. For finite $H > 0$ and each $n \in \aleph$ put

$$X_n(H) := -H\mathbf{1}_{(-\infty,-H)}(X_n) + X_n\mathbf{1}_{[-H,H]}(X_n) + H\mathbf{1}_{(H,\infty)}(X_n).$$

This censored version of X_n being uniformly bounded and therefore uniformly integrable, we have

$$\lim_{n\to\infty} E\,|X_n^r| \geq \lim_{n\to\infty} E\,|X_n(H)^r| = \int |x(H)^r| \cdot dF,$$

and the conclusion follows on pushing $H \uparrow \infty$. $\square$

Example 6.11 has already shown that inequality (6.13) *can* be strict for some sequences and that $\lim_{n\to\infty} E\,|X_n^r|$ may be infinite even when $E\,|X^r| < \infty$. The next example makes the point again as it demonstrates how to check for uniform integrability.

Example 6.17. Since uniform integrability is necessary and sufficient for limits of moments and moments of limiting distributions to coincide, it should be true that $|x^r|$ ($r > 0$) is u.i. with respect to the sequence of distributions in Example 6.11 if and only if $\theta < 1/r$. To see that this is so, note that $\int |x^r| \cdot dF_n = n^{\theta r - 1}$ and that

$$\int_{|x^r| \geq h} |x^r| \cdot dF_n = \begin{cases} 0, & n^{\theta r} < h \\ n^{\theta r - 1}, & n^{\theta r} \geq h \end{cases}.$$

If $\theta < 1/r$ then $r\theta - 1 < 0$ and $\sup_{n \in \aleph} n^{\theta r - 1}$ is attained at $n = 1$, so

$$\sup_{n \in \aleph} \int_{|x^r| \geq h} |x^r| \cdot dF_n = 0$$

for all $h > 1$ and $E\,|X_n|^r \to E\,|X|^r = 0$. On the other hand when $\theta \geq 1/r$ we have

$$\sup_{n \in \aleph} \int_{|x^r| \geq h} |x^r| \cdot dF_n = \sup_{n \in \aleph} n^{\theta r - 1} = \begin{cases} 1, & \theta = r^{-1} \\ +\infty, & \theta > r^{-1} \end{cases}$$

for every h, and $\lim_{n\to\infty} E\,|X_n|^r$ equals unity and $+\infty$, respectively, in the two cases.

6.6   Laws of Large Numbers

Laws of large numbers are theorems that state conditions under which sample moments converge to their population counterparts. *Weak* laws of large numbers give conditions under which they converge in probability, whereas *strong* laws govern almost-sure convergence—that is, convergence with probability one. Such statements about sample moments may appear rather narrow, but we shall see that they apply to all other statistics that are sample averages of well behaved functions of the data. When combined with the continuity theorem, laws of large numbers become extremely powerful tools. Indeed, they are the rigorous basis for the usual (frequentist) interpretation of probabilities as "long-run relative frequencies".

6.6.1 *Some Important Theorems*

Since almost-sure convergence implies convergence in probability, one may ask why "weak" laws of large numbers are of interest when theorems are available that guarantee a.s. convergence of sample means. The reason is that stronger conditions on the moments of the data may be required in order to guarantee the stronger form of convergence. This is true, for example, when the data are independent but not identically distributed, and when they are dependent but have identical marginal distributions. However, the limiting behavior of means of *random samples* is completely settled by one powerful theorem, *Kolmogorov's strong law of large numbers (s.l.l.n.)*, which delivers a.s. convergence under the weakest possible moment condition.

Theorem 6.10. *(Kolmogorov) If* $\{X_j\}_{j=1}^{\infty}$ *is a sequence of i.i.d. r.v.s with* $E|X_1| < \infty$ *and* $EX_1 = \mu$, *then* $\bar{X} := n^{-1}\sum_{j=1}^{n} X_j$ *converges almost surely to* μ *as* $n \to \infty$.

The power of this result comes from the fact that it delivers strong convergence to μ without any requirement on higher moments. The implication is that if we keep gathering data there is some sample size beyond which $\bar{X}$ will almost surely never deviate from μ by more than any given error tolerance. Another way to put this is that the discrepancies $\{X_j - \mu\}_{j=1}^{\infty}$ will eventually average out, so that $n^{-1}\sum_{j=1}^{n}(X_j - \mu) \to 0$. The standard (and lengthy) proof relies on a truncation argument in which each X_j is replaced by $X_j 1_{[-j,j]}(X_j)$ or some other bounded version.[11]

Example 6.18. The (standard) *Wiener* or *Brownian motion* process $\{W_t\}_{t\geq 0}$ was defined in Section 4.3.7 as a stochastic process adapted to an information structure (filtration) $\{\mathcal{F}_t\}_{t\geq 0}$ that (i) evolves in continuous time from $W_0 = 0$; (ii) has independent increments $\{W_{t_j} - W_{t_{j-1}}\}$ over

---

[11]For the proof consult any advanced text on probability; e.g., Billingsley (1986, 1995), Chung (1974), Williams (1991).

nonoverlapping intervals $\{[t_{j-1}, t_j)\}$; and (iii) is such that $W_{t_j} - W_{t_{j-1}} \sim N(0, t_j - t_{j-1})$. These defining properties imply that $\{W_t\}_{t \geq 0}$ is a.s. continuous but nowhere differentiable. With the aid of the s.l.l.n. we can now deduce another remarkable feature.

The *quadratic variation (q.v.)* of a real-valued function g on a finite interval $[0, t]$ is defined as $\langle g \rangle_t := \sup \sum_{j=1}^{n} [g(t_j) - g(t_{j-1})]^2$, where the supremum is taken over all finite partitions of $[0, t]$. If function g is *differentiable* throughout the interval and has bounded derivative $|g'| \leq B$, then $|g(t_j) - g(t_{j-1})| = |g'(t_j^*)| (t_j - t_{j-1})$ for some $t_j^* \in [t_{j-1}, t_j]$. Allowing the intervals to shrink in such a way that $\max\{t_j - t_{j-1}\}_{j=1}^{n} = n^{-1}t + o(n^{-1})$, it follows that

$$\langle g \rangle_t = \lim_{n \to \infty} \sum_{j=1}^{n} |g'(t_j^*)|^2 (t_j - t_{j-1})^2 \leq B^2 \lim_{n \to \infty} \sum_{j=1}^{n} (t_j - t_{j-1})^2 = 0.$$

Thus, the ordinary real-valued functions with which we are most familiar have quadratic variations equal to zero.

What can be said about the quadratic variation of a *nondifferentiable* Brownian motion? Since $\left\{ W_{t_j} - W_{t_{j-1}} \sim Z_j \sqrt{t_j - t_{j-1}} \right\}_{j=1}^{n}$ for independent standard normals $\{Z_j\}$, and since $EZ_j^2 = 1 < \infty$, the s.l.l.n. implies that

$$\langle W \rangle_t = \lim_{n \to \infty} \sum_{j=1}^{n} \left(W_{t_j} - W_{t_{j-1}} \right)^2 = \lim_{n \to \infty} \sum_{j=1}^{n} Z_j^2 (t_j - t_{j-1})$$

$$= \lim_{n \to \infty} \frac{t}{n} \sum_{j=1}^{n} Z_j^2 + o(1) = t$$

with probability one. Thus, not only does the q.v. on $[0, t]$ not vanish, it is perfectly *predictable* given initial (trivial) information $\mathcal{F}_0$—even though we have no idea what track the process will follow from $t = 0$ forward!

Although Kolmogorov's strong law is extremely powerful, we sometimes need to deal with averages of independent r.v.s that are not identically distributed. The next theorem covers that case but does require a condition on the second moment.[12]

Theorem 6.11. *If $\{X_j\}_{j=1}^{\infty}$ is a sequence of independent r.v.s with zero means and variances $\{\sigma_j^2\}_{j=1}^{\infty}$ such that $\sum_{j=1}^{\infty} \sigma_j^2/j^2 < \infty$, then $\bar{X} = n^{-1} \sum_{j=1}^{n} X_j$ converges almost surely to zero as $n \to \infty$.*

---

[12]For the proof see Rao (1973).

Besides the obvious fact that it restricts the variances to be finite, the condition $\sum_{j=1}^{\infty} \sigma_j^2/j^2 < \infty$ avoids situations in which variances increase in such a way that adding more data adds enough noise to overwhelm the averaging-out process. The restriction that the means equal zero is unimportant, for if $\{Y_j\}_{j=1}^{\infty}$ obey the other conditions but have means $\{\mu_j\}_{j=1}^{\infty}$ whose running averages $\left\{n^{-1}\sum_{j=1}^{n}\mu_j\right\}_{n=1}^{\infty}$ approach a finite limit μ, then $n^{-1}\sum_{j=1}^{n}(Y_j - \mu_j) \to^{a.s.} 0$, and so

$$n^{-1}\sum_{j=1}^{n} Y_j = n^{-1}\sum_{j=1}^{n}(Y_j - \mu_j) + n^{-1}\sum_{j=1}^{n}\mu_j \to^{a.s.} \mu.$$

Before Kolmogorov proved that the finiteness of μ alone suffices for $\bar{X} \to^{a.s.} \mu$ in the i.i.d. case, Y.A. Khintchine had obtained the weaker result that $\bar{X} \to^{p} \mu$ under the same condition. Khintchine's theorem is, for i.i.d. random variables, the best possible version of the *weak* law of large numbers. A proof based on characteristic functions is easy and greatly instructive, for it demonstrates the great theoretical power of those functions.

Theorem 6.12. *(Khintchine) If $\{X_j\}_{j=1}^{\infty}$ is a sequence of i.i.d. r.v.s with $E|X_1| < \infty$ and $EX_1 = \mu$, then $\bar{X} = n^{-1}\sum_{j=1}^{n} X_j$ converges in probability to μ as $n \to \infty$.*

Proof. Let $\phi(t)$ be the c.f. of each X_j. Under the hypothesis of the theorem we know about ϕ only that it is differentiable at least once at the origin and that $\phi'(0) = i\mu$. But with (6.6) this implies that $\phi(t) = 1 + it\mu + o(t)$ as $t \to 0$ and hence that $\phi_{\bar{X}}(t) = \phi(t/n)^n = [1 + i\mu t/n + o(n^{-1})]^n \to e^{it\mu}$ as $n \to \infty$. That $\bar{X} \to^{p} \mu$ then follows from the continuity theorem for c.f.s. $\qquad\square$

Khintchine's theorem can be strengthened by relaxing the requirement that the $\{X_j\}$ be identically distributed, so long as they are independent and have the same mean μ; for in that case it is still true that

$$\phi_{\bar{X}}(t) = \prod_{j=1}^{n}\phi_{X_j}\left(\frac{t}{n}\right) = \prod_{j=1}^{n}\left[1 + \frac{it\mu}{n} + o_j(n^{-1})\right] \to e^{it\mu}. \qquad (6.14)$$

Again, it is not even necessary that there be a common mean, for if $\{Y_j\}_{j=1}^{\infty}$ are independent with means such that $\left|n^{-1}\sum_{j=1}^{n}\mu_j\right| \to |\mu| < \infty$, then $n^{-1}\sum_{j=1}^{n}(Y_j - \mu_j) \to^{p} 0$ and $\bar{Y} \to^{p} \mu$.[13] Notice that for this weaker

---

[13]The restriction $\lim_{n\to\infty}\left|n^{-1}\sum_{j=1}^{n}\mu_j\right| = |\mu| < \infty$ rules out situations in which the means increase with n in such a way that the remainder terms in (6.14) are not $o(n^{-1})$. For example, $\bar{Y}$ would clearly not converge if $\{\mu_j = j\}_{j=1}^{\infty}$.

conclusion of convergence in probability the $\{Y_j\}_{j=1}^{\infty}$ need not even have finite variances.

When variances *are* finite, demonstrating convergence in mean square affords simple proofs of weak laws of large numbers, as seen in Example 6.9. The method is sufficiently powerful that it extends weak laws to sequences of r.v.s that are neither identically distributed nor independent, provided that their variances and covariances are suitably restricted.

Theorem 6.13. (*Chebychev*) *Let* $S_n := \sum_{j=1}^{n} X_j$, *where* $\{X_j\}_{j=1}^{\infty}$ *are such that* $V\left(n^{-1}S_n\right) \to 0$ *as* $n \to \infty$ *and that* $n^{-1}\sum_{j=1}^{n} EX_j$ *approaches a limit* μ. *Then* $\bar{X} = n^{-1}S_n \longrightarrow^p \mu$.

Proof. $\bar{X} = n^{-1}\left(S_n - ES_n\right) + n^{-1}ES_n = n^{-1}\left(S_n - ES_n\right) + E\bar{X}$, so $E\left(\bar{X} - E\bar{X}\right)^2 = V\left(n^{-1}S_n\right) \to 0$, and the result follows from Chebychev's inequality. □

As an example of when the condition of the theorem would hold, put

$$\sigma_{ij} := Cov\left(X_i, X_j\right) = \sigma_i\sigma_j\rho_{ij}$$

and write

$$V\left[n^{-1}\left(S_n - ES_n\right)\right] = \frac{1}{n^2}\sum_{i=1}^{n}\sigma_i^2 + \frac{1}{n}\sum_{i=1}^{n}\sigma_i\left(\frac{1}{n}\sum_{j\neq i}\sigma_j\rho_{ij}\right). \quad (6.15)$$

Now suppose the variances are bounded, as $\sigma_i^2 \leq b < \infty$. Under that condition the first term on the right vanishes as $n \to \infty$, and the absolute value of the second term is bounded by $bn^{-1}\sum_{i=1}^{n}\left(n^{-1}\sum_{j\neq i}|\rho_{ij}|\right)$. Now if the sum of correlations $\sum_{j\neq i}|\rho_{ij}|$ converges for each i, then the average is $o(1)$ and, likewise, the final term in (6.15). This condition amounts to a sufficiently rapid weakening of the dependence as the distance between observations increases. If the $\{X_j\}$ are uncorrelated, then of course the boundedness of the variances is sufficient.

Remarkably, for uncorrelated r.v.s the boundedness of the $\{\sigma_j^2\}$ even delivers *strong* convergence.

Theorem 6.14. (*Rajchman*) *If uncorrelated r.v.s* $\{X_j\}_{j=1}^{\infty}$ *have means* $\{\mu_j\}$ *such that* $\lim_{n\to\infty}\left|n^{-1}\sum_{j=1}^{n}\mu_j\right| = |\mu| < \infty$ *and variances* $\{\sigma_j^2 \leq b < \infty\}$, *then* $\bar{X} = n^{-1}S_n \longrightarrow^{a.s.} \mu$.

Chung (1974, pp. 103-4) gives an instructive proof, first applying Borel–Cantelli to show that the subsequence $\left\{n^{-2}\left(S_{n^2} - ES_{n^2}\right)\right\}_{n=1}^{\infty} = \left\{n^{-2}\sum_{j=1}^{n^2}\left(X_j - \mu_j\right)\right\}_{n=1}^{\infty}$ converges a.s., then showing that arbitrary elements of $\left\{n^{-1}\left(S_n - ES_n\right)\right\}_{n=1}^{\infty}$ do not differ much from near members of the subsequence.

Exercise 6.16. $\{X_j\}_{j=1}^{\infty}$ *are nonnegative and i.i.d. with* $E \log X_j = \log \gamma$ *for each j. Find the probability limit of the sample geometric mean,* $G_n := \sqrt[n]{X_1 X_2 \cdots \cdots X_n}$.

Exercise 6.17. $\{X_j\}_{j=1}^{\infty}$ *have identical marginal distributions with zero mean and unit variance, but they are dependent, with* $Cov\left(X_j, X_k\right) = \rho \neq 0$ *for each* $j \neq k$. *Determine whether* $\bar{X} := n^{-1}\sum_{j=1}^{n} X_j$ *converges in mean square to zero.*

6.6.2 *Interpretations*

Laws of large numbers are the foundation of statistical inference. They tell us that we should eventually come close to the truth if we just keep collecting data. The intuition is that in large random samples the unrepresentative values that might be picked up will ultimately either be offset by others that err in the opposite direction, or else will be dominated by the representatives. In other words, the mistakes will eventually average out. However, this averaging process may well be very slow if the variance of the population is large or infinite. Moreover, if the tails of the distribution are so thick that not even the mean exists, the bad values may *never* average out. This was demonstrated dramatically in Example 3.49, which showed that the sample mean of i.i.d. Cauchy variates has precisely the same (Cauchy) distribution for all n. The explanation is that the Cauchy tails are so thick, and highly deviant outliers so likely, that increasing n raises—at just the right rate!—the odds of getting extreme, dominating values. This is illustrated in Fig. (6.2), which plots for $n \in (1, 2, ..., 1000)$ sample means $\left\{\bar{X}_n\right\}$ of i.i.d. pseudorandom (computer-generated) standard normal and standard Cauchy variates.

If the variances of the $\{X_j\}_{j=1}^{\infty}$ are finite as well as the means, then there is a way of bounding the rate of convergence of $\bar{X}$. This remarkable result, due also to Kolmogorov in the form given below, is known as the *law of the iterated logarithm.*

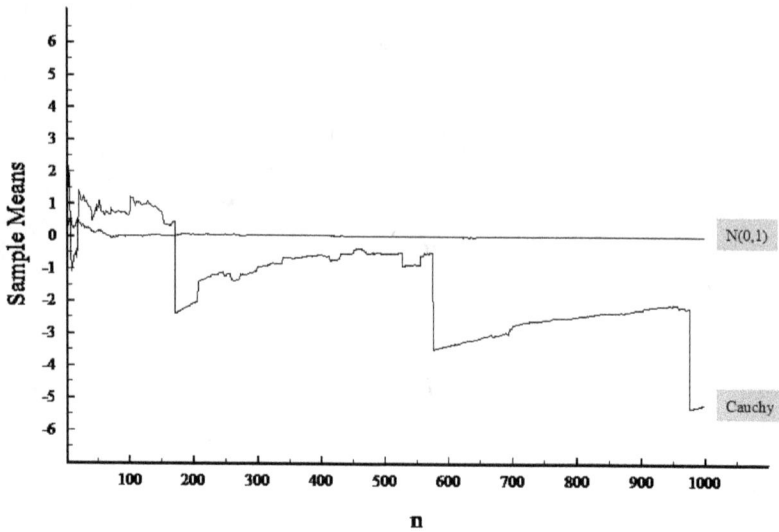

Fig. 6.2   Sample means of i.i.d. standard normals and standard Cauchys *vs.* sample size, n.

Theorem 6.15. *If $\{Y_j\}_{n=1}^{\infty}$ are i.i.d. with $EY_j = 0$ and $VY_j = 1$, and $\bar{Y} := n^{-1}\sum_{j=1}^{n} Y_j$, then with probability one*

$$\limsup_n \left[\frac{\sqrt{n}\bar{Y}}{\sqrt{2\log(\log n)}} \right] = 1. \tag{6.16}$$

Note that since $\{-Y_j\}_{j=1}^{\infty}$ have the same properties as those assumed for $\{Y_j\}_{n=1}^{\infty}$, the same limit holds also for $|\bar{Y}|$. To get a better sense of what (6.16) implies, suppose $\{X_j\}_{j=1}^{\infty}$ are i.i.d. with mean μ and finite variance σ^2. Then, for some sample size n, $(\bar{X} - \mu)/\sigma = n^{-1}\sum_{j=1}^{n}(X_j - \mu)/\sigma$ corresponds to $\bar{Y}$ in the theorem. Thus, roughly speaking, the theorem tells us that when n is large we are not apt to observe values of $|\bar{X} - \mu|$ larger than $n^{-1/2}\sigma\sqrt{2\log(\log n)}$. These bounds shrink just *slightly* more slowly than $1/\sqrt{n}$ since the "iterated" logarithm is very slowly increasing.

6.6.3   *Extensions and Applications*

Laws of large numbers are more general than they may seem, since they apply to sample averages of all functions of r.v.s $\{X_j\}_{j=1}^{\infty}$ that retain the properties needed for convergence of means of the r.v.s themselves. For example, if $\{X_j\}_{j=1}^{\infty}$ are i.i.d., Kolmogorov's strong law guarantees that $\bar{g} := n^{-1} \sum_{j=1}^{n} g(X_j) \to^{a.s.} Eg(X)$ so long as $g : \Re \to \Re$ is such that $E|g(X)| < \infty$. This is (trivially) true because $\{g(X_j)\}_{j=1}^{\infty}$ are themselves just i.i.d. r.v.s with finite means.

Restricting attention to the case of random samples, here are examples of some specific functions of the data to which the strong law may be applied.

(i) If $g(X) := X^k$, then $\bar{g} = M_k'$, the kth sample moment about the origin. Thus, the kth sample moment about the origin converges a.s. to μ_k' if the corresponding absolute moment $\nu_k' = E|X^k|$ is finite.

(ii) With B as any Borel set, put $g(X) := 1_B(X)$. Observing the n independent realizations, the sample mean, $\overline{1_B} := n^{-1} \sum_{j=1}^{n} 1_B(X_j)$, is just the relative frequency of occurrences of the event $X \in B$. Since $E|1_B(X)| = E1_B(X) = \mathbb{P}_X(B) = \Pr(X \in B)$ is clearly finite, the strong law implies that $\overline{1_B} \to^{a.s.} \Pr(X \in B)$. This is the rigorous meaning of the statement that $\mathbb{P}_X(B)$ is the "long-run relative frequency" of the event $X \in B$.

(iii) In particular, taking $B = (-\infty, x]$ for any $x \in \Re$, $\overline{1_{(-\infty,x]}} =: F_n(x)$ gives the value at x of the *empirical* distribution function, and so F_n converges a.s. (pointwise) to population c.d.f. F.

(iv) For any $t \in \Re$ put $g(X; t) := e^{itX}$. The sample mean, $\bar{g} = \phi_n(t) := n^{-1} \sum_{j=1}^{n} e^{itX_j}$, is the *empirical* c.f., which by the strong law converges a.s. pointwise to the population c.f. Corresponding results hold for empirical m.g.f.s, p.g.f.s, etc. when their population counterparts exist.

Notice that the convergence of empirical c.d.f.s and c.f.s in the last two examples requires no moment condition on the data at all. For example, empirical c.d.f.s and c.f.s of Cauchy variates do converge a.s. to their population counterparts.

Not only do sample means of appropriate functions of the data converge to the expected values, it is also true that appropriate *functions* of the means converge to the corresponding functions of the expectations— a joint implication of laws of large numbers and the continuity theorem for stochastic convergence. Specifically, assuming that $E|g(X)| < \infty$, put $Eg(X) =: \gamma$ and let $h : \Re \to \Re$ be continuous at γ. Then, for a random sample $\{X_j\}_{j=1}^{n}$ the strong law and continuity theorem imply that $h(\bar{g}) \to^{a.s.} h(\gamma)$. This all carries through when g, X, and even h are vector-valued and/or complex.

Example 6.19. Take $g(X) := X$ and $h(g) = g^2$, and assume that $E\,|X| < \infty$. Then $\bar{X}^2 \to^{a.s.} \mu^2$.

Example 6.20. Take $\mathbf{g}(X) := (g_1(X), g_2(X))' = (X, X^2)'$ and $h(\mathbf{g}) := g_2 - g_1^2$, and assume that $EX^2 < \infty$. Then $\bar{\mathbf{g}} = (\bar{X}, M_2')' \to^{a.s.} (\mu, \mu_2')'$,[14] so the continuity theorem implies

$$h(\bar{\mathbf{g}}) = M_2' - \bar{X}^2 = M_2 \to^{a.s.} \mu_2' - \mu^2 = \sigma^2.$$

Thus, the second sample moment about the sample mean converges a.s. to the population variance. Since $S^2 = \frac{n}{n-1} M_2$ and $n/(n-1) \to 1$, this holds for the sample variance as well. Moreover, since $h(x) = +\sqrt{x}$ is continuous for $x \geq 0$, $S := +\sqrt{S^2} \to^{a.s.} +\sqrt{\sigma^2} =: \sigma$.

Example 6.21. Put

$$\mathbf{g}(X) := (g_0(X), g_1(X), ..., g_k(X))' = (1, X, ..., X^{k-1}, X^k)'$$

for positive integer k such that $E\,|X^k| < \infty$, and let

$$h(\mathbf{g}) = \sum_{j=0}^{k} \binom{k}{j} g_{k-j}(X)\,[-g_1(X)]^j.$$

Then $\bar{\mathbf{g}} = (1, \bar{X}, M_2', ..., M_k')'$ and

$$h(\bar{\mathbf{g}}) = \sum_{j=0}^{k} \binom{k}{j} M_{k-j}'(-\bar{X})^j \to^{a.s.} \sum_{j=0}^{k} \binom{k}{j} \mu_{k-j}'(-\mu)^j = E(X - \mu)^k = \mu_k.$$

Thus, any (positive-integer-valued) sample moment about the sample mean converges a.s. to the corresponding population central moment.

Example 6.22. The famous Weierstrass approximation theorem asserts that any real function g that is continuous on a closed interval $[a, b]$ can be approximated "arbitrarily well" by some polynomial; that is, given $\varepsilon > 0$ there are an integer n and real constants $\{\alpha_j\}_{j=0}^{n}$ such that $\sup_{x \in [a,b]} \left| g(x) - \sum_{j=0}^{n} \alpha_j x^j \right| \leq \varepsilon$. A simple proof can be developed from the law of large numbers, as follows. Putting $p := (x - a)/(b - a)$ and $h(p) := g[a + (b - a)p]$, we will show that h can be approximated by a

---

[14]Note that $\bar{X} \to^{a.s.} \mu$ and $M_2' \to^{a.s.} \mu_2'$ imply that each of the sets of exceptional outcomes on which these sequences do not converge—$\mathfrak{N}_1$ and $\mathfrak{N}_2$, say—are $\mathbb{P}$-null. Thus, the set on which either $\bar{X}$ or M_2' does not converge has measure $\mathbb{P}(\mathfrak{N}_1 \cup \mathfrak{N}_2) \leq \mathbb{P}(\mathfrak{N}_1) + \mathbb{P}(\mathfrak{N}_2) = 0$. The joint conditions $\bar{X} \to^{a.s.} \mu$ and $M_2' \to^{a.s.} \mu_2'$ therefore imply that $(\bar{X}, M_2') \to^{a.s.} (\mu, \mu_2')$.

polynomial on $[0, 1]$. The like assertion for g on $[a, b]$ then follows. For any $p \in [0, 1]$ define a (Bernstein) polynomial of degree n as

$$B_n(p) := \sum_{y=0}^{n} \binom{n}{y} p^y (1-p)^{n-y} h\left(\frac{y}{n}\right),$$

noting that $B_n(0) = h(0)$ and $B_n(1) = h(1)$ for each $n \in \aleph$. For each fixed $p \in (0, 1)$ we observe that $B_n(p) = Eh(n^{-1}Y_{n,p})$, where $Y_{n,p} \sim B(n, p)$ (binomial distribution). Recognizing $n^{-1}Y_{n,p}$ as the mean of i.i.d. Bernoulli r.v.s with expected value p, one sees that $Y_{n,p} \longrightarrow^{a.s.} p$ as $n \to \infty$. Since h (being continuous on $[0, 1]$) is bounded, dominated convergence then implies that $B_n(p) = Eh(n^{-1}Y_{n,p}) \to Eh(p) = h(p)$ as $n \to \infty$. This establishes *pointwise* convergence, but convergence is actually *uniform*. To see this, observe first that $|B_n(p) - h(p)| \leq E|h(n^{-1}Y_{n,p}) - h(p)|$ by the absolute-value inequality. Next, letting $\mathbf{1}_{[0,\delta]}$ and $\mathbf{1}_{(\delta,\infty)}$ be indicators of the sets on which $|h(n^{-1}Y_{n,p}) - h(p)|$ is less than/greater than some $\delta > 0$, we have

$$|B_n(p) - h(p)| \leq E|h(n^{-1}Y_{n,p}) - h(p)|\left(\mathbf{1}_{[0,\delta]} + \mathbf{1}_{(\delta,\infty)}\right). \qquad (6.17)$$

We can now show that a uniform bound applies to the term corresponding to each indicator. Since h is continuous on a closed interval, it is *uniformly* continuous there; thus, given $\varepsilon > 0$, there is a δ such that $|h(n^{-1}Y_{n,p}) - h(p)| < \varepsilon/2$ when $|n^{-1}Y_{n,p} - p| \leq \delta$. The first term in (6.17) can then be bounded uniformly as

$$E|h(n^{-1}Y_{n,p}) - h(p)|\mathbf{1}_{[0,\delta]} \leq \frac{\varepsilon}{2} \cdot \Pr\left(|n^{-1}Y_{n,p} - p| \leq \delta\right) \leq \frac{\varepsilon}{2}.$$

The second term can be bounded via Chebychev, as follows. Letting $H := \sup_{p \in [0,1]} |h(p)|$, we have

$$E|h(n^{-1}Y_{n,p}) - h(p)|\mathbf{1}_{(\delta,\infty)} \leq 2H \cdot \Pr\left(|n^{-1}Y_{n,p} - p| > \delta\right)$$

$$\leq 2H \cdot \frac{VY_{n,p}}{n^2\delta^2} = 2H \cdot \frac{p(1-p)}{n\delta^2}$$

$$\leq \frac{2H}{4n\delta^2} \text{ (since } p(1-p) \leq \frac{1}{4} \text{ on } [0, 1])$$

This is made less than $\varepsilon/2$ by taking $n > H/(\varepsilon\delta^2)$, and for such n we have $|B_n(p) - h(p)| < \varepsilon$ uniformly in p, so that

$$\lim_{n \to \infty} \sup_{p \in [0,1]} |h(p) - B_n(p)| = 0.$$

Exercise 6.18. *Show for empirical c.f.* $\phi_n(t)$ *that*

$$-\left.\frac{d^2 \log \phi_n(t)}{dt^2}\right|_{t=0} \to^{a.s.} \sigma^2$$

if the population has finite variance.

6.7 Central Limit Theorems

In their most intuitive interpretation, laws of large numbers imply that sample moments are "essentially the same as" population moments when n is sufficiently large. Were this literally true, a moment statistic would fall within in any Borel set with either probability one or probability zero, depending on whether the set contained the corresponding population moment. In reality, of course, for any large but finite n such probabilities might just be *close* to zero or to unity. Central limit theorems (c.l.t.s) provide sharper estimates of such probabilities for arbitrary Borel sets by setting forth conditions under which standardized values of statistics have limiting normal distributions. There are many such theorems, some applying when samples are i.i.d., others when the data are independent but not identically distributed, and others that allow various forms of dependence. Here we focus on the first two classes, as c.l.t.s under dependence are more relevant for time series analysis or the study of stochastic processes. The existence and relevance of broad sets of conditions for "asymptotic" normality help account for the importance of the Gaussian law in applied statistics.

6.7.1 *Lindeberg–Lévy and Lyapounov Theorems*

The weakest conditions under which standardized means of i.i.d. r.v.s converge in distribution to the normal are given by the Lindeberg–Lévy theorem. The proof is a simple application of the continuity theorem for c.f.s. Lyapounov's theorem, presented next, is perhaps the most useful for the independent, non-i.d. case. It should not surprise that central limit theorems require stronger moment conditions than are needed for laws of large numbers.

Theorem 6.16. (*Lindeberg–Lévy*). *If* $\{X_j\}_{j=1}^{\infty}$ *are i.i.d. with mean* μ *and positive, finite variance* σ^2*, then the sequence of standardized sample means* $\left\{\bar{X} := n^{-1}\sum_{j=1}^{n} X_j\right\}$ *converges in distribution to standard normal as* $n \to \infty$*. That is,*

$$\frac{\bar{X} - \mu}{\sigma/\sqrt{n}} \rightsquigarrow N(0, 1).$$

Proof. The standardized mean is

$$Z_n := \frac{\sqrt{n}\,(\bar{X} - \mu)}{\sigma} = n^{-1/2}\sum_{j=1}^{n} Y_j,$$

where $\{Y_j := (X_j - \mu)/\sigma\}_{j=1}^{\infty}$ are the standardized observations ($EY_j = 0$, $VY_j = 1$). Then the c.f. of Z_n is

$$\phi_{Z_n}(t) = \phi_Y\left(\frac{t}{\sqrt{n}}\right)^n$$

$$= \left[1 + \phi_Y'(0)\frac{t}{\sqrt{n}} + \phi_Y''(0)\frac{t^2}{2n} + o(n^{-1})\right]^n$$

$$= \left[1 - \frac{t^2}{2n} + o(n^{-1})\right]^n.$$

This converges to $e^{-t^2/2}$ as $n \to \infty$, so the conclusion follows from the continuity theorem for c.f.s. $\qquad\square$

Although the theorem is stated in terms of standardized sample means, it applies also to standardized sums, $S_n := n\bar{X} = \sum_{j=1}^{n} X_j$, since the two standardized quantities have exactly the same realizations; i.e.,

$$Z_n := \frac{\bar{X} - E\bar{X}}{\sqrt{V\bar{X}}} = \frac{n\left(\bar{X} - E\bar{X}\right)}{n\sqrt{V\bar{X}}} = \frac{S_n - n\mu}{\sigma\sqrt{n}} = \frac{S_n - ES_n}{\sqrt{VS_n}}.$$

Thus, standardized sample sums of i.i.d. r.v.s with finite variance are asymptotically normally distributed. Of course, if

$$Z_n = \left(\frac{S_n - n\mu}{\sigma\sqrt{n}}\right) = \frac{\sqrt{n}\left(\bar{X} - \mu\right)}{\sigma} \rightsquigarrow N(0,1),$$

then it is also true that $\sigma Z_n = \sqrt{n}\left(\bar{X} - \mu\right) \rightsquigarrow N\left(0, \sigma^2\right)$.[15]

Example 6.23. Bricks produced by a certain manufacturing process average 20 cm. in length, with standard deviation 1 cm. A project requires enough bricks to produce a row 19.8 meters long when laid end to end. Choosing bricks at random, what is the probability that more than 100 will be needed? More than 100 will be needed if $S_{100} := \sum_{j=1}^{100} X_j < 19.8$, where $\{X_j\}_{j=1}^{100}$ are the lengths of the individual bricks. Regarding these as i.i.d. with mean $\mu = .20$ m. and standard deviation $\sigma = .01$ m.,

$$\Pr(S_{100} < 19.8) = \Pr\left(\frac{S_{100} - 100\mu}{\sigma\sqrt{100}} < \frac{19.8 - 20}{.01(10)}\right)$$

$$\doteq \Phi(-2.0) \doteq .02.$$

---

[15]Obviously, however, it is *not* true that $\sqrt{n}\bar{X} \rightsquigarrow N\left(\mu, \sigma^2\right)$ unless $\mu = 0$. Remarkably, one sometimes encounters in applied papers such absurdities as "$\bar{X} \rightsquigarrow N\left(\mu, \sigma^2/n\right)$". While this is indeed absurd (how could "n" appear in an expression representing a limit as $n \to \infty$?), it is perfectly correct to state that $N\left(\mu, \sigma^2/n\right)$ *approximates* the distribution of $\bar{X}$ when n is large. It is to obtain such approximations that we make use of asymptotic theory.

Example 6.8 proved the De Moivre–Laplace theorem, which states that

$$\frac{Y_n - n\theta}{\sqrt{n}} \rightsquigarrow N\left[0, \theta\left(1 - \theta\right)\right]$$

if $Y_n \sim B(n, \theta)$. This implies that $(Y_n - n\theta)/\sqrt{n\theta(1-\theta)}$ is asymptotically standard normal. Noting that binomial variates are sums of i.i.d. Bernoulli r.v.s makes clear that De Moivre–Laplace is just a special case of Lindeberg–Lévy.

Example 6.24. A gambler fears that a pair of dice are "loaded" to increase the frequency of "sevens"—the event that the numbers on the upmost faces add to 7. Rolling the dice 120 times as a test and observing 28 sevens, the gambler wants to know the chance of getting so many with *fair* dice. The exact probability is

$$\Pr\left(N_7 \geq 28\right) = 1 - \Pr\left(N_7 < 28\right) = 1 - \sum_{j=0}^{27} \binom{120}{j} \left(\frac{1}{6}\right)^j \left(\frac{5}{6}\right)^{120-j},$$

but the tedious computation can be avoided by approximating using the c.l.t. Letting $N_7 := \sum_{j=1}^{120} X_j$, where the $\{X_j\}$ are i.i.d. as Bernoulli with $\theta = 1/6$, we have $EN_7 = 120\,(1/6) = 20$, $VN_7 = 120(1/6)(5/6) = 100/6$, and

$$\Pr\left(N_7 \geq 28\right) = \Pr\left(\frac{N_7 - EN_7}{\sqrt{VN_7}} \geq \frac{28 - 20}{10\sqrt{1/6}}\right)$$

$$\doteq 1 - \Phi\left(1.96\right) \doteq .025.$$

Accuracy is improved somewhat with a continuity correction that spreads the probability mass on the integer 28 over the interval $[27.5, 28.5]$. Thus,

$$\Pr\left(N_7 \geq 28\right) \doteq \Pr\left(\frac{N_7 - EN_7}{\sqrt{VN_7}} \geq \frac{27.5 - 20}{10\sqrt{1/6}}\right)$$

$$\doteq 1 - \Phi\left(1.84\right) \doteq .033.$$

Example 6.25. We saw in Example 6.18 that the Brownian motion process $\{W_t\}_{t \geq 0}$ has positive and predictable quadratic variation $\langle W \rangle_t = t$ over $[0, t]$. Another of its bizarre properties pertains to its *first-order* variation. The (first-order) *variation* of a function $g : \Re \to \Re$ over $[0, t]$ is the supremum over all finite partitions of $\sum_{j=1}^{n} |g\left(t_j\right) - g\left(t_{j-1}\right)|$. If g is differentiable throughout $[0, t]$ and $|g'| < B < \infty$, then

$$Var_{[0,t]}g = \lim_{n \to \infty} \sum_{j=1}^{n} \left|g'\left(t_j^*\right)\right| \left(t_j - t_{j-1}\right) \leq B \lim_{n \to \infty} \sum_{j=1}^{n} \left(t_j - t_{j-1}\right) = B.$$

Thus, all our ordinary functions have *finite* variation over finite intervals, but what of Brownian motions? To simplify we can take $t = 1$ and $t_j - t_{j-1} = 1/n$ for each j and put $S_n := \sum_{j=1}^n \left| W_{t_j} - W_{t_{j-1}} \right| \sim n^{-1/2} \sum_{j=1}^n |Z_j|$, so that $ES_n = \sqrt{n} E |Z| = \sqrt{2n/\pi}$ and $VS_n = V |Z| = EZ^2 - (E |Z|)^2 = 1 - 2/\pi$. Then for any finite B we have

$$\Pr\left(Var_{[0,1]} W > B\right) = \lim_{n \to \infty} \Pr\left(S_n > B\right)$$

$$= \lim_{n \to \infty} \Pr\left(\frac{S_n - ES_n}{\sqrt{VS_n}} > \frac{B - \sqrt{2n/\pi}}{\sqrt{1 - 2/\pi}}\right).$$

The r.v. on the left of the inequality converges in distribution to $N(0,1)$ by Lindeberg–Lévy, and the right side approaches $-\infty$, so $\Pr\left(Var_{[0,1]} W > B\right) = 1$ for any finite B. The variation of $\{W_t\}$ over $[0,1]$—indeed, over any open set of $\Re$—is, therefore, *infinite*. This implies that any portion of the track of a Brownian motion (such as the approximation depicted in Fig. 4.14) is infinitely long, so that one who actually tried to trace such a track in exquisite detail with pen and ink would always run out of ink! Putting any portion of the track under arbitrarily high magnification would reveal something having all the roughness and chaotic behavior of the original image. Such "self-similar" curves are often called *fractals*, a term popularized by Mandelbrot (1983).

Bear in mind that the conclusion of Lindeberg–Lévy applies only to sums and averages of i.i.d. r.v.s with finite variance. Necessary and sufficient conditions for asymptotic normality of standardized sums of independent but non-i.d. r.v.s are given in a theorem by Lindeberg and Feller, but these are hard to verify except in special cases. The following result is narrower but often more useful in practice.

Theorem 6.17. (*Lyapounov*) *Let* $S_n := \sum_{j=1}^n X_j$, *where* $\{X_j\}_{j=1}^\infty$ *are independent r.v.s with means* $\{\mu_j\}_{j=1}^\infty$, *variances* $\{\sigma_j^2\}_{j=1}^\infty$, *and finite* $(2 + \delta)th$ *absolute moments,* $\{E|X_j|^{2+\delta}\}_{j=1}^\infty$, *for some* $\delta > 0$. *If*

$$\lim_{n \to \infty} \frac{\sum_{j=1}^n E|X_j|^{2+\delta}}{\left(\sum_{j=1}^n \sigma_j^2\right)^{1+\delta/2}} = 0, \tag{6.18}$$

then as $n \to \infty$

$$\frac{S_n - ES_n}{\sqrt{VS_n}} \rightsquigarrow N(0,1).$$

Exercise 6.19. $\{X_j\}_{j=1}^{\infty}$ *are independent r.v.s with* $X_j \sim B(1, \theta_j)$ *(Bernoulli), where the* $\{\theta_j\}_{j=1}^{\infty}$ *are bounded away from zero and unity. (That is, there exists* $\varepsilon > 0$ *such that* $\varepsilon \leq \theta_j \leq 1 - \varepsilon$ *for each j.) Checking that the conditions for Lyapounov hold, apply the theorem to approximate the distribution of* $S_n := \sum_{j=1}^{n} X_j$ *for large n.*

Knowing that the distribution of a standardized sample mean or sample sum approaches the normal leaves open the question "How large must n be in practice?" Of course, there can be no general answer, since it depends on how the approximation will be used. However, a famous theorem due to Berry (1941) and Esseen (1945) provides a bound that helps one calculate a minimal n for particular applications. Here is a special case of the result that applies to means or sums of random samples—i.e., the i.i.d. case.

Theorem 6.18. *(Berry–Esseen)* $\{X_j\}_{j=1}^{\infty}$ *are i.i.d. with mean* μ, *variance* $\sigma^2 > 0$, *and third absolute central moment* $\nu_3 := E|X - \mu|^3 < \infty$. *Let*

$$Z_n := \frac{\sum_{j=1}^{n} X_j - n\mu}{\sigma\sqrt{n}}$$

be the standardized partial sum and $F_n(z) := \Pr(Z_n \leq z)$ *its c.d.f. Then*

$$|F_n(z) - \Phi(z)| \leq \frac{C\nu_3}{\sigma^3\sqrt{n}}, \tag{6.19}$$

where Φ *is the standard normal c.d.f. and C is a positive constant.*

The smallest value of C for which (6.19) is true is unknown, but the inequality is known to hold when $C = .7975$. In any case, we learn that the greatest difference between the two c.d.f.s diminishes at the rate $n^{-1/2}$ and that the maximal difference for any n depends on the thickness of the tails of the population, as governed by ν_3. When absolute central moments are difficult to calculate explicitly, one can make use of the upper bound $\nu_3 \leq \nu_4^{3/4} \equiv \mu_4^{3/4}$ that was established in Example 3.33.

Exercise 6.20. *Express the Berry–Esseen bound for* $|F_n(z) - \Phi(z)|$ *if* $\{X_j\}_{j=1}^{\infty}$ *are i.i.d. as Student's* $t(4)$.

6.7.2   Limit Theorems for Sample Quantiles

The $p \in (0,1)$ quantile of a r.v. X was defined in Section 2.2.4 as a number x_p such that $\Pr(X \leq x_p) \geq p$ and $\Pr(X \geq x_p) \geq 1 - p$. The *sample quantile* of a sample $\{X_j\}_{j=1}^{n}$ can be defined analogously as a value $X_{p,n}$

such that the proportion of sample elements less than or equal to $X_{p,n}$ is at least p and the proportion greater than or equal to $X_{p,n}$ is at least $1 - p$. Since this does not always specify a unique value, and since our interest here is in asymptotic behavior, we will simply take $X_{p,n}$ to be the $[np]$th order statistic, $X_{([np],n)}$. (Recall that $[np] := \max\{j \in \mathbb{N}_0 : j \le np\}$, so that $[np]/n \to p$ as $n \to \infty$.) We will show that a normalization of $X_{p,n}$ is distributed asymptotically as normal when $0 < p < 1$ and when $\{X_j\}_{j=1}^n$ are i.i.d. with a c.d.f. F that is strictly increasing and differentiable on an open interval $\mathbb{X}$ with $\mathbb{P}_X(\mathbb{X}) = 1$. Note that under these conditions the p quantile of each X_j is uniquely determined, with $x_p = F^{-1}(p)$ and $F'(x_p) = f(x_p) > 0$. We think of central limit theorems as applying to *sums* of r.v.s, and we will in fact be able to relate events pertaining to $X_{p,n}$ to events involving a sum of i.i.d. indicator functions—i.e., to a binomially distributed r.v. This is the key to what follows.

To proceed, put $Y_n := \sqrt{n}(X_{p,n} - x_p)$ as a normalization, so that event $Y_n \le y$ corresponds to $X_{p,n} \le y/\sqrt{n} + x_p$. Clearly, this event occurs if and only if (realizations of) at least $[np]$ of the $\{X_j\}$ are no greater than $y/\sqrt{n} + x_p$, in which case *at most* $n - [pn]$ *of the* $\{X_j\}$ *exceed* $y/\sqrt{n} + x_p$. The number of these "large" values—$N(y, p_n)$, say—is distributed as $B(n, p_n)$ (binomial), where $p_n := 1 - F(y/\sqrt{n} + x_p)$. Then

$$\Pr(Y_n \le y) = \Pr\{N(y, p_n) \le n - [pn]\}$$

$$= \Pr\left\{ \frac{N(y, p_n) - np_n}{\sqrt{np_n(1 - p_n)}} \le \frac{\sqrt{n}(1 - p_n - [pn]/n)}{\sqrt{p_n(1 - p_n)}} \right\}.$$

Noting that $p_n \to F(x_p) = p$ as $n \to \infty$, on the left side of the inequality we have

$$\frac{N(y, p_n) - np_n}{\sqrt{np_n(1 - p_n)}} = \frac{N(y, p) - np}{\sqrt{np(1 - p)}} + o(1) \rightsquigarrow N(0, 1)$$

by De Moivre–Laplace.[16] On the right side, $[pn]/n = p + O(n^{-1}) = F(x_p) + O(n^{-1})$, so that the numerator is $\sqrt{n}[F(y/\sqrt{n} + x_p) - F(x_p)] + o(1)$; while the denominator is $p(1 - p) + o(1)$. Thus,

$$\lim_{n \to \infty} \frac{\sqrt{n}(1 - p_n - [pn]/n)}{\sqrt{p_n(1 - p_n)}}$$

$$= \frac{y}{\lim_{n \to \infty}\sqrt{p_n(1 - p_n)}} \lim_{n \to \infty} \frac{F(y/\sqrt{n} + x_p) - F(x_p)}{y/\sqrt{n}}$$

$$= \frac{yf(x_p)}{\sqrt{p(1 - p)}}.$$

---

[16] By putting $\theta_n = \theta + o(1)$ in Example 6.8 it is easy to see that the dependence of p_n on n does not alter the conclusion of De Moivre–Laplace.

Applying Corollary 6.1, we conclude that

$$\Pr\left(Y_n \leq y\right) \to \Phi\left(\frac{yf\left(x_p\right)}{\sqrt{p(1-p)}}\right)$$

and hence that

$$\sqrt{n}\left(X_{p,n} - x_p\right) \rightsquigarrow N\left[0, \frac{p\left(1-p\right)}{f\left(x_p\right)^2}\right].$$

In particular, with $p = 1/2$ and $x_{.5}$ as the unique median of each X_j,

$$\sqrt{n}\left(X_{.5,n} - x_{.5}\right) \rightsquigarrow N\left[0, \frac{1}{4f\left(x_{.5}\right)^2}\right]. \tag{6.20}$$

For example, if $\{X_j\}_{j=1}^n$ are i.i.d. as $U\left(0,1\right)$, $X_{.5,n}$ is distributed approximately as $N\left(\frac{1}{2}, \frac{1}{4n}\right)$ for large n. Notice that if density $f\left(x_{.5}\right)$ is close to zero the asymptotic variance of $X_{.5,n}$ can be very large, even for large n. This would occur, for example, if $f\left(x\right)$ were symmetrical and U-shaped, as for certain beta distributions. By contrast, the sample median is very well behaved for peaked, unimodal distributions such as the Laplace.

When n is odd, we have $[.5\left(n+1\right)] \equiv .5\left(n+1\right)$, and it is conventional to regard the unique *central* order statistic, $X_{(.5(n+1):n)}$ as the sample median. Of course, whether n is even or odd, the limiting distributions of normalized versions of $X_{(.5(n+1):n)}$ and $X_{([.5(n+1)]:n)}$ are just the same.

6.7.3 *Approximations to Higher Order*

There are several ways to improve the normal approximation to the distribution of a sample sum or other asymptotically normal statistic. Generally, these take account of higher moments than the second, assuming that these exist, and for practical purposes one usually limits these to moments up to the fourth.

The *Gram–Charlier series* of order four approximates the density f of a standardized r.v X as the product of standard normal density f_Z and a fourth-order polynomial:

$$\hat{f}\left(x\right) = f_Z(x)\left[1 + \frac{\kappa_3}{3!} H_3\left(x\right) + \frac{\kappa_4}{4!} H_4\left(x\right)\right]. \tag{6.21}$$

Here κ_3 and κ_4 are the third and fourth cumulants of X, and the $\{H_j\}$ are the corresponding *Hermite* polynomials. The Hermite polynomials are generated by repeated differentiation of f_Z, as

$$H_j\left(x\right) = \left(-1\right)^j \frac{f_Z^{(j)}\left(x\right)}{f_Z\left(x\right)}, \tag{6.22}$$

with $f^{(0)} \equiv f$. Thus, $H_0(x) = 1$, while

$$H_1(x) = -\frac{df_Z(x)/dx}{f_Z(x)} = -\frac{d\log f_Z(x)}{dx} = x$$

$$H_2(x) = x^2 - 1$$

$$H_3(x) = x^3 - 3x$$

$$H_4(x) = x^4 - 6x^2 + 3.$$

The collection $\{H_j\}_{j=0}^{\infty}$ constitutes an orthogonal sequence with respect to f_Z, in that

$$\int_{-\infty}^{\infty} H_j(x) H_k(x) f_Z(x) \cdot dx = \begin{cases} 0, & k \neq j \\ j!, & k = j \end{cases}.$$

One basis for approximation (6.21) lies in the theory of orthogonal functions. Any $\mathcal{L}_2$ function $f : \Re \to \Re$ (i.e., one such that $\int_{-\infty}^{\infty} f(x)^2 \cdot dx < \infty$) can be approximated by a linear function of Hermite polynomials, in the special sense that for any $\varepsilon > 0$ there are a positive integer $n(\varepsilon)$ and coefficients $\{c_j\}_{j=0}^{n(\varepsilon)}$ such that

$$\int_{-\infty}^{\infty} \left| f(x) - f_Z(x) \sum_{j=0}^{n(\varepsilon)} c_j H_j(x) \right|^2 \cdot dx \leq \varepsilon.$$

Writing

$$f(x) = f_Z(x)\left[c_0 + c_1 H_1(x) + c_2 H_2(x) + \cdots + c_j H_j(x) + \cdots \right],$$

multiplying by $H_j(x)$, and integrating show that $c_j = EH_j(X)/j!$ when $E|H_j(X)| < \infty$. Thus, $c_0 = 1$ and, when X is in standard form,

$$c_1 = c_2 = 0$$

$$c_3 = \frac{1}{3!} E\left(X^3 - X\right) = \frac{\mu_3}{3!} = \frac{\kappa_3}{3!}$$

$$c_4 = \frac{1}{4!} E\left(X^4 - 6X^2 + 3\right) = \frac{\mu_4 - 3}{4!} = \frac{\kappa_4}{4!},$$

as per (6.21).[17]

---

[17]Here is a more direct approach to (6.21). With r.v. X in standard from, express the logarithm of its c.f. in terms of its cumulants, as in (3.49):

$$\log \phi(t) = -\frac{t^2}{2} + \frac{\kappa_3 (it)^3}{3!} + \frac{\kappa_4 (it)^4}{4!} + \dots.$$

Exponentiate and expand the exponential in the terms involving κ_3, κ_4 to get

$$\phi(t) \doteq e^{-t^2/2}\left[1 + \frac{\kappa_3 (it)^3}{3!} + \frac{\kappa_4 (it)^4}{4!} \right]. \quad \text{(continued)}$$

Of course, the real issue is whether an expansion based on the first few moments provides a good fit. Fortunately, it often does. Moreover, the approximation is easy to use in practical settings because it leads directly to an expression for the c.d.f. $\hat{F}$, as follows. The definition (6.22) implies that

$$\int_{-\infty}^{x} H_j(t) f_Z(t) \cdot dt = (-1)^j \int_{-\infty}^{x} f_Z^{(j)}(t) \cdot dt$$
$$= -(-1)^{j-1} f_Z^{(j-1)}(x)$$
$$= -H_{j-1}(x) f_Z(x),$$

so that the fourth-order approximant of the c.d.f. is

$$\hat{F}(x) = \Phi(x) - f_Z(x) \left[\frac{\kappa_3}{3!} H_2(x) + \frac{\kappa_4}{4!} H_3(x) \right]. \tag{6.23}$$

Fisher and Cornish (1960) provide a table of quantiles based on such approximations—that is, numbers x_p such that $\hat{F}(x_p) = p$.

Example 6.26. Recall from Section 4.3.2 that the exponential distributions,

$$\left\{ f(x) = \theta^{-1} e^{-x/\theta} \mathbf{1}_{(0,\infty)}(x) : \theta > 0 \right\},$$

exhibit constant hazard rate, $h(x) = f(x) / [1 - F(x)]$, equal to the inverse of scale parameter θ. If elements of $\mathbf{X} = (X_1, X_2, ..., X_n)'$ are i.i.d. as exponential, the scale-invariant statistic

$$C_n := \sqrt{48n} \left(n^{-1} \sum_{j=1}^{n} e^{-X_j/\bar{X}} - \frac{1}{2} \right)$$

converges in distribution to $N(0,1)$. On the other hand, it can be shown that $C_n \longrightarrow^p -\infty$ or $+\infty$ if the data come from a distribution with increasing or decreasing hazard, respectively. Accordingly, a test that $\{X_j\}_{j=1}^{n}$ are i.i.d. as exponential can be carried out by comparing the realization of C_n with its quantiles, $\{c_{n,p}\}$. Asymptotically, these approach the standard normal quantiles $\{z_p\}$; however, convergence to normality happens to be slow, so that applying the test successfully with small samples requires more accurate estimates. To this end, by putting $Y_j := e^{-X_j/\bar{X}}$ one can see that

---

Now apply inversion formula (3.52) to approximate $f(x)$ as

$$f_Z(x) \int_{-\infty}^{\infty} f_Z(t + ix) \left[1 + \frac{\kappa_3 (it)^3}{3!} + \frac{\kappa_4 (it)^4}{4!} \right] \cdot dt.$$

Setting $z := t + ix$ and carrying out the integration lead to (6.21).

C_n can be expressed in terms of $\{Y_j\}_{j=1}^{n-1}$ alone and can show that the joint distribution of $\{Y_j\}_{j=1}^{n-1}$ is

$$f(y_1, ..., y_{n-1}) = \frac{n!}{n^n y_1 y_2 \cdots \cdots y_{n-1}}$$

for each $y_j \leq 1$ and $\prod_{j=1}^{n-1} y_j \geq e^{-n}$. From this (and tedious algebra) it is possible to work out the first few cumulants of C_n. These and the Fisher–Cornish tables lead to the following estimates of quantiles $\{c_{n,p}\}$ for various n, which can be compared with the asymptotic values $\{z_p\}$ in the last row:[18]

$n \setminus p$	.010	.025	.050	.950	.975	.990
10	-2.06	-1.85	-1.66	1.38	1.77	2.25
20	-2.16	-1.91	-1.67	1.50	1.88	2.33
50	-2.24	-1.94	-1.68	1.57	1.93	2.35
100	-2.27	-1.95	-1.67	1.60	1.94	2.35
∞	-2.33	-1.96	-1.64	1.64	1.96	2.33

While the Gram–Charlier procedure is reasonably straightforward and often gives excellent results, a significant deficiency is that the approximant $\hat{f}$ is not everywhere nonnegative for certain values of the $\{\kappa_j\}$. An alternative approach that avoids this defect is to approximate f using some flexible *system* of distributions. The best known of these is the *Pearson system*,[19] whose members are solutions to a differential equation involving a rational function of polynomials. By altering the coefficients of these polynomials one generates distributions of many forms, including most of the standard models; e.g., normal, beta, gamma, and Student. Tables are available that present selected quantiles as functions of skewness and kurtosis, $\alpha_3 = \kappa_3$ and $\alpha_4 = \kappa_4 + 3$; for example, Table 32 in Volume 2 of Pearson and Hartley (1976).

Example 6.27. Let $\{X_j\}_{j=1}^n$ be i.i.d. as uniform on $(0, 1)$ and put $Y_n := \sum_{j=1}^n X_j^2$ and $Z_n := (Y_n - EY_n)/\sqrt{VY_n}$. We will approximate the distribution of Z_n using (6.23) and also as the member of the Pearson class that corresponds to its skewness and kurtosis. From $EX^k = \int_0^1 x^k \cdot dx = (k+1)^{-1}$

---

[18]For details and proofs see Epps and Pulley (1986).
[19]After Karl Pearson, whose original papers on the subject are reprinted in Pearson (1948). For a detailed discussion of the Pearson system, see Ord (1985).

for $k \geq 0$ we find $EY_n = n/3$, $VY_n = 4n/45$, and, after some calculation,

$$\alpha_3(n) := EZ_n^3 = \frac{2\sqrt{5}}{7\sqrt{n}}$$

$$\alpha_4(n) := EZ_n^4 = 3 - \frac{6}{7n}.$$

The first two rows of the table show the approximate values of these for various n. The last six rows show estimates of upper .05 and .01 quantiles, $z_n^{.05}$ and $z_n^{.01}$. Those marked (PH) are by linear interpolation from Pearson and Hartley's Table 32; those marked (GC) were found by solving $\hat{F}(z_n^p) = 1 - p$ with $\hat{F}$ as in (6.23); and those marked (sim) were found by simulation in 200,000 trials. The "∞" column presents the standard normal skewness, kurtosis, and quantiles $\{z^p\}$.

n :	10	20	50	100	∞
$\alpha_3(n)$	.2020	.1429	.0904	.0639	0.000
$\alpha_4(n)$	2.914	2.957	2.983	2.991	3.000
$z_n^{.05}(PH)$	1.706	1.694	1.671	1.663	1.645
$z_n^{.05}(GC)$	1.707	1.688	1.672	1.664	1.645
$z_n^{.05}(sim)$	1.712	1.685	1.679	1.664	1.645
$z_n^{.01}(PH)$	2.440	2.413	2.385	2.369	2.326
$z_n^{.01}(GC)$	2.448	2.419	2.388	2.371	2.326
$z_n^{.01}(sim)$	2.451	2.416	2.392	2.371	2.326

6.7.4   *Applications of Mann–Wald: The Delta Method*

Previously, we used the laws of large numbers and the continuity theorem to find probability limits of various functions of statistics that converge stochastically to constants. Adding the c.l.t.s and the Mann–Wald theorem to our tool kit lets us treat cases in which limiting distributions are not degenerate.

Example 6.28. Take $\mathbf{g}(X) := (g_1(X), g_2(X))' := (X, X^2)'$ and

$$h(\mathbf{g}) := \frac{\sqrt{n}(g_1 - \mu)}{\sqrt{g_2 - g_1^2}}.$$

Then for a random sample of size n

$$h(\bar{\mathbf{g}}) = \sqrt{n} \frac{\bar{X} - \mu}{\sqrt{M_2' - \bar{X}^2}} = \frac{\sqrt{n}(\bar{X} - \mu)/\sigma}{\sqrt{(M_2' - \bar{X}^2)/\sigma^2}}.$$

The numerator converges in distribution to standard normal by Lindeberg–Lévy, while the denominator converges in probability to unity by the laws of large numbers and the continuity theorem, since $M_2' - \bar{X}^2 \to^{a.s.} \mu_2' - \mu^2 = \sigma^2$. Mann–Wald thus implies that $h(\bar{\mathbf{g}}) \rightsquigarrow N(0, 1)$. Since $S^2 = \frac{n}{n-1}(M_2' - \bar{X}^2)$ and $\frac{n}{n-1} \to 1$, it follows also that

$$\frac{\bar{X} - \mu}{S/\sqrt{n}} \rightsquigarrow N(0, 1). \tag{6.24}$$

We shall see in Chapters 7 and 8 that this fundamental result is of great importance for statistical inference about population means when no parametric family is assumed for the population.

Example 6.29. Let $\{X_j\}_{j=1}^{n}$ and $\{Y_j\}_{j=1}^{m}$ be independent random samples from two populations, and let S_X and S_Y be the two sample standard deviations. Then as $n \to \infty$ and $m \to \infty$ with n/m tending to some positive constant (to keep one sample from dominating asymptotically), the foregoing result and another application of Mann–Wald imply that

$$\left(\frac{\bar{X} - \mu_X}{S_X/\sqrt{n}}\right)^2 + \left(\frac{\bar{Y} - \mu_Y}{S_Y/\sqrt{m}}\right)^2 \rightsquigarrow \chi^2(2).$$

Independence of the samples is crucial for this result, for two reasons: (i) The limiting normals themselves must be independent in order to get the chi-squared result; and (ii) independence lets us infer the limiting joint distribution of the studentized means from the known limiting behavior of their marginals.

We have seen that the continuity theorem delivers probability limits of continuous functions of certain statistics. The *delta method* provides higher-order approximations for the distributions of such functions, just as the c.l.t.s give higher-order approximations for the statistics themselves. This does come at a price, however, since it requires a stronger condition on the functions than continuity. Let $\{T_n\}$ be a sequence of r.v.s; for example, a sequence of statistics indexed by the sample size, n. Suppose that $T_n \to^p \theta$, that $\sqrt{n}(T_n - \theta)/\sigma_T \rightsquigarrow N(0, 1)$ for some positive constant σ_T, that $g : \Re \to \Re$ is *continuously differentiable* (i.e., g' exists and is continuous) in a neighborhood of θ, and that $g'(\theta) := g'(t)|_{t=\theta} \neq 0$. Then the expansion $g(T_n) = g(\theta) + g'(\theta_n)(T_n - \theta)$ leads to

$$\sqrt{n}[g(T_n) - g(\theta)] = g'(\theta_n)\sqrt{n}(T_n - \theta).$$

Here θ_n is a r.v. taking values between T_n and θ; i.e., $|\theta_n - \theta| < |T_n - \theta|$. Now $T_n \to^p \theta$ implies $\theta_n \to^p \theta$, and therefore $g'(\theta_n) \to^p g'(\theta)$ by the continuity theorem. Since $\sqrt{n}(T_n - \theta) \rightsquigarrow N(0, \sigma_T^2)$, Mann–Wald yields

$$\sqrt{n}\left[g(T_n) - g(\theta)\right] \rightsquigarrow N\left[0, \sigma_T^2 g'(\theta)^2\right], \ g'(\theta) \neq 0. \tag{6.25}$$

For large n the distribution of $g(T_n)$ is thus *approximated* as $N\left[g(\theta), \sigma_T^2 g'(\theta)^2/n\right]$ whenever $g'(\theta) \neq 0$. The delta method gets its name because it links $\Delta g \equiv g(T_n) - g(\theta)$ to $\Delta T_n \equiv T_n - \theta$, quantities that are sometimes called "statistical differentials". In this way the method supplies an affine approximation to the nonlinear g and thereby exploits the closure of the normal family under such transformations.

No such normal approximation exists for a θ_0 at which $g'(\theta_0) = 0$; but if g'' exists and is continuous near there, and if $g''(\theta_0) \neq 0$, there is a chi-squared approximation instead. This is based on a second-order expansion, as

$$g(T_n) = g(\theta_0) + \frac{g''(\theta_n)}{2}\left(T_n - \theta_0\right)^2,$$

and the fact that $n(T_n - \theta_0)^2/\sigma_T^2 \rightsquigarrow \chi^2(1)$ when $\theta = \theta_0$.

Example 6.30. If $\bar{X}$ is the mean of a random sample of size n from a population with mean μ and finite variance σ^2, then $\sqrt{n}\left(\bar{X} - \mu\right) \rightsquigarrow N\left(0, \sigma^2\right)$ by Lindeberg–Lévy. Applying the delta method with $T_n = \bar{X}$ and $g(\bar{X}) = \bar{X}e^{-\bar{X}}$, we have $\bar{X}e^{-\bar{X}} = \mu e^{-\mu} - (\mu_n - 1)e^{-\mu_n}(\bar{X} - \mu)$ and

$$\sqrt{n}\left(\bar{X}e^{-\bar{X}} - \mu e^{-\mu}\right) \rightsquigarrow N\left[0, \sigma^2(\mu - 1)^2 e^{-2\mu}\right], \mu \neq 1.$$

Since $S^2 \longrightarrow^p \sigma^2$ and $S \longrightarrow^p \sigma$, Mann–Wald implies, as well, that

$$\frac{\sqrt{n}\left(\bar{X}e^{-\bar{X}} - \mu e^{-\mu}\right)}{S} \rightsquigarrow N\left[0, (\mu - 1)^2 e^{-2\mu}\right], \mu \neq 1.$$

If $\mu = 1$ take the expansion out to the next term, as

$$\bar{X}e^{-\bar{X}} = \mu e^{-\mu} - (\mu - 1)e^{-\mu}(\bar{X} - \mu) + \frac{1}{2}(\mu_n - 2)e^{-\mu_n}(\bar{X} - \mu)^2,$$

and use the fact that $n(\bar{X} - \mu)/S^2 \rightsquigarrow N(0,1)$ to get $n\left(\bar{X}e^{-\bar{X}} - e^{-1}\right)/S^2 \rightsquigarrow -\frac{1}{2}e^{-1}\chi^2(1)$.

Example 6.31. An exercise in an astronomy class involves estimating the radius of the earth by measuring the apparent shift of location of the planet Venus against the background of stars when viewed at intervals of 12 hours. 100 independent measurements are taken with identical apparatus, and the sample mean and sample standard deviation are calculated as $\bar{R} = 4,000$ miles and $S_R = 200$ miles, respectively. An estimate of the volume of the earth (regarded as a perfect sphere) is then $\hat{V} := 4\pi\bar{R}^3/3$. Let us now estimate the probability that $\hat{V}$ differs from the true volume by more than 10^{10} cubic miles.

Assuming that the estimates are i.i.d., Lindeberg–Lévy implies that $\sqrt{n}(\bar{R}-\rho) \rightsquigarrow N(0,\sigma_R^2)$, where ρ is the true (unknown) radius and σ_R^2 is the variance of the measurement error. Let $V := 4\pi\rho^3/3$ be the true volume. Taking $T_n = \bar{R}$ and $g(\bar{R}) = \hat{V} = 4\pi\bar{R}^3/3$, we have

$$\sqrt{n}(\hat{V} - V) \rightsquigarrow N(0, 16\pi^2\rho^4\sigma_R^2).$$

Since $S_R^2 \rightarrow^p \sigma_R^2$ and $\bar{R} \rightarrow^p \rho$, it also follows by Mann–Wald that

$$\sqrt{n}\frac{(\hat{V} - V)}{4\pi\bar{R}^2 S_R} = \sqrt{n}\frac{(\hat{V} - V)}{4\pi\rho^2\sigma_R} \cdot \frac{\rho^2\sigma_R}{\bar{R}^2 S_R} \rightsquigarrow N(0,1) \cdot 1.$$

Thus, we can approximate

$$\Pr(|\hat{V} - V| > 10^{10}) = \Pr\left(\sqrt{100}\frac{|\hat{V} - V|}{4\pi\bar{R}^2 S_R} > \frac{10 \cdot 10^{10}}{4\pi(16 \times 10^6)200}\right)$$

as $\Pr(|Z| > 2.49) \doteq .0128$.

Example 6.32. Letting $\{X_j\}_{j=1}^\infty$ be i.i.d. with positive variance and finite fourth moment, let us work out the limiting distribution of $\sqrt{n}(\sqrt{M_2} - \sigma)$, where $M_2 = M_2' - \bar{X}^2 = n^{-1}\sum_{j=1}^n X_j^2 - \bar{X}^2$. The delta method with $g(M_2) = \sqrt{M_2}$ gives $\sqrt{M_2} = \sqrt{\sigma^2} + (2\sigma_n)^{-1}(M_2 - \sigma^2)$, where $|\sigma_n^2 - \sigma^2| < |M_2 - \sigma^2|$. Thus,

$$\sqrt{n}\left(\sqrt{M_2} - \sigma\right) = (2\sigma_n)^{-1}\sqrt{n}\left(M_2 - \sigma^2\right). \qquad (6.26)$$

To proceed, we must determine the limiting distribution of

$$\sqrt{n}\left(M_2 - \sigma^2\right) = \sqrt{n}\left(M_2' - \bar{X}^2 - \sigma^2\right) = \sqrt{n}\left(M_2' - \sigma^2\right) - \bar{X}\sqrt{n}\bar{X}.$$

M_2 being invariant with respect to μ, set $\mu = 0$ without loss of generality. Now $\bar{X} \rightarrow^p 0$ and $\sqrt{n}\bar{X} \rightsquigarrow N\left(0,\sigma^2\right)$, so $\bar{X}\sqrt{n}\bar{X} \rightarrow^p 0$ by Mann–Wald, indicating that the limiting distributions of $\sqrt{n}\left(M_2 - \sigma^2\right)$ and $\sqrt{n}\left(M_2' - \sigma^2\right)$ are the same. Since the $\{X_j\}_{j=1}^n$ have finite fourth moment, it follows that $VM_2' = V\left(n^{-1}\sum_{j=1}^n X_j^2\right) = n^{-1}\left(\mu_4 - \sigma^4\right) < \infty$ ($\mu_4 = \mu_4'$ when $\mu = 0$). The Lindeberg–Lévy c.l.t. then applies to give

$$\frac{M_2' - EM_2'}{\sqrt{VM_2'}} = \sqrt{n}\frac{(M_2' - \frac{n-1}{n}\sigma^2)}{\sqrt{\mu_4 - \sigma^4}} \rightsquigarrow N(0,1)$$

or, equivalently, $\sqrt{n}\left(M'_2 - \sigma^2\right) \rightsquigarrow N\left(0, \mu_4 - \sigma^4\right)$. Returning now to (6.26) and noting that $\sigma_n \to^{a.s.} \sigma$, we have the final result:

$$\sqrt{n}\left(\sqrt{M_2} - \sigma\right) \rightsquigarrow N\left(0, \frac{\mu_4 - \sigma^4}{4\sigma^2}\right).$$

Note, too, that since $S^2 - M_2 \to^{a.s.} 0$, we also have

$$\sqrt{n}\left(S^2 - \sigma^2\right) \rightsquigarrow N\left(0, \mu_4 - \sigma^4\right) \qquad (6.27)$$

$$\sqrt{n}\left(S - \sigma\right) \rightsquigarrow N\left(0, \frac{\mu_4 - \sigma^4}{4\sigma^2}\right).$$

The delta method applies also to vector-valued statistics. Suppose that

$$\mathbf{T}_n := (T_{1n}, T_{2n}, ..., T_{kn})' \to \boldsymbol{\theta} := (\theta_1, \theta_2, ..., \theta_k)'$$

and that $\sqrt{n}\left(\mathbf{T}_n - \boldsymbol{\theta}\right) \rightsquigarrow N\left(\mathbf{0}, \boldsymbol{\Sigma_T}\right)$, and let $g : \Re^k \to \Re$ have continuous first derivatives in a neighborhood of $\boldsymbol{\theta}$. Represent the gradient vector as $\mathbf{g_T} := (\partial g/\partial T_{1n}, \partial g/\partial T_{2n}, ..., \partial g/\partial T_{kn})'$. Then the multivariate version of Taylor's formula gives, upon multiplying by $\sqrt{n}$,

$$\sqrt{n}\left[g\left(\mathbf{T}_n\right) - g\left(\boldsymbol{\theta}\right)\right] = \mathbf{g_T}\left(\boldsymbol{\theta}_n\right)' \sqrt{n}\left(\mathbf{T}_n - \boldsymbol{\theta}\right), \qquad (6.28)$$

where $\boldsymbol{\theta}_n$ lies on the line segment connecting $\mathbf{T}_n$ and $\boldsymbol{\theta}$. Depending on the circumstances, there are two ways to proceed from here.

First, suppose the components of $\mathbf{T}_n$ are all expressible as sample means of functions of the data, as $\left\{T_{in} = n^{-1}\sum_{j=1}^{n} h_i\left(X_j\right)\right\}_{i=1}^{k}$. Then we can take the inner product in (6.28) and express $g\left(\mathbf{T}_n\right) - g\left(\boldsymbol{\theta}\right)$ as the sample mean of a linear combination of scalar functions; i.e., as $n^{-1}\sum_{j=1}^{n} Y_j\left(\boldsymbol{\theta}_n\right)$, where $Y_j\left(\boldsymbol{\theta}_n\right)$ is given by

$$\frac{\partial g\left(\boldsymbol{\theta}_n\right)}{\partial T_{1n}}\left[h_1\left(X_j\right) - \theta_1\right] + \frac{\partial g\left(\boldsymbol{\theta}_n\right)}{\partial T_{2n}}\left[h_2\left(X_j\right) - \theta_2\right] + \cdots + \frac{\partial g\left(\boldsymbol{\theta}_n\right)}{\partial T_{kn}}\left[h_k\left(X_j\right) - \theta_k\right].$$

Since $\mathbf{T}_n \to^p \boldsymbol{\theta}$ implies $\boldsymbol{\theta}_n \to \boldsymbol{\theta}$, the limiting distribution of the above is the same if the derivatives are evaluated at $\boldsymbol{\theta}$ rather than $\boldsymbol{\theta}_n$. Thus, the Lindeberg–Lévy c.l.t. leads to

$$\sqrt{n}\left[g(\mathbf{T}_n) - g(\boldsymbol{\theta})\right] \doteq n^{-1/2}\sum_{j=1}^{n} Y_j\left(\boldsymbol{\theta}\right) \rightsquigarrow N\left(0, \sigma_Y^2\right).$$

Working out σ_Y^2 requires finding the variance of the linear combination of the $h_i\left(X\right)$.

Example 6.33. Assuming that the population has finite fourth moment, let us work out the limiting distribution of a standardized version of the sample *coefficient of variation*, $\sqrt{M_2}/\bar{X}$. Taking $\mathbf{T}_n := (M_2, \bar{X})'$ and $\boldsymbol{\theta} := (\sigma^2, \mu)'$, we have $g(M_2, \bar{X}) = \sqrt{M_2}/\bar{X}$ with gradient vector

$$\mathbf{g_T}(\mathbf{T}_n) = \left(\frac{\partial g}{\partial M_2}, \frac{\partial g}{\partial \bar{X}} \right)' = \left[\frac{1}{2\bar{X}\sqrt{M_2}}, -\frac{\sqrt{M_2}}{\bar{X}^2} \right]'$$

and $\mathbf{g_T}(\boldsymbol{\theta}_n) = [1/(2\mu_n\sigma_n), -\sigma_n/\mu_n^2]'$, where $\boldsymbol{\theta}_n = (\sigma_n^2, \mu_n)'$ lies on a line connecting $(M_2, \bar{X})$ and (σ^2, μ). Corresponding to (6.28), then,

$$\sqrt{n}\left(\frac{\sqrt{M_2}}{\bar{X}} - \frac{\sigma}{\mu} \right) = \left(\frac{1}{2\mu_n\sigma_n}, -\frac{\sigma_n}{\mu_n^2} \right) \sqrt{n}\left(\begin{matrix} M_2 - \sigma^2 \\ \bar{X} - \mu \end{matrix} \right).$$

Now $M_2 = n^{-1}\sum_{j=1}^n (X_j - \bar{X})^2 = n^{-1}\sum_{j=1}^n (Y_j - \bar{Y})^2 = n^{-1}\sum_{j=1}^n Y_j^2 - \bar{Y}^2$, where $Y_j := X_j - \mu$. Thus,

$$\sqrt{n}\left(M_2 - \sigma^2, \bar{X} - \mu \right) = \left(n^{-1/2}\sum_{j=1}^n (Y_j^2 - \sigma^2) - \sqrt{n}\bar{Y}^2, \sqrt{n}\bar{Y} \right).$$

But $\bar{Y} \to^p 0$ and $\sqrt{n}\bar{Y} \rightsquigarrow N(0, \sigma^2)$, so $\sqrt{n}\bar{Y}^2 \to^p 0$. Thus, the above is asymptotically equivalent to

$$n^{-1/2}\sum_{j=1}^n (Y_j^2 - \sigma^2, Y_j)$$

(in the sense that the difference between them converges in probability to the zero vector). Moreover,

$$\mathbf{g_T}(\boldsymbol{\theta}_n) = \left(\frac{1}{2\mu_n\sigma_n}, -\frac{\sigma_n}{\mu_n^2} \right)' \to^p \mathbf{g_T}(\boldsymbol{\theta}) = \left(\frac{1}{2\mu\sigma}, -\frac{\sigma}{\mu^2} \right)'$$

implies that these two expressions are asymptotically equivalent as well. (In the terminology and notation of Definition 6.6 on page 393 we would say that $|\mathbf{g_T}(\boldsymbol{\theta}_n) - \mathbf{g_T}(\boldsymbol{\theta})| = o_P(1)$.) Replacing $\mathbf{g_T}(\boldsymbol{\theta}_n)$ and $\sqrt{n}(M_2 - \sigma^2, \bar{X} - \mu)$ by their asymptotic equivalents and taking the inner product, we see that $\sqrt{n}(\sqrt{M_2}/\bar{X} - \sigma/\mu)$ has the same limiting distribution as

$$n^{-1/2}\sum_{j=1}^n \left(\frac{Y_j^2 - \sigma^2}{2\mu\sigma} - \frac{\sigma}{\mu^2}Y_j \right) =: n^{-1/2}\sum_{j=1}^n W_j,$$

say. Now the $\{W_j\}_{j=1}^n$ are i.i.d. with $EW = 0$ and finite variance $VW = EW^2$ given by

$$EW^2 = \frac{EY^4 - 2\sigma^2 EY^2 + \sigma^4}{4\mu^2\sigma^2} - \frac{1}{\mu^3}EY^3 + \frac{\sigma^2}{\mu^4}EY^2$$

$$= \frac{\mu_4 - \sigma^4}{4\mu^2\sigma^2} - \frac{\mu_3}{\mu^3} + \frac{\sigma^4}{\mu^4}.$$

Therefore, the Lindeberg–Lévy c.l.t. implies that $\sqrt{n}\left(\sqrt{M_2}/\bar{X} - \sigma/\mu\right)$ converges in distribution to $N\left(0, EW^2\right)$.

The second way of extending the delta method to the multivariate case is more general, in that the components of $\mathbf{T}_n$ need not (but *may*) all be sample means. However, we must be able to establish in some way that $\sqrt{n}\left(\mathbf{T}_n - \boldsymbol{\theta}\right)$ has a limiting multivariate normal distribution. If so, then (6.28) leads immediately to the multivariate version of (6.25):

$$\sqrt{n}\left[g(\mathbf{T}_n) - g(\boldsymbol{\theta})\right] \rightsquigarrow N\left[0, \mathbf{g}'_{\mathbf{T}}\left(\boldsymbol{\theta}\right)\boldsymbol{\Sigma}_{\mathbf{T}}\mathbf{g}_{\mathbf{T}}\left(\boldsymbol{\theta}\right)\right]. \tag{6.29}$$

To show that $\sqrt{n}\left(\mathbf{T}_n - \boldsymbol{\theta}\right)$ converges in distribution to multivariate normal is easy if it happens that the components of vector $\mathbf{T}_n$ are independent, since in that case one need show only that each of the marginal distributions is asymptotically normal. When the components are not independent, a multivariate central limit theorem is needed. Here is a multivariate version of Lindeberg–Lévy.

Theorem 6.19. *(Multivariate Central Limit) Let $\{\mathbf{X}_j\}_{j=1}^n$ be i.i.d. k-vector-valued r.v.s with mean $\boldsymbol{\mu} = (\mu_1, \mu_2, ..., \mu_k)'$, finite variances, and nonsingular covariance matrix $\boldsymbol{\Sigma}$. Then, if $\boldsymbol{\Sigma}^{-1/2}$ is the unique symmetric square-root of $\boldsymbol{\Sigma}^{-1}$, the standardized sum*

$$\mathbf{Z}_n := n^{-1/2}\boldsymbol{\Sigma}^{-1/2}\sum_{j=1}^n\left(\mathbf{X}_j - \boldsymbol{\mu}\right)$$

converges in distribution to $N\left(\mathbf{0}, \mathbf{I}_k\right)$, where $\mathbf{I}_k$ is the $k \times k$ identity matrix. Equivalently, $n^{-1/2}\sum_{j=1}^n\left(\mathbf{X}_j - \boldsymbol{\mu}\right) \rightsquigarrow N\left(\mathbf{0}, \boldsymbol{\Sigma}\right)$.

Proof. Let $\mathbf{a} := (a_1, a_2, ..., a_k)'$ be any vector of constants, not all zero. The scalar r.v.s $\left\{Y_j = \mathbf{a}'\boldsymbol{\Sigma}^{-1/2}\left(\mathbf{X}_j - \boldsymbol{\mu}\right)\right\}_{j=1}^n$ are i.i.d. with mean zero and variance

$$\sigma_Y^2 = \mathbf{a}'\boldsymbol{\Sigma}^{-1/2}\,\boldsymbol{\Sigma}\,\boldsymbol{\Sigma}^{-1/2}\mathbf{a} = \mathbf{a}'\mathbf{a} > 0,$$

so $n^{-1/2}\sum_{j=1}^n Y_j/\sigma_Y \rightsquigarrow N(0,1)$ by Lindeberg–Lévy and thus $\mathbf{Z}_n \rightsquigarrow N\left(\mathbf{0}, \mathbf{I}_k\right)$ by Cramér–Wold. $\qquad\square$

Example 6.34. The vector methods can be applied to the problem posed in Example 6.33. The same arguments show that $\sqrt{n}\left(\sqrt{M_2}/\bar{X} - \sigma/\mu\right)$ is asymptotically equivalent to the inner product

$$\left(\frac{1}{2\mu\sigma}, -\frac{\sigma}{\mu^2}\right)n^{-1/2}\sum_{j=1}^n\left(\frac{Y_j^2 - \sigma^2}{Y_j}\right).$$

Each of the i.i.d. vector summands having covariance matrix

$$\Sigma = \begin{pmatrix} \mu_4 - \sigma^4 & \mu_3 \\ \mu_3 & \sigma^2 \end{pmatrix},$$

the multivariate c.l.t. implies that $\sqrt{n}\,(M_2 - \sigma^2, \bar{X} - \mu)$ is asymptotically bivariate normal with mean $(0,0)$ and covariance matrix Σ. Therefore,

$$\sqrt{n}\left(\frac{\sqrt{M_2}}{\bar{X}} - \frac{\sigma}{\mu}\right) \rightsquigarrow N\left[0, \left(\frac{1}{2\mu\sigma}, -\frac{\sigma}{\mu^2}\right)\begin{pmatrix} \mu_4 - \sigma^4 & \mu_3 \\ \mu_3 & \sigma^2 \end{pmatrix}\begin{pmatrix} \frac{1}{2\mu\sigma} \\ -\frac{\sigma}{\mu^2} \end{pmatrix}\right]$$

$$= N\left(0, \frac{\mu_4 - \sigma^4}{4\mu^2\sigma^2} - \frac{\mu_3}{\mu^3} + \frac{\sigma^4}{\mu^4}\right).$$

Note that the asymptotic equivalence of S^2 and M_2 implies that $\sqrt{n}\,(S/\bar{X} - \sigma/\mu)$ has the same limiting distribution.

Exercise 6.21. $\{X_j\}_{j=1}^{n_X}$ and $\{Y_j\}_{j=1}^{n_Y}$ *are random samples of sizes* n_X *and* n_Y *that are drawn—independently of each other—from populations with means* μ_X, μ_Y *and finite variances* σ_X^2, σ_Y^2. $\bar{X}, \bar{Y}$ *and* S_X^2, S_Y^2 *are the sample means and variances. If the ratio of sample sizes* n_Y/n_X *approaches a finite, positive number* ν *as* $n_X \to \infty$, *show that*

$$\bar{D}_{n_X, n_Y} := \frac{(\bar{X} - \bar{Y}) - (\mu_X - \mu_Y)}{\sqrt{S_X^2/n_X + S_Y^2/n_Y}} \rightsquigarrow N(0,1)$$

as $n_X \to \infty$. *(Hint: Write* $\bar{D}_{n_X, n_Y}$ *as*

$$\left(\frac{\bar{X} - \mu_X}{\sqrt{S_X^2/n_X}}\right)\frac{\sqrt{S_X^2/n_X}}{\sqrt{S_X^2/n_X + S_Y^2/n_Y}} - \left(\frac{\bar{Y} - \mu_Y}{\sqrt{S_Y^2/n_Y}}\right)\frac{\sqrt{S_Y^2/n_Y}}{\sqrt{S_X^2/n_X + S_Y^2/n_Y}}.)$$

Exercise 6.22. *The sample covariance of a bivariate sample* $\{(X_j, Y_j)\}_{j=1}^{n}$ *is*

$$S_{XY} := \frac{1}{n-1}\sum_{j=1}^{n}(X_j - \bar{X})(Y_j - \bar{Y}) = \frac{1}{n-1}\left(\sum_{j=1}^{n}X_j Y_j - n\bar{X}\bar{Y}\right),$$

and the sample coefficient of correlation is

$$R_{XY} := \begin{cases} 0, & S_{XY} = 0 \\ \frac{S_{XY}}{S_X S_Y}, & S_{XY} \neq 0 \end{cases}.$$

If the population is bivariate normal with zero correlation (i.e., independent normals), show that $\sqrt{n}R_{XY} \rightsquigarrow N(0,1)$. *(Hint: Show that* R_{XY} *is invariant under affine transformations of* X *and* Y, *so that nothing is lost by regarding* X *and* Y *as independent standard normals. Then show that* *the terms involving products and squares of $\bar{X}$ and $\bar{Y}$ are asymptotically negligible, so that $\sqrt{n}R_{XY}$ can be approximated as $\sqrt{n}M_{XY}/\sqrt{M_{XX}M_{YY}}$, where the M's are sample second moments and product moment about the origin. Linearize using the delta method to represent this as a linear function of means of products and second powers of observations X_j and Y_j, then apply the univariate c.l.t.)*

6.8   U Statistics

Recall (Section 5.2) that the sample variance can be calculated in either of the following ways:

$$S^2 = \frac{1}{n-1}\sum_{j=1}^{n}(X_j - \bar{X})^2 = \frac{1}{\binom{n}{2}}\sum_{i<j}\frac{(X_i - X_j)^2}{2}. \tag{6.30}$$

Both expressions involve sums of identically distributed r.v.s, but these are clearly not *independent*. ($\bar{X}$ is in each term of the first sum, and X_j appears in $j-1$ terms of the double summation.) Thus, the Lindeberg–Lévy central limit theorem, of itself, does not imply the asymptotic normality of the (standardized) sample variance. Nevertheless, $\sqrt{n}\left(S^2 - \sigma^2\right)$ is indeed asymptotically normal when the $\{X_j\}$ are i.i.d. with positive variance and finite fourth moment. This was shown in Example (6.32) by appealing to the law of large numbers and the Mann–Wald theorem, as well as Lindeberg–Lévy. We now describe another way to establish this result, and one that applies to many other functions of r.v.s for which alternative methods are not always at hand.

To this end, consider statistics of the form

$$U_{m,n} := \frac{1}{\binom{n}{m}}\sum_{\mathbf{j}\in\mathcal{J}_{m,n}}K(X_{j_1}, X_{j_2}, ..., X_{j_m}), \tag{6.31}$$

where (i) $m \in \{1, 2, ..., n\}$; (ii) $\mathcal{J}_{m,n}$ represents the $\binom{n}{m}$ subsets of $\{1, 2, ..., n\}$ containing m distinct integers without regard to order; (iii) $\mathbf{j} : = (j_1, j_2, ..., j_m)$ represents any such subset; and (iv) function $K : \Re^m \to \Re$ is *symmetric* in its arguments—i.e., invariant under permutations of $X_{j_1}, X_{j_2}, ..., X_{j_m}$. We refer to K as a *symmetric kernel* and consider versions of these that have positive, finite variance and expectation $EK(X_{j_1}, X_{j_2}, ..., X_{j_m}) =: \kappa$, say. It is then clear that $EU_{m,n} = \kappa$, as well. Statistics $\{U_{m,n}\}$ with these specifications are known as *U statistics*. An example is right at hand, for the second expression in (6.30) presents S^2

in the form (6.31) with $m = 2$, $K(X_i, X_j) = (X_i - X_j)^2 / 2$, and $\kappa = \sigma^2$. We will show using a *projection* argument that $\sqrt{n}(U_{m,n} - \kappa)$ is, for all members of the class, distributed asymptotically as normal—a result due to Hoeffding (1948).[20]

To proceed requires some preliminary facts about kernel functions. The projection theorem will involve conditional expectations of the form

$$K_c(X_1, ..., X_c) := E[K(X_1, ..., X_c, X_{c+1}, ..., X_m) \mid X_1, ..., X_c],$$

where $c \in \{1, 2, ..., m\}$. Because K is symmetric and the $\{X_j\}$ are i.i.d., this expectation is the same regardless of which set of c variables are conditioned upon; we take the *first* c just for convenience. To simplify notation, let $\mathcal{X}_c$, $\mathcal{X}_m$ and $\mathcal{X}_m \backslash \mathcal{X}_c$ represent, respectively, the sets $\{X_1, ..., X_c\}$, $\{X_1, ..., X_c, X_{c+1}, ..., X_m\}$, and $\{X_{c+1}, ..., X_m\}$, so that the above becomes

$$K_c(\mathcal{X}_c) = E[K(\mathcal{X}_m) \mid \mathcal{X}_c].$$

We shall also need a second set of m variables $\mathcal{X}_m' := \{X_1, ..., X_c, X_{m+1}, ..., X_{2m-c}\}$, for which $\mathcal{X}_m \cap \mathcal{X}_m' = \mathcal{X}_c$ and for which $\mathcal{X}_m \backslash \mathcal{X}_c$ and $\mathcal{X}_m' \backslash \mathcal{X}_c$ are i.i.d.

The first two moments of $K_c(\mathcal{X}_c)$ can now be determined. The first moment comes directly via the tower property of conditional expectation: $EK_c(\mathcal{X}_c) = EK(\mathcal{X}_m) = \kappa$. Finding the variance is a little more complicated. Here are the steps; explanations follow.

$$
\begin{aligned}
VK_c(\mathcal{X}_c) &= E\{E[K(\mathcal{X}_m) \mid \mathcal{X}_c]\}^2 - \kappa^2 \\
&= E\{E[K(\mathcal{X}_m) \mid \mathcal{X}_c] \cdot E[K(\mathcal{X}_m') \mid \mathcal{X}_c]\} - \kappa^2 \\
&= E\{E[K(\mathcal{X}_m) \cdot K(\mathcal{X}_m') \mid \mathcal{X}_c]\} - \kappa^2 \\
&= EK(\mathcal{X}_m) K(\mathcal{X}_m') - \kappa^2 \\
&= Cov[K(\mathcal{X}_m), K(\mathcal{X}_m')] \\
&=: \xi_c.
\end{aligned}
$$

The second equality holds because $\mathcal{X}_m \backslash \mathcal{X}_c$ and $\mathcal{X}_m' \backslash \mathcal{X}_c$ are identically distributed; the third, because they are independent; and the fourth, by the tower property. Notice that the result shows that kernel functions with any $c \in \{1, 2, ..., m\}$ arguments in common are *positively* correlated. Also, we can set $\xi_0 = 0$ since functions with *no* arguments in common are independent (and since $K_0(\varnothing) \equiv EK(\mathcal{X}_m)$ is a constant).

---

[20]For more details on one-sample U statistics and extensions to the two-sample case, see Randles and Wolfe (1979) and Serfling (1980).

As example, with $K(X_1, X_2) = (X_1 - X_2)^2 / 2$ and

$$K_1(X_1) = \frac{1}{2} E\left(X_1^2 - 2X_1 X_2 + X_2^2 \mid X_1\right)$$

$$= \frac{1}{2}\left(X_1^2 - 2X_1 \mu + \sigma^2 + \mu^2\right)$$

$$= \frac{1}{2}\left[(X_1 - \mu)^2 + \sigma^2\right],$$

we have

$$EK_1(X_1) = \sigma^2$$

$$VK_1(X_1) = \frac{1}{4} E\left[(X_1 - \mu)^2 + \sigma^2\right]^2 - \sigma^4$$

$$= \frac{\mu_4 - \sigma^4}{4}$$

$$\xi_1 = \frac{1}{4} E(X_1 - X_2)^2(X_1 - X_3)^2 - \sigma^4$$

$$= \frac{\mu_4 - \sigma^4}{4}$$

$$\xi_2 = \frac{1}{4} E(X_1 - X_2)^4 - \sigma^4$$

$$= \frac{\mu_4 + \sigma^4}{2}.$$

The variance of $U_{m,n}$ can now be expressed in terms of the covariances $\{\xi_c\}$, working from the following expressions:

$$VU_{m,n} = \frac{1}{\binom{n}{m}^2} \sum_{\mathbf{i} \in \mathcal{J}_{m,n}} \sum_{\mathbf{j} \in \mathcal{J}_{m,n}} \left[EK\left(X_{i_1}, ..., X_{i_m}\right) K\left(X_{j_1}, ..., X_{j_m}\right) - \kappa^2\right]$$

$$= \frac{1}{\binom{n}{m}^2} \sum_{\mathbf{i} \in \mathcal{J}_{m,n}} \sum_{\mathbf{j} \in \mathcal{J}_{m,n}} Cov\left[K\left(X_{i_1}, ..., X_{i_m}\right), K\left(X_{j_1}, ..., X_{j_m}\right)\right].$$

Here, the two kernels in some of the $\binom{n}{m}^2$ terms of the summation have sample elements in common, and some do not. If particular sets $\mathbf{i}, \mathbf{j}$ have c common elements, then the covariance of the kernels is ξ_c, with $\xi_0 = 0$ if there are no common elements. Now for $c \in \{1, 2, ..., m\}$ there are $\binom{n}{m}$ ways to choose m variables from the n and then $\binom{m}{c}\binom{n-m}{m-c}$ ways to choose from those m precisely c common variables. Thus,

$$VU_{m,n} = \frac{1}{\binom{n}{m}^2} \sum_{c=1}^{m} \binom{n}{m}\binom{m}{c}\binom{n-m}{m-c} \xi_c$$

$$= \frac{1}{\binom{n}{m}} \sum_{c=1}^{m} \binom{m}{c}\binom{n-m}{m-c} \xi_c.$$

As example, with $K(X_1, X_2) = (X_1 - X_2)^2/2$ and $U_{2,n} \equiv S^2$ we have

$$VS^2 = \frac{1}{\binom{n}{2}}\left[\binom{2}{1}\binom{n-1}{2-1}\frac{\mu_4 - \sigma^4}{4} + \binom{2}{2}\binom{n-2}{0}\frac{\mu_4 + \sigma^4}{2}\right]$$

$$= \frac{n-2}{n(n-1)}(\mu_4 - \sigma^4) + \frac{1}{n(n-1)}(\mu_4 + \sigma^2)$$

$$= \frac{\mu_4 - \sigma^4}{n} + \frac{2\sigma^4}{n(n-1)},$$

which corresponds to direct solution (5.4).

Finally, we define the auxiliary quantity

$$U^*_{m,n} - \kappa := \sum_{i=1}^{n} E(U_{m,n} - \kappa \mid X_i)$$

$$= \sum_{i=1}^{n}\left\{\frac{1}{\binom{n}{m}}\sum_{j \in \mathcal{J}_{m,n}} E[K(X_{j_1}, X_{j_2}, ..., X_{j_m}) \mid X_i]\right\} - n\kappa.$$

Conditional expectation $E(U_{m,n} - \kappa \mid X_i)$ is regarded as the *projection* of $U_{m,n} - \kappa$ onto X_i, and $U^*_{m,n} - \kappa$ thus comprises projections onto the entire sample. To express $U^*_{m,n} - \kappa$ more simply, note that if i belongs to a particular set $\mathbf{j}$ that specifies the arguments of K, then

$$E[K(X_{j_1}, X_{j_2}, ..., X_{j_m}) \mid X_i] = K_1(X_i).$$

Otherwise, if i is *not* included in $\mathbf{j}$, then $E[K(X_{j_1}, X_{j_2}, ..., X_{j_m}) \mid X_i] = \kappa$. Since there are $\binom{1}{1}\binom{n-1}{m-1}$ sets $\mathbf{j}$ that include a given i and $\binom{1}{0}\binom{n-1}{m}$ sets that do not, we have

$$U^*_{m,n} - \kappa = \frac{1}{\binom{n}{m}}\sum_{i=1}^{n}\left[\binom{n-1}{m-1}K_1(X_i) + \binom{n-1}{m}\kappa\right] - n\kappa$$

$$= \frac{m}{n}\sum_{i=1}^{n}[K_1(X_i) - \kappa].$$

Observe, then, that $U^*_{m,n} - \kappa$ is simply the sample mean of i.i.d. r.v.s. Its expected value equals zero, and its variance is

$$VU^*_{m,n} = \frac{m^2}{n}VK_1(X_1) = \frac{m^2\xi_1}{n}.$$

To express its covariance with $U_{m,n}$, we have

$$Cov(U^*_{m,n}, U_{m,n}) = EU_{m,n}(U^*_{m,n} - \kappa)$$

$$= \frac{1}{\binom{n}{m}}\frac{m}{n}E\left\{\sum_{j \in \mathcal{J}_{m,n}}\sum_{i=1}^{n}K(X_{j_1}, ..., X_{j_m})[K_1(X_i) - \kappa]\right\}.$$

Here, for each $i \in \{1, 2, ..., n\}$ there are $\binom{n-1}{m-1}$ terms in the summation over **j** in which X_i appears among the arguments of K, and there are $\binom{n}{m-1}$ terms in which X_i does not so appear. When **j** does include i the term's expected value is

$$E\left[K(X_{j_1}, ..., X_{j_m})K_1(X_i)\right] - \kappa E K(X_{j_1}, ..., X_{j_m})$$
$$= E\left\{K(X_{j_1}, ..., X_{j_m})E\left[K(X_{j_1}, ..., X_{j_m}) \mid X_i\right]\right\} - \kappa^2$$
$$= E\left\{E\left[K(X_{j_1}, ..., X_{j_m}) \mid X_i\right]\right\}^2 - \kappa^2$$
$$= V K_1(X_i)$$
$$= \xi_1;$$

and when **j** does not include i the expectation is

$$E K(X_{j_1}, X_{j_2}, ..., X_{j_m}) E\left[K_1(X_i) - \kappa\right] = 0$$

by independence. Thus,

$$Cov\left(U^*_{m,n}, U_{m,n}\right) = \frac{1}{\binom{n}{m}} \frac{m}{n} \left[n \binom{n-1}{m-1}\xi_1 + n\binom{n-1}{m}0\right]$$
$$= \frac{m^2 \xi_1}{n}$$
$$= V U^*_{m,n}.$$

With these facts established, it is now easy to prove the following U-statistics theorem.

Theorem 6.20. *When* $\{X_j\}_{j=1}^n$ *are i.i.d. and* $K(X_1, ..., X_m)$ *(m* $\leq$ *n) is a symmetric function with* $E K(X_1, ..., X_m) = \kappa$, $0 < V K(X_1, ..., X_m) < \infty$, *and*

$$Cov\left[K(X_1, X_2, ..., X_m), K(X_1, X_3, ..., X_{m+1})\right] =: \xi_1 > 0,$$

then $\sqrt{n}(U_{m,n} - \kappa) \rightsquigarrow N(0, m^2\xi_1)$ *as* $n \to \infty$.

Proof. Since $U^*_{m,n} - \kappa$ is the arithmetic mean of n i.i.d. r.v.s having finite variance, Lindeberg–Lévy gives

$$\sqrt{n}\left(U^*_{m,n} - \kappa\right) \rightsquigarrow N\left[0, m^2 V K_1(X_1)\right] = N\left(0, m^2\xi_1\right).$$

The conclusion will follow if we can show that $\sqrt{n}\left|U_{m,n} - U^*_{m,n}\right|$ converges to zero in mean square—hence in probability, hence in distribution. We have

$$nE\left(U_{m,n} - U^*_{m,n}\right)^2 = nV U_{m,n} - 2nCov\left(U_{m,n}, U^*_{m,n}\right) + nV U^*_{m,n}$$
$$= nV U_{m,n} - nV U^*_{m,n}$$
$$= \sum_{c=1}^{m} \frac{n}{\binom{n}{m}}\binom{m}{c}\binom{n-m}{m-c}\xi_c - m^2\xi_1.$$

Writing out the binomials and simplifying give for a generic term of the sum

$$c! \binom{m}{c}^2 \left[\frac{(n-m)(n-m-1) \cdots (n-2m+c+1)}{(n-1)(n-2) \cdots (n-m+1)} \right] \xi_c$$

$$= c! \binom{m}{c}^2 \frac{n^{m-c}[1+o(1)]}{n^{m-1}[1+o(1)]} \xi_c$$

$$= c! \binom{m}{c}^2 \left[n^{1-c} + o(1) \right] \xi_c.$$

For $c > 1$ this vanishes as $n \to \infty$, and for $c = 1$ it approaches $m^2 \xi_1$. Thus,

$$\lim_{n \to \infty} E \left(U_{m,n} - U_{m,n}^* \right)^2 = m^2 \xi_1 - m^2 \xi_1 = 0. \qquad \square$$

Example 6.35. With $K(X_1, X_2) = (X_1 - X_2)^2 / 2$ and $U_{2,n} = S^2$ we have

$$\sqrt{n} \left(S^2 - \sigma^2 \right) \rightsquigarrow N \left(0, 2^2 \cdot \frac{\mu_4 - \sigma^4}{4} \right) = N \left(0, \mu_4 - \sigma^4 \right).$$

If in addition $\{X_j\}_{j=1}^n$ are i.i.d. as $N(\mu, \sigma^2)$, this becomes

$$\sqrt{n} \left(S^2 - \sigma^2 \right) \rightsquigarrow N \left(0, 2\sigma^4 \right).$$

Exercise 6.23. $\{X_j\}_{j=1}^n$ *are i.i.d. as* $N(\mu, \sigma^2)$. *Find the limiting distribution of*

$$\sqrt{n} \left(U_{2,n} - \kappa \right) := \sqrt{n} \left[\frac{1}{\binom{n}{2}} \sum_{i<j} e^{-(X_i - X_j)^2 / 2} - E e^{-(X_1 - X_2)^2 / 2} \right]$$

by the following steps:

(i) Show that $\kappa := E e^{-(X_1 - X_2)^2 / 2} = \left(1 + 2\sigma^2 \right)^{-1/2}$. *(Hint:* $(X_1 - X_2) / \sqrt{2} \sim N(0, \sigma^2)$.*)*

(ii) Show that

$$K_1(X_1) := E \left[e^{-(X_1 - X_2)^2 / 2} \mid X_1 \right] = \frac{1}{\sqrt{1 + \sigma^2}} \exp \left[-\frac{X_1^2}{2(1 + \sigma^2)} \right].$$

(iii) Show that

$$\xi_1 := Cov \left[e^{-(X_1 - X_2)^2 / 2}, e^{-(X_1 - X_3)^2 / 2} \right]$$

$$= \frac{1}{\sqrt{1 + 4\sigma^2 + 3\sigma^4}} - \frac{1}{1 + 2\sigma^2}.$$

(Hint: $\xi_1 = V K_1(X_1)$.*)*

Exercise 6.24. *For i.i.d. r.v.s $\{X_j\}_{j=1}^n$ put*

$$\mathfrak{G} := \frac{1}{\binom{n}{2}} \sum_{i<j} |X_i - X_j|,$$

which is a sample counterpart of Gini's mean difference, $\mathfrak{g} := E\,|X_1 - X_2|$. If $\{X_j\}_{j=1}^n$ are i.i.d. as $N\left(\mu, \sigma^2\right)$, find $V\mathfrak{G}$ and the asymptotic distribution of $\sqrt{n}\,(\mathfrak{G} - \mathfrak{g})$. (This builds on Exercise 4.52.)

6.9 Solutions to Exercises

1. Taking $g\left(t\right) = f^{(m+1)}\left(t\right)$ and $h\left(t\right) = \left(t - b\right)^m / m!$, the theorem implies that there exists $\hat{b}$ between b and x such that

$$\int_b^x \frac{(t - b)^m}{m!} f^{(m+1)}(t) \cdot dt = f^{(m+1)}\left(\hat{b}\right) \int_b^x \frac{(t-b)^m}{m!}$$

$$= \frac{f^{(m+1)}\left(\hat{b}\right)}{(m+1)!}(x - b)^{m+1}.$$

2. By De Moivre–Laplace $Z_n := \left(Y_n - n\theta\right)/\sqrt{n\theta\left(1 - \theta\right)} \rightsquigarrow N\left(0, 1\right)$ so for each $z \in \Re$ $F_n\left(z\right) = \Pr\left(Z_n \leq z\right) \to \Phi\left(z\right)$. Since $\left(x - n\theta\right)/\sqrt{n\theta\left(1 - \theta\right)} \to -\infty$ for each finite x, we have

$$\lim_{n\to\infty} \Pr\left(\frac{a - n\theta}{\sqrt{n\theta\left(1 - \theta\right)}} < Z_n < \frac{b - n\theta}{\sqrt{n\theta\left(1 - \theta\right)}}\right)$$

$$= \lim_{n\to\infty} F_n\left(\frac{b - n\theta}{\sqrt{n\theta\left(1 - \theta\right)}}\right) - F_n\left(\frac{a - n\theta}{\sqrt{n\theta\left(1 - \theta\right)}}-\right)$$

$$= \Phi\left(-\infty\right) - \Phi\left(-\infty\right) = 0.$$

3. $\phi_n(t) = Ee^{itX_n} = \phi_Y(t/n)^n = (1 - i\beta t/n)^{-\alpha n}$ converges as $n \to \infty$ to $e^{i\alpha\beta t}$, which is the c.f. of a degenerate r.v. equal to $\alpha\beta$ with probability one

4. $\phi_n(t) = Ee^{itX_n} = \phi_{Y_n}(t/n) = \exp\left[n\left(e^{it/n} - 1\right)\right]$. Expanding $e^{it/n}$ about $t = 0$ gives $\phi_n(t) = e^{n[it/n + o(1/n)]}$, which converges to $\phi_\infty(t) = e^{it}$, the c.f. of a degenerate r.v. equal to unity with probability one.

5. We have

$$\phi_n(t) = Ee^{it(Y_n - n)/\sqrt{n}}$$

$$= e^{-it\sqrt{n}} Ee^{it\sqrt{n}Y_n}$$

$$= e^{-it\sqrt{n}} \exp\left[n\left(e^{it/\sqrt{n}} - 1\right)\right].$$

Expanding $e^{it/\sqrt{n}}$ to order n^{-1} gives

$$\phi_n(t) = e^{-it\sqrt{n}}e^{n\left[it/\sqrt{n}-t^2/2n+o\left(n^{-1}\right)\right]} = e^{-t^2/2+o(1)},$$

which converges to $e^{-t^2/2}$. Thus, $X_n \rightsquigarrow N(0,1)$.

6. The c.d.f. of X_n is

$$F_n(x) = \left\{ \begin{array}{ll} 0, & x < 0 \\ [nx]/n, & 0 \le x < 1 \\ 1, & x \ge 1 \end{array} \right\} = \frac{[nx]}{n}\mathbf{1}_{[0,1)}(x) + \mathbf{1}_{[1,\infty)}(x),$$

where $[nx]$ is the greatest integer contained in nx. Since $0 \le x - [nx]/n < 1/n$ it follows that $x - [nx]/n \to 0$ as $n \to \infty$. Therefore,

$$F_n(x) \to F_\infty(x) := x\mathbf{1}_{[0,1)}(x) + \mathbf{1}_{[1,\infty)}(x),$$

which is the c.d.f. of a (continuous) r.v. distributed as $U(0,1)$ (uniform on $[0,1]$).

One can also find the limiting c.f., as follows. Multiplying $\phi_n(t) = n^{-1}\sum_{j=1}^{n} e^{itj/n}$ by $e^{it/(2n)} - e^{-it/(2n)}$ collapses the summation, as

$$\frac{1}{n}\left(-e^{-\frac{it}{2n}}+e^{\frac{it}{2n}}\right)\left(e^{\frac{it}{n}}+e^{\frac{2it}{n}}+e^{\frac{3it}{n}}+\cdots+e^{it}\right)$$

$$=\frac{1}{n}\left(-e^{\frac{it}{2n}}+e^{\frac{3it}{2n}}-e^{\frac{3it}{2n}}+e^{\frac{5it}{2n}}-e^{\frac{5it}{2n}}+-\cdots+e^{\frac{(2n+1)it}{2n}}\right)$$

$$=\frac{1}{n}\left(e^{\frac{(2n+1)it}{2n}}-e^{\frac{it}{2n}}\right).$$

Thus, for $t \ne 0$

$$\phi_n(t) = \frac{e^{\frac{(2n+1)it}{2n}}-e^{\frac{it}{2n}}}{n\left(e^{\frac{it}{2n}}-e^{-\frac{it}{2n}}\right)} = \frac{e^{\frac{(2n+1)it}{2n}}-e^{\frac{it}{2n}}}{n\left[\frac{it}{n}+o\left(n^{-1}\right)\right]}$$

$$\to \phi_\infty(t) := \frac{e^{it}-1}{it},$$

which is the c.f. of $U(0,1)$. Of course, $\phi_n(0) = 1$ for each n implies $\phi_\infty(0) = 1$, which corresponds with the above if defined at $t = 0$ by continuity.

7. To prove necessity write

$$\Pr(|X_n - c| \ge \varepsilon) = [F_n(c - \varepsilon)] + [1 - F_n(c + \varepsilon-)]$$

and note that $X_n \to^p c$ implies that $F_n(c-\varepsilon) \to 0$ and $F_n(c+\varepsilon-) \to 1$ for any $\varepsilon > 0$. For sufficiency, the fact that both bracketed terms are nonnegative means that each must converge to zero if $\Pr(|X_n - c| \ge \varepsilon)$ does. This being so for all $\varepsilon > 0$, it follows that F_n converges pointwise to $F(x) = \mathbf{1}_{[c,\infty)}$ except perhaps at $x = c$.

8. We have

$$\lim_{n\to\infty} \Pr\left(|X_n| \geq \varepsilon\right) = \lim_{n\to\infty} \Pr\left(Y \geq n\varepsilon\right)$$

$$= \lim_{n\to\infty} \min\left\{1, (n\varepsilon)^{-1}\right\} = 0.$$

9. Since

$$E\left(X_n - c\right)^2 = E\left[(X_n - \mu_n) + (\mu_n - c)\right]^2$$
$$= E\left(X_n - \mu_n\right)^2 + 2\left(\mu_n - c\right)E\left(X_n - \mu_n\right) + (\mu_n - c)^2$$
$$= \sigma_n^2 + (\mu_n - c)^2,$$

and since both terms are nonnegative, $E\left(X_n - c\right)^2 \to 0$ if and only if each term converges to zero.

10. Sequence $\{\cap_{m=n}^{\infty} S_m\left(\varepsilon\right)\}_{n=1}^{\infty}$ is monotone increasing, so

$$\lim_{n\to\infty} \mathbb{P}\left[\bigcap_{m=n}^{\infty} S_m\left(\varepsilon\right)\right] = \mathbb{P}\left[\lim_{n\to\infty} \bigcap_{m=n}^{\infty} S_m\left(\varepsilon\right)\right]$$

$$= \mathbb{P}\left[\bigcup_{n=1}^{\infty} \bigcap_{m=n}^{\infty} S_m\left(\varepsilon\right)\right]$$

$$= \mathbb{P}\left[\liminf_n S_n\left(\varepsilon\right)\right].$$

11. $\Pr\left(|X_n| \geq \varepsilon\right) = 1 - \Pr\left(|X_n| < \varepsilon\right) = 1 - \left[F_n\left(\varepsilon-\right) - F_n\left(-\varepsilon\right)\right]$. But $F_n\left(-\varepsilon\right) = 0$ and $F_n\left(\varepsilon-\right) = \max\left\{0, 1 - n^{-2}/\varepsilon\right\}$, so $\Pr\left(|X_n| \geq \varepsilon\right) \leq n^{-2}/\varepsilon$. Since $\sum_{n=1}^{\infty} n^{-2} = \zeta\left(2\right) = \pi^2/6 < \infty$, it follows that $X_n \to^{a.s.} 0$. $F_n\left(x\right)$ is continuous and differentiable except at n^{-2}, with p.d.f. $f_n\left(x\right) = \left(n^2 x^2\right)^{-1} \mathbf{1}_{(n^{-2}, \infty)}\left(x\right)$. Thus,

$$EX_n^2 = \int_{n^{-2}}^{\infty} \frac{x^2}{n^2 x^2} \cdot dx = +\infty$$

for each n.

12. Clearly, $\Pr\left(|X_n| \geq \varepsilon\right) = 0$ when $\varepsilon > 1$ and $\Pr\left(|X_n| \geq \varepsilon\right) = \Pr\left(X_n \geq \varepsilon\right) = n^{-1}$ for $0 < \varepsilon \leq 1$. Thus, $X_n \to^p 0$. However, by the divergence part of Borel–Cantelli, $\Pr\left(|X_n| \geq \varepsilon \text{ i.o.}\right) = 1$ when $\varepsilon \leq 1$, since $\sum_{n=1}^{\infty} n^{-1} = \infty$.

13. $\Pr\left(|X_n| > \varepsilon\right) = \Pr\left(|Y_n| > n\varepsilon\right) = \Pr\left(|Y_1| > n\varepsilon\right) \to 0$ as $n \to \infty$, so $X_n \to^p 0$ as claimed. However, applying (3.18) shows that

$$\frac{E|Y_1|}{\varepsilon} - 1 \leq \sum_{n=1}^{\infty} \Pr\left(|X_n| > \varepsilon\right) \leq \frac{E|Y_1|}{\varepsilon}.$$

The divergence part of Borel–Cantelli and the first inequality show that $E|Y_1| < \infty$ is necessary for $X_n \to^{a.s.} 0$, and the convergence part of B-C and the second inequality show that $E|Y_1| < \infty$ is sufficient.

14. $EX_n^2 = \int_{-\infty}^{\infty} x^2 \cdot dF_n(x) = \int_{n^{-1/3}}^{\infty} 3n^{-1}/x^2 \cdot dx = 3n^{-2/3} \to 0$, so $X_n \to^{m.s.} 0$. However, since $\Pr(|X_n| \geq \varepsilon) = n^{-1}\varepsilon^{-3}$ when $n \geq \varepsilon^{-3}$ and $\sum_{n=1}^{\infty} n^{-1} = \infty$, the divergence part of Borel–Cantelli implies $\Pr(|X_n| \geq \varepsilon \text{ i.o.}) = 1$, so $X_n \not\to^{a.s.} 0$.

15. This follows trivially from Chebychev and (6.12).

16. $G_n = \exp\left(n^{-1} \sum_{j=1}^{n} \log X_j\right) \to^{a.s.} e^{\log \gamma} = \gamma$ by the strong law of large numbers. Therefore, convergence is also i.p.

17. It does not so converge, since

$$EX_n^2 = n^{-2} \sum_{j,k=1}^{n} EX_j X_k = n^{-2} \left(\sum_{j=1}^{n} EX_j^2 + \sum_{j \neq k} EX_j X_k\right)$$

$$= n^{-2} \left[n + (n^2 - n)\rho\right] = n^{-1}(1-\rho) + \rho \to \rho.$$

18. With $\phi_n(t) = n^{-1} \sum_{j=1}^{n} e^{itX_j}$, we have $\phi_n'(t) = in^{-1} \sum_{j=1}^{n} X_j e^{itX_j}$ and $\phi_n''(t) = -n^{-1} \sum_{j=1}^{n} X_j^2 e^{itX_j}$, so that

$$-\left.\frac{d^2 \log \phi_n(t)}{dt^2}\right|_{t=0} = -\frac{\phi_n''(0) - \phi_n'(0)^2}{\phi_n(0)^2}$$

$$= M_2 - \bar{X}^2 \to^{a.s.} \sigma^2.$$

19. $\sigma_j^2 = VX_j = \theta_j - \theta_j^2$ and $EX_j^{2+\delta} = \theta_j$ for any $\delta > 0$. Since $0 < \varepsilon \leq \theta_j \leq 1 - \varepsilon$ implies $\sigma_j^2 \geq \varepsilon(1-\varepsilon)$, and since $\theta_j \leq 1 - \varepsilon$, it follows that

$$0 \leq \lim_{n \to \infty} \frac{\sum_{j=1}^{n} EX_j^{2+\delta}}{\left(\sum_{j=1}^{n} \sigma_j^2\right)^{1+\delta/2}} \leq \lim_{n \to \infty} \frac{n(1-\varepsilon)}{n^{1+\delta/2}\left[\varepsilon(1-\varepsilon)\right]^{1+\delta/2}} = 0,$$

verifying that the condition for Lyapounov does hold. Thus, $(S_n - ES_n)/\sqrt{VS_n} \rightsquigarrow N(0,1)$, so that the distribution of S_n is approximately normal with mean $ES_n = \sum_{j=1}^{n} \theta_j$ and variance $VS_n = \sum_{j=1}^{n} \theta_j(1 - \theta_j)$ when n is large.

20. For $k < \nu/2$, the $2k$th central moment of the Student distribution with ν d.f. is

$$\mu_{2k} = \frac{\nu^k \Gamma(\nu/2 - k) \Gamma(k + 1/2)}{\Gamma(1/2) \Gamma(\nu/2)}.$$

The same expression gives the mth *absolute* central moment, $\nu_m := E|T|^m$, upon setting $k = m/2$. With $\nu = 4$ and $k = 3/2$ we have

$$\nu_3 = \frac{4^{3/2} \Gamma(1/2) \Gamma(2)}{\Gamma(1/2) \Gamma(2)} = 2^3,$$

while the standard deviation of $T \sim t(4)$ is

$$\sigma = \sqrt{\mu_2} = \sqrt{\frac{4\Gamma(1)\Gamma(3/2)}{\Gamma(1/2)\Gamma(2)}} = \sqrt{2}.$$

Thus, $\nu_3/\sigma^3 = 2^{3/2} = \sqrt{8}$ and $|F_n(z) - \Phi(z)| \leq C\sqrt{8/n}$.

21. Write $\bar{D}_{n_X, n_Y}$ as

$$\bar{D}_{n_X, n_Y} = \left(\frac{\bar{X} - \mu_X}{\sqrt{S_X^2/n_X}}\right) \frac{\sqrt{S_X^2/n_X}}{\sqrt{S_X^2/n_X + S_Y^2/n_Y}}$$
$$- \left(\frac{\bar{Y} - \mu_Y}{\sqrt{S_Y^2/n_Y}}\right) \frac{\sqrt{S_Y^2/n_Y}}{\sqrt{S_X^2/n_X + S_Y^2/n_Y}}.$$

Condition $\lim_{n_X \to \infty} n_Y/n_X = \nu > 0$ implies that $n_Y \to \infty$ as $n_X \to \infty$. Lindeberg–Lévy and Mann–Wald therefore imply that the two expressions in parentheses each converge in distribution to $N(0,1)$ as $n_X \to \infty$. Since the expressions are independent, there exist independent standard normal r.v.s Z_X and Z_Y to which they converge in probability. Now $S_X^2 \to^{a.s.} \sigma_X^2$ as $n_X \to \infty$ and $S_Y^2 \to^{a.s.} \sigma_Y^2$ as $n_Y \to \infty$, so as $n_X \to \infty$ we have

$$\frac{\sqrt{S_X^2/n_X}}{\sqrt{S_X^2/n_X + S_Y^2/n_Y}} = \frac{\sqrt{S_X^2 n_Y/n_X}}{\sqrt{S_X^2 n_Y/n_X + S_Y^2}} \to^{a.s.} \sqrt{\frac{\sigma_X^2 \nu}{\sigma_X^2 \nu + \sigma_Y^2}}$$

$$\frac{\sqrt{S_Y^2/n_Y}}{\sqrt{S_X^2/n_X + S_Y^2/n_Y}} \to^{a.s.} \sqrt{\frac{\sigma_Y^2/\nu}{\sigma_X^2 + \sigma_Y^2/\nu}} = \sqrt{\frac{\sigma_Y^2}{\sigma_X^2 \nu + \sigma_Y^2}}$$

and

$$\bar{D}_{n_X, n_Y} \to^p Z_X \sqrt{\frac{\sigma_X^2 \nu}{\sigma_X^2 \nu + \sigma_Y^2}} + Z_Y \sqrt{\frac{\sigma_Y^2}{\sigma_X^2 \nu + \sigma_Y^2}} \sim N(0,1).$$

22. The problem simplifies if we note two things. The first is that the sample correlation coefficient is affine invariant. That is, transforming each X_j as $X_j' = a + bX_j$ and each Y_j as $Y_j' = c + dY_j$, with $bd > 0$ leaves R_{XY} unchanged. The specific choice $a = -\mu_X/\sigma_X$, $b = 1/\sigma_X$, $c = -\mu_Y/\sigma_Y$, $d = 1/\sigma_Y$ lets us work with a function of standard normals with no loss of generality. The second thing to note is that when the population means are zero the components of R_{XY} that are products and squares of sample means are asymptotically negligible relative to the remaining terms. (To see this, suppose we have a differentiable

function $g\left(\bar{X}^2\right)$ of $\bar{X}^2$ and that $\mu = 0$, $\sigma^2 = 1$. Then applying the delta method and differentiating with respect to $\bar{X}^2$ give

$$\sqrt{n}\left[g\left(\bar{X}^2\right) - g\left(0\right)\right] = \sqrt{n}g'\left(\mu_n^2\right)\bar{X}^2 = g'\left(\mu_n^2\right)\bar{X}\left(\sqrt{n}\bar{X}\right).$$

Since $g'\left(\mu_n^2\right) \to g'\left(0\right)$, $\sqrt{n}\bar{X} \rightsquigarrow N\left(0,1\right)$ and $\bar{X} \to^{a.s.} 0$, the right side vanishes, leaving $g(0)$ as the $O\left(n^{-1/2}\right)$ approximation to $g\left(\bar{X}^2\right)$. Alternatively, applying the delta method by differentiating with respect to $\bar{X}$ gives

$$\sqrt{n}\left[g\left(\bar{X}^2\right) - g\left(0\right)\right] = g'\left(\mu_n^2\right)\left(2\mu_n\right)\sqrt{n}\bar{X},$$

which leads to the same conclusion.) This implies that the sample means can just be set to zero, since they enter R_{XY} only as squares or as products. Thus, letting M_{XX}, M_{YY}, and M_{XY} be the second sample moments about the origin (e.g., $M_{XY} = n^{-1}\sum_{j=1}^{n} X_j Y_j$), the sample correlation is approximated to $O\left(n^{-1/2}\right)$ as

$$R_{XY} = g\left(M_{XY}, M_{XX}, M_{YY}, \bar{X}, \bar{Y}\right)$$
$$\doteq g\left(M_{XY}, M_{XX}, M_{YY}, 0, 0\right)$$
$$= M_{XY}M_{XX}^{-1/2}M_{YY}^{-1/2}.$$

Applying the delta method and expanding using the multivariate version of Taylor's formula give for $\sqrt{n}R_{XY}$

$$\frac{1}{\sigma_{X,n}\sigma_{Y,n}}\sqrt{n}M_{XY} - \frac{\sigma_{XY,n}}{\left(\sigma_{X,n}\right)^3 \sigma_{Y,n}}\sqrt{n}\left(M_{XX} - 1\right)$$
$$- \frac{\sigma_{XY,n}}{\sigma_{X,n}\left(\sigma_{Y,n}\right)^3}\sqrt{n}\left(M_{YY} - 1\right),$$

where $\left(\sigma_{XY,n}, \sigma_{XX,n}, \sigma_{YY,n}\right)$ is somewhere on a line from $\left(\sigma_{XY}, \sigma_{XX}, \sigma_{YY}\right) = \left(0, 1, 1\right)$ to $\left(M_{XY}, M_{XX}, M_{YY}\right)$, and where we use the facts $\sigma_{XY} = 0$ and $\sigma_{XX} = \sigma_{YY} = 1$. Since the sample moments converge a.s. to the population moments, so do the n-subscripted σ's. In particular, $\sigma_{XY,n} \to 0$, and $\sigma_{X,n}$ and $\sigma_{Y,n}$ converge to unity. Thus, $\sqrt{n}R_{XY}$ has the same limiting distribution as

$$\sqrt{n}M_{XY} = \frac{1}{\sqrt{n}}\sum_{j=1}^{n} X_j Y_j.$$

The terms of the sum being i.i.d. with variance $E\left(X^2Y^2\right) = EX^2 \cdot EY^2 = 1$, the conclusion $\sqrt{n}R_{XY} \rightsquigarrow N\left(0,1\right)$ follows from Lindeberg–Lévy.

23. (i) The hint implies that $\left(X_1 - X_2\right)^2/2 \sim \Gamma(\frac{1}{2}, 2\sigma^2)$, so $Ee^{-(X_1-X_2)^2/2} = \left(1+2\sigma^2\right)^{-1/2}$.

(ii)

$$K_1\left(X_1\right) = \int \frac{1}{\sqrt{2\pi\sigma^2}} \exp\left[-\frac{1}{2}\left(y-X_1\right)^2 - \frac{y^2}{2\sigma^2}\right] \cdot dy.$$

Complete the square in the exponent and integrate out y to get

$$K_1\left(X_1\right) = \frac{1}{\sqrt{1+\sigma^2}} \exp\left[-\frac{X_1^2}{2\left(1+\sigma^2\right)}\right].$$

(iii)

$$\begin{aligned}
\xi_1 &= EK_1\left(X_1\right)^2 - \kappa^2 \\
&= \frac{1}{1+\sigma^2}E\exp\left[-X_1^2/\left(1+\sigma^2\right)\right] - \frac{1}{1+2\sigma^2} \\
&= \frac{1}{1+\sigma^2}\left(1+\frac{2\sigma^2}{1+\sigma^2}\right)^{-1/2} - \frac{1}{1+2\sigma^2} \\
&= \frac{1}{\sqrt{1+4\sigma^2+3\sigma^4}} - \frac{1}{1+2\sigma^2}.
\end{aligned}$$

Theorem 6.20 then implies that $\sqrt{n}\left(U_{2,n} - \kappa\right) \rightsquigarrow N\left(0, 4\xi_1\right)$.

24. Clearly, $E\mathfrak{G} = \mathfrak{g}$, $V\mathfrak{G} = \binom{n}{2}^{-1}\sum_{c=1}^2 \binom{2}{c}\binom{n-2}{2-c}\xi_c$, and $\sqrt{n}\left(\mathfrak{G} - \mathfrak{g}\right) \rightsquigarrow N\left(0, 2^2\xi_1\right)$ by Theorem 6.20. Exercise 4.52 showed that $\mathfrak{g} = 2\sigma/\sqrt{\pi}$, and so $\xi_2 = E\left(X_1 - X_2\right)^2 - \mathfrak{g}^2 = 2\sigma^2\left(1 - 2/\pi\right)$. The only remaining task is to find $\xi_1 \equiv \sigma^2 E\left|X - Y\right|\left|X - Z\right| - \mathfrak{g}^2$ when X, Y, Z are i.i.d. as $N\left(0,1\right)$.

With Φ as the standard normal c.d.f., we have

$$\begin{aligned}
E\left|X - Y\right|\left|X - Z\right| &= \int\int\int \left|x - y\right|\left|x - z\right| \cdot d\Phi\left(z\right)\, d\Phi\left(y\right)\, d\Phi\left(x\right) \\
&= \int\left[\int \left|x - y\right| \cdot d\Phi\left(y\right)\right]\left[\int \left|x - z\right| \cdot d\Phi\left(z\right)\right] \cdot d\Phi\left(x\right) \\
&= \int\left[\int \left|x - y\right| \cdot d\Phi\left(y\right)\right]^2 \cdot d\Phi\left(x\right).
\end{aligned}$$

But

$$\int_{-\infty}^{\infty} |x - y| \cdot d\Phi(y) = \int_{-\infty}^{x} (x - y) \cdot d\Phi(y) - \int_{x}^{\infty} (x - y) \cdot d\Phi(y)$$

$$= 2 \int_{-\infty}^{x} (x - y) \cdot d\Phi(y) - \int_{-\infty}^{\infty} (x - y) \cdot d\Phi(y)$$

$$= 2 \int_{-\infty}^{x} (x - y) \cdot d\Phi(y) - x$$

$$= 2 \left[x\Phi(x) - \int_{-\infty}^{x} y \cdot d\Phi(y) \right] - x$$

$$= 2 \left[x\Phi(x) + f_Z(x) \right] - x,$$

where the last step follows from $-\int_{-\infty}^{x} y \cdot d\Phi(y) = -\int_{-\infty}^{x} y f_Z(y) \cdot dy = \int_{-\infty}^{x} df_Z(y)$ with f_Z as the $N(0,1)$ p.d.f. Thus,

$$E|X - Y||X - Z| = E\{2[X\Phi(X) + f_Z(X)] - X\}^2.$$

Squaring leads to terms involving expected values of $X^2\Phi(X)^2$, $X\Phi(X)f_Z(X)$, $f_Z(X)^2$, $X^2\Phi(X)$, and $Xf_Z(X)$. These can be found by applying Stein's lemma (page 282) and the facts that $E\Phi(X)^{j-1} = j^{-1}$ and $Ef_Z(X)^j = (2\pi)^{-j/2} Ee^{-jX^2/2} = (2\pi)^{-j/2}(1+j)^{-1/2}$ for $j \in \aleph$ and $X \sim N(0,1)$. Assembling the results gives

$$E|X - Y||X - Z| = \frac{1}{3} + \frac{2\sqrt{3}}{\pi}$$

and

$$\xi_1 = \sigma^2 \left[\frac{1}{3} - \frac{2}{\pi}\left(2 - \sqrt{3}\right) \right].$$

Chapter 7

Estimation

Chapters 5 and 6 have shown how to determine exact and approximate distributions of various sample statistics, such as $\bar{X}$ and S^2, and also of various functions of statistics and parameters, such as $\sqrt{n}\left(\bar{X} - \mu\right)/S$. We now begin to apply this knowledge of sampling distributions to the task of drawing intelligent inferences about populations. This and the next two chapters present techniques for statistical inference within the "frequentist" paradigm, which bases inferences on how functions of sample data would behave in hypothetical meta-experiments involving indefinitely many repeated samples. Inference within the alternative Bayesian approach to statistics is described in Chapter 10. The specific goal of the present chapter is to see how to use sample information efficiently to deduce good estimates of various quantities that describe the population. There are two general settings in which this takes place.

In the first setting a parametric family of models has been adopted to represent the distribution of the population—and therefore the marginal distribution of a generic element X of the random sample. For example, based on theory, past experience, or statistical procedures one may have decided to model the distribution of X as a member of the normal family, or gamma family, or Poisson family, etc. The statistical problem is then to determine the specific member of the family that characterizes the data at hand. Thus, for the normal family one would need to estimate the mean and variance, μ and σ^2; for the gamma family, the shape and scale parameters, α and β; for the Poisson family, the "intensity" parameter, λ; and so forth. More generally, for a parametric family $\{F(x;\theta) : \theta \in \Theta\}$, where $\Theta \subset \Re^k$ is the parameter *space*, one would want to estimate θ. A still more general goal is to estimate the value of some measurable function, $\tau(\theta)$.

443

The second setting for estimation problems is not so highly structured. One wishes to infer certain general features of the data-generating process without assuming a specific parametric family. For example, one might want to estimate one or more population moments—μ, σ^2, $\{\mu'_j\}_{j=2}^m$, etc.—or selected values of the c.d.f., $\{F(x_j)\}_{j=1}^m$. Each of these particular features of interest is the mathematical expectation of some function of X; namely, X^j or $(X - \mu)^j$ for moments and $\mathbf{1}_{(-\infty, x_j]}(X)$ for the c.d.f. Alternatively, one might want to estimate a measurable function of such quantities, such as the coefficient of variation, σ/μ, or the distribution's p quantile, $F^{-1}(p)$, or even the value of the p.d.f. at one or more points, $f(x_j) = F'(x_j) \doteq [F(x_j + \delta/2) - F(x_j - \delta/2)]/\delta$. By adopting some parametric family $\{F(x; \boldsymbol{\theta}) : \boldsymbol{\theta} \in \boldsymbol{\Theta}\}$ *a priori*, these various characteristics of the population could all be expressed as functions of $\boldsymbol{\theta}$, but the quality of the resulting estimates could depend critically on having chosen the right family of models. While inference that does not depend on parametric assumptions is more robust in this sense, it is still usually the case that *some* unverifiable thing has to be assumed; e.g., the finiteness of population moments or the continuity of the c.d.f. or p.d.f.

In both the structured and unstructured settings the process of obtaining numerical estimates of population characteristics is referred to as *point estimation*. The theory and specific techniques are described in the next two sections of the chapter. As we shall emphasize, point estimates are sample realizations of specialized statistics called *estimators*; hence, they are realizations of random variables. For example, by calculating the mean of sample data as a point estimate of the population mean one obtains the value $\bar{x} = n^{-1} \sum_{j=1}^n x_j = n^{-1} \sum_{j=1}^n X_j(\omega) = \bar{X}(\omega)$ corresponding to the observed outcome of the sampling experiment. Were the sampling experiment to be repeated—this being a step in the "meta-experiment" referred to above—a different realization of $\{X_j\}_{j=1}^n$ and a different $\bar{x}$ would likely be obtained. A particular sample might, by chance, be highly unrepresentative of the population and produce a realization $\bar{x}$ that differed greatly from the true μ. Recognizing this inherent variability of point estimators, one typically wants a sense of how close the point estimate is apt to be to the unknown population value—i.e., how much confidence to place in the estimate. The usual way to ascertain this is to build a *confidence interval* for the population value. The final section of the chapter discusses this process of *interval estimation*.

7.1 Criteria for Point Estimators

As just described, a point *estimator* is a sample statistic—some function $T(\mathbf{X})$ of the random sample—that is used to produce an *estimate* of an unknown population quantity. Of course, infinitely many such functions could be considered as possible estimators of a given population characteristic. The first task in studying statistical estimation is to develop criteria by which alternative estimators can be judged. Since they are r.v.s, the only way we can compare estimators is in terms of their probability distributions. Broadly speaking, it is clear that a good estimator should have a distribution that is highly concentrated about the value of the unknown population quantity. Of course, the distributions of potential estimators typically depend on the sample size—a fact that we emphasize by attaching subscripts, as T_n. Comparisons of estimators can be based on their distributions for arbitrary, finite n when these can be worked out, and also on their asymptotic properties as $n \to \infty$. We consider finite-sample and asymptotic criteria separately. Of these, we shall see that the latter are usually more compelling.

In what follows θ is a scalar that represents some generic, unobservable characteristic of the population about which an inference is to be drawn. In the unstructured setting in which no parametric family is assumed θ could stand for one of the features previously discussed, such as a moment, a function of moments, or a quantile. In the parametric setting we usually take θ—or $\boldsymbol{\theta}$ in the multiparameter case—to represent the quantity that defines a specific member of the parametric family, $\{F(x;\theta) : \theta \in \Theta\}$ or $\{F(x;\boldsymbol{\theta}) : \boldsymbol{\theta} \in \boldsymbol{\Theta}\}$. $\tau(\boldsymbol{\theta})$ would then represent some measurable function of the parameter(s). In both settings we often refer to θ—loosely or strictly, according to the setting—as a "parameter". Symbols $T_n = T_n(\mathbf{X}), T_n^* = T_n^*(\mathbf{X}), T_n' = T_n'(\mathbf{X}), T_n'' = T_n''(\mathbf{X}), \dots$ represent various members of a set $\mathcal{T}_n$ of feasible estimators of θ from sample $\mathbf{X} := (X_1, X_2, \dots, X_n)'$, and symbols $t_n = T_n(\mathbf{x}), t_n^* = T_n^*(\mathbf{x}), t_n' = T_n'(\mathbf{x}), t_n'' = T_n''(\mathbf{x}), \dots$ represent their generic realizations.

7.1.1 *Finite-Sample Properties*

In estimating some population characteristic θ one may have in mind a specific course of action that depends on the estimate t_n. For example, a pharmaceutical company may decide whether to produce a certain drug based on patients' average response to the medication in a clinical trial;

or an investor may decide whether to invest in a corporation's long-term debt based on an estimate of the probability of default. In such practical applications it may be possible to specify a *loss function* $\mathfrak{L}(t_n, \theta)$ that assigns to a particular t_n, θ pair a measure of the cost associated with the estimation error. For the drug company or investor $\mathfrak{L}(t_n, \theta)$ would represent the cost of acting under the belief that t_n was the population characteristic when the value was actually θ. Any reasonable loss function would have the property that $\mathfrak{L}(\theta, \theta) = 0$, but otherwise the functional form would be specific to the particular situation and to the particular entity that bears the loss. Given a specification of $\mathfrak{L}$, a plausible sounding strategy would be to choose among alternative estimators the one that minimizes the *expected* loss $E\mathfrak{L}(T_n, \theta)$—and that does so whatever the true, unknown value of θ happens to be. In symbols, the goal would be to find the T_n that solves

$$\min_{T_n \in \mathcal{T}_n} \{E\mathfrak{L}(T_n, \theta)\}, \text{ each } \theta \in \Theta. \tag{7.1}$$

This expected loss, a mapping from Θ into $\mathfrak{R}$, is often called the *risk function*. We interpret it as the average loss if realizations t_n of T_n were observed in indefinitely many repeated samples.

Unfortunately, there are three objections to this minimum-risk approach as a general guide to choosing estimators. The most serious problem is that there simply is no solution to (7.1) unless the choice set $\mathcal{T}_n$ is appropriately restricted. To see this, suppose $\mathcal{T}_n$ is allowed to include degenerate statistics—constants that do not depend at all on the sample realization. Then the choice $T_n = \theta_0 \in \Theta$ produces loss $\mathfrak{L}(\theta_0, \theta)$ for sure, which is therefore the expected loss. While this could be very large for most $\theta \in \Theta$, it would be zero should θ_0 happen to be the true value of θ. Clearly, this is a result that no nondegenerate statistic could match. Only by excluding such degenerate statistics from $\mathcal{T}_n$ could there be an estimator that minimizes expected loss for *each* $\theta \in \Theta$.

A second difficulty is that in many circumstances it is hard to know how to attach a specific loss to a given t_n, θ pair. For example, how would a scientist attempting to measure the distance to a celestial object from independent observations quantify the cost of a particular estimation error? The essential arbitrariness of such a specification should be apparent.

This leads to the final point. Estimates based on the expected-loss principle are necessarily determined subjectively, in that they depend on the particular loss function that the analyst has adopted. Different researchers with different loss functions might come to very different conclusions and might remain skeptical of findings based on others' subjective criteria.[1]

7.1.1.1 *The Mean Squared Error Criterion*

The last two points suggest adopting a universal class of loss functions. A plausible suggestion is to restrict $\mathfrak{L}(T_n, \theta)$ to quadratic functions of the absolute estimation error $|T_n - \theta|$, such as

$$\mathfrak{L}(T_n, \theta) = |T_n - \theta| + \eta (T_n - \theta)^2.$$

Here η would be a "taste" parameter that, depending on sign, could allow loss to grow more or less than in proportion to the absolute size of error. Attaching an arbitrary scale factor, such a specification could serve as a second-order Taylor-series approximation to more general smooth functions of $|T_n - \theta|$. However, there are still two objections. While $E(T_n - \theta)^2$ can be expressed in terms of θ and the moments of T_n only, evaluating $E|T_n - \theta|$ usually requires knowing the exact distribution of T_n, which rules out applications in which a parametric family for the data has not been assumed. The second problem is that the presence of the taste parameter still leaves an element of subjectivity, in that the costs attached to errors would be case- and researcher-dependent.

Both of these problems are eliminated if the $|T_n - \theta|$ term is dropped and η is restricted just to be positive. In that case all with loss functions in this restricted class would choose among some class of estimators the one that minimizes the *mean squared error (m.s.e.)*. That is, they would solve

$$\min_{T_n \in \mathcal{T}_n'} \left\{ E(T_n - \theta)^2 \right\}, \text{ each } \theta \in \Theta,$$

where $\mathcal{T}_n'$ is a restricted class of estimators (if such exists) for which a solution can be found.

To see in what situations such solutions do exist, it is helpful to decompose the m.s.e. as

$$E(T_n - \theta)^2 = VT_n + (ET_n - \theta)^2. \tag{7.2}$$

Here, the second term represents the square of the *bias* of T_n, the difference between its expected value and the value of the quantity being estimated. In other words, an estimator's bias is the average of the (signed) estimation errors that would be generated in indefinitely many replications of the sampling experiment. Identity (7.2) shows that an estimator that minimizes

---

[1] This, of course, is the frequentist statistician's standard objection to Bayesian methods, in which such subjective considerations figure prominently.

mean squared error has the smallest possible sum of variance and squared bias.

Exercise 7.1. *Verify the identity (7.2).*

Example 7.1. For parametric model $f(x; \theta) = \theta^{-1} 1_{[0,\theta]}(x), \theta > 0$, consider the class of estimators $T'_n = \{\varphi \bar{X} : \varphi > 0\}$. Since $E(\varphi \bar{X}) = \varphi \theta / 2$ and $V(\varphi \bar{X}) = \varphi^2 \theta^2 / (12n)$,

$$E(\varphi \bar{X} - \theta)^2 = \left(\frac{\varphi}{2} - 1\right)^2 \theta^2 + \frac{\varphi^2 \theta^2}{12n}$$

$$= \theta^2 \left[\left(\frac{\varphi}{2} - 1\right)^2 + \frac{\varphi^2}{12n}\right].$$

The minimum for all $\theta > 0$ is at $\varphi = 6n/(3n+1)$. Thus, estimator $T_n := 6 \sum_{j=1}^{n} X_j / (3n+1)$ is the minimum-m.s.e. estimator in the class T'_n.

Exercise 7.2. *For parametric model $X \sim N(\mu, \sigma^2), \mu \in \Re, \sigma^2 > 0$, find the minimum-m.s.e. estimator of σ^2 in the class $T'_n = \{\varphi S^2 : \varphi > 0\}$, where S^2 is the sample variance. (Hint: Use the result of Exercise 5.12.)*

The example and exercise show that there do exist estimation problems and classes of estimators—albeit narrow classes—for which the minimum-m.s.e. criterion is feasible. However, it is more typical that the function that minimizes m.s.e. turns out to depend on θ itself and/or on other unobserved characteristics of the population. Such functions of sample data and unobservables are not *statistics*, and in such cases the minimum-m.s.e. criterion is infeasible.

Example 7.2. For parametric model $X \sim \Gamma(\alpha, \beta), \alpha > 0, \beta > 0$, consider the class $T'_n = \{\varphi \bar{X} : \varphi > 0\}$ of estimators of $\mu = \alpha\beta$. The bias of $T_n \in T'_n$ is $\varphi E \bar{X} - \mu = (\varphi - 1)\mu$, the variance is $VT_n = \varphi^2 \alpha \beta^2 / n = \varphi^2 \mu \beta / n$, and the m.s.e. is

$$E(\varphi \bar{X} - \mu)^2 = (\varphi - 1)^2 \mu^2 + \frac{\varphi^2 \mu \beta}{n}.$$

The minimizing value, $\varphi = \alpha/(\alpha + 1/n)$, depends on the unknown α, so the class T'_n contains no feasible minimum-m.s.e. estimator.

7.1.1.2   *Minimum-Variance Unbiased Estimators*

The m.s.e. criterion often does become feasible if class T_n is restricted to *uniformly unbiased* estimators; that is, to estimators T_n such that $ET_n = \theta$ for all $\theta \in \Theta$. Since the m.s.e. is then just the variance, the problem reduces to finding a member of the unbiased class that has the least possible variance. As we shall see below, this program can sometimes be carried out in parametric settings to obtain "globally efficient" or "minimum-variance unbiased" (m.v.u.) estimators. In a more restricted form it is feasible in broader settings when θ represents the mathematical expectation of some function $g(X)$ with finite second moment.

A further restriction for unstructured applications is that T_n comprise just *linear* unbiased estimators—i.e., estimators $T_n = \sum_{j=1}^{n} a_j g(X_j)$ that are linear functions of the quantities whose expected values are being estimated. For such a statistic to be unbiased it is necessary and sufficient that $\sum_{j=1}^{n} a_j = 1$. Then, if $Vg(X) = \gamma^2$, say, we have

$$VT_n = \sum_{j=1}^{n} a_j^2 Vg(X_j) = \gamma^2 \sum_{j=1}^{n} a_j^2. \tag{7.3}$$

This attains a minimum subject to $\sum_{j=1}^{n} a_j = 1$ when $a_j = n^{-1}$ for each j, in which case $VT_n = \gamma^2/n$. Thus, $T_n = \bar{g} = n^{-1} \sum_{j=1}^{n} g(X_j)$ is the minimum-variance estimator in this linear, unbiased class. As such, it is often referred to as the "best" linear unbiased estimator, or the "b.l.u.e."[2] Note that for $n^{-1} \sum_{j=1}^{n} g(X_j)$ to be b.l.u., it is not *necessary* that $\{X_j\}_{j=1}^{n}$ be i.i.d.; it is enough that all the terms have the same mean and (finite) variance and that they be uncorrelated. However, if $\{X_j\}_{j=1}^{n}$ are i.i.d. and the requisite moments are finite, it follows that all of $\bar{X}, M_2', ..., M_k', ..., n^{-1} \sum_{j=1}^{n} \mathbf{1}_B(X_j), n^{-1} \sum_{j=1}^{n} e^{itX_j}, ...$ are, respectively, the b.l.u.e.s of $\mu, \mu_2', ..., \mu_k', ..., \mathbb{P}_X(B), \phi(t), ... $.[3]

Exercise 7.3. *Show that* $\{a_j = n^{-1}\}_{j=1}^{n}$ *is the solution to* $\min_{\{a_j\}_{j=1}^{n}} \left\{ \sum_{j=1}^{n} a_j^2 \right\}$ *subject to* $\sum_{j=1}^{n} a_j = 1$.

An unbiased estimator whose variance is smaller than that of another unbiased estimator is said to be *relatively efficient*. To see why this is a sensible term, let us contrast the b.l.u.e. $T_n = n^{-1} \sum_{j=1}^{n} g(X_j)$ with the

---

[2]If T_n is restricted not to linear estimators but to statistics that are invariant under permutations of i.i.d. $\{X_j\}$ with $Eg(X_j)^2 < \infty$, then the linear form with $a_j = n^{-1}$ emerges naturally as the m.v.u. estimator. This arises from the fact that permutation-invariant statistics can be expressed as functions of the order statistics $\{X_{(j:n)}\}_{j=1}^{n}$, which are the (only) complete sufficient statistics for the universal family of *all* distributions. For discussion and references see Lehmann and Casella (1998, Sect. 2.4). (The concepts of completeness and sufficiency are discussed below beginning on p. 460.)

[3]$Vg(X)$ would be defined for complex-valued g as $Eg(X)\overline{g(X)} - Eg(X)\overline{Eg(X)}$, where $\bar{g}$ is the complex conjugate.

following rather strange looking alternative:

$$T_n' = 2n^{-1}g\left(X_1\right) + \frac{n-2}{n-1}n^{-1}\sum_{j=2}^{n}g\left(X_j\right).$$

Although T_n' gives the first observation added weight, it is easily seen to be unbiased. However, a short calculation shows the variance to be $VT_n' = \gamma^2/\left(n-1\right)$, as contrasted with $VT_n = \gamma^2/n$. Thus, (as it was specifically designed to do for this illustration) T_n' delivers estimates with precisely the same precision as T_{n-1}, effectively—and inefficiently—wasting one observation. The relative efficiency of T_n' vs. T_n can thus be expressed as the ratio of the variances: $VT_n/VT_n' = \left(n-1\right)/n$. Alternatively—and in this case *equivalently*—it could be expressed as the ratio of sample sizes needed to attain a given variance.

Exercise 7.4. *Consider two alternative estimators for the mean μ of a population with some positive and finite—but unknown—variance σ^2: (i) $\bar{X}_{n+1}$, the mean of a random sample of fixed size $n+1$ ($n \geq 1$), and (ii) $\bar{X}_{N_n+1}$, the mean of a random sample of* random *size, where $N_n \sim P\left(n\right)$ (Poisson). (Note that the* expected *sample sizes are the same.) Show that estimator $\bar{X}_{N_n+1}$ is unbiased, and determine its efficiency relative to $\bar{X}_{n+1}$.*

7.1.1.3 *"Global" Efficiency*

Although sample moments are efficient relative to other *linear* unbiased estimators of the population moments, there may be an estimator outside the linear class with even lower variance; however, there is no way of discovering whether this is so without confining models for the data to some specific class. In the parametric case, where θ represents either a parameter or some function of parameters (a moment, a quantile, the value of the c.d.f. at some x), it is sometimes possible to find the estimator with the least variance in the entire unbiased class. Such a minimum-variance unbiased estimator is said to be *globally efficient* or, to be brief, simply *efficient*. There are two ways to find efficient estimators. One is based on the *information inequality*; the other, on the concept of *sufficiency*. Each takes a bit of explaining.

The Information Inequality Let $\{f(\cdot\,;\theta) : \theta \in \Theta\}$ represent the parametric model for a r.v. X, where f is a density with respect to some dominating measure $\mathcal{M}$—typically either counting measure or Lebesgue measure—with $f\left(x;\theta\right) > 0$ on a support $\mathbb{X}$. Thus, for an integrable g and a r.v. X of any type we can express $Eg(X)$ in terms of the density as $\int g(x) f(x; \theta) \mathcal{M}(dx)$.[4] (The advantage of such a representation will become apparent as we proceed.) For now, take θ to be a scalar. Since we deal with random samples, the joint p.m.f./p.d.f. of $\mathbf{X} := (X_1, X_2, ..., X_n)'$ is $f(\mathbf{x}; \theta) = \prod_{j=1}^{n} f(x_j; \theta)$. The results to be presented apply not to all parametric families but to the so-called "regular" distributions.

Definition 7.1. The distribution $\mathbb{P}_X = \mathbb{P}_X(\cdot; \theta)$ of a r.v. X is *regular* if for each θ in the interior of Θ and any integrable function g (not depending on θ)

$$\frac{d}{d\theta} \int g(x) f(x; \theta) \mathcal{M}(dx) = \int g(x) \frac{df(x; \theta)}{d\theta} \mathcal{M}(dx). \qquad (7.4)$$

Thus, regularity implies that

$$\frac{d}{d\theta} Eg(X) = \int g(x) \frac{d\log f(x; \theta)}{d\theta} f(x; \theta) \mathcal{M}(dx) = E\left[g(X) \frac{d\log f(X; \theta)}{d\theta}\right].$$

Likewise, when random sample $\mathbf{X}$ is from a regular distribution and g : $\Re^n \to \Re$ is integrable, we also have (*via* Fubini's theorem)

$$\frac{dEg(\mathbf{X})}{d\theta} = E\left[g(\mathbf{X}) \frac{d\log f(\mathbf{X}; \theta)}{d\theta}\right], \qquad (7.5)$$

a result that will be used frequently in the sequel.

How does one determine in practice whether a distribution is regular? Condition (7.4) obviously requires density $f(x; \theta)$ to be a.e. differentiable, but more than that is needed. To see when the condition might fail, recall Leibnitz' formula for differentiating a definite integral with respect to a parameter that appears in (either or both) the limits of integration and the integrand:

$$\frac{d}{d\theta} \int_{\ell(\theta)}^{u(\theta)} h(x; \theta) \cdot dx = h(u; \theta) u'(\theta) - h(\ell; \theta) \ell'(\theta) + \int_{\ell(\theta)}^{u(\theta)} \frac{dh(x; \theta)}{d\theta} \cdot dx.$$

Applied here in the continuous case with $h(x; \theta) = g(x) f(x; \theta)$ and $\mathbb{X} = [\ell(\theta), u(\theta)]$, this is

$$\frac{d}{d\theta} \int_{\ell}^{u} g(x) f(x; \theta) \cdot dx = g(u) f(u; \theta) u' - g(\ell) f(\ell; \theta) \ell'$$

$$+ \int_{\ell}^{u} g(x) \frac{df(x; \theta)}{d\theta} \cdot dx.$$

---

[4]Recalling the discussion in Section 3.3, Lebesgue measure λ dominates when X is continuous, and counting measure N dominates when X is supported on a countable set. Since N and λ are mutually singular, taking $\mathcal{M} = N + \lambda$ gives a general density representation in discrete, continuous, and mixed cases alike, as $\int g(x) f(x; \theta) N d(x) + \int g(x) f(x; \theta) \lambda(dx) = \sum_{x \in \mathcal{D}} g(x) [F(x; \theta) - F(x-; \theta)] + \int g(x) f(x; \theta) \cdot dx = \int g(x) \cdot dF(x; \theta)$, where $\mathcal{D}$ is the countable set of atoms.

Condition (7.4) requires the first two terms to vanish for all integrable g. Unless $\ell'(\theta) = u'(\theta) = 0$, this imposes the impossible condition that $g(u)/g(\ell) = [f(\ell;\theta)\,\ell']/[f(u;\theta)\,u']$ for all integrable g. Thus, f cannot be regular if support $\mathbb{X}$ depends on θ. This is the requirement that is most apt to be violated by models used in applied work; for example, it is violated by uniform model $f(x;\theta) = \theta^{-1}\mathbf{1}_{[0,\theta]}(x)$ and by the location-shifted exponential, $f(x;\theta_1,\theta_2) = \theta_2^{-1}e^{-(x-\theta_1)/\theta_2}\mathbf{1}_{[\theta_1,\infty)}(x)$. The validity of Leibnitz' formula itself requires further conditions on h—which is to say, on $g \cdot f$— that *are* met by most standard models. These can be checked in each case by applying Theorem 3.2 (Lebesgue dominated convergence).[5]

Fortunately, no checks of regularity are needed for common models in the *exponential class*. As described in Section 4.1, the exponential class comprises models with densities of the form $f(x;\boldsymbol{\eta}) = a(\boldsymbol{\eta})\,c(x)\,e^{\boldsymbol{\eta}(\boldsymbol{\theta})'T(x)}$, where elements of $\boldsymbol{\eta}(\boldsymbol{\theta})$ are the distribution's "natural" parameters ($\boldsymbol{\eta}$ and $\boldsymbol{\theta}$ typically being of the same dimension) and $\mathbf{H} := \left\{\boldsymbol{\eta} : \int c(x)\,e^{\boldsymbol{\eta}'\mathbf{T}(x)}\mathcal{M}(dx) < \infty\right\}$ is the *natural parameter space*. Since for $p \in [0,1]$

$$\int c(x)\,e^{[p\boldsymbol{\eta}_1 + (1-p)\boldsymbol{\eta}_2]'\mathbf{T}(x)}\mathcal{M}(dx)$$

$$\leq \int c(x)\left[pe^{p\boldsymbol{\eta}_1'\mathbf{T}(x)} + (1-p)\,e^{p\boldsymbol{\eta}_2'\mathbf{T}(x)}\right]\mathcal{M}(dx) < \infty$$

whenever $\boldsymbol{\eta}_1, \boldsymbol{\eta}_2 \in \mathbf{H}$, the natural parameter space is a convex set (i.e., an *interval* when η is scalar-valued). Distributions in the exponential class are regular when (i) $\boldsymbol{\eta}$ is differentiable, (ii) $\{\boldsymbol{\eta}(\boldsymbol{\theta}) : \boldsymbol{\theta} \in \boldsymbol{\Theta}\} = \mathbf{H}$, and (iii) $\mathbf{H}$ is an open set (an open interval in the scalar case). All of the exponential-class models (or specified submodels) in Tables 4.1 and 4.2 (marked by asterisks) are regular except the inverse Gaussian, where (Exercise 4.68) $\mathbf{H}$ is not open. Nevertheless, even in the latter case, regularity holds for $\boldsymbol{\theta}$ such that $\boldsymbol{\eta}(\boldsymbol{\theta})$ is in the interior of $\mathbf{H}$. In fact, (7.4) extends to *all* derivatives $d^r(\cdot)/d\theta^r, r \in \mathbb{N}$ when η is likewise differentiable.[6]

We are now ready to put the concept of regularity to work. The following theorem establishes for regular distributions a lower bound for the variance of any unbiased estimator of a *function* of θ (which, for the time being, we continue to regard as a scalar).

---

[5]Applying Theorem 3.2, the interchange of differentiation and integration is valid if $df(x;\theta)/d\theta = \lim_{m\to\infty} m\left[f(x;\theta + m^{-1}) - f(x;\theta)\right]$ exists and if there is an integrable function q such that $|g(x)|\,m|f(x;\theta + m^{-1}) - f(x;\theta)| \leq q(x)$ for each $m \in \mathbb{N}$.

[6]For further discussion and proofs of these features of members of the exponential class see Barndorff-Nielsen (1982), Lehmann and Casella (1998, Section 1.5), and Lehmann and Romano (2005, Section 2.7).

Theorem 7.1. *Let X be a r.v. with density in the regular family $\{f(x; \theta) : \theta \in \Theta\}$, and let $\mathbf{X}$ be a random sample of size n (i.i.d. replicates of X). If statistic $T_n := T(\mathbf{X})$ is such that $ET_n =: \tau(\theta)$ and $VT_n < \infty$, then $\tau(\theta)$ is differentiable, and*

$$VT_n \geq \frac{\tau'(\theta)^2}{n\mathcal{I}(\theta)}, \tag{7.6}$$

where $\mathcal{I}(\theta) := E\left[d\log f(X; \theta)/d\theta\right]^2.$

Proof. Random sample $\mathbf{X}$ has joint density $f(\mathbf{x}; \theta) = \prod_{j=1}^{n} f(x_j; \theta)$, so that for all $\theta \in \Theta$

$$1 = \int_{\Re^n} f(\mathbf{x}; \theta) \mathcal{M}(d\mathbf{x}) \text{ and}$$

$$\tau(\theta) = \int_{\Re^n} T(\mathbf{x}) f(\mathbf{x}; \theta) \mathcal{M}(d\mathbf{x}).$$

Differentiate each of these identities with respect to θ and apply (7.4) and (7.5) to obtain

$$0 = \int_{\Re^n} \frac{d\log f(\mathbf{x}; \theta)}{d\theta} f(\mathbf{x}; \theta) \mathcal{M}(d\mathbf{x}) = E\left[\frac{d\log f(\mathbf{X}; \theta)}{d\theta}\right] \tag{7.7}$$

$$\tau'(\theta) = \int_{\Re^n} T(\mathbf{x}) \frac{d\log f(\mathbf{x}; \theta)}{d\theta} f(\mathbf{x}; \theta) \mathcal{M}(d\mathbf{x})$$

$$= E\left[T_n \cdot \frac{d\log f(\mathbf{X}; \theta)}{d\theta}\right]. \tag{7.8}$$

Note that the differentiability of $\tau(\theta)$ in the second line follows from (7.4). Now (7.7) implies that the expectation in (7.8) equals the covariance of the two factors in brackets. Schwarz' inequality then implies $\tau'(\theta)^2 \leq VT_n \cdot E\left[d\log f(\mathbf{X}; \theta)/d\theta\right]^2$ and hence

$$VT_n \geq \frac{\tau'(\theta)^2}{E\left[d\log f(\mathbf{X}; \theta)/d\theta\right]^2}.$$

By virtue of (7.7) the denominator is

$$V\left[\frac{d\log f(\mathbf{X}; \theta)}{d\theta}\right] = V\sum_{j=1}^{n}\left[\frac{d\log f(X_j; \theta)}{d\theta}\right]$$

$$= \sum_{j=1}^{n} E\left[\frac{d\log f(X_j; \theta)}{d\theta}\right]^2$$

$$= nE\left[\frac{d\log f(X; \theta)}{d\theta}\right]^2$$

$$= n\mathcal{I}(\theta).$$

Here, the second equality follows from independence and the fact that

$$E\left[\frac{d\log f(\mathbf{X};\theta)}{d\theta}\right] = \sum_{j=1}^{n} E\left[\frac{d\log f(X_j;\theta)}{d\theta}\right] = 0$$

for each n implies that $Ed\log f(X_j;\theta)/d\theta = 0$ for each j. $\qquad\square$

Corollary 7.1. *(i) When T_n is a biased estimator of some $\varphi(\theta)$, with $ET_n = \tau(\theta) = \varphi(\theta) + \beta(\theta)$, then*

$$VT_n \geq \frac{[\varphi'(\theta) + \beta'(\theta)]^2}{n\mathcal{I}(\theta)}.$$

(ii) When T_n is an unbiased estimator of θ itself,

$$VT_n \geq \frac{1}{n\mathcal{I}(\theta)}. \tag{7.9}$$

The crucial quantity

$$n\mathcal{I}(\theta) := nE\left[\frac{d\log f(X;\theta)}{d\theta}\right]^2 = E\left[\frac{d\log f(\mathbf{X};\theta)}{d\theta}\right]^2,$$

called the *Fisher information* of the sample,[7] is interpreted below. Inequalities (7.6) and (7.9) are known as the *information inequalities,* and the lower bounds themselves are referred to as the *Cramér–Rao bounds.*[8] Any unbiased estimator from a sample of size n whose variance actually attains the bound $[n\mathcal{I}(\theta)]^{-1}$ is said to be *uniformly* (meaning for all $\theta \in \Theta$) *minimum-variance unbiased (m.v.u.).* It is therefore the *efficient* estimator.

Example 7.3. Elements of $\mathbf{X} := (X_1, X_2, ..., X_n)'$ are i.i.d. as $P(\mu)$ (Poisson), with $f(x;\mu) = (\mu^x e^{-\mu}/x!)\mathbf{1}_{\aleph_0}(x)$. Then $d\log f(X;\mu)/d\mu = (X-\mu)/\mu$, and

$$\mathcal{I} = \frac{E(X-\mu)^2}{\mu^2} = \frac{1}{\mu}.$$

The Cramér–Rao bound for an unbiased estimator of μ is therefore $(n\mathcal{I})^{-1} = \mu/n$, while the bound for an unbiased estimator of e^μ is $e^{2\mu}\mu/n$.

Example 7.4. Elements of $\mathbf{X} := (X_1, X_2, ..., X_n)'$ are i.i.d. as exponential, with $f(x;\theta) = \theta^{-1}e^{-x/\theta}\mathbf{1}_{[0,\infty)}(x)$. Then $d\log f(X;\theta)/d\theta = (X-\theta)/\theta^2$, and

$$\mathcal{I} = \frac{E(X-\theta)^2}{\theta^4} = \frac{1}{\theta^2}.$$

---

[7]After R.A. Fisher (1890-1962), often regarded as the "father" of modern statistics in recognition of his pioneering work during the first half of the 20th century.
[8]After independent discoverers Harald Cramér (1893-1985) and C.R. Rao (1920–).

The Cramér–Rao bound for an unbiased estimator of θ is therefore $(n\mathcal{I})^{-1} = \theta^2/n$, while the bound for an unbiased estimator of $\log \theta$ is simply n^{-1}.

Example 7.5. Elements of $\mathbf{X} := (X_1, X_2, ..., X_n)'$ are i.i.d. as $N\left(0, \sigma^2\right)$, with $f\left(x; \sigma^2\right) = e^{-x^2/(2\sigma^2)}/\sqrt{2\pi\sigma^2}$. Then

$$\frac{d \log f\left(X; \sigma^2\right)}{d\sigma^2} = \frac{X^2 - \sigma^2}{2\sigma^4},$$

and

$$\mathcal{I} = \frac{E(X^2 - \sigma^2)^2}{4\sigma^8} = \frac{1}{2\sigma^4}.$$

The Cramér–Rao bound for an unbiased estimator of the variance of a normal population is therefore $(n\mathcal{I})^{-1} = 2\sigma^4/n$, while the bound for an unbiased estimator of the *inverse* of the variance, $\tau\left(\sigma^2\right) := 1/\sigma^2$, is $2/\left(n\sigma^4\right)$.

There is an easier calculation for $\mathcal{I}(\theta)$ if the parametric family is such that (7.4) extends to the second derivative, as it does for members of the exponential class whose natural parameters are differentiable and contained in an open set. In this case, using (7.7),

$$0 = \frac{d}{d\theta} \int \frac{d \log f\left(\mathbf{x}; \theta\right)}{d\theta} f\left(\mathbf{x}; \theta\right) \mathcal{M}\left(d\mathbf{x}\right)$$

$$= \int \left[\frac{d^2 \log f\left(\mathbf{x}; \theta\right)}{d\theta^2} f\left(\mathbf{x}; \theta\right) + \frac{d \log f\left(\mathbf{x}; \theta\right)}{d\theta} \frac{df\left(\mathbf{x}; \theta\right)}{d\theta} \right] \mathcal{M}\left(d\mathbf{x}\right)$$

$$= \int \frac{d^2 \log f\left(x; \theta\right)}{d\theta^2} f\left(\mathbf{x}; \theta\right) \mathcal{M}\left(d\mathbf{x}\right) + \int \left[\frac{d \log f\left(\mathbf{x}; \theta\right)}{d\theta} \right]^2 f\left(\mathbf{x}; \theta\right) \mathcal{M}\left(d\mathbf{x}\right)$$

$$= E\left[\frac{d^2 \log f\left(\mathbf{X}; \theta\right)}{d\theta^2} \right] + E\left[\frac{d \log f\left(\mathbf{X}; \theta\right)}{d\theta} \right]^2,$$

and so

$$n\mathcal{I} = -E\left[\frac{d^2 \log f\left(\mathbf{X}; \theta\right)}{d\theta^2} \right] \tag{7.10}$$

$$= -nE\left[\frac{d^2 \log f\left(X; \theta\right)}{d\theta^2} \right].$$

This is usually easier to evaluate because $d^2 \log f\left(X; \theta\right)/d\theta^2$ is usually a simpler function of X than is $[d \log f\left(X; \theta\right)/d\theta]^2$.

Example 7.6. Applying (7.10) in Example 7.4 gives

$$-E\left[\frac{d^2 \log f\left(X; \theta\right)}{d\theta^2} \right] = -E\left(\frac{1}{\theta^2} - \frac{2X}{\theta^3} \right) = \frac{1}{\theta^2}.$$

Example 7.7. Applying (7.10) in Example 7.5 with $\theta = \sigma^2$ gives

$$-E\left[\frac{d^2 \log f(X;\theta)}{d\theta^2}\right] = -E\left(-\frac{X^2}{\theta^3} + \frac{1}{2\theta^2}\right) = \frac{1}{2\theta^2} = \frac{1}{2\sigma^4}.$$

To interpret the all-important quantity $\mathcal{I}$ as "information" requires some sense of how sampling provides information about θ. For this it is helpful momentarily to adopt the Bayesian viewpoint that θ itself is an uncertain quantity about which we have some *prior* conception—that is, a conception based on information that was available prior to viewing the sample. This prior belief is represented explicitly through a *prior distribution*, $_*P(\cdot)$ (the "$*$" *preceding* the "P" reminding us that $_*P$ is "prior"). In the Bayesian view, which is discussed at length in Chapter 10, $_*P(B) = \int_B d_*P(\theta)$ represents the *subjective* probability that $\theta \in B$ (some Borel set) *before* acquiring sample information. In this setting the customary symbol $f(\cdot;\theta)$ for the density of X represents a *conditional* distribution for X, equivalent to $f(\cdot \mid \theta)$ in the usual notation.

With this background, we now can ask how obtaining one specific sample realization, $X(\omega) = x$ say, alters the extent of uncertainty about θ. Formally, this change can be expressed via Bayes' rule in terms of a *posterior* distribution $P_*(\cdot \mid x)$, which combines the sample and prior information in a manner consistent with the rules of probability:

$$dP_*(\theta \mid x) = \frac{f(x;\theta) \cdot d_*P(\theta)}{\int_\Theta f(x;\theta) \cdot d_*P(\theta)} = \frac{f(x;\theta) \cdot d_*P(\theta)}{f_X(x)}.$$

Here, the numerator of the last expression represents the joint distribution of X and θ, and the denominator is the marginal for X after θ is integrated out.

Now the sample datum can be considered informative about θ only to the extent that $P_*(\cdot \mid x)$ is less dispersed—more concentrated—than prior $_*P(\cdot)$. In that case, assuming now that $f(x;\cdot)$ and $d_*P(\cdot)$ are continuous on Θ, there will be some $\theta^* \in \Theta$ at which $r(\theta \mid x) := dP_*(\theta \mid x)/d_*P(\theta) = f(x;\theta)/f_X(x)$ achieves a maximum value. Assuming that θ^* is in the interior of Θ, we will have $d \log r(\theta \mid x)/d\theta|_{\theta^*} = d \log f(x;\theta)/d\theta|_{\theta^*} = 0$, and in the vicinity of θ^* we will have

$$-\frac{d^2 \log r(\theta \mid x)}{d\theta^2} = -\frac{d^2 \log f(x;\theta)}{d\theta^2} > 0.$$

The greater the relative concentration of $P_*(\cdot \mid x)$ the greater the curvature at the maximum, and the larger will be $-d^2 \log f(x;\theta)/d\theta^2$. Thus, $-d^2 \log f(x;\theta)/d\theta^2$ indicates the amount of information contributed by one particular sample observation. Averaging this information content over the support of X gives the Fisher information of a single observation:

$$\mathcal{I}(\theta) = -\int_{\mathbb{X}} \frac{d^2 \log f(x;\theta)}{d\theta^2} f(x;\theta) \mathcal{M}(dx) = -E\left[\frac{d^2 \log f(X;\theta)}{d\theta^2}\right].$$

With a random sample of size n the joint density of $\mathbf{X} = (X_1, X_2, ..., X_n)'$ at sample point $\mathbf{x} = (x_1, x_2, ..., x_n)'$ is $f(\mathbf{x};\theta) = \prod_{j=1}^{n} f(x_j;\theta)$, so the information content of $\mathbf{x}$ is $-d^2 \log f(\mathbf{x};\theta)/d\theta^2 = -\sum_{j=1}^{n} d^2 \log f(x_j;\theta)/d\theta^2$, and the average value is

$$-E\left[\frac{d^2 \log f(\mathbf{X};\theta)}{d\theta^2}\right] = -\sum_{j=1}^{n} E\left[\frac{d^2 \log f(X_j;\theta)}{d\theta^2}\right]$$

$$= -nE\left[\frac{d^2 \log f(X;\theta)}{d\theta^2}\right]$$

$$= n\mathcal{I}(\theta).$$

We have the pleasing result that the information content of a random sample—that is, of a set of i.i.d. random variables—is proportional to its size.

Exercise 7.5. *As in Example 7.5 elements of $\mathbf{X} = (X_1, X_2, ..., X_n)'$ are i.i.d. as $N(0, \sigma^2)$. Taking $\tau(\sigma^2) = \sigma := +\sqrt{\sigma^2}$, find the Cramér–Rao bound for the variance of an unbiased estimator of σ.*

Exercise 7.6. *In random samples from normal populations the facts that*

$$C_{n-1} := \frac{(n-1)S^2}{\sigma^2} \sim \chi^2(n-1)$$

and that $EC_{n-1} = n-1$ imply that sample variance S^2 is an unbiased estimator of σ^2. Use these facts to show, more generally, that $q(n,r)S^r$ is unbiased for σ^r, where

$$q(n,r) := \left(\frac{n-1}{2}\right)^{r/2} \frac{\Gamma\left(\frac{n-1}{2}\right)}{\Gamma\left(\frac{r+n-1}{2}\right)}, r > -(n-1).$$

Conclude that $q(n,1)S$ is an unbiased estimator of σ, and compare its variance with the variance bound obtained in Exercise 7.5.

We can now extend the information inequality to allow $\theta := (\theta_1, \theta_2, ..., \theta_k)'$ to be a vector of parameters in a space $\Theta \subset \Re^k$ and $\tau(\theta) \in \Re^m$ a vector of $m \leq k$ functions of θ. Let $\mathbf{T}_n := (T_{1n}, T_{2n}, ..., T_{mn})'$ be an m-vector of statistics with $E\mathbf{T}_n = \tau(\theta)$, and let $V\mathbf{T}_n$ stand for the $m \times m$ covariance matrix. Finally, for square matrices $\mathbf{A}$ and $\mathbf{B}$ interpret "$\mathbf{A} \geq \mathbf{B}$" to mean that $\mathbf{A} - \mathbf{B}$ is positive semidefinite, in which case $\mathbf{a}'(\mathbf{A} - \mathbf{B})\mathbf{a} \geq 0$ for all conformable vectors $\mathbf{a}$. Then the Cramér–Rao bound for the covariance matrix of unbiased estimators of $\boldsymbol{\tau}$ is

$$V\mathbf{T}_n \geq \frac{\partial \boldsymbol{\tau}(\boldsymbol{\theta})}{\partial \boldsymbol{\theta}} \left[nE\left(\frac{\partial \log f(X;\boldsymbol{\theta})}{\partial \boldsymbol{\theta}} \frac{\partial \log f(X;\boldsymbol{\theta})}{\partial \boldsymbol{\theta}'} \right) \right]^{-1} \frac{\partial \boldsymbol{\tau}(\boldsymbol{\theta})}{\partial \boldsymbol{\theta}'}$$

$$=: \frac{\partial \boldsymbol{\tau}(\boldsymbol{\theta})}{\partial \boldsymbol{\theta}} (n\mathcal{I})^{-1} \frac{\partial \boldsymbol{\tau}(\boldsymbol{\theta})}{\partial \boldsymbol{\theta}'}.$$

Here $\partial \boldsymbol{\tau}(\boldsymbol{\theta})/\partial \boldsymbol{\theta}$ is the $m \times k$ matrix with $\partial \tau_i(\boldsymbol{\theta})/\partial d\theta_j$ in the ith row and jth column, and $\partial \log f(X;\boldsymbol{\theta})/\partial \boldsymbol{\theta}$ is a k-vector of partial derivatives—i.e., gradient of f with respect to $\boldsymbol{\theta}$. The product of $\partial \log f(X;\boldsymbol{\theta})/\partial \boldsymbol{\theta}$ and its transpose is a $k \times k$ symmetric matrix with $[\partial \log f(X;\boldsymbol{\theta})/\partial \theta_i][\partial \log f(X;\boldsymbol{\theta})/\partial \theta_j]$ in row i and column j. The matrix $\mathcal{I}$ comprising the expected values of these elements is the *Fisher information matrix*. Note that it is the *inverse* of this matrix that enters into the minimum-variance bound. When the regularity condition extends to second derivatives, $\mathcal{I}$ is also given by the expectation of $-\partial^2 \log f(X;\boldsymbol{\theta})/\partial \boldsymbol{\theta} \partial \boldsymbol{\theta}'$, which is the (negative of the) $k \times k$ (Hessian) matrix of second derivatives of $\log f$. In this case, we also have

$$V\mathbf{T}_n \geq \frac{\partial \boldsymbol{\tau}(\boldsymbol{\theta})}{\partial \boldsymbol{\theta}} \left[-nE\left(\frac{\partial^2 \log f(X;\boldsymbol{\theta})}{\partial \boldsymbol{\theta} \partial \boldsymbol{\theta}'} \right) \right]^{-1} \frac{\partial \boldsymbol{\tau}(\boldsymbol{\theta})}{\partial \boldsymbol{\theta}'}.$$

When $\boldsymbol{\tau}(\boldsymbol{\theta}) = \boldsymbol{\theta}$, matrix $\partial \boldsymbol{\tau}(\boldsymbol{\theta})/\partial \boldsymbol{\theta}$ is just the identity matrix, and these expressions reduce to $V\mathbf{T}_n \geq (n\mathcal{I})^{-1}$.

If $\mathbf{a} \in \Re^m$ is a vector of known constants not all equal to zero, then the scalar statistic $\mathbf{a}'\mathbf{T}_n$ is an unbiased estimator of $\mathbf{a}'\boldsymbol{\tau}(\boldsymbol{\theta})$ when $\mathbf{T}_n$ itself is unbiased for $\boldsymbol{\tau}$. In this case, $V(\mathbf{a}'\mathbf{T}_n) = \mathbf{a}'(V\mathbf{T}_n)\mathbf{a}$, so that

$$V(\mathbf{a}'\mathbf{T}_n) - \mathbf{a}' \frac{\partial \boldsymbol{\tau}(\boldsymbol{\theta})}{\partial \boldsymbol{\theta}} (n\mathcal{I})^{-1} \frac{\partial \boldsymbol{\tau}(\boldsymbol{\theta})}{\partial \boldsymbol{\theta}'} \mathbf{a}$$

$$= \mathbf{a}' \left[V\mathbf{T}_n - \frac{\partial \boldsymbol{\tau}(\boldsymbol{\theta})}{\partial \boldsymbol{\theta}} (n\mathcal{I})^{-1} \frac{\partial \boldsymbol{\tau}(\boldsymbol{\theta})}{\partial \boldsymbol{\theta}'} \right] \mathbf{a} \geq 0.$$

The information inequality thus gives a minimum-variance bound for an unbiased estimator of $\mathbf{a}'\boldsymbol{\tau}(\boldsymbol{\theta})$. In particular, taking $\mathbf{a}$ to consist entirely of zeroes except for a unit in any position $j \in \{1, 2, ..., m\}$, we have

$$V(\mathbf{a}'\mathbf{T}_n) = VT_{jn} \geq n^{-1} \frac{\partial \tau_j(\boldsymbol{\theta})}{\partial \boldsymbol{\theta}} \mathcal{I}^{-1} \frac{\partial \tau_j(\boldsymbol{\theta})}{\partial \boldsymbol{\theta}'}$$

and, when $\boldsymbol{\tau}(\boldsymbol{\theta}) = \boldsymbol{\theta}$,

$$VT_{jn} \geq n^{-1} \mathcal{I}^{jj}$$

where $\mathcal{I}^{jj}$ is the jth element on the principal diagonal of $\mathcal{I}^{-1}$.

Example 7.8. With $X \sim N(\mu, \sigma^2)$ and $\tau(\boldsymbol{\theta}) = \boldsymbol{\theta} = (\mu, \sigma^2)'$,

$$\log f(X; \boldsymbol{\theta}) = -\frac{1}{2}\left[\log(2\pi) + \log\sigma^2 + \frac{(X-\mu)^2}{\sigma^2}\right]$$

$$\frac{\partial \log f(X; \boldsymbol{\theta})}{\partial \boldsymbol{\theta}'} = \left(\frac{X-\mu}{\sigma^2}, -\frac{1}{2\sigma^2} + \frac{(X-\mu)^2}{2\sigma^4}\right)$$

$$-\frac{\partial^2 \log f(X; \boldsymbol{\theta})}{\partial \boldsymbol{\theta}\partial \boldsymbol{\theta}'} = \begin{pmatrix} \frac{1}{\sigma^2} & \frac{X-\mu}{\sigma^4} \\ \frac{X-\mu}{\sigma^4} & -\frac{1}{2\sigma^4} + \frac{(X-\mu)^2}{\sigma^6} \end{pmatrix} \qquad (7.11)$$

$$-E\left[\frac{\partial^2 \log f(X; \boldsymbol{\theta})}{\partial \boldsymbol{\theta}\partial \boldsymbol{\theta}'}\right] = \begin{pmatrix} \frac{1}{\sigma^2} & 0 \\ 0 & \frac{1}{2\sigma^4} \end{pmatrix}.$$

Multiplying by n and inverting, the lower bound for the covariance matrix of an unbiased estimator of (μ, σ^2) is

$$V\mathbf{T}_n \geq (n\mathcal{I})^{-1} = \begin{pmatrix} \frac{\sigma^2}{n} & 0 \\ 0 & \frac{2\sigma^4}{n} \end{pmatrix}.$$

Thus, for given constants a_1 and a_2, $a_1^2\sigma^2/n + 2a_2^2\sigma^4/n$ is the minimum variance bound for an unbiased estimator of $a_1\mu + a_2\sigma^2$. In particular, taking $\mathbf{a} = (1,0)'$ and then $\mathbf{a} = (0,1)'$ shows that σ^2/n and $2\sigma^4/n$ are the lower bounds for unbiased estimators of μ and σ^2.

Exercise 7.7. $\{\mathbf{Z}_j = (X_j, Y_j)\}_{j=1}^n$ *is a (bivariate) random sample from a population distributed as (bivariate) Bernoulli, having joint p.m.f.*

$$f_{\mathbf{Z}}(x, y; \boldsymbol{\theta}) = \theta_1^x \theta_2^y (1 - \theta_1 - \theta_2)^{1-x-y}$$

for $x, y \in \{0, 1\}$, $x + y \in \{0, 1\}$, $\theta_1, \theta_2 \in (0, 1)$, $\theta_1 + \theta_2 \in (0, 1)$.

(i) *Find the lower bound for the covariance matrix of an unbiased estimator of* $\boldsymbol{\theta} = (\theta_1, \theta_2)'$.

(ii) *Find the lower bound for the variance of an unbiased estimator of* $\theta_1 + \theta_2$.

(iii) *Find a minimum variance unbiased estimator for* $\theta_1 + \theta_2$ *from sample* $\{\mathbf{Z}_j\}_{j=1}^n$.

As we have seen, $\bar{X}$ does attain the lower bound for variances of unbiased estimators of μ when the population is $N(\mu, \sigma^2)$. On the other hand, unbiased sample variance S^2 has variance $2\sigma^4/(n-1)$, which exceeds the bound of $2\sigma^4/n$ for estimators of σ^2. The same bound for the variance also applies in the one-parameter case that μ is known, as was seen in Example 7.5, where we took $\mu = 0$. When μ has any known value, the bound of $2\sigma^4/n$ is certainly attainable, since that is the variance of unbiased statistic $n^{-1}\sum_{j=1}^{n}(X_j - \mu)^2$. When μ is not known, it turns out that the bound $2\sigma^4/n$ is simply *unattainable*, in that there is no unbiased estimator with a variance smaller than $2\sigma^4/(n-1) = VS^2$. This will be seen in the discussion of *sufficient* statistics, which follows. While the possible unattainability of the Cramér–Rao bound is a drawback, we will see later that the difficulty vanishes asymptotically as $n \to \infty$ and that the information inequality has particular relevance in large-sample theory.

'Sufficient' Statistics   A statistic from a sample $\mathbf{X}$ can convey information about constants $\boldsymbol{\tau}(\boldsymbol{\theta})$ only to the extent that its distribution depends on $\boldsymbol{\theta}$. There are two polar cases.

Definition 7.2. A statistic $\mathbf{U}_n$ is *ancillary* for $\boldsymbol{\theta}$ if (and only if) the distribution of $\mathbf{U}_n$ does not depend on $\boldsymbol{\theta}$.

Definition 7.3. A statistic $\mathbf{T}_n$ is *sufficient* for $\boldsymbol{\theta}$ if (and only if) the conditional distribution of $\mathbf{X}$ given $\mathbf{T}_n$ does not depend on $\boldsymbol{\theta}$.[9]

Alternatively, we could say that $\mathbf{T}_n$ is sufficient for $\boldsymbol{\theta}$ if and only if any statistic that is not a function of $\mathbf{T}_n$ alone (i.e., not $\sigma(\mathbf{T}_n)$ measurable) is ancillary conditional on $\mathbf{T}_n$. By itself, an ancillary statistic conveys no information about $\boldsymbol{\theta}$, although it may tell us something about the precision with which some other statistic estimates $\boldsymbol{\theta}$ or some function thereof. As an example of ancillarity, in sampling from a normal population the sample variance is ancillary for the mean, since $(n-1)S^2/\sigma^2$ is distributed as $\chi^2(n-1)$ for all $\mu \in \Re$. On the other hand, the sample mean is *not* ancillary for σ^2, for realizations of $\bar{X} \sim N(\mu, \sigma^2/n)$ certainly convey information about the variance. We will see later that $\bar{X}$ is sufficient for μ when σ^2 is known and that $\bar{X}, S^2$ together are jointly sufficient for μ, σ^2.

It is clear that when a sufficient statistic for $\boldsymbol{\theta}$ has been found, there is no point in looking elsewhere for more sample information. It is therefore not surprising that if a minimum-variance unbiased (m.v.u.) estimator for some $\boldsymbol{\tau}(\boldsymbol{\theta})$ exists—whether or not it attains the Cramér-Rao bound—it must be a function of a sufficient statistic. This is a simple consequence of the following theorem:

---

[9]That is, "if ... there is a *version* of the conditional distribution...". Recall (p. 99) that distributions conditional on continuous r.v.s must be thought of as equivalence classes that generate the same unconditional probabilities.

Theorem 7.2. *(Rao–Blackwell) The risk $E\mathcal{L}\left(\mathbf{S}_n, \boldsymbol{\tau}\left(\boldsymbol{\theta}\right)\right)$ associated with any integrable statistic $\mathbf{S}_n$ is at least as great as the risk of $\boldsymbol{\xi}_n\left(\mathbf{T}_n\right) :=$ $E\left(\mathbf{S}_n \mid \mathbf{T}_n\right)$ when (i) $\mathbf{T}_n$ is a sufficient statistic for $\boldsymbol{\theta}$ and (ii) loss function $\mathcal{L}$ is convex in its first argument; i.e.,*

$$E\mathcal{L}\left(\mathbf{S}_n, \boldsymbol{\tau}\left(\boldsymbol{\theta}\right)\right) \geq E\mathcal{L}\left(\boldsymbol{\xi}_n, \boldsymbol{\tau}\left(\boldsymbol{\theta}\right)\right). \tag{7.12}$$

The inequality is strict if $\mathcal{L}$ is strictly convex, $E\mathcal{L}\left(\mathbf{S}_n, \boldsymbol{\tau}\left(\boldsymbol{\theta}\right)\right) < \infty$, and $\Pr\left(\mathbf{S}_n = \mathbf{T}_n\right) \neq 1$.

Proof. Since $\mathbf{T}_n$ is sufficient, $\boldsymbol{\xi}_n$ does not depend on $\boldsymbol{\theta}$ and is therefore a statistic. (7.12) then follows from Jensen's inequality. $\qquad\square$

Corollary 7.2. *If $\boldsymbol{\xi}_n$ is a m.v.u. estimator of $\boldsymbol{\tau}\left(\boldsymbol{\theta}\right)$, then $\boldsymbol{\xi}_n = E\left(\mathbf{S}_n \mid \mathbf{T}_n\right)$ for some unbiased statistic $\mathbf{S}_n$.*

The Rao–Blackwell theorem limits the search for an m.v.u. statistic to those that are functions of some sufficient $\mathbf{T}_n$. To progress in the search we must first see how to find a sufficient statistic, then how to determine whether the one we have found generates the m.v.u. estimator.

Since any statistic depends just on known constants and sample data $\mathbf{X} = \left(X_1, X_2, ..., X_n\right)'$, the condition for the sufficiency of a statistic $\mathbf{T}_n$ can be stated as $F_{\mathbf{X}\mid\mathbf{T}_n}\left(\mathbf{x} \mid \mathbf{T}_n = \mathbf{T}_n\left(\mathbf{x}\right)\right) = H'(\mathbf{x})$ for all $\mathbf{x} \in \Re^n$ and all $\boldsymbol{\theta} \in \Theta$, where H' is some function of the data alone and $F_{\mathbf{X}\mid\mathbf{T}_n}$ is (some version of) the conditional c.d.f. This just expresses formally that the distribution of the data—and therefore of any function of the data—conditional on $\mathbf{T}_n$ is the same for all $\boldsymbol{\theta}$. Now, the joint c.d.f. of the sample itself can be expressed as the product of the conditional c.d.f. and the marginal c.d.f. of $\mathbf{T}_n$:

$$F_{\mathbf{X}}\left(\mathbf{x}; \boldsymbol{\theta}\right) = F_{\mathbf{X}\mid\mathbf{T}_n}\left(\mathbf{x} \mid \mathbf{T}_n = \mathbf{t}\right) F_{\mathbf{T}_n}\left(\mathbf{t}; \boldsymbol{\theta}\right).$$

It follows that a necessary and sufficient condition for the sufficiency of $\mathbf{T}_n$ is that the joint c.d.f. of the sample factor for all $\boldsymbol{\theta}$ and $\mathbf{x}$ as

$$F_{\mathbf{X}}\left(\mathbf{x}; \boldsymbol{\theta}\right) = F_{\mathbf{T}_n}\left(\mathbf{T}_n\left(\mathbf{x}\right); \boldsymbol{\theta}\right) H'(\mathbf{x}), \tag{7.13}$$

where $F_{\mathbf{T}_n}\left(\mathbf{T}_n\left(\mathbf{x}\right); \boldsymbol{\theta}\right)$ is the marginal c.d.f. and where $H'(\mathbf{x})$ does not depend on $\boldsymbol{\theta}$.

While (7.13) does serve to verify that a given $\mathbf{T}_n$ is sufficient, it has two shortcomings if the ultimate purpose is to find an m.v.u. estimator of some $\boldsymbol{\tau}\left(\boldsymbol{\theta}\right)$. First, one would have had to guess that $\mathbf{T}_n$ *might* be sufficient in order to check, and, second, one would have to work out its c.d.f. However, both the identification and the verification can be accomplished together if one can express $F_{\mathbf{X}}\left(\mathbf{x}; \boldsymbol{\theta}\right)$ as the product of two nonnegative functions on

$\Re^n$, one depending on $\boldsymbol{\theta}$ and some statistic $\mathbf{T}_n$ and the other being free of $\boldsymbol{\theta}$ entirely, as

$$F_{\mathbf{X}}(\mathbf{x}; \boldsymbol{\theta}) = G(\mathbf{T}_n(\mathbf{x}); \boldsymbol{\theta}) H(\mathbf{x}).$$

A factorization of this sort both identifies $\mathbf{T}_n$ as being sufficient and confirms that an expression in the form (7.13) is possible, for $H(\mathbf{x}) = H'(\mathbf{x}) F_{\mathbf{T}_n}(\mathbf{T}_n; \boldsymbol{\theta}) / G(\mathbf{T}_n; \boldsymbol{\theta})$ just absorbs any component of $F_{\mathbf{T}_n}(\cdot; \boldsymbol{\theta})$ that does not depend on $\mathbf{T}_n$ or $\boldsymbol{\theta}$. In the usual case that X is purely discrete or purely continuous, it is more convenient to work with the *density*—i.e., the p.d.f. or p.m.f., for which the corresponding factorization is

$$f_{\mathbf{X}}(\mathbf{x}; \boldsymbol{\theta}) = g(\mathbf{T}_n(\mathbf{x}); \boldsymbol{\theta}) h(\mathbf{x}). \tag{7.14}$$

When $\boldsymbol{\theta}$ is vector-valued, a statistic T_{n1} (not necessarily a scalar) is sufficient for the single element θ_1, say, if the joint density can be factored as $f_{\mathbf{X}}(\mathbf{x}; \boldsymbol{\theta}) = g(t_1(\mathbf{x}); \theta_1) h(\mathbf{x})$, where $h(\mathbf{x})$ can now depend on $\theta_2, \theta_3, ..., \theta_k$ but not on θ_1.

Example 7.9. The n elements of $\mathbf{X}$ are i.i.d. as Poisson with joint p.m.f.

$$f_{\mathbf{X}}(\mathbf{x}; \theta) = \frac{\theta^{x_1 + x_2 + \cdots + x_n} e^{-n\theta}}{x_1! x_2! \cdots \cdot x_n!}$$
$$= \frac{\left(\theta^{\bar{x}} e^{-\theta}\right)^n}{x_1! x_2! \cdots \cdot x_n!}$$

when each $x_j \in \aleph_0$. This has the form (7.14) with $T_n(\mathbf{x}) = \bar{x}$, $g(T_n; \theta) = \left(\theta^{\bar{x}} e^{-\theta}\right)^n$ and $h(\mathbf{x}) = (x_1! x_2! \cdots \cdot x_n!)^{-1}$. Thus, $\bar{X}$ is a sufficient statistic for θ in the Poisson model.

Example 7.10. The n elements of $\mathbf{X}$ are i.i.d. as exponential with joint p.d.f.

$$f_{\mathbf{X}}(\mathbf{x}; \theta) = \theta^{-n} e^{-(x_1 + x_2 + \cdots + x_n)/\theta}$$
$$= \theta^{-n} e^{-n\bar{x}/\theta}.$$

This has the form (7.14) with $T_n = \bar{x}$, $g(T_n; \theta) = \left(\theta e^{\bar{x}/\theta}\right)^{-n}$ and $h(\mathbf{x}) = 1$. Thus, $\bar{X}$ is a sufficient statistic for θ in the exponential model.

In both these examples a single statistic was found to be sufficient for the single parameter θ. However, sufficient statistics are never unique. For one thing, it is clear that any one-to-one function of a sufficient statistic is also sufficient; for if $\mathbf{S}_n = \mathbf{q}(\mathbf{T}_n)$ is such a function, then (7.14) implies also that

$$f_{\mathbf{X}}(\mathbf{x}; \boldsymbol{\theta}) = g\left[\mathbf{q}^{-1}(\mathbf{S}_n); \boldsymbol{\theta}\right] h(\mathbf{x}),$$

which is of the same form. Moreover, if $\mathbf{T}_n$ is sufficient, then $(\mathbf{T}'_n, \mathbf{U}'_n)$ is also sufficient, where $\mathbf{U}_n$ is any quantity that does not involve unknown parameters. As an extreme example, the entire sample $\mathbf{X}$ is a sufficient statistic, since its joint density trivially "factors" as $f_\mathbf{X}(\mathbf{x}; \boldsymbol{\theta}) = g(\mathbf{x}; \boldsymbol{\theta}) h(\mathbf{x})$ with $g(\mathbf{x}; \boldsymbol{\theta}) = f_\mathbf{X}(\mathbf{x}; \boldsymbol{\theta})$ and $h(\mathbf{x}) = 1$. While this fact is not often of *practical* use, it does force us to recognize that the relevant goal is to achieve the most economical or succinct way to package the data so as to extract all relevant information about $\boldsymbol{\theta}$. Doing this requires finding what is called a *minimal-sufficient statistic*. Briefly, statistic $\mathbf{T}_n$ is minimal-sufficient if it can be expressed as a function of any other sufficient statistic. For example, $\mathbf{T}_n = \mathbf{T}(\mathbf{X})$ is a function of the sufficient collection $\mathbf{X}$. Of course, if T_n is a single (scalar) sufficient statistic for the scalar parameter θ, then it is obviously minimal-sufficient.

A result by Lehmann and Scheffé (1950) makes it relatively easy to find a minimal $\mathbf{T}_n$. Specifically, $\mathbf{T}_n$ is minimal-sufficient for $\boldsymbol{\theta}$ if and only if (there is a version of $f_\mathbf{X}$ such that) the value of $f_\mathbf{X}(\mathbf{x}; \boldsymbol{\theta}) / f_\mathbf{X}(\mathbf{y}; \boldsymbol{\theta})$ at each $\mathbf{x}$ and each $\mathbf{y}$ for which $f_\mathbf{X}(\cdot; \boldsymbol{\theta}) > 0$ is invariant with respect to $\boldsymbol{\theta}$ when $\mathbf{T}_n(\mathbf{x}) = \mathbf{T}_n(\mathbf{y})$. In practice, since the logarithms of densities are typically of simpler form, it is usually easier to check for invariance of $\log f_\mathbf{X}(\mathbf{x}; \boldsymbol{\theta}) - \log f_\mathbf{X}(\mathbf{y}; \boldsymbol{\theta})$. Some examples will show how to put this into practice.

Example 7.11. With elements of $\mathbf{X}$ i.i.d. as

$$f_X(x; \theta) = \frac{\theta^x e^{-\theta}}{x!} \mathbf{1}_{\aleph_0}(x) = \frac{e^{x \log \theta - \theta}}{x!} \mathbf{1}_{\aleph_0}(x)$$

we have (when each element of $\mathbf{x}$ and each element of $\mathbf{y}$ is in $\aleph_0 :=$ $\{0, 1, 2, ...\}$)

$$\log\left[\frac{f_\mathbf{X}(\mathbf{x}; \theta)}{f_\mathbf{X}(\mathbf{y}; \theta)}\right] = n(\bar{x} - \bar{y}) \log \theta - \sum_{j=1}^n \log x_j! + \sum_{j=1}^n \log y_j!.$$

This is independent of θ if and only if $\bar{x} = \bar{y}$, confirming that $\bar{X}$ (or any one-to-one function thereof) is minimal-sufficient for the Poisson parameter.

Example 7.12. With $\boldsymbol{\theta} = (\alpha, \beta)'$ and elements of $\mathbf{X}$ i.i.d. as

$$f_X(x; \alpha, \beta) = \Gamma(\alpha)^{-1} \beta^{-\alpha} x^{\alpha-1} e^{-x/\beta} \mathbf{1}_{(0,\infty)}(x)$$
$$= \Gamma(\alpha)^{-1} \beta^{-\alpha} \exp\left[(\alpha - 1) \log x - \beta^{-1} x\right] \mathbf{1}_{(0,\infty)}(x)$$

we have (when each element of $\mathbf{x}$ and each element of $\mathbf{y}$ is positive)

$$\log\left[\frac{f_\mathbf{X}(\mathbf{x}; \alpha, \beta)}{f_\mathbf{X}(\mathbf{y}; \alpha, \beta)}\right] = (\alpha - 1) \sum_{j=1}^n (\log x_j - \log y_j) - \beta^{-1} \sum_{j=1}^n (x_j - y_j).$$

This is invariant with respect to (α, β) if and only if $\sum_{j=1}^{n} \log x_j = \sum_{j=1}^{n} \log y_j$ and $\sum_{j=1}^{n} x_j = \sum_{j=1}^{n} y_j$. Thus,

$$\mathbf{T}_n\left(\mathbf{X}\right) = \left(\sum_{j=1}^{n} \log X_j, \sum_{j=1}^{n} X_j\right)'$$

is minimal-sufficient for the parameters of the gamma distribution. Also, when β has known value β_0, say, putting $h(\mathbf{x}) = \exp\left(-\beta_0^{-1} \sum_{j=1}^{n} x_j\right)$ and factoring $f_{\mathbf{X}}\left(\mathbf{x};\alpha, \beta_0\right)$ as

$$f_{\mathbf{X}}\left(\mathbf{x};\alpha, \beta_0\right) = \Gamma\left(\alpha\right)^{-n} \beta_0^{-n\alpha} \exp\left[(\alpha - 1) \sum_{j=1}^{n} \log x_j\right] h(\mathbf{x})$$

shows that $\sum_{j=1}^{n} \log X_j$ is sufficient for α. Likewise, $\sum_{j=1}^{n} X_j$ is sufficient for β when α is known.

Exercise 7.8. *Show that* $\mathbf{T}_n\left(\mathbf{X}\right) = \left(\sum_{j=1}^{n} X_j, \sum_{j=1}^{n} X_j^{-1}\right)'$ *is a minimal-sufficient statistic for* $\boldsymbol{\theta} = (\mu, \lambda)'$ *when* $\{X_j\}_{j=1}^{n}$ *are i.i.d. as* $IG\left(\mu, \lambda\right)$ *(inverse Gaussian).*

In the two examples and the exercise dimensions of $\boldsymbol{\theta}$ and minimal-sufficient statistic $\mathbf{T}_n$ were the same, but even in such cases it is not *necessarily* true that each component of $\mathbf{T}_n$ is individually sufficient for a component of $\boldsymbol{\theta}$.

Example 7.13. With $\{X_j\}_{j=1}^{n}$ i.i.d. as $N\left(\mu, \sigma^2\right)$ we have

$$\log\left[\frac{f_{\mathbf{X}}\left(\mathbf{x};\mu, \sigma^2\right)}{f_{\mathbf{X}}\left(\mathbf{y};\mu, \sigma^2\right)}\right] = \frac{1}{2\sigma^2}\left[\sum_{j=1}^{n} (y_j - \mu)^2 - \sum_{j=1}^{n} (x_j - \mu)^2\right]$$

$$= \frac{1}{2\sigma^2} \sum_{j=1}^{n} [(y_j - \bar{y}) + (\bar{y} - \mu)]^2$$

$$- \frac{1}{2\sigma^2} \sum_{j=1}^{n} [(x_j - \bar{x}) + (\bar{x} - \mu)]^2 .$$

Now

$$\sum_{j=1}^{n} [(x_j - \bar{x}) + (\bar{x} - \mu)]^2 = \sum_{j=1}^{n} (x_j - \bar{x})^2 + n (\bar{x} - \mu)^2$$

$$= (n - 1) s_X^2 + n (\bar{x} - \mu)^2$$

since $\sum_{j=1}^{n} (x_j - \bar{x}) = 0$, and so

$$\log\left[\frac{f_{\mathbf{X}}(\mathbf{x}; \mu, \sigma^2)}{f_{\mathbf{X}}(\mathbf{y}; \mu, \sigma^2)}\right] = \frac{1}{2\sigma^2}\left\{(n-1)\left(s_Y^2 - s_X^2\right) + n\left[(\bar{y} - \mu)^2 - (\bar{x} - \mu)^2\right]\right\}.$$

Thus, $(\bar{X}, S^2)$ are jointly minimal-sufficient for (μ, σ^2). $\bar{X}$ is also individually sufficient for μ when σ^2 is known, since the joint p.d.f. can be factored as

$$f\left(\mathbf{x}; \mu, \sigma^2\right) = \exp\left[-\frac{1}{2}\left(\frac{\bar{x} - \mu}{\sigma/\sqrt{n}}\right)^2\right] h(\mathbf{x})$$

with $h(\mathbf{x}) = \left(2\pi\sigma^2\right)^{-n/2} e^{-(n-1)s^2/(2\sigma^2)}$. However, S^2 is not individually sufficient for σ^2 even if μ is known, because there is no factorization of the form $f(\mathbf{x}; \mu, \sigma^2) = g(s^2; \sigma^2) h(\mathbf{x})$ in which $g(s^2; \sigma^2)$ depends just on s^2 and $h(\mathbf{x})$ is invariant with respect to σ^2. Instead, when $\mu = \mu_0$ the sufficient statistic for σ^2 is $\sum_{j=1}^{n} (X_j - \mu_0)^2$.

Exercise 7.9. *Continuing Exercise 7.8, is $\sum_{j=1}^{n} X_j$ alone sufficient for inverse Gaussian parameter μ when λ is known? Is $\sum_{j=1}^{n} X_j^{-1}$ by itself sufficient for λ when μ is known?*

Although one hopes to find a sufficient statistic of smaller dimension than the full sample $\mathbf{X}$, this cannot always be done.

Example 7.14. If $\{X_j\}_{j=1}^{n}$ are i.i.d. as $f(x; \theta) = \pi^{-1}\left[1 + (x - \theta)^2\right]^{-1}$ (Cauchy with location parameter θ), then

$$\log\left[\frac{f(\mathbf{x}; \theta)}{f(\mathbf{y}; \theta)}\right] = \sum_{j=1}^{n} \log\left[\frac{1 + (y_j - \theta)^2}{1 + (x_j - \theta)^2}\right]$$

has the same value for all θ only if $x_j = y_j$ for each j. Thus, the entire sample $\mathbf{T}_n = \mathbf{X}$ is minimal-sufficient for the single scalar parameter. Of course, any permutation of the sample, such as the set of order statistics $\{X_{(j:n)}\}_{j=1}^{n}$, is also sufficient.

Exercise 7.10. $\{X_j\}_{j=1}^{n}$ *are i.i.d. as Weibull, with*

$$f(x; \alpha, \beta) = \alpha\beta^{-1} x^{\alpha-1} e^{-x^\alpha/\beta} \mathbf{1}_{(0,\infty)}(x).$$

Show that

(i) $\mathbf{X}$ itself is minimal-sufficient for (α, β).
(ii) $\sum_{j=1}^{n} X_j^\alpha$ is sufficient for β when α is known.

Example 7.15. If $\{X_j\}_{j=1}^n$ are i.i.d. as $f(x;\theta) = \theta^{-1}\mathbf{1}_{[0,\theta]}(x)$ the joint density is $f(\mathbf{x};\theta) = \theta^{-n}\prod_{j=1}^n \mathbf{1}_{[0,\theta]}(x_j)$. The product equals unity if and only if $x_{(1:n)} := \min\{x_1,...,x_n\} \geq 0$ and $x_{(n:n)} := \max\{x_1,...,x_n\} \leq \theta$, so the density can be written as

$$f(\mathbf{x};\theta) = \theta^{-n}\mathbf{1}_{[0,\infty)}\left(x_{(1:n)}\right)\mathbf{1}_{(-\infty,\theta]}\left(x_{(n:n)}\right).$$

Since

$$\frac{f(\mathbf{x};\theta)}{f(\mathbf{y};\theta)} = \frac{\mathbf{1}_{[0,\infty)}\left(x_{(1:n)}\right)\mathbf{1}_{(-\infty,\theta]}\left(x_{(n:n)}\right)}{\mathbf{1}_{[0,\infty)}\left(y_{(1:n)}\right)\mathbf{1}_{(-\infty,\theta]}\left(y_{(n:n)}\right)}$$

is invariant with θ if and only if $x_{(n:n)} = y_{(n:n)}$, the sample maximum $X_{(n:n)}$ is minimal sufficient.

It happens that the only *regular* distributions that admit minimal-sufficient statistics are in the *exponential class*. For these the joint density of the sample will have some "minimal" representation as

$$f(\mathbf{x};\boldsymbol{\theta}) = a(\boldsymbol{\eta})^n \prod_{j=1}^n c(x_j)\exp\left[\boldsymbol{\eta}(\boldsymbol{\theta})'\sum_{j=1}^n \mathbf{T}(x_j)\right], \qquad (7.15)$$

in which the vector $\boldsymbol{\eta}$ of natural parameters has the smallest dimension.[10] Such densities are of the form $g(\mathbf{T}_n(\mathbf{x});\boldsymbol{\theta})h(\mathbf{x})$ in (7.14) with

$$\mathbf{T}_n(\mathbf{x}) = \sum_{j=1}^n \mathbf{T}(x_j),$$

$$g(\mathbf{T}_n(\mathbf{x});\boldsymbol{\theta}) = a(\boldsymbol{\eta})^n\, e^{\boldsymbol{\eta}(\boldsymbol{\theta})'\mathbf{T}_n(\mathbf{x})},$$

$$h(\mathbf{x}) = \prod_{j=1}^n c(x_j).$$

Thus, $\mathbf{T}_n(\mathbf{X}) = \sum_{j=1}^n \mathbf{T}(X_j)$ is sufficient for $\boldsymbol{\theta}$. It is also *minimally sufficient*, since

$$\log\left[\frac{f(\mathbf{x};\boldsymbol{\theta})}{f(\mathbf{y};\boldsymbol{\theta})}\right] = \boldsymbol{\eta}(\boldsymbol{\theta})'\left[\sum_{j=1}^n \mathbf{T}(x_j) - \sum_{j=1}^n \mathbf{T}(y_j)\right] + \sum_{j=1}^n [\log c(x_j) - \log c(y_j)]$$

is parameter-free if and only if $\mathbf{T}_n(\mathbf{x}) = \mathbf{T}_n(\mathbf{y})$. Thus, it is easy to find minimal-sufficient statistics when using a model in the exponential class.

Now let us proceed to see how to apply the sufficiency concept to find the m.v.u.e. of some $\tau(\boldsymbol{\theta})$. To recap a bit at this point, we have seen

---

[10]It would not be minimal if elements of $\boldsymbol{\eta}$ satisfied a linear constraint, such as $\mathbf{a}'\boldsymbol{\eta}(\boldsymbol{\theta}) = b$, since any η_j could then be expressed in terms of the others.

how to find sufficient statistics, including the minimal ones that extract information in the most economical way. We also know from the Rao–Blackwell theorem that any m.v.u. estimator of $\tau(\boldsymbol{\theta})$ must be a function of a statistic $\mathbf{T}_n$ that is sufficient for $\boldsymbol{\theta}$. Such a function can be obtained from any unbiased estimator $\mathbf{S}_n$ by conditioning, as $\boldsymbol{\xi}_n := E(\mathbf{S}_n \mid \mathbf{T}_n)$. Alternatively, without going though such a procedure, it might be easy to show from first principles that some statistic known to be sufficient is also unbiased; as for example, estimator $\xi_n := \bar{X} + S^2$ of $\tau(\mu, \sigma^2) = \mu + \sigma^2$ for the normal distribution. In either case a critical question remains: Will the particular function we have obtained produce the *efficient* estimator— the one with minimum variance? Clearly, $\boldsymbol{\xi}_n$ will have minimum variance among unbiased estimators if it attains the Cramér-Rao bound; but, if it does not, it might simply be that the bound is unattainable. The one piece of the puzzle that is lacking is the concept of *completeness*.

Note: Random vector $\boldsymbol{\xi}_n \in \Re^m$ will be said to have "lower variance" than some $\boldsymbol{\zeta}_n \in \Re^m$ if the difference between their covariance matrices, $V\boldsymbol{\zeta}_n - V\boldsymbol{\xi}_n$, is positive definite. In symbols, this is $V\boldsymbol{\xi}_n < V\boldsymbol{\zeta}_n$, with "$\le$" indicating semi-definiteness.

Definition 7.4. A statistic $\mathbf{T}_n$ is *complete* if and only if

$$Eg(\mathbf{T}_n) = \int g(\mathbf{t}) f_{\mathbf{T}_n}(\mathbf{t}; \boldsymbol{\theta}) \mathcal{M}(d\mathbf{t}) = 0 \text{ for all } \boldsymbol{\theta} \text{ implies } g(\mathbf{T}_n) = 0 \text{ a.s.}$$

$$(7.16)$$

To explain the origin of the term, observe that the definition requires that no function other than $g \equiv 0$ be "orthogonal" to all members of the family $\{f_{\mathbf{T}_n}(\mathbf{t}; \boldsymbol{\theta}), \boldsymbol{\theta} \in \Theta\}$. In linear algebra a collection $\mathcal{Z} = \{\mathbf{z}_j\}$ of orthogonal vectors in an inner product space is said to be complete if there is no other vector that is orthogonal to all members of $\mathcal{Z}$. In such a case $\mathcal{Z}$ *spans* the space. Before giving an example of a complete statistic, let us see how this rather odd condition fits in and delivers our m.v.u. estimator. Suppose we have a $\boldsymbol{\xi}_n \equiv \boldsymbol{\xi}_n(\mathbf{T}_n) := E(\mathbf{S}_n \mid \mathbf{T}_n)$ for which $E\boldsymbol{\xi}_n = \tau(\boldsymbol{\theta})$, but we discover that another unbiased function $\boldsymbol{\zeta}_n \equiv \boldsymbol{\zeta}_n(\mathbf{T}_n) := E(\mathbf{U}_n \mid \mathbf{T}_n)$ is available as well. Might $\boldsymbol{\zeta}_n$ do better (in the sense that $V\boldsymbol{\zeta}_n < V\boldsymbol{\xi}_n$) or, more generally, might some convex combination $(1-p)\boldsymbol{\xi}_n + p\boldsymbol{\zeta}_n = \boldsymbol{\xi}_n + p(\boldsymbol{\zeta}_n - \boldsymbol{\xi}_n)$ do better? (Note that $E(\boldsymbol{\zeta}_n - \boldsymbol{\xi}_n) = \mathbf{0}$ since both are unbiased.) Clearly, the combination might do better for some p depending on the variances and covariances of the elements of $\boldsymbol{\xi}_n$ and $\boldsymbol{\zeta}_n$. But if sufficient statistic $\mathbf{T}_n$ were known to be complete, the only function $p(\boldsymbol{\zeta}_n - \boldsymbol{\xi}_n) =: g(\mathbf{T}_n)$ with $Eg(\mathbf{T}_n) = 0$ for all $\boldsymbol{\theta}$

would be the zero function—that is, one equal to zero for all realizations of $\mathbf{T}_n$, and in that case $\Pr(\boldsymbol{\zeta}_n = \boldsymbol{\xi}_n) = 1$. Thus, a uniformly unbiased function of a complete sufficient statistic is *unique*.[11] Together, then, Rao–Blackwell and completeness imply that there is no hope of reducing the variance of $\boldsymbol{\xi}_n$ by adding anything to it, nor is there any more general transformation that would reduce its variance while leaving it uniformly unbiased. We have reached the following conclusion:

Theorem 7.3. *A uniformly unbiased function of a complete sufficient statistic is the unique minimum-variance unbiased estimator.*

Of course, one immediately wants to know how to determine whether a given sufficient statistic $\mathbf{T}_n$ is complete. First, the *bad* news: Because condition (7.16) is so general, verifying it *can* require some nontrivial and inventive analysis, specific to each case.

Example 7.16. When $\{X_j\}_{j=1}^n$ are i.i.d. as uniform on $[0,\theta]$, Example 7.15 showed that sample maximum $X_{(n:n)}$ is minimal sufficient for θ. The c.d.f. and p.d.f. of $T_n := X_{(n:n)}$ are $F_{T_n}(t;\theta) = \theta^{-n} t^n \mathbf{1}_{[0,\theta]}(t) + \mathbf{1}_{(\theta,\infty)}(t)$ and $f_{T_n}(t;\theta) = n\theta^{-n} t^{n-1} \mathbf{1}_{[0,\theta]}(t)$. Any g that satisfies (7.16) must of course be integrable in the Lebesgue sense, so that each of $Eg^+(T_n) := E\max\{g(T_n),0\}$ and $Eg^-(T_n) := E\max\{-g(T_n),0\}$ is finite. Thus $Eg(T_n) = 0$ for all $\theta > 0$ implies that

$$\int_0^\theta g^+(t) f_{T_n}(t;\theta) \cdot dt - \int_0^\theta g^-(t) f_{T_n}(t;\theta) \cdot dt = 0$$

for all $\theta > 0$. Since the sets $\{[0,\theta] : \theta > 0\}$ generate the Borel sets $\mathcal{B}_+$ of $\Re_+$, it follows that $\int_B [g^+(t) - g^-(t)] f_{T_n}(t;\theta) \cdot dt = 0$ for all $B \in \mathcal{B}_+$ and hence that $g = g^+ - g^- = 0$ a.s.

Fortunately, for the broad range of models in the exponential class there are general results that eliminate the need to verify completeness of the sufficient statistics. The minimal sufficient statistic $\mathbf{T}_n \in \Re^m$ from a distribution in exponential class (7.15) is, in fact, complete if the natural parameter space $\mathbf{H} = \left\{ \boldsymbol{\eta} : \int c(x) e^{\boldsymbol{\eta}(\theta)' \mathbf{T}(x)} \mathcal{M}(dx) < \infty \right\}$ ($\mathcal{M}(\cdot)$ being the relevant dominating measure) is of dimension m.[12] To verify dimensionality, it is enough to check that *some* m-dimensional rectangle can be embedded in

---

[11] $\mathbf{T}_n$, and therefore $\boldsymbol{\xi}_n$, would also be minimal, for if some other unbiased sufficient statistic $\boldsymbol{\zeta}_n$ were minimal it would have to be a function of $\mathbf{T}_n$ (else, it would not be minimal), in which case $\boldsymbol{\zeta}_n = \boldsymbol{\xi}_n$, as just seen.

[12] For a proof see pp. 116-117 in Lehmann and Romano (2005).

H—that is, to check that **H** can contain some such rectangle. When might this condition fail? Of itself, the minimality of $\mathbf{T}_n$ excludes *affine* relations among the $\{\eta_j\}$, such as $\mathbf{a}'\boldsymbol{\eta} = b$. Thus, **H** will be of dimension m unless some *a priori* restriction on the elements of $\boldsymbol{\theta}$ imposes a *nonlinear* relation. An instance of this appears in Example 7.18(ii) below. In the absence of such a restriction, **H** is typically a product space in $\Re^m$ when $m > 1$; and when $m = 1$, H is typically an interval—a 1-dimensional "rectangle". Thus, the dimensionality condition is usually trivial to check.

Here is how these general results for the exponential class apply to two models we have already considered on their own.

Example 7.17. The Poisson family
$$\left\{ f(x;\theta) = e^{-\theta} \frac{e^{x\log\theta}}{x!} \mathbf{1}_{\aleph_0}(x) : \theta > 0 \right\}$$
has minimal representation (7.15) with $c(x) = 1/x!$, $\eta = \log\theta$, and $T(x) = x$. Thus, $\sum_{j=1}^{n} T(X_j) = n\bar{X}$ (and therefore $\bar{X}$ itself) is a minimal-sufficient statistic from a sample of size n. With $N(\cdot)$ as counting measure, the integral $\int_{\aleph_0} (x!)^{-1} e^{x\log\theta} N(dx) \equiv \sum_{x=0}^{\infty} (x!)^{-1} e^{x\log\theta}$ is finite for $\eta \in H = \Re$, so $\bar{X}$ is complete; and $\tau(\bar{X})$, if unbiased, is the unique m.v.u. estimator of $\tau(\theta)$.

Example 7.18. (i) The normal family
$$\left\{ f(x;\mu,\sigma^2) = \frac{1}{\sqrt{2\pi\sigma^2}} e^{-\mu^2/(2\sigma^2)} \exp\left(-\frac{x^2}{2\sigma^2} + \frac{\mu x}{\sigma^2} \right) : \mu \in \Re, \sigma^2 > 0 \right\}$$
has minimal representation (7.15) with $\eta_1 = -1/(2\sigma^2)$, $\eta_2 = \mu/\sigma^2$, and $\mathbf{T}(x)' = (x^2, x)$. Integral $\int e^{\eta_1 x^2 + \eta_2 x} \cdot dx$ is finite for (η_1, η_2) in product space $\mathbf{H} = (-\infty, 0) \times \Re$, so $\mathbf{T}_n(\mathbf{X})' = \left(\sum_{j=1}^{n} X_j^2, \sum_{j=1}^{n} X_j \right)$ is complete-sufficient. Thus,
$$\boldsymbol{\tau}(\mathbf{T}_n) := \begin{pmatrix} n^{-1} \sum_{j=1}^{n} X_j \\ \frac{1}{n-1} \sum_{j=1}^{n} X_j^2 - \frac{1}{n(n-1)} \left(\sum_{j=1}^{n} X_j \right)^2 \end{pmatrix} = \begin{pmatrix} \bar{X} \\ S^2 \end{pmatrix}$$
is m.v.u. for $(\mu, \sigma^2)'$.

(ii) The subfamily of normals with $\mu = \sigma > 0$ and $\boldsymbol{\eta}' = \left(-1/(2\sigma^2), 1/\sigma \right)$ has $\eta_2 = \sqrt{-2\eta_1}$, the locus **H** of which for $\sigma > 0$ is a rectifiable curve in $\Re^2$—a one-dimensional subspace. For this subfamily $\mathbf{T}_n$ is *not* complete, as can be seen directly by the fact that $E\left[\bar{X}^2 - (n+1) S^2/n \right] = 0$ for all $\sigma > 0$.[13]

---

[13] See Lehmann and Casella (1998) for discussion and further references pertaining to estimation in such so-called "curved" exponential families.

Exercise 7.11. *Find the m.v.u. estimator of $EX = \mu$ when $X \sim \Gamma(\alpha, \beta)$.*

We can now finally resolve the issue of whether the sample variance is the m.v.u. estimator of the parameter σ^2 of the normal distribution when μ is unknown. That $\boldsymbol{\xi}_n := \left(\bar{X}, S^2\right)'$ is m.v.u. for $\boldsymbol{\theta} = \left(\mu, \sigma^2\right)'$ implies that $\mathbf{a}'\boldsymbol{\xi}_n$ is m.v.u. for $\mathbf{a}'\boldsymbol{\theta}$, any $\mathbf{a} \in \Re^2$. In particular, setting $a_1 = 0$ and $a_2 = 1$ shows that S^2 is m.v.u. for σ^2. Thus, even though S^2 does not attain the Cramér–Rao bound and is not *individually* sufficient, it is in fact the efficient estimator. Likewise, for $r > -(n-1)$ the estimator $q(n,r) S^r$ of Exercise 7.6 is m.v.u. for σ^r when the population is normal, with

$$q(n,r) := \left(\frac{n-1}{2}\right)^{r/2} \frac{\Gamma\left(\frac{n-1}{2}\right)}{\Gamma\left(\frac{r+n-1}{2}\right)}.$$

Exercise 7.12. *Find the m.v.u. estimator of $\mu^2 + \sigma^2$ when $\{X_j\}_{j=1}^n$ are i.i.d. as $N\left(\mu, \sigma^2\right)$.*

The concepts of sufficiency, ancillarity, and completeness are basic to the following result, which can greatly simplify distribution theory from parametric models:

Theorem 7.4. *(Basu) If $\mathbf{T}_n(\mathbf{X})$ is a complete sufficient statistic for family $\{F(\mathbf{x}; \boldsymbol{\theta}), \boldsymbol{\theta} \in \boldsymbol{\Theta}\}$, and if $\mathbf{S}_n(\mathbf{X}) \in \Re^m$ is ancillary, then $\mathbf{T}_n$ and $\mathbf{S}_n$ are independent.*

Proof. Put $\mathbb{P}_{\mathbf{S}_n}(B) := E\mathbf{1}_B(\mathbf{S}_n)$ and $\mathbb{P}_{\mathbf{S}_n|\mathbf{T}_n}(B \mid \mathbf{T}_n) := E\left[\mathbf{1}_B(\mathbf{S}_n) \mid \mathbf{T}_n\right]$ for arbitrary Borel set $B \in \mathcal{B}^m$. Neither $\mathbb{P}_{\mathbf{S}_n}(B)$ nor $\mathbb{P}_{\mathbf{S}_n|\mathbf{T}_n}(B \mid \mathbf{T}_n)$ depends on $\boldsymbol{\theta}$ by the definitions of ancillarity and sufficiency. Since $\mathbf{T}_n$ is complete and $E\left[\mathbb{P}_{\mathbf{S}_n}(B) - \mathbb{P}_{\mathbf{S}_n|\mathbf{T}_n}(B \mid \mathbf{T}_n)\right] = 0$ for all $\boldsymbol{\theta}$, it follows that $\mathbb{P}_{\mathbf{S}_n}(B) = \mathbb{P}_{\mathbf{S}_n|\mathbf{T}_n}(B \mid \mathbf{T}_n)$ a.s.; hence, $\mathbf{T}_n$ and $\mathbf{S}_n$ are independent. $\square$

In normal samples $\bar{X}$ is a complete sufficient statistic for μ, and S^2 is ancillary. Thus, the independence of $\bar{X}$ and S^2 follows immediately from Basu's theorem—a conclusion that was arrived at by brute force in Section 5.3. Since $\bar{X}$ is unbiased for μ and $q(n,r) S^r$ is unbiased for σ^r, independence implies that $\bar{X}q(n,r) S^r$ is m.v.u. for $\mu\sigma^r$. The implications of sufficiency, ancillarity, and completeness are thus very far reaching.

We conclude the discussion of efficient statistics with a proposition whose conclusion is weaker than that of Basu's theorem but which holds under weaker conditions. It, too, is often extremely useful in applications.

Proposition 7.1. *If T_n is an efficient (i.e., m.v.u.) estimator of θ and U_n is any other unbiased estimator with finite variance, then $Cov\,(T_n, U_n - T_n) = 0$.*

The conditions here are weaker than for Basu's theorem, because T_n is neither necessarily *individually* sufficient nor complete, nor is $U_n - T_n$ necessarily ancillary. The equality here is just another way of expressing the fact that another unbiased estimator cannot be used to increase the efficiency of an efficient T_n. The proof is very direct: Construct an alternative unbiased estimator $Q_n\,(p) = (1 - p)T_n + pU_n = T_n + pD_n$, where $D_n := U_n - T_n$. Then $VQ_n(p) = VT_n + 2pCov(T_n, D_n) + p^2 VD_n$ attains a minimum value of

$$VQ_n^* = VT_n - \frac{Cov(T_n, D_n)^2}{VD_n}$$

at $p^* := -Cov(T_n, D_n)/VD_n$. But that T_n is efficient implies $VQ_n^* \geq VT_n$, which in turn implies that $Cov(T_n, D_n) = 0$.

Exercise 7.13. *If $ET_n = EU_n = \theta$ and T_n is m.v.u., show that $Cov(T_n, U_n) = VT_n$. Conclude that $V\,(U_n - T_n) = VU_n - VT_n$.*

Exercise 7.14. *$\{X_j\}_{j=1}^n$ are i.i.d. as $P\,(\lambda)$ (Poisson). Find $Cov\,(S^2, \bar{X})$.*

Exercise 7.15. *If $X \sim P\,(\lambda)$, the probability generating function (p.g.f.) at $t \in \Re$ is $p\,(t; \lambda) := Et^X = e^{\lambda(t-1)}$. With $\{X_j\}_{j=1}^n$ i.i.d. as $P\,(\lambda)$, put $P_n\,(t) := n^{-1} \sum_{j=1}^n t^{X_j}$ (the empirical p.g.f.) and*

$$\hat{p}_n\,(t) := \left(1 + \frac{t-1}{n}\right)^{n\bar{X}}.$$

(i) Show that both $P_n\,(t)$ and $\hat{p}_n\,(t)$ are unbiased estimators of $p\,(t; \lambda)$.
(ii) Prove that $\hat{p}_n\,(t)$ is the m.v.u. estimator.
(iii) Express $VP_n\,(t)$ and $V\hat{p}_n\,(t)$ and verify that $VP_n\,(t) \geq V\hat{p}_n\,(t)$ with strict inequality when $t \neq 1$.
(iv) $p\,(t; \bar{X}) = e^{\bar{X}(t-1)}$ is another estimator of $p\,(t; \lambda)$, a function of the complete sufficient statistic $\bar{X}$. Is it also an m.v.u.e.?

7.1.2 *Asymptotic Properties*

Despite the elegance of the supporting theory, the goal of finding uniformly m.v.u. estimators does have its shortcomings. For one, it is simply not attainable in many complicated estimation problems, where not even unbiased estimators—much less globally efficient ones—can be found. Even when such are available, their appeal diminishes on recognizing that the m.v.u. property is not sustained under nonlinear transformations. Thus, finding a statistic $\mathbf{T}_n$ that is m.v.u. for $\boldsymbol{\theta}$ does not automatically solve the problem of finding one for a function $g(\boldsymbol{\theta})$ unless $g = a + \mathbf{b}'\boldsymbol{\theta}$. For example, although $\bar{X}$ and S^2 are m.v.u. for μ and σ^2 when the population is normal (and are *unbiased* generally), $S/\bar{X}$ is not unbiased for σ/μ. This limitation of the m.v.u. criterion becomes even more apparent when one considers that the conventional parameterizations of models for distributions are essentially arbitrary. That is, any one-to-one transformation that takes parameter space $\boldsymbol{\Theta} \subset \Re^k$ onto another subset $\boldsymbol{\Theta}'$ of $\Re^k$ would represent the probability distribution equally well.[14] By contrast, it is usually much easier to find estimators with good *asymptotic* properties as $n \to \infty$, and these good properties are preserved under all continuous transformations.

7.1.2.1 *Consistent Estimators*

The most basic large-sample criterion is that of *consistency*.

Definition 7.5. A statistic T_n is a strongly (weakly) *consistent* estimator of a constant θ if and only if $T_n \to^{a.s.(p)} \theta$.

Thus, by using a consistent estimator the risk of getting an estimate that differs much from θ can be made arbitrarily small by taking n to be sufficiently large. The term "consistent" applies because inference with T_n from the sample would, in the limit, be consistent with what could be deduced from the entire population. Of course, consistency is no guarantee of good results for any finite n—after all, $T_n + 10^{23}/\log(\log n)$ is consistent if T_n is! On the other hand, one could hardly be satisfied with an estimator that did *not* have this potential of ultimately producing the right answer.

Chapter 6 has already supplied many examples of consistent estimators. The strong law of large numbers (s.l.l.n.) tells us that sample moments about the origin are strongly consistent for the corresponding population moments. Together, the s.l.l.n. and continuity theorem imply strong consistency of moments about the mean as well. In general, the continuity theorem implies that $g(T_n) \to^{a.s.(p)} g(\theta)$ whenever $T_n \to^{a.s.(p)} \theta$, so long as g is continuous at θ. Thus, the property of consistency is indeed preserved

---

[14]Indeed, one suspects that it was the availability of unbiased estimators for some forms that made those parameterizations traditional.

even under continuous nonaffine transformations. For example, $S^2 \to^{a.s.} \sigma^2$ implies $S \to^{a.s.} \sigma$, $e^{-S} \to^{a.s.} e^{-\sigma}$, and so forth.

7.1.2.2 *Consistent and Asymptotically Normal Estimators*

When a statistic T_n is at least weakly consistent for some θ, there is often a positive ϵ such that the scaled deviation $n^{\epsilon/2}(T_n - \theta)$ converges in distribution to normal with mean zero and some positive variance—σ_T^2, say, which generally depends on one or more parameters of the population. By far the most common case, of which we have seen many examples, is that $\epsilon = 1$. Thus, for populations with moments through order four, $n^{1/2}(\bar{X} - \mu) \rightsquigarrow N(0, \sigma^2)$ by the Lindeberg–Lévy c.l.t. and $n^{1/2}(S - \sigma) \rightsquigarrow N\left[0, (\mu_4 - \sigma^4)/(4\sigma^2)\right]$ by the c.l.t. and delta method (Example 6.32). An estimator with the property that $n^{\epsilon/2}(T_n - \theta) \rightsquigarrow N(0, \sigma_T^2)$ for some $\epsilon > 0$ is said to be *consistent and asymptotically normal*, or *c.a.n.* for short.

Since $n^{\epsilon/2}(T_n - \theta) =: Z_n$ is distributed approximately as $N(0, \sigma_T^2)$ for large n when T_n is c.a.n., it follows that $T_n = \theta + Z_n/n^{\epsilon/2}$ is itself distributed approximately as $N(\theta, \sigma_T^2/n^\epsilon)$. The variance of this *approximating* distribution has a special name.

Definition 7.6. If $n^{\epsilon/2}(T_n - \theta) \rightsquigarrow N(0, \sigma_T^2)$, then σ_T^2/n^ϵ is called the *asymptotic variance* of T_n.

This terminology, although convenient and widely used, is apt to mislead. As a function of n, the asymptotic variance of a statistic T_n is most emphatically *not* the limiting value of VT_n, which could take any value in $[0, +\infty]$; nor is it necessarily equal to finite-sample variance VT_n at any specific n. Rather, the asymptotic variance is the variance of the distribution that is used to *approximate* the distribution of T_n when n is large. Such approximations are needed when exact finite-sample distributions are either unknown, difficult to work out, or cumbersome to use.

Example 7.19. **X** is a random sample of size n from a population with mean μ and finite variance σ^2 but no presumed parametric form. We want to estimate $\theta = e^{-\mu}$. Now $T_n = e^{-\bar{X}}$ is a strongly consistent estimator by the s.l.l.n. and continuity theorem, and an application of the delta method shows that $\sqrt{n}(T_n - \theta) = -e^{-\mu_n^*}\sqrt{n}(\bar{X} - \mu)$ converges in distribution to $N(0, e^{-2\mu}\sigma^2) = N(0, \theta^2\sigma^2)$, where $|\mu_n^* - \mu| < |\bar{X} - \mu|$. Thus, the asymptotic variance of $T_n = e^{-\bar{X}}$ is $\theta^2\sigma^2/n$. The exact, finite-sample variance simply cannot be determined, since no parametric family of models for the population is available.

Example 7.20. Continuing Example 7.19, if the decision is made to model the population as $N\left(\mu,\sigma^2\right)$, then the asymptotic variance of T_n remains $\theta^2\sigma^2/n$, but the finite-sample distribution and variance can now be worked out explicitly for comparison. With $\bar{X} \sim N\left(\mu,\sigma^2/n\right)$ we have $T_n \sim LN\left(-\mu,\sigma^2/n\right)$, so that $ET_n = e^{-\mu}e^{\sigma^2/(2n)} = \theta + O\left(n^{-1}\right)$ and $VT_n = \theta^2 e^{\sigma^2/n}\left(e^{\sigma^2/n}-1\right) = \theta^2\sigma^2/n + O(n^{-2})$. In this case there is no need for an asymptotic approximation, since the exact distribution is both known and easy to use.

Here is one final example that, while rather contrived, makes quite dramatic the distinction between an estimator's finite-sample and asymptotic variances.

Example 7.21. R.v. Y has unknown mean $EY = \theta$ and some finite variance σ^2. We try to observe n i.i.d. replicates $\{Y_j\}_{j=1}^n$ through an experiment of n independent trials, but our observations are subject to errors that tend to diminish as we progress through the trials. Specifically, the error at step j is distributed as $j^{-\delta}\mathcal{E}_j$, where $\delta > 1$ and $\{\mathcal{E}_j\}_{j=1}^n$ are i.i.d. as Cauchy, independent of the $\{Y_j\}$. Thus, the observation at trial j is $X_j := Y_j + \mathcal{E}_j/j^\delta$. Taking $T_n := \bar{X}$ as estimator of θ, let $Z_n := \sqrt{n}\left(T_n - \theta\right) = \sqrt{n}\left(\bar{Y} - \theta\right) + n^{-1/2}\sum_{j=1}^n \mathcal{E}_j/j^\delta$. The c.f. of Z_n is

$$\phi_{Z_n}\left(t\right) = \phi_{\sqrt{n}(\bar{Y}-\theta)}\left(t\right) \prod_{j=1}^n \exp\left(-|t|n^{-1/2}j^{-\delta}\right)$$

$$= \phi_{\sqrt{n}(\bar{Y}-\theta)}\left(t\right)\exp\left(-|t|n^{-1/2}\sum_{j=1}^n j^{-\delta}\right).$$

Now since $\delta > 1$ the sum in the exponent converges as $n \to \infty$, and so (because of the factor $n^{-1/2}$) the exponent converges to zero. We conclude that $|\phi_{Z_n}\left(t\right) - \phi_{\sqrt{n}(\bar{Y}-\theta)}\left(t\right)| \to 0$ and hence that the limiting distributions of $\sqrt{n}\left(T_n - \theta\right)$ and $\sqrt{n}\left(\bar{Y} - \theta\right)$ are precisely the same. By Lindeberg–Lévy that common distribution is $N\left(0,\sigma^2\right)$. Accordingly, T_n has asymptotic variance σ^2/n, whereas $VT_n = +\infty$ for any finite n.

Exercise 7.16. *For some Borel set B take $\theta = E\mathbf{1}_B(X)$ and $T_n := n^{-1}\sum_{j=1}^n \mathbf{1}_B(X_j)$. Determine the exact distribution of T_n and its asymptotic approximation, and compare the finite-sample and asymptotic variances.*

Let us back up for a moment and consider how to regard at a practical level the distinction between asymptotic moments and those from finite samples. We have some family of models for our data—a parametric family such as the normal, exponential, or gamma, or else one of a broad class with general properties—continuous, finite moments, etc. We also have chosen a statistic T_n that is supposed to tell us something about an unknown feature of the generating process, as represented by this model—a parameter, a quantile, a moment, etc. If we have taken our sample, we can simply observe the realized value of T_n; apart from computational error there is no uncertainty at all about its value. The real question is how it relates to the unknown quantity of interest, and we may as well face the fact that this question can never be answered with certainty. Following the frequentist approach to statistics, the best we can do is to consider how our statistic would behave (given our model) if we had new data; in particular, if we were able to replicate the sampling experiment indefinitely and observe new realizations *ad infinitum*. Since we cannot actually carry out such a fantastic procedure, we must rely on theory to tell us what might happen. Applying theory to the model, we can *sometimes* work out and estimate the exact moments of T_n—mean, variance, etc., corresponding to the actual sample size n, whereas getting the exact finite-sample *distribution* is usually difficult, if not impossible. It is for this reason that we rely on asymptotic theory when the sample seems large enough to justify it. The asymptotic distribution is supposed to tell us—at least *roughly*—what we might see were we to carry out the indefinite replication.

Now what do we conclude if the asymptotic variance, say, differs considerably from its theoretically derived finite-sample counterpart? Just understanding intuitively how such disagreement can arise is a first step to drawing conclusions. The theoretically derived finite-sample variance is simply the estimate of the variance of realizations of T_n in our hypothetical metaexperiment. The asymptotic variance—σ_T^2/n if, as usual, $n^{1/2}(T_n - \theta) \rightsquigarrow N(0, \sigma_T^2)$—is the variance of the $N(\theta, \sigma_T^2/n)$ distribution that we take to represent the totality of results. At the intuitive level, how is it that they can differ? The first thing to say is that obtaining an appreciable difference would be unusual with models that are apt to be good descriptors of real data, although, as we have seen, one can devise examples to show the possibility. Indeed, we saw in Section 6.5.2 and in Example 7.21 that nVT_n can increase without bound as $n \to \infty$ or even be infinite at each n even though $\sigma_T^2 = \lim_{n\to\infty} n(\sigma_T^2/n)$ is finite. If we do observe $VT_n >> \sigma_T^2/n$, what it means is that there is some probability mass in the extreme tails of our finite-sample distribution which, for large enough n, has little effect on the probability of large error, even though it exerts a large influence on the second moment. If we are genuinely operating under squared-error loss, the large second moment is obviously of concern. If, however, we are willing simply to ignore rare events—and take the consequences—then we could be comfortable with the asymptotic approximation.[15] Of course, neither VT_n nor σ_T^2/n is of much relevance unless we are reasonably confident in the model!

7.1.2.3 *Asymptotically Efficient Estimators*

Two different c.a.n. estimators of the same constant θ may have different asymptotic variances. All else equal, one clearly prefers the one that delivers greater precision. If we are relying strictly on asymptotic properties, we would like, if possible, to find the one estimator with smallest asymptotic variance among all the possible c.a.n. estimators. It turns out that under condition (7.4) the Cramér–Rao bound applies to the asymptotic variance as well.[16]

Definition 7.7. Let θ be a parameter or function of parameters of a model that obeys condition (7.4), and let T_n be a c.a.n. estimator of θ. T_n is said to be *asymptotically efficient* if its asymptotic variance attains the Cramér–Rao bound.

Example 7.22. Elements of $\mathbf{X} := (X_1, X_2, ..., X_n)'$ are i.i.d. as $P(\mu)$ (Poisson), where $\mu > 0$. Set $\theta := e^{-\mu} < 1$, so that $f(x; \mu) = (\mu^x e^{-\mu}/x!) \mathbf{1}_{\aleph_0}(x) = (\log \theta^{-1})^x (\theta/x!) \mathbf{1}_{\aleph_0}(x)$. The Fisher information is

$$\mathcal{I} = -E\left[\frac{d^2 \log f\left(X; \log \theta^{-1}\right)}{d\theta^2}\right] = -\frac{1}{\theta^2 \log \theta},$$

and the Cramér–Rao bound is $(n\mathcal{I})^{-1} = -n^{-1}\theta^2 \log \theta = n^{-1}\theta^2\mu$. The asymptotic variance of $T_n = e^{-\bar{X}}$ was shown in Example 7.19 to be

---

[15]There are poignant examples of unforeseen—and large—consequences of ignoring small risks. The financial meltdowns of 1987 and 2008 can be attributed in large part to reliance on Gaussian asymptotics to model prices of financial assets. Such models work just fine *most* of the time, but improbable events can and do occur, and in today's interconnected markets shocks in one sector can destabilize others and feed back on themselves.

[16]There are examples of "superefficient" estimators that beat the Cramér-Rao bound for certain point values of θ by asymptotically "shrinking" the statistic toward those values. For example, if $\mu = 0$, $\bar{X}\left[.51_{[0,n^{-1/4})}\left(|\bar{X}|\right) + \mathbf{1}_{[n^{-1/4},\infty)}\left(|\bar{X}|\right)\right]$ would still converge a.s. to μ and have smaller asymptotic variance than $\bar{X}$ itself. Since one would need to know the true θ to design such an estimator, they are of little practical interest. For examples and discussion, see Cox and Hinkley (1974) or Lehmann and Casella (1998).

$n^{-1}\theta^2\sigma^2$. Since $\sigma^2 = \mu$ for the Poisson model, we conclude that T_n is an asymptotically efficient estimator of θ.

Asymptotically efficient estimators have a nice invariance property. Suppose statistic M_n is consistent, asymptotically normal, and asymptotically efficient for some parameter μ (not necessarily a first moment) that can take values in an open set $\mathcal{S} \subset \Re$; and let $\theta = g(\mu)$, where g is continuously differentiable in $\mathcal{S}$. Thus, the asymptotic variance of M_n is the inverse of the Fisher information, $n\mathcal{I}(\mu) = nE[d\log f(X;\mu)/d\mu]^2$. Likewise, the Cramér–Rao bound for the asymptotic variance for a c.a.n. estimator of θ is the inverse of

$$n\mathcal{I}(\theta) = nE\left[\frac{d\log f(X;\mu)}{d\theta}\right]^2$$

$$= nE\left[\frac{d\log f(X;\mu)}{d\mu} \cdot \frac{d\mu}{d\theta}\right]^2 = n\mathcal{I}(\mu)\,g'(\mu)^{-2}. \qquad (7.17)$$

Now $g(M_n)$ is consistent for θ by the continuity theorem, and an application of the delta method shows it to be asymptotically normal with asymptotic variance $g'(\mu)^2[n\mathcal{I}(\mu)]^{-1}$ (when $g'(\mu) \neq 0$). It follows that a continuously differentiable function of an asymptotically efficient estimator of a parameter is itself an asymptotically efficient estimator of the function of that parameter.

Exercise 7.17. *Use relation (7.17) and the result of Example 7.3 to show directly that the variance bound for estimators of $\theta = e^{-\mu}$ in the Poisson(μ) model is as given in Example 7.22.*

Example 7.23. Continuing Example 7.22, consider the estimator $T_n' := n^{-1}\sum_{j=1}^{n}\mathbf{1}_{\{0\}}(X_j)$, which is the relative frequency of zeros in the sample. T_n' converges a.s. to $E\mathbf{1}_{\{0\}}(X) = \Pr(X = 0) = e^{-\mu} = \theta$ by the strong law of large numbers. Exercise 7.16 shows that T_n' is also c.a.n., with asymptotic variance $n^{-1}\theta(1-\theta)$. But $n^{-1}\theta(1-\theta) > -n^{-1}\theta^2\log\theta$ for $\theta \in (0,1)$, so T_n' is not asymptotically efficient.

Example 7.24. By Example 7.8, when the population is distributed as $N(\mu,\sigma^2)$ the Cramér–Rao bound for the variance of an unbiased estimator of σ^2 is $2\sigma^4/n$, which is smaller than $VS^2 = 2\sigma^4/(n-1)$. On the other hand, the *asymptotic* variance of S^2 does attain the bound, as we can see by working out the asymptotic distribution of $\sqrt{n}\left(S^2 - \sigma^2\right)$. Now $(n-1)S^2/\sigma^2$ has the same $\chi^2(n-1)$ distribution as the sum of $n-1$ squared, independent standard normals, $\sum_{j=1}^{n-1} Z_j^2$. Letting

$$Y_n := \frac{\sum_{j=1}^{n-1} Z_j^2 - E\sum_{j=1}^{n-1} Z_j^2}{\sqrt{V\sum_{j=1}^{n-1} Z_j^2}} = \frac{\sum_{j=1}^{n-1} Z_j^2 - (n-1)}{\sqrt{2(n-1)}}$$

be the standardized sum, $\sqrt{n}\left(S^2 - \sigma^2\right)$ thus has the same distribution as $\sqrt{2n/(n-1)}\sigma^2 Y_n$. Since $Y_n \rightsquigarrow N(0,1)$ and $n/(n-1) \to 1$ as $n \to \infty$, we conclude that $\sqrt{n}\left(S^2 - \sigma^2\right) \rightsquigarrow N\left(0, 2\sigma^4\right)$ and hence that $2\sigma^4/n$ is the asymptotic variance of S^2. Therefore, S^2 is an asymptotically efficient estimator of the variance of a normal population.

Exercise 7.18. *Compare the asymptotic variances of $\sqrt{\pi}\mathfrak{G}/2$ and $S = +\sqrt{S^2}$ when $\{X_j\}_{j=1}^n$ are i.i.d. as $N\left(\mu, \sigma^2\right)$, where $\mathfrak{G} := \binom{n}{2}^{-1} \sum_{i<j} |X_i - X_j|$. (Hint: Recall Exercises 4.52 and 6.24.)*

7.2   Classical Methods of Estimation

There are many systematic approaches to obtaining estimates of population quantities. We describe in detail just the two most important ones: the method of moments and the method of maximum likelihood.

7.2.1   *The Method of Moments*

We have seen that there are two relevant settings for estimation: (i) a structured setting in which the data are modeled according to a specific parametric family, and (ii) an unstructured setting in which minimal assumptions are made about the data-generating process. In the latter situation it is usually feasible to estimate only quantities (or functions of quantities) that can be expressed as expectations. Thus, for r.v. X defined on a population we might want to infer the mean, the variance, some higher moment, or the probability associated with some Borel set. In each such case our interest is in $Eg(X)$ for some real-valued (or possibly complex-valued) function g; that is, our interest is in the first moment of $g(X)$ itself. The obvious estimator is the corresponding *sample* moment, $\bar{g}_n = n^{-1}\sum_{j=1}^n g(X_j)$, which is (i) unbiased, (ii) strongly consistent, and (iii) asymptotically normal if $Eg(X)^2 < \infty$. Thus, in the unstructured setting virtually all estimation can be considered a "method of moments", and all estimators can be classified as *method-of-moments* (m.o.m.) estimators.

The term "method of moments" is, however, usually applied to a certain class of estimators within the structured setting, where the data are assumed to have come from some particular family $\left\{F\left(x;\boldsymbol{\theta}\right):\boldsymbol{\theta}\in\boldsymbol{\Theta}\subset\Re^k\right\}$. In this case the interest is in estimating some or all of the elements of $\boldsymbol{\theta}$, for once a consistent estimator $\tilde{\boldsymbol{\theta}}_n$ is found we automatically have a consistent estimator of any continuous $h\left(\boldsymbol{\theta}\right)$. The method of moments often provides an easy way to obtain such a $\tilde{\boldsymbol{\theta}}_n$.

Here is how it works. To begin, choose k functions of X whose expectations are continuous functions of $\boldsymbol{\theta}$; i.e., functions $\{g_m\left(X\right)\}_{m=1}^k$ such that $E\left|g_m\left(X\right)\right|<\infty$ and $Eg_m\left(X\right)=\gamma_m\left(\boldsymbol{\theta}\right)$. The resulting multivariate transformation $\boldsymbol{\gamma}:\boldsymbol{\Theta}\to\boldsymbol{\Gamma}\subset\Re^k$ must also be one to one. Having estimated each $Eg_m\left(X\right)$ by its strongly consistent sample counterpart, $\bar{g}_{nm}=n^{-1}\sum_{j=1}^n g_m\left(X_j\right)$, now equate each $\bar{g}_{nm}$ to $\gamma_m(\tilde{\boldsymbol{\theta}}_n)$ and solve the resulting set of k equations for $\tilde{\boldsymbol{\theta}}_n$. Here, the notation $\tilde{\boldsymbol{\theta}}_n$ (as with $\mathbf{T}_n$ earlier) indicates that we now have an "estimator" as opposed to a parameter. Thus, writing out the k equations in vector form as

$$\bar{\mathbf{g}}_n=\left(\bar{g}_{n1},\bar{g}_{n2},...,\bar{g}_{nk}\right)'=\boldsymbol{\gamma}(\tilde{\boldsymbol{\theta}}_n)=\left(\gamma_1(\tilde{\boldsymbol{\theta}}_n),\gamma_2(\tilde{\boldsymbol{\theta}}_n),...,\gamma_k(\tilde{\boldsymbol{\theta}}_n)\right)',$$

one solves to express $\tilde{\boldsymbol{\theta}}_n$ as $\boldsymbol{\gamma}^{-1}(\bar{\mathbf{g}}_n)$. Given that the transformation $\boldsymbol{\gamma}:\boldsymbol{\Theta}\to\boldsymbol{\Gamma}$ is one to one and that the inverse $\boldsymbol{\gamma}^{-1}$ is continuous at $\left(\gamma_1\left(\boldsymbol{\theta}\right),\gamma_2\left(\boldsymbol{\theta}\right),...,\gamma_k\left(\boldsymbol{\theta}\right)\right)'$, it is clear that the m.o.m. estimators thus constructed are strongly consistent. If $\boldsymbol{\gamma}^{-1}$ is continuously differentiable as well (and with nonzero derivatives), then the delta method also delivers asymptotic normality.

Example 7.25. Modeling X as gamma($\boldsymbol{\theta}$), where $\boldsymbol{\theta}=\left(\theta_1,\theta_2\right)'$ comprises shape and scale parameters, respectively, take $g_1(X)=X$ and $g_2\left(X\right)=X^2$. Then $\gamma_1\left(\boldsymbol{\theta}\right)=EX=\theta_1\theta_2$, $\gamma_2\left(\boldsymbol{\theta}\right)=EX^2=VX+\left(EX\right)^2=\theta_1\theta_2^2+\theta_1^2\theta_2^2$, and $\bar{g}_{n1}=\bar{X}$ and $\bar{g}_{n2}=M_2'=n^{-1}\sum_{j=1}^n X_j^2$ are strongly consistent estimators. Solving

$$\bar{\mathbf{g}}_n=\begin{pmatrix}\bar{X}\\M_2'\end{pmatrix}=\boldsymbol{\gamma}(\tilde{\boldsymbol{\theta}}_n)=\begin{pmatrix}\tilde{\theta}_{n1}\tilde{\theta}_{n2}\\\tilde{\theta}_{n1}\tilde{\theta}_{n2}^2+\tilde{\theta}_{n1}^2\tilde{\theta}_{n2}^2\end{pmatrix}$$

gives as the m.o.m. estimators

$$\begin{pmatrix}\tilde{\theta}_{n1}\\\tilde{\theta}_{n2}\end{pmatrix}=\begin{pmatrix}\frac{\bar{X}^2}{M_2'-\bar{X}^2}\\\frac{M_2'-\bar{X}^2}{\bar{X}}\end{pmatrix}=\begin{pmatrix}\frac{\bar{X}^2}{M_2}\\\frac{M_2}{\bar{X}}\end{pmatrix}.$$

As the example indicates, one is free to choose both moments about the origin and moments about the mean as the functions that provide a link to the parameters. Lower-order moments are usually better than higher-order moments, because the corresponding sample estimators are typically more precise. Thus, $\bar{X}$ typically has lower variance than M_2, which in turn is more precise than M_3, etc. However, other features of the distribution besides integer-valued moments can also be used: (i) fractional or negative moments if X is a.s. positive; (ii) values of the m.g.f. or c.f., $M(t)$ or $\phi(t)$, at various values of t; (iii) values of c.d.f. $F(x)$ at various x; or (iv) probabilities associated with other Borel sets than $(-\infty, x]$. For instance, taking $g(X) = \mathbf{1}_{\{0\}}(X)$ and $\gamma(\mu) = e^{-\mu}$, Example 7.23 shows that a feasible m.o.m. estimator of the intensity parameter μ of the Poisson model is $\tilde{\mu}_n = -\log \bar{g}_n$, where $\bar{g}_n = T_n' = n^{-1} \sum_{j=1}^n \mathbf{1}_{\{0\}}(X_j)$. Of course, while this flexibility in designing consistent estimators has its advantages, there clearly can be no guarantee that such a flexible approach will deliver an asymptotically efficient estimator.

Exercise 7.19. *Elements of* $\mathbf{X} := (X_1, X_2, ..., X_n)'$ *are i.i.d. as* $P(\mu)$ *(Poisson). Use the delta method to find the asymptotic variance of*

$$\tilde{\mu}_n := -\log\left(\frac{1}{n}\sum_{j=1}^n \mathbf{1}_{\{0\}}(X_j)\right).$$

Comparing with the Cramér–Rao bound worked out in Example 7.3, show that this particular m.o.m. estimator is not asymptotically efficient. Then find a m.o.m. estimator that does attain the minimum-variance bound.

An important tool in econometric applications is an extension of the m.o.m. procedure known as the *generalized method of moments* (*g.m.m.*).[17] The idea is to increase asymptotic efficiency by bringing to bear more sample information than is strictly needed to obtain consistent estimates. To see the essence of the idea, consider estimating just a scalar parameter θ, and suppose one can find $m > 1$ functions $\{g_i(X)\}_{i=1}^m$ with finite expectations $\{Eg_i(X) = \gamma_i(\theta)\}_{i=1}^m$ such that each transformation $\gamma_i : \Theta \to \Re$ is one to one. Any one element of $\boldsymbol{\gamma}(\theta) = (\gamma_1(\theta), \gamma_2(\theta), ..., \gamma_m(\theta))'$ *identifies* θ, in the sense that the equation $g_i = \gamma_i(\theta)$ could be solved uniquely for θ. The remaining $m - 1$ conditions are then *overidentifying* restrictions that would be redundant if values of any of $\{\gamma_i(\theta)\}_{i=1}^m$ were known. However, they are *not* known in fact, and the system of equations $\bar{\mathbf{g}}_n - \boldsymbol{\gamma}(\theta) = \mathbf{0}$ is generally inconsistent, in the algebraic sense that no single value of θ

---

[17]For details of this procedure see Hansen's (1982) seminal paper and/or subsequent treatments by Davidson and MacKinnon (1993) or Green (1997).

satisfies all of them for almost all realizations of estimators $\{\bar{g}_{ni}\}_{i=1}^{m}$. On the other hand, given some positive-definite matrix $\mathbf{M}$, there is usually a unique $\tilde{\theta}_n$ that minimizes the quadratic form

$$Q_n = (\bar{\mathbf{g}}_n - \boldsymbol{\gamma}(\theta))' \mathbf{M} (\bar{\mathbf{g}}_n - \boldsymbol{\gamma}(\theta)), \qquad (7.18)$$

even though there is usually no θ that equates Q_n to zero. The g.m.m. estimator is such a minimizer. It is at least weakly consistent if $\bar{\mathbf{g}}_n \to^p \boldsymbol{\gamma}(\theta)$ and elements of $\boldsymbol{\gamma}(\theta)$ are one to one. Given a particular choice of $\bar{\mathbf{g}}_n$ with nonsingular covariance matrix

$$\mathbf{G}_n(\theta) = E (\bar{\mathbf{g}}_n - \boldsymbol{\gamma}(\theta)) (\bar{\mathbf{g}}_n - \boldsymbol{\gamma}(\theta))'$$

the best asymptotic efficiency is achieved by taking $\mathbf{M} = \mathbf{G}_n(\theta)^{-1}$ or a consistent estimator thereof.

G.m.m. is particularly useful in econometric applications when a structural model either is incompletely specified or is sufficiently complicated that the method of maximum likelihood (our next topic) cannot be easily be applied. For example, suppose we have a multivariate sample $\{Y_j, \mathbf{X}_j\}_{j=1}^{n}$, where $\mathbf{X}_j = \{X_{j\ell}\}_{l=1}^{m}$ is the jth observation on m "covariates". Take the model to be $Y_j = h(\mathbf{X}_j; \boldsymbol{\theta}) + U_j, j \in \{1, 2, ..., n\}$, where $\boldsymbol{\theta}$ is a k-vector of unknown parameters, $h : \Re^m \times \Re^k \to \Re$ is a known function, and $\{U_j\}_{j=1}^{n}$ are unobserved errors. Each realization of U_j could arise from measurement errors in Y_j or from the effects of other influences besides $\mathbf{X}_j$. While it would be hard to formulate a specific model for $\{U_j\}_{j=1}^{n}$, it might at least be assumed that they have zero expected values and are *orthogonal to* (uncorrelated with) the observed covariates; i.e., that $EU_j = 0$ and $EU_j\mathbf{X}_j = \mathbf{0}$ for each j. If m (the number of variables) exceeds k (the number of parameters) the orthogonality conditions $EU_j\mathbf{X}_j = E[Y_j - h(\mathbf{X}_j; \boldsymbol{\theta})]\mathbf{X}_j = \mathbf{0}$ provide $m - k$ overidentifying restrictions on $\boldsymbol{\theta}$.[18] The sample counterpart of these conditions is $\left\{\bar{v}_{n\ell}(\boldsymbol{\theta}) := n^{-1} \sum_{j=1}^{n} [Y_j - h(\mathbf{X}_j; \boldsymbol{\theta})] X_{j\ell} = 0\right\}_{\ell=1}^{m}$. To set up for g.m.m. estimation, the m-vector $\bar{\mathbf{v}}_n(\boldsymbol{\theta}) = (\bar{v}_{n1}(\boldsymbol{\theta}), \bar{v}_{n2}(\boldsymbol{\theta}), ..., \bar{v}_{nm}(\boldsymbol{\theta}))'$ of sample orthogonality conditions would take the place of $\bar{\mathbf{g}}_n - \boldsymbol{\gamma}(\theta)$ in a quadratic form such as (7.18). An initial consistent estimate $\tilde{\boldsymbol{\theta}}_n^{(1)}$ of $\boldsymbol{\theta}$ could be found by taking $\mathbf{M} = \mathbf{I}_m$ (the identity matrix) and minimizing $Q_n^{(1)} = \bar{\mathbf{v}}_n(\boldsymbol{\theta})' \bar{\mathbf{v}}_n(\boldsymbol{\theta})$. This would yield a consistent estimate of the $m \times m$ covariance matrix of $\bar{\mathbf{v}}_n(\boldsymbol{\theta})$, as

$$\tilde{\mathbf{V}}_n = \left\{n^{-1} \sum_{j=1}^{n} \left[Y_j - h\left(\mathbf{X}_j; \tilde{\boldsymbol{\theta}}_n^{(1)}\right)\right] X_{j\ell} \left[Y_j - h\left(\mathbf{X}_j; \tilde{\boldsymbol{\theta}}_n^{(1)}\right)\right] X_{jq}\right\}_{\ell,q=1}^{m},$$

---

[18] Assuming that h is such that any subset of k of the equations can be solved uniquely for—and thus identifies—$\boldsymbol{\theta}$.

which would in turn produce a second-stage estimate $\tilde{\boldsymbol{\theta}}_n^{(2)}$ as minimizer of

$$Q_n^{(2)} = \bar{\mathbf{v}}_n(\boldsymbol{\theta})'\, \tilde{\mathbf{V}}_n^{-1} \bar{\mathbf{v}}_n(\boldsymbol{\theta}).$$

7.2.2   The Method of Maximum Likelihood

Although m.o.m. and its g.m.m. extension find many useful applications in statistics and econometrics, the method of maximum likelihood remains the standard tool for point estimation when a parametric model is available. As an introductory example, consider again the problem of estimating the parameter μ for the Poisson model, $f(x; \mu) = (\mu^x e^{-\mu}/x!)\, \mathbf{1}_{\aleph_0}(x)$. The joint p.m.f. of random sample $\mathbf{X} := (X_1, X_2, ..., X_n)'$ is

$$f(\mathbf{x}; \mu) = \frac{\mu^{\sum_{j=1}^n x_j} e^{-n\mu}}{x_1! x_2! \cdots x_n!} \mathbf{1}_{\aleph_0^n}(\mathbf{x}) = \frac{\mu^{n\bar{x}} e^{-n\mu}}{x_1! x_2! \cdots x_n!} \mathbf{1}_{\aleph_0^n}(\mathbf{x}).$$

If μ is known, $f(\mathbf{x}; \mu)$ tells us the probability of getting the particular sample realization $\mathbf{x}$. Clearly, for a given μ some samples are more likely than others. Conversely, the probability of obtaining any *particular* sample varies with μ. Figure 7.1 shows plots of $f(\mathbf{x}; \mu)$ versus μ for two actual (pseudo-random) samples of $n = 10$ from the Poisson law with $\mu = 5$: $\mathbf{x} = (4, 5, 3, 4, 5, 5, 4, 8, 4, 8)'$ (sample 1) and $\mathbf{x} = (2, 7, 4, 3, 5, 2, 5, 7, 4, 4)'$.

Regarding things from a statistical perspective, the sample $\mathbf{x}$ itself is known once the sampling experiment has been completed. What is *not* known is the value of the parameter that pins down the particular member of the parametric family that generated the data. Let us then adopt this point of view and regard $f(\mathbf{x}; \mu)$ not as a function of the known $\mathbf{x}$ but of the unknown μ. Having done so, we now refer to $f(\mathbf{x}; \mu)$ as the *likelihood function* of the sample, and denote it $L(\mathbf{x}; \mu)$. The method of maximum likelihood (m.l.) has us take as the estimate of μ the value that maximizes the likelihood function. In other words, the m.l. estimate of μ is the value that maximizes the probability of obtaining the sample that was actually observed. Obviously, this estimate will turn out to be a function of sample data $\mathbf{x}$. *Ex ante*, the same function of an as-yet-unobserved sample $\mathbf{X}$ is called the m.l. *estimator (m.l.e.)*.

Generalizing from the Poisson example, let the parametric model for the population be given by the family of densities $\{f(x; \boldsymbol{\theta}) : \boldsymbol{\theta} \in \boldsymbol{\Theta} \subset \Re^k\}$ with respect to dominating measure $\mathcal{M}$. Then the likelihood function corresponding to the realization $\mathbf{x}$ of a random sample of size n is

$$L(\mathbf{x}; \boldsymbol{\theta}) = \prod_{j=1}^n f(x_j; \boldsymbol{\theta}).$$

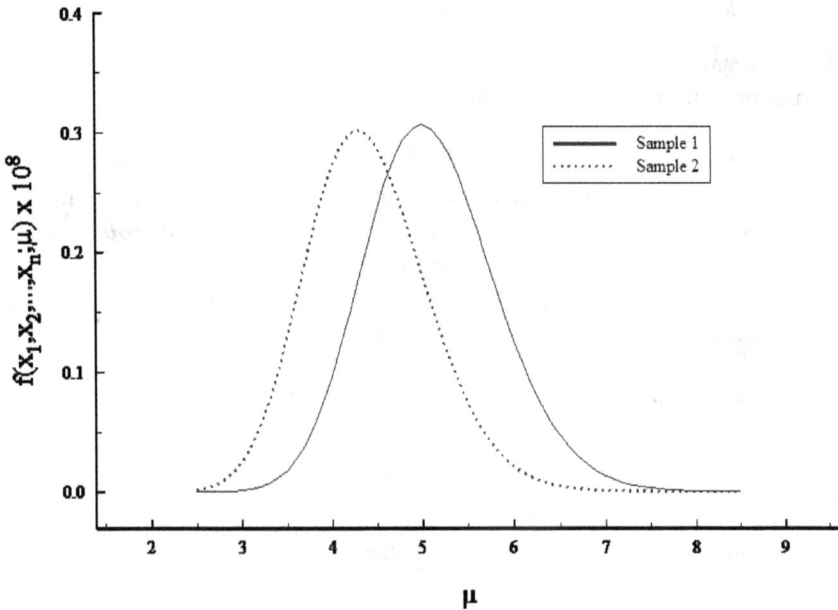

Fig. 7.1   Two Poisson(μ) likelihood functions *vs.* μ.

When r.v. X is purely discrete, as in the Poisson case, $L(\mathbf{x}; \boldsymbol{\theta})$ does represent the actual probability associated with sample point $\mathbf{x} \in \Re^n$. When X is purely (absolutely) continuous, we think of $L(\mathbf{x}; \boldsymbol{\theta})$ as being *proportional* to the probability mass in a "small" open hypercube $\mathbb{C}_\varepsilon := \times_{j=1}^n (x_j - \varepsilon, x_j + \varepsilon)$ about $\mathbf{x}$, as $L(\mathbf{x}; \boldsymbol{\theta}) \propto \int_{\mathbb{C}_\varepsilon} \mathbb{P}_\mathbf{X}(d\mathbf{u})$. In either case the m.l. estimate, denoted $\hat{\boldsymbol{\theta}}_n$, is the solution (if one exists) to $\sup_{\boldsymbol{\theta} \in \Theta} L(\mathbf{x}; \boldsymbol{\theta})$. That is, $\hat{\boldsymbol{\theta}}_n$ is a point within or on the boundary of parameter space Θ that satisfies $L\left(\mathbf{x}; \hat{\boldsymbol{\theta}}_n\right) \geq L(\mathbf{x}; \boldsymbol{\theta})$ for all $\boldsymbol{\theta} \in \Theta$.[19] *Ex ante*, before the sample realization is observed, we regard the m.l. *estimator* as the random variable $\hat{\boldsymbol{\theta}}_n$ that solves $\sup_{\boldsymbol{\theta} \in \Theta} L(\mathbf{X}; \boldsymbol{\theta})$.

Since the natural log function is one-to-one and increasing, solutions to $\sup_{\boldsymbol{\theta} \in \Theta} L(\mathbf{X}; \boldsymbol{\theta})$ and $\sup_{\boldsymbol{\theta} \in \Theta} \log L(\mathbf{X}; \boldsymbol{\theta})$ coincide. As a practical matter, the log-likelihood function is usually easier to optimize.

---

[19]The m.l. estimate is often defined as the solution to $\max_{\boldsymbol{\theta} \in \Theta} L(\mathbf{x}; \boldsymbol{\theta})$ rather than to $\sup_{\boldsymbol{\theta} \in \Theta} L(\mathbf{x}; \boldsymbol{\theta})$. The two solutions are usually the same, but the sup definition does yield reasonable m.l. estimates in anomalous cases where there is otherwise no solution, as in Example 7.33 and Exercise 7.22 below.

7.2.2.1 *Finding M.L. Estimates*

There are various ways to find m.l. estimates, when they exist. The following indicate the range of situations encountered.

(i) The typical circumstance is that the likelihood function is differentiable with respect to θ and has a unique maximum at a point in the interior of Θ, at which the k first-order conditions $\partial \log L(\mathbf{x}; \theta) / \partial \theta = \mathbf{0}$ are satisfied. In this context these first-order conditions are called the *likelihood equations*. There is often an explicit solution that gives $\hat{\theta}_n$ as a function of observables (data and n).

Example 7.26. With $X \sim P(\mu)$, the log-likelihood is

$$\log L(\mathbf{x}; \mu) = n\bar{x} \log \mu - n\mu - \sum_{j=1}^{n} \log x_j!$$

provided each $x_j \in \aleph_0$. The single likelihood equation,

$$\frac{\partial \log L(\mathbf{x}; \mu)}{\partial \mu} = \frac{n\bar{x}}{\mu} - n = 0,$$

has as its solution $\bar{x}$. The second-order condition is satisfied since $\log L(\mathbf{x}; \mu)$ is globally concave, so the m.l. estimate is $\hat{\mu}_n = \bar{x}$. The m.l. estimates for the two samples in the introductory example are $\hat{\mu}_{10} = 5.0$ and 4.3, respectively, corresponding to the peaks of the likelihood plots in Fig. 7.1.

Example 7.27. If $X \sim N(\mu, \sigma^2)$, $\mu \in \Re, \sigma^2 > 0$, the log-likelihood is

$$\log L(\mathbf{x}; \mu, \sigma^2) = -\frac{n}{2} \log(2\pi) - \frac{n}{2} \log \sigma^2 - \frac{1}{2\sigma^2} \sum_{j=1}^{n} (x_j - \mu)^2,$$

and the likelihood equations are

$$\frac{\partial \log L(\mathbf{x}; \mu, \sigma^2)}{\partial \mu} = \frac{1}{\sigma^2} \sum_{j=1}^{n} (x_j - \mu) = 0$$

$$\frac{\partial \log L(\mathbf{x}; \mu, \sigma^2)}{\partial \sigma^2} = -\frac{n}{2\sigma^2} + \frac{1}{2\sigma^4} \sum_{j=1}^{n} (x_j - \mu)^2 = 0.$$

Provided $x_1, x_2, ..., x_n$ do not all have the same value (a situation considered in Exercise 7.22 below), the unique interior solution is $\hat{\mu}_n = \bar{x}$,

$\hat{\sigma}_n^2 = n^{-1} \sum_{j=1}^{n} (x_j - \bar{x})^2 = m_2$. Checking the second-order conditions, the matrix of second derivatives evaluated at $\bar{x}, m_2$,

$$\frac{\partial^2 \log L(X; \mu, \sigma^2)}{\partial (\mu, \sigma^2)' \, \partial (\mu, \sigma^2)} \bigg|_{\mu = \bar{x}, \sigma^2 = m_2} = \begin{pmatrix} -\frac{n}{m_2} & 0 \\ 0 & -\frac{n}{2m_2^2} \end{pmatrix},$$

is seen to be negative definite (first principal minor negative, second principal minor positive). Thus, $(\hat{\mu}_n, \hat{\sigma}_n^2) = (\bar{x}, m_2)$ is the m.l. estimate.

Exercise 7.20. *If* $X \sim IG(\mu, \lambda), \mu >, \lambda > 0$ *(inverse Gaussian), the log-likelihood is*

$$\log L(\mathbf{x}; \mu, \lambda) = -\frac{n}{2} \log (2\pi) + \frac{n}{2} \log \lambda - \frac{3}{2} \sum_{j=1}^{n} \log x_j - \frac{\lambda}{2\mu^2} \sum_{j=1}^{n} \frac{(x_j - \mu)^2}{x_j},$$

(7.19)

provided all the $\{x_j\}$ are positive. Find the m.l. estimates $\hat{\mu}_n$ and $\hat{\lambda}_n$. What happens if all the $\{x_j\}$ have the same (positive) value?

Exercise 7.21. *Continuing Exercise 7.20, show that $\hat{\mu}_n$ and $\hat{\lambda}_n (n - 3) / n$ are minimum-variance unbiased estimators of μ and λ. (Hint: Use the results of Exercises 5.13 and 7.8.)*

(ii) In many applications the likelihood equations are highly nonlinear in the parameters, and solutions must be found by numerical methods. When there are many parameters it is usually easier to use a numerical procedure to optimize $L(\mathbf{x}; \boldsymbol{\theta})$ or $\log L(\mathbf{x}; \boldsymbol{\theta})$ directly than to solve a system of nonlinear likelihood equations. However, in some cases it is helpful to "concentrate" the likelihood by replacing one or more parameters with implicit solutions derived from one or more likelihood equations. Method-of-moments estimators are often used as starting values in numerical search routines.

Example 7.28. If $X \sim \Gamma(\alpha, \beta)$, the log-likelihood is

$$\log L(\mathbf{x}; \alpha, \beta) = -n \log \Gamma(\alpha) - n\alpha \log \beta + (\alpha - 1) \sum_{j=1}^{n} \log x_j - \beta^{-1} \sum_{j=1}^{n} x_j$$

provided all the $\{x_j\}$ are positive, and the likelihood equations are then

$$\frac{\partial \log L(\mathbf{x}; \alpha, \beta)}{\partial \alpha} = -n\Psi(\alpha) - n \log \beta + \sum_{j=1}^{n} \log x_j = 0 \quad (7.20)$$

$$\frac{\partial \log L(\mathbf{x}; \alpha, \beta)}{\partial \beta} = -\frac{n\alpha}{\beta} + \beta^{-2} \sum_{j=1}^{n} x_j = 0. \quad (7.21)$$

Here $\Psi(\alpha) := d \log \Gamma(\alpha) / d\alpha$ is the *digamma function*, values of which can approximated by series expansions.[20] Obviously, the solutions must be found numerically. One approach is to use (7.21) to express β implicitly in terms of α, as $\beta = \bar{x}/\alpha$, then plug this into (7.20) to obtain a single nonlinear equation in α. The root of this equation must be found numerically—e.g., by Newton's method. Alternatively, $\bar{x}/\alpha$ can be substituted for β in $\log L(\mathbf{x};\alpha,\beta)$, and then the maximizer $\hat{\alpha}_n$ of the resulting *concentrated* log likelihood can be found by a hill-climbing technique or a "golden-section" search. Either way, an initial guess is needed to start the numerical search, and the m.o.m. estimate $\tilde{\alpha}_n = \bar{x}^2/m_2$ is a reasonable choice (provided $\{x_j\}_{j=1}^n$ are not all equal). Second-order conditions for a maximum of $\log L$ are satisfied for $\hat{\beta}_n = \bar{x}/\hat{\alpha}_n$ and any $\hat{\alpha}_n > 0$.

(iii) Sometimes—notably, when the support of X depends on $\boldsymbol{\theta}$—the likelihood function has a maximum in the interior of $\boldsymbol{\Theta}$, but the maximizing value of $\hat{\boldsymbol{\theta}}_n$ does not satisfy the likelihood equations.

Example 7.29. For $f(x;\theta) = e^{-(x-\theta)}\mathbf{1}_{[\theta,\infty)}(x)$, $\theta \in \Re$, the log-likelihood is

$$\log L(\mathbf{x};\theta) = -\sum_{j=1}^n (x_j - \theta) = -n\bar{x} + n\theta$$

provided each $x_j \geq \theta$. This increases linearly with θ up to $\theta = x_{(1:n)}$, the (realization of the) first *order statistic* of the sample (the sample minimum), beyond which $L(\mathbf{x};\theta) = 0$. The m.l. estimate is therefore $\hat{\theta}_n = x_{(1:n)}$.

Example 7.30. X has p.d.f. $f(x;\theta) = \theta^{-1}\mathbf{1}_{[0,\theta]}(x)$, $\theta > 0$. The log-likelihood is $\log L(\mathbf{x};\theta) = -n\log\theta$ provided that each $x_j \leq \theta$. Provided also that not all elements of $\mathbf{x}$ are zero, the log likelihood increases monotonically as θ declines to $\theta = x_{(n:n)} > 0$, which is (the realization of) the nth order statistic (the sample maximum), below which $L(\mathbf{x};\theta) = 0$. Thus, $\hat{\theta}_n = x_{(n:n)}$ is the m.l. estimate.

---

[20]See, for example, Abramowitz and Stegun (1972, pp. 258 ff.). Good approximations can also be made numerically using routines that calculate the log of the gamma function, as

$$\Psi(\alpha) \doteq \frac{\log \Gamma(\alpha + h) - \log \Gamma(\alpha - h)}{2h}$$

for some small h.

(iv) In some cases the likelihood is not everywhere differentiable, and so the likelihood equations do not exist for all values of $\boldsymbol{\theta}$.

Example 7.31. X is distributed as $B(m, \theta)$ (binomial) with unknown values of both $m \in \aleph$ and $\theta \in (0, 1)$. Likelihood function

$$L(\mathbf{x}; m, \theta) = \prod_{j=1}^{n} \binom{m}{x_j} \theta^{x_j} (1 - \theta)^{m - x_j}$$

is obviously not differentiable with respect to the integer-valued m; however, it is differentiable with respect to θ, and likelihood equation

$$\frac{\partial \log L(\mathbf{x}; m, \theta)}{\partial \theta} = \frac{n\bar{x}}{\theta} - \frac{n(m - \bar{x})}{1 - \theta} = 0$$

yields the implicit solution $\theta = \bar{x}/m$. By using this to concentrate the likelihood one can search numerically for the maximizing value $\hat{m}_n$.[21]

Example 7.32. If X has p.d.f.

$$f(x; \phi, \psi) = \frac{1}{2\psi} e^{-|x - \phi|/\psi}, \phi \in \Re, \psi > 0,$$

(Laplace with scale parameter ψ, centered at ϕ), then

$$\log L(\mathbf{x}; \phi, \psi) = -n \log 2 - n \log \psi - \psi^{-1} \sum_{j=1}^{n} |x_j - \phi|.$$

This is differentiable with respect to ψ but it is not differentiable with respect to ϕ at the sample points $\{x_j\}_{j=1}^{n}$. However, for any given (positive) value of ψ, $\log L(\mathbf{x}; \phi, \psi)$ attains its maximum at the value of ϕ that minimizes $\sum_{j=1}^{n} |x_j - \phi|$. The m.l. estimate of ϕ is therefore any number that qualifies as a median of the (discrete, *empirical*) distribution of sample data $\{x_j\}_{j=1}^{n}$.[22] If either (i) n is odd or (ii) n is even and realizations $x_{(n/2:n)}$ and $x_{(n/2+1:n)}$ of order statistics $X_{(n/2:n)}$ and $X_{(n/2+1:n)}$ are equal, then the m.l. estimate $\hat{\phi}_n$ is the (unique) sample median, but in neither case does $\partial \log L(\mathbf{x}; \phi, \psi)/\partial \phi$ exist at $\hat{\phi}_n$. If n is even and $x_{(n/2:n)} \neq x_{(n/2+1:n)}$, then $L(\mathbf{x}; \phi, \psi)$ attains its maximum

---

[21] A maximum can usually be found by evaluating $L(\mathbf{x}; m, \bar{x}/m)$ at a sequence of integer values of m beginning with $x_{(n:n)}$ (the sample maximum); however, in samples for which $\bar{x} - s^2$ is near zero or negative the likelihood often seems to increase monotonically. Obviously, the standard m.o.m. estimates $\hat{m}_n = \bar{x}^2/(\bar{x} - s^2)$ and $\hat{\theta}_n = \bar{x}/\hat{m}_n$ are also useless in this situation. Simulations with $\theta > .2$ seldom produce such errant samples, and m.l. estimates then seem to have excellent properties. For further discussion see Johnson *et al.* (1992, pp. 125-129) and the references cited there.

[22] Recall Proposition 3.4.1.1 on page 168.

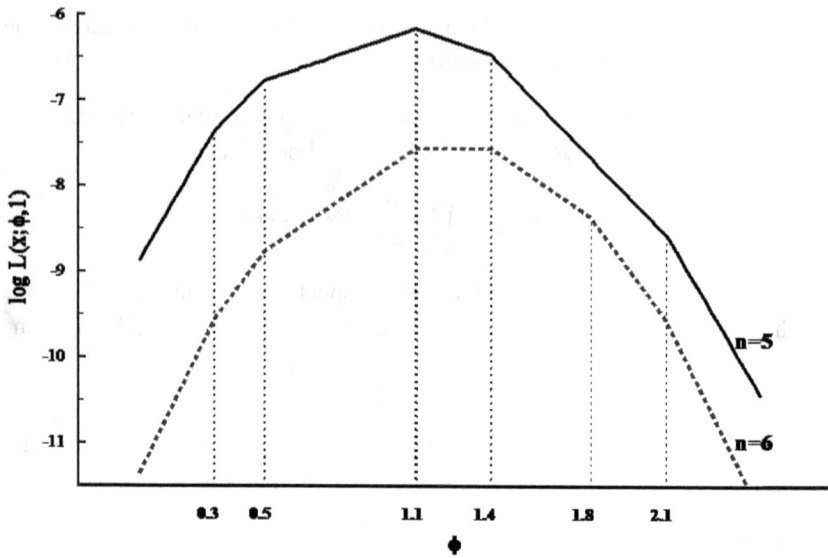

Fig. 7.2   Laplace log likelioods *vs.* ϕ with $\mathbf{x} = \{.3, .5, 1.1, 1.4, 2, 1\}$ ($n = 5$) and $\mathbf{x} = \{.3, .5, 1.1, 1.4, 1.8, 2, 1\}$ ($n = 6$).

and $\partial \log L(\mathbf{x}; \phi, \psi) / \partial \phi = 0$ for all $\phi \in \left[x_{(n/2:n)}, x_{(n/2+1:n)} \right]$, and so $\hat{\phi}_n$ is not uniquely determined. (Figure 7.2 illustrates with plots of Laplace log likelihoods for a sample of $n = 5$ and for another sample with one added observation.) However, provided that the $\{x_j\}_{j=1}^n$ are not all equal, the m.l.e. of ψ is uniquely determined from the corresponding likelihood equation as $\hat{\psi}_n = n^{-1} \sum_{j=1}^n |x_j - \hat{\phi}_n|$—this quantity being the same for all values of median $\hat{\phi}_n$. If the $\{x_j\}_{j=1}^n$ *are* all equal, then we have an instance of the following case.

(v) Sometimes there is no interior maximum and $\sup_{\boldsymbol{\theta} \in \Theta} L(\mathbf{x}; \boldsymbol{\theta})$ is attained on the boundary of Θ.

Example 7.33. With $X \sim P(\mu), \mu > 0$, as in Example 7.26, suppose $\mathbf{x} = \mathbf{0}$ (the sample realization comprising zero values only). The likelihood function is then proportional to $e^{-n\mu}$ and $\sup_{\mu > 0} L(\mathbf{0}; \mu)$ occurs at $\mu = 0$, which is the m.l. estimate. (Note, however, that this agrees with the solution $\hat{\mu}_n = \bar{x}$ that applies so long as all elements of $\mathbf{x}$ are nonnegative.) Likewise, if $\mathbf{x} = \mathbf{0}$ in Example 7.30 and if $\mathbf{x} = x\mathbf{1} = (x, x, ..., x)'$ in Example 7.32, then $\sup_{\theta > 0} L(\mathbf{0}; \theta)$ and

$\sup_{\phi \in \Re, \psi > 0} L(x1; \phi, \psi)$ occur at $\theta = 0$ and at $\phi = x, \psi = 0$, respectively (which again agree in these particular cases with the solutions $\hat{\theta}_n$, and $\hat{\phi}_n, \hat{\psi}_n$ that otherwise apply).

Exercise 7.22. *With $X \sim N(\mu, \sigma^2)$ suppose the sample realization satisfies $\mathbf{x} = x_1 \mathbf{1}$. Show that $\sup_{\mu \in \Re, \sigma^2 > 0} L(x_1 \mathbf{1}; \mu, \sigma^2)$ occurs at $\hat{\mu}_n = x_1$ and $\hat{\sigma}_n^2 = 0$. Explain why this "boundary" problem is less apt to occur than the one in Example 7.33.*

(vi) Sometimes, notably in finite *mixture* models, the likelihood function is unbounded at more than one point on the boundary of Θ, there being then no unique solution to $\sup_{\theta \in \Theta} L(\mathbf{x}; \theta)$. In this case the m.l. estimate, as we have defined it, does not exist.[23]

Example 7.34. X is distributed as $N(\mu, \sigma^{2Y})$ conditional on Y, where Y is distributed as $B(1, 1/2)$ (Bernoulli with "success" probability $1/2$). That is, $X(\omega)$ can be either the realization of $W_\sigma \sim N(\mu, \sigma^2), \mu \in \Re, \sigma^2 > 0$, or of $W_1 \sim N(\mu, 1), \mu \in \Re$, each with probability $1/2$. The p.d.f. is

$$f(x; \mu, \sigma^2) = \frac{1}{2\sqrt{2\pi\sigma^2}} e^{-(x-\mu)^2/(2\sigma^2)} + \frac{1}{2\sqrt{2\pi}} e^{-(x-\mu)^2/2}.$$

For each $\mathbf{x} \in \Re^n$ likelihood $L(\mathbf{x}; \mu, \sigma^2)$ is proportional to

$$\prod_{j=1}^{n} \left[\frac{1}{\sigma} e^{-(x_j - \mu)^2/(2\sigma^2)} + e^{-(x_j - \mu)^2/2} \right],$$

and satisfies $\lim_{\sigma^2 \to 0} L(\mathbf{x}; \mu, \sigma^2) = +\infty$ for each $\mu \in \{x_1, x_2, ..., x_n\}$. For example, when $n = 2$,

$$L(\mathbf{x}; \mu, \sigma^2) \propto \left(\frac{1}{\sigma} + 1 \right) \left[\frac{1}{\sigma} e^{-(x_2 - x_1)^2/(2\sigma^2)} + e^{-(x_2 - x_1)^2/2} \right]$$

for $\mu \in \{x_1, x_2\}$.

(vii) If the data are realizations of a r.v. that is neither purely discrete nor purely absolutely continuous, then the likelihood function would be expressed as

$$L(\mathbf{x}; \theta) = \prod_{\{j: x_j \in \mathcal{D}\}} [F(x_j; \theta) - F(x_j-; \theta)] \prod_{\{j: x_j \in \mathcal{C}\}} F'(x_j; \theta),$$

---

[23] However, in certain mixture problems the estimator corresponding to the largest *finite* value of $L(\mathbf{X}; \theta)$ has been found to have good properties. See Everitt and Hand (1981, pp. 8-9).

where $\mathcal{D}$ is the set of discontinuity points of F and $\mathcal{C}$ comprises the points at which F is differentiable (any remaining continuity points necessarily forming a set of measure zero). As in the purely continuous case, we think of this as being proportional to the probability mass in a neighborhood of $\mathbf{x}$, as $L(\mathbf{x};\boldsymbol{\theta}) \propto \prod_{j=1}^{n} dF(x_j;\boldsymbol{\theta})$, where $dF(x;\boldsymbol{\theta}) := [F(x;\boldsymbol{\theta}) - F(x-;\boldsymbol{\theta})]\, 1_{\mathcal{D}}(x) + F'(x;\boldsymbol{\theta})\, 1_{\mathcal{C}}(x) \cdot dx.$[24]

Example 7.35. X has absolutely continuous c.d.f. F_X and p.d.f. f_X, but its realizations are *censored* to give observations of $W := a1_{(-\infty,a]}(X) + X1_{(a,b)}(X) + b1_{[b,+\infty)}(X)$. (Recall Exercise 4.56.) Thus,

$$dF_W(w) = F_X(a)\, 1_{\{a\}}(w) + f_X(w)\, 1_{(a,b)}(w) \cdot dw + [1 - F_X(b)]\, 1_{\{b\}}(w),$$

and

$$L(\mathbf{w};\boldsymbol{\theta}) = F_X(a)^{n_a}\,[1 - F_X(b)]^{n_b}\,\prod_{\{j:a<w_j<b\}} f_X(w_j),$$

where n_a, n_b have the obvious interpretation.

7.2.2.2 Properties of M.L. Estimators

When they do exist, m.l. estimators have many attractive properties.

Invariance and Sufficiency   If maximum likelihood is used to estimate σ^2 from normal data, as $\hat{\sigma}_n^2 = m_2$, then it is natural to take $\sqrt{m_2}$ as the estimate of σ. Is this (square-root) function of the estimate also the m.l. estimate of the (square-root) function of σ^2? It is, in fact, and this is a particular example of what is called the *invariance* property of maximum likelihood estimators. Suppose, more generally, that $\hat{\boldsymbol{\theta}}_n$ is the m.l. estimator in a parametric model $\{f(x;\boldsymbol{\theta}) : \boldsymbol{\theta} \in \boldsymbol{\Theta} \subset \Re^k\}$, and let $\mathbf{g} : \boldsymbol{\Theta} \to \boldsymbol{\Gamma} \subset \Re^m$ be a vector-valued function of the parameters, with elements $\{\gamma_j = g_j(\boldsymbol{\theta})\}_{j=1}^{m}$. Then the m.l. estimator of $\boldsymbol{\gamma}$ is $\hat{\boldsymbol{\gamma}}_n = \mathbf{g}(\hat{\boldsymbol{\theta}}_n)$. This is easily seen to be true in the case that $k = m = 1$ and g is one to one. The general result can be established as follows.[25] Let $\boldsymbol{\Theta}_{\boldsymbol{\gamma}}$ be the set of values of $\boldsymbol{\theta}$ satisfying the constraint $\mathbf{g}(\boldsymbol{\theta}) = \boldsymbol{\gamma}$; i.e., $\boldsymbol{\Theta}_{\boldsymbol{\gamma}} = \{\boldsymbol{\theta} : \mathbf{g}(\boldsymbol{\theta}) = \boldsymbol{\gamma}\}$. If $\mathbf{g}$ is not one to one,

---

[24]Specifically, if n_0 of $\{x_j\}_{j=1}^{n}$ are in $\mathcal{C}$ and $n - n_0$ in $\mathcal{D}$, then $L(\mathbf{x};\boldsymbol{\theta})$ is proportional to the product of (i) the probability mass associated with the points in $\mathcal{D}$ and (ii) the mass in a "small" n_0-dimensional subspace containing $\mathbf{x}$.

Note that if we take $N(\cdot)$ to be counting measure on the measurable space comprising $\mathcal{D}$ and all it subsets and $\lambda(\cdot)$ to be Lebesgue measure on $(\Re, \mathcal{B})$, then, since N and λ are mutually singular, we could write $L(\mathbf{x};\boldsymbol{\theta})$ in the usual way as $\prod_{j=1}^{n} f(x_j;\boldsymbol{\theta})$, where f is the density with respect to measure $\mathcal{M} := N + \lambda$.

[25]The argument follows that of Casella and Berger (1990, p. 294).

then the likelihood is not necessarily a function of γ, because more than one value of L can correspond to a given γ. To define a proper function of γ, set $L^*(\mathbf{x};\gamma) = \sup_{\theta \in \Theta_\gamma} L(\mathbf{x};\theta)$. This is the *largest* value (actually, the supremum) of the likelihood subject to the constraint $\mathbf{g}(\theta) = \gamma$, which is clearly single-valued. Let us take the m.l. estimate of γ to be the value $\hat{\gamma}_n$ such that $L^*(\mathbf{x};\hat{\gamma}_n) = \sup_{\gamma \in \Gamma} L^*(\mathbf{x};\gamma)$. Then

$$L^*(\mathbf{x};\hat{\gamma}_n) = \sup_{\gamma \in \Gamma} \left\{ \sup_{\theta \in \Theta_\gamma} L(\mathbf{x};\theta) \right\} = \sup_{\theta \in \Theta} L(\mathbf{x};\theta) = L\left(\mathbf{x};\hat{\theta}_n\right).$$

But $L\left(\mathbf{x};\hat{\theta}_n\right) = \sup_{\theta \in \Theta_{\mathbf{g}(\hat{\theta}_n)}} L(\mathbf{x};\theta)$, since if the largest value of the likelihood occurs at $\hat{\theta}_n$ it is also the largest value consistent with $g(\theta) = \mathbf{g}(\hat{\theta}_n)$. Finally, since $\sup_{\theta \in \Theta_{\mathbf{g}(\hat{\theta}_n)}} L(\mathbf{x};\theta) = L^*\left[\mathbf{x};\mathbf{g}(\hat{\theta}_n)\right]$, we conclude that $\hat{\gamma}_n = \mathbf{g}(\hat{\theta}_n)$.

Exercise 7.23. *Suppose that θ is scalar valued and that $g : \Theta \to \Gamma$ is one to one. If $\hat{\theta}_n$ satisfies the first-order condition $\partial \log L(\mathbf{x};\theta)/\partial\theta = 0$, and if $g'\left(\hat{\theta}_n\right)$ exists and is not equal to zero, show that $\hat{\gamma}_n = g(\hat{\theta}_n)$ is the m.l. estimate of $\gamma = g(\theta)$.*

Example 7.27, in which $(\bar{X}, M_2) = (\hat{\mu}, \hat{\sigma}^2)$ was shown to be the m.l. estimator of (μ, σ^2) in the normal model, demonstrates that m.l. estimators are not necessarily unbiased, for $EM_2 = E(n-1)S^2/n = (n-1)\sigma^2/n$. Obviously, therefore, m.l. estimators are not necessarily *minimum-variance* unbiased (m.v.u.). However, suppose that the model $\{f(x;\theta) : \theta \in \Re^k\}$, is such that there is a collection $\mathbf{T}_n \in \Re^k$ of complete minimal-sufficient statistics for θ. We saw in Section 7.1.1.3 that any function of such a collection that is unbiased for θ is also m.v.u. We also saw that a sufficient collection exists if the joint p.d.f./p.m.f. of the sample factors as $f_\mathbf{X}(\mathbf{x};\theta) = g(\mathbf{t};\theta)h(\mathbf{x})$, where $\mathbf{t} = \mathbf{T}_n(\mathbf{x})$ represents a realization of $\mathbf{T}_n$. In the present context we express this condition in terms of the likelihood function:

$$L(\mathbf{x};\theta) = g(\mathbf{t};\theta)h(\mathbf{x}). \tag{7.22}$$

It is now clear that maximizing $L(\mathbf{x};\theta)$ with respect to θ is equivalent to maximizing $g(\mathbf{t};\theta)$, so that the m.l. estimator $\hat{\theta}_n$ is necessarily a function of the jointly sufficient (vector-valued) statistic $\mathbf{T}_n$. Consequently, when the factorization (7.22) holds and $\mathbf{T}_n$ is complete, a function of the m.l. estimator $\hat{\theta}_n$ that is unbiased (if such exists) is also m.v.u. For instance, in the normal case we have already seen that $(\hat{\mu}, n\hat{\sigma}^2/(n-1)) = (\bar{X}, S^2)$ is unbiased and complete sufficient for (μ, σ^2) and therefore m.v.u. More generally, for all r.v.s with densities in the exponential class,

$$f(x; \boldsymbol{\theta}) = b(\boldsymbol{\theta}) c(x) e^{\boldsymbol{\eta}(\boldsymbol{\theta})' \mathbf{T}(x)},$$

unbiased functions of the m.l. estimators $\hat{\boldsymbol{\theta}}_n$ are m.v.u. and therefore globally efficient when $\boldsymbol{\eta}$ lies in a space of the required dimensionality.

Consistency, Asymptotic Normality, and Efficiency Various sets of sufficient conditions are known under which m.l. estimators from random samples converge either strongly or weakly to the true parameters. To introduce the ideas, let $\{f(x; \boldsymbol{\theta}) : \boldsymbol{\theta} \in \boldsymbol{\Theta} \subset \Re^k\}$ be the parametric family of models, with $\mathbb{P}_X(\cdot; \boldsymbol{\theta})$ the corresponding induced probability measure and $\mathbb{X} \subset \Re$ a support, with $f(x; \boldsymbol{\theta}) > 0$ for $x \in \mathbb{X}$. Let $\boldsymbol{\theta}_0 \in \boldsymbol{\Theta}$ be the true value of $\boldsymbol{\theta}$ that defines the particular model in the class from which the data are generated. An obvious requirement for the consistency of any estimator is that $\boldsymbol{\theta}_0$ be *identified*, in the sense that there be no $\boldsymbol{\theta}_1$ such that $\mathbb{P}_X(B; \boldsymbol{\theta}_1) = \mathbb{P}_X(B; \boldsymbol{\theta}_0)$ for all Borel sets B. Were such a $\boldsymbol{\theta}_1$ to exist, there would clearly be no way for the data to distinguish $\boldsymbol{\theta}_0$ from $\boldsymbol{\theta}_1$.[26] Assuming that parameter space $\boldsymbol{\Theta}$ is a compact subset of $\Re^k$ (e.g., a closed, bounded rectangle), that X is either discrete for all $\boldsymbol{\theta}$ or else continuous for all $\boldsymbol{\theta}$, and that $f(x; \boldsymbol{\theta})$ meets certain other technical conditions, Wald (1949) proves *strong* consistency (almost-sure convergence) of the m.l.e. using the following lemma.[27]

Lemma 7.1. *Let X have density f with respect to measure $\mathcal{M}$ (e.g., counting measure or Lebesgue measure) and induced probability measure $\mathbb{P}_X$, with $f(x) > 0$ for $x \in \mathbb{X}$, where $\mathbb{P}_X(\mathbb{X}) = 1$ and such that $E|\log f(X)| < \infty$. If g is another density (relative to $\mathcal{M}$) then*

$$E \log g(X) \le E \log f(X), \tag{7.23}$$

with equality if and only if $g = f$ almost everywhere with respect to $\mathbb{P}_X$.

---

[26] For example, if X were modeled as $N(\theta_1 + \theta_2, \sigma^2)$, then $\mathbb{P}_X(B; \boldsymbol{\theta}, \sigma^2)$ would be the same for all $\boldsymbol{\theta}$ such that $\theta_1 + \theta_2 = \mu$.

[27] Wolfowitz (1949) gives a simpler proof that $\hat{\boldsymbol{\theta}}_n \to \boldsymbol{\theta}$ in probability.

Proof. Put $E \log [g(X)/f(X)] = -\infty$ if $g = 0$ on any set of positive $\mathbb{P}_X$ measure. Otherwise, write

$$E \log \left[\frac{g(X)}{f(X)} \right] \leq \log E \left[\frac{g(X)}{f(X)} \right] \quad \text{(Jensen)}$$

$$= \log \int_{\mathbb{X}} \frac{g(x)}{f(x)} f(x) \mathcal{M}(dx)$$

$$= \log \int_{\mathbb{X}} g(x) \mathcal{M}(dx)$$

$$= \log 1$$

$$= 0.$$

Equality clearly holds if $g = f$ a.e. $\mathbb{P}_X$. On the other hand the concavity of the log function implies strict inequality whenever $g \neq f$ on any set $\mathcal{U}$ such that $\mathbb{P}_X(\mathcal{U}) > 0$. $\qquad\square$

Wald's proof of the strong consistency of the m.l.e. is lengthy and technical, but the following argument[28] gives the essence of the idea. The log-likelihood function for i.i.d. sample $\mathbf{X} = (X_1, X_2, ..., X_n)'$ is $\log L(\mathbf{X}; \boldsymbol{\theta}) = \sum_{j=1}^{n} \log f(X_j; \boldsymbol{\theta})$. Now $n^{-1} \log L(\mathbf{X}; \boldsymbol{\theta}) \to^{a.s.} E \log L(\mathbf{X}; \boldsymbol{\theta})$ for any fixed $\boldsymbol{\theta} \in \boldsymbol{\Theta}$ by the strong law of large numbers, and thus (7.23) implies that $\Pr\left[n^{-1} \log L(\mathbf{X}; \boldsymbol{\theta}) < n^{-1} \log L(\mathbf{X}; \boldsymbol{\theta}_0) \right] \to 1$ as $n \to \infty$ for any $\boldsymbol{\theta} \neq \boldsymbol{\theta}_0$. But m.l. estimator $\hat{\boldsymbol{\theta}}_n$ satisfies $\log L(\mathbf{X}; \hat{\boldsymbol{\theta}}_n) = \sup_{\boldsymbol{\theta} \in \boldsymbol{\Theta}} \log L(\mathbf{X}; \boldsymbol{\theta})$, and so $\log L(\mathbf{X}; \hat{\boldsymbol{\theta}}_n) \geq \log L(\mathbf{X}; \boldsymbol{\theta}_0)$ for each n. Thus, assuming that $\boldsymbol{\theta}_0$ is identified, $\hat{\boldsymbol{\theta}}_n$ cannot converge to any point other than $\boldsymbol{\theta}_0$.

Wald's condition that $\boldsymbol{\Theta}$ be compact is violated in many of our standard models. For example, if $X \sim N(\mu, \sigma^2)$ then μ can take any real value and σ^2 any positive value. However, building on the ideas of Cramér (1946, pp. 499-504), it is possible to prove weak consistency even when $\boldsymbol{\Theta}$ is unbounded, provided that the likelihood is sufficiently smooth in $\boldsymbol{\theta}$.[29] Specifically, under conditions that often do hold in applications there is a consistent root $\hat{\boldsymbol{\theta}}_n$ of the likelihood equations $\partial \log L(\mathbf{X}; \boldsymbol{\theta}) / \partial \boldsymbol{\theta} = \mathbf{0}$. If there is only one such root, then solving the likelihood equations delivers the consistent estimator. If there are multiple roots, one would need to find another consistent estimate (e.g., by the method of moments) and take the root nearest to it. Fortunately, there is at most one root in models

---

[28] Due to Kendall and Stuart (1973, p. 42).

[29] Scholz (1985) surveys the extensive literature on properties of m.l. estimators. The books by Lehmann and Casella (1998) and Cox and Hinkley (1974) provide more detail. The consistency proof given below follows the argument of Lehmann and Casella.

of the exponential class, and so the multiplicity problem is often avoided. An advantage of Cramér's approach is that the consistent root $\hat{\boldsymbol{\theta}}_n$ is also asymptotically normal and asymptotically efficient when the likelihood is sufficiently well behaved. These facts about the estimators' approximate large-n sampling distributions will be used in Section 7.3 and Chapter 8 to build confidence intervals and to test parametric hypotheses.

Cramér's conditions can be stated as follows:

(i) True value $\boldsymbol{\theta}_0$ lies within an open set $\mathcal{S}_{\boldsymbol{\theta}}$ contained in $\boldsymbol{\Theta}$.
(ii) $E \partial \log f(X; \boldsymbol{\theta}_0) / \partial \boldsymbol{\theta} := E \, \partial \log f(X; \boldsymbol{\theta}) / \partial \boldsymbol{\theta}|_{\boldsymbol{\theta} = \boldsymbol{\theta}_0} = \mathbf{0}$.
(iii)

$$
\begin{aligned}
\mathcal{I}(\boldsymbol{\theta}_0) \; : \; = E & \left[\frac{\partial \log f(X; \boldsymbol{\theta}_0)}{\partial \boldsymbol{\theta}} \right] \left[\frac{\partial \log f(X; \boldsymbol{\theta}_0)}{\partial \boldsymbol{\theta}'} \right] \\
= -E & \left[\frac{\partial^2 \log f(X; \boldsymbol{\theta}_0)}{\partial \boldsymbol{\theta} \partial \boldsymbol{\theta}'} \right],
\end{aligned}
$$

and $\mathcal{I}(\boldsymbol{\theta}_0)$ is positive definite.
(iv) Elements of Hessian matrix $\partial^2 \log f(\mathbf{X}; \boldsymbol{\theta}) / \partial \boldsymbol{\theta} \partial \boldsymbol{\theta}'$ are themselves differentiable, and the derivatives are bounded within $\mathcal{S}_{\boldsymbol{\theta}}$ by functions not depending on $\boldsymbol{\theta}$ that have finite expectations.

These conditions will be referred to henceforth as the *regularity conditions for m.l. estimation*. As we saw in Section 7.1.1 in connection with global efficiency, conditions (ii) and (iii) rule out models in which the support depends on $\boldsymbol{\theta}$. (Of course, they also imply the *existence* of the first two derivatives almost everywhere with respect to $\mathbb{P}_X(\cdot; \boldsymbol{\theta}_0)$.) Here is an example that shows how to check the conditions.

Example 7.36. If $X \sim W(\theta, 1)$ (Weibull with unit scale), $\log f(x; \theta) = \log \theta + (\theta - 1) \log x - x^\theta$ for $x > 0$ and $\theta \in \Theta = (0, \infty)$.

(i) There is obviously an open set $\mathcal{S}_\theta$ within Θ that contains any $\theta_0 > 0$.
(ii) We have $Ed \log f(X; \theta) / d\theta = \theta^{-1} + E \log X - E(X^\theta \log X)$. From $E e^{t \log X} = EX^t = \Gamma(1 + t/\theta)$ we find that $E \log X = \theta^{-1} \Gamma'(1) = \theta^{-1} \Psi(1)$, where Ψ is the digamma function; and one can determine from a good table of definite integrals (e.g., Gradshteyn *et al.* (2000)) that $E(X^\theta \log X) = \theta^{-1} \Psi(2) = \theta^{-1}[1 + \Psi(1)]$. Thus, condition (ii) is satisfied.
(iii) Using $E\left[X^{k\theta}(\log X)^2\right] = \theta^{-2}\left[\Psi(k)^2 + \zeta(2, k)\right]$ and $\Psi(k + 1) = k^{-1} + \Psi(k)$ for $k \in \{1, 2, 3\}$, where $\zeta(2, k) := \sum_{j=0}^{\infty}(j + k)^{-2}$ and $\zeta(2, 0) :=$

$\sum_{j=1}^{\infty} j^{-2} = \zeta(2) = \pi^2/6$ (the Riemann zeta function), one can show that

$$-E\left[\frac{d^2 \log f(X;\theta)}{d\theta^2}\right] = \theta^{-2}\left\{1 + [1 + \Psi(1)]^2 + \zeta(2)\right\}$$

$$= E\left[\frac{d \log f(X;\theta)}{d\theta}\right]^2 > 0.$$

(iv) Finally, third derivative $\left|d^3 \log f(X;\theta)/d\theta^3\right| = \left|2\theta^{-3} + X^\theta (\log X)^3\right|$ is bounded by $2\theta^{-3} + \left|(\log X)^3\right|$ for $X < 1$ and by $2\theta^{-3} + X^{3+\theta}$ for $X \geq 1$, both of which are integrable.

Conveniently, the conditions do not have to be checked for models in the exponential class. From a minimal canonical representation

$$f(x;\boldsymbol{\eta}) = a(\boldsymbol{\eta}) c(x) e^{\boldsymbol{\eta}' \mathbf{T}} \tag{7.24}$$

with $\boldsymbol{\eta} = \boldsymbol{\eta}(\boldsymbol{\theta})$ a one-to-one, differentiable function of the original parameters, we have

$$\frac{\partial \log f(x;\boldsymbol{\eta})}{\partial \boldsymbol{\eta}} = \frac{\partial \log a(\boldsymbol{\eta})}{\partial \boldsymbol{\eta}} + \mathbf{T}.$$

Thus, recalling (4.6), $E\partial \log f(X;\boldsymbol{\eta})/\partial \boldsymbol{\eta} = \mathbf{0}$, so that condition (ii) is met. Moreover,

$$E\left(\frac{\partial \log f(X;\boldsymbol{\eta})}{\partial \boldsymbol{\eta}}\right)\left(\frac{\partial \log f(X;\boldsymbol{\eta})}{\partial \boldsymbol{\eta}'}\right) = E(\mathbf{T}-E\mathbf{T})(\mathbf{T}-E\mathbf{T})' = V\mathbf{T}$$

which is positive definite given that the representation (7.24) is minimal (i.e., $\mathbf{T}$ satisfies no affine constraint). Also, applying (4.7) shows that $V\mathbf{T} = \partial^2 \log a(\boldsymbol{\eta})/\partial \boldsymbol{\eta}\partial \boldsymbol{\eta}'$, verifying condition (iii). Finally, condition (iv) is satisfied because the third derivatives are simply constants. Since $\boldsymbol{\eta}(\boldsymbol{\theta})$ is one to one, the conditions in terms of derivatives with respect to $\boldsymbol{\theta}$ are met as well.

We can now sketch a proof that the likelihood equations have a consistent, asymptotically normal and asymptotically efficient root. With $\boldsymbol{\theta}_0$ as the true value let $\mathcal{S}_\delta := \{\boldsymbol{\theta} : |\boldsymbol{\theta} - \boldsymbol{\theta}_0| = \delta\}$, where $\delta > 0$ is small enough that $\mathcal{S}_\delta$ lies within the open subset $\mathcal{S}_\theta$ of Θ specified by condition (i). Thus, $\mathcal{S}_\delta$ is the set of parameter values that lie at a distance δ from $\boldsymbol{\theta}_0$. Now put

$$\mathbb{X}_\delta := \{\mathbf{x} : L(\mathbf{x};\boldsymbol{\theta}_0) > L(\mathbf{x};\boldsymbol{\theta}), \boldsymbol{\theta} \in \mathcal{S}_\delta\}.$$

$\mathbb{X}_\delta$ is thus the set of realizations of $\mathbf{X}$ at which the likelihood of the sample is larger at $\boldsymbol{\theta}_0$ than at any point on the "sphere" $\mathcal{S}_\delta$ that we have chosen to contain it. ($\mathbb{X}_\delta$ might well be empty for certain n and δ, but we will see that its measure under $\mathbb{P}_\mathbf{X}$ for any $\delta > 0$ approaches unity as $n \rightarrow \infty$.) For any $\mathbf{x}$ in this set the likelihood has a local maximum somewhere *within* the sphere $\mathcal{S}_\delta$, and at this local maximum the likelihood equations will be satisfied. For such $\mathbf{x} \in \mathbb{X}_\delta$ let $\hat{\boldsymbol{\theta}}_n(\delta; \mathbf{x})$ be the point of local maximum, so that $\partial L\left(\mathbf{x}; \hat{\boldsymbol{\theta}}_n(\delta; \mathbf{x})\right)/\partial\boldsymbol{\theta} := \partial L(\mathbf{x}; \boldsymbol{\theta})/\partial\boldsymbol{\theta}\Big|_{\boldsymbol{\theta}=\hat{\boldsymbol{\theta}}_n(\delta;\mathbf{x})} = \mathbf{0}$. Of course, this implies that $\left|\hat{\boldsymbol{\theta}}_n(\delta; \mathbf{x}) - \boldsymbol{\theta}_0\right| < \delta$ for $\mathbf{x} \in \mathbb{X}_\delta$. (For $\mathbf{x} \in \mathbb{X}\backslash\mathbb{X}_\delta$, $\hat{\boldsymbol{\theta}}_n(\delta; \mathbf{x})$ can be defined arbitrarily.) But Lemma 7.1 implies that $\lim_{n\to\infty} \Pr(\mathbf{X} \in \mathbb{X}_\delta) = \lim_{n\to\infty} E\mathbf{1}_{\mathbb{X}_\delta}(\mathbf{X}) = 1$ for any $\delta > 0$, since $E \log L(\mathbf{X}; \boldsymbol{\theta})$ is greatest at $\boldsymbol{\theta}_0$. It follows that $\lim_{n\to\infty} \Pr\left[\left|\hat{\boldsymbol{\theta}}_n(\delta; \mathbf{X}) - \boldsymbol{\theta}_0\right| < \delta\right] \to 1$, and since this is true for any $\delta > 0$ we conclude that there is a root of the likelihood equations that converges in probability to $\boldsymbol{\theta}_0$.

To prove asymptotic normality, begin by expanding $\partial \log L(\mathbf{X}; \hat{\boldsymbol{\theta}}_n)/d\boldsymbol{\theta} \equiv \partial \log L(\mathbf{X}; \boldsymbol{\theta})/d\boldsymbol{\theta}\big|_{\boldsymbol{\theta}=\hat{\boldsymbol{\theta}}_n}$ about $\boldsymbol{\theta}_0$ to get

$$\mathbf{0} = \frac{\partial \log L(\mathbf{X}; \boldsymbol{\theta}_0)}{\partial\boldsymbol{\theta}} + \frac{\partial^2 \log L(\mathbf{X}; \boldsymbol{\theta}_0)}{\partial\boldsymbol{\theta}\partial\boldsymbol{\theta}'}\left(\hat{\boldsymbol{\theta}}_n - \boldsymbol{\theta}_0\right) + \frac{1}{2}\left(\hat{\boldsymbol{\theta}}_n - \boldsymbol{\theta}_0\right)'\mathbf{Q}_n\left(\hat{\boldsymbol{\theta}}_n - \boldsymbol{\theta}_0\right),$$

where $k \times k \times k$ array $\mathbf{Q}_n$ involves the third derivatives evaluated at an intermediate point between $\hat{\boldsymbol{\theta}}_n$ and $\boldsymbol{\theta}_0$.[30] Then $\sqrt{n}\left(\hat{\boldsymbol{\theta}}_n - \boldsymbol{\theta}_0\right)$ equals

$$\left[-\frac{1}{n}\frac{\partial^2 \log L(\mathbf{X}; \boldsymbol{\theta}_0)}{\partial\boldsymbol{\theta}\partial\boldsymbol{\theta}'} - \frac{1}{2n}\left(\hat{\boldsymbol{\theta}}_n - \boldsymbol{\theta}_0\right)'\mathbf{Q}_n\right]^{-1} \cdot n^{-1/2}\frac{\partial \log L(\mathbf{X}; \boldsymbol{\theta}_0)}{\partial\boldsymbol{\theta}}.$$

In the bracketed factor the term

$$-\frac{1}{n}\frac{\partial^2 \log L(\mathbf{X}; \boldsymbol{\theta}_0)}{\partial\boldsymbol{\theta}\partial\boldsymbol{\theta}'} = -\frac{1}{n}\sum_{j=1}^{n}\frac{\partial^2 \log f(X_j; \boldsymbol{\theta}_0)}{\partial\boldsymbol{\theta}\partial\boldsymbol{\theta}'}$$

converges to $-E\left[\partial^2 \log f(X; \boldsymbol{\theta}_0)/\partial\boldsymbol{\theta}\partial\boldsymbol{\theta}'\right] = \mathcal{I}(\boldsymbol{\theta}_0)$ by condition (iii) and the s.l.l.n. The boundedness of third derivatives $\partial^3 f(X; \boldsymbol{\theta})/\partial\theta_i\partial\theta_j\partial\theta_k$ by integrable functions under condition (iv), together with the consistency of $\hat{\boldsymbol{\theta}}_n$, imply that the term involving $\mathbf{Q}_n$ converges in probability to zero. Finally, by conditions (i) and (ii) and the central limit theorem we have

$$n^{-1/2}\frac{\partial \log L(\mathbf{X}; \boldsymbol{\theta}_0)}{\partial\boldsymbol{\theta}} = n^{-1/2}\sum_{j=1}^{n}\frac{\partial \log f(X_j; \boldsymbol{\theta}_0)}{\partial\boldsymbol{\theta}} \rightsquigarrow N(\mathbf{0}, \mathcal{I}(\boldsymbol{\theta}_0)).$$

---

[30]To visualize the array operations in the last term, think of $\mathbf{Q}_n$ as a "stack" of $k \times k$ matrices. Then $\left(\hat{\boldsymbol{\theta}}_n - \boldsymbol{\theta}_0\right)'\mathbf{Q}_n$ is a stack of $1 \times k$ vectors (a $k \times k$ matrix), and post-multiplying by $\left(\hat{\boldsymbol{\theta}}_n - \boldsymbol{\theta}_0\right)$ yields a stack of k *scalars*—i.e., a $k \times 1$ vector.

Thus, by Mann–Wald it follows that

$$\sqrt{n}\left(\hat{\boldsymbol{\theta}}_n - \boldsymbol{\theta}_0\right) \rightsquigarrow N\left(\mathbf{0}, \mathcal{I}\left(\boldsymbol{\theta}_0\right)^{-1}\right)$$

and hence that $\hat{\boldsymbol{\theta}}_n$ has asymptotic variance equal to Cramér–Rao bound $n^{-1}\mathcal{I}\left(\boldsymbol{\theta}_0\right)^{-1}$.

Example 7.37. For the exponential-class model $X \sim \Gamma\left(\alpha, \beta\right)$ the likelihood equations are

$$\frac{\partial \log L\left(\mathbf{x}; \alpha, \beta\right)}{\partial \alpha} = -n\Psi\left(\alpha\right) - n\log\beta + \sum_{j=1}^{n} \log x_j = 0$$

$$\frac{\partial \log L\left(\mathbf{x}; \alpha, \beta\right)}{\partial \beta} = -\frac{n\alpha}{\beta} + \beta^{-2}\sum_{j=1}^{n} x_j = 0,$$

with unique solutions for $\hat{\alpha}_n$ and $\hat{\beta}_n$. The Fisher information is

$$n\mathcal{I}\left(\alpha, \beta\right) = n\begin{pmatrix} \Psi'\left(\alpha\right) & \frac{1}{\beta} \\ \frac{1}{\beta} & \frac{\alpha}{\beta^2} \end{pmatrix},$$

where $\Psi'\left(\alpha\right) := d^2 \log \Gamma\left(\alpha\right)/d\alpha^2$ is the *trigamma function*.[31] The gamma family being in the exponential class, we have

$$\sqrt{n}\begin{pmatrix} \hat{\alpha}_n - \alpha \\ \hat{\beta}_n - \beta \end{pmatrix} \rightsquigarrow N\left(\mathbf{0}, \mathcal{I}\left(\alpha, \beta\right)^{-1}\right).$$

The asymptotic covariance matrix of the m.l. estimators is

$$
\begin{aligned}
n^{-1}\mathcal{I}\left(\alpha, \beta\right)^{-1} &= n^{-1}\begin{pmatrix} \frac{\alpha}{\beta^2|\mathcal{I}(\alpha,\beta)|} & -\frac{1}{\beta|\mathcal{I}(\alpha,\beta)|} \\ -\frac{1}{\beta|\mathcal{I}(\alpha,\beta)|} & \frac{\Psi'(\alpha)}{|\mathcal{I}(\alpha,\beta)|} \end{pmatrix} \\
&= \begin{pmatrix} \frac{\alpha}{n[\alpha\Psi'(\alpha)-1]} & -\frac{\beta}{n[\alpha\Psi'(\alpha)-1]} \\ -\frac{\beta}{n[\alpha\Psi'(\alpha)-1]} & \frac{\beta^2\Psi'(\alpha)}{n[\alpha\Psi'(\alpha)-1]} \end{pmatrix}.
\end{aligned}
\tag{7.25}
$$

Exercise 7.24. *This extends Exercise 7.20. Verify that Cramér's conditions hold for the inverse Gaussian distribution, and find the asymptotic covariance matrix of m.l. estimators $\hat{\mu}_n, \hat{\lambda}_n$ when $\{X_j\}_{j=1}^{n}$ are i.i.d. as $IG\left(\mu, \lambda\right)$.*

---

[31] Tables of values are widely available; e.g., in Abramowitz and Stegun (1972). Numerical approximations can be made as

$$\Psi'\left(\alpha\right) \doteq \frac{\log\Gamma\left(\alpha + h\right) + \log\Gamma\left(\alpha - h\right) - 2\log\Gamma\left(\alpha\right)}{h^2}$$

for some small h.

Exercise 7.25. $\{X_j, Y_j\}_{j=1}^n$ *is a random sample from a bivariate normal distribution. Each of X and Y has mean μ and variance equal to unity, and their coefficient of correlation is a known constant ρ, with $0 < |\rho| < 1$.*

(i) Find m.l. estimator $\hat{\mu}_n$.

(ii) Find the Fisher information for μ and determine whether $\hat{\mu}_n$ is m.v.u.

(iii) Compare the information content of bivariate sample $\{X_j, Y_j\}_{j=1}^n$ with that of an i.i.d. sample $\{X_j\}_{j=1}^{2n}$ that comprises $2n$ observations on the first variable alone.

(iv) How do the information content of $\{X_j, Y_j\}_{j=1}^n$ and the variance of $\hat{\mu}_n$ behave as $\rho \to -1$? Give an intuitive explanation.

The next example and the following exercises apply maximum likelihood theory to *regression models*, which are treated at length in Chapter 11.

Example 7.38. **U** is an n-vector of unobservable r.v.s distributed as multivariate normal with mean **0** and nonsingular covariance matrix $\boldsymbol{\Sigma}$, elements of which are known constants. The n-vector of observable r.v.s **Y** is modeled as $\mathbf{Y} = \mathbf{X}\boldsymbol{\beta} + \mathbf{U}$, where **X** is an $n \times k$ matrix of known constants having rank k, and $\boldsymbol{\beta}$ is a k-vector of unknown constants, to be estimated. The log likelihood for **Y** is then

$$\log L\left(\mathbf{y}; \boldsymbol{\beta}\right) = -\frac{n}{2}\log\left(2\pi\right) - \frac{1}{2}\log|\boldsymbol{\Sigma}| - \frac{1}{2}\left(\mathbf{y} - \mathbf{X}\boldsymbol{\beta}\right)'\boldsymbol{\Sigma}^{-1}\left(\mathbf{y} - \mathbf{X}\boldsymbol{\beta}\right),$$

and the m.l. estimator for $\boldsymbol{\beta}$ is

$$\hat{\boldsymbol{\beta}} = \left(\mathbf{X}'\boldsymbol{\Sigma}^{-1}\mathbf{X}\right)^{-1}\mathbf{X}'\boldsymbol{\Sigma}^{-1}\mathbf{Y}.$$

Substituting for **Y** gives

$$\hat{\boldsymbol{\beta}} = \boldsymbol{\beta} + \left(\mathbf{X}'\boldsymbol{\Sigma}^{-1}\mathbf{X}\right)^{-1}\mathbf{X}'\boldsymbol{\Sigma}^{-1}\mathbf{U},$$

from which we see directly that $E\hat{\boldsymbol{\beta}} = \boldsymbol{\beta}$ and that

$$\begin{aligned}V\hat{\boldsymbol{\beta}} &= E\left(\hat{\boldsymbol{\beta}} - \boldsymbol{\beta}\right)\left(\hat{\boldsymbol{\beta}} - \boldsymbol{\beta}\right)' \\ &= \left(\mathbf{X}'\boldsymbol{\Sigma}^{-1}\mathbf{X}\right)^{-1}\mathbf{X}'\boldsymbol{\Sigma}^{-1}\boldsymbol{\Sigma}\boldsymbol{\Sigma}^{-1}\mathbf{X}\left(\mathbf{X}'\boldsymbol{\Sigma}^{-1}\mathbf{X}\right)^{-1} \\ &= \left(\mathbf{X}'\boldsymbol{\Sigma}^{-1}\mathbf{X}\right)^{-1},\end{aligned}$$

an exact result.

Exercise 7.26. *Show that the expression for $V\hat{\boldsymbol{\beta}}$ in Example 7.38 equals the inverse of the Fisher information and conclude that $\hat{\boldsymbol{\beta}}$ is the m.v.u. estimator.*

Exercise 7.27. *Put* $\mathbf{Y} = \mathbf{X}\boldsymbol{\beta} + \mathbf{U}$ *with* $E\mathbf{U} = \mathbf{0}$ *and* $V\mathbf{U} = \boldsymbol{\Sigma}$ *as before, but no longer assume* $\mathbf{U}$ *to be multivariate normal. Show that* $\hat{\boldsymbol{\beta}}$ *is still the best* linear *unbiased estimator of* $\boldsymbol{\beta}$; *that is, the minimum-variance estimator of the form* $\mathbf{CY}$, *where* $\mathbf{C}$ *is a* $k \times n$ *matrix of rank* k. *(Hint: Set* $\mathbf{C} := \left(\mathbf{X}'\boldsymbol{\Sigma}^{-1}\mathbf{X} \right)^{-1} \mathbf{X}'\boldsymbol{\Sigma}^{-1} + \mathbf{D}$. *What condition must* $\mathbf{D}$ *satisfy if* $\mathbf{CY}$ *is to be unbiased?)*

7.2.2.3 *Estimating Asymptotic Variances and Covariances*

As functions of the Fisher information under regularity conditions, asymptotic variances and covariances of m.l. estimators are themselves typically functions of some or all of the parameters they are used to estimate. Using m.l. estimators to build confidence intervals for the parameters, as we do in Section 7.3, and to carry out tests of parametric hypotheses, as described in Chapter 8, typically requires having a consistent estimate of information matrix $\mathcal{I}(\boldsymbol{\theta})$. How can this be found? There are two general situations.

(i) If $\partial^2 \log f(X; \boldsymbol{\theta})/\partial\boldsymbol{\theta}\partial\boldsymbol{\theta}'$, the Hessian of the log p.d.f./p.m.f., is a function of the sample whose mathematical expectation can be worked out analytically, then the elements of $\mathcal{I}(\boldsymbol{\theta}) = -E\partial^2 \log f(X; \boldsymbol{\theta})/\partial\boldsymbol{\theta}\partial\boldsymbol{\theta}'$ will appear as explicit functions of the model's parameters. The regularity conditions require these to be continuous, so the consistency of m.l. estimator $\hat{\boldsymbol{\theta}}_n$ and the continuity theorem imply that $\mathcal{I}(\hat{\boldsymbol{\theta}}_n)$ is a consistent estimator of $\mathcal{I}(\boldsymbol{\theta})$. Indeed, the invariance property of m.l.e.s implies that $\mathcal{I}(\hat{\boldsymbol{\theta}}_n)$ is the m.l. estimator of $\mathcal{I}(\boldsymbol{\theta})$. Moreover, because elements of the inverse of any invertible matrix $\mathbf{A}$ are continuous functions of the elements of $\mathbf{A}$, it is also true that $\mathcal{I}(\hat{\boldsymbol{\theta}}_n)^{-1}$ is consistent for $\mathcal{I}(\boldsymbol{\theta})^{-1}$. Thus, in Example 7.37 the matrix-valued statistics

$$\mathcal{I}\left(\hat{\alpha}_n, \hat{\beta}_n \right) = \begin{pmatrix} \Psi'(\hat{\alpha}_n) & \frac{1}{\hat{\beta}_n} \\ \frac{1}{\hat{\beta}_n} & \frac{\hat{\alpha}_n}{\hat{\beta}_n^2} \end{pmatrix} \tag{7.26}$$

and

$$\mathcal{I}\left(\hat{\alpha}_n, \hat{\beta}_n \right)^{-1} = \begin{pmatrix} \frac{\hat{\alpha}_n}{\hat{\alpha}_n \Psi'(\hat{\alpha}_n) - 1} & -\frac{\hat{\beta}_n}{\hat{\alpha}_n \Psi'(\hat{\alpha}_n) - 1} \\ -\frac{\hat{\beta}_n}{\hat{\alpha}_n \Psi'(\hat{\alpha}_n) - 1} & \frac{\hat{\beta}_n^2 \Psi'(\hat{\alpha}_n)}{\hat{\alpha}_n \Psi'(\hat{\alpha}_n) - 1} \end{pmatrix}$$

are weakly consistent estimators of $\mathcal{I}(\alpha, \beta)$ and $\mathcal{I}(\alpha, \beta)^{-1}$.

(ii) If $\partial^2 \log f(X; \boldsymbol{\theta})/\partial\boldsymbol{\theta}\partial\boldsymbol{\theta}'$ is not a function of X whose expectation can be found explicitly, then we must appeal to the law of large numbers, noting that

$$-\frac{1}{n}\frac{\partial^2 \log L\left(\mathbf{X};\boldsymbol{\theta}\right)}{\partial\boldsymbol{\theta}\partial\boldsymbol{\theta}'} = -\frac{1}{n}\sum_{j=1}^{n}\frac{\partial^2 \log f\left(X_j;\boldsymbol{\theta}\right)}{\partial\boldsymbol{\theta}\partial\boldsymbol{\theta}'} \xrightarrow{a.s.} \mathcal{I}\left(\boldsymbol{\theta}\right).$$

Thus, $-n^{-1}$ times the Hessian of the log likelihood is itself a consistent estimator of the Fisher information, just as the mean of any random sample estimates consistently its own (finite) expected value. Moreover, given the continuity of second derivatives $\partial^2 \log f\left(X_j;\boldsymbol{\theta}\right)/\partial\boldsymbol{\theta}\partial\boldsymbol{\theta}'$, we can again replace the parameters by the m.l. estimators and remain confident that

$$-\frac{1}{n}\frac{\partial^2 \log L(\mathbf{X};\hat{\boldsymbol{\theta}}_n)}{\partial\boldsymbol{\theta}\partial\boldsymbol{\theta}'} := -\frac{1}{n}\frac{\partial^2 \log L\left(\mathbf{X};\boldsymbol{\theta}\right)}{\partial\boldsymbol{\theta}\partial\boldsymbol{\theta}'}\bigg|_{\boldsymbol{\theta}=\hat{\boldsymbol{\theta}}_n}$$

consistently estimates $\mathcal{I}\left(\boldsymbol{\theta}\right)$ as well.

Example 7.39. For Weibull family

$$\left\{f\left(x;\alpha,\beta\right) = \frac{\alpha}{\beta}x^{\alpha-1}\exp\left(-\frac{x^\alpha}{\beta}\right)\mathbf{1}_{(0,\infty)}\left(x\right):\alpha>0,\beta>0\right\},\qquad (7.27)$$

the Fisher information is

$$\mathcal{I}\left(\alpha,\beta\right) = \begin{pmatrix} \frac{1}{\alpha^2}+\frac{1}{\beta}E\left[X^\alpha\left(\log X\right)^2\right] & -\frac{1}{\beta^2}E\left(X^\alpha\log X\right) \\ -\frac{1}{\beta^2}E\left(X^\alpha\log X\right) & -\frac{1}{\beta^2}+\frac{2}{\beta^3}EX^\alpha \end{pmatrix}.$$

The second diagonal element reduces to β^{-2} since $EX^\alpha = \beta$ (recall that X^α has the exponential distribution with scale β), but the other expectations are not so easy to find. However, a consistent estimator of $\mathcal{I}\left(\alpha,\beta\right)$ is

$$\tilde{\mathcal{I}}\left(\hat{\alpha}_n,\hat{\beta}_n\right) = \begin{pmatrix} \frac{1}{\hat{\alpha}_n^2}+\frac{1}{\hat{\beta}_n n}\sum_{j=1}^{n}\left[X_j^{\hat{\alpha}_n}\left(\log X_j\right)^2\right] & -\frac{1}{\hat{\beta}_n^2 n}\sum_{j=1}^{n}\left(X_j^{\hat{\alpha}_n}\log X_j\right) \\ -\frac{1}{\hat{\beta}_n^2 n}\sum_{j=1}^{n}\left(X_j^{\hat{\alpha}_n}\log X_j\right) & \frac{1}{\hat{\beta}_n^2} \end{pmatrix}.$$

$$(7.28)$$

Here are two exercises that apply the m.l. theory under regularity. Both require some computational resource.

Exercise 7.28. *A random sample of* $n = 10$ *from Weibull distribution* (7.27) *yields*

$$\mathbf{x}' = (.262, 2.023, 3.013, 2.098, 1.526, .619, .351, .401, 1.747, 1.270).$$

As practice with this small sample, find

(i) The m.l.e.s $\hat{\alpha}_{10}, \hat{\beta}_{10}$

(ii) The estimate $\widetilde{\mathcal{I}}\left(\hat{\alpha}_{10}, \hat{\beta}_{10}\right)$ of the information matrix

(iii) The estimate $10^{-1}\widetilde{\mathcal{I}}\left(\hat{\alpha}_{10}, \hat{\beta}_{10}\right)^{-1}$ of the asymptotic covariance matrix. (The results of this exercise will be used again in Exercises 7.43 and 8.14.)

Exercise 7.29. *The table presents, in absolute frequency form, a random sample of size $n = 100$ from a population modeled as*

$$f(x; \varphi, \theta) = \frac{\Gamma(\varphi + x)}{x! \Gamma(\varphi)} \theta^{\varphi} (1 - \theta)^{x} \mathbf{1}_{\aleph_0}(x).$$

x:	0	1	2	3	4	5	6	7	8	9	10	11	12	13
$N(\{x\})$:	1	19	16	10	14	9	10	4	4	3	4	2	2	2

(i) *Find method of moments estimates $\tilde{\varphi}_{100}$ and $\tilde{\theta}_{100}$ of φ and θ based on the realizations of sample moments $\bar{X}$ and M_2. (Given the form of the data, it is easiest to calculate moments as $100^{-1} \sum_{x=1}^{13} x^k N(\{x\})$.)*

(ii) *Write out likelihood equations $\partial \log L(\mathbf{x}; \varphi, \theta)/\partial\theta = 0$ and $\partial \log L(\mathbf{x}; \varphi, \theta)/\partial\varphi = 0$ and obtain from the second of these an implicit solution for $\hat{\varphi}_n$ in terms of $\hat{\theta}_n$.*

(iii) *Find the m.l. estimates by either substituting the expression for $\hat{\varphi}_n$ in the log likelihood function and maximizing with respect to $\hat{\theta}_n$, or else substituting in the first likelihood equation and finding the root $\hat{\theta}_n$. In either case use m.o.m. estimate $\tilde{\theta}_{100}$ as starting value for the algorithm used to find the numerical solution. (You will need the appropriate software for computing values of digamma function $\Psi(\cdot)$.)*

(iv) *Express the Fisher information matrix for the full sample,*

$$n\mathcal{I}(\varphi, \theta) := \begin{pmatrix} -\frac{E\partial \log L(\mathbf{X}; \varphi, \theta)}{\partial \varphi^2} & -\frac{E\partial \log L(\mathbf{X}; \varphi, \theta)}{\partial \varphi \partial \theta} \\ -\frac{E\partial \log L(\mathbf{X}; \varphi, \theta)}{\partial \varphi \partial \theta} & -\frac{E\partial \log L(\mathbf{X}; \varphi, \theta)}{\partial \theta^2} \end{pmatrix},$$

and then calculate the estimate $100\widetilde{\mathcal{I}}\left(\hat{\varphi}_{100}, \hat{\theta}_{100}\right)$ using $\sum_{x=0}^{13} \Psi'(\hat{\varphi}_{100} + x_j) N(\{x\})$ as estimate of $\sum_{j=1}^{100} E\Psi'(\varphi + X_j)$. (Either use software that computes trigamma function $\Psi'(\cdot)$ directly, or else approximate as $\Psi'(x) \doteq [\Psi(x + h) - \Psi(x - h)]/(2h)$ for some small h.)

(v) *Obtain a numerical estimate of the asymptotic covariance matrix of $\hat{\varphi}_n, \hat{\theta}_n$.*

7.2.2.4　*Asymptotics When the Regularity Conditions Do not Hold*

When the regularity conditions for m.l. estimation are not met, one must go back to first principles in order to ascertain the asymptotic properties of m.l.e.s. Provided that the m.l.e. can be expressed explicitly as a function of the data, it is often possible to discover its essential properties using the basic tools of asymptotic theory laid out in Chapter 6.

Example 7.40. $\{X_j\}_{j=1}^n$ are i.i.d. with p.d.f. $f_X(x) = e^{-(x-\theta)}1_{[\theta,\infty)}(x)$, which corresponds to the (displaced) exponential with origin at some $\theta \in \Re$. Example 7.29 showed that the m.l. estimator is $\hat{\theta}_n = X_{(1:n)}$, the sample minimum. The regularity conditions fail because the support depends on θ; however, it is not hard to determine the essential properties of the m.l.e., as follows. First, note that the marginal c.d.f. of each X_j is $F_X(x) = \left[1 - e^{-(x-\theta)}\right]1_{[\theta,\infty)}(x)$ and that $\Pr(X_j > x) = 1_{(-\infty,\theta)}(x) + e^{-(x-\theta)}1_{[\theta,\infty)}(x)$. Letting $F_{(1:n)}(x) := \Pr\left(X_{(1:n)} \le x\right)$, we have

$$
\begin{aligned}
F_{(1:n)}(x) &= 1 - \Pr\left(X_{(1:n)} > x\right) \\
&= 1 - \Pr(X_1 > x) \cdot \Pr(X_2 > x) \cdots \Pr(X_n > x) \\
&= 1 - \left[1_{(-\infty,\theta)}(x) + e^{-(x-\theta)}1_{[\theta,\infty)}(x)\right]^n \\
&= \left[1 - e^{-n(x-\theta)}\right]1_{[\theta,\infty)}(x).
\end{aligned}
$$

Applying Expression (3.16) to find the expected value gives

$$
\begin{aligned}
E(\hat{\theta}_n - \theta) &= \int_0^\infty \Pr\left(X_{(1:n)} - \theta > x\right) \cdot dx \\
&= \int_0^\infty \left[1 - F_{(1:n)}(\theta + x)\right] \cdot dx \\
&= \int_0^\infty e^{-nx} \cdot dx \\
&= n^{-1}.
\end{aligned}
$$

Thus, we find (no surprise!) that $\hat{\theta}_n$ has positive bias. However, since

$$
\lim_{n\to\infty} F_{(1:n)}(x) = \lim_{n\to\infty} \left[1 - e^{-n(x-\theta)}\right]1_{[\theta,\infty)}(x) = 1_{[\theta,\infty)}(x),
$$

we see (Definition 6.2 on page 385) that $\hat{\theta}_n$ is weakly consistent; i.e., that $\hat{\theta}_n \to^p \theta$. Moreover, for any $\varepsilon > 0$

$$
\sum_{n=1}^\infty \Pr\left(X_{(1:n)} > \theta + \varepsilon\right) = \sum_{n=1}^\infty e^{-n\varepsilon} < \infty,
$$

so the convergence part of Borel–Cantelli (page 26) confirms that

$$\Pr\left(X_{(1:n)} > \theta \text{ for infinitely many } n\right) = 0$$

and therefore, since $\Pr\left(X_{(1:n)} \geq \theta\right) = 1$ for each n, that $\hat{\theta}_n \to^{a.s.} \theta$. Finally, although there is little need for asymptotic approximation since the finite-sample distribution of $\hat{\theta}_n$ is known and has a simple form, it takes but a moment to understand the asymptotics of the normalized statistic $T_n = n(\hat{\theta}_n - \theta)$. For fixed n the c.d.f. of T_n is $F_{T_n}(t) = F_{(1:n)}(\theta + t/n) = (1 - e^{-t})\mathbf{1}_{[0,\infty)}(t)$. This being independent of n, it is also the limiting distribution. Note that with the usual normalizing factor $\sqrt{n}$, the limiting distribution of $\sqrt{n}\left(X_{(1:n)} - \theta\right)$ is still of the degenerate form $\mathbf{1}_{[\theta,\infty)}(x)$.

Exercise 7.30. *When sample elements $\{X_j\}_{j=1}^n$ are i.i.d. with p.d.f.*

$$f(x) = \theta^{-1}\mathbf{1}_{[0,\theta]}(x), \theta > 0,$$

Examples 7.30 and 7.33 show that the m.l. estimator of θ is $\hat{\theta}_n = X_{(n:n)}$, the sample maximum.

(i) *Show that $\hat{\theta}_n$ is weakly consistent for θ; i.e., that $\hat{\theta}_n \to^p \theta$ as $n \to \infty$.*
(ii) *Show that $\hat{\theta}_n$ is strongly consistent for θ; i.e., that $\hat{\theta}_n \to^{a.s} \theta$ as $n \to \infty$.*
(iii) *Determine the unique value of α such that normalized m.l.e. $T_n(\alpha) := -n^\alpha(\hat{\theta}_n - \theta)$ has a nondegenerate limiting distribution as $n \to \infty$ (i.e., such that $T_n(\alpha)$ converges in probability to a r.v. that is not a constant); and express this limiting form.*

Example 7.41. When n is odd and $\{X_j\}_{j=1}^n$ are i.i.d. as Laplace, with p.d.f. $f(x; \phi, \psi) = \frac{1}{2\psi}e^{-|x-\phi|/\psi}, \phi \in \Re, \psi > 0$, Example 7.32 shows that sample median $\hat{\phi}_n = X_{(.5(n+1):n)}$ is the the m.l. estimator of ϕ. The regularity conditions do not hold because the log likelihood fails to be differentiable for $\phi \in \{x_1, x_2, ..., x_n\}$, but the asymptotic behavior of $T_n := \sqrt{n}\left(\hat{\phi}_n - \phi\right)$ can be determined as follows. It is shown on page 609 that the p.d.f. of the jth order statistic in a random sample of size n from a distribution with absolutely continuous c.d.f. F and p.d.f. f is

$$f_{(j:n)}(x) = j\binom{n}{j}f(x)F(x)^{j-1}[1 - F(x)]^{n-j}.$$

Restricting to odd values of n by setting $n = 2m + 1$ for some $m \in \aleph_0$, the p.d.f. of $\hat{\phi}_n = X_{(m+1:n)}$ from the Laplace sample is seen to be

$$f_{(m+1:n)}(x) = \frac{n!}{(m!)^2}f(x; \phi, \psi)F(x; \phi, \psi)^m[1 - F(x; \phi, \psi)]^m,$$

where $F(x; \phi, \psi) = \frac{1}{2}e^{(x-\phi)/\psi}1_{(-\infty,\phi)}(x) + \left[1 - \frac{1}{2}e^{-(x-\phi)/\psi}\right]1_{[\phi,\infty)}(x)$.
The change-of-variable formula and a little algebra give as the p.d.f. of T_n itself

$$f_n(t) = \frac{1}{\psi\sqrt{n}}\frac{n!}{(m!)^2}\left(\frac{1}{2}e^{-|t|/(\psi\sqrt{n})}\right)^{m+1}\left(1 - \frac{1}{2}e^{-|t|/(\psi\sqrt{n})}\right)^m$$

$$= \frac{1}{\psi\sqrt{n}}\frac{n!}{(m!)^2}\frac{1}{2^{2m+1}}\left(e^{-|t|/(\psi\sqrt{n})}\right)^{m+1}\left(2 - e^{-|t|/(\psi\sqrt{n})}\right)^m$$

$$= \frac{1}{\psi}\left[\frac{(2m+1)!}{\sqrt{2m+1}(m!)^2}\frac{1}{2^{2m+1}}\right]e^{-|t|/(\psi\sqrt{n})} \tag{7.29}$$

$$\times \left(2e^{-|t|/(\psi\sqrt{n})} - e^{-2|t|/(\psi\sqrt{n})}\right)^m.$$

By expanding the exponentials in the last factor one finds that

$$\lim_{m\to\infty}\left(2e^{-|t|/(\psi\sqrt{n})} - e^{-2|t|/(\psi\sqrt{n})}\right)^m = \lim_{m\to\infty}\left(1 - \frac{t^2}{\psi^2 n} + o(n^{-1})\right)^m$$

$$= \lim_{m\to\infty}\left(1 - \frac{t^2}{2\psi^2 m} + o(m^{-1})\right)^m$$

$$= e^{-t^2/(2\psi^2)}.$$

Finally, Stirling's approximation (Expression (4.12)) and algebra give for the limit of the bracketed factor in (7.29)

$$\lim_{n\to\infty}\frac{1}{\sqrt{2\pi}}\left[1 + \frac{1}{n} + o\left(n^{-1}\right)\right]^n e^{-1} + o(1) = \frac{1}{\sqrt{2\pi}}.$$

Thus, we have $f_n(t) \to \frac{1}{\psi\sqrt{2\pi}}e^{-t^2/(2\psi^2)}$ at each $t \in \Re$. Using the Stirling bounds in (4.11) one can show that for each fixed t the integrand in

$$F_n(t) = \int_{-\infty}^{t} f_n(y) \cdot dy$$

is bounded for all n by an integrable function.[32] From this it follows (via dominated convergence) that

$$F_n(t) \to \int_{-\infty}^{t}\frac{1}{\psi\sqrt{2\pi}}e^{-y^2/(2\psi^2)} \cdot dy$$

for each $t \in \Re$ and hence that $T_n = \sqrt{n}\left(\hat{\phi}_n - \phi\right) \rightsquigarrow N(0, \psi^2)$. Of course, this in turn implies that $\hat{\phi}_n \to^P \phi$ as $n \to \infty$.

---

[32]The last factor in (7.29) attains a maximum value of unity at $t = 0$, so that $f_n(y) \le \psi^{-1}B(m)e^{-|y|/(\psi\sqrt{n})}$, where $B(m)$ is the bracketed factor. The Stirling bounds then give the (very conservative) upper bound $B(m) < 2\left(1 + \frac{1}{2m}\right)^{2m}e^{-1} < 2$ (the last following because $\left(1 + \frac{1}{2m}\right)^{2m} \uparrow e$). Thus, $f_n(y) < 2\psi^{-1}e^{-|y|/(\psi\sqrt{n})} = 4f(y; 0, \psi)$.

Exercise 7.31. *As in Section 6.7.2 we could simply* define *the sample median as the* $[.5n]$*th order statistic,* $X_{([.5n]:n)}$ *("*$[x]$*" signifying "greatest integer* $\leq x$*"), a quantity that is uniquely determined for all* $n \geq 2$*. Noting that the limiting distributions of normalized versions of* $X_{([.5n]:n)}$ *and* $X_{([.5(n+1)]:n)}$ *are just the same, apply the results of that section to verify that the normalized Laplace m.l.e. converges in distribution to* $N(0, \psi^2)$*.*

7.3 Confidence Intervals and Regions

Having obtained a point estimate of some constant that characterizes the data-generating process, such as a parameter, a moment, or the probability associated with some Borel set, one naturally wants to assess the reliability of the result. One way to gauge and indicate accuracy is simply to calculate and present the standard deviation of the estimator—usually referred to in this context as its *standard error*. When an exact standard error cannot be found, an asymptotic standard error sometimes can be. However, a more formal—and more conventional—way to assess accuracy is to construct an *interval estimate* or *confidence interval*.

Constructing a confidence interval for some constant θ requires finding statistics $L_{p\alpha}(\mathbf{X})$ and $U_{(1-p)\alpha}(\mathbf{X})$ that bracket θ with some prespecified probability $1-\alpha \in (0,1)$. That is, $L_{p\alpha}$ and $U_{(1-p)\alpha}$ are functions of the data such that $\Pr\left[L_{p\alpha}(\mathbf{X}) \leq \theta \leq U_{(1-p)\alpha}(\mathbf{X})\right]$ equals, or at least approximates, $1 - \alpha$. Typically, one picks some small value for α, such as $.10, .05,$ or $.01$, so that the coverage probability—the probability of bracketing θ—is large. The constant $p \in (0,1)$ serves to apportion the residual probability, $\alpha = 1 - (1 - \alpha)$, over the extremities of the space Θ of possible values. That is, p is such that $\Pr\left[L_{p\alpha}(\mathbf{X}) > \theta\right] = p\alpha$ and $\Pr\left[U_{(1-p)\alpha}(\mathbf{X}) < \theta\right] = (1-p)\alpha$. It is common practice just to take $p = .5$ for an even split, but we will see that other choices sometimes give better results. Having chosen $\alpha, p,$ and statistics $L_{p\alpha}(\mathbf{X})$ and $U_{(1-p)\alpha}(\mathbf{X})$, one presents the sample realizations, $l_{p\alpha} := L_{p\alpha}(\mathbf{x})$ and $u_{(1-p)\alpha} := U_{(1-p)\alpha}(\mathbf{x})$ as *indications* of how well θ can be determined by the data. The interval $\left[l_{p\alpha}, u_{(1-p)\alpha}\right]$ is then called a $1 - \alpha$ *confidence interval* (c.i.) for θ.[33] Although we cannot know whether a specific realization $\left[l_{p\alpha}, u_{(1-p)\alpha}\right] = \left[L_{p\alpha}(\mathbf{x}), U_{(1-p)\alpha}(\mathbf{x})\right]$ does contain θ, we do know that a proportion of such intervals equal to or approximating $1 - \alpha$ would contain θ were the sampling experiment to be repeated indefinitely. While $\left[L_{p\alpha}(\mathbf{X}), U_{(1-p)\alpha}(\mathbf{X})\right]$ could appropriately be

---

[33]This is often referred to as a "$100(1 - \alpha)\%$" c.i., but we use the simpler expression.

called a $1 - \alpha$ "probability interval" for θ, we apply the term "confidence interval" to the *ex post* realization $[l_{p\alpha}, u_{(1-p)\alpha}]$. More generally, if $\boldsymbol{\theta} \in \Re^k$ is a vector of parameters, we might wish to find some $\mathbb{R}_{1-\alpha}(\mathbf{X}) \subset \Re^k$ that contains $\boldsymbol{\theta}$ with probability $1 - \alpha$ and a corresponding *confidence region (c.r.)*, $\mathbb{R}_{1-\alpha}(\mathbf{x})$. The more general term *confidence set* applies to intervals and regions alike.

We stress that under the frequentist interpretation of probability the statement "θ lies between $l_{p\alpha}$ and $u_{(1-p)\alpha}$ with probability $1 - \alpha$" is simply *incorrect*. Although it is unknown to us, θ is a fixed quantity, as are the known values $l_{p\alpha}$ and $u_{(1-p)\alpha}$. An omniscient observer would know whether or not θ lies between $l_{p\alpha}$ and $u_{(1-p)\alpha}$, and such an observer— whether trained as a frequentist or as a Bayesian—would assign to the statement $\theta \in [l_{p\alpha}, u_{(1-p)\alpha}]$ a probability of either zero (if the statement were false) or unity (if true). The *non*omniscient frequentist could say only that the probability has one or the other of these two values, and such completely uninformative statements are never made. (We will see what the Bayesian has to say in Chapter 10.)

All else equal, we would like confidence sets to place the tightest possible bounds on the parameters of interest. Of course, there is a trade-off between precision—tightness of the bounds—and levels of "confidence". Obviously, we can be completely confident that a real, finite constant θ lies on $(-\infty, \infty)$ and that a collection of k constants $\boldsymbol{\theta}$ lies somewhere in $\Re^k$, but such "bounds" have no precision at all. In general, contracting (expanding) any (properly constructed) confidence set lowers (raises) our confidence that it contains the parameters of interest. Thus, in comparing different ways of constructing confidence sets, we must control (i.e., fix) the coverage probability $1 - \alpha$. Given α, a sensible goal in building a c.i. for a scalar θ is to choose p, $L_{p\alpha}(\mathbf{X})$, and $U_{(1-p)\alpha}(\mathbf{X})$ so as to minimize the average length, $E\left[U_{(1-p)\alpha}(\mathbf{X}) - L_{p\alpha}(\mathbf{X})\right]$. Likewise, among alternative regions $\left\{\mathbb{R}_{1-\alpha}^{(j)}(\mathbf{X})\right\}_{j=1}^{\infty}$ such that $\Pr\left(\boldsymbol{\theta} \in \mathbb{R}_{1-\alpha}^{(j)}(\mathbf{X})\right) = 1 - \alpha$, we would want the one with least average volume. However, finding such minimal sets is not always possible. Moreover, there are other considerations besides precision; in particular, (i) the degree to which actual coverage probability approximates nominal value $1 - \alpha$ and (ii) computational difficulty.

The following subsections present two approaches to building confidence sets. Both are based on what are called *pivotal variables*, *pivotal quantities*, or simply *pivots*.

Definition 7.8. A pivotal quantity obtained from a sample $\mathbf{X} = (X_1, X_2, ..., X_n)'$ with distribution $F_{\mathbf{X}}(\mathbf{x}; \boldsymbol{\theta}), \boldsymbol{\theta} \in \boldsymbol{\Theta}$, is a function $\Pi : \Re^n \times \boldsymbol{\Theta} \to \Re$ whose distribution is independent of $\boldsymbol{\theta}$. $\Pi(\mathbf{X}, \boldsymbol{\theta})$ is *asymptotically* pivotal if its limiting distribution exists as $n \to \infty$ and has this invariance property.

In other words, $\Pi(\mathbf{X}, \boldsymbol{\theta})$ is pivotal if and only if $F_{\Pi}(c) := \Pr[\Pi(\mathbf{X}, \boldsymbol{\theta}) \leq c]$ is, for each real c, the same for each $\boldsymbol{\theta} \in \boldsymbol{\Theta}$. As examples, $\sqrt{n}(\bar{X} - \mu) \sim N(0, \sigma^2)$ is pivotal in random samples of size n from $N(\mu, \sigma^2)$ when σ has any known value, and $\sqrt{n}(\bar{X} - \mu)/\sigma \sim N(0, 1)$ and $\sqrt{n}(\bar{X} - \mu)/S \sim t(n-1)$ are always pivotal; likewise, $(n-1)S^2/\sigma^2 \sim \chi^2(n-1)$. More generally, if $\hat{\boldsymbol{\theta}}_n$ is the maximum likelihood estimator from a distribution that meets the regularity conditions for m.l. estimation, then we will see later that $n\left(\hat{\boldsymbol{\theta}}_n - \boldsymbol{\theta}\right)' \mathcal{I}(\boldsymbol{\theta})\left(\hat{\boldsymbol{\theta}}_n - \boldsymbol{\theta}\right)$ and $n\left(\hat{\boldsymbol{\theta}}_n - \boldsymbol{\theta}\right)' \mathcal{I}\left(\hat{\boldsymbol{\theta}}_n\right)\left(\hat{\boldsymbol{\theta}}_n - \boldsymbol{\theta}\right)$ are both asymptotically pivotal, both converging in distribution to $\chi^2(k)$ for each $\boldsymbol{\theta} \in \boldsymbol{\Theta} \subset \Re^k$.

In our first approach to building confidence sets we start with a point estimator $T_n(\mathbf{X})$ of a scalar parameter θ and then look for a pivotal function involving θ and T_n (and perhaps other functions of the data). This is a natural way to proceed, since it is often the accuracy of a particular point estimate that one wants to assess. The method is usually easy to apply, and very versatile. Approximate confidence regions for a vector $\boldsymbol{\theta}$ can be constructed from confidence intervals for its separate components or from a pivotal quantity for $\boldsymbol{\theta}$ itself. Even when there is no parametric model for the data, one can in this way build confidence sets for quantities that can be expressed as expectations, such as moments and the probabilities of arbitrary Borel sets. The disadvantage is that there is often some arbitrariness in the choice of estimator and pivotal quantity and thus no assurance that the resulting confidence set is tight or has accurate coverage probability.

In our second approach we derive confidence sets from a pivotal variable constructed from the likelihood function. Naturally, this requires having a fully specified parametric model, and it often involves significant computation. The positive aspects are that there is just one choice of pivotal quantity, that the confidence sets are usually small, and that they often have accurate coverage probabilities even for vector-valued parameters. Moreover, many find the underlying logic and systematic nature of the approach to be extremely compelling. Finally, some tables are available for standard applications that make the computations trivial.

7.3.1 *Confidence Sets from Pivots Based on Estimators*

Let us now see how to build confidence sets for one or more parameters from some appropriate estimator. To start with the simplest case, we will construct an interval for a scalar-valued θ using an estimator $T_n = T_n(\mathbf{X})$. This involves four steps.

(i) Find a pivotal quantity $\Pi(\mathbf{X}, \theta)$ and a distribution $F_\Pi(\cdot)$ that characterizes, or at least approximates, its stochastic behavior. Since we are describing a method based on estimators, the pivot necessarily will depend directly on T_n, but it may depend on the data in other ways as well. It makes sense to try to develop a pivot based on a minimal sufficient statistic for θ when such exists, since this embodies all the relevant information in the sample in the most compact way.

(ii) Knowing the function $F_\Pi(\cdot)$ and the desired confidence level $1 - \alpha$, find or approximate the $(1-p)\alpha$ quantile $c_{(1-p)\alpha}$ and the upper-$p\alpha$ quantile $c^{p\alpha}$ for some $p \in (0, 1)$. (How to choose p is explained in the next step.) In many applications the distribution (or approximating distribution) $F_\Pi(\cdot)$ is continuous and strictly increasing, so that the quantiles are uniquely determined. In such applications the event $\Pi(\mathbf{X}, \theta) \in [c_{(1-p)\alpha}, c^{p\alpha}]$ occurs with probability $1 - \alpha$. When $F_\Pi(\cdot)$ is not continuous or is known only approximately, one has to settle for an interval that contains $\Pi(\mathbf{X}, \theta)$ with probability close to $1 - \alpha$.

(iii) Having found the probability interval $[c_{(1-p)\alpha}, c^{p\alpha}]$, manipulate algebraically the inequality $c_{(1-p)\alpha} \leq \Pi(\mathbf{X}, \theta) \leq c^{p\alpha}$ to find an equivalent event of the form $L_{p\alpha}(\mathbf{X}) \leq \theta \leq U_{(1-p)\alpha}(\mathbf{X})$. "Equivalent" means specifically that (a) $\Pi(\mathbf{X}, \theta) \in [c_{(1-p)\alpha}, c^{p\alpha}]$ if and only if (b) $L_{p\alpha}(\mathbf{X}) \leq \theta \leq U_{(1-p)\alpha}(\mathbf{X})$, in which case the two events occur with the same probability. When $\Pi(\mathbf{X}, \theta)$ is monotone in θ, as it typically is, it is usually easy to find an equivalent event of this interval form. We refer to the manipulation that transforms (a) into (b) as the *inversion process*. When it is feasible to do so, choose p so as to minimize the expected length of the interval, $E\left[U_{(1-p)\alpha}(\mathbf{X}) - L_{p\alpha}(\mathbf{X})\right]$. We will see that $p = .5$ is often optimal, and it is common just to use this value by default for simplicity.

(iv) Finally, once sample realization $\mathbf{x}$ is obtained, calculate the desired $1 - \alpha$ c.i. as $\left[l_{p\alpha}, u_{(1-p)\alpha}\right] := \left[L_{p\alpha}(\mathbf{x}), U_{(1-p)\alpha}(\mathbf{x})\right]$.

The next subsection gives several realistic applications of the procedure, but let us first look at an especially simple case that illustrates the process step by step.

Example 7.42. $\mathbf{X} = (X_1, X_2, ..., X_n)'$ is a random sample from the exponential distribution with mean (scale parameter) θ:

$$f(x; \theta) = \theta^{-1} e^{-x/\theta} \mathbf{1}_{(0,\infty)}(x), \theta \in \Theta = (0, \infty).$$

Given $\alpha \in (0, 1)$, here are the steps for building a $1 - \alpha$ c.i. using a pivot involving minimal-sufficient statistic $\bar{X}$.

(i) We know that $n\bar{X} = \sum_{j=1}^{n} X_j \sim \Gamma(n, \theta)$, and so the rescaled quantity $n\bar{X}/\theta \sim \Gamma(n, 1)$ is pivotal. However, a more convenient choice is $\Pi(\mathbf{X}, \boldsymbol{\theta}) = 2n\bar{X}/\theta \sim \chi^2(2n)$, since the quantiles can be found in standard chi-squared tables, such as that in Appendix C.3.

(ii) Taking $p = .5$ for simplicity, we find the unique quantiles $c_{\alpha/2}, c^{\alpha/2}$ of $\chi^2(2n)$ such that $\Pr\left[c_{\alpha/2} \leq 2n\bar{X}/\theta \leq c^{\alpha/2}\right] = 1 - \alpha$.

(iii) Now *literally* inverting this monotonic pivotal quantity, we find a $1 - \alpha$ probability interval for θ as

$$1 - \alpha = \Pr\left(\frac{1}{c_{\alpha/2}} \geq \frac{\theta}{2n\bar{X}} \geq \frac{1}{c^{\alpha/2}}\right) = \Pr\left(\frac{2n\bar{X}}{c^{\alpha/2}} \leq \theta \leq \frac{2n\bar{X}}{c_{\alpha/2}}\right).$$

(iv) A random sample of $n = 20$ from an exponential distribution with $\theta = 1$ yields the realization $2n\bar{x} = 37.89$. For a .95 c.i. we use $\chi^2(40)$ quantiles $c_{.025} = 24.43$ and $c^{.025} = 59.34$ to obtain $[37.89/59.34, 37.89/24.43] \doteq [.638, 1.551]$. Trading precision for confidence, the .99 c.i. would be $[37.89/66.77, 37.89/20.71] \doteq [.567, 1.830]$; and, going the other way, the .90 c.i. would be $[.680, 1.429]$.

As it happens, choosing $p = .5$ and allocating residual probability α equally to the two tails of $\chi^2(2n)$ does not produce, on average, the shortest c.i.s in this application. We will see later how to find a better value. The following proposition is often helpful for this purpose.

Proposition 7.2. *Let Y be a continuous r.v. with p.d.f. f that is continuous and differentiable a.e. on a support $\mathbb{Y}$, and let $y_{p\alpha}$ and $y^{(1-p)\alpha}$ be the $p\alpha$ quantile and upper $(1-p)\alpha$ quantile, where $\alpha \in (0, 1)$ and $p \in (0, 1)$. Then if $f'(y) < 0$ for $y \geq y^\alpha$ and $f'(y) > 0$ for $y \leq y_\alpha$, the difference $y^{(1-p)\alpha} - y^{p\alpha}$ is minimized at a value of p such that $f\left(y^{(1-p)\alpha}\right) = f(y_{p\alpha})$.*

Proof. Differentiating implicitly with respect to p the relations $F\left(y^{(1-p)\alpha}\right) = 1 - (1-p)\alpha$ and $F(y_{p\alpha}) = p\alpha$ gives $dy^{(1-p)\alpha}/dp =$

$\alpha / f\left(y^{(1-p)\alpha}\right)$ and $dy_{p\alpha}/dp = \alpha / f\left(y_{p\alpha}\right)$. The first-order condition for minimum of $y^{(1-p)\alpha} - y_{p\alpha}$ is then

$$\frac{d}{dp}\left[y^{(1-p)\alpha} - y_{p\alpha}\right] = \alpha\left[\frac{1}{f\left(y^{(1-p)\alpha}\right)} - \frac{1}{f\left(y_{p\alpha}\right)}\right] = 0,$$

which holds if and only if the densities are equal at the two points. The second-order condition is met since

$$\frac{d^2}{dp^2}\left[y^{(1-p)\alpha} - y_{p\alpha}\right] = \alpha^2\left[-\frac{f'\left(y^{(1-p)\alpha}\right)}{f\left(y^{(1-p)\alpha}\right)^3} + \frac{f'\left(y_{p\alpha}\right)}{f\left(y_{p\alpha}\right)^3}\right] > 0$$

when $f'(y) < 0$ for $y \geq y^\alpha$ and $f'(y) > 0$ for $y \leq y_\alpha$ $\square$

7.3.1.1   Confidence Intervals from Estimators: Applications

Example 7.43. With $\{X_j\}_{j=1}^n$ i.i.d. as $N\left(\mu, \sigma^2\right)$ and σ^2 unknown, we build a .95 c.i. for μ based on $\bar{X}$. Random variable $\sqrt{n}(\bar{X}-\mu)/S \sim t\,(n-1)$ is pivotal. With $\alpha = .05$ and $t_{.05(1-p)} = -t^{.05(1-p)}$ and $t^{.05p}$ as the $.05\,(1-p)$ and upper-$.05p$ quantiles of the symmetric Student's $t\,(n-1)$, we have

$$0.95 = \Pr\left[-t^{.05(1-p)} \leq \frac{\bar{X}-\mu}{S/\sqrt{n}} \leq t^{.05p}\right]$$

for any $p \in (0,1)$. To invert the inequality, multiply by $S/\sqrt{n}$, subtract $\bar{X}$, and change signs and directions of the inequalities. Here is the process, step by step:

$$0.95 = \Pr\left[-t^{.05(1-p)}\frac{S}{\sqrt{n}} \leq \bar{X} - \mu \leq t^{.05p}\frac{S}{\sqrt{n}}\right]$$

$$= \Pr\left[-\bar{X} - t^{.05(1-p)}\frac{S}{\sqrt{n}} \leq -\mu \leq -\bar{X} + t^{.05p}\frac{S}{\sqrt{n}}\right]$$

$$= \Pr\left[\bar{X} - t^{.05p}\frac{S}{\sqrt{n}} \leq \mu \leq \bar{X} + t^{.05(1-p)}\frac{S}{\sqrt{n}}\right].$$

The expected length of the probability interval is proportional to $t^{.05(1-p)} + t^{.05p} = t^{.05(1-p)} - t_{.05p}$. Being symmetric about the origin, differentiable, and unimodal, the Student distribution satisfies the conditions of Proposition 7.2 for any $\alpha < .5$. The expected length of any $1 - \alpha$ interval is therefore least when the densities at $t^{.05(1-p)}$ and $t_{.05p} = -t^{.05p}$ are the same. By symmetry this requires $p = .5$. Thus, the shortest .95 c.i. based on pivot $\sqrt{n}(\bar{X} - \mu)/S$ is $\left[\bar{x} - t^{.025}s/\sqrt{n}, \bar{x} + t^{.025}s/\sqrt{n}\right]$.

Example 7.44. Assuming only that $\{X_j\}_{j=1}^n$ are i.i.d. with finite variance, an approximate .90 c.i. for $\mu = EX$ can be found as follows. $\bar{X}$ is the natural estimator since it is best *linear* unbiased, and Lindeberg–Lévy and Mann–Wald imply that $\sqrt{n}(\bar{X} - \mu)/S \rightsquigarrow N(0,1)$ is asymptotically pivotal as $n \to \infty$. With $z^{.10p}$ and $z_{.10(1-p)} = -z^{.10(1-p)}$ as the upper and lower quantiles of $N(0,1)$, we have, for sufficiently large n,

$$0.90 \doteq \Pr\left[-z^{.10(1-p)} \le \frac{\bar{X} - \mu}{S/\sqrt{n}} \le z^{.10p}\right]$$

$$= \Pr\left[\bar{X} - z^{.10p}\frac{S}{\sqrt{n}} \le \mu \le \bar{X} + z^{.10(1-p)}\frac{S}{\sqrt{n}}\right].$$

The $N(0,1)$ density is symmetric about the origin, differentiable, and unimodal, and so Proposition 7.2 again implies that residual probability $\alpha = .10$ should be divided equally between the tails. Based on the large-sample approximations, the shortest .90 c.i. is therefore

$$\left[\bar{x} - z^{.05}\frac{s}{\sqrt{n}}, \bar{x} + z^{.05}\frac{s}{\sqrt{n}}\right] = \left[\bar{x} - 1.645\frac{s}{\sqrt{n}}, \bar{x} + 1.645\frac{s}{\sqrt{n}}\right].$$

Example 7.45. With $\{X_j\}_{j=1}^n$ i.i.d. as $\Gamma(\alpha, \beta)$ an approximate large-sample .99 c.i. for α is found as follows. Maximum likelihood estimator $\hat{\boldsymbol{\theta}}_n = \left(\hat{\alpha}_n, \hat{\beta}_n\right)'$ is weakly consistent and asymptotically normal with asymptotic covariance matrix given in Example 7.37. The asymptotic variance of $\hat{\alpha}_n$ is $n^{-1}\alpha/\left[\alpha\Psi'(\alpha) - 1\right]$ (the first element on the principal diagonal of $n^{-1}\mathcal{I}(\alpha, \beta)^{-1}$ in (7.25)). Thus, $\sqrt{n}(\hat{\alpha}_n - \alpha)\sqrt{\Psi'(\alpha) - \alpha^{-1}} \rightsquigarrow N(0,1)$ is asymptotically pivotal; however, the inversion step would require a numerical procedure. Replacing $\mathcal{I}(\alpha, \beta)^{-1}$ by the consistent estimator $\mathcal{I}(\hat{\alpha}_n, \hat{\beta}_n)^{-1}$, as in (7.26), yields another asymptotically standard normal pivot that is easy to invert: $\sqrt{n}(\hat{\alpha}_n - \alpha)\sqrt{\Psi'(\hat{\alpha}_n) - \hat{\alpha}_n^{-1}}$. Proposition 7.2 again justifies dividing the residual .01 probability equally between the two tails of $N(0,1)$, and the resulting approximate .99 c.i. is

$$\left[l_{.005} = \hat{\alpha}_n - \frac{2.576}{\sqrt{n\left[\Psi'(\hat{\alpha}_n) - \hat{\alpha}_n^{-1}\right]}}, u_{.005} = \hat{\alpha}_n + \frac{2.576}{\sqrt{n\left[\Psi'(\hat{\alpha}_n) - \hat{\alpha}_n^{-1}\right]}}\right].$$

Note that the interval $[l, u]$ based on such an asymptotic approximation may not be wholly contained in the allowed parameter space Θ. In such cases the relevant c.i. is $[l, u] \cap \Theta$. Thus, if $l_{.005} \le 0$ in Example 7.45, the c.i. for the shape parameter of the gamma density would be $(0, u]$.

In the above examples it was optimal to divide residual probability α equally between the two tails of the pivotal distributions because those distributions were symmetric and because the pivots depended linearly on the relevant parameter. When these conditions do not hold, other values of p may produce shorter average c.i.s, but a numerical procedure is typically needed to find them.

Example 7.46. With $\{X_j\}_{j=1}^n$ i.i.d. as $N\left(\mu, \sigma^2\right)$, we build a $1 - \alpha$ c.i. for σ^2 from unbiased sample variance S^2. Pivot $(n-1)\,S^2/\sigma^2$ is distributed as $\chi^2(n-1)$, which is supported on $(0, \infty)$ and right-skewed. With $c_{(1-p)\alpha}$ and $c^{p\alpha}$ as the $(1-p)\alpha$ and upper-$p\alpha$ quantiles

$$1 - \alpha = \Pr\left[c_{(1-p)\alpha} \leq \frac{(n-1)\,S^2}{\sigma^2} \leq c^{p\alpha}\right]$$

$$= \Pr\left[\frac{(n-1)\,S^2}{c^{p\alpha}} \leq \sigma^2 \leq \frac{(n-1)\,S^2}{c_{(1-p)\alpha}}\right].$$

Proposition 7.2 implies that $c^{p\alpha} - c_{(1-p)\alpha}$ is minimized at p such that $f\left(c^{p\alpha}\right) = f\left(c_{(1-p)\alpha}\right)$, where f is the chi-squared density. However, since the pivot depends inversely on σ^2, the length of the probability interval for σ^2 is proportional to $\left(c_{(1-p)\alpha}\right)^{-1} - \left(c^{p\alpha}\right)^{-1}$. Minimizing this requires that p satisfy

$$\left(c_{(1-p)\alpha}\right)^2 f\left(c_{(1-p)\alpha}\right) = \left(c^{p\alpha}\right)^2 f\left(c^{p\alpha}\right). \tag{7.30}$$

Quantiles $c_{(1-p^*)\alpha}$ and $c^{p^*\alpha}$ that solve (7.30) for various α and n can be found numerically, and tables are available.[34] However, since c.i.s for the variance of a normal population are so often required, it would be nice if things could be made a little easier. Instead of having the quantiles $c_{(1-p^*)\alpha}$ and $c^{p^*\alpha}$ it would be better to have values of $v_\alpha := (n-1)/c^{p^*\alpha}$ and $v^\alpha := (n-1)/c_{(1-p^*)\alpha}$, for these give the $1 - \alpha$ c.i. directly, as $[v_\alpha s^2, v^\alpha s^2]$. Table 7.1 presents the bounds v_α, v^α for selected α, n. However, for $n > 30$ these "shortest" c.i.s do not differ much from those based on the equal-probability quantiles $c_{\alpha/2}, c^{\alpha/2}$.

Example 7.47. If $n = 100$ and $s^2 = 41.781$ the shortest .90 c.i. based on pivot $(n-1)\,S^2/\sigma^2$ is

$$[.781\,(41.781)\,, 1.254\,(41.781)] \doteq [32.63, 52.39]\,,$$

but the c.i. using equal-probability quantiles is less than two per cent longer:

$$\left[\frac{99\,(41.781)}{123.225}, \frac{99\,(41.781)}{77.046}\right] \doteq [33.57, 53.69]$$

---

[34]E.g., Table A.5 of Bartosyznski and Niewiadomska-Bugaj (1996).

Table 7.1   Lower/upper values by which to multiply sample variance to find shortest c.i. for variance of normal population.

n	$1 - \alpha = .90$		$1 - \alpha = .95$		$1 - \alpha = .99$	
	v_α	v^α	v_α	v^α	v_α	v^α
5	0.221	3.788	0.190	5.648	0.146	13.474
6	0.264	3.137	0.229	4.389	0.178	9.035
7	0.302	2.759	0.264	3.696	0.208	6.896
8	0.334	2.511	0.294	3.260	0.234	5.668
9	0.363	2.335	0.321	2.960	0.258	4.879
10	0.388	2.204	0.345	2.741	0.279	4.332
11	0.411	2.102	0.367	2.574	0.299	3.932
12	0.431	2.020	0.386	2.441	0.317	3.626
13	0.449	1.952	0.404	2.334	0.333	3.385
14	0.466	1.896	0.420	2.245	0.349	3.191
15	0.481	1.847	0.435	2.170	0.363	3.030
16	0.495	1.806	0.449	2.106	0.376	2.895
17	0.508	1.769	0.462	2.050	0.389	2.780
18	0.520	1.737	0.474	2.001	0.400	2.680
19	0.531	1.708	0.485	1.958	0.411	2.594
20	0.541	1.682	0.496	1.919	0.422	2.517
22	0.560	1.638	0.515	1.853	0.440	2.389
24	0.576	1.600	0.532	1.799	0.458	2.285
26	0.591	1.569	0.547	1.753	0.473	2.198
28	0.605	1.542	0.561	1.713	0.487	2.125
30	0.617	1.518	0.573	1.679	0.500	2.063
35	0.643	1.469	0.600	1.610	0.528	1.940
40	0.664	1.432	0.623	1.558	0.552	1.849
45	0.681	1.402	0.641	1.517	0.572	1.778
50	0.697	1.377	0.658	1.483	0.590	1.721
60	0.721	1.339	0.684	1.431	0.619	1.635
70	0.741	1.310	0.705	1.392	0.642	1.572
80	0.756	1.288	0.722	1.362	0.661	1.524
90	0.770	1.269	0.737	1.338	0.678	1.486
100	0.781	1.254	0.749	1.317	0.692	1.454
150	0.819	1.203	0.792	1.252	0.741	1.354
200	0.842	1.174	0.818	1.215	0.772	1.299

Exercise 7.32. *Independent random samples of sizes n_1 and n_2 are drawn from two normal populations with unknown variances, σ_1^2 and σ_2^2. Find a pivot that involves the sample variances S_1^2 and S_2^2, and show how to obtain $1 - \alpha$ c.i.s for the ratio, σ_1^2/σ_2^2. Will it be optimal to use equal-probability quantiles?*

Exercise 7.33. *A single observation is drawn from exponential distribution*

$$f(x; \theta) = \theta^{-1} e^{-x/\theta} \mathbf{1}_{[0,\infty)}(x).$$

Taking X/θ as the pivot, show that the shortest $1 - \alpha$ c.i. for θ is $[0, -x/\log \alpha]$.

Exercise 7.34. *Use the result of Exercise 6.21 to find an approximate large-sample .95 c.i. for the difference in means of two finite-variance r.v.s X and Y based on independent samples of sizes n_X, n_Y.*

7.3.1.2 Confidence Regions from Estimators of $\boldsymbol{\theta} \in \Re^k$

Just as the statistics $L_{p\alpha}(\mathbf{X})$ and $U_{(1-p)\alpha}(\mathbf{X})$ define a stochastic interval in $\Re$ that contains some scalar-valued θ with probability $1 - \alpha$, it is often possible in the vector case to construct a stochastic region $\mathbb{R}_{1-\alpha}(\mathbf{X}) \subset \Re^k$ such that $\Pr(\boldsymbol{\theta} \in \mathbb{R}_{1-\alpha}(\mathbf{X}))$ approximates the specified $1 - \alpha$. When evaluated *ex post*, $\mathbb{R}_{1-\alpha}(\mathbf{x})$ is called a *joint confidence region* (c.r.) for $\boldsymbol{\theta}$. Note that it is the *region* $\mathbb{R}_{1-\alpha}(\mathbf{X})$ here that is stochastic, and not the constant $\boldsymbol{\theta}$; thus, one should interpret the symbolic statement "$\boldsymbol{\theta} \in \mathbb{R}_{1-\alpha}(\mathbf{X})$" as "the event that $\mathbb{R}_{1-\alpha}(\mathbf{X})$ contains $\boldsymbol{\theta}$". Two approaches based on pivots from estimators are commonly used to find confidence regions, but both only approximate the desired coverage probability. In neither case is $\mathbb{R}_{1-\alpha}(\mathbf{x})$ necessarily contained in parameter space $\boldsymbol{\Theta}$, so we always take $\mathbb{R}_{1-\alpha}(\mathbf{x}) \cap \boldsymbol{\Theta}$ as the c.r.

The first approach is to construct a rectangular region from confidence intervals for the separate elements of $\boldsymbol{\theta}$, using an appropriate pivotal quantity for each. Thus, for $\boldsymbol{\theta} \in \Re^k$ one finds k pairs of statistics $\left\{L_{p_j\alpha_k}(\mathbf{X}), U_{(1-p_j)\alpha_k}(\mathbf{X})\right\}_{j=1}^k$ such that

$$\Pr\left[L_{p_j\alpha_k}(\mathbf{X}) \leq \theta_j \leq U_{(1-p_j)\alpha_k}(\mathbf{X})\right] = 1 - \alpha_k$$

for each j. The confidence region is then the product set

$$\mathbb{R}_{1-\alpha} = \times_{j=1}^k \left[L_{p_j\alpha_k}(\mathbf{x}), U_{(1-p_j)\alpha_k}(\mathbf{x})\right].$$

Coverage probability $1 - \alpha_k$ for each interval is chosen so as to make the overall coverage probability close to $1-\alpha$. As usual, the $\{p_j\}_{j=1}^k$ apportion the residual α_k of probability beyond the two extremities of each interval. Were events $\left\{L_{p_j\alpha_k}(\mathbf{X}) \leq \theta_j \leq U_{(1-p_j)\alpha_k}(\mathbf{X})\right\}_{j=1}^k$ mutually independent, taking $1-\alpha_k = \sqrt[k]{1-\alpha}$ would produce an exact $1-\alpha$ coverage probability; however, the usual situation is that the coverage probability can only be bounded. Letting $A_j(\alpha_k)$ represent the event $L_{p_j\alpha_k} \leq \theta_j \leq U_{(1-p_j)\alpha_k}$,

Bonferroni's inequality (page 25) implies that

$$\mathbb{P}\left(\bigcap_{j=1}^{k} A_j\left(\alpha_k\right)\right) \geq \sum_{j=1}^{k} \mathbb{P}(A_j\left(\alpha_k\right)) - (k-1)$$
$$= k(1-\alpha_k) - (k-1)$$
$$= 1 - k\alpha_k.$$

Taking $\alpha_k = \alpha/k$ therefore ensures an overall coverage probability of at least $1 - \alpha$.

Example 7.48. A random sample is taken from $N(\mu, \sigma^2)$. Stochastic intervals that cover each of μ and σ^2 separately with probability $1 - \alpha_2$ are

$$\bar{X} - t^{\alpha_2/2}\frac{S}{\sqrt{n}} \leq \mu \leq \bar{X} + t^{\alpha_2/2}\frac{S}{\sqrt{n}}$$
$$(n-1)\frac{S^2}{c^{p\alpha_2}} \leq \sigma^2 \leq (n-1)\frac{S^2}{c_{(1-p)\alpha_2}},$$

where $p \in (0,1)$ and the t's and c's are the indicated quantiles of $t(n-1)$ and $\chi^2(n-1)$, respectively. With $\alpha_2 = \alpha/2$ and with $\mathbb{R}_{1-\alpha}$ as the Cartesian product of the intervals, we have $\Pr\left[(\mu, \sigma^2) \in \mathbb{R}_{1-\alpha}\right] \geq 1 - \alpha$. Thus, the rectangle

$$\left[\bar{x} - t^{\alpha/4}\frac{S}{\sqrt{n}}, \bar{x} + t^{\alpha/4}\frac{S}{\sqrt{n}}\right] \times \left[(n-1)\frac{s^2}{c^{p\alpha/2}}, (n-1)\frac{s^2}{c_{(1-p)\alpha/2}}\right]$$

constitutes a minimal $1 - \alpha$ joint c.r. for μ and σ^2. The rectangular region in Fig. 7.3 depicts a .95 c.r. from a sample of size $n = 100$ from $N(5, 49)$, with $\bar{x} = 5.021$, $s^2 = 41.718$, and $p = .5$.

A second approach to building joint c.r.s from an estimator $\mathbf{T}_n$ of $\boldsymbol{\theta}$ involves finding a function of these that is jointly pivotal for the entire vector $\boldsymbol{\theta}$. Usually, this requires n to be large enough to justify the use of asymptotic approximations. Thus, with $\mathbf{T}_n$ as a k-vector of estimators of $\boldsymbol{\theta}$, suppose we have determined that $\sqrt{n}\left(\mathbf{T}_n - \boldsymbol{\theta}\right) \rightsquigarrow N\left(\mathbf{0}, \boldsymbol{\Sigma}_{\mathbf{T}}\left(\boldsymbol{\theta}\right)\right)$, where covariance matrix $\boldsymbol{\Sigma}_{\mathbf{T}}\left(\boldsymbol{\theta}\right)$ is nonsingular with elements that are continuous functions of $\boldsymbol{\theta}$. For example, if $\mathbf{T}_n$ is the maximum likelihood estimator of parameters of a distribution satisfying the regularity conditions for m.l. estimation (page 494), then $n^{-1}\boldsymbol{\Sigma}_{\mathbf{T}}\left(\boldsymbol{\theta}\right) = n^{-1}\mathcal{I}\left(\boldsymbol{\theta}\right)^{-1}$, the inverse of the Fisher information from the entire sample. Theorem 5.1 establishes that a quadratic form in "standardized" multivariate normals has a chi-squared distribution. With Theorem 6.7 (Mann–Wald) this implies that

$$\Pi_1\left(\mathbf{X}, \boldsymbol{\theta}\right) := n\left(\mathbf{T}_n - \boldsymbol{\theta}\right)' \boldsymbol{\Sigma}_{\mathbf{T}}\left(\boldsymbol{\theta}\right)^{-1}\left(\mathbf{T}_n - \boldsymbol{\theta}\right) \rightsquigarrow \chi^2\left(k\right)$$

as $n \to \infty$. Moreover, since $\mathbf{T}_n$ is at least weakly consistent, the quantity

$$\Pi_2\left(\mathbf{X}, \boldsymbol{\theta}\right) := n\left(\mathbf{T}_n - \boldsymbol{\theta}\right)' \boldsymbol{\Sigma_T}\left(\mathbf{T}_n\right)^{-1}\left(\mathbf{T}_n - \boldsymbol{\theta}\right)$$

is also asymptotically pivotal, with the same limiting distribution. With c^α as the upper-α quantile of $\chi^2\left(k\right)$, it follows that

$$\lim_{n \to \infty} \Pr\left[\Pi_j\left(\mathbf{X}, \boldsymbol{\theta}\right) \le c^\alpha\right] = 1 - \alpha$$

for each $\Pi_j, j \in \{1, 2\}$. Taking $\mathbf{t}_n$ as the sample realization, the sets

$$\mathbb{R}_{1,1-\alpha} := \left\{\boldsymbol{\theta} : n\left(\mathbf{t}_n - \boldsymbol{\theta}\right)' \boldsymbol{\Sigma_T}\left(\boldsymbol{\theta}\right)^{-1}\left(\mathbf{t}_n - \boldsymbol{\theta}\right) \le c^\alpha\right\}$$

$$\mathbb{R}_{2,1-\alpha} := \left\{\boldsymbol{\theta} : n\left(\mathbf{t}_n - \boldsymbol{\theta}\right)' \boldsymbol{\Sigma_T}\left(\mathbf{t}_n\right)^{-1}\left(\mathbf{t}_n - \boldsymbol{\theta}\right) \le c^\alpha\right\}$$

thus form approximate $1 - \alpha$ c.r.s for $\boldsymbol{\theta}$ when n is large. When $\mathbf{T}_n = \hat{\boldsymbol{\theta}}_n$ (the m.l. estimator), these are

$$\mathbb{R}_{1,1-\alpha} = \left\{\boldsymbol{\theta} : n\left(\hat{\boldsymbol{\theta}}_n - \boldsymbol{\theta}\right)' \mathcal{I}\left(\boldsymbol{\theta}\right)\left(\hat{\boldsymbol{\theta}}_n - \boldsymbol{\theta}\right) \le c^\alpha\right\}$$

$$\mathbb{R}_{2,1-\alpha} = \left\{\boldsymbol{\theta} : n\left(\hat{\boldsymbol{\theta}}_n - \boldsymbol{\theta}\right)' \mathcal{I}\left(\hat{\boldsymbol{\theta}}_n\right)\left(\hat{\boldsymbol{\theta}}_n - \boldsymbol{\theta}\right) \le c^\alpha\right\}. \qquad (7.31)$$

The geometry of $\mathbb{R}_{1,1-\alpha}$ is usually complicated, but $\mathbb{R}_{2,1-\alpha}$ is always a simple ellipse or ellipsoid. Again, nothing guarantees that either region lies wholly within admissible space $\boldsymbol{\Theta}$, so we always restrict to the intersection.

Example 7.49. Sampling from a normal population with $\boldsymbol{\theta} = \left(\mu, \sigma^2\right)'$, m.l. estimator $\hat{\boldsymbol{\theta}}_n = \left(\bar{X}, M_2\right)'$ has inverse asymptotic covariance matrix

$$n\mathcal{I}\left(\mu, \sigma^2\right) = \begin{pmatrix} \frac{n}{\sigma^2} & 0 \\ 0 & \frac{n}{2\sigma^4} \end{pmatrix}.$$

This happens to involve just the one parameter σ^2, of which M_2 is a consistent estimator; thus, by Mann–Wald and continuity, quadratic forms

$$\Pi_1\left(\mathbf{X}, \boldsymbol{\theta}\right) := \frac{n\left(\bar{X} - \mu\right)^2}{\sigma^2} + \frac{n\left(M_2 - \sigma^2\right)^2}{2\sigma^4}$$

$$\Pi_2\left(\mathbf{X}, \boldsymbol{\theta}\right) := \frac{n\left(\bar{X} - \mu\right)^2}{M_2} + \frac{n\left(M_2 - \sigma^2\right)^2}{2M_2^2}$$

both converge in distribution to $\chi^2\left(2\right)$. With c^α as the upper-α quantile and $\bar{x}$ and m_2 as sample realizations, asymptotic $1 - \alpha$ joint c.r.s are

$$\mathbb{R}_{1,1-\alpha} = \left\{\left(\mu, \sigma^2\right) : \frac{n\left(\bar{x} - \mu\right)^2}{\sigma^2} + \frac{n\left(m_2 - \sigma^2\right)^2}{2\sigma^4} \le c^\alpha\right\}$$

$$\mathbb{R}_{2,1-\alpha} = \left\{\left(\mu, \sigma^2\right) : \frac{n\left(\bar{x} - \mu\right)^2}{m_2} + \frac{n\left(m_2 - \sigma^2\right)^2}{2m_2^2} \le c^\alpha\right\}.$$

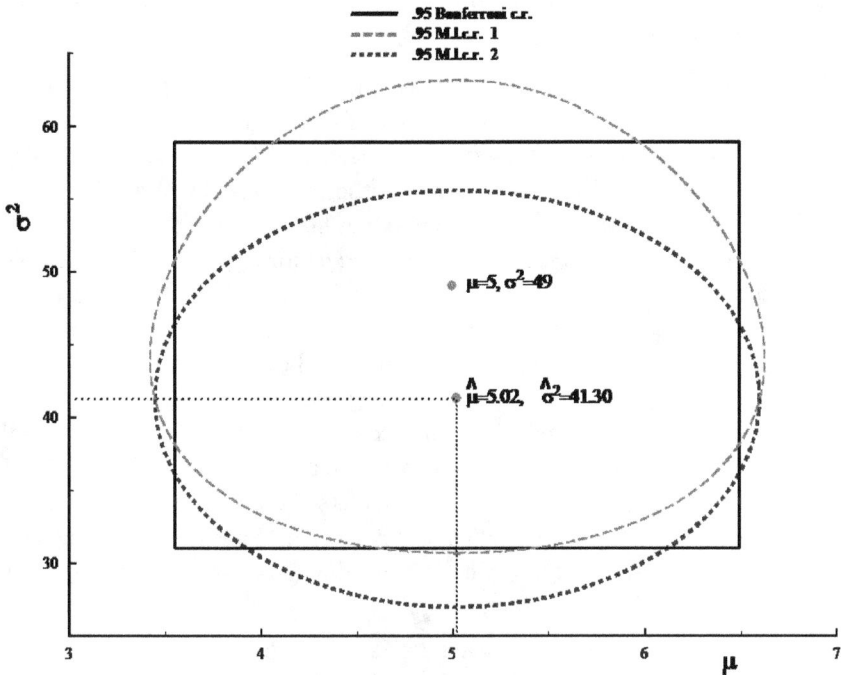

Fig. 7.3 Joint .95 c.r.s for μ, σ^2 from Bonferroni's inequality and joint asymptotic distribution of m.l.e.s.

The dashed curves in Figure 7.3 outline the m.l.-based .95 c.r.s for the data of Example 7.48. Major axes of all the regions are parallel to the coordinate axes because $\bar{X}$ and M_2 are independent. Notice that $\mathbb{R}_{1,.95}$ and the rectangular Bonferroni c.r. do a better job of allowing for the right skewness of M_2 than does the elliptical $\mathbb{R}_{2,.95}$.

Exercise 7.35. $\left\{ \mathbf{Z}_j = (X_j, Y_j)' \right\}_{j=1}^n$ *is a bivariate random sample from a population distributed as bivariate Bernoulli, having joint p.m.f.* $f_{\mathbf{Z}}(x, y; \boldsymbol{\theta}) = \theta_1^x \theta_2^y (1 - \theta_1 - \theta_2)^{1-x-y}$ *for* $x + y \in \{0, 1\}$, $\theta_1, \theta_2 \in (0, 1)$, $\theta_1 + \theta_2 \in (0, 1)$. *A sample of* $n = 40$ *yields* $\sum_{j=1}^n x_j = 15, \sum_{j=1}^n y_j = 10$. *Find the asymptotic c.r. corresponding to* $\mathbb{R}_{2,.95}$ *in (7.31). (Hint: This problem is an extension of Exercise 7.7 on page 459.)*

7.3.2   Confidence Sets from Pivots Based on Likelihood Functions

We introduced maximum likelihood estimation in Section 7.2.2 by depicting likelihood functions $L(\mathbf{x}_1; \mu)$ and $L(\mathbf{x}_2; \mu)$ for two random samples $\mathbf{x}_1$ and $\mathbf{x}_2$ of size $n = 10$ from a Poisson distribution with parameter $\mu = 5$. The functions, plotted in Fig. 7.1, show how the probabilities of getting these sample realizations vary with μ, the maxima being attained at the m.l. estimates $\bar{x}_1 = 5.0, \bar{x}_2 = 4.3$. If we divide $L(\mathbf{x}; \mu)$ for a given realized sample $\mathbf{x}$ by its maximum value $L(\mathbf{x}; \hat{\mu}_n = \bar{x})$, we have what is called the *likelihood ratio* or *relative likelihood*, $\Lambda(\mathbf{x}; \mu) := L(\mathbf{x}; \mu)/L(\mathbf{x}; \hat{\mu}_n)$. This function clearly lies on $[0, 1]$ and attains the upper bound of unity at $\mu = \hat{\mu}_n$. Thus, to any $\lambda \in (0, 1)$ there corresponds a nonempty set of values of μ whose relative likelihood is *at least* λ. When $\Lambda(\mathbf{x}; \mu)$ is a continuous and unimodal function of μ, as it is in the Poisson case, such a set $\{\mu > 0 : \Lambda(\mathbf{x}; \mu) \geq \lambda\}$ is an interval, $[\mu_\lambda, \mu^\lambda]$, say. Figure 7.4 depicts the likelihood ratios for the two Poisson samples with $\bar{x}_1 = 5.0$ and $\bar{x}_2 = 4.3$, along with the "highest-likelihood" intervals corresponding to the value $\lambda = .1465$. Intervals of this

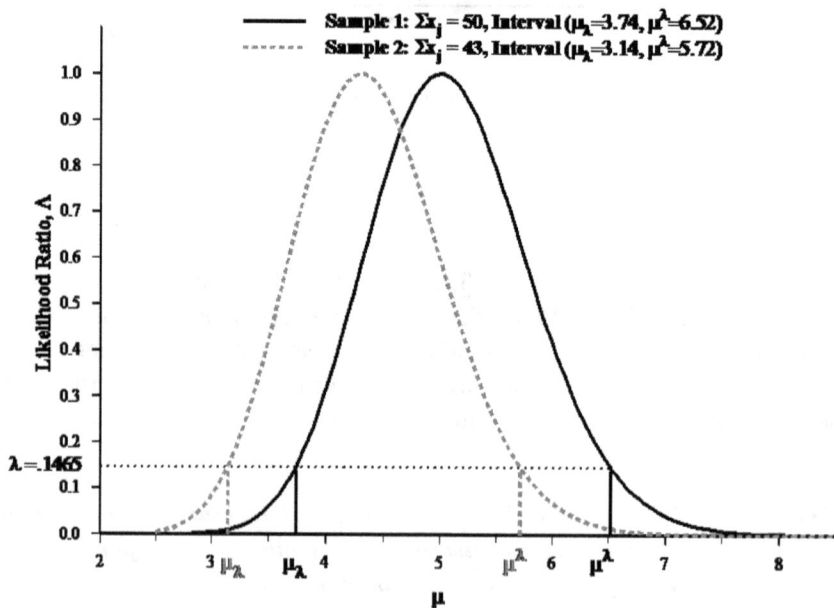

Fig. 7.4   Highest-likelihood c.i.s for two Poisson samples of $n = 10$.

sort represent the values of μ that are "most consistent" with the sample data, and so they seem ideally suited to represent confidence intervals. But how can one choose the critical λ so as to attain a given desired confidence level $1 - \alpha$?

To answer the question, let us first consider a generic model represented by a density $f(x; \theta)$ with scalar parameter $\theta \in \Theta \subset \Re$. Working with *ex post* random sample $\mathbf{x} = (x_1, x_2, \ldots, x_n)'$, the likelihood ratio is

$$\Lambda(\mathbf{x}; \theta) = \frac{L(\mathbf{x}; \theta)}{L\left(\mathbf{x}; \hat{\theta}_n\right)} = \frac{\prod_{j=1}^{n} f(x_j; \theta)}{\prod_{j=1}^{n} f\left(x_j; \hat{\theta}_n\right)},$$

where $\hat{\theta}_n$ is the m.l.e. *Ex ante* (before the sample is observed), $\Lambda(\mathbf{X}; \theta)$ is, for each $\theta \in \Theta$, a random variable supported on $[0, 1]$. In fact, we shall see that for a broad class of models f the likelihood ratio is either exactly or asymptotically *pivotal*. Its c.d.f. (or approximating, large-sample c.d.f.) F_Λ is continuous, strictly increasing, and—since Λ is pivotal—the same *for all* $\theta \in \Theta$. Confining attention to this class of models, and given a desired coverage probability $1 - \alpha$, we can find a unique quantile λ_α such that $\Pr[\Lambda(\mathbf{X}; \theta) \geq \lambda_\alpha] = 1 - F_\Lambda(\lambda_\alpha) = 1 - \alpha$ holds for all $\theta \in \Theta$—either *exactly* or approximately in large samples. To such λ_α there corresponds a set $\{\theta \in \Theta : \Lambda(\mathbf{x}; \theta) \geq \lambda_\alpha\}$, and since Λ will be unimodal in the models we consider, the resulting set of θ's will be an interval. We take this interval to be the $1 - \alpha$ *highest-likelihood confidence interval (h.l.c.i.)* for θ. It is an exact such interval or an approximate one depending on the status of the operative λ_α. Notice that the h.l.c.i. is always contained in parameter space Θ, so that no *ad hoc* restriction is ever needed.

In seeking to find the critical threshold λ_α one encounters two common situations. The first, more felicitous situation is that $\Lambda(\mathbf{X}; \theta)$ is a function $g[h(\mathbf{X}, \theta)]$ of a simpler pivotal quantity h whose exact distribution is known, where g depends on $\mathbf{X}$ and θ only through $h(\mathbf{X}, \theta)$. In this case, although Λ itself is exactly pivotal, it will usually be easier to express the event $\Lambda(\mathbf{X}; \theta) \geq \lambda_\alpha$ in terms of the simpler $h(\mathbf{X}, \theta)$. The second, more common situation is that event $\Lambda(\mathbf{X}; \theta) \geq \lambda_\alpha$ *cannot* be expressed in terms of a pivotal quantity whose exact distribution is known. In this case we will need an asymptotic approximation for F_Λ. The following result, which we present for the k-parameter case $\boldsymbol{\theta} \in \Re^k$, applies to all models $f(x; \boldsymbol{\theta})$ that satisfy the regularity conditions for m.l. estimation that were given on page 494.

Theorem 7.5. *If* $\mathbf{X} = (X_1, X_2, ..., X_n)'$ *is a random sample from a regular distribution* $f(x; \boldsymbol{\theta})$ *with* $\boldsymbol{\theta} \in \Theta \subset \Re^k$, *and if* $\hat{\boldsymbol{\theta}}_n$ *is the m.l.e., then for each fixed* $\boldsymbol{\theta} \in \Theta$ *the r.v.*

$$-2 \log \Lambda(\mathbf{X}; \boldsymbol{\theta}) = 2 \left[\log L\left(\mathbf{X}; \hat{\boldsymbol{\theta}}_n \right) - \log L(\mathbf{X}; \boldsymbol{\theta}) \right]$$

converges in distribution to $\chi^2(k)$ *as* $n \to \infty$.

Proof. For brevity, put $\mathcal{L}(\boldsymbol{\theta}) := \log L(\mathbf{X}; \boldsymbol{\theta})$ and let $\mathcal{L}_{\boldsymbol{\theta}}(\mathbf{t}) := \partial \mathcal{L}(\boldsymbol{\theta}) / \partial \boldsymbol{\theta}|_{\boldsymbol{\theta}=\mathbf{t}}$ and $\mathcal{L}_{\boldsymbol{\theta}\boldsymbol{\theta}'}(\mathbf{t}) := \partial^2 \mathcal{L}(\boldsymbol{\theta}) / \partial \boldsymbol{\theta} \partial \boldsymbol{\theta}'|_{\boldsymbol{\theta}=\mathbf{t}}$ represent, respectively, the k-vector of partial derivatives and the $k \times k$ Hessian matrix, each evaluated at some $\mathbf{t} \in \Theta$. Expanding $\mathcal{L}(\boldsymbol{\theta})$ about $\hat{\boldsymbol{\theta}}_n$ gives

$$\begin{aligned}
-2 \log \Lambda(\mathbf{X}; \boldsymbol{\theta}) &= 2 \left[-\mathcal{L}_{\boldsymbol{\theta}}\left(\hat{\boldsymbol{\theta}}_n \right)\left(\boldsymbol{\theta} - \hat{\boldsymbol{\theta}}_n \right) - \frac{1}{2}\left(\boldsymbol{\theta} - \hat{\boldsymbol{\theta}}_n \right)' \mathcal{L}_{\boldsymbol{\theta}\boldsymbol{\theta}'}\left(\bar{\boldsymbol{\theta}}_n \right)\left(\boldsymbol{\theta} - \hat{\boldsymbol{\theta}}_n \right) \right] \\
&= -\left(\hat{\boldsymbol{\theta}}_n - \boldsymbol{\theta} \right)' \mathcal{L}_{\boldsymbol{\theta}\boldsymbol{\theta}'}\left(\bar{\boldsymbol{\theta}}_n \right)\left(\hat{\boldsymbol{\theta}}_n - \boldsymbol{\theta} \right) \\
&= \sqrt{n}\left(\hat{\boldsymbol{\theta}}_n - \boldsymbol{\theta} \right)' \left[-n^{-1} \mathcal{L}_{\boldsymbol{\theta}\boldsymbol{\theta}'}\left(\bar{\boldsymbol{\theta}}_n \right) \right] \sqrt{n}\left(\hat{\boldsymbol{\theta}}_n - \boldsymbol{\theta} \right).
\end{aligned}$$

Here, $\bar{\boldsymbol{\theta}}_n$ is a r.v. with realizations intermediate between those of $\hat{\boldsymbol{\theta}}_n$ and $\boldsymbol{\theta}$, and the second line follows from the fact that $\mathcal{L}_{\boldsymbol{\theta}}\left(\hat{\boldsymbol{\theta}}_n \right) = \mathbf{0}$. Now $\sqrt{n}\left(\hat{\boldsymbol{\theta}}_n - \boldsymbol{\theta} \right) \rightsquigarrow N\left(\mathbf{0}, \mathcal{I}^{-1} \right)$ for each fixed $\boldsymbol{\theta}$, where $\mathcal{I} \equiv \mathcal{I}(\boldsymbol{\theta}) = -E \partial^2 \log f(X; \boldsymbol{\theta}) / \partial \boldsymbol{\theta} \partial \boldsymbol{\theta}'$ is the Fisher information from a single observation, and

$$-n^{-1} \mathcal{L}_{\boldsymbol{\theta}\boldsymbol{\theta}'}(\boldsymbol{\theta}) = -n^{-1} \sum_{j=1}^{n} \frac{\partial^2 \log f(X_j; \boldsymbol{\theta})}{\partial \boldsymbol{\theta} \partial \boldsymbol{\theta}'} \xrightarrow{a.s.} \mathcal{I}$$

by the strong law of large numbers. Also, the weak consistency of $\bar{\boldsymbol{\theta}}_n$ follows from the like property of $\hat{\boldsymbol{\theta}}_n$, and so the continuity theorem implies that $-n^{-1} \mathcal{L}_{\boldsymbol{\theta}\boldsymbol{\theta}'}\left(\bar{\boldsymbol{\theta}}_n \right) \xrightarrow{p} \mathcal{I}$. Applications of Mann–Wald and Theorem 5.1 now deliver the asserted conclusion. □

Thus, the likelihood ratio itself is always either pivotal or asymptotically pivotal under the regularity conditions. In the scalar case, and still with regular $f(x; \boldsymbol{\theta})$, the theorem implies that

$$\lim_{n \to \infty} \Pr[\Lambda(\mathbf{X}; \boldsymbol{\theta}) \geq \lambda_\alpha] = \lim_{n \to \infty} \Pr[-2 \log \Lambda(\mathbf{X}; \boldsymbol{\theta}) \leq -2 \log \lambda_\alpha] = 1 - \alpha$$

when $\lambda_\alpha = \exp(-c_{1-\alpha}/2)$ and $c_{1-\alpha}$ is the $1 - \alpha$ quantile of $\chi^2(1)$.

7.3.2.1 *Highest-Likelihood C.I.s: Applications*

We now present some examples of highest-likelihood c.i.s.[35] These both illustrate the use of large-n approximations and show how to find exact results when $\Lambda(\mathbf{X};\theta)$ is a function of a simpler exact pivot.

Example 7.50. Consider again the Poisson case, with (regular) likelihood

$$L\left(\mathbf{X};\mu\right) = \frac{\mu^{n\bar{X}}e^{-n\mu}}{X_1!\cdots X_n!}$$

for $\mu > 0$, m.l.e. $\hat{\mu}_n = \bar{X}$, and likelihood ratio

$$\Lambda(\mathbf{X};\mu) = \frac{\mu^{n\bar{X}}e^{-n\mu}}{\bar{X}^{n\bar{X}}e^{-n\bar{X}}} = \left(\frac{\mu}{\bar{X}}\right)^{n\bar{X}}e^{-n(\mu-\bar{X})}.$$

This is not expressible in terms of a simpler pivotal quantity, and the exact distribution depends on μ, so we must rely on the asymptotic approximation. To build an approximate $1-\alpha$ h.l.c.i., take $\lambda_\alpha = \exp\left(-c_{1-\alpha}/2\right) = \exp\left(-c^\alpha/2\right)$, where $c_{1-\alpha} = c^\alpha$ are the appropriate quantiles of $\chi^2(1)$. Now Λ is unimodal in μ, with maximum value unity at $\mu = \bar{X}$. Provided sample realization $\mathbf{x} \neq \mathbf{0}$, to any $\lambda_\alpha \in (0,1)$ there correspond two roots $0 < \mu_\lambda < \mu^\lambda < 1$ such that $\Lambda(\mathbf{x};\mu_\lambda) = \Lambda(\mathbf{x};\mu^\lambda) = \lambda_\alpha$ and such that $\Lambda(\mathbf{x};\mu) \geq \lambda_\alpha$ for $\mu \in [\mu_\lambda,\mu^\lambda]$. The interval $[\mu_\lambda,\mu^\lambda]$ represents the $1-\alpha$ h.l.c.i. The roots may be found by a numerical procedure, such as bisection. For the two samples in Fig. 7.4 with $n = 10$, $\bar{x}_1 = 5.0$, and $\bar{x}_2 = 4.3$ we need $\lambda_{.05} = \exp\left(-3.841/2\right) \doteq .1465$ for .95 c.i.s. This yields the intervals $[3.739, 6.516]$ and $[3.140, 5.716]$ shown in the figure.

Alternatively, as in Section 7.3.1, we could develop approximate c.i.s using either of the m.l.e.-based asymptotically $N(0,1)$ pivots $\sqrt{n}\left(\bar{X}-\mu\right)/\sqrt{\mu}$ or $\sqrt{n}\left(\bar{X}-\mu\right)/\sqrt{\bar{X}}$, which we refer to as "m.l.1" and "m.l.2", respectively. These c.i.s have the form $\left[\max\left\{0,l_{\alpha/2}\right\},u_{\alpha/2}\right]$, where (recalling that $z^{\alpha/2} = \sqrt{c_{1-\alpha}}$)

$$l_{\alpha/2} := \begin{cases} \bar{x} + \frac{c_{1-\alpha}}{2n} - \sqrt{\frac{\bar{x}c_{1-\alpha}}{n} + \frac{c_{1-\alpha}^2}{4n^2}}, & \text{for m.l.1} \\ \bar{x} - z^{\alpha/2}n^{-1/2}\sqrt{\bar{x}}, & \text{for m.l.2} \end{cases}$$

$$u_{\alpha/2} := \begin{cases} \bar{x} + \frac{c_{1-\alpha}}{2n} + \sqrt{\frac{\bar{x}c_{1-\alpha}}{n} + \frac{c_{1-\alpha}^2}{4n^2}}, & \text{for m.l.1} \\ \bar{x} + z^{\alpha/2}n^{-1/2}\sqrt{\bar{x}}, & \text{for m.l.2} \end{cases}.$$

---

[35]Boyles (2008) gives other examples of h.l.c.i.s and some persuasive arguments for using them.

With the data for the two samples in Fig. 7.4 these are $[3.793, 6.591]$ and $[3.193, 5.792]$ for the m.l.1 pivot and $[3.614, 6.386]$ and $[3.015, 5.585]$ for m.l.2. Simulations with various $\lambda \geq 1$ and $\alpha \in \{.10, .05, .01\}$ indicate that all three c.i.s have accurate average coverage rates and similar average lengths for n as small as 10.

Example 7.51. A random sample of size n from a Bernoulli distribution with parameter $\theta \in (0, 1)$ yields a total number of "successes" $X \sim B(n, \theta)$ (binomial). With $\hat{\theta}_n = X/n$, the relative likelihood is

$$\Lambda(X; \theta) = \frac{\theta^X (1 - \theta)^{n-X}}{\hat{\theta}_n^X \left(1 - \hat{\theta}_n\right)^{n-X}}.$$

Again, this is not exactly pivotal, and so an asymptotic approximation is required. $\Lambda(x; \theta)$ is unimodal in θ, and when $x \in \{1, 2, ..., n-1\}$ and $\lambda \in (0, 1)$ there are two roots $0 < \theta_\lambda < \theta^\lambda < 1$ satisfying $\Lambda(x; \theta_\lambda) = \Lambda(x; \theta^\lambda) = \lambda$. Setting $\lambda_\alpha = \exp(-c_{1-\alpha}/2)$ gives $[\theta_\lambda, \theta^\lambda]$ as the approximate $1 - \alpha$ h.l.c.i.

Alternatively, since both $\sqrt{n}(\hat{\theta}_n - \theta)/\sqrt{\theta(1 - \theta)}$ (m.l.1) and $\sqrt{n}(\hat{\theta}_n - \theta)/\sqrt{\hat{\theta}_n \left(1 - \hat{\theta}_n\right)}$ (m.l.2) are asymptotically standard normal, the interval $[\max\{0, l_{\alpha/2}\}, \min\{1, u_{\alpha/2}\}]$ is an approximate c.i. based on the m.l.e., where

$$l_{\alpha/2} := \begin{cases} \dfrac{\hat{\theta}_n + c_{1-\alpha}/(2n) - \sqrt{\hat{\theta}_n \left(1 - \hat{\theta}_n\right) c_{1-\alpha}/n + c_{1-\alpha}^2/(4n^2)}}{1 + c_{1-\alpha}/n}, & \text{for m.l.1} \\[2ex] \hat{\theta}_n - z^{\alpha/2} n^{-1/2} \sqrt{\hat{\theta}_n \left(1 - \hat{\theta}_n\right)}, & \text{for m.l.2} \end{cases}$$

$$u_{\alpha/2} := \begin{cases} \dfrac{\hat{\theta}_n + c_{1-\alpha}/(2n) + \sqrt{\hat{\theta}_n \left(1 - \hat{\theta}_n\right) c_{1-\alpha}/n + c_{1-\alpha}^2/(4n^2)}}{1 + c_{1-\alpha}/n}, & \text{for m.l.1} \\[2ex] \hat{\theta}_n + z^{\alpha/2} n^{-1/2} \sqrt{\hat{\theta}_n \left(1 - \hat{\theta}_n\right)}, & \text{for m.l.2} \end{cases}$$

As example, with $x = 3$ from a sample of $n = 20$ with $\theta = .2$, the approximate .95 h.l.c.i. is $[.0396, .3444]$, while the corresponding c.i.s from pivots m.l.1 and m.l.2 are $[.0524, .3604]$ and $[0, .3065]$, respectively. Simulations indicate that h.l.c.i.s are typically more accurate and shorter than those based on the m.l.e. Not surprisingly, all the methods require larger samples for good accuracy as θ approaches the boundary points $\{0, 1\}$, but the h.l.c.i.s have very accurate coverage probabilities for $\alpha \in \{.01, .05, .10\}$ when $n \geq 20$ and $\theta \in [.2, .8]$.

Exercise 7.36. *(i) In example 7.50 the event* **X** = **0** *occurs with probability* $e^{-n\mu} > 0$. *For given* $1 - \alpha$ *what would be the h.l.c.i. in this case? What c.i.s, if any, would be implied by the pivotal quantities based on the m.l.e.?*

(ii) In example 7.51 the events $X = 0$ *and* $X = n$ *also occur with positive probability for any finite* n. *For given* $1 - \alpha$ *what would be the h.l.c.i.s in these cases? What c.i.s, if any, would be implied by the pivotal quantities based on the m.l.e.?*

Exercise 7.37. *A random sample* $\{x_j\}_{j=1}^{100}$ *from a population modeled as* $\mathfrak{Be}\,(\theta, 1)$ *(beta distribution) yields values such that* $\sum_{j=1}^{100} \log x_j = -15.1837$. *Find (i) the sample realization of m.l. estimator* $\hat{\theta}_{100}$ *and the estimate of its asymptotic variance; (ii) a .95 m.l.1 c.i. for* θ; *(iii) a .95 m.l.2 c.i. for* θ; *and (iv) the .95 h.l.c.i., using the asymptotic distribution of the relative likelihood.*

Example 7.52. A random sample of size n from $f(x; \theta) = \theta^{-1} e^{-x/\theta} \mathbf{1}_{(0,\infty)}(x)$, $\theta \in \Theta = (0, \infty)$, yields likelihood ratio

$$\Lambda(\mathbf{X}; \theta) = \left(\frac{\hat{\theta}_n}{\theta}\right)^n \exp\left[-n\left(\frac{\hat{\theta}_n}{\theta} - 1\right)\right],$$

where $\hat{\theta}_n = \bar{X}$. Λ is expressible in terms of the exact pivotal quantity $Y := \bar{X}/\theta$, with $nY = \sum_{j=1}^n X_j/\theta \sim \Gamma(n, 1)$. For any $\lambda_\alpha \in (0, 1)$ the equation $y^n e^{-n(y-1)} = \lambda_\alpha$ has roots $0 < y_\lambda < 1 < y^\lambda$, and values of y between the roots correspond to large values of Λ. The exact h.l.c.i. is found by the following steps.

(i) Given an arbitrary $\lambda \in (0, 1)$, find the roots y_λ, y^λ and evaluate $\Pr(\Lambda \geq \lambda)$ as $F(ny^\lambda) - F(ny_\lambda)$, where $F(x) = \int_0^x \Gamma(n)^{-1} y^{n-1} e^{-y} \cdot dy$ (the $\Gamma(n, 1)$ c.d.f.). (ii) Search numerically for the unique λ_α such that $\Pr(\Lambda \geq \lambda_\alpha) = 1 - \alpha$. (iii) Invert the inequality $y_{\lambda_\alpha} \leq \hat{\theta}_n/\theta \leq y^{\lambda_\alpha}$ to get the h.l.c.i.: $\left[\hat{\theta}_n/y^{\lambda_\alpha}, \hat{\theta}_n/y_{\lambda_\alpha}\right]$.

In this application there are several alternative constructions of c.i.s. First, we could use the asymptotic approximation $-2 \log \Lambda(\mathbf{X}; \theta) \rightsquigarrow \chi^2(1)$ and set $\lambda_\alpha = \exp(-c_{1-\alpha}/2)$ without having to search. Second, as in Example 7.42, we could get an exact c.i. based directly on the marginal distribution of pivotal quantity nY, without explicit reference to the sample likelihood. Since $2nY \sim \Gamma(n, 2) \equiv \chi^2(2n)$, we could use tabulated values of shortest-interval quantiles of $\chi^2(2n)$, $c_{(1-p^*)\alpha}$ and $c^{p^*\alpha}$, to get an exact c.i. as $\left[2n\hat{\theta}_n/c^{p^*\alpha}, 2n\hat{\theta}_n/c_{(1-p^*)\alpha}\right]$. Third, we could just take

$p = .5$ and divide residual probability α equally to get the exact interval $\left[2n\hat{\theta}_n/c^{\alpha/2}, 2n\hat{\theta}_n/c_{\alpha/2} \right]$. Finally, we could rely on the asymptotic normality of pivots $\sqrt{n}\left(\hat{\theta}_n - \theta \right)/\theta$ (m.l.1) and $\sqrt{n}\left(\hat{\theta}_n - \theta \right)/\hat{\theta}_n$ (m.l.2) to develop approximate c.i.s from the m.l.e., as

$$\left[\bar{x}\left(1 + z^{\alpha/2}/\sqrt{n}\right)^{-1}, \bar{x}\left(1 - z^{\alpha/2}/\sqrt{n}\right)^{-1} \right]$$

$$\left[\bar{x}\left(1 - z^{\alpha/2}/\sqrt{n}\right), \bar{x}\left(1 + z^{\alpha/2}/\sqrt{n}\right) \right],$$

respectively. As examples, here are the various .95 versions from a sample of $n = 20$ with $\bar{x} = .8353$:

Method	$\lambda_{.05}$	θ_l	θ^u	$\theta^u - \theta_l$
Exact highest likelihood	.1442	.5543	1.3429	.7886
Asymptotic highest likelihood	.1465	.5552	1.3401	.7849
Exact pivot: shortest interval	-	.5234	1.2938	.7704
Exact pivot: equal–prob. interval	-	.5630	1.3674	.8044
Asymptotic pivot m.l.1	-	.5808	1.4869	.9061
Asymptotic pivot m.l.2	-	.4692	1.2013	.7321

Simulations indicate that the asymptotic h.l. and m.l.1 c.i.s have average coverage rates close to nominal levels for n as small as 20, although the likelihood-based intervals are much shorter on average. The m.l.2 c.i.s are usually the shortest of all, but their coverage rates are uniformly low. Of course, the rates for the other c.i.s are exact, and their average lengths are similar.

Exercise 7.38. *Express the p.d.f. of $Y : -\log X$ when $X \sim \mathfrak{Be}\,(\theta, 1)$ (beta distribution). (i) Using the sample values of $\sum_{j=1}^{100} y_j \equiv -\sum_{j=1}^{100} \log x_j$ from Exercise 7.37, work out the exact .95 h.l.c.i. for θ using the fact that relative likelihood $\Lambda\,(\mathbf{Y}; \theta)$ can be expressed in terms of the pivotal quantity $\theta \bar{Y}_{100}$. (This involves some significant numerical computation, as in Example 7.52.) (ii) Compare the exact h.l.c.i. with the exact "shortest" and "equal–probability" intervals based on the distribution of $200\theta \bar{Y}_{100}$.*

7.3.2.2 *Joint and "Profile" Confidence Regions when* $\boldsymbol{\theta} \in \Re^k$

The pivotal nature of the likelihood ratio makes it possible to build confidence regions for $\boldsymbol{\theta} \in \boldsymbol{\Theta} \subset \Re^k$, as well as c.i.s for the individual components of $\boldsymbol{\theta}$ and joint c.r.s for subsets of $\{\theta_1, ..., \theta_k\}$. To find the c.r. for all k elements of $\boldsymbol{\theta}$, we need a λ_α such that $\Pr[\Lambda(\mathbf{X}; \boldsymbol{\theta}) \le \lambda_\alpha]$ equals or approximates α for each fixed $\boldsymbol{\theta} \in \boldsymbol{\Theta}$. For a given sample realization $\mathbf{x}$ the set $\{\boldsymbol{\theta} \in \boldsymbol{\Theta} : \Lambda(\mathbf{x}; \boldsymbol{\theta}) \ge \lambda_\alpha\}$ then constitutes an exact or approximate $1 - \alpha$ c.r. To find a c.r. for a subset $\{\theta_1, ..., \theta_{k_0}\}$ of k_0 elements of $\boldsymbol{\theta}$ (taking the *first* k_0 without loss of generality), let $\boldsymbol{\phi} \in \boldsymbol{\Phi} \subset \Re^{k_0}$ be the vector of these components and put $\boldsymbol{\theta} = (\boldsymbol{\phi}', \boldsymbol{\psi}')'$, where $\boldsymbol{\psi} \in \boldsymbol{\Psi} \subset \Re^{k-k_0}$ and $\boldsymbol{\Theta} = \boldsymbol{\Phi} \times \boldsymbol{\Psi}$. Then the *profile* likelihood ratio, $\Lambda(\mathbf{X}; \boldsymbol{\phi}) = L\left(\mathbf{X}; \boldsymbol{\phi}, \check{\boldsymbol{\psi}}_n\right) / L\left(\mathbf{X}; \hat{\boldsymbol{\theta}}_n\right)$, is a function of $\boldsymbol{\phi}$ alone. Here $\hat{\boldsymbol{\theta}}_n = \left(\hat{\boldsymbol{\phi}}_n', \hat{\boldsymbol{\psi}}_n'\right)'$ is the usual m.l.e. of all k parameters, while $\check{\boldsymbol{\psi}}_n$ is the constrained m.l.e. that solves $\max_{\boldsymbol{\theta} \in \boldsymbol{\Theta}} \{L(\mathbf{X}; \boldsymbol{\theta})\}$ subject to the restriction $(\theta_1, ..., \theta_{k_0})' = \boldsymbol{\phi}$. Obviously, we still have $\Lambda(\mathbf{X}; \boldsymbol{\phi}) \in [0, 1]$ a.s., so for $\lambda_\alpha \in (0, 1)$ such that $\Pr(\Lambda \le \lambda_\alpha) = \alpha$ the set $\{\boldsymbol{\phi} \in \boldsymbol{\Phi} : \Lambda(\mathbf{x}; \boldsymbol{\phi}) \ge \lambda_\alpha\}$ represents a $1 - \alpha$ confidence set for $\boldsymbol{\phi}$. When $k_0 = 1$ (ϕ a scalar), then in standard cases this set is a closed interval $[\phi_\lambda, \phi^\lambda]$, as in the one-parameter case.

In constructing such confidence sets we can sometimes express $\Lambda(\mathbf{X}; \boldsymbol{\phi})$ in terms of an exact pivotal quantity and find an exact λ_α. Otherwise, an approximant can be found in regular cases from the fact that $-2 \log \Lambda(\mathbf{X}; \boldsymbol{\phi}) \rightsquigarrow \chi^2(k_0)$. This was established in Theorem 7.5 for the case $k_0 = k$ (i.e., the case $\boldsymbol{\Phi} = \boldsymbol{\Theta}$, $\boldsymbol{\Psi} = \varnothing$). The result for $k_0 \in \{1, 2, ..., k-1\}$ is harder to establish because the likelihood ratio depends on estimators $\check{\boldsymbol{\psi}}_n$ of the "nuisance" parameters $\boldsymbol{\psi}$. It follows from Theorem 8.1, which we present in connection with tests of statistical hypotheses.[36]

7.3.2.3 *Multiparameter Applications*

Our first two applications make use of the asymptotic approximations just described. At this point it may be helpful to review Examples 7.28 and 7.37 (pages 485 and 497). Shape and scale parameters α and β there correspond to ϕ and ψ here.

---

[36] In the testing context Theorem 8.1 establishes the limiting behavior of likelihood ratio *statistic* $\Lambda_n(\mathbf{X}; \boldsymbol{\phi}_0)$ when $\boldsymbol{\theta} = (\boldsymbol{\phi}_0', \boldsymbol{\psi}')'$ contains nuisance parameters not restricted by null hypothesis $\boldsymbol{\phi} = \boldsymbol{\phi}_0$. The theorem applies here because the relative likelihood at any fixed $\boldsymbol{\phi} \in \boldsymbol{\Phi}$ coincides with statistic $\Lambda_n(\mathbf{X}; \boldsymbol{\phi}_0)$ when $\boldsymbol{\phi}_0 = \boldsymbol{\phi}$.

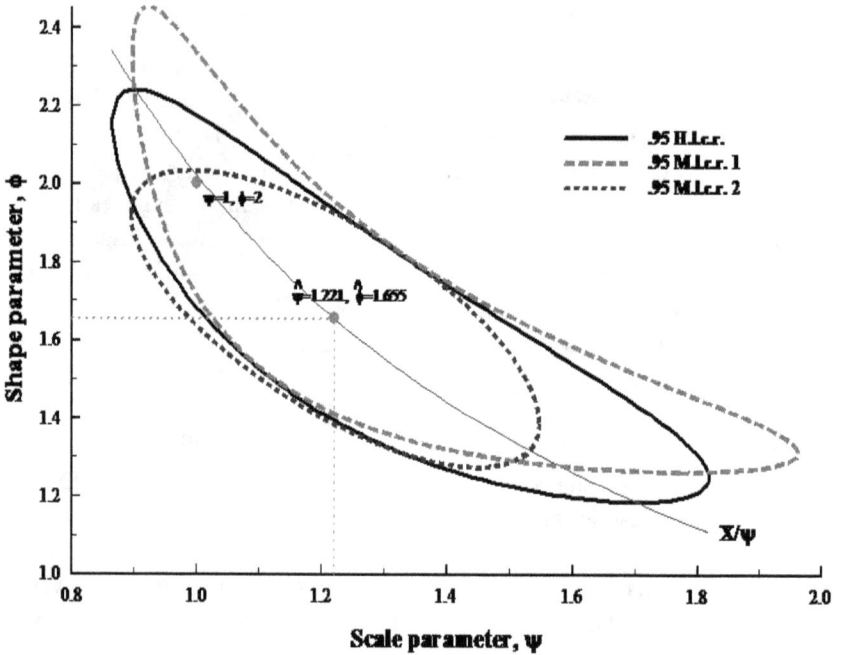

Fig. 7.5   Joint .95 c.r.s for parameters of $\Gamma(\phi, \psi)$ from likelihood-ratio and m.l.e-based pivots.

Example 7.53. With $\{X_j\}_{j=1}^n$ i.i.d. as $\Gamma(\phi, \psi)$ we build an approximate large-sample $1 - \alpha$ c.r. for $\boldsymbol{\theta} := (\phi, \psi)'$ (there being here no nuisance parameter at all). The likelihood ratio is

$$\Lambda(\mathbf{X}; \boldsymbol{\theta}) = \left[\frac{\Gamma\left(\hat{\phi}_n\right)\hat{\psi}_n^{\hat{\phi}_n}}{\Gamma(\phi)\psi^\phi}\right]^n \exp\left[\left(\phi - \hat{\phi}_n\right)\sum_{j=1}^n \log X_j - \left(\frac{1}{\psi} - \frac{1}{\hat{\psi}_n}\right)\sum_{j=1}^n X_j\right],$$

where $\hat{\phi}_n$ and $\hat{\psi}_n$ are functions of jointly sufficient statistics $\sum_{j=1}^n \log X_j$ and $\sum_{j=1}^n X_j$, and $\hat{\psi}_n = \bar{X}/\hat{\phi}_n$. With $\lambda_\alpha = e^{-c_{1-\alpha}/2}$, where $c_{1-\alpha}$ is the $1 - \alpha$ quantile of $\chi^2(2)$, the locus of points $\{(\phi, \psi) : \Lambda(\mathbf{x}; \boldsymbol{\theta}) = \lambda_\alpha\}$ defines a closed curve in $\Re^2$ that bounds an approximate $1 - \alpha$ c.r. for (ϕ, ψ). From a (pseudo)random sample of $n = 100$ from $\Gamma(2, 1)$ we obtain $\sum_{j=1}^n x_j = 202.020$, $\sum_{j=1}^n \log x_j = 37.1453$, and m.l.e.s $\hat{\phi}_n \doteq 1.6546, \hat{\psi}_n \doteq 1.2208$. With $c_{.95} = 5.991$, the approximate .95 c.r. appears as the solid curve in Fig. 7.5. For comparison, the dashed curve labeled m.l.c.r. 1 is an approximate c.r. based on the limiting $\chi^2(2)$ distribution of m.l.e.-based pivot

$$n\left(\hat{\phi}_n - \phi, \hat{\psi}_n - \psi\right)\mathcal{I}(\boldsymbol{\theta})\begin{pmatrix}\hat{\phi}_n - \phi \\ \hat{\psi}_n - \psi\end{pmatrix}.$$

The elliptical m.l.c.r. 2 uses the same $\chi^2(2)$ approximation but replaces $\mathcal{I}(\boldsymbol{\theta})$ with $\mathcal{I}\left(\hat{\boldsymbol{\theta}}_n\right)$. The thin open curve is a section of the hyperbola that is the locus $\bar{x} = \phi\psi$. Note that all three c.r.s from this particular sample do happen to cover the true values $\psi = 1, \phi = 2$. Simulations with these same values indicate that actual average coverage rates for the h.l.c.r. are slightly conservative but very close to nominal levels for $n \geq 20$. M.l.c.r. 2 is usually more accurate than m.l.c.r. 1, but less accurate than h.l.c.r.

Exercise 7.39. *The three regions depicted in Fig. 7.5 differ greatly in area and overlap substantially, and yet we find that coverage rates of regions constructed by the three methods are all very close to 95% in samples of this size. Explain how this is possible.*

Example 7.54. To build a c.i. for ϕ alone from a random sample from $\Gamma(\phi, \psi)$, we use the profile likelihood ratio $\Lambda(\mathbf{X}; \phi) = L\left(\mathbf{X}; \phi, \check{\psi}\right)/L\left(\mathbf{X}; \hat{\boldsymbol{\theta}}_n\right)$, where $\check{\psi}_n = \bar{x}/\phi$ is obtained by maximizing $L(\mathbf{X}; \boldsymbol{\theta})$ subject to the constraint $\theta_1 = \phi$. An approximate $1 - \alpha$ c.i. comprises $\left\{\phi : \Lambda(\mathbf{X}; \phi) \geq e^{-c_{1-\alpha}/2}\right\}$, where now $c_{1-\alpha}$ is the $1 - \alpha$ quantile of $\chi^2(1)$. For the same pseudorandom sample of $n = 100$ as above and with $c_{.95} = 3.841$, the approximate .95 h.l.c.i. for ϕ is $[1.271, 2.114]$. For comparison, the c.i.s based on the asymptotic distribution of (normalized, unconstrained) m.l.e. $\hat{\phi}_n$ are (i) $[1.140, 2.169]$ when asymptotic variance is expressed in terms of ϕ and (ii) $[1.234, 2.075]$ when variance is based on $\hat{\phi}_n$. Simulations indicate that coverage rates for the m.l.e.-based c.i.s can be quite erratic, sometimes conservative and sometimes excessive depending on α, even for n as large as 100. By contrast the h.l.c.i. is always slightly conservative but very accurate for $n \geq 50$.

Exercise 7.40. *With the data for Example 7.53 verify that the .95 c.i. based on m.l.e. $\hat{\phi}_n$ is $[1.234, 2.075]$ when asymptotic variance is computed from $\hat{\phi}_n$ (Hint: Use $\hat{\phi}_n = 1.65459$ and $\Psi'\left(\hat{\phi}_n\right) = .82163$.)*

Exercise 7.41. *The log-likelihood of a random sample* $\mathbf{x} = (x_1, x_2, ..., x_n)'$ *from* $IG\left(\mu, \lambda\right)$ *(inverse Gaussian) can be written as*

$$\log L\left(\mathbf{x}; \mu, \lambda\right) = -\frac{n}{2}\log\left(2\pi\right) - \frac{3}{2}\sum_{j=1}^{n}\log x_j + K\left(\mathbf{x}; \mu, \lambda\right),$$

where

$$K\left(\mathbf{x}; \mu, \lambda\right) := \frac{n}{2}\log\lambda - \frac{\lambda}{2\mu^2}\sum_{j=1}^{n}\frac{\left(x_j - \mu\right)^2}{x_j}$$

is the component of $\log L\left(\mathbf{x}; \mu, \lambda\right)$ *that depends on the parameters. With* $\bar{x}$ *as the sample mean,* $\bar{y} := n^{-1}\sum_{j=1}^{n} x_j^{-1}$, *and* $c_{1-\alpha}$ *as the* $1 - \alpha$ *quantile of* $\chi^2\left(2\right)$, *show that the locus of points* $\left(\mu, \lambda\right)$ *whose interior constitutes a large-sample* $1 - \alpha$ *h.l.c.r. is given explicitly by*

$$\mu\left(\lambda\right)^{\pm} = \frac{\bar{x}}{1 \mp \sqrt{1 - \bar{x}[\bar{y} - q\left(\lambda, \mathbf{x}, \alpha\right)]}} \quad for \ \{\lambda : \bar{y} - q\left(\lambda, \mathbf{x}, \alpha\right) \leq 1/\bar{x}\},$$

$$(7.32)$$

where

$$q\left(\lambda, \mathbf{x}, \alpha\right) := \frac{\log\lambda + c_{1-\alpha}/n - 2K\left(\mathbf{x}; \hat{\mu}_n, \hat{\lambda}_n\right)/n}{\lambda}.$$

A (pseudo-random) sample of $n = 100$ *from* $IG\left(1, 1\right)$ *yields* $\bar{x} = 0.9718$ *and* $\bar{y} = 1.8513$. *Plot the* $1 - \alpha = 0.95$ *h.l.c.r. for these data. (Hint: By numerical search or other means find the two roots* $\lambda^{\pm}$ *of* $\bar{y} - q\left(\lambda, \mathbf{x}, .05\right) - 1/\bar{x} = 0$, *which define the applicable range of* λ. *Then generate the pair* $\mu^-\left(\lambda\right), \mu^+\left(\lambda\right)$ *via (7.32) for an array of values* $\lambda \in [\lambda^-, \lambda^+]$ *and graph the two series. Note that* $\mu^{\pm}\left(\lambda^-\right) = \mu^{\pm}\left(\lambda^+\right) = \hat{\lambda}_n$.*

Exercise 7.42. *Using the data from Exercise 7.41, find .95 c.i.s for inverse Gaussian parameter* λ *based on (i) the profile likelihood ratio, (ii) the m.l. estimator* $\hat{\lambda}_n$ *with asymptotic variance expressed in terms of* λ, *and (iii) the m.l. estimator* $\hat{\lambda}_n$ *with asymptotic variance estimated in terms of* $\hat{\lambda}_n$.

In the important case of inferences about the normal distribution, the likelihood ratio can be expressed in terms of functions that are jointly pivotal for μ and σ^2. This makes it possible to find exact h.l.c.r.s, but it does take a bit of work. Exact profile c.i.s for μ and σ^2 separately are also available, although we shall see that the c.i. for μ corresponds precisely to that based on the usual t-distributed pivot $\sqrt{n}\left(\bar{X} - \mu\right)/S$.

Example 7.55. With $\hat{\mu}_n = \bar{X}$ and $\hat{\sigma}_n^2 = M_2 = n^{-1}\sum_{j=1}^n (X_j - \bar{X})^2$, the relative likelihood for a random sample of size n from $N(\mu, \sigma^2)$ is

$$\Lambda(\mathbf{X}; \boldsymbol{\theta}) = \left(\frac{\hat{\sigma}_n^2}{\sigma^2}\right)^{n/2} \exp\left\{\frac{n}{2}\left[1 - \frac{\hat{\sigma}_n^2}{\sigma^2} - \frac{(\bar{X}-\mu)^2}{\sigma^2}\right]\right\}, \tag{7.33}$$

where $\boldsymbol{\theta} = (\mu, \sigma^2)'$. Putting $W_n := \hat{\sigma}_n^2/\sigma^2$ and $Y_n := n(\bar{X}-\mu)^2/\sigma^2$, we have for $\lambda \in (0,1)$

$$\Pr[\Lambda(\mathbf{X}; \boldsymbol{\theta}) \geq \lambda] = \Pr\left(W_n^{n/2} e^{-n(W_n-1)/2 - Y_n/2} \geq \lambda\right)$$

$$= \Pr\left(W_n^{n/2} e^{-n(W_n-1)/2} \geq \lambda e^{Y_n/2}\right)$$

$$= E\Pr\left(W_n^{n/2} e^{-n(W_n-1)/2} \geq \lambda e^{Y_n/2} \mid Y_n\right),$$

where the last follows from the law of total probability. The function of W_n on the left of the inequality attains a maximum value of unity at $W_n = 1$. For given $\lambda \in (0,1)$ and any $y \in [0, -2\log\lambda)$ the equation $w^{n/2} e^{-n(w-1)/2} = \lambda e^{y/2}$ will have two roots, $0 \leq w_\lambda(y) < 1 < w^\lambda(y)$, say. Now W_n and Y_n are independently distributed as $nW_n \sim \chi^2(n-1)$ and $Y_n \sim \chi^2(1)$. If we put $U_n := nW_n$, $u_\lambda(Y_n) := nw_\lambda(Y_n)$, and $u^\lambda(Y_n) := w^\lambda(Y_n)$, then

$$\Pr\left(W_n^{n/2} e^{-n(W_n-1)/2} \geq \lambda e^{Y_n/2} \mid Y_n\right) = \Pr\left[u_\lambda(Y_n) \leq U_n \leq u^\lambda(Y_n) \mid Y_n\right]$$

$$= F_U\left[u^\lambda(Y_n)\right] - F_U\left[u_\lambda(Y_n)\right],$$

where F_U is the $\chi^2(n-1)$ c.d.f. Then

$$\Pr(\Lambda \geq \lambda) = \int_0^{-2\log\lambda} \left\{F_U\left[u^\lambda(y)\right] - F_U\left[u_\lambda(y)\right]\right\} f_Y(y) \cdot dy,$$

where $f_Y(y) = (2\pi y)^{-1/2} e^{-y/2} \mathbf{1}_{[0,\infty)}(y)$ is the $\chi^2(1)$ p.d.f. We must now find the root λ_α of $1 - \alpha - \Pr(\Lambda \geq \lambda) = 0$ by numerical search, using numerical integration to evaluate $\Pr(\Lambda \geq \lambda)$ for each trial value of λ. Figure 7.6 compares the exact .95 h.l.c.r. from the data of Example 7.48 with the m.l.e.-based versions depicted in Fig. 7.3.

Assuming that the critical quantile λ_α is determined accurately, the h.l.c.r. does have the exact coverage rate $1 - \alpha$, while the two c.r.s based on m.l.e.s are known only to be asymptotically correct. Table 7.2 shows the average coverage rates of all three c.r.s in a simulation of 10^6 samples of various sizes from $N(0,1)$. (All three pivots are location/scale invariant, so the results would be the same for all μ, σ^2.) Notice that evaluating

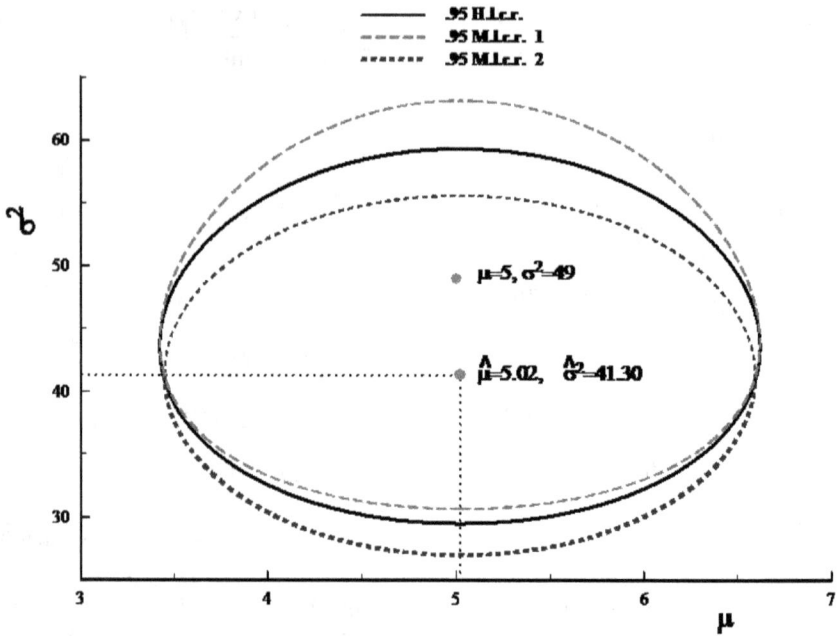

Fig. 7.6   Joint .95 c.r.s for parameters of $N(\mu, \sigma^2)$ from likelihood-ratio and m.l.e.-based pivots.

Table 7.2   Average coverage rates for h.l.c.r. and two c.r.s based on the m.l.e.s, via simulations.

n	$1 - \alpha = .90$			$1 - \alpha = .95$			$1 - \alpha = .99$		
	H.l.	M.l. 1	M.l. 2	H.l.	M.l. 1	M.l. 2	H.l.	M.l. 1	M.l. 2
20	.900	.910	.817	.950	.955	.865	.990	.989	.924
50	.900	.904	.865	.949	.952	.912	.990	.990	.962
100	.901	.903	.882	.952	.952	.931	.990	.990	.976
200	.900	.901	.891	.950	.950	.940	.990	.990	.983

the asymptotic covariance matrix of $\bar{X}, M_2$ at σ^2, as in m.l.c.r. 1, greatly improves accuracy for this range of n.

The one drawback of the highest-likelihood procedure in this application is the effort involved in finding the critical λ_α's. Fortunately, it is possible to calculate these once and for all for various α and n. Table 7.3 presents some selected values.

Table 7.3 Critical values of relative likelihood for exact h.l.c.r.s of normal mean and variance.

n	$\lambda_{.10}$	$\lambda_{.05}$	$\lambda_{.01}$
10	.0789	.0367	.00619
12	.0824	.0388	.00677
14	.0849	.0404	.00719
16	.0868	.0416	.00752
18	.0883	.0425	.00778
20	.0894	.0432	.00799
25	.0915	.0446	.00838
30	.0930	.0455	.00864
35	.0940	.0461	.00883
40	.0947	.0466	.00897
45	.0953	.0470	.00908
50	.0958	.0473	.00917
60	.0965	.0477	.00931
70	.0970	.0480	.00940
80	.0974	.0483	.00948
90	.0977	.0485	.00954
100	.0979	.0486	.00958
125	.0983	.0489	.00966
150	.0986	.0491	.00972
200	.0989	.0493	.00979
∞	.1000	.0500	.01000

Example 7.56. A c.i. for μ alone comes from $\Lambda(\mathbf{X}; \mu) = L(\mathbf{X}; \mu, \check{\sigma}_n^2)/L(\mathbf{X}; \hat{\mu}_n, \hat{\sigma}_n^2)$. With $\check{\sigma}_n^2 = n^{-1}\sum_{j=1}^{n}(X_j - \mu)^2$ this is

$$\Lambda(\mathbf{X}; \mu) = \left(\frac{\check{\sigma}_n^2}{\hat{\sigma}_n^2}\right)^{-n/2}$$

$$= \left[\frac{\sum_{j=1}^{n}(X_j - \bar{X})^2 + n(\bar{X} - \mu)^2}{\sum_{j=1}^{n}(X_j - \bar{X})^2}\right]^{-n/2}$$

$$= \left[1 + \frac{1}{n-1}\left(\frac{\bar{X} - \mu}{S/\sqrt{n}}\right)^2\right]^{-n/2}.$$

This depends on μ only through pivotal quantity $\sqrt{n}(\bar{X} - \mu)/S \sim t(n-1)$. Since large values of Λ correspond to small values of $\left|\sqrt{n}(\bar{X} - \mu)/S\right|$, the $1 - \alpha$ h.l.c.i. for μ is the usual one based on the pivot: $[\bar{x} - t^{\alpha/2}s/\sqrt{n}, \bar{x} + t^{\alpha/2}s/\sqrt{n}]$.

Example 7.57. A c.i. for σ^2 alone comes from $\Lambda(\mathbf{X}; \sigma^2) = L(\mathbf{X}; \check{\mu}_n, \sigma^2)/L(\mathbf{X}; \hat{\mu}_n, \hat{\sigma}_n^2)$, where $\check{\mu}_n = \bar{X}$ does not depend on σ^2. Putting $W_n := \hat{\sigma}_n^2/\sigma^2$, we have

$$\Pr\left[\Lambda\left(\mathbf{X};\sigma^2\right) \geq \lambda\right] = \Pr\left(W_n^{n/2} e^{-n(W_n-1)/2} \geq \lambda\right).$$

For $\lambda \in (0,1)$ the equation $w^{n/2} e^{-n(w-1)/2} - \lambda = 0$ has two roots, $0 < w_\lambda < 1 < w^\lambda$, and so

$$\Pr\left(\Lambda \geq \lambda\right) = \Pr\left(w_\lambda \leq W_n \leq w^\lambda\right).$$

Since $nW_n \sim \chi^2(n-1)$, the probability can be calculated exactly for given w_λ, w^λ, but a numerical procedure is needed to find the root λ_α of $1 - \alpha - \Pr(\Lambda \geq \lambda) = 0$. Given the corresponding w_λ, w^λ, the $1 - \alpha$ c.i. for σ^2 is found from the following equivalences:

$$w_\lambda \leq \frac{\hat{\sigma}_n^2}{\sigma^2} \leq w^\lambda$$

$$\Longleftrightarrow \frac{n}{n-1} w_\lambda \leq \frac{s^2}{\sigma^2} \leq \frac{n}{n-1} w^\lambda$$

$$\Longleftrightarrow \frac{n-1}{n} \frac{s^2}{w^\lambda} \leq \sigma^2 \leq \frac{n-1}{n} \frac{s^2}{w_\lambda},$$

where s^2 is the realization of the unbiased sample variance. For example, with $n = 100$, $99/(100w^\lambda) = .790$, and $99/(100w_\lambda) = 1.264$. With $s^2 = 41.781$ the .90 h.l.c.i. is

$$[.790\,(41.781), 1.264\,(41.781)] \doteq [33.01, 52.81].$$

This differs from the interval $[32.63, 52.39]$ in Example 7.47, which is the shortest interval based on pivot $(n-1)S^2/\sigma^2$.

The slightly greater length of the likelihood-based c.i.s for σ^2—0.2% greater in this example—is clearly a disadvantage. Still, some may find it advantageous to follow a systematic, general approach, even if other methods work a little better in certain cases. Those who do want to construct h.l.c.i.s for the variance of a normal population will need the roots w_λ and w^λ for various n and α. To make things even easier, we provide in Table 7.4 the values of $h_\alpha := (n-1)/(nw^\lambda)$ and $h^\alpha := (n-1)/(nw_\lambda)$ for selected α, n. These can be compared with the slightly shorter intervals $[v_\alpha, v^\alpha]$ in Table 7.1. Given a realization s^2 of the unbiased sample variance, the $1-\alpha$ h.l.c.i. for σ^2 is simply $[h_\alpha s^2, h^\alpha s^2]$.

Exercise 7.43. *Using the data and results from Exercise 7.28, find .95 c.i.s for Weibull shape parameter α (i) from the m.l.e. and (ii) via the highest-likelihood approach.*

Table 7.4 Lower/upper values by which to multiply sample
variance to find h.l.c.i. for variance of normal population.

n	$1 - \alpha = .90$		$1 - \alpha = .95$		$1 - \alpha = .90$	
	h_α	h^α	h_α	h^α	h^α	h^α
5	0.282	3.921	0.241	5.782	0.183	13.618
6	0.326	3.257	0.282	4.508	0.218	9.157
7	0.362	2.867	0.316	3.803	0.247	7.004
8	0.392	2.609	0.345	3.357	0.273	5.764
9	0.419	2.426	0.370	3.049	0.297	4.967
10	0.442	2.288	0.393	2.824	0.317	4.413
11	0.462	2.179	0.413	2.651	0.336	4.007
12	0.480	2.092	0.431	2.513	0.353	3.696
13	0.496	2.020	0.447	2.402	0.369	3.451
14	0.510	1.960	0.461	2.309	0.383	3.253
15	0.524	1.908	0.475	2.230	0.396	3.089
16	0.536	1.863	0.487	2.163	0.409	2.950
17	0.547	1.823	0.499	2.104	0.420	2.833
18	0.558	1.789	0.510	2.053	0.431	2.731
19	0.567	1.757	0.520	2.007	0.441	2.642
20	0.576	1.729	0.529	1.966	0.450	2.563
22	0.593	1.681	0.546	1.896	0.468	2.431
24	0.607	1.641	0.561	1.839	0.483	2.324
26	0.620	1.606	0.575	1.790	0.498	2.235
28	0.632	1.576	0.587	1.748	0.511	2.160
30	0.643	1.550	0.598	1.712	0.522	2.095
35	0.665	1.497	0.622	1.639	0.548	1.968
40	0.684	1.457	0.642	1.583	0.570	1.874
45	0.700	1.424	0.659	1.539	0.589	1.800
50	0.714	1.398	0.674	1.503	0.605	1.741
60	0.736	1.356	0.698	1.448	0.632	1.652
70	0.753	1.325	0.717	1.407	0.654	1.587
80	0.768	1.300	0.733	1.375	0.672	1.537
90	0.780	1.280	0.746	1.349	0.687	1.497
100	0.790	1.264	0.758	1.328	0.700	1.465
150	0.825	1.210	0.798	1.259	0.747	1.361
200	0.847	1.179	0.822	1.220	0.776	1.304

7.4 Solutions to Exercises

1. We have

$$
\begin{aligned}
E\,(T_n - \theta)^2 &= E\,[(T_n - ET_n) + (ET_n - \theta)]^2 \\
&= E\,(T_n - ET_n)^2 + 2\,(ET_n - \theta)\,E\,(T_n - ET_n) + (ET_n - \theta)^2 \\
&= VT_n + (ET_n - \theta)^2\,.
\end{aligned}
$$

2. Since $E\left(\varphi S^2 - \sigma^2\right) = (\varphi - 1)\sigma^2$ and $V\left(\varphi S^2\right) = \varphi^2 \cdot 2\sigma^4 / (n-1)$, the m.s.e. is

$$E\left(\varphi S^2 - \sigma^2\right)^2 = \sigma^4 \left[(\varphi - 1)^2 + \frac{2\varphi^2}{n-1}\right].$$

For each $\sigma^2 > 0$ this is minimized at $\varphi = (n-1)/(n+1)$. Thus, in random samples from normal populations the minimum-m.s.e., $\mathcal{T}_n'$-class estimator of σ^2 is

$$\frac{n-1}{n+1} \cdot S^2 = \frac{1}{n+1} \sum_{j=1}^{n} \left(X_j - \bar{X}\right)^2.$$

3. Imposing the constraint, we must solve

$$\min_{\{a_1, a_2, \ldots, a_{n-1}\}} \left\{ \sum_{j=1}^{n-1} a_j^2 + \left(1 - \sum_{j=1}^{n-1} a_j\right)^2 \right\}.$$

Differentiating with respect to each a_j and equating to zero give $2a_j = 2\left(1 - \sum_{i=1}^{n-1} a_i\right) = 2a_n$ for $j \in \{1, 2, \ldots, n-1\}$, so $a_j = n^{-1}$ for each j. At this stationary point $\sum_{j=1}^{n} a_j^2 = \sum_{j=1}^{n} n^{-2} = n^{-1}$. The sum of squares being quadratic, there is only one extremum, which is either a minimum or a maximum. The solution is a minimum since $\{a_1, a_2, \ldots, a_n\} = \{1, 0, \ldots, 0\}$ satisfies the constraint and gives $\sum_{j=1}^{n} a_j^2 = 1 > n^{-1}$ when $n > 1$.

4. $\bar{X}_{N_n+1}$ is unbiased since $E\bar{X}_{N_n+1} = E\left[E\left(\bar{X}_{N_n+1} \mid N_n\right)\right] = E\mu = \mu$. The variance is

$$V\bar{X}_{N_n+1} = E\left[V\left(\bar{X}_{N_n+1} \mid N_n\right)\right] + V\left[E\left(\bar{X}_{N_n+1} \mid N_n\right)\right]$$

$$= E\left(\frac{\sigma^2}{N_n+1}\right) + 0 = \sigma^2 \sum_{j=0}^{\infty} \frac{1}{j+1} \frac{n^j e^{-n}}{j!}$$

$$= \frac{\sigma^2}{n} \sum_{j=0}^{\infty} \frac{n^{j+1} e^{-n}}{(j+1)!} = \frac{\sigma^2}{n} \sum_{k=1}^{\infty} \frac{n^k e^{-n}}{k!}$$

$$= \frac{\sigma^2}{n} \left(1 - e^{-n}\right).$$

The efficiency relative to $\bar{X}_{n+1}$ is

$$\frac{V\bar{X}_{n+1}}{V\bar{X}_{N_n+1}} = \frac{n}{(n+1)\left(1 - e^{-n}\right)}.$$

This is less than unity if and only if $0 < 1 - (n+1)e^{-n}$, which follows from $e^n > 1 + n$ for $n > 0$.

5. The Fisher information for σ^2 in a sample of size n is $n\mathcal{I}\left(\sigma^2\right) = n/\left(2\sigma^4\right)$, so the m.v. bound for an unbiased estimator of σ is

$$\left(\frac{d\sqrt{\sigma^2}}{d\sigma^2}\right)^2 n^{-1}\mathcal{I}\left(\sigma^2\right)^{-1} = \left(\frac{1}{2\sigma}\right)^2 \left(\frac{2\sigma^4}{n}\right) = \frac{\sigma^2}{2n}.$$

6. $S^r = \sigma^r C_{n-1}^{r/2}/\left(n-1\right)^{r/2}$, and, since $C_{n-1} \sim \chi^2\left(n-1\right) = \Gamma\left(\frac{n-1}{2}, 2\right)$,

$$
\begin{aligned}
ES^r &= \frac{\sigma^r}{\left(n-1\right)^{r/2}} \int_0^\infty \frac{1}{\Gamma\left(\frac{n-1}{2}\right) 2^{(n-1)/2}} c^{(r+n-1)/2-1} e^{-c/2} \cdot dc \\
&= \frac{\sigma^r}{\left(n-1\right)^{r/2}} \frac{\Gamma\left(\frac{r+n-1}{2}\right) 2^{(r+n-1)/2}}{\Gamma\left(\frac{n-1}{2}\right) 2^{(n-1)/2}} \\
&\quad \cdot \int_0^\infty \frac{1}{\Gamma\left(\frac{r+n-1}{2}\right) 2^{(r+n-1)/2}} c^{(r+n-1)/2-1} e^{-c/2} \cdot dc \\
&= \frac{\sigma^r}{\left(n-1\right)^{r/2}} \frac{\Gamma\left(\frac{r+n-1}{2}\right) 2^{r/2}}{\Gamma\left(\frac{n-1}{2}\right)}, r > -\left(n-1\right).
\end{aligned}
$$

Thus,

$$q\left(n,r\right) := \left(\frac{n-1}{2}\right)^{r/2} \frac{\Gamma\left(\frac{n-1}{2}\right)}{\Gamma\left(\frac{r+n-1}{2}\right)}, r > -\left(n-1\right).$$

In particular, $q\left(n,1\right) S$ is unbiased for σ, and

$$
\begin{aligned}
Vq\left(n,1\right) S &= q\left(n,1\right)^2 VS = q\left(n,1\right)^2 \left[ES^2 - \left(ES\right)^2\right] = \sigma^2 \left[q\left(n,1\right)^2 - 1\right] \\
&= \sigma^2 \left[\frac{n-1}{2} \frac{\Gamma\left(\frac{n-1}{2}\right)^2}{\Gamma\left(\frac{n}{2}\right)^2} - 1\right].
\end{aligned}
$$

This is larger than the minimum variance bound, but the relative difference declines rapidly with n, as shown in Fig. 7.7. Moreover, as we shall see in the discussion of complete sufficient statistics, $q\left(n,1\right) S$ is actually the m.v.u. estimator, the m.v. bound here being unattainable.

7. It is helpful first to note that the marginals of the bivariate Bernoulli distribution are themselves (univariate) Bernoullis; i.e.,

$$f_X\left(x; \theta_1\right) = \sum_{y=0}^{1-x} f_{\mathbf{Z}}\left(x, y; \boldsymbol{\theta}\right) = \theta_1^x \left(1 - \theta_1\right)^{1-x} \mathbf{1}_{\{0,1\}}\left(x\right)$$

$$f_Y\left(y; \theta_2\right) = \sum_{x=0}^{1-y} f_{\mathbf{Z}}\left(x, y; \boldsymbol{\theta}\right) = \theta_2^y \left(1 - \theta_2\right)^{1-y} \mathbf{1}_{\{0,1\}}\left(y\right).$$

Thus, $EX = \theta_1$ and $EY = \theta_2$.

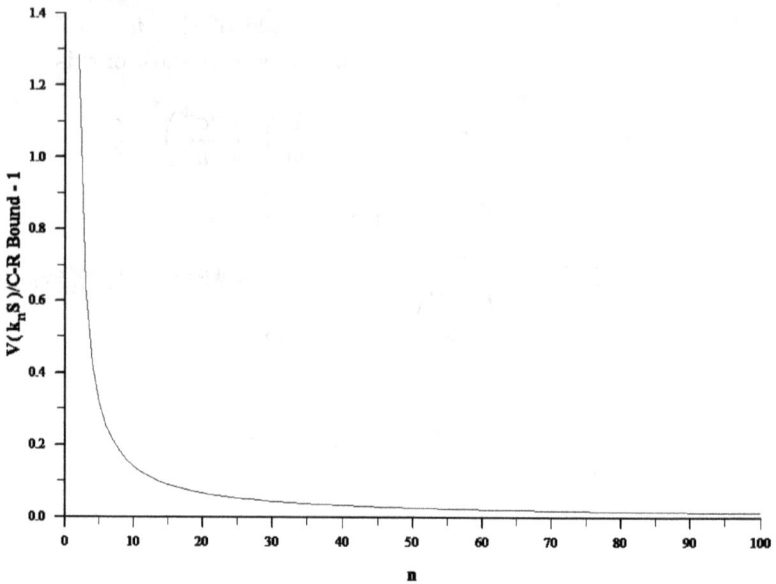

Fig. 7.7 Relative difference between $V\left(q\left(n,1\right)S\right)$ and Cramér–Rao bound for variance of an unbiased estimator of σ.

(i) Proceeding to find the Cramér–Rao bound, we have

$$\log f_{\mathbf{Z}}\left(x,y;\boldsymbol{\theta}\right) = x\log\theta_1 + y\log\theta_2$$
$$+ \left(1-x-y\right)\log\left(1-\theta_1-\theta_2\right)$$

$$\frac{\partial \log f_{\mathbf{Z}}\left(x,y;\boldsymbol{\theta}\right)}{\partial\boldsymbol{\theta}'} = \left(\frac{x}{\theta_1} - \frac{1-x-y}{1-\theta_1-\theta_2}, \frac{y}{\theta_2} - \frac{1-x-y}{1-\theta_1-\theta_2}\right)$$

$$\frac{\partial^2 \log f_{\mathbf{Z}}\left(x,y;\boldsymbol{\theta}\right)}{\partial\boldsymbol{\theta}\partial\boldsymbol{\theta}'} = \begin{pmatrix} -\frac{x}{\theta_1^2} - \frac{1-x-y}{(1-\theta_1-\theta_2)^2} & -\frac{1-x-y}{(1-\theta_1-\theta_2)^2} \\ -\frac{1-x-y}{(1-\theta_1-\theta_2)^2} & -\frac{y}{\theta_2^2} - \frac{1-x-y}{(1-\theta_1-\theta_2)^2} \end{pmatrix}$$

$$-E\left[\frac{\partial^2 \log f_{\mathbf{Z}}\left(X,Y;\boldsymbol{\theta}\right)}{\partial\boldsymbol{\theta}\partial\boldsymbol{\theta}'}\right] = \frac{1}{1-\theta_1-\theta_2}\begin{pmatrix} \frac{1-\theta_2}{\theta_1} & 1 \\ 1 & \frac{1-\theta_1}{\theta_2} \end{pmatrix}.$$

Finally, inverting the matrix and simplifying lead to

$$n^{-1}\mathcal{I}\left(\boldsymbol{\theta}\right)^{-1} = \frac{1}{n}\begin{pmatrix} \theta_1\left(1-\theta_1\right) & -\theta_1\theta_2 \\ -\theta_1\theta_2 & \theta_2\left(1-\theta_2\right) \end{pmatrix}.$$

(ii) Letting $\mathbf{T}_n$ be an m.v.u. estimator (m.v.u.e.) for $\boldsymbol{\theta}$, the m.v.u.e. for $\theta_1 + \theta_2 = (1,1)\,\boldsymbol{\theta}$ is $(1,1)\,\mathbf{T}_n$, and its variance is

$$n^{-1}\,(1,1)\,\mathcal{I}\,(\boldsymbol{\theta})^{-1}\,(1,1)' = n^{-1}\,(\theta_1 + \theta_2)\,(1 - \theta_1 - \theta_2)\,.$$

(iii) Let $\bar{\mathbf{Z}}_n := n^{-1}\sum_{j=1}^{n}(X_j, Y_j)' = (\bar{X}_n, \bar{Y}_n)'$. Since $\{\mathbf{Z}_j = (X_j, Y_j)\}_{j=1}^{n}$ are i.i.d., we have $VX_j = \theta_1(1 - \theta_1)$, $VY_j = \theta_2(1 - \theta_2)$, and

$$Cov\,(X_j, Y_j) = EX_jY_j - \theta_1\theta_2 = -\theta_1\theta_2$$

for each $j \in \{1, 2, ..., n\}$ and

$$V\bar{\mathbf{Z}}_n = n^{-2}\sum_{j=1}^{n} V\,(X_j, Y_j)' = n^{-2}\begin{pmatrix} n\theta_1(1 - \theta_1) & -n\theta_1\theta_2 \\ -n\theta_1\theta_2 & n\theta_2(1 - \theta_2) \end{pmatrix} = (n\mathcal{I})^{-1},$$

from which we see that $\bar{\mathbf{Z}}_n$ is the m.v.u.e. of $\boldsymbol{\theta}$ and that $\bar{X}_n + \bar{Y}_n$ is the m.v.u.e. of $\theta_1 + \theta_2$.

8. Apart from terms free of parameters, the logarithm of $f_{\mathbf{X}}\,(\mathbf{x}; \mu, \lambda)\,/f_{\mathbf{X}}\,(\mathbf{y}; \mu, \lambda)$ is

$$\frac{\lambda}{2\mu^2}\left[\sum_{j=1}^{n}\frac{(y_j - \mu)^2}{y_j} - \sum_{j=1}^{n}\frac{(x_j - \mu)^2}{x_j}\right]$$

$$= \frac{\lambda}{2}\left[\frac{1}{\mu^2}\sum_{j=1}^{n}(y_j - x_j) + \sum_{j=1}^{n}(y_j^{-1} - x_j^{-1})\right].$$

This is independent of μ and λ if and only if $\sum_{j=1}^{n} y_j = \sum_{j=1}^{n} x_j$ and $\sum_{j=1}^{n} y_j^{-1} = \sum_{j=1}^{n} x_j^{-1}$, so $\sum_{j=1}^{n} X_j$ and $\sum_{j=1}^{n} X_j^{-1}$ are jointly minimal-sufficient.

9. When $\lambda = \lambda_0$ joint density $f\,(\mathbf{x}; \mu, \lambda)$ factors as $g\,(\bar{x}; \mu)\,h\,(\mathbf{x})$, with

$$g\,(\bar{x}; \mu) = \exp\left[-\frac{n\lambda_0}{2\mu}\left(\frac{\bar{x}}{\mu} - 2\right)\right]$$

$$h\,(\mathbf{x}) = \left(\frac{\lambda_0}{2\pi}\right)^{n/2}\prod_{j=1}^{n} x_j^{-3/2}\exp\left(-\frac{\lambda_0}{2}\sum_{j=1}^{n} x_j^{-1}\right).$$

Thus, $\bar{X}$ (or, equivalently, $\sum_{j=1}^{n} X_j$) is sufficient for μ when λ is known, but there is no factorization $g\left(\sum_{j=1}^{n} x_j^{-1}; \lambda\right)h\,(\mathbf{x})$ when $\mu = \mu_0$, so $\sum_{j=1}^{n} X_j^{-1}$ alone is not sufficient for λ.

10. The logarithm of $f_{\mathbf{X}}\left(\mathbf{x}; \alpha, \beta\right) / f_{\mathbf{X}}\left(\mathbf{y}; \alpha, \beta\right)$ is

$$(\alpha - 1) \left[\sum_{j=1}^{n} \log x_j - \sum_{j=1}^{n} \log y_j \right] - \beta^{-1} \left[\sum_{j=1}^{n} x_j^\alpha - \sum_{j=1}^{n} y_j^\alpha \right].$$

The form of the second term is such that the entire expression can be parameter-free only if $x_j = y_j$ for all j. However, if $\alpha = \alpha_0$ is known the factorization

$$f_{\mathbf{X}}\left(\mathbf{x}; \alpha_0, \beta\right) = \beta^{-n} \exp\left(-\beta^{-1} \sum_{j=1}^{n} x_j^{\alpha_0} \right) h(\mathbf{x})$$

with $h(\mathbf{x}) = \alpha_0^n \exp\left[(\alpha_0 - 1) \sum_{j=1}^{n} \log x_j \right]$ shows that $\sum_{j=1}^{n} X_j^{\alpha_0}$ is sufficient for β.

11. As shown in Example 7.12, the gamma family is in the exponential class, and natural statistics $T_n\left(\mathbf{X}\right) = \left(\sum_{j=1}^{n} \log X_j, \sum_{j=1}^{n} X_j \right)'$ from random sample $\mathbf{X}$ are minimal sufficient for $\boldsymbol{\theta} = (\alpha, \beta)'$, with natural parameters $\boldsymbol{\eta} = \left(\alpha - 1, -\beta^{-1}\right)'$. Natural parameter space $\mathbf{H} = (-1, \infty) \times (-\infty, 0)$ contains a 2-dimensional rectangle, so $T_n\left(\mathbf{X}\right)$ is complete. The sample mean $\bar{X}$ is unbiased for $\mu = \alpha\beta$ and, as the function $n^{-1} (0, 1) \cdot T_n\left(\mathbf{X}\right)$ of complete sufficient statistics, is the m.v.u. (efficient) estimator.

12. $E\left(\bar{X}^2 + S^2\right) = \mu^2 + \sigma^2/n + \sigma^2$, so $\bar{X}^2 + S^2 (n - 1)/n = \bar{X}^2 + M_2$ is an unbiased function of complete sufficient statistics and is therefore m.v.u.

13. We can set $\theta = 0$ without loss of generality since variances and covariances involve expectations of deviations from the mean. Then

$$\begin{aligned}
Cov\left(T_n, U_n\right) &= Cov\left[T_n\left(T_n + U_n - T_n\right) \right] \\
&= E T_n^2 + E\left[T_n\left(U_n - T_n\right) \right] \\
&= V T_n + 0
\end{aligned}$$

and

$$\begin{aligned}
V\left(U_n - T_n\right) &= V U_n - 2 Cov(T_n, U_n) + V T_n \\
&= V U_n - V T_n.
\end{aligned}$$

14. For the Poisson sample $E S^2 = E \bar{X} = \lambda$ and $\bar{X}$ is efficient, so $Cov\left(S^2, \bar{X}\right) = V \bar{X} = \lambda/n$ by the previous exercise. (To work it out directly takes half a page of scribbling!)

15. (i) $EP_n(t) = n^{-1} \sum_{j=1}^n Et^{X_j} = p(t; \lambda)$. Since $n\bar{X} \sim P(n\lambda)$, $E\hat{p}_n(t) = p\left(1 + \frac{t-1}{n}; n\lambda\right) = p(t; \lambda)$.

(ii) $\bar{X}$ is a complete sufficient statistic for λ, and so $\hat{p}_n(t)$ is m.v.u.

(iii)

$$VP_n(t) = n^{-1}Vt^X = n^{-1}\left[p(t^2; \lambda) - p(t; \lambda)^2\right]$$

$$= p(t; \lambda)^2 n^{-1}\left[e^{\lambda(t-1)^2} - 1\right]$$

$$V\hat{p}_n(t) = p\left[\left(1 + \frac{t-1}{n}\right)^2; n\lambda\right] - p(t; \lambda)^2$$

$$= p(t; \lambda)^2\left[e^{\lambda(t-1)^2/n} - 1\right].$$

Put $s := \lambda(t-1)^2$ and $q(s) := e^s - 1$ for $s \geq 0$. Then $VP_n(t) > V\hat{p}_n(t)$ for $t \neq 1$ if and only if $n^{-1}q(s) > q(s/n)$ for all $s > 0$, which follows from the convexity of q. Of course, $VP_n(1) = V\hat{p}_n(1) = 0$.

(iv) Evidently not, since the m.v.u.e. is unique. Moreover, $Ee^{\bar{X}(t-1)} > e^{E\bar{X}(t-1)} = p(t; \lambda)$ by Jensen, so $p(t; \bar{X})$ is biased.

16. $nT_n \sim B(n, \theta)$, so $ET_n = \theta$ and $VT_n = n^{-1}\theta(1-\theta)$. Lindeberg–Lévy (or De Moivre–Laplace) implies $\sqrt{n}(T_n - \theta) \rightsquigarrow N[0, \theta(1-\theta)]$, so the asymptotic and finite-sample variances of T_n are the same.

17. By 7.3 the bound for the asymptotic variance of a c.a.n. estimator of Poisson parameter μ is μ/n. Relation (7.17) shows that the bound for estimators of $\theta = g(\mu) = e^{-\mu}$ is $n^{-1}\mu(de^{-\mu}/du)^2 = n^{-1}\mu e^{-2\mu} = n^{-1}\mu\theta^2$.

18. Exercise 6.24 established that $\sqrt{n}(\mathfrak{G} - \mathfrak{g})$ is asymptotically normal with zero mean and variance

$$4\sigma^2\left[\frac{1}{3} - \frac{2}{\pi}\left(2 - \sqrt{3}\right)\right],$$

so that $\sqrt{\pi}\mathfrak{G}/2$ (which is an unbiased estimator of σ) has asymptotic variance

$$\frac{\sigma^2}{n}\left[\frac{\pi}{3} - 2\left(2 - \sqrt{3}\right)\right]. \tag{7.34}$$

The delta method gives

$$S = \sigma + \frac{1}{2\sigma_n^*}\left(S^2 - \sigma^2\right),$$

where σ_n^* is between S and σ. Hence, the asymptotic variance of S is $(2\sigma)^{-2}$ times that of S^2, or

$$\frac{1}{4\sigma^2} \frac{2\sigma^4}{n} = \frac{\sigma^2}{2n}.$$

As an estimator of σ, the asymptotic efficiency of $\sqrt{\pi}\mathfrak{S}/2$ relative to S is thus

$$\frac{1/2}{\pi/3 - 2\left(2 - \sqrt{3}\right)} \doteq .9779.$$

19. With $\gamma(\mu) = e^{-\mu}$ the delta method gives

$$\sqrt{n}\left(\tilde{\mu}_n - \mu\right) = \sqrt{n}\left(-\log \bar{g}_n + \log \gamma\right) = -\hat{\gamma}_n^{-1}\sqrt{n}\left(\bar{g}_n - \gamma\right),$$

where $\hat{\gamma}_n$ is between $\bar{g}_n$ and γ. Since $\bar{g}_n \to^{a.s.} \gamma$ and $V 1_{\{0\}}(X_j) = \gamma(1 - \gamma)$, the right side converges in distribution to that of $-\gamma^{-1}$ times a normal r.v. with mean zero and variance $\gamma(1 - \gamma) = e^{-\mu}\left(1 - e^{-\mu}\right)$. Thus,

$$\sqrt{n}\left(\tilde{\mu}_n - \mu\right) \rightsquigarrow N\left(0, e^{\mu} - 1\right).$$

The asymptotic variance, $n^{-1}\left(e^{\mu} - 1\right)$, exceeds μ/n for all $\mu > 0$. On the other hand, $\tilde{\mu}_n = \bar{X}$ is another m.o.m. estimator that happens to be consistent and asymptotically efficient (as well as m.v.u. in finite samples).

20. The likelihood equations are

$$\frac{\partial \log L\left(\mathbf{x}; \mu, \lambda\right)}{\partial \mu} = \frac{\lambda}{\mu^3} \sum_{j=1}^{n} \left(x_j - \mu\right) = 0$$

$$\frac{\partial \log L\left(\mathbf{x}; \mu, \lambda\right)}{\partial \lambda} = \frac{n}{2\lambda} - \frac{1}{2\mu^2} \sum_{j=1}^{n} \frac{\left(x_j - \mu\right)^2}{x_j} = 0.$$

The first equation yields $\hat{\mu}_n = \bar{x}$, and the second gives $\hat{\lambda}_n = \left(n^{-1}\sum_{j=1}^{n} x_j^{-1} - \bar{x}^{-1}\right)^{-1}$, provided not all $\{x_j\}$ are the same. In that case $\hat{\lambda}_n$ exists and is positive by part (ii) of Example 3.34. The second-order condition is satisfied since

$$\left(\begin{array}{cc} \frac{\partial^2 \log L}{\partial \mu^2} & \frac{\partial^2 \log L}{\partial \lambda \partial \mu} \\ \frac{\partial^2 \log L}{\partial \mu \partial \lambda} & \frac{\partial^2 \log L}{\partial \lambda^2} \end{array}\right)\Bigg|_{\hat{\mu}, \hat{\lambda}} = \left(\begin{array}{cc} -\frac{n\hat{\lambda}}{\hat{\mu}^3} & 0 \\ 0 & -\frac{n}{2\hat{\lambda}^2} \end{array}\right)$$

is negative definite.

21. First, that $E\hat{\mu}_n = E\bar{X} = \mu$ follows from the fact that $EX = \mu$ when $X \sim IG(\mu, \lambda)$. Second, Exercise 5.13 shows that $n\lambda(\bar{Y} + \bar{X}^{-1}) = n\lambda/\hat{\lambda}_n \sim \chi^2(n-1)$, where $\bar{Y} := n^{-1}\sum_{j=1}^n X_j^{-1}$; and $E\hat{\lambda}_n = \lambda n/(n-3)$ then follows from (4.32). Finally, the joint minimal sufficiency of $\bar{X}$ and $\bar{Y}$ (Exercise 7.8) imply that $\hat{\mu}_n$ and $\hat{\lambda}_n(n-3)/n$ are the m.v.u.e.s.

22. If $x_j = x_1$ for each j, the normal likelihood evaluated at $\mu = x_1$ is proportional to σ^{-n}. Since $\sup_{\mu \in \mathfrak{R}, \sigma^2 > 0} L(\mathbf{x}; \mu, \sigma^2)$ occurs on the boundary at $\sigma = 0$, we take $\hat{\sigma}_n^2 = 0$ as the m.l. estimate. This is consistent with the result $\hat{\sigma}_n^2 = m_2$ that holds when $m_2 > 0$. Unlike the Poisson case, however, the event $M_2 = 0$ is a null set apart from the effects of rounding.

23. In the one-to-one case write $\theta = g^{-1}(\gamma)$ and put $L^*(\mathbf{x}; \gamma) := L(\mathbf{x}; g^{-1}(\gamma))$. Then the value $\hat{\gamma}_n$ that solves

$$
\begin{aligned}
0 &= \frac{\partial L^*(\mathbf{x}; \gamma)}{\partial \gamma} \\
&= \frac{dL(\mathbf{x}; g^{-1}(\gamma))}{dg^{-1}(\gamma)} \frac{dg^{-1}(\gamma)}{d\gamma} \\
&= \frac{dL(\mathbf{x}; \theta)}{d\theta} \frac{dg^{-1}(\gamma)}{d\gamma} \\
&= \frac{dL(\mathbf{x}; \theta)}{d\theta} \div \frac{dg(\theta)}{d\theta}
\end{aligned}
$$

is the value of $g(\theta)$ evaluated at the $\hat{\theta}_n$ that solves $dL(\mathbf{x}; \theta)/d\theta = 0$.

24. It is enough to observe that the inverse Gaussian family is in the exponential class; however, we can verify conditions (ii) and (iii) directly. With

$$
\log f(x; \mu, \lambda) = \frac{1}{2}\left[\log \lambda - \log(2\pi) - \frac{\lambda(x-\mu)^2}{\mu^2 x}\right] \mathbf{1}_{(0,\infty)}(x)
$$

and the result $EX^{-1} = \lambda^{-1} + \mu^{-1}$ from Exercise 4.66 we have

$$
E\left(\frac{\partial \log f}{\partial \mu}\right) = \frac{\lambda}{\mu^3} E(X - \mu) = 0
$$

$$
\begin{aligned}
E\left(\frac{\partial \log f}{\partial \lambda}\right) &= \frac{1}{2\lambda} - \frac{1}{2\mu^2} E(X - 2\mu + \mu^2 X^{-1}) \\
&= \frac{1}{2\lambda} - \frac{1}{2\mu^2} \frac{\mu^2}{\lambda} = 0,
\end{aligned}
$$

Thus, condition (ii) is met. Then with the result $EX^{-2} = \mu^{-2} + 3\mu^{-1}\lambda^{-1} + 3\lambda^{-2}$ from Exercise 4.66 we have

$$E\left(\frac{\partial \log f}{\partial \mu}\right)^2 = \frac{\lambda^2}{\mu^6} E\left(X - \mu\right)^2 = \frac{\lambda^2}{\mu^6} \frac{\mu^3}{\lambda} = \frac{\lambda}{\mu^3}$$

$$-E\frac{\partial^2 \log f}{\partial \mu^2} = \lambda\left(\frac{3EX}{\mu^4} - \frac{2}{\mu^3}\right) = \frac{\lambda}{\mu^3}$$

$$E\left(\frac{\partial \log f}{\partial \mu}\right)\left(\frac{\partial \log f}{\partial \lambda}\right) = -\frac{\lambda}{2\mu^5}\left[E\left(X - \mu\right)^2 - \mu E\left(X - 2\mu + \mu^2 X^{-1}\right)\right]$$

$$= -\frac{\lambda}{2\mu^5}\left[\frac{\mu^3}{\lambda} - \mu\left(\frac{\mu^2}{\lambda}\right)\right] = 0$$

$$E\left(\frac{\partial^2 \log f}{\partial \mu \partial \lambda}\right) = \frac{E\left(X - \mu\right)}{\mu^3} = 0$$

$$E\left(\frac{\partial \log f}{\partial \lambda}\right)^2 = \frac{1}{4\lambda^2} - \frac{1}{2\lambda\mu^2} E\left(X - 2\mu + \mu^2 X^{-1}\right)$$

$$+ \frac{1}{4\mu^4} E\left(X - 2\mu + \mu^2 X^{-1}\right)^2$$

$$= \frac{1}{2\lambda^2}$$

$$-E\frac{\partial^2 \log f}{\partial \lambda^2} = \frac{1}{2\lambda^2}.$$

Thus, condition (iii) is met. The asymptotic covariance matrix is then

$$n^{-1}\mathcal{I}\left(\mu, \lambda\right)^{-1} = \begin{pmatrix} \frac{n\lambda}{\mu^3} & 0 \\ 0 & \frac{n}{2\lambda^2} \end{pmatrix}^{-1} = \begin{pmatrix} \frac{\mu^3}{n\lambda} & 0 \\ 0 & \frac{2\lambda^2}{n} \end{pmatrix},$$

so that

$$\sqrt{n}\begin{pmatrix} \hat{\mu}_n - \mu \\ \hat{\lambda}_n - \lambda \end{pmatrix} \rightsquigarrow N\left[\begin{pmatrix} 0 \\ 0 \end{pmatrix}, \begin{pmatrix} \frac{\mu^3}{\lambda} & 0 \\ 0 & 2\lambda^2 \end{pmatrix}\right].$$

Hence, $\hat{\mu}_n, \hat{\lambda}_n$ are *asymptotically* independent (as, indeed, they are in finite samples also).

25. (i) The log likelihood is

$$-n\log\left(2\pi\right) - \frac{n}{2}\log\left(1 - \rho^2\right)$$

$$-\frac{\sum_{j=1}^{n}\left[(X_j - \mu)^2 - 2\rho\left(X_j - \mu\right)\left(Y_j - \mu\right) + (Y_j - \mu)^2\right]}{2\left(1 - \rho^2\right)},$$

and $\hat{\mu}_n$ satisfies likelihood equation

$$0 = \frac{2 \sum_{j=1}^{n} \left[(X_j - \mu) - \rho(X_j - \mu) - \rho(Y_j - \mu) + (Y_j - \mu) \right]}{2(1 - \rho^2)}$$

$$= \frac{\sum_{j=1}^{n}(X_j + Y_j - 2\mu)}{1 + \rho}.$$

Thus, $\hat{\mu}_n = (2n)^{-1} \sum_{j=1}^{n}(X_j + Y_j)$.

(ii) The Fisher information for the full sample is

$$n\mathcal{I} = -E\left(\frac{d^2 \log L}{d\mu^2}\right) = \frac{2n}{1 + \rho}.$$

Since $E\hat{\mu}_n = \mu$ and $V\hat{\mu}_n = (1 + \rho)/(2n) = (n\mathcal{I})^{-1}$, it follows that the m.l.e. is m.v.u.

(iii) The information content of sample $\{X_j\}_{j=1}^{2n}$ is simply $2n$. Thus, estimation from the bivariate sample is more or less efficient according as ρ is less than zero or greater than zero, respectively.

(iv) Clearly, $n\mathcal{I} \to +\infty$ and $V\hat{\mu}_n \to 0$ as $\rho \to -1$. When X and Y have the same mean and variance and have correlation near -1, realizations $(x - \mu, y - \mu)$ of $(X - \mu, Y - \mu)$ will, with probability approaching unity, be of equal magnitude but opposite sign, in which case the value of μ will be close to $(x + y)/2$.

26. The matrix of second derivatives of the log likelihood is $-\mathbf{X}'\boldsymbol{\Sigma}^{-1}\mathbf{X}$, elements of which are known constants. Therefore, $(n\mathcal{I})^{-1} = (\mathbf{X}'\boldsymbol{\Sigma}^{-1}\mathbf{X})^{-1}$.

27. Putting

$$\mathbf{C} = (\mathbf{X}'\boldsymbol{\Sigma}^{-1}\mathbf{X})^{-1}\mathbf{X}'\boldsymbol{\Sigma}^{-1} + \mathbf{D}$$

for some $k \times n$ matrix of constants $\mathbf{D}$, we have

$$\mathbf{CY} = \mathbf{C}(\mathbf{X}\boldsymbol{\beta} + \mathbf{U})$$
$$= (\mathbf{X}'\boldsymbol{\Sigma}^{-1}\mathbf{X})^{-1}\mathbf{X}'\boldsymbol{\Sigma}^{-1}\mathbf{X}\boldsymbol{\beta} + \mathbf{DX}\boldsymbol{\beta} + \mathbf{CU}$$
$$= \boldsymbol{\beta} + \mathbf{DX}\boldsymbol{\beta} + \mathbf{CU},$$

so unbiasedness requires $\mathbf{DX} = \mathbf{0}$. Then

$$V\mathbf{CY} = \mathbf{C}(E\mathbf{UU}')\mathbf{C}'$$
$$= \left[(\mathbf{X}'\boldsymbol{\Sigma}^{-1}\mathbf{X})^{-1}\mathbf{X}'\boldsymbol{\Sigma}^{-1} + \mathbf{D}\right]\boldsymbol{\Sigma}\left[(\mathbf{X}'\boldsymbol{\Sigma}^{-1}\mathbf{X})^{-1}\mathbf{X}'\boldsymbol{\Sigma}^{-1} + \mathbf{D}\right]'$$
$$= (\mathbf{X}'\boldsymbol{\Sigma}^{-1}\mathbf{X})^{-1} + \mathbf{D}\boldsymbol{\Sigma}\mathbf{D}'.$$

But since $\boldsymbol{\Sigma}$ is nonsingular and therefore (as a covariance matrix) positive definite, we have $V\mathbf{CY} \geq (\mathbf{X}'\boldsymbol{\Sigma}^{-1}\mathbf{X})^{-1}$ with equality if and only if $\mathbf{D} = \mathbf{0}$.

28. The log likelihood is

$$\log L\left(\mathbf{x}; \alpha, \beta\right) = 10\log\left(\frac{\alpha}{\beta}\right) + (\alpha - 1)\sum_{j=1}^{10}\log x_j - \frac{1}{\beta}\sum_{j=1}^{10} x_j^\alpha,$$

and the likelihood equations are

$$\frac{\partial \log L}{\partial \alpha} = \frac{10}{\alpha} + \sum_{j=1}^{10}\log x_j - \frac{1}{\beta}\sum_{j=1}^{10} x_j^\alpha \log x_j = 0 \qquad (7.35)$$

$$\frac{\partial \log L}{\partial \beta} = -\frac{10}{\beta} + \frac{1}{\beta^2}\sum_{j=1}^{10} x_j^\alpha = 0. \qquad (7.36)$$

Replacing β in (7.35) with the implicit solution from (7.36) gives the nonlinear equation

$$\frac{10}{\alpha} + \sum_{j=1}^{10}\log x_j - \frac{10\sum_{j=1}^{10} x_j^\alpha \log x_j}{\sum_{j=1}^{10} x_j^\alpha} = 0.$$

The solution with the given sample is $\hat{\alpha}_{10} = 1.52330$, which with (7.36) yields $\hat{\beta}_{10} = 1.81282$. Plugging data and m.l.e.s into (7.28) gives

$$\tilde{\mathcal{I}}\left(\hat{\alpha}_{10}, \hat{\beta}_{10}\right) = \begin{pmatrix} 1.07093 & -.361477 \\ -.361477 & .304291 \end{pmatrix}$$

and

$$10^{-1}\tilde{\mathcal{I}}\left(\hat{\alpha}_{10}, \hat{\beta}_{10}\right)^{-1} = \begin{pmatrix} .15588 & .18517 \\ .18517 & .54861 \end{pmatrix}.$$

29. (i) Exercise 4.9 showed the negative binomial mean and variance to be $\mu = \varphi(1-\theta)/\theta$ and $\sigma^2 = \varphi(1-\theta)/\theta^2$. The simplest solutions from these are $\theta = \mu/\sigma^2$ and $\varphi = \mu^2/(\sigma^2 - \mu)$, and the corresponding m.o.m. estimators are $\tilde{\theta}_n = \bar{X}/M_2$ and $\tilde{\varphi}_n = \bar{X}^2/(M_2 - \bar{X})$. Sample values $\bar{x} = 4.4100$ and $m_2 \doteq 10.0019$ yield estimates $\tilde{\theta}_{100} \doteq .4409$ and $\tilde{\varphi}_{100} \doteq 3.4779$.

(ii) With $\Psi\left(\cdot\right) := d\log\Gamma\left(\cdot\right)/d\left(\cdot\right)$ we have

$$\frac{\partial \log L\left(\mathbf{x}; \varphi, \theta\right)}{\partial \varphi} = \sum_{j=1}^{n}\Psi\left(\varphi + x_j\right) - n\Psi\left(\varphi\right) + n\log\theta = 0\,(7.37)$$

$$\frac{\partial \log L\left(\mathbf{x}; \varphi, \theta\right)}{\partial \theta} = \frac{n\varphi}{\theta} - \frac{n\bar{x}}{1-\theta} = 0,$$

and the second equation yields implicit solution $\hat{\varphi}_n = \bar{x}\hat{\theta}_n/\left(1 - \hat{\theta}_n\right)$.

(iii) Plugging the implicit solution for $\hat{\varphi}_n$ into (7.37) and starting a simple bisection search at $\tilde{\theta}_{100} \doteq .4409$ give $\hat{\theta}_{100} \doteq .44599$ as the root and $\hat{\varphi}_{100} \doteq 3.5501$ as the corresponding estimate of φ.

(iv) The sample information matrix is

$$n\mathcal{I}\left(\varphi, \theta\right) = \begin{pmatrix} n\Psi'\left(\varphi\right) - E\sum_{j=1}^{n}\Psi'\left(\varphi + X_j\right) & -\frac{n}{\theta} \\ -\frac{n}{\theta} & \frac{n\varphi}{\theta^2(1-\theta)} \end{pmatrix};$$

the estimated value is

$$100\widetilde{\mathcal{I}}\left(\hat{\varphi}_{100}, \hat{\theta}_{100}\right) \doteq \begin{pmatrix} .16785 & -2.2422 \\ -2.2422 & 32.216 \end{pmatrix};$$

and

(v) The estimated asymptotic covariance matrix is

$$100^{-1}\widetilde{\mathcal{I}}\left(\hat{\varphi}_{100}, \hat{\theta}_{100}\right)^{-1} \doteq \begin{pmatrix} .8478 & .0590 \\ .0590 & .0044 \end{pmatrix}.$$

30. (i) The c.d.f. is

$$\begin{aligned} F_{(n:n)}\left(x\right) &= \Pr\left(X_{(n:n)} \leq x\right) \\ &= \Pr\left(X_1 \leq x, X_2 \leq x, ..., X_n \leq x\right) \\ &= \Pr\left(X_1 \leq x\right)\Pr\left(X_2 \leq x\right) \cdots \Pr\left(X_n \leq x\right) \\ &= \left(\frac{x}{\theta}\right)^n \mathbf{1}_{[0,\theta]}\left(x\right) + \mathbf{1}_{(\theta,\infty)}\left(x\right). \end{aligned}$$

Taking limits, we have $F_\infty\left(x\right) := \lim_{n\to\infty} F_{(n:n)}\left(x\right) = 0$ for $x < \theta$ and $F_\infty\left(x\right) = 1$ for $x \geq \theta$. Thus, the limiting c.d.f. is that of a degenerate r.v. that takes the value θ with probability one, and so $X_{(n:n)} \to^p \theta$.

(ii) Pick any $\varepsilon \in (0, 1)$. Then $\Pr\left(X_{(n:n)} \leq \theta - \varepsilon\theta\right) = F_{(n:n)}\left(\theta - \varepsilon\theta\right) = (1 - \varepsilon)^n$ and $\sum_{n=1}^{\infty}(1 - \varepsilon)^n = [1 - (1 - \varepsilon)]^{-1} = 1/\varepsilon < \infty$. Since ε can be arbitrarily close to zero, the convergence part of the Borel–Cantelli lemma implies that $\Pr(X_{(n:n)} < \theta \text{ i.o.}) = 0$, and since $\Pr\left(X_{(n:n)} \leq \theta\right) = 1$ for each n, we conclude that $X_{(n:n)} \to^{a.s.} \theta$.

(iii) Put $T_n = -n\left(X_{(n:n)} - \theta\right)$. (The minus sign makes the right side positive, since $X_{(n:n)}$ can be no greater than θ.) Then

$$\begin{aligned} F_{T_n}\left(t\right) &= \Pr\left[n\left(X_{(n:n)} - \theta\right) \geq -t\right] \\ &= 1 - \Pr\left(X_{(n:n)} < \theta - tn^{-1}\right) \\ &= \mathbf{1}_{(-\infty,\infty)}\left(t\right) - \left[\left(1 - \frac{t}{\theta n}\right)^n \mathbf{1}_{[0,n\theta]}\left(t\right) + \mathbf{1}_{(-\infty,0)}\left(t\right)\right] \\ &= \mathbf{1}_{[0,\infty)}\left(t\right) - \left(1 - \frac{t}{\theta n}\right)^n \mathbf{1}_{[0,n\theta]}\left(t\right). \end{aligned}$$

The second term term on the right goes to $e^{-t/\theta} \mathbf{1}_{[0,\infty)}(t)$, so we have

$$\lim_{n \to \infty} F_{T_n}(t) = \left(1 - e^{-t/\theta}\right) \mathbf{1}_{[0,\infty)}(t).$$

This is the c.d.f. of an exponential distribution, which (modulo the change of sign of $n\left(X_{(n:n)} - \theta\right)$) belongs to one of the three classes of the "extreme-value" distributions that comprise the limiting forms of all appropriately normalized sample maxima.

31. Take $f(x_{.5}) = f(\phi; \phi, \psi) = \frac{1}{2\psi}$ in expression (6.20).

32. Since $(n_j - 1)S_j^2/\sigma_j^2 \sim \chi^2(n_j - 1)$ for $n = 1, 2$, and since these are independent, the ratio $R := S_1^2/\sigma_1^2 \div S_2^2/\sigma_2^2$ has Snedecor's F distribution with $n_1 - 1$ and $n_2 - 1$ d.f. Therefore, for $p \in (0, 1)$

$$\Pr\left[r_{(1-p)\alpha} \le \frac{S_1^2/\sigma_1^2}{S_2^2/\sigma_2^2} \le r^{p\alpha}\right] = 1 - \alpha,$$

where $r_{(1-p)\alpha}$ and $r^{p\alpha}$ are the $(1-p)\alpha$ and upper-$p\alpha$ quantiles. Inverting yields

$$\Pr\left[\frac{S_1^2}{r^{p\alpha}S_2^2} \le \frac{\sigma_1^2}{\sigma_2^2} \le \frac{S_1^2}{r_{(1-p)\alpha}S_2^2}\right] = 1 - \alpha$$

and $1 - \alpha$ c.i.

$$\left[\frac{s_1^2}{r^{p\alpha}s_2^2}, \frac{s_1^2}{r_{(1-p)\alpha}s_2^2}\right].$$

Since the F distribution is right-skewed, shorter c.i.s can be obtained for some values p above 0.5, but the precise optimal value must be found numerically.

33. Taking $Y := \Pi(X; \theta) = X/\theta$ as pivot, the change-of-variable formula gives $f_Y(y) = e^{-y}\mathbf{1}_{[0,\infty)}(y)$, so $F_Y(y) = (1 - e^{-y})\mathbf{1}_{[0,\infty)}(y)$. This has unique inverse $F_Y^{-1}(\alpha) = -\log(1 - \alpha)$ for $\alpha \in [0, 1)$. With $\alpha \in (0, 1)$ and $y_{(1-p)\alpha}$ and $y^{p\alpha}$ as $(1-p)\alpha$ quantile and upper-$p\alpha$ quantile we have $1 - \alpha = \Pr\left[y_{(1-p)\alpha} \le X/\theta \le y^{p\alpha}\right] = \Pr\left[X/y^{p\alpha} \le \theta \le X/y_{(1-p)\alpha}\right]$ for any $p \in (0, 1)$. The length of this interval is proportional to

$$\frac{1}{y_{(1-p)\alpha}} - \frac{1}{y^{p\alpha}} = \frac{1}{F_Y^{-1}(\alpha - p\alpha)} - \frac{1}{F_Y^{-1}(1 - p\alpha)}$$

$$= \frac{1}{\log(p\alpha)} - \frac{1}{\log(1 - \alpha + p\alpha)}.$$

As p increases from zero to unity this increases monotonically from $-1/\log(1 - \alpha)$ to $+\infty$. Therefore, for each α the optimal value of p is zero, and the shortest c.i. given the realization x is

$$\left[0, \frac{x}{y_\alpha}\right] = \left[0, -\frac{x}{\log(1 - \alpha)}\right].$$

34. Exercise 6.21 shows that

$$\bar{D}_{n_X,n_Y} := \frac{(\bar{X} - \bar{Y}) - (\mu_X - \mu_Y)}{\sqrt{S_X^2/n_X + S_Y^2/n_Y}} \rightsquigarrow N(0,1)$$

as $n_X \uparrow$ and n_Y/n_X does not vanish. $\bar{D}_{n_X,n_Y}$ is therefore a pivotal quantity, and one for which it is optimal to divide the residual .05 probability equally between the two tails. Inverting $-z^{.025} \le \bar{D}_{n_X,n_Y} \le z^{.025}$, the shortest approximate .95 c.i. is

$$\left[(\bar{x} - \bar{y}) - 1.96\sqrt{\frac{s_X^2}{n_X} + \frac{s_Y^2}{n_Y}}, (\bar{x} - \bar{y}) + 1.96\sqrt{\frac{s_X^2}{n_X} + \frac{s_Y^2}{n_Y}} \right].$$

35. With $c_{.95} = 5.991$ as the .95 quantile of $\chi^2(2)$, we have that

$$n\left(\hat{\boldsymbol{\theta}}_n - \boldsymbol{\theta}\right)' \mathcal{I}\left(\hat{\boldsymbol{\theta}}_n\right) \left(\hat{\boldsymbol{\theta}}_n - \boldsymbol{\theta}\right) \le 5.991$$

with probability approaching .95 as $n \to \infty$. For the relevant data with $n = 40$ we have $\hat{\boldsymbol{\theta}}_{40} = (\bar{x}, \bar{y})' = (.375, .250)'$ and

$$40\mathcal{I}\left(\hat{\boldsymbol{\theta}}_{40}\right) = \frac{40}{1 - \bar{x} - \bar{y}} \left[\begin{array}{cc} \frac{1-\bar{y}}{\bar{x}} & 1 \\ 1 & \frac{1-\bar{x}}{\bar{y}} \end{array} \right] = \left[\begin{array}{cc} \frac{640}{3} & \frac{320}{3} \\ \frac{320}{3} & \frac{800}{3} \end{array} \right].$$

Thus, the relevant c.r., $\mathbb{R}_{2,.95}$, is

$$\left\{ \boldsymbol{\theta} : \frac{640}{3}(\theta_1 - .375)^2 + \frac{640}{3}(\theta_1 - .375)(\theta_2 - .25) + \frac{800}{3}(\theta_2 - .25)^2 \le 5.991 \right\}$$

Figure 7.8 depicts the region.

36. (i) If $\mathbf{X} = \mathbf{0}$, then $\hat{\mu}_n = 0$, and likelihood ratio $\Lambda(\mathbf{0}; \mu) = e^{-n\mu}$ is monotone decreasing in μ. (Note that 0^0 is defined by continuity as $\lim_{n \to \infty} (2^{-n})^{2^{-n}} = 1$.) In this case $e^{-n\mu} = \lambda_\alpha = e^{-c_{1-\alpha}/2}$ has but one root, and the $1 - \alpha$ h.l.c.i. is $\{\mu > 0 : e^{-n\mu} \ge e^{-c_{1-\alpha}/2}\} = (0, c_{1-\alpha}/(2n)]$. With both the pivotal quantities m.l.1 and m.l.2 we have $l_{\alpha/2} = u_{\alpha/2} = 0$, so neither yields a c.i.

(ii) If $\mathbf{X} = \mathbf{0}$ we have $\Lambda(\mathbf{0}; \theta) = (1 - \theta)^n$, and if $\mathbf{X} = \mathbf{n}$ we have $\Lambda(\mathbf{n}; \theta) = \theta^n$, both being monotonic functions of θ. The h.l.c.i.s are

$$\left\{ \theta \in (0,1) : (1 - \theta)^n \le e^{-c_{1-\alpha}/2} \right\} = (0, 1 - e^{-c_{1-\alpha}/(2n)}]$$

and

$$\left\{ \theta \in (0,1) : \theta^n \le e^{-c_{1-\alpha}/2} \right\} = [e^{-c_{1-\alpha}/(2n)}, 1),$$

respectively. Again, the pivotal quantities based on the m.l.e. yield no c.i.

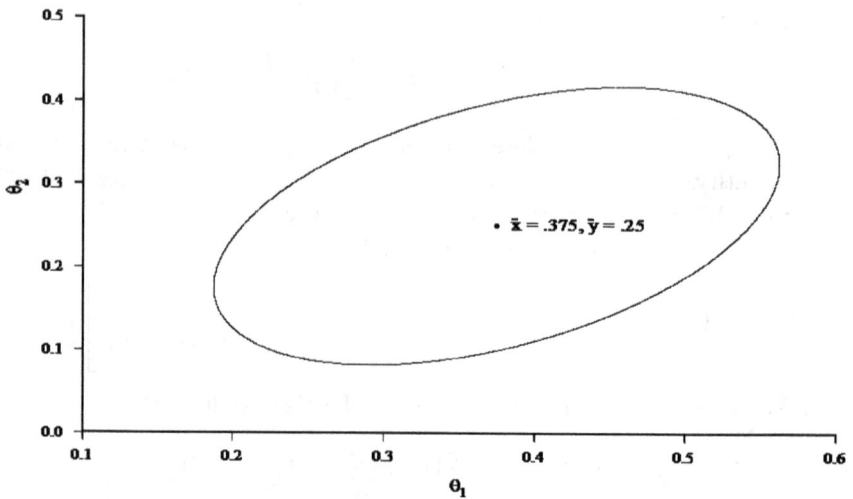

Fig. 7.8   .95 c.r. for parameters of bivariate Bernoulli distribution.

37. (i) The p.d.f. of $\mathfrak{Be}\left(\theta, 1\right)$ is

$$f\left(x\right) = \mathfrak{B}\left(\theta, 1\right)^{-1} x^{\theta}\left(1 - x\right)^{0} \mathbf{1}_{(0,1)}\left(x\right) = \theta x^{\theta - 1} \mathbf{1}_{(0,1)}\left(x\right),$$

and the log likelihood for the sample of $n = 100$ is $\log L\left(\mathbf{x}; \theta\right) = 100 \log \theta + \left(\theta - 1\right) \sum_{j=1}^{100} \log x_j$. The solution of likelihood equation $d \log L\left(\mathbf{x}; \theta\right) / d\theta = 100/\theta + \sum_{j=1}^{100} \log x_j = 0$ is $\hat{\theta}_{100} = -100 / \sum_{j=1}^{100} \log x_j \doteq 6.586$. The sample Fisher information is $-Ed^2 \log L\left(\mathbf{x}; \theta\right) / d\theta^2 = 100/\theta^2$, so the estimate of asymptotic variance is $\hat{\theta}_{100}^2 / 100 \doteq 0.43375$.

(ii) Since $\sqrt{n}\left(\hat{\theta}_n - \theta\right) / \theta \rightsquigarrow N\left(0, 1\right)$, we have

$$.95 \doteq \Pr\left[-1.96 \le 10\left(\frac{\hat{\theta}_{100}}{\theta} - 1\right) \le 1.96\right]$$

$$= \Pr\left(1 - .196 \le \frac{\hat{\theta}_{100}}{\theta} \le 1 + .196\right)$$

$$= \Pr\left(\frac{\hat{\theta}_{100}}{1.196} \le \theta \le \frac{\hat{\theta}_{100}}{.804}\right),$$

so that the .95 m.l.l c.i. is approximately $\left[5.507, 8.192\right]$.

(iii) Since $\sqrt{n}\left(\hat{\theta}_n - \theta\right)/\hat{\theta}_n \rightsquigarrow N(0,1)$, we have .

$$95 \doteq \Pr\left(-1.96\frac{\hat{\theta}_{100}}{10} \le \hat{\theta}_{100} - \theta \le 1.96\frac{\hat{\theta}_{100}}{10}\right)$$

$$= \Pr\left(.804\hat{\theta}_{100} \le \theta \le 1.196\hat{\theta}_{100}\right),$$

so that the .95 m.l.2 c.i. is approximately $[5.295, 7.877]$.

(iv) The relative likelihood is

$$\Lambda(\mathbf{x};\theta) = \left(\frac{\theta}{\hat{\theta}_{100}}\right)^{100}\prod_{j=1}^{100} x_j^{\theta-\hat{\theta}_{100}}.$$

With 3.841 as the .95 quantile of $\chi^2(1)$, the h.l.c.i. comprises values of θ between the roots of $\Lambda(\mathbf{x};\theta) - e^{-3.841/2} = 0$ or, equivalently, of $\log\Lambda(\mathbf{x};\theta) = 100\log(\theta/6.856) - (\theta - 6.856)\,15.1837 + 3.841/2 = 0$. Solving numerically gives $[5.378, 7.962]$ as the c.i.

38. With $f_X(x) = \theta x^{\theta-1}\mathbf{1}_{(0,1)}(x)$ and $Y = -\log X$ the change-of-variable formula gives $f_Y(y) = \theta e^{-\theta y}\mathbf{1}_{(0,\infty)}(y)$, so that Y is distributed as exponential with mean θ^{-1}; that is, as $\Gamma\left(1,\theta^{-1}\right)$.

(i) The m.l.e. is $\hat{\theta}_{100} = 1/\bar{y} = -100/\sum_{j=1}^{100}\log x_j \doteq 6.586$, as before, and the relative likelihood is

$$\Lambda(\mathbf{y};\theta) = \frac{\theta^{100}\exp\left(-\theta\sum_{j=1}^{100} y_j\right)}{\bar{y}^{-100}\exp(-100)} = (\theta\bar{y})^{100}\, e^{-100(\theta\bar{y}-1)}.$$

Put $q := \theta\bar{y}$ and $Q := \theta\bar{Y}$, so that $100Q \sim \Gamma(100,1)$. One must develop a program that finds the roots q_λ, q^λ of $q^{100}e^{-100(q-1)} - \lambda = 0$ for arbitrary $\lambda \in (0,1)$, then search for the specific value $\lambda_{.05}$ such that

$$\Pr\left[Q^{100}e^{-100(Q-1)} \ge \lambda_{.05}\right] = F\left(100q^{\lambda_{.05}}\right) - F\left(100q_{\lambda_{.05}}\right) = .95,$$

where $F(\cdot)$ is the c.d.f. of $\Gamma(100,1)$. The value of $\lambda_{.05}$ turns out to be approximately .14603, and the corresponding roots are $q_{\lambda_{.05}} \doteq .8165$ and $q^{\lambda_{.05}} \doteq 1.2092$. Values of $q = \theta\bar{y}$ *between* these roots correspond to large values of $\Lambda(\mathbf{y};\theta)$, so the exact .95 h.l.c.i. comprises values of θ satisfying $\theta\bar{y} \in [.8165, 1.2092]$ or $\theta \in [5.377, 7.964]$. Notice that with this relatively large sample the "exact" critical value $\lambda_{.05} = .14603$ for the relative likelihood is very close to that based on the asymptotic chi-squared approximation, $e^{-3.841/2} = .14653$. Accordingly, the c.i. is almost identical to that from Exercise 7.37.

(ii) Since $200\theta\bar{Y} \sim \chi^2(200)$, we have

$$1 - \alpha = \Pr\left(c_{(1-p)\alpha} \le 200\theta\bar{Y} \le c^{p\alpha}\right) = \Pr\left(\frac{c_{(1-p)\alpha}}{200\bar{Y}} \le \theta \le \frac{c^{p\alpha}}{200\bar{Y}}\right)$$

for any $\alpha \in (0,1), p \in (0,1)$, where the c's are the $\chi^2 (200)$ quantiles. Taking p as "optimal" value p^* gives $c_{(1-p^*)(.05)} = 200/v^{.05} = 200/1.215 \doteq 164.61$ and $c^{p^*(.05)} = 200/v_{.05} = 200/.818 \doteq 244.50$, where the v's come from Table 7.1. The shortest .95 c.i. based on pivotal quantity $200\theta\bar{Y}$ is then $\left[\frac{164.61}{2(15.1837)}, \frac{244.50}{2(15.1837)} \right] \doteq$ $[5.421, 8.051]$. Alternatively, taking $p = .5$ and upper and lower quantiles $c_{.025} = 162.72, c^{.025} = 241.06$ yields the equal-probability interval $[5.358, 7.938]$.

39. The regions are based on different functions of the data. That is, for any given value of $\boldsymbol{\theta} = (\phi', \psi')' \in \boldsymbol{\Theta}$, the functions $\Lambda (\mathbf{X}; \boldsymbol{\theta})$, $n \left(\hat{\boldsymbol{\theta}}_n - \boldsymbol{\theta} \right)' \mathcal{I}(\boldsymbol{\theta}) \left(\hat{\boldsymbol{\theta}}_n - \boldsymbol{\theta} \right)$, and $n \left(\hat{\boldsymbol{\theta}}_n - \boldsymbol{\theta} \right)' \mathcal{I}(\hat{\boldsymbol{\theta}}_n) \left(\hat{\boldsymbol{\theta}}_n - \boldsymbol{\theta} \right)$ have different distributions, some being more concentrated about $\boldsymbol{\theta}$ than others. Of course, all else equal (computational ease, accuracy of coverage probability), regions that place tighter bounds on $\boldsymbol{\theta}$ are to be preferred.

40. From expression (7.25), the asymptotic variance of $\hat{\phi}_n$ is $n^{-1}\phi/[\phi\Psi'(\phi) - 1]$. Since $\hat{\phi}_n$ is consistent, the asymptotic theory for m.l.e.s implies that

$$n \left(\hat{\phi}_n - \phi \right)^2 \left[\frac{\hat{\phi}_n \Psi' \left(\hat{\phi}_n \right) - 1}{\hat{\phi}_n} \right] \rightsquigarrow \chi^2 (1)$$

and hence that

$$\Pr \left\{ n \left(\hat{\phi}_n - \phi \right)^2 \left[\frac{\hat{\phi}_n \Psi' \left(\hat{\phi}_n \right) - 1}{\hat{\phi}_n} \right] \leq c_{.95} \right\} \to .95$$

as $n \to \infty$, where $c_{.95} = 3.841$ is the .95 quantile. Equivalently,

$$\Pr \left\{ \hat{\phi}_n - z^{.025} \sqrt{\frac{n^{-1}\hat{\phi}_n}{\hat{\phi}_n \Psi' \left(\hat{\phi}_n \right) - 1}} \leq \phi \leq \hat{\phi}_n + z^{.025} \sqrt{\frac{n^{-1}\hat{\phi}_n}{\hat{\phi}_n \Psi' \left(\hat{\phi}_n \right) - 1}} \right\}$$
$$\to .95,$$

where $z^{.025} = \sqrt{c_{.95}} \doteq 1.96$ is the upper-.025 quantile of $N(0,1)$. Plugging in the given values of $\hat{\phi}_n, \Psi' \left(\hat{\phi}_n \right)$, and n gives $[1.234, 2.075]$ as the c.i.

41. The log of the likelihood ratio is

$$\log \Lambda (\mathbf{x}; \mu, \lambda) = \frac{n}{2} \left[\log \lambda - \lambda \left(\frac{\bar{x}}{\mu^2} - \frac{2}{\mu} + \bar{y} \right) \right] - K \left(\mathbf{x}; \hat{\mu}_n, \hat{\lambda}_n \right),$$

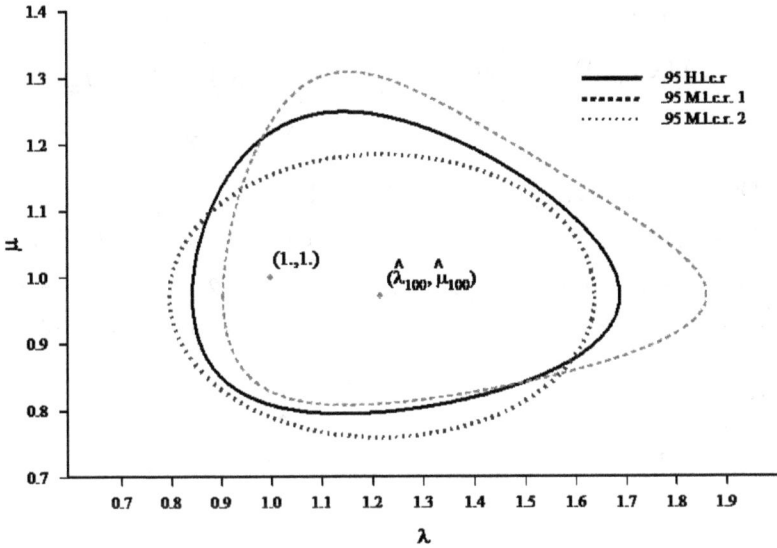

Fig. 7.9   A .95 h.l.c.r. for parameters of $IG(\mu, \lambda)$ with sample of $n = 100$ from $IG(1, 1)$.

and this is less than or equal to $-c_{1-\alpha}/2$ if and only if

$$\frac{\bar{x}}{\mu^2} - \frac{2}{\mu} + \bar{y} \geq \frac{\log \lambda + c_{1-\alpha}/n - 2K\left(\mathbf{x}; \hat{\mu}_n, \hat{\lambda}_n\right)/n}{\lambda} =: q(\lambda, \mathbf{x}, \alpha).$$

For $\{\lambda : 1 - \bar{x}[\bar{y} - q(\lambda, \mathbf{x}, \alpha)] \geq 0\}$ the quadratic formula gives (7.32) as the (real-valued) roots for which this holds as an equality. The given data yield $\hat{\mu}_{100} = \bar{x} = .9718$, $\hat{\lambda}_{100} = (\bar{y} - \bar{x}^{-1})^{-1} = 1.2161$, and $K\left(\mathbf{x}; \hat{\mu}_{100}, \hat{\lambda}_{100}\right) = -40.2173$. The minimum and maximum values of λ satisfying $1 - \bar{x}[\bar{y} - q(\lambda, \mathbf{x}, \alpha)] \geq 0$ are $\lambda^- \doteq .8423$ and $\lambda^+ \doteq 1.6870$. Figure 7.9 shows plots of $\mu(\lambda)^{\pm}$ between these values. The two c.r.s based on the m.l. estimators are shown for comparison. (Note that all three loci satisfy $\mu(\lambda)^{\pm} = \mu(\lambda)^{\pm} = \hat{\mu}_{100} = \bar{x}$ at the applicable bounds $\lambda^{\pm}$.)

42. (i) The log of the profile likelihood satisfies

$$\lim_{n \to \infty} \Pr\left[\log \Lambda(\mathbf{X}; \breve{\mu}_n, \lambda) \leq -\frac{c_{1-\alpha}}{2}\right] = 1 - \alpha,$$

where $c_{1-\alpha}$ is the $1 - \alpha$ quantile of $\chi^2(1)$. The constrained and unconstrained estimates $\breve{\mu}_n, \hat{\mu}_n$ being both equal to $\bar{x}$, we have for sample realization $\mathbf{x}$

$$
\begin{aligned}
\log \Lambda\left(\mathbf{x} ; \breve{\mu}_n, \lambda\right) &= \frac{n}{2}\left[\log \lambda - \lambda\left(\frac{\bar{x}}{\breve{\mu}_n^2} - \frac{2}{\breve{\mu}_n} + \bar{y}\right)\right] - K\left(\mathbf{x} ; \hat{\mu}_n, \hat{\lambda}_n\right) \\
&= \frac{n}{2}\left[\log \lambda - \lambda\left(\bar{y} - \bar{x}^{-1}\right)\right] - K\left(\mathbf{x} ; \hat{\mu}_n, \hat{\lambda}_n\right).
\end{aligned}
$$

Thus, $\log \Lambda\left(\mathbf{x} ; \breve{\mu}_n, \lambda\right) \leq -c_{1-\alpha}/2$ if and only if $\log \lambda - \lambda\left(\bar{y} - \bar{x}^{-1}\right) - 2K\left(\mathbf{x} ; \hat{\mu}_n, \hat{\lambda}_n\right)/n + c_{1-\alpha}/n \leq 0$. With $n = 100$, $\bar{x} = 0.9718$, $\bar{y} = 1.8513$, $\hat{\lambda}_{100} = 1.2161$, $K\left(\mathbf{x} ; \hat{\mu}_{100}, \hat{\lambda}_{100}\right) = -40.2173$, and $c_{95} = 3.841$ a numerical search for the roots shows that the inequality holds for $\lambda \in [.909, 1.585]$.

(ii) The asymptotic variance of $\hat{\lambda}_n$ is $2\lambda^2/n$, so

$$
\begin{aligned}
1 - \alpha &\doteq \operatorname{Pr}\left(-z_{\alpha/2} \leq \sqrt{\frac{n}{2}}\left(\frac{\hat{\lambda}_n}{\lambda} - 1\right) \leq z_{\alpha/2}\right) \\
&= \operatorname{Pr}\left(\frac{\hat{\lambda}_n}{1 + z_{\alpha/2}\sqrt{2/n}} \leq \lambda \leq \frac{\hat{\lambda}_n}{1 - z_{\alpha/2}\sqrt{2/n}}\right).
\end{aligned}
$$

With $\alpha = .05$, $n = 100$, and $\hat{\lambda}_{100} = 1.2161$ the approximate .95 c.i. is $[.952, 1.682]$.

(iii) Evaluating the asymptotic variance at $\hat{\lambda}_n$ gives

$$
\begin{aligned}
1 - \alpha &\doteq \operatorname{Pr}\left[-z_{\alpha/2} \leq \sqrt{\frac{n}{2}}\left(\frac{\lambda}{\hat{\lambda}_n} - 1\right) \leq z_{\alpha/2}\right] \\
&= \operatorname{Pr}\left[\hat{\lambda}_n\left(1 - z_{\alpha/2}\sqrt{2/n}\right) \leq \lambda \leq \hat{\lambda}_n\left(1 + z_{\alpha/2}\sqrt{2/n}\right)\right]
\end{aligned}
$$

and $[.879, 1.553]$ as the approximate c.i.

43. (i) From the solution for $10^{-1}\tilde{\mathcal{I}}\left(\hat{\alpha}_{10}, \hat{\beta}_{10}\right)^{-1}$ we obtain the estimate $\sqrt{.15588} \doteq .39482$ for the asymptotic standard error of $\hat{\alpha}_{10}$. With m.l.e. $\hat{\alpha}_{10} = 1.52330$ this yields the .95 c.i.

$$
[1.52330 - 1.96\,(.39482), 1.5233 + 1.96\,(.39482)] = [.7495, 2.2971].
$$

(ii) Imposing the restriction $\breve{\beta}_{10} = 10^{-1}\sum_{j=1}^{10} x_j^\alpha$, the relative likelihood is

$$
\Lambda\left(\mathbf{x} ; \alpha\right) = \left(\frac{\alpha \sum_{j=1}^n x_j^{\hat{\alpha}_{10}}}{\hat{\alpha}_{10} \sum_{j=1}^n x_j^\alpha}\right)^{10} \prod_{j=1}^{10} x_j^{\alpha - \hat{\alpha}_{10}}.
$$

Finding the two roots of $\Lambda\left(\mathbf{x} ; \alpha\right) - \exp\left(-c_{.95}/2\right)$ (where $c_{.95}$ is the .95 quantile of $\chi^2\left(1\right)$) gives for the h.l.c.i. $[.8620, 2.4170]$.

Chapter 8

Tests of Parametric Hypotheses

8.1 Introduction to Hypothesis Testing

In statistical inference one uses a sample to make educated guesses about the data-generating process (d.g.p.) that produced the data in hand and that would generate additional observations in like experiments. Point estimation, discussed in Chapter 7, is one type of statistical inference. There, one may make an assumption about the family of models for the d.g.p., such as that sample $\mathbf{x} = (x_1, x_2, ..., x_n)'$ is the realization of i.i.d. r.v.s, each with marginal distribution in family $\{F(x; \boldsymbol{\theta}) : \boldsymbol{\theta} \in \boldsymbol{\Theta}\}$. In this case inference consists of getting an estimate of $\boldsymbol{\theta}$, or of one or more scalar components $\boldsymbol{\theta}$, so as to restrict the applicable set of models for F. Alternatively, without imposing the structure of a parametric family, one may wish just to estimate some numerical quantity θ that describes the population, such as a moment or a quantile. Whatever the setting, nothing in the *frequentist* estimation procedures discussed thus far (as contrasted with the Bayesian procedures treated in Chapter 10) depends on having any prior beliefs or guesses about the value of θ. Sometimes, however, one does have a preconceived hypothesis about θ and wants to see whether sample data are at least broadly consistent with it. One may even want to check whether the proposed parametric family adequately characterizes the d.g.p., or, for that matter, whether the data are really i.i.d. Methods for doing these things comprise the branch of statistical inference known as *hypothesis testing*.

This chapter deals with statistical techniques for testing what are loosely called "parametric" hypotheses. These are beliefs or assertions about values of certain constants—either the formal parameters of some family of models or else moments or other quantitative features that describe the population. There are also *nonparametric* hypotheses, which we take up in Chapter 9.

This introductory section explains the distinction between these types and presents fundamental concepts that apply to both.

A statistical hypothesis—whether parametric or nonparametric—might be based on some theoretical consideration ("This pill *ought* to be more effective than a placebo", or "The Poisson family *ought* to be a good model for the number of typographic errors in a page of text") or on hopes or fears ("The mean number of defects in our products *should* be no greater than before the strike", or "Since it makes the analysis easier, I'm *hoping* the normal model represents the data adequately"). The test may be intended to lead either to a definite decision and a course of action ("Given these results, we'll go ahead and market the product") or to a provisional judgment ("The data are consistent with the predictions of my theory"). In either case the hypothesis must be some meaningful statement whose truth or falsity would be apparent if the d.g.p. were actually known; in other words, it must be a statement that could, in principle, be objectively refuted. We adopt the following formal definition:

Definition 8.1. A *statistical hypothesis* is a refutable statement about a d.g.p.

8.1.1   *Types of Statistical Hypotheses*

Here are examples of some of the statistical hypotheses to be considered in this chapter and the next.

(i) Our data are realizations of two independent random samples,

$$\mathbf{X} := (X_1, X_2, ..., X_{n_X})' \text{ and } \mathbf{Y} := (Y_1, Y_2, ..., Y_{n_Y})',$$

and the (marginal) distributions of each X and each Y are just the same.

(ii) Our data are realizations of two independent samples from populations that may have different medians but that are otherwise identical.

(iii) Our data are i.i.d. draws from normal (or gamma, or Poisson, ...) distributions with unspecified parameter(s).

(iv) Our sample data are i.i.d. as normal with mean $\mu = 13$ and unspecified variance.

(v) Our data are i.i.d. as Poisson with parameter $\theta > 13$.

These examples serve to illustrate several conventional ways of classifying statistical hypotheses.

8.1.1.1 *Maintained* vs. *Tested Hypotheses*

Notice that the statement in each example imposes more than one restriction on the d.g.p. In effect, each hypothesis comprises several distinct subhypotheses. Thus, the assertion in Example (i) includes (a) that members of each sample are i.i.d. and (b) that the marginal distributions of X_1 and Y_1 are just the same. Hypothesis (ii) includes the i.i.d. restriction for each sample and adds a specific statement about how the marginal distributions of X_1 and Y_1 may differ. Besides stating that members of the single sample are i.i.d., hypothesis (iii) actually specifies the family of their marginal distributions. Typically, in each of these cases one or more of the stated conditions is simply assumed and is not subjected to test. For example, when working with a simple random sample, the i.i.d. assumption is certainly justified. Sometimes, as in Example (iv), one goes farther and assumes a special parametric family of distributions. This may be justified because of one's own prior experience with such data, because other researchers have adopted the same model, or possibly just because it makes the subsequent analysis tractable. Such assumptions about the data are referred to as *maintained hypotheses.* They are restrictions that one simply accepts without question, either because they are thought to be reasonable approximations to reality or because statistical tools requiring weaker restrictions are unavailable or difficult to use.

The *tested* hypotheses are the ones about which the data will help us decide. How to design the tests and come to a decision usually depends on the maintained hypotheses. Thus, in Example (i) the techniques used to look for some unspecified difference in the marginal distributions might be appropriate only for independent samples. In Example (ii), that the X's and Y's be alike except for median might be the weakest possible assumption consistent with existing ways of testing for differences in location. Maintaining particular distributional forms, as in Examples (iv) and (v), might support specialized techniques that are highly informative about the parameters of those specific models.

Formulating and testing statistical hypotheses thus require making a choice about what to maintain and what to test, and this involves some trade-offs. Typically, the stronger the *a priori* restrictions on the d.g.p., the sharper and more reliable will be the inferences about what is not known—provided those *a priori* restrictions are correct. The last section of the chapter is devoted to *distribution-free* tests that require relatively few assumptions about the d.g.p. Some examples there show what one may have to sacrifice to gain the robustness these tests afford.

8.1.1.2   *Parametric vs. Nonparametric Hypotheses*

The distinction between parametric and nonparametric hypotheses has become blurred in the statistical literature, because the terms are sometimes used imprecisely to refer to different statistical methods rather than to different types of hypotheses. All would agree that the tested hypotheses in Examples (iv) and (v) above are parametric, since they are statements about parameters of families of models. All would agree as well that the tested hypotheses in Example (i) (that the populations are the same) and Example (iii) (that the data are drawn from a specific family of distributions) are nonparametric, since they do not refer to any specific constants and since nothing is *assumed* about a parametric family. On the other hand, some writers might describe the hypothesis in Example (ii) as nonparametric, because no parametric model is maintained. As we use the term here, it is a parametric hypothesis since it refers to constants (medians) that describe populations, but it is one to which a distribution-free test would be applied. It is helpful to keep the terms "nonparametric" and "distribution-free" separate, as some hypotheses of each type can be tested by methods that require restrictive maintained hypotheses and some can be tested by methods that do not. In this chapter we apply both types of methods to test parametric hypotheses. Nonparametric hypotheses are the subject of Chapter 9.

8.1.1.3   *Null and Alternative Hypotheses*

Accompanying any tested hypothesis about the d.g.p. is a counter-hypothesis that would be adopted, at least tentatively, should the data not support the original statement. In classical hypothesis testing one of these statements is treated as a "strawman" that one expects to show is untenable, the test being structured in such a way that the dummy is knocked down only when the contrary evidence is sufficiently strong. The strawman, denoted H_0, is called the *null* hypothesis, since it typically negates the conclusion that one expects to draw. By "rejecting" the null, we are disposed to favor the opposing, or "alternate" hypothesis, denoted H_1. Since our prior beliefs about the d.g.p. are usually imprecise, the statement H_1 is usually broad, while H_0 is often more specific. Thus, in Example (i) the specific null hypothesis that the populations are the same would be set up to oppose the broader alternative that they differ. In Example (ii) the null H_0: $x_{.5} = y_{.5}$ (equality of medians) would oppose the more general H_1: $x_{.5} \neq y_{.5}$, and in Example (iv) H_0: $\mu = 13$ would be contrasted with H_1: $\mu \neq 13$. Numbers (ii) and (iv) are examples of "two-sided" parametric alternatives, since the signs of the differences, $x_{.5} - y_{.5}$ and $\mu - 13$, are not specified. In Example (v) the one-sided hypothesis $\theta > 13$ would typically correspond to our prior belief. In this case the opposing hypothesis, $H_0 : \theta \leq 13$, is no more specific than H_1.[1]

8.1.1.4 *Simple* vs. *Composite Hypotheses*

Hypotheses that fully specify the d.g.p. are called *simple*. Those that do not are called *composite*. Thus, if we were to take H_0: $\theta = 13$ in Example (v) (ruling out $\theta < 13$ *a priori*) then this combined with the maintained hypothesis that the data are i.i.d. as Poisson completely specify the model, whereas the alternative H_1: $\theta > 13$ does not. In Example (iv) neither H_0: $\mu = 13$ nor H_1: $\mu \neq 13$ is simple because the variance of the normal distribution is not specified. Nonparametric hypotheses such as those in Examples (i)-(iii) are almost always composite, so the distinction usually is made only in the parametric case.

8.1.2 *An Overview of Classical Hypothesis Testing*

The problem posed in classical hypothesis testing is to use sample data to decide between two exclusive and collectively exhaustive hypotheses: a "null" hypothesis H_0 that often puts relatively narrow restrictions on the d.g.p., and an "alternative" or "alternate" hypothesis H_1 that is typically broader and less restrictive. Since H_0 and H_1 are exclusive, only one *can* be true, and since they are exhaustive, one *must* be true. Classical tests focus on the null and lead the decision maker either to reject H_0 or to accept H_0. While this classification seems to leave no room for doubt, it is important not to take the words "reject" and "accept" too literally. They should in fact be regarded as provisional conclusions that might well be reversed upon acquiring additional data or upon reassessing the validity of the maintained hypothesis. More will be said about this later.

---

[1] Example (iii) is a bit different. There, the researcher probably hopes the data will be consistent with the specified distributional family in order to justify using specific tools for estimation. Even so, there is no other option but to take the specified family as the null, leaving as the alternative the broad statement "H_0 is false." These "goodness-of-fit" tests are discussed in Chapter 9.

8.1.2.1   *The General Framework*

In the most general terms testing a null hypothesis amounts to judging whether the data are sufficiently inconsistent with H_0 as to warrant rejecting it. How can this be done in a systematic and logical manner? Recognize that realization $\mathbf{x} := (x_1, x_2, ..., x_n)'$ of the sampling experiment is simply a point in the n-dimensional sample space $\Re^n$. To make a formal decision about H_0 requires establishing a specific region of $\Re^n$—call it $\mathcal{R}$—such that H_0 is rejected when $\mathbf{x} \in \mathcal{R}$ and is accepted otherwise. The prospect of carving out regions of n-dimensional space sounds daunting but is not as difficult as it sounds. We almost always work with regions of $\Re^n$ that are defined implicitly in terms of the values of scalar test statistics. That is, for each specific test there is a scalar-valued statistic $T = T(\mathbf{X})$ that maps $\Re^n$ to $\Re$, and rejection and acceptance regions are defined in terms of values of T. For example, a test of H_0: $\theta \leq 13$ *vs.* H_1: $\theta > 13$ in Example (v) could be based on $T = \bar{X}$, the sample mean, with $\mathcal{R} = \{\mathbf{x} : \bar{x} \geq c\}$, where c is some specific "critical" value of $\bar{x}$. The choice of rejection region—sometimes called the "critical region" of the test—thus involves both a choice of test statistic T and a choice of the critical value or values of T that lead to rejection. The obvious question is, "How does one make these choices that determine $\mathcal{R}$?" Providing a cogent answer to this question requires introducing two more terms.

8.1.2.2   *Significance Levels and Powers of Tests*

In statistical inference one must invariably draw conclusions from limited information, and the conclusion drawn in any specific application may well be wrong. Since making an error is always possible, the goal must be to minimize or at least control the risk of error, which involves both the consequences of mistakes and the probabilities of making them. As we have seen, from the frequentist point of view that underlies classical (as opposed to Bayesian) statistics, "probability" refers to the relative frequency with which an event would occur were the chance experiment to be repeated indefinitely. In statistical inference specifically, one imagines taking not just the one sample that is available but infinitely many such samples, drawing the required inferences from each of them, then assessing the frequency and nature of error. One chooses among inference procedures based on the potential outcomes of this hypothetical meta-experiment, as determined from theory, simulation, or a combination of the two. In Chapter 7 we considered the goal of minimizing the expected loss associated with estimation errors in repeated samples and used statistical theory to show that certain estimators minimize expected loss (within certain categories of loss functions and estimators). In the context of hypothesis testing losses are associated with wrong decisions—rejecting a true null hypothesis or not rejecting a false one. Statisticians of the frequentist school judge inference procedures according to the probabilities that they will produce such errors.

There is a standard—albeit highly "unmnemonic"— terminology for the two types of errors that can be made.

I. A "type-I" error is made when a *true* null hypothesis is rejected.

II. A "type-II" error is made when a *false* null hypothesis is accepted.

Thus, a type-I error would be made in testing H_0: $\theta \leq 13$ *vs.* H_1: $\theta > 13$ if H_0 was rejected even though the true value of θ was 12.1, while a type-II error would be made if θ was really 13.5 and H_0 was not rejected. Of course, in applied work one never knows the true value of θ and cannot tell whether one's decision is right or wrong. Nevertheless, theory and simulation can tell us how the choices of test statistic and rejection/acceptance regions affect the probabilities of error—the relative frequencies in the hypothetical meta-experiment.

It is customary in the statistical literature to let "α" and "β" represent probabilities of type-I and type-II errors, respectively. Obviously, if there is probability β of accepting a false H_0, there is chance $1 - \beta$ of making the correct decision to reject a false H_0. The probability of rejecting a false null is called the *power* of the test. Ideally, one wants to make α small and $1 - \beta$ large, but there are limitations on the degree to which both can be achieved simultaneously. To make things more concrete, let us see how α and $1 - \beta$ can actually be calculated in a particular application, given two possible choices of rejection region. We can begin to see how to choose such a region once the relation between α and $1 - \beta$ becomes clear.

Example 8.1. Maintaining that the data are i.i.d. as $P(\theta)$ (Poisson), we test H_0: $\theta \leq 13$ *vs.* H_1: $\theta > 13$ using $T(\mathbf{X}) = \bar{X}$ from a sample of size $n = 100$. Consider the rejection region $\mathcal{R} = \{\mathbf{x} : \bar{x} > 13\}$, such that H_0 is rejected if sample realization $\bar{x}$ exceeds 13.0 and is not rejected otherwise. To judge the effect of this choice let us see how the probability of rejecting H_0 depends on the true value of θ. For this we construct the *power function*

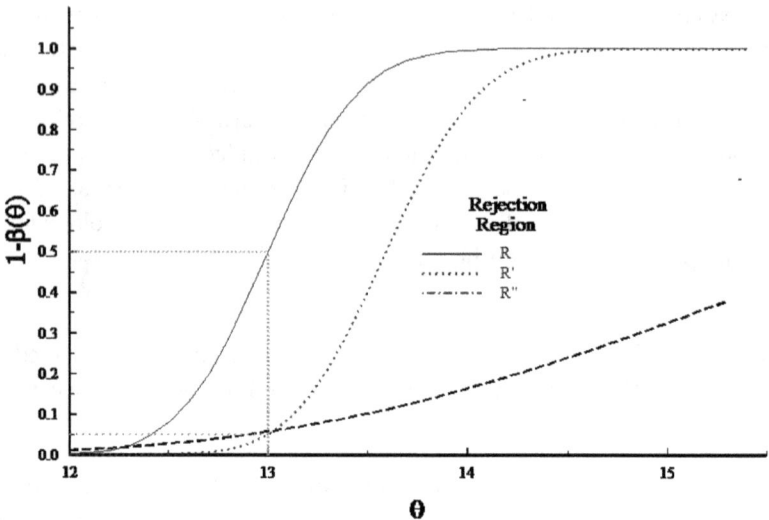

Fig. 8.1   Probability of Rejecting H_0: $\theta \leq 13$ for $\mathcal{R} = \{x : \bar{x} \geq 13\}$, $\mathcal{R}' = \{x : \bar{x} \geq 13.593\}$, and $\mathcal{R}'' = \{x : s^2 \geq 16.082\}$.

of the test,

$$\Pr\left(\mathbf{X} \in \mathcal{R}; \theta\right) = \Pr\left(\bar{X} > 13; \theta\right) = \Pr\left(\sum_{j=1}^{100} X_j > 1300; \theta\right).$$

Since $\sum_{j=1}^{100} X_j$ is distributed as Poisson with parameter 100θ, the last probability could be computed exactly as $1 - \sum_{s=0}^{1300} (100\theta)^s\, e^{-100\theta}/s!$. Fortunately, this tedious calculation can be avoided by applying asymptotic theory. Exercise 6.5 shows that $n\bar{X} = \sum_{j=1}^{n} X_j$ is distributed approximately as $N(n\theta, n\theta)$ for large n, so for $n = 100$

$$\Pr\left(\mathbf{X} \in \mathcal{R}; \theta\right) = \Pr\left(\frac{100\bar{X} - 100\theta}{\sqrt{100\theta}} > \frac{1300 - 100\theta}{\sqrt{100\theta}}\right)$$

$$\doteq \Pr\left(Z > \frac{130}{\sqrt{\theta}} - 10\sqrt{\theta}\right).$$

(As usual, "Z" represents a generic standard-normal r.v.) The second ($\mathcal{R}$) column of Table 8.1 shows such asymptotic estimates of rejection probabilities for selected values of θ. The solid curve in Fig. 8.1 is a plot on a finer and more extensive grid. For values of θ less than or equal to 13.0 these are probabilities of type-I error—of rejecting a true H_0. For values of θ above

Table 8.1   Rejection   probabilities
for three critical regions.

θ	$1 - \beta(\theta)$		
	$\mathcal{R}$	$\mathcal{R}'$	$\mathcal{R}''$
12.4	.0442	.0004	.0198
12.6	.1299	.0026	.0277
12.8	.2881	.0133	.0376
13.0	.5000	.0500	.0500
13.2	.7090	.1397	.0648
13.6	.9481	.5076	.1025
13.8	.9844	.7113	.1253
14.0	.9962	.8617	.1507
14.2	.9993	.9464	.1785
14.4	.9999	.9833	.2084
14.8	1.0000	.9991	.2735
15.2	1.0000	1.0000	.3432

13.0 they are powers. Similar calculations with $\mathcal{R}' = \{\mathbf{x} : \bar{x} \geq 13.593\}$ produce the figures in the third column of the table, which are plotted as the dotted curve in the figure. This region delivers lower probabilities of type-I error, but lower powers also. (The relevance of $\mathcal{R}''$ and the dashed curve will be explained later.)

The two regions $\mathcal{R}$ and $\mathcal{R}'$ in the example illustrate the usual trade-off between probability of type-I error and power. Specifically, *given* a (sensible) design of test statistic and geometry for $\mathcal{R}$, shrinking the critical region reduces both α and $1 - \beta$.[2] That is, reducing the probability of type-I error in this way increases the chance of making an error of type II. Of course, since the true population characteristic θ is unknown in applications, the actual error probabilities one faces cannot be determined. However, one *can* find or approximate the largest possible probability of type-I error associated with a particular rejection region. In the setup of the example the largest such probability corresponds to the boundary value $\theta = 13.0$ that separates H_0 and H_1. For all values less than or equal to 13.0 rejecting H_0 is an error, while for all values greater than this it is a correct decision. For region $\mathcal{R}$ in the example the largest probability of type-I error is .50, but this maximum value could be set at any desired value between zero and unity by adjusting the critical value of $\bar{x}$. For example, the error probability

---

[2]The proviso "sensible" is needed because really bad designs, such as regions $\mathcal{R}^*$ and $\mathcal{R}''$ described in Section 8.2, are capable of improvement on both counts. Strictly speaking, such "Pareto improvements" are always possible *in principle* except in the few situations that "optimal" tests are available.

could be limited to .05 or less by finding the critical value $c\,(.05)$ such that $\max_{0<\theta\le13}\left\{\Pr\left(\bar{X}\ge c\,(.05)\,;\theta\right)\right\}\le.05$. Since the maximum occurs at $\theta=13.0$, $c\,(.05)$ satisfies

$$.05 = \Pr\left(\frac{100\bar{X}-1300}{\sqrt{1300}}\ge\frac{100c\,(.05)-1300}{\sqrt{1300}}\right)$$
$$\doteq \Pr\left(Z\ge\frac{10c\,(.05)}{\sqrt{13}}-10\sqrt{13}\right).$$

The upper-.05 quantile of the standard normal is $z^{.05}=1.645$, so solving

$$1.645 = \frac{10c\,(.05)}{\sqrt{13}}-10\sqrt{13}$$

gives $c\,(.05)\doteq13.593$, corresponding to rejection region $\mathcal{R}'$ in the example.

A test that limits the maximum probability of type-I error to some predetermined value α is said to be a test at *significance level* α. The significance level is thus the maximum *size* of the test, where "size" refers to the *probability* measure of $\mathcal{R}$ when H_0 is true (as opposed to its Lebesgue measure). As we shrink the size of the rejection region and reduce α, we also make it harder to reject H_0 when it is false and therefore reduce power. Rejections at low values of α are therefore considered statistically more "significant", in that they are less apt to be type-I errors. The traditional approach in classical statistics is to choose a significance level for the test based on some notion about the cost of making a type-I error, then to look for a rejection region of maximum size α—$\mathcal{R}_\alpha$, say—that delivers high power against alternatives. We shall see next that in certain special cases it is possible to conduct truly optimal tests of parametric hypotheses—tests in which, given the choice of α and a sample size n, power is maximized for each of the possible departures from H_0 that are included in H_1.

8.2 Narrow Optimality: The Neyman–Pearson Lemma

Example 8.1 showed that with $n=100$ a test of $H_0:\theta\le13$ vs. $H_1:\theta>13$ using $\mathcal{R}'=\{\mathbf{x}:\bar{x}\ge13.593\}$ delivers type-I error of at most $\alpha=.05$, based on the asymptotic approximation for the distribution of $\bar{X}$. Of course, with the same test statistic there would be many other rejection regions of the same size; for example, the region $\mathcal{R}^*=\{\mathbf{x}:13.0\le\bar{x}\le13.045\}$. This region is clearly a bad choice because the very large values of $\bar{X}$ that are least consistent with H_0 would not lead to rejections. One expects the power of $\mathcal{R}^*$ to be lower than that of $\mathcal{R}'$ for a large range of values of θ.

Exercise 8.1. *Verify that the region* $\mathcal{R}^* = \{\mathbf{x} : 13.0 \leq \bar{x} \leq 13.045\}$ *has maximum size of about .05 in testing* $H_0 : \theta \leq 13$ *vs.* $H_1 : \theta > 13$ *when the data are i.i.d. as* $P(\theta)$ *(Poisson), and calculate the powers for the values of* θ *shown in the table in Example 8.1.*

Another possibility is to base the rejection region on a different statistic entirely. Since the Poisson distribution's mean and variance both equal θ, one might consider using sample variance S^2 as test statistic. Example 6.32 showed that normalized sample second central moment $\sqrt{n}(M_2 - \sigma^2)$ converges in distribution to $N(0, \mu_4 - \sigma^4)$ for any population with finite fourth moment, and the limiting distribution of $\sqrt{n}(S^2 - \sigma^2)$ is just the same. In the Poisson case with $\sigma^2 = \theta$ and $\mu_4 = \theta + 3\theta^2$ we have $\sqrt{n}(S^2 - \sigma^2) \rightsquigarrow N(0, \theta + 2\theta^2)$. Using this approximation, the region $\mathcal{R}'' = \{\mathbf{x} : s^2 \geq 16.082\}$ limits the maximum probability of type-I error to about .05. One can verify that its approximate power function is that shown in the last column of Table 8.1 and as the dashed line in Fig. 8.1. By this comparison, the test based on $\bar{X}$ seems by far the better procedure. But calculating power functions is just one way to compare tests; another is to determine the sample sizes that would give the tests the same power at any given θ.

Exercise 8.2. *Referring to tests of* $H_0 : \theta \leq 13$ *vs.* $H_1 : \theta > 13$ *with random* $P(\theta)$ *(Poisson) samples,*

(i) *Use the asymptotic approximation* $\sqrt{n}(\bar{X} - \theta) \dot{\sim} N(0, \theta)$ *to find the smallest integer* $n_{\bar{X}}$ *such that a test based on* $\bar{X}$ *at significance level* $\alpha = .05$ *would have at least .95 power for* $\theta = 14.0$.

(ii) *Use the asymptotic approximation* $\sqrt{n}(S^2 - \theta) \dot{\sim} N(0, \theta + 2\theta^2)$ *to find the smallest integer* n_{S^2} *such that a test based on* S^2 *at significance level* $\alpha = .05$ *would have at least .95 power for* $\theta = 14.0$.
(Answers: $n_{\bar{X}} = 146$, $n_{S^2} = 4,092$.*)*

Clearly, basing a test of the Poisson parameter on the sample variance instead of the sample mean is a bad idea. Is there any way to understand why? Is there any reason to think that the sample mean is the best statistic to use? A celebrated contribution by Neyman and Pearson (1933)[3] shows how to find the "best" (i.e., most powerful) rejection region of a given size

---

[3]Contemporaries (and often professional antagonists) of R.A. Fisher, Jerzey Neyman (1894-1981) and Egon S. Pearson (1885-1980) were among the most renowned statisticians of the 20th century. They are best remembered today for this joint work on tests of statistical hypotheses.

in testing a *simple* null against a *simple* alternative hypothesis. (Recall that simple hypotheses are those that completely specify the underlying distribution.) We will present their argument for this narrow case of simple hypotheses and then extend it to cover the case at hand, where both H_0 and H_1 are composite.

Take as maintained hypothesis that the sample elements are i.i.d. with a density relative to a dominating measure $\mathcal{M}$ (typically counting measure or Lebesgue measure) that belongs to parametric family $\{f(x; \theta) : \theta \in \Theta \subset \Re\}$. Consider hypotheses $H_0 : \theta = \theta_0$ vs. $H_1 : \theta = \theta_1$, where θ_0 and θ_1 are specific points in Θ. *Ex ante*, the likelihood function corresponding to sample $\mathbf{X} := (X_1, X_2, ..., X_n)'$ under H_0 is $L(\mathbf{X}; \theta_0) = \Pi_{j=1}^n f(X_j; \theta_0)$, and under H_1 it is $L(\mathbf{X}; \theta_1) = \Pi_{j=1}^n f(X_j; \theta_1)$. The best rejection region of size $\alpha \in (0, 1)$, if one exists, is the one region of that size that has the highest chance of containing the sample point when H_1 is true. That is, it is a region $\mathcal{R}_\alpha \in \Re^n$ that solves

$$\max_{\mathcal{R}} \{\Pr(\mathbf{X} \in \mathcal{R} \mid H_1)\} = \max_{\mathcal{R}} \left\{ \int_{\mathcal{R}} L(\mathbf{x}; \theta_1) \mathcal{M}(d\mathbf{x}) \right\} \qquad (8.1)$$

subject to $\int_{\mathcal{R}} L(\mathbf{x}; \theta_0) \mathcal{M}(d\mathbf{x}) = \alpha$. Momentarily confining the search for the best $\mathcal{R}_\alpha$ to regions of $\Re^n$ in which $L(\mathbf{x}; \theta_0) > 0$, the objective function in (8.1) can be written

$$\int_{\mathcal{R}} L(\mathbf{x}; \theta_1) \mathcal{M}(d\mathbf{x}) = \int_{\mathcal{R}} \frac{L(\mathbf{x}; \theta_1)}{L(\mathbf{x}; \theta_0)} L(\mathbf{x}; \theta_0) \mathcal{M}(d\mathbf{x}).$$

It is now apparent that maximum power is attained on a region where likelihood ratio $L(\mathbf{x}; \theta_1)/L(\mathbf{x}; \theta_0)$ is as large as possible subject to the size constraint. In fact, it is helpful instead to deal with the *inverse* ratio, $L(\mathbf{x}; \theta_0)/L(\mathbf{x}; \theta_1)$, which should be made small on the desired critical region. Doing this lets us add to $\mathcal{R}_\alpha$ any points with $L(\mathbf{x}; \theta_0) = 0$ and $L(\mathbf{x}; \theta_1) > 0$, which augment power with no consequence for type-I error. Consider, then, regions of points that satisfy the inequality $L(\mathbf{x}; \theta_0)/L(\mathbf{x}; \theta_1) \leq \lambda$ for some $\lambda \geq 0$; that is, regions

$$\mathcal{R}(\lambda) = \left\{ \mathbf{x} : \frac{L(\mathbf{x}; \theta_0)}{L(\mathbf{x}; \theta_1)} \leq \lambda \right\}.$$

Since $L(\mathbf{x}; \theta_0) > 0$ on the support, $\Pr(\mathbf{X} \in \mathcal{R}(\lambda) \mid H_0) = 0$ at $\lambda = 0$ and increases from there as $\lambda \uparrow$. If a specific positive number—λ_α, say—can be found such that $\Pr(\mathbf{X} \in \mathcal{R}(\lambda_\alpha) \mid H_0) = \alpha$ then the test based on $\mathcal{R}(\lambda_\alpha)$ will have power at least as great as that of any other test of size α. If an exact such λ_α cannot be found, as is typically the case when X is discrete, it may be possible to find a $\lambda_{\alpha'}$ such that α' is less than but as close as possible to α. In that case the test based on $\mathcal{R}(\lambda_{\alpha'})$ will the the most powerful test of size no greater than α. This result is the conclusion of what is referred to as the *Neyman–Pearson lemma*.

Example 8.2. Maintaining that the sample of size n is i.i.d. as $P(\theta)$ (Poisson), let us apply the Neyman–Pearson lemma to find the best size-α rejection region for $H_0 : \theta = \theta_0$ vs. $H_1 : \theta = \theta_1 > \theta_0$. For $h \in \{0, 1\}$ $L(\mathbf{X}; \theta_h) = \prod_{j=1}^{n} \theta_h^{X_j} e^{-\theta_h} / X_j!$, so

$$\frac{L(\mathbf{X}; \theta_0)}{L(\mathbf{X}; \theta_1)} = \left(\frac{\theta_0}{\theta_1}\right)^{\sum_{j=1}^{n} X_j} e^{n(\theta_1 - \theta_0)}.$$

The likelihood ratio is less than or equal to some $\lambda > 0$ if and only if

$$-\log(\theta_1/\theta_0) \sum_{j=1}^{n} X_j + n(\theta_1 - \theta_0) \leq \log \lambda$$

or

$$\sum_{j=1}^{n} X_j \geq \frac{n(\theta_1 - \theta_0) - \log \lambda}{\log(\theta_1/\theta_0)}. \tag{8.2}$$

The best rejection region of size no greater than α can therefore be stated just in terms of the sample sum—or any one-to-one function thereof, such as the sample mean. Specifically, it is the set $R_\alpha = \left\{\mathbf{x} : \sum_{j=1}^{n} x_j \geq c^\alpha\right\}$, where c^α is the smallest number such that $\Pr\left(\sum_{j=1}^{n} X_j \geq c^\alpha \mid H_0\right) \leq \alpha$. Since $\sum_{j=1}^{n} X_j$ is distributed as Poisson($n\theta_0$) under H_0, this condition is met by the smallest *integer* c^α such that $\sum_{x=c^\alpha}^{\infty} (n\theta_0)^x e^{-n\theta_0} / x! \leq \alpha$. In this case, as is typical with discrete populations, achieving an exact size of α is possible only for countably many values of α. However, using the normal approximation $\sum_{j=1}^{n} X_j \sim N(n\theta_0, n\theta_0)$, an approximate size-$\alpha$ test is obtained for arbitrary $\alpha \in (0, 1)$ by choosing c^α to solve

$$\Pr\left(\frac{\sum_{j=1}^{n} X_j - n\theta_0}{\sqrt{n\theta_0}} \geq \frac{c^\alpha - n\theta_0}{\sqrt{n\theta_0}} = z^\alpha\right) = \alpha, \tag{8.3}$$

where z^α is the upper-α quantile of the standard normal distribution. Thus, for particular values $\theta_0 = 9$, $n = 100$, and $\alpha = .05$ we would have $c^\alpha = 900 + 1.645\sqrt{900} = 949.35$.

As can be inferred from the example, implementing the Neyman–Pearson procedure does not ordinarily require calculating the constant λ_α.

Generally, one tries to find an event equivalent to $\{L(\mathbf{X}; \theta_0)/L(\mathbf{X}; \theta_1) \leq \lambda_\alpha\}$ that involves a statistic $T(\mathbf{X})$ whose distribution is known or can be well approximated under H_0. Finding test statistic $T(\mathbf{X})$ and the equivalent event is a matter of algebraic manipulation, as in the steps leading to (8.2). Once a candidate $T(\mathbf{X})$ is identified, one can choose whatever one-to-one function of it is most convenient to use.

Example 8.3. Using the asymptotic approximation in Example 8.2, it would have been simpler to take as test statistic not $\sum_{j=1}^{n} X_j$ itself, but its standardized form under H_0; i.e.,

$$T(\mathbf{X}) = \frac{\sum_{j=1}^{n} X_j - n\theta_0}{\sqrt{n\theta_0}} = \frac{\bar{X} - \theta_0}{\sqrt{\theta_0/n}}.$$

Doing this makes computing c^α unnecessary, since the rejection region now just comprises realizations equal to z^α or larger; i.e., $\mathcal{R}_\alpha = \{\mathbf{x} : T(\mathbf{x}) \geq z^\alpha\}$.

Exercise 8.3. *Modify Example 8.2 to handle the case that $\theta_1 < \theta_0$, showing that the best rejection region is of the form $\mathcal{R}_\alpha = \left\{\mathbf{x} : 0 \leq \sum_{j=1}^{n} x_j \leq c_\alpha\right\}$ for some c_α. Using the normal approximation to the distribution of the sum, show that the best test can be based on $T(\mathbf{X}) = (\bar{X} - \theta_0)/\sqrt{\theta_0/n}$ and the (approximately) size-α region $\mathcal{R}'_\alpha = \{\mathbf{x} : T(\mathbf{x}) \leq -z^\alpha\}$. Why is $\mathcal{R}_\alpha$ bounded below and $\mathcal{R}'_\alpha$ not?*

Exercise 8.4. *In Example 8.2 let $\alpha \in (0,1)$ be some fixed value such that there is a positive integer $c^{\alpha'}$ with $\sum_{x=c^{\alpha'}}^{\infty} (n\theta_0)^x e^{-n\theta_0}/x! = \alpha' < \alpha$ and $\sum_{x=c^{\alpha'}-1}^{\infty}(n\theta_0)^x e^{-n\theta_0}/x! = \alpha'' > \alpha$. Thus, α is such that the sizes of the feasible critical regions $\mathcal{R}_{\alpha'} = \left\{\mathbf{x} : \sum_{j=1}^{n} x_j \geq c^{\alpha'}\right\}$ and $\mathcal{R}_{\alpha''} = \left\{\mathbf{x} : \sum_{j=1}^{n} x_j \geq c^{\alpha'} - 1\right\}$ merely bracket α. An experiment is performed that yields a realization of the r.v. $Q \sim U(0,1)$ (uniform), independently of Poisson sample $\mathbf{X}$. The following decision rule is then applied: If $Q \leq q$, reject H_0 if and only if $\mathbf{X} \in \mathcal{R}_{\alpha'}$; otherwise, reject H_0 if and only if $\mathbf{X} \in \mathcal{R}_{\alpha''}$. Find the value of q such that resulting randomized test has exact size α.[4]*

---

[4]To be sure, such randomized procedures can scarcely be recommended: They allow different researchers with a common maintained hypothesis and identical sample data to come to different conclusions! While such disputes do sometimes arise among *Bayesians* (as we will see in chapter 10), they at least are attributable to something other than chance.

8.3   Extending to Classes of Simple Hypotheses

8.3.1   *Uniformly Most Powerful Tests*

When a best critical region $\mathcal{R}_\alpha$ for $H_0 : \theta = \theta_0$ vs. $H_1 : \theta = \theta_1$ depends only on the sign of $\theta_1 - \theta_0$ and not on the specific value of θ_1, as in Example 8.2 and Exercise 8.3, then $\mathcal{R}_\alpha$ is most powerful for *all* values of θ_1 such that $\theta_1 - \theta_0$ has the same sign. Thus, in the Poisson example $\left\{ \mathbf{x} : \sum_{j=1}^n x_j \geq c^\alpha \right\}$ is most powerful for all $\theta_1 > \theta_0$, while $\left\{ \mathbf{x} : 0 \leq \sum_{j=1}^n x_j \leq c_\alpha \right\}$ is most powerful for all $\theta_1 < \theta_0$. In such a case the test of $H_0 : \theta = \theta_0$ is said to be *uniformly most powerful (u.m.p.)* for the entire class of hypotheses—each *individually* simple—that put θ on one specific side of θ_0. In other words, the test is u.m.p. against the pertinent one-sided composite hypothesis—either $H_1 : \theta > \theta_0$ or $H_1 : \theta < \theta_0$.

The conditions for the existence of such u.m.p. tests are satisfied for distributions having a *monotone* likelihood ratio, meaning that there is some function $T(\mathbf{x})$ such that $L(\mathbf{x}; \theta_0)/L(\mathbf{x}; \theta)$ is strictly increasing or strictly decreasing in $T(\mathbf{x})$ when $\theta \neq \theta_0$.[5] As a prominent example, likelihoods of distributions in the one-parameter exponential class, $f_X(x; \theta) = b(\theta) c(x) e^{\eta(\theta) T(x)}$, are monotone so long as $\eta(\theta)$ is a monotone function. The likelihood ratio for random sample $\mathbf{X}$ is then proportional to $\exp\left\{ [\eta(\theta_0) - \eta(\theta)] \sum_{j=1}^n T(x_j) \right\}$. If η is monotone increasing and $\theta > \theta_0$, then small values of the likelihood ratio correspond to large values of $T(\mathbf{x}) := \sum_{j=1}^n T(x_j)$, which then constitute the u.m.p. critical region for $H_0 : \theta = \theta_0$ vs. $H_1 : \theta > \theta_0$. Conversely, when $\eta(\theta)$ is strictly decreasing H_0 is rejected for small values of $\sum_{j=1}^n T(x_j)$. Thus, in the Poisson example with $f_X(x; \theta) = e^{-\theta + x \log \theta} \mathbf{1}_{\aleph_0}(x)/x!$ and $\eta(\theta) = \log \theta$ the u.m.p. test rejects H_0 in favor of $H_1 : \theta > \theta_0$ for large values of $\sum_{j=1}^n x_j$, as we have already seen.

When the null is also composite but one-sided, as for $H_0 : \theta \leq \theta_0$ vs. $H_1 : \theta > \theta_0$, and when the likelihood has the above monotone property, then a u.m.p. test of type-I error *at most* α corresponds to the best critical region for the member of H_0 that is closest to H_1—i.e., for the boundary value θ_0. Thus, based on the normal approximation for $\sum_{j=1}^n X_j$, region $\mathcal{R}' = \{\mathbf{x} : \bar{x} \geq 13.593\}$ in Example 8.1 is the u.m.p. .05-level test of $H_0 : \theta \leq 13.0$ vs. $H_1 : \theta > 13.0$ in Poisson samples of $n = 100$.

---

[5]The statement is also correct if "strictly" is replaced by "weakly", but cases involving weak monotonicity are rarely encountered.

Exercise 8.5. X *is a random sample of size n from a population with p.d.f.*
$f(x; \theta) = \theta^{-1} e^{-x/\theta} \mathbf{1}_{(0,\infty)}(x)$. *Find the u.m.p. size-$\alpha$ test of $H_0 : \theta = \theta_0$ vs.*
$H_1 : \theta > \theta_0$. *If $n = 12$, $\theta_0 = 2$, and $x_1 + x_2 + \cdots + x_{12} = 37$, test H_0 at*
level $\alpha = .05$. (Hint: Quantiles of the distribution of $\sum_{j=1}^{n} X_j$ under H_0
are in one of the tables in Appendix B.)

8.3.2   Unbiased Tests

One sometimes wants to test a simple null against a *two-sided* class of simple
alternatives, such as $H_0 : \theta = \theta_0$ *vs.* $H_1 : \theta \neq \theta_0$ in the Poisson case. An
obvious idea is to use the Neyman–Pearson lemma to find separate, disjoint
rejection regions whose sizes add to α, as $\mathcal{R}_{\alpha p}^+$ for the case $\theta > \theta_0$ and
$\mathcal{R}_{\alpha(1-p)}^-$ for $\theta < \theta_0$, where p is some number on $(0, 1)$. The rejection region
would then be $\mathcal{R}_\alpha = \mathcal{R}_{\alpha p}^+ \cup \mathcal{R}_{\alpha(1-p)}^-$. Unfortunately, no matter what the
choice of p such a test is no longer uniformly most powerful, since for $\theta > \theta_0$
the larger size-α one-sided region $\mathcal{R}_\alpha^+$ would be optimal, while $\mathcal{R}_\alpha^-$ would
be best for $\theta < \theta_0$. Still, using separate regions is clearly the only sensible
thing to do. For one thing, no u.m.p. test can exist when alternatives are
two-sided. In addition, using separate regions with an appropriate choice of
p does produce a u.m.p. test within the subclass of tests that are *unbiased*.

Definition 8.2. A size-α test based on rejection region $\mathcal{R}_\alpha$ is *unbiased* if
and only if

$$\Pr(\mathbf{X} \in \mathcal{R}_\alpha \mid H_1) \geq \alpha.$$

In other words, an unbiased test is one that is at least as likely to reject H_0
when H_0 is false as when H_0 is true.

To find a u.m.p. unbiased region $\mathcal{R}_\alpha = \mathcal{R}_{\alpha p}^+ \cup \mathcal{R}_{\alpha(1-p)}^-$, one first uses the
Neyman–Pearson lemma to characterize the best one-sided tests of sizes αp
and $\alpha(1-p)$ for any arbitrary $p \in (0, 1)$. These regions $\mathcal{R}_{\alpha p}^+$ and $\mathcal{R}_{\alpha(1-p)}^-$
will comprise values of a test statistic $T(\mathbf{X})$ lying beyond critical values
$c_{\alpha p}$ and $c^{\alpha(1-p)}$, which are lower and upper quantiles of the distribution
of $T(\mathbf{X})$ under H_0. For example, a large-sample, size-$\alpha(1-p)$ test of
$H_0 : \theta \leq \theta_0$ *vs.* $H_1 : \theta > \theta_0$ with Poisson data would comprise values of
$T(\mathbf{x}) = (\bar{x} - \theta_0)/\sqrt{\theta_0/n}$ exceeding the upper-$\alpha(1-p)$ quantile $z^{\alpha(1-p)}$ of
$N(0, 1)$, while the best critical region for $H_0 : \theta \geq \theta_0$ *vs.* $H_1 : \theta < \theta_0$ would
comprise values below $z_{\alpha p}$. Once the general form of each one-sided region
is thus determined, p must be chosen so as to make the test unbiased. When
the test statistic has a unimodal density f_T, this amounts to positioning

$c_{\alpha p}$ and $c^{\alpha(1-p)}$ so that the density at any point between them exceeds that at any other point. Doing this is equivalent to choosing p so as to minimize the distance between the quantiles, $c^{\alpha(1-p)} - c_{\alpha p}$. When f_T is differentiable and unimodal, Proposition 7.2 shows that the densities at the two critical points are equal at the optimal p; i.e., $f_T(c_{\alpha p^*}) = f_T(c^{\alpha(1-p^*)})$. In many cases this optimal p must be found numerically. However, when f_T is also symmetric about the origin, then $c_{\alpha p} = -c^{\alpha p}$, and equality of the densities requires $p = 1/2$. In many standard applications distributions of test statistics are indeed represented by differentiable, symmetric, unimodal densities. The optimal critical points are then the upper- and lower-$\alpha/2$ quantiles of $T(\mathbf{X})$, $c^{\alpha/2}$ and $c_{\alpha/2} = -c^{\alpha/2}$, which divide the maximum probability of type-I error equally.

Example 8.4. Maintaining that elements of sample $\mathbf{X} := (X_1, X_2, ..., X_n)'$ are i.i.d. as $P(\theta)$ (Poisson), the u.m.p. unbiased size-α test of H_0 : $\theta = \theta_0$ *vs.* $H_1 : \theta \neq \theta_0$ using the normal approximation to $T(\mathbf{X}) = (\bar{X} - \theta_0)/\sqrt{\theta_0/n}$ has critical region

$$\mathcal{R}_\alpha = \left\{ \mathbf{x} : \frac{|\bar{x} - \theta_0|}{\sqrt{\theta_0/n}} \geq z^{\alpha/2} \right\}.$$

Exercise 8.6. *Modify the result of Exercise 8.5 to find the rejection region that constitutes a u.m.p. unbiased size-α test of $H_0 : \theta = \theta_0$ vs. $H_1 : \theta \neq \theta_0$ for the parameter of the exponential distribution. Use a large-sample approximation to the distribution of $\sum_{j=1}^{n} X_j$.*

8.4 Nuisance Parameters and "Similar" Tests

To be applied directly, the Neyman–Pearson recipe for optimal tests requires the null and alternative to be either *simple* hypotheses or *classes* of simple hypotheses—those such that the value of $L(\mathbf{x}; \theta)$ is known at each particular value of θ within the sets specified by H_0 and H_1. These are the cases we have dealt with up to this point. However, when the parametric family depends on more than one unknown parameter, hypotheses that do not specify *all* of them are intrinsically composite. In such cases the direct application of the Neyman–Pearson lemma fails to determine an operable rejection region. Indeed, the ratio of likelihoods under H_0 and H_1 is no longer even a *statistic*, since it depends on the unknown parameters. Because they are the source of this difficulty, the unrestricted parameters are often referred to as "nuisance parameters".

Example 8.5. Maintaining that elements of $\mathbf{X}$ are i.i.d. as $N\left(\mu, \sigma^2\right)$, we wish to test at level α that $\mu = \mu_0$ *vs.* the two-sided alternative that $\mu \neq \mu_0$, no restriction being placed on the variance. Since the parameter space for $N\left(\mu, \sigma^2\right)$ is $\Re \times \Re_+ = (-\infty, \infty) \times (0, \infty)$, the full statement of the null and alternative hypotheses is $H_0 : \mu = \mu_0, \sigma^2 > 0$ and $H_1 : \mu \neq \mu_0, \sigma^2 > 0$. Let us see why the Neyman–Pearson principle fails to produce a feasible u.m.p. unbiased test. The likelihood ratio for the sample is

$$\frac{L(\mathbf{X}; \mu_0, \sigma^2)}{L(\mathbf{X}; \mu, \sigma^2)} = \exp\left\{\frac{n}{2\sigma^2}\left[(\mu^2 - \mu_0^2) - 2\bar{X}(\mu - \mu_0)\right]\right\}.$$

This is less than or equal to some positive λ if and only if

$$\bar{X}(\mu - \mu_0) \geq \frac{\mu^2 - \mu_0^2}{2} - \frac{\sigma^2}{n}\log \lambda.$$

Rejection would thus occur for values of $\bar{X}$ greater than or equal to some critical $c^{\alpha/2}$ if $\mu > \mu_0$ or less than some $c_{\alpha/2}$ if $\mu < \mu_0$. However, these critical values cannot be found, because σ is unknown. For example, the $c^{\alpha/2}$ that satisfies

$$\frac{\alpha}{2} = \Pr\left(\bar{X} \geq c^{\alpha/2} \mid H_0\right) = \Pr\left(\frac{\bar{X} - \mu_0}{\sigma/\sqrt{n}} \geq \frac{c^{\alpha/2} - \mu_0}{\sigma/\sqrt{n}}\right)$$

is $c^{\alpha/2} = \mu_0 + z^{\alpha/2}\sigma/\sqrt{n}$.

While the *direct* application of Neyman–Pearson is infeasible in the presence of nuisance parameters, the lemma can nevertheless be a guide to uniformly most powerful tests when there exists a *boundedly complete* sufficient statistic for those parameters. Explaining how this works requires a little background. To set things up for a parametric family $\{f(x; \boldsymbol{\theta})\}$ under the maintained hypothesis, let us suppose that $\boldsymbol{\theta} = (\phi, \psi)'$ with $(\phi, \psi) \in \Phi \times \Psi$ and that we aim to test either $H_0 : \phi \leq \phi_0, \psi \in \Psi$ or $H_0 : \phi \geq \phi_0, \psi \in \Psi$ versus the corresponding one-sided alternative. Here ψ represents the nuisance parameter, and we consider cases in which there there is a complete sufficient statistic $T = T(\mathbf{X})$ for ψ when ϕ equals the boundary value ϕ_0. Recalling Definition 7.4 on page 467, a statistic T with c.d.f. $F(t; \phi_0, \psi)$ is complete if the function $g \equiv 0$ is the only one that satisfies

$$\int g(t) \cdot dF(t; \phi_0, \psi) \equiv \int g(t) f(t; \phi_0, \psi) \mathcal{M}(dt) = 0 \qquad (8.4)$$

for all $\psi \in \Psi$. T is said to be *boundedly* complete if such a g is the only such *bounded* function—and this weaker condition is enough for what follows.

Now since T is sufficient for ψ, the conditional distribution of any other statistic given T does not depend on ψ. In particular, if there is a region $\mathcal{R}_\alpha$ of $\Re^n$ such that $E\left[1_{\mathcal{R}_\alpha}(\mathbf{X}) \mid T\right] = \alpha$ when $\phi = \phi_0$, then $E1_{\mathcal{R}_\alpha}(\mathbf{X}) = E\left\{E\left[1_{\mathcal{R}_\alpha}(\mathbf{X}) \mid T\right]\right\} = \alpha$ as well. And since $E\left[1_{\mathcal{R}_\alpha}(\mathbf{X}) \mid T\right]$ does not depend on ψ, the region $\mathcal{R}_\alpha$ has size α for all values of ψ. $\mathcal{R}_\alpha$ is then called a "similar" critical region, and the corresponding test is said to be a *similar* test.[6] Now here is where completeness comes in. If sufficient statistic T is at least boundedly complete, then the $\mathcal{R}_\alpha$ that satisfies $E\left[1_{\mathcal{R}_\alpha}(\mathbf{X}) \mid T\right] = \alpha$ is (essentially) the *only* such similar critical region; for if $\mathcal{R}'_\alpha$ were asserted to be another such region, the fact that $E\left[1_{\mathcal{R}_\alpha}(\mathbf{X}) - 1_{\mathcal{R}'_\alpha}(\mathbf{X}) \mid T\right] = \alpha - \alpha = 0$ for all $\psi \in \Psi$ implies $\mathbb{P}_{\mathbf{X}}\left[(\mathcal{R}_\alpha \backslash \mathcal{R}'_\alpha) \cup (\mathcal{R}'_\alpha \backslash \mathcal{R}_\alpha)\right] = 0$, in which case $\mathcal{R}_\alpha$ and $\mathcal{R}'_\alpha$ could differ only on null sets.

The best critical region $\mathcal{R}_\alpha$ that applies to any particular ϕ allowed by H_1 can be found via the Neyman–Pearson prescription as the region comprising small values of the ratio of *conditional* likelihoods. While the conditions for all this are rather special—particularly when ϕ and ψ are vectors—the theory has been used to show that the standard t, chi-squared, and F tests for parameters of normal distributions are uniformly most powerful among unbiased tests. We develop these standard tests in Section 8.6.1 as examples of exact likelihood-ratio tests.[7]

8.5 Asymptotic Properties of Tests

Unfortunately, finding optimal tests by any of the principles just described is often infeasible. Clearly, there can be no basis for judging the optimality of tests about moments or other characteristics of a data-generating process when no parametric family is specified. In such cases the distribution-free methods described in Section 8.8 are often helpful. In parametric settings Bayesian methods provide an alternative approach, which we discuss briefly in Section 8.7.1 and in more detail in Chapter 10. There are also tests that invoke the principle of invariance, which holds that decisions should be invariant under transformations of the data that do not change a model's basic structure—e.g., sample space, parameter space, and family of distri-

---

[6]$\mathcal{R}_\alpha$ is "similar" to the entire sample space $\Re^n$, in the sense that $\mathbb{P}_{\mathbf{X}}(\mathcal{R}_\alpha \mid T)$ and $\mathbb{P}_{\mathbf{X}}(\Re^n \mid T)$ are both invariant with respect to ψ.

[7]Details of the theory of similar tests and their application to normal distributions are given in Chapters 4 and 5 of Lehmann and Romano (2005).

butions.[8] Of broader application are *asymptotic* tests, whose characteristics are judged only by their limiting behavior as $n \to \infty$. These methods are often the only feasible ones in nonparametric settings and with complicated parametric models that involve many parameters, such as are often encountered in applied disciplines.

Let us now see what criteria have been proposed for judging the large-sample properties of tests. The principal ones are (i) asymptotic *validity*, (ii) *consistency*, and (iii) asymptotic *efficiency*. These constitute a hierarchy of increasingly stringent conditions.

Definition 8.3. A test with rejection region $\mathcal{R}$ and stated significance level α is *valid* if $\Pr(\mathbf{X} \in \mathcal{R} \mid H_0) \leq \alpha$ for all cases contained in H_0. A test based on a sample of size n and rejection region $\mathcal{R}$ is (uniformly) *asymptotically valid* if

$$\lim_{n \to \infty} \sup \Pr(\mathbf{X} \in \mathcal{R} \mid H_0) \leq \alpha,$$

where the supremum is over all distributions consistent with H_0.

Thus, a valid test is one whose probability of type-I error does not exceed the stated significance level. The distinction between "valid" and "asymptotically valid" has already arisen implicitly in several examples. In testing $H_0 : \theta = \theta_0$ vs. $H_1 : \theta > \theta_0$ for Poisson data in Example 8.2 we saw that the discreteness of the sample mean usually makes it impossible to find a critical upper quantile c^α such that $\mathcal{R} = \left\{\mathbf{x} : \sum_{j=1}^{n} x_j \geq c^\alpha\right\}$ has size exactly equal to α. To obtain a valid test requires finding a value of $c^{\alpha'}$ such that $\Pr\left(\sum_{j=1}^{n} X_j \geq c^{\alpha'} \mid H_0\right) = \alpha' \leq \alpha$. The alternative is to find an approximate solution for c^α using the normal approximation for $\bar{X}$, as in (8.3). In finite samples it may well be that the value of c^α so determined produces type-I error larger than α; however, the central limit theorem does assure the *asymptotic* validity of the test. In most unstructured parametric settings asymptotic procedures are the only option, because there is no test statistic whose exact distribution under H_0 is known.

Once at least rough control over type-I error is assured, the logical succeeding goal in test design is to assure that a *false* null would ultimately be rejected if given an abundance of data. This brings us to the criterion of *consistency*.

---

[8]For a description of invariant tests see Chapter 6 of Lehmann and Romano (2005).

Definition 8.4. A test based on a sample of size n and rejection region $\mathcal{R}_\alpha$ is *consistent* if

$$\lim_{n \to \infty} \inf \Pr\left(\mathbf{X} \in \mathcal{R}_\alpha \mid H_1\right) = 1,$$

where the infimum is over all distributions consistent with H_1.

A consistent test, therefore, has power that tends to unity as the sample size grows. With a consistent test, the probability of type-II error can be made arbitrarily small by taking n sufficiently large, regardless of what specific condition under H_1 actually holds.

It is usually easy to design asymptotically valid and consistent tests, and several alternative procedures are often available. When possible, it is useful to compare and select among these on the basis of a third criterion: relative asymptotic *efficiency*. We have seen that the relative efficiency of unbiased estimators can be indicated by the ratio of sample sizes that would give them equal variances. In the same way the efficiency of tests of the same size could be measured by the ratio of sample sizes at which they have the same power. Thus, Exercise 8.2 shows that for samples from $P(\theta)$ (Poisson) the efficiency of a .05-level test of $H_0 : \theta \leq 13$ vs. $H_1 : \theta > 13$ based on $\bar{X}$ relative to a test based on S^2 is $n_{S^2}/n_{\bar{X}} \doteq 4,092 \div 146$, or about 28 times, given that both tests are to deliver power of .95 at $\theta = 14$. However, this measure of relative efficiency has the disadvantage of depending on the particular values of α, of power, and of θ. Thus, in the Poisson example changing any of the conditions $\alpha = .05, 1 - \beta = .95$, or $\theta = 14$ would change this measure of relative efficiency.

To develop a summary measure of greater generality requires looking at the asymptotic properties of tests as $n \to \infty$. Of two prominent measures in the literature the one due to Pitman (1948) and Noether (1955) is most commonly used. The obvious difficulty in making asymptotic comparisons is that powers of consistent tests against fixed alternatives all approach unity. Pitman's idea was to compare powers against a *sequence* of alternatives that get closer to H_0 as n increases, and at rates that keep power below unity. These are often called "local" alternatives. For example, consistent tests of $H_0 : \theta = \theta_0$ vs. $H_1 : \theta > \theta_0$ would be compared against alternatives of the form $\left\{H_n : \theta_n = \theta_0 + c/n^\delta\right\}_{n=1}^\infty$ for some constants $c, \delta > 0$. Remarkably, the comparison typically turns out not to depend on c or δ or on the specific choice of α.

Definition 8.5. The (Pitman) *asymptotic efficiency* of test T_1 relative to T_2 is the limiting value of n_2/n_1 as $n_1 \to \infty$, where n_j is the sample size used in test T_j and where n_2 is adjusted so that T_2 and T_1 have the same limiting power against alternatives $\{H_n\}_{n=1}^{\infty}$.

For tests based on statistics whose standardized values approach normality the constant δ in H_n turns out to be the power of n to which the statistic's asymptotic standard error is proportional—usually $1/2$. Not surprisingly, in standard cases asymptotic relative efficiency depends on the ratio of the slopes of the tests' power functions at θ_0, with T_1 being more efficient if its power increases more rapidly as θ departs from θ_0.

Here is an illustration of the Pitman criterion in a simple setting without nuisance parameters.

Example 8.6. Testing $H_0 : \theta \leq \theta_0$ *vs.* $H_1 : \theta > \theta_0$ in Poisson samples of size n, let us compare the asymptotic efficiencies of tests based on the limiting theory for $\bar{X}_n$ and S_n^2; namely,

$$\frac{\bar{X}_n - \theta}{\sqrt{\theta/n}} \rightsquigarrow N(0,1) \text{ and } \frac{S_n^2 - \theta}{\sqrt{(\theta + 2\theta^2)/n}} \rightsquigarrow N(0,1).$$

With $\sigma_{\bar{X}_n}(\theta) = \sqrt{\theta/n}$ and $\sigma_{S_n^2}(\theta) = \sqrt{(\theta + 2\theta^2)/n}$ as the asymptotic standard errors (square roots of the asymptotic variances), the size-α rejection regions are

$$\mathcal{R}_\alpha = \left\{ \mathbf{x} : \frac{\bar{x} - \theta_0}{\sigma_{\bar{X}_n}(\theta_0)} \geq z^\alpha \right\} = \left\{ \mathbf{x} : \bar{x} \geq \theta_0 + z^\alpha \sigma_{\bar{X}_n}(\theta_0) \right\}$$

and

$$\mathcal{R}_\alpha'' = \left\{ \mathbf{x} : \frac{s^2 - \theta_0}{\sigma_{S_n^2}(\theta_0)} \geq z^\alpha \right\} = \left\{ \mathbf{x} : s^2 \geq \theta_0 + z^\alpha \sigma_{S_n^2}(\theta_0) \right\},$$

where z^α is the upper-α quantile of $N(0,1)$. For large n the powers at any specific θ are approximately

$$\mathfrak{P}_{\bar{X}_n}(\theta) := \Pr\left[\bar{X}_n \geq \theta_0 + z^\alpha \sigma_{\bar{X}_n}(\theta_0) \mid \theta \right]$$

$$= \Pr\left[\frac{\bar{X}_n - \theta}{\sigma_{\bar{X}_n}(\theta)} \geq \frac{\theta_0 - \theta + z^\alpha \sigma_{\bar{X}_n}(\theta_0)}{\sigma_{\bar{X}_n}(\theta)} \mid \theta \right]$$

$$= 1 - \Phi\left[\frac{\theta_0 - \theta + z^\alpha \sigma_{\bar{X}_n}(\theta_0)}{\sigma_{\bar{X}_n}(\theta)} \right]$$

and

$$\mathfrak{P}_{S_n^2}(\theta) := 1 - \Phi\left[\frac{\theta_0 - \theta + z^\alpha \sigma_{S_n^2}(\theta_0)}{\sigma_{S_n^2}(\theta)} \right],$$

where Φ is the standard normal c.d.f. For each test, using $\delta = 1/2$ in each case, we choose a *sequence* of alternatives converging to H_0, as $H_{\bar{X}_n} : \theta = \theta_0 + c_{\bar{X}}/\sqrt{n_{\bar{X}}}$ and $H_{S_n^2} : \theta = \theta_0 + c_{S^2}/\sqrt{n_{S^2}}$. The goals are to choose constants $c_{\bar{X}}, c_{S^2}$ and sample sizes $n_{\bar{X}}, n_{S^2}$ so that (i) the alternatives themselves become indistinguishable for large samples and (ii) the powers of the tests approach the same value. Goal (i) is met by requiring

$$\lim_{n_{\bar{X}} \to \infty} \frac{n_{S^2}}{n_{\bar{X}}} = \left(\frac{c_{S^2}}{c_{\bar{X}}} \right)^2. \tag{8.5}$$

The limiting powers are

$$\lim_{n_{\bar{X}} \to \infty} \mathfrak{P}_{\bar{X}_n} \left(\theta_0 + \frac{c_{\bar{X}}}{\sqrt{n_{\bar{X}}}} \right) = 1 - \lim_{n_{\bar{X}} \to \infty} \Phi \left[\frac{-\frac{c_{\bar{X}}}{\sqrt{n_{\bar{X}}}} + z^\alpha \sigma_{\bar{X}_n}(\theta_0)}{\sigma_{\bar{X}_n} \left(\theta_0 + \frac{c_{\bar{X}}}{\sqrt{n_{\bar{X}}}} \right)} \right]$$

$$= 1 - \lim_{n_{\bar{X}} \to \infty} \Phi \left[\frac{-c_{\bar{X}} + z^\alpha \sqrt{\theta_0}}{\sqrt{\theta_0 + \frac{c_{\bar{X}}}{\sqrt{n_{\bar{X}}}}}} \right]$$

$$= 1 - \Phi \left[-\frac{c_{\bar{X}}}{\sqrt{\theta_0}} + z^\alpha \right]$$

and

$$\lim_{n_{S^2} \to \infty} \mathfrak{P}_{S_n^2} \left(\theta_0 + \frac{c_{S^2}}{\sqrt{n_{S^2}}} \right) = 1 - \lim_{n_{S^2} \to \infty} \Phi \left[\frac{-c_{S^2} + z^\alpha \sqrt{\theta_0 + 2\theta_0^2}}{\sqrt{\left(\theta_0 + \frac{c_{S^2}}{\sqrt{n_{S^2}}} \right) + 2 \left(\theta_0 + \frac{c_{S^2}}{\sqrt{n_{S^2}}} \right)^2}} \right]$$

$$= 1 - \Phi \left[-\frac{c_{S^2}}{\sqrt{\theta_0 + 2\theta_0^2}} + z^\alpha \right].$$

Thus, goal (ii) (matching the limiting powers) requires

$$\frac{c_{\bar{X}}}{\sqrt{\theta_0}} = \frac{c_{S^2}}{\sqrt{\theta_0 + 2\theta_0^2}}$$

or

$$\frac{c_{S^2}}{c_{\bar{X}}} = \frac{\sqrt{\theta_0 + 2\theta_0^2}}{\sqrt{\theta_0}} = \sqrt{1 + 2\theta_0};$$

and combining with (8.5) gives

$$\lim_{n_{\bar{X}} \to \infty} \frac{n_{S^2}}{n_{\bar{X}}} = 1 + 2\theta_0$$

as the asymptotic relative efficiency of the test based on $\bar{X}$ to that based on S^2.

Exercise 8.7. *Show that the asymptotic efficiency of the test of* $H_0 : \theta \leq \theta_0$ *vs.* $H_1 : \theta > \theta_0$ *based on* $\bar{X}$ *relative to that based on* S^2 *equals the square of the limit of the relative slopes of the power functions at* θ_0; *i.e., that*

$$1 + 2\theta_0 = \lim_{n \to \infty} \left[\frac{\mathfrak{P}'_{\bar{X}_n}(\theta)\,|_{\theta=\theta_0}}{\mathfrak{P}'_{S_n^2}(\theta)\,|_{\theta=\theta_0}} \right]^2 .$$

8.6   Three Classical Asymptotic Tests

We now examine three specific parametric tests that do have good asymptotic properties when nuisance parameters are present: the likelihood-ratio test, the Lagrange-multiplier or "score" test, and the Wald test. The notation here is the same as that introduced with confidence sets in Section 7.3.2 (p. 525); that is, the parametric family for the data under the maintained hypothesis is $\left\{ f(x; \boldsymbol{\theta}) : \boldsymbol{\theta} \in \boldsymbol{\Theta} \subset \Re^k \right\}$, and is $\boldsymbol{\theta}$ broken down into subvectors $\boldsymbol{\phi} \in \boldsymbol{\Phi} \subset \Re^{k_0}$ and $\boldsymbol{\psi} \in \boldsymbol{\Psi} \subset \Re^{k-k_0}$, as $\boldsymbol{\theta}' = (\boldsymbol{\phi}', \boldsymbol{\psi}')$. We consider null hypotheses that limit $\boldsymbol{\phi}$ alone and leave nuisance parameters $\boldsymbol{\psi}$ unrestricted. Specifically, with $\boldsymbol{\phi}_0$ a k_0-vector of known constants in the subspace $\boldsymbol{\Phi}$ and $\boldsymbol{\Theta}_0 = \{\boldsymbol{\phi}_0\} \times \boldsymbol{\Psi}$, the goal is to test $H_0 : \boldsymbol{\theta} \in \boldsymbol{\Theta}_0$ *vs.* $H_1 : \boldsymbol{\theta} \in \boldsymbol{\Theta} \backslash \boldsymbol{\Theta}_0$.

8.6.1   *The Likelihood-Ratio Test*

The likelihood-ratio (l.r.) test is a natural extension of the Neyman–Pearson scheme. The idea is to replace the nuisance parameters in the likelihood function by their maximum likelihood estimators (m.l.e.s). This again makes the likelihood ratio a *statistic*, as it was for simple hypotheses. Letting $L(\mathbf{X}; \boldsymbol{\theta}) = \prod_{j=1}^{n} f(X_j; \boldsymbol{\theta})$ be the likelihood function of the random sample, *two* sets of m.l.e.s of $\boldsymbol{\theta}$ are found by solving (i) $\sup_{\boldsymbol{\theta} \in \boldsymbol{\Theta}_0} \{L(\mathbf{X}; \boldsymbol{\theta})\}$ and (ii) $\sup_{\boldsymbol{\theta} \in \boldsymbol{\Theta}} \{L(\mathbf{X}; \boldsymbol{\theta})\}$. The first problem is equivalent to $\sup_{\boldsymbol{\psi} \in \boldsymbol{\Psi}} \{L(\mathbf{X}; \boldsymbol{\phi}_0, \boldsymbol{\psi})\}$, in which $\boldsymbol{\phi}$ is fixed at the value specified by H_0 and maximization is carried out with respect to the $k - k_0$ nuisance parameters. Letting $\breve{\boldsymbol{\psi}}_n$ represent these restricted m.l.e.s, which are often functions of $\boldsymbol{\phi}_0$, we have $\sup_{\boldsymbol{\psi} \in \boldsymbol{\Psi}} \{L(\mathbf{X}; \boldsymbol{\phi}_0, \boldsymbol{\psi})\} = L(\mathbf{X}; \boldsymbol{\phi}_0, \breve{\boldsymbol{\psi}}_n)$. Problem (ii) is the conventional one of finding unrestricted m.l.e.s of all k elements of $\boldsymbol{\theta}$, namely $\hat{\boldsymbol{\theta}}'_n = (\hat{\boldsymbol{\phi}}'_n, \hat{\boldsymbol{\psi}}'_n)$. Thus, $\sup_{\boldsymbol{\theta} \in \boldsymbol{\Theta}} \{L(\mathbf{X}; \boldsymbol{\theta})\} = L(\mathbf{X}; \hat{\boldsymbol{\theta}}_n)$. The *likelihood-ratio statistic* is simply the ratio of these two expressions for the likelihood—restricted over unrestricted:

$$\Lambda\left(\mathbf{X};\phi_0\right) := \frac{L(\mathbf{X};\phi_0,\check{\psi}_n)}{L\left(\mathbf{X};\hat{\phi}_n,\hat{\psi}_n\right)}. \tag{8.6}$$

In chapter 7 we used such a ratio of likelihoods evaluated at an arbitrary $\phi \in \boldsymbol{\Phi}$ as a pivotal quantity for building confidence sets. Evaluating the ratio now at the known value ϕ_0, we will use the resulting pivotal *statistic* to test $H_0 : \phi = \phi_0$. Since the numerator in (8.6) is the solution to a restricted maximization problem that confines m.l.e.s to the subspace $\boldsymbol{\Theta}_0 \subset \boldsymbol{\Theta}$, it is clear that $\Lambda\left(\mathbf{X};\phi_0\right)$ cannot exceed unity. Of course, it is also nonnegative. Obtaining a realization close to unity indicates that the data are relatively consistent with the restriction imposed by H_0, while values close to zero indicate the converse. A plausible idea, therefore, is to choose some critical value of the ratio—λ_α, say—and reject H_0 whenever $\Lambda\left(\mathbf{X};\phi_0\right) \leq \lambda_\alpha$. That is, the rejection region would be of the form $\mathcal{R}_\alpha = \{\mathbf{x} : \Lambda\left(\mathbf{x};\phi_0\right) \leq \lambda_\alpha\}$, where λ_α would be chosen so that $\Pr\left[\Lambda\left(\mathbf{X};\phi_0\right) \leq \lambda_\alpha \mid H_0\right] = \alpha$ in order to control type-I error. Critical value λ_α would thus represent the α quantile of the distribution of the l.r. statistic. (Observe the notational distinction between the *statistic* $\Lambda\left(\mathbf{X};\phi_0\right)$, which is a random variable, and its realization $\Lambda\left(\mathbf{x};\phi_0\right)$ from an observed sample.)

Although the complexity of the l.r. statistic in most applications makes finding its quantiles no easy matter, there are nevertheless two general ways in which the idea can be implemented. These correspond precisely to the two ways of building confidence sets from the likelihood ratio, as described in Section 7.3.2. First, as in the Neyman–Pearson setting with simple hypotheses, it is sometimes possible to restate the inequality $\Lambda\left(\mathbf{X};\phi_0\right) \leq \lambda_\alpha$ in terms of another statistic whose exact distribution under H_0 is known. In these cases the l.r. principle yields valid finite-sample tests of H_0, some of which even turn out to be uniformly most powerful (u.m.p.). When no such restatement is possible, we must rely on large-sample approximations to the distribution of $\Lambda\left(\mathbf{X};\phi_0\right)$ to produce tests with desirable asymptotic properties. We consider the two cases in turn.

8.6.1.1 *Exact Tests Based on the L.R. Principle*

Some of the standard tests that are encountered in a first course in statistics turn out to be l.r. tests. Here are some examples and exercises. In each case inequality $\Lambda\left(\mathbf{X};\phi_0\right) \leq \lambda_\alpha$ is shown to be equivalent to an inequality that bounds a simpler (pivotal) function of the data whose distribution is known.

Example 8.7. Maintaining that the data are i.i.d. as $N(\mu, \sigma^2)$, we test $H_0 : \mu = \mu_0$ vs. $H_1 : \mu \neq \mu_0$. Here μ, σ^2 correspond to ϕ, ψ, with parameter space $\Theta = \Re \times \Re_+$. The restricted m.l.e. for nuisance parameter σ^2 is $\breve{\sigma}_n^2 = n^{-1} \sum_{j=1}^n (X_j - \mu_0)^2$, while the unrestricted m.l.e.s are $\hat{\mu}_n = \bar{X}$ and $\hat{\sigma}_n^2 = M_2$. The l.r. statistic is

$$\Lambda(\mathbf{X}; \mu_0) = \frac{(2\pi\breve{\sigma}_n^2)^{-n/2} \exp\left[-\frac{1}{2\breve{\sigma}_n^2} \sum_{j=1}^n (X_j - \mu_0)^2\right]}{(2\pi\hat{\sigma}_n^2)^{-n/2} \exp\left[-\frac{1}{2\hat{\sigma}_n^2} \sum_{j=1}^n (X_j - \hat{\mu}_n)^2\right]} = \left(\frac{\breve{\sigma}_n^2}{\hat{\sigma}_n^2}\right)^{-n/2},$$

so that $\Lambda(\mathbf{X}; \mu_0) \leq \lambda_\alpha$ is equivalent to $\lambda_\alpha^{-2/n} \leq \breve{\sigma}_n^2/\hat{\sigma}_n^2$. Thus, H_0 is to be rejected if

$$\lambda_\alpha^{-2/n} \leq \frac{\sum_{j=1}^n (X_j - \mu_0)^2}{\sum_{j=1}^n (X_j - \bar{X})^2}$$

$$= \frac{\sum_{j=1}^n [(X_j - \bar{X}) + (\bar{X} - \mu_0)]^2}{\sum_{j=1}^n (X_j - \bar{X})^2}$$

$$= \frac{\sum_{j=1}^n (X_j - \bar{X})^2 + n(\bar{X} - \mu_0)^2}{\sum_{j=1}^n (X_j - \bar{X})^2}.$$

Noting that $\sum_{j=1}^n (X_j - \bar{X})^2 = (n-1) S^2$, where S^2 is the unbiased sample variance, further manipulation reduces the inequality to

$$\left|\frac{\bar{X} - \mu_0}{S/\sqrt{n}}\right| \geq \sqrt{(n-1)\left(\lambda_\alpha^{-2/n} - 1\right)}. \tag{8.7}$$

The particular form of the expression on the right side of the inequality and the value of λ_α itself are now completely irrelevant. What matters is that H_0 should be rejected for large values of the statistic on the left. Since $T(\mathbf{X}) := (\bar{X} - \mu_0)/(S/\sqrt{n})$ is distributed as Student's t with $n-1$ degrees of freedom under H_0, an exact size-α test involves the rejection region

$$\mathcal{R}_\alpha = \left\{\mathbf{x} : |T(\mathbf{x})| \geq t^{\alpha/2}\right\},$$

where $t^{\alpha/2}$ is the upper-$\alpha/2$ quantile of Student's t. This is the familiar "t test" for means of normal distributions. In testing H_0 vs. a *one-sided* alternative, $H_1^+ : \mu > \mu_0$ or $H_1^- : \mu < \mu_0$, a one-sided rejection region would be used:

$$\mathcal{R}_\alpha^+ = \{\mathbf{x} : T(\mathbf{x}) \geq t^\alpha\} \text{ for } H_1^+,$$
$$\mathcal{R}_\alpha^- = \{\mathbf{x} : T(\mathbf{x}) \leq -t^\alpha\} \text{ for } H_1^-.$$

From the theory of similar tests (Section 8.4) it can be shown that the tests are u.m.p.-unbiased.[9]

Exercise 8.8. *Maintaining that the data are i.i.d. as $N\left(\mu, \sigma^2\right)$, show that exact size-$\alpha$ tests of $H_0 : \sigma^2 = \sigma_0^2$ vs. (i) $H_1^+ : \sigma^2 > \sigma_0^2$ (ii) $H_1^- : \sigma^2 < \sigma_0^2$, and (iii) $H_1 : \sigma^2 \neq \sigma_0^2$ can be based on the rejection regions $\mathcal{R}_\alpha^+ = \{\mathbf{x} : C\left(\mathbf{x}\right) \geq c^\alpha\}$, $\mathcal{R}_\alpha^- = \{\mathbf{x} : C\left(\mathbf{x}\right) \leq c_\alpha\}$, and $\mathcal{R}_\alpha = \mathcal{R}_{(1-p)\alpha}^+ \cup \mathcal{R}_{p\alpha}^-$, respectively, where (i) $C\left(\mathbf{X}\right) := (n-1)S^2/\sigma_0^2$ (S^2 being the unbiased sample variance), (ii) c_α and c^α are α and upper-α quantiles of $\chi^2\left(n-1\right)$ (the chi-squared distribution with $n-1$ degrees of freedom), and (iii) $p \in (0,1)$. With the appropriate choice of p these standard chi-squared tests for hypotheses about the variance of a normal population are u.m.p.-unbiased.*

Exercise 8.9. *Maintaining that the data are i.i.d. as $IG\left(\mu, \lambda\right)$ (inverse Gaussian), show that an exact size-α l.r. test of $H_0 : \lambda = \lambda_0$ vs. $H_1 : \lambda > \lambda_0$ can be based on rejection region $\mathcal{R}_\alpha = \{\mathbf{x} : C\left(\mathbf{x}\right) \leq c_\alpha\}$, where $C\left(\mathbf{x}\right) = n\lambda_0/\hat{\lambda}_n$ ($\hat{\lambda}_n$ being the m.l.e.) and c_α is the α quantile of $\chi^2\left(n-1\right)$. Carry out a .05-level test for $\lambda_0 = 2.5$ if a sample of $n = 101$ yields $\sum_{j=1}^{101} x_j = 209.306$ and $\sum_{j=1}^{101} x_j^{-1} = 77.7896$. (Hint: Use the results of Exercises 7.20 and 5.13.)*

Example 8.8. The data consist of independent random samples $\{X_j\}_{j=1}^{n_X}$ and $\{Y_j\}_{j=1}^{n_Y}$ from different populations distributed as $N\left(\mu_X, \sigma^2\right)$ and $N\left(\mu_Y, \sigma^2\right)$. Notice that the variances of the two populations are assumed to be the same, but that the common value is unspecified. We test $H_0 : \mu_X = \mu_Y$ against one- and two-sided alternatives, σ^2 being a nuisance parameter. The likelihood function for the two independent samples is proportional to

$$\left(\sigma^2\right)^{-n/2} \exp\left\{-\frac{1}{2\sigma^2}\left[\sum_{j=1}^{n_X}(X_j - \mu_X)^2 + \sum_{j=1}^{n_Y}(Y_j - \mu_Y)^2\right]\right\},$$

where $n := n_X + n_Y$. Unrestricted m.l.e.s are $\hat{\mu}_{n_X} = \bar{X}, \hat{\mu}_{n_Y} = \bar{Y}$, and

$$\hat{\sigma}_n^2 = n^{-1}\left[\sum_{j=1}^{n_X}\left(X_j - \bar{X}\right)^2 + \sum_{j=1}^{n_Y}\left(Y_j - \bar{Y}\right)^2\right]$$

---

[9] For proofs of properties of this and other tests for normal distributions see chapter 5 of Lehmann and Romano (2005).

(the "pooled-sample" variance). Imposing $H_0 : \mu_X - \mu_Y = 0$, the restricted estimators are

$$\breve{\mu}_n = n^{-1}\left(\sum_{j=1}^{n_X} X_j + \sum_{j=1}^{n_Y} Y_j\right) = \nu_X \bar{X} + \nu_Y \bar{Y},$$

$$\breve{\sigma}_n^2 = n^{-1}\left[\sum_{j=1}^{n_X}(X_j - \breve{\mu}_n)^2 + \sum_{j=1}^{n_Y}(Y_j - \breve{\mu}_n)^2\right],$$

where $\nu_X := n_X/n$ and $\nu_Y := n_Y/n$. Substituting and simplifying, the l.r. statistic is

$$\Lambda_n(\mathbf{X};0) = \left(\frac{\breve{\sigma}_n^2}{\hat{\sigma}_n^2}\right)^{-n/2}.$$

Small values (values less than or equal to some λ_α) correspond to large values of

$$\frac{\breve{\sigma}_n^2}{\hat{\sigma}_n^2} = \frac{\sum_{j=1}^{n_X}(X_j - \nu_X\bar{X} - \nu_Y\bar{Y})^2 + \sum_{j=1}^{n_Y}(Y_j - \nu_X\bar{X} - \nu_Y\bar{Y})^2}{n\hat{\sigma}_n^2}$$

$$= \frac{\sum_{j=1}^{n_X}\left[(X_j - \bar{X}) + \nu_Y(\bar{X} - \bar{Y})\right]^2 + \sum_{j=1}^{n_Y}\left[(Y_j - \bar{Y}) + \nu_X(\bar{Y} - \bar{X})\right]^2}{n\hat{\sigma}_n^2}$$

$$= \frac{\sum_{j=1}^{n_X}(X_j - \bar{X})^2 + \sum_{j=1}^{n_Y}(Y_j - \bar{Y})^2 + \left(n_X\nu_Y^2 + n_Y\nu_X^2\right)(\bar{X} - \bar{Y})^2}{n\hat{\sigma}_n^2}$$

$$= 1 + \frac{\nu_X\nu_Y(\bar{X} - \bar{Y})^2}{\hat{\sigma}_n^2},$$

and therefore to large *absolute* values of

$$T(\mathbf{X},\mathbf{Y}) := \sqrt{n-2}\sqrt{\nu_X\nu_Y}\frac{(\bar{X} - \bar{Y})}{\hat{\sigma}_n}.$$

We can write this alternatively as

$$T(\mathbf{X},\mathbf{Y}) = \left[\frac{\bar{X} - \bar{Y}}{\sigma/\sqrt{n\nu_X\nu_Y}}\right] \div \left[\frac{\sqrt{n\hat{\sigma}_n^2/\sigma^2}}{\sqrt{n-2}}\right] \qquad (8.8)$$

Under the null hypothesis of equal means and the maintained hypothesis of normality with common variance

$$\bar{X} - \bar{Y} \sim N\left[0, \sigma^2\left(n_X^{-1} + n_Y^{-1}\right)\right] = N\left(0, \frac{\sigma^2}{n\nu_X\nu_Y}\right),$$

so that the first bracketed quantity in (8.8) is distributed as $N(0,1)$ under H_0. The square of the second bracketed quantity can be expressed as

$$\frac{n\hat{\sigma}_n^2}{\sigma^2(n-2)} = \frac{(n_X-1)S_X^2/\sigma^2 + (n_Y-1)S_Y^2/\sigma^2}{n-2}.$$

Here the numerator is the sum of independent chi-squared variates with $n_X - 1$ and $n_Y - 1$ d.f., which is distributed (independently of $\bar{X} - \bar{Y}$) as $\chi^2(n-2)$. Therefore, under the null $T(\mathbf{X}, \mathbf{Y})$ is distributed as Student's t with $n - 2 = n_X + n_Y - 2$ d.f. Rejection regions

$$\mathcal{R}_\alpha^+ = \{(\mathbf{x}, \mathbf{y}) : T(\mathbf{x}, \mathbf{y}) \geq t^\alpha\}$$

$$\mathcal{R}_\alpha^- = \{(\mathbf{x}, \mathbf{y}) : T(\mathbf{x}, \mathbf{y}) \leq -t^\alpha\}$$

$$\mathcal{R}_\alpha = \left\{(\mathbf{x}, \mathbf{y}) : |T(\mathbf{x}, \mathbf{y})| \geq t^{\alpha/2}\right\}$$

would be used for alternatives $H_1^+ : \mu_X > \mu_Y$, $H_1^- : \mu_X < \mu_Y$, and $H_1 : \mu_X \neq \mu_Y$, respectively. These tests are known to be u.m.p.-unbiased.

Example 8.9. Data $\{x_j, y_j\}_{j=1}^n$ were obtained by sampling at random from a *bivariate normal* population with mean vector $\boldsymbol{\mu} = (\mu_X, \mu_Y)'$ and covariance matrix

$$\boldsymbol{\Sigma} = \begin{pmatrix} \sigma_X^2 & \sigma_{XY} \\ \sigma_{XY} & \sigma_Y^2 \end{pmatrix}.$$

We test $H_0 : \mu_X - \mu_Y = 0$ *vs.* $H_1 : \mu_X - \mu_Y \neq 0$ with $\boldsymbol{\Sigma}$, unspecified, as nuisance parameter. This is referred to as a "paired-sample" test of the difference in means. The setup differs from that in Example 8.8 in that (i) there are the same number of observations on X and Y and (ii) the X, Y pairs may be dependent. The situation often arises when X_j and Y_j refer to characteristics of the *same* entity before and after some event; e.g., to (logs of) blood cholesterol levels of individual j before and after some treatment. Alternatively, X_j and Y_j could represent continuously compounded rates of return of two stocks in a bivariate sample $\{X_j, Y_j\}_{j=1}^n$ over n time steps, assumed to be i.i.d. Defining new variables $\{D_j := X_j - Y_j\}_{j=1}^n$, which are i.i.d. as

$$N\left(\mu_D := \mu_X - \mu_Y, \sigma_D^2 := \sigma_X^2 - 2\sigma_{XY} + \sigma_Y^2\right),$$

a likelihood-ratio test of H_0 can be carried out as a simple one-sample "t" test of $H_0 : \mu_D = 0$. The test statistic is $T(\mathbf{X}, \mathbf{Y}) = \sqrt{n}\bar{D}/S_D$, distributed as Student's $t(n-1)$, and the rejection region is

$$\mathcal{R}_\alpha = \left\{(\mathbf{x}, \mathbf{y}) : |T(\mathbf{x}, \mathbf{y})| \geq t^{\alpha/2}\right\},$$

where $\bar{D} := \bar{X} - \bar{Y}$ and $S_D^2 := (n-1)^{-1}\sum_{j=1}^n (D_j - \bar{D})^2$. The test is u.m.p.-unbiased, as are the corresponding one-sided tests.

Exercise 8.10. *The data consist of two random samples, $\{X_j\}_{j=1}^{n_X}$ and $\{Y_j\}_{j=1}^{n_Y}$, from different populations distributed independently as $N\left(\mu_X, \sigma_X^2\right)$ and $N\left(\mu_Y, \sigma_Y^2\right)$. Show that an exact size-α l.r. test of H_0 : $\sigma_X^2 = \sigma_Y^2$ vs. $H_1^+ : \sigma_X^2 > \sigma_Y^2$ (without restriction on the means) is based on rejection region $\mathcal{R}_\alpha = \{(\mathbf{x}, \mathbf{y}) : R(\mathbf{x}, \mathbf{y}) \geq f^\alpha\}$, where $R\left(\mathbf{X}, \mathbf{Y}\right) := S_X^2 / S_Y^2$ is the ratio of sample variances and f^α is the upper-α quantile of the F distribution with $n_X - 1$ and $n_Y - 1$ degrees of freedom. Likewise, show that an exact size-α test of H_0 against the two-sided alternative $H_1 : \sigma_X^2 \neq \sigma_Y^2$ is based on*

$$\mathcal{R}_\alpha = \{(\mathbf{x}, \mathbf{y}) : R(\mathbf{x}, \mathbf{y}) \leq f_{p\alpha}\} \cup \left\{(\mathbf{x}, \mathbf{y}) : R(\mathbf{x}, \mathbf{y}) \geq f^{(1-p)\alpha}\right\}$$

for some $p \in (0, 1)$. With the appropriate choice of p these well known F tests for equality of variances are in fact u.m.p.-unbiased.

8.6.1.2 *Asymptotic Tests Based on the L.R. Principle*

When event $\Lambda\left(\mathbf{X}; \phi_0\right) \leq \lambda_\alpha$ cannot be restated in terms of a statistic with known distribution, appeal can often be made to asymptotic theory. For this the maintained hypothesis will be restricted to parametric families that satisfy the regularity conditions for m.l. estimation (page 494). Let $\{f(x; \boldsymbol{\theta}) : \boldsymbol{\theta} \in \boldsymbol{\Theta} \subset \Re^k\}$ be such a family, with $\boldsymbol{\theta}' = \left(\boldsymbol{\phi}', \boldsymbol{\psi}'\right)$ comprising k_0-vector $\boldsymbol{\phi} \in \boldsymbol{\Phi}$ and $(k - k_0)$-vector $\boldsymbol{\psi} \in \boldsymbol{\Psi}$ of restricted and nuisance parameters, respectively. Null hypothesis $H_0 : \boldsymbol{\theta} \in \boldsymbol{\Theta}_0 = \{\boldsymbol{\phi}_0\} \times \boldsymbol{\Psi}$ equates elements of $\boldsymbol{\phi}$ to known constants.[10] The alternative is simply that H_0 is false; i.e., $H_1 : \boldsymbol{\theta} \in \boldsymbol{\Theta}\backslash\boldsymbol{\Theta}_0 = (\boldsymbol{\Phi}\backslash\{\boldsymbol{\phi}_0\}) \times \boldsymbol{\Psi}$. If there are no nuisance parameters, then $k = k_0$ and $\boldsymbol{\Theta}_0 = \{\boldsymbol{\phi}_0\}$. We will prove the following.

Theorem 8.1.

(i) *Under the conditions referred to above and when H_0 is true, the statistic $-2 \log \Lambda\left(\mathbf{X}; \phi_0\right)$ converges in distribution to chi-squared with k_0 degrees of freedom, where k_0 is the number of restrictions imposed by H_0.*

(ii) *When H_1 is true $-2 \log \Lambda\left(\mathbf{X}; \phi_0\right)$ diverges stochastically, in the sense that*

$$\lim_{n\to\infty} \Pr\left[-2 \log \Lambda\left(\mathbf{X}; \phi_0\right) > b\right] = 1$$

for each $b \in \Re$.

---

[10] Hypotheses that impose certain other types of restrictions can be fit into this setup by reparameterizing. For example in the two-sample problem the null $H_0 : \mu_Y = \mu_X$ can be handled by setting $\mu_Y = \mu_X + \Delta$ and testing $\Delta = 0$. μ_X then becomes a nuisance parameter. Tests of these and other sorts of parametric restrictions are discussed in Section 8.6.5.

It follows from the two assertions of the theorem that a size-α test of H_0 *vs.* H_1 with rejection region $\mathcal{R}_\alpha = \{\mathbf{x} : -2\log\Lambda\left(\mathbf{x}; \phi_0\right) \geq c^\alpha\}$ is asymptotically valid and consistent, where c^α is the upper-α quantile of the chi-squared distribution with k_0 degrees of freedom.

When there are no nuisance parameters (the case $k_0 = k$), the fact that $-2\log\Lambda\left(\mathbf{X}; \phi_0\right) \equiv -2\log\Lambda\left(\mathbf{X}; \theta_0\right) \rightsquigarrow \chi^2\left(k_0\right)$ under H_0 follows from Theorem 7.5. There we showed in the context of interval estimation that $-2\log\Lambda\left(\mathbf{X}; \theta\right) \rightsquigarrow \chi^2\left(k\right)$ for each fixed $\theta \in \Theta$. When nuisance parameters *are* present, developing the limiting distribution under H_0 is considerably more complicated, but we give the main thread of the argument below.[11]

First, some abbreviated notation and terminology. In what follows (i) $\mathcal{L}\left(\theta\right) := \log L\left(\mathbf{X}; \theta\right)$ represents the log-likelihood function; (ii) $\hat{\theta}_n = \left(\hat{\phi}_n', \hat{\psi}_n'\right)'$ and $\check{\theta}_n = \left(\phi_0', \check{\psi}_n'\right)'$ are the k-vectors of unrestricted and restricted maximum likelihood estimators (m.l.e.s); and (iii) $\theta_0 := \left(\phi_0', \psi'\right)'$, where ψ is the true but unknown value of the unrestricted parameter vector. The gradient vector (vector of partial derivatives) and Hessian matrix (matrix of second derivatives) of $\mathcal{L}\left(\theta\right)$ evaluated at some vector $\mathbf{t}$ are expressed as

$$\frac{\partial \log L\left(\mathbf{X}; \theta\right)}{\partial\theta}\Big|_{\theta=\mathbf{t}} =: \mathcal{L}_\theta\left(\mathbf{t}\right)$$

$$\frac{\partial^2 \log L\left(\mathbf{X}; \theta\right)}{\partial\theta\partial\theta'}\Big|_{\theta=\mathbf{t}} =: \mathcal{L}_{\theta\theta'}\left(\mathbf{t}\right).$$

Note that "$\mathcal{L}_\theta$" and "$\mathcal{L}_{\theta'}$" denote a column vector and a row vector, respectively, and that $\mathcal{L}_{\theta\theta'}$ is the matrix obtained by replacing each element of $\mathcal{L}_\theta$ by the row vector comprising its derivatives with respect to elements of θ. It is customary to refer to gradient $\mathcal{L}_\theta$ as the *score* function. Equating the score to the zero vector gives the usual first-order conditions or likelihood equations, whose solutions in regular cases are the m.l.e.s; thus, $\mathcal{L}_\theta(\hat{\theta}_n) \equiv \mathbf{0}$. In the same way, the restricted m.l.e.s satisfy the $k - k_0$ equations $\mathcal{L}_\psi(\check{\theta}_n) = \mathbf{0}$. We are now ready to prove the theorem.

Proof. (i) Putting

$$-2\log\Lambda\left(\mathbf{X}; \phi_0\right) = 2\left[\mathcal{L}\left(\hat{\theta}_n\right) - \mathcal{L}\left(\check{\theta}_n\right)\right], \tag{8.9}$$

expand $\mathcal{L}\left(\check{\theta}_n\right)$ about $\hat{\theta}_n$ as

$$\mathcal{L}\left(\check{\theta}_n\right) = \mathcal{L}\left(\hat{\theta}_n\right) + \mathcal{L}_{\theta'}\left(\hat{\theta}_n\right)\left(\check{\theta}_n - \hat{\theta}_n\right) + \frac{1}{2}\left(\check{\theta}_n - \hat{\theta}_n\right)'\mathcal{L}_{\theta\theta'}\left(\bar{\theta}_n\right)\left(\check{\theta}_n - \hat{\theta}_n\right)$$

$$= \mathcal{L}\left(\hat{\theta}_n\right) + \frac{1}{2}\left(\hat{\theta}_n - \check{\theta}_n\right)'\mathcal{L}_{\theta\theta'}\left(\bar{\theta}_n\right)\left(\hat{\theta}_n - \check{\theta}_n\right),$$

---

[11] One can consult Wilks (1962) and Cox and Hinkley (1974) for different approaches.

where $\bar{\boldsymbol{\theta}}_n$ is a point on the line segment connecting $\hat{\boldsymbol{\theta}}_n$ and $\check{\boldsymbol{\theta}}_n$, and where the second equality follows from $\mathcal{L}_{\boldsymbol{\theta}'}(\hat{\boldsymbol{\theta}}_n) \equiv \mathbf{0}'$. Substituting in (8.9) and writing out the vectors and Hessian matrix give for $-2\log\Lambda(\mathbf{X};\boldsymbol{\phi}_0)$

$$-\left(\hat{\boldsymbol{\phi}}_n'-\boldsymbol{\phi}_0', \hat{\boldsymbol{\psi}}_n'-\check{\boldsymbol{\psi}}_n'\right)\begin{pmatrix}\mathcal{L}_{\boldsymbol{\phi}\boldsymbol{\phi}'}(\bar{\boldsymbol{\theta}}_n) & \mathcal{L}_{\boldsymbol{\phi}\boldsymbol{\psi}'}(\bar{\boldsymbol{\theta}}_n) \\ \mathcal{L}_{\boldsymbol{\psi}\boldsymbol{\phi}'}(\bar{\boldsymbol{\theta}}_n) & \mathcal{L}_{\boldsymbol{\psi}\boldsymbol{\psi}'}(\bar{\boldsymbol{\theta}}_n)\end{pmatrix}\begin{pmatrix}\hat{\boldsymbol{\phi}}_n-\boldsymbol{\phi}_0 \\ \hat{\boldsymbol{\psi}}_n-\check{\boldsymbol{\psi}}_n\end{pmatrix}$$

$$=-\left(\hat{\boldsymbol{\phi}}_n-\boldsymbol{\phi}_0\right)'\mathcal{L}_{\boldsymbol{\phi}\boldsymbol{\phi}'}(\bar{\boldsymbol{\theta}}_n)\left(\hat{\boldsymbol{\phi}}_n-\boldsymbol{\phi}_0\right)-\left(\hat{\boldsymbol{\psi}}_n-\check{\boldsymbol{\psi}}_n\right)'\mathcal{L}_{\boldsymbol{\psi}\boldsymbol{\phi}'}(\bar{\boldsymbol{\theta}}_n)\left(\hat{\boldsymbol{\phi}}_n-\boldsymbol{\phi}_0\right)$$

$$-\left(\hat{\boldsymbol{\phi}}_n-\boldsymbol{\phi}_0\right)'\mathcal{L}_{\boldsymbol{\phi}\boldsymbol{\psi}'}(\bar{\boldsymbol{\theta}}_n)\left(\hat{\boldsymbol{\psi}}_n-\check{\boldsymbol{\psi}}_n\right)-\left(\hat{\boldsymbol{\psi}}_n-\check{\boldsymbol{\psi}}_n\right)'\mathcal{L}_{\boldsymbol{\psi}\boldsymbol{\psi}'}(\bar{\boldsymbol{\theta}}_n)\left(\hat{\boldsymbol{\psi}}_n-\check{\boldsymbol{\psi}}_n\right).$$

Since $-n^{-1}\mathcal{L}_{\boldsymbol{\theta}\boldsymbol{\theta}'}(\boldsymbol{\theta}_0) \to \mathcal{I}(\boldsymbol{\theta}_0)$ and $\bar{\boldsymbol{\theta}}_n \to \boldsymbol{\theta}_0$ in probability under H_0, and since the second derivatives are a.s. continuous, it follows that $-2\log\Lambda(\mathbf{X};\boldsymbol{\phi}_0)$ has the same limiting distribution as that of

$$\sqrt{n}\left(\hat{\boldsymbol{\phi}}_n - \boldsymbol{\phi}_0\right)'\mathcal{I}_{\boldsymbol{\phi}\boldsymbol{\phi}'}\sqrt{n}\left(\hat{\boldsymbol{\phi}}_n - \boldsymbol{\phi}_0\right)$$

$$+\sqrt{n}\left(\hat{\boldsymbol{\psi}}_n - \check{\boldsymbol{\psi}}_n\right)'\mathcal{I}_{\boldsymbol{\psi}\boldsymbol{\phi}'}\sqrt{n}\left(\hat{\boldsymbol{\phi}}_n - \boldsymbol{\phi}_0\right)$$

$$+\sqrt{n}\left(\hat{\boldsymbol{\phi}}_n - \boldsymbol{\phi}_0\right)'\mathcal{I}_{\boldsymbol{\phi}\boldsymbol{\psi}'}\sqrt{n}\left(\hat{\boldsymbol{\psi}}_n - \check{\boldsymbol{\psi}}_n\right)$$

$$+\sqrt{n}\left(\hat{\boldsymbol{\psi}}_n - \check{\boldsymbol{\psi}}_n\right)'\mathcal{I}_{\boldsymbol{\psi}\boldsymbol{\psi}'}\sqrt{n}\left(\hat{\boldsymbol{\psi}}_n - \check{\boldsymbol{\psi}}_n\right), \tag{8.10}$$

where $\mathcal{I}_{\boldsymbol{\phi}\boldsymbol{\phi}'}$, $\mathcal{I}_{\boldsymbol{\psi}\boldsymbol{\phi}'}$, etc. are submatrices of the Fisher information from a single observation; that is,

$$\mathcal{I}=\begin{pmatrix}\mathcal{I}_{\boldsymbol{\phi}\boldsymbol{\phi}'} & \mathcal{I}_{\boldsymbol{\phi}\boldsymbol{\psi}'} \\ \mathcal{I}_{\boldsymbol{\psi}\boldsymbol{\phi}'} & \mathcal{I}_{\boldsymbol{\psi}\boldsymbol{\psi}'}\end{pmatrix}.$$

For later reference, the inverse can be represented as[12]

$$\mathcal{I}^{-1}=\begin{pmatrix}\mathcal{I}^{\boldsymbol{\phi}\boldsymbol{\phi}'} & \mathcal{I}^{\boldsymbol{\phi}\boldsymbol{\psi}'} \\ \mathcal{I}^{\boldsymbol{\psi}\boldsymbol{\phi}'} & \mathcal{I}^{\boldsymbol{\psi}\boldsymbol{\psi}'}\end{pmatrix},$$

with

$$\mathcal{I}^{\boldsymbol{\phi}\boldsymbol{\phi}'} = \left(\mathcal{I}_{\boldsymbol{\phi}\boldsymbol{\phi}'} - \mathcal{I}_{\boldsymbol{\phi}\boldsymbol{\psi}'}\mathcal{I}_{\boldsymbol{\psi}\boldsymbol{\psi}'}^{-1}\mathcal{I}_{\boldsymbol{\psi}\boldsymbol{\phi}'}\right)^{-1}. \tag{8.11}$$

An expression for $\sqrt{n}\left(\hat{\boldsymbol{\psi}}_n - \check{\boldsymbol{\psi}}_n\right)$ can now be developed as follows. Expanding $\mathcal{L}_{\boldsymbol{\theta}}(\boldsymbol{\theta}_0)$ about $\hat{\boldsymbol{\theta}}_n$ and using $\mathcal{L}_{\boldsymbol{\theta}}\left(\hat{\boldsymbol{\theta}}_n\right) = \mathbf{0}$ give $\mathcal{L}_{\boldsymbol{\theta}}(\boldsymbol{\theta}_0) =$

---

[12]Formulas for inverses of partitioned matrices are given in many texts on matrix algebra; e.g., Schott (1997).

$\mathcal{L}_{\theta\theta'}\left(\tilde{\boldsymbol{\theta}}_n\right)\left(\boldsymbol{\theta}_0 - \hat{\boldsymbol{\theta}}_n\right)$, where $\tilde{\boldsymbol{\theta}}_n$ is intermediate between $\boldsymbol{\theta}_0$ and $\hat{\boldsymbol{\theta}}_n$. Written out, this is

$$\begin{pmatrix} \mathcal{L}_\phi\left(\boldsymbol{\theta}_0\right) \\ \mathcal{L}_\psi\left(\boldsymbol{\theta}_0\right) \end{pmatrix} = \begin{pmatrix} \mathcal{L}_{\phi\phi'}\left(\tilde{\boldsymbol{\theta}}_n\right)\left(\boldsymbol{\phi}_0 - \hat{\boldsymbol{\phi}}_n\right) + \mathcal{L}_{\phi\psi'}\left(\tilde{\boldsymbol{\theta}}_n\right)\left(\boldsymbol{\psi} - \hat{\boldsymbol{\psi}}_n\right) \\ \mathcal{L}_{\psi\phi'}\left(\tilde{\boldsymbol{\theta}}_n\right)\left(\boldsymbol{\phi}_0 - \hat{\boldsymbol{\phi}}_n\right) + \mathcal{L}_{\psi\psi'}\left(\tilde{\boldsymbol{\theta}}_n\right)\left(\boldsymbol{\psi} - \hat{\boldsymbol{\psi}}_n\right) \end{pmatrix},$$

and the second member of this pair of vector equations yields the following expression for unrestricted estimator $\hat{\boldsymbol{\psi}}_n$:

$$\hat{\boldsymbol{\psi}}_n = \boldsymbol{\psi} - \mathcal{L}_{\psi\psi'}\left(\tilde{\boldsymbol{\theta}}_n\right)^{-1}\mathcal{L}_\psi\left(\boldsymbol{\theta}_0\right) - \mathcal{L}_{\psi\psi'}\left(\tilde{\boldsymbol{\theta}}_n\right)^{-1}\mathcal{L}_{\psi\phi'}\left(\tilde{\boldsymbol{\theta}}_n\right)\left(\hat{\boldsymbol{\phi}}_n - \boldsymbol{\phi}_0\right). \tag{8.12}$$

Next, we get the following expression for restricted estimator $\check{\boldsymbol{\psi}}_n$ by expanding $\mathcal{L}_\psi\left(\boldsymbol{\theta}_0\right)$ about $\check{\boldsymbol{\theta}}_n$ and using $\mathcal{L}_\psi(\check{\boldsymbol{\theta}}_n) \equiv \mathbf{0}$:

$$\check{\boldsymbol{\psi}}_n = \boldsymbol{\psi} - \mathcal{L}_{\psi\psi'}\left(\check{\boldsymbol{\theta}}_n\right)^{-1}\mathcal{L}_\psi\left(\boldsymbol{\theta}_0\right). \tag{8.13}$$

Here $\check{\boldsymbol{\theta}}_n$ is another intermediate point. Finally, subtracting (8.13) from (8.12) and multiplying by $\sqrt{n}$ give

$$\sqrt{n}\left(\hat{\boldsymbol{\psi}}_n - \check{\boldsymbol{\psi}}_n\right) = \left[n\mathcal{L}_{\psi\psi'}\left(\check{\boldsymbol{\theta}}_n\right)^{-1} - n\mathcal{L}_{\psi\psi'}\left(\tilde{\boldsymbol{\theta}}_n\right)^{-1}\right]n^{-1/2}\mathcal{L}_\psi\left(\boldsymbol{\theta}_0\right)$$
$$- n\mathcal{L}_{\psi\psi'}\left(\tilde{\boldsymbol{\theta}}_n\right)^{-1}n^{-1}\mathcal{L}_{\psi\phi'}\left(\tilde{\boldsymbol{\theta}}_n\right)\sqrt{n}\left(\hat{\boldsymbol{\phi}}_n - \boldsymbol{\phi}_0\right).$$

The first term converges in probability to zero since both $n\mathcal{L}_{\psi\psi'}\left(\check{\boldsymbol{\theta}}_n\right)^{-1} = \left[n^{-1}\mathcal{L}_{\psi\psi'}\left(\check{\boldsymbol{\theta}}_n\right)\right]^{-1}$ and $n\mathcal{L}_{\psi\psi'}\left(\tilde{\boldsymbol{\theta}}_n\right)^{-1}$ converge in probability to $-\mathcal{I}_{\psi\psi'}^{-1}$, and since

$$n^{-1/2}\mathcal{L}_\psi\left(\boldsymbol{\theta}_0\right) = n^{-1/2}\sum_{j=1}^n \frac{\partial f\left(X_j;\boldsymbol{\theta}_0\right)}{\partial\psi} \rightsquigarrow N\left(0,\mathcal{I}_{\psi\psi'}\right).$$

Thus, $\sqrt{n}\left(\hat{\boldsymbol{\psi}}_n - \check{\boldsymbol{\psi}}_n\right) \approx -\mathcal{I}_{\psi\psi'}^{-1}\mathcal{I}_{\psi\phi'}\sqrt{n}\left(\hat{\boldsymbol{\phi}}_n - \boldsymbol{\phi}_0\right)$, where "$\approx$" indicates asymptotic equivalence, meaning that the difference between the two sides is $o_P\left(1\right)$ and thus converges in probability to zero. Now plugging into (8.10) and simplifying, we find that, in the same asymptotic sense,

$$-2\log\Lambda\left(\mathbf{X};\boldsymbol{\phi}_0\right) \approx \sqrt{n}\left(\hat{\boldsymbol{\phi}}_n - \boldsymbol{\phi}_0\right)'\left(\mathcal{I}_{\phi\phi'} - \mathcal{I}_{\phi\psi'}\mathcal{I}_{\psi\psi'}^{-1}\mathcal{I}_{\psi\phi'}\right)\sqrt{n}\left(\hat{\boldsymbol{\phi}}_n - \boldsymbol{\phi}_0\right)$$
$$= \sqrt{n}\left(\hat{\boldsymbol{\phi}}_n - \boldsymbol{\phi}_0\right)'\left(\mathcal{I}^{\phi\phi'}\right)^{-1}\sqrt{n}\left(\hat{\boldsymbol{\phi}}_n - \boldsymbol{\phi}_0\right). \tag{8.14}$$

But the theory of m.l. estimators tells us that $\sqrt{n}\left(\hat{\boldsymbol{\theta}}_n - \boldsymbol{\theta}_0\right) \rightsquigarrow N\left(0,\mathcal{I}^{-1}\right)$ under H_0, so that, in particular, $\sqrt{n}\left(\hat{\boldsymbol{\phi}}_n - \boldsymbol{\phi}_0\right) \rightsquigarrow N\left(0,\mathcal{I}^{\phi\phi'}\right)$. Thus, it follows from Mann–Wald and Theorem 5.1 that $-2\log\Lambda\left(\mathbf{X};\boldsymbol{\phi}_0\right)\rightsquigarrow\chi^2\left(k_0\right)$ as $n\to\infty$ under the null in the case that H_0 restricts $0 < k_0 < k$ parameters.

(ii) Discovering the limiting behavior of $-2\log\Lambda\left(\mathbf{X};\boldsymbol{\phi}_0\right)$ under H_1 is more straightforward. If $k_0 = k$, so that there are no nuisance parameters, then $\hat{\boldsymbol{\theta}}_n \equiv \hat{\boldsymbol{\phi}}_n \to^p \boldsymbol{\phi} \neq \boldsymbol{\phi}_0 \equiv \boldsymbol{\theta}_0$, and it is clear from (8.14) that $-2\log\Lambda\left(\mathbf{X};\boldsymbol{\phi}_0\right)\to +\infty$. If $k_0 < k$ the unrestricted m.l.e. $\hat{\boldsymbol{\theta}}_n$ still converges in probability to true value $\boldsymbol{\theta} = \left(\boldsymbol{\phi}',\boldsymbol{\psi}'\right)'$, but what happens under H_1 to the restricted m.l.e.s , $\check{\boldsymbol{\theta}}_n = \left(\boldsymbol{\phi}_0',\check{\boldsymbol{\psi}}_n'\right)'$, is more problematic. Suppose first that the m.l.e.s do exist for all sufficiently large n and converge to some $\boldsymbol{\theta}_1 := (\boldsymbol{\phi}_0',\boldsymbol{\psi}_1')'$ in the interior of $\boldsymbol{\Theta} = \boldsymbol{\Phi}\times\boldsymbol{\Psi}$. That is, suppose we can find $\check{\boldsymbol{\psi}}_n \in \boldsymbol{\Psi}$ that satisfies $\mathcal{L}_{\boldsymbol{\psi}}(\check{\boldsymbol{\theta}}_n) := \mathcal{L}_{\boldsymbol{\psi}}\left(\boldsymbol{\phi}_0,\check{\boldsymbol{\psi}}_n\right) = \mathbf{0}$ with negative-definite Hessian $\mathcal{L}_{\boldsymbol{\psi}\boldsymbol{\psi}'}\left(\boldsymbol{\phi}_0,\check{\boldsymbol{\psi}}_n\right)$, and that $\check{\boldsymbol{\psi}}_n \to \boldsymbol{\psi}_1$ in probability, where $\boldsymbol{\psi}_1$ is some point in the interior of $\boldsymbol{\Psi}$. Then

$$-n^{-1}\log\Lambda\left(\mathbf{X};\boldsymbol{\phi}_0\right) = n^{-1}\sum_{j=1}^{n}\left[\log f(X_j;\hat{\boldsymbol{\theta}}_n) - \log f(X_j;\check{\boldsymbol{\theta}}_n)\right]$$

converges in probability under H_1 to

$$E[\log f(X;\boldsymbol{\theta}) - \log f(X;\boldsymbol{\theta}_1)] = \int [\log f(x;\boldsymbol{\theta}) - \log f(x;\boldsymbol{\theta}_1)]\cdot dF(x;\boldsymbol{\theta}),$$

and Equation (7.23) shows this quantity to be positive. Thus, $-2\log\Lambda\left(\mathbf{X};\boldsymbol{\phi}_0\right)\to +\infty$ a.s. under H_1. Alternatively, if for some n the problem $\max_{\boldsymbol{\psi}\in\boldsymbol{\Psi}}\left\{L\left(\mathbf{X};\boldsymbol{\phi}_0,\boldsymbol{\psi}\right)\right\}$ has no interior solution, we simply set $-2\log\Lambda\left(\mathbf{X};\boldsymbol{\phi}_0\right) = +\infty$. In either case the l.r. test with $\mathcal{R}_\alpha = \left\{\Lambda_n\left(\mathbf{x}\right) : -2\log\Lambda\left(\mathbf{x};\boldsymbol{\phi}_0\right)\geq c^\alpha\right\}$ is indeed consistent.[13] □

Section 8.6.4 gives an example that compares the l.r. test with the other two asymptotic procedures, Lagrange-multiplier and Wald, to which we now turn.

8.6.2 *The Lagrange-multiplier Test*

As we have seen, performing the likelihood-ratio test requires finding both the unrestricted and restricted maxima of the likelihood function. Finding

---

[13]There remains the possibility that an interior solution exists for each sufficiently large n but that the sequence $\{\check{\boldsymbol{\psi}}_n\}$ of restricted estimates does not converge. Consistency could still be established under a stronger identification condition; e.g., that $E\left[\log f\left(X;\boldsymbol{\phi},\boldsymbol{\psi}\right) - \log f\left(X;\boldsymbol{\phi}_0,\boldsymbol{\psi}\right)\right]$ be bounded away from zero for all $\boldsymbol{\psi}\in\boldsymbol{\Psi}$.

the maximum under the restriction $\phi = \phi_0$ can be posed as a Lagrange-multiplier problem, as follows. Take as the Lagrangian function

$$\mathbb{L} := \log L\left(\mathbf{X}; \boldsymbol{\theta}\right) - \boldsymbol{\lambda}'_n\left(\boldsymbol{\phi} - \boldsymbol{\phi}_0\right),$$

where $\boldsymbol{\lambda}_n$ is a k_0-vector of undetermined multipliers (indexed by sample size n) and $\boldsymbol{\theta}' = \left(\boldsymbol{\phi}', \boldsymbol{\psi}'\right)$ is the vector of k_0 restricted and $k - k_0$ nuisance parameters. The idea behind the test is that the "size" of $\boldsymbol{\lambda}_n$ (e.g., $|\boldsymbol{\lambda}_n| = \sqrt{\boldsymbol{\lambda}'_n \boldsymbol{\lambda}_n}$), once solved for as a function of the data, should indicate the degree to which the constraint imposed by H_0 is binding and therefore the degree to which the data are consistent with it. Remarkably, developing the distribution theory for $\boldsymbol{\lambda}_n = \boldsymbol{\lambda}_n\left(\mathbf{X}\right)$ under H_0 turns out to be relatively easy when (as we continue to assume) our sample is from one of the parametric families that satisfy the regularity conditions for m.l. estimation.

Recalling notation $\mathcal{L}\left(\boldsymbol{\theta}\right) := \log L\left(\mathbf{X}; \boldsymbol{\theta}\right)$ and $\mathcal{L}_{\phi}\left(\boldsymbol{\theta}\right), \mathcal{L}_{\psi}\left(\boldsymbol{\theta}\right)$ for the derivatives, the first-order conditions for a maximum of $\mathbb{L}$ are

$$\frac{\partial \mathbb{L}}{\partial \boldsymbol{\phi}} = \mathcal{L}_{\phi}\left(\boldsymbol{\theta}\right) - \boldsymbol{\lambda}_n = \mathbf{0}$$

$$\frac{\partial \mathbb{L}}{\partial \boldsymbol{\psi}} = \mathcal{L}_{\psi}\left(\boldsymbol{\theta}\right) = \mathbf{0}$$

$$\frac{\partial \mathbb{L}}{\partial \boldsymbol{\lambda}_n} = \boldsymbol{\phi} - \boldsymbol{\phi}_0 = \mathbf{0}.$$

Substituting $\boldsymbol{\phi} = \boldsymbol{\phi}_0$ into the second vector equation presents us with $k - k_0$ equations in as many unknowns: $\mathcal{L}_{\psi}\left(\boldsymbol{\phi}_0, \boldsymbol{\psi}\right) = \mathbf{0}$. In regular cases these can be solved for $\breve{\boldsymbol{\psi}}_n$, a restricted estimator of the nuisance parameters. Evaluating $\mathcal{L}_{\phi}\left(\boldsymbol{\theta}\right)$ in the first equation at $\breve{\boldsymbol{\theta}}_n := \left(\boldsymbol{\phi}'_0, \breve{\boldsymbol{\psi}}'_n\right)'$, then gives a solution for $\boldsymbol{\lambda}_n$ as

$$\boldsymbol{\lambda}_n\left(\mathbf{X}\right) = \mathcal{L}_{\phi}\left(\breve{\boldsymbol{\theta}}_n\right) = \sum_{j=1}^{n} \frac{\partial}{\partial \boldsymbol{\phi}} \log f\left(X_j; \breve{\boldsymbol{\theta}}_n\right).$$

Thus, in tests of $H_0 : \boldsymbol{\phi} = \boldsymbol{\phi}_0$ the Lagrange multiplier turns out to be nothing more than the score with respect to $\boldsymbol{\phi}$, evaluated at the restricted estimators. A test of $H_0 : \boldsymbol{\phi} = \boldsymbol{\phi}_0$ based on $\boldsymbol{\lambda}_n\left(\mathbf{X}\right)$ is therefore merely a test based on the score function, and so the procedure is often referred to as the "score" test.

Our interpretation of Lagrange multipliers affords an immediate sense of why $\boldsymbol{\lambda}_n\left(\mathbf{X}\right)$ should be informative about H_0. Is there, independently of the logical connection, a compelling intuition for the informativeness of the score as well? Indeed, there is. In working with regular parametric families we know that $\mathcal{L}_\phi$ evaluated at unrestricted m.l.e. $\hat{\boldsymbol{\theta}}_n$ would just be a vector of zeroes, since the likelihood function attains its maximum there. By contrast, the "steepness" of the likelihood at the restricted estimate $\check{\boldsymbol{\theta}}_n$—the "size" of vector $\mathcal{L}_\phi\left(\check{\boldsymbol{\theta}}_n\right)$, or its Euclidean distance from vector $\mathbf{0}$— indicates how much is to be gained by moving away from $\check{\boldsymbol{\theta}}_n$, and therefore the degree to which the data are consistent with H_0. Indeed, if we consider *local* alternatives of the form $\boldsymbol{\phi}_\varepsilon = \boldsymbol{\phi}_0 + \boldsymbol{\varepsilon}$ for some small $|\varepsilon|$, then the logarithm of the likelihood ratio is approximately proportional to the score, since

$$\mathcal{L}\left(\boldsymbol{\phi}_0, \check{\boldsymbol{\psi}}_n\right) - \mathcal{L}\left(\boldsymbol{\phi}_\varepsilon, \check{\boldsymbol{\psi}}_n\right) = \mathcal{L}\left(\check{\boldsymbol{\theta}}_n\right) - \left[\mathcal{L}\left(\check{\boldsymbol{\theta}}_n\right) + \mathcal{L}_{\phi'}\left(\check{\boldsymbol{\theta}}_n\right)\boldsymbol{\varepsilon} + o\left(|\varepsilon|\right)\right]$$
$$= -\mathcal{L}_{\phi'}\left(\check{\boldsymbol{\theta}}_n\right)\boldsymbol{\varepsilon} + o\left(|\varepsilon|\right).$$

Of course, developing a formal, valid test requires adopting a metric based on probability rather than on mere Euclidean distance. That is, we need to know how likely it is that various discrepant scores would be observed when H_0 is true, and we require at least a qualitative sense of their behavior under H_1. Thus, we require a distribution theory for $\mathcal{L}_\phi\left(\check{\boldsymbol{\theta}}_n\right)$.

For this we have a good place to start, since the theory of maximum likelihood estimators has already told us a great deal about the limiting distribution of $\mathcal{L}_\phi$. Specifically, when the full $k \times 1$ score vector $\mathcal{L}_\theta$ is evaluated at the true parameters, then $E\mathcal{L}_\theta\left(\boldsymbol{\theta}\right) = \left(E\mathcal{L}_{\phi'}\left(\boldsymbol{\theta}\right), E\mathcal{L}_{\psi'}\left(\boldsymbol{\theta}\right)\right)' = \mathbf{0}$ under the regularity conditions and

$$E\mathcal{L}_\theta\left(\boldsymbol{\theta}\right)\mathcal{L}_{\theta'}\left(\boldsymbol{\theta}\right) = \begin{pmatrix} E\mathcal{L}_\phi\left(\boldsymbol{\theta}\right)\mathcal{L}_{\phi'}\left(\boldsymbol{\theta}\right) & E\mathcal{L}_\phi\left(\boldsymbol{\theta}\right)\mathcal{L}_{\psi'}\left(\boldsymbol{\theta}\right) \\ E\mathcal{L}_\psi\left(\boldsymbol{\theta}\right)\mathcal{L}_{\phi'}\left(\boldsymbol{\theta}\right) & E\mathcal{L}_\psi\left(\boldsymbol{\theta}\right)\mathcal{L}_{\psi'}\left(\boldsymbol{\theta}\right) \end{pmatrix}$$
$$= n\begin{pmatrix} \mathcal{I}_{\phi\phi'}\left(\boldsymbol{\theta}\right) & \mathcal{I}_{\phi\psi'}\left(\boldsymbol{\theta}\right) \\ \mathcal{I}_{\psi\phi'}\left(\boldsymbol{\theta}\right) & \mathcal{I}_{\psi\psi'}\left(\boldsymbol{\theta}\right) \end{pmatrix}$$
$$= n\mathcal{I}\left(\boldsymbol{\theta}\right),$$

where $\mathcal{I}\left(\boldsymbol{\theta}\right)$ is the Fisher information from a single observation. Moreover, by the central limit theorem

$$n^{-1/2}\mathcal{L}_\theta\left(\boldsymbol{\theta}\right) = n^{-1/2}\sum_{j=1}^{n}\frac{\partial}{\partial\boldsymbol{\theta}}\log f\left(X_j; \boldsymbol{\theta}\right) \rightsquigarrow N\left[\mathbf{0}, \mathcal{I}\left(\boldsymbol{\theta}\right)\right].$$

Thus, focusing just on the subvector of $\mathcal{L}_\theta\left(\boldsymbol{\theta}\right)$ corresponding to the restricted parameters, we have under H_0 that $n^{-1/2}\mathcal{L}_\phi\left(\boldsymbol{\theta}\right) \rightsquigarrow N\left[\mathbf{0}, \mathcal{I}_{\phi\phi'}\left(\boldsymbol{\phi}_0, \boldsymbol{\psi}\right)\right]$.

Unfortunately, this is indeed just a place to *start*. We must determine the limiting distribution not of $n^{-1/2}\mathcal{L}_\phi\left(\boldsymbol{\theta}\right) = n^{-1/2}\mathcal{L}_\phi\left(\boldsymbol{\phi}_0, \boldsymbol{\psi}\right)$ but of $n^{-1/2}\mathcal{L}_\phi(\breve{\boldsymbol{\theta}}_n) = n^{-1/2}\mathcal{L}_\phi(\phi_0, \breve{\psi}_n)$, which is a function of the restricted estimator (and r.v.) $\breve{\psi}_n$. This takes a bit more work, as follows.

Expanding $\mathcal{L}_\phi(\phi_0, \breve{\psi}_n)$ about (ϕ_0, ψ) gives

$$\mathcal{L}_\phi(\phi_0, \breve{\psi}_n) = \mathcal{L}_\phi(\phi_0, \psi) + \mathcal{L}_{\phi\psi'}(\phi_0, \psi_n^*)(\breve{\psi}_n - \psi), \qquad (8.15)$$

where $|\psi_n^* - \psi| < |\breve{\psi}_n - \psi|$. Likewise, expanding $\mathcal{L}_\psi(\phi_0, \psi)$ about $(\phi_0, \breve{\psi}_n)$ gives

$$\mathcal{L}_\psi(\phi_0, \psi) = \mathcal{L}_\psi(\phi_0, \breve{\psi}_n) - \mathcal{L}_{\psi\psi'}(\phi_0, \psi_n^{**})(\breve{\psi}_n - \psi).$$

The first vector on the right being identically equal to $\mathbf{0}$, we have

$$\breve{\psi}_n - \psi = -\mathcal{L}_{\psi\psi'}(\phi_0, \psi_n^{**})^{-1}\mathcal{L}_\psi(\phi_0, \psi),$$

assuming that the inverse exists. Moreover, it *does* exist for sufficiently large n if the log likelihood is twice continuously differentiable and if $\mathcal{I}(\boldsymbol{\theta})$ is nonsingular throughout $\boldsymbol{\Theta} = \boldsymbol{\Phi} \times \boldsymbol{\Psi}$. Substituting in (8.15) gives for $n^{-1/2}\mathcal{L}_\phi(\phi_0, \breve{\psi}_n)$ the expression

$$\frac{1}{\sqrt{n}}\mathcal{L}_\phi(\phi_0, \psi) - \left[\frac{1}{n}\mathcal{L}_{\phi\psi'}(\phi_0, \psi_n^*)\right]\left[\frac{1}{n}\mathcal{L}_{\psi\psi'}(\phi_0, \psi_n^{**})\right]^{-1}\frac{1}{\sqrt{n}}\mathcal{L}_\psi(\phi_0, \psi).$$

Now since $\psi_n^* \to^p \psi$ and $\psi_n^{**} \to^p \psi$ under H_0, the continuity theorem and law of large numbers imply that the right side is asymptotically equivalent to

$$n^{-1/2}\mathcal{L}_\phi(\phi_0, \psi) - \mathcal{I}_{\phi\psi'}\mathcal{I}_{\psi\psi'}^{-1}n^{-1/2}\mathcal{L}_\psi(\phi_0, \psi)$$

$$= n^{-1/2}\sum_{j=1}^{n}\left[\frac{\partial \log f(X_j; \phi_0, \psi)}{\partial \phi} - \mathcal{I}_{\phi\psi'}\mathcal{I}_{\psi\psi'}^{-1}\frac{\partial \log f(X_j; \phi_0, \psi)}{\partial \psi}\right].$$

Terms of the sum are i.i.d. with mean $\mathbf{0}$ under H_0, so $n^{-1/2}\mathcal{L}_\phi(\phi_0, \breve{\psi}_n)$ is asymptotically multivariate normal with mean vector $\mathbf{0}$ and covariance matrix given by

$$E\left[\frac{\partial \log f}{\partial \phi} - \mathcal{I}_{\phi\psi'}\mathcal{I}_{\psi\psi'}^{-1}\frac{\partial \log f}{\partial \psi}\right]\left[\frac{\partial \log f}{\partial \phi} - \mathcal{I}_{\phi\psi'}\mathcal{I}_{\psi\psi'}^{-1}\frac{\partial \log f}{\partial \psi}\right]'.$$

Multiplying and taking expectations term by term, this is

$$\mathcal{I}_{\phi\phi'} - \mathcal{I}_{\phi\psi'}\mathcal{I}_{\psi\psi'}^{-1}\mathcal{I}_{\psi\phi'} - \mathcal{I}_{\phi\psi'}\mathcal{I}_{\psi\psi'}^{-1}\mathcal{I}_{\psi\phi'} + \mathcal{I}_{\phi\psi'}\mathcal{I}_{\psi\psi'}^{-1}\mathcal{I}_{\psi\psi'}\mathcal{I}_{\psi\psi'}^{-1}\mathcal{I}_{\psi\phi'},$$

which (recalling (8.11)) reduces to

$$\mathcal{I}_{\phi\phi'}(\boldsymbol{\theta}) - \mathcal{I}_{\phi\psi'}(\boldsymbol{\theta})\mathcal{I}_{\psi\psi'}(\boldsymbol{\theta})^{-1}\mathcal{I}_{\psi\phi'}(\boldsymbol{\theta}) \equiv \mathcal{I}^{\phi\phi'}(\boldsymbol{\theta})^{-1}. \qquad (8.16)$$

Now since $\breve{\boldsymbol{\theta}}_n \to^p \boldsymbol{\theta}_0 = (\boldsymbol{\phi}_0', \boldsymbol{\psi}')'$ when H_0 is true, the continuity theorem implies that $\mathcal{I}^{\phi\phi'}(\breve{\boldsymbol{\theta}}_n) \to^p \mathcal{I}^{\phi\phi'}(\boldsymbol{\theta}_0)$ under the null. Mann–Wald then confirms that

$$\mathcal{S}_n(\mathbf{X}) := n^{-1} \mathcal{L}_{\phi'}(\breve{\boldsymbol{\theta}}_n) \mathcal{I}^{\phi\phi'}(\breve{\boldsymbol{\theta}}_n) \mathcal{L}_\phi(\breve{\boldsymbol{\theta}}_n) \rightsquigarrow \chi^2(k_0). \tag{8.17}$$

Quadratic form $\mathcal{S}_n(\mathbf{X})$ is the score statistic that is the basis for the L.m. test.

Under the alternative hypothesis that $\boldsymbol{\phi} \neq \boldsymbol{\phi}_0$ suppose first that $\breve{\boldsymbol{\theta}}_n \to^p \boldsymbol{\theta}_1 := (\boldsymbol{\phi}_0', \boldsymbol{\psi}_1')'$ for some $\boldsymbol{\psi}_1$ in the interior of $\boldsymbol{\Psi}$. Then

$$n^{-1}\mathcal{L}_\phi(\breve{\boldsymbol{\theta}}_n) = n^{-1} \sum_{j=1}^n \frac{\partial}{\partial \boldsymbol{\phi}} \log f(X_j; \breve{\boldsymbol{\theta}}_n)$$

$$\to^p E\left[\frac{\partial}{\partial \boldsymbol{\phi}} \log f(X; \boldsymbol{\theta}_1)\right] =: \boldsymbol{\delta}_1 \neq \mathbf{0}$$

and

$$\mathcal{S}_n(\mathbf{X}) = n\left[n^{-1}\mathcal{L}_{\phi'}(\breve{\boldsymbol{\theta}}_n)\right] \mathcal{I}^{\phi\phi'}(\breve{\boldsymbol{\theta}}_n) \left[n^{-1}\mathcal{L}_\phi(\breve{\boldsymbol{\theta}}_n)\right] \to^p +\infty.$$

Alternatively, if for some n the problem $\max_{\psi \in \Psi}\{L(\mathbf{X}; \boldsymbol{\phi}_0, \boldsymbol{\psi})\}$ has no interior solution, we set $\mathcal{S}_n(\mathbf{X}) = +\infty$. Accordingly, with c^α as the upper-α quantile of $\chi^2(k_0)$, a test based on rejection region $\mathcal{R}_\alpha = \{\mathbf{x} : \mathcal{S}_n(\mathbf{x}) \geq c^\alpha\}$ is asymptotically of size α and is consistent.

As we have seen, in regular cases Fisher information

$$n\mathcal{I}(\boldsymbol{\theta}) = E\mathcal{L}_\theta(\boldsymbol{\theta})\mathcal{L}_{\theta'}(\boldsymbol{\theta})$$

can also be evaluated as

$$n\mathcal{I}(\boldsymbol{\theta}) = -E\mathcal{L}_{\theta\theta'}(\boldsymbol{\theta})$$

$$= -E\left[\frac{\partial^2 \log L(\mathbf{X}; \boldsymbol{\theta})}{\partial \boldsymbol{\theta} \partial \boldsymbol{\theta}'}\right]$$

$$= -nE\left[\frac{\partial^2 \log f(X; \boldsymbol{\theta})}{\partial \boldsymbol{\theta} \partial \boldsymbol{\theta}'}\right].$$

However, it often happens that neither $E\mathcal{L}_\theta(\boldsymbol{\theta})\mathcal{L}_{\theta'}(\boldsymbol{\theta})$ nor $-E\mathcal{L}_{\theta\theta'}(\boldsymbol{\theta})$ can be expressed explicitly. Recalling the discussion on page 499, $\mathcal{I}(\boldsymbol{\theta})$ can be estimated consistently as

$$\widetilde{\mathcal{I}}(\boldsymbol{\theta}) := -n^{-1} \sum_{j=1}^n \frac{\partial^2 \log f(X_j; \boldsymbol{\theta})}{\partial \boldsymbol{\theta} \partial \boldsymbol{\theta}'}$$

$$= \begin{pmatrix} -n^{-1}\sum_{j=1}^n \frac{\partial^2 \log f(X_j; \boldsymbol{\theta})}{\partial \boldsymbol{\phi} \partial \boldsymbol{\phi}'} & -n^{-1}\sum_{j=1}^n \frac{\partial^2 \log f(X_j; \boldsymbol{\theta})}{\partial \boldsymbol{\phi} \partial \boldsymbol{\psi}'} \\ -n^{-1}\sum_{j=1}^n \frac{\partial^2 \log f(X_j; \boldsymbol{\theta})}{\partial \boldsymbol{\psi} \partial \boldsymbol{\phi}'} & -n^{-1}\sum_{j=1}^n \frac{\partial^2 \log f(X_j; \boldsymbol{\theta})}{\partial \boldsymbol{\psi} \partial \boldsymbol{\psi}'} \end{pmatrix}$$

$$=: \begin{pmatrix} \widetilde{\mathcal{I}}_{\phi\phi'}(\boldsymbol{\theta}) & \widetilde{\mathcal{I}}_{\phi\psi'}(\boldsymbol{\theta}) \\ \widetilde{\mathcal{I}}_{\psi\phi'}(\boldsymbol{\theta}) & \widetilde{\mathcal{I}}_{\psi\psi'}(\boldsymbol{\theta}) \end{pmatrix}.$$

Putting $\widetilde{\mathcal{I}}^{\phi\phi'}(\boldsymbol{\theta}) := \left[\widetilde{\mathcal{I}}_{\phi\phi'}(\boldsymbol{\theta}) - \widetilde{\mathcal{I}}_{\phi\psi'}(\boldsymbol{\theta})\widetilde{\mathcal{I}}_{\psi\psi'}(\boldsymbol{\theta})^{-1}_{\psi\phi'}\widetilde{\mathcal{I}}(\boldsymbol{\theta})\right]^{-1}$ and evaluating at $\check{\boldsymbol{\theta}}_n$, the score statistic can then be evaluated as

$$\mathcal{S}_n(\mathbf{X}) = n^{-1}\mathcal{L}_{\phi'}(\check{\boldsymbol{\theta}}_n)\widetilde{\mathcal{I}}^{\phi\phi'}(\check{\boldsymbol{\theta}}_n)\mathcal{L}_\phi(\check{\boldsymbol{\theta}}_n)$$

without compromising the test's asymptotic validity or its consistency.

8.6.3 The Wald Test

Notice that the L.m. test requires finding only the *restricted* estimates of parameters, whereas both the restricted and the unrestricted estimates, $\check{\boldsymbol{\theta}}_n$ and $\hat{\boldsymbol{\theta}}_n$, are needed for the likelihood-ratio procedure. The third possibility is a test that requires only unrestricted estimates $\hat{\boldsymbol{\theta}}_n$. The Wald test is such a procedure. In this test departures from $H_0 : \phi = \phi_0$ are judged entirely by the discrepancy between ϕ_0 and $\hat{\phi}_n$, the unrestricted m.l.e.

The idea and the distribution theory are very simple under the regularity conditions for m.l. estimation. From the theory for m.l. estimators that then applies we know that

$$\sqrt{n}\left(\hat{\boldsymbol{\theta}}_n - \boldsymbol{\theta}\right) = \left[\sqrt{n}\left(\hat{\phi}_n - \phi\right)', \sqrt{n}\left(\hat{\psi}_n - \psi\right)'\right]' \rightsquigarrow N\left[\mathbf{0}, \mathcal{I}(\boldsymbol{\theta})^{-1}\right].$$

Again writing out the Fisher information and its inverse in partitioned form as[14]

$$\mathcal{I}(\boldsymbol{\theta}) = \begin{pmatrix} \mathcal{I}_{\phi\phi'}(\boldsymbol{\theta}) & \mathcal{I}_{\phi\psi'}(\boldsymbol{\theta}) \\ \mathcal{I}_{\psi\phi'}(\boldsymbol{\theta}) & \mathcal{I}_{\psi\psi'}(\boldsymbol{\theta}) \end{pmatrix}, \ \mathcal{I}(\boldsymbol{\theta})^{-1} = \begin{pmatrix} \mathcal{I}^{\phi\phi'}(\boldsymbol{\theta}) & \mathcal{I}^{\phi\psi'}(\boldsymbol{\theta}) \\ \mathcal{I}^{\psi\phi'}(\boldsymbol{\theta}) & \mathcal{I}^{\psi\psi'}(\boldsymbol{\theta}) \end{pmatrix},$$

one sees that $\sqrt{n}\left(\hat{\phi}_n - \phi_0\right) \rightsquigarrow N\left[\mathbf{0}, \mathcal{I}^{\phi\phi'}(\boldsymbol{\theta}_0)\right]$ under the null. Evaluating $\mathcal{I}^{\phi\phi'}$ at unrestricted m.l.e. $\hat{\boldsymbol{\theta}}_n$ and applying the continuity and Mann–Wald theorems, we conclude that when H_0 is true

$$\mathcal{W}_n(\mathbf{X}) := n\left(\hat{\phi}_n - \phi_0\right)'\mathcal{I}^{\phi\phi'}\left(\hat{\boldsymbol{\theta}}_n\right)^{-1}\left(\hat{\phi}_n - \phi_0\right) \rightsquigarrow \chi^2(k_0).$$

Statistic $\mathcal{W}_n(\mathbf{X})$ is the basis for the Wald test.

Under H_1, when $\phi = \phi_1 := \phi_0 + \delta_1$ and $\hat{\boldsymbol{\theta}}_n \rightarrow^p (\phi_1', \psi_1')'$ say, we have $\mathcal{W}_n(\mathbf{X}) \rightarrow^{a.s.} +\infty$. Thus, with c^α as the upper-α quantile of $\chi^2(k_0)$, a test based on rejection region

$$\mathcal{R}_\alpha = \{\mathbf{x} : \mathcal{W}_n(\mathbf{x}) \geq c^\alpha\} \tag{8.18}$$

---

[14]Notice that the matrices on the principal diagonal of $\mathcal{I}(\boldsymbol{\theta})^{-1}$ are not the inverses of the corresponding components of $\mathcal{I}(\boldsymbol{\theta})$ unless the latter is block diagonal—that is, unless $\mathcal{I}^{\psi\phi'}(\boldsymbol{\theta})$ is a matrix of zeroes. Thus, finding $\mathcal{I}^{\phi\phi'}(\boldsymbol{\theta})$ usually requires inverting the full matrix $\mathcal{I}(\boldsymbol{\theta})$. Its expression in terms of the components of $\mathcal{I}(\boldsymbol{\theta})$ is given in (8.11).

is asymptotically valid and consistent.

As with the score test, in regular cases $\mathcal{I}(\boldsymbol{\theta})$ can be replaced not just by $\mathcal{I}\left(\hat{\boldsymbol{\theta}}_n\right)$ but also by the alternative consistent estimator

$$\widetilde{\mathcal{I}}\left(\hat{\boldsymbol{\theta}}_n\right) = -n^{-1} \sum_{j=1}^{n} \frac{\partial^2 \log f(X_j; \hat{\boldsymbol{\theta}}_n)}{\partial \boldsymbol{\theta} \partial \boldsymbol{\theta}'}.$$

Although the above development is for parametric models that obey the regularity conditions for maximum likelihood estimation, tests based on the Wald principle are, in fact, not limited to such cases. The minimum requirement is that the limiting distribution of $\sqrt{n}\left(\hat{\boldsymbol{\phi}}_n - \boldsymbol{\phi}_0\right)$ be normal with a covariance matrix that can be estimated consistently. When the regularity conditions do not hold, one must try to verify these conditions from first principles.

Example 8.10. Recall from Example 7.32 that when n is odd and the elements of sample $\mathbf{X} = (X_1, X_2, ..., X_n)'$ are i.i.d. with (Laplace) p.d.f.

$$f(x; \phi, \psi) = \frac{1}{2\psi} e^{-|x-\phi|/\psi}, \phi \in \Re, \psi > 0,$$

then the m.l. estimators of ϕ and ψ are $\hat{\phi}_n = X_{((n+1)/2:n)}$ (the unique sample median) and $\hat{\psi}_n = n^{-1} \sum_{j=1}^{n} |X_j - \hat{\phi}_n|$. Example 7.41 showed further that normalized sample median $T_n := \sqrt{n}\left(\hat{\phi}_n - \phi\right)$ converges in distribution to $N\left(0, \psi^2\right)$. Since $\hat{\psi}_n$ is a continuous (although not a differentiable) function of the consistent estimator $\hat{\phi}_n$, one can see (strong law of large numbers and continuity theorem) that $\hat{\psi}_n \to^P \psi$. Thus, by Mann–Wald we have that

$$\mathcal{W}_n(\mathbf{X}) = \frac{n\left(\hat{\phi}_n - \phi_0\right)^2}{\hat{\psi}_n^2} \rightsquigarrow \chi^2(1)$$

when $H_0 : \phi = \phi_0$ is true. It is easy to see also that $\mathcal{W}_n(\mathbf{X}) \to^{a.s.} +\infty$ when H_0 is false. Thus, the test based on rejection region (8.18) (with $k_0 = 1$) is asymptotically valid and consistent in this application even though the regularity conditions fail to hold. Likewise, for one-sided hypotheses $H_0 : \phi \leq \phi_0$ vs. $H_1 : \phi > \phi_0$ and $H_0 : \phi \geq \phi_0$ vs. $H_1 : \phi < \phi_0$ rejection regions $\mathcal{R}_\alpha^+ = \left\{\mathbf{x} : \sqrt{n}\left(\hat{\phi}_n - \phi_0\right)/\hat{\psi}_n \geq z^\alpha\right\}$ and $\mathcal{R}_\alpha^- = \left\{\mathbf{x} : \sqrt{n}\left(\hat{\phi}_n - \phi_0\right)/\hat{\psi}_n \leq -z^\alpha\right\}$, respectively, have these same asymptotic properties, where z^α is the upper-α quantile of $N(0, 1)$.

8.6.4 *Applying the Three Tests: An Illustration*

Here are examples of how to apply the three classical asymptotic tests for a model that does satisfy the regularity conditions for m.l. estimation and which contains a nuisance parameter. Maintaining that the n elements of $\mathbf{X}$ are i.i.d. as gamma with shape parameter ϕ and scale parameter ψ, we test $H_0 : \phi = 1$ *vs.* $H_1 : \phi \neq 1$ without restriction on ψ. Thus, the maintained hypothesis is that the data are from the family

$$\left\{ f(x; \phi, \psi) = \Gamma(\phi)^{-1} \psi^{-\phi} x^{\phi-1} e^{-x/\psi} \mathbf{1}_{(0,\infty)}(x) : \phi > 0, \psi > 0 \right\},$$

and this reduces under H_0 to the simpler subfamily of exponential distributions:

$$\left\{ f(x; \psi) = \psi^{-1} e^{-x/\psi} \mathbf{1}_{(0,\infty)}(x) : \psi > 0 \right\}.$$

Numerical examples will be based on two data sets: (i) a sample of size $n = 10$,

$$\mathbf{x}' = (2.9, 0.7, 1.7, 0.9, 1.0, 1.7, 4.3, 0.3, 5.1, 0.3),$$

for which $\bar{x} = 1.890$, $\log \bar{x} \doteq .637$, and $10^{-1} \sum_{j=1}^{10} \log x_j \doteq .234$; and (ii) a sample of size $n = 100$ with $\bar{x} \doteq 2.02020$, $\log \bar{x} \doteq .703199$, and $100^{-1} \sum_{j=1}^{100} \log x_j \doteq .371453$. The samples are in fact pseudorandom gamma deviates with $\phi = 2.0$, $\psi = 1.0$. Thus, in both cases the null hypothesis is actually false. To make it easier to check the numbers with a calculator, elements of the smaller sample have been rounded to one decimal place. Of course, using asymptotic procedures with such a small sample is appropriate merely as an illustration.

(i) Applying the likelihood-ratio test, we construct the l.r. statistic

$$-2 \log \Lambda(\mathbf{X}; 1) = -2 \left[\log L\left(\mathbf{X}; 1, \check{\psi}_n\right) - \log L\left(\mathbf{X}; \hat{\phi}_n, \hat{\psi}_n\right) \right] \quad (8.19)$$

by evaluating at the unrestricted and restricted m.l.e.s the log-likelihood function,

$$\log L(\mathbf{X}; \phi, \psi) = -n \log \Gamma(\phi) - n\phi \log \psi + (\phi - 1) \sum_{j=1}^{n} \log X_j - \frac{n\bar{X}}{\psi}.$$

Example 7.28 showed that the unrestricted m.l.e.s satisfy the nonlinear likelihood equations

$$0 = \mathcal{L}_\phi\left(\hat{\phi}_n, \hat{\psi}_n\right) = -n\Psi\left(\hat{\phi}_n\right) - n \log \hat{\psi}_n + \sum_{j=1}^{n} \log X_j \quad (8.20)$$

$$0 = \mathcal{L}_\psi\left(\hat{\phi}_n, \hat{\psi}_n\right) = -\frac{n\hat{\phi}_n}{\hat{\psi}_n} + \frac{n\bar{X}}{\hat{\psi}_n^2}, \quad (8.21)$$

where $\Psi(\phi) := d\log\Gamma(\phi)/d\phi$. To solve, use (8.21) to get $\hat{\psi}_n = \bar{X}/\hat{\phi}_n$ and substitute in (8.20) to produce the following equation, which must be solved numerically for $\hat{\phi}_n$:

$$0 = \Psi\left(\hat{\phi}_n\right) - \log\hat{\phi}_n + \log\bar{X} - n^{-1}\sum_{j=1}^{n}\log X_j. \qquad (8.22)$$

Setting $\phi = 1$ for the restricted case, Equation (8.21) yields the explicit solution $\check{\psi}_n = \bar{X}$. Plugging these into (8.19) gives for $-2\log\Lambda_n(\mathbf{X})$ the factor $2n$ times the quantity

$$1 - \log\Gamma\left(\hat{\phi}_n\right) + \hat{\phi}_n\left(\log\hat{\phi}_n - 1\right) + \left(\hat{\phi}_n - 1\right)\left(\frac{1}{n}\sum_{j=1}^{n}\log X_j - \log\bar{X}\right).$$

Since this is distributed asymptotically as $\chi^2(1)$ under H_0, the rejection region is $\mathcal{R}_\alpha = \{\mathbf{x} : -2\log\Lambda_n(\mathbf{x}) \geq c^\alpha\}$, where c^α as the upper-α quantile.

    Let us carry out the test at level $\alpha = .05$ with the example data. First, with the sample of size $n = 10$ we find from (8.22) and (8.21) that $\hat{\phi}_n \doteq 1.386$ and $\hat{\psi}_n \doteq 1.364$, so that $\Gamma\left(\hat{\phi}_n\right) \doteq 0.888$, $\log\Gamma\left(\hat{\phi}_n\right) \doteq -0.119$, $\log\hat{\phi}_n \doteq 0.326$, and $-2\log\Lambda_{10}(\mathbf{x}) \doteq 0.586$. Since this is less than critical value $c^{.05} = 3.84$, the l.r. test does not reject H_0 at level .05 for this sample. Next, with the sample of size $n = 100$ we find $\hat{\phi}_n \doteq 1.65459$, $\hat{\psi}_n \doteq 1.22084$, and $-2\log\Lambda_{100}(\mathbf{x}) \doteq 13.175 > 3.84$, so with the larger sample we do reject H_0 at level .05.

(ii) Applying the Lagrange-multiplier test, we must compute the score statistic $S_n(\mathbf{X}) = n^{-1}\mathcal{L}_{\phi'}\left(\check{\boldsymbol{\theta}}_n\right)\mathcal{I}^{\phi\phi'}\left(\check{\boldsymbol{\theta}}_n\right)\mathcal{L}_\phi\left(\check{\boldsymbol{\theta}}_n\right)$ with $\mathcal{I}^{\phi\phi'} = \left(\mathcal{I}_{\phi\phi'} - \mathcal{I}_{\phi\psi'}\mathcal{I}_{\psi\psi'}^{-1}\mathcal{I}_{\psi\phi'}\right)^{-1}$. Since ϕ and ψ are scalars now, this is simply

$$S_n(\mathbf{X}) = \frac{\mathcal{L}_\phi^2}{n\left(\mathcal{I}_{\phi\phi} - \mathcal{I}_{\phi\psi}^2/\mathcal{I}_{\psi\psi}\right)}$$

with $\mathcal{L}$ and $\mathcal{I}$ evaluated at $(1, \check{\psi}_n)$. The restricted estimator is $\check{\psi}_n = \bar{X}$, so

$$\mathcal{L}_\phi\left(1, \check{\psi}_n\right) = n\left[-\Psi(1) - \log\bar{X} + n^{-1}\sum_{j=1}^{n}\log X_j\right].$$

From (7.25) in Example 7.37 the information matrix is

$$\mathcal{I}(\phi, \psi) = \begin{pmatrix} \mathcal{I}_{\phi\phi} & \mathcal{I}_{\phi\psi} \\ \mathcal{I}_{\phi\psi} & \mathcal{I}_{\psi\psi} \end{pmatrix} = \begin{pmatrix} \Psi'(\phi) & \psi^{-1} \\ \psi^{-1} & \phi\psi^{-2} \end{pmatrix},$$

where $\Psi'(\phi) := d^2 \log \Gamma(\phi)/d\phi^2$ (the trigamma function). Thus,

$$\mathcal{I}^{\phi\phi'}\left(1, \check{\psi}_n\right) = \left[\mathcal{I}_{\phi\phi}\left(1, \check{\psi}_n\right) - \frac{\mathcal{I}_{\phi\psi}^2\left(1, \check{\psi}_n\right)}{\mathcal{I}_{\psi\psi}\left(1, \check{\psi}_n\right)}\right]^{-1} = \frac{1}{\Psi'(1) - 1},$$

and

$$\mathcal{S}_n(\mathbf{X}) = \frac{n}{\Psi'(1) - 1}\left[\Psi(1) + \log \bar{X} - n^{-1}\sum_{j=1}^n \log X_j\right]^2.$$

Again, H_0 is rejected at level α when $\mathcal{S}_n(\mathbf{x})$ is equal to or greater than the upper-α quantile of $\chi^2(1)$.

Testing at level $\alpha = .05$ with the example data, we obtain from a table of digamma and trigamma functions (or by numerical approximation) $\Psi(1) \doteq -.577$, $\Psi'(1) \doteq 1.645$. The sample with $n = 10$ then gives $\mathcal{S}_{10}(\mathbf{x}) \doteq 0.469 < 3.84$, so H_0 is not rejected at level .05. We do reject with the larger sample, since $\mathcal{S}_{100}(\mathbf{x}) \doteq 9.343 > 3.84$.

(iii) Applying the Wald test, we must calculate the statistic

$$\mathcal{W}_n(\mathbf{X}) = \frac{n\left(\hat{\phi}_n - 1\right)^2}{\mathcal{I}^{\phi\phi}\left(\hat{\phi}_n, \hat{\psi}_n\right)}.$$

As before, the unrestricted m.l.e. for ϕ is found by solving (8.22). Equation (7.25) in Example 7.37 shows the first diagonal element of the inverse of the information matrix to be $\mathcal{I}^{\phi\phi}(\phi, \psi) = \phi/[\phi\Psi'(\phi) - 1]$, so that

$$\mathcal{W}_n(\mathbf{X}) = n\left(\hat{\phi}_n - 1\right)^2\left[\Psi'\left(\hat{\phi}_n\right) - \hat{\phi}_n^{-1}\right].$$

As for the l.r. and L.m. tests H_0 is rejected when $\mathcal{W}_n(\mathbf{x})$ is at or above the upper-α quantile of $\chi^2(1)$.

Testing at level .05, we find for the sample of $n = 10$ that $\hat{\phi}_n \doteq 1.386$, $\Psi'\left(\hat{\phi}_n\right) \doteq 1.039$, and $\mathcal{W}_{10}(\mathbf{x}) \doteq 0.473 < 3.84$, so that the Wald test fails to reject H_0. The larger sample yields $\hat{\phi}_{100} \doteq 1.65459$ and $\mathcal{W}_{100}(\mathbf{x}) \doteq 9.308 > 3.84$, indicating rejection at level .05.

Exercise 8.11. *Adapt the Wald procedure to test $H_0 : \phi = \phi_0$ against one-sided alternative $H_1 : \phi > \phi_0$ under the maintained hypothesis that the sample is i.i.d. as $\Gamma(\phi, \psi)$.*

Exercise 8.12. *Relying on the asymptotic normality of the pivotal quantity*

$$\sqrt{n}\left(\hat{\phi}_n - \phi\right)\sqrt{\Psi'\left(\hat{\phi}_n\right) - \hat{\phi}_n^{-1}}$$

and recalling Example 7.45, an approximate $1 - \alpha$ large-sample confidence interval for the shape parameter of the gamma distribution is

$$\left[\hat{\phi}_n - \frac{z^{\alpha/2}}{\sqrt{n\left[\Psi'\left(\hat{\phi}_n\right) - \hat{\phi}_n^{-1}\right]}}, \hat{\phi}_n + \frac{z^{\alpha/2}}{\sqrt{n\left[\Psi'\left(\hat{\phi}_n\right) - \hat{\phi}_n^{-1}\right]}}\right].$$

Show that the corresponding open *interval represents the set of values of ϕ_0 for which the Wald test would not reject $H_0 : \phi = \phi_0$ in favor of the two-sided alternative. That is, show that the $1 - \alpha$ c.i. represents the set of hypotheses acceptable at level α.*

Exercise 8.13. *Consider testing $H_0 : \phi = 1$ vs. $H_1 : \phi \neq 1$ for the Weibull model:*

$$f\left(x; \phi, \psi\right) = \phi\psi^{-1}x^{\phi-1}\exp\left(-\frac{x^\phi}{\psi}\right)\mathbf{1}_{(0,\infty)}\left(x\right), \phi > 0, \psi > 0.$$

Choose among the three asymptotic tests the one that can be carried out with only a hand calculator, and apply it (as practice) to the data

$$\mathbf{x}' = \left(0.4, \ 1.2, \ 3.4, \ 4.8, \ 8.6\right).$$

Exercise 8.14. *Using the data and results for Exercises 7.28 and 7.43, test at the .10 level $H_0 : \alpha = 1$ vs. $H_1 : \alpha \neq 1$ in the Weibull model,*

$$f\left(x; \alpha, \beta\right) = \alpha\beta^{-1}x^{\alpha-1}\exp\left(-\frac{x^\alpha}{\beta}\right)\mathbf{1}_{(0,\infty)}\left(x\right), \alpha > 0, \beta > 0,$$

using the (i) likelihood-ratio, (ii) score, and (iii) Wald tests.

Exercise 8.15. *A random sample of $n = 100$ from $IG\left(\mu, \lambda\right)$ (inverse Gaussian) yields $\bar{x} = 0.9718$ and $\bar{y} := n^{-1}\sum_{j=1}^{n} x_j^{-1} = 1.8513$. Test $H_0 : \mu = \lambda = 1$ at the .05 level using the (i) likelihood-ratio, (ii) score, and (iii) Wald tests. (Hint: Refer to the results of Exercises 7.20, 7.24 and 7.41.)*

8.6.5 Comparison of Tests

We have seen that all three asymptotic tests are asymptotically valid and consistent. Moreover, in standard (regular) cases they have the same Pitman asymptotic efficiencies—meaning that each test has unitary asymptotic efficiency relative to each of the others. However, the numerical examples illustrate that the test statistics can take quite different values against real (non-local) alternatives, even though their distributions under the null and under local alternatives are the same asymptotically. It is unsettling that different tests may lead to different decisions about rejection. While not uncommon, this often goes unnoticed, since applied researchers typically just choose the one test that is easiest to apply. If unrestricted m.l.e.s and estimates of their standard errors have already been obtained, then the Wald test of an hypothesis like $H_0 : \phi = \phi_0$ comes essentially for free. For this reason, it is the most popular of the three in such applications. However, it does have some drawbacks: (i) The statistic is not invariant under different parameterizations of the model (e.g., do we take ψ or ψ^{-1} as the scale parameter of the gamma family?); (ii) inversion of the information matrix is required in the multiparameter case; and (iii) in small samples, and when H_0 is badly at variance with the data, $\mathcal{I}^{\phi\phi'}(\hat{\boldsymbol{\theta}}_n)$ may fail to provide a good approximation to the variance of $\hat{\phi}$.[15] Also, it is sometimes much easier to find restricted estimates than to find unrestricted estimates. As in the gamma example, where nuisance scale parameter ψ was trivial to estimate when shape parameter ϕ was known, this situation favors the L.m. test. Of course, the l.r. test has the disadvantage of requiring both sets of estimates, but it has the decided advantage of not requiring the calculation of covariance matrices. This is particularly relevant in testing linear or nonlinear restrictions among the parameters; that is, hypotheses such as $H_0 : \phi - 2\psi = 0$ or $H_0 : \phi - \psi^2 = 0$. Although the Wald test works well enough in linear cases, testing a nonlinear relation requires making a linear approximation in working out the asymptotics, and the value of the test statistic will depend on how this is posed; for example, in testing $\phi - \psi^2 = 0$ one could base the Wald statistic on either $\hat{\phi}_n - \hat{\psi}_n^2$ or $\sqrt{\hat{\phi}_n} - \hat{\psi}_n$. In any case, both the Wald and L.m. procedures require asymptotic variances to be worked out from scratch for each specific hypothesis.

Example 8.11. Consider testing $H_0 : \phi = 2\psi$ against $H_1 : \phi \neq 2\psi$ under the maintained hypothesis that $\{X_j\}_{j=1}^n$ are i.i.d. as $\Gamma(\phi, \psi)$.

---

[15] See Vaeth (1985) and Mantel (1987) for further discussion of Wald procedures.

(i) The l.r. test is very straightforward: One obtains unrestricted m.l.e.s $\hat{\phi}_n, \hat{\psi}_n$ in the usual way, then gets restricted estimates $\check{\phi}_n$ and $\check{\psi}_n = \check{\phi}_n/2$ by expressing the likelihood in terms of one parameter or the other, then maximizing. With the data for $n = 100$ from the examples in Section 8.6.4 ($\bar{x} = 2.02020$, $100^{-1}\sum_{j=1}^{100} \log x_j = .371453$) one obtains in this way $\check{\phi}_{100} \doteq 1.981$ and $-2\log\Lambda(\mathbf{x}) \doteq 2.207$. Since H_0 imposes just a single restriction (it is expressed with just one "="!), the asymptotic distribution under the null is still $\chi^2(1)$, and so the test fails to reject at level .05.

(ii) The Wald test is also reasonably straightforward: One forms the test statistic by dividing the square of $\hat{\phi}_n - 2\hat{\psi}_n$ by its estimated asymptotic variance,

$$n^{-1}\left[\mathcal{I}^{\phi\phi}\left(\hat{\phi}_n, \hat{\psi}_n\right) - 4\mathcal{I}^{\phi\psi}\left(\hat{\phi}_n, \hat{\psi}_n\right) + 4\mathcal{I}^{\psi\psi}\left(\hat{\phi}_n, \hat{\psi}_n\right)\right].$$

The result with the given data is $\mathcal{W}_{100}(\mathbf{x}) \doteq 1.947$, which also fails to reject. Of course, because the relation is linear, a test based on $\hat{\phi}_n/2 - \hat{\psi}_n$ instead of $\hat{\phi}_n - 2\hat{\psi}_n$ would yield precisely the same result.

(iii) For the L.m. test the first-order conditions for maximizing $\mathbb{L} = \mathcal{L}(\phi, \psi) - \lambda(\phi - 2\psi)$ yield *two* alternative score expressions for λ: $\mathcal{L}_\phi(\phi, \phi/2)$ and $-\frac{1}{2}\mathcal{L}_\psi(\phi, \phi/2)$. From these come an equation that can be solved for ϕ:

$$0 = \mathcal{L}_\phi\left(\phi, \frac{\phi}{2}\right) + \frac{1}{2}\mathcal{L}_\psi\left(\phi, \frac{\phi}{2}\right)$$
$$= -n\Psi(\phi) - n\log\left(\frac{\phi}{2}\right) + \sum_{j=1}^n \log x_j - n + \frac{2n\bar{x}}{\phi^2}.$$

Naturally, the solution is the same as that obtained in the l.r. test. Putting $n^{-1/2}\check{\lambda} = n^{-1/2}\mathcal{L}_\phi(\check{\phi}, \check{\phi}/2)$ and expanding about ϕ, lengthy calculation shows that the normalized multiplier statistic is asymptotically normal with zero mean and variance V_λ, say, given by

$$\frac{\mathcal{I}_{\phi\phi}\left(\mathcal{I}_{\phi\psi} + \frac{\mathcal{I}_{\psi\psi}}{2}\right)^2 - 2\mathcal{I}_{\phi\psi}\left(\mathcal{I}_{\phi\psi} + \frac{\mathcal{I}_{\psi\psi}}{2}\right)\left(\mathcal{I}_{\phi\phi} + \frac{\mathcal{I}_{\phi\psi}}{2}\right) + \mathcal{I}_{\psi\psi}\left(\mathcal{I}_{\phi\phi} + \frac{\mathcal{I}_{\phi\psi}}{2}\right)^2}{4(\mathcal{I}_{\phi\phi} + \mathcal{I}_{\phi\psi} + \mathcal{I}_{\psi\psi}/4)^2}.$$

Hence, $\mathcal{S}_n(\mathbf{X}) = n^{-1}\check{\lambda}^2/V_\lambda$ is distributed asymptotically as $\chi^2(1)$ under H_0. The sample data give $V_\lambda \doteq 7.46564$ and $\mathcal{S}_{100}(\mathbf{x}) \doteq 1.167 < 3.841$, so that $H_0 : \phi = 2\psi$ is again not rejected. The same value for $\mathcal{S}_{100}(\mathbf{x})$ would have been obtained had we worked instead from $n^{-1/2}\check{\lambda} = -\frac{1}{2}\mathcal{L}_\psi(\check{\phi}, \check{\phi}/2)$ or, since the relation is linear, from $-\frac{1}{2}\mathcal{L}_\psi(2\check{\psi}, \check{\psi})$.

Exercise 8.16. *Verify the numerical results for the l.r. and Wald tests in Example 8.11.*

Exercise 8.17. *Use the same data as in Example 8.11 to perform an l.r. test of $H_0 : \phi = \psi^2$ at level .05.*

Exercise 8.18. *Use the same data as in Example 8.11 to perform two versions of the Wald test of $H_0 : \phi = \psi^2$ at level .05: one based on $\hat{\phi}_n - \hat{\psi}_n^2$ and another on $\sqrt{\hat{\phi}_n} - \hat{\psi}_n$.*

Exercise 8.19. *Independent random samples $\{X_j\}_{j=1}^{n}$, $\{Y_j\}_{j=1}^{m}$ are maintained to be i.i.d. as $N(\phi, 1)$ and $N(\psi, 1)$, respectively. If samples of $n = m = 100$ yield $\sum_{j=1}^{100} x_j = 54.85$, $\sum_{j=1}^{100} x_j^2 = 138.01$, $\sum_{j=1}^{100} y_j = 59.92$, and $\sum_{j=1}^{100} y_j^2 = 129.50$, test $H_0 : \phi^2 + \psi^2 = 1$ at level .10. (Hint: Polar coordinates.)*

When more than one asymptotic test is feasible in a given application, one can choose among them by using Monte Carlo simulation to compare estimates of type-I errors and powers. This would be done with parametric bootstrapping. For example, in testing $H_0 : \phi = \phi_0, \psi \in \Psi$ one would first calculate restricted and unrestricted m.l.e.s $(\phi_0, \check{\psi}_n)$ and $(\hat{\phi}_n, \hat{\psi}_n)$ from the actual sample $\mathbf{x}$ of size n, then do the following. To estimate type-I errors (i) take S repeated pseudorandom samples $\{\mathbf{x}_s\}_{s=1}^{S}$ of size n from the distribution with $\phi = \phi_0$ and $\psi = \check{\psi}_n$; (ii) apply all three tests to each pseudorandom sample $\mathbf{x}_s$; and (iii) compare the proportions of the S samples in which the test rejects the null. This exercise will indicate whether critical values based on the asymptotic χ^2 approximation yield valid tests at the given n. Then, to estimate the tests' power functions, first construct a set of values $\{\phi_m\}_{m=1}^{M}$ of the restricted parameter(s) having a range that includes ϕ_0 and $\hat{\phi}_n$. For each ϕ_m (i) take S repeated pseudorandom samples $\{\mathbf{x}_s\}_{s=1}^{S}$ of size n from the distribution with $\phi = \phi_m$ and $\psi = \check{\psi}_n$; then perform steps (ii) and (iii) above. This exercise will help determine which of three tests has the best power in the relevant part of the parameter space.

8.7 Criticisms and Alternative Procedures

The procedures considered thus far for testing parametric hypotheses all follow the general approach advocated by J. Neyman and E. Pearson in the 1930s. These are the "classical" tests that have long been the standard tools in applied statistics. Distilled to its essence, the Neyman–Pearson approach involves just two steps:

α. Choose a significance level α that represents the largest "acceptable" probability of type-I error.

β. Choose among available tests with probability of type-I error $\leq \alpha$ the one with highest power $1 - \beta$, if such exists.

As we have seen, some compromises usually have to be made at each step. In many cases—and often when there are nuisance parameters—the finite-sample distributions of reasonable test statistics are unknown even under H_0, and asymptotic approximations must be used. Thus, one is often obliged to settle for asymptotic validity. Next, except in very special cases there is no test that delivers highest power uniformly—i.e., for all possibilities allowed under H_1. We often have to choose on the basis of "local" power—power against alternatives "close" to H_0—or simply to settle for a test that is known to be asymptotically valid and consistent. Of course, these properties depend on asymptotic theory that may not be a good guide unless n is very large.

While critics who cite the need for such compromises may themselves be criticized for counseling perfection, there are indeed valid concerns about the suitability of the Neyman–Pearson approach in applied work. Here are some of them.

(i) The classical setup forces one to take far too rigid a position about whether hypotheses are true or false. A knife-edge hypothesis like $H_0 : \theta = \theta_0$ would almost never be *precisely* true. Therefore, a consistent test of H_0 *vs.* $H_1 : \theta \neq \theta_0$ at fixed significance level α would reject H_0 with probability arbitrarily close to unity if the sample were sufficiently large. Knowing this, one may ask (as many have) "What is the point of carrying out the test?"

(ii) A related point is that users of statistical research—especially the lay public, but even some experienced researchers—are apt to be misled by reports that a result has high *statistical* significance; i.e., that a certain hypothesis can be rejected at a very small α. In large enough samples, tiny discrepancies that have no practical importance can nevertheless be found to be highly statistically significant. Conversely, the *lack* of significance at conventional levels is by no means an assurance that H_1

is false or irrelevant. Issues of power aside, the *consequences* of wrongly ignoring such a possibility must also be considered.[16]

(iii) Although we can always put at least an asymptotic bound on the probability of type-I error, there is usually no way to estimate a test's actual power in a particular situation, because the alternative hypothesis usually embraces a large class of possibilities. For example, even when a uniformly most powerful test is available, as when testing $H_0 : \theta \geq \theta_0$ against a one-sided class $H_1 : \theta < \theta_0$, one would have to know the actual value of θ in order to estimate power at any particular significance level α. Why is this a problem? We know that increasing significance level α increases power $1 - \beta$, whatever be the actual θ; but there is no way to quantify the effect so as to assess the trade-off between risks of the two types of error. Thus, the choice of α seems essentially arbitrary (hence, the quotation marks about "acceptable" in step α. above). If the test will really lead to some specific action (Do we market this drug that is supposed to lower cholesterol levels, or not?), how can one begin to make a rational decision?

(iv) In scientific studies statistical tests are typically used to assess the validity—or at least the *usefulness*—of some theory, rather than to decide between two specific courses of action. Since a theory can never be *proven* empirically, the most one can do is to determine whether the data *dis*confirm it—in the sense of being substantially and consistently (over repeated experiments) at odds with it. Now in parametric settings a theory commonly places only vague restrictions on a parameter. For examples, an epidemiologist might want to test that of two populations with different dietary habits the one with more fat intake has higher incidence of colon cancer; or an economist might want to test that an institutional change promoting dealer competition on the NASDAQ exchange has increased the mean rate of stock trading. In both examples, the theory might be posed as the hypothesis $H_1 : \theta = \mu_1 - \mu_2 > 0$. In such cases, as explained in the introduction to this chapter, the null $H_0 : \theta \leq 0$ would be considered just a contrary statement that the data were expected to disconfirm. The usual approach would be

---

[16]Bialik (2007) reports the finding by two psychologists that, of 15,000 business school students, those whose names began with C or D had below-average grades. The finding was widely circulated in the media, yet the "highly significant" difference amounted to two one-hundredths of a grade point.

On the other side of the coin, Ziliak (2011) describes a case decided by the U.S. Supreme Court (Matrixx Initiatives *v.* Siracusano, March 22, 2011) in which a company cited statistical insignificance as defense for marketing a drug with occasional severe side effects. (To their credit, the justices were *not* misled: They found the company liable.)

to focus on the specific contrary hypothesis $\theta = 0$ that is closest to H_1 and set up a rejection region with low probability of type-I error, such as $\alpha \in \{.10, .05, .01\}$. By thus making small the probability of rejecting a true null, empirical findings that did lead to rejection could be proclaimed statistically "significant". On the other hand, since the maximum probability of type-I is deliberately made low, failure to reject cannot be interpreted as offering support for H_0. This is why it is customary to hedge by using the phrase "fail to reject" rather than the term "accept". Now while the Neyman-Pearson framework serves well enough in this setting (apart from the other objections), one occasionally encounters theories that do place sharp restrictions on parameters. For example, a cosmologist working from a theory of the "big bang" might have a fairly precise prediction of the temperature, θ_0, of the residual background thermal radiation and want to test this from repeated measurements subject to independent observational errors. This presents a problem. Since the relevant theory proposes no specific alternative to θ_0, it seems appropriate to test $H_1 : \theta = \theta_0$ with $H_0 : \theta \neq \theta_0$ as the opposing statement that the theorist hopes to reject. But the Neyman–Pearson prescription simply fails in this case, for any unbiased rejection region that limits the probability of type-I error to α limits the power to α as well. If, instead, the prediction of the theory is taken as the null, setting $H_0 : \theta = \theta_0$, then it needs to be *easy* to reject H_0 in order to claim statistical support for the theory. But this requires setting a value of α close to unity, in which case a *true* hypothesis would likely be rejected no matter how large the sample. In short, the Neyman–Pearson theory just does not apply unless one can set up a sharp null hypothesis to oppose the alternative that is believed to be true.

(v) Finally, there is the objection that the classical framework relies excessively on unverifiable (or at least typically *unverified*) assumptions about the data. Specifically, the properties of each test we have considered depend crucially on the truth of the applicable maintained hypothesis—in particular, that the data were generated by a model from a particular parametric family. Rarely is there more than conjecture to support such a maintained hypothesis, and all too often the choice is made just to simplify the resulting analysis. While the "goodness-of-fit" tests treated in Chapter 9 do provide a statistical means of assessing such a choice, they can only *disconfirm* it. Moreover, they are subject to the objection raised in point (iv).

8.7.1 *The Bayesian Approach*

The methods of Bayesian statistics do offer a framework for hypothesis testing that overcomes the first three criticisms. In brief, Bayesians take the view that probability just measures one's degree of belief and is therefore inherently subjective. In this view it is reasonable to make probabilistic statements about the value of some parameter, as if the parameter were a r.v. Thus, a Bayesian would feel justified in assigning a probability to an "event" such as $\theta \in B \in \mathcal{B}$, whereas the frequentist would say that the constant either lies in the set or does not, and that whether it does or not is simply unknown. In the Bayesian view, then, it is meaningful to assign probabilities to hypotheses of the form $H_0 : \theta \in \Theta_0$ and $H_1 : \theta \in \Theta_1$, and a decision between these might be made on the basis of these probabilities and an assessment of the relative losses from incorrect decisions. We provide details of this approach in Chapter 10.

Within the frequentist school some ways have emerged of mitigating some of the difficulties associated with the Neyman–Pearson scheme. The first, calculating and presenting "P" values, is a straightforward extension of the classical methods. The second, using "distribution-free" procedures, involves many new ideas and is treated in Section 8.8.

8.7.2 *P Values*

Here is a simple procedure that follows the Neyman–Pearson prescription in part but avoids the more or less arbitrary choice of α—and with it the rigid accept/reject form of the conclusion. First, choose a test that is thought to be suitable for the problem at hand—a u.m.p. or u.m.p.-unbiased test if one is available, or one of the classical asymptotic tests—and calculate the test statistic's sample realization $T(\mathbf{x})$. Finally, calculate the probability under H_0 that *statistic* $T(\mathbf{X})$ would take a value beyond the observed $T(\mathbf{x})$ in the direction (or directions) that support H_1. For example, if the rejection region $\mathcal{R}_\alpha$ comprises algebraically large values of $T(\mathbf{x})$, one would find

$$p(\mathbf{x}) := \Pr[T(\mathbf{X}) \geq T(\mathbf{x}) \mid H_0].$$

Alternatively, if $\mathcal{R}_\alpha$ comprises large absolute values of $T(\mathbf{x})$, then

$$p(\mathbf{x}) := \Pr[|T(\mathbf{X})| \geq |T(\mathbf{x})| \mid H_0].$$

The quantity $p(\mathbf{x})$, called variously the test's "P value", "prob value", or "marginal significance level", is just the probability of type-I error for a rejection region whose critical boundary coincides with $T(\mathbf{x})$. The statistician merely presents this probability value as a summary of the findings, leaving users to interpret it as they wish. If willing to accept a chance of type-I error greater than or equal to $p(\mathbf{x})$, one would reject H_0; otherwise, not.

The P-value approach is now widely used in applied research. Besides obviating the arbitrary choice of significance level, presenting a P value conveys precisely the extent to which the observed data are consistent with H_0. There is, however, an essential *caveat*. The precision with which the rejection probability can be calculated may make one forget that $p(\mathbf{x})$ is itself just the realization of a *random variable*. To explain, while the construction of the P value takes into account the sampling variability of $T(\mathbf{X})$ *under the null*, the calculated value $p(\mathbf{x})$ depends on the realized sample (which—quite obviously—is from a population for which H_0 is *not* known to be true). Thus, the P value must be regarded as the realization of a statistic $p(\mathbf{X})$ that has a sampling variability of its own. In principal this variability could be observed through repeated sampling, and in actual practice the dispersion of $p(\mathbf{X})$ can be estimated by nonparametric bootstrapping from the realized $\mathbf{x}$. Since the variability is commonly high relative to the mean, there is really no point in reporting values of $p(\mathbf{x})$ to more than one or two significant digits.[17] Since most statistical tables present only selected quantiles of the distributions of standard test statistics, P values typically must be estimated either by interpolation or with computer software that evaluates c.d.f.s.

Example 8.12. Maintaining that $n = 25$ observations are i.i.d. as $N(\mu, \sigma^2)$, we test $H_0 : \mu \leq 10$ *vs.* $H_1 : \mu > 10$ without restriction on the variance. The sample data yield $\bar{x} = 12$, $s^2 = 9$. For this situation the exact version of the likelihood-ratio test is appropriate, small values of the likelihood ratio being equivalent to large values of $T(\mathbf{X}) := (\bar{X} - 10)/(S/\sqrt{25})$, which is distributed as Student's $t(24)$ under H_0. The observed value is $T(\mathbf{x}) = 10/3 \doteq 3.33$. Since the u.m.p. rejection region for the one-sided alternative comprises just positive values of T we have

$$p(\mathbf{x}) = \Pr[T(\mathbf{X}) \geq 3.33 \mid H_0].$$

The table in Appendix C.4 shows 3.33 to lie between the upper-0.005 and upper-0.001 quantiles of $t(24)$. A more precise estimate is $p(\mathbf{x}) \doteq 0.0014$.

---

[17]See Boos and Stefanski (2011) for an elaboration of this argument and for some distribution theory for $p(\mathbf{X})$.

The P value for two-sided alternative $H_1 : \mu \neq 10$ would be exactly twice this: $p(\mathbf{x}) = \Pr\left[|T(\mathbf{X})| \geq 3.33 \mid H_0\right] \doteq 0.0028$.

Exercise 8.20. *Maintaining that $n = 100$ observations are i.i.d. as $\Gamma(\phi, \psi)$ (gamma distribution), a Wald test of $H_0 : \phi = 1$ vs. $H_1 : \phi \neq 1$ yields $\hat{\phi} \doteq 1.65417$ and $\mathcal{W}_n(\mathbf{x}) \doteq 9.301$. Find an approximate P value.*

8.8 Some Distribution-Free Tests

We now look at some methods of testing that require minimal assumptions about the underlying data-generating process. The term "distribution-free" indicates their broad applicability.[18] Since no parametric family for the population is assumed, such tests typically deal with features such as moments or quantiles. We begin by describing a simple but very useful approach that relies just on good intuition, moderate assumptions, and in most cases no more than a cursory knowledge of asymptotic theory.

8.8.1 *Asymptotic Procedures*

In constructing confidence intervals for a parameter θ we began by searching for a pivotal quantity $\Pi(\mathbf{X}; \theta)$, whose distribution was known at least asymptotically and depended on no unknown parameters besides θ. These pivotal quantities can be used to construct valid and consistent tests of hypotheses in parametric settings even when nuisance parameters are present. When no parametric model is assumed, such statistics can often be used to construct asymptotically valid and consistent tests of hypotheses of the form $H_0 : \gamma = \gamma_0$, where $\gamma = Eg(X)$ for some (real or complex) function g.

Here is how. The first step is to find a consistent estimator of γ, $G_n(\mathbf{X})$ say. Often this will be just the sample counterpart of $Eg(X)$, or $n^{-1}\sum_{j=1}^{n} g(X_j)$. Next, verify that there is some $\delta > 0$ (typically $\delta = 1$) such that $n^{\delta/2}[G_n(\mathbf{X}) - \gamma]$ is asymptotically normal with a variance—$\sigma_G^2 > 0$, say—that can be estimated consistently by some statistic S_G^2 under both H_0 and H_1. Applying either or both of the continuity theorem and Mann–Wald, it will follow that the pivotal quantity $n^{\delta/2}[G_n(\mathbf{X}) - \gamma]/S_G$ converges in distribution to $N(0, 1)$. In this case

$$T_n(\mathbf{X}) := \frac{n^{\delta/2}[G_n(\mathbf{X}) - \gamma_0]}{S_G} \tag{8.23}$$

---

[18] As explained in the introduction to this chapter, distribution-free methods are sometimes called "nonparametric" because they do not require a particular parametric family to be maintained; however, we reserve the term "nonparametric" to describe *hypotheses* rather than *procedures*.

is a plausible test statistic for $H_0 : \gamma = \gamma_0$. Since $G_n(\mathbf{X}) \to^p \gamma \neq \gamma_0$ and $S_G \to^p \sigma_G$ under H_1, $|T_n(\mathbf{X})| \to^p +\infty$ when H_0 is false. Rejection regions $\mathcal{R}_\alpha = \{\mathbf{x} : T_n(\mathbf{x}) \geq z^\alpha\}$ and $\mathcal{R}_\alpha = \{\mathbf{x} : |T_n(\mathbf{x})| \geq z^{\alpha/2}\}$ would then afford asymptotically valid, consistent tests against alternatives $H_1 : \gamma > \gamma_0$ and $H_1 : \gamma \neq \gamma_0$, respectively. Of course, the test comes with no guarantee of good small-sample properties, but without a structured, parametric model for the data there is no way that an optimal test could be found.

Here are examples of this *ad hoc* method and two exercises that apply it.

Example 8.13. Assuming only that elements of sample $\mathbf{X} :=$ $(X_1, X_2, ..., X_n)'$ are i.i.d. with finite, positive (unknown) variance σ^2, we devise an asymptotically size-α test of $H_0 : EX = \mu = \mu_0$ vs. $H_1 : \mu \neq \mu_0$. With $\bar{X}$ and S^2 as the sample mean and variance, the quantity $\sqrt{n}(\bar{X} - \mu)/S$ is asymptotically standard normal and pivotal, so $T_n(\mathbf{X}) := \sqrt{n}(\bar{X} - \mu_0)/S$ is a potential test statistic for H_0. Since $\bar{X} \to^{a.s.} \mu$ (the true value of the mean), we expect either large positive or large negative values of $T_n(\mathbf{X})$ according as $\mu - \mu_0 > 0$ or $\mu - \mu_0 < 0$. Moreover,

$$|T_n(\mathbf{X})| = \left| \frac{\bar{X} - \mu}{S/\sqrt{n}} + \frac{\mu - \mu_0}{S/\sqrt{n}} \right|$$

$$\approx \left| Z + \frac{\sqrt{n}(\mu - \mu_0)}{\sigma} \right| \to^{a.s.} +\infty$$

when $\mu \neq \mu_0$. Thus, an α-level test based on $\mathcal{R}_\alpha = \{\mathbf{x} : |T_n(\mathbf{x})| \geq z^{\alpha/2}\}$ is asymptotically valid and consistent. The probability of type-I error in small samples being unknown, applied researchers commonly compare $T_n(\mathbf{x})$ with quantiles of Student's t instead of $N(0,1)$ to make this "asymptotic t test" more conservative.

Example 8.14. Assuming only that elements of sample $\mathbf{X} :=$ $(X_1, X_2, ..., X_n)'$ are i.i.d., we devise an approximate large-sample, size-α test of $H_0 : E\mathbf{1}_B(X) \leq \pi_0(B)$ vs. $H_1 : E\mathbf{1}_B(X) > \pi_0(B)$, where B is a Borel set on which X lies with some unknown probability $\pi(B)$. Letting $P_n(B) := n^{-1} \sum_{j=1}^n \mathbf{1}_B(X_j)$ be the sample relative frequency, the asymptotic variance is $n^{-1}\pi(B)[1 - \pi(B)]$. Since $P_n(B) \to^{as} \pi(B)$,

$$\frac{P_n(B) - \pi(B)}{\sqrt{P_n(B)[1 - P_n(B)]/n}} \rightsquigarrow N(0, 1)$$

and is pivotal. As usual, to control the test's size, we focus on the particular value of $\pi(B)$ that is at the boundary between H_0 and H_1 and take

$$T_n(\mathbf{X}) := \frac{P_n(B) - \pi_0(B)}{\sqrt{P_n(B)\left[1 - P_n(B)\right]/n}}$$

as potential test statistic. Since $P_n(B) \to^{a.s.} \pi(B)$ (the true value), we expect large positive values of $T_n(\mathbf{X})$ under H_1. Moreover, since

$$T_n(\mathbf{X}) = \frac{P_n(B) - \pi(B)}{\sqrt{P_n(B)\left[1 - P_n(B)\right]/n}} + \frac{\pi(B) - \pi_0(B)}{\sqrt{P_n(B)\left[1 - P_n(B)\right]/n}}$$

$$\approx Z + \frac{\sqrt{n}\left[\pi(B) - \pi_0(B)\right]}{\sqrt{\pi(B)\left[1 - \pi(B)\right]}} \to^{a.s.} +\infty$$

when $\pi(B) > \pi_0(B)$, an α-level test based on $\mathcal{R}_\alpha = \{\mathbf{x} : T_n(\mathbf{x}) \geq z^\alpha\}$ is asymptotically valid and consistent. Convergence can be very slow, however, when $\pi(B)$ is close to zero or unity. Replacing $P_n(B)$ in the denominator with $\pi_0(B)$ preserves consistency and usually yields more accurate type-I error in small samples.

Exercise 8.21. *Use the result of Exercise 6.21 to conduct an asymptotically valid and consistent test of $H_0 : \mu_X \leq \mu_Y$ vs. $H_1 : \mu_X > \mu_Y$ from independent random samples of sizes n_X and n_Y from populations with finite but unknown variances σ_X^2 and σ_Y^2.*

A large-sample, distribution-free paired-sample test can be fashioned along the lines of the parametric test in Example 8.9.

Exercise 8.22. *Construct an asymptotically valid and consistent test of $H_0 : \mu_X \leq \mu_Y$ vs. $H_1 : \mu_X > \mu_Y$ when $\{X_j, Y_j\}_{j=1}^n$ is a bivariate sample from a distribution with finite second moments.*

Although one may reasonably expect these asymptotic tests to perform satisfactorily in large samples, there are special situations in which they do have very bad properties. For example, the next exercise shows that the asymptotic, α-level t test in Example 8.13 can yield type-I errors with near-unit frequency in repeated trials, even when n is arbitrarily large and when sufficient conditions for the asymptotic theory are met.

Exercise 8.23. *Let $\{X_j\}_{j=1}^n$ be i.i.d. as*

$$\Pr(X = x) = \begin{cases} 1 - \delta, & x = (\delta - 1)^{-1} \\ \delta, & x = \delta^{-1} \end{cases}$$

for some $\delta \in (0,1)$. Thus, $EX = \mu = 0$, and $VX = \delta^{-1}(1-\delta)^{-1}$ is positive and finite. The asymptotic t test in Example 8.13 rejects the true $H_0 : \mu = 0$ in favor of $H_0 : \mu \neq 0$ at level α when $\left|\sqrt{n}\bar{X}\right| \geq z^{\alpha/2}S$. Show that for any $\varepsilon \in (0,1)$ and any $n \in \aleph$ there are values of δ such that H_0 is rejected with probability exceeding $1 - \varepsilon$.

Fortunately, one is unlikely to encounter such situations in applied work. Indeed, for this particular problem even a cursory inspection of the sample would be apt to send up the necessary red flags. However, the fact remains that the t test—and most such *ad hoc* asymptotic procedures— can behave very differently in practice than large-sample theory leads us to expect. Nonparametric bootstrapping—i.e., resampling—can give at least some assurance that the procedure performs satisfactorily in the case at hand. Indeed, it is almost always worthwhile to back up theory with simulation—advice that applies much more generally than just to these asymptotic tests.

8.8.2   *Small-Sample Procedures*

Various distribution-free tests are available for hypotheses about "location" (mean or median) in one- and two-sample situations when samples are small. We give here just a brief overview of some of the more useful procedures.[19] While the validity of these depends on no specific parametric model, some *a priori* restriction on the population is inevitably required, such as continuity and/or symmetry. The common procedures depend on one or more of the following sample statistics:

(i) The *order statistics*, $\mathbf{X}_{(\cdot:n)} := \left(X_{(1:n)}, X_{(2:n)}, ..., X_{(n:n)}\right)'$, which are the sample observations arrayed in increasing order;

(ii) The *signs* of the data (or of some transform of them),

$$\mathbf{S} := (S(X_1), S(X_2), ..., S(X_n))',$$

where

$$S(x) := \begin{cases} \frac{x}{|x|}, & x \neq 0 \\ 1, & x = 0 \end{cases}; \qquad (8.24)$$

and

---

[19]References for the theory and practical implementation of these procedures are, respectively, Randles and Wolfe (1979) and Hollander and Wolfe (1973). Maritz (1995) provides a balanced blend of theory and practice. Good (1994) gives a lucid account of permutation tests.

(iii) The *ranks* of the observations, $\mathbf{R} := (R(X_1), R(X_2), ..., R(X_n))'$, where $R(X_j) \in \{1, 2, ..., n\}$ is the number of observations no larger than X_j.

Some distribution theory for these statistics is required before we take up specific techniques. Since the main applications are for continuous r.v.s, we limit attention to that case.

8.8.2.1 *Preliminaries*

Theorem 8.2. *(Joint distribution of order statistics) If* $\{X_j\}_{j=1}^n$ *are i.i.d., continuous r.v.s with p.d.f.* f, *then the joint p.d.f. of order statistics* $\mathbf{X}_{(\cdot:n)}$ *is*

$$f_{(\cdot:n)}(\mathbf{x}_{(\cdot:n)}) = n! \prod_{j=1}^n f\left(x_{(j:n)}\right) \qquad (8.25)$$

for each $\mathbf{x}_{(\cdot:n)} \in \Re^n$ *such that* $x_{(1:n)} < x_{(2:n)} < \cdots < x_{(n:n)}$.

Proof. Let $\mathbf{i} := (i_1, i_2, ..., i_n)'$ be one of the $n!$ permutations of the first n integers, and set $\mathbf{X}_{\mathbf{i}} := (X_{i_1}, X_{i_2}, ..., X_{i_n})'$. For any ordered $\mathbf{x}_{(\cdot:n)} \in \Re^n$ the joint p.d.f. of $\mathbf{X}_{\mathbf{i}}$ at $\mathbf{x}_{(\cdot:n)}$ is $\prod_{j=1}^n f\left(x_{(j:n)}\right)$. Since order statistics are invariant under permutations of the sample, we have $\Pr\left(\mathbf{X}_{(\cdot:n)} \in A\right) = n! \Pr\left(\mathbf{X}_{\mathbf{i}} \in A\right)$ for any measurable set $A \in \Re^n$, so that

$$\int_A f_{(\cdot:n)}(\mathbf{x}_{(\cdot:n)}) \cdot d\mathbf{x}_{(\cdot:n)} = \int_A n! \prod_{j=1}^n f\left(x_{(j:n)}\right) \cdot dx_{(j:n)}. \qquad \square$$

Theorem 8.3. *(Marginal distribution of an order statistic) The marginal c.d.f. and p.d.f. of the* j*th order statistic from independent, continuous r.v.s.* $\{X_j\}_{j=1}^n$ *with c.d.f.* F *and p.d.f.* f *are, respectively,*[20]

$$F_{(j:n)}(x) = \sum_{i=j}^n \binom{n}{i} F(x)^i [1 - F(x)]^{n-i} \qquad (8.26)$$

$$f_{(j:n)}(x) = j \binom{n}{j} F(x)^{j-1} [1 - F(x)]^{n-j} f(x). \qquad (8.27)$$

Proof. Event $X_{(j:n)} \leq x$ occurs if and only if at least j observations are no greater than x. For $i \in \{j, j+1, ..., n\}$ the probability that precisely i observations are no greater than x is $\binom{n}{i} F(x)^i [1 - F(x)]^{n-i}$. Expression (8.26) follows on summing over i.

---

[20] The c.d.f.s of $X_{(1:n)}$ and $X_{(n:n)}$, the sample minimum and sample maximum, were worked out previously in Example 7.40 and Exercise 7.30.

For the p.d.f., let $I(j, n; p) := j\binom{n}{j} \int_0^p t^{j-1}(1-t)^{n-j} \cdot dt$. Then $I(n, n; p) = p^n$, and integration by parts shows that

$$I(j, n; p) = \binom{n}{j} p^j (1-p)^{n-j} + I(j+1, n; p)$$

for $j \in \{1, 2, ..., n-1\}$. Thus,

$$I(j, n; p) = \sum_{i=j}^{n} \binom{n}{i} p^i (1-p)^{n-i}, \tag{8.28}$$

and $I(j, n; F(x)) = F_{(j:n)}(x)$. Differentiating gives (8.27). $\square$

Theorem 8.4. *(Joint distribution of two order statistics) The joint p.d.f. of the jth and kth order statistics ($j < k$) corresponding to independent, continuous r.v.s. $\{X_j\}_{j=1}^n$ with c.d.f. F and p.d.f. f is*

$$f_{(j:n),(k:n)}(x, y) = C_{k,j-1}^n F(x)^{j-1}[F(y)-F(x)]^{k-j-1}[1-F(y)]^{n-k} f(x)f(y) \tag{8.29}$$

for $x < y$, where

$$C_{k,j-1}^n := \frac{n!}{(j-1)!(k-j-1)!(n-k)!}.$$

Proof. For $h, h' > 0$ the event that (i) $X_\ell \leq x$ for $\ell \in \{1, 2, ..., j-1\}$, (ii) $x < X_j \leq x + h$, (iii) $x + h < X_\ell \leq y$ for $\ell \in \{j+1, j+2, ..., k\}$, (iv) $y < X_{k+1} \leq y + h'$, and (v) $X_\ell > y + h'$ for $\ell \in \{k+1, k+2, ..., n\}$ has probability

$$F(x)^{j-1}[F(x+h) - F(x)][F(y) - F(x+h)]^{k-j-1} \tag{8.30}$$
$$\cdot [F(y+h') - F(y)][1 - F(y+h')]^{n-k}.$$

There are $C_{k,j-1}^n$ ways in which the observations could be assigned to such sets. The result follows on multiplying (8.30) by $C_{k,j-1}^n$, dividing by hh', and letting h and h' approach 0. $\square$

Theorem 8.5. *(Independence of signs and absolute values of continuous r.v.s symmetric about the origin.) Let X be a continuous r.v. with c.d.f. F such that $F(-x) = 1 - F(x)$. Then $S(X)$ is independent of $|X|$.*

Proof. The conditions imply that $\Pr[S(X) = (-1)^j] = \frac{1}{2}$ for $j \in \{0, 1\}$. Thus, for each such j

$$\Pr[S(X) = (-1)^j, |X| \leq x] = \Pr[0 < (-1)^j X \leq x]$$
$$= \frac{1}{2} \Pr(-x \leq X \leq x)$$
$$= \Pr\left[S(X) = (-1)^j\right] \Pr(|X| \leq x). \qquad \square$$

Theorem 8.6. *(Joint distribution of rank statistics) If* $\{X_j\}_{j=1}^{n}$ *is a random sample from a continuous population, then the rank statistics*

$$\mathbf{R} := (R(X_1), R(X_2), ..., R(X_n))'$$

have joint distribution

$$\Pr(\mathbf{R} = \mathbf{i}) = \frac{1}{n!},$$

where $\mathbf{i} := (i_1, i_2, ..., i_n)'$ *is any of the* $n!$ *permutations of the first* n *integers.*

Proof. The $n!$ permutations are all equally likely. □

Corollary 8.1. *(Marginal and bivariate distributions of rank statistics)*

(i) The distribution of the k*th rank statistic,* $k \in \{1, 2, ..., n\}$, *is*

$$\Pr\left[R\left(X_k\right) = i\right] = n^{-1}, i \in \{1, 2, ..., n\}.$$

(ii) The joint distribution of the j*th and* k*th rank statistics,* $j \neq k$, *is*

$$\Pr\left[R(X_j) = i, R\left(X_k\right) = i'\right] = [n\left(n-1\right)]^{-1}, i \neq i' \in \{1, 2, ..., n\}.$$

Proof. (i) Fix $i \in \{1, 2, ..., n\}$. Of the $n!$ permutations of the first n integers there are $(n-1)!$ members in which integer i appears in the kth position. Each being equally likely, we have $\Pr\left[R\left(X_k\right) = i\right] = (n-1)!/n! = n^{-1}$.

(ii) Fix i and $i' \neq i$ among the first n integers. Of the $n!$ permutations of the first n integers there are $(n-2)!$ in which i appears in position j and i' appears in position k, so $\Pr\left[R(X_j) = i, R\left(X_k\right) = i'\right] = (n-2)!/n! = [n\left(n-1\right)]^{-1}$. □

Corollary 8.2. *(Moments of rank statistics)*

(i) $ER\left(X_j\right) = (n+1)/2$
(ii) $VR\left(X_j\right) = \left(n^2 - 1\right)/12$
(iii) $Cov\left[R(X_j), R(X_k)\right] = -\left(n+1\right)/12$ *for* $j, k \in \{1, 2, ..., n\}$ *and* $k \neq j$.

Proof. (i) $ER\left(X_j\right) = n^{-1}\sum_{i=1}^{n} i = (n+1)/2$.
(ii) $ER\left(X_j\right)^2 = n^{-1}\sum_{i=1}^{n} i^2 = (n+1)(2n+1)/6$, so

$$VR\left(X_j\right) = (n+1)\left[\frac{2n+1}{6} - \frac{n+1}{4}\right] = \frac{n^2 - 1}{12}.$$

(iii)

$$Cov\left[R(X_j), R(X_k)\right] = ER\left(X_j\right)R\left(X_k\right) - ER\left(X_j\right) \cdot ER\left(X_k\right)$$

$$= [n(n-1)]^{-1} \sum_{i \neq i'=1}^{n} ii' - \frac{(n+1)^2}{4}$$

$$= [n(n-1]^{-1} \left[\left(\sum_{i=1}^{n} i\right)^2 - \sum_{i=1}^{n} i^2\right] - \frac{(n+1)^2}{4}$$

$$= \frac{n+1}{n-1}\left(\frac{3n^2-n-2}{12}\right) - \frac{(n+1)^2}{4}$$

$$= -\frac{n+1}{12}. \qquad \qquad \square$$

Theorem 8.7. *(Independence of rank and order statistics) If $\{X_j\}_{j=1}^{n}$ is a random sample from a continuous population, then the vector $\mathbf{R}$ of rank statistics is stochastically independent of the vector $\mathbf{X}_{(\cdot:n)}$ of order statistics.*

Proof. Condition on the event $\mathbf{R} = \mathbf{i} = (i_1, i_2, ..., i_n)'$, which is any of the $n!$ permutations of the first n integers. Then since the rank of the jth observation is i_j, we have $X_j = X_{(i_j:n)}$ and $X_{p_j} = X_{(j:n)}$, where p_j is the number of the observation that has rank j. Therefore, for each $\mathbf{x}_{(\cdot:n)} \in \Re^n$ and each $\mathbf{i}$

$$\Pr\left(\mathbf{X}_{(\cdot:n)} \leq \mathbf{x}_{(\cdot:n)} | \mathbf{R} = \mathbf{i}\right) = \frac{\Pr(X_{(1:n)} \leq x_{(1:n)}, ..., X_{(n:n)} \leq x_{(n:n)})}{\Pr\left(\mathbf{R} = \mathbf{i}\right)}$$

$$= n! \Pr\left(X_{p_1} \leq x_{(1:n)}, ..., X_{p_n} \leq x_{(n:n)}\right)$$

$$= n! F(x_{(1:n)}) F(x_{(2:n)}) \cdot \cdots \cdot F\left(x_{(n:n)}\right).$$

By Theorem 8.2 this coincides with the unconditional joint c.d.f. of $\mathbf{X}_{(\cdot)}$, establishing independence. $\qquad \square$

Exercise 8.24. *Develop a direct argument for Expression (8.27) as in the proof of Theorem 8.4.*

8.8.2.2   *Conditional Procedures: Permutation Tests*

Modern computing capabilities have made feasible in many applications some *conditional* test procedures in which rejection regions are determined from the observed sample. These computationally intensive "randomization" or "permutation" tests are conducted in the following way. First, a statistic $T(\mathbf{X})$ is chosen that is thought to be powerful in distinguishing

H_0 from H_1 and whose qualitative behavior under H_1 is understood; for example, large values of T might be expected under the alternative. Next, the statistic is calculated for all permutations of the observations that are equally likely conditional on some feature of the actual data, such as the order statistics or the absolute values. These N (say) values of T are ranked in order of consistency with H_1, from most consistent to least. The pth-ranked value is then chosen as critical value for a test at level α, where p is the largest integer such that $p/N \leq \alpha$. While the actual critical region is conditional on the data and would vary in repeated samples, we shall see that the procedure does limit to α or less the unconditional probability of rejecting a true null hypothesis. The method is illustrated with two applications and accompanying numerical examples.

First, given a random sample $\mathbf{X} := (X_1, X_2, ..., X_n)'$ from a continuous, symmetric population, let us see how to test at level $\alpha \in (0,1)$ an hypothesis that specifies the population median. Without loss of generality take the null to be $H_0 : x_{.5} = 0$. (Any nonzero value can be subtracted from the observations.) Choose a statistic $T(\mathbf{X})$, such as $\bar{X}$, that will be informative about central tendency. According to Theorem 8.5 signs and absolute values of the data are independent under the stated conditions of continuity and symmetry, and under H_0 each X_j can be positive with probability 0.5 and negative with probability 0.5. Thus, the value of $T(\mathbf{x})$ that is associated with any of the 2^n possible configurations of signs, $\mathbf{S} := (S(X_1), S(X_2), ..., S(X_n))'$, can occur with probability 2^{-n}. Ranking these values of $T(\mathbf{x})$ from most favorable to H_1 to least favorable and taking the pth-ranked one as critical value defines a conditional rejection region of size α or less, where p is the greatest integer such that $2^{-n}p \leq \alpha$. Note that although the rejection region depends on observed sample $\mathbf{x}$, the fact that α bounds the test size conditional on *each* sample realization means that it is also bounds the unconditional probability of rejecting a true null hypothesis. While either the sample mean or sample sum, $\sum_{j=1}^{n} x_j$, could serve as $T(\mathbf{x})$, it simplifies calculation to use just the sum of *positives*: $\sum_{j=1}^{n} x_j^+$, where $x_j^+ = x_j \mathbf{1}_{(0,\infty)}(x_j)$. Since $\sum_{j=1}^{n} x_j^+ = .5\left(\sum_{j=1}^{n} x_j + \sum_{j=1}^{n} |x_j|\right)$, and since $\sum_{j=1}^{n} |x_j|$ is the same for all permutations of signs, the rankings of $\sum_{j=1}^{n} x_j^+$ and $\bar{x}$ would be identical.

Example 8.15. We have a single sample of size $n = 4$, $\mathbf{x} = (-3, -1, 4, 9)'$, and wish to test $H_0 : x_{.5} \leq 0$ versus $H_1 : x_{.5} > 0$ at level $\alpha = .10$. The population is assumed to be continuous and symmetrically distributed about $x_{.5}$. We will work with the sum of positives, $T(\mathbf{x}) = \sum_{j=1}^{4} x_j^+$,

Table 8.2   Ordered sums of positives for 1-sample permutation test.

$S(X_1)$	$S(X_2)$	$S(X_3)$	$S(X_4)$	$T(\mathbf{X})$
+	+	+	+	17
+	−	+	+	16
−	+	+	+	14
+	+	−	+	13
−	−	+	+	13
+	−	−	+	12
−	+	−	+	10
−	−	−	+	9
+	+	+	−	8
+	−	+	−	7
−	+	+	−	5
+	+	−	−	4
−	−	+	−	4
+	−	−	−	3
−	+	−	−	1
−	−	−	−	0

which takes the value 13 for these data. Clearly, larger values of $T(\mathbf{X})$ are expected under H_1 than under H_0. The values for the $2^n = 16$ possible arrangements of + and − signs are displayed in decreasing order in Table 8.2. Under the boundary null, each has conditional probability $1/16 = .0625$. The appropriate rejection region that limits the chance of type-I error to $\alpha = .10$ would be $\{T(\mathbf{x}) : T(\mathbf{x}) \geq 17\}$, so H_0 would not be rejected at level .10.

For the next application, suppose we have independent random samples
$$\mathbf{X}_p := (X_{1p}, X_{2p}, ..., X_{n_p p})', p \in \{1, 2\}$$
from two populations that are assumed to be continuous and identically distributed up to a shift parameter δ; i.e., $X_{k2} \sim X_{j1} + \delta$ for each $j \in \{1, 2, ..., n_1\}$ and each $k \in \{1, 2, ..., n_2\}$. We wish to test $H_0 : \delta = 0$ against $H_0 : \delta \neq 0$ at level $\alpha \in (0, 1)$. The two samples are merged into an $n_1 + n_2 = n$-vector $\mathbf{Y} := (\mathbf{X}_1', \mathbf{X}_2')'$, with $Y_i = X_{i1}$ for $i \in \{1, 2, ..., n_1\}$ and $Y_{i+n_1} = X_{i2}$ for $i \in \{1, 2, ..., n_2\}$. An informative statistic about difference in location is chosen. Let $\mathbf{R}$ be the n-vector of ranks of the merged samples, R_i representing the number of elements of $\mathbf{Y}$ no larger than Y_i, $i \in \{1, 2, ..., n\}$; and let $\mathbf{Y}_{(\cdot : n)}$ be the vector of order statistics, with $\mathbf{y}_{(\cdot : n)}$ as the observed values. Rearrangements (permutations) of the elements of $\mathbf{Y}$ change the ranks but leave the order statistics unchanged. To any such arrangement there correspond a mean of the first n_1 elements and a mean

of the last n_2 elements and hence a value of $T(\mathbf{Y})$. Theorem 8.7 tells us that $\mathbf{R}$ and $\mathbf{Y}_{(\cdot)}$ are independent under H_0, and this combined with Theorem 8.6 tells us that $\Pr\left(\mathbf{R} = \mathbf{i} \mid \mathbf{Y}_{(\cdot)} = \mathbf{y}_{(\cdot)}\right) = 1/n!$ for any vector $\mathbf{i}$ that is a permutation of the first n integers. Thus, conditional on $\mathbf{Y}_{(\cdot:n)} = \mathbf{y}_{(\cdot:n)}$ each value of $T(\mathbf{y})$ corresponding to a given permutation of the data has probability $1/n!$. However, not all $n!$ permutations will alter $T(\mathbf{y})$, since it is unaffected by rearrangements of the n_1 elements of $\mathbf{x}_1$ alone or the n_2 elements of $\mathbf{x}_2$. This means that we can consider just the $\binom{n}{n_1}$ (conditionally) equally likely values of $T(\mathbf{y})$ corresponding to the ways that n_1 positions can be chosen for the elements of $\mathbf{X}_1$. Calculating $T(\mathbf{y})$ for each of these and ranking by absolute value, we choose as critical value the pth largest, where p is the greatest integer such that $p/\binom{n}{n_1} \leq \alpha$. Again, while the critical value is data dependent, the procedure limits to α the unconditional probability of type-I error.

As the "informative statistic" one could choose the difference in sample sums $\sum_{i=1}^{n_1} Y_i - \sum_{i=n_1+1}^{n} Y_i$, but it is easier just to use the sum of the values of the smaller sample, the ordering of which would be the same.

Exercise 8.25. *Show that $\sum_{i=1}^{n_1} y_i - \sum_{i=n_1+1}^{n} y_i$ and $\sum_{i=1}^{n_1} y_i$ are identically ordered among the $\binom{n}{n_1}$ possible ways of assigning the data to the two samples.*

Example 8.16. With sample data $\mathbf{x}_1 = (2,4)'$ and $\mathbf{x}_2 = (3,6,9)'$, and supposing $X_{2k} \sim X_{1j} + \delta$ for $j \in \{1,2\}$ and $k \in \{1,2,3\}$, we test $H_0 : \delta = 0$ versus $H_1 : \delta > 0$ at level $\alpha = .10$. The merged sample is $\mathbf{y} = (2,4,3,6,9)'$ and the sum of the elements of the smaller sample is $T(\mathbf{y}) = 2 + 4 = 6$. Small values of $T(\mathbf{y})$ are expected under H_1. Table 8.3 presents in increasing order the $\binom{5}{2} = 10$ values of $T(\mathbf{y})$ corresponding to the positions chosen for $\mathbf{x}_1$. The conditional rejection region of size .10 comprises values of $T(\mathbf{y})$ less than or equal to 5, so H_0 is not rejected. The P value of the test is 0.20.

8.8.2.3 *Unconditional Tests Based on Signs and Ranks*

For a permutation test the conditional distribution of the test statistic under H_0 has to be worked out by calculating its value for each of a set of (conditionally) equally likely permutations of the data. We now look at some distribution-free one- and two-sample location tests for which rejection regions can be determined without reference to the observed sample.

Table 8.3   Ordered absolute differences in means for 2-sample permutation test.

y_1	y_2	y_3	y_4	y_5	$T(\mathbf{y})$
2	3	4	6	9	5
2	4	3	6	9	6
3	4	2	6	9	7
2	6	3	4	9	8
3	6	2	4	9	9
6	4	3	2	9	10
2	9	3	6	4	11
3	9	2	6	4	12
9	4	3	6	1	13
6	9	3	2	4	15

These are really just permutation tests in which the observations are represented not by their numerical values but by either their signs or their
ranks. Since the numerical values are not involved, the distributions of
the test statistics can be worked out and tabulated once and for all, just
as for conventional tests based on parametric models. Although typically
less powerful than permutation tests, these sign and rank methods are less
demanding computationally.

The Sign Test   A simple test for the one-sample location problem can
be based just on the signs of the data. Given a random sample $\mathbf{X} :=$
$(X_1, X_2, ..., X_n)'$, we test $H_0 : x_{.5} = 0$ *vs.* a one- or two-sided alternative,
where $x_{.5}$ is the population median. (If the interest is in $H_0 : x_{.5} = \delta_0$,
subtract δ_0 from each observation.) The population is assumed to be distributed symmetrically about $x_{.5}$ and, for full generality, to be continuous.[21]
With $S(x)$ as defined at (8.24) we construct an indicator function for the
sign as

$$I(x) := \frac{S(x) + 1}{2}.$$

Thus, in our usual but less concise notation, $I(x) \equiv \mathbf{1}_{[0,\infty)}(x)$; i.e., $I(x) = 1$
if $x \geq 0$ and $I(x) = 0$ otherwise. Given symmetry and continuity, r.v.s
$\{I(X_j)\}_{j=1}^n$ are i.i.d. as Bernoulli, with $\theta := \Pr\left[I(X_1) = 1\right] \gtreqless .5$ according
as $x_{.5} \gtreqless 0$. Statistic $T(\mathbf{X}) := \sum_{j=1}^n I(X_j)$ is therefore distributed as
$B(n, \theta)$ (binomial). The appropriate size-α rejection region for testing H_0 :

---

[21]For the sign test it is enough that there be no probability mass at the specific point
δ_0.

$x_{.5} \leq 0$ *vs.* $H_1 : x_{.5} > 0$ would be $\mathcal{R}_\alpha = \{\mathbf{x} : T(\mathbf{x}) \geq b^\alpha\}$, where b^α is the smallest integer such that

$$2^{-n} \sum_{j=b_\alpha}^{n} \binom{n}{j} \leq \alpha.$$

For $H_0 : x_{.5} = 0$ *vs.* $H_1 : x_{.5} \neq 0$ the rejection region would comprise values of $T(\mathbf{x})$ lying outside the open interval $(n - b^{\alpha/2}, b^{\alpha/2})$. Although these regions depend on n, they do not have to be determined from the sample data as they did for permutation tests.

Wilcoxon's Signed-Rank Test The sign test is indeed almost distribution free, in that it is valid for all continuous, symmetric distributions, but it extracts very little information from the sample. For example, $T(\mathbf{x})$ would take the same value for $\mathbf{x} = (-2, 1, 1)'$ as for $\mathbf{x} = (-2, 1, 10)$, even though the latter is clearly stronger evidence against the hypothesis $x_{.5} = 0$. Of course, a statistic that depends on the numerical values would not be distribution free, but the sample does contain other information that preserves this feature; namely, the *ranks* of the observations. The *signed-rank test* of $H_0 : x_{.5} = \delta_0$, due to Wilcoxon (1945), incorporates this additional information.

To develop the Wilcoxon test, we assume that δ_0 has been subtracted from the data, so that $x_{.5} = 0$ under the null. Continuity and symmetry are again maintained. To each observation X_j we associate both the indicator of its sign $I(X_j)$ and its rank among the absolute values, $R(|X_j|)$. For brevity we refer to $R(|X_j|)$ as the "absolute rank" of X_j; it equals the number of elements of the sample whose absolute values are no greater than $|X_j|$. Now letting

$$\mathbf{I} := (I(X_1), I(X_2), ..., I(X_n))'$$
$$\mathbf{R} := (R(|X_1|), R(|X_2|), ..., R(|X_n|))'$$

be vectors of these statistics, the signed-rank statistic is given by

$$W^+ := \mathbf{I'R} = \sum_{j=1}^{n} I(X_j) R(|X_j|).$$

Thus, W^+ equals the sum of the absolute ranks of the positive observations. Clearly, the range of W^+ comprises all the integers from zero (when all observations are nonpositive) to $n(n+1)/2$, which is the sum of the first n positive integers. Since positives and negatives are equally likely under H_0, we guess at once that the distribution is symmetric about $n(n+1)/4$

under the null and that more than half the probability mass would lie to the right (left) of this point if $x_{.5} > 0$ (< 0). Thus, large (small) values of W^+ would favor alternatives $H_1 : x_{.5} > 0$ (< 0).

Determining critical values for tests of one- or two-sided alternatives requires working out explicitly the distribution of W^+ under the null. The starting point is the following result, which follows easily from Theorems 8.5 and 8.6:

Proposition 8.1. *The joint distribution of* $\mathbf{I}$ *and* $\mathbf{R}$ *is given by*

$$\Pr\left(\mathbf{I} = \mathbf{s}, \mathbf{R} = \mathbf{i}\right) = \frac{2^{-n}}{n!} \tag{8.31}$$

for each n-vector $\mathbf{s}$ *comprising zeroes and ones and each vector* $\mathbf{i}$ *that is a permutation of the first n positive integers.*

Exercise 8.26. *Prove Proposition 8.1.*

The event $W^+ = \mathbf{I}'\mathbf{R} = w$ for some $w \in \{0, 1, 2, ..., n(n+1)/2\}$ is the event that the sum of the absolute ranks of the positive values is w. Let $N_n(w)$ be the number of subsets of the first n integers whose sum is w, counting the empty set as one when $w = 0$. Corresponding to any one such subset there is a number of positive signs $p \in \{0, 1, ..., n\}$ and a set of p integers corresponding to the ranks of the positive values. For each p there are $\binom{n}{p}$ ways to choose p positives, and for each of these the p positive and $n - p$ negative observations can be rearranged in $p!(n-p)!$ ways. Thus, there is a total of $\binom{n}{p}p!(n-p)! = n!$ equally likely ways of picking p positives. Applying (8.31), the probability of any one configuration of positives and ranks whose sum is j is therefore 2^{-n}, and so the distribution of W^+ is given by

$$f_{W^+}(w) = N_n(w)2^{-n}\mathbf{1}_{\{0,1,2,...,n(n+1)/2\}}(w).$$

For example, with $n = 3$ we have $f_{W^+}(w) = 1/8$ for $w \in \{0, 1, 2, 4, 5, 6\}$ and $f_{W^+}(3) = 1/4$.

Tables of the distribution for various values of n are available in many statistics texts,[22] and the Wilcoxon test is a standard feature in statistical software. Upper-tail probabilities $\Pr\left(W^+ \geq j\right) = \sum_{w=j}^{n(n+1)/2} N_n(w)2^{-n}$ are required for testing $H_0 : x_{.5} \leq 0$ vs. $H_1 : x_{.5} > 0$, the critical value w^α for a test at level $\alpha \in (0,1)$ being the smallest integer w^α such that $\Pr\left(W^+ \geq w^\alpha\right) \leq \alpha$. Exploiting the symmetry of f_{W^+} about $n(n+1)/4$, the lower-tail critical value for testing $H_0 : x_{.5} \geq 0$ vs. $H_1 : x_{.5} < 0$ is

[22]E.g., Hollander and Wolfe (1973).

$w_\alpha = n(n+1)/2 - w^\alpha$, and the rejection region for a two-sided test at level α is

$$\left\{ w^+ : \left| w^+ - \frac{n(n+1)}{4} \right| \geq w^{\alpha/2} - \frac{n(n+1)}{4} \right\}.$$

Example 8.17. The first column of the table below contains ten pseudo-random observations distributed as $N(1,1)$, rounded to three decimal places and ranked by absolute value. The second column contains the signed ranks. Eight values are positive, and the probability of obtaining eight or more positive observations under $H_0 : x_{.5} \leq 0$ is at most

$$1 - \left[\binom{10}{0} + \binom{10}{1} + \binom{10}{2} \right] 2^{-10} \doteq .0547.$$

Thus, the sign test does not reject H_0 in favor of $H_1 : x_{.5} > 0$ at the .05 level. However, the sum of the signed ranks is $W^+ = (10 + 9 + \cdots + 3) = 52$, and (using statistical software) $\Pr(W^+ \geq 52 \mid H_0)$ is at most 0.007. Thus, the signed-rank test does reject this (known-to-be-false) null at the .05 level.

X	$I(X)R(|X|)$
2.572	10
2.363	9
2.113	8
1.508	7
1.502	6
0.991	5
0.965	4
0.590	3
-0.475	0
-0.292	0

The example suggests that the signed-rank test has greater power than the sign test, even though its validity depends on no stronger maintained hypothesis. Since the data in the example were actually distributed as normal, a test based on Student's t would also have been appropriate—and, as pointed out in Example 8.7, uniformly most powerful. This illustrates the trade-off that one faces in choosing test procedures. While distribution-free methods are valid under more general conditions, parametric procedures tend to have greater power when their more restrictive conditions actually hold.

Exercise 8.27. *Using statistical software, calculate the minimum P value for the t test of $H_0 : \mu \leq 0$ vs. $H_1 : \mu > 0$ using the data of Example 8.17.*

The Wilcoxon–Mann–Whitney (W-M-W) Rank-Sum Test We conclude the discussion of distribution-free tests with an unconditional procedure for the two-sample location problem.[23] Again we have independent random samples $\mathbf{X}_1 := (X_{11}, X_{21}, ..., X_{n_1 1})'$ and $\mathbf{X}_2 := (X_{12}, X_{22}, ..., X_{n_2 2})'$ from two populations that are assumed to be continuous and identically distributed up to a (possibly nonzero) shift parameter δ; i.e., $X_{k2} \sim X_{j1} + \delta$ for each $j \in \{1, 2, ..., n_1\}$ and each $k \in \{1, 2, ..., n_2\}$. Note that symmetry is not maintained. We wish to test $H_0 : \delta = 0$ against a one- or two-sided alternative at level $\alpha \in (0, 1)$. The two samples are merged into an $n_1 + n_2 = n$-vector $\mathbf{Y} := (\mathbf{X}'_1, \mathbf{X}'_2)'$ with $Y_j = X_{j1}$ for $j \in \{1, 2, ..., n_1\}$ and $Y_{j+n_1} = X_{j2}$ for $j \in \{1, 2, ..., n_2\}$, and ranks $\{R(Y_j)\}_{j=1}^{n}$ are recorded. Put $W := \sum_{j=1}^{n_2} R(Y_{j+n_1})$, which is the sum of the ranks of the $\mathbf{X}_2$ sample. Then the range of W is the set

$$\mathbb{W} = \left\{ \frac{n_2(n_2 + 1)}{2}, \frac{n_2(n_2 + 1)}{2} + 1, ..., \frac{n_2(n_1 + n + 1)}{2} \right\},$$

extending from the sum of the first n_2 integers to the sum of the integers from $n_1 + 1$ through n. Under $H_0 : \delta = 0$ one expects the elements of $\mathbf{X}_2$ to be distributed more or less uniformly among the n observations of the merged samples, and one guesses that the distribution W is symmetric about $n_2(n + 1)/2$, the median of $\mathbb{W}$. If, however, $\delta > 0$ (< 0) one expects W to take on a value nearer the upper (lower) end of the range.

Determining critical values of W requires working out the distribution under $H_0 : \delta = 0$. The value of W is determined once n_2 of the $n = n_1 + n_2$ positions is selected for the elements of $\mathbf{X}_2$. Under H_0 all of these $\binom{n}{n_2} = \binom{n}{n_1}$ arrangements are equally likely. For $w \in \mathbb{W}$ let $N_{n_1, n_2}(w)$ be the number of subsets comprising n_2 of the first $n_1 + n_2$ integers whose elements sum to w. Then the distribution of W is given by

$$f_W(w) = \frac{N_{n_1, n_2}(w)}{\binom{n}{n_2}}, w \in \mathbb{W}.$$

Upper-tail probabilities $\Pr(W \geq j) = \sum_{w=j}^{n_2(n_1 + n + 1)/2} f_W(w)$, tabulated in many statistics texts[24], are used to find critical values w^α for level-α tests against $H_1 : \delta > 0$. The symmetry of the distribution implies that the corresponding w_α for tests against $H_1 : \delta < 0$ is $n_2(n + 1) - w^\alpha$. Finally, the rejection region for the two-sided alternative $H_1 : \delta \neq 0$ is

---

[23] A test based on the W statistic was first proposed by Wilcoxon (1945). The statistic proposed by Mann and Whitney (1947) can be expressed as a one-to-one function of W.
[24] E.g., Hollander and Wolfe (1973).

Table 8.4 Possible ranks of elements of second sample and values of W-M-W statistic.

$R(y_3)$	$R(y_4)$	$R(y_5)$	W
3	4	5	12
2	4	5	11
2	3	5	10
1	4	5	10
2	3	4	9
1	3	5	9
1	3	4	8
1	2	5	8
1	2	4	7
1	2	3	6

$\left\{ w : |w - n_2(n+1)/2| \geq w^{\alpha/2} - n_2(n+1)/2 \right\}$. Statistical software packages commonly calculate exact P values for the W-M-W test.

Exercise 8.28. *Show that the W-M-W statistic is distributed symmetrically about the mean value $n_2(n+1)/2$. (Hint: Consider how the numerical value of W is changed when the signs of all members of the combined sample are reversed.)*

Example 8.18. With sample data $\mathbf{x_1} = (2,4)'$, $\mathbf{x_2} = (3,6,9)'$ we test $H_0 : \delta = 0$ versus $H_1 : \delta > 0$ at level $\alpha = .10$. The merged sample is $\mathbf{y} = (2,4,3,6,9)'$, and the realization of $W = \sum_{j=3}^{n} R(Y_j)$ is $3+4+5 = 11$. Table 8.4 shows the possible ranks of the elements of $\mathbf{x_2}$ and the corresponding values of W. For $n = 5, n_2 = 3$ the distribution of W is

$$f_W(w) = \begin{cases} \frac{1}{10}, & w = 6, 7, 11, 12 \\ \frac{2}{10}, & w = 8, 9, 10 \end{cases}.$$

Thus, $\Pr(W \geq 11) = 0.20$, so that H_0 cannot be rejected at the .10 level.

8.9 Pitfalls and Red Flags

Both the producer and the consumer of statistical research need to be aware of common practices that, either by design or through inadvertence, can lead to erroneous conclusions. Five prominent examples are described below. The first three, drawn from the penetrating critical article by Young and Karr (2011), pertain to hypothesis tests on "observational" data; that is, data obtained by observing "natural" rather than controlled experiments. The last two abuses are more general, applying to estimation and testing alike and to data of both sorts.

(i) Multiple testing. A researcher asks a group of pregnant women a series of questions about their personal characteristics and behavior: marital status, age, overall health, dietary and smoking habits, amounts of exercise, etc. Then, after they have given birth, the women are queried again to assess the frequencies of certain events or characteristics. Based on the responses the researcher issues the startling finding— "confirmed" by a high level of statistical significance—that children of women who had eaten nonfat yogurt during pregnancy were 1.6 times more likely to develop asthma.[25]

Findings of this sort appear continually in the media; and the more implausible or surprising they seem, the more attention they receive. While they do sometimes reveal genuine, previously unsuspected causal relations, they are often not confirmed in controlled experiments—e.g., a randomized trial in which pregnant women are assigned to various yogurt regimens. It is not hard to see why such results often fail to replicate. Represent the outcome of the test as $\omega \in A$ or $\omega \in A^c$ (child is asthmatic, or not) and the presence or absence of some supposed causal factor as $\omega \in C$ or $\omega \in C^c$. The assessment of dichotomous outcomes splits the sample into two groups, as does the presence or absence of the supposed causal condition. A significant relation $C \implies A$ corresponds to a sizeable overlap of the C and A groups, in the sense that the frequency of $C \cap A$ relative to C exceeds the relative frequency of A in the entire sample. While an inference from one such comparison might be perfectly valid, a proper judgment requires more information about the experimental design. In particular, one needs to know how many questions were asked initially; that is, how many *potential* causal factors were there? If another 49 questions had been asked about the women's habits besides the preference for yogurt, the responses would have created a total of 100 subsets, $\left\{C_j, C_j^c\right\}_{j=1}^{50}$. It would be a great surprise if at least one of these subsets did not contain an above-average proportion of women with an asthmatic child.

---

[25] A finding of this general nature was actually reported in various media outlets in September 2011; e.g., http://junkscience.com/2011/09/17/low-fat-yogurt-during-pregnancy-causes-asthma-hay-fever-in-kids/. Nothing about the validity of this particular study is to be inferred from the hypothetical example presented here.

Moreover, the publicized dichotomy A, A^c might have been just one of many conditions about which the women were queried after they gave birth: was the child a male, did it have brown eyes, low birth weight, superior intelligence, twelve toes, etc. If enough comparisons are made it is almost certain that the association between *some* initial factor and *some* supposed resultant will achieve an impressive P value. Of course, it is *that* association which will be reported and—if sufficiently provocative—trumpeted in the media. To assess such findings intelligently, one must know the full scope of the study; and the researcher who fails to make this known is simply behaving unethically.

(ii) Selection bias. This arises when a sample is chosen in a way that skews the result of a test. Sometimes this is done deliberately to obfuscate or to twist the outcome, as when observations are deleted from or added to a sample until a desired result is obtained. Here are two less blatant examples, which could result either from a researcher's bad intentions or from mere ignorance.

A medical researcher measures the blood pressure of a group of individuals and classifies each person as "hypertensive" or "normal". The former group is given a medication designed to lower blood pressure and retested at some later time, with the finding that there was a "significant" reduction. Although the medication may indeed have been effective, the statistical design would most likely have overstated the effect, owing to a phenomenon known as "regression to the mean". An individual's blood pressure typically fluctuates considerably over the course of a day and in response to exercise, food intake, stress, etc. Those assigned initially to the hypertensive group likely included some individuals whose pressures happened to be unusually high at the moment of testing but which would later have reverted to more normal levels—that is, regressed to the mean—even without the medication. An investigator might defend such a research design on the grounds that it would be improper to treat asymptomatic individuals, but that issue could be avoided just by monitoring a control group that receives no treatment regardless of symptoms. The change in pressure of that group's hypertensives could then be compared with the change for those who were treated.[26]

Practitioners in finance commonly evaluate investment strategies by what is known as "back testing". While the general approach is sound,

---

[26]The example is taken from Senn (2011), who uses it to show that regression to the mean can contribute to the "placebo effect" that is often observed in drug trials.

many follow a sampling procedure that overstates performance. For example, suppose one takes as the sample the 500 stocks in the current S&P 500 index and simulates the gain from a multi-year trading strategy using historical records of their prices. The return from the strategy is then compared with that of a passive portfolio that simply tracks the S&P index over the period. The fallacy here is that the composition of the S&P 500 changes over time, the least successful companies being replaced periodically by others. While the declining share prices of the dropouts would have been reflected in the contemporary values of the index, those companies would have been excluded from a back test on companies in the *current* S&P, thus biasing the results in favor of the active strategy. Clearly, the proper approach is to apply the investment strategy to the S&P as it was constituted when each simulated trade was made, adopting a protocol for dropouts and replacements that could actually be followed in real time.

(iii) Multiple modeling. This is a fallacy encountered most commonly in regression analyses, but it can arise in other contexts as well. As we will see in Chapter 11, the goal of regression analysis is to assess how the conditional mean of a variate Y responds to a specified set of "covariates" $\mathbf{x} = (x_1, x_2, ..., x_k)$, as $E(Y \mid \mathbf{x}) = g(\mathbf{x}; \boldsymbol{\theta})$. As a maintained hypothesis, g is specified up to a set of unknown parameters $\boldsymbol{\theta}$, which govern the sensitivity of the response variable to elements of $\mathbf{x}$. A sample $\{Y_j, \mathbf{x}_j\}_{j=1}^n$ would be used to estimate these parameters and/or to test various hypotheses. The outcome of a test of the influence of some particular variate $x_i \in \mathbf{x}$ typically depends on what other covariates are assumed to enter the model and on the functional form of g. As in the multiple testing fallacy, searching over many specifications enhances the chance of statistical support for any given preconception. Such possibilities for deception exist even in univariate settings. For example, in testing $H_0 : \mu = 0$ from a random sample, one could plausibly adopt any of several models for the population having mean μ. When multiple models are considered, the proper course is to make this known and to report the test results for each specification. One should be suspicious when no such report accompanies the findings.

(iv) Multiple statistical procedures. There are usually many ways to attack a particular inferential problem, even when a model for the data has been selected. For example, in a large-sample parametric setting one could test $H_0 : \boldsymbol{\theta} = \boldsymbol{\theta}_0$ using any of the three classical asymptotic tests—likelihood ratio, Lagrange multiplier, or Wald; and one could estimate θ using the basic or generalized method of moments, maximum likelihood, or any of several robust procedures that reduce the influence of outliers. The creative but ethically disadvantaged researcher with a vested interest in the outcome can offer a plausible-seeming justification for almost any such choice.

(v) Selective reporting. Here is a headline that no one will ever see: "Statistical tests show no link between a child's asthma and mom's choice of yogurt." Negative results are rarely of general interest unless they overturn some established view. For that reason they are seldom reported even in academic journals, much less in the public media. Of more concern, one who can control the publication of results has an obvious incentive to filter them for personal benefit. Thus, a journal editor who had accepted a scientific paper may be reluctant later on to publish conflicting evidence, and companies whose products fail to pass in-house tests for safety or efficacy have a clear incentive to suppress negative findings.[27]

8.10   Solutions to Exercises

1. Using the normal approximation for $\bar{X}$,

$$\Pr\left(13.0 \le \bar{X} \le 13.045 \mid \theta_0\right) = \Pr\left(0 \le \frac{\bar{X} - 13}{\sqrt{13/100}} \le \frac{.045}{\sqrt{13/100}}\right)$$

$$\doteq \Pr\left(0 \le Z \le .1248\right) \doteq .05.$$

For powers we have

$$\Pr\left(13.0 \le \bar{X} \le 13.045 \mid \theta\right) \doteq \Pr\left(\frac{13 - \theta}{\sqrt{\theta/100}} \le Z \le \frac{13.045 - \theta}{\sqrt{\theta/100}}\right).$$

To illustrate, when $\theta = 12.4$ this is

$$\Pr\left(\frac{13 - 12.4}{\sqrt{12.4/100}} \le Z \le \frac{13.045 - 12.4}{\sqrt{12.4/100}}\right) = \Pr\left(1.70 \le Z \le 1.83\right)$$

$$\doteq .011,$$

and when $\theta = 15.2$ it is

$$\Pr\left(\frac{13 - 15.2}{\sqrt{15.2/100}} \le Z \le \frac{13.045 - 15.2}{\sqrt{15.2/100}}\right) = \Pr\left(-5.64 \le Z \le -5.53\right)$$

$$\doteq .000.$$

---

[27]Ince (2011) presents a compelling account of editorial stonewalling of attempts to discredit fraudulent cancer research. For accounts of some especially sordid practices by drug companies see the article by Hsu (2010).

2. In each case we must find a critical value of the test statistic, denoted q_n, such that a test that rejects $H_0 : \theta = 13$ for values greater than q_n has probability of type-I error no larger than .05 and has power at least .95 at $\theta = 14$.

(i) Assuring size no greater than $\alpha = .05$ for the $\bar{X}$ test requires

$$\Pr\left(\frac{\bar{X} - 13}{\sqrt{13/n}} \geq \frac{q_n - 13}{\sqrt{13/n}} \mid H_0\right) \doteq \Pr\left(Z \geq \frac{q_n - 13}{\sqrt{13/n}}\right) \leq .05,$$

whence $q_n \geq 13 + 1.645\sqrt{13/n}$. Assuring power $1 - \beta \geq .95$ at $\theta = 14$ requires

$$\Pr\left(\frac{\bar{X} - 14}{\sqrt{14/n}} \geq \frac{q_n - 14}{\sqrt{14/n}} \mid \theta = 14\right) \doteq \Pr\left(Z \geq \frac{q_n - 14}{\sqrt{14/n}}\right) \geq .95,$$

which requires that $q_n \geq 14 - 1.645\sqrt{14/n}$. Equating the two expressions for q_n and solving for n give $n_{\bar{X}} = \left[1.645\left(\sqrt{13} + \sqrt{14}\right)\right]^2 \doteq 146.075$ or about 146.

(ii) Assuring size no greater than $\alpha = .05$ for the S^2 test requires

$$\Pr\left(\frac{S^2 - 13}{\sqrt{(13 + 2 \cdot 13^2)/n}} \geq \frac{q_n - 13}{\sqrt{(13 + 2 \cdot 13^2)/n}}\right) \doteq \Pr\left(Z \geq \frac{q_n - 13}{\sqrt{351/n}}\right)$$
$$\leq .05,$$

whence $q_n \geq 13 + 1.645\sqrt{351/n}$. Assuring power $1 - \beta \geq .95$ at $\theta = 14$ requires

$$\Pr\left(\frac{S^2 - 14}{\sqrt{(14 + 2 \cdot 14^2)/n}} \geq \frac{q_n - 14}{\sqrt{(14 + 2 \cdot 14^2)/n}}\right) \doteq \Pr\left(Z \geq \frac{q_n - 14}{\sqrt{406/n}}\right)$$
$$\geq .95,$$

whence $q_n = 14 - 1.645\sqrt{406/n}$. Equating and solving give

$$n_{S^2} = \left[1.645\left(\sqrt{351} + \sqrt{406}\right)\right]^2 \doteq 4091.508$$

or about 4,092.

3. We have

$$\frac{L(\mathbf{X}; \theta_0)}{L(\mathbf{X}; \theta_1)} = \left(\frac{\theta_0}{\theta_1}\right)^{\sum_{j=1}^{n} X_j} e^{n(\theta_1 - \theta_0)} \leq \lambda$$

if and only if

$$\log\left(\frac{\theta_0}{\theta_1}\right) \sum_{j=1}^{n} X_j \leq n(\theta_0 - \theta_1) + \log \lambda.$$

$\log(\theta_0/\theta_1)$ being positive under H_1, the above is equivalent to

$$\sum_{j=1}^{n} X_j \le \frac{n(\theta_0 - \theta_1) + \log \lambda}{\log(\theta_0/\theta_1)}.$$

Since $\sum_{j=1}^{n} X_j$ is necessarily nonnegative, the rejection region is of the form

$$\mathcal{R}_\alpha = \left\{ \mathbf{x} : 0 \le \sum_{j=1}^{n} x_j \le c_\alpha \right\},$$

where c_α is the largest number such that $\Pr\left(\sum_{j=1}^{n} X_j \le c_\alpha \mid H_0\right) \le \alpha$. Because $T(\mathbf{X}) = (\bar{X} - \theta_0) / \sqrt{\theta_0/n}$ is asymptotically standard normal (of which all of $\Re$ is the support), the test can be based also on $\mathcal{R}'_\alpha = \{\mathbf{x} : T(\mathbf{x}) \le -z^\alpha\}$. (Obviously, however, obtaining a sample containing an $X_j < 0$ would lead one to abandon the *maintained* hypothesis that $X \sim P(\theta)$.)

4. Let $\mathcal{R}_\alpha := \mathcal{R}_{\alpha'} \mathbf{1}_{(-\infty, q]}(Q) + \mathcal{R}_{\alpha''} \mathbf{1}_{(q, \infty)}(Q)$ represent the random(ized) critical region. Then for $q \in (0, 1)$ the size of the randomized test is

$$\Pr(\mathbf{X} \in \mathcal{R}_\alpha \mid H_0) = \Pr(\mathbf{X} \in \mathcal{R}_{\alpha'} \mid H_0) \Pr(\mathcal{R}_\alpha = \mathcal{R}_{\alpha'})$$
$$+ \Pr(\mathbf{X} \in \mathcal{R}_{\alpha''} \mid H_0) \Pr(\mathcal{R}_\alpha = \mathcal{R}_{\alpha''})$$
$$= \alpha' q + \alpha''(1 - q),$$

and putting $q := (\alpha'' - \alpha)/(\alpha'' - \alpha')$ gives $\Pr(\mathbf{X} \in \mathcal{R}_\alpha \mid H_0) = \alpha$.

5. The likelihood ratio is

$$\frac{L(\mathbf{X}; \theta_0)}{L(\mathbf{X}; \theta)} = \left(\frac{\theta}{\theta_0}\right)^n \exp\left[n\bar{X}\left(\theta^{-1} - \theta_0^{-1}\right)\right].$$

For any $\theta > \theta_0$ this is less than or equal to some positive λ if and only if

$$n\bar{X} \ge \frac{n\log(\theta/\theta_0) - \log \lambda}{\theta_0^{-1} - \theta^{-1}} =: c > 0.$$

In the specific case $\theta_0 = 2$ it is easy to see that $12\bar{X} = \sum_{j=1}^{12} X_j$ is distributed as $\chi^2(24)$. (The c.f. of $12\bar{X}$ under H_0 is $(1 - 2it)^{12}$.) Thus, the desired value of c for a test at level $\alpha = .05$ is the upper-$.05$ quantile of $\chi^2(24)$, or $c^{.05} = 36.415$. Since $12\bar{x} = 37$ does fall into the rejection region, we reject H_0 at level $.05$.

6. If $\theta < \theta_0$, small values of $L(\mathbf{X}; \theta_0)/L(\mathbf{X}; \theta)$ correspond to small values of $n\bar{X}$, whereas when $\theta > \theta_0$ small values of the likelihood ratio imply large values of $n\bar{X}$. Since $EX = \theta_0$ and $VX = \theta_0^2$ under H_0, the statistic $T(\mathbf{X}) = (n\bar{X} - n\theta_0)/\sqrt{n\theta_0^2}$ is approximately normal when n is large. The normal density being unimodal, differentiable, and symmetric about the origin, an approximate u.m.p. unbiased test of size α is based on

$$\mathcal{R}_\alpha = \left\{ \mathbf{x} : \left| \frac{n\bar{x} - n\theta_0}{\sqrt{\theta_0^2 n}} \right| \geq z^{\alpha/2} \right\} = \left\{ \mathbf{x} : \left| \frac{\bar{x} - \theta_0}{\sqrt{\theta_0^2/n}} \right| \geq z^{\alpha/2} \right\}.$$

7. Differentiating

$$\mathfrak{P}_{\bar{X}_n}(\theta) = 1 - \Phi \left[\frac{\theta_0 - \theta + z^\alpha \sigma_{\bar{X}_n}(\theta_0)}{\sigma_{\bar{X}_n}(\theta)} \right]$$

$$\mathfrak{P}_{S_n^2}(\theta) = 1 - \Phi \left[\frac{\theta_0 - \theta + z^\alpha \sigma_{S_n^2}(\theta_0)}{\sigma_{S_n^2}(\theta)} \right]$$

with respect to θ and evaluating at θ_0 give

$$\mathfrak{P}'_{\bar{X}_n}(\theta)|_{\theta=\theta_0} = f_Z(z^\alpha)\sqrt{n} \left[\frac{1}{\sqrt{\theta_0}} + \frac{z^\alpha \sigma_{\bar{X}_n}(\theta_0)}{2\theta_0^{3/2}} \right]$$

$$= f_Z(z^\alpha)\sqrt{n} \left[\frac{1}{\sqrt{\theta_0}} + \frac{z^\alpha}{2\theta_0\sqrt{n}} \right],$$

$$\mathfrak{P}'_{S_n^2}(\theta)|_{\theta=\theta_0} = f_Z(z^\alpha)\sqrt{n} \left[\frac{1}{\sqrt{\theta_0 + 2\theta_0^2}} + \frac{z^\alpha \sigma_{S_n^2}(\theta_0)(1 + 4\theta_0)}{2(\theta_0 + 2\theta_0^2)^{3/2}} \right]$$

$$= f_Z(z^\alpha)\sqrt{n} \left[\frac{1}{\sqrt{\theta_0 + 2\theta_0^2}} + \frac{z^\alpha(1 + 4\theta_0)}{(\theta_0 + 2\theta_0^2)\sqrt{n}} \right],$$

where f_Z represents the standard normal p.d.f. Thus,

$$\lim_{n\to\infty} \left[\frac{\mathfrak{P}'_{\bar{X}_n}(\theta)|_{\theta=\theta_0}}{\mathfrak{P}'_{S_n^2}(\theta)|_{\theta=\theta_0}} \right]^2 = \left[\lim_{n\to\infty} \frac{\mathfrak{P}'_{\bar{X}_n}(\theta)|_{\theta=\theta_0}}{\mathfrak{P}'_{S_n^2}(\theta)|_{\theta=\theta_0}} \right]^2 = 1 + 2\theta.$$

8. Now σ^2, μ correspond to ϕ, ψ, with parameter space $\Theta = \Re_+ \times \Re$. The unrestricted m.l.e. for σ^2 is still $\hat{\sigma}_n^2 = M_2 = (n-1)S^2/n$, while the restricted and unrestricted m.l.e.s for μ are $\hat{\mu}_n = \check{\mu}_n = \bar{X}$. The l.r. statistic is thus

$$\Lambda(\mathbf{X}; \sigma_0^2) = \frac{(2\pi\sigma_0^2)^{-n/2} \exp\left[-\frac{1}{2\sigma_0^2} \sum_{j=1}^n (X_j - \bar{X})^2 \right]}{(2\pi\hat{\sigma}_n^2)^{-n/2} \exp\left[-\frac{1}{2\hat{\sigma}_n^2} \sum_{j=1}^n (X_j - \bar{X})^2 \right]}.$$

Algebra reduces this to $\Lambda(\mathbf{X}; \sigma_0^2) = (e/n)^{n/2} C(\mathbf{X})^{n/2} e^{-C(\mathbf{X})/2}$, where $C(\mathbf{X}) := (n-1) S^2/\sigma_0^2$ is distributed as $\chi^2(n-1)$ under H_0. Noting that $\Lambda(\mathbf{X}; \sigma_0^2)$ increases from zero to a maximum of unity as $C(\mathbf{X})$ increases from zero to n, then declines monotonically toward zero beyond, we see that $\Lambda(\mathbf{X}; \sigma_0^2) \le \lambda_\alpha$ holds whenever C is either sufficiently small or sufficiently large. Clearly, large values of S^2 should be taken as favoring H_1^+, so the one-sided rejection region $\mathcal{R}_\alpha^+$ is appropriate in this case, whereas $\mathcal{R}_\alpha^-$ is appropriate for H_1^-. An "equal-tails" test for H_1 (not unbiased) would use rejection region

$$\mathcal{R}_\alpha = \left\{ \mathbf{x} : C(\mathbf{x}) \le c_{\alpha/2}, C(\mathbf{x}) \ge c^{\alpha/2} \right\}.$$

Alternatively, one could find the $p \in (0,1)$ that minimizes $c^{(1-p)\alpha} - c_{p\alpha}$ and choose

$$\mathcal{R}_\alpha = \left\{ \mathbf{x} : C(\mathbf{x}) \le c_{p\alpha}, C(\mathbf{x}) \ge c^{(1-p)\alpha} \right\}.$$

9. The unrestricted and restricted m.l.e.s for μ are $\hat{\mu}_n = \breve{\mu}_n = \bar{X}$, , while the m.l.e. for λ is $\hat{\lambda}_n = \left(\bar{Y} - \bar{X}^{-1} \right)^{-1}$, where $\bar{Y} := n^{-1} \sum_{j=1}^n X_j^{-1}$. The l.r. statistic is

$$
\begin{aligned}
\Lambda(\mathbf{X}; \lambda_0) &= \frac{\lambda_0^{n/2} \exp\left[-\frac{\lambda_0}{2\bar{X}^2} \sum_{j=1}^n (X_j - \bar{X})^2 / X_j \right]}{\hat{\lambda}_n^{n/2} \exp\left[-\frac{\hat{\lambda}_n}{2\bar{X}^2} \sum_{j=1}^n (X_j - \bar{X})^2 / X_j \right]} \\
&= \left(\frac{\lambda_0}{\hat{\lambda}_n} \right)^{n/2} \exp\left[(\hat{\lambda}_n - \lambda_0) \sum_{j=1}^n \left(\frac{X_j - 2\bar{X} + \bar{X}^2 X_j^{-1}}{2\bar{X}^2} \right) \right] \\
&= \left(\frac{\lambda_0}{\hat{\lambda}_n} \right)^{n/2} \exp\left[(\hat{\lambda}_n - \lambda_0) \frac{n}{2} (\bar{Y} - \bar{X}^{-1}) \right] \\
&= \left(\frac{\lambda_0}{\hat{\lambda}_n} \right)^{n/2} \exp\left[\frac{n}{2} \left(1 - \frac{\lambda_0}{\hat{\lambda}_n} \right) \right] \\
&\propto C(\mathbf{X})^{n/2} e^{-C(\mathbf{X})/2},
\end{aligned}
$$

where $C(\mathbf{X}) := n\lambda_0/\hat{\lambda}_n$ is distributed as $\chi^2(n-1)$ under H_0. Small values of $\Lambda(\mathbf{X}; \lambda_0)$ correspond to both small and large values of $C(\mathbf{X})$, but small values of $C(\mathbf{X})$ favor H_1. Thus, with $n = 101$ an exact .05-level rejection region is $\mathcal{R}_{.05} = \{ \mathbf{x} : C(\mathbf{x}) \le c_{.05} = 77.929 \}$. The given data yield $\hat{\lambda}_{101} \doteq 3.4765$ and $C(\mathbf{x}) \doteq 72.63$, so that H_0 is rejected at the .05 level.

10. $L\left(\mathbf{X},\mathbf{Y};\mu_X,\sigma_X^2,\mu_Y,\sigma_Y^2\right)$, the likelihood function for the pair of independent samples, is proportional to

$$\frac{1}{(\sigma_X^2)^{n_X/2}(\sigma_Y^2)^{n_Y/2}}\exp\left[-\frac{1}{2\sigma_X^2}\sum_{j=1}^{n_X}(X_j-\mu_X)^2-\frac{1}{2\sigma_Y^2}\sum_{j=1}^{n_Y}(Y_j-\mu_Y)^2\right].$$

Unrestricted m.l.e.s are

$$\hat{\mu}_{n_X}=\bar{X},$$
$$\hat{\mu}_{n_Y}=\bar{Y},$$
$$\hat{\sigma}_{n_X}^2=\frac{(n_X-1)S_X^2}{n_X},$$
$$\hat{\sigma}_{n_Y}^2=\frac{(n_Y-1)S_Y^2}{n_Y},$$

while under restriction $\sigma_X^2-\sigma_Y^2=0$ they are $\check{\mu}_{n_X}=\bar{X},\check{\mu}_{n_Y}=\bar{Y}$, and (with $n:=n_X+n_Y$)

$$\check{\sigma}_n^2=n^{-1}\left[\sum_{j=1}^{n_X}\left(X_j-\bar{X}\right)^2+\sum_{j=1}^{n_Y}\left(Y_j-\bar{Y}\right)^2\right]$$
$$=\frac{n_X-1}{n}S_X^2+\frac{n_Y-1}{n}S_Y^2$$

(the "pooled-sample" variance). Substituting, the likelihood ratio simplifies to

$$\Lambda\left(\mathbf{X},\mathbf{Y};0\right)=\frac{\left(\check{\sigma}_n^2\right)^{-n/2}}{\left(\hat{\sigma}_{n_X}^2\right)^{-n_X/2}\left(\hat{\sigma}_{n_Y}^2\right)^{-n_Y/2}}$$
$$=\frac{\left[\frac{n_X-1}{n}S_X^2+\frac{n_Y-1}{n}S_Y^2\right]^{-n/2}}{\left[\frac{n_X-1}{n_X}S_X^2\right]^{-n_X/2}\left[\frac{n_Y-1}{n_Y}S_Y^2\right]^{-n_Y/2}}.$$

This is equal to a positive constant times

$$\left(1+\frac{n_Y-1}{n_X-1}R\left(\mathbf{X},\mathbf{Y}\right)^{-1}\right)^{-n_X/2}\left(1+\frac{n_X-1}{n_Y-1}R(\mathbf{X},\mathbf{Y})\right)^{-n_Y/2},$$

where $R(\mathbf{X},\mathbf{Y}):=S_X^2/S_Y^2$ is the ratio of unbiased sample variances. The derivative of Λ with respect to R is proportional to $R^{-1}-\frac{n_Y/(n_Y-1)}{n_X/(n_X-1)}$, which is negative for $R>\frac{n_X/(n_X-1)}{n_Y/(n_Y-1)}$, zero for $R=\frac{n_X/(n_X-1)}{n_Y/(n_Y-1)}$, and positive for smaller values of R. Thus, the rejection region for H_1^+ comprises large values of R, while the rejection region for the two-sided H_1 comprises both small and large values of R. Recall that $(n_X - 1) S_X^2/\sigma_X^2$ and $(n_Y - 1) S_Y^2/\sigma_Y^2$ are independent chi-squared variates with $n_X - 1$ and $n_Y - 1$ degrees of freedom, respectively. Therefore, under $H_0 : \sigma_X^2 = \sigma_Y^2$ the variance ratio R is the ratio of independent chi-squared variates, each divided by its d.f., and so is distributed as $F(n_X - 1, n_Y - 1)$.

11. When the null restricts a single parameter, the Wald statistic,

$$W_n(\mathbf{X}) = n \frac{\left(\hat{\phi}_n - \phi_0\right)^2}{\mathcal{I}^{\phi\phi}\left(\hat{\phi}_n, \hat{\psi}_n\right)} = n\left(\hat{\phi}_n - 1\right)^2 \left[\Psi'\left(\hat{\phi}_n\right) - \hat{\phi}_n^{-1}\right],$$

is simply the square of the "Studentized" unrestricted m.l.e.; that is, it is the square of the difference between $\hat{\phi}_n$ and ϕ_0 divided by its estimated asymptotic standard deviation. The Studentized m.l.e. itself is asymptotically distributed as standard normal under the null. Power against a one-sided alternative $\phi > \phi_0$ is enhanced by restricting the rejection region to large values of this quantity—e.g., to values of $\sqrt{n}\left(\hat{\phi}_n - 1\right)\sqrt{\Psi'\left(\hat{\phi}_n\right) - \hat{\phi}_n^{-1}}$ that exceed or equal the upper-α quantile of the standard normal.

12. The Wald test of $H_0 : \phi = \phi_0$ fails to reject at level α if and only if $n\left(\hat{\phi}_n - \phi_0\right)^2 \left[\Psi'\left(\hat{\phi}_n\right) - \hat{\phi}_n^{-1}\right] < c^\alpha$, where c^α is the upper-α quantile of $\chi^2(1)$. Since $Z^2 \sim \chi^2(1)$ when $Z \sim N(0, 1)$, this is equivalent to the condition

$$z^{\alpha/2} > \sqrt{n}\left|\hat{\phi}_n - \phi_0\right|\sqrt{\Psi'\left(\hat{\phi}_n\right) - \hat{\phi}_n^{-1}},$$

or

$$\hat{\phi}_n - \frac{z^{\alpha/2}}{\sqrt{n\left[\Psi'\left(\hat{\phi}_n\right) - \hat{\phi}_n^{-1}\right]}} < \phi_0 < \hat{\phi}_n + \frac{z^{\alpha/2}}{\sqrt{n\left[\Psi'\left(\hat{\phi}_n\right) - \hat{\phi}_n^{-1}\right]}}.$$

13. With $\boldsymbol{\theta} := (\phi, \psi)'$ the log likelihood for a sample of size n is

$$\mathcal{L}(\boldsymbol{\theta}) = n \log \phi - n \log \psi + (\phi - 1)\sum_{j=1}^{n} \log X_j - \psi^{-1}\sum_{j=1}^{n} X_j^\phi,$$

and the likelihood equations are

$$\mathcal{L}_\phi(\boldsymbol{\theta}) = n\phi^{-1} + \sum_{j=1}^{n} \log X_j - \psi^{-1}\sum_{j=1}^{n} X_j^\phi \log X_j = 0 \quad (8.32)$$

$$\mathcal{L}_\psi(\boldsymbol{\theta}) = -n\psi^{-1} + \psi^{-2}\sum_{j=1}^{n} X_j^\phi = 0. \quad (8.33)$$

The unrestricted m.l.e.s must be found by numerical methods, but under the restriction $\phi = 1$ the Weibull family specializes to the exponential, and (8.33) yields $\check{\psi}_n = \bar{X}$. Thus, the L.m. test will be the easiest to carry out. Evaluating the score function in (8.32) at $\check{\boldsymbol{\theta}}_n := (1, \bar{X})'$ gives

$$\mathcal{L}_\phi\left(\check{\boldsymbol{\theta}}_n\right) = n + \sum_{j=1}^n \log X_j - \bar{X}^{-1} \sum_{j=1}^n X_j \log X_j.$$

The components of the Fisher information are,

$$\mathcal{I}_{\phi\phi}\left(\boldsymbol{\theta}\right) = \phi^{-2} + \psi^{-1} E\left[X^\phi \left(\log X\right)^2\right],$$
$$\mathcal{I}_{\phi\psi}\left(\boldsymbol{\theta}\right) = -\psi^{-2} E\left(X^\phi \log X\right),$$
$$\mathcal{I}_{\psi\psi}\left(\boldsymbol{\theta}\right) = -\psi^{-2} + 2\psi^{-3} E X^\phi.$$

Under H_0 these are estimated consistently by

$$\widetilde{\mathcal{I}}_{\phi\phi}\left(\check{\boldsymbol{\theta}}_n\right) = 1 + \bar{X}^{-1} n^{-1} \sum_{j=1}^n X_j \left(\log X_j\right)^2,$$

$$\widetilde{\mathcal{I}}_{\phi\psi}\left(\check{\boldsymbol{\theta}}_n\right) = -\bar{X}^{-2} n^{-1} \sum_{j=1}^n X_j \log X_j,$$

$$\widetilde{\mathcal{I}}_{\psi\psi}\left(\check{\boldsymbol{\theta}}_n\right) = \bar{X}^{-2}$$

so the score statistic is

$$\mathcal{S}_n\left(\mathbf{X}\right) = \frac{\mathcal{L}_\phi\left(\check{\boldsymbol{\theta}}_n\right)^2}{n\left(\widetilde{\mathcal{I}}_{\phi\phi} - \widetilde{\mathcal{I}}_{\phi\psi}^2 / \widetilde{\mathcal{I}}_{\psi\psi}\right)}.$$

With the example data we have $\check{\psi}_n = \bar{x} = 3.68$, $\mathcal{L}_\phi\left(\check{\boldsymbol{\theta}}_n\right) \doteq 1.045$, $\widetilde{\mathcal{I}}_{\phi\phi} \doteq 4.103$, $\widetilde{\mathcal{I}}_{\phi\psi} \doteq -.444$, $\widetilde{\mathcal{I}}_{\psi\psi} \doteq .0738$, and $\Sigma\left(\check{\boldsymbol{\theta}}_n\right) \doteq 1.432$, giving $\mathcal{S}_n\left(\mathbf{x}\right) \doteq 0.153$.

14. (i) The solution to Exercise 7.43 gives the likelihood ratio as

$$\Lambda\left(\mathbf{x}; \alpha\right) = \left(\frac{\alpha \sum_{j=1}^n x_j^{\hat{\alpha}_{10}}}{\hat{\alpha}_{10} \sum_{j=1}^n x_j^\alpha}\right)^{10} \prod_{j=1}^{10} x_j^{\alpha - \hat{\alpha}_{10}},$$

from which we find $-2 \log \Lambda\left(\mathbf{x}; 1\right) = 2.2262$.

(ii) Imposing $\alpha = 1$, the restricted estimate of β is simply $\bar{x} = 1.331$. The score for α evaluated at $\alpha = 1, \beta = \bar{x}$ is

$$\mathcal{L}_\alpha(1, \bar{x}) = 10 + \sum_{j=1}^{10} \log x_j - \frac{1}{\bar{x}} \sum_{j=1}^{10} x_j \log x_j = 4.8459,$$

and the estimate of asymptotic variance is

$$\tilde{\mathcal{I}}^{\alpha\alpha}\left(1,\bar{x}\right) = \left[\tilde{\mathcal{I}}_{\alpha\alpha}\left(1,\bar{x}\right) - \frac{\tilde{\mathcal{I}}_{\alpha\beta}\left(1,\bar{x}\right)^2}{\tilde{\mathcal{I}}_{\beta\beta}\left(1,\bar{x}\right)}\right]^{-1}$$

$$= \left[1.6042 - \frac{(-.3864)^2}{5645}\right]^{-1}$$

$$= .7464$$

yielding $S_{10} = 10^{-1}\mathcal{L}_\alpha(1,\bar{x})^2\tilde{\mathcal{I}}^{\alpha\alpha}\left(1,\bar{x}\right) = 1.7528$.

(iii) From the m.l.e. $\hat{\alpha}_{10} = 1.52330$ and the estimate .15588 of asymptotic variance we obtain Wald statistic

$$\mathcal{W}_{10}\left(\mathbf{x}\right) = \frac{(1.52330 - 1)^2}{.15588} = 1.7568.$$

Comparing with upper-.10 quantile $c^{.10} = 2.706$ of $\chi^2\left(1\right)$, none of the tests rejects H_0 at the .10 level.

15. (i) The logarithm of the likelihood ratio can be written as $K\left(\mathbf{x}; \mu_0, \lambda_0\right) - K\left(\mathbf{x}; \hat{\mu}_n, \hat{\lambda}_n\right)$, where

$$K\left(\mathbf{x}; \mu, \lambda\right) := \frac{n}{2}\left[\log\lambda - \lambda\left(\frac{\bar{x}}{\mu^2} - \frac{2}{\mu} + \bar{y}\right)\right].$$

With $\mu_0 = \lambda_0 = 1$, $n = 100$, $\hat{\mu}_n = \bar{x} = .9718$, $\bar{y} := n^{-1}\sum_{j=1}^{n}x_j^{-1} = 1.8513$, $\hat{\lambda}_n = \left(\bar{y} - \bar{x}^{-1}\right)^{-1} = 1.2161$, and $K\left(\mathbf{x}; \hat{\mu}_n, \hat{\lambda}_n\right) = -40.2173$ we have

$$-2\left[K\left(\mathbf{x}; 1, 1\right) - K\left(\mathbf{x}; \hat{\mu}_n, \hat{\lambda}_n\right)\right] \doteq 1.88 < c^{.05} = 5.991,$$

where $c^{.05}$ is the upper-.05 quantile of $\chi^2\left(2\right)$. Thus, H_0 is not rejected at level .05.

(ii) There being no nuisance parameters, the score statistic is just

$$S_n\left(\mathbf{x}\right) = n^{-1}\mathcal{L}_{\theta'}\left(\boldsymbol{\theta}_0\right)\mathcal{I}^{\theta\theta'}\left(\boldsymbol{\theta}_0\right)\mathcal{L}_\theta\left(\boldsymbol{\theta}_0\right),$$

where $\boldsymbol{\theta}_0 = \left(\mu_0, \lambda_0\right)' = \left(1, 1\right)'$ and

$$\mathcal{L}_{\theta'}\left(\boldsymbol{\theta}_0\right) = \left\{n\left(\bar{x} - 1\right), \frac{n}{2}\left[1 - (\bar{x} - 2 + \bar{y})\right]\right\}$$

$$\mathcal{I}_{\theta\theta'}\left(\boldsymbol{\theta}_0\right) = \begin{pmatrix} \mu_0^3/\lambda_0 & 0 \\ 0 & 2\lambda_0^2 \end{pmatrix} = \begin{pmatrix} 1 & 0 \\ 0 & 2 \end{pmatrix}.$$

The calculation yields $S_{100}\left(\mathbf{x}\right) \doteq 1.64 < c^{.05}$, so that H_0 is not rejected.

(iii) The Wald statistic is $W_n(\mathbf{x}) = \left(\hat{\boldsymbol{\theta}}_n - \boldsymbol{\theta}_0\right)' \mathcal{I}_{\boldsymbol{\theta}\boldsymbol{\theta}'}\left(\hat{\boldsymbol{\theta}}_n\right)^{-1}\left(\hat{\boldsymbol{\theta}}_n - \boldsymbol{\theta}_0\right)$.
With $\hat{\boldsymbol{\theta}}'_n = \left[\bar{x}, (\bar{y} - \bar{x}^{-1})^{-1}\right]$ and

$$\mathcal{I}_{\boldsymbol{\theta}\boldsymbol{\theta}'}\left(\hat{\boldsymbol{\theta}}_n\right)^{-1} = \begin{pmatrix} \hat{\lambda}_n/\hat{\mu}_n^3 & 0 \\ 0 & 1/(2\hat{\lambda}_n^2) \end{pmatrix}$$

the calculation yields $W_{100}(\mathbf{x}) \doteq 1.68 < c^{.05}$, so that H_0 is not rejected. (In fact, the data were generated with $\mu = \lambda = 1$, so H_0 is actually true.)

16. (The data for the exercise were generated as $\Gamma(2,1)$, so $H_0 : \phi = 2\psi$ is in fact true.) Imposing $\psi = \phi/2$, the log likelihood is

$$\log L\left(\mathbf{x} : \phi, \frac{\phi}{2}\right) = -n\log\Gamma(\phi) - n\phi\log\left(\frac{\phi}{2}\right) + (\phi-1)\sum_{j=1}^{n}\log x_j - \frac{2n\bar{x}}{\phi}.$$

With the given data for $\bar{x}$ and $\sum_{j=1}^{100}\log x_j$, the root $\check{\phi}_{100}$ of

$$0 = \frac{d}{d\phi}\log L = 100\left[-\Psi(\phi) - \log\left(\frac{\phi}{2}\right) - 1 + \frac{1}{100}\sum_{j=1}^{100}\log x_j + \frac{2\bar{x}}{\phi^2}\right]$$

is approximately 1.981, while the unrestricted m.l.e.s are $\hat{\phi}_{100} \doteq 1.655, \hat{\psi}_{100} \doteq 1.221$. The maxima of the restricted and unrestricted log likelihoods are approximately -164.836 and -163.732, so that $-2\log\Lambda(\mathbf{x}) \doteq 2(164.836 - 163.732) = 2.208$. For the Wald test, the above unrestricted estimates give $\hat{\phi}_{100} - 2\hat{\psi}_{100} \doteq -.787$, and the estimated asymptotic variance is

$$\frac{\hat{\phi}_{100} + 4\hat{\psi}_{100} + 4\hat{\psi}_{100}^2\Psi'\left(\hat{\phi}_{100}\right)}{100\left[\hat{\phi}_{100}\Psi'\left(\hat{\phi}_{100}\right) - 1\right]} \doteq 0.31817,$$

so that $W_{100}(\mathbf{x}) \doteq (-.787)^2/.31817 \doteq 1.947$.

17. The restricted log likelihood and its derivative with respect to ϕ are, respectively,

$$-n\log\Gamma(\phi) - n\phi\log\sqrt{\phi} + (\phi-1)\sum_{j=1}^{n}\log x_j - n\bar{x}\phi^{-1/2},$$

$$-n\Psi(\phi) - \frac{n}{2}(\log\phi + 1) + \sum_{j=1}^{n}\log x_j + \frac{n\bar{x}}{2}\phi^{-3/2}.$$

With the given data the root of the likelihood equation is $\check{\phi}_{100} \doteq$ 1.6058, and so $\check{\psi}_{100} = \sqrt{\check{\phi}_{100}} \doteq 1.2672$ and $\log L\left(\mathbf{x}; \check{\phi}_{100}, \check{\psi}_{100}\right) \doteq$ -163.762. With unrestricted estimates $\hat{\phi}_{100} \doteq 1.6546$ and $\hat{\psi}_{100} \doteq$ 1.2208 we have $\log L\left(\mathbf{x}; \hat{\phi}_{100}, \hat{\psi}_{100}\right) \doteq -163.732$, yielding $\Lambda\left(\mathbf{x}\right) \doteq$ $-2\left(-163.762 + 163.732\right) \doteq 0.060$. The observed data are thus highly consistent with H_0.

18. For the version based on $\hat{\phi}_n - \hat{\psi}_n^2$ the delta method gives

$$\sqrt{n}\left[\left(\hat{\phi}_n - \hat{\psi}_n^2\right) - \left(\phi - \psi^2\right)\right] \rightsquigarrow N\left(0, \mathcal{I}^{\phi\phi} - 4\psi\mathcal{I}^{\phi\psi} + 4\psi^2\mathcal{I}^{\psi\psi}\right),$$

with $\mathcal{I}^{\psi\psi} \equiv \mathcal{I}^{\psi\psi}\left(\phi, \psi\right) = \Psi'\left(\phi\right)\psi^2/\left[\Psi'\left(\phi\right)\phi - 1\right]$, $\mathcal{I}^{\phi\psi} = -\psi/\left[\Psi'\left(\phi\right)\phi - 1\right]$, and $\mathcal{I}^{\phi\phi} = \phi/\left[\Psi'\left(\phi\right)\phi - 1\right]$. The corresponding Wald statistic is

$$\mathcal{W}_n\left(\mathbf{X}\right) = \frac{n\left(\hat{\phi}_n - \hat{\psi}_n^2\right)^2}{\mathcal{I}^{\phi\phi} - 4\psi\mathcal{I}^{\phi\psi} + 4\psi^2\mathcal{I}^{\psi\psi}},$$

with $\mathcal{W}_{100}\left(\mathbf{x}\right) = 0.077$ from the given data.

For the version based on $\sqrt{\hat{\phi}_n} - \hat{\psi}_n$ the delta method gives

$$\sqrt{n}\left[\left(\sqrt{\hat{\phi}_n} - \hat{\psi}_n\right) - \sqrt{\phi} - \psi\right] \rightsquigarrow N\left(0, \frac{\mathcal{I}^{\phi\phi}}{4\phi} - \frac{\mathcal{I}^{\phi\psi}}{\sqrt{\phi}} + \mathcal{I}^{\psi\psi}\right).$$

The corresponding Wald statistic is

$$\mathcal{W}_n\left(\mathbf{X}\right) = \frac{n\left(\sqrt{\hat{\phi}_n} - \hat{\psi}_n\right)^2}{\mathcal{I}^{\phi\phi}/\left(4\hat{\phi}_n\right) - \mathcal{I}^{\phi\psi}/\sqrt{\hat{\phi}_n} + \mathcal{I}^{\psi\psi}},$$

with $\mathcal{W}_{100}\left(\mathbf{x}\right) = 0.064$ from the given data.

19. Imposing $\phi^2 + \psi^2 = 1$ by setting $\phi = \cos\theta$ and $\psi = \sin\theta$, the log likelihood is

$$\log L\left(\mathbf{x}, \mathbf{y}; \phi, \psi\right) = c - \frac{1}{2}\sum_{j=1}^{n}\left(x_j - \cos\theta\right)^2 - \frac{1}{2}\sum_{j=1}^{m}\left(y_j - \sin\theta\right)^2,$$

where irrelevant constant $c := -\left(n + m\right)\log\sqrt{2\pi}$; and the likelihood equation for θ is

$$0 = \frac{d}{d\theta}\log L\left(\mathbf{x}, \mathbf{y}; \theta\right) = \cos\theta\sum_{j=1}^{m}y_j - \sin\theta\sum_{j=1}^{n}x_j + \left(n - m\right)\cos\theta\sin\theta.$$

With $n = m = 100$ and the given data the solution is $\check{\theta}_{100} = \tan^{-1}\left(\sum_{j=1}^{100} y_j / \sum_{j=1}^{100} x_j\right) \doteq 0.8295$. The restricted and unrestricted estimates of ϕ and ψ are $\check{\phi}_{100} \doteq 0.6752$, $\check{\psi}_{100} \doteq 0.7376$ and $\hat{\phi}_{100} = \bar{x} = 0.5485$, $\hat{\psi}_{100} = \bar{y} = 0.5992$. The restricted and unrestricted values of the log likelihood are $L\left(\mathbf{x}, \mathbf{y}; \check{\phi}_{100}, \check{\psi}_{100}\right) \doteq c - 102.521$, $L\left(\mathbf{x}, \mathbf{y}; \hat{\phi}_{100}, \hat{\psi}_{100}\right) \doteq c - 100.760$; hence, $-2\log\Lambda\left(\mathbf{x}, \mathbf{y}\right) \doteq 3.522$. This lies between the upper .10 and upper .05 quantiles of $\chi^2\left(1\right)$, so H_0 is rejected at level .10.

20. $W_n(\mathbf{X})$ is distributed asymptotically as $\chi^2\left(1\right)$ under H_0, while under H_1 one expects large values. Thus, as a large-sample approximation, $p(\mathbf{x}) \doteq \Pr\left(C \geq 9.301 \mid H_0\right)$, where $C \sim \chi^2\left(1\right)$. A standard chi-squared table shows this to lie above the upper-0.005 quantile. Since the square of a standard normal variate is distributed as $\chi^2\left(1\right)$, a sharper estimate can be got from the normal table, as $\Pr\left(|Z| \geq \sqrt{9.301} \doteq 3.05\right) \doteq 0.0011$.

21. Under the null statistic

$$T_n\left(\mathbf{X}, \mathbf{Y}\right) = \frac{\bar{X} - \bar{Y}}{\sqrt{S_X^2/n_X + S_Y^2/n_Y}}$$

is asymptotically standard normal, and under H_1 we expect to see positive values. Since

$$T_n\left(\mathbf{X}, \mathbf{Y}\right) = \frac{(\bar{X} - \bar{Y}) - (\mu_X - \mu_Y)}{\sqrt{S_X^2/n_X + S_Y^2/n_Y}} + \frac{\mu_X - \mu_Y}{\sqrt{S_X^2/n_X + S_Y^2/n_Y}}$$

$$\approx Z + \sqrt{n_X}\frac{\mu_X - \mu_Y}{\sqrt{\sigma_X^2 + \sigma_Y^2 n_X/n_Y}} \to^{a.s.} +\infty$$

under H_1, a level-α test based on $\mathcal{R}_\alpha = \{T_n\left(\mathbf{x}, \mathbf{y}\right) : T_n\left(\mathbf{x}, \mathbf{y}\right) \geq z^\alpha\}$ is asymptotically valid and consistent.

22. Letting $D_j := X_j - Y_j$, we have $\mu_D \leq 0$ under H_0 and positive under H_1, while $\sigma_D^2 = \sigma_X^2 - 2\sigma_{XY} + \sigma_Y^2$ is finite under the maintained hypothesis and can be estimated consistently by $S_D^2 := (n-1)\sum_{j=1}^n (D_j - \bar{D})^2$. Accordingly, $\sqrt{n}(\bar{D} - \mu_D)/S_D$ is asymptotically normal. Taking $T_n\left(\mathbf{X}, \mathbf{Y}\right) = \sqrt{n}\bar{D}/S_D = \sqrt{n}(\bar{D} - \mu_D)/S_D + \sqrt{n}\mu_D/S_D$, it follows that $T_n\left(\mathbf{X}, \mathbf{Y}\right) \rightsquigarrow N\left(0, 1\right)$ when $\mu_D = 0$, while

$$T_n\left(\mathbf{X}, \mathbf{Y}\right) \to^{a.s.} \begin{cases} -\infty, & \mu_D < 0 \\ +\infty, & \mu_D > 0 \end{cases}.$$

Accordingly, $\Pr[T_n(\mathbf{X}, \mathbf{Y}) \geq z^\alpha]$ has limiting value no larger than α when H_0 is true and approaches unity under H_1, so that a test with

$$\mathcal{R}_\alpha = \{(\mathbf{x}, \mathbf{y}) : T_n(\mathbf{x}, \mathbf{y}) \geq z^\alpha\}$$

is asymptotically valid and consistent.

23. The $\{X_j\}_{j=1}^n$ are i.i.d. with zero mean and finite, positive variance, and so $\lim_{n \to \infty} \Pr\left(\left|\sqrt{n}\bar{X}\right| \geq z^{\alpha/2} S\right) = \alpha$. Thus, the t test of $H_0 : \mu = 0$ is indeed asymptotically valid. However, all members of the sample take on the same value with probability $(1 - \delta)^n + \delta^n$, in which case $S = 0$. Should that event occur, H_0 would be rejected even at level $\alpha = 1$. Thus,

$$\Pr\left(\left|\sqrt{n}\bar{X}\right| \geq z^{\alpha/2} S\right) \geq (1 - \delta)^n + \delta^n.$$

In the trivial case $n = 1$ the right side equals unity for all δ. For $n = 2, 3, \ldots$ the right side exceeds $\max\{(1 - \delta)^n, \delta^n\}$ and so exceeds $1 - \varepsilon$ if either $\delta \geq \sqrt[n]{1 - \varepsilon}$ or $\delta \leq 1 - \sqrt[n]{1 - \varepsilon}$.

24. The probability that the first $j - 1$ sample elements are less than or equal to x, that $x < X_j \leq x + h$, and that the remaining $n - 1$ elements exceed $x + h$ is

$$F(x)^{j-1} [F(x + h) - F(x)] [1 - F(x + h)]^{n-j}.$$

There are $\binom{n}{n-j} = \binom{n}{j}$ ways to choose $n - j$ elements to exceed $x + h$, and j ways to choose one of j elements to lie in $(x, x + h]$. Multiplying the above expression by $j\binom{n}{j}$, dividing by h, and letting $h \to 0$ give the result.

25. The sum of the merged samples, $\sum_{i=1}^n y_i$, is invariant over the $\binom{n}{n_1}$ assignments, and $\sum_{i=1}^{n_1} y_i = .5\left(\sum_{i=1}^{n_1} y_i - \sum_{i=n_1+1}^n y_i + \sum_{i=1}^n y_i\right)$.

26. $\mathbf{I}$ and $\mathbf{R}$ are functions of the signs and absolute values of the independent observations, and the independence of signs and absolute values of continuous, symmetric r.v.s with median zero implies that the $2n$ elements of $\mathbf{I}$ and $\mathbf{R}$ are mutually independent r.v.s. Each element of $\mathbf{I}$ is distributed as Bernoulli with parameter $1/2$; and for each of the $n!$ permutations $\mathbf{i}$ of integers the events $\mathbf{R} = \mathbf{i}$ are equally likely.

27. The sample mean $\bar{x} = 1.184$ and sample standard deviation $s = 1.041$ yield for the test of $H_0 : \mu = 0$ the realization $\bar{x}/\left(s/\sqrt{10}\right) = 3.597$. The probability that a Student's t r.v. with 9 d.f. exceeds 3.597 is approximately 0.0029.

28. Reversing the signs reverses the ranks of the observations, so that $R(-Y_j) = n + 1 - R(Y_j)$ for each $j \in \{1, 2, \ldots, n\}$. Letting W' be

the sum of the ranks of the n_2 elements of $-\mathbf{X}_2$, we have $W' = n_2(n+1) - W$ and so

$$W' - \frac{n_2(n+1)}{2} = \frac{n_2(n+1)}{2} - W.$$

But W' and W have the same distribution under $H_0 : \delta = 0$, so

$$\Pr\left[W - \frac{n_2(n+1)}{2} = w\right] = \Pr\left[\frac{n_2(n+1)}{2} - W = w\right]$$

for each w, indicating that the distribution of W is symmetric about $n_2(n+1)/2$. Since the range of W is bounded, the mean exists and equals the median.

Chapter 9

Tests of Nonparametric Hypotheses

This chapter deals with tests about aspects of data-generating processes other than constants that define parametric subfamilies or that describe characteristics such as location or dispersion. We begin with a short treatment of tests for independence of components of a bivariate sample, then undertake a detailed survey of tests of goodness of fit. These important procedures afford ways to assess the validity of assumptions required for model-based inference, such as maximum likelihood estimation, exact likelihood-ratio tests, the three classical asymptotic tests, and Bayesian procedures generally.

9.1 Tests for Independence

Given the bivariate random sample $\{(X_j, Y_j)\}_{j=1}^{n}$, we wish to test whether the X and Y observations are independent. Now a parametric test of independence could well be designed under appropriate additional maintained hypotheses. For example, if bivariate normality were thought to be tenable, one could simply test $H_0 : \rho = 0$ vs. $H_1 : \rho \neq 0$. using any of the three classical asymptotic tests of Section 8.6. However, an *exact* Wald test is feasible, since the distribution of sample correlation coefficient (and maximum likelihood estimator)

$$R_{XY} := \frac{S_{XY}}{S_X S_Y} = \frac{\sum_{j=1}^{n} (X_j - \bar{X}) (Y_j - \bar{Y})}{\sqrt{\sum_{j=1}^{n} (X_j - \bar{X})^2 \sum_{j=1}^{n} (Y_j - \bar{Y})^2}}$$

is known exactly for finite n. When $H_0 : \rho = 0$ is true, this has the simple form

$$f_{R_{XY}} (r) = \frac{\left(1 - r^2\right)^{n/2 - 2}}{\mathcal{B} (1/2, n/2 - 1)} \mathbf{1}_{(-1,1)} (r) .$$

Moreover, it is not hard to show that

$$T_n(R_{XY}) := \sqrt{n-2} \frac{R_{XY}}{\sqrt{1-R_{XY}^2}} \sim t(n-2) \tag{9.1}$$

under H_0, so that rejection region $\mathcal{R}_\alpha = \{T_n : |T_n| \geq t^{\alpha/2}\}$ also yields an exact level-α test. Even easier for moderate-to-large n is to use the fact (Exercise 6.22) that $\sqrt{n}R_{XY} \rightsquigarrow N(0,1)$ under the null. Indeed, there are transformations that approach normality even faster, such as Fisher's "z" statistic,

$$Z_n(R_{XY}) := \frac{1}{2} \log \left(\frac{1+R_{XY}}{1-R_{XY}} \right), \tag{9.2}$$

for which $\sqrt{n-1}Z_n(R_{XY}) \rightsquigarrow N(0,1)$ when $\rho = 0$. However, the present task is to treat the independence problem nonparametrically. For this we look at just one well known procedure that requires minimal assumptions about the data-generating process: Spearman's rank correlation test.

A test based on the correlation coefficient can be made distribution-free by correlating not the data themselves but the *ranks* of the data within the respective vectors $\mathbf{X}, \mathbf{Y}$. Let $\{R(X_j), R(Y_j)\}_{j=1}^n$ be the ranks of these observations. Each set of ranks is just a permutation of the first n integers, so that

$$\sum_{j=1}^n R(X_j) = \sum_{j=1}^n R(Y_j) = \sum_{j=1}^n j = \frac{n(n+1)}{2}$$

and

$$\sum_{j=1}^n R(X_j)^2 = \sum_{j=1}^n R(Y_j)^2 = \sum_{j=1}^n j^2 = \frac{n(n+1)(2n+1)}{6}.$$

The sample correlation coefficient of the ranks is therefore

$$\begin{aligned} C_{XY} :&= \frac{\sum_{j=1}^n R(X_j)R(Y_j) - n\bar{R}(X)\bar{R}(Y)}{\sqrt{\left[\sum_{j=1}^n R(X_j)^2 - n\bar{R}(X)^2\right]\left[\sum_{j=1}^n R(Y_j)^2 - n\bar{R}(Y)^2\right]}} \\ &= \frac{\sum_{j=1}^n R(X_j)R(Y_j) - n(n+1)^2/4}{n(n+1)(2n+1)/6 - n(n+1)^2/4} \\ &= \frac{\sum_{j=1}^n R(X_j)R(Y_j) - n(n+1)^2/4}{n(n^2-1)/12}. \end{aligned} \tag{9.3}$$

In fact, this can be expressed in a still simpler way as

$$C_{XY} = 1 - \frac{6\sum_{j=1}^n [R(X_j) - R(Y_j)]^2}{n(n^2-1)}. \tag{9.4}$$

Exercise 9.1. *Verify Expression (9.4).*

Statistic $\mathcal{C}_{XY}$ should take values near zero if the null hypothesis of independence is true, while large absolute values would be expected if there were either positive or negative association between X and Y. Developing an operational test requires determining or approximating the distribution of $\mathcal{C}_{XY}$ under the null, but since $\mathcal{C}_{XY}$ equals $\mathcal{S}_{XY} := \sum_{j=1}^{n} R(X_j) R(Y_j)$ apart from a change in location and scale, we can focus on the distribution of $\mathcal{S}_{XY}$. Unfortunately, the limiting theory for $\mathcal{S}_{XY}$ is complicated by the dependence in ranks of different observations, but it has been shown[1] that under the null hypothesis of independence

$$\mathcal{Z}_{XY} := \frac{\mathcal{S}_{XY} - E\mathcal{S}_{XY}}{\sqrt{V\mathcal{S}_{XY}}} \rightsquigarrow N(0,1).$$

Thus, an asymptotically valid test of H_0 vs. either positive or negative association rejects at level α if $|\mathcal{Z}_{XY}| \geq z^{\alpha/2}$, the upper-$\alpha/2$ quantile of $N(0,1)$.

To make use of this fact requires working out expressions for the mean and variance of $\mathcal{S}_{XY}$. For this we can rely on the results of corollaries 8.1 and 8.2 pertaining to distributions of rank statistics. Getting the mean of $\mathcal{S}_{XY}$ is easy.

Exercise 9.2. *Show that $E\mathcal{S}_{XY} = n(n+1)^2/4$ under H_0.*

To attack the variance, we have

$$V\mathcal{S}_{XY} = \sum_{i=1}^{n} V[R(X_i) R(Y_i)] + 2 \sum_{i=2}^{n} \sum_{j=1}^{i-1} Cov[R(X_i) R(Y_i), R(X_j) R(Y_j)].$$

Since the observations are i.i.d., the first term is just $nV[R(X_1) R(Y_1)]$ and the second is $n(n-1) Cov[R(X_1) R(Y_1), R(X_2) R(Y_2)]$. Taking these one at a time,

$$V[R(X_1) R(Y_1)] = \left[ER(X_1)^2\right]^2 - [ER(X_1)]^4,$$

since $R(X_1)$ and $R(Y_1)$ are independent and identically distributed under H_0. Using

$$ER(X_1)^2 = VR(X_1) + [ER(X_1)]^2$$

and the expressions for the moments in corollary 8.2, we get

$$V[R(X_1) R(Y_1)] = \left(\frac{n^2-1}{12}\right)^2 + 2\left(\frac{n^2-1}{12}\right)\left(\frac{n+1}{2}\right)^2.$$

---

[1] See Randles and Wolfe (1979, Ch. 8).

For the covariance expression, $Cov[R(X_1)R(Y_1), R(X_2)R(Y_2)]$, again using independence under H_0 and the fact that the observations are i.i.d., we have, letting $R_j := R(X_j)$ for brevity,

$$[E(R_1R_2)]^2 - (ER_1)^4 = \left[Cov(R_1, R_2) + (ER_1)^2\right]^2 - (ER_1)^4$$
$$= \left(\frac{n+1}{12}\right)^2 - 2\left(\frac{n+1}{12}\right)\left(\frac{n+1}{2}\right)^2.$$

Finally, combining the variance and covariance terms and simplifying give

$$V\mathcal{S}_{XY} = \frac{n^2(n+1)^2(n-1)}{144}.$$

The expressions for $E\mathcal{S}_{XY}$ and $V\mathcal{S}_{XY}$ reveal a nice connection between the standardized value of $\mathcal{S}_{XY}$ and the correlation statistic $\mathcal{C}_{XY}$:

$$\frac{\mathcal{S}_{XY} - E\mathcal{S}_{XY}}{\sqrt{V\mathcal{S}_{XY}}} = \frac{\mathcal{S}_{XY} - n(n+1)^2/4}{[n(n+1)\sqrt{n-1}]/12}$$
$$= \sqrt{n-1}\frac{\mathcal{S}_{XY} - n(n+1)^2/4}{n(n^2-1)/12}$$
$$= \sqrt{n-1}\mathcal{C}_{XY}.$$

Thus, a large-sample, size-α test of H_0: "X, Y independent" uses the rejection region $\{\mathcal{C}_{XY} : |\mathcal{C}_{XY}| \geq z^{\alpha/2}/\sqrt{n-1}\}$. Tables of critical values for small n are widely available; e.g., in Zar (1972) and Neave (1978).[2]

Example 9.1. To illustrate the exact and asymptotic tests based on correlation and rank correlation, ten pseudorandom standard bivariate normal observations were generated to have correlation $\rho = .5$. The data are

$$\mathbf{X}: \ .076, .949, -2.149, 1.053, .457,$$
$$.925, -.025, .081, 1.198, -.512$$
$$\mathbf{Y}: \ -.807, 1.461, -.640, 1.937, 1.962,$$
$$-.724, -.280, -.321, 1.300, -.504.$$

The realization of sample correlation R_{XY} is $r_{XY} \doteq .577$. The exact t test in (9.1) based on $T_{10}(r_{XY}) \doteq 1.998$ rejects H_0 at the .10 level, as does the asymptotic test (9.2) based on $Z_{10}(r_{XY}) \doteq 1.974$. However, the Spearman statistic $\sqrt{9}\mathcal{C}_{XY}$ takes the value 1.618, which is not significant at the .10 level.

---

[2] Tables and software for calculation are also available on the internet; e.g., at http://udel.edu/~mcdonald/statspearman.html.

Not surprisingly, the results of this one test suggest that the rank correlation test is less powerful than that based on R_{XY} when the data are bivariate normal. In fact, the efficiency is only about 10% less—meaning that a 10% larger sample for the rank test is needed to equate powers. Of course, the rank test's greater generality is a significant attribute. On the other hand, except when the data are normal *no* test based on correlation— whether of the actual data or their ranks—will be consistent against all alternatives to independence. Specifically, they will not be good at picking up nonaffine associations, such as a positive relation between Y and X^2.

Example 9.2. With X observations as in the previous example, Y observations were generated as $Y = X^2 + Z$, where $Z \sim N(0, 1)$ is independent of X. The new data for Y are

$$\mathbf{Y}: \quad -.970, 2.040, 5.120, 2.737, 2.211,$$
$$-.514, -.309, -.411, 2.244, -.024.$$

The sample correlation is now $r_{XY} \doteq -.314$, with $T_n(r_{XY}) \doteq -.935$, $Z_n(r_{XY}) \doteq -.975$, and $\mathcal{C}_{XY} \doteq .491$; thus, none of the tests gives statistically meaningful indication of dependence.

9.2 Tests of Goodness of Fit

Tests of fit are statistical procedures to infer whether specific parametric families of models adequately represent the data in hand. Thus, if F represents the common c.d.f. of the (possibly vector-valued) independent observations, the null hypothesis is of the form

$$H_0 : F(x) = F_0(x; \boldsymbol{\theta}), \boldsymbol{\theta} \in \boldsymbol{\Theta}_0.$$

Here, $\{F_0(x; \boldsymbol{\theta}) : \boldsymbol{\theta} \in \boldsymbol{\Theta}\}$ is some family of distributions indexed by the parameter vector $\boldsymbol{\theta}$, and $\boldsymbol{\Theta}_0$ is some subspace of the k-dimensional parameter space $\boldsymbol{\Theta}$. Possible specifications of $\boldsymbol{\Theta}_0$ range from a single point in $\Re^k$ to the entire parameter space. In the former case, which rarely arises, distribution F is fully specified and H_0 is a simple hypothesis. An example would be in testing that the data are distributed as uniform on $[0, 1]$. At the other extreme, F is proposed merely to be one of infinitely many members of a parametric family. An example—and an important and common application—is to test whether the data are from a normal population with unspecified mean and variance. Usually, the alternative hypothesis for tests of fit is not narrowly specified. The common goal is simply to test H_0

against the alternative that H_0 is false. While there do exist procedures for testing marginal distributions of time-series data, we consider only tests with random samples. Thus, the maintained hypothesis is in each instance that the elements of the sample are i.i.d. Apart from describing one test for multivariate normality, we confine attention to the univariate case.

All tests of fit are based on statistics that measure the degree to which some aspect of the sample matches the corresponding aspect of the proposed family of distributions. Tests differ according to the features of sample and population that are compared and the ways of measuring propinquity. Given the multitude of aspects of data and model that could be compared and the variety of ways of making such comparisons, it is not surprising that there is an immense literature on tests of fit, comprising hundreds of procedures of varying degrees of generality. Of these we describe only a few whose generality, ease of use, and/or reputation for power recommend them for applied work,[3] classifying them according to the aspects of model and sample that are compared:

(i) Expected frequencies on various subsets of the support
(ii) C.d.f.s
(iii) Population moments
(iv) Expected values of order statistics
(v) Generating functions—c.f.s, m.g.f.s, and p.g.f.s.

9.2.1   *A Test Based on Frequencies: Pearson's Chi-Squared*

The most venerable and still most widely applicable test of fit is the *chi-squared* test, proposed in 1900 by Karl Pearson. While testing the fit of univariate distributions is the most common application, the procedure can be used also for multivariate models. It is especially well adapted to categorical data; for example, observations classified by sex, race, or marital status. However, since our focus throughout this section is on quantifiable, univariate data, we confine the discussion here to that case. We first derive the usual computational form of the Pearson statistic from its more theoretically insightful expression as a quadratic form in sample frequencies, then develop its limiting distribution for *simple* null hypotheses that specify the parameters as well as the parametric model, then extend

---

[3]For an extensive survey and practical guide to tests of fit see D'Agostino and Stephens (1986). Unfortunately, the survey gives only passing reference to the newer tests based on generating functions.

to composite hypotheses that require parameters to be estimated. Finally, we discuss practical issues for implementation and some not-too-obvious limitations of this procedure.[4]

9.2.1.1 *Construction of the Pearson Statistic*

Take as the null hypothesis $H_0 : F(x) = F_0(x; \boldsymbol{\theta}_0)$ with $\boldsymbol{\theta}_0 \in \Re^k$ known, and let $\mathbb{P}_X$ be the corresponding measure, with $\mathbb{X} \subset \Re$ as support. There is no restriction on F_0; that is, the distribution can be discrete, continuous, or mixed. Construction of the Pearson statistic begins by partitioning $\mathbb{X}$ into $m + 1 \geq 2$ disjoint intervals or bins, $\{\mathbb{X}_j := (x_{j-1}, x_j]\}_{j=1}^{m+1}$, such that $\mathbb{P}_X(\mathbb{X}_j) =: p_j(\boldsymbol{\theta}_0) > 0$ for each j. One sets $x_0 = -\infty$ and/or $x_{m+1} = +\infty$ if $\mathbb{X}$ is unbounded below and/or above. For now we take m as given (any positive integer); criteria for choosing it are discussed later. Given a random sample $\{X_i\}_{i=1}^n$, let $\mathbf{1}_{\mathbb{X}_j}(X_i)$ be the indicator for the event $\{\omega : X_i(\omega) \in \mathbb{X}_j\}$, and let $N_j := \sum_{i=1}^n \mathbf{1}_{\mathbb{X}_j}(X_i)$ be the sample frequency— the number of observations in the jth bin. As the sum of Bernoulli variates, each N_j has binomial marginal distribution $B(n, p_j)$. Further, setting $\mathbf{p} := (p_1, p_2, ..., p_m)'$ and $\mathbf{1} := (1, 1, ..., 1)'$, the joint distribution of counts in the first m bins, $\mathbf{N} := (N_1, N_2, ..., N_m)'$, is *multinomial*, with

$$\Pr(\mathbf{N} = \mathbf{r}) = \binom{n}{\mathbf{r}} p_1^{r_1} p_2^{r_2} \cdots \cdots p_m^{r_m} (1 - \mathbf{p}'\mathbf{1})^{n - \mathbf{r}'\mathbf{1}}$$

for each collection of nonnegative integers $\mathbf{r}$ such that $\mathbf{r}'\mathbf{1} \leq n$.[5]

The mean vector and covariance matrix of $\mathbf{N}$ are $E\mathbf{N} = n\mathbf{p}$ and $V\mathbf{N} = n\mathbb{V}$, where

$$\mathbb{V} := \begin{pmatrix} p_1(1 - p_1) & -p_1 p_2 & \cdots & -p_1 p_m \\ -p_1 p_2 & p_2(1 - p_2) & \cdots & -p_2 p_m \\ \cdots & \cdots & \ddots & \cdots \\ -p_1 p_m & -p_2 p_m & \cdots & p_m(1 - p_m) \end{pmatrix}. \tag{9.5}$$

Letting $\mathbf{P}$ be a diagonal matrix with elements of $\mathbf{p}$ on the principal diagonal, $\mathbb{V}$ can be written (as in Expression (4.17)) $\mathbb{V} = \mathbf{P} - \mathbf{pp}'$. This matrix has full rank m and inverse

$$\mathbb{V}^{-1} = \begin{pmatrix} p_1^{-1} + p_{m+1}^{-1} & p_{m+1}^{-1} & \cdots & p_{m+1}^{-1} \\ p_{m+1}^{-1} & p_2^{-1} + p_{m+1}^{-1} & \cdots & p_{m+1}^{-1} \\ \cdots & \cdots & \ddots & \cdots \\ p_{m+1}^{-1} & p_{m+1}^{-1} & \cdots & p_m^{-1} + p_{m+1}^{-1} \end{pmatrix}.$$

---

[4]Excellent sources for the theory of chi-squared tests are Moore (1978) and the chapter by David Moore in D'Agostino and Stephens (1986).

[5]Recall the definition of vector combinatoric $\binom{n}{\mathbf{r}}$ on page 245.

The Pearson statistic, denoted $P_{n,m}(\boldsymbol{\theta}_0)$, is given by the quadratic form

$$P_{n,m}(\boldsymbol{\theta}_0) := n^{-1} (\mathbf{N} - n\mathbf{p})' \, \mathbb{V}^{-1} \, (\mathbf{N} - n\mathbf{p}),$$

which reduces to the following simple expression:

$$P_{n,m}(\boldsymbol{\theta}_0) = \sum_{j=1}^{m+1} \frac{(N_j - np_j)^2}{np_j}. \tag{9.6}$$

Exercise 9.3. *Verify that the expression for $\mathbb{V}^{-1}$ is indeed the inverse of $\mathbb{V}$ and show that quadratic form $n^{-1} (\mathbf{N} - n\mathbf{p})' \, \mathbb{V}^{-1} \, (\mathbf{N} - n\mathbf{p})$ can be expressed as (9.6).*

9.2.1.2 The Asymptotic Distribution of the Pearson Statistic

The multivariate central limit and Mann–Wald theorems (Theorems 6.19 and 6.7) make it easy to establish the following result, which justifies the label "chi-squared test" for the Pearson procedure.

Theorem 9.1. *Under $H_0 : F(\cdot) = F_0(\cdot; \boldsymbol{\theta}_0)$ with $\boldsymbol{\theta}_0$ a known vector of parameters, the statistic $P_{n,m}(\boldsymbol{\theta}_0)$ converges in distribution to $\chi^2(m)$ as $n \to \infty$.*

Exercise 9.4. *Prove Theorem 9.1. (Hint: Write $P_{n,m}(\boldsymbol{\theta}_0) = \mathbf{Z}_n' \mathbf{Z}_n$, where*

$$\mathbf{Z}_n := n^{-1/2} \mathbb{V}^{-1/2} [\mathbf{N} - n\mathbf{p}(\boldsymbol{\theta}_0)],$$

and show that $\mathbf{Z}_n \rightsquigarrow N(\mathbf{0}, \mathbf{I}_m)$, where $\mathbf{I}_m$ is the $m \times m$ identity matrix.)

Exercise 9.5. *Let F_0 be a discrete distribution supported on the $m + 1$ distinct real numbers $\{x_j\}_{j=1}^{m+1}$; and take $F_0(x_j) - F_0(x_j-) =: p_j$ for $j \in \{1, 2, ..., m+1\}$, where $\sum_{j=1}^{m+1} p_j = 1$.*

(i) *Show that the maximum likelihood estimators from sample $\{X_i\}_{i=1}^n$ are $\hat{p}_j = N_j/n$, where $N_j := \sum_{i=1}^n \mathbf{1}_{\{x_j\}}(X_i)$.*

(ii) *Show that, under H_0, minus twice the log of the likelihood-ratio statistic is asymptotically equivalent to Pearson statistic $P_{n,m}$, in the sense that $|-2 \log \Lambda - P_{n,m}| \to^p 0$ as $n \to \infty$.*

(iii) *Apply these results and the asymptotic theory of likelihood-ratio tests to obtain an alternative proof of Theorem 9.1 for discrete distributions with bounded support.*

Consider now a sample from a distribution alternative to $F_0(x; \boldsymbol{\theta}_0)$, at least one of whose bin probabilities differ from $\mathbf{p}$, so that $E\mathbf{N} = n\mathbf{p}^*$, say, with $|\mathbf{p} - \mathbf{p}^*| > 0$. Assuming that the bins are such that covariance matrix $V\mathbf{N} = n\mathbb{V}^*$ still has rank $m - 1$, we can write

$$
\begin{aligned}
P_{n,m}(\boldsymbol{\theta}_0) &= n^{-1} \left[(\mathbf{N} - n\mathbf{p}^*) + n (\mathbf{p}^* - \mathbf{p}) \right]' \mathbb{V}^{*-1} \left[(\mathbf{N} - n\mathbf{p}^*) + n (\mathbf{p}^* - \mathbf{p}) \right] \\
&= n^{-1} (\mathbf{N} - n\mathbf{p}^*)' \mathbb{V}^{*-1} (\mathbf{N} - n\mathbf{p}^*) + 2 (\mathbf{p}^* - \mathbf{p})' \mathbb{V}^{*-1} (\mathbf{N} - n\mathbf{p}^*) \\
&\quad + n (\mathbf{p}^* - \mathbf{p})' \mathbb{V}^{*-1} (\mathbf{p}^* - \mathbf{p}) .
\end{aligned}
$$

The first term now has the same limiting χ^2 distribution that $P_{n,m}(\boldsymbol{\theta}_0)$ does when H_0 is true, while the second term is[6] $O_P\left(n^{1/2}\right)$ and the third term is $O(n)$. Since the third term is positive whenever $|\mathbf{p} - \mathbf{p}^*| > 0$, we would have $P_{n,m}(\boldsymbol{\theta}_0) \to +\infty$ under alternatives whose bin probabilities differ from those under H_0. This justifies the usual practice of using upper-tailed rejection regions for the Pearson test, rejecting H_0 at level α if $P_{n,m}(\boldsymbol{\theta}_0) \geq c^{\alpha}$, where c^{α} is the upper-α quantile of $\chi^2(m)$. It also shows that such a test is consistent against all alternatives to H_0 that have different bin probabilities. Under *local* (Pitman) alternatives with $\mathbf{p}_n^* := \mathbf{p} + n^{-1/2}\Delta$, say, the Pearson statistic does have a proper limiting distribution, with $P_{n,m}(\boldsymbol{\theta}_0) \rightsquigarrow \chi^2(m) + \Delta'\mathbb{V}^{-1}\Delta$. Thus, for simple null hypotheses the asymptotic theory of the Pearson test is straightforward and supports its effectiveness as a test of fit.

Unfortunately, the distribution theory of the Pearson statistic changes drastically when parameters $\boldsymbol{\theta}$ are not specified under H_0, the limiting distribution no longer being $\chi^2(m)$ and the precise form depending on the choice of estimators. The simplest result applies for *minimum-chi-squared (m.c.s.)* estimators; i.e., the estimator $\breve{\boldsymbol{\theta}}$ obtained by minimizing the Pearson statistic with respect to $\boldsymbol{\theta}$. We will first show that this estimator is consistent under H_0 (and under mild conditions on $F_0(\cdot : \boldsymbol{\theta})$ and the choice of bins), then will establish that the limiting distribution of $P_{n,m}(\breve{\boldsymbol{\theta}})$ is $\chi^2(m - k)$. Thus, when m.c.s. estimators are used there is a loss of one degree of freedom for each parameter that is estimated. The same result does not hold for maximum likelihood estimators $\hat{\boldsymbol{\theta}}$. In that case it has been established (see Moore (1978)) that the limiting distribution of $P_{n,m}(\hat{\boldsymbol{\theta}})$ is not independent of the true $\boldsymbol{\theta}$, although the upper quantiles are bounded between those of $\chi^2(m - k)$ and $\chi^2(m)$.

Focusing on the m.c.s. estimators, let us first establish consistency under the condition that there are more bins than there are parameters to be estimated.

---

[6] Recalling Definition 6.6, this means that the product with $n^{-1/2}$ converges in probability to a (possibly degenerate) r.v. (but not to zero). To see this, express the second term as $2n^{1/2} (\mathbf{p}^* - \mathbf{p})' \mathbb{V}^{*-1/2}\mathbf{Z}_n$, where $\mathbf{Z}_n = n^{-1/2}\mathbb{V}^{*-1/2} (\mathbf{N} - n\mathbf{p}^*) \rightsquigarrow N(\mathbf{0}, \mathbf{I}_m)$.

Theorem 9.2. *Fixing $m \geq k$, suppose the function $\mathbf{p}(\boldsymbol{\theta}) : \Re^k \to \Re^m$ has a continuous inverse, so that $\boldsymbol{\theta}$ can be expressed uniquely as a continuous function of the bin probabilities. Then under $H_0 : F(\cdot) = F_0(\cdot; \boldsymbol{\theta}_0)$ the statistic $\breve{\boldsymbol{\theta}}$ that satisfies $P_{n,m}(\breve{\boldsymbol{\theta}}) = \min_{\boldsymbol{\theta} \in \Theta} \{P_{n,m}(\boldsymbol{\theta})\}$ converges in probability to $\boldsymbol{\theta}_0$.*

Proof. $n^{-1} P_{n,m}(\boldsymbol{\theta}) = \sum_{j=1}^{m+1} [N_j/n - p_j(\boldsymbol{\theta})]^2 \to^p 0$ (by the law of large numbers) and $0 \leq P_{n,m}(\breve{\boldsymbol{\theta}}) \leq P_{n,m}(\boldsymbol{\theta})$ for all $\boldsymbol{\theta} \in \Theta$ (since $\breve{\boldsymbol{\theta}}$ is the minimizer) imply that $n^{-1} P_{n,m}(\breve{\boldsymbol{\theta}}) \to^p 0$ also. Therefore (with "$P\lim_{n\to\infty}$" denoting "probability limit"),

$$
0 = P \lim_{n\to\infty} \left[n^{-1} P_{n,m}(\breve{\boldsymbol{\theta}}) - n^{-1} P_{n,m}(\boldsymbol{\theta}) \right]
$$

$$
= P \lim_{n\to\infty} \sum_{j=1}^{m+1} \left\{ \frac{2N_j}{n} \left[p_j(\boldsymbol{\theta}) - p_j(\breve{\boldsymbol{\theta}}) \right] + p_j(\breve{\boldsymbol{\theta}})^2 - p_j(\boldsymbol{\theta})^2 \right\}
$$

$$
= P \lim_{n\to\infty} \sum_{j=1}^{m+1} \left[2p_j(\boldsymbol{\theta})^2 - 2p_j(\boldsymbol{\theta}) p_j(\breve{\boldsymbol{\theta}}) + p_j(\breve{\boldsymbol{\theta}})^2 - p_j(\boldsymbol{\theta})^2 \right]
$$

$$
= P \lim_{n\to\infty} \sum_{j=1}^{m+1} \left[p_j(\breve{\boldsymbol{\theta}}) - p_j(\boldsymbol{\theta}) \right]^2
$$

$$
= P \lim_{n\to\infty} \left| \mathbf{p}(\breve{\boldsymbol{\theta}}) - \mathbf{p}(\boldsymbol{\theta}) \right|^2.
$$

Since there is a unique, continuous inverse, the result follows by the continuity theorem. □

Theorem 9.3. *Add to the conditions of Theorem 9.2 that $\boldsymbol{\theta}_0$ (the true value of $\boldsymbol{\theta}$) lies in the interior of parameter space Θ, that $\mathbf{p}(\boldsymbol{\theta})$ is continuously differentiable, and that $m > k$. Then $P_{n,m}(\breve{\boldsymbol{\theta}}) \rightsquigarrow \chi^2(m-k)$.*

Proof. Since $\breve{\boldsymbol{\theta}}$ is consistent and $\boldsymbol{\theta}_0$ lies in the interior of Θ, the minimizer satisfies

$$
0 = \frac{\partial P_{n,m}(\breve{\boldsymbol{\theta}})}{\partial \boldsymbol{\theta}} := \frac{\partial P_{n,m}(\boldsymbol{\theta})}{\partial \boldsymbol{\theta}} \bigg|_{\boldsymbol{\theta}=\breve{\boldsymbol{\theta}}} \tag{9.7}
$$

$$
= -2n^{-1} \mathbf{p}_\theta(\breve{\boldsymbol{\theta}})' \mathbf{V}^{-1} [\mathbf{N} - n\mathbf{p}(\breve{\boldsymbol{\theta}})]
$$

$$
= -2n^{-1} \mathbf{p}_\theta(\breve{\boldsymbol{\theta}})' \mathbf{V}^{-1} [\mathbf{N} - n\mathbf{p}(\boldsymbol{\theta}_0)] + 2\mathbf{p}_\theta(\breve{\boldsymbol{\theta}})' \mathbf{V}^{-1} [\mathbf{p}(\breve{\boldsymbol{\theta}}) - \mathbf{p}(\boldsymbol{\theta}_0)]
$$

with probability arbitrarily close to unity for large enough n, where $\mathbf{p}_\theta(\breve{\boldsymbol{\theta}}) := \partial \mathbf{p}(\boldsymbol{\theta}) / \partial \boldsymbol{\theta}|_{\boldsymbol{\theta}=\breve{\boldsymbol{\theta}}}$. Since $\mathbf{p}$ is differentiable, the mean value theorem implies $\mathbf{p}(\breve{\boldsymbol{\theta}}) - \mathbf{p}(\boldsymbol{\theta}_0) = \mathbf{p}_\theta(\boldsymbol{\theta}^*)(\breve{\boldsymbol{\theta}} - \boldsymbol{\theta}_0)$ for some intermediate point

$\boldsymbol{\theta}^*$. Inserting in the last equation and solving for $\check{\boldsymbol{\theta}} - \boldsymbol{\theta}_0$ give

$$\check{\boldsymbol{\theta}} - \boldsymbol{\theta}_0 = n^{-1} \left[\mathbf{p}_{\boldsymbol{\theta}} \left(\check{\boldsymbol{\theta}} \right)' \mathbb{V}^{-1} \mathbf{p}_{\boldsymbol{\theta}} \left(\boldsymbol{\theta}^* \right) \right]^{-1} \mathbf{p}_{\boldsymbol{\theta}} \left(\check{\boldsymbol{\theta}} \right)' \mathbb{V}^{-1} \left[\mathbf{N} - n\mathbf{p} \left(\boldsymbol{\theta}_0 \right) \right]. \quad (9.8)$$

Now expand $P_{n,m}(\boldsymbol{\theta}_0)$ about $\check{\boldsymbol{\theta}}$ as

$$\begin{aligned} P_{n,m}(\boldsymbol{\theta}_0) &= P_{n,m} \left(\check{\boldsymbol{\theta}} \right) + \frac{1}{2} \left(\check{\boldsymbol{\theta}} - \boldsymbol{\theta}_0 \right)' \frac{\partial^2 P_{n,m} \left(\boldsymbol{\theta}^{\cdot} \right)}{\partial \boldsymbol{\theta} \partial \boldsymbol{\theta}'} \left(\check{\boldsymbol{\theta}} - \boldsymbol{\theta}_0 \right) \\ &= P_{n,m} \left(\check{\boldsymbol{\theta}} \right) + n \left(\check{\boldsymbol{\theta}} - \boldsymbol{\theta}_0 \right)' \mathbf{p}_{\boldsymbol{\theta}} \left(\boldsymbol{\theta}^{\cdot} \right)' \mathbb{V}^{-1} \mathbf{p}_{\boldsymbol{\theta}} \left(\boldsymbol{\theta}^{\cdot} \right) \left(\check{\boldsymbol{\theta}} - \boldsymbol{\theta}_0 \right), \end{aligned}$$

noting that the first-order term in $\check{\boldsymbol{\theta}} - \boldsymbol{\theta}_0$ vanishes because of (9.7). Solving for $P_{n,m} \left(\check{\boldsymbol{\theta}} \right)$, substituting for $\check{\boldsymbol{\theta}} - \boldsymbol{\theta}_0$ from (9.8), and simplifying give, apart from terms converging in probability to zero,

$$P_{n,m} \left(\check{\boldsymbol{\theta}} \right) = \mathbf{Z}_n' \mathbf{D}_n \mathbf{Z}_n, \quad (9.9)$$

where

$$\mathbf{Z}_n := n^{-1/2} \mathbb{V}^{-1/2} \left[\mathbf{N} - n\mathbf{p} \left(\boldsymbol{\theta}_0 \right) \right]$$

$$\mathbf{D}_n := \mathbf{I}_m - \mathbb{V}^{-1/2} \mathbf{p}_{\boldsymbol{\theta}} \left(\boldsymbol{\theta}^* \right) \left[\mathbf{p}_{\boldsymbol{\theta}} \left(\check{\boldsymbol{\theta}} \right)' \mathbb{V}^{-1} \mathbf{p}_{\boldsymbol{\theta}} \left(\boldsymbol{\theta}^* \right) \right]^{-1} \mathbf{p}_{\boldsymbol{\theta}} \left(\check{\boldsymbol{\theta}} \right)' \mathbb{V}^{-1/2}.$$

Now $\mathbf{Z}_n \rightsquigarrow N(\mathbf{0}, \mathbf{I}_m)$ by Theorem 6.19, and matrix $\mathbf{D}_n$ is idempotent with rank equal to the trace of

$$\mathbf{I}_m - \left[\mathbf{p}_{\boldsymbol{\theta}} \left(\check{\boldsymbol{\theta}} \right)' \mathbb{V}^{-1} \mathbf{p}_{\boldsymbol{\theta}} \left(\boldsymbol{\theta}^* \right) \right]^{-1} \mathbf{p}_{\boldsymbol{\theta}} \left(\check{\boldsymbol{\theta}} \right)' \mathbb{V}^{-1} \mathbf{p}_{\boldsymbol{\theta}} \left(\boldsymbol{\theta}^* \right) = \mathbf{I}_m - \mathbf{I}_k,$$

which is $m - k$. The conclusion now follows by Theorem (5.2). $\qquad \square$

9.2.1.3 *Implementing the Test*

The main considerations in applying the Pearson test are the choice of estimators for $\boldsymbol{\theta}$ and the number and sizes of the bins. Minimum-chi-squared estimators can ordinarily be found only through a numerical optimization, which requires special programming for each model. Moreover, since m.c.s. estimators use only the bin counts rather than the numerical values of the data, they are not asymptotically efficient. Thus, in applied work one would likely want to report maximum likelihood estimates in any case. Note that the previous remark about upper quantiles of $P_{n,m} \left(\hat{\boldsymbol{\theta}} \right)$ being bounded between those of $\chi^2 \left(m - k \right)$ and $\chi^2 \left(m \right)$ for m.l.e.s implies that the use of $\chi^2 \left(m \right)$ quantiles affords a conservative test, in that the (limiting) probability of rejecting a true null may be lower than the stated significance level. Thus, the use of m.l.e.s with upper-α quantiles of $\chi^2 \left(m \right)$ is a practice that trades power for convenience.

The choices of the number of bins, $m + 1$, and of their positions have implications both for power and for the distribution of $P_{n,m}$ under the null. For count data, where the support of $F_0(x; \boldsymbol{\theta})$ comprises the nonnegative integers, it is natural to associate bins with the integers themselves; e.g., to take $\mathbb{X}_1 = [0, J]$, $\mathbb{X}_{j+1} = \{J + j\}$ for $j \in \{1, 2, ..., m - 1\}$ and $\mathbb{X}_{m+1} = [J + m, \infty)$, where J and m would be set so that most of the probability mass under H_0 is thought to lie in bins 2 through m.

Example 9.3. We test $H_0 : X \sim P(\theta)$, $\theta > 0$ (Poisson) with the following sample of $n = 10$ pseudo-random observations distributed as $B(10, .5)$:

$$\mathbf{X} = (3, 5, 6, 6, 6, 5, 4, 6, 2, 7)'.$$

Choosing $J = 0$ and $m = 9$ (i.e., 10 bins), a numerical search shows that $P_{n,9}(\theta)$ attains a minimum value of about 5.42 at about $\check{\theta} = 5.548$. Actual and predicted relative bin frequencies with this value of θ are as follows:

$\mathbb{X}(j)$	N_j/n	$p_j(\check{\theta})$
$\{0\}$	.000	.004
$(0, 1]$	.000	.022
$(1, 2]$	.100	.060
$(2, 3]$	.100	.111
$(3, 4]$	.100	.154
$(4, 5]$	.200	.171
$(5, 6]$	.400	.158
$(6, 7]$	.100	.125
$(7, 8]$	.000	.087
$(8, \infty)$	.000	.110

With the approximation $P_{n,9}(\check{\theta}) \sim \chi^2(8)$ under H_0, the test's approximate P-value is 0.712, so H_0 is not rejected with these data.

With continuous data the selecting of bins is more problematical. The usual recommendation is to choose them so that the bin probabilities under H_0 are about the same; i.e., so that

$$p_j(\boldsymbol{\theta}) = \Pr(X \in \mathbb{X}_j \mid H_0) \doteq (m + 1)^{-1}$$

and to choose the number of bins such that $np_j(\boldsymbol{\theta}) \geq 1$ for each j.

Example 9.4. We test $H_0 : X \sim \Gamma(1, \theta)$, $\theta > 0$ (exponential with scale θ), with the following sample of $n = 10$ pseudo-random observations from $\Gamma(5, 1)$:

$$\mathbf{x} = (3.78, 9.48, 4.88, 3.60, 5.01, 8.07, 5.48, 3.01, 7.60, 6.67)'.$$

Since $F_0(x;\theta) = \left(1 - e^{-x/\theta}\right) \mathbf{1}_{[0,\infty)}(x)$ under H_0, bins

$$\mathbb{X}_j = \left[-\theta\log\left(1 - \frac{j-1)}{m+1}\right), -\theta\log\left(1 - \frac{j}{m+1}\right)\right), j \in \{1, 2, ..., m\}$$

$$\mathbb{X}_{m+1} = \left[-\theta\log\left(1 - \frac{j}{m+1}\right), \infty\right)$$

are equally likely. Choosing $m = 9$, a numerical search shows that $P_{n,9}(\theta)$ attains a minimum value of about 12.00 at about $\check{\theta} = 5.728$. Actual relative bin frequencies with this value of θ are as follows:

$\mathbb{X}(j)$	N_j/n
$[0, .60)$	.00
$[.60, 1.28)$	.00
$[1.28, 2.04)$	.00
$[2.04, 2.93)$	.00
$[2.93, 3.97)$	.30
$[3.97, 5.25)$	.20
$[5.25, 6.90)$	.20
$[6.90, 9.22)$	.20
$[9.22, 13.2)$	.10
$[13.2, \infty)$	.00

With the approximation $P_{n,9}(\check{\theta}) \sim \chi^2(8)$ under H_0, the test's approximate P-value is 0.151.

Some *caveats* apply to the usual recommendations for implementing the Pearson test. When H_0 does not fully specify $F(\cdot)$ and parameters have to be estimated, the attempt to make cell frequencies equally probable under the null makes the bin boundaries *data dependent*. (For example, for the test of exponentiality above the minimizing value $\check{\theta}$ was used to define the $\mathbb{X}_j$.) Fortunately, although this dependence greatly complicates the asymptotic theory, it turns out that the limiting distribution of the test statistic remains the same.[7] Of more concern is that the number of cells $m + 1$ may depend on the data and on n. Common practice is to let m grow in some relation to n in the hope that finer partitions of the support will enhance power; indeed, following the recommendations that bins be equally likely under H_0 and that $np_j(\theta) \geq 1$ for each j clearly requires that

---

[7] For the proof, which involves concepts of weak convergence on metric spaces, see Section 6 of Moore (1978).

m depend on n. In this case, assuming that m is not also data-dependent, the large-n distribution of $P_{n,m}(\boldsymbol{\theta}_0)$ is better approximated as normal with $\mu = m$ and variance $\sigma^2 = 2\mu$ when $\boldsymbol{\theta}_0$ is known. What happens then in the usual case that $\boldsymbol{\theta}$ must also be estimated is unclear. Of equal concern is the possibility that m will in some way depend on the sample data. This can happen with count data, for example, if unit bin widths are chosen by default and m is made large enough that most of the observed data fall in the first m bins. When m does depend on the sample, one might guess that $P_{n,m}(\boldsymbol{\theta}_0)$ converges weakly to some mixture of chi-squareds, but precise analysis would depend on which of the many possible forms of data dependence is actually present. Thus, the wide applicability of the Pearson test comes at a substantial price, and for most applications there are newer and better alternatives.

9.2.2   *Tests Based on Empirical Distribution Functions*

Here we look at one member of a large family of *empirical distribution function (e.d.f.)* tests, which are based on statistics that measure the closeness of the empirical c.d.f. to that of a proposed model. Given a random sample $\mathbf{X} := (X_1, X_2, ..., X_n)'$ the empirical c.d.f. was defined in Section 6.6.3 as

$$F_n(x) := n^{-1} \sum_{j=1}^{n} \mathbf{1}_{(-\infty, x]}(X_j), \, x \in \Re.$$

Thus, the value of F_n at any real x is the proportion of sample elements whose values are no greater than x. F_n is clearly a nondecreasing, right-continuous step function with $F_n(-\infty) = 0$ and $F_n(+\infty) = 1$ and is therefore a legitimate c.d.f. It corresponds to the "empirical" measure that places probability mass n^{-1} on each of the order statistics of the sample. We have already seen, as a simple consequence of the law of large numbers, that $F_n(x) \to^{a.s.} F(x)$ at each $x \in \Re$, where $F(x)$ is the common c.d.f. of the elements of the random sample. Recall that almost-sure convergence means that the set of outcomes ω on which the sequence $\{F_n(x)(\omega)\}_{n=1}^{\infty}$ fails to converge is a set of zero $\mathbb{P}$ measure—i.e., a $\mathbb{P}$-null set. By the countable additivity of measures, it follows that $F_n(x_j) \to F(x_j)$ with probability one for each x_j in any countable set; in other words, the set of outcomes on which this fails is still $\mathbb{P}$-null. However, convergence even on a countable, dense set such as the rationals still does not guarantee that the convergence of F_n to F is *uniform*, which is to say that the supremum of the difference between F_n and F approaches zero.[8] That F_n does in fact converge to F uniformly is the conclusion of the following famous theorem:

Theorem 9.4. *(Glivenko–Cantelli) If F_n is the empirical c.d.f. corresponding to a random sample from a population with c.d.f. F, then*

$$\sup_{x \in \Re} |F_n(x) - F(x)| \to^{a.s.} 0. \tag{9.10}$$

Proof. The plan is to show that (i) with probability one F_n and F can be made uniformly arbitrarily close for all large n on any sufficiently short interval $\mathcal{I} := [t, t+\delta)$; (ii) there is a finite interval $[t_-, t_+)$ *outside of which* F_n and F are uniformly close; and (iii) $[t_-, t_+)$ can be partitioned into finitely many subintervals such as $\mathcal{I}$. That F_n and F are right continuous, nondecreasing, and bounded are the crucial facts from which these follow.

(i) Since F is right continuous and nondecreasing, there exists for each $t \in \Re$ and each $\varepsilon > 0$ a $\delta = \delta(t, \varepsilon) > 0$ such that $\sup_{x \in \mathcal{I} := [t, t+\delta)} F(x) - F(t) \le F(t+\delta) - F(t) \le \varepsilon/4$. Since $F_n(x) \to^{a.s.} F(x)$ at countably many x, there is an $N(t, \delta, \varepsilon)$ such that $|F_n(t) - F(t)| \vee |F_n(t+\delta) - F(t+\delta)| \le \varepsilon/4$ with probability one for all $n > N(t, \delta, \varepsilon)$. Then, also with probability one for such n,

$$
\begin{aligned}
\sup_{x \in \mathcal{I}} F_n(x) - F_n(t) &\le F_n(t+\delta) - F_n(t) \\
&= [F_n(t+\delta) - F(t+\delta)] + [F(t+\delta) - F(t)] \\
&\quad - [F_n(t) - F(t)] \\
&\le |F_n(t+\delta) - F(t+\delta)| + [F(t+\delta) - F(t)] \\
&\quad + |F_n(t) - F(t)| \\
&\le \frac{3\varepsilon}{4}.
\end{aligned}
$$

---

[8] As an extreme example of nonuniform pointwise convergence (from Billingsley, 1968) consider the following sequence of functions on $(0, 1]$:

$$
g_n(x) = \begin{cases} nx, & 0 < x \le n^{-1} \\ 2 - nx, & n^{-1} < x \le 2n^{-1} \\ 0, & 2n^{-1} < x \le 1 \end{cases},
$$

for $n \in \{2, 3, \ldots\}$. While $g_n(x) \to 0$ for each x in the domain, we have $\sup |g_n(x) - 0| = 1$ for all n.

It follows that

$$\sup_{x \in \mathcal{I}} |F_n(x) - F(x)| = \sup_{x \in \mathcal{I}} |[F_n(x) - F_n(t)] + [F_n(t) - F(t)] - [F(x) - F(t)]|$$

$$\leq \left[\sup_{x \in \mathcal{I}} F_n(x) - F_n(t) \right] + |F_n(t) - F(t)|$$

$$\leq \frac{3\varepsilon}{4} + \frac{\varepsilon}{4} = \varepsilon$$

for all $n > N(t, \delta, \varepsilon)$.

(ii) Pick x_-, x_+, and N such that

$$\max \{ F(x_-), F_n(x_-), 1 - F(x_+), 1 - F_n(x_+) \} \leq \varepsilon$$

with probability one for all $n \geq N$. For such n it follows (from monotonicity and boundedness of F and F_n) that

$$\sup_{x \in (-\infty, x_-)} |F_n(x) - F(x)| \leq F_n(x_-) \vee F(x_-) \leq \varepsilon$$

and

$$\sup_{x \in [x_+, +\infty)} |F_n(x) - F(x)| = \sup_{x \in [x_+, +\infty)} |[1 - F(x)] - [1 - F_n(x)]|$$

$$\leq [1 - F_n(x_+)] \vee [1 - F(x_+)] \leq \varepsilon.$$

(iii) By the argument in (i) $[x_-, x_+)$ can be partitioned into finitely many intervals $\{ \mathcal{I}_i := [t_i, t_i + \delta(t_i, \varepsilon)) \}_{i=1}^m$ such that, for each i, $\sup_{x \in \mathcal{I}_i} |F_n(x) - F(x)| \leq \varepsilon$ for all n greater than some $N_i := N(t_i, \delta(t_i, \varepsilon), \varepsilon)$. Accordingly, there is a finite $M := \max_{i \in \{1,2,\dots,m\}} \{N_i\}$ such that $\sup_{x \in [x_-, x_+)} |F_n(x) - F(x)| \leq \varepsilon$ with probability one for all $n > M$. Then (deleting the arguments for brevity)

$$\sup_{\Re} |F_n - F| = \max \left\{ \sup_{(-\infty, x_-)} |F_n - F|, \sup_{[x_-, x_+)} |F_n - F|, \sup_{[x_+, +\infty)} |F_n - F| \right\}$$

$$\leq \varepsilon$$

with probability one for all $n > N \vee M$. (9.10) now follows since ε is arbitrary. $\qquad\square$

There are many statistical applications of Glivenko–Cantelli. One is the well-known Kolmogorov-Smirnov test for goodness of fit of continuous distributions. This is based on the statistic $\sup_{x \in \Re} |F_n(x) - F_0(x; \boldsymbol{\theta})|$, which can be expressed as an elementary function of the order statistics. While its elegance and simplicity commend it, the test is notorious for its low power.[9] A more promising application is the class of *Cramér–von Mises* statistics, which measure closeness of empirical to model c.d.f.s. as

$$n \int_{\Re} [F_n(x) - F_0(x; \boldsymbol{\theta})]^2 \, w(x) \cdot dF_0(x; \boldsymbol{\theta}).$$

Here $w(\cdot)$ is a positive function on support $\mathbb{X}$ that governs the weight given to deviations of F_n from F_0 at various places on the line. Clearly, when $H_0 : F(x) = F_0(x; \boldsymbol{\theta})$ is true (and when $w(\cdot)$ is consistent with integrability), Glivenko–Cantelli implies that the integral in the above expression converges a.s. to zero under the null, whereas it takes positive values whenever $F(\cdot)$ differs from $F_0(\cdot; \boldsymbol{\theta})$. Our focus will be on the member of this class that uses the specific weight function $w_{AD}(x) := \{F_0(x; \boldsymbol{\theta}) [1 - F_0(x; \boldsymbol{\theta})]\}^{-1}$. The resulting *Anderson–Darling (A-D)* test is thus based on the statistic

$$A^2 := n \int_{\Re} \frac{[F_n(x) - F_0(x; \boldsymbol{\theta})]^2}{F_0(x; \boldsymbol{\theta}) [1 - F_0(x; \boldsymbol{\theta})]} \cdot dF_0(x; \boldsymbol{\theta}). \tag{9.11}$$

Since $w_{AD}(x) \to +\infty$ as $|x| \to \infty$, this function accentuates discrepancies in the tails, where they often have the largest influence on parametric inference. This feature gives the test high power in detecting "contamination"—e.g., data errors or inhomogeneities in the population. Clearly, large values of the statistic are expected under alternatives to H_0, and $A^2 \to \infty$ as $n \to \infty$ in such cases. Thus, a test that rejects H_0 for large values of A^2 is consistent against *all* alternatives. Note that this vital property does not always hold for the Pearson test—and *never* in the case of continuous distributions, since there are infinitely many distinct distributions with the same set of bin probabilities.

When model F_0 is absolutely continuous, the A-D statistic has a relatively simple representation in terms of the order statistics, $\{X_{(j:n)}\}_{j=1}^n$.

Theorem 9.5. *For an absolutely continuous distribution F the A-D statistic is given by*

$$A^2 = -\sum_{j=1}^n \frac{2j-1}{n} \left[\log F_{(j:n)} + \log \left(1 - F_{(n-j+1:n)}\right) \right] - n, \tag{9.12}$$

where $F_{(j:n)} := F_0\left(X_{(j:n)}; \boldsymbol{\theta}\right)$.

---

[9] Somehow this fact has not percolated down to producers of statistical software, who invariably include it among the handful of tests they feature.

Proof. Since $F_n\left(X_{(j:n)}\right) = j/n$, the integral form (9.11) can be expressed in terms of order statistics as

$$n \sum_{j=0}^{n-1} \int_{X_{(j:n)}}^{X_{(j+1:n)}} \frac{(j/n - F_0)^2}{F_0(1 - F_0)} \cdot dF_0 + n \int_{X_{(n:n)}}^{\infty} \frac{1 - F_0}{F_0} \cdot dF_0,$$

where we set $X_{(0:n)} := -\infty$. Now $F_0(X)$ is distributed as uniform on $[0, 1]$, so changing variables as $F_0(x) = t$ yields

$$A^2 = n \sum_{j=0}^{n-1} \int_{F_{(j:n)}}^{F_{(j+1:n)}} \frac{(j/n - t)^2}{t(1 - t)} \cdot dt + n \int_{F_{(n:n)}}^{1} \frac{1 - t}{t} \cdot dt.$$

The indefinite forms of the integrals are

$$\int \frac{(j/n - t)^2}{t(1 - t)} \cdot dt = -\left(\frac{j}{n} - 1\right)^2 \log(1 - t) + \left(\frac{j}{n}\right)^2 \log t - t$$

$$\int \frac{1 - t}{t} \cdot dt = \log t - t.$$

Using these one obtains for A^2

$$\sum_{j=0}^{n-1} \left\{ -\frac{(n-j)^2}{n} \log\left(\frac{1 - F_{(j+1:n)}}{1 - F_{(j:n)}}\right) + \frac{j^2}{n} \log\left(\frac{F_{(j+1:n)}}{F_{(j:n)}}\right) \right\}$$
$$-n\left[\log F_{(n:n)} + 1\right], \tag{9.13}$$

from which (9.12) can be seen to follow on writing out and collecting the terms. $\square$

Exercise 9.6. *Write out and collect terms in (9.13) to verify (9.12).*

When the null hypothesis is simple, $\boldsymbol{\theta}$ being specified as well as parametric family $\{F_0(\cdot; \boldsymbol{\theta}) : \boldsymbol{\theta} \in \boldsymbol{\Theta}\}$, the A-D statistic is a function just of order statistics of uniform r.v.s. In this case the distribution of A^2 is the same for all parametric families. The limiting distribution, to which A^2 converges rapidly, is known,[10] and tables of upper quantiles are widely available. However, in the almost universal situation that $\boldsymbol{\theta}$ is *not* specified under the null the distribution changes from one parametric family to the next. Two classes of such cases are relevant for the applicability of the A-D test.

---

[10] Anderson and Darling (1952, 1954).

(i) The model proposed under H_0 is a two-parameter family,

$$\{F\left(\cdot;\theta_1,\theta_2\right):(\theta_1,\theta_2)\in\boldsymbol{\Theta}=\Re\times\Re_+\},$$

such that if X has c.d.f. $F\left(\cdot;\theta_1,\theta_2\right)$ then $Z:=(X-\theta_1)/\theta_2$ has c.d.f. $F\left(\cdot;0,1\right)$. Such a family of distributions is called a *location–scale* family since its parameters govern only these properties. Examples are the uniform, normal, logistic, and Laplace distributions; also, the subfamilies of three-parameter gammas and Weibulls with common shape parameters but arbitrary origin and scale. The exponential distributions are one such subfamily of each of these last two. The lognormal is not location–scale but can be made so simply by taking logs and transforming to normal. For testing location–scale families, one can estimate θ_1,θ_2 by maximum likelihood and apply the A-D test with transformed data $\left\{\hat{Z}_j=\left(X_j-\hat{\theta}_1\right)/\hat{\theta}_2\right\}$. In this case the limiting distribution of A^2 for any *particular* location–scale family is the same for all (θ_1,θ_2), and tables of its quantiles for many such particular families have been developed. Table 9.1[11] presents approximate upper-α quantiles of modified versions of A^2 for the normal and one-parameter exponential families.[12] The modifications to A^2 make the same critical values work reasonably well for all sample sizes.

(ii) The proposed model is not a location–scale family. Examples include the gamma and Weibull families with unspecified shape parameters, the beta distributions, and Student's t. In such cases the distribution of A^2 depends, even asymptotically, on the parameters themselves. This makes general tabulation impossible and leaves parametric bootstrapping as the only feasible way to apply the test. That is, an estimate $\hat{\boldsymbol{\theta}}$ would be found from the sample data, then the distribution of A^2 would be inferred by repeatedly drawing pseudo-random samples from $F_0\left(\cdot;\hat{\boldsymbol{\theta}}\right)$. Not surprisingly, the A-D test is less often used in such cases.

---

[11] Adapted from Stephens (1983).

[12] That is, the family $F_0\left(x;\theta\right)=[1-\exp\left(-x/\theta\right)]\mathbf{1}_{(0,\infty)}\left(x\right),\theta>0$. For the two-parameter case,

$$F_0\left(x;\theta_1,\theta_2\right)=\left[1-\exp\left(-\frac{x-\theta_1}{\theta_2}\right)\right]\mathbf{1}_{(\theta_1,\infty)}\left(x\right),\theta_2>0,$$

the one-parameter quantiles are appropriate if A^2 is applied to the reduced sample of $n-1$ deviations of the order statistics from the sample minimum, $\left\{Y_j:=X_{(j:n)}-X_{(1:n)}\right\}_{j=2}^n$. This follows from the fact that the $\{Y_j\}$ are i.i.d. as $\Pr\left(Y\le y\right)=[1-\exp\left(-y/\theta_2\right)]\mathbf{1}_{(0,\infty)}\left(y\right),\theta_2>0$.

Table 9.1   Upper quantiles for A-D test for normal and exponential distributions with estimated parameters.

Distribution	Modified A^2	$\alpha = .10$	$\alpha = .05$	$\alpha = .01$
Normal	$A^2\left(1 + .75/n + 2.25/n^2\right)$	0.631	0.752	1.035
Exponential	$A^2\left(1 + .3/n\right)$	1.062	1.321	1.959

Example 9.5. We apply the A-D procedure to test

$$H_0 : f(x) = \theta^{-1} e^{-x/\theta} \mathbf{1}_{(0,\infty)}(x), \theta > 0,$$

using the ordered $\Gamma(5, 1)$ sample from Example 9.4:

$$\mathbf{x} = (3.01, 3.60, 3.78, 4.88, 5.01, 5.48, 6.67, 7.60, 8.07, 9.48)'.$$

The m.l.e. of θ is $\bar{x} = 5.758$, and the values of the ordered c.d.f.s at $\left\{x_{(j:10)}/\bar{x}\right\}_{j=1}^{10}$ are about

$$0.407, 0.465, 0.481, 0.572, 0.581, 0.614, 0.686, 0.733, 0.754, 0.807.$$

The resulting A-D statistic and its modified value are $A^2 = 1.958$ and $A^2(1.03) = 2.017$. Comparison with upper quantiles in Table 9.1 shows that H_0 is rejected at the .01 level.

9.2.3   *Tests Based on Moments*

In many cases the relation between sample and population moments gives a good indication of a model's adequacy to represent the data at hand. Test statistics for these moment procedures are often simple to compute (if not always to derive), and they have the additional advantage of showing what particular features of the model are inconsistent with the data whenever H_0 is rejected. On the other hand, since not all distributions are uniquely defined by their moments, no test of this sort can be consistent against all alternatives. We look briefly at a specialized test for univariate normality, then in greater detail at the much broader class of "smooth" tests.

9.2.3.1   *Skewness–Kurtosis Tests*

The coefficients of skewness and kurtosis were defined in (3.27) and (3.28) as $\alpha_3 := \mu_3/\sigma^3$ and $\alpha_4 := \mu_4/\sigma^4$, respectively. The strong consistency of the sample moments and the continuity theorem imply that sample statistics $A_3 := M_3/S^3$ and $A_4 := M_4/S^4$ converge a.s. to their population counterparts, which suggests a role for these quantities in inference about the distribution of the population. In the case of normal populations, we saw in Exercise 4.54 that α_3 and α_4 take the specific values 0.0 and 3.0. These by no means characterize normality, in that distributions in many other families can be found with the same coefficients of skewness and kurtosis, but the ease of computing these statistics has made them popular as rough-and-ready indicators. Moreover, the sample values of A_3 and A_4 convey some useful information about *how* the data depart from normality. What is needed for a formal test is a distribution theory for (A_3, A_4) under H_0 : "population is normal". It turns out that standardized versions of these statistics are distributed asymptotically as independent normals under H_0. Specifically, we have the following result:

Theorem 9.6. *For random samples from a normal population*

$$\sqrt{n} \begin{pmatrix} A_3 \\ A_4 - 3 \end{pmatrix} \rightsquigarrow N \left[\begin{pmatrix} 0 \\ 0 \end{pmatrix}, \begin{pmatrix} 6 & 0 \\ 0 & 24 \end{pmatrix} \right].$$

Thus, A_3 and A_4 have asymptotic variance $6/n$ and $24/n$, respectively.

Proof. We first develop some asymptotically equivalent expressions for A_3 and A_4. Since these are invariant under changes in location and scale of the data (μ and σ), we can work with samples from a standard normal population. Applying (6.29) to A_3 with $\mathbf{T}_n = (M_3, S^2)'$, $\boldsymbol{\theta} = (0, 1)'$, $g(\boldsymbol{\theta}) = 0$, $g(\mathbf{T}_n) = M_3 / (S^2)^{3/2}$, and $\mathbf{g}_{\mathbf{T}}'(\boldsymbol{\theta}) = (1, 0)$ shows that $\sqrt{n}A_3$ has the same limiting distribution as

$$\sqrt{n}(1, 0) \cdot \begin{pmatrix} M_3 \\ S^2 - 1 \end{pmatrix} = \sqrt{n}M_3.$$

But

$$\sqrt{n}M_3 = n^{-1/2} \sum_{j=1}^{n} Z_j^3 - 3\bar{Z}n^{-1/2} \sum_{j=1}^{n} Z_j^2 + 2\sqrt{n}\bar{Z}^3$$

$$= n^{-1/2} \sum_{j=1}^{n} Z_j^3 - 3\sqrt{n}\bar{Z} + 3\bar{Z}n^{-1/2} \sum_{j=1}^{n} (Z_j^2 - 1) + 2\sqrt{n}\bar{Z}^3,$$

of which the last two terms converge a.s. to zero. Thus, asymptotically,

$$\sqrt{n}A_3 \sim n^{-1/2} \sum_{j=1}^{n} (Z_j^3 - 3Z_j).$$

Working out an asymptotic equivalence for A_4 in the same manner and combining the two expressions give

$$\sqrt{n} \begin{pmatrix} A_3 \\ A_4 - 3 \end{pmatrix} \sim n^{-1/2} \sum_{j=1}^{n} \begin{pmatrix} Z_j^3 - 3Z_j \\ Z_j^4 - 6Z_j^2 + 3 \end{pmatrix},$$

which by the multivariate central limit theorem converges in distribution to bivariate normal with the stated mean vector and covariance matrix. $\square$

Exercise 9.7. *Show that* $\sqrt{n}\,(A_4 - 3)$ *and* $n^{-1/2} \sum_{j=1}^{n} \left(Z_j^4 - 6Z_j^2 + 3 \right)$ *have the same limiting distribution, and verify the expression for the asymptotic covariance matrix of A_3 and A_4.*

Corollary 9.1. $K_n^2 := n \left[A_3^2/6 + (A_4 - 3)^2 /24 \right] \rightsquigarrow \chi^2 (2)$ *as $n \to \infty$ when the random sample is from a normal population.*

Proof. Apply Mann–Wald. $\square$

While it is not implausible that A_3 and A_4 should be given equal weight relative to standard error, it is true that the statistic K_n^2 is just one of infinitely many ways of combining these into one statistic. One might prefer to give more weight to A_3, for example, if the alternative distribution were thought likely to be asymmetric or if asymmetry was of more relevance for subsequent analysis.[13] However, K_n^2 is not so arbitrary as it seems, since it turns out to be a score or Lagrange-multiplier statistic when the normal is regarded as a subfamily of the Pearson class of distributions. These distributions, which we have already encountered in Section 6.7.3, have p.d.f.s that satisfy a certain class of ordinary differential equations. The Pearson system groups distributions according to the particular subclasses of o.d.e.s that they satisfy.[14] Since $K_n^2 \to \infty$ whenever $\alpha_3 \neq 0$ or $\alpha_4 - 3 \neq 0$, a test that rejects H_0 : "population is normal" at level α when K_n^2 equals or exceeds the upper-α quantile of $\chi^2 (2)$ is asymptotically valid and consistent against all alternatives whose skewness and kurtosis differ from those of the normal. Moreover, the test inherits the local efficiency characteristics of score tests when the data come from any other family in the Pearson class.

While K_n^2 thus provides a test with desirable large-sample properties, it turns out that the very slow convergence of $\sqrt{n}\,(A_3, A_4 - 3)$ renders the chi-squared approximation unacceptably inaccurate unless the sample is extremely large. However, it is an easy matter to estimate appropriate rejection regions for various n by simulation. Table 9.2[15] presents such

---

[13]Bowman and Shenton (1975, 1986) and Pearson *et al.* (1977) describe other ways of using these two statistics in testing normality.

[14]For details see Ord (1985).

[15]From Jarque and Bera (1987), who (to my knowledge) were first to interpret K_n^2 as a score test.

Table 9.2 Empirical upper quantiles of Jarque-Bera statistic.

$n \backslash \alpha$	0.10	0.05
20	2.13	3.26
30	2.49	3.71
40	2.70	3.99
50	2.90	4.26
75	3.09	4.27
100	3.14	4.29
125	3.31	4.34
150	3.43	4.39
200	3.48	4.43
250	3.54	4.51
300	3.68	4.60
400	3.76	4.74
500	3.91	4.82
800	4.32	5.46
∞	4.61	5.99

simulation-based estimates of critical values of K_n^2 for samples ranging from $n = 20$ to $n = 800$. The corresponding $\chi^2(2)$ values are in the last row.

9.2.3.2 *"Smooth" Tests*

We now examine a procedure that is not limited to normality but that can be applied to a wide variety of distributions having finite moments—both discrete and continuous. Like the test for normality based on K_n^2, smooth tests are asymptotic score tests applied to a parametric class of distributions that converge to the null family as certain parameters converge to zero. The name arises from this convergence property.[16] In describing the procedure we use the notation from the general discussion of score tests in Section 8.6.2.

Construction. Suppose H_0 specifies that elements of sample $\mathbf{X}$ are i.i.d. as some member of family $\left\{ F_0\left(\cdot; \psi\right) : \psi \in \mathbf{\Psi} \subset \Re^k \right\}$, with corresponding p.d.f./p.m.f. $f_0\left(\cdot; \psi\right)$. The idea is to embed null or "core" model f_0 in a broader, augmented parametric family

$$\left\{ f_m\left(\cdot; \phi, \psi\right) : \psi \in \mathbf{\Psi}, \phi \in \mathbf{\Phi} \subset \Re^m \right\}$$

---

[16]The original idea appeared in Neyman (1937) as a test of the simple hypothesis of uniformity, but it has been extensively developed and broadened in a series of papers by Rayner and Best, culminating in their 1989 monograph.

and to define f_m so that it can capture the main features of realistic alternatives that might be encountered in practice. It will turn out that the features involved are, in fact, the first m moments. To construct the augmented family f_m, the p.d.f./p.m.f. of the null model is multiplied by a function of polynomials, $\{p_i(x; \psi)\}_{i=1}^m$, as

$$f_m(x; \theta) = K(\theta) \exp\left(\sum_{i=1}^m \phi_i p_i(x, \psi)\right) f_0(x; \psi). \tag{9.14}$$

Here $\theta' := (\phi', \psi')$ is the composite vector of augmented and core parameters, and $K(\theta)$ is a normalizing constant that makes f_m integrate to unity. The polynomials have the following properties: (i) each p_i is of order x^i; (ii) $E_0 p_i(X; \psi) = 0$; and (iii)

$$E_0 p_i(X; \psi) p_{i'}(X; \psi) = \int p_i(x; \psi) p_{i'}(x; \psi) \cdot dF_0(x; \psi) = \begin{cases} 1, & i = i' \\ 0, & i \neq i' \end{cases}.$$

Property (ii) indicates that the polynomials have zero mean under the core model, and so are orthogonal to a units vector. Property (iii) indicates that they are also mutually orthogonal with respect to f_0 and have unit variance. Thus, $\{p_i(x; \psi)\}_{i=1}^m$ are referred to as "f_0-orthonormal". There are systems of such orthonormal polynomials corresponding to many standard models f_0.[17] With this specification the null family corresponds to the parametric hypothesis $H_0 : \phi = 0$, which could be tested against $H_1 : \phi \neq 0$ with any of the three classical asymptotic procedures discussed in Section 8.6: the Lagrange-multiplier or "score" test, the likelihood-ratio test, or the Wald test. Of these, the score test offers the marked advantage that only the nuisance parameters ψ of the core model need be estimated—and these would likely be desired in any case. The nice feature of constructing f_m from orthogonal polynomials is that the score statistic turns out to be of very simple—and often very illuminating—form.

To develop the smooth test we begin with the log likelihood of the augmented model,

$$\mathcal{L}(\theta) = n \log K(\theta) + \sum_{i=1}^m \left[\phi_i \sum_{j=1}^n p_i(X_j; \psi)\right] + \sum_{j=1}^n \log f_0(X_j; \psi),$$

and the general expression for the score statistic as

$$S_{n,m}(\mathbf{X}) := n^{-1} \mathcal{L}_{\phi'}(\breve{\theta}_n) \mathcal{I}^{\phi\phi'}(\breve{\theta}_n) \mathcal{L}_\phi(\breve{\theta}_n). \tag{9.15}$$

---

[17]See Ch. 22 of Abramowitz and Stegun (1972) and Griffiths (1985) for lists of common systems. Three of these appear below in examples of common applications of the smooth test.

Here $\mathcal{L}_\phi$ is the m-vector of partial derivatives with respect to the augmented parameters (the "score") and $n\mathcal{I}^{\phi\phi'}(\boldsymbol{\theta})^{-1}$ is their covariance matrix. Recalling Expression (8.16), we have

$$\mathcal{I}^{\phi\phi'}(\boldsymbol{\theta})^{-1} = \mathcal{I}_{\phi\phi'}(\boldsymbol{\theta}) - \mathcal{I}_{\phi\psi'}(\boldsymbol{\theta})\mathcal{I}_{\psi\psi'}(\boldsymbol{\theta})^{-1}\mathcal{I}_{\psi\phi'}(\boldsymbol{\theta}),$$

where matrices $\mathcal{I}_{\phi\phi'}, \mathcal{I}_{\phi\psi'}, \mathcal{I}_{\psi\psi'}$ are (apart from factor n) the blocks of the Fisher information matrix under f_m:

$$-E_m\mathcal{L}_\theta(\boldsymbol{\theta})\mathcal{L}_{\theta'}(\boldsymbol{\theta}) = n\begin{pmatrix} \mathcal{I}_{\phi\phi'}(\boldsymbol{\theta}) & \mathcal{I}_{\phi\psi'}(\boldsymbol{\theta}) \\ \mathcal{I}_{\psi\phi'}(\boldsymbol{\theta}) & \mathcal{I}_{\psi\psi'}(\boldsymbol{\theta}) \end{pmatrix}.$$

Here "E_m" denotes expectation with respect to the augmented model. Also, in (9.15) $\breve{\boldsymbol{\theta}}_n := \left(\mathbf{0}, \hat{\boldsymbol{\psi}}_n\right)'$ represents the estimates of the parameters under the restriction of H_0. As the usual "hat" notation signifies, the restricted m.l.e. of $\boldsymbol{\psi}$ is just the ordinary m.l.e. for the core model.

Constructing $\mathcal{S}_{n,m}(\mathbf{X})$ requires working out the derivatives of the log likelihood, which depend on the derivatives of normalizing constant $K(\boldsymbol{\theta})$. These have nice interpretations when f_0 satisfies the usual regularity conditions for m.l. estimation, which is henceforth assumed. Under the regularity conditions derivatives with respect to $\boldsymbol{\psi}$ of expected values can be equated with expected values of derivatives. The form of (9.14) assures that this interchange of operations is valid also for derivatives with respect to $\boldsymbol{\phi}$. Using this feature to differentiate $\int dF_m(x; \boldsymbol{\theta})$ under the integral sign gives

$$0 = \int \frac{\partial \log f_m(x; \boldsymbol{\theta})}{\partial \boldsymbol{\theta}} \cdot dF_m(x; \boldsymbol{\theta}) = E_m\left[\frac{\partial \log f_m(X; \boldsymbol{\theta})}{\partial \boldsymbol{\theta}}\right].$$

For the m elements of $\boldsymbol{\phi}$ we have $0 = \partial \log K(\boldsymbol{\theta})/\partial \phi_i + E_m p_i(X; \boldsymbol{\psi})$ and hence

$$\frac{\partial \log K(\boldsymbol{\theta})}{\partial \phi_i} = -E_m p_i(X; \boldsymbol{\psi}); \qquad (9.16)$$

and for the k elements of $\boldsymbol{\psi}$,

$$\frac{\partial \log K(\boldsymbol{\theta})}{\partial \psi_\ell} = -\sum_{i=1}^{m} \phi_i E_m\left[\frac{\partial p_i(X; \boldsymbol{\psi})}{\partial \psi_\ell}\right] - E_m\left[\frac{\partial \log f_0(X; \boldsymbol{\psi})}{\partial \psi_\ell}\right]. \qquad (9.17)$$

With these expressions the derivatives of the log likelihood become

$$\partial \mathcal{L}(\boldsymbol{\theta})/\partial \phi_i = \sum_{j=1}^{n}(1 - E_m)p_i(X_j; \boldsymbol{\psi})$$

$$\partial \mathcal{L}(\boldsymbol{\theta})/\partial \psi_\ell = \sum_{j=1}^{n}\left[(1-E_m)\frac{\partial \log f_0(X_j; \boldsymbol{\psi})}{\partial \psi_\ell} + \sum_{i=1}^{m}\phi_i(1-E_m)\frac{\partial p_i(X_j; \boldsymbol{\psi})}{\partial \psi_\ell}\right],$$

where operator $(1 - E_m)(\cdot) := (\cdot) - E_m(\cdot)$ creates deviations from expected value under f_m.

As the last two expressions indicate, finding second derivatives of $\mathcal{L}(\boldsymbol{\theta})$ requires finding derivatives of expectations of various quantities: $p_i(X; \boldsymbol{\psi})$, $\partial \log f_0(X; \boldsymbol{\psi}) / \partial \psi_\ell$, etc. Letting $Q(X)$ be some generic integrable function, we can use (9.16) to express $\partial E_m Q(X) / \partial \phi_i$ as

$$
\begin{aligned}
\partial E_m Q(X) / \partial \phi_i &= \int Q(x) \frac{\partial \log f_m(x; \boldsymbol{\theta})}{\partial \phi_i} \cdot dF_m(x; \boldsymbol{\theta}) \\
&= \int Q(x) \left[\frac{\partial \log K(\boldsymbol{\theta})}{\partial \phi_i} + p_i(x; \boldsymbol{\psi}) \right] \cdot dF_m(x; \boldsymbol{\theta}) \\
&= -E_m Q(X) \cdot E_m p_i(X; \boldsymbol{\psi}) + E_m[Q(X) p_i(X; \boldsymbol{\psi})] \\
&= Cov_m[Q(X), p_i(X; \boldsymbol{\psi})].
\end{aligned}
$$

Similarly, for $\partial E_m Q(X) / \partial \psi_\ell$ we use (9.17) to get

$$
\int \frac{\partial Q(x)}{\partial \psi_\ell} \cdot dF_m(x; \boldsymbol{\theta}) + \int Q(x) \frac{\partial \log f_m(x; \boldsymbol{\theta})}{\partial \psi_\ell} \cdot dF_m(x; \boldsymbol{\theta}).
$$

The first term is just $E_m \partial Q(X) / \partial \psi_\ell$, and the second is

$$
\begin{aligned}
\int Q(x) &\left[\frac{\partial \log K(\boldsymbol{\theta})}{\partial \psi_\ell} + \sum_{i=1}^{m} \phi_i \frac{\partial p_i(x; \boldsymbol{\psi})}{\partial \psi_\ell} + \frac{\partial \log f_0(x; \boldsymbol{\psi})}{\partial \psi_\ell} \right] \cdot dF_m(x; \boldsymbol{\theta}) \\
&= \sum_{i=1}^{m} \phi_i Cov_m \left[Q(X), \frac{\partial p_i(X; \boldsymbol{\psi})}{\partial \psi_\ell} \right] + Cov_m \left[Q(X), \frac{\partial \log f_0(X; \boldsymbol{\psi})}{\partial \psi_\ell} \right].
\end{aligned}
$$

From these general expressions the second derivatives of $\mathcal{L}(\boldsymbol{\theta})$ are found to be

$$
\begin{aligned}
\frac{\partial^2 \mathcal{L}(\boldsymbol{\theta})}{\partial \phi_i \partial \phi_{i'}} &= -\sum_{j=1}^{n} \frac{\partial E_m p_i(X_j; \boldsymbol{\psi})}{\partial \phi_{i'}} \\
&= -n Cov_m[p_i(X; \boldsymbol{\psi}), p_{i'}(X; \boldsymbol{\psi})]
\end{aligned}
$$

$$
\begin{aligned}
\frac{\partial^2 \mathcal{L}(\boldsymbol{\theta})}{\partial \phi_i \partial \psi_\ell} &= \sum_{j=1}^{n} \left[(1 - E_m) \frac{\partial p_i(X_j; \boldsymbol{\psi})}{\partial \psi_\ell} \right] \\
&\quad - n \sum_{i'=1}^{m} \phi_{i'} Cov_m \left[p_i(X; \boldsymbol{\psi}), \frac{\partial p_{i'}(X; \boldsymbol{\psi})}{\partial \psi_\ell} \right] \\
&\quad - n Cov_m \left[p_i(X; \boldsymbol{\psi}), \frac{\partial \log f_0(X; \boldsymbol{\psi})}{\partial \psi_\ell} \right]
\end{aligned}
$$

$$\frac{\partial^2 \mathcal{L}(\boldsymbol{\theta})}{\partial \psi_\ell \partial \psi_{\ell'}} = \sum_{j=1}^{n} \left[(1 - E_m) \frac{\partial^2 \log f_0 (X_j; \boldsymbol{\psi})}{\partial \psi_\ell \partial \psi_{\ell'}} \right]$$

$$- n Cov_m \left[\frac{\partial \log f_0 (X; \boldsymbol{\psi})}{\partial \psi_\ell}, \frac{\partial \log f_0 (X; \boldsymbol{\psi})}{\partial \psi_{\ell'}} \right]$$

$$- n \sum_{i=1}^{m} \phi_i Cov_m \left[\frac{\partial \log f (X; \boldsymbol{\psi})}{\partial \psi_\ell}, \frac{\partial p_i (X; \boldsymbol{\psi})}{\partial \psi_{\ell'}} \right]$$

$$+ \sum_{j=1}^{n} \sum_{i=1}^{m} \phi_i \frac{\partial}{\partial \psi_{\ell'}} \left[(1 - E_m) \frac{\partial p_i (X_j; \boldsymbol{\psi})}{\partial \psi_\ell} \right].$$

Component blocks of the information matrix are now given by expectations of the negatives of these under the constraint $H_0 : \boldsymbol{\phi} = \mathbf{0}$. Thus,

$$n \mathcal{I}_{\boldsymbol{\phi}\boldsymbol{\phi}'} (\mathbf{0}, \boldsymbol{\psi}) = \left\{ -E_0 \frac{\partial^2 \mathcal{L}(\boldsymbol{\theta})}{\partial \phi_i \partial \phi_{i'}} \right\}_{i,i'=1}^{m}$$

$$= n \left\{ Cov_0 \left[p_i (X; \boldsymbol{\psi}), p_{i'} (X; \boldsymbol{\psi}) \right] \right\} = n \mathbf{I}_m$$

$$n \mathcal{I}_{\boldsymbol{\phi}\boldsymbol{\psi}'} (\mathbf{0}, \boldsymbol{\psi}) = \left\{ -E_0 \frac{\partial^2 \mathcal{L}(\boldsymbol{\theta})}{\partial \phi_i \partial \psi_\ell} \right\}_{i=1,\ell=1}^{m,k}$$

$$= n \left\{ Cov_0 \left[p_i (X; \boldsymbol{\psi}), \frac{\partial \log f_0 (X; \boldsymbol{\psi})}{\partial \psi_\ell} \right] \right\}$$

$$n \mathcal{I}_{\boldsymbol{\psi}\boldsymbol{\psi}'} (\mathbf{0}, \boldsymbol{\psi}) = \left\{ -E_0 \frac{\partial^2 \mathcal{L}(\boldsymbol{\theta})}{\partial \psi_\ell \partial \psi_{\ell'}} \right\}_{\ell,\ell'=1}^{k}$$

$$= n \left\{ Cov_0 \left[\frac{\partial \log f_0 (X; \boldsymbol{\psi})}{\partial \psi_\ell}, \frac{\partial \log f_0 (X; \boldsymbol{\psi})}{\partial \psi_{\ell'}} \right] \right\},$$

where the first expression follows from orthonormality of the $\{p_i\}$ under f_0. Note in particular that $\mathcal{I}_{\boldsymbol{\phi}\boldsymbol{\phi}'} (\mathbf{0}, \boldsymbol{\psi})$ is just the $m \times m$ identity matrix. The remaining components are evaluated at the restricted m.l.e. $\hat{\boldsymbol{\psi}}_n$, giving

$$\mathcal{I}^{\boldsymbol{\phi}\boldsymbol{\phi}'} \left(\mathbf{0}, \hat{\boldsymbol{\psi}}_n \right)^{-1} = \mathbf{I}_m - \mathcal{I}_{\boldsymbol{\phi}\boldsymbol{\psi}'} \left(\mathbf{0}, \hat{\boldsymbol{\psi}}_n \right) \mathcal{I}_{\boldsymbol{\psi}\boldsymbol{\psi}'} \left(\mathbf{0}, \hat{\boldsymbol{\psi}}_n \right)^{-1} \mathcal{I}_{\boldsymbol{\psi}\boldsymbol{\phi}'} \left(\mathbf{0}, \hat{\boldsymbol{\psi}}_n \right).$$

To complete the construction of the test statistic, we need the score vector under the restriction of H_0. This is simply the vector of sample sums of the orthonormal polynomials,

$$\mathcal{L}_{\boldsymbol{\phi}} \left(\mathbf{0}, \hat{\boldsymbol{\psi}}_n \right) = \left(\sum_{j=1}^{n} p_1 \left(X_j; \hat{\boldsymbol{\psi}}_n \right), \sum_{j=1}^{n} p_2 \left(X_j; \hat{\boldsymbol{\psi}}_n \right), ..., \sum_{j=1}^{n} p_m \left(X_j; \hat{\boldsymbol{\psi}}_n \right) \right)'.$$

Provided that $\Sigma\left(\mathbf{0}, \hat{\boldsymbol{\psi}}_n\right)$ has rank m, we can now construct the score statistic as the quadratic form

$$\mathcal{S}_{n,m}\left(\mathbf{X}\right) = n\mathcal{L}_{\phi'}\left(\mathbf{0}, \hat{\boldsymbol{\psi}}_n\right) \Sigma\left(\mathbf{0}, \hat{\boldsymbol{\psi}}_n\right)^{-1} \mathcal{L}_{\phi}\left(\mathbf{0}, \hat{\boldsymbol{\psi}}_n\right).$$

The general asymptotic properties of score tests are such that $\mathcal{S}_{n,m}\left(\mathbf{X}\right)$ converges in distribution to $\chi^2\left(m\right)$ under H_0 as $n \to \infty$ and that $\mathcal{S}_{n,m}\left(\mathbf{X}\right) \to^{a.s.} +\infty$ under $H_1 : \boldsymbol{\phi} \neq \mathbf{0}$. Thus, rejecting H_0 when $\mathcal{S}_{n,m}\left(\mathbf{X}\right)$ exceeds the upper-α quantile of $\chi^2\left(m\right)$ gives an asymptotically valid and consistent test. Since expected values of the $\{p_i\}_{i=1}^m$ under H_1 are just functions of the population moments, the test amounts to checking agreement between the first m moments of model and data.

There is, however, one complication. Covariance matrix $\Sigma\left(\mathbf{0}, \hat{\boldsymbol{\psi}}_n\right)$ will be singular whenever f_0 is such that any of the derivatives $\{\partial \log f_0\left(X; \boldsymbol{\psi}\right) / \partial \psi_\ell\}_{\ell=1}^k$ is itself an orthogonal polynomial of order m or less—and therefore proportional to one of the $\{p_i\left(X; \boldsymbol{\psi}\right)\}_{i=1}^m$. For example, suppose $k = 1$ (i.e., one nuisance parameter) and $\partial \log f_0\left(X; \boldsymbol{\psi}\right) / \partial \psi_1 = \beta p_1\left(X; \boldsymbol{\psi}\right)$. Then $\mathcal{I}_{\phi\psi'} = \left(\beta, 0, ..., 0\right)'$ and $\mathcal{I}_{\psi\psi'}\left(\mathbf{0}, \boldsymbol{\psi}\right)^{-1} = \beta^{-2}$, so that

$$\Sigma\left(\mathbf{0}, \boldsymbol{\psi}\right) = \mathbf{I}_m - \left(\beta, 0, ..., 0\right)' \beta^{-2}\left(\beta, 0, ..., 0\right)$$

$$= \mathbf{I}_m - \begin{pmatrix} 1 & 0 & \cdots & 0 \\ 0 & 0 & \cdots & 0 \\ \vdots & \vdots & \ddots & \vdots \\ 0 & 0 & \cdots & 0 \end{pmatrix} = \begin{pmatrix} 0 & 0 & \cdots & 0 \\ 0 & 1 & \cdots & 0 \\ \vdots & \vdots & \ddots & \vdots \\ 0 & 0 & \cdots & 1 \end{pmatrix},$$

which has rank $m - 1$. In such cases it is simplest just to replace any orthogonal polynomial to which some $\partial \log f_0\left(X; \boldsymbol{\psi}\right) / \partial \psi_\ell$ is proportional with one of higher order. We shall see two examples below in which such proportionality exists between all k derivatives $\{\partial \log f_0\left(X; \boldsymbol{\psi}\right) / \partial \psi_\ell\}_{\ell=1}^k$ and the first k orthogonal polynomials. In such cases f_m can be represented in terms of $\{p_i\}_{i=k+1}^{m+k}$ as

$$f_m\left(x; \boldsymbol{\psi}\right) = K\left(\boldsymbol{\theta}\right) \exp\left[\sum_{i=1}^m p_{i+k}\left(x; \boldsymbol{\psi}\right)\right] f_0\left(x; \boldsymbol{\psi}\right).$$

Discarding the first k polynomials causes no loss of information about the shape of f under H_1, because the scores with respect to $\boldsymbol{\psi}$—and therefore the first k polynomials—are zero when evaluated at $\hat{\boldsymbol{\psi}}_n$. Once these are excluded, the matrix $\mathcal{I}_{\phi\psi'}$ of covariances between the remaining p_i and the $\partial \log f_0 / \partial \psi_\ell$ consists entirely of zeros and $\Sigma\left(\mathbf{0}, \hat{\boldsymbol{\psi}}_n\right)$ reduces to identity matrix $\mathbf{I}_m$. The score statistic then becomes just the sum of squared sample sums of the m orthogonal polynomials, and the magnitudes of the individual sums convey some idea of how the data depart from f_0.

We now derive specific versions of the test and give numerical examples for three prominent cases, of which the second and third illustrate the singularity of $\mathbf{\Sigma}\left(\mathbf{0}, \hat{\psi}_n\right)$.[18]

Testing Uniformity. Many contributors to the goodness-of-fit literature regard the uniform as the canonical model, because any fully specified model can be converted to uniform *via* the probability integral transform.[19] To apply the smooth test to the simple null hypothesis $H_0 : f_0\left(x\right) = \mathbf{1}_{[0,1]}\left(x\right)$, one begins by transforming the data as $Y = 2X - 1$. This takes us to uniform on $[-1, 1]$, for which the *Legendre* polynomials $\{L_j\left(\cdot\right)\}_{j=1}^{\infty}$ are an orthogonal set. Starting with $L_1\left(y\right) = y$ and $L_2\left(y\right) = \left(3y^2 - 1\right)/2$, higher orders are generated by the recurrence relation

$$nL_n\left(y\right) = \left(2n - 1\right)yL_{n-1}\left(y\right) - \left(n - 1\right)L_{n-1}\left(y\right).$$

These must be rescaled to have unit variance and thus be orthonormal. In terms of the original X data the first four orthonormal polynomials for the test are

$$p_1\left(x\right) = \sqrt{3}\left(2x - 1\right)$$
$$p_2\left(x\right) = \sqrt{5}\left(6x^2 - 6x + 1\right)$$
$$p_3\left(x\right) = \sqrt{7}\left(20x^3 - 30x^2 + 12x - 1\right)$$
$$p_4\left(x\right) = \sqrt{9}\left(70x^4 - 140x^3 + 90x^2 - 20x + 1\right).$$

Since no parameters are estimated, we have $\mathbf{\Sigma}\left(\breve{\theta}\right) = \mathbf{\Sigma}\left(\mathbf{0}\right) = \mathbf{I}_m$ and

$$\mathcal{S}_{n,m}\left(\mathbf{X}\right) = n^{-1}\sum_{i=1}^{m}\left[\sum_{j=1}^{n}p_i\left(X_j\right)\right]^2$$

Example 9.6. Merely to illustrate this asymptotic procedure, a pseudo-random sample of $n = 5$ *squared* uniform r.v.s on $[0, 1]$ was generated. These appear in the first column of the table with values of p_1, p_2, p_3, p_4

---

[18]Rayner and Best (1989) give applications to other specific null hypotheses, including several discrete families. They conduct Monte Carlo studies to investigate how the test's power depends on m and conclude that $m = 4$ is usually a good choice.

[19]One might reasonably question the preeminence of the uniform model on considering that a second *inverse* probability transform would take us to any other distribution we care to name—normal, gamma, Poisson... .

in the adjacent columns. The data give $\mathcal{S}_{5,4}(\mathbf{x}) \doteq 4.75$, which (taken as a realization of $\chi^2(4)$) has P-value of about 0.31.

x_j	$p_1(x_j)$	$p_2(x_j)$	$p_3(x_j)$	$p_4(x_j)$
.012	−1.692	2.082	−2.288	2.340
.281	−0.757	−0.477	1.182	−0.545
.521	0.072	−1.112	−0.166	1.105
.469	−0.109	−1.105	0.247	1.081
.285	−0.745	−0.497	1.181	−0.509
$\Sigma p_i(x_j)$	−3.230	−1.109	0.157	3.472

Testing Exponentiality. The *Laguerre* polynomials,

$$p_i(x) := \sum_{r=1}^{i} \binom{i}{r} \frac{(-x/\psi)^r}{r!}$$

are orthonormal for exponential densities $f_0(x;\psi) = \psi^{-1} e^{-x/\psi} 1_{[0,\infty)}(x)$. With $y = x/\psi$, the first five are

$$p_1(y) = 1 - y$$

$$p_2(y) = 1 - 2y + y^2$$

$$p_3(y) = 1 - 3y + \frac{3y^2}{2} - \frac{y^3}{6}$$

$$p_4(y) = 1 - 4y + 3y^2 - \frac{2y^3}{3} + \frac{y^4}{24}$$

$$p_5(y) = 1 - 5y + 5y^2 - \frac{5y^3}{3} + \frac{5y^4}{24} - \frac{y^5}{120}.$$

Note that

$$\frac{\partial \log f_0(x;\psi)}{\partial \psi} = \psi^{-1} \left(\frac{x}{\psi} - 1 \right)$$

is proportional to p_1 and therefore orthogonal to the remaining p_i, causing the matrix $\boldsymbol{\Sigma}\left(0, \hat{\psi}_n\right)$ to be singular. Dropping p_1 and using $p_2, p_3, ..., p_{m+1}$ leaves $\boldsymbol{\Sigma}\left(0, \hat{\psi}_n\right)$ as the identity matrix $\mathbf{I}_m$. With $\hat{\psi}_n = \bar{X}$ as the m.l.e. the score statistic is

$$\mathcal{S}_{n,m}(\mathbf{X}) = n^{-1} \sum_{i=1}^{m} \left[\sum_{j=1}^{n} p_{i+1}\left(\frac{X_j}{\bar{X}} \right) \right]^2.$$

To increase accuracy of the test for small n, the upper-α quantiles of $\chi^2(m)$ are multiplied by $(1 - 1.5/\sqrt{n})$ and $(1 - 1.8/\sqrt{n})$ for $\alpha = .05$ and $\alpha = .10$, respectively.

Example 9.7. The ten pseudorandom $\Gamma(5,1)$ observations from Examples 9.4 and 9.5 are

$$\mathbf{x} = (3.78, 9.48, 4.88, 3.60, 5.01, 8.07, 5.48, 3.01, 7.60, 6.67)',$$

with mean $\bar{x} = 5.758$. With $m = 4$ the score statistic takes the value $\mathcal{S}_{10,4}(\mathbf{x}) = 6.520$. Scaling the upper-.05 quantile of $\chi^2(4)$ (9.49) by $\left(1 - 1.5/\sqrt{10}\right)$ gives the value 4.99, indicating rejection of H_0 at the $\alpha = .05$ level.

Testing Normality. Orthogonal polynomials for normal family

$$\left\{ f_0(x; \psi_1, \psi_2) = (2\pi\psi_2)^{-1/2} \exp\left[-\frac{(x - \psi_1)^2}{2\psi_2} \right] : \psi_1 \in \Re, \psi_2 > 0 \right\}$$

are the *Hermite* polynomials. Dividing the ith-order member of these by $\sqrt{i!}$ makes them orthonormal. With $z := (x - \psi_1)/\sqrt{\psi_2}$ these scaled Hermite polynomials are given explicitly by the formula

$$p_i(z) := \sqrt{i!} \sum_{r=0}^{[i/2]} \frac{(-1/2)^r}{r!(i - 2r)!} z^{i-2r},$$

where "$[\cdot]$" signifies "greatest integer contained in $\cdot$". The first six orthonormal polynomials are

$$p_1(z) = z$$

$$p_2(z) = \frac{z^2 - 1}{\sqrt{2}}$$

$$p_3(z) = \frac{z^3 - 3z}{\sqrt{6}}$$

$$p_4(z) = \frac{z^4 - 6z^2 + 3}{\sqrt{24}}$$

$$p_5(z) = \frac{z^5 - 10z^3 + 15z}{\sqrt{120}}$$

$$p_6(z) = \frac{z^6 - 15z^4 + 45z^2 - 15}{\sqrt{720}}.$$

Since

$$\frac{\partial \log f_0(x; \boldsymbol{\psi})}{\partial \psi_1} = \frac{x - \psi_1}{\psi_2}$$

$$\frac{\partial \log f_0(x; \boldsymbol{\psi})}{\partial \psi_2} = \frac{(x - \psi_1)^2}{2\psi_2^2} - \frac{1}{2\psi_2},$$

the scores for the two nuisance parameters are proportional to p_1 and p_2 and therefore orthogonal to $\{p_i\}_{i\geq3}$. Excluding polynomials p_1 and p_2 and using $\{p_{i+2}\}_{i=1}^m$, covariance matrix $\Sigma\left(0,\hat{\psi}_n\right)$ becomes the identity matrix $\mathbf{I}_m$. With $\hat{\psi}_{1n} = \bar{X}$ and $\hat{\psi}_{2n} = S^2$ (disregarding the distinction between M_2 and S^2 for this asymptotic test) the score statistic is

$$S_{n,m}\left(\mathbf{X}\right) = n^{-1}\sum_{i=1}^m\left\{\sum_{j=1}^n p_{i+2}\left(\frac{X_j - \bar{X}}{S}\right)\right\}^2.$$

To increase accuracy for small samples, the upper-α quantiles of $\chi^2\left(m\right)$ should be multiplied by $(1 - 1.6/\sqrt{n})$ and $(1 - 1.8/\sqrt{n})$ for $\alpha = .05$ and $\alpha = .10$, respectively. Like the skewness–kurtosis test based on K_n^2, the test with $m = 2$ responds just to the first four moments, while with $m = 4$ there is sensitivity to the next two higher moments also. Monte Carlo experiments indicate that the power with $m = 4$ is never much lower than that for $m = 2$ and is sometimes much higher.

Example 9.8. The skewness–kurtosis and smooth tests are applied to the $n = 20$ pseudo-random observations shown below, which were generated from the Laplace (double-exponential) distribution with p.d.f. $f\left(x\right) = \frac{1}{2}e^{-|x|}, x \in \Re$. This has mean and skewness zero, variance 2.0, and coefficient of kurtosis $\alpha_4 = 6.0$.

$\mathbf{x} = (-.634, .379, 1.033, .501, .618, -.444, -3.020, .303, -.774, .332,$

$\qquad -.527, .055, .294, -2.306, .627, -3.664, -2.223, 1.542, -.269, -2.129)'.$

The corresponding sample statistics are

$$\bar{x} = -.515, s^2 = 2.025, a_3 = -.743, a_4 = 2.358.$$

The value of skewness–kurtosis statistic K_{20}^2 is 2.185, and values of the score statistics with $m = 2$ and $m = 4$ are $S_{20,2}\left(\mathbf{x}\right) = 1.939$ and $S_{20,4}\left(\mathbf{x}\right) = 4.638$, respectively. Comparing K_{20}^2 with the empirical quantiles in Table 9.2 shows that the test just rejects at level $\alpha = .10$, while $S_{20,2}\left(\mathbf{x}\right)$ and $S_{20,4}\left(\mathbf{x}\right)$ are below the adjusted upper quantiles of $\chi^2\left(2\right)$ and $\chi^2\left(4\right)$: 2.75 = $4.61\left(1 - 1.8/\sqrt{20}\right)$ and 4.65 = $7.78\left(1 - 1.8/\sqrt{20}\right)$.

9.2.4   *Tests Based on Order Statistics*

Let $(X_1, X_2, ..., X_n)$ be a random sample from location–scale family $\{F\left(x; \mu, \sigma\right) : \mu \in \Re, \sigma > 0\}$, with $\mathbf{X}_{(\cdot:n)} := \left(X_{(1:n)}, X_{(2:n)}, ..., X_{(n:n)}\right)'$ as the vector of order statistics. The $\{X_{(j:n)}\}$ will be related to the order statistics $\{Z_{(j:n)}\}$ of standardized variables $\{Z_j := (X_j - \mu)/\sigma\}$ as $X_{(j:n)} = \mu + \sigma Z_{(j:n)}$. If $H_0 : F(x) = F_0(x; \mu, \sigma)$ is true, then, a plot of sample values $\{X_{(j:n)}\}$ against the *expected values* of the standardized order statistics—$m_j := E_0 Z_{(j:n)}$, say—should be, apart from sampling error, a straight line with slope equal to scale parameter σ. Therefore, under H_0 this slope estimator should roughly correspond to another (model-independent) consistent estimator, such as the sample standard deviation, S. Equivalently, under H_0 one expects the sample order statistics and their expected values to be highly correlated. These are the intuitive bases for the very powerful Shapiro–Wilk (1965,1972) and Shapiro–Francia (1972) tests for normality and a related correlation test for the exponential distribution with unspecified origin and scale.

9.2.4.1 *The Shapiro–Wilk/Francia Tests for Normality*

The linearity of the relation between raw and standardized order statistics suggests the linear model $X_{(j:n)} = \mu + \sigma m_j + U_j$, where $U_j := \sigma\left(Z_{(j:n)} - m_j\right)$. In vector form this is $\mathbf{X}_{(\cdot:n)} = \mathbf{M}\boldsymbol{\beta} + \mathbf{U}$, where $\boldsymbol{\beta} := (\mu, \sigma)'$ and $\mathbf{M} := (\mathbf{1}, \mathbf{m})$ ($\mathbf{1}$ being a units vector). The unobserved deviations $\{U_j\}$ have zero mean but are correlated, with covariance matrix $\mathbf{V} := E_0 \mathbf{U}\mathbf{U}'$ which is the same as the covariance matrix of $\mathbf{X}_{(\cdot:n)}$ itself. Recalling Exercise 7.27, the best linear unbiased estimator of $\boldsymbol{\beta}$ is $\left(\mathbf{M}'\mathbf{V}^{-1}\mathbf{M}\right)^{-1}\left(\mathbf{M}'\mathbf{V}^{-1}\mathbf{X}_{(\cdot:n)}\right)$. When the null distribution is symmetric, as it is for the normal family, the b.l.u.e. estimator of component σ has the simple form

$$\tilde{\sigma} = \frac{\mathbf{m}'\mathbf{V}^{-1}\mathbf{X}_{(\cdot, n)}}{\mathbf{m}'\mathbf{V}^{-1}\mathbf{m}},$$

which is a weighted sum of the elements of $\mathbf{X}_{(\cdot:n)}$. Since vector $\mathbf{m}'\mathbf{V}^{-1}$ has length $\left|\mathbf{m}'\mathbf{V}^{-1}\right| = \sqrt{\mathbf{m}'\mathbf{V}^{-1}\mathbf{V}^{-1}\mathbf{m}}$, the normalized statistic

$$\tilde{\sigma}\frac{\mathbf{m}'\mathbf{V}^{-1}\mathbf{m}}{\left|\mathbf{m}'\mathbf{V}^{-1}\right|} = \frac{\mathbf{m}'\mathbf{V}^{-1}}{\left|\mathbf{m}'\mathbf{V}^{-1}\right|}\mathbf{X}_{(\cdot:n)} =: \mathbf{a}'\mathbf{X}_{(\cdot:n)}$$

is a weighted sum, $\sum_{j=1}^{n} a_j X_{(j:n)}$, with weights satisfying $\sum_{j=1}^{n} a_j^2 = 1$. The Shapiro–Wilk (S-W) test of normality is based on the statistic

$$W := \frac{\left(\sum_{j=1}^{n} a_j X_{(j:n)}\right)^2}{(n-1) S^2},$$

where S^2 is the unbiased sample variance.

Unfortunately, calculating the $\{a_j\}$ is no simple matter, as moments of standard normal order statistics have to be obtained by numerical integration. However, values of $a_1, a_2, ..., a_{[n/2]}$ for n ranging from 2 to 50 have been tabulated, along with critical values for W based on simulations.[20] The $\{a_j\}$ are symmetric, with $a_j = -a_{n-j}$ and $a_{[n/2]+1} = 0$ when n is odd. The symmetry of the weights makes evident that W is the square of the ratio of two alternative measures of scale. *Small* values of W are considered to oppose H_0.

The difficulty of having to rely on tabulated coefficients to calculate W makes some simplification desirable. A modified test that approximates **V** as the identity matrix has been found to retain good properties. This reduces the weighting constants to normalized expected values of standard normal order statistics, $a'_j = m_j/\sqrt{\mathbf{m'm}}$. The resulting *Shapiro–Francia* (S-F) test is usually denoted W':

$$W' = \frac{\left(\sum_{j=1}^{n} a'_j X_{(j:n)}\right)^2}{(n-1)\, S^2}. \tag{9.18}$$

While tables of the $\{m_j\}$ have been published in various sources, a good approximation is[21]

$$m'_j = \Phi^{-1}\left(\frac{j - 3/8}{n + 1/4}\right), \tag{9.19}$$

where Φ^{-1} is the inverse of the standard normal c.d.f. Computer software for Φ^{-1} is available. Table 9.3 (from Stephens (1986)) gives simulation-based upper quantiles of the transformation $W_S = n\,(1 - W')$ of W' using the m'_j approximation. H_0 is rejected for *large* values of W_S. Both the S-W and S-F tests have been shown to be consistent against all alternatives to normality having finite second moment.[22]

Example 9.9. The modified Shapiro–Francia test is applied to the $n = 20$ Laplace deviates of Example 9.8. The order statistics are

$$\mathbf{x}_{(\cdot:20)} = (-3.664, -3.020, -2.306, -2.223, -2.129, -.774, -.634, -.527, -.444,$$
$$-.269, .055, .294, .303, .332, .379, .501, .618, .627, 1.033, 1.542)'.$$

---

[20] Shapiro and Wilk (1965).

[21] For tables see Harter (1961) or Pearson and Hartley (1976). Weisberg and Bingham (1972) and Stephens (1986) discuss modified tests using $\left\{a'_j\right\}$ and $\left\{m'_j\right\}$. Stephens' expression for m'_j incorrectly gives the denominator of the argument of Φ^{-1} as $n + 1/8$.

[22] Sarkadi (1975) and Leslie *et al.* (1986).

Table 9.3 Upper quantiles
of modified Shapiro-Francia
statistic for normality.

$n \setminus \alpha$	0.10	0.05	0.01
10	1.26	1.58	2.27
20	1.56	1.91	2.81
40	1.83	2.23	3.30
60	1.96	2.39	3.43
80	2.05	2.49	3.54
100	2.11	2.56	3.61
400	2.45	2.89	3.95
600	2.52	2.98	4.04
1000	2.60	3.11	4.25

The values of $\left\{-m'_{10-j+1} = m'_{10+j}\right\}_{j=1}^{10}$ are

$$.062, .1868, .3146, .4478, .5895, .7441, .9191, 1, 128, 1.403, 1.868$$

with $\sum_{j=1}^{n} \left(m'_j\right)^2 = 17.634$. The values of $\left\{-a'_{10-j+1} = a'_{10+j}\right\}_{j=1}^{10}$ are

$$.015, .044, .075, .107, .140, .177, .219, .269, .334, .445.$$

With $s^2 = 2.025$ the statistic W_S has the value 1.839, which is significant at the $\alpha = .10$ level.

9.2.4.2 *Testing Exponentiality with Unknown Origin and Scale*

Consider the statistic

$$R_{Xm}^2 := \frac{\left[\sum_{j=1}^{n}(m_j - \bar{m})\left(X_{(j:n)} - \bar{X}\right)\right]^2}{\sum_{j=1}^{n}(m_j - \bar{m})^2 \sum_{j=1}^{n}(X_j - \bar{X})^2},$$

which is the squared correlation between sample order statistics and their expected values. In testing normality (or other symmetric distribution) the mean of the expected standardized order statistics is zero, so R_{Xm}^2 reduces to

$$R_{Xm}^2 = \frac{\sum_{j=1}^{n} m_j X_{(j:n)}}{\sum_{j=1}^{n} m_j^2 \cdot (n-1) S^2} = W',$$

the Shapiro–Francia statistic. The correlation test can also be used to test that the data have the exponential distribution with unknown origin and scale:

$$H_0 : f(x) = \sigma^{-1} e^{-(x-\mu)/\sigma} \mathbf{1}_{(0,\infty)}(x).$$

However, since the exponential distribution is not symmetric, the $\{m_j\}$ no longer sum to zero and $R^2_{Xm} \neq W'$ in this case. However, expressing R^2_{Xm} is easy because explicit expressions for expected order statistics are known for the exponential model.

Theorem 9.7. *When $\{X_j\}_{j=1}^n$ are i.i.d. as standard exponential ($\mu = 0, \sigma = 1$), the order statistics have expected value*

$$m_j = \sum_{i=0}^{j-1} (n-i)^{-1}, j \in \{1, 2, ..., n\}.$$

Proof. From Theorem 8.3 the marginal c.d.f. of $X_{(j:n)}$ is

$$F_{(j:n)}(x) = \sum_{i=j}^n \binom{n}{i} (1 - e^{-x})^i (e^{-x})^{n-i}$$

$$= 1 - \sum_{i=0}^{j-1} \binom{n}{i} (1 - e^{-x})^i (e^{-x})^{n-i}.$$

Therefore, using (3.16),

$$EX_{(j:n)} = \int_0^\infty \left[1 - F_{(j:n)}(x)\right] \cdot dx$$

$$= \sum_{i=0}^{j-1} \binom{n}{i} \int_0^\infty (1 - e^{-x})^i (e^{-x})^{n-i} \cdot dx$$

$$= \sum_{i=0}^{j-1} \binom{n}{i} \int_0^1 (1 - t)^i t^{n-i-1} \cdot dt$$

$$= \sum_{i=0}^{j-1} \binom{n}{i} \frac{\Gamma(i+1)\Gamma(n-i)}{\Gamma(n+1)}$$

$$= \sum_{i=0}^{j-1} (n-i)^{-1}.$$

$\square$

Exercise 9.8. *Show that $\sum_{j=1}^n m_j = n$ for the standard exponential distribution.*

The conclusions of the theorem and exercise give the form of the correlation statistic in the exponential case as

$$R^2_X = \frac{\left[\sum_{j=1}^n m_j \left(X_{(j:n)} - \bar{X}\right)\right]^2}{\left(\sum_{j=1}^n m_j^2 - n\right)(n-1)S^2}.$$

Table 9.4 Upper Quantiles of Modified Shapiro-Francia Statistic for Exponentiality.

$n\backslash\alpha$	0.10	0.05	0.01
10	1.56	1.92	2.67
20	2.20	2.77	4.10
40	3.05	3.94	6.42
60	3.52	4.67	8.01
80	3.96	5.25	9.33
100	4.30	5.70	10.35

Since the correlation approaches unity under the null as $n \to \infty$, small values of R_X^2 disconfirm H_0, but transforming as $W_S = n\left(1 - R_X^2\right)$ makes an upper-tailed rejection region appropriate. Selected upper quantiles of W_S are given in Table 9.4 (from Stephens (1986)).

Example 9.10. The correlation test based on W_S is applied to the $n = 10$ ordered pseudorandom $\Gamma\left(5, 1\right)$ observations from Examples 9.4, 9.5, and 9.7:

$$\mathbf{x}_{(\cdot:10)} = (3.01, 3.60, 3.78, 4.88, 5.01, 5.48, 6.67, 7.60, 8.07, 9.48)'.$$

These give $\bar{x} = 5.758$ and $(n-1)\,s^2 = 40.952$. The expected values of the order statistics are

$$\mathbf{m} = (.100, .211, .336, .479, .646, .846, 1.096, 1.429, 1.929, 2.929)',$$

with $\mathbf{m'm} - n = 7.071$, $\sum_{j=1}^{10} m_j\left(x_{(j:10)} - \bar{x}\right) = 16.459$. These yield $R_X^2 = 0.9355$ and $W_S = 0.645$, which is not significant at the .10 level.[23]

Exercise 9.9. *Show that m.g.f. of the jth order statistic of the standard exponential distribution is*

$$M_{X_{(j:n)}}\left(\zeta\right) = \frac{n!\,\Gamma\left(n - j + 1 - \zeta\right)}{(n - j)!\,\Gamma\left(n + 1 - \zeta\right)}.$$

Use the m.g.f. to

(i) Verify the conclusion of Theorem 9.7.
(ii) Show that the variance of $X_{(j:n)}$ is

$$v_j = \sum_{i=0}^{j-1} (n - i)^{-2}.$$

---

[23] In comparison with examples 9.4, 9.5, and 9.7 note that the $\Gamma\left(5, 1\right)$ data are more nearly compatible with the exponential model when H_0 does not specify the origin to be 0.0.

The following relations for the di- and trigamma functions may be used:

$$\Psi(n+1) = \Psi(1) + \sum_{i=1}^{n} i^{-1}$$

$$\Psi'(n+1) = \Psi'(1) - \sum_{i=1}^{n} i^{-2}.$$

9.2.5　Tests Based On Generating Functions

Section 3.4.2 introduced several integral transforms of c.d.f.s, which are real- or complex-valued functions of the form

$$\gamma(t) := Eg(X;t) = \int g(x;t) \cdot dF(x)$$

on some domain $\mathbb{T}$. We have already seen many applications of three of these:

(i) Moment generating functions (m.g.f.s): $M(t) := Ee^{tX}$,
(ii) Probability generating functions (p.g.f.s): $p(t) := Et^{X}$, and
(iii) Characteristic functions (c.f.s): $\phi(t) := Ee^{itX}, i = \sqrt{-1}$.

Each of these functions has its sample or empirical counterpart, as

$$\gamma_n(t) := n^{-1} \sum_{j=1}^{n} g(X_j;t).$$

While all three have found application in statistical inference—and especially in testing goodness of fit, the empirical c.f. has been the most useful. We now look at a family of tests known as "i.c.f." tests (for "integrated c.f.") that are based on the discrepancy between c.f.s of model and sample. Some theory is needed before looking at applications to specific models.

9.2.5.1　Construction and Theory of ICF Tests

The empirical c.f.,

$$\phi_n(t) := n^{-1} \sum_{j=1}^{n} e^{itX_j} = n^{-1} \sum_{j=1}^{n} \cos tX_j + i \cdot n^{-1} \sum_{j=1}^{n} \sin tX_j,$$

was introduced in Section 6.6.3, where it was pointed out that $\phi_n(t) \to^{a.s.}$ $\phi(t)$ for each $t \in \Re$. Letting $\mathcal{N}(t) \subset \Omega$ be the null set of outcomes on which convergence fails at a specific t, the countable additivity of probability measures assures that $\mathbb{P}\{\cup_{k=1}^{\infty}\mathcal{N}(t_k)\} = 0$ for each countable collection $\{t_k\}_{k=1}^{\infty}$. Thus, ϕ_n converges a.s. to ϕ on any countable set, such as the rationals. Indeed, more is true. Specifically, it is true that $\sup_{t \in \mathbb{T}} |\phi_n(t) - \phi(t)| \to 0$ holds generally for any *bounded* set $\mathbb{T} \subset \Re$ and that $\sup_{t \in \Re} |\phi_n(t) - \phi(t)| \to 0$ when X is a discrete r.v.[24] Note that this is not quite as strong as the result delivered by Glivenko–Cantelli (Theorem 9.4) for the empirical c.d.f., which converges uniformly on the entire real line for samples from continuous as well as discrete populations. Uniform convergence of empirical c.f.s of continuous distributions is confined to a limited range, because each sample realization of ϕ_n—the sum of sines and cosines of the data—is periodic with periods that lengthen as t moves away from the origin, whereas the population c.f. ϕ approaches zero as $|t| \to \infty$. Nevertheless, we shall see that the limiting behavior of ϕ_n is adequate to assure its usefulness in statistical inference.

I.c.f. tests resemble Cramér–von Mises and Anderson–Darling tests, but with c.f.s replacing c.d.f.s. Like those "e.d.f." tests based on the empirical distribution function, i.c.f. tests are most practical for continuous distributions that depend just on location and scale parameters. Whereas the e.d.f. statistics are weighted integrals of discrepancies between empirical and model c.d.f.s, i.c.f. statistics are weighted integrals of discrepancies between the characteristic functions. More specifically, consider the test of $H_0 : F(\cdot) = F_0(\cdot; \boldsymbol{\theta})$, where $\boldsymbol{\theta} := (\theta_1, \theta_2)'$ comprises location and/or scale parameters, respectively. Let $F_0(\cdot) := F_0(\cdot; 0, 1)$ represent the standard form, with $\phi_0(\cdot)$ as the corresponding c.f. Also, let

$$\hat{\phi}_n(t) := n^{-1} \sum_{j=1}^{n} \exp\left[it \left(\frac{X_j - \hat{\theta}_1}{\hat{\theta}_2} \right) \right] =: n^{-1} \sum_{j=1}^{n} e^{it\hat{Z}_j}$$

be the empirical c.f. constructed from the standardized observations, where standardization is with m.l.e.s or other consistent estimators. (Of course, should one or both of the parameters be known, just remove the "hats".)

Besides consistency, the estimators should have another property, known as *equivariance*. To understand the meaning of this term, notice that any location–scale family of distributions is closed under affine transformations, in the sense that if $F_0(\cdot; \theta_1, \theta_2)$ is the c.d.f. of X with $(\theta_1, \theta_2) \in \Re \times \Re_+$, then $F_0(\cdot; a + b\theta_1, b\theta_2)$ is the c.d.f. of $a + bX$ for any real a and positive b. Such an affine transformation could result from a change of units, as from Fahrenheit to Celsius, US dollars to Euros, or miles to kilometers. One would clearly want any test statistic for $H_0 : F = F_0$ to be invariant under such transformations, since they do not take us out of the family of distributions specified by the null. For affine invariance to hold, the

---

[24] For proofs see Csörgo (1981), Feuerverger and Mureika (1977).

estimators $\hat{\theta}_1, \hat{\theta}_2$ must themselves be equivariant with respect to location–scale transformations. This means that if $\hat{\theta}_1, \hat{\theta}_2$ are estimates from sample $\{x_j\}_{j=1}^n$, then the corresponding estimates from sample $\{a + bx_j\}_{j=1}^n$ should be $a + b\hat{\theta}_1, b\hat{\theta}_2$. Maximum likelihood estimators do have this attribute, a consequence of their general invariance property. (See Section 7.2.2.) Equivariance of the estimators confers invariance on the empirical c.f. of standardized observations $\hat{\phi}_n$ and ultimately on the i.c.f. statistic, which, as we see next, depends on the data through $\hat{\phi}_n$ alone.

I.c.f. statistics have the form

$$I_n := n \int_{-\infty}^{\infty} \left| \hat{\phi}_n(t) - \phi_0(t) \right|^2 w(t) \cdot dt, \tag{9.20}$$

where w is a suitable weight function. The rationale for the abbreviation "i.c.f." (*integrated* ...) thus becomes clear. Note that $\hat{\phi}_n(t)$ and $\phi_0(t)$ are, in general, complex valued, and that "$|\cdot|$" represents the modulus, which of course is real.[25] To be "suitable" a weight function must meet several requirements:

- It must be real valued if we are to avoid dealing with complex-valued test statistics.
- It must approach zero rapidly enough as $|t| \to \infty$ to make the integral converge. (Recall that $\hat{\phi}_n(t)$ is periodic, whereas $|\phi_0(t)| \to 0$ for continuous distributions.)
- To emphasize discrepancies in the tails of the distribution, w should put most of the weight near the origin. (Recall that the degree of smoothness of c.f.s near the origin indicates the degree of thickness of the tails.)
- Finally, to facilitate computation, w should give the integral a closed form.

Weight functions of the class $\left\{ K \cdot |\phi_0(t\eta)|^2 : \eta > 0, K > 0 \right\}$ typically satisfy these requirements. Obviously, $|\phi_0(t\eta)|^2$ is real, and it does approach zero as $|t| \to \infty$, since it is the c.f. of a continuous, symmetric distribution—specifically, of $\eta(Z - Z')$, where Z, Z' are independent copies of $(X - \theta_1)/\theta_2$. Of course, the maximum value of $|\phi_0(t\eta)|^2$ (unity) is attained at $t = 0$, and the concentration of weight near the origin can be made arbitrarily high by making η sufficiently large. While η can be chosen to

---

[25]Recall that the squared modulus of a complex number is the product of the number and its complex conjugate; i.e., $|z|^2 = z \cdot \bar{z}$. For c.f.s in particular we have $\bar{\phi}(t) = \phi(-t)$, so that $|\phi(t)|^2 = \phi(t)\phi(-t)$.

enhance power against specific types of alternatives, $\eta = 1$ is a natural choice. Also, since $\left|\hat{\phi}_n(t) - \phi_0(t)\right|^2 \leq 4$, the integral in (9.20) is finite so long as $|\phi_0(t\eta)|^2$ itself is integrable.

Exercise 9.10. *Show the following:*

(i) $\int_{-\infty}^{\infty} |\phi(t\eta)| \cdot dt < \infty$ implies $\int_{-\infty}^{\infty} |\phi(t\eta)|^p \cdot dt < \infty$ for all $p \geq 1$.

(ii) If ϕ is absolutely integrable, then $\int_{-\infty}^{\infty} |\phi(t\eta)|^2 \cdot dt = 2\pi f_{\eta(Z-Z')}(0)$, where Z, Z' are independent r.v.s with c.f. ϕ. (Hint: Apply Theorem 3.5.)

Adopting the specific weight function $w(t) = |\phi_0(t)|^2 / \int_{-\infty}^{\infty} |\phi_0(u)|^2 \cdot du$, we can obtain an expression for I_n in closed form. For this, expand $\left|\hat{\phi}_n(t) - \phi_0(t)\right|^2$ and write out $\hat{\phi}_n(t)$ in terms of standardized data $\left\{\hat{Z}_j = \left(X_j - \hat{\theta}_1\right)/\hat{\theta}_2\right\}_{j=1}^{n}$ as

$$\frac{n}{K} \int_{-\infty}^{\infty} \left[n^{-1} \sum_{j=1}^{n} e^{it\hat{Z}_j} - \phi_0(t)\right] \left[n^{-1} \sum_{j=1}^{n} e^{-it\hat{Z}_j} - \phi_0(-t)\right] |\phi_0(t)|^2 \cdot dt,$$

where $K := \int_{-\infty}^{\infty} |\phi_0(u)|^2 \cdot du$. Multiplying and reversing the order of integration and (finite) sums, the integral becomes

$$\frac{1}{n^2} \sum_{j,k=1}^{n} \int_{-\infty}^{\infty} e^{it(\hat{Z}_j - \hat{Z}_k)} |\phi_0(t)|^2 \cdot dt - \frac{1}{n} \sum_{j=1}^{n} \int_{-\infty}^{\infty} e^{it\hat{Z}_j} \phi_0(-t) |\phi_0(t)|^2 \cdot dt$$

$$-\frac{1}{n} \sum_{j=1}^{n} \int_{-\infty}^{\infty} e^{-it\hat{Z}_j} \phi_0(t) |\phi_0(t)|^2 \cdot dt + \int_{-\infty}^{\infty} |\phi_0(t) \phi_0(t)|^2 \cdot dt.$$

The integral in the first term equals $\int_{-\infty}^{\infty} \cos\left[t(\hat{Z}_j - \hat{Z}_k)\right] |\phi_0(t)|^2 \cdot dt$, since the imaginary part is an odd function of t and integrates to zero. Similarly the integrals in the second and third terms are just the same, since they have the same real parts and the imaginary parts vanish. Thus, multiplying by n/K, we have the simpler expression

$$\begin{aligned}
I_n = {} & \frac{1}{nK} \sum_{j,k=1}^{n} \int_{-\infty}^{\infty} \cos[t\left(\hat{Z}_j - \hat{Z}_k\right)] |\phi_0(t)|^2 \cdot dt \\
& - \frac{2}{K} \sum_{j=1}^{n} \int_{-\infty}^{\infty} e^{it\hat{Z}_j} \phi_0(-t) |\phi_0(t)|^2 \cdot dt \qquad (9.21) \\
& + \frac{n}{K} \int_{-\infty}^{\infty} |\phi_0(t) \phi_0(t)|^2 \cdot dt
\end{aligned}$$

The integrals must be worked out for the specific version of ϕ_0 implied by the null hypothesis in order to obtain an operational test statistic.

The three integrals in I_n have an interpretation that can help in expressing them and thereby producing practical computational formulas for specific null hypotheses. Recall (Theorem 3.5) that, when c.f. ϕ of r.v. Z is absolutely integrable, the p.d.f. at any real z can be recovered as

$$f_Z(z) = \frac{1}{2\pi} \int_{-\infty}^{\infty} e^{-itz} \phi(t) \cdot dt$$

or, when the c.f. is real, as

$$f_Z(z) = \frac{1}{2\pi} \int_{-\infty}^{\infty} \cos(tz) \phi(t) \cdot dt.$$

Thus, in the first term of (9.21), using the fact (Exercise 9.10) that $|\phi_0(t)|^2$ is the (real-valued) c.f. of $Z - Z'$, we have

$$\int_{-\infty}^{\infty} \cos[t\left(\hat{Z}_j - \hat{Z}_k\right)] |\phi_0(t)|^2 \cdot dt = 2\pi f_{Z-Z'}(\hat{Z}_j - \hat{Z}_k),$$

where $f_{Z-Z'}$ is the p.d.f. of $Z - Z'$. Using similar notation for the other integrals and noting that $K = 2\pi f_{Z-Z'}(0)$, we can represent the i.c.f. statistic in closed form as

$$n^{-1} \sum_{j,k=1}^{n} \frac{f_{(Z-Z')}(\hat{Z}_j - \hat{Z}_k)}{f_{Z-Z'}(0)} - 2 \sum_{j=1}^{n} \frac{f_{(Z-Z')-Z''}\left(\hat{Z}_j\right)}{f_{Z-Z'}(0)} + n \frac{f_{(Z-Z')+(Z''-Z''')}(0)}{f_{Z-Z'}(0)}.$$

We shall see below what forms these density expressions take in some specific applications.

Apart from a common factor the three terms of (9.21) can also be understood as c.f.s of convolutions of various distributions. The first term is the c.f. of the convolution of the model distribution of $Z - Z'$ with the empirical distribution of $Z - Z'$; the second term is minus twice the c.f. of the convolution of the model distribution of $Z - Z'$, the empirical distribution of Z, and the model distribution of $-Z'$; and the third term is the c.f. of the convolution of the model distribution of $Z - Z'$ with the model distribution of $Z'' - Z'''$. One expects each of these c.f.s to approach the same value in the limit if H_0 is true, which suggests that $n^{-1}I_n$ should approach zero as $n \to \infty$. This is in fact true and can be stated formally as follows.

Theorem 9.8. *If I_n is constructed from a random sample from a location–scale family with $\int_{-\infty}^{\infty} |\phi_0(t)| \cdot dt < \infty$, elements of which are standardized using strongly consistent estimates of location and scale, then $n^{-1}I_n \to^{a.s.} 0$ when H_0 is true.*

Exercise 9.11. *Prove the theorem. (Hint: Use the triangle-inequality bound*

$$\left|\hat{\phi}_n(t) - \phi_0(t)\right|^2 \le \left|\hat{\phi}_n(t) - \phi_n(t)\right|^2 + \left|\phi_n(t) - \phi_0(t)\right|^2$$

and consider the limiting behavior of each component on a domain $\mathbb{T}$ *for which* $\int_{\mathbb{T}^c} |\phi_0(t)|^2 \cdot dt$ *is arbitrarily small.)*

While $n^{-1}I_n$ approaches zero under the null, the facts that c.f.s are continuous and that $\phi_1(t) - \phi_0(t) \ne 0$ for some t if $F_1(x) \ne F_0(x)$ for some x suggest that I_n itself should approach $+\infty$ under *any* alternative H_1. If proper allowance is made for the possible misbehavior of estimators $\hat{\theta}_1$ and $\hat{\theta}_2$ as $n \to \infty$, this can indeed generally be demonstrated.[26] It indicates that an upper-tailed rejection region is appropriate and that a test that rejects for large values of I_n is consistent against all alternatives to H_0. Of course, it is necessary to determine the distribution of I_n under H_0 in order to specify a rejection region having a given size $\alpha \in (0,1)$. As it was for Anderson–Darling tests, this is the requirement that limits the practical application of i.c.f. tests to location–scale families. Since the standardized values $\left\{ \left(X_j - \hat{\theta}_1 \right) / \hat{\theta}_2 \right\}_{j=1}^n$ are then independent of the parameters when $\hat{\theta}_1, \hat{\theta}_2$ are equivariant, so is the distribution of I_n. This makes it possible to find a unique upper quantile I_n^α for each n such that a test with rejection region $\mathcal{R}_\alpha = \{\mathbf{x} : I_n \ge I_n^\alpha\}$ is of size α for each member of the family $\{F_0(x; \boldsymbol{\theta}) : \boldsymbol{\theta} \in \boldsymbol{\Theta}\}$. As explained in Section 8.4, rejection regions with this property are called *similar* regions. While some theoretical results about the limiting distribution of I_n under the null have been obtained for specific location–scale families,[27] they are not of practical use in finding the needed upper quantiles—even for large n. However, simulation affords an easy way to make the tests operational for any n.

We now look at applications to the standard location–scale models. In each case formulas are provided from which empirical upper quantiles can be determined for various sample sizes. These are based on simulations of

---

[26] When the behavior is anomalous, as for example when the sample is such that scale estimate $\hat{\theta}_2 = 0$, we can just *define* I_n as equal to its maximum value. Since $\left|\hat{\phi}_n(t) - \phi_0(t)\right| \le 2$, the maximum value is $4n$, so this alteration assures the consistency of the test.

[27] Baringhaus and Henze (1989), Henze (1990), Henze and Wagner (1997), Gürtler and Henze (2000).

50,000 replications for sample sizes[28]

$$n \in \{10(2)32(4)52(6)70(10)120(20)200(30)290, 310(30)400\}.$$

Except where a different support is indicated, the formulas apply for values of n between 10 and 400.

9.2.5.2   Testing Uniformity

The composite null hypothesis is

$$H_0 : f(x) = (2\theta_2)^{-1} \mathbf{1}_{[\theta_1 - \theta_2, \theta_1 + \theta_2]}(x), \theta_1 \in \Re, \theta_2 > 0,$$

with both parameters unknown. The m.l.e.s are $\hat{\theta}_1 = \left(X_{(1:n)} + X_{(n:n)}\right)/2$ and $\hat{\theta}_2 = (X_{(n:n)} - X_{(1:n)})/2$, where $X_{(1:n)}$ and $X_{(n:n)}$ are the sample minimum and maximum, respectively. Transforming as $Z := (X - \theta_1)/\theta_2$ makes Z uniform on $[-1, 1]$. The c.f. of Z is $\phi_0(t) = t^{-1}\sin t$ for $t \neq 0$ and $\phi_0(0) = 1$, with $\int_{-\infty}^{\infty} \phi_0(t)^2 \cdot dt = \pi$. With $\hat{Z}_j = \left(X_j - \hat{\theta}_1\right)/\hat{\theta}_2$ the i.c.f. statistic is (with $0^{-1}\sin 0 := 1$)

$$I_n = \pi^{-1} n \int_{-\infty}^{\infty} \left| n^{-1} \sum_{j=1}^{n} e^{it\hat{Z}_j} - \frac{\sin t}{t} \right|^2 \frac{\sin^2(t)}{t^2} \cdot dt$$

$$= \pi^{-1} n^{-1} \sum_{j=1}^{n} \sum_{k=1}^{n} \int_{-\infty}^{\infty} e^{it(\hat{Z}_j - \hat{Z}_k)} \frac{\sin^2(t)}{t^2} \cdot dt$$

$$-2\pi^{-1} \sum_{j=1}^{n} \int_{-\infty}^{\infty} \frac{\cos t\hat{Z}_j \sin t \sin^2(t)}{t^3} \cdot dt$$

$$+\pi^{-1} n \int_{-\infty}^{\infty} \frac{\sin^2 t \sin^2(t)}{t^4} \cdot dt.$$

Now $\sin^2(t)/t^2$ is the c.f. of $Y := Z - Z'$, which is the difference between independent copies of r.v.s distributed as uniform on $[-1, 1]$. Since this has p.d.f.

$$f_Y(y) = \frac{1}{2}\left(1 - \frac{|y|}{2}\right) \mathbf{1}_{[-2,2]}(y),$$

we see that

$$\int_{-\infty}^{\infty} e^{it(\hat{Z}_j - \hat{Z}_k)} \frac{\sin^2(t)}{t^2} \cdot dt = 2\pi f_Y\left(\hat{Z}_j - \hat{Z}_k\right) = \pi\left(1 - \frac{\left|\hat{Z}_j - \hat{Z}_k\right|}{2}\right).$$

---

[28]Notation $m(j)n$ signifies the set $\{m, m+j, m+2j, ..., n-j, n\}$; e.g., $10(2)32 = \{10, 12, ..., 30, 32\}$.

With similar manipulations in the second and third terms I_n can ultimately be reduced to the simple form

$$I_n = -\frac{1}{n}\sum_{j<k}\left|\hat{Z}_j - \hat{Z}_k\right| + \frac{1}{2}\sum_{j=1}^{n}\hat{Z}_j^2 + \frac{n}{6}. \tag{9.22}$$

The following expressions are used for finding approximate upper-α quantiles of I_n for values of n between 10 and 400:

α	I_n^{α}
.10	$.701 - .064n^{-1/2} + .084n^{-1} - n^{-3/2}$
.05	$.929 + .030n^{-1/2} - .83n^{-1} + .44n^{-3/2}$

Thus, for $n = 10$ one rejects H_0 at level $\alpha = .05$ if I_n is at least .869.

Example 9.11. The following pseudorandom sample of $n = 10$ was generated as *squares* of uniforms on $[0, 1]$:

$$\mathbf{x} = (.005, .003, .862, .953, .189, .541, .136, .108, .372, .116)'.$$

The minimum and maximum are $x_{(1:10)} = .003$ and $x_{(10:10)} = .953$, and the m.l. estimates are $\hat{\theta}_1 = \left(x_{(1:10)} + x_{(10:10)}\right)/2 = .478$ and $\hat{\theta}_2 = (x_{(10:10)} - x_{(1:10)})/2 = .475$. These give as standardized values

$$\hat{\mathbf{z}} = (-.996, -1.000, .808, 1.000, -.609, .133, -.719, -.779, -.224, -.762)'.$$

Formula (9.22) gives the value $I_{10}(1) = 0.869$, which just attains significance at the 0.05 level.

9.2.5.3 *Testing Exponentiality*

We have the composite null hypothesis H_0 : $f(x) = \theta_2^{-1}e^{-(x-\theta_1)/\theta_2}\mathbf{1}_{[\theta_1,\infty)}(x)$ for $\theta_1 \in \Re$ and $\theta_2 > 0$, in which scale parameter θ_2 is unknown and location parameter θ_1 may be known or may be unknown. We can define m.l. estimators for both cases as

$$\hat{\theta}_1 = \begin{cases} X_{(1:n)}, & \theta_1 \text{ unknown} \\ \theta_1, & \theta_1 \text{ known} \end{cases}$$

$$\hat{\theta}_2 = \bar{X} - \hat{\theta}_1.$$

The p.d.f. and c.f. of $Z := (X - \theta_1)/\theta_2$ are $f_0(z) = e^{-z}\mathbf{1}_{[0,\infty)}(x)$ and $\phi_0(t) = (1 - it)^{-1}$, respectively, with $\int |\phi_0(\tau)|^2 d\tau = \pi$. Putting $\hat{Z}_j := (X_j - \hat{\theta}_1)/\hat{\theta}_2$ the i.c.f. statistic is

$$I_n = \pi^{-1} n \int_{-\infty}^{\infty} \left| n^{-1} \sum_{j=1}^{n} e^{it\hat{Z}_j} - \frac{1}{1 - it} \right|^2 \frac{1}{1 + t^2} \cdot dt.$$

With algebra and integration this reduces to

$$1 + \frac{2}{n} \sum_{j<k} e^{-|\hat{Z}_j - \hat{Z}_k|} - \sum_{j=1}^{n} \left(1 + 2\hat{Z}_j\right) e^{-\hat{z}_j} + \frac{n}{2}.$$

Simulation gives the following expressions from which approximate upper-α quantiles of I_n may be determined:

α	I_n^α: θ_1 known	I_n^α: θ_1 unknown
.10	.635	$.635 + .022n^{-1/2} - .36n^{-1} + .75n^{-3/2}$
.05	.765	$.785 - .290n^{-1/2} + 1.3n^{-1} - 2.1n^{-3/2}$

Thus, for $n = 10$ and θ_1 unknown we would reject H_0 at level .05 if I_n exceeds .757.

Example 9.12. The test is applied to the $n = 10$ pseudorandom $\Gamma(5,1)$ observations from Examples 9.10, 9.7, 9.5, and 9.4, ordered here just for display:

$$\mathbf{x}_{(\cdot:10)} = (3.01, 3.60, 3.78, 4.88, 5.01, 5.48, 6.67, 7.60, 8.07, 9.48)'.$$

These give $\bar{x} = 5.758$ and $x_{(1:10)} = 3.01$. For the cases θ_1 known to be zero and θ_1 unknown, the i.c.f. statistics are, respectively, $I_{10} = 1.189$ (significant at the 0.05 level) and $I_{10} = .238$ (not significant).

9.2.5.4   Testing H_0: Laplace

The composite null hypothesis is

$$H_0 : f(x) = \frac{1}{2\theta_2} \exp\left(- |x - \theta_1| /\theta_2\right), \theta_1 \in \Re, \theta_2 > 0.$$

Equivariant estimators of θ_1 and θ_2 are

$$\hat{\theta}_1 = \begin{cases} X_{((n+1)/2:n)}, & n \text{ odd} \\ \frac{X_{(n/2:n)} + X_{(n/2+1:n)}}{2}, & n \text{ even} \end{cases}$$

and $\hat{\theta}_2 = n^{-1} \sum_{j=1}^{n} \left| X_j - \hat{\theta}_1 \right|$. (When n is odd, these are the unique m.l. estimators.) The p.d.f. and c.f. of $Z := (X - \theta_1)/\theta_2$ under H_0 are $f_0(z) =$

$\frac{1}{2}e^{-|z|}$ and $\phi_0(t) = \left(1+t^2\right)^{-1}$, respectively, with $\int |\phi_0(\tau)|^2 \, d\tau = \pi/2$. Putting $\hat{Z}_j := (X_j - \hat{\theta}_1)/\hat{\theta}_2$, the i.c.f. statistic is

$$I_n = 2\pi^{-1} n \int_{-\infty}^{\infty} \left| n^{-1} \sum_{j=1}^{n} e^{it\hat{Z}_j} - \left(1+t^2\right)^{-1} \right|^2 \left(1+t^2\right)^{-2} \cdot dt$$

$$= 1 + \frac{2}{n} \sum_{j<k} \left(1 + \left|\hat{Z}_j - \hat{Z}_k\right|\right) e^{-\left|\hat{Z}_j - \hat{Z}_k\right|}$$

$$- \frac{1}{2} \sum_{j=1}^{n} \left(3 + 3\left|\hat{Z}_j\right| + \hat{Z}_j^2\right) e^{-\left|\hat{Z}_j\right|} + \frac{5n}{8}.$$

Expressions for interpolating empirical upper-α quantiles for I_n are given below.

α	I_n^α
.10	$.283 + .06n^{-1/2} - 1.4n^{-1} + 3n^{-3/2}$
.05	$.348 - .04n^{-1/2} - n^{-1} + 2.3n^{-3/2}$

For example, with $n = 10$ the approximate upper-.10 quantile is about .257.

Example 9.13. An ordered sample of $n = 10$ Cauchy variates is

$$\mathbf{x} = (-23.81, -2.614, -1.634, -.247, -.177, .654, .771, 1.987, 6.404, 7.535)'.$$

These give $\hat{\theta}_1 = .238$, $\hat{\theta}_2 = 4.58$, and $I_{10} = .263$, which is significant at the .10 level.

9.2.5.5 Testing H_0: Logistic

The standard form of the logistic model is represented by c.d.f., p.d.f., and c.f., respectively, as

$$F(z) = \left(1 + e^{-z}\right)^{-1}$$

$$f(z) = \frac{e^{-z}}{(1+e^{-z})^2} = \frac{1}{4\cosh^2(z/2)}$$

$$\phi(t) = \frac{\pi t}{\sinh(\pi t)}$$

for $z \in \Re$ and $t \in \Re$. Introducing unknown location and scale parameters and putting $X := \theta_1 + \theta_2 Z$, we wish to test the composite null hypothesis

$$H_0 : f(x) = \theta_2^{-1} \frac{e^{-(x-\theta_1)/\theta_2}}{\left(1 + e^{-(x-\theta_1)/\theta_2}\right)^2} = \left[4\theta_2 \cosh^2\left(\frac{x-\theta_1}{2\theta_2}\right)\right]^{-1}$$

for $\theta_1 \in \Re$, and $\theta_2 > 0$. The mean and variance are $\mu = \theta_1$ and $\sigma^2 = \pi^2 \theta_2^2/3$. Since finding m.l. estimates requires solving two simultaneous nonlinear equations, it is easier to use method-of-moments estimators $\tilde{\theta}_1 = \bar{X}$ and $\tilde{\theta}_2 = \sqrt{3}S/\pi$, where S is the sample standard deviation. These are consistent, equivariant, and not much less efficient than m.l. estimators. Taking $\tilde{Z}_j := (X_j - \tilde{\theta}_1)/\tilde{\theta}_2$ and determining that $\int |\phi_0(\tau)|^2 \cdot d\tau = \pi/3$, the i.c.f. statistic can be shown to have the following computational form:[29]

$$1 + \frac{12}{n} \sum_{j<k} \left[g(d_{jk})(d_{jk} + 2) - g(d_{jk})^2 (3d_{jk} + 2) + g(d_{jk})^3 (2d_{jk}) \right]$$

$$-6 \sum_{j=1}^{n} \left[h(\tilde{Z}_j) \left(6 + \pi^2 + 6\tilde{Z}_j + \tilde{Z}_j^2 \right) - h(\tilde{Z}_j)^2 \left(6 + 7\pi^2 + 18\tilde{Z}_j + 7\tilde{Z}_j^2 \right) \right]$$

$$-36 \sum_{j=1}^{n} \left[2h(\tilde{Z}_j)^3 \left(\pi^2 + \tilde{Z}_j + \tilde{Z}_j^2 \right) - h(\tilde{Z}_j)^4 \left(\pi^2 + \tilde{Z}_j^2 \right) \right] + 0.6840n,$$

where

$$d_{jk} := \tilde{Z}_j - \tilde{Z}_k$$

$$g(d) := \frac{e^d}{e^d - 1}, d \neq 0$$

$$h(z) := \frac{e^z}{e^z + 1}.$$

Although $g(d_{jk})$ is undefined at $d_{jk} = 0$, the summand in the first line of the expression for I_n may be defined there by continuity as its limiting value, $1/6$. In practice, since the calculations are unstable for small $|d_{jk}|$, it is helpful to use the $O\left(d_{jk}^2\right)$ approximation $1/6 - d_{jk}^2/60$ when $|d_{jk}| < .01$.

Below are expressions for empirical upper-α quantiles. For example, $I_{20}^{.05} \doteq .262$.

α	I_n^α
.10	$.246 + .153n^{-1/2} - 2.925n^{-1} + 8.01n^{-3/2} - 9.36n^{-2}$
.05	$.315 + .345n^{-1/2} - 5.67n^{-1} + 19.65n^{-3/2} - 26.64n^{-2}$

Example 9.14. Here is an ordered pseudorandom sample of $n = 20$ Laplace variates:

$$-3.664, -3.020, -2.306, -2.223, -2.129, -.774, -.634, -.527, -.444,$$
$$-.269, .055, .294, .303, .332, .379, .501, .618, .627, 1.033, 1.542.$$

---

[29]Expressions for densities of sums of i.i.d. logistic variates can be deduced from George and Mudholkar (1983).

These data give $\bar{x} = \hat{\theta}_1 = -.515$, $s^2 = 2.025$, $\hat{\theta}_2 = .785$, and $I_{20}(1) = .087$, significant at the .05 level.

9.2.5.6 Testing H_0: Extreme-Value and H_0: Weibull

Limiting distributions of normalized sample maxima can take one of three forms, all of which are called "extreme-value" distributions. We consider here the form that applies for r.v.s with unbounded support. For this the composite null hypothesis is

$$H_0 : F(x) = \exp\left[-e^{-(x-\theta_1)/\theta_2}\right]$$

for $x \in \Re$, $\theta_1 \in \Re$, and $\theta_2 > 0$. The p.d.f. and c.f. of $Z := (X - \theta_1)/\theta_2$ under H_0 are $f_0(z) = \exp(-z - e^{-z})$ and $\phi_0(t) = \Gamma(1 - it)$. Also,

$$|\phi_0(t)|^2 = \Gamma(1 - it)\Gamma(1 + it) = \frac{\pi t}{\sinh(\pi t)}$$

and $\int |\phi_0(\tau)|^2 \cdot d\tau = \pi/2$. M.l. estimates can be found using the result of the following exercise.

Exercise 9.12. *Show that maximum likelihood estimates of location and scale parameters in the extreme-value model can be found by solving*

$$\hat{\theta}_2 - \bar{X} + \frac{\sum_{j=1}^n X_j e^{-X_j/\hat{\theta}_2}}{\sum_{j=1}^n e^{-X_j/\hat{\theta}_2}} = 0 \tag{9.23}$$

numerically for $\hat{\theta}_2$ and then using

$$\hat{\theta}_1 = -\hat{\theta}_2 \log\left(n^{-1} \sum_{j=1}^n e^{-X_j/\hat{\theta}_2}\right). \tag{9.24}$$

Taking $\hat{Z}_j := (X_j - \hat{\theta}_1)/\hat{\theta}_2$, the i.c.f. statistic is

$$I_n = n \int_{-\infty}^{\infty} \left| n^{-1} \sum_{j=1}^n e^{it\hat{Z}_j} - \Gamma(1 - it) \right|^2 \frac{\pi t}{\sinh(\pi t)} \cdot dt.$$

With $K = \pi^{-1}$ as normalizing constant, I_n reduces to

$$\frac{1}{n} \sum_{j,k=1}^n \frac{1}{\cosh^2\left(\frac{\hat{Z}_j - \hat{Z}_k}{2}\right)} + 8 \sum_{j=1}^n e^{-\hat{Z}_j} \left[\left(e^{-\hat{Z}_j} + 1\right) e^{e^{-\hat{Z}_j}} \operatorname{Ei}\left(-e^{-\hat{Z}_j}\right) + 1\right] + \frac{2n}{3},$$

where $\text{Ei}\,(\cdot)$ is the exponential integral function, defined for arguments $x > 0$ as

$$\text{Ei}\,(-x) = -\int_x^\infty \zeta^{-1} e^{-\zeta} \cdot d\zeta.$$

Software for computing this function is widely available, just as for the normal c.d.f. Setting $\widehat{W}_j := e^{-\hat{Z}_j}$, an equivalent expression for I_n is

$$1 + \frac{8}{n} \sum_{j<k} \frac{\widehat{W}_j \widehat{W}_k}{\left(\widehat{W}_j + \widehat{W}_k\right)^2} + 8 \sum_{j=1}^n \widehat{W}_j \left[\left(\widehat{W}_j + 1\right) e^{\widehat{W}_j}\, \text{Ei}\left(-\widehat{W}_j\right) + 1\right] + \frac{2n}{3}.$$

$$(9.25)$$

Empirical upper-α quantiles for $n \in \{10, 11, ..., 400\}$ are

α	I_n^α
.10	$.226 - .106n^{-1/2} + .42n^{-1} - .8n^{-3/2}$.
.05	$.2946 - .2n^{-1/2} + .8n^{-1} - 1.4n^{-3/2}$

Example 9.15. A pseudorandom sample of $n = 10$ from $N(0,1)$ is

$$\mathbf{X} = (.52, -.95, .68, 1.23, .08, -2.10, -.98, .44, -.21, .36)'.$$

These yield $\hat{\theta}_1 = -.593$, $\hat{\theta}_2 = 1.023$, and $I_{10} = .218$, significant at the .10 level.

The Weibull family of distributions, denoted $W(\alpha, \beta, \gamma)$, is represented by the c.d.f.

$$F(w) = \left\{1 - \exp\left[-\left(\frac{w - \gamma}{\alpha}\right)^\beta\right]\right\} \mathbf{1}_{(\gamma, \infty)}(w),$$

where γ is a location parameter, $\alpha > 0$ governs the scale, and $\beta > 0$ controls the shape. This is not a location-scale family, but suppose we restrict to the subfamily with $\gamma = 0$ and make the transformation $X = -\log W$. Then for any real x we have

$$F_X(x) = \Pr\left(W \geq e^{-x}\right)$$

$$= 1 - \left\{1 - \exp\left[-\left(\frac{e^{-x}}{\alpha}\right)^\beta\right]\right\}$$

$$= \exp\left[-e^{-(x-\theta_1)/\theta_2}\right],$$

where $\theta_1 := -\log \alpha$ and $\theta_2 := \beta^{-1}$. Thus, if $W \sim W(\alpha, \beta, 0)$ then $-\log W$ has the extreme-value distribution. Given a random sample $\{W_j\}_{j=1}^n$, the i.c.f. test of H_0: Weibull can be performed by transforming the data as $X_j = -\log W_j$ and testing the composite hypothesis that the $\{X_j\}$ are distributed as extreme value with unspecified location and scale. Indeed, since

$$\widehat{W}_j := e^{-\hat{Z}_j} = \exp\left(-\frac{X_j - \hat{\theta}_1}{\hat{\theta}_2}\right) = \frac{W_j}{\hat{\alpha}},$$

(9.25) gives the i.c.f. statistic directly in terms of empirically standardized Weibull data.

9.2.5.7 *Testing H_0: Cauchy*

The composite null hypothesis is

$$H_0 : f(x) = (\theta_2 \pi)^{-1}\left[1 + \left(\frac{x - \theta_1}{\theta_2}\right)^2\right]^{-1}, \theta_1 \in \Re, \theta_2 > 0.$$

Consistent estimators of θ_1 and θ_2 are, respectively,[30]

$$\check{\theta}_1 = n^{-1}\sum_{j=1}^{n} X_{(j:n)} \frac{\sin\left[4\pi(\frac{j}{n+1} - \frac{1}{2})\right]}{\tan\left[\pi(\frac{j}{n+1} - \frac{1}{2})\right]},$$

$$\check{\theta}_2 = 8n^{-1}\sum_{j=1}^{n} X_{(j:n)} \frac{\tan\left[\pi(\frac{j}{n+1} - \frac{1}{2})\right]}{\sec^4\left[\pi(\frac{j}{n+1} - \frac{1}{2})\right]},$$

for *even* n. To achieve equivariance, we replace $\check{\theta}_1$ by $\tilde{\theta}_1 = n\check{\theta}_1/(n+1)$. The p.d.f. and c.f. of $Z := (X - \theta_1)/\theta_2$ under H_0 are $f_0(z) = \pi^{-1}(1 + z^2)^{-1}$ and $\phi_0(t) = e^{-|t|}$, respectively, with $\int |\phi_0(\tau)|^2 d\tau = 1$. Putting $\hat{Z}_j := (X_j - \tilde{\theta}_1)/\check{\theta}_2$ the i.c.f. statistic is

$$I_n = n \int_{-\infty}^{\infty} \left| n^{-1}\sum_{j=1}^{n} e^{it\hat{Z}_j} - e^{-|t|} \right|^2 e^{-2|t|} \cdot dt$$

$$= 1 + \frac{1}{n}\sum_{j<k} \frac{8}{4 + \left(\tilde{Z}_j - \tilde{Z}_k\right)^2} - \sum_{j=1}^{n} \frac{12}{9 + \tilde{Z}_j^2} + \frac{n}{2}.$$

---

[30]Chernoff *et al.* (1967).

Expressions for interpolating empirical upper-α quantiles for I_n are given below.

α	Range of n	I_n^α
.10	$\{14, 16, ..., 400\}$	$.342 + 1.90n^{-1/2} - 25.3n^{-1} + 199n^{-3/2} - 403n^{-2}$
.05	$\{16, 18, ..., 400\}$	$.151 + 12.56n^{-1/2} - 177.5n^{-1} + 1092n^{-3/2} - 2020n^{-2}$

For example, with $n = 30$ the upper-.10 quantile is about .609.

Example 9.16. The following is an ordered pseudorandom sample of $n = 30$ from $N(0, 1)$:

$$-2.34, -2.10, -1.87, -1.42, -1.12, -.98, -.95, -.77, -.76, -.69,$$
$$-.60, -.59, -.56, -.50, -.21, .08, .15, .16, .21, .36.$$
$$.40, .44, .48, .52, .52, .63, .68, .81, 1.23, 1.27$$

The location and scale estimates are $\hat{\theta}_1 = -.13$ and $\hat{\theta}_2 = .74$. The standardized data give $I_{30} = .740$, which is significant at the .10 level.

9.2.5.8 *Testing Normality*

The composite null hypothesis is

$$H_0 : f(x) = \left(2\pi\theta_2^2\right)^{-1/2} \exp\left[-\frac{1}{2}\left(\frac{x - \theta_1}{\theta_2}\right)^2\right], \theta_1 \in \Re, \theta_2 > 0.$$

We use estimators $\hat{\theta}_1 = \bar{X}$ and $\tilde{\theta}_2 = S^2$ to produce standardized observations $\left\{\hat{Z}_j := (X_j - \bar{X})/S\right\}$. The c.f. of the standard normal being $\phi_0(t) = e^{-t^2/2}$ with $\int |\phi_0(t)|^2 \cdot dt = \sqrt{\pi}$, the i.c.f. statistic is

$$I_n = \frac{n}{\sqrt{\pi}} \int_{-\infty}^{\infty} \left| n^{-1} \sum_{j=1}^{n} e^{it\hat{Z}_j} - e^{-t^2/2} \right|^2 e^{-t^2} \cdot dt. \tag{9.26}$$

This can be reduced to the computational formula

$$1 + 2n^{-1} \sum_{j<k} e^{-(\hat{Z}_j - \hat{Z}_k)^2/4} - 2\sqrt{\frac{2}{3}} \sum_{j=1}^{n} e^{-\hat{Z}_j^2/6} + \frac{n}{\sqrt{2}}. \tag{9.27}$$

Exercise 9.13. *Show that (9.26) reduces to formula (9.27). (Hint: Write out the squared modulus in (9.26) and reverse orders of summation and integration. Use the inversion theorem to represent the integrals as 2π times p.d.f.s of certain normally distributed r.v.s evaluated at specific points.)*

The properties of I_n have been extensively studied.[31] The test is known to have high power against both long- and short-tailed alternatives and to be very competitive with the Anderson–Darling and the Shapiro–Wilk/Francia procedures. Below are expressions for determining empirical upper quantiles of I_n. For example, with $n = 20$ the null is rejected at level .05 if $I_n \geq .138$.

α	I_n^α
.10	$.121 - .18n^{-1/2} + .85n^{-1} - 2n^{-3/2}$
.05	$.161 - .25n^{-1/2} + 1.41n^{-1} - 3.4n^{-3/2}$

Example 9.17. The test based on I_n is applied to the $n = 20$ Laplace deviates of Examples 9.8 and 9.9, for which $\bar{x} = -.515$, $s^2 = 2.025$, and $I_n = .150$, significant at the .05 level.

9.2.5.9 *Testing Multivariate Normality*

The i.c.f. procedure generalizes easily to the multivariate case. Moreover, in the multinormal case it provides the only test now available that is consistent against all alternatives and (as explained below) is invariant under affine transformations. Introducing notation, take the null hypothesis to be

$$H_0 : f(\mathbf{x}) = (2\pi)^{-d/2} |\Theta|^{-d/2} \exp\left[-\frac{1}{2}(\mathbf{x} - \boldsymbol{\theta})' \Theta^{-1} (\mathbf{x} - \boldsymbol{\theta})\right],$$

where $d \in \{1, 2, ...\}$ is the dimension of vector-valued r.v. $\mathbf{X}$, $\boldsymbol{\theta} := (\theta_1, \theta_2, ..., \theta_d)'$ is the vector of means, and Θ is a nonsingular $d \times d$ covariance matrix. The p.d.f. and c.f. of $\mathbf{Z} := \Theta^{-1/2}(\mathbf{X} - \boldsymbol{\theta})$ under H_0 are

$$f_0(\mathbf{z}) = (2\pi)^{-d/2} \exp(-\mathbf{z}'\mathbf{z}/2)$$
$$\phi_0(\mathbf{t}) = e^{-\mathbf{t}'\mathbf{t}/2}, \mathbf{t} \in \Re^d,$$

with $\int_{\Re^d} |\phi_0(\mathbf{t})|^2 \cdot d\mathbf{t} = \pi^{d/2}$. Parameters $\boldsymbol{\theta}$ and Θ are estimated consistently in a sample of size n by the sample mean vector and sample covariance matrix,

$$\bar{\mathbf{X}} = n^{-1} \sum_{j=1}^{n} \mathbf{X}_j, \mathbf{S}^2 = (n-1)^{-1} \sum_{j=1}^{n} (\mathbf{X}_j - \bar{\mathbf{X}})(\mathbf{X}_j - \bar{\mathbf{X}})',$$

where $\mathbf{X}_j$ is the jth column of the $d \times n$ data array (the jth observation of $\mathbf{X}$). Letting $\mathbf{S}^{-1}$ be the square-root matrix of the inverse of $\mathbf{S}^2$, we have,

---

[31]Epps and Pulley (1983), Baringhaus *et al.* (1989), Henze (1990,1997), Henze and Wagner (1997), Epps (1999).

corresponding to the jth univariate standardized sample element $\hat{Z}_j$, the jth d-variate standardized element $\hat{\mathbf{Z}}_j = \mathbf{S}^{-1}\left(\mathbf{X}_j - \bar{\mathbf{X}}\right)$. The affine transformation from $\mathbf{X}_j$ to $\mathbf{a} + \mathbf{B}\mathbf{X}_j$, where $\mathbf{a}$ is a d-vector and $\mathbf{B}$ is a $d \times d$ nonsingular matrix, takes $\bar{\mathbf{X}}$ to $\mathbf{a} + \mathbf{B}\bar{\mathbf{X}}$ and $\mathbf{S}^{-1}$ to $\mathbf{S}^{-1}\mathbf{B}^{-1}$, leaving $\hat{\mathbf{Z}}_j$ unchanged for each j. Thus, a test statistic that depends on the data through $\left\{\hat{\mathbf{Z}}_j\right\}_{j=1}^{n}$ alone is indeed affine invariant. With $K = (2\pi)^{-d/2}$ as normalizing factor, the generalized version of (9.26) is

$$I_{n,d} = \frac{n}{\pi^{d/2}} \int_{\Re^d} \left| n^{-1} \sum_{j=1}^{n} e^{it'\hat{\mathbf{Z}}_j} - e^{-t't/2} \right| e^{-t't} \cdot dt,$$

which reduces to

$$1 + 2n^{-1} \sum_{j<k} e^{-(\hat{\mathbf{Z}}_j - \hat{\mathbf{Z}}_k)'(\hat{\mathbf{Z}}_j - \hat{\mathbf{Z}}_k)/4} - 2\left(\frac{2}{3}\right)^{d/2} \sum_{j=1}^{n} e^{-\hat{\mathbf{Z}}_j'\hat{\mathbf{Z}}_j/6} + \frac{n}{2^{d/2}}. \quad (9.28)$$

Calculating this is easier than it looks. Square-root matrix $\mathbf{S}^{-1}$ does not have to be found since

$$\left(\hat{\mathbf{Z}}_j - \hat{\mathbf{Z}}_k\right)'\left(\hat{\mathbf{Z}}_j - \hat{\mathbf{Z}}_k\right) = (\mathbf{X}_j - \mathbf{X}_k)'\mathbf{S}^{-2}(\mathbf{X}_j - \mathbf{X}_k)$$

and

$$\hat{\mathbf{Z}}_j'\hat{\mathbf{Z}}_j = (\mathbf{X}_j - \bar{\mathbf{X}})'\mathbf{S}^{-2}(\mathbf{X}_j - \bar{\mathbf{X}}),$$

where $\mathbf{S}^{-2}$ is the inverse of $\mathbf{S}^2$.

The test in dimensions $d = 2$ and 5 has been shown to have high power against various alternatives to bivariate and 5-variate normal.[32] Below are expressions from which empirical upper quantiles of $\{I_{n,d}\}_{d=2}^{5}$ can be inferred for various n. For example, for $d = 2$ and $n = 10$ the null would be rejected at level .10 if $I_{10,2} \geq .175$.

	$\alpha = .10$	$\alpha = .05$
$I_{n,2}^{\alpha}$, $n \in \{10, 11, ..., 400\}$	$.224 - .376n^{-1/2} + 3.80n^{-1}$ $-18.6n^{-3/2} + 27.8n^{-2}$	$.270 - .556n^{-1/2} + 5.8n^{-1}$ $-27.6n^{-3/2} + 41.6n^{-2}$
$I_{n,3}^{\alpha}$, $n \in \{12, 13, ..., 400\}$	$.314 - .06n^{-1/2} - .06n^{-1}$ $-1.4n^{-3/2}$	$.351 - .06n^{-1/2} + .08n^{-1}$ $-2n^{-3/2}$
$I_{n,4}^{\alpha}$, $n \in \{12, 13, ..., 400\}$	$.412 - .12n^{-1/2} + .28n^{-1}$ $-2.4n^{-3/2}$	$.448 - .12n^{-1/2} + .12n^{-1}$ $-2.24n^{-3/2}$
$I_{n,5}^{\alpha}$, $n \in \{12, 13, ..., 400\}$	$.498 - .057n^{-1/2} - .107n^{-1}$ $-1.58n^{-3/2}$	$.531 - .079n^{-1/2} - .11n^{-1}$ $-1.7n^{-3/2}$

---

[32]Henze and Zirkler (1990).

Example 9.18. The following are independent, pseudorandom Laplace samples of size $n = 10$:

$$\mathbf{x}_1 = (2.228, 0.326, -0.628, 0.409, 0.401, 1.235, -1.319, 0.037, -4.311, 0.012)'$$
$$\mathbf{x}_2 = (3.480, 0.292, 0.406, -1.287, 0.072, -2.718, -4.012, -0.234, 1.661, -2.130)'.$$

The means are

$$\left(\bar{X}_1, \bar{X}_2\right) = (-.161, -.447),$$

and the sample covariance matrix and its inverse are

$$\mathbf{S}^2 = \begin{pmatrix} 3.043 & 0.124 \\ 0.124 & 4.763 \end{pmatrix}$$

$$\mathbf{S}^{-2} = \begin{pmatrix} 0.329 & -0.009 \\ -0.009 & 0.210 \end{pmatrix}.$$

(9.28) gives $I_{10,2}(1) = 0.212$, significant at the .10 level.

9.3 Solutions to Exercises

1. Starting with

$$c_{XY} = \frac{\sum_{j=1}^{n} R(X_j) R(Y_j) - (n+1)^2 / 4}{n(n^2 - 1)/12},$$

write

$$
\begin{aligned}
\sum_{j=1}^{n} R(X_j) R(Y_j) &= \frac{1}{2} \sum_{j=1}^{n} R(X_j)^2 + \frac{1}{2} \sum_{j=1}^{n} R(Y_j)^2 - \frac{1}{2} \sum_{j=1}^{n} [R(X_j) - R(Y_j)]^2 \\
&= \sum_{j=1}^{n} j^2 - \frac{1}{2} \sum_{j=1}^{n} [R(X_j) - R(Y_j)]^2 \\
&= \frac{n(n+1)(2n+1)}{6} - \frac{1}{2} \sum_{j=1}^{n} [R(X_j) - R(Y_j)]^2.
\end{aligned}
$$

Thus,

$$
\begin{aligned}
c_{XY} &= \frac{-\frac{1}{2} \sum_{j=1}^{n} [R(X_j) - R(Y_j)]^2 + n(n+1)(2n+1)/6 - (n+1)^2/4}{n(n^2-1)/12} \\
&= \frac{-\frac{1}{2} \sum_{j=1}^{n} [R(X_j) - R(Y_j)]^2 + n(n^2-1)/12}{n(n^2-1)/12},
\end{aligned}
$$

which simplifies to the desired expression.

2. $ES_{XY} = \sum_{j=1}^{n} E\left[R\left(X_j\right)R\left(Y_j\right)\right] = \sum_{j=1}^{n} ER\left(X_j\right) \cdot ER\left(Y_j\right) = n\left(\frac{n+1}{2}\right)^2$.

3. The jth row of $\mathbb{V}$ and the ℓth column of the asserted $\mathbb{V}^{-1}$ are, respectively,

$$\left(-p_1 p_j, -p_2 p_j, ..., -p_{j-1} p_j, p_j\left(1-p_j\right), -p_{j+1} p_j, ..., -p_m p_j\right)$$
$$\left(p_{m+1}^{-1}, p_{m+1}^{-1}, ..., p_{m+1}^{-1}, p_j^{-1} + p_{m+1}^{-1}, p_{m+1}^{-1}, ..., p_{m+1}^{-1}\right).$$

The scalar products are

$$-p_j p_{m+1}^{-1}\left(\sum_{i=1}^{m-1} p_i\right) + 1 - p_j + p_j p_m^{-1} = 1, j = \ell$$

$$-p_j p_m^{-1}\left(\sum_{i=1}^{m} p_i - p_j + p_j p_{m+1}^{-1}\right) = 0, j \neq \ell.$$

We have for $n^{-1}\left(\mathbf{N} - n\mathbf{p}\right)' \mathbb{V}^{-1}$

$$n^{-1}\left(N_1 - np_1, ..., N_m - np_m\right)\begin{pmatrix} p_1^{-1} + p_{m+1}^{-1} & p_{m+1}^{-1} & \cdots & p_{m+1}^{-1} \\ p_{m+1}^{-1} & p_2^{-1} + p_{m+1}^{-1} & \cdots & p_{m+1}^{-1} \\ \vdots & \vdots & \ddots & \vdots \\ p_{m+1}^{-1} & p_{m+1}^{-1} & \cdots & p_m^{-1} + p_{m+1}^{-1} \end{pmatrix}.$$

Apart from factor n^{-1} the jth term of the m elements of the product vector is

$$p_j^{-1}\left(N_j - np_j\right) + p_{m+1}^{-1}\sum_{\ell=1}^{m}\left(N_\ell - np_\ell\right) = p_j^{-1}\left(N_j - np_j\right) - \left(N_{m+1} - np_{m+1}\right)$$

since $\sum_{\ell=1}^{m+1}\left(N_\ell - np_\ell\right) = \sum_{\ell=1}^{m}\left(N_\ell - np_\ell\right) + \left(N_{m+1} - np_{m+1}\right) = 0$.
Thus,

$$n^{-1}\left(\mathbf{N} - n\mathbf{p}\right)' \mathbb{V}^{-1}\left(\mathbf{N} - n\mathbf{p}\right) = \sum_{j=1}^{m} \frac{\left(N_j - np_j\right)^2}{np_j}$$

$$- \frac{N_{m+1} - np_{m+1}}{p_{m+1}}\sum_{j=1}^{m}\left(N_j - np_j\right)$$

$$= \sum_{j=1}^{m+1} \frac{\left(N_j - np_j\right)^2}{np_j}.$$

4. Recognizing that

$$\mathbf{N} - n\mathbf{p} = \sum_{i=1}^{n}\left(1_{\mathbb{X}_1}\left(X_i\right) - p_1, 1_{\mathbb{X}_2}\left(X_i\right) - p_2, ..., 1_{\mathbb{X}_m}\left(X_i\right) - p_m\right)$$

is the sum of independent m-vector-valued r.v.s with nonsingular covariance matrix $n\mathbb{V}$, we conclude from the multivariate c.l.t. that $\mathbf{Z}_n := n^{-1/2}\mathbb{V}^{-1/2}(\mathbf{N} - n\mathbf{p}) \leadsto N(\mathbf{0}, \mathbf{I}_m)$ and thence from the Mann–Wald theorem that

$$P_{n,m}(\boldsymbol{\theta}_0) = \mathbf{Z}_n'\mathbf{Z}_n = n^{-1}(\mathbf{N} - n\mathbf{p})'\mathbb{V}^{-1}(\mathbf{N} - n\mathbf{p}) \leadsto \chi^2(m).$$

5. (i) The log-likelihood function is

$$\mathcal{L}(p_1, p_2, ..., p_{m+1}) = \sum_{j=1}^{m+1} N_j \log p_j.$$

Maximizing subject to $\sum_{j=1}^{m+1} p_j = 1$ gives $\hat{p}_j = N_j/n$.
(ii) We have

$$
\begin{aligned}
-2\log\Lambda &= 2\sum_{j=1}^{m+1} N_j \log\left(\frac{\hat{p}_j}{p_j}\right) \\
&= 2n\sum_{j=1}^{m+1} \hat{p}_j \log\left(1 + \frac{\hat{p}_j - p_j}{p_j}\right) \\
&= 2n\sum_{j=1}^{m+1} \hat{p}_j\left[\hat{r}_j - \frac{1}{2}\hat{r}_j^2 + \frac{1}{3}\frac{\hat{r}_j^3}{(1 + \theta_{j,n})^3}\right],
\end{aligned}
$$

where $\hat{r}_j := (\hat{p}_j - p_j)/p_j$ and $|\theta_{j,n}| < |\hat{r}_j|$. Now $\sqrt{n}\hat{r}_j$ has for each j a limiting normal distribution as $n \to \infty$, and $\theta_{j,n} \to^P 0$ since $\hat{p}_j \to^P p_j$. It follows that

$$n\hat{r}_j^3 (1 + \theta_{j,n})^{-3} = n^{-1/2}\left(\sqrt{n}\hat{r}_j\right)^3 (1 + \theta_{j,n})^{-3} \to^P 0.$$

Thus, apart from terms that are $o_P(1)$ (converge to zero in probability),

$$
\begin{aligned}
-2\log\Lambda &= 2n\sum_{j=1}^{m+1} \frac{\hat{p}_j}{p_j}\left[(\hat{p}_j - p_j) - \frac{(\hat{p}_j - p_j)^2}{2p_j}\right] \\
&= 2n\sum_{j=1}^{m+1}\left(1 + \frac{\hat{p}_j - p_j}{p_j}\right)\left[(\hat{p}_j - p_j) - \frac{(\hat{p}_j - p_j)^2}{2p_j}\right] \\
&= 2n\sum_{j=1}^{m+1}(\hat{p}_j - p_j) + n\sum_{j=1}^{m+1}\frac{(\hat{p}_j - p_j)^2}{p_j} - n\sum_{j=1}^{m+1} p_j\hat{r}_j^3.
\end{aligned}
$$

The first term vanishes, since probabilities and relative frequencies sum to zero; and the last term converges in probability to zero. Therefore,

$$-2 \log \Lambda = n \sum_{j=1}^{m+1} \frac{(\hat{p}_j - p_j)^2}{p_j} + o_P(1)$$

$$= \sum_{j=1}^{m+1} \frac{(N_j - np_j)^2}{np_j} + o_P(1).$$

(iii) The model for $F_0\left(x; \{p_j\}_{j=1}^{m+1}\right)$ satisfies the regularity conditions for maximum likelihood estimation. It follows that $-2 \log \Lambda \rightsquigarrow \chi^2(m)$, since H_0 places m constraints on the $m+1$ parameters. The asymptotic equivalence of likelihood-ratio and Pearson statistics thus implies the conclusion of Theorem 9.1 for discrete distributions with bounded support.

6. Writing out the terms, we have for $-nA^2$

$$n^2 \log\left(1 - F_{(1:n)}\right) + (n-1)^2 \log\left(\frac{1 - F_{(2:n)}}{1 - F_{(1:n)}}\right) + \cdots + \log\left(\frac{1 - F_{(n:n)}}{1 - F_{(n-1:n)}}\right)$$

$$- \log\left(\frac{F_{(2:n)}}{F_{(1:n)}}\right) - 2^2 \log\left(\frac{F_{(3:n)}}{F_{(2:n)}}\right) - \cdots - (n-1)^2 \log\left(\frac{F_{(n:n)}}{F_{(n-1:n)}}\right)$$

$$+ n^2 \log F_{(n-1:n)} + n^2$$

$$= \sum_{j=1}^{n} \left[j^2 - (j-1)^2\right] \left[\log F_{(j:n)} + \log\left(1 - F_{(n-j+1:n)}\right)\right] + n^2$$

$$= \sum_{j=1}^{n} (2j-1) \left[\log F_j + \log F_{n-j+1}\right] + n^2$$

7. Applying the delta method to A_4 with $\mathbf{T}_n = (M_4, S^2)'$, $\boldsymbol{\theta} = (3,1)'$, $g(\boldsymbol{\theta}) = 3$, $g(\mathbf{T}_n) = M_4/(S^2)^2$, and $\mathbf{g}_{\mathbf{T}}'(\boldsymbol{\theta}) = (1, -6)$ shows that $\sqrt{n}(A_4 - 3)$ has the same limiting distribution as

$$\sqrt{n}(1, -6)\begin{pmatrix} M_4 - 3 \\ S^2 - 1 \end{pmatrix} = \sqrt{n}\left[(M_4 - 3) - 6(S^2 - 1)\right].$$

But

$$\sqrt{n}(M_4 - 3) = n^{-1/2} \sum_{j=1}^{n} Z_j^4 - 4\bar{Z}n^{-1/2} \sum_{j=1}^{n} Z_j^3 + 6\bar{Z}^2 n^{-1/2} \sum_{j=1}^{n} Z_j^2 - 3\sqrt{n}\bar{Z}^4,$$

of which all terms but the first converge a.s. to zero. Similarly,

$$\sqrt{n}\left(S^2 - 1\right) \sim n^{-1/2} \sum_{j=1}^{n} \left(Z_j^2 - 1\right),$$

asymptotically, so that

$$\sqrt{n}\left(A_4 - 3\right) \sim n^{-1/2} \sum_{j=1}^{n} \left(Z_j^4 - 6Z_j^2 + 3\right).$$

Next, the variances of the asymptotically equivalent expressions for $\sqrt{n}A_3$ and $\sqrt{n}\left(A_4 - 3\right)$ are, respectively,

$$\begin{aligned}
E\left(Z^3 - 3Z\right)^2 &= EZ^6 - 6EZ^4 + 9EZ^2 \\
&= 15 - 6 \cdot 3 + 9 \\
&= 6
\end{aligned}$$

and

$$\begin{aligned}
E\left(Z^4 - 6Z^2 + 3\right)^2 &= EZ^8 - 12EZ^6 + 6EZ^4 + 36EZ^4 - 36EZ^2 + 9 \\
&= 105 - 12 \cdot 15 + 42 \cdot 3 - 36 + 9 \\
&= 24.
\end{aligned}$$

(Use the relation $EZ^{2k} = (2k-1)(2k-3) \cdots \cdots 1$ for the higher moments.) Finally, since $\left(Z^3 - 3Z\right)\left(Z^4 - 6Z^2 + 3\right)$ is a polynomial in odd powers of Z, the covariance is zero.

8. We have

$$\begin{aligned}
\sum_{j=1}^{n} \sum_{i=0}^{j-1} (n-i)^{-1} &= \sum_{j=1}^{n} \sum_{i=1}^{j} (n-i+1)^{-1} \\
&= \sum_{i=1}^{n} \sum_{j=i}^{n} (n-i+1)^{-1} \\
&= \sum_{i=1}^{n} \frac{(n-i+1)}{(n-i+1)} \\
&= n.
\end{aligned}$$

9. The marginal p.d.f. of $X_{(j:n)}$ for an exponential sample is

$$f_{(j:n)}(x) = \frac{n!}{(j-1)!(n-j)!} \left(1 - e^{-x}\right)^{j-1} e^{-x(n-j+1)} \mathbf{1}_{[0,\infty)}(x).$$

Therefore,

$$M_{X_{(j:n)}}(\zeta) = \frac{n!}{(j-1)!\,(n-j)!} \int_0^\infty \left(1 - e^{-x}\right)^{j-1} e^{-x(n-j+1-\zeta)} \cdot dx$$

$$= \frac{n!}{(j-1)!\,(n-j)!} \int_0^1 \left(1 - t\right)^{j-1} t^{n-j-\zeta} \cdot dt$$

$$= \frac{n!}{(j-1)!\,(n-j)!} \frac{\Gamma(j)\,\Gamma(n-j+1-\zeta)}{\Gamma(n+1-\zeta)}$$

$$= \frac{n!\,\Gamma(n-j+1-\zeta)}{(n-j)!\,\Gamma(n+1-\zeta)}.$$

(i) Differentiating $\log M_{X_{(j:n)}}(\zeta)$ and simplifying give

$$\frac{M'_{X_{(j:n)}}(\zeta)}{M_{X_{(j:n)}}(\zeta)} = \Psi(n+1-\zeta) - \Psi(n-j+1-\zeta),$$

where $\Psi(\cdot) = d \log \Gamma(\cdot) / d(\cdot)$ is the digamma function. Evaluating at $\zeta = 0$ gives

$$m_j = \Psi(n+1) - \Psi(n-j+1)$$

$$= \left[\Psi(1) + \sum_{i=1}^{n} i^{-1}\right] - \left[\Psi(1) + \sum_{i=1}^{n-j} i^{-1}\right]$$

$$= \sum_{i=n-j+1}^{n} i^{-1}$$

$$= \sum_{i=0}^{j-1} (n-i)^{-1}.$$

(ii) Differentiating $\log M_{X_{(j:n)}}(\zeta)$ a second time and evaluating at $\zeta = 0$ give

$$v_j = \Psi'(n-j+1) - \Psi'(n+1)$$

$$= \left[\Psi'(1) - \sum_{i=1}^{n-j} i^{-2}\right] - \left[\Psi'(1) - \sum_{i=1}^{n} i^{-2}\right]$$

$$= \sum_{i=n-j+1}^{n} i^{-2}$$

$$= \sum_{i=0}^{j-1} (n-i)^{-2}.$$

10. (i) Follows from $|\phi(t\eta)| \leq 1$ and the dominance property of the integral.

(ii) $|\phi(t\eta)|^2 = \phi(t\eta)\phi(-t\eta) = Ee^{it\eta Z}Ee^{-it\eta Z'} = Ee^{it\eta(Z-Z')}$ when Z, Z' are i.i.d. with c.f. ϕ. Integrability of $|\phi(t\eta)|^2$ follows from part (i), and the inversion theorem implies that $f_{\eta(Z-Z')}(z) = \frac{1}{2\pi}\int_{-\infty}^{\infty} e^{itz}|\phi(t\eta)|^2 \cdot dt$ for $z \in \Re$. Set $z = 0$ to conclude.

11. Since $|\phi_0(\cdot)|^2$ is integrable, there is a bounded interval $\mathbb{T}$ such that, for any $\varepsilon > 0$, $\int_{\mathbb{T}^c}|\phi_0(t)|^2 \cdot dt \leq \varepsilon/4$ and therefore

$$\int_{\mathbb{T}^c}\left|\hat{\phi}_n(t) - \phi_0(t)\right|^2 |\phi_0(t)|^2 \cdot dt \leq \frac{\varepsilon}{2}.$$

Since $\sup_{t\in\mathbb{T}}|\phi_n(t) - \phi_0(t)|^2 \to^{a.s.} 0$ and since $\hat{\theta}_1, \hat{\theta}_2$ are strongly consistent, there is an $N(\varepsilon)$ such that for $n > N(\varepsilon)$ we have simultaneously

$$\sup_{t\in\mathbb{T}}|\phi_n(t) - \phi_0(t)|^2 < \frac{\varepsilon}{4\lambda(\mathbb{T})}$$

and

$$\sup_{t\in\mathbb{T}}\left|\hat{\phi}_n(t) - \phi_n(t)\right| < \frac{\varepsilon}{4\lambda(\mathbb{T})}$$

for almost all ω, where $\lambda(\mathbb{T})$ is the Lebesgue measure of $\mathbb{T}$. Therefore, for large enough n, almost surely

$$\frac{I_n}{nK} = \int_{\mathbb{T}}\left|\hat{\phi}_n(t) - \phi_0(t)\right|^2 |\phi_0(t)|^2 \cdot dt + \int_{\mathbb{T}^c}\left|\hat{\phi}_n(t) - \phi_0(t)\right|^2 |\phi_0(t)|^2 \cdot dt$$

$$\leq \int_{\mathbb{T}}\left[\left|\hat{\phi}_n(t) - \phi_0(t)\right|^2 + |\phi_n(t) - \phi_0(t)|^2\right]|\phi_0(t)|^2 \cdot dt + \frac{\varepsilon}{2}$$

$$\leq \lambda(\mathbb{T})\left[\frac{\varepsilon}{4\lambda(\mathbb{T})} + \frac{\varepsilon}{4\lambda(\mathbb{T})}\right] + \frac{\varepsilon}{2} = \varepsilon$$

Since ε is arbitrary, the conclusion follows.

12. The extreme-value p.d.f. is

$$f(x) = \theta_2^{-1}\exp\left[-\left(\frac{x-\theta_1}{\theta_2}\right) - e^{-(x-\theta_1)/\theta_2}\right]$$

for $x \in \Re$, and the log-likelihood function for sample $\mathbf{x} = (x_1, x_2, ..., x_n)'$ is

$$\log L(\mathbf{x}; \boldsymbol{\theta}) = -n\log\theta_2 - \sum_{j=1}^{n}\left(\frac{x_j - \theta_1}{\theta_2}\right) - \sum_{j=1}^{n}e^{-(x_j-\theta_1)/\theta_2}.$$

The likelihood equations are

$$\frac{\partial \log L}{\partial \theta_1}\bigg|_{\theta=\hat{\theta}} = \hat{\theta}_2^{-1}\left[n - \sum_{j=1}^{n}e^{-(x_j-\hat{\theta}_1)/\hat{\theta}_2}\right] = 0$$

$$\frac{\partial \log L}{\partial \theta_2}\bigg|_{\theta=\hat{\theta}} = -n\hat{\theta}_2^{-1} + \hat{\theta}_2^{-2}\sum_{j=1}^{n}(x_j - \hat{\theta}_1)\left[1 - e^{-(x_j-\hat{\theta}_1)/\hat{\theta}_2}\right] = 0.$$

The first equation gives

$$1 - e^{-\hat{\theta}_1/\hat{\theta}_2} n^{-1} \sum_{j=1}^{n} e^{-x_j/\hat{\theta}_2} = 0, \tag{9.29}$$

from which one obtains (9.24) on solving for $\hat{\theta}_1$. The second likelihood equation gives

$$n\hat{\theta}_2 = \sum_{j=1}^{n} x_j \left[1 - e^{-(x_j - \hat{\theta}_1)/\hat{\theta}_2}\right] - \hat{\theta}_1 \sum_{j=1}^{n} \left[1 - e^{-(x_j - \hat{\theta}_1)/\hat{\theta}_2}\right]$$

$$= \sum_{j=1}^{n} x_j \left[1 - e^{-(x_j - \hat{\theta}_1)/\hat{\theta}_2}\right],$$

since the second term in the first line vanishes by virtue of (9.29). Expression (9.23) follows on substituting for $\hat{\theta}_1$ from (9.24) and solving for $\hat{\theta}_2$.

13. We have

$$I_n = \frac{n}{\sqrt{\pi}} \int_{-\infty}^{\infty} \left| n^{-1} \sum_{j=1}^{n} e^{it\hat{Z}_j} - e^{-t^2/2} \right|^2 e^{-t^2} \cdot dt$$

Expanding the integrand and reversing the order of summation and integration in the first three terms give for the integral itself

$$n^{-2} \sum_{j,k} \int_{-\infty}^{\infty} e^{it(\hat{Z}_j - \hat{Z}_k) - t^2} \cdot dt - n^{-1} \sum_{j=1}^{n} \int_{-\infty}^{\infty} e^{it\hat{Z}_j - 3t^2/2} \cdot dt$$

$$-n^{-1} \sum_{j=1}^{n} \int_{-\infty}^{\infty} e^{-it\hat{Z}_j - 3t^2/2} \cdot dt + \int_{-\infty}^{\infty} e^{-2t^2} \cdot dt.$$

Since e^{-t^2} is the c.f. of $N(0,2)$, the first integral is 2π times the corresponding p.d.f. evaluated at $\hat{Z}_j - \hat{Z}_k$, or

$$\sqrt{\pi} e^{-(\hat{Z}_j - \hat{Z}_k)^2/4}.$$

Similarly, the second and third integrals are each 2π times the p.d.f. of $N(0,3)$ evaluated at $\pm\hat{Z}_j$, or

$$\frac{\sqrt{2\pi}}{\sqrt{3}} e^{-\hat{Z}_j^2/6}.$$

Finally, $\int_{-\infty}^{\infty} e^{-2t^2} \cdot dt$ is 2π times the p.d.f. of $N(0,4)$ evaluated at 0, or $\sqrt{\pi/2}$. Combine the terms and multiply by $n/\sqrt{\pi}$ to get computational form

$$I_n = 1 + 2n^{-1} \sum_{j<k} e^{-(\hat{Z}_j - \hat{Z}_k)^2/4} - 2\sqrt{\frac{2}{3}} \sum_{j=1}^{n} e^{-\hat{Z}_j^2/6} + \frac{n}{\sqrt{2}}.$$

Chapter 10

A Survey of Bayesian Methods

10.1 Introduction

Thus far we have dealt almost entirely with the frequentist or objectivist interpretation of probability and with the techniques for statistical inference that are based on that interpretation. Probabilities have heretofore been connected strictly with repeatable chance experiments whose outcomes are not fully predictable. In this context the intuitive concept of a probability as a "long-run relative frequency" takes its substantive meaning from the law of large numbers. Within the axiomatic framework set up by Kolmogorov, we know that relative frequencies of events do in fact converge to numbers assigned by measures that obey the various axioms.

While the objectivist conception is thus internally consistent and—as we should fully perceive by now—immensely powerful, there are nonetheless some important limitations. Is it meaningful to ask the probability that my stock will earn more than a 10% rate of return in the coming year? That the home team will win the homecoming football game? That my latest tax return will be audited? None of these situations quite fits the conditions assumed by the objectivist. Can we ever control the factors we realize to be relevant and repeat the marketplace "experiment" that will determine what investors are willing to pay for my stock a year from now? Will the known conditions of the players, their experience thus far in the season, the condition of the playing field, the anticipated attendance, and so on ever be just the same as they are for the coming game? Although we might be able to determine the proportion of *past* audits among taxpayers with similar levels and sources of income, the taxing authorities this year might well use different criteria. These and many other such common situations do not fit the objectivist's conditions, yet we all intuitively and subjectively judge the likelihood of such events and even base our actions on the subjective beliefs.

For everyday purposes it may suffice to express our subjective judgments about future events in rough, intuitive terms, but truly important decisions need better inputs. Fortunately, it is possible to develop axiomatic theories that give subjective beliefs much the same formal properties that Kolmogorov assumed for probability measures,[1] and actions based on such formal structures can be shown to be consistent with defensible notions of what constitutes rational behavior. Moreover, the result of thinking of probabilities as subjective constructs—strengths of belief—does more than just rationalize our everyday practices. It opens up an entirely new approach to statistical inference, known as *Bayesian statistics*. This chapter serves as an introduction to Bayesian methods.

10.2   A Motivating Example

Since objective probability is meaningful only in settings of repeatable chance experiments, the frequentist approach to statistical inference must be based on hypothetical replications of the sampling experiment. That is, we imagine what would happen were the experiment carried out many times, leading to indefinitely many estimates, confidence intervals, and hypothesis tests, then pick the estimators and test statistics that—according to our model—would generate results to our liking. This concept of a meta-experiment underlies not only inference from genuine random samples, as in pristine cross-sectional data, but also inference from time series, where the "innovations" that make current observations unpredictable from the past are assumed to behave as independent random draws. By contrast, the Bayesian approach bases inferences just on (i) the family of models that is believed to have generated the data, (ii) a set of prior beliefs about the unknown parameters that define the particular member of the family, and (iii) the single sample that is actually observed.

To see the contrast between the two approaches, let us consider a simple coin-tossing experiment. Both the frequentist (F) and the Bayesian (B) want to judge the frequency of heads in future flips of a particular coin; both propose the same Bernoulli family of models, $\left\{ f\left(x;\theta\right) = \theta^x \left(1-\theta\right)^{1-x} \mathbf{1}_{\{0,1\}}\left(x\right) : \theta \in (0,1) \right\}$ (with $X = 1$ for heads, say); and both expect to have a sample $\mathbf{X} = (X_1, X_2, ..., X_n)'$ with which

---

[1] As done, for example, by Savage (1972), Pratt *et al.* (1964), and Rényi (1970).

to work once they flip the coin n times. Once they observe the sample how will they then proceed to draw an inference about θ? We already know what F will do. F will try to find a statistic that would have desirable properties in the hypothetical meta-experiment; for example, one that has small mean squared error. It is an easy choice, since $\bar{X} = n^{-1} \sum_{j=1}^{n} X_j$ is a complete sufficient statistic, unbiased and efficient, and with all the desirable large-sample properties besides. Accordingly, once the sample is observed F takes $\bar{x}$ as the best guess of θ, perhaps supplementing the point estimate with a confidence interval that would contain θ in a specified proportion of those hypothetical samples. To test $H_0 : \theta = .5$, F would rely on the Neyman-Pearson theory to construct a u.m.p.-unbiased test that would reject a false H_0 in the largest fraction of samples.

Statistician B, on the other hand, gives no thought to hypothetical samples but does have to make some conjectures about θ itself. It may be that B has never seen this coin before and has had no experience with it before conducting the sampling experiment. Nevertheless B has seen many coins that look more or less the same and have seemed more or less fair, so, prior to observing the sample, B guesses that θ for this coin is "close to" 0.5. To carry out formal inference, this intuitive belief must be expressed in a very precise way. This is done by imposing on θ a *subjective* probability distribution that puts most of the mass close to where it would *all* lie for an idealized fair coin. The subjective distribution that applies before the sample is observed is known as the *prior distribution*, or simply as the "prior". The task then is to determine logically, in ways consistent with the axioms of probability theory, how to adjust the prior belief in the face of the experimental data to be obtained.

To get a quick glimpse of the procedure, represent B's prior as

$$ _*P(\{\theta\}) = \begin{cases} .25, \ \theta = .49 \\ .50, \ \theta = .50 \\ .25, \ \theta = .51 \end{cases} . $$

The *support* of $_*P$—that is, $\Theta = \{.49, 50, .51\} \subset (0, 1)$—effectively becomes the parameter *space* in the ensuing analysis. We will see later that this is a very strange and unsatisfactory prior, but it makes for a simple example. (The "$*$" that appears *prior* to the "P" both reminds us of what $_*P$ represents and distinguishes it from *posterior* probability P_*, now to be defined.) With prior in hand, B now observes the sample and calculates the likelihood function, which we represent as $L(\mathbf{x} \mid \theta) = \theta^{n\bar{x}} (1 - \theta)^{n - n\bar{x}}$. Here, the conditioning notation emphasizes that the probability of obtaining sample $\mathbf{x}$ is conditional on the uncertain value of θ; that is, $L(\mathbf{x} \mid \theta) = \Pr(\mathbf{X} = \mathbf{x} \mid \theta)$. B's next step is to use Bayes' rule to reverse the roles of $\mathbf{x}$ and θ and thus determine a *posterior distribution* of θ conditional on the data, as

$$P_*(\{\theta\} \mid \mathbf{x}) = \frac{L(\mathbf{x} \mid \theta)_* P(\{\theta\})}{L(\mathbf{x})}, \theta \in \Theta.$$

Here, the "unconditional" probability $L(\mathbf{x}) := \Pr(\mathbf{X} = \mathbf{x})$ is simply the average of conditional likelihoods weighted by the prior probabilities of θ, as

$$L(\mathbf{x}) = \int L(\mathbf{x} \mid \theta) \cdot d_* P(\theta) = \sum_{\theta \in \Theta} L(\mathbf{x} \mid \theta)_* P(\{\theta\})$$

$$= L(\mathbf{x} \mid .49)(.25) + L(\mathbf{x} \mid .50)(.50) + L(\mathbf{x} \mid .51)(.25).$$

Obviously, $L(\mathbf{x})$ just serves as a norming factor that makes the posterior probabilities add up to unity. For example, with $n = 5$, $\omega = \{H, T, T, H, H\}$, and $\sum_{j=1}^{5} x_j = 3$, we have

$$L(\mathbf{x}) = .25(.49)^3(.51)^2 + .50(.5)^5 + .25(.49)^2(.51)^3 \doteq .0312$$

and

$$P_*(\{\theta\} \mid \mathbf{x}) = \begin{cases} .2449, \theta = .49 \\ .5002, \theta = .50 \\ .2549, \theta = .51 \end{cases}$$

as the posterior distribution. With this distribution B could determine a "best guess" of θ in various ways, depending on how the costs of errors are reckoned. For example, with squared-error loss, B would use the conditional mean:

$$E(\theta \mid \mathbf{x}) = (.2449)(.49) + (.5002)(.50) + (.2549)(.51) \doteq .5001.$$

Compared with F's estimate $\bar{x} = 3/5 = .60$, this seems a much more reasonable inference from this tiny sample. On the other hand, if there were $x = 30$ heads in $n = 50$ flips, F's estimate of .60 might seem more reasonable than B's estimate of .501; indeed, with this restricted prior B's estimate could get no larger than .51 even if there were 5,000 heads in 5,000 trials. Having been thus counseled, however, B would presumably recognize that the precise description of prior belief should not build in too much certainty and should allow a preponderance of sample evidence to swamp the prior belief. We shall return to this example later and suggest a prior that affords such flexibility.

10.3 The Scope of Bayesian Inference

We now begin a more formal overview of Bayesian methods. As does the frequentist, the Bayesian starts with a family of models for the data that have been or will be generated. Whereas the frequentist's model could be rather vague—for example, that the data are i.i.d. with finite variance—the Bayesian's model must be explicit. Specifically, the Bayesian must be able to express sample likelihood $L(\mathbf{x} \mid \boldsymbol{\theta})$ in terms of a *finite* number of parameters $\boldsymbol{\theta} = (\theta_1, \theta_2, ..., \theta_k)' \in \boldsymbol{\Theta} \subset \Re^k$. There must also be an explicit prior distribution $_*P$ that gives the prior probability that $\boldsymbol{\theta}$ is in any k-dimensional Borel set B_k as

$$_*P(B_k) = \int_{B_k} d_*P(\boldsymbol{\theta}).$$

Given L and $_*P$, Bayes' rule is then applied to form the posterior distribution P_*, as

$$dP_*(\boldsymbol{\theta} \mid \mathbf{x}) = \frac{L(\mathbf{x} \mid \boldsymbol{\theta}) \cdot d_*P(\boldsymbol{\theta})}{\int_{\Re^k} L(\mathbf{x} \mid \boldsymbol{\theta}) \cdot d_*P(\boldsymbol{\theta})} = \frac{L(\mathbf{x} \mid \boldsymbol{\theta}) \cdot d_*P(\boldsymbol{\theta})}{L(\mathbf{x})}.$$

This is a density with respect to whatever measure dominates the prior $_*P$—e.g., Lebesgue measure if $_*P$ is continuous or counting measure if, as in the introductory example, $_*P$ is discrete. As we have seen, the quantity $L(\mathbf{x})$ in the denominator is just a norming constant that makes $\int dP_*(\boldsymbol{\theta} \mid \mathbf{x}) = 1$.

If $L(\mathbf{x} \mid \boldsymbol{\theta})$ is put roughly into words as "the probability of data given $\boldsymbol{\theta}$", then $dP_*(\boldsymbol{\theta} \mid \mathbf{x})$ can be rendered correspondingly as "the probability of $\boldsymbol{\theta}$ given data". P_* now encompasses both the prior belief and the experimental evidence, and it does so in a way that is consistent with the rules that apply to objective probability. If $\boldsymbol{\theta}$ is a vector and one wants to express beliefs about a single component θ_1, then the superfluous $\{\theta_j\}_{j=2}^k$ are simply integrated out. Thus, the marginal distribution for θ_1 is

$$dP_*(\theta_1 \mid \mathbf{x}) = \int_{(\theta_2, ..., \theta_k) \in \Re^{k-1}} dP_*(\theta_1, \theta_2, ..., \theta_k \mid \mathbf{x}).$$

Should more information become available in the form of a new sample $\mathbf{y} \in \Re^m$, say, then posterior $dP_*(\boldsymbol{\theta} \mid \mathbf{x})$ becomes the new prior, yielding an updated posterior

$$dP_*(\boldsymbol{\theta} \mid \mathbf{x}, \mathbf{y}) \propto L(\mathbf{y} \mid \boldsymbol{\theta}) \cdot dP_*(\boldsymbol{\theta} \mid \mathbf{x}).$$

Note that all these operations are consistent with how we would calculate frequentist probabilities if $\boldsymbol{\theta}$ were regarded as a (vector-valued) r.v. rather than as an unknown constant.

Now, once the posterior for some parameter θ is derived, what does one do with it? As in the frequentist case we can break down the goals as point estimation, interval estimation, and tests of hypotheses. However, the Bayesian approaches these goals in a very different way.

10.3.1   *Point Estimation*

Suppose first that one has an explicit loss function $\mathfrak{L} : \Theta \times \Theta \to \mathfrak{R}$ that quantifies the effect of the error $\theta^* - \theta$ associated with a point estimate θ^*. Taking $\mathfrak{L}(\theta, \theta) = 0$ without loss of generality, a single point estimate θ^* can be chosen to minimize expected loss. That is, θ^* satisfies

$$\min_{\theta^* \in \Theta} \{ E\left[\mathfrak{L}(\theta^*, \theta) \mid \mathbf{x}\right] \} = \min_{\theta^* \in \Theta} \left\{ \int \mathfrak{L}(\theta^*, \theta) \cdot dP_*(\theta \mid \mathbf{x}) \right\}.$$

If the loss is proportional to the squared error, then this yields $\theta^* = E(\theta \mid \mathbf{x})$. If loss is proportional to absolute value, then it yields $\theta^* = \theta_{.5}$, a median. Alternatively, one who is vague about the loss function might just take the modal value of θ.

10.3.2   *Interval Estimation*

In the frequentist framework one must continually suppress the inclination to think of a confidence interval as containing θ with some stated probability. In the Bayesian view, however, it is quite proper to speak of the probability that θ lies within some interval or that vector $\boldsymbol{\theta}$ lies in some measurable subset of $\mathfrak{R}^k$. Any interval $[\theta_\ell, \theta_u]$ contains the scalar θ with some prior probability $\int_{[\theta_\ell, \theta_u]} d_* P(\theta) = {}_* P([\theta_\ell, \theta_u])$ and with some posterior probability $P_*([\theta_\ell, \theta_u] \mid \mathbf{x})$. If the distribution is continuous and unimodal, then the shortest interval that delivers a given posterior probability $1 - \alpha$ is the highest-likelihood interval $[\theta_\ell, \theta_u]$ such that $dP_*(\theta_\ell \mid \mathbf{x}) = dP_*(\theta_u \mid \mathbf{x})$ and $P_*([\theta_\ell, \theta_u] \mid \mathbf{x}) = 1 - \alpha$. In the vector case one wants the region $\boldsymbol{\Theta}_{1-\alpha}$ such that $dP_*(\boldsymbol{\theta} \mid \mathbf{x}) \geq dP_*(\mathring{\boldsymbol{\theta}} \mid \mathbf{x})$ for all $\boldsymbol{\theta} \in \boldsymbol{\Theta}_{1-\alpha}$ and $\mathring{\boldsymbol{\theta}} \in \boldsymbol{\Theta} \backslash \boldsymbol{\Theta}_{1-\alpha}$ and such that $P_*(\boldsymbol{\Theta}_{1-\alpha} \mid \mathbf{x}) = 1 - \alpha$.

10.3.3   *Tests of Hypotheses*

Testing $H_0 : \boldsymbol{\theta} \in \boldsymbol{\Theta}_0$ vs. $H_1 : \boldsymbol{\theta} \in \boldsymbol{\Theta} \backslash \boldsymbol{\Theta}_0$ involves calculating the posterior probabilities $P_*(\boldsymbol{\Theta}_0 \mid \mathbf{x}) =: P_*(H_0 \mid \mathbf{x})$ and $P_*(\boldsymbol{\Theta} \backslash \boldsymbol{\Theta}_0 \mid \mathbf{x}) = 1 - P_*(\boldsymbol{\Theta}_0 \mid \mathbf{x}) =: P_*(H_1 \mid \mathbf{x})$. The ratio of these, $P_*(H_0 \mid \mathbf{x}) / P_*(H_1 \mid \mathbf{x})$, is known as the *posterior odds ratio* in favor of H_0. If Lebesgue measure $\lambda(\Theta_0) = 0$, as for the "simple" hypothesis $H_0 : \theta = \theta_0$, then it is clear that prior $_*P$ must place some positive probability mass on θ_0 if the comparison is to be meaningful. However, unlike the frequentist methods based on the Neyman-Pearson theory, the Bayesian approach does not *oblige* one to consider such knife-edge hypotheses.

Once the probabilities have been calculated, one can factor in the consequences of wrong decisions. If there is a well-defined loss function for these, then one can pick the hypothesis that minimizes expected loss. For example, if $\mathfrak{L}(H, H')$ is the loss associated with accepting H when H' is true, then one accepts H_0 if

$$\mathfrak{L}(H_0, H_1) P_*(H_1 \mid \mathbf{x}) < \mathfrak{L}(H_1, H_0) P_*(H_0 \mid \mathbf{x}).$$

It is possible in this way to deal with more than two exclusive and exhaustive hypotheses $\{H, H', H'', ...\}$. For example, if there are three such hypotheses, one accepts H_0 if the first of the three expected losses

$$\mathfrak{L}(H_0, H_1) P_*(H_1 \mid \mathbf{x}) + \mathfrak{L}(H_0, H_2) P_*(H_2 \mid \mathbf{x}),$$
$$\mathfrak{L}(H_1, H_0) P_*(H_0 \mid \mathbf{x}) + \mathfrak{L}(H_1, H_2) P_*(H_2 \mid \mathbf{x}),$$
$$\mathfrak{L}(H_2, H_0) P_*(H_0 \mid \mathbf{x}) + \mathfrak{L}(H_2, H_1) P_*(H_1 \mid \mathbf{x})$$

is also the least. One who is vague about the relative losses can just pick the hypothesis with the greatest posterior probability.

10.3.4 *Prediction*

Often, one cares to know the value of a parameter not for its own sake, but merely because it tells us which specific model of a given family applies to future observations. For example, suppose what one really wants to know is $\mathbb{P}_X(B) = \Pr(X \in B)$ for some Borel set B. Having assumed that the data are governed by some member of family $\{f(x \mid \theta) : \theta \in \Theta\}$, inference on θ is carried out just to make this calculation. In the frequentist setup, one has no choice but to calculate this probability as $\int_B dF\left(x \mid \hat{\theta}_n\right)$, using the single point estimate $\hat{\theta}_n$ obtained from the sample. This ignores the fact that the specific $\hat{\theta}_n$ is most unlikely to equal the true θ, and of itself takes no account of the degree of uncertainty associated with $\hat{\theta}_n$.[2] The Bayesian approach, on the other hand, deals with the matter in a fully

---

[2]Of course, one could in principle derive the sampling distribution of $\int_B dF\left(x \mid \hat{\theta}_n\right)$ to judge its behavior in the frequentist's hypothetical meta-experiment.

coherent way. The posterior distribution $P_*\left(\boldsymbol{\theta} \mid \mathbf{x}\right)$ expressly characterizes the uncertainty about $\boldsymbol{\theta}$ and makes it possible to construct a *predictive distribution*, $f\left(x \mid \mathbf{x}\right)$, that does explicitly take this into account. Of course, the construction of $f\left(x \mid \mathbf{x}\right)$ is based on the usual rules for calculating conditional probabilities, as[3]

$$f\left(x \mid \mathbf{x}\right) = \int_{\boldsymbol{\theta} \in \Theta} f\left(x \mid \boldsymbol{\theta}\right) \cdot dP_*\left(\boldsymbol{\theta} \mid \mathbf{x}\right). \tag{10.1}$$

If the effort had not been made to conduct the sampling experiment and to calculate the posterior distribution, predictions would have had to be made from the prior alone, as

$$f\left(x\right) = \int_{\boldsymbol{\theta} \in \Theta} f\left(x \mid \boldsymbol{\theta}\right) \cdot d_*P\left(\boldsymbol{\theta}\right). \tag{10.2}$$

By comparing the two predictive distributions in an appropriate way, a decision maker may be able to place a subjective value on the information provided by the sample. We will now see how this can be done.

10.3.5   *Valuing Information*

Let us suppose that the individual has a complete, transitive preference ordering over "states of the world". We can think of a "state" loosely just as a detailed description of the individual's circumstances at a moment in time. The preference ordering is complete in the sense that the individual can always say of any two states $s, s' \in \mathbb{S}$ (the state space) either that $s \succ s'$ (s preferred) or that $s' \succ s$ or that $s' \sim s$ (indifference). Transitivity means that $s'' \succsim s'$ and $s' \succsim s$ imply $s'' \succsim s$, the last preference being strong if and only if at least one of the others is. Moreover, we assume that the individual's choices among uncertain *prospects* over states—i.e., among different probability distributions of future states—satisfy certain other rationality conditions. It can then be shown that there exists a "utility" function $u : \mathbb{S} \to \Re$ such that different prospects over states are ranked according to the individual's preference by their *expected* utilities. Note that the formal axiomatic development of utility provides a rigorous underpinning for the loss functions $\mathfrak{L}$ that we have considered up to now; that is, the loss associated with an incorrect inference is simply the difference in utilities of states corresponding to correct and incorrect inferences. We have already encountered expected-utility theory in Section 3.4.1, where we showed how

---

[3]Note that $f\left(x \mid \boldsymbol{\theta}, \mathbf{x}\right) \equiv f\left(x \mid \boldsymbol{\theta}\right)$ since the model is defined without reference to the sample.

Jensen's inequality can be used in modeling financial decisions. Here we will apply the theory in a broader context.

For this, suppose that the individual must make a decision in today's current state s_0, the consequence of which for tomorrow's uncertain state S_1 will depend on the realization of r.v. (or random vector) X. For example, the decision might be whether to invest in a particular stock whose price tomorrow will be X; or it might be whether to reserve a cottage at the beach, the enjoyment of which will depend on tomorrow's temperature X_1, rainfall X_2, etc. In making the decision the individual has available a model $f(x \mid \boldsymbol{\theta})$ and a prior $_*P$ over the Borel sets of $\boldsymbol{\Theta}$. He also has the option of procuring a sample $\mathbf{X}_n := (X_1, X_2, ..., X_n)'$. (Why the sample size is indicated in symbol "$\mathbf{X}_n$" will become apparent.) From these he could construct predictive distributions $f(x \mid \mathbf{x}_n)$ and $f(x)$ based, respectively, on posterior P_* and prior $_*P$, as in (10.1) and (10.2). Of course, there are other uncertainties about the future state besides X, but we can focus on X by considering the expectation of utility conditional on $X = x$, $Eu(S_1 \mid s_0, x)$. We will suppose that the individual makes an optimal— i.e., utility-maximizing—decision conditional on whatever information is available. With prior information alone the result would be $U(s_0) := \max\{\int Eu(S_1 \mid s_0, x) f(x) \cdot dx\}$, while information $\mathbf{x}_n$ would produce expected utility $U(s_0, \mathbf{x}_n) := \max\{\int Eu(S_1 \mid s_0, x) f(x \mid \mathbf{x}_n) \cdot dx\}$. But

$$
\begin{aligned}
U(s_0) &= \max\left\{\int \left[\int Eu(S_1 \mid s_0, x) f(x \mid \mathbf{x}_n) \cdot dx\right] f(\mathbf{x}_n) \cdot d\mathbf{x}_n\right\} \\
&\leq \int \max\left\{\int Eu(S_1 \mid s_0, x) f(x \mid \mathbf{x}_n) \cdot dx\right\} f(\mathbf{x}_n) \cdot d\mathbf{x}_n \\
&= \int U(s_0, \mathbf{x}_n) f(\mathbf{x}_n) \cdot d\mathbf{x}_n \\
&= EU(s_0, \mathbf{X}_n).
\end{aligned}
$$

Thus, on average there will be a gain from using the additional information.[4] Moreover, extending the argument shows that the gain is nondecreasing in the sample size n. Can we now assign a subjective value to any such sample?

In fact, we can assign an explicit *monetary* value, as follows. The description of the current state s_0 will include the individual's current wealth w_0, so we can write $s_0 \equiv s_0(w_0)$ to emphasize the connection. Presuming that greater wealth today provides opportunities for more desirable states tomorrow, we suppose that $U[s_0(w_0), \mathbf{x}_n]$ increases strictly with w_0 for

---

[4]To understand the inequality in the second line, think of n students who each took m exams. Compare the best of the n *average* scores with the average of the m *best* scores.

each $\mathbf{x}_n$. Assuming also that preferences are "smooth", there will be some amount of cash v_n that delivers the precise equality

$$EU\left[s_0\left(w_0 - v_n\right), \mathbf{X}_n\right] = U\left[s_0\left(w_0\right)\right].$$

v_n represents the greatest sum that the individual would pay to obtain information $\mathbf{X}_n$, and therefore its subjective valuation.

10.3.6 *Optimal Sampling*

It is now but a short step to see that it is possible to determine a sample size that is *optimal*, based on the prior information embodied in $_*P$. Given a potential sample $\mathbf{X}_n$, one could calculate its prior informational value v_n as above. Suppose a certain cost c_n is associated with obtaining and processing such a sample. Then the expected utility to be attained from a sample of size n is $EU\left[s_0\left(w_0 - c_n\right), \mathbf{X}_n\right]$, with $U\left[s_0\left(w_0\right)\right]$ corresponding to $n = 0$. Procuring a sample of size n will be preferable to using the prior alone so long as $v_n > c_n$; and an optimal sample size is any integer $n^* \geq 0$ such that $EU\left[s_0\left(w_0 - c_{n^*}\right), \mathbf{X}_{n^*}\right] \geq EU\left[s_0\left(w_0 - c_n\right), \mathbf{X}_n\right]$ for all $n \in \aleph_0$.

10.4 Implementing the Bayes Appoach

The above examples illustrate the versatility of Bayesian methods. Unfortunately, applying them is not always easy. Aside from the problem of formulating a specific prior, the main challenge is a computational one. Finding the posterior distribution

$$dP_*\left(\boldsymbol{\theta} \mid \mathbf{x}\right) = \frac{L\left(\mathbf{x} \mid \boldsymbol{\theta}\right) \cdot d_*P\left(\boldsymbol{\theta}\right)}{L\left(\mathbf{x}\right)} = \frac{L\left(\mathbf{x} \mid \boldsymbol{\theta}\right) \cdot d_*P\left(\boldsymbol{\theta}\right)}{\int_{\Re^k} L\left(\mathbf{x} \mid \boldsymbol{\theta}\right) \cdot d_*P\left(\boldsymbol{\theta}\right)}$$

requires carrying out the integration that produces the norming constant $L\left(\mathbf{x}\right)$.[5] Updating in response to new information $\mathbf{y}$ requires another integration with respect to the previous posterior $P_*\left(\boldsymbol{\theta} \mid \mathbf{x}\right)$, and still more such operations are needed to calculate predictive distributions, expected losses, and the value of information. Several ways have been found to make all this calculation feasible: (i) the use of special "conjugate" priors $_*P$ designed to facilitate analytical calculation for specific models $f\left(\cdot \mid \boldsymbol{\theta}\right)$; (ii) the use of various large-n approximations for likelihood functions; and (iii) the application of numerical and Monte Carlo integration in high-dimensional

---

[5]Note, however, that this step is not required in calculating posterior odds ratios for testing hypotheses, since $L\left(\mathbf{x}\right)^{-1}$ is a common factor to numerator and denominator.

cases $\theta \in \Re^k$ and when analytical calculations are otherwise infeasible. A discussion of approximations and numerical methods is beyond the scope of this general survey,[6] but below we present and illustrate the use of conjugate priors for several standard models. These have the flexibility to accommodate diverse prior beliefs, and they facilitate the analytics because they produce posteriors that are in the same parametric family. Posteriors updated with new information retain the same general form as well, and expressions for predictive distributions developed from the posteriors are usually manageable.

In each case below **x** represents the realization of a random sample of size n. In working through these examples it may be helpful to refer to the summaries of models for distributions on pages 256 and 305.

(i) Bernoulli–beta: $X \sim B(1, \theta), \theta \sim \mathfrak{Be}(\alpha, \beta)$. Assuming that $x_i \in \{0, 1\}$ for each element i of the sample, the likelihood is

$$L(\mathbf{x} \mid \theta) = \theta^{n\hat{\theta}}(1 - \theta)^{n(1-\hat{\theta}_n)},$$

where $\hat{\theta}_n = n^{-1}\sum_{i=1}^{n} x_i$ is the m.l. estimate (and minimal-sufficient statistic). The prior is

$$d_*P(\theta) = \frac{1}{\mathfrak{B}(\alpha, \beta)}\theta^{\alpha-1}(1 - \theta)^{\beta-1}\mathbf{1}_{(0,1)}(\theta) \cdot d\theta, \alpha > 0, \beta > 0,$$

where (beta function) $\mathfrak{B}(\alpha, \beta)$ is defined as

$$\mathfrak{B}(\alpha, \beta) := \int_0^1 t^{\alpha-1}(1 - t)^{\beta-1} \cdot dt = \frac{\Gamma(\alpha + \beta)}{\Gamma(\alpha)\Gamma(\beta)}.$$

The posterior is then

$$dP_*(\theta \mid \mathbf{x}) \propto \theta^{\alpha+n\hat{\theta}_n-1}(1 - \theta)^{\beta+n(1-\hat{\theta}_n)-1}\mathbf{1}_{(0,1)}(\theta) \cdot d\theta$$

and $\mathfrak{B}\left[\alpha + n\hat{\theta}_n, \beta + n(1 - \hat{\theta}_n)\right]^{-1}$ is the appropriate normalizing constant. Thus, conditional on **x**, $\theta \sim \mathfrak{Be}\left[\alpha + n\hat{\theta}_n, \beta + n(1 - \hat{\theta}_n)\right]$. If $\alpha + n\hat{\theta}_n > 1$ and $\beta + n(1 - \hat{\theta}_n) > 1$, then the posterior has a single mode at

$$\frac{\alpha + n\hat{\theta}_n - 1}{\alpha + \beta + n - 2} \in (0, 1).$$

---

[6]References for such discussion include Lindley (1985), Press (1989), Chen *et al.* (2000), Gelman *et al.* (2004), Calvetti and Somersalo (2007) and Albert (2009).

(ii) Negative binomial–beta: $X \sim NB(n, \theta)$, $\theta \sim \mathfrak{Be}(\alpha, \beta)$. With $L(x \mid \theta)$ proportional to $\theta^n (1 - \theta)^x \mathbf{1}_{\aleph_0}(x)$ and $d_* P(\theta) \propto \theta^{\alpha-1} (1 - \theta)^{\beta-1} \mathbf{1}_{(0,1)}(\theta) \cdot d\theta$ we have

$$dP_*(\theta \mid x) \propto \theta^{\alpha+n-1} (1 - \theta)^{\beta+x-1} \mathbf{1}_{(0,1)}(\theta) \cdot d\theta,$$

with $\mathfrak{B}(\alpha + n, \beta + x)^{-1}$ as normalizing constant.   Thus, $\theta \sim \mathfrak{Be}(\alpha + n, \beta + x)$ conditional on x.

(iii) Uniform–power: $X \sim U(0, \theta)$, $f(\theta) \propto \theta^{-k}$, $k > 1$. With $f(x \mid \theta) = \theta^{-1} \mathbf{1}_{[0,\theta)}(x)$, we have $L(\mathbf{x} \mid \theta) = \theta^{-n}$ assuming that each $x_i \in (0, \theta)$. Taking $d_* P(\theta) \propto \theta^{-1-\alpha} \mathbf{1}_{[\beta,\infty]}(\theta) \cdot d\theta$ for $\alpha > 0$ and $\beta > 0$ as the prior gives

$$dP_*(\theta \mid \mathbf{x}) \propto \frac{1}{\theta^{\alpha+n+1}} \mathbf{1}_{[\beta \vee \hat{\theta}_n, \infty)}(\theta) \cdot d\theta,$$

where $\hat{\theta}_n = x_{(n:n)}$ (the nth order statistic or sample maximum) and $a \vee b := \max\{a, b\}$. This is again of the power form, with normalizing constant equal to the inverse of

$$\int_{\beta \vee \hat{\theta}_n}^{\infty} \frac{1}{\theta^{\alpha+n+1}} \cdot d\theta = (\alpha + n)^{-1} (\beta \vee \hat{\theta}_n)^{-(\alpha+n)}.$$

(iv) Exponential–gamma: $X \sim \Gamma(1, \theta^{-1})$, $\theta \sim \Gamma(\alpha, \beta)$. For samples all of whose elements are positive the likelihood is

$$L(\mathbf{x} \mid \theta) = \theta^n \exp\left(-\theta \sum_{i=1}^{n} x_i\right) = \theta^n \exp\left(-\frac{n\theta}{\hat{\theta}_n}\right),$$

where sufficient statistic $\hat{\theta}_n$ is the m.l. estimate. With prior $d_* P(\theta) \propto \theta^{\alpha-1} e^{-\theta/\beta} \cdot d\theta$ for $\alpha, \beta > 0$, we have

$$dP_*(\theta \mid \mathbf{x}) \propto \theta^{\alpha+n-1} \exp\left[-\theta \left(\frac{1}{\beta} + \frac{n}{\hat{\theta}_n}\right)\right] \mathbf{1}_{(0,\infty)}(\theta) \cdot d\theta$$

with normalizing constant

$$\frac{\left(\beta^{-1} + n/\hat{\theta}_n\right)^{\alpha+n}}{\Gamma(\alpha + n)}.$$

Thus, conditional on $\mathbf{x}$, $\theta \sim \Gamma\left[\alpha + n, \left(\beta^{-1} + n/\hat{\theta}_n\right)^{-1}\right]$.

(v) Poisson–gamma: $X \sim P(\theta)$, $\theta \sim \Gamma(\alpha, \beta)$. With

$$L(\mathbf{x} \mid \theta) \propto \theta^{n\bar{x}} e^{-n\theta} = \theta^{n\hat{\theta}_n} e^{-n\theta}$$

and $d_* P(\theta) \propto \theta^{\alpha-1} e^{-\theta/\beta} \cdot d\theta$ for $\alpha, \beta > 0$, we have

$$dP_*(\theta \mid \mathbf{x}) \propto \theta^{\alpha+n\hat{\theta}_n-1} \exp\left[-\theta\left(\beta^{-1}+n\right)\right] \mathbf{1}_{(0,\infty)}(\theta) \cdot d\theta$$

with normalizing constant

$$\Gamma\left(\alpha+n\hat{\theta}_n\right)^{-1} (\beta^{-1}+n)^{\alpha+n\hat{\theta}_n}.$$

Thus, conditional on $\mathbf{x}$, $\theta \sim \Gamma\left[\alpha+n\hat{\theta}_n, (\beta^{-1}+n)^{-1}\right]$.

(vi) Normal–normal mean: $X \sim N\left(\theta, \phi^{-1}\right)$, $\theta \sim N\left(\mu, \delta^{-1}\phi^{-1}\right)$, ϕ known.
The likelihood is

$$L(\mathbf{x} \mid \theta, \phi) \propto \phi^{n/2} \exp\left[-\frac{\phi}{2} \sum_{i=1}^{n} (x_i - \theta)^2\right]$$

$$= \phi^{n/2} \exp\left[-\frac{\phi}{2} \sum_{i=1}^{n} (x_i - \bar{x})^2\right] \exp\left[-\frac{n\phi}{2} (\bar{x} - \theta)^2\right]$$

$$\propto \exp\left[-\frac{n\phi}{2} \left(\hat{\theta}_n - \theta\right)^2\right],$$

where $\hat{\theta}_n = \bar{x}$ is the sufficient statistic and m.l. estimate. With

$$d_* P(\theta) \propto \exp\left[-\frac{\phi}{2} \delta (\theta - \mu)^2\right] \cdot d\theta$$

we have

$$dP_*(\theta \mid \mathbf{x}) \propto \exp\left\{-\frac{\phi}{2}\left[n\left(\hat{\theta}_n - \theta\right)^2 + \delta(\theta-\mu)^2\right]\right\} \cdot d\theta$$

$$\propto \exp\left\{-\frac{\phi}{2}(n+\delta)\left[\theta - \left(\frac{n\hat{\theta}_n + \delta\mu}{n+\delta}\right)\right]^2\right\} \cdot d\theta,$$

with normalizing constant $\sqrt{\phi(n+\delta)/(2\pi)}$. Thus, conditional on $\mathbf{x}$,

$$\theta \sim N\left[\left(\frac{n\hat{\theta}_n + \delta\mu}{n+\delta}\right), \frac{1}{\phi(n+\delta)}\right].$$

Note that the posterior mean is a weighted average of the m.l.e. and the mean of the prior, with weights proportional to the *precision* (inverse variance) of $\hat{\theta}_n$ and the precision of θ in the prior.

(vii) Normal–gamma precision: $X \sim N\left(\theta, \frac{1}{\phi}\right)$, $\phi \sim \Gamma\left(\frac{\alpha}{2}, 2\beta\right)$, θ known.
The likelihood is

$$L(\mathbf{x} \mid \theta, \phi) \propto \phi^{n/2} \exp\left[-\frac{\phi}{2} \sum_{i=1}^{n} (x_i - \theta)^2\right] \tag{10.3}$$

and $d_* P(\phi) \propto \phi^{\alpha/2-1} \exp\left(-\phi\beta^{-1}/2\right) \cdot d\phi$. Thus,

$$dP_*(\phi \mid \mathbf{x}) \propto \phi^{(n+\alpha)/2-1} \exp\left[-\frac{\phi}{2}\left(\frac{1}{\beta} + \frac{n}{\hat{\phi}_n}\right)\right] \mathbf{1}_{(0,\infty)}(\phi) \cdot d\phi,$$

where $\hat{\phi}_n = n/\sum_{i=1}^{n}(x_i - \theta)^2$ is the m.l.e., and the normalizing constant is

$$\Gamma\left(\frac{n+\alpha}{2}\right)^{-1} 2^{-(n+\alpha)/2} \left(\frac{1}{\beta} + \frac{n}{\hat{\phi}_n}\right)^{(n+\alpha)/2}.$$

Thus, given $\mathbf{x}$ we have

$$\phi \sim \Gamma\left[\frac{n+\alpha}{2}, \frac{1}{2}\left(\frac{1}{\beta} + \frac{n}{\hat{\phi}_n}\right)^{-1}\right].$$

(viii) Normal–normal mean, gamma precision: $X \sim N\left(\theta, \frac{1}{\phi}\right)$, $\theta \sim N\left(\mu, \frac{1}{\delta\phi}\right)$, $\phi \sim \Gamma\left(\frac{\alpha}{2}, 2\beta\right)$. With $L(\mathbf{x} \mid \theta, \phi)$ as in (10.3) and

$$d_* P(\theta, \phi) \propto \phi^{1/2} \exp\left[-\frac{\phi}{2}\delta(\theta - \mu)^2\right] \cdot \phi^{\alpha/2-1} \exp\left(-\frac{\phi}{2\beta}\right) \cdot d\theta d\phi \tag{10.4}$$

we have

$$dP_*(\theta, \phi \mid \mathbf{x}) \propto \phi^{1/2} \exp\left\{-\frac{\phi}{2}\left[n\left(\hat{\theta}_n - \theta\right)^2 + \delta(\theta - \mu)^2\right]\right\}$$

$$\cdot \phi^{(\alpha+n)/2-1} \exp\left[-\frac{\phi}{2}\left(\frac{1}{\beta} + \frac{n}{\hat{\phi}_n}\right)\right] \mathbf{1}_{(0,\infty)}(\phi) \cdot d\theta d\phi,$$

where $\hat{\theta}_n = \bar{x}$ and $\hat{\phi}_n = n/\sum_{i=1}^{n}(x_i - \bar{x})^2$ are the jointly sufficient statistics and m.l.e.s. With some algebra the right side reduces to

$$\phi^{1/2} \exp\left\{-\frac{\phi}{2}(n+\delta)\left(\theta - \frac{n\hat{\theta}_n + \delta\mu}{n+\delta}\right)^2\right\} \cdot d\theta$$

$$\cdot \phi^{(\alpha+n)/2-1} \exp\left\{-\frac{\phi}{2}\left[\frac{1}{\beta} + \frac{n}{\hat{\phi}_n} + \frac{n\delta\left(\mu - \hat{\theta}_n\right)^2}{n+\delta}\right]\right\} \mathbf{1}_{(0,\infty)}(\phi) \cdot d\phi.$$

Apart from a normalizing constant, this corresponds to the product of independent normal and gamma r.v.s:

$$N\left(\frac{n\hat{\theta}_n + \delta\mu}{n+\delta}, \frac{1}{\phi(n+\delta)}\right) \cdot \Gamma\left(\frac{\alpha+n}{2}, 2\left[\frac{1}{\beta} + \frac{n}{\hat{\phi}_n} + \frac{n\delta\left(\mu - \hat{\theta}_n\right)^2}{\delta+n}\right]^{-1}\right),$$

so the appropriate constant is

$$
\frac{2^{-(\alpha+n)/2}}{\sqrt{2\pi\,(n+\delta)}\,\Gamma\left(\frac{\alpha+n}{2}\right)}\left[\frac{1}{\beta}+\frac{n}{\hat{\phi}_n}+\frac{n\delta\left(\mu-\hat{\theta}_n\right)^2}{\delta+n}\right]^{(\alpha+n)/2}.
$$

In the case of the normal with both θ and ϕ unknown we will require the marginal posteriors for the individual parameters in order to draw inferences about them separately. These are, of course, found by integrations, as $dP_*\left(\theta\mid\mathbf{x}\right)=\int_{\phi\in(0,\infty)}dP_*\left(\theta,\phi\mid\mathbf{x}\right)$ and $dP_*\left(\phi\mid\mathbf{x}\right)=\int_{\theta\in\Re}dP_*\left(\theta,\phi\mid\mathbf{x}\right)$. To see what form these take, let us simplify notation by working out the marginals that correspond to the joint *prior*. Those for the posterior can then be found from the correspondences below:

Prior	Posterior
μ	$\mu_{\mathbf{x}}:=\frac{n\hat{\theta}_n+\delta\mu}{n+\delta}$
δ	$\delta_{\mathbf{x}}:=n+\delta$
α	$\alpha_{\mathbf{x}}:=\alpha+n$
β	$\beta_{\mathbf{x}}:=\left[\frac{1}{\beta}+\frac{n}{\hat{\phi}_n}+\frac{n\delta\left(\mu-\hat{\theta}_n\right)^2}{n+\delta}\right]^{-1}$

$$(10.5)$$

(ix) Marginal for θ. We have for $d_*P\left(\theta,\phi\right)$, apart from a constant norming factor,

$$
\phi^{1/2}\exp\left[-\frac{\phi}{2}\delta\left(\theta-\mu\right)^2\right]\phi^{\alpha/2-1}\exp\left(-\frac{\phi}{2\beta}\right)\cdot d\theta d\phi \quad (10.6)
$$

$$
=\phi^{(\alpha+1)/2-1}\exp\left\{-\frac{\phi}{2}\left[\beta^{-1}+\delta\left(\theta-\mu\right)^2\right]\right\}\cdot d\theta d\phi
$$

and

$$
d_*P\left(\theta\right)\propto d\theta\int_0^\infty\phi^{(\alpha+1)/2-1}\exp\left\{-\frac{\phi}{2}\left[\beta^{-1}+\delta\left(\theta-\mu\right)^2\right]\right\}\cdot d\phi
$$

$$
=\Gamma\left(\frac{\alpha+1}{2}\right)2^{(\alpha+1)/2}\left[\beta^{-1}+\delta\left(\theta-\mu\right)^2\right]^{-(\alpha+1)/2}\cdot d\theta
$$

$$
\propto\frac{\Gamma\left(\frac{\alpha+1}{2}\right)}{\Gamma\left(\frac{\alpha}{2}\right)\Gamma\left(\frac{1}{2}\right)}\sqrt{\beta\delta}\left[1+\beta\delta\left(\theta-\mu\right)^2\right]^{-(\alpha+1)/2}\cdot d\theta
$$

$$
=\frac{\sqrt{\alpha\beta\delta}}{\mathfrak{B}\left(\frac{1}{2},\frac{\alpha}{2}\right)\sqrt{\alpha}}\left[1+\frac{\alpha\beta\delta\left(\theta-\mu\right)^2}{\alpha}\right]^{-(\alpha+1)/2}\cdot d\theta. \quad (10.7)
$$

Putting $T := (\theta - \mu)\sqrt{\alpha\beta\delta}$ and applying the change-of-variable formula give

$$f_T(t) = \frac{1}{\mathfrak{B}\left(\frac{1}{2}, \frac{\alpha}{2}\right)\sqrt{\alpha}}\left(1 + \frac{t^2}{\alpha}\right)^{-(\alpha+1)/2},$$

which is the p.d.f. of Student's $t(\alpha)$. Expression (10.7) thus integrates to unity without further normalization and shows that θ is distributed as a rescaled, recentered Student variate. Notice that the uncertainty about ϕ leads to a marginal for θ that has thicker tails than the normal form that appears in the joint prior; however, α is replaced by $\alpha + n$ in the marginal *posterior* distribution, which therefore becomes more nearly normal as n increases.

(x) Marginal for ϕ. Integrating out θ in (10.6) shows immediately that

$$d_*P(\phi) \propto \phi^{\alpha/2-1}\exp\left(-\frac{\phi}{2\beta}\right)\mathbf{1}_{(0,\infty)}(\phi)\cdot d\phi$$

and hence that ϕ remains distributed as $\Gamma(\alpha/2, 2\beta)$, as it was in the construction of $d_*P(\theta, \phi)$.

Here now are some exercises that give practice in applying the conjugate priors, and some lessons that can be drawn from them.

Exercise 10.1. *The prior for the probability θ of turning up "sevens" in the roll of a pair of dice is $\mathfrak{Be}(2,6)$. In $n = 20$ throws of the dice $x = 5$ "sevens" are observed. Find the posterior distribution P_*. What is the best estimate of θ under squared-error loss? What is the most likely value (in the sense of highest posterior density)?*

Exercise 10.2. *Occurrence number $m = 5$ of "sevens" in throws of a pair of dice takes place on the $m + y = 20$th throw, where y is the number of throws resulting in "failures". Taking the prior for θ (the probability of sevens) as in the previous exercise, find the posterior P_*, the best estimate under squared-error loss, and the modal value.*

The results of these two exercises illustrate that Bayesian methods conform to what is called the *strong likelihood principle*. This asserts that the inference to be drawn from the data "ought" to be the same, regardless of the experiment that generated them, if the likelihoods are proportional. In this case, the principle asserts that it should be irrelevant that the binomial experiment involved a fixed number of trials whereas the negative-binomial experiment involved a fixed number of "successes". The validity of the strong likelihood principle is by no means universally acknowledged and is often violated in the application of frequentist methods.[7] For example, the likelihoods in these binomial and negative-binomial experiments are proportional to $\theta^x (1-\theta)^{n-x}$ and $\theta^m (1-\theta)^y$, respectively. Although these quantities are equal for the given values of n, x, m, and y, they would not always be so in *repeated* samples where r.v.s X and Y have different realizations. Hence, the tests and confidence intervals obtained by frequentist methods would not generally coincide.

The next exercise gives a somewhat more surprising example, since the change in how the "experiment" is viewed involves a shift from a continuous model to a discrete. Note that the likelihoods are indeed proportional despite this difference in interpretation.

Exercise 10.3. *The fifth trade in stock XYZ occurred 20 seconds after the opening of today's trading on the exchange.*

(i) *Modeling the time between trades as exponential with mean θ^{-1} and taking $\theta \sim \Gamma\left(1, \frac{1}{2}\right)$ as prior, find the expected value of θ given the data.*

(ii) *Modeling the number of trades per second as $P(\theta)$ and taking $\theta \sim \Gamma\left(1, \frac{1}{2}\right)$ as prior, find the expected value of θ given the data.*

Finally, here are two exercises on inference with normal data.

Exercise 10.4. *The data are modeled as $X \sim N(\theta, 1)$ and the prior is $\theta \sim N(0, 11)$. A sample of $n = 25$ produces $\bar{x} = .20$. Find the posterior odds associated with $H_0 : \theta > 0$ vs. $H_1 : \theta \le 0$.*

Exercise 10.5. *The data are modeled as $X \sim N\left(\theta, \frac{1}{\phi}\right)$, and the priors for θ and ϕ are $\theta \sim N\left(\mu, \frac{1}{\delta\phi}\right)$, $\phi \sim \Gamma(\alpha/2, 2\beta)$ with $\mu = .25$, $\delta = 1$, $\alpha = .5$, $\beta = 1$. A sample of $n = 100$ yields $\hat{\theta}_{100} = .5$, $\hat{\phi}_{100} = 2$. If the loss is squared error, find the best estimates of θ and ϕ.*

10.5 Illustrations and an Application

As a first illustration of how Bayesian methods are applied, let us return to the introductory example of inferring the probability of obtaining heads in flipping a coin. Replacing the discrete prior used in the example with

---

[7] For further discussion of this controversial principle see Cox and Hinkley (1974) and Casella and Berger (1990).

$\mathfrak{Be}\,(\alpha,\beta)$ gives as the posterior

$$dP_*\,(\theta \mid \mathbf{x}) = \frac{\theta^{\alpha+n\hat{\theta}_n-1}\,(1-\theta)^{\beta+n(1-\hat{\theta}_n)-1}}{\mathfrak{B}\left[\alpha+n\hat{\theta}_n,\beta+n(1-\hat{\theta}_n)\right]}\mathbf{1}_{(0,1)}\,(\theta)\cdot d\theta,$$

where $\hat{\theta}_n$ is the proportion of heads in the sample of size n. With squared-error loss, this implies that the best point estimate of θ is

$$E\,(\theta \mid \mathbf{x}) = \frac{\alpha+n\hat{\theta}_n}{\alpha+\beta+n}.$$

Notice that this approaches the frequentist estimate $\hat{\theta}_n$ as n increases,

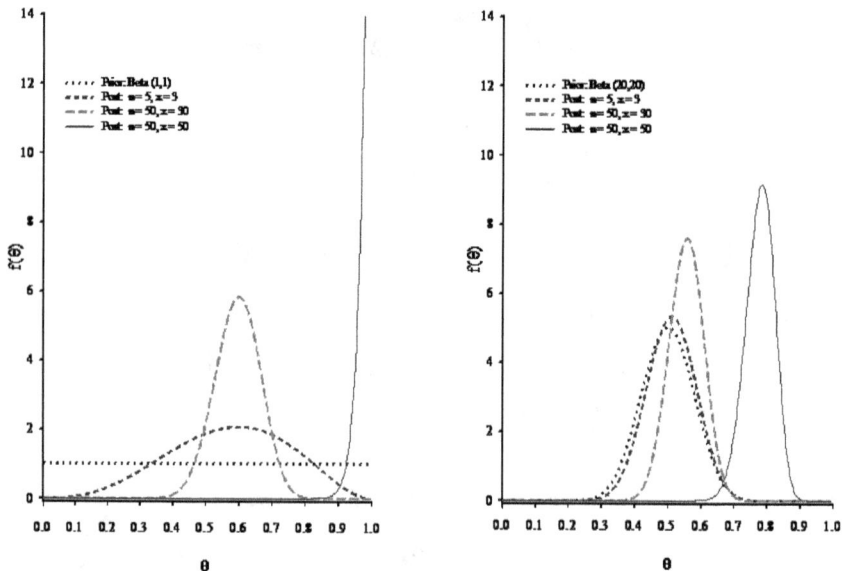

Fig. 10.1   Inference for probability of heads with diffuse priors.

so that sufficient data can always swamp the prior belief $E\theta = \alpha/(\alpha+\beta)$. We consider a sequence of priors that illustrate different degrees of belief that θ is fair, ranging from the diffuse ("vague") prior $\mathfrak{Be}\,(1,1) \equiv U\,(0,1)$ to the highly concentrated $\mathfrak{Be}(100,100)$. Figures 10.1 and 10.2 show how the posterior corresponding to each prior changes with sample outcomes: $\left(n,n\hat{\theta}_n\right) \in \{(5,3)\,,(50,30)\,,(50,50)\}$. Notice that the more diffuse is the prior the more responsive is the posterior to the sample information. In

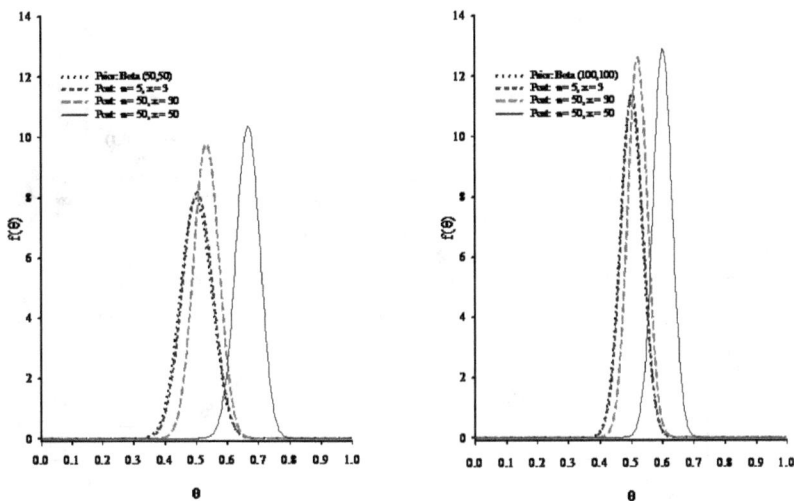

Fig. 10.2 Inference for the probability of heads with concentrated priors.

choosing a prior it is a good idea to think what it would seem reasonable to believe under various extreme sample outcomes. (One would then be apt to regard the $\mathfrak{Be}(100, 100)$ prior as too restrictive here. After getting 50 heads in 50 flips, would $E(\theta \mid \mathbf{x}) = 0.6$ really seem to be the best estimate of $\mathbb{P}(\{H\})$?)

For the next example we develop the predictive distribution for a r.v. $X \sim N\left(\theta, \phi^{-1}\right)$ based on the normal–gamma prior for θ and ϕ. Again, to simplify the notation we work out the predictive distribution $f(x)$ based on the prior alone. The correspondences in (10.5) then yield the distribution $f(x \mid \mathbf{x})$ that takes into account sample data $\mathbf{x}$. With

$$f(x \mid \theta, \phi) \propto \phi^{1/2} \exp\left[-\frac{\phi}{2}(x - \theta)^2\right]$$

and $d_* P(\theta, \phi)$ as in (10.4), we have

$$f(x \mid \theta, \phi) \cdot d_* P(\theta, \phi) \propto \phi^{1/2} \exp\left\{-\frac{\phi}{2}\left[(x - \theta)^2 + \delta(\theta - \mu)^2\right]\right\}$$

$$\cdot \phi^{(\alpha-1)/2} \exp\left(-\frac{\phi}{2\beta}\right).$$

Some algebra puts the right-hand side in the form

$$\phi^{1/2} \exp\left[-\frac{\phi}{2}\left(\theta - \frac{x + \delta\mu}{1 + \delta}\right)^2\right] \cdot \phi^{(\alpha-1)/2} \exp\left\{-\frac{\phi}{2\beta}\left[1 + \frac{\beta\delta(x - \mu)^2}{1 + \delta}\right]\right\},$$

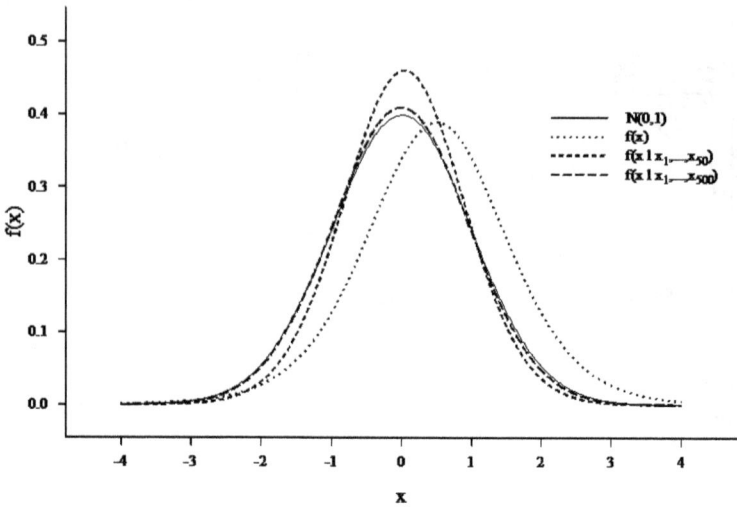

Fig. 10.3   Actual and predictive distributions of normal data with normal–gamma prior.

and integrating with respect to θ gives

$$f\left(x \mid \phi\right) \propto \phi^{(\alpha-1)/2} \exp\left\{-\frac{\phi}{2\beta}\left[1+\frac{\beta\delta\left(x-\mu\right)^2}{1+\delta}\right]\right\}.$$

This has the gamma form with shape parameter $(\alpha+1)/2$ and scale $2\beta\left[1+\frac{\beta\delta}{1+\delta}(x-\mu)^2\right]^{-1}$, and so

$$f\left(x\right) = \int_0^\infty \int_{-\infty}^\infty f\left(x \mid \theta, \phi\right) \cdot d_* P\left(\theta, \phi\right) \propto \left[1+\frac{\alpha\beta\delta}{1+\delta}\frac{\left(x-\mu\right)^2}{\alpha}\right]^{-(\alpha+1)/2}.$$

Finally, recognizing the Student form with α d.f., we can supply the necessary norming constant and write the predictive p.d.f. as

$$f\left(x\right) = \frac{\sqrt{\beta\delta/\left(1+\delta\right)}}{\mathcal{B}\left(\frac{1}{2}, \frac{\alpha}{2}\right)}\left[1+\frac{\alpha\beta\delta}{1+\delta}\frac{\left(x-\mu\right)^2}{\alpha}\right]^{-(\alpha+1)/2}. \qquad (10.8)$$

To illustrate how sample information affects the result, we generate pseudo-random samples of size $n \in \{50, 500\}$ from $N\left(\theta = 0, \phi = 1\right)$ and apply the prior based on $\theta \sim N\left(\mu = .5, \delta\phi = 1.\right)$ and $\phi \sim \Gamma\left(\alpha/2 = 5, 2\beta = .4\right)$. Although the predictive distribution $f\left(x\right)$ based on prior alone is badly centered, Fig. 10.3 shows that a sample of even modest size brings $f\left(x \mid \mathbf{x}\right)$ close to that of the data.

As an application, let us consider the following investment problem. As of $t = 0$ an individual has a sum of money that is to provide for retirement at a distant date T. The plan is to invest the funds in some combination of a stock index fund and certificates of deposit (CDs), making an initial allocation now, updating at the beginning of each succeeding month $t \in \{1, 2, ..., T - 1\}$, and then cashing out at T. The monthly updating is done both to rebalance because of changes in the assets' relative values and to incorporate new information. It is assumed, however, that no new funds are added and none withdrawn, so purchases of one asset must be financed by sales of the other.

Let R_{t+1} represent the stock fund's continuously compounded rate of return over the month ending at $t + 1$. This is defined as $R_{t+1} :=$ $\log(P_{t+1}/P_t)$, where the terminal price per share, P_{t+1}, is adjusted to take account of any dividend received during the month, as if it were immediately reinvested in stock. We let r_t be the CD rate that will apply over $(t, t + 1]$. While this is known at t, the stock's return R_{t+1} is, of course, a random variable based on time-t information. Despite this uncertainty, the individual must somehow decide at t how to allocate funds between the two assets. If proportion p_t is placed in the stock fund at t and proportion $1 - p_t$ in CDs, then the *portfolio's* continuously compounded rate over $(t, t + 1]$ is

$$R_{p_t, t+1} = \log\left[(1 - p_t)\, e^{r_t} + p_t e^{R_{t+1}}\right]$$

$$= \log\left[e^{r_t} + p_t\left(e^{R_{t+1}} - e^{r_t}\right)\right]$$

$$= r_t + \log\left[1 + p_t\left(e^{X_{t+1}} - 1\right)\right],$$

where $X_{t+1} := R_{t+1} - r_t$ is the stock fund's "excess" rate of return. What sort of criterion might the individual use to choose p_t?

Let us suppose that the individual reasons as follows. By starting at $t = 0$ with funds W_0, pursuing an investment policy $p_0, p_1, ..., p_{T-1}$ through month $T - 1$, and neither adding nor withdrawing funds in the interim, the terminal value of the portfolio would be

$$W_T = W_0 \prod_{t=0}^{T-1} \exp\left(R_{p_t, t+1}\right) = W_0 \exp\left(\sum_{t=0}^{T-1} R_{p_t, t+1}\right).$$

Obviously, one would like this quantity to be as large as possible. Unfortunately, the future returns are *uncertain*, and there is simply no way to "maximize" a r.v. However, consider the *average* continuously compounded growth rate of wealth from this policy. We can write this as

$$\log \sqrt[T]{\frac{W_T}{W_0}} = T^{-1} \sum_{t=0}^{T-1} R_{p_t,t+1}$$

$$= T^{-1} \sum_{t=0}^{T-1} E(R_{p_t,t+1} \mid \mathcal{F}_t) + T^{-1} \sum_{t=0}^{T-1} [R_{p_t,t+1} - E(R_{p_t,t+1} \mid \mathcal{F}_t)],$$

where the conditioning at each date t is on the information (i.e., the σ field) $\mathcal{F}_t$ that is then available. Now there are a number of reasonable conditions under which the last term would converge to zero as $T \to \infty$. Certainly it would be true if the $\{R_{p_t,t+1}\}$ were i.i.d. with finite mean, but that is too restrictive for the present application. On the other hand, if the returns were merely uncorrelated and the variances bounded, Theorem 6.14 would deliver strong convergence. Even if the rates were correlated, Theorem 6.13 would certify weak convergence if the dependence decayed rapidly enough over separations of time and if the variances were more or less stable. Under such reasonable conditions the "long-run" average growth rate in wealth could indeed be maximized simply by choosing p_t in each period so as to maximize $E(R_{p_t,t+1} \mid \mathcal{F}_t)$. In other words, the optimal *long-term* policy would be the "myopic" one of simply maximizing at each t the expected log of the one-period growth rate or, equivalently, maximizing the expectation of $\log [1 + p_t (e^{X_{t+1}} - 1)]$. Of course, this asymptotic result provides no assurance of good results for any finite T, but it does at least constitute a definite, well-defined goal, and one that seems intuitively reasonable.[8]

Unfortunately, to attain even this modest-seeming goal requires having a good model for the stock fund's excess rate of return, X_{t+1}. We will use Bayesian methods to help the investor to attack this problem. Here is the plan. Adopting the model $N(\theta, \phi^{-1})$ for excess rates of return, we will use the normal-gamma prior (10.4) and the resulting Student forms (10.8) for predictive distributions $f(x)$ and $f(x \mid \mathbf{x})$ based on the prior alone and on prior and sample. Sample $\mathbf{x}$ will comprise monthly continuously compounded returns on the Standard & Poors 500 (S&P) index and CD rates for the period May 1969–April 2009, a total of $n = 480$ months. Then, assuming that we are standing at time $t = 0$, we will try to determine the optimal investment proportion p_0 for the initial decision. To do this, we

---

[8]This "growth-optimal" policy has a long history in finance, beginning with the pioneering work of Kelly (1956), Latané (1959), and Breiman (1961). For an application of Bayesian inference to optimal portfolio choice in *continuous* time, see Brown and Whitt (1996).

will use simulation to estimate

$$E\left\{\log\left[1+p_0\left(e^{X_1}-1\right)\right] \mid \mathbf{x}\right\} = \int \log\left[1+p_0\left(e^x-1\right)\right] f\left(x \mid \mathbf{x}\right) \cdot dx$$

for fixed p_0, then optimize over p_0. To see how the sample data affect the result, we will carry out the same optimization with

$$E\log\left[1+p_0\left(e^{X_1}-1\right)\right] = \int \log\left[1+p_0\left(e^x-1\right)\right] f\left(x\right) \cdot dx,$$

as estimated from the prior alone.

To carry out the plan, we first calculate the sample mean and variance of the excess returns over the $n = 480$ months of the sample, obtaining $\hat{\theta}_n = -.00073$ and $\hat{\phi}_n^{-1} \doteq .00207$.[9] Now we must formulate the priors. To get some guidance on the $\Gamma\left(\alpha/2, 2\beta\right)$ prior for ϕ, we might think of fitting a gamma distribution to the inverse squares of excess returns over the period. However, the resulting highly skewed, backwards-J-shaped distribution implies a value of shape parameter α on the order of 0.17, and this leads to a predictive distribution based on the prior alone for which $\log\left[1+p_0\left(e^{X_1}-1\right)\right]$ is not integrable w.r.t. $\mathbb{P}_{X_1}$ when $p_0 > 0$. As a compromise that assures a finite expectation and yet retains some connection to past experience, we shall take $\alpha = 2$ and use the value β implied by the inverse squared returns: $n^{-1}\sum_{t=1}^n X_t^{-2}/\alpha \doteq 58,700/2$.[10] For the prior on θ we must now choose μ and δ. The mean value of X in the posterior predictive distribution will be a weighted average of μ and $\hat{\theta}_n$ with weights δ and n, respectively. It seems reasonable to take δ of the same order as n so as to give the prior belief and the sample information comparable weight. Since an exact value must be chosen, we take $\delta = n$. We will then consider a span of μ values ranging from pessimistic to optimistic and find the corresponding "optimal" p_0 for each.

The two plots in Fig. 10.4 depict the results of the experiment. These show how the "optimal" proportion p_0 varies with μ when expectations are estimated from predictive distributions $f\left(x\right)$ and $f\left(x \mid \mathbf{x}\right)$. Not surprisingly, information obtained from the sample makes the results far less sensitive to the prior belief about the mean of θ. Notice that the two assets (CDs and stock fund) get equal weight at $\mu = 0$ when predictions are based on the prior alone, whereas the equal split occurs at $\mu = .00073 = -\hat{\theta}_n$ when sample data are taken into account. These correspond to the values of μ at

---

[9] Yes, the average *excess* rate of return for the S&P over this 40-year period is in fact *negative!*
[10] To see that $E\left|\log\left[1+p_0\left(e^{X_1}-1\right)\right]\right| < \infty$ when $X_1 \sim t\left(2\right)$ recall Exercise 5.16 on page 356.

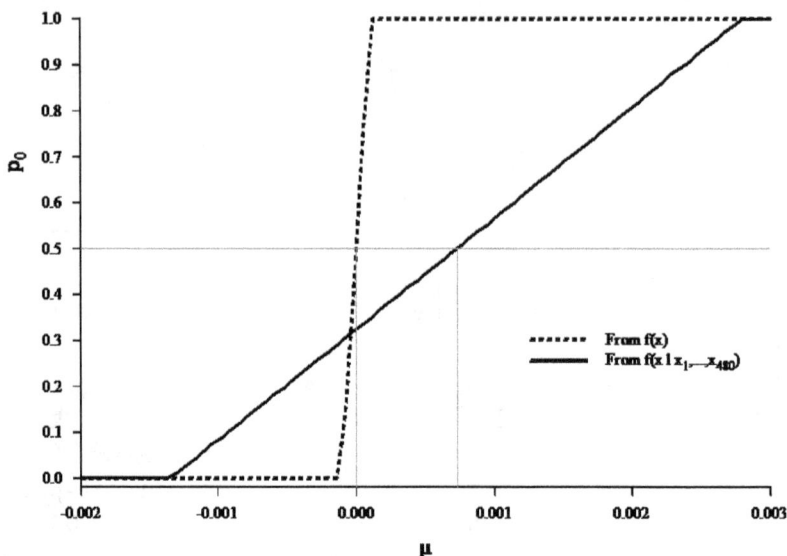

Fig. 10.4   "Optimal" investment proportion *vs.* mean of prior distribution for mean rate of return.

which the two predictive distributions have zero mean. The next exercise explains this regularity.

Exercise 10.6. *X has a distribution that is symmetric about the origin and such that* $E \left| \log \left[1 + p \left(e^X - 1 \right) \right] \right| < \infty$ *for each* $p \in [0, 1]$. *Show that* $E \log \left[1 + p \left(e^X - 1 \right) \right]$ *attains its maximum value at* $p = .5$.

10.6   A Brief Appraisal

Given the elegance of the Bayesian approach, its flawless internal consistency or "coherence", and the ease with which it allows one to tackle intricate decision problems, one may ask why not all statisticians are Bayesians and why we have devoted most of Part II to frequentist methods. There are several reasons, which is why many professional statisticians still argue heatedly about the proper approach to statistical inference.

The standard objection to Bayesian methods is that the inferences are too subjective, being skewed one way or the other by the prior distribution imposed by the researcher. Obviously, individuals B and F who saw the same data but had different priors would come to different conclusions, and why should F be influenced by B's preconceptions? To this query B would respond that if the prior is disclosed along with the posterior F is free to form his own conclusions. And to this F would counter that B could just keep his prior and posterior and show us the data—and on and on the argument would go. The intemperance of the debate and the intransigence of the debaters reminds one of squabbles between adherents of opposing religious sects.

Some other drawbacks to the Bayesian approach have already been alluded to. Besides the computational difficulty there is the matter that the Bayesian must commit in advance to a specific parametric model for the data, whereas the frequentist often has the option of adopting small-sample distribution-free methods or relying on asymptotic theory. As well, the not-too-fastidious frequentist who does want to use an explicit model is free to do some preliminary tests of fit or use formal techniques for model selection. The *purist* Bayesian would reject this opportunity because of its inconsistency with the strong likelihood principle and the evident fact that posterior inferences would depend in complicated ways on the results of the pretest. The fastidious frequentist might worry a bit about this as well, but since the implications for the ultimate conclusions are less explicit, the matter might well be swept under the rug. Of course, one who makes use of the statistical analysis might well be—and *should* be!—grateful for some confirmatory evidence for the underlying model.

Finally, there is the matter of the ambiguity in selecting a prior. Do we take a $\Gamma(2,5)$ prior for ϕ in the $X \sim N\left(\theta, \phi^{-1}\right)$ model or a $\Gamma(2.5, 4.0)$? If this is a toss-up and the one leads to acceptance of $H_0 : \theta \leq 0$ *vs.* $H_1 : \theta > 0$ and the other leads to rejection, what do we do? More worrisome is the possibility that an unethical researcher—or *any* researcher with enough personal stake in the outcome—might deliberately configure the prior so as to support some favored conclusion. On the other hand, the frequentist has many such opportunities for cooking the data as well, and they are easy to hide: "Outliers" can be discarded on plausible-seeming grounds, samples can be lengthened or truncated until desired results are achieved, and seemingly objective, defensible choices can be made among a variety of point and interval estimators and tests. Accordingly, one must acknowledge the merit of Bayesians' argument that frequentist methods are no less subjective, merely that the subjective influences are subtler and harder to detect by those who see only the results and not the data. Still, Bayesians have as much opportunity as do frequentists to stretch and truncate samples and to toss outliers—and they can manipulate their priors besides.

While the debate continues, a growing number of statisticians now acknowledge the force of arguments on both sides and take the pragmatic course of choosing among frequentist and Bayesian approaches the one that best suits the purpose at hand. Clearly, Bayesian methods are ideal for *personal* decision problems with well defined loss functions and in applications to small samples when one does have some reasonable preconceptions—as in the introductory coin-tossing example. On the other hand, frequentist methods seem more appropriate for answering purely scientific questions— "how does the world work?" *vs.* "what do I do now?"[11] Obviously, whatever methods were applied in carrying out a statistical analysis, the *user* must rely upon the researcher's high ethical standards—but should never take for granted that they were adhered to.

10.7   Solutions to Exercises

1. The prior puts the modal value of θ at $1/6$, which would be the objective probability for fair dice. The likelihood of the sample for this binomial experiment of n trials is $L\left(x \mid \theta\right) \propto \theta^x \left(1-\theta\right)^{n-x} = \theta^5 \left(1-\theta\right)^{15}$, and the posterior is

$$dP_*\left(\theta \mid x\right) = \frac{\theta^6 \left(1-\theta\right)^{20}}{\mathcal{B}\left(7, 21\right)} \mathbf{1}_{(0,1)}\left(\theta\right).$$

The best point estimate with squared-error loss is $E\left(\theta \mid \mathbf{x}\right) = 7/28 = 1/4$, with modal value $6/26 = 3/13$.

2. For this negative-binomial experiment we have

$$L\left(y \mid \theta\right) \propto \theta^m \left(1-\theta\right)^y = \theta^5 \left(1-\theta\right)^{15},$$

since there were $m = 5$ "successes" and $y = 15$ "failures" prior to the fifth success. Thus, $L\left(y \mid \theta\right)$ for this experiment is proportional to that from the binomial experiment; and since there is a common prior, it follows that the posteriors are the same—and hence the resulting estimates.

3. (i) The likelihood is $L\left(\mathbf{x} \mid \theta\right) = \theta^5 e^{-20\theta}$ and the prior is $d_*P\left(\theta\right) \propto e^{-2\theta}$, so $dP_*\left(\theta \mid \mathbf{x}\right) \propto \theta^5 e^{-22\theta}$ and $E\left(\theta \mid \mathbf{x}\right) = 3/11$.
   (ii) The likelihood is $L\left(\mathbf{x} \mid \theta\right) \propto \theta^5 e^{-20\theta}$ and the prior is $d_*P\left(\theta\right) \propto e^{-2\theta}$, so $dP_*\left(\theta \mid \mathbf{x}\right) \propto \theta^5 e^{-22\theta}$ and $E\left(\theta \mid \mathbf{x}\right) = 3/11$.

---

[11] Bayesian techniques can also be used to develop and improve frequentist estimators; for example, to find "minimax" estimators that minimize the maximum expected loss. For details, see Chapters 4 and 5 of Lehmann and Casella (1998).

4. Given that $\hat{\theta}_{25} = \bar{x} = .20$, the posterior distribution of θ is

$$N\left[\left(\frac{25\,(.20) + 11\,(0)}{25 + 11}\right), \frac{1}{25 + 11}\right] = N\left(\frac{5}{36}, \frac{1}{36}\right),$$

so $P_*\,((0, \infty) \mid \mathbf{x}) = \Pr\left(Z > -\frac{5}{6}\right) \doteq .7977$, and

$$\frac{P_*\,(H_0 \mid \mathbf{x})}{P_*\,(H_1 \mid \mathbf{x})} \doteq \frac{.7977}{.2023} \doteq 3.943.$$

5. Given $\mathbf{x}$, the marginal distribution of θ is that of $\mu_{\mathbf{x}} + \sqrt{\alpha_{\mathbf{x}}\beta_{\mathbf{x}}\delta_{\mathbf{x}}}T_{\alpha_{\mathbf{x}}}$, where $T_{\alpha_{\mathbf{x}}} \sim t\,(\alpha_{\mathbf{x}})$ and the parameters are as in (10.5). But $ET_{\alpha_{\mathbf{x}}} = 0$ since $\alpha_{\mathbf{x}} = \alpha + n = .5 + 100 > 1$, and so

$$E\,(\theta \mid \mathbf{x}) = \mu_{\mathbf{x}} = \frac{100\,(.5) + 1\,(.25)}{100 + 1} \doteq .4975.$$

Given $\mathbf{x}$, the inverse variance ϕ is distributed as $\Gamma\,(\alpha_{\mathbf{x}}/2, 2\beta_{\mathbf{x}})$, so

$$E\,(\phi \mid \mathbf{x}) = \alpha_{\mathbf{x}}\beta_{\mathbf{x}} = \frac{.5 + 100}{\left[\frac{1}{1} + \frac{100}{2} + \frac{100(1)(.25 - .5)^2}{100 + 1}\right]} \doteq 1.968.$$

6. Write $E\log\left[1 + p(e^X - 1)\right] = \int \log\left[1 + p(e^x - 1)\right] \cdot dF(x)$ and break the range of integration into $(-\infty, 0)$ and $(0, \infty)$. (Note that the integrand is zero at $x = 0$ even if $F(0) - F(0-) > 0$.) Changing variables as $x \longrightarrow -x$ in the first integral and using symmetry,

$$E\log\left[1 + p(e^X - 1)\right] = \int_{(0,\infty)} \log\left\{\left[1 + p(e^x - 1)\right]\left[1 + p(e^{-x} + 1)\right]\right\} \cdot dF(x)$$

$$= \int_{(0,\infty)} \log\left\{1 + 2p(1 - p)\left[\cosh(x) - 1\right]\right\} \cdot dF(x).$$

Since $\cosh(x) > 1$ on $(0, \infty)$, the integrand is greatest for *each* x at $p = \frac{1}{2}$, and likewise the integral.

Chapter 11

Regression Analysis

11.1 Introduction

We have already seen several ways of applying statistical inference in multi-variate models, including point and interval estimation, tests of parametric hypotheses, and nonparametric tests of independence and of goodness of fit; however, all of these have just been extensions of corresponding procedures for univariate models. This chapter deals with inferences about a feature of data-generating processes that is present only in the multivariate case; specifically, the aim is to infer how the conditional mean of one variable depends on the realizations of one or more others. There are several settings in which such inference might take place, and Figs. 11.1–11.4 illustrate these with bivariate data.

Figure 11.1 is a scatter plot of annual consumption expenditures *vs.* incomes for a sample of 1,759 US households, drawn from the 2002 Bureau of Labor Statistics' (BLS) Consumer Expenditure Survey. The sample is not really random, in that the BLS imposed various conditions for inclusion; also, for display purposes the figure limits to households with incomes less than $150,000. Nevertheless, we can think of these data as representative of a (truncated) bivariate random sample and as an example of *cross-sectional* data—i.e., data pertaining to different entities at the same point in time or, as in this case, over the same period of time. The plot supports a conclusion that we could have inferred from common experience and observation; namely, that expenditure and income are positively related in the cross section. Indeed, the sample coefficient of correlation is $r = +0.58$.

Figure 11.2 is a similar scatter plot, but now of *time-series* rather than cross-sectional data; namely, contemporaneous monthly simple rates of return of IBM common stock and of the Standard and Poors 500-stock index

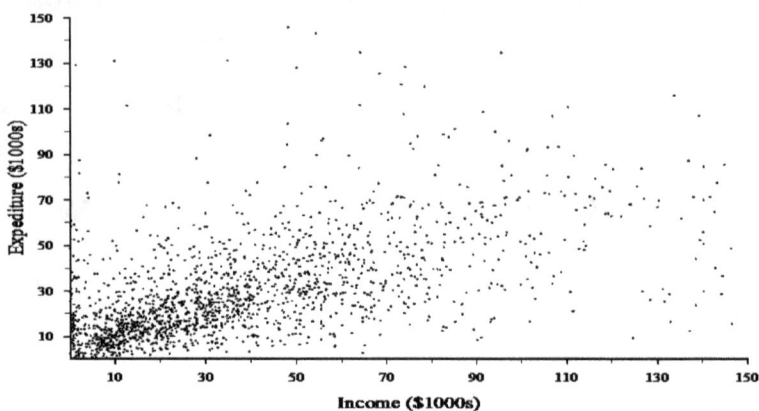

Fig. 11.1   Annual expenditure *vs.* income of sample of US households.

(S&P). Again, both the visual display and the sample correlation of +0.60 suggest some positive relationship, which (one presumes) exists mainly because investors regard the values of IBM and of other stocks in the S&P as subject to common influences. Still, as for the cross-sectional data on expenditure and income, it is apparent that neither coordinate of any sample point could be perfectly predicted from values of the remaining data.

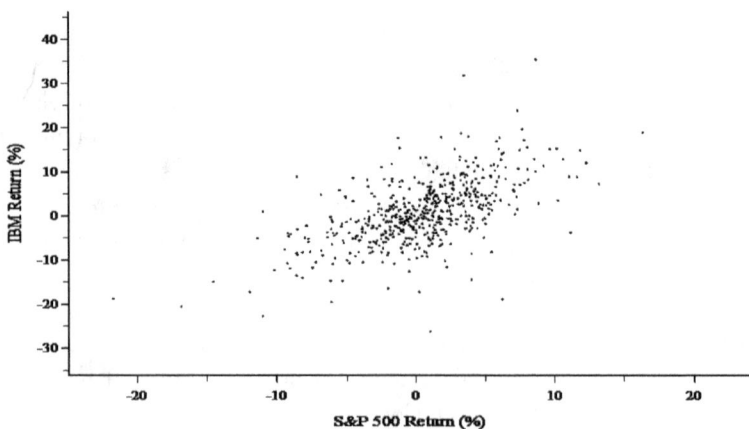

Fig. 11.2   Monthly rates of return, IBM and S&P, June 1962-February 2009.

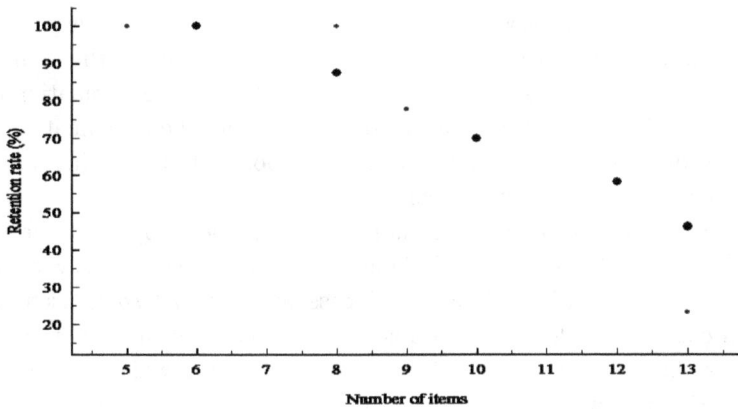

Fig. 11.3 Proportion of symbols in random strings that were correctly ordered, *vs.* length of string.

Figure 11.3 illustrates a contrary case in which the variable on the horizontal axis was actually *controlled*, and therefore perfectly predictable. Here the data represent the accuracy with which an experimental subject could recall and recite various strings of numbers, letters, and words whose lengths were chosen by the experimenter. (The larger dots in the figure denote repeated values.) Both the graphical view and the sample correla-

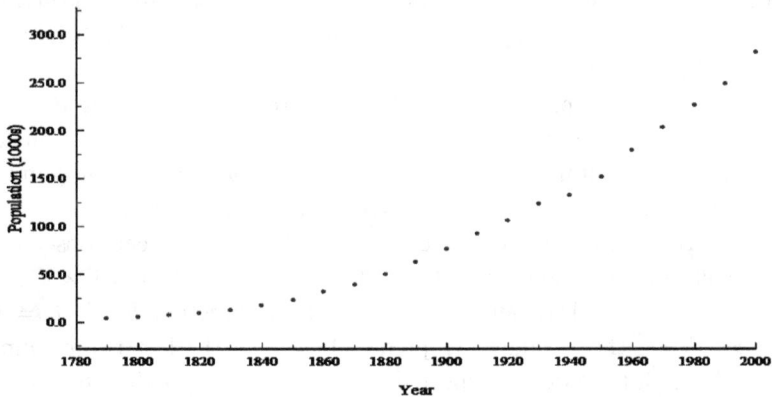

Fig. 11.4 Census counts of US population, 1790-2000.

tion of -0.94 suggest (as we might have surmised from personal experience) that retention rate is negatively related to the complexity of the stimulus.

As a final example, Fig. 11.4 tracks the size of the US population over the period 1790-2000 at intervals of ten years, from the decennial census. The relation here is between a variable that is not perfectly predictable and one that is predictable but *not* controlled.

The common feature of these examples is that there appears to be some relation between the variables. Although each relation is obviously stochastic, the appearance of the data in each case suggests that *some* useful prediction could be made of the variable on one axis if the value on the other axis were known. The choice of which to predict and which to regard as predictor is clear when one variable is actually under control or is otherwise deterministic. Indeed, experiments like that in Fig. 11.3 are *designed* to see how one variable is affected by others that the experimenter chooses. In the case of Fig. 11.4 the interest might be in predicting the population at a future date or in inferring its value between past censuses. When neither variable is controlled or predictable, as in Figs. 11.1 and 11.2, we may still have in mind some theoretical causal link that makes one variable *depend* in some way on the other. Thus, we would think of household consumption as being influenced by income rather than the other way around (although families' ability to save, to borrow, and to spend from past savings makes the connection between *contemporaneous* values far less rigid). In the case of stock prices, we might think of S&P returns as proxying for general economic influences that affect the valuations of IBM. Thus, although mere contemporaneous statistical relationships can never *establish* causality, theoretical notions in that regard often make us want to link the behavior of one variable to one or more others.

In the four examples we would thus want to draw inferences about the expected value of the variable on the vertical axis *given* a value assumed by that on the horizontal. We refer to these as the *dependent variable* and the *explanatory variable*, respectively, denoted for the present as Y and x. When neither variable is predictable, as in the first two examples, each observation $(Y_i, \breve{x}_i)$ is, *ex ante*, a random vector with support contained in $\mathbb{Y} \times \mathbb{X} \subset \Re^2$, and the goal is to draw an inference about the function $E(Y \mid \breve{x}_i = x) : \mathbb{X} \to \mathbb{Y}$. When the x values *are* predictable, as in the last two examples, the goal is still to infer the expected value of Y corresponding to values of x in some set $\mathbb{X}$. In each case $E(Y \mid x) : \mathbb{X} \to \mathbb{Y}$ is called the *regression function* of Y on x. More generally, our data may be multivariate, with *ex post* observations $\{(y_i, \mathbf{x}_i)\}_{i=1}^n$, where $\mathbf{x}_i := (x_{i1}, x_{i2}, ..., x_{im})$. Again, the goal is to estimate $E(Y \mid \mathbf{x})$ for $\mathbf{x} \in \mathbb{X} \subset \Re^m$. When we perform such an estimation, we are said to be "regressing" Y on $\mathbf{x}$. In common (albeit rather perverse) usage the explanatory variables $\mathbf{x}$ are often referred to as the *regressors* or *covariates*, and "dependent" variable Y is called the *regressand*.

Just as regression analysis has its own special jargon, some special notation is needed as well, owing to the fact that matrix notation is essential for this topic. Notice that $\mathbf{x}_i \in \Re^m$ was defined above as a *row* vector, contrary to our usual convention. Below, we will stack the n row vectors $\{\mathbf{x}_i\}_{i=1}^n$ into an $n \times m$ matrix, denoted by $\mathbf{X}$, but will use the symbol $\mathbf{Y}$ in the customary way to represent the column vector $(Y_1, Y_2, ..., Y_n)'$. Since the upper case for "$\mathbf{X}$" now signifies "matrix", we use the lower-case $\check{\mathbf{x}}$ when it is necessary to refer specifically to vectors of stochastic explanatory variables. Symbols $\{Y_i\}$ and $\{y_i\}$ are used to distinguish r.v.s and their *ex post* realizations as we have done heretofore, and realizations of $\{\check{\mathbf{x}}_i\}$ are indicated just as $\{\mathbf{x}_i\}$.

Unless $\mathbb{X}$ is a finite set, estimating the *function* $E(Y \mid \mathbf{x}) : \mathbb{X} \to \mathbb{Y}$ from a finite sample clearly requires imposing some structure on the functional form. If $\{(Y_i, \check{\mathbf{x}}_i)\}_{i=1}^n$ are r.v.s, as in the first two examples, and if a parametric model $f(y_1, \mathbf{x}_1, ..., y_n, \mathbf{x}_n; \boldsymbol{\theta})$ has been adopted, then $E(Y \mid \check{\mathbf{x}} = \mathbf{x}) =: \xi(\mathbf{x}; \boldsymbol{\theta})$, say, will be known up to the vector of parameters $\boldsymbol{\theta}$, and we will have $\xi\left(\mathbf{x}; \hat{\boldsymbol{\theta}}_n\right)$ as the maximum likelihood estimate at each $\mathbf{x} \in \mathbb{X}$. However, we most often deal with situations in which there is no parametric model for random vectors $\{(y_i, \check{\mathbf{x}}_i)\}_{i=1}^n$ or in which the explanatory variables $\mathbf{x}$ are not stochastic, as for the examples in Figs. 11.3 and 11.4. In these situations some functional form for $E(Y \mid \mathbf{x})$ has to be proposed *a priori*, and we are led to consider models of the form $Y_i = \xi(\mathbf{x}_i; \boldsymbol{\beta}) + U_i$. Here $U_i := Y_i - E(Y_i \mid \mathbf{x}_i)$ is an "error" or "disturbance" term that comprises stochastic influences on the realization of Y_i for which the model does not explicitly account, and $\boldsymbol{\beta}$ is a k-vector of parameters that govern just the conditional mean and not the errors. (Note that k does not necessarily equal the number of explanatory variables m.)

How we go about doing inference on $\boldsymbol{\beta}$ depends on two general considerations: (i) the assumptions made about the joint distribution of the errors $\{U_i\}_{i=1}^n$ and (ii) how parameters $\boldsymbol{\beta}$ enter regression function ξ. If the errors represent the net effect of numerous small, independently acting influences, it may be appropriate to treat the $\{U_i\}$ as i.i.d. as $N(0, \sigma^2)$ conditionally on the $\{\mathbf{x}_i\}$, where σ^2 is a positive, unknown parameter. Of course, that $E(U_i \mid \mathbf{x}_i) = 0$ follows from the definition, but that the

$\{U_i\}$ are independent, normally distributed, and have the same variance are strong conditions that are at best working approximations. The attraction of imposing them is that one then has available powerful tools for drawing inferences about $\boldsymbol{\beta}$, often without having to rely on asymptotic theory. The next section treats inference under this strong assumption. In the final section we dispense with normality and develop techniques for confidence intervals and tests of hypotheses based on asymptotic theory. Unfortunately, the errors often fail even to be *approximately* i.i.d. in time series data and with cross sections having wide variability in the explanatory variables. When they are not i.i.d. conditionally on the regressors and when the explanatory variables are stochastic, there is also the possibility of dependence between errors and regressors. Any of these conditions can seriously affect the quality of inferences using the standard methods we discuss. The many specialized techniques that can be used under such conditions are beyond our present scope but are the principal focus of books and courses in the field of econometrics.

Within each of the next two sections we treat separately models that are *linear* in the parameters and those in which at least one component of $\boldsymbol{\beta}$ enters *non*linearly. For example, (i) $\xi(x_i; \boldsymbol{\beta}) = \beta_1 + \beta_2 x_i$, (ii) $\xi(x_i, \boldsymbol{\beta}) = \beta_1 + \beta_2 x_i + \beta_3 x_i^2$, and (iii) $\xi(\mathbf{x}_i; \boldsymbol{\beta}) = \beta_1 + \beta_2 \log x_{i1} + \beta_3 e^{x_{i2}}$ are all considered linear models, even though they are not all linear in the explanatory variables; whereas $\xi(x_i; \boldsymbol{\beta}) = \beta_1 + \beta_2 x_i^{\beta_3}$ is genuinely nonlinear. Generally speaking, one would adopt a linear form for the *statistical* model unless it is underlain by some deep theoretical structure, while allowing enough freedom in the specification of regressors to give a satisfactory fit to the sample data. On the other hand, theoretical considerations do sometimes suggest a particular nonlinear form.[1]

Inference for linear models is much more straightforward and provides a foundation for treating the nonlinear case, and so we take up the two cases in that order. Also, we shall see that linear regression affords a simple way to motivate and perform tests for equality of mean effects of categorical or unquantifiable "treatments"—tests that are usually described under the rubric "analysis of variance" (ANOVA).

---

[1]For example, in Epps (1976) a model of stock market transactions derived from primitive assumptions about investors and markets implies the following specific form for the expected number of shares that trade during $(t, t+1]$: $\xi(c_t, p_t; \boldsymbol{\beta}) = \frac{\beta_1}{4\beta_2\beta_3} p_t - \frac{\beta_1}{\beta_2} c_t + \frac{2\beta_1\beta_3}{\beta_2} \frac{c_t^2}{p_t} - \frac{4\beta_1\beta_3^2}{3\beta_2} \frac{c_t^3}{p_t^2}$. Here c_t and p_t are the dollar cost of transacting and the price of the stock at t, and the β's characterize investors' taste for risk and the rate and nature of the information shocks that generate trades.

11.2   Regression with Normal Errors

11.2.1   The Linear Model

11.2.1.1   Point Estimation

Taking $\xi\left(\mathbf{x},\boldsymbol{\beta}\right) = \mathbf{x}\boldsymbol{\beta}$, we consider the model $Y_i = \mathbf{x}_i\boldsymbol{\beta} + U_i$ with $\{U_i\}_{i=1}^n$ i.i.d. as $N\left(0,\sigma^2\right)$, conditional on the explanatory variables. Since we always condition on $\check{\mathbf{x}} = \mathbf{x}$ when explanatory variables $\check{\mathbf{x}}$ are stochastic, and since we also allow those variables *not* to be stochastic, we henceforth omit reference to conditionality except when special emphasis is needed. Thus, when we refer to distributions, moments, etc., it will be understood that these are the conditional versions when that is appropriate.

Adopting matrix notation, we put

$$\mathbf{Y} := \begin{pmatrix} Y_1 \\ Y_2 \\ \vdots \\ Y_n \end{pmatrix}, \mathbf{X} := \begin{pmatrix} \mathbf{x}_1 \\ \mathbf{x}_2 \\ \vdots \\ \mathbf{x}_n \end{pmatrix} = \begin{pmatrix} x_{11} & x_{12} & \cdots & x_{1k} \\ x_{21} & x_{22} & \cdots & x_{2k} \\ \vdots & \vdots & \ddots & \vdots \\ x_{n1} & x_{n2} & \cdots & x_{nk} \end{pmatrix}, \mathbf{U} := \begin{pmatrix} U_1 \\ U_2 \\ \vdots \\ U_n \end{pmatrix}.$$

Thus, $\mathbf{X}$ is an $n \times k$ matrix with k-vector $\mathbf{x}_i$ in row i and scalar x_{ij} in row i and column j. Also, $\boldsymbol{\beta} := \left(\beta_1, \beta_2, ..., \beta_k\right)'$ is a column vector with the same number of elements as there are columns in $\mathbf{X}$. Note that the k elements of $\mathbf{x}$ might well be functions of $m < k$ distinct variables, as $\mathbf{x} = \left(1, x, x^2, e^x\right)$ in the model $Y_i = \beta_1 + \beta_2 x_i + \beta_3 x_i^2 + \beta_4 e^{x_i} + U_i$, but there is no need to distinguish this from the case that $m = k$. In particular, as illustrated here and explained further below, we commonly take the first column of $\mathbf{X}$ to be an n-vector of units, $\mathbf{1} = \left(1, 1, ..., 1\right)'$, thus giving the linear model an intercept term. We do always assume, however, that $n > k$ and that $\mathbf{X}$ has full column rank k. This amounts to a condition for in-sample *identification* of $\boldsymbol{\beta}$; for if the rank were less than k there would exist (infinitely many) vectors $\mathbf{a} = \left(a_1, a_2, ..., a_k\right)' \neq \mathbf{0}$ such that $\mathbf{X}\mathbf{a} = \mathbf{0}$, in which case $\mathbf{Y} = \mathbf{X}\boldsymbol{\beta} + \mathbf{U} \equiv \mathbf{X}(\mathbf{a} + \boldsymbol{\beta}) + \mathbf{U}$. Parameter vector $\boldsymbol{\beta}$ lies in some subspace $\mathbb{B} \subset \Re^k$, and of course we assume that $\sigma^2 \in (0, \infty)$.[2]

Matrix $\mathbf{X}$ is often referred to as the "*design* matrix", because in controlled experiments the choice of regressors and their values *define* the experimental design. For example, in a horticultural experiment to ascertain

---

[2]If $\sigma^2 = 0$ there is no longer a statistical problem, as each of the (up to) $\binom{n}{k}$ sets of k observations for which abbreviated matrix $\mathbf{X}_{(k)}$ is of full rank would give $\boldsymbol{\beta} = \mathbf{X}_{(k)}^{-1}\mathbf{Y}$ uniquely. Finding such determinism in the data likely means that one has written down an *identity* rather than a model.

the effects of moisture and fertilizer on the growth of plants, $\mathbf{x}_i = (x_{i1}, x_{i2})$ would represent the quantities of these factors applied to the ith plant in a sample of n. Choosing the applications such that the columns of $\mathbf{X}$ (i.e., the two explanatory variables) are uncorrelated would make it easier to discern—*identify*—the separate effect of each. For example, if moisture (x_1) was applied in either one or two units and fertilizer (x_2) in either one, two, or three units, an optimal design for a sample of size $n = 6$ would be

$$\mathbf{X}' = \begin{pmatrix} 1\ 1\ 1\ 2\ 2\ 2 \\ 1\ 2\ 3\ 1\ 2\ 3 \end{pmatrix}.$$

Under the normality assumption we have that $\mathbf{U} \sim N\left(\mathbf{0}, \sigma^2 \mathbf{I}_n\right)$, where $\mathbf{I}_n$ is the $n \times n$ identity matrix. Thus, the $\{Y_i\}$ are themselves independent (remember that conditioning on the $\{\mathbf{x}_i\}$ is understood!) with $Y_i \sim N\left(\mathbf{x}_i\boldsymbol{\beta}, \sigma^2\right)$. We can thus write down the log-likelihood function, $\mathcal{L}\left(\boldsymbol{\beta}, \sigma^2\right) := \log L\left(\mathbf{Y}; \boldsymbol{\beta}, \sigma^2\right)$, as

$$\mathcal{L}\left(\boldsymbol{\beta}, \sigma^2\right) \; : \; = -\frac{n}{2} \log\left(2\pi\sigma^2\right) - \frac{1}{2\sigma^2} \sum_{i=1}^{n} (Y_i - \mathbf{x}_i\boldsymbol{\beta})^2$$

$$= -\frac{n}{2} \log\left(2\pi\sigma^2\right) - \frac{1}{2\sigma^2} (\mathbf{Y} - \mathbf{X}\boldsymbol{\beta})' (\mathbf{Y} - \mathbf{X}\boldsymbol{\beta}).$$

Clearly, the $\hat{\boldsymbol{\beta}}_n$ that satisfies $\sup_{\boldsymbol{\beta} \in \mathbb{B}} \mathcal{L}\left(\boldsymbol{\beta}, \sigma^2\right)$ also satisfies

$$\inf_{\boldsymbol{\beta} \in \mathbb{B}} (\mathbf{Y} - \mathbf{X}\boldsymbol{\beta})' (\mathbf{Y} - \mathbf{X}\boldsymbol{\beta}), \tag{11.1}$$

independently of σ^2. (Note that, following our usual convention for m.l. estimators, the subscript "n" pertains to the sample size, not to the *dimension* of $\boldsymbol{\beta}$.) The objective function in (11.1) is quadratic and strictly convex in $\boldsymbol{\beta}$. Thus, if $\mathbb{B}$ is not restricted—not a proper subset of $\mathfrak{R}^k$—then $\hat{\boldsymbol{\beta}}_n$ is the unique solution to

$$\frac{\partial}{\partial \boldsymbol{\beta}} (\mathbf{Y} - \mathbf{X}\boldsymbol{\beta})' (\mathbf{Y} - \mathbf{X}\boldsymbol{\beta}) \bigg|_{\boldsymbol{\beta} = \hat{\boldsymbol{\beta}}_n} = -2\mathbf{X}'\left(\mathbf{Y} - \mathbf{X}\hat{\boldsymbol{\beta}}_n\right) = \mathbf{0} \tag{11.2}$$

and has the explicit form

$$\hat{\boldsymbol{\beta}}_n = (\mathbf{X}'\mathbf{X})^{-1} \mathbf{X}'\mathbf{Y}. \tag{11.3}$$

Note that $k \times k$ matrix $\mathbf{X}'\mathbf{X}$ is nonsingular on the condition that $\mathbf{X}$ have rank k.

The explicit form (11.3) does not apply if $\mathbb{B}$ is subject to inequality restrictions, since the solution to (11.1) might then happen to lie on a boundary. For example, this could occur if one refused to let the data overturn an *a priori* belief that $\partial E\left(Y \mid \mathbf{x}\right)/\partial x_j = \beta_j > 0$ for some regressor x_j. Such would not be an issue if there were just one or more linear *equality* restrictions on $\boldsymbol{\beta}$, as $\mathbf{a}'\boldsymbol{\beta} = 0$ for some constant vector $\mathbf{a}$. In that case $\mathbb{B}$ would be confined to a linear subspace of $\Re^k$, and the columns of $\mathbf{X}$ could be reconfigured to produce an unrestricted model of some dimension $\ell < k$ with explicit solution $\hat{\boldsymbol{\beta}}_n \in \Re^\ell$. If there are *nonlinear* restrictions, then the reduced model becomes nonlinear in parameters and the methods of Section 11.2.2 apply.

Elements of vector $\mathbf{e} := \mathbf{Y} - \mathbf{X}\hat{\boldsymbol{\beta}}_n$ are known as the regression *residuals*, and the n elements of $\hat{\mathbf{Y}} := \mathbf{X}\hat{\boldsymbol{\beta}}_n$ are the *fitted values*—estimators of $\{E\left(Y_i \mid \mathbf{x}_i\right)\}_{i=1}^{n}$. Likelihood equation (11.2) implies that $\mathbf{e}$ is *orthogonal* to each column of $\mathbf{X}$ when $\mathbb{B}$ is unrestricted, and so the residuals are (within the sample) uncorrelated with each explanatory variable and, therefore, with the fitted values themselves. There is a geometric interpretation here that helps build the intuition. Since it is a linear combination of the columns of $\mathbf{X}$, vector $\hat{\mathbf{Y}} = \mathbf{X}\hat{\boldsymbol{\beta}}_n$ clearly lies within the k-dimensional subspace of $\Re^n$ that is spanned by those columns. Writing $\hat{\mathbf{Y}}$ as $\mathbf{X}\left(\mathbf{X}'\mathbf{X}\right)^{-1}\mathbf{X}'\mathbf{Y}$ shows that $n \times n$ matrix $\mathbf{H} := \mathbf{X}\left(\mathbf{X}'\mathbf{X}\right)^{-1}\mathbf{X}'$ *projects* $\mathbf{Y}$ into this subspace. (The symmetric, idempotent matrix $\mathbf{H}$ is sometimes referred to as the "hat matrix", since $\mathbf{HY}$ puts the "hat" on $\mathbf{Y}$.) Writing residual vector $\mathbf{e}$ as $\mathbf{MY}$, where $\mathbf{M} := \mathbf{I}_n - \mathbf{H}$ (another symmetric, idempotent matrix), we see that $\mathbf{M}$ projects into the $(n-k)$-dimensional subspace of $\Re^n$ that is *orthogonal* to that spanned by the columns of $\mathbf{X}$. Note that if $\mathbf{w}$ is any vector in the subspace spanned by $\mathbf{X}$, then $\mathbf{Hw} = \mathbf{w}$. Thus, $\mathbf{H}$ projects $\mathbf{w}$ onto itself, while $\mathbf{Mw} = \mathbf{0}$. It follows, in particular, that

- $\mathbf{HX} = \mathbf{X}$
- $\mathbf{MX} = \mathbf{0}$
- $\mathbf{MH} = \mathbf{HM} = \mathbf{0}$.

Since $\hat{\boldsymbol{\beta}}_n$ minimizes the sum of squared residuals, $\sum_{i=1}^{n} e_i^2 = \mathbf{e}'\mathbf{e}$, our m.l. estimator under normality is also a *least-squares* estimator. It is easy to see that the m.l. estimator of σ^2 is $\hat{\sigma}_n^2 = n^{-1}\mathbf{e}'\mathbf{e}$, the average of the sum of squared residuals. This makes intuitive sense, in that $\mathbf{e}$ is the m.l. estimator of the vector of unobserved errors $\mathbf{U}$, and $\sigma^2 = EU_i^2 = n^{-1}E\mathbf{U}'\mathbf{U}$.

Exercise 11.1. *For the model* $Y_i = \beta_1 + \beta_2 x_i + U_i$ *with* $\boldsymbol{\beta} = (\beta_1, \beta_2)' \in \mathbb{B} = \Re^2$ *and* $\mathbf{x}_i = (1, x_i)$ *show that*

$$\hat{\beta}_{n2} = \frac{\sum_{i=1}^{n}\left(x_i - \bar{x}\right)Y_i}{\sum_{i=1}^{n}\left(x_i - \bar{x}\right)^2} \tag{11.4}$$

and that $\hat{\beta}_{n1} = \bar{Y} - \hat{\beta}_{n2}\bar{x}$.

Example 11.1. When $Y_i = \beta_1 + \beta_2 x_i + U_i$ and β_1 and β_2 are unrestricted, there is a simple solution for $\hat{\beta}_n$, as follows. Put $\alpha := \beta_1 + \beta_2\bar{x}$ and write

$$\sum_{i=1}^{n} (Y_i - \beta_1 - \beta_2 x_i)^2 = \sum_{i=1}^{n} [Y_i - \alpha - \beta_2 (x_i - \bar{x})]^2.$$

The likelihood equations for α and β_2 are

$$\frac{\partial \mathcal{L}(\beta,\sigma^2)}{\partial \alpha} = \frac{1}{\sigma^2} \sum_{i=1}^{n} [Y_i - \alpha - \beta_2 (x_i - \bar{x})]$$

$$= \frac{1}{\sigma^2} \left(\sum_{i=1}^{n} Y_i - n\alpha \right) = 0$$

$$\frac{\partial \mathcal{L}(\beta,\sigma^2)}{\partial \beta_2} = \frac{1}{\sigma^2} \sum_{i=1}^{n} [Y_i - \alpha - \beta_2 (x_i - \bar{x})] (x_i - \bar{x})$$

$$= \frac{1}{\sigma^2} \left[\sum_{i=1}^{n} (x_i - \bar{x}) Y_i - \beta_2 \sum_{i=1}^{n} (x_i - \bar{x})^2 \right] = 0.$$

The second equation gives $\hat{\beta}_{n2}$ as in (11.4), and the first gives $\hat{\alpha}_n = \bar{Y}$ and $\hat{\beta}_{n1} = \hat{\alpha}_n - \hat{\beta}_{n2}\bar{x} = \bar{Y} - \hat{\beta}_{n2}\bar{x}$.

The statistical properties of $\hat{\beta}_n$ depend on the nature of parameter space $\mathbb{B}$. As already stated, if there are inequality restrictions on elements of β, then the solution to $\inf_{\beta \in \mathbb{B}} (\mathbf{Y} - \mathbf{X}\beta)' (\mathbf{Y} - \mathbf{X}\beta)$ may lie on a boundary, in which case (11.2) need not be satisfied. Clearly, under such a restriction $\hat{\beta}_n$ cannot be unbiased, and not much else can be said about its properties in finite samples; therefore, we confine attention to the unrestricted case $\mathbb{B} = \Re^k$. Unbiasedness then follows at once from the explicit solution (11.3), since

$$E\hat{\beta}_n = (\mathbf{X}'\mathbf{X})^{-1} \mathbf{X}'E\mathbf{Y} = (\mathbf{X}'\mathbf{X})^{-1} \mathbf{X}'\mathbf{X}\beta = \beta. \tag{11.5}$$

The covariance matrix of $\hat{\beta}_n$, denoted $\Sigma_{\hat{\beta}}$, is then

$$\begin{aligned}
\Sigma_{\hat{\beta}} : &= E \left(\hat{\beta}_n - \beta \right) \left(\hat{\beta}_n - \beta \right)' \\
&= (\mathbf{X}'\mathbf{X})^{-1} \mathbf{X}' \left(E\mathbf{U}\mathbf{U}' \right) \mathbf{X} (\mathbf{X}'\mathbf{X})^{-1} \\
&= \sigma^2 (\mathbf{X}'\mathbf{X})^{-1} \mathbf{X}'\mathbf{X} (\mathbf{X}'\mathbf{X})^{-1} \\
&= \sigma^2 (\mathbf{X}'\mathbf{X})^{-1}.
\end{aligned} \tag{11.6}$$

Moreover, since (11.3) represents $\hat{\beta}_n$ as a linear function of the jointly normal components of $\mathbf{Y}$, it follows that $\hat{\beta}_n \sim N\left(\beta, \Sigma_{\hat{\beta}}\right)$. While all these properties of the unrestricted estimates apply *conditionally* on $\mathbf{X}$, the fact that $E\hat{\beta}_n = E\left[E\left(\hat{\beta}_n \mid \mathbf{X}\right)\right]$ indicates that $\hat{\beta}_n$ is also unbiased *unconditionally*. On the other hand, if explanatory variables $\check{\mathbf{x}}$ are stochastic, an explicit model for $\mathbf{X} \equiv (\mathbf{x}_1, \mathbf{x}_2, ..., \mathbf{x}_n)'$ is needed to determine the unconditional versions of $\Sigma_{\hat{\beta}}$ and the distribution of $\hat{\beta}_n$.

Exercise 11.2. *Express covariance matrix* $\Sigma_{\hat{\beta}} = \sigma^2 (\mathbf{X}'\mathbf{X})^{-1}$*for the bivariate model* $Y_i = \beta_1 + \beta_2 x_i + U_i$. *Under what condition is* $\mathbf{X}'\mathbf{X}$ *"nearly" singular in this case? Explain intuitively why* $\hat{\beta}_{n1}$ *and* $\hat{\beta}_{n2}$ *would then have high variance in repeated samples with fixed design matrix* $\mathbf{X}$.

Now consider the properties of $\hat{\sigma}_n^2$. As seen above, we can write residual vector $\mathbf{e}$ as $\mathbf{MY}$, where $\mathbf{M} := \mathbf{I}_n - \mathbf{H} := \mathbf{I}_n - \mathbf{X}(\mathbf{X}'\mathbf{X})^{-1}\mathbf{X}'$ is a symmetric, idempotent matrix with $\mathbf{MX} = \mathbf{0}$. Thus, $\mathbf{e} = \mathbf{M}(\mathbf{X}\beta + \mathbf{U}) = \mathbf{MU}$ is jointly normal with mean $\mathbf{0}$ and covariance matrix

$$E\mathbf{ee}' = \mathbf{M}\left(E\mathbf{UU}'\right)\mathbf{M}' = \sigma^2\mathbf{M}.$$

$\mathbf{M}$ not being an identity matrix, the residuals themselves are correlated and do not all have the same variance even when the unobserved errors are i.i.d. Indeed, $n \times n$ matrix $\mathbf{M}$ is not even of full rank. Since it is idempotent, its rank equals its *trace*—the sum of elements of its principal diagonal, and so

$$\text{rank}(\mathbf{M}) = \text{tr}\left[\mathbf{I}_n - \mathbf{X}(\mathbf{X}'\mathbf{X})^{-1}\mathbf{X}'\right]$$
$$= \text{tr}(\mathbf{I}_n) - \text{tr}\left[\mathbf{X}'\mathbf{X}(\mathbf{X}'\mathbf{X})^{-1}\right]$$
$$= \text{tr}(\mathbf{I}_n) - \text{tr}(\mathbf{I}_k) = n - k.$$

Their covariance matrix being not positive definite, the residuals are thus more properly said to have a *singular* normal distribution.

Having worked out the trace of $\mathbf{M}$, it is now easy to find the expectation of inner (scalar) product $\mathbf{e}'\mathbf{e}$. Since the trace of the product of two vectors or matrices is invariant under permutation of the factors, and since a scalar equals its own trace, we have

$$E\mathbf{e}'\mathbf{e} = tr\left(E\mathbf{e}'\mathbf{e}\right) = \text{tr}\left(E\mathbf{e}\mathbf{e}'\right) = \sigma^2\text{tr}(\mathbf{M}) = \sigma^2(n - k).$$

M.l. estimator $\hat{\sigma}_n^2 = n^{-1}\mathbf{e}'\mathbf{e}$ is therefore biased, but a simple rescaling gives at once an unbiased alternative:

$$s_n^2 := (n - k)^{-1}\mathbf{e}'\mathbf{e}.$$

Just as estimating the mean of a univariate sample takes away one degree of freedom in the deviations used to estimate the variance, k degrees of freedom are lost in estimating the components of β—a fact that explains why $E\mathbf{ee}'$ has rank $n-k$. Using s_n^2 as the estimator of σ^2, the corresponding unbiased estimator of covariance matrix $\Sigma_{\hat{\beta}}$ is $\mathbf{S}_{\hat{\beta}} := s_n^2 \left(\mathbf{X}'\mathbf{X}\right)^{-1}$.

Applying the Lehmann-Scheffé and Rao-Blackwell theorems, it is easy to see that $\hat{\beta}_n$ and s_n^2 are jointly complete and minimal-sufficient for β, σ^2 when $\mathbb{B}$ is restricted; hence, they are the unique minimum-variance unbiased statistics.

Exercise 11.3. *Show that $\hat{\beta}_n, s_n^2$ are jointly minimal-sufficient for β, σ^2 if $\mathbf{U} \sim N(\mathbf{0}, \sigma^2 \mathbf{I}_n)$.*

Exercise 11.4. *Suppose the errors $\{U_i\}_{i=1}^n$ in the model $\mathbf{Y} = \mathbf{X}\beta + \mathbf{U}$ are not i.i.d., but rather $\mathbf{U} \sim N\left(\mathbf{0}, \Sigma\right)$ for some matrix Σ that is not proportional to identity matrix $\mathbf{I}_n$. Is the least-squares estimator $\hat{\beta}_n$ still unbiased when β is unrestricted? If so, is it still minimum-variance unbiased? (Hint: Recall Example 7.38 and Exercise 7.26.)*

To complete the basic distribution theory, we can now show, as the first steps, that $\hat{\beta}_n - \beta$ and $\mathbf{e}$ are independent and that $\mathbf{e}'\mathbf{e}/\sigma^2 = (n-k) s_n^2/\sigma^2 \sim \chi^2 (n-k)$. Independence holds since both $\hat{\beta}_n - \beta = (\mathbf{I}_n - \mathbf{M}) \mathbf{U} = \mathbf{H}\mathbf{U}$ and $\mathbf{e} = \mathbf{M}\mathbf{U}$ are jointly normal, and since

$$E\left(\hat{\beta}_n - \beta\right) \mathbf{e}' = E\mathbf{H}\mathbf{U}\mathbf{U}'\mathbf{M}' = \mathbf{H}\left(\sigma^2 \mathbf{I}_n\right)\mathbf{M}' = \sigma^2 \mathbf{H}\mathbf{M}' = \mathbf{0}.$$

Next, $\mathbf{e}'\mathbf{e}/\sigma^2 = \mathbf{U}'\mathbf{M}\mathbf{U}/\sigma^2 \sim \chi^2 (n-k)$ by Theorem 5.2, since $\sigma^{-1}\mathbf{U} \sim N(\mathbf{0}, \mathbf{I}_n)$ and since the rank of idempotent matrix $\mathbf{M}$ equals its trace, $n-k$. From these results and the fact that $\hat{\beta}_n - \beta \sim N\left[\mathbf{0}, \sigma^2 \left(\mathbf{X}'\mathbf{X}\right)^{-1}\right]$ we see that

$$\frac{\left(\hat{\beta}_n - \beta\right)' \mathbf{X}'\mathbf{X} \left(\hat{\beta}_n - \beta\right) / k}{\mathbf{e}'\mathbf{e}/(n-k)} \sim F\left(k, n-k\right),$$

(the F distribution with k and $n-k$ d.f.). Again with $\mathbf{S}_{\hat{\beta}} := s_n^2 \left(\mathbf{X}'\mathbf{X}\right)^{-1}$, the above is

$$\frac{1}{k}\left(\hat{\beta}_n - \beta\right)' \mathbf{S}_{\hat{\beta}}^{-1} \left(\hat{\beta}_n - \beta\right) \sim F\left(k, n-k\right). \tag{11.7}$$

Finally, taking $\hat{\beta}_{nj}$ as the jth element of jointly normal $\hat{\beta}_n$, we have that $\hat{\beta}_{nj} \sim N\left(\beta_j, \sigma_{\hat{\beta}_j}^2\right)$, where variance $\sigma_{\hat{\beta}_j}^2$ equals σ^2 times the jth diagonal element of $(\mathbf{X}'\mathbf{X})^{-1}$. Now let $S^2_{\hat{\beta}_j}$ be the sample counterpart of $\sigma^2_{\hat{\beta}_j}$; i.e., s^2_n times the jth diagonal element of $(\mathbf{X}'\mathbf{X})^{-1}$. Then

$$\frac{\hat{\beta}_j - \beta_j}{S_{\hat{\beta}_j}} = \frac{\left(\hat{\beta}_j - \beta_j\right)/\sigma_{\hat{\beta}_j}}{\sqrt{S^2_{\hat{\beta}_j}/\sigma^2_{\hat{\beta}_j}}} = \frac{\left(\hat{\beta}_j - \beta_j\right)/\sigma_{\hat{\beta}_j}}{\sqrt{s^2_n/\sigma^2}}$$

is the ratio of a standard normal to the square root of an independent chi-squared r.v. divided by its d.f.,[3] so that

$$\frac{\hat{\beta}_j - \beta_j}{S_{\hat{\beta}_j}} \sim t(n - k).$$

The basic facts needed for inference about $\boldsymbol{\beta}$ are now in place, but before applying them to interval estimation and testing we consider two issues pertaining to the specification and assessment of the model.

The linear specification $E\left(Y_i \mid \mathbf{x}_i\right) = \mathbf{x}_i\boldsymbol{\beta} = \beta_1 x_{i1} + \beta_2 x_{i2} + \ldots + \beta_k x_{ik}$ clearly implies that $E\left(Y_i \mid \mathbf{x}_i\right) = 0$ when all the regressors take the value zero. Recognizing the need for greater flexibility, one typically wants to allow an intercept term. The standard convention for this is to include as the *first* regressor in $\mathbf{X}$ a column of units, setting $x_{i1} = 1$ for $i = 1, 2, \ldots, n$. In this way $E\left(Y_i \mid x_{i2} = \cdots = x_{ik} = 0\right) = \beta_1$ becomes the desired intercept.

Under this condition that one column of $\mathbf{X}$ is a units vector, there are two standard measures of the goodness of fit of the regression model, denoted R^2 and $\bar{R}^2$. To explain, let $\mathbf{y}$ be the sample realization of vector $\mathbf{Y}$ and $\hat{\mathbf{y}} := \mathbf{X}\hat{\boldsymbol{\beta}}_n$ be the vector of observed fitted values, with $\mathbf{y} = \hat{\mathbf{y}} + \mathbf{e}$. The sample means of $\mathbf{y}$ and $\hat{\mathbf{y}}$ are the same when $\mathbf{X}$ contains a units column, since $\mathbf{e}$ is orthogonal to each column of $\mathbf{X}$; that is,

$$\bar{y} = n^{-1}\mathbf{1}'\mathbf{y} = n^{-1}\mathbf{1}'\left(\hat{\mathbf{y}} + \mathbf{e}\right) = n^{-1}\mathbf{1}'\hat{\mathbf{y}} =: \bar{\hat{\mathbf{y}}}.$$

In other words, when $\mathbf{X}$ contains a units column, the sample sum (and, therefore, the mean) of residuals $\{e_i\}$ equals zero. The sum of squared deviations of $\{y_i\}$ from $\bar{y}$ is then

$$\sum_{i=1}^n \left(y_i - \bar{y}\right)^2 = \sum_{i=1}^n \left[(y_i - \hat{y}_i) + (\hat{y}_i - \bar{y})\right]^2 = \sum_{i=1}^n \left[e_i + (\hat{y}_i - \bar{y})\right]^2.$$

Squaring, the cross product vanishes because of the orthogonality of $\mathbf{e}$ and $\mathbf{X}$, and we are left with

$$\sum_{i=1}^n \left(y_i - \bar{y}\right)^2 = \sum_{i=1}^n (\hat{y}_i - \bar{y})^2 + \sum_{i=1}^n e_i^2.$$

---

[3] That is, $s_n^2/\sigma^2 = \left[(n-k)\,s_n^2/\sigma^2\right] \div (n-k)$.

On the left we have what is referred to as the *total sum of squares*, which measures the variation in $\{y_i\}$ over the sample, relative to the sample mean. On the right are, first, the variation that is "explained" by the model—that is, by the variation in the $\{x_i\}$—and, second, the part that is "unexplained". The proportion that is explained is

$$R^2 := \frac{\sum_{i=1}^{n}(\hat{y}_i - \bar{y})^2}{\sum_{i=1}^{n}(y_i - \bar{y})^2} = 1 - \frac{\sum_{i=1}^{n} e_i^2}{\sum_{i=1}^{n}(y_i - \bar{y})^2}.$$

This is referred to as the *coefficient of determination*, or simply as the *R-squared*. It is not hard to see that R^2 is in fact the square of the sample correlation between $\{y_i\}$ and $\{\hat{y}_i\}$.

Now, by adding to the model *any* $n - k$ new explanatory variables such that the n columns of $\mathbf{X}$ remain linearly independent, one would obtain a perfect fit to the n observations of Y, reducing the sum of squared residuals to zero and bringing R^2 up to unity. That is, the n columns of $\mathbf{X}$ would span the Euclidean space $\Re^n$ in which $\mathbf{y}$ resides, and therefore any realization $\mathbf{y} = \mathbf{Y}(\omega)$ could be expressed as a linear combination of those columns. This is true regardless of whether the new variables have any conceivable theoretical relation to the dependent variable. To counter tendencies to boost the R-squared by adding superfluous variables, it is common to report the *adjusted R-squared*,

$$\bar{R}^2 := 1 - \frac{\sum_{i=1}^{n} e_i^2 / (n - k)}{\sum_{i=1}^{n}(y_i - \bar{y})^2 / (n - 1)}.$$

Here, dividing by $n - k$ acts as an *ad hoc* way of penalizing overparameterized models.

Exercise 11.5. *Show that R^2 equals the square of the sample correlation between $\{y_i\}$ and $\{\hat{y}_i\}$.*

Exercise 11.6. *With $\mathbf{X}$ as an $n \times k$ design matrix on k regressors ($n \gg k$) you estimate the model $\mathbf{Y} = \mathbf{X}\beta + \mathbf{U}$ and obtain an estimate $\hat{\beta}$. You then consider revising the model to add another regressor z that is, within the sample, uncorrelated with the regressors in $\mathbf{X}$. Thus, with $\mathbf{z}$ as the (column) vector $(z_1, z_2, ..., z_n)'$, we have $\mathbf{X}'\mathbf{z} = \mathbf{0}$. How would the estimates $\hat{\beta}^*$ from the augmented model $\mathbf{Y} = \mathbf{X}\beta + \mathbf{z}\gamma + \mathbf{U}^*$ compare with those from the original model?*

Exercise 11.7. *With $\mathbf{X}$ as an $n \times k$ design matrix on k regressors ($n \gg k$) you estimate the trial model $\mathbf{Y} = \mathbf{X}\beta + \mathbf{U}$ and obtain a residual vector $\mathbf{e}$, from which you calculate the R^2. Finding that this is very close to zero and* *being discouraged by the poor fit, you consider estimating the augmented model* $\mathbf{Y} = \mathbf{X}\boldsymbol{\beta} + \mathbf{e}\gamma + \mathbf{U}^*$. *What would the* R^2 *and* $\bar{R}^2$ *from the augmented model turn out to be? Should you be pleased?*

11.2.1.2   Confidence Intervals and Regions

Section 7.3 described two methods of building confidence intervals and confidence regions (c.i.s. and c.r.s) for parameters, one involving a pivotal quantity constructed directly from an estimator, the other using the likelihood ratio as pivot. Let us consider first how to build confidence sets for components of $\boldsymbol{\beta}$ using pivots based on the distribution of $\hat{\boldsymbol{\beta}}_n$.

To introduce the concepts, decompose model $\mathbf{Y} = \mathbf{X}\boldsymbol{\beta} + \mathbf{U}$ as $\mathbf{Y} = \mathbf{X}_1\boldsymbol{\beta}_1 + \mathbf{X}_2\boldsymbol{\beta}_2 + \mathbf{U}$, where $\mathbf{U} \sim N\left(\mathbf{0}, \sigma^2 \mathbf{I}_n\right)$, $\boldsymbol{\beta}_1 \in \Re^{k_1}$, $\boldsymbol{\beta}_2 \in \Re^{k_2}$, and $\mathbf{X}_1$, $\mathbf{X}_2$, and $\mathbf{X} = (\mathbf{X}_1, \mathbf{X}_2)$ have full column ranks (k_1, k_2, and $k_1 + k_2 = k$, respectively). The goal is to build a confidence set for $\boldsymbol{\beta}_2$, obtaining an expression that applies for all $k_2 \in \{1, 2, ..., k\}$. (When $k_2 = 1$ the set is a c.i. for β_k; otherwise, it is a c.r. for $\beta_{k_1+1}, ..., \beta_k$. Of course, the columns of $\mathbf{X}$ and corresponding elements of $\boldsymbol{\beta}$ can be ordered arbitrarily.) Partitioning the m.l. estimator and its estimated covariance matrix, $\mathbf{S}_{\hat{\boldsymbol{\beta}}} = s_n^2 \left(\mathbf{X}'\mathbf{X}\right)^{-1} = (n-k)^{-1} \mathbf{e}'\mathbf{e} \cdot \boldsymbol{\Sigma}_{\hat{\boldsymbol{\beta}}}/\sigma^2$, as

$$\hat{\boldsymbol{\beta}}_n = \begin{pmatrix} \hat{\boldsymbol{\beta}}_{1n} \\ \hat{\boldsymbol{\beta}}_{2n} \end{pmatrix}, \mathbf{S}_{\hat{\boldsymbol{\beta}}} = \begin{pmatrix} \mathbf{S}_{\hat{\boldsymbol{\beta}}_1} & \mathbf{S}_{\hat{\boldsymbol{\beta}}_1\hat{\boldsymbol{\beta}}_2} \\ \mathbf{S}_{\hat{\boldsymbol{\beta}}_2\hat{\boldsymbol{\beta}}_1} & \mathbf{S}_{\hat{\boldsymbol{\beta}}_2} \end{pmatrix},$$

we have $\left(\hat{\boldsymbol{\beta}}_{2n} - \boldsymbol{\beta}_2\right)' \boldsymbol{\Sigma}_{\hat{\boldsymbol{\beta}}_2}^{-1} \left(\hat{\boldsymbol{\beta}}_{2n} - \boldsymbol{\beta}_2\right) \sim \chi^2\left(k_2\right)$, independent of $\mathbf{e}'\mathbf{e}/\sigma^2 \sim \chi^2\left(n-k\right)$, and

$$\frac{1}{k_2}\left(\hat{\boldsymbol{\beta}}_{2n} - \boldsymbol{\beta}_2\right)' \mathbf{S}_{\hat{\boldsymbol{\beta}}_2}^{-1} \left(\hat{\boldsymbol{\beta}}_{2n} - \boldsymbol{\beta}_2\right) = \frac{\left(\hat{\boldsymbol{\beta}}_{2n} - \boldsymbol{\beta}_2\right)' \boldsymbol{\Sigma}_{\hat{\boldsymbol{\beta}}_2}^{-1} \left(\hat{\boldsymbol{\beta}}_{2n} - \boldsymbol{\beta}_2\right)}{k_2 (n-k)^{-1} \mathbf{e}'\mathbf{e}/\sigma^2}$$

$$\sim F(k_2, n-k).$$

Of course, this corresponds to (11.7) when $k_2 = k$. With $f_{k_2, n-k}^\alpha$ as the upper-α quantile of the F distribution with k_2 and $n-k$ d.f., the inequality

$$\frac{1}{k_2}\left(\hat{\boldsymbol{\beta}}_{2n} - \boldsymbol{\beta}_2\right)' \mathbf{S}_{\hat{\boldsymbol{\beta}}_2}^{-1} \left(\hat{\boldsymbol{\beta}}_{2n} - \boldsymbol{\beta}_2\right) \leq f_{k_2, n-k}^\alpha \tag{11.8}$$

holds with probability $1 - \alpha$, and substituting the realizations for $\hat{\boldsymbol{\beta}}_{2n}$ and $\mathbf{S}_{\hat{\boldsymbol{\beta}}_2}$ gives a $1 - \alpha$ c.r. for $\boldsymbol{\beta}_2$. In the case $k_2 = 1$, $\mathbf{S}_{\hat{\boldsymbol{\beta}}_2}$ is the scalar $S_{\hat{\beta}_k}^2$, and $\left(\hat{\beta}_{kn} - \beta_k\right)^2 / S_{\hat{\beta}_k}^2$ is distributed as the square of a Student's t r.v. with $n - k$ d.f., so a $1 - \alpha$ c.i. is $\left[\hat{\beta}_k - t_{n-k}^{\alpha/2} S_{\hat{\beta}_k}, \hat{\beta}_k + t_{n-k}^{\alpha/2} S_{\hat{\beta}_k}\right]$.

The foregoing results can be generalized as follows. Letting $\mathbf{A}$ be an $\ell \times k$ matrix of constants having rank $\ell \leq k$, we have $\mathbf{A}\left(\hat{\boldsymbol{\beta}}_n - \boldsymbol{\beta}\right) \sim N\left(\mathbf{0}, \mathbf{A}\boldsymbol{\Sigma}_{\hat{\boldsymbol{\beta}}}\mathbf{A}'\right)$, and

$$\left(\hat{\boldsymbol{\beta}}_n - \boldsymbol{\beta}\right)' \mathbf{A}' \left(\mathbf{A}\boldsymbol{\Sigma}_{\hat{\boldsymbol{\beta}}}\mathbf{A}'\right)^{-1} \mathbf{A}\left(\hat{\boldsymbol{\beta}}_n - \boldsymbol{\beta}\right) \sim \chi^2\left(\ell\right)$$

independently of $\mathbf{e}'\mathbf{e}/\sigma^2 \sim \chi^2\left(n-k\right)$. Since $\mathbf{S}_{\hat{\boldsymbol{\beta}}} = \left(n-k\right)^{-1}\mathbf{e}'\mathbf{e}\boldsymbol{\Sigma}_{\hat{\boldsymbol{\beta}}}/\sigma^2$, the estimators $\hat{\boldsymbol{\beta}}_n, \mathbf{S}_{\hat{\boldsymbol{\beta}}}$ satisfy

$$\frac{1}{\ell}\left(\hat{\boldsymbol{\beta}}_n - \boldsymbol{\beta}\right)' \mathbf{A}' \left(\mathbf{A}\mathbf{S}_{\hat{\boldsymbol{\beta}}}\mathbf{A}'\right)^{-1} \mathbf{A}\left(\hat{\boldsymbol{\beta}}_n - \boldsymbol{\beta}\right) \leq f^{\alpha}_{\ell, n-k} \qquad (11.9)$$

with probability $1 - \alpha$. Thus, expressed in terms of *estimates* $\hat{\boldsymbol{\beta}}_n$ and $\mathbf{S}_{\hat{\boldsymbol{\beta}}}$, the left side of (11.9) is a $1-\alpha$ c.r. for $\mathbf{A}\boldsymbol{\beta}$. Taking $\mathbf{A}$ as the $k_2 \times k$ matrix $(\mathbf{0}, \mathbf{I}_{k_2})$ gives (11.8).

Example 11.2. A sample of $n = 20$ for the model $Y_i = \beta_1 + \beta_2 x_{i2} + \beta_3 x_{i3} + \beta_4 x_{i4} + U_i$, $\{U_i\}$ i.i.d. as $N\left(0, \sigma^2\right)$, yields the estimates

$$\hat{\boldsymbol{\beta}}_{20} = \begin{pmatrix} 1.395 \\ 1.851 \\ 3.044 \\ 2.995 \end{pmatrix}$$

and

$$\mathbf{S}_{\hat{\boldsymbol{\beta}}} = \begin{pmatrix} .23683 & -.00842 & -.00547 & -.00652 \\ -.00842 & .00212 & -.00053 & .00017 \\ -.00547 & -.00053 & .00091 & -.00002 \\ -.00652 & .00017 & -.00002 & .00031 \end{pmatrix}.$$

A .90 c.i. for β_2 is

$$\left[\hat{\beta}_{2,20} - t^{.05}_{16}s_{\hat{\beta}_2}, \hat{\beta}_{2,20} + t^{.05}_{16}s_{\hat{\beta}_2}\right] = \left[1.851 - 1.746\sqrt{.00212}, 1.851 + 1.746\sqrt{.00212}\right]$$
$$\doteq \left[1.77, 1.93\right].$$

To build a .95 c.i. for $4\beta_4 - 3\beta_3$, take $\mathbf{A} = \left(0, 0, -3, 4\right)$, so that

$$\mathbf{A}\left(\hat{\boldsymbol{\beta}}_{20} - \boldsymbol{\beta}\right) = -3\left(3.044\right) + 4\left(2.995\right) + \left(4\beta_4 - 3\beta_3\right) = 2.848 + \left(4\beta_4 - 3\beta_3\right)$$

and

$$\mathbf{A}\mathbf{S}_{\hat{\boldsymbol{\beta}}}\mathbf{A}' = \left(-3\right)^2\left(.00091\right) + 2\left(-3\right)\left(4\right)\left(-.00002\right) + \left(4\right)^2\left(.00031\right) \doteq .117^2.$$

The left side of (11.9) is then $\left[2.848 + \left(4\beta_4 - 3\beta_3\right)\right]^2/.117^2$, and the right side is $f^{.05}_{1,16} = \left(t^{.025}_{16}\right)^2 = 2.12^2$, so the .95 c.i. is $2.848 \pm \left(2.12\right)\left(.117\right) \doteq \left[2.60, 3.10\right]$.

Now suppose that one wants a c.i. for $E(Y \mid \mathbf{x}_0) = \mathbf{x}_0\boldsymbol{\beta}$, where $\mathbf{x}_0 :=$ $(x_{01}, x_{02}, ..., x_{0k})$ represents some set of specific values of the explanatory variables. This could be the same as one of the existing rows of $\mathbf{X}$, but it could be a new set of values not previously observed. Taking $\mathbf{A}$ as the $1 \times k$ vector $\mathbf{x}_0$ in (11.9) gives as the $1 - \alpha$ c.i.

$$\left(\mathbf{x}_0\hat{\boldsymbol{\beta}}_n - \mathbf{x}_0\boldsymbol{\beta}\right)' \left(\mathbf{x}_0\mathbf{S}_{\hat{\beta}}\mathbf{x}_0'\right)^{-1} \left(\mathbf{x}_0\hat{\boldsymbol{\beta}}_n - \mathbf{x}_0\boldsymbol{\beta}\right) \leq f_{1,n-k}^{\alpha}.$$

When $\mathbf{X} =: (\mathbf{1}, \mathbf{Z})$ (the first column being units) and $\mathbf{x}_0 =: (1, \mathbf{z}_0)$, the variance of $\hat{E}(Y \mid \mathbf{x}_0) = \mathbf{x}_0\hat{\boldsymbol{\beta}}_n$ is

$$\mathbf{x}_0\boldsymbol{\Sigma}_{\hat{\beta}}\mathbf{x}_0' = \sigma^2\mathbf{x}_0(\mathbf{X}'\mathbf{X})^{-1}\mathbf{x}_0' = \sigma^2 n^{-1}\left[1 + (\mathbf{z}_0 - \bar{\mathbf{z}})\,\mathbf{S}_{\mathbf{Z}}^{-1}(\mathbf{z}_0 - \bar{\mathbf{z}})'\right], \quad (11.10)$$

where $\bar{\mathbf{z}} := n^{-1}\mathbf{1}'\mathbf{Z}$ and $\mathbf{S}_{\mathbf{Z}} := n^{-1}\mathbf{Z}'\mathbf{Z} - \bar{\mathbf{z}}\bar{\mathbf{z}}'$. This shows that estimates of $E(Y \mid \mathbf{x}_0)$ that extrapolate far beyond the center of the data cloud (where $|\mathbf{z}_0 - \bar{\mathbf{z}}| \gg 0$) can be extremely noisy.[4]

Now suppose $Y_0 = \mathbf{x}_0\boldsymbol{\beta} + U_0$ is a prospective new observation corresponding to $\mathbf{x}_0$, with $U_0 \sim N(0, \sigma^2)$ independently of $\{U_i\}_{i=1}^n$. If we use $\hat{Y}_0 = \mathbf{x}_0\hat{\boldsymbol{\beta}}_n$ to predict Y_0, then the prediction error is $Y_0 - \hat{Y}_0 = -\mathbf{x}_0\left(\hat{\boldsymbol{\beta}}_n - \boldsymbol{\beta}\right) + U_0$. This is distributed as normal with mean zero and variance $\sigma^2 + \mathbf{x}_0\boldsymbol{\Sigma}_{\hat{\beta}}\mathbf{x}_0'$. Thus,

$$\frac{Y_0 - \hat{Y}_0}{\sqrt{s_n^2 + \mathbf{x}_0\mathbf{S}_{\hat{\beta}}\mathbf{x}_0'}} \sim t(n - k),$$

and so a $1 - a$ *prediction* interval for Y_0 is $\hat{y}_0 \pm t_{n-k}^{\alpha/2}\sqrt{s_n^2 + \mathbf{x}_0\mathbf{S}_{\hat{\beta}}\mathbf{x}_0'}$.

We saw in Section 7.3 that c.i.s. for the mean of a normal distribution based on pivotal quantity $\sqrt{n}(\bar{X} - \mu)/S$ and on the relative likelihood

---

[4]Verifying (11.10) requires working with partitioned matrices. Putting $\mathbf{Q} := n^{-1}\mathbf{Z}'\mathbf{Z} = \mathbf{S}_{\mathbf{Z}} + \bar{\mathbf{z}}\bar{\mathbf{z}}'$, the formula for the inverse of a partitioned matrix [e.g., Schott (1997, pp. 247-8)] gives

$$(\mathbf{X}'\mathbf{X})^{-1} = n^{-1}\begin{pmatrix} 1 & \bar{\mathbf{z}} \\ \bar{\mathbf{z}}' & \mathbf{Q} \end{pmatrix}^{-1} = n^{-1}\begin{pmatrix} 1 + \bar{\mathbf{z}}\mathbf{S}_{\mathbf{Z}}^{-1}\bar{\mathbf{z}}' & -\bar{\mathbf{z}}\mathbf{S}_{\mathbf{Z}}^{-1} \\ -\mathbf{Q}^{-1}\bar{\mathbf{z}}'\left(1 + \bar{\mathbf{z}}\mathbf{S}_{\mathbf{Z}}^{-1}\bar{\mathbf{z}}'\right) & \mathbf{S}_{\mathbf{Z}}^{-1} \end{pmatrix}.$$

Multiplying and simplifying give

$$\mathbf{x}_0(\mathbf{X}'\mathbf{X})^{-1}\mathbf{x}_0' = n^{-1}\left[1 + \bar{\mathbf{z}}\mathbf{S}_{\mathbf{Z}}^{-1}\bar{\mathbf{z}}' - \mathbf{z}_0\mathbf{Q}^{-1}\left(1 + \bar{\mathbf{z}}'\bar{\mathbf{z}}\mathbf{S}_{\mathbf{Z}}^{-1}\right)\bar{\mathbf{z}}' + (\mathbf{z}_0 - \bar{\mathbf{z}})\,\mathbf{S}_{\mathbf{Z}}^{-1}\bar{\mathbf{z}}_0'\right].$$

Now $\mathbf{Q}^{-1}\left(1 + \bar{\mathbf{z}}'\bar{\mathbf{z}}\mathbf{S}_{\mathbf{Z}}^{-1}\right) = \mathbf{Q}^{-1}(\mathbf{S}_{\mathbf{Z}} + \bar{\mathbf{z}}'\bar{\mathbf{z}})\mathbf{S}_{\mathbf{Z}}^{-1} = \mathbf{S}_{\mathbf{Z}}^{-1}\mathbf{S}_{\mathbf{Z}}^{-1}$, and so the above reduces to

$$n^{-1}\left[1 + \bar{\mathbf{z}}\mathbf{S}_{\mathbf{Z}}^{-1}\bar{\mathbf{z}}' - \mathbf{z}_0\mathbf{S}_{\mathbf{Z}}^{-1}\bar{\mathbf{z}}' + (\mathbf{z}_0 - \bar{\mathbf{z}})\,\mathbf{S}_{\mathbf{Z}}^{-1}\bar{\mathbf{z}}'\right] = n^{-1}\left[1 + (\mathbf{z}_0 - \bar{\mathbf{z}})\,\mathbf{S}_{\mathbf{Z}}^{-1}(\mathbf{z}_0 - \bar{\mathbf{z}})'\right].$$

turned out to be just the same, since the relative likelihood depends on the data and parameters only through this pivot. It happens that the two methods also yield the same exact confidence sets for components of β in the linear model with normal errors. To see this, we will set up the "profile" likelihood ratio $\Lambda\left(\mathbf{Y};\beta_2\right) = L\left(\mathbf{Y};\check{\beta}_{1n},\beta_2,\check{\sigma}_n^2\right)/L\left(\mathbf{Y};\hat{\beta}_n,\hat{\sigma}_n^2\right)$ to find a highest-likelihood confidence set for β_2 in the model $\mathbf{Y} = \mathbf{X}_1\beta_1+\mathbf{X}_2\beta_2+\mathbf{U}$ with $\beta_1 \in \Re^{k_1}$, $\beta_2 \in \Re^{k_2}$ and $k_1 + k_2 = k$. Here the restricted estimator $\check{\beta}_{1n}$ would be found by regressing $\{Y_i - \mathbf{x}_{i2}\beta_2\}$ on $\{\mathbf{x}_{i1}\}$; thus, $\check{\beta}_{1n} = \left(\mathbf{X}_1'\mathbf{X}_1\right)^{-1}\mathbf{X}_1'\left(\mathbf{Y} - \mathbf{X}_2\beta_2\right)$, with residual vector

$$\check{\mathbf{e}} = \mathbf{M}_1\left(\mathbf{Y} - \mathbf{X}_2\beta_2\right) = \mathbf{M}_1\left(\mathbf{X}_1\beta_1 + \mathbf{U}\right) = \mathbf{M}_1\mathbf{U},$$

where symmetric, idempotent matrix $\mathbf{M}_1 = \mathbf{I}_n - \mathbf{X}_1\left(\mathbf{X}_1'\mathbf{X}_1\right)^{-1}\mathbf{X}_1'$ projects into the subspace orthogonal to that spanned by the columns of $\mathbf{X}_1$. With $\mathbf{e} = \mathbf{M}\mathbf{Y} = \mathbf{M}\mathbf{U}$ as the residuals from the full model, it is easy to see that the likelihood ratio (a function of the unknown elements of β_2) is

$$\Lambda\left(\mathbf{Y};\beta_2\right) = \left(\frac{\check{\sigma}_n^2}{\hat{\sigma}_n^2}\right)^{-n/2} = \left(\frac{\check{\mathbf{e}}'\check{\mathbf{e}}}{\mathbf{e}'\mathbf{e}}\right)^{-n/2} = \left[1 + \frac{\check{\mathbf{e}}'\check{\mathbf{e}} - \mathbf{e}'\mathbf{e}}{\mathbf{e}'\mathbf{e}}\right]^{-n/2}.$$

Large values of $\Lambda\left(\mathbf{X};\beta_2\right)$ correspond to small values of

$$\frac{\left(\check{\mathbf{e}}'\check{\mathbf{e}} - \mathbf{e}'\mathbf{e}\right)/k_2}{\mathbf{e}'\mathbf{e}/\left(n - k\right)} = \frac{\mathbf{U}'\left(\mathbf{M}_1 - \mathbf{M}\right)\mathbf{U}/k_2}{\mathbf{U}'\mathbf{M}\mathbf{U}/\left(n - k\right)}. \tag{11.11}$$

The rank of $\mathbf{M}_1 - \mathbf{M}$ is $tr\left(\mathbf{M}_1\right) - tr\left(\mathbf{M}\right) = \left(n - k_1\right) - \left(n - k\right) = k_2$ and

$$\left(\mathbf{M}_1 - \mathbf{M}\right)\mathbf{M} = \left[\mathbf{I}_n - \mathbf{X}_1\left(\mathbf{X}_1'\mathbf{X}_1\right)^{-1}\mathbf{X}_1'\right]\mathbf{M} - \mathbf{M}$$
$$= \left(\mathbf{M} - \mathbf{0}\right) - \mathbf{M} = \mathbf{0},$$

since $\mathbf{M}$ projects into the subspace orthogonal to that spanned by $\mathbf{X} = \left(\mathbf{X}_1,\mathbf{X}_2\right)$. Thus, (11.11) is the ratio of independent chi-squareds relative to their degrees of freedom, and so

$$\frac{\left(\check{\mathbf{e}}'\check{\mathbf{e}} - \mathbf{e}'\mathbf{e}\right)/k_2}{\left(\mathbf{e}'\mathbf{e}\right)/\left(n - k\right)} = \frac{\check{\mathbf{e}}'\check{\mathbf{e}} - \mathbf{e}'\mathbf{e}}{k_2 s_n^2} \tag{11.12}$$

has the same $F\left(k_2, n - k\right)$ distribution as the pivotal quantity in (11.8).

In fact the two quantities in (11.12) and (11.8) are numerically *identical* for all realizations of $\mathbf{Y}$. We can see this quickly—albeit, in a somewhat backhanded way—as follows. From $\check{\mathbf{e}} = \mathbf{M}_1\left(\mathbf{Y} - \beta_2\mathbf{X}_2\right)$ and $\mathbf{Y} = \mathbf{X}_1\hat{\beta}_{1n} + \mathbf{X}_2\hat{\beta}_{2n} + \mathbf{e}$ we have

$$\check{\mathbf{e}} = \mathbf{M}_1\left(\mathbf{X}_1\hat{\beta}_{1n} + \mathbf{X}_2\hat{\beta}_{2n} + \mathbf{e} - \mathbf{X}_2\beta_2\right).$$

But $M_1 X_1 = 0$ and $M_1 e = e$, since e is orthogonal to X_1 (and X_2) and M_1 projects orthogonally to X_1. Thus, $\check{e} = M_1 X_2 \left(\hat{\beta}_{2n} - \beta_2 \right) + e$ and

$$\frac{\check{e}'\check{e} - e'e}{k_2 s_n^2} = \frac{\left(\hat{\beta}_{2n} - \beta_2 \right)' X_2' M_1 X_2 \left(\hat{\beta}_{2n} - \beta_2 \right)}{k_2 s_n^2}.$$

For *each* fixed X of full column rank this has the same distribution as pivotal quantity $k_2^{-1} \left(\hat{\beta}_{2n} - \beta_2 \right)' S_{\hat{\beta}_2}^{-1} \left(\hat{\beta}_{2n} - \beta_2 \right)$, and so the two quantities must be numerically equal for each realization of Y.

Exercise 11.8. *We wish to build a c.i. for β_2 in model $Y_i = \beta_1 + \beta_2 x_i + U_i$. In this case (where $k_1 = k_2 = 1$, $x_{i1} = 1, x_{i2} = x_i$) show algebraically that*

$$\frac{\check{e}'\check{e} - e'e}{s_n^2} = \frac{\left(\hat{\beta}_{2n} - \beta_2 \right)^2}{s_n^2} \sum_{i=1}^n (x_i - \bar{x})^2 = \left(\frac{\hat{\beta}_{2n} - \beta_2}{S_{\hat{\beta}_2}} \right)^2.$$

Example 11.3. The data for the regression in Example 11.2 ($n = 20, k = 4$) yield $s_{20}^2 = .441$ as the estimate of σ^2. We can find a $1 - \alpha = .95$ highest-likelihood confidence interval for error variance σ^2 by appropriately choosing and adjusting the entries in Table 7.4. With $\hat{\sigma}_n^2 = (n-k)s_n^2/n$ and $W_n := \hat{\sigma}_n^2/\sigma^2$ the relative likelihood is $\Lambda(Y; \sigma^2) = W_n^{n/2} e^{-n(W_n-1)/2}$, and the equation $\Lambda(y; \sigma^2) - \lambda = 0$ has roots w_λ, w^λ. Hence,

$$\Lambda(y; \sigma^2) \geq \lambda$$

$$\Longleftrightarrow w_\lambda \leq \frac{\hat{\sigma}_n^2}{\sigma^2} \leq w^\lambda$$

$$\Longleftrightarrow n w_\lambda \leq \frac{(n-k) s_n^2}{\sigma^2} \leq n w^\lambda$$

$$\Longleftrightarrow \frac{n-k}{n} \frac{s_n^2}{w^\lambda} \leq \sigma^2 \leq \frac{n-k}{n} \frac{s_n^2}{w_\lambda}.$$

The tabulated entries for h_α, h^α in row n of Table 7.4 correspond to $(n-1)/(nw^\lambda)$ and $(n-1)/(nw_\lambda)$ where $n-1$ is the d.f. of the unbiased sample variance in a univariate model. In the regression context $(n-k)s_n^2/\sigma^2 \sim \chi^2(n-k)$, so we must take the multipliers from the $n-k+1$ row of the table. These were scaled by $(n-k)/(n-k+1)$, whereas we want a scaling of $(n-k)/n$, so we must rescale by factor $(n-k+1)/n$. Here, with $n-k = 16$ and $1 - \alpha = .95$ we have $h_{.05}^* = h_{.05}(17/20) = .499(17/20)$ and $h^{.05*} = 2.104(17/20)$, so the (exact) .95 c.i. is $\left[.441 h_{.05}^*, .441 h^{.05*} \right] \doteq [.187, .789]$. This is substantially shorter than the interval $\left[\frac{16(.441)}{28.845}, \frac{16(.441)}{6.908}\right] = [.245, 1.021]$ based on pivot $(n-k)\, s_n^2/\sigma^2 = 16 s_{20}^2/\sigma^2$ with residual probability .05 divided evenly between the tails.

Exercise 11.9. *Use the appropriate entries in Table 7.1 with the data from Example 11.3 to find the shortest .95 c.i. for σ^2 based on pivotal quantity $(n-k)\, s_n^2/\sigma^2 = 16 s_{20}^2/\sigma^2$.*

11.2.1.3 Tests of Hypotheses about β

In the model $\mathbf{Y} = \mathbf{X}_1 \boldsymbol{\beta}_1 + \mathbf{X}_2 \boldsymbol{\beta}_2 + \mathbf{U}$ with $\mathbf{U} \sim N\left(\mathbf{0}, \sigma^2 \mathbf{I}_n\right)$, we have seen that

$$k_2^{-1}\left(\hat{\boldsymbol{\beta}}_{2n} - \boldsymbol{\beta}_2\right)' \mathbf{S}_{\hat{\boldsymbol{\beta}}_2}^{-1}\left(\hat{\boldsymbol{\beta}}_{2n} - \boldsymbol{\beta}_2\right) \sim F\left(k_2, n-k\right),$$

where k_2 and $k_1 = k - k_2$ are the numbers of components of $\boldsymbol{\beta}_2$ and $\boldsymbol{\beta}_1$, respectively. Thus, under the null hypothesis $\boldsymbol{\beta}_2 = \boldsymbol{\beta}_2^0$ for some $\boldsymbol{\beta}_2^0 \in \Re^{k_2}$ the region

$$\mathcal{R}_\alpha = \left\{\mathbf{y} : k_2^{-1}\left(\hat{\boldsymbol{\beta}}_{2n} - \boldsymbol{\beta}_2^0\right)' \mathbf{S}_{\hat{\boldsymbol{\beta}}_2}^{-1}\left(\hat{\boldsymbol{\beta}}_{2n} - \boldsymbol{\beta}_2^0\right) \geq f_{k_2, n-k}^\alpha\right\} \tag{11.13}$$

constitutes an exact size-α rejection region for $H_0 : \boldsymbol{\beta}_2 = \boldsymbol{\beta}_2^0$ *vs.* $H_1 : \boldsymbol{\beta}_2 \neq \boldsymbol{\beta}_2^0$. This is essentially a Wald test, although it is based on the unbiased estimator $\mathbf{S}_{\hat{\boldsymbol{\beta}}_2}$ of the covariance matrix rather than on the m.l. estimator $\hat{\boldsymbol{\Sigma}}_{\hat{\boldsymbol{\beta}}_2}$. Alternatively, applying the likelihood-ratio principal, the exact size-α rejection region would be

$$\mathcal{R}_\alpha = \left\{\mathbf{y} : \frac{(\breve{\mathbf{e}}'\breve{\mathbf{e}} - \mathbf{e}'\mathbf{e})/k_2}{(\mathbf{e}'\mathbf{e})/(n-k)} \geq f_{k_2, n-k}^\alpha\right\}, \tag{11.14}$$

where residuals $\breve{\mathbf{e}} = \mathbf{M}_1\left(\mathbf{Y} - \boldsymbol{\beta}_2^0 \mathbf{X}_2\right)$ are obtained by regressing $\{Y_i - \boldsymbol{\beta}_2^0 \mathbf{x}_{i2}\}$ on $\{\mathbf{x}_{i1}\}$. As we have seen, realizations of the two statistics are a.s. equal, and so the tests are precisely the same.

Likewise, if $\mathbf{A}$ and $\mathbf{B}$ are an $\ell \times k$ matrix of rank ℓ and an ℓ-vector of constants, respectively, an α-level test of $H_0 : \mathbf{A}\boldsymbol{\beta} = \mathbf{B}$ *vs.* $H_1 : \mathbf{A}\boldsymbol{\beta} \neq \mathbf{B}$ would be based on

$$\mathcal{R}_\alpha = \left\{\mathbf{y} : \frac{1}{\ell}\left(\mathbf{A}\hat{\boldsymbol{\beta}}_n - \mathbf{B}\right)'\left(\mathbf{A}\mathbf{S}_{\hat{\boldsymbol{\beta}}}\mathbf{A}'\right)^{-1}\left(\mathbf{A}\hat{\boldsymbol{\beta}}_n - \mathbf{B}\right) \geq f_{\ell, n-k}^\alpha\right\}$$

$$\equiv \left\{\mathbf{y} : \frac{(\breve{\mathbf{e}}'\breve{\mathbf{e}} - \mathbf{e}'\mathbf{e})/\ell}{(\mathbf{e}'\mathbf{e})/(n-k)} \geq f_{\ell, n-k}^\alpha\right\}.$$

Here, the restricted residuals are $\left\{\breve{e}_i = Y_i - \mathbf{z}_i\hat{\mathbf{B}}\right\}$, where $\hat{\mathbf{B}}$ is the least-squares estimator in the restricted model $Y_i = \mathbf{z}_i\mathbf{B} + U_i = \mathbf{x}_i\mathbf{A}'\mathbf{A}\boldsymbol{\beta} + U_i$.

Thus, $\check{\mathbf{e}} = \mathbf{M}_\mathbf{Z}\mathbf{Y}$, with $\mathbf{Z} = \mathbf{X}\mathbf{A}'$. Taking $\mathbf{A} = (0, ..., 0, 1)$ and $\mathbf{B} = \beta_k^0$ (a scalar constant) gives the standard t test that rejects $H_0 : \beta_k = \beta_k^0$ at level α in favor of $H_1 : \beta_k \neq \beta_k^0$ when $\left| \hat{\beta}_k - \beta_k^0 \right| / S_{\hat{\beta}_k} \geq t_{n-k}^{\alpha/2}$, with corresponding one-sided rejection regions $\left\{ \mathbf{y} : \left(\hat{\beta}_k - \beta_k^0 \right) / S_{\hat{\beta}_k} \geq t_{n-k}^{\alpha} \right\}$ for $H_1 : \beta_k > \beta_k^0$ and $\left\{ \mathbf{y} : \left(\hat{\beta}_k - \beta_k^0 \right) / S_{\hat{\beta}_k} \leq -t_{n-k}^{\alpha} \right\}$ for $H_1 : \beta_k < \beta_k^0$.

11.2.1.4 *ANOVA in the Regression Context*

The regression framework applies even in cases that one or more explanatory variables are purely qualitative. For example, an experiment might be proposed to see how different types of fertilizer, different soils, or different planting dates affect the growth rates of plants. A test that the mean growth rates under all treatments are the same as that of a standard control group could be carried out as follows. If there are $k - 1$ "treatments", one would create $k - 1$ "dummy" variables $x_2, ..., x_k$, with $x_{ij} = 1$ for plant $i \in \{1, 2, ..., n\}$ if it had treatment j and $x_{ij} = 0$ otherwise. With Y_i as the growth rate to be observed, we would have the linear model

$$Y_i = \beta_1 + \beta_2 x_{i2} + \cdots + \beta_k x_{ik} + U_i = \mathbf{x}_i \boldsymbol{\beta} + U_i.$$

Here β_1 would represent the mean growth of the controls, while β_j for $j > 1$ would represent the *difference* in mean growths of group j and controls. Assuming that the unobserved errors $\{U_i\}$ are i.i.d. as $N\left(0, \sigma^2\right)$, an F test of $H_0 : \beta_2 = \beta_3 = \cdots = \beta_k = 0$ would be carried out via (11.14) with $k_2 = k - 1$. Here the restricted estimator $\check{\beta}_1$ would just be the sample mean $\bar{y}$, and the restricted residuals $\{\check{e}_i\}$ would be the deviations $\{y_i - \bar{y}\}$. This corresponds to the traditional ANOVA one-way layout test of equality of means, familiar from a first course in statistics. Of course, tests and confidence regions for subsets and linear combinations of the $\{\beta_j\}$ could be carried out in the usual way as well, and additional terms could be added to look for interactions among two or more treatments, as $\beta_{k+1} x_{i2} x_{i3}$, and/or *quantifiable* influences.

11.2.2 *Nonlinear Models*

11.2.2.1 *Point Estimation*

We now extend regression theory from the benchmark linear case to the nonlinear model $Y_i = \xi_i(\boldsymbol{\beta}) + U_i$, where $\xi_i(\boldsymbol{\beta}) \equiv \xi\left(\mathbf{x}_i; \boldsymbol{\beta}\right) := E\left(Y_i \mid \mathbf{x}_i\right)$ is a differentiable function of $\boldsymbol{\beta}$ and the $\{U_i\}_{i=1}^n$ are i.i.d. as $N\left(0, \sigma^2\right)$. In vector form the model is $\mathbf{Y} = \boldsymbol{\xi}(\boldsymbol{\beta}) + \mathbf{U}$. Again, we take $\boldsymbol{\beta} \in \mathbb{B} \subset \Re^k$ so that there are k parameters, but the vector of explanatory variables $\mathbf{x}_i \in \Re^m$ can now be of any dimension $m \geq 1$. For example, $k = 1 < m = 2$ when $\xi_i(\boldsymbol{\beta}) = \beta_1 x_{i1} + x_{i2}^{\beta_1}$; $k = m = 2$ when $\xi_i(\boldsymbol{\beta}) = \beta_1 x_{i1} + \beta_2 x_{i2}^{\beta_1}$; and $k = 3 > m = 2$ for $\xi_i(\boldsymbol{\beta}) = \beta_1 x_{i1} + \beta_2 x_{i2}^{\beta_3}$.

In the linear model we had $\{\partial E(Y_i \mid \mathbf{x}_i)/\partial\boldsymbol{\beta}' = \mathbf{x}_i\}_{i=1}^n$ and $\mathbf{X}$ as the $n \times k$ matrix of these. What corresponds to $\mathbf{X}$ in the nonlinear setup is the $n \times k$ matrix of partial derivatives $\boldsymbol{\Xi}(\boldsymbol{\beta}) := \partial\boldsymbol{\xi}(\boldsymbol{\beta})/\partial\boldsymbol{\beta}' \equiv \{\partial\xi_i(\boldsymbol{\beta})/\partial\boldsymbol{\beta}'\}_{i=1}^n$. This matrix arises naturally when we differentiate to try to find the unrestricted maximum of log-likelihood function,

$$\mathcal{L}(\boldsymbol{\beta}, \sigma^2) : = -\frac{n}{2}\log(2\pi\sigma^2) - \frac{1}{2\sigma^2}\sum_{i=1}^n [Y_i - \xi_i(\boldsymbol{\beta})]^2$$

$$= -\frac{n}{2}\log(2\pi\sigma^2) - \frac{1}{2\sigma^2}[\mathbf{Y} - \boldsymbol{\xi}(\boldsymbol{\beta})]'[\mathbf{Y} - \boldsymbol{\xi}(\boldsymbol{\beta})].$$

The likelihood equations for $\boldsymbol{\beta}, \sigma^2$ are then

$$\frac{\partial\mathcal{L}(\boldsymbol{\beta}, \sigma^2)}{\partial\boldsymbol{\beta}} = \frac{1}{\sigma^2}\sum_{i=1}^n [Y_i - \xi_i(\boldsymbol{\beta})]\frac{\partial\xi_i(\boldsymbol{\beta})}{\partial\boldsymbol{\beta}}$$

$$= \frac{1}{\sigma^2}\boldsymbol{\Xi}(\boldsymbol{\beta})'[\mathbf{Y} - \boldsymbol{\xi}(\boldsymbol{\beta})] = \mathbf{0} \tag{11.15}$$

$$\frac{\partial\mathcal{L}(\boldsymbol{\beta}, \sigma^2)}{\partial\sigma^2} = -\frac{n}{2\sigma^2} + \frac{1}{2\sigma^4}\sum_{i=1}^n [Y_i - \xi_i(\boldsymbol{\beta})]^2 = 0. \tag{11.16}$$

Note that equations (11.15) correspond to $\sigma^{-2}\mathbf{X}'(\mathbf{Y} - \mathbf{X}\boldsymbol{\beta}) = \mathbf{0}$ in the linear model. In the unrestricted case $\mathbb{B} = \Re^k$, these are satisfied at the value $\hat{\boldsymbol{\beta}}_n$ at which $[\mathbf{Y} - \boldsymbol{\xi}(\boldsymbol{\beta})]'[\mathbf{Y} - \boldsymbol{\xi}(\boldsymbol{\beta})]$ attains its minimum (and $\mathcal{L}(\boldsymbol{\beta}, \sigma^2)$ attains its maximum). Again, the solution can be found independently of σ^2 (assuming that no functional relation between σ^2 and $\boldsymbol{\beta}$ is specified). Just as $\mathbf{X}$ was required to be of rank k to identify the k elements of $\boldsymbol{\beta}$ in the linear model, $\boldsymbol{\xi}(\boldsymbol{\beta})$ must be such that (11.15) yields a unique minimizing solution $\hat{\boldsymbol{\beta}}_n = \hat{\boldsymbol{\beta}}_n(\mathbf{Y})$ for almost all realizations $\mathbf{Y}(\omega)$. With $\mathbf{e} := \mathbf{Y} - \boldsymbol{\xi}(\hat{\boldsymbol{\beta}}_n)$ as the vector of residuals, (11.15) implies the orthogonality relation $\boldsymbol{\Xi}(\hat{\boldsymbol{\beta}}_n)'\mathbf{e} = \mathbf{0}$, corresponding to $\mathbf{X}'\mathbf{e} = \mathbf{0}$ in the linear model. As in that model, (11.16) gives $\hat{\sigma}_n^2 = n^{-1}\mathbf{e}'\mathbf{e}$. for the m.l. estimator of σ^2.

Unlike the linear case, there will rarely be an *explicit* solution for $\hat{\boldsymbol{\beta}}_n$, but there are several general ways of finding numerical approximations. One is to obtain a numerical solution to the k nonlinear equations $\boldsymbol{\Xi}(\boldsymbol{\beta})'[\mathbf{y} - \boldsymbol{\xi}(\boldsymbol{\beta})] = \mathbf{0}$; however, there may be no such interior solution in the case that $\mathbb{B}$ is restricted. Another approach is a "downhill" search procedure to look for a point at which $[\mathbf{y} - \boldsymbol{\xi}(\boldsymbol{\beta})]'\,[\mathbf{y} - \boldsymbol{\xi}(\boldsymbol{\beta})]$ attains a local minimum, which might well occur on a boundary of $\mathbb{B}$. There are many numerical algorithms for each of these procedures.[5] Also, a wide assortment of statistical and econometric software packages are available that provide black-box answers—although not necessarily *solutions*—without requiring any understanding of the process.

All numerical methods require some initial guess $\boldsymbol{\beta}^{(0)}$ at which the algorithm is to begin. For those who like to do things themselves or to check the black boxes, good results can often be found by applying linear regression iteratively to the pseudo model $\mathbf{Y} = \boldsymbol{\xi}\left(\boldsymbol{\beta}^{(q)}\right) + \boldsymbol{\Xi}\left(\boldsymbol{\beta}^{(q)}\right)\left(\boldsymbol{\beta} - \boldsymbol{\beta}^{(q)}\right) + \boldsymbol{\varepsilon}$ that amounts to a first-order linear approximation in the neighborhood of a point $\boldsymbol{\beta}^{(q)}$. For this, one starts at $q = 0$ with an initial guess $\boldsymbol{\beta}^{(0)}$ and performs a linear regression of $\left\{Y_i - \xi_i\left(\boldsymbol{\beta}^{(0)}\right)\right\}$ on $\left\{\partial\xi_i(\boldsymbol{\beta})/\partial\boldsymbol{\beta}\big|_{\boldsymbol{\beta}=\boldsymbol{\beta}^{(0)}}\right\}$ to estimate $\boldsymbol{\delta}^{(1)} := \boldsymbol{\beta} - \boldsymbol{\beta}^{(0)}$. This gives a new approximation $\boldsymbol{\beta}^{(1)} = \boldsymbol{\beta}^{(0)} + \boldsymbol{\delta}^{(1)}$ that sets up a second regression that produces $\boldsymbol{\delta}^{(2)}$ and $\boldsymbol{\beta}^{(2)}$. The process is continued until the distance $\left|\boldsymbol{\beta}^{(q)} - \boldsymbol{\beta}^{(q-1)}\right|$ between successive estimates falls below some tolerance level—or until computation time (or patience!) hits a limit. Whether such convergence occurs depends on the nature of the regression function and the quality of the initial guess.

Whatever method is used, there is always the danger that the successive estimates will drift to some local extremum or other stationary point that is not the global minimum. In low-dimensional cases—$k = 2$ or, at most, $k = 3$—it may be feasible to plot $[\mathbf{y} - \boldsymbol{\xi}(\boldsymbol{\beta})]'\,[\mathbf{y} - \boldsymbol{\xi}(\boldsymbol{\beta})]$ over a grid of points $\{\boldsymbol{\beta}\}$ and get a direct feel for the nature of the surface. When it can be done, this is the surest way to find the minimum. When it cannot be done, one is obliged to restart the selected algorithm repeatedly at various initial points $\boldsymbol{\beta}^{(0)}$ and choose the solution that yields the smallest sum of squares. Algorithms employing "simulated annealing" methods (Kirkpatrick *et al.* (1983)) do this automatically, but there is no guarantee that any finite number of restarts will produce the desired m.l. estimate.

Given these computational challenges, one should always do some checking of the putative solutions of a nonlinear problem. First, if other than a direct numerical maximization of $\mathcal{L}\left(\boldsymbol{\beta}, \sigma^2\right)$ is employed, the value of the

---

[5] Press *et al.* (2007) (and earlier editions) describe various algorithms and provide computer code in C++, Fortran, and other languages. The IMSL® Fortran Numerical Library has an extensive collection of such routines.

function at $\left(\hat{\boldsymbol{\beta}}_n, \hat{\sigma}_n^2\right)$ should be compared with values at a sampling of nearby points to ensure that the routine has found at least a *local* maximum. Second, no matter how the estimates were obtained, the likelihood estimate $\Xi\left(\hat{\boldsymbol{\beta}}_n\right)$ of $\Xi\left(\boldsymbol{\beta}\right)$ should (in the unrestricted case) be orthogonal to the residuals, $\mathbf{e} = \mathbf{Y} - \boldsymbol{\xi}\left(\hat{\boldsymbol{\beta}}_n\right)$. A simple way to check this is to perform a supplementary *linear* regression of $\mathbf{e}$ on the columns of $\Xi\left(\hat{\boldsymbol{\beta}}_n\right)$, as in the artificial model $\mathbf{e} = \Xi\left(\hat{\boldsymbol{\beta}}_n\right)\boldsymbol{\alpha} + \boldsymbol{\varepsilon}$. The resulting estimates of $\boldsymbol{\alpha}$ should be indistinguishable from zero. If $\xi\left(\boldsymbol{\beta}\right)$ contains a constant term (as in $\xi\left(\boldsymbol{\beta}\right) = \beta_1 + \beta_2 x^{\beta_3}$), then the R^2 coefficient from the artificial regression has the usual interpretation and should also be close to zero.

Since neither $\hat{\boldsymbol{\beta}}_n$ nor any simple transformation of $\hat{\sigma}_n^2$ will be unbiased in the nonlinear model, whatever good properties these have will pertain to asymptotic behavior. Asymptotic theory, even under normality, is complicated by the fact that the $\{Y_i\}$ are usually not (unconditionally) identically distributed in the regression setting. This requires that something be assumed about the progressive behavior of conditional means $\{\xi_i(\boldsymbol{\beta}) \equiv \xi\left(\mathbf{x}_i; \boldsymbol{\beta}\right)\}_{i=1}^n$ as n increases. When explanatory variables $\{\mathbf{x}_i\}$ are (realizations of) time series, when they contain linear trends, or when they are fixed values chosen by researchers in controlled experiments, the asymptotic theory for $\hat{\boldsymbol{\beta}}_n$ and $\hat{\sigma}_n^2$ depends critically on how the data would behave were the sample indefinitely augmented.

To see the problem, consider trying to apply Wald's (1949) proof of consistency of m.l. estimators in the i.i.d. case, as sketched out on page 493. Letting $\boldsymbol{\beta}_0$ represent the true value of the parameters that enter the conditional mean and $\boldsymbol{\beta} \in \mathbb{B}$ be arbitrary, we have for the (rescaled) log likelihood

$$n^{-1}\mathcal{L}\left(\boldsymbol{\beta}, \sigma^2\right) = -\frac{1}{2}\log\left(2\pi\sigma^2\right) - \frac{1}{2\sigma^2 n}\sum_{i=1}^n [\xi_i(\boldsymbol{\beta}_0) - \xi_i(\boldsymbol{\beta}) + U_i]^2.$$

Apart from the factor $-\left(2\sigma^2\right)^{-1}$ the critical component involving $\boldsymbol{\beta}$ is

$$\frac{1}{n}\sum_{i=1}^n [\xi_i(\boldsymbol{\beta}_0) - \xi_i(\boldsymbol{\beta})]^2 + \frac{2}{n}\sum_{i=1}^n [\xi_i(\boldsymbol{\beta}_0) - \xi_i(\boldsymbol{\beta})] U_i + \frac{1}{n}\sum_{i=1}^n U_i^2. \quad (11.17)$$

From Lemma 7.1 we know that $E\mathcal{L}\left(\boldsymbol{\beta}, \sigma^2\right)$ has its maximum at $\boldsymbol{\beta}_0$, but we need to know under what conditions the *sample* mean of the likelihood elements converges to the expectation, because it is this that we maximize to find $\hat{\boldsymbol{\beta}}_n$ at each n. That $\{\xi(\boldsymbol{\beta}) \equiv \xi\left(\mathbf{x}; \boldsymbol{\beta}\right)\}$ be uniformly bounded for all $\boldsymbol{\beta} \in \mathbb{B}$ and all $\mathbf{x}$ in the relevant space would assure that the second term in (11.17) converged in probability to zero, for then its mean would vanish and its variance would approach zero as $n \to \infty$. We would then require as well that $n^{-1} \sum_{i=1}^{n} [\xi_i(\boldsymbol{\beta}_0) - \xi_i(\boldsymbol{\beta})]^2$ converge to some $g(\boldsymbol{\beta}_0, \boldsymbol{\beta})$ with $g(\boldsymbol{\beta}_0, \boldsymbol{\beta}_0) = 0$. However, the uniform boundedness condition is extremely restrictive; for example, even if $\mathbb{B}$ were known to be a compact set, it would rule out the linear model $\xi(\boldsymbol{\beta}) = \mathbf{x}\boldsymbol{\beta}$ unless the $\mathbf{x}$'s were appropriately restricted as well.

There are, however, two common scenarios under which consistency is easily established. First, the data may come from controlled experiments in which the observations $\{\mathbf{x}_i\}$ are predetermined. If the actual sample of size n contains n_0 distinct values of these with $N_0 = n/n_0$ observations of Y from each, then we can consider these n_0 values to be fixed and think of n increasing as $n = Nn_0$, $N \in \{N_0, N_0 + 1, ...\}$. In this way we would merely observe new observations of Y for each member of the fixed set $\{\mathbf{x}_i\}$. In this case the $\{\xi_i(\boldsymbol{\beta})\}$ are obviously uniformly bounded for each $\boldsymbol{\beta} \in \mathbb{B}$, and the strong consistency of $\hat{\boldsymbol{\beta}}_n$ would follow as in Wald's (1949) proof, provided $\boldsymbol{\beta}$ is identified and $\mathbb{B}$ is a compact set. We will give details of such an argument in Section 11.3. The second scenario is that $\{\mathbf{x}_i\}_{i=1}^{n}$ are realizations of i.i.d. random vectors $\{\mathbf{\breve{x}}_i\}_{i=1}^{n}$ independent of $\{U_i\}_{i=1}^{n}$, and that $E\xi(\boldsymbol{\beta})^2 < \infty$. In that case expression (11.17) converges a.s. to $E[\xi(\boldsymbol{\beta}_0) - \xi(\boldsymbol{\beta})]^2 + \sigma^2$, and again Wald's proof would apply given identification and compactness. Of course, appealing to either of these scenarios amounts just to telling a *story* that would justify using asymptotic approximations for our model with *given* n.

Another approach with broader implications is applicable under both (virtual) sampling scenarios if the true value $\boldsymbol{\beta}_0$ is in the interior of an open set contained in parameter space $\mathbb{B}$. Under that condition we can follow Cramér (1946) and attempt to show that likelihood equations (11.15) have a consistent root. This is feasible, since the terms $\{[Y_i - \xi_i(\boldsymbol{\beta})] \partial\xi_i(\boldsymbol{\beta})/\partial\boldsymbol{\beta}\}_{i=1}^{n}$ are again i.i.d. when the $\{\mathbf{x}_i\}_{i=1}^{n}$ are realizations of i.i.d. r.v.s., and they are just ensembles of i.i.d. r.v.s in the experimental setting. Also, regularity conditions (i)-(iii) for m.l. estimation (p. 494) are easily verified if $\boldsymbol{\xi}(\boldsymbol{\beta})$ is at least twice differentiable. In particular, the Fisher information $n\mathcal{I}$ from the full sample is

$$
E \begin{bmatrix} \frac{\partial\mathcal{L}(\boldsymbol{\beta},\sigma^2)}{\partial\boldsymbol{\beta}} \frac{\partial\mathcal{L}(\boldsymbol{\beta},\sigma^2)}{\partial\boldsymbol{\beta}'} & \frac{\partial\mathcal{L}(\boldsymbol{\beta},\sigma^2)}{\partial\boldsymbol{\beta}} \frac{\partial\mathcal{L}(\boldsymbol{\beta},\sigma^2)}{\partial\sigma^2} \\ \frac{\partial\mathcal{L}(\boldsymbol{\beta},\sigma^2)}{\partial\sigma^2} \frac{\partial\mathcal{L}(\boldsymbol{\beta},\sigma^2)}{\partial\boldsymbol{\beta}'} & \frac{\partial^2\mathcal{L}(\boldsymbol{\beta},\sigma^2)}{\partial(\sigma^2)^2} \end{bmatrix} = -E \begin{bmatrix} \frac{\partial^2\mathcal{L}(\boldsymbol{\beta},\sigma^2)}{\partial\boldsymbol{\beta}\partial\boldsymbol{\beta}'} & \frac{\partial^2\mathcal{L}(\boldsymbol{\beta},\sigma^2)}{\partial\boldsymbol{\beta}\partial\sigma^2} \\ \frac{\partial^2\mathcal{L}(\boldsymbol{\beta},\sigma^2)}{\partial\sigma^2\partial\boldsymbol{\beta}'} & \frac{\partial^2\mathcal{L}(\boldsymbol{\beta},\sigma^2)}{\partial\sigma^2\partial\sigma^2} \end{bmatrix},
$$

or

$$n\mathcal{I} = \frac{n}{\sigma^2} \begin{bmatrix} E\frac{\partial\xi(\beta)}{\partial\beta}\frac{\partial\xi(\beta)}{\partial\beta'} & \mathbf{0} \\ \mathbf{0}' & \frac{1}{2\sigma^2} \end{bmatrix}. \tag{11.18}$$

If the specific $\xi(\beta)$ is such that the third derivatives can be bounded by integrable functions within some neighborhood of (β_0, σ_0^2) (condition (iv)), and assuming the finiteness of the relevant moments, then the proof of consistency, asymptotic normality, and asymptotic efficiency of $\left(\hat{\beta}_n, \hat{\sigma}_n^2\right)$ goes through as outlined on page 495. Specifically, we have

$$\sqrt{n}\left(\hat{\beta}_n - \beta_0\right) \rightsquigarrow N\left\{0, \sigma^2\left[E\frac{\partial\xi(\beta)}{\partial\beta}\frac{\partial\xi(\beta)}{\partial\beta'}\right]^{-1}\right\}$$

$$\sqrt{n}\left(\hat{\sigma}_n^2 - \sigma_0^2\right) \rightsquigarrow N\left(0, 2\sigma^4\right).$$

Estimators $\hat{\beta}_n$ and $\hat{\sigma}_n^2$ are independent, just as are the m.l. estimators of mean and variance in the univariate normal model. A consistent estimator of the asymptotic covariance matrix of $\hat{\beta}_n$ is

$$\hat{\boldsymbol{\Sigma}}_n\left(\hat{\beta}_n\right) \equiv \begin{pmatrix} \hat{\sigma}^2_{\hat{\beta}_{1n}} & \hat{\sigma}_{\hat{\beta}_{1n}\hat{\beta}_{2n}} & \cdots & \hat{\sigma}_{\hat{\beta}_{1n}\hat{\beta}_{kn}} \\ \hat{\sigma}_{\hat{\beta}_{1n}\hat{\beta}_{2n}} & \hat{\sigma}^2_{\hat{\beta}_{2n}} & \cdots & \hat{\sigma}_{\hat{\beta}_{2n}\hat{\beta}_{kn}} \\ \vdots & \vdots & \ddots & \vdots \\ \hat{\sigma}_{\hat{\beta}_{1n}\hat{\beta}_{kn}} & \hat{\sigma}_{\hat{\beta}_{2n}\hat{\beta}_{kn}} & \cdots & \hat{\sigma}^2_{\hat{\beta}_{kn}} \end{pmatrix} := \hat{\sigma}_n^2\left[\boldsymbol{\Xi}(\hat{\beta}_n)'\boldsymbol{\Xi}(\hat{\beta}_n)\right]^{-1}. \tag{11.19}$$

Exercise 11.10. *Verify the equalities in (11.18).*

Example 11.4. The model $Y_i = \beta_1 + \beta_2 x_i^{\beta_3} + U_i$ was estimated from the $n = 20$ observations shown in the first two columns of Table 11.1 with the results

$$\hat{\beta}_{20} = (.548, 2.534, .406)'$$

$$\hat{\sigma}_{20}^2 = 0.033$$

$$\hat{\boldsymbol{\Sigma}}_{20}\left(\hat{\beta}_{20}\right) = \hat{\sigma}_{20}^2\left[\boldsymbol{\Xi}(\hat{\beta}_{20})'\boldsymbol{\Xi}(\hat{\beta}_{20})\right]^{-1} \tag{11.20}$$

$$= \begin{pmatrix} .1083 & -.1223 & .0191 \\ -.1223 & .1402 & -.0217 \\ .0191 & -.0217 & .0036 \end{pmatrix}.$$

Fitted values $\{\hat{y}_i\}$ and residuals $\{e_i\}$ are shown in the next two columns. The residuals sum to zero in this case because the model has a constant term, making **e** orthogonal to a units column. The last three columns contain the derivatives $\partial\xi_i(\beta)/\partial\beta_j|_{\beta=\hat{\beta}_n}$. An artificial regression of the residuals on the derivatives yields slope and R^2 coefficients equal to 0.000.

Table 11.1 Observation matrix, fitted values, residuals, and derivatives from NLS regression.

x	y	$\hat{y}$	e	$\frac{\partial \boldsymbol{\xi}(\hat{\boldsymbol{\beta}}_n)}{\partial \beta_1}$	$\frac{\partial \boldsymbol{\xi}(\hat{\boldsymbol{\beta}}_n)}{\partial \beta_2}$	$\frac{\partial \boldsymbol{\xi}(\hat{\boldsymbol{\beta}}_n)}{\partial \beta_3}$
.106	1.455	1.565	-.110	1.000	.401	-2.285
1.000	3.228	3.082	.146	1.000	1.000	.001
4.718	5.445	5.307	.138	1.000	1.878	7.383
1.209	3.525	3.285	.240	1.000	1.080	.520
.309	2.513	2.121	.391	1.000	.621	-1.846
.956	2.681	3.036	-.355	1.000	.982	-.113
.101	1.572	1.545	.028	1.000	.393	-2.289
.107	1.569	1.568	.001	1.000	.403	-2.284
1.535	3.640	3.564	.076	1.000	1.190	1.292
.363	2.147	2.226	-.079	1.000	.662	-1.702
.104	1.797	1.557	.240	1.000	.398	-2.287
.499	2.541	2.459	.083	1.000	.754	-1.328
.445	2.196	2.371	-.176	1.000	.720	-1.477
.105	1.483	1.563	-.080	1.000	.401	-2.285
1.081	3.088	3.163	-0.75	1.000	1.032	.203
.100	1.468	1.543	-.074	1.000	.392	-2.289
4.105	4.862	5.046	-.184	1.000	1.775	6.351
.668	2.644	2.699	-.055	1.000	.849	-.868
.372	2.169	2.245	-.076	1.000	.669	-1.675
.110	1.506	1.583	-.077	1.000	.408	-2.281

11.2.2.2 *Confidence Sets and Tests of Hypotheses*

The inference procedures worked out for the linear model extend to the nonlinear case, with two changes: (i) Procedures for the linear model that were based on the F distribution now must be based on the chi-squared, and (ii) the resulting confidence sets and tests are not exact but approach the nominal coverage probabilities and test sizes only asymptotically. For example, if we partition $\boldsymbol{\beta}$ as $(\boldsymbol{\beta}_1, \boldsymbol{\beta}_2)'$, where $\boldsymbol{\beta}_2 \in \Re^{k_2}$, and then break down $\hat{\boldsymbol{\beta}}_n$ and estimated covariance matrix $\hat{\boldsymbol{\Sigma}}_n \left(\hat{\boldsymbol{\beta}}_n \right)$ in the corresponding way, we have the asymptotic result $\left(\hat{\boldsymbol{\beta}}_{2n} - \boldsymbol{\beta}_2 \right)' \hat{\boldsymbol{\Sigma}}_{\hat{\boldsymbol{\beta}}_{2n}}^{-1} \left(\hat{\boldsymbol{\beta}}_{2n} - \boldsymbol{\beta}_2 \right) \rightsquigarrow \chi^2 (k_2)$. Accordingly, the set

$$\left\{ \boldsymbol{\beta}_2 : \left(\hat{\boldsymbol{\beta}}_{2n} - \boldsymbol{\beta}_2 \right)' \hat{\boldsymbol{\Sigma}}_{\hat{\boldsymbol{\beta}}_{2n}}^{-1} \left(\hat{\boldsymbol{\beta}}_{2n} - \boldsymbol{\beta}_2 \right) \leq c_{1-\alpha} \right\}$$

constitutes an approximate $1 - \alpha$ confidence region (c.r.) for $\boldsymbol{\beta}_2$ in large samples, where $c_{1-\alpha}$ is the $1 - \alpha$ quantile of $\chi^2 (k_2)$. Also, with $\hat{\sigma}^2_{\hat{\beta}_{jn}}$ as the estimated variance of $\hat{\beta}_{jn}$ (the jth member of $\hat{\boldsymbol{\beta}}_n$), an approximate $1 - \alpha$

c.i. for the single parameter β_j is

$$\left[\hat{\beta}_{jn} - z^{\alpha/2}\hat{\sigma}_{\hat{\beta}_{jn}}, \hat{\beta}_{jn} + z^{\alpha/2}\hat{\sigma}_{\hat{\beta}_{jn}}\right].$$

The $1 - \alpha$ confidence regions and intervals enclose the parameter values that constitute acceptable hypotheses at the α level. Thus, we would reject $H_0 : \beta_2 = \beta_2^0$ at level α in favor of $H_1 : \beta_2 \neq \beta_2^0$ if sample point $\mathbf{y}$ falls in

$$\mathcal{R}_\alpha = \left\{\mathbf{y} : \left(\hat{\beta}_{2n} - \beta_2^0\right)' \hat{\mathbf{\Sigma}}_{\hat{\beta}_{2n}}^{-1} \left(\hat{\beta}_{2n} - \beta_2^0\right) \geq c_{1-\alpha}\right\}.$$

Likewise, $H_0 : \beta_j = \beta_j^0$ would be rejected in favor of $H_1 : \beta_j \neq \beta_j^0$ if $\left|\hat{\beta}_j - \beta_j^0\right|/\hat{\sigma}_{\hat{\beta}_{jn}} \geq z^{\alpha/2}$ or, equivalently, if $\left(\hat{\beta}_j - \beta_j^0\right)^2/\hat{\sigma}_{\hat{\beta}_{jn}}^2 \geq c^\alpha$, where c^α is the upper-α quantile of $\chi^2(1)$. For a one-sided test of $H_0 : \beta_j \leq \beta_j^0$ vs. $H_1 : \beta_j > \beta_j^0$ the rejection region would be $\left\{\mathbf{y} : \left(\hat{\beta}_j - \beta_j^0\right)/\hat{\sigma}_{\hat{\beta}_{jn}} \geq z^\alpha\right\}$. These, of course, are Wald tests.

Exercise 11.11. *Use the estimates in Example 11.4 to test at level $\alpha = .10$ (i) $H_0 : (\beta_1, \beta_2) = (1.0, 2.0)$ vs. $H_0 : (\beta_1, \beta_2) \neq (1.0, 2.0)$ and (ii) $H_0 : \beta_3 \geq .5$ vs. $H_1 : \beta_3 < .5$.*

The Wald procedure offers the most direct way to test more general *linear* restrictions of the form $H_0 : \mathbf{A}\boldsymbol{\beta} = \mathbf{B}$, where $\mathbf{A}, \mathbf{B}$ are an $\ell \times k$ matrix of rank $\ell \leq k$ and an ℓ-vector of constants, respectively. The procedure corresponds again to that in the linear model; that is, H_0 is rejected at level α if

$$\left(\mathbf{A}\hat{\boldsymbol{\beta}}_n - \mathbf{B}\right)' \left(\mathbf{A}\hat{\mathbf{\Sigma}}_{\hat{\boldsymbol{\beta}}_n}\mathbf{A}'\right)^{-1} \left(\mathbf{A}\hat{\boldsymbol{\beta}}_n - \mathbf{B}\right) \geq c_{1-\alpha},$$

where $c_{1-\alpha}$ is the $1-\alpha$ quantile of $\chi^2(\ell)$. For *nonlinear* restrictions $\mathbf{g}(\boldsymbol{\beta}) = \boldsymbol{\gamma}$, where $\mathbf{g} : \Re^k \to \Re^\ell$, one would fall back on either the likelihood-ratio test or the Lagrange multiplier test.

11.2.3   *Testing Normality*

Our discussion of linear and nonlinear regression has thus far been limited to models in which the errors $\{U_i\}_{i=1}^n$ are i.i.d. as $N(0, \sigma^2)$. Testing this crucial assumption is obviously complicated by the fact that the errors are not directly observable. What we can observe are the regression residuals $\{e_i\}_{i=1}^n$, but these do not preserve all the properties of the disturbances. In the linear model, with residual and error vectors related linearly as $\mathbf{e} = \mathbf{MY} = \mathbf{MU}$, the residuals are obviously unbiased estimators of the errors, in the sense that $E\left(\mathbf{e} - \mathbf{U}\right) = \mathbf{0}$. In fact, they are "best" linear unbiased, in the sense that for any other linear unbiased estimator $\mathbf{e}^*$ of $\mathbf{U}$ the covariance matrix of $\mathbf{e}^* - \mathbf{U}$ exceeds that of $\mathbf{e} - \mathbf{U}$ by a positive semi-definite matrix. Moreover, the joint (singular) normality of $\mathbf{e}$ does follow from that of $\mathbf{U}$. The difficulty is that the covariance matrix of $\mathbf{e}$ is not of the "scalar" form $\sigma^2 \mathbf{I}_n$ that is supposed to pertain to $\mathbf{U}$; indeed $E\mathbf{ee}'$ is not even of full rank. Thus, the standard tests for univariate normality from i.i.d. observations are simply invalid. Likewise, the $\{e_i\}_{i=1}^n$ do not of themselves support exact tests of the basic assumption that the $\{U_i\}$ are i.i.d.[6]

There is, however, a way to construct *modified* residuals that support valid, exact tests in the linear model $\mathbf{Y} = \mathbf{X}\beta + \mathbf{U}$. Theil (1971, Ch. 5) has shown how to replace $\mathbf{e} = \mathbf{MY} \in \Re^n$ with a new vector $\hat{\mathbf{e}} \in \Re^{n-k}$ that is multivariate normal with a scalar covariance matrix (when this is true of $\mathbf{U}$). This requires arranging and partitioning observation matrix $(\mathbf{Y}, \mathbf{X})$ into components with k and $n - k$ observations in such a way that the $k \times k$ submatrix of $\mathbf{X}$ has rank k. The modified residuals are then constructed as $\hat{\mathbf{e}} = \mathbf{CY} = \mathbf{Ce}$, where $\mathbf{C}$ is a certain $(n - k) \times n$ matrix whose rows are orthogonal to $\mathbf{X}$, so that $\mathbf{HC}' = \mathbf{X}\left(\mathbf{X}'\mathbf{X}\right)^{-1}\mathbf{X}'\mathbf{C}' = \mathbf{0}$ and $\mathbf{MC}' = \left(\mathbf{I}_n - \mathbf{H}\right)\mathbf{C}' = \mathbf{C}'$. Specifically, matrix $\mathbf{C}$ comprises latent vectors corresponding to the $n - k$ unit roots of $\mathbf{M}$, these being selected so that the covariance matrix of $\hat{\mathbf{e}} - \mathbf{U}_{(n-k)} := (\hat{e}_1 - U_1, \hat{e}_2 - U_2, ..., \hat{e}_{n-k} - U_{n-k})'$ is a minimum. Given the particular partitioning of observations, the resulting "BLUS" residuals are then "Best Linear Unbiased with Scalar covariance matrix". With these modified residuals any of the conventional tests for normality discussed in Section 9.2 would be appropriate. The obvious dis-advantage of the procedure is that one must choose arbitrarily some par-ticular one of the (up to) $\binom{n}{k}$ ways of partitioning the observations. In any case the procedure applies only to linear models.[7]

---

[6] To be sure, tests of the i.i.d. assumption do exist. The well-known Goldfeld–Quandt (1965) test of equal variances uses independent sets of residuals from separate regres-sions with subsets of the n observations. This and other procedures in the econometrics literature have substantial power only against restricted classes of alternatives to H_0. Likewise, in testing for independence the need to narrow the possible alternatives ordi-narily limits the procedures to time-series models. The venerable Dubwin–Watson test (1950, 1951) for first-order *auto*correlation of the errors (i.e., $EU_i U_{i+1} \neq 0$) does use the residuals from a single regression, but the theory of the test puts only an upper bound on type-I error.

[7] The development of BLUS residuals was undeniably a *tour de force* and a significant contribution to the theory of linear models. Still, as stressed in Section 8.9, users of statistical results should be wary of findings based on procedures that can be executed

The only alternatives to such exact tests in linear models—and the only options at all for nonlinear models—are asymptotic procedures and simulation. Under conditions such that the estimators $\hat{\boldsymbol{\beta}}_n$ are consistent when $H_0 : \mathbf{U} \sim N\left(\mathbf{0}, \sigma^2 \mathbf{I}_n\right)$ is true, the residuals $\mathbf{e} = \mathbf{y} - \hat{\mathbf{y}}$ are also consistent estimators of the errors, and this is true in linear and nonlinear models alike. Thus, a conventional i.i.d. test for univariate normality should at least be reliable in very large samples. Even in small samples the normality test that is applied may not be highly sensitive to moderate departures from the maintained hypothesis that the data are i.i.d. Either to gauge the test's sensitivity or to improve its accuracy, one could carry out a parametric bootstrapping procedure to infer a critical point for the test statistic that is appropriate for the model and data at hand. For this purpose, one would calculate K replicates $\left\{T\left(\breve{\mathbf{e}}^{(j)}\right)\right\}_{j=1}^{K}$ of desired test statistic T from n-vectors of *pseudo-residuals* $\left\{\breve{\mathbf{e}}^{(j)}\right\}_{j=1}^{K}$. Each such set $\breve{\mathbf{e}}^{(j)}$ would be obtained by estimating the regression model with a set of pseudo-*observations* $\breve{\mathbf{y}}^{(j)}$, constructed as $\left\{\breve{y}_i^{(j)} = \xi\left(\mathbf{x}_i; \hat{\boldsymbol{\beta}}_n\right) + \hat{\sigma}_n z_i^{(j)}\right\}_{i=1}^{n}$. Here, $\hat{\boldsymbol{\beta}}_n, \hat{\sigma}_n$ are the m.l. estimates from original sample $\{y_i, \mathbf{x}_i\}_{i=1}^{n}$, and $\left\{z_i^{(j)}\right\}_{i=1, j=1}^{n, K}$ are pseudo-random standard normals. For an α-level test that rejects for large values of test statistic $T\left(\mathbf{e}\right)$, one would choose as critical value the numerical value of the replicate whose rank among the $\left\{T\left(\breve{\mathbf{e}}^{(j)}\right)\right\}$ is the smallest integer greater than or equal to $K(1 - \alpha)$. An approximate P value for the test in the real sample would be based on the rank of $T\left(\mathbf{e}\right)$ among the K replicates.

What is one to do if the normality test rejects the null? Of course, at a minimum such an occurrence would call into question the validity of the normality-based confidence intervals and tests on β, particularly in small samples; however, there is a more serious possibility. While it is certainly possible that the errors of an otherwise correctly specified model are highly nonnormal, a test outcome that is *grossly* inconsistent with normality may well signal a seriously incorrect specification of conditional mean $E\left(Y \mid \mathbf{x}\right)$. The error could be in the functional form, in the choice of explanatory variables, or in both. A severe rejection might also signal an

---

in so many ways. For example, if asked to demonstrate the appropriateness of tests based on normal errors, an author or an applicant for a grant could report just the most favorable of the (up to) $\binom{n}{k}$ results based on BLUS residuals. Being mindful of such potential abuses, the careful researcher should take pains to describe fully how such procedures were applied. One who is *really* careful might ask a disinterested colleague to supply a specific protocol for the procedure and to certify—*in writing*—that it was carried out.

error in the data themselves, as through improper recording. As an illustration, consider the data in Example 11.4, which were actually generated as $Y_i = \beta_1 + \beta_2 x_i^{\beta_3} + U_i$ with $\beta_1 = 1.0$, $\beta_2 = 2.0$, $\beta_3 = 0.5$, and $\{U_i\}$ i.i.d. as (pseudo-random) $N\left(0, .2^2\right)$. Applied to the residuals in column four of Table 11.1, the i.c.f. test for normality (Section 9.2.5) yields test statistic $I_{20} \doteq 0.025$, which is well below the upper-.10 quantile $I_{20}^{.10} = .072$. Despite the small sample, the test does at least give no false indication that the data are inconsistent with H_0. By contrast, when applied to the residuals $\{e_i\}$ of the incorrectly specified *linear* model $Y_i = \beta_1 + \beta_2 x_i + U_i$, the test yields $I_{20} \doteq 1.05$, which far exceeds upper-.05 quantile $I_{20}^{.05} = .096$. In low-dimensional models, such as this one, a plot of residuals *vs.* explanatory variables may show the nature of the problem. For example, Fig. 11.5 indicates a pronounced nonlinear relation. (Notice the points corresponding to $x = 4.105$ and $x = 4.718$.)

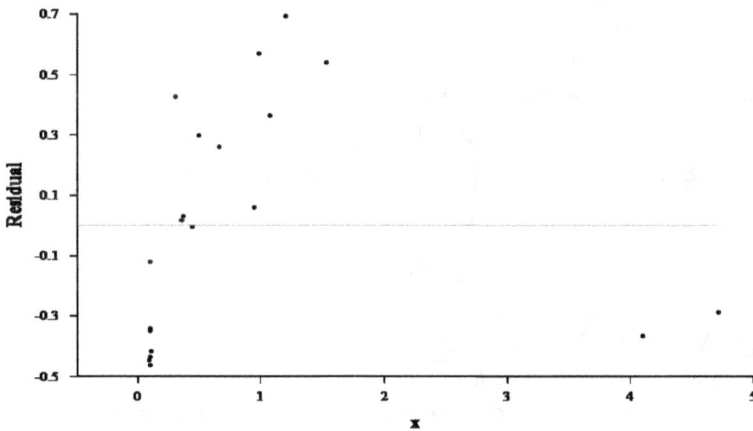

Fig. 11.5 Residuals $\{e_i\}$ from incorrectly specified linear model *vs.* explanatory observations $\{x_i\}$.

11.3 Regression without Normality

In principle, likelihood methods could be applied to the regression model when errors $\{U_i\}_{i=1}^n$ have any specified joint distribution. However, the benchmark case of normal errors that we have thus far considered is the one of greatest interest, and in any case it serves to illustrate the general procedure. We now consider inference for linear and nonlinear models when nothing is assumed about the errors except that they are i.i.d. with zero mean and finite, positive variance σ^2. Since nothing is really ever *known* about the errors, it is of particular interest to know what happens under these far weaker conditions.

11.3.1   *The Linear Model*

11.3.1.1   *Point Estimation*

Since likelihood methods are now unavailable, let us try an approach to estimation based on the method of moments (m.o.m.). From the form of the model as $\mathbf{Y} = \mathbf{X}\boldsymbol{\beta} + \mathbf{U}$, it follows that

$$E\mathbf{X}'\mathbf{Y} = \mathbf{X}'\mathbf{X}\boldsymbol{\beta} + \mathbf{X}'E\mathbf{U} = \mathbf{X}'\mathbf{X}\boldsymbol{\beta},$$

where conditioning is again understood if the explanatory variables are stochastic. Assuming that $\mathbf{X}$ has full column rank k, this implies that $\boldsymbol{\beta} = (\mathbf{X}'\mathbf{X})^{-1} E\mathbf{X}'\mathbf{Y}$. The corresponding expression in terms of sample averages gives as the m.o.m. estimator

$$\breve{\boldsymbol{\beta}}_n = \left(n^{-1}\mathbf{X}'\mathbf{X}\right)^{-1} \begin{pmatrix} n^{-1}\sum_{i=1}^n x_{i1}Y_i \\ \vdots \\ n^{-1}\sum_{i=1}^n x_{ik}Y_i \end{pmatrix} = (\mathbf{X}'\mathbf{X})^{-1} \mathbf{X}'\mathbf{Y}.$$

Of course, this is identical in form to the m.l. estimator $\hat{\boldsymbol{\beta}}_n$ in the linear model with normal errors. Like $\hat{\boldsymbol{\beta}}_n$, it is also the solution to $\inf_{\boldsymbol{\beta}\in\mathbb{B}} (\mathbf{Y} - \mathbf{X}\boldsymbol{\beta})' (\mathbf{Y} - \mathbf{X}\boldsymbol{\beta})$ when allowed space $\mathbb{B}$ is unrestricted, and so $\breve{\boldsymbol{\beta}}_n$ is again in that case a *least-squares* estimator. With $\mathbf{e} := \mathbf{Y} - \mathbf{X}\breve{\boldsymbol{\beta}}_n$ again representing the residual vector, the corresponding m.o.m. estimator for σ^2 is $\tilde{\sigma}_n^2 := n^{-1}\mathbf{e}'\mathbf{e}$; however, we still choose to work with the rescaled version $s_n^2 = \mathbf{e}'\mathbf{e}/(n-k)$.

What are the properties of $\breve{\boldsymbol{\beta}}_n$ and s_n^2? That $\breve{\boldsymbol{\beta}}_n$ is unbiased when $\mathbb{B}$ is unrestricted is seen immediately, for

$$E\breve{\boldsymbol{\beta}}_n = E (\mathbf{X}'\mathbf{X})^{-1} \mathbf{X}' (\mathbf{X}\boldsymbol{\beta} + \mathbf{U}) = \boldsymbol{\beta} + (\mathbf{X}'\mathbf{X})^{-1} \mathbf{X}'E\mathbf{U} = \boldsymbol{\beta}.$$

Also, covariance matrix $\boldsymbol{\Sigma}_{\breve{\boldsymbol{\beta}}_n} := E\left(\breve{\boldsymbol{\beta}}_n - \boldsymbol{\beta}\right)\left(\breve{\boldsymbol{\beta}}_n - \boldsymbol{\beta}\right)'$ is still

$$\boldsymbol{\Sigma}_{\breve{\boldsymbol{\beta}}_n} = (\mathbf{X}'\mathbf{X})^{-1} \mathbf{X}' (E\mathbf{U}\mathbf{U}') \mathbf{X} (\mathbf{X}'\mathbf{X})^{-1}$$
$$= \sigma^2 (\mathbf{X}'\mathbf{X})^{-1}.$$

Finally, as when the errors are normal, we have $Ee'e = (n - k)\sigma^2$, and so $s_n^2 = e'e/(n - k)$ remains unbiased for σ^2.

Of course, lacking specific knowledge of the distribution of the $\{U_i\}$, we cannot expect these estimators to be fully efficient, yet there is a strong result nonetheless. Clearly, $\breve{\beta}_n$ belongs to the class of *linear* estimators— linear in the sense that each element of $\breve{\beta}_n$ is a linear function of observations $\{Y_i\}_{i=1}^n$. Let $\mathbf{CY}$ be a representative member of the class of such linear estimators, where $\mathbf{C}$ is any $k \times n$ matrix of observables having rank k, and put $\mathbf{C} =: (\mathbf{X'X})^{-1}\mathbf{X'} + \mathbf{D}$, implicitly defining $k \times n$ matrix $\mathbf{D}$. If we consider now only those matrices $\mathbf{C}$ such that $E\mathbf{CY} = \beta$ for all β in a k-dimensional space $\mathbb{B}$, it is clear that we must have $E\mathbf{DY} = \mathbf{DX}\beta + \mathbf{DEU} = \mathbf{DX}\beta = 0$ for all such β, and this can hold only if $\mathbf{DX} = 0$. Let us now consider the covariance matrix of members of this linear unbiased subclass. We have

$$\mathbf{CY} - \beta = \left[(\mathbf{X'X})^{-1}\mathbf{X'} + \mathbf{D}\right](\mathbf{X}\beta + \mathbf{U}) - \beta = \left[(\mathbf{X'X})^{-1}\mathbf{X'} + \mathbf{D}\right]\mathbf{U},$$

so that

$$E(\mathbf{CY} - \beta)(\mathbf{CY} - \beta)' = \sigma^2\left[(\mathbf{X'X})^{-1} + 2\mathbf{DX}(\mathbf{X'X})^{-1} + \mathbf{DD'}\right]$$

$$= \sigma^2\left[(\mathbf{X'X})^{-1} + \mathbf{DD'}\right].$$

This differs from the covariance matrix of $\breve{\beta}_n$ by the positive semi-definite matrix $\mathbf{DD'}$, which tells us that $\breve{\beta}_n$ is the efficient estimator within the linear unbiased class—and so is the "best" linear unbiased estimator. We have already encountered this result in a more general form in Exercise 7.27. It is the conclusion of what is referred to as the *Gauss–Markov theorem*.

Exercise 11.12. *Devise an alternative proof of Gauss–Markov, as follows. Putting* $\mathbf{B} := \mathbf{CY}$, *where* $\mathbf{C}$ *is a* $k \times n$ *matrix of rank* k, *show that*
  (i) Estimator $\mathbf{B}$ *is unbiased if and only if* $\mathbf{CX} = \mathbf{I}_k$.
  (ii) For such $\mathbf{C}$ *the covariance matrix of* $\mathbf{B}$ *is* $\Sigma_{\mathbf{B}} = \sigma^2\mathbf{CC'}$.
  (iii) For any $\mathbf{a} \in \Re^k$ *with* $\mathbf{a'a} > 0$, $V(\mathbf{a'B})$ *attains its minimum value subject to* $\mathbf{CX} = \mathbf{I}_k$ *when* $\mathbf{C} = (\mathbf{X'X})^{-1}\mathbf{X'}$.

We turn now to the large-sample properties of $\breve{\beta}_n$ and s_n^2. Again, the asymptotic behaviors depend on our conception of how explanatory variables $\mathbf{x}$ would behave if the sample were augmented. If the original sample of n arises from an experiment with n_0 distinct predetermined values of $\mathbf{x}$ and $N_0 = n/n_0$ observations of Y for each, then, as argued previously, we may think of expanding the sample as $n = Nn_0$, $N \in \{N_0 + 1, N_0 + 2, ...\}$. In this case the $n_0 \times k$ matrix $\mathbf{X}_0$ of distinct observations $\{\mathbf{x}_i\}_{i=1}^{n_0}$ would just be replicated, and we would simply observe additional observations $\{Y_i\}$ resulting from new realizations of independent errors $\{U_i\}$. It is then clear that $(Nn_0)^{-1}\mathbf{X}'\mathbf{X} = n_0^{-1}\mathbf{X}_0'\mathbf{X}_0 =: \boldsymbol{\Sigma}_{\mathbf{x}}$, say, would be for each N a finite, positive-definite matrix of rank k, and hence that $\lim_{n\to\infty} n^{-1}\mathbf{X}'\mathbf{X} = \boldsymbol{\Sigma}_{\mathbf{x}}$ would have the same properties. A common alternative setting is one in which the $\{\mathbf{x}_i\}$ are realizations of i.i.d. random (row) vectors $\{\breve{\mathbf{x}}_i\}$, independent of $\{U_i\}$ and such that $n^{-1}\sum_{i=1}^n \breve{\mathbf{x}}_i\breve{\mathbf{x}}_i'$ converges a.s. to a finite, positive-definite matrix $\boldsymbol{\Sigma}_{\mathbf{x}} := E\breve{\mathbf{x}}\breve{\mathbf{x}}'$ of rank k. Under such conditions it is straightforward to work out the limiting properties of $\breve{\boldsymbol{\beta}}_n$ and s_n^2, as follows.

The weak consistency of $\breve{\boldsymbol{\beta}}_n$ follows at once from the facts that $E\breve{\boldsymbol{\beta}}_n = \boldsymbol{\beta}$ and that $V\breve{\boldsymbol{\beta}}_n = \boldsymbol{\Sigma}_{\breve{\boldsymbol{\beta}}_n} = \sigma^2(\mathbf{X}'\mathbf{X})^{-1}$, for then

$$\lim_{n\to\infty} V\breve{\boldsymbol{\beta}}_n = \lim_{n\to\infty} \left(n^{-1}\sigma^2\right)\left(n^{-1}\mathbf{X}'\mathbf{X}\right)^{-1} = 0 \cdot \boldsymbol{\Sigma}_{\mathbf{x}}^{-1}.$$

For $s_n^2 = \mathbf{e}'\mathbf{e}/(n-k)$ it suffices to consider the biased but asymptotically equivalent $\tilde{\sigma}_n^2 := n^{-1}\mathbf{e}'\mathbf{e} = s_n^2(n-k)/n$:

$$\begin{aligned}
\tilde{\sigma}_n^2 &= n^{-1}\left(\mathbf{Y}-\mathbf{X}\breve{\boldsymbol{\beta}}_n\right)'\left(\mathbf{Y}-\mathbf{X}\breve{\boldsymbol{\beta}}_n\right) \\
&= n^{-1}\left[\mathbf{Y}-\mathbf{X}\boldsymbol{\beta}-\mathbf{X}\left(\breve{\boldsymbol{\beta}}_n-\boldsymbol{\beta}\right)\right]'\left[\mathbf{Y}-\mathbf{X}\boldsymbol{\beta}-\mathbf{X}\left(\breve{\boldsymbol{\beta}}_n-\boldsymbol{\beta}\right)\right] \\
&= n^{-1}\mathbf{U}'\mathbf{U} - 2\left(\breve{\boldsymbol{\beta}}_n-\boldsymbol{\beta}\right)'n^{-1}\mathbf{X}'\mathbf{U} + \left(\breve{\boldsymbol{\beta}}_n-\boldsymbol{\beta}\right)'\left(n^{-1}\mathbf{X}'\mathbf{X}\right)\left(\breve{\boldsymbol{\beta}}_n-\boldsymbol{\beta}\right).
\end{aligned}$$

The first term converges a.s. to $EU^2 = \sigma^2$, and the third term converges in probability to zero since $n^{-1}\mathbf{X}'\mathbf{X} \to \boldsymbol{\Sigma}_{\mathbf{x}}$ and $\left(\breve{\boldsymbol{\beta}}_n-\boldsymbol{\beta}\right) \longrightarrow^p \mathbf{0}$. If rows $\{\breve{\mathbf{x}}_i = \breve{x}_{i1}, ..., \breve{x}_{ik}\}_{i=1}^n$ of $\mathbf{X}$ are i.i.d. random vectors, independent of the errors and with finite moments, then in the second term

$$n^{-1}\mathbf{X}'\mathbf{U} = \begin{pmatrix} n^{-1}\sum_{i=1}^n \breve{x}_{i1}U_i \\ n^{-1}\sum_{i=1}^n \breve{x}_{i2}U_i \\ \vdots \\ n^{-1}\sum_{i=1}^n \breve{x}_{ik}U_i \end{pmatrix} \to^{a.s.} \begin{pmatrix} E\breve{x}_1 U = E\breve{x}_1 EU \\ E\breve{x}_2 U = E\breve{x}_2 EU \\ \vdots \\ E\breve{x}_k U = E\breve{x}_k EU \end{pmatrix} = \mathbf{0},$$

and hence this second term vanishes as well. The result is the same in the case of added samples with the same deterministic $\mathbf{X}_0$. To see this, let $\mathbf{x}' = (x_1, ..., x_n)$ represent an arbitrary one of the k rows of $\mathbf{X}'$—i.e., the n-vector of observations of any one of the k explanatory variables, and write out $\mathbf{x}'\mathbf{U} = \sum_{i=1}^{n} x_i U_i$ as

$$\mathbf{x}'\mathbf{U} = x_1 U_1 + x_2 U_2 + \cdots + x_{n_0} U_{n_0} +$$
$$x_1 U_{n_0+1} + x_2 U_{n_0+2} + \cdots + x_{n_0} U_{2n_0} +$$
$$\cdots +$$
$$x_1 U_{(N-1)n_0+1} + x_2 U_{(N-1)n_0+2} + \cdots + x_{n_0} U_{Nn_0}$$
$$= x_1 \sum_{\ell=1}^{N} U_{(\ell-1)n_0+1} + x_2 \sum_{\ell=1}^{N} U_{(\ell-1)n_0+2} + \cdots + x_{n_0} \sum_{\ell=1}^{N} U_{\ell n_0}.$$

Then as $N \to \infty$

$$n^{-1}\mathbf{x}'\mathbf{U} = \frac{1}{n_0} \sum_{i=1}^{n_0} \left[\frac{x_i}{N} \sum_{\ell=1}^{N} U_{(\ell-1)n_0+i} \right] \tag{11.21}$$

$$\xrightarrow{a.s.} \frac{1}{n_0} (x_1 + x_2 + \cdots + x_{n_0}) EU = 0,$$

so that $\lim_{n\to\infty} n^{-1}\mathbf{X}'\mathbf{U} = \mathbf{0}$. Thus, $s_n^2 \xrightarrow{p} \sigma^2$ under either of the sampling scenarios.

The asymptotic joint normality of $\breve{\boldsymbol{\beta}}_n$ follows under these two scenarios as well. Writing

$$\sqrt{n}\left(\breve{\boldsymbol{\beta}}_n - \boldsymbol{\beta}\right) = \left(n^{-1}\mathbf{X}'\mathbf{X}\right)^{-1} n^{-1/2}\mathbf{X}'\mathbf{U} \tag{11.22}$$

and noting that $n^{-1}\mathbf{X}'\mathbf{X} \to \boldsymbol{\Sigma}_\mathbf{x}$ in both cases, it is evident that we can focus on the behavior of $n^{-1/2}\mathbf{X}'\mathbf{U}$. Suppose first that rows $\{\breve{\mathbf{x}}_i\}_{i=1}^{n}$ are i.i.d. random k-vectors—still having finite second moments and being independent of the $\{U_i\}_{i=1}^{n}$. Letting $\mathbf{a}$ be any k-vector of constants not all equal to zero, we have

$$n^{-1/2}\mathbf{a}'\mathbf{X}'\mathbf{U} = n^{-1/2} \sum_{j=1}^{k} a_j \sum_{i=1}^{n} \breve{x}_{ij} U_i$$

$$= n^{-1/2} \sum_{i=1}^{n} \left(\sum_{j=1}^{k} a_j \breve{x}_{ij} \right) U_i$$

$$= n^{-1/2} \sum_{i=1}^{n} \mathbf{a}'\breve{\mathbf{x}}_i U_i.$$

The n terms of the sum being i.i.d. r.v.s with zero mean and finite second moment, the expression converges in distribution to normal with mean zero and variance

$$E\left(\mathbf{a}'\breve{\mathbf{x}}U\right)^2 = \sigma^2 \mathbf{a}' \left(E\breve{\mathbf{x}}\breve{\mathbf{x}}'\right) \mathbf{a} = \sigma^2 \mathbf{a}'\boldsymbol{\Sigma}_\mathbf{x}\mathbf{a}.$$

Theorem 4.1 (Cramér–Wold) thus implies that $n^{-1/2}\mathbf{X}'\mathbf{U} \rightsquigarrow N\left(\mathbf{0}, \sigma^2 \mathbf{\Sigma_x}\right)$.

Once again, the same result follows when $\mathbf{X}$ comprises replicates of a deterministic $\mathbf{X}_0$. Referring to expression (11.21), one sees that $n^{-1/2}$ times the product of the jth row of $\mathbf{X}'$ and $\mathbf{U}$ is

$$\frac{1}{\sqrt{n_0}}\left[\frac{x_{1j}}{\sqrt{N}}\sum_{\ell=1}^{N}U_{(\ell-1)n_0+1} + \frac{x_{2j}}{\sqrt{N}}\sum_{\ell=1}^{N}U_{(\ell-1)n_0+2} + \cdots + \frac{x_{n_0 j}}{\sqrt{N}}\sum_{\ell=1}^{N}U_{\ell n_0}\right].$$

The limiting distribution of normalized sums $\left\{N^{-1/2}\sum_{\ell=1}^{N}U_{(\ell-1)n_0+i}\right\}_{i=1}^{n_0}$ is the same as that of $\{\sigma Z_i\}_{i=1}^{n_0}$, where the $\{Z_i\}$ are independent standard normals. Thus, the limiting distribution of $n^{-1/2}\mathbf{a}'\mathbf{X}'\mathbf{U}$ is that of

$$\frac{\sigma}{\sqrt{n_0}}\sum_{j=1}^{k}a_j\sum_{i=1}^{n_0}x_{ij}Z_i = \frac{\sigma}{\sqrt{n_0}}\sum_{i=1}^{n_0}\left(\sum_{j=1}^{k}a_j x_{ij}\right)Z_i$$

$$= \frac{\sigma}{\sqrt{n_0}}\sum_{i=1}^{n_0}\mathbf{a}'\mathbf{x}_{0i}Z_i,$$

where $\mathbf{x}_{0i}$ is the ith row of sample matrix $\mathbf{X}_0$. This has mean zero and variance

$$\frac{\sigma^2}{n_0}\sum_{i=1}^{n_0}\left(\mathbf{a}'\mathbf{x}_{0i}\right)^2 = \frac{\sigma^2}{n_0}\mathbf{a}'\left(\sum_{i=1}^{n_0}\mathbf{x}_{0i}\mathbf{x}_{0i}'\right)\mathbf{a} = \frac{\sigma^2}{n_0}\mathbf{a}'\mathbf{X}_0'\mathbf{X}_0\mathbf{a} = \sigma^2\mathbf{a}'\mathbf{\Sigma_x}\mathbf{a},$$

as in the case of stochastic $\{\breve{\mathbf{x}}_i\}$, so again we have $n^{-1/2}\mathbf{X}'\mathbf{U} \rightsquigarrow N\left(\mathbf{0}, \sigma^2\mathbf{\Sigma_x}\right)$.

Finally, referring to expression (11.22) and recalling that $n^{-1}\mathbf{X}'\mathbf{X} \to \mathbf{\Sigma_x}$, we see that the limiting distribution of $\sqrt{n}\left(\breve{\boldsymbol{\beta}}_n - \boldsymbol{\beta}\right)$ is, in both scenarios, $N\left(\mathbf{0}, \sigma^2\mathbf{\Sigma_x}^{-1}\right)$ and hence that the asymptotic covariance matrix of $\breve{\boldsymbol{\beta}}_n$ is $n^{-1}\sigma^2\mathbf{\Sigma_x}^{-1}$. In both cases this would be estimated consistently from the sample as $\mathbf{S}_{\breve{\boldsymbol{\beta}}} = s_n^2\left(\mathbf{X}'\mathbf{X}\right)^{-1}$. It follows then by Mann–Wald that

$$\left(\breve{\boldsymbol{\beta}}_n - \boldsymbol{\beta}\right)'\mathbf{S}_{\breve{\boldsymbol{\beta}}}^{-1}\left(\breve{\boldsymbol{\beta}}_n - \boldsymbol{\beta}\right) \rightsquigarrow \chi^2\left(k\right). \tag{11.23}$$

11.3.1.2 *Confidence Sets and Tests of Hypotheses*

The asymptotic result (11.23) when $\{U_i\}_{i=1}^{n}$ are i.i.d. with zero means and equal variances corresponds to the *exact* results

$$\left(\hat{\boldsymbol{\beta}}_n - \boldsymbol{\beta}\right)'\mathbf{\Sigma}_{\hat{\boldsymbol{\beta}}}^{-1}\left(\hat{\boldsymbol{\beta}}_n - \boldsymbol{\beta}\right) \sim \chi^2\left(k\right)$$

$$k^{-1}\left(\hat{\boldsymbol{\beta}}_n - \boldsymbol{\beta}\right)'\mathbf{S}_{\hat{\boldsymbol{\beta}}}^{-1}\left(\hat{\boldsymbol{\beta}}_n - \boldsymbol{\beta}\right) \sim F\left(k, n-k\right)$$

when the errors are i.i.d. as $N\left(0, \sigma^{2}\right)$. The development of asymptotic confidence sets and tests of hypotheses in the present case thus parallels that for normal errors, except that these are now based on chi-squared approximations rather than exact F distributions. Specifically, if $\boldsymbol{\beta} \in \Re^{k}$ is decomposed into vectors $\boldsymbol{\beta}_{1} \in \Re^{k_{1}}$ and $\boldsymbol{\beta}_{2} \in \Re^{k_{2}}$, if $\mathbf{S}_{\breve{\boldsymbol{\beta}}}$ is broken down conformably, and if $c_{k_{2}}^{\alpha}$ is the upper-α quantile of $\chi^{2}\left(k_{2}\right)$, then corresponding to (11.8) in the normal case, the set

$$\left\{\boldsymbol{\beta}_{2}:\left(\breve{\boldsymbol{\beta}}_{2n}-\boldsymbol{\beta}_{2}\right)' \mathbf{S}_{\breve{\boldsymbol{\beta}}_{2}}^{-1}\left(\hat{\boldsymbol{\beta}}_{2n}-\boldsymbol{\beta}_{2}\right) \leq c_{k_{2}}^{\alpha}\right\}$$

constitutes an approximate $1-\alpha$ confidence region for $\boldsymbol{\beta}_{2}$. Likewise, corresponding to (11.13) the rejection region

$$\mathcal{R}_{\alpha}=\left\{\mathbf{y}:\left(\breve{\boldsymbol{\beta}}_{2n}-\boldsymbol{\beta}_{2}^{0}\right)' \mathbf{S}_{\breve{\boldsymbol{\beta}}_{2}}^{-1}\left(\breve{\boldsymbol{\beta}}_{2n}-\boldsymbol{\beta}_{2}^{0}\right) \geq c_{k_{2}}^{\alpha}\right\}$$

yields an asymptotically valid and consistent test of $H_{0}: \boldsymbol{\beta}_{2}=\boldsymbol{\beta}_{2}^{0}$ vs. $H_{1}: \boldsymbol{\beta}_{2} \neq \boldsymbol{\beta}_{2}^{0}$. More generally, when $\mathbf{A}$ is an $\ell \times k$ matrix of rank ℓ and c_{ℓ}^{α} is the upper-α quantile of $\chi^{2}(\ell)$,

$$\left\{\mathbf{A}\boldsymbol{\beta}:\left(\mathbf{A}\breve{\boldsymbol{\beta}}_{n}-\mathbf{A}\boldsymbol{\beta}\right)'\left(\mathbf{A}\mathbf{S}_{\breve{\boldsymbol{\beta}}}\mathbf{A}'\right)^{-1}\left(\mathbf{A}\breve{\boldsymbol{\beta}}_{n}-\mathbf{A}\boldsymbol{\beta}\right) \leq c_{\ell}^{\alpha}\right\}$$

and

$$\mathcal{R}_{\alpha}=\left\{\mathbf{y}:\left(\mathbf{A}\breve{\boldsymbol{\beta}}_{n}-\mathbf{B}\right)'\left(\mathbf{A}\mathbf{S}_{\breve{\boldsymbol{\beta}}}\mathbf{A}'\right)^{-1}\left(\mathbf{A}\breve{\boldsymbol{\beta}}_{n}-\mathbf{B}\right) \geq c_{\ell}^{\alpha}\right\} \tag{11.24}$$

are an asymptotically correct confidence region for $\mathbf{A}\boldsymbol{\beta}$ and an asymptotically valid critical region for $H_{0}: \mathbf{A}\boldsymbol{\beta}=\mathbf{B}$, respectively. In particular, taking $\mathbf{A}=(0, \ldots, 0,1)$ and $\mathbf{B}=\beta_{k}^{0}$ (a scalar constant) in (11.24) shows that $\left\{\mathbf{y}:\left|\left(\tilde{\beta}_{k}-\beta_{k}^{0}\right) / S_{\tilde{\beta}_{k}}\right| \geq z^{\alpha / 2}\right\}$ is an asymptotically valid α-level test of $H_{0}: \beta_{k}=\beta_{k}^{0}$ vs. $H_{1}: \beta_{k} \neq \beta_{k}^{0}$. (Of course, upper-$\alpha/2$ quantile $z^{\alpha / 2}$ of $N(0,1)$ equals $\sqrt{c_{1}^{\alpha}}$.) Since this test has only *asymptotic* validity, it is common in applied work to impose the more stringent requirement that $\left|\left(\tilde{\beta}_{k}-\beta_{k}^{0}\right) / S_{\tilde{\beta}_{k}}\right| \geq t_{n-k}^{\alpha / 2}$, where $t_{n-k}^{\alpha / 2}>z^{\alpha / 2}$ is the upper-$\alpha/2$ quantile of the Student distribution with $n-k$ d.f. Although this practice is purely *ad hoc*, it does have the effect of making the test more conservative.

11.3.2 Nonlinear Models

Under conditions like those applied to the linear model we can show that the nonlinear least-squares estimator has similar asymptotic properties, provided that $E(Y_i \mid \mathbf{x}_i)$ meets certain conditions. Thus, in the model $Y_i = \xi_i(\boldsymbol{\beta}) + U_i$ with $E(Y_i \mid \mathbf{x}_i) = \xi(\mathbf{x}_i; \boldsymbol{\beta}) =: \xi_i(\boldsymbol{\beta})$ a differentiable function, we consider the estimator $\breve{\boldsymbol{\beta}}_n$ that solves $\inf_{\boldsymbol{\beta} \in \mathbb{B}} n^{-1} \sum_{i=1}^n [Y_i - \xi_i(\boldsymbol{\beta})]^2$ or, in vector form, $\inf_{\boldsymbol{\beta} \in \mathbb{B}} [\mathbf{Y} - \boldsymbol{\xi}(\boldsymbol{\beta})]' [\mathbf{Y} - \boldsymbol{\xi}(\boldsymbol{\beta})]$. With $\boldsymbol{\beta}_0$ as the true value of $\boldsymbol{\beta}$, we have

$$Y_i - \xi_i(\boldsymbol{\beta}) = [Y_i - \xi_i(\boldsymbol{\beta}_0)] - [\xi_i(\boldsymbol{\beta}) - \xi_i(\boldsymbol{\beta}_0)] = U_i - [\xi_i(\boldsymbol{\beta}) - \xi_i(\boldsymbol{\beta}_0)]$$

and

$$\frac{1}{n} \sum_{i=1}^n [Y_i - \xi_i(\boldsymbol{\beta})]^2 = \frac{1}{n} \sum_{i=1}^n U_i^2 - \frac{2}{n} \sum_{i=1}^n U_i [\xi_i(\boldsymbol{\beta}) - \xi_i(\boldsymbol{\beta}_0)]$$
$$+ \frac{1}{n} \sum_{i=1}^n [\xi_i(\boldsymbol{\beta}) - \xi_i(\boldsymbol{\beta}_0)]^2.$$

Clearly, the first term converges a.s. to σ^2 when the $\{U_i\}$ are i.i.d. with $EU_i = 0$ and $VU_i = \sigma^2$. We must now consider the limiting behavior of the remaining terms under the two sampling scenarios previously considered. If there is random sampling and $E\xi(\boldsymbol{\beta})^2 < \infty$ for $\boldsymbol{\beta} \in \mathbb{B}$, then the third term converges a.s. to $E[\xi(\boldsymbol{\beta}) - \xi(\boldsymbol{\beta}_0)]^2 \geq 0$. In the experimental setting where there are N_0 observations of Y for each of the n_0 distinct $\{\mathbf{x}_i\}$ and the sample grows as $n = Nn_0$, we have $n^{-1} \sum_{i=1}^n [\xi_i(\boldsymbol{\beta}) - \xi_i(\boldsymbol{\beta}_0)]^2 \equiv n_0^{-1} \sum_{i=1}^{n_0} [\xi_i(\boldsymbol{\beta}) - \xi_i(\boldsymbol{\beta}_0)]^2 \geq 0$ for each N. Thus, only the second term is at issue. Given that $E\xi(\boldsymbol{\beta})^2 < \infty$ it is clear that $n^{-1} \sum_{i=1}^n U_i [\xi_i(\boldsymbol{\beta}) - \xi_i(\boldsymbol{\beta}_0)] \longrightarrow^{a.s.} EU[\xi(\boldsymbol{\beta}) - \xi(\boldsymbol{\beta}_0)] = 0$ in the case of random sampling. Likewise, with experimental data

$$\frac{1}{n} \sum_{i=1}^n U_i [\xi_i(\boldsymbol{\beta}) - \xi_i(\boldsymbol{\beta}_0)] = \frac{1}{n_0} \sum_{i=1}^{n_0} \left\{ [\xi_i(\boldsymbol{\beta}) - \xi_i(\boldsymbol{\beta}_0)] \frac{1}{N} \sum_{\ell=1}^N U_{(\ell-1)n_0+i} \right\},$$

which converges to zero as well. Thus, in both cases we have

$$\frac{1}{n} \sum_{i=1}^n [Y_i - \xi_i(\boldsymbol{\beta})]^2 \longrightarrow^{a.s.} \sigma^2 + E[\xi(\boldsymbol{\beta}) - \xi(\boldsymbol{\beta}_0)]^2,$$

which is minimized at $\boldsymbol{\beta} = \boldsymbol{\beta}_0$. Assuming that $\boldsymbol{\beta}_0$ is identified and that $\mathbb{B}$ is compact, the sequence of minimizers $\left\{ \breve{\boldsymbol{\beta}}_n \right\}$ converges a.s. to $\boldsymbol{\beta}_0$.

We can now look for some further conditions under which $\sqrt{n} \left(\breve{\boldsymbol{\beta}}_n - \boldsymbol{\beta}_0 \right)$ is also asymptotically normal. We require to begin with that $\xi(\boldsymbol{\beta})$ be at least twice continuously differentiable with finite second moment (when the explanatory variables are stochastic) and that $\boldsymbol{\beta}_0$ lie in the interior of $\mathbb{B}$. We start with the first-order condition for the minimization of

$\sum_{i=1}^{n} \left[Y_i - \xi_i \left(\boldsymbol{\beta} \right) \right]^2$. Since $\boldsymbol{\beta}_0$ is in the interior of $\mathbb{B}$, consistent estimator $\breve{\boldsymbol{\beta}}_n$ must (for large enough n) satisfy

$$0 = \frac{\partial}{\partial \boldsymbol{\beta}} \sum_{i=1}^{n} \left[Y_i - \xi_i \left(\boldsymbol{\beta} \right) \right]^2 \bigg|_{\breve{\boldsymbol{\beta}}_n} = -2 \sum_{i=1}^{n} \frac{\partial \xi_i \left(\breve{\boldsymbol{\beta}}_n \right)}{\partial \boldsymbol{\beta}} \left[Y_i - \xi_i \left(\breve{\boldsymbol{\beta}}_n \right) \right], \quad (11.25)$$

with

$$\sum_{i=1}^{n} \mathbf{q}_i \left(\breve{\boldsymbol{\beta}}_n \right) := \sum_{i=1}^{n} \frac{\partial^2}{\partial \boldsymbol{\beta} \partial \boldsymbol{\beta}'} \left[Y_i - \xi_i \left(\boldsymbol{\beta} \right) \right]^2 \bigg|_{\breve{\boldsymbol{\beta}}_n} > 0.$$

Dividing (11.25) by $-2\sqrt{n}$ and expanding about $\boldsymbol{\beta}_0$ give

$$0 = \left[\frac{1}{\sqrt{n}} \sum_{i=1}^{n} \frac{\partial \xi_i \left(\boldsymbol{\beta}_0 \right)}{\partial \boldsymbol{\beta}} U_i \right] - \left[\frac{1}{2n} \sum_{i=1}^{n} \mathbf{q}_i \left(\boldsymbol{\beta}_n^* \right) \right] \sqrt{n} \left(\breve{\boldsymbol{\beta}}_n - \boldsymbol{\beta}_0 \right), \quad (11.26)$$

where $\left| \boldsymbol{\beta}_n^* - \boldsymbol{\beta}_0 \right| < \left| \breve{\boldsymbol{\beta}}_n - \boldsymbol{\beta}_0 \right|$. Clearly, the limiting behavior of $\sqrt{n} \left(\breve{\boldsymbol{\beta}}_n - \boldsymbol{\beta}_0 \right)$ is governed entirely by the behaviors of the two expressions in brackets.

What happens to the first expression is easy to see. If stochastic regressors $\{ \breve{\mathbf{x}}_i \}$ are i.i.d. and independent of the errors, and if

$$\mathbf{Q} \left(\boldsymbol{\beta}_0 \right) := E \left[\frac{\partial \xi \left(\boldsymbol{\beta} \right)}{\partial \boldsymbol{\beta}} \frac{\partial \xi \left(\boldsymbol{\beta} \right)}{\partial \boldsymbol{\beta}'} \right] \bigg|_{\boldsymbol{\beta}_0}$$

is a finite, positive-definite matrix, then for any $\mathbf{a} \in \Re^k$ with $\mathbf{a}'\mathbf{a} > 0$ the Lindeberg–Lévy central limit theorem implies that

$$\frac{1}{\sqrt{n}} \sum_{i=1}^{n} \mathbf{a}' \frac{\partial \xi_i \left(\boldsymbol{\beta}_0 \right)}{\partial \boldsymbol{\beta}} U_i \rightsquigarrow N \left[0, \sigma^2 \mathbf{a}' \mathbf{Q} \left(\boldsymbol{\beta}_0 \right) \mathbf{a} \right].$$

Alternatively, if the regressors are deterministic and distinct observations $\{ \mathbf{x}_i \}_{i=1}^{n_0}$ of the original sample are repeated as $n = N n_0$ increases, then

$$\frac{1}{\sqrt{n}} \sum_{i=1}^{n} \mathbf{a}' \frac{\partial \xi_i \left(\boldsymbol{\beta}_0 \right)}{\partial \boldsymbol{\beta}} U_i = \frac{1}{\sqrt{n_0}} \sum_{i=1}^{n_0} \left[\mathbf{a}' \frac{\partial \xi_i \left(\boldsymbol{\beta}_0 \right)}{\partial \boldsymbol{\beta}} \left(\frac{1}{\sqrt{N}} \sum_{\ell=1}^{N} U_{(\ell-1)n_0+i} \right) \right]$$

is again distributed asymptotically as $N \left[0, \sigma^2 \mathbf{a}' \mathbf{Q} \left(\boldsymbol{\beta}_0 \right) \mathbf{a} \right]$, where now

$$\mathbf{Q} \left(\boldsymbol{\beta}_0 \right) := n_0^{-1} \sum_{i=1}^{n} \frac{\partial \xi_i \left(\boldsymbol{\beta}_0 \right)}{\partial \boldsymbol{\beta}} \frac{\partial \xi_i \left(\boldsymbol{\beta}_0 \right)}{\partial \boldsymbol{\beta}'}.$$

Thus, for samples of both kinds the first bracketed term in (11.26) converges in distribution to multivariate normal with zero mean and covariance matrix $\mathbf{Q}(\boldsymbol{\beta}_0)$.

Turning to the second bracketed expression in (11.26), we have

$$
\begin{aligned}
\frac{1}{2}\mathbf{q}_i(\boldsymbol{\beta}) &= -\frac{\partial}{\partial\boldsymbol{\beta}'}\left\{\frac{\partial\xi_i(\boldsymbol{\beta})}{\partial\boldsymbol{\beta}}\,[Y_i - \xi_i(\boldsymbol{\beta})]\right\} \\
&= -\frac{\partial^2\xi_i(\boldsymbol{\beta})}{\partial\boldsymbol{\beta}\partial\boldsymbol{\beta}'}\,[Y_i - \xi_i(\boldsymbol{\beta})] + \frac{\partial\xi_i(\boldsymbol{\beta})}{\partial\boldsymbol{\beta}}\frac{\partial\xi_i(\boldsymbol{\beta})}{\partial\boldsymbol{\beta}'} \\
&= -\frac{\partial^2\xi_i(\boldsymbol{\beta})}{\partial\boldsymbol{\beta}\partial\boldsymbol{\beta}'}\,\{U_i - [\xi_i(\boldsymbol{\beta}) - \xi_i(\boldsymbol{\beta}_0)]\} + \frac{\partial\xi_i(\boldsymbol{\beta})}{\partial\boldsymbol{\beta}}\frac{\partial\xi_i(\boldsymbol{\beta})}{\partial\boldsymbol{\beta}'}.
\end{aligned}
$$

Now under either of the sampling scenarios

$$
\frac{1}{2n}\sum_{i=1}^{n}\mathbf{q}_i(\boldsymbol{\beta}) \longrightarrow^{a.s.} E\frac{\partial^2\xi(\boldsymbol{\beta})}{\partial\boldsymbol{\beta}\partial\boldsymbol{\beta}'}\,[\xi(\boldsymbol{\beta}) - \xi(\boldsymbol{\beta}_0)] + \mathbf{Q}(\boldsymbol{\beta}) =: \mathbf{H}(\boldsymbol{\beta}),
$$

say, with $\mathbf{H}(\boldsymbol{\beta}_0) = \mathbf{Q}(\boldsymbol{\beta}_0)$. However, we must determine the limiting value when the summands $\{\mathbf{q}_i\}$ are evaluated at the random vector $\boldsymbol{\beta}_n^*$, and the mere fact that $\boldsymbol{\beta}_n^* \longrightarrow^{a.s.} \boldsymbol{\beta}_0$ does not guarantee that $(2n)^{-1}\sum_{i=1}^{n}\mathbf{q}_i(\boldsymbol{\beta}_n^*) \to \mathbf{Q}(\boldsymbol{\beta}_0)$, even in probability. However, if it can be determined that $(2n)^{-1}\sum_{i=1}^{n}\mathbf{q}_i(\boldsymbol{\beta})$ converges *uniformly* for $\boldsymbol{\beta}$ in some open ball $|\boldsymbol{\beta} - \boldsymbol{\beta}_0| < \delta$, and if the limiting $\mathbf{H}(\boldsymbol{\beta})$ is continuous in that neighborhood, then the desired limiting result does follow. In this case we get from (11.26) the result that

$$
\sqrt{n}\left(\breve{\boldsymbol{\beta}}_n - \boldsymbol{\beta}_0\right) \rightsquigarrow N\left[\mathbf{0}, \mathbf{Q}(\boldsymbol{\beta}_0)^{-1}\right],
$$

and we can estimate $\mathbf{Q}(\boldsymbol{\beta}_0)$ consistently by

$$
\breve{\mathbf{Q}}(\boldsymbol{\beta}_n^*) := n^{-1}\sum_{i=1}^{n}\frac{\partial\xi_i(\boldsymbol{\beta}_n^*)}{\partial\boldsymbol{\beta}}\frac{\partial\xi_i(\boldsymbol{\beta}_n^*)}{\partial\boldsymbol{\beta}'}.
$$

Unfortunately, whether the sufficient conditions do hold depends on the nature of the particular regression function ξ and even on the unknown $\boldsymbol{\beta}_0$. Because the plausibility of the conditions is so difficult to judge, the result is often just taken for granted. Since the results are only asymptotic in any case, one is doubly advised to check their adequacy in the problem at hand by simulation. We conclude with an example that illustrates such a process.

Example 11.5. We conduct a "stress test" of the adequacy of asymptotic theory for the model in Example 11.4 as follows, using samples

Table 11.2   Simulated nonlinear least squares estimates.

Actual $\boldsymbol{\beta}$	Uniform Errors			Student Errors		
	1.000	2.000	.500	1.000	2.000	.500
Mean $\left\{ \breve{\boldsymbol{\beta}}_n^{(\ell)} \right\}$	.998	2.014	.500	.999	2.013	.500
	.0028	-.0090	.0008	.0028	-.0093	.0008
Mean $\left\{ \tilde{\mathbf{Q}}^{(\ell)} (\boldsymbol{\beta}_n^*)^{-1} \right\}$	-.0090	.0301	-.0028	-.0093	.0320	-.0028
	.0008	-.0028	.0004	.0008	-.0028	.0004
	.0028	-.0090	.0008	.0028	-.0091	.0008
Covariance of $\left\{ \breve{\boldsymbol{\beta}}_n^{(\ell)} \right\}$	-.0090	.0297	-.0028	-.0091	.0304	-.0028
	.0008	-.0028	.0004	.0008	-.0028	.0004
% Rejections:						
$\quad \alpha = .10$	10.4	10.1	10.0	10.5	10.1	10.4
$\quad \alpha = .05$	5.1	5.1	5.2	5.4	5.2	5.3
$\quad \alpha = .01$	1.1	1.3	1.1	1.4	1.5	1.2

of more appropriate size $n = 100$. We generate $10,000$ samples as $\left\{ Y_i = 1 + 2x_i^{.5} + U_i \right\}_{i=1}^n$, each having the same $\{x_i\}$ and with pseudo-random $\{U_i\}$. For each sample $\ell \in \left\{ 1, 2, ..., 100^2 \right\}$ we compute coefficient estimate $\breve{\boldsymbol{\beta}}_n^{(\ell)}$ and covariance estimate $\breve{\mathbf{Q}}^{(\ell)} (\boldsymbol{\beta}_n^*)^{-1}$. We then compare the mean of the $\left\{ \breve{\boldsymbol{\beta}}_n^{(\ell)} \right\}$ with the actual $\boldsymbol{\beta}_0 = (1.0, 2.0, 0.5)'$ and the covariance matrix of the $\left\{ \breve{\boldsymbol{\beta}}_n^{(\ell)} \right\}$ with the mean of matrices $\left\{ \breve{\mathbf{Q}}^{(\ell)} (\boldsymbol{\beta}_n^*)^{-1} \right\}$. We also compare with nominal significance levels $\alpha \in \{.10, .05, .01\}$ the proportion of rejections of $H_0 : \beta_j = \beta_{j0}$ using critical regions $\left| \tilde{\beta}_{jn} - \beta_{j0} \right| / \tilde{\sigma}_{\tilde{\beta}_{jn}} \geq z^{\alpha/2}$. The "stress" is applied by generating pseudorandom errors from radically different distributions: first the thin-tailed uniform $(-1, 1)$, then the thick-tailed Student's $t(3)$. In each case the $\{U_i\}$ are scaled to have the same standard deviation, $\sigma = 0.20$. (In a real application, one would generally simulate with σ equal to the standard deviation of the sample residuals and, if testing hypotheses, using the $\boldsymbol{\beta}_0$ specified under H_0.) For this model, parameters, and sample size the results shown in Table 11.2 indicate good agreement with the theory, although the standard tests do seem slightly excessive, particularly with Student errors. (Standard errors of the percentage rejections in 100^2 *valid* tests are $\sqrt{(.10)(.90)} = 0.30$ for $\alpha = .10$, $\sqrt{(.05)(.95)} \doteq 0.22$ for $\alpha = .05$, and $\sqrt{(.01)(.99)} \doteq .099$ for $\alpha = .01$.)

11.4   Solutions to Exercises

1. We have

$$\mathbf{X}'\mathbf{X} = \begin{pmatrix} n & n\bar{x} \\ n\bar{x} & \sum_{i=1}^{n} x_i^2 \end{pmatrix}$$

$$|\mathbf{X}'\mathbf{X}| = n \sum_{i=1}^{n} x_i^2 - n^2 \bar{x}^2 = n \sum_{i=1}^{n} (x_i - \bar{x})^2$$

$$(\mathbf{X}'\mathbf{X})^{-1} = |\mathbf{X}'\mathbf{X}|^{-1} \begin{pmatrix} \sum_{i=1}^{n} x_i^2 & -n\bar{x} \\ -n\bar{x} & n \end{pmatrix}$$

$$\mathbf{X}'\mathbf{Y} = \begin{pmatrix} \sum_{i=1}^{n} Y_i \\ \sum_{i=1}^{n} x_i Y_i \end{pmatrix}$$

and

$$\hat{\beta}_n = \begin{pmatrix} \dfrac{\sum_{i=1}^{n} x_i^2 \sum_{i=1}^{n} Y_i - n\bar{x} \sum_{i=1}^{n} x_i Y_i}{n \sum_{i=1}^{n} (x_i - \bar{x})^2} \\ \dfrac{-n\bar{x} \sum_{i=1}^{n} Y_i + n \sum_{i=1}^{n} x_i Y_i}{n \sum_{i=1}^{n} (x_i - \bar{x})^2} \end{pmatrix}.$$

Simplifying, the second component is

$$\hat{\beta}_{n2} = \frac{\sum_{i=1}^{n} (x_i - \bar{x}) Y_i}{\sum_{i=1}^{n} (x_i - \bar{x})^2},$$

and the first is

$$\hat{\beta}_{n1} = \frac{\left(\sum_{i=1}^{n} x_i^2 - n\bar{x}^2\right) n^{-1} \sum_{i=1}^{n} Y_i - \bar{x} \sum_{i=1}^{n} x_i Y_i + \bar{x}^2 \sum_{i=1}^{n} Y_i}{\sum_{i=1}^{n} (x_i - \bar{x})^2}$$

$$= \bar{Y} - \bar{x} \frac{\sum_{i=1}^{n} (x_i - \bar{x}) Y_i}{\sum_{i=1}^{n} (x_i - \bar{x})^2} = \bar{Y} - \hat{\beta}_{n2}\bar{x}.$$

2. With $\{\mathbf{x}_i := (1, x_i)\}_{i=1}^{n}$ we have

$$\mathbf{X}'\mathbf{X} = \begin{pmatrix} n & \sum_{i=1}^{n} x_i \\ \sum_{i=1}^{n} x_i & \sum_{i=1}^{n} x_i^2 \end{pmatrix},$$

$$|\mathbf{X}'\mathbf{X}| = n \sum_{i=1}^{n} x_i^2 - n^2 \bar{x}^2$$
$$= n \sum_{i=1}^{n} (x_i - \bar{x})^2$$
$$= n m_2$$

(n times the sample second moment about the mean), and

$$(\mathbf{X}'\mathbf{X})^{-1} = \frac{1}{n m_2} \begin{pmatrix} \sum_{i=1}^{n} x_i^2 & -n\bar{x} \\ -n\bar{x} & n \end{pmatrix}.$$

Near singularity implies that $m_2 \doteq 0$, meaning that there is little sample variation in regressor x. (Of course, there is *no* variation in regressor 1!) In that case $\sigma^2_{\hat{\beta}_{n2}} = \sigma^2/m_2 = \sigma^2/\sum_{i=1}^n (x_i - \bar{x})^2$ and $\sigma^2_{\hat{\beta}_{n1}} = m'_2/m_2$ are large, relative to when the $\{x_i\}$ are dispersed. The intuition is that there is no way to pin down the *function* $\xi(x) := E(Y \mid x)$ if we do not observe variation in x. (Think of trying to fit a straight line to a narrow vertical strip of the data cloud in either Fig. 11.1 or Fig. 11.2. Minor variations in the $\{y_i\}$ correspond to the observed $\{x_i\}$ would have large effects on slope and intercept.)

3. For arbitrary sample points $\mathbf{y}, \mathbf{w} \in \Re^n$ we must show that likelihood ratio $L(\mathbf{y}; \beta, \sigma^2)/L(\mathbf{w}; \beta, \sigma^2)$ does not depend on β or σ^2 when $\hat{\beta}_n(\mathbf{y}) := (\mathbf{X}'\mathbf{X})^{-1}\mathbf{X}'\mathbf{y}$ equals $\hat{\beta}_n(\mathbf{w}) := (\mathbf{X}'\mathbf{X})^{-1}\mathbf{X}'\mathbf{w}$ and

$$s_n^2(\mathbf{y}) = (n-k)^{-1} \left[\mathbf{y} - \mathbf{X}\hat{\beta}_n(\mathbf{y})\right]' \left[\mathbf{y} - \mathbf{X}\hat{\beta}_n(\mathbf{y})\right]$$
$$= (n-k)^{-1} \mathbf{e}(\mathbf{y})' \mathbf{e}(\mathbf{y})$$

equals $s_n^2(\mathbf{w}) = (n-k)^{-1} \mathbf{e}(\mathbf{w})' \mathbf{e}(\mathbf{w})$. With a little algebra we can express $\log L(\mathbf{y}; \beta, \sigma^2) = -n \log(2\pi\sigma^2) - (\mathbf{y} - \mathbf{X}\beta)'(\mathbf{y} - \mathbf{X}\beta)/(2\sigma^2)$ as

$$-n \log(2\pi\sigma^2) - \frac{1}{2\sigma^2} \left\{\left[\hat{\beta}_n(\mathbf{y}) - \beta\right]' \mathbf{X}'\mathbf{X}\left[\hat{\beta}_n(\mathbf{y}) - \beta\right] + (n-k) s_n^2(\mathbf{y})\right\}$$

and similarly for $\log L(\mathbf{w}; \beta, \sigma^2)$. Then the difference between the log likelihoods at $\mathbf{y}$ and $\mathbf{w}$ equals $(2\sigma^2)^{-1}$ times

$$\left[\hat{\beta}_n(\mathbf{w}) - \beta\right]' \mathbf{X}'\mathbf{X} \left[\hat{\beta}_n(\mathbf{w}) - \beta\right] - \left[\hat{\beta}_n(\mathbf{y}) - \beta\right]' \mathbf{X}'\mathbf{X} \left[\hat{\beta}_n(\mathbf{y}) - \beta\right]$$
$$+ (n-k) \left[s_n^2(\mathbf{w}) - s_n^2(\mathbf{y})\right].$$

All the terms vanish, and thus become independent of β and σ^2, when $s_n^2(\mathbf{w}) = s_n^2(\mathbf{y})$ and $\hat{\beta}_n(\mathbf{w}) = \hat{\beta}_n(\mathbf{y})$.

4. $\hat{\beta}_n$ is still unbiased, since $E\mathbf{U} = \mathbf{0}$, but it is not m.v.u., since by Example 7.38 and Exercise 7.26 the m.v.u. estimator is $(\mathbf{X}'\mathbf{\Sigma}^{-1}\mathbf{X})^{-1} \mathbf{X}'\mathbf{\Sigma}^{-1}\mathbf{Y}$.

5. Noting that $\bar{y} \equiv \bar{\hat{y}}$ when $\mathbf{X}$ contains a units column and that residuals $\mathbf{e}$ are orthogonal to the columns of $\mathbf{X}$ (and therefore to $\hat{\mathbf{y}}$), we have

$$R^2 = \frac{\sum_{i=1}^n (\hat{y}_i - \bar{y})(y_i - e_i - \bar{y})}{\sum_{i=1}^n (y_i - \bar{y})^2} = \frac{\sum_{i=1}^n (\hat{y}_i - \bar{\hat{y}})(y_i - \bar{y})}{\sum_{i=1}^n (y_i - \bar{y})^2}$$

$$= \frac{\sum_{i=1}^n (\hat{y}_i - \bar{\hat{y}})(y_i - \bar{y})}{\sqrt{\sum_{i=1}^n (\hat{y}_i - \bar{\hat{y}})^2 \sum_{i=1}^n (y_i - \bar{y})^2}} \sqrt{\frac{\sum_{i=1}^n (\hat{y}_i - \bar{\hat{y}})^2}{\sum_{i=1}^n (y_i - \bar{y})^2}}$$

$$= r_{y\hat{y}} \cdot R.$$

6. We have

$$\begin{pmatrix} \hat{\boldsymbol{\beta}}^* \\ \hat{\gamma} \end{pmatrix} = \left[(\mathbf{X}, \mathbf{z})' \, (\mathbf{X}, \mathbf{z}) \right]^{-1} (\mathbf{X}, \mathbf{z})' \, \mathbf{Y}$$

$$= \begin{pmatrix} \mathbf{X}'\mathbf{X} & 0 \\ 0' & \mathbf{z}'\mathbf{z} \end{pmatrix}^{-1} \begin{pmatrix} \mathbf{X}' \\ \mathbf{z}' \end{pmatrix} \mathbf{Y}$$

$$= \begin{bmatrix} (\mathbf{X}'\mathbf{X})^{-1} & 0 \\ 0' & (\mathbf{z}'\mathbf{z})^{-1} \end{bmatrix} \begin{pmatrix} \mathbf{X}'\mathbf{Y} \\ \mathbf{z}'\mathbf{Y} \end{pmatrix}$$

$$= \begin{bmatrix} (\mathbf{X}'\mathbf{X})^{-1} \mathbf{X}'\mathbf{Y} \\ \dfrac{\mathbf{z}'\mathbf{Y}}{\mathbf{z}'\mathbf{z}} \end{bmatrix} = \begin{pmatrix} \hat{\boldsymbol{\beta}} \\ \hat{\gamma} \end{pmatrix}.$$

Thus, adding to the linear model a regressor that is uncorrelated within sample does not alter the estimates of the other parameters.

7. Since $\mathbf{e}$ and $\mathbf{X}$ are orthogonal, we have $\hat{\boldsymbol{\beta}}^* = \hat{\boldsymbol{\beta}}$,

$$\hat{\gamma} = \frac{\mathbf{e}'\mathbf{Y}}{\mathbf{e}'\mathbf{e}} = \frac{\mathbf{e}'\left(\hat{\mathbf{Y}} + \mathbf{e}\right)}{\mathbf{e}'\mathbf{e}} = \frac{\mathbf{e}'\mathbf{e}}{\mathbf{e}'\mathbf{e}} = 1,$$

and $\hat{\mathbf{Y}}^* = \mathbf{X}\hat{\boldsymbol{\beta}} + 1 \cdot \mathbf{e} = \mathbf{X}\hat{\boldsymbol{\beta}} + \left(\mathbf{Y} - \hat{\mathbf{Y}}\right) = \mathbf{X}\hat{\boldsymbol{\beta}} + \left(\mathbf{Y} - \mathbf{X}\hat{\boldsymbol{\beta}}\right) = \mathbf{Y}$. You would thus achieve a *perfect fit*, with $R^2 = \bar{R}^2 = 1$. You should be pleased if you enjoy using the computer to discover identities.

8. Restricted estimator $\check{\beta}_{1n}$ would be found by regressing $\{Y_i - x_i\beta_2\}$ on $\{1\}$. The estimate from sample realization $\{y_i\}_{i=1}^{n}$ is just the sample mean, $\bar{y} - \bar{x}\beta_2$, and so the restricted residual is $\check{e}_i = y_i - \bar{y} - \beta_2 (x_i - \bar{x})$. Now $y_i = \hat{y}_i + e_i = \hat{\beta}_{1n} + \hat{\beta}_{2n} x_i + e_i$, where $\hat{\beta}_{1n}, \hat{\beta}_{2n}$ are the unrestricted estimates, and

$$y_i - \bar{y} = y_i - \bar{\hat{y}} = \hat{\beta}_{2n}(x_i - \bar{x}) + e_i.$$

Thus, $\check{e}_i = \left(\hat{\beta}_{2n} - \beta_2\right)(x_i - \bar{x}) + e_i$ and

$$\frac{\check{\mathbf{e}}'\check{\mathbf{e}} - \mathbf{e}'\mathbf{e}}{s_n^2} = \frac{\left(\hat{\beta}_{2n} - \beta_2\right)^2}{s_n^2} \sum_{i=1}^{n}(x_i - \bar{x})^2.$$

9. Row n in Table 7.1 pertains to pivotal quantity $(n-1)S^2/\sigma^2 \sim \chi^2 (n-1)$ from the univariate normal model. In the regression context our pivot $(n-k)s_n^2/\sigma^2$ is distributed as $\chi^2 (n-k)$, so we simply take the multipliers from row $n-k+1$. With $n-k = 16$, $1-\alpha = .95$, and $s_{20}^2 = .441$, entries $v_\alpha = .462, v^\alpha = 2.050$ in row 17 give as the c.i. $[.441 \, (.462), .441 \, (2.050)] \doteq [.204, .904]$.

10. The errors $\{U_i\}_{i=1}^n$ being i.i.d. with variance σ^2 and independent of $\{\xi_i(\beta)\}_{i=1}^n$, we have

$$E\left[\frac{\partial \mathcal{L}(\beta,\sigma^2)}{\partial \beta}\frac{\partial \mathcal{L}(\beta,\sigma^2)}{\partial \beta'}\right] = \frac{1}{\sigma^4}\sum_{i=1}^n\sum_{j=1}^n E\left[U_iU_j\frac{\partial \xi_i(\beta)}{\partial \beta}\frac{\partial \xi_j(\beta)}{\partial \beta'}\right]$$

$$= \frac{1}{\sigma^2}\sum_{i=1}^n E\left[\frac{\partial \xi_i(\beta)}{\partial \beta}\frac{\partial \xi_i(\beta)}{\partial \beta'}\right]$$

$$= \frac{n}{\sigma^2}E\left[\frac{\partial \xi(\beta)}{\partial \beta}\frac{\partial \xi(\beta)}{\partial \beta'}\right]$$

$$E\left[\frac{\partial \mathcal{L}(\beta,\sigma^2)}{\partial \beta}\frac{\partial \mathcal{L}(\beta,\sigma^2)}{\partial \sigma^2}\right] = -\frac{n}{2\sigma^4}\sum_{i=1}^n E\left[U_i\frac{\partial \xi_i(\beta)}{\partial \beta}\right]$$

$$+\frac{1}{2\sigma^6}\sum_{i=1}^n\sum_{j=1}^n E\left[U_iU_j^2\frac{\partial \xi_j(\beta)}{\partial \beta'}\right] = 0$$

$$E\left[\frac{\partial \mathcal{L}(\beta,\sigma^2)}{\partial \sigma^2}\right]^2 = E\left[-\frac{n}{2\sigma^2}+\frac{1}{2\sigma^4}\sum_{i=1}^n U_i^2\right]^2 = \frac{n}{2\sigma^4}$$

and

$$E\frac{\partial^2 \mathcal{L}(\beta,\sigma^2)}{\partial \beta \partial \beta'} = \frac{1}{\sigma^2}\sum_{i=1}^n E\left[U_i\frac{\partial^2 \xi_i(\beta)}{\partial \beta \partial \beta'}\right] - \frac{1}{\sigma^2}\sum_{i=1}^n E\left[\frac{\partial \xi_i(\beta)}{\partial \beta}\frac{\partial \xi_i(\beta)}{\partial \beta'}\right]$$

$$= -\frac{1}{\sigma^2}\sum_{i=1}^n E\left[\frac{\partial \xi_i(\beta)}{\partial \beta}\frac{\partial \xi_i(\beta)}{\partial \beta'}\right]$$

$$E\frac{\partial^2 \mathcal{L}(\beta,\sigma^2)}{\partial \beta \partial \sigma^2} = 0$$

$$E\frac{\partial^2 \mathcal{L}(\beta,\sigma^2)}{\partial (\sigma^2)^2} = \frac{n}{2\sigma^4}-\frac{1}{\sigma^6}E\sum_{i=1}^n U_i^2 = -\frac{n}{2\sigma^4}.$$

11. (i) The asymptotic covariance matrix of $\left(\hat{\beta}_{n1},\hat{\beta}_{n2}\right)$ is the first 2×2 block in (11.20), and its inverse is

$$\hat{\Sigma}_{\hat{\beta}_{n1},\hat{\beta}_{n2}}^{-1} = \begin{pmatrix} .1083 & -.1223 \\ -.1223 & .1402 \end{pmatrix}^{-1}$$

$$\doteq \begin{pmatrix} 619.340 & 540.266 \\ 540.266 & 478.420 \end{pmatrix}.$$

The test statistic is

$$\left(\hat{\beta}_{n1} - 1.0, \hat{\beta}_{n2} - 2.0\right) \hat{\Sigma}^{-1}_{\hat{\beta}_{n1}, \hat{\beta}_{n2}} \begin{pmatrix} \hat{\beta}_{n1} - 1.0 \\ \hat{\beta}_{n2} - 2.0 \end{pmatrix}$$

$$= (-.452, .534) \begin{pmatrix} 619.340 & 540.266 \\ 540.266 & 478.420 \end{pmatrix} \begin{pmatrix} -.452 \\ .534 \end{pmatrix}$$

$$\doteq 2.152.$$

This is less than $c_{.90} \equiv c^{10} = 4.605$, the upper-.10 quantile of $\chi^2(2)$, so that H_0 cannot be rejected at the .10 level.

(ii) $\left(\hat{\beta}_{n3} - .5\right)/\hat{\sigma}^2_{\hat{\beta}_{n3}} = -.094/\sqrt{.0036} = -1.567 < -1.282 = -z^{.10}$, so H_0 is rejected at level .10.

12. (i) $ECY = E(CX\beta + CU) = CX\beta$.

(ii) $\Sigma_{\mathbf{B}} = E(CX\beta + CU - \beta)(CX\beta + CU - \beta)' = ECUU'C' = \sigma^2 CC'$.

(iii) We have $V(\mathbf{a}'\beta) = \mathbf{a}'\Sigma_{\mathbf{B}}\mathbf{a} = \sigma^2 \mathbf{a}'CC'\mathbf{a}$. Differentiating Lagrangian $\mathbb{L} := \frac{1}{2}\mathbf{a}'CC'\mathbf{a} - \mathbf{a}'(CX - I_k)\lambda$ (where $\lambda \in \Re^k$) yields

$$\frac{\partial \mathbb{L}}{\partial C'\mathbf{a}} = C'\mathbf{a} - X\lambda = 0$$

$$\frac{\partial \mathbb{L}}{\partial \lambda} = (I_k - X'C')\mathbf{a} = 0.$$

Premultiplying the first equation by X' and applying $X'C' = I_k$ give $\lambda = (X'X)^{-1}\mathbf{a}$ and hence $C'\mathbf{a} = X(X'X)^{-1}\mathbf{a}$.

Appendix A

Abbreviations

Abbreviation	Meaning	Page
a.c.	absolutely continuous	63
a.e.	almost everywhere	23
a.s.	almost surely	23
b.l.u.e.	best linear unbiased estimator	449
c.a.n.	consistent & asymptotically normal	473
c.d.f.	cumulative distribution function	54
c.f.	characteristic function	194
c.g.f.	cumulant generating function	201
c.i.	confidence interval	505
c.l.t.	central limit theorem	410
c.o.v.	change of variable	353
c.r.	confidence region	506
d.f.	degrees of freedom	349
d.g.p.	data-generating process	335
e.d.f.	empirical distribution function	652
g.m.m.	generalized method of moments	480
h.l.c.i.	highest-likelihood c.i.	519
i.c.f.	integrated c.f.	676
i.i.d.	independent & identically distributed	112

Abbreviation	Meaning	Page
i.o.	infinitely often	12
L.m.	Lagrange multiplier	590
l.r.	likelihood ratio	576
m.c.s.	minimum chi-squared	647
m.g.f.	moment generating function	190
m.l.	maximum likelihood	482
m.l.e.	m.l. estimator/estimate	482
m.o.m.	method of moments	478
m.s.e.	mean squared error	447
m.v.u.	minimum-variance unbiased	449
m.v.u.e.	m.v.u. estimator/estimate	537
p.d.f.	probability density function	65
p.g.f.	probability generating function	238
p.m.f.	probability mass function	65
q.v.	quadratic variation	402
r.v.	random variable	48
s.l.l.n.	strong law of large numbers	401
u.i.	uniformly integrable	397
u.m.p.	uniformly most powerful	567
w.r.t.	with respect to	64

Appendix B

Glossary of Symbols

Symbol	Meaning	Page
$\Re$	set of real numbers	5
$\Re_+$	nonnegative reals	5
$\aleph$	positive integers	5
$\aleph_0$	nonnegative integers	5
$\mathbb{Q}$	rational numbers	5
$\mathcal{C}$	complex numbers	5
$S_1 \times S_2$	Cartesian product of 2 sets	5
$S^k, k \in \aleph$	k-fold product of S with itself	5
$\Re^k, k \in \aleph$	k-dimensional reals	5
Ω	sample space	6
ω	a generic outcome	6
$\varnothing$	empty set	8
$\cup, \cap, \backslash, {}^c$	union, intersect, difference, complement of sets	9
$\subset, \supset, =$	subset, superset, equality of sets	9
$\{A_j\} \uparrow, \{A_j\} \downarrow$	increasing, decreasing sequences of sets	10
$\liminf_n A_n, \limsup_n A_n$	limits inferior and superior of sets	11
$\mathcal{F}$	field or σ field of sets	14
$\mathcal{B}, \mathcal{B}_{[0,1]}$	Borel sets of $\Re$, of $[0,1]$	17
$(\Omega, \mathcal{F})$	measurable space	19
$\mathcal{M}$	a generic measure	19
$(\Omega, \mathcal{F}, \mathcal{M})$	measure space	19
$N(A)$	counting measure of set A	19
$\lambda(A)$	Lebesgue measure of set A	19
$\mathbb{P}(A)$	probability measure of set A	21
$(\Omega, \mathcal{F}, \mathbb{P})$	probability space	22
P_k^n	permutations of k of n objects	29
$C_k^n = \binom{n}{k}$	combinations of k of n objects	30
$\mathbb{P}(A \mid \cdot)$	conditional probability of A given $(\cdot)$	31
$X : \Omega \to \Re$	random variable (r.v.)	48
$X(\Omega)$	range of random variable	48

Symbol	Meaning	Page
$X^{-1}(B)$	inverse image of set B under r.v. X	49
$\mathbb{P}_X(B)$	probability measure of B induced by r.v. X	52
$(\Re, \mathcal{B}, \mathbb{P}_X)$	probability space induced by r.v. X	53
$F(x)$	cumulative distribution function (c.d.f.) at x	54
$\mathbf{1}_A(s)$	indicator function of set A at s	54
$[x]$	greatest integer $\leq x$	56
$F(x+), F(x-)$	limit of F at x from above, below	59
$\mathbb{X}$	canonical support of a r.v.	61
$f(x)$	probability mass/density function (p.m.f./p.d.f.) at x	64
x_p	p quantile of r.v. X	69
x^p	upper p quantile of r.v. X	70
$\mathcal{B}^k$	Borel sets of $\Re^k$	82
$F(x,y), f(x,y)$	joint c.d.f., p.m.f. or p.d.f. at (x,y)	109
$f_{Y|X}(y \mid x)$	conditional p.m.f./p.d.f.	97
$J_{\mathbf{h}}$	Jacobian of transformation $\mathbf{h}$	106
EX	expected value of r.v. X	127
$\mathfrak{F}$	generic space of distributions	161
μ'_k	kth moment of r.v. about the origin	162
ν'_k	kth absolute moment about the origin	162
μ	first moment or mean of r.v.	163
μ_k	kth central moment of r.v.	170
$\sigma^2 = VX$	variance or 2d central moment	170
$\mathfrak{g}$	Gini's mean difference	171
α_3, α_4	coefficients of skewness & kurtosis	173
η	harmonic mean	178
γ	geometric mean	178
$\sigma_{XY} = Cov(X,Y)$	covariance between X and Y	181
ρ_{XY}	coefficient of correlation	183
$\mathbf{\Sigma_X}$	covariance matrix	185
$\mathbf{R_X}$	correlation matrix	186
$M(t)$	moment generating function at t	190
$\phi(t)$	characteristic function at t	194
κ_k	kth cumulant of a r.v.	201
$E(Y|x)$	conditional expectation of Y given $X = x$	211
$E(Y|\mathcal{G})$	conditional expectation given σ field $\mathcal{G}$	212
$\sigma(X)$	σ field generated by r.v. X	212
$V(Y|x)$	conditional variance of Y given $X = x$	216
$p(t)$	probability generating function at t	238
$\mu'_{[k]}$	kth descending factorial moment	239
$\sim$	is distributed as	240
$B(n, \theta)$	binomial distribution with parameters n, θ	241
$\binom{n}{\mathbf{y}}$	vector or multinomial combinatoric	245
$NB(n, \theta)$	negative binomial distribution	247
$\Gamma(x)$	gamma function at x	249
$HG(S, F, n)$	hypergeometric distribution with parameters S, F, n	250

Symbol	Meaning	Page
$P(\lambda)$	Poisson distribution with parameter λ	254
$m(t)$	central moment generating function at t	255
$U(\alpha,\beta)$	uniform distribution on (α,β)	258
$\Gamma(\alpha,\beta)$	gamma distribution with parameters α,β	265
$h(t)$	hazard function at t	267
$\mathfrak{B}(\alpha,\beta)$	beta function at α,β	270
$\mathfrak{Be}(\alpha,\beta)$	beta distribution with parameters α,β	271
$W(\alpha,\beta)$	Weibull distribution with parameters α,β	273
$L(\alpha,\beta)$	Laplace distribution with parameters α,β	275
$N(0,1)$	standard normal distribution	277
$\Phi(z)$	standard normal c.d.f. at z	278
$N(\mu,\sigma^2)$	normal distribution with parameters μ,σ^2	284
$\mathbf{I}_k, k\in\aleph$	$k\times k$ identity matrix	286
$N(\boldsymbol{\mu},\boldsymbol{\Sigma})$	multivariate normal distribution with parameters $\boldsymbol{\mu},\boldsymbol{\Sigma}$	287
$\{W_t\}_{t=0}^{\infty}$	Wiener or Brownian motion process	296
$LN(\mu,\sigma^2)$	lognormal distribution with parameters μ,σ^2	299
$IG(\mu,\lambda)$	inverse Gaussian distribution with parameters μ,λ	303
$(X)^+ = X\vee 0$	greater of X and zero	302
$X_{(j:n)}$	jth order statistic of sample of n	306
$\mathbb{P}^n$	probability measure on $(\Omega^n,\mathcal{F}^n)$	336
M_k'	kth sample moment about the origin	341
$\bar{X}$	mean of sample $(X_1,X_2,...,X_n)$	341
M_k	kth sample moment about the mean	341
S^2, S	sample variance, standard deviation	342
$\chi^2(\nu)$	chi-squared distribution with ν degrees of freedom	349
$t(\nu)$	Student's t distribution with ν degrees of freedom	354
$F(\nu_1,\nu_2)$	F distribution with ν_1,ν_2 degrees of freedom	358
$O(t), o(t)$	of order t, of order less than t as $t\to 0$	371
$X_n\rightsquigarrow$	convergence of $\{X_n\}$ in distribution	377
$X_n\to^p$	convergence of $\{X_n\}$ in probability	385
$X_n\to^{m.s.}$	convergence of $\{X_n\}$ in mean square	386
$X_n\to^{a.s.}$	almost-sure convergence of $\{X_n\}$	387
$O_P(n^q)$	of order n^q in probability	393
S_{XY}	sample covariance between X and Y	427
R_{XY}	sample correlation between X and Y	427
$U_{m,n}$	U statistic of order m from sample of n	429
$\Theta\subset\Re^k$	parameter space	443
$\mathfrak{L}(t_n,\theta)$	loss function at estimate t_n and parameter θ	446
$\mathcal{I}(\theta)$	Fisher information at parameter θ	454
$_*P(\theta)$	prior distribution for θ	456
$P_*(\theta|x)$	posterior distribution for θ given datum x	456
$\tilde{\theta}_n$	method-of-moments estimator of θ from sample of n	478
$L(\mathbf{x};\boldsymbol{\theta})$	likelihood function for θ at sample $\mathbf{x}$	483
$\hat{\theta}_n$	maximum likelihood estimator of θ from sample of n	483

Symbol	Meaning	Page
$\Psi(\alpha)$	digamma function at α	269
$\Psi'(\alpha)$	trigamma function at α	497
$\tilde{\mathcal{I}}\left(\hat{\theta}_n\right)$	sample estimate of Fisher information at $\hat{\theta}_n$	500
$\left[L_{p\alpha}(\mathbf{x}), U_{(1-p)\alpha}(\mathbf{x})\right]$	$1 - \alpha$ confidence interval at sample $\mathbf{x}$	505
$\mathbb{R}_{1-\alpha}(\mathbf{x})$	$1 - \alpha$ confidence region at sample $\mathbf{x}$	506
$\Pi(\mathbf{x}, \boldsymbol{\theta})$	pivotal quantity for $\boldsymbol{\theta}$ at sample $\mathbf{x}$	507
$\Lambda(\mathbf{x}; \theta)$	likelihood ratio for θ at sample $\mathbf{x}$	519
$\mathcal{L}(\boldsymbol{\theta})$	logarithm of likelihood function at $\boldsymbol{\theta}$	520
$\check{\psi}_n$	restricted maximum likelihood estimator	525
H_0, H_1	null, alternate hypotheses	556
$\mathcal{R}_\alpha$	rejection region for H_0 of size α	562
$\mathcal{S}_n$	score statistic for Lagrange-multiplier test	590
$\mathcal{W}_n$	statistic for Wald test	591
$p(\mathbf{x})$	P value or "prob value" of statistical test	603
$\mathbf{S}, \mathbf{R}$	sign, rank statistics for distribution-free tests	609
W^+	Wilcoxon statistic	617
W	Wilcoxon–Mann–Whitney statistic	620
$P_{n,m}$	Pearson chi-squared statistic	646
A^2	Anderson-Darling statistic	655
A_3, A_4	sample skewness, kurtosis	658
$\phi_n(t)$	empirical characteristic function at t	676
I_n	I.c.f. statistic for goodness of fit	678
$P_*(H_0|\mathbf{x})/P_*(H_1|\mathbf{x})$	posterior odds ratio	707
$f(x \mid \mathbf{x})$	predictive distribution at x from sample $\mathbf{x}$	708
$\{E(Y|x) : x \in \mathbb{X}\}$	regression function of r.v. Y on x	732
$\xi(\mathbf{x}; \boldsymbol{\theta})$	regression function at data $\mathbf{x}$, parameters $\boldsymbol{\theta}$	733
$\hat{\mathbf{Y}}$	fitted values $\mathbf{X}\hat{\boldsymbol{\beta}}_n$ from regression	737
$\mathbf{H}$	"hat" matrix, projecting to space spanned by $\mathbf{X}$	737
$\mathbf{M}$	matrix projecting to space orthogonal to $\mathbf{X}$	737
R^2	coefficient of determination	742
$\bar{R}^2$	adjusted R^2	742
$\boldsymbol{\Xi}(\boldsymbol{\beta})$	partial derivatives of nonlinear regression function	750

Appendix C

Statistical Tables

C.1 $N(0,1)$ C.D.F.

$$\Phi(z) = \int_{-\infty}^{z} \frac{1}{\sqrt{2\pi}} e^{-x^2/2} \cdot dx$$

z	0.00	0.01	0.02	0.03	0.04	0.05	0.06	0.07	0.08	0.09
0.0	0.5000	0.5040	0.5080	0.5120	0.5160	0.5199	0.5239	0.5279	0.5319	0.5359
0.1	0.5398	0.5438	0.5478	0.5517	0.5557	0.5596	0.5636	0.5675	0.5714	0.5753
0.2	0.5793	0.5832	0.5871	0.5910	0.5948	0.5987	0.6026	0.6064	0.6103	0.6141
0.3	0.6179	0.6217	0.6255	0.6293	0.6331	0.6368	0.6406	0.6443	0.6480	0.6517
0.4	0.6554	0.6591	0.6628	0.6664	0.6700	0.6736	0.6772	0.6808	0.6844	0.6879
0.5	0.6915	0.6950	0.6985	0.7019	0.7054	0.7088	0.7123	0.7157	0.7190	0.7224
0.6	0.7257	0.7291	0.7324	0.7357	0.7389	0.7422	0.7454	0.7486	0.7517	0.7549
0.7	0.7580	0.7611	0.7642	0.7673	0.7704	0.7734	0.7764	0.7794	0.7823	0.7852
0.8	0.7881	0.7910	0.7939	0.7967	0.7995	0.8023	0.8051	0.8078	0.8106	0.8133
0.9	0.8159	0.8186	0.8212	0.8238	0.8264	0.8289	0.8315	0.8340	0.8365	0.8389
1.0	0.8413	0.8438	0.8461	0.8485	0.8508	0.8531	0.8554	0.8577	0.8599	0.8621
1.1	0.8643	0.8665	0.8686	0.8708	0.8729	0.8749	0.8770	0.8790	0.8810	0.8830
1.2	0.8849	0.8869	0.8888	0.8907	0.8925	0.8944	0.8962	0.8980	0.8997	0.9015
1.3	0.9032	0.9049	0.9066	0.9082	0.9099	0.9115	0.9131	0.9147	0.9162	0.9177
1.4	0.9192	0.9207	0.9222	0.9236	0.9251	0.9265	0.9279	0.9292	0.9306	0.9319
1.5	0.9332	0.9345	0.9357	0.9370	0.9382	0.9394	0.9406	0.9418	0.9429	0.9441
1.6	0.9452	0.9463	0.9474	0.9484	0.9495	0.9505	0.9515	0.9525	0.9535	0.9545
1.7	0.9554	0.9564	0.9573	0.9582	0.9591	0.9599	0.9608	0.9616	0.9625	0.9633
1.8	0.9641	0.9649	0.9656	0.9664	0.9671	0.9678	0.9686	0.9693	0.9699	0.9706
1.9	0.9713	0.9719	0.9726	0.9732	0.9738	0.9744	0.9750	0.9756	0.9761	0.9767
2.0	0.9772	0.9778	0.9783	0.9788	0.9793	0.9798	0.9803	0.9808	0.9812	0.9817
2.1	0.9821	0.9826	0.9830	0.9834	0.9838	0.9842	0.9846	0.9850	0.9854	0.9857
2.2	0.9861	0.9864	0.9868	0.9871	0.9875	0.9878	0.9881	0.9884	0.9887	0.9890
2.3	0.9893	0.9896	0.9898	0.9901	0.9904	0.9906	0.9909	0.9911	0.9913	0.9916
2.4	0.9918	0.9920	0.9922	0.9925	0.9927	0.9929	0.9931	0.9932	0.9934	0.9936
2.5	0.9938	0.9940	0.9941	0.9943	0.9945	0.9946	0.9948	0.9949	0.9951	0.9952
2.6	0.9953	0.9955	0.9956	0.9957	0.9959	0.9960	0.9961	0.9962	0.9963	0.9964
2.7	0.9965	0.9966	0.9967	0.9968	0.9969	0.9970	0.9971	0.9972	0.9973	0.9974
2.8	0.9974	0.9975	0.9976	0.9977	0.9977	0.9978	0.9979	0.9979	0.9980	0.9981
2.9	0.9981	0.9982	0.9982	0.9983	0.9984	0.9984	0.9985	0.9985	0.9986	0.9986
3.0	0.9987	0.9987	0.9987	0.9988	0.9988	0.9989	0.9989	0.9989	0.9990	0.9990

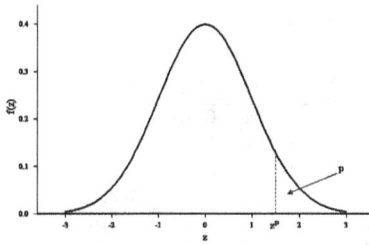

Fig. C.1

C.2 Upper $N(0,1)$ Quantiles z^p

p	$z^p = \Phi^{-1}(1-p)$
0.50000	0.00000
0.45000	0.12566
0.40000	0.25335
0.35000	0.38532
0.30000	0.52440
0.25000	0.67449
0.20000	0.84162
0.15000	1.03643
0.10000	1.28155
0.09000	1.34075
0.08000	1.40507
0.07000	1.47579
0.06000	1.55477
0.05000	1.64485
0.04000	1.75069
0.03000	1.88079
0.02500	1.95996
0.02000	2.05375
0.01000	2.32635
0.00500	2.57583
0.00400	2.65207
0.00300	2.74778
0.00250	2.80703
0.00200	2.87816
0.00100	3.09023
0.00050	3.29053
0.00040	3.35279
0.00030	3.43161
0.00025	3.48076
0.00020	3.54008
0.00010	3.71902
0.00005	3.89059

C.3 Upper $\chi^2(\nu)$ Quantiles c_ν^p

ν	p :0.995	0.99	0.975	0.95	0.9	0.1	0.05	0.025	0.01	0.005
1	0.000	0.000	0.001	0.004	0.016	2.706	3.841	5.024	6.635	7.880
2	0.010	0.020	0.051	0.103	0.211	4.605	5.991	7.378	9.210	10.597
3	0.072	0.115	0.216	0.352	0.584	6.251	7.815	9.349	11.345	12.838
4	0.207	0.297	0.484	0.711	1.064	7.779	9.488	11.143	13.277	14.861
5	0.412	0.554	0.831	1.145	1.610	9.236	11.071	12.833	15.086	16.750
6	0.676	0.872	1.237	1.635	2.204	10.645	12.592	14.449	16.812	18.548
7	0.989	1.239	1.690	2.167	2.833	12.017	14.067	16.013	18.476	20.279
8	1.344	1.646	2.180	2.733	3.490	13.362	15.507	17.535	20.090	21.956
9	1.735	2.088	2.700	3.325	4.168	14.684	16.919	19.023	21.666	23.590
10	2.156	2.558	3.247	3.940	4.865	15.987	18.307	20.483	23.210	25.189
11	2.603	3.053	3.816	4.575	5.578	17.275	19.675	21.920	24.725	26.757
12	3.074	3.571	4.404	5.226	6.304	18.549	21.026	23.337	26.217	28.300
13	3.565	4.107	5.009	5.892	7.042	19.812	22.362	24.736	27.688	29.819
14	4.075	4.660	5.629	6.571	7.790	21.064	23.685	26.119	29.141	31.319
15	4.601	5.229	6.262	7.261	8.547	22.307	24.996	27.488	30.578	32.801
16	5.142	5.812	6.908	7.962	9.312	23.542	26.296	28.845	32.000	34.267
17	5.697	6.408	7.564	8.672	10.085	24.769	27.587	30.191	33.409	35.718
18	6.265	7.015	8.231	9.390	10.865	25.989	28.869	31.526	34.805	37.156
19	6.844	7.633	8.907	10.117	11.651	27.204	30.144	32.852	36.191	38.582
20	7.434	8.260	9.591	10.851	12.443	28.412	31.410	34.170	37.566	39.997
21	8.034	8.897	10.283	11.591	13.240	29.615	32.671	35.479	38.932	41.401
22	8.643	9.542	10.982	12.338	14.041	30.813	33.924	36.781	40.289	42.796
23	9.260	10.196	11.689	13.091	14.848	32.007	35.172	38.076	41.638	44.181
24	9.886	10.856	12.401	13.848	15.659	33.196	36.415	39.364	42.980	45.559
25	10.520	11.524	13.120	14.611	16.473	34.382	37.652	40.646	44.314	46.928
26	11.160	12.198	13.844	15.379	17.292	35.563	38.885	41.923	45.642	48.290
27	11.808	12.879	14.573	16.151	18.114	36.741	40.113	43.195	46.963	49.645
28	12.461	13.565	15.308	16.928	18.939	37.916	41.337	44.461	48.278	50.993
29	13.121	14.256	16.047	17.708	19.768	39.087	42.557	45.722	49.588	52.336
30	13.787	14.953	16.791	18.493	20.599	40.256	43.773	46.979	50.892	53.672
40	20.707	22.164	24.433	26.509	29.050	51.805	55.758	59.342	63.691	66.766
50	27.990	29.707	32.357	34.764	37.689	63.167	67.505	71.420	76.154	79.490
60	35.533	37.484	40.481	43.188	46.459	74.397	79.082	83.298	88.379	91.952
70	43.246	45.423	48.750	51.739	55.333	85.521	90.528	95.026	100.44	104.24
80	51.145	53.523	57.147	60.391	64.282	96.572	101.88	106.63	112.34	116.35
90	59.171	61.738	65.641	69.126	73.295	107.56	113.14	118.14	124.13	128.32
100	67.303	70.049	74.216	77.929	82.362	118.49	124.34	129.56	135.82	140.19
150	109.12	112.66	117.98	122.69	128.28	172.58	179.58	185.80	193.22	198.38
200	152.22	156.42	162.72	168.28	174.84	226.02	233.99	241.06	249.46	255.28

C.4　Upper $t(\nu)$ Quantiles t_ν^p

ν	p :0.400	0.300	0.200	0.100	0.050	0.025	0.010	0.005	0.001
1	0.325	0.727	1.376	3.078	6.314	12.706	31.821	63.657	318.309
2	0.289	0.617	1.061	1.886	2.920	4.303	6.965	9.925	22.327
3	0.277	0.584	0.978	1.638	2.353	3.182	4.541	5.841	10.215
4	0.271	0.569	0.941	1.533	2.132	2.776	3.747	4.604	7.173
5	0.267	0.559	0.920	1.476	2.015	2.571	3.365	4.032	5.893
6	0.265	0.553	0.906	1.440	1.943	2.447	3.143	3.707	5.208
7	0.263	0.549	0.896	1.415	1.895	2.365	2.998	3.499	4.785
8	0.262	0.546	0.889	1.397	1.860	2.306	2.896	3.355	4.501
9	0.261	0.543	0.883	1.383	1.833	2.262	2.821	3.250	4.297
10	0.260	0.542	0.879	1.372	1.812	2.228	2.764	3.169	4.144
11	0.260	0.540	0.876	1.363	1.796	2.201	2.718	3.106	4.025
12	0.259	0.539	0.873	1.356	1.782	2.179	2.681	3.055	3.930
13	0.259	0.538	0.870	1.350	1.771	2.160	2.650	3.012	3.852
14	0.258	0.537	0.868	1.345	1.761	2.145	2.624	2.977	3.787
15	0.258	0.536	0.866	1.341	1.753	2.131	2.602	2.947	3.733
16	0.258	0.535	0.865	1.337	1.746	2.120	2.583	2.921	3.686
17	0.257	0.534	0.863	1.333	1.740	2.110	2.567	2.898	3.646
18	0.257	0.534	0.862	1.330	1.734	2.101	2.552	2.878	3.610
19	0.257	0.533	0.861	1.328	1.729	2.093	2.539	2.861	3.579
20	0.257	0.533	0.860	1.325	1.725	2.086	2.528	2.845	3.552
21	0.257	0.532	0.859	1.323	1.721	2.080	2.518	2.831	3.527
22	0.256	0.532	0.858	1.321	1.717	2.074	2.508	2.819	3.505
23	0.256	0.532	0.858	1.319	1.714	2.069	2.500	2.807	3.485
24	0.256	0.531	0.857	1.318	1.711	2.064	2.492	2.797	3.467
25	0.256	0.531	0.856	1.316	1.708	2.060	2.485	2.787	3.450
26	0.256	0.531	0.856	1.315	1.706	2.056	2.479	2.779	3.435
27	0.256	0.531	0.855	1.314	1.703	2.052	2.473	2.771	3.421
28	0.256	0.530	0.855	1.313	1.701	2.048	2.467	2.763	3.408
29	0.256	0.530	0.854	1.311	1.699	2.045	2.462	2.756	3.396
30	0.256	0.530	0.854	1.310	1.697	2.042	2.457	2.750	3.385
40	0.255	0.529	0.851	1.303	1.684	2.021	2.423	2.704	3.307
50	0.255	0.528	0.849	1.299	1.676	2.009	2.403	2.678	3.261
60	0.254	0.527	0.848	1.296	1.671	2.000	2.390	2.660	3.232
70	0.254	0.527	0.847	1.294	1.667	1.994	2.381	2.648	3.211
80	0.254	0.526	0.846	1.292	1.664	1.990	2.374	2.639	3.195
90	0.254	0.526	0.846	1.291	1.662	1.987	2.368	2.632	3.183
100	0.254	0.526	0.845	1.290	1.660	1.984	2.364	2.626	3.174
150	0.254	0.526	0.844	1.287	1.655	1.976	2.351	2.609	3.145
200	0.254	0.525	0.843	1.286	1.653	1.972	2.345	2.601	3.131
∞	0.253	0.524	0.842	1.282	1.645	1.960	2.326	2.576	3.090

C.5　Upper $F(\nu_1, \nu_2)$ Quantiles $f^p_{\nu_1, \nu_2}$

$$p = .10$$

ν_2	ν_1 :1	2	3	4	5	6	7	8	9	10
1	39.86	49.50	53.59	55.83	57.24	58.20	58.91	59.44	59.86	60.19
2	8.53	9.00	9.16	9.24	9.29	9.33	9.35	9.37	9.38	9.39
3	5.54	5.46	5.39	5.34	5.31	5.28	5.27	5.25	5.24	5.23
4	4.54	4.32	4.19	4.11	4.05	4.01	3.98	3.95	3.94	3.92
5	4.06	3.78	3.62	3.52	3.45	3.40	3.37	3.34	3.32	3.30
6	3.78	3.46	3.29	3.18	3.11	3.05	3.01	2.98	2.96	2.94
7	3.59	3.26	3.07	2.96	2.88	2.83	2.78	2.75	2.72	2.70
8	3.46	3.11	2.92	2.81	2.73	2.67	2.62	2.59	2.56	2.54
9	3.36	3.01	2.81	2.69	2.61	2.55	2.51	2.47	2.44	2.42
10	3.29	2.92	2.73	2.61	2.52	2.46	2.41	2.38	2.35	2.32
11	3.23	2.86	2.66	2.54	2.45	2.39	2.34	2.30	2.27	2.25
12	3.18	2.81	2.61	2.48	2.39	2.33	2.28	2.24	2.21	2.19
13	3.14	2.76	2.56	2.43	2.35	2.28	2.23	2.20	2.16	2.14
14	3.10	2.73	2.52	2.39	2.31	2.24	2.19	2.15	2.12	2.10
15	3.07	2.70	2.49	2.36	2.27	2.21	2.16	2.12	2.09	2.06
16	3.05	2.67	2.46	2.33	2.24	2.18	2.13	2.09	2.06	2.03
17	3.03	2.64	2.44	2.31	2.22	2.15	2.10	2.06	2.03	2.00
18	3.01	2.62	2.42	2.29	2.20	2.13	2.08	2.04	2.00	1.98
19	2.99	2.61	2.40	2.27	2.18	2.11	2.06	2.02	1.98	1.96
20	2.97	2.59	2.38	2.25	2.16	2.09	2.04	2.00	1.96	1.94
21	2.96	2.57	2.36	2.23	2.14	2.08	2.02	1.98	1.95	1.92
22	2.95	2.56	2.35	2.22	2.13	2.06	2.01	1.97	1.93	1.90
23	2.94	2.55	2.34	2.21	2.11	2.05	1.99	1.95	1.92	1.89
24	2.93	2.54	2.33	2.19	2.10	2.04	1.98	1.94	1.91	1.88
25	2.92	2.53	2.32	2.18	2.09	2.02	1.97	1.93	1.89	1.87
26	2.91	2.52	2.31	2.17	2.08	2.01	1.96	1.92	1.88	1.86
27	2.90	2.51	2.30	2.17	2.07	2.00	1.95	1.91	1.87	1.85
28	2.89	2.50	2.29	2.16	2.06	2.00	1.94	1.90	1.87	1.84
29	2.89	2.50	2.28	2.15	2.06	1.99	1.93	1.89	1.86	1.83
30	2.88	2.49	2.28	2.14	2.05	1.98	1.93	1.88	1.85	1.82
40	2.84	2.44	2.23	2.09	2.00	1.93	1.87	1.83	1.79	1.76
50	2.81	2.41	2.20	2.06	1.97	1.90	1.84	1.80	1.76	1.73
60	2.79	2.39	2.18	2.04	1.95	1.87	1.82	1.77	1.74	1.71
70	2.78	2.38	2.16	2.03	1.93	1.86	1.80	1.76	1.72	1.69
80	2.77	2.37	2.15	2.02	1.92	1.85	1.79	1.75	1.71	1.68
90	2.76	2.36	2.15	2.01	1.91	1.84	1.78	1.74	1.70	1.67
100	2.76	2.36	2.14	2.00	1.91	1.83	1.78	1.73	1.69	1.66
150	2.74	2.34	2.12	1.98	1.89	1.81	1.76	1.71	1.67	1.64
200	2.73	2.33	2.11	1.97	1.88	1.80	1.75	1.70	1.66	1.63

$p = .10$ (continued)

ν_2	ν_1 :12	14	17	20	25	30	40	60	100	150
1	60.71	61.07	61.46	61.74	62.05	62.26	62.53	62.79	63.01	63.11
2	9.41	9.42	9.43	9.44	9.45	9.46	9.47	9.47	9.48	9.48
3	5.22	5.20	5.19	5.18	5.17	5.17	5.16	5.15	5.14	5.14
4	3.90	3.88	3.86	3.84	3.83	3.82	3.80	3.79	3.78	3.77
5	3.27	3.25	3.22	3.21	3.19	3.17	3.16	3.14	3.13	3.12
6	2.90	2.88	2.85	2.84	2.81	2.80	2.78	2.76	2.75	2.74
7	2.67	2.64	2.61	2.59	2.57	2.56	2.54	2.51	2.50	2.49
8	2.50	2.48	2.45	2.42	2.40	2.38	2.36	2.34	2.32	2.31
9	2.38	2.35	2.32	2.30	2.27	2.25	2.23	2.21	2.19	2.18
10	2.28	2.26	2.22	2.20	2.17	2.16	2.13	2.11	2.09	2.08
11	2.21	2.18	2.15	2.12	2.10	2.08	2.05	2.03	2.01	1.99
12	2.15	2.12	2.08	2.06	2.03	2.01	1.99	1.96	1.94	1.93
13	2.10	2.07	2.03	2.01	1.98	1.96	1.93	1.90	1.88	1.87
14	2.05	2.02	1.99	1.96	1.93	1.91	1.89	1.86	1.83	1.82
15	2.02	1.99	1.95	1.92	1.89	1.87	1.85	1.82	1.79	1.78
16	1.99	1.95	1.92	1.89	1.86	1.84	1.81	1.78	1.76	1.74
17	1.96	1.93	1.89	1.86	1.83	1.81	1.78	1.75	1.73	1.71
18	1.93	1.90	1.86	1.84	1.80	1.78	1.75	1.72	1.70	1.68
19	1.91	1.88	1.84	1.81	1.78	1.76	1.73	1.70	1.67	1.66
20	1.89	1.86	1.82	1.79	1.76	1.74	1.71	1.68	1.65	1.64
21	1.87	1.84	1.80	1.78	1.74	1.72	1.69	1.66	1.63	1.62
22	1.86	1.83	1.79	1.76	1.73	1.70	1.67	1.64	1.61	1.60
23	1.84	1.81	1.77	1.74	1.71	1.69	1.66	1.62	1.59	1.58
24	1.83	1.80	1.76	1.73	1.70	1.67	1.64	1.61	1.58	1.56
25	1.82	1.79	1.75	1.72	1.68	1.66	1.63	1.59	1.56	1.55
26	1.81	1.77	1.73	1.71	1.67	1.65	1.61	1.58	1.55	1.54
27	1.80	1.76	1.72	1.70	1.66	1.64	1.60	1.57	1.54	1.52
28	1.79	1.75	1.71	1.69	1.65	1.63	1.59	1.56	1.53	1.51
29	1.78	1.75	1.71	1.68	1.64	1.62	1.58	1.55	1.52	1.50
30	1.77	1.74	1.70	1.67	1.63	1.61	1.57	1.54	1.51	1.49
40	1.71	1.68	1.64	1.61	1.57	1.54	1.51	1.47	1.43	1.42
50	1.68	1.64	1.60	1.57	1.53	1.50	1.46	1.42	1.39	1.37
60	1.66	1.62	1.58	1.54	1.50	1.48	1.44	1.40	1.36	1.34
70	1.64	1.60	1.56	1.53	1.49	1.46	1.42	1.37	1.34	1.31
80	1.63	1.59	1.55	1.51	1.47	1.44	1.40	1.36	1.32	1.30
90	1.62	1.58	1.54	1.50	1.46	1.43	1.39	1.35	1.30	1.28
100	1.61	1.57	1.53	1.49	1.45	1.42	1.38	1.34	1.29	1.27
150	1.59	1.55	1.50	1.47	1.43	1.40	1.35	1.30	1.26	1.23
200	1.58	1.54	1.49	1.46	1.41	1.38	1.34	1.29	1.24	1.21

$p = .05$

ν_2	ν_1 :1	2	3	4	5	6	7	8	9	10
1	161.45	199.50	215.71	224.58	230.16	233.99	236.77	238.88	240.54	241.88
2	18.51	19.00	19.16	19.25	19.30	19.33	19.35	19.37	19.38	19.40
3	10.13	9.55	9.28	9.12	9.01	8.94	8.89	8.85	8.81	8.79
4	7.71	6.94	6.59	6.39	6.26	6.16	6.09	6.04	6.00	5.96
5	6.61	5.79	5.41	5.19	5.05	4.95	4.88	4.82	4.77	4.74
6	5.99	5.14	4.76	4.53	4.39	4.28	4.21	4.15	4.10	4.06
7	5.59	4.74	4.35	4.12	3.97	3.87	3.79	3.73	3.68	3.64
8	5.32	4.46	4.07	3.84	3.69	3.58	3.50	3.44	3.39	3.35
9	5.12	4.26	3.86	3.63	3.48	3.37	3.29	3.23	3.18	3.14
10	4.96	4.10	3.71	3.48	3.33	3.22	3.14	3.07	3.02	2.98
11	4.84	3.98	3.59	3.36	3.20	3.09	3.01	2.95	2.90	2.85
12	4.75	3.89	3.49	3.26	3.11	3.00	2.91	2.85	2.80	2.75
13	4.67	3.81	3.41	3.18	3.03	2.92	2.83	2.77	2.71	2.67
14	4.60	3.74	3.34	3.11	2.96	2.85	2.76	2.70	2.65	2.60
15	4.54	3.68	3.29	3.06	2.90	2.79	2.71	2.64	2.59	2.54
16	4.49	3.63	3.24	3.01	2.85	2.74	2.66	2.59	2.54	2.49
17	4.45	3.59	3.20	2.96	2.81	2.70	2.61	2.55	2.49	2.45
18	4.41	3.55	3.16	2.93	2.77	2.66	2.58	2.51	2.46	2.41
19	4.38	3.52	3.13	2.90	2.74	2.63	2.54	2.48	2.42	2.38
20	4.35	3.49	3.10	2.87	2.71	2.60	2.51	2.45	2.39	2.35
21	4.32	3.47	3.07	2.84	2.68	2.57	2.49	2.42	2.37	2.32
22	4.30	3.44	3.05	2.82	2.66	2.55	2.46	2.40	2.34	2.30
23	4.28	3.42	3.03	2.80	2.64	2.53	2.44	2.37	2.32	2.27
24	4.26	3.40	3.01	2.78	2.62	2.51	2.42	2.36	2.30	2.25
25	4.24	3.39	2.99	2.76	2.60	2.49	2.40	2.34	2.28	2.24
26	4.23	3.37	2.98	2.74	2.59	2.47	2.39	2.32	2.27	2.22
27	4.21	3.35	2.96	2.73	2.57	2.46	2.37	2.31	2.25	2.20
28	4.20	3.34	2.95	2.71	2.56	2.45	2.36	2.29	2.24	2.19
29	4.18	3.33	2.93	2.70	2.55	2.43	2.35	2.28	2.22	2.18
30	4.17	3.32	2.92	2.69	2.53	2.42	2.33	2.27	2.21	2.16
40	4.08	3.23	2.84	2.61	2.45	2.34	2.25	2.18	2.12	2.08
50	4.03	3.18	2.79	2.56	2.40	2.29	2.20	2.13	2.07	2.03
60	4.00	3.15	2.76	2.53	2.37	2.25	2.17	2.10	2.04	1.99
70	3.98	3.13	2.74	2.50	2.35	2.23	2.14	2.07	2.02	1.97
80	3.96	3.11	2.72	2.49	2.33	2.21	2.13	2.06	2.00	1.95
90	3.95	3.10	2.71	2.47	2.32	2.20	2.11	2.04	1.99	1.94
100	3.94	3.09	2.70	2.46	2.31	2.19	2.10	2.03	1.97	1.93
150	3.90	3.06	2.66	2.43	2.27	2.16	2.07	2.00	1.94	1.89
200	3.89	3.04	2.65	2.42	2.26	2.14	2.06	1.98	1.93	1.88

$p = .05$ (continued)

ν_2	ν_1 :12	14	17	20	25	30	40	60	100	150
1	243.91	245.36	246.92	248.01	249.26	250.10	251.14	252.20	253.04	253.46
2	19.41	19.42	19.44	19.45	19.46	19.46	19.47	19.48	19.49	19.49
3	8.74	8.71	8.68	8.66	8.63	8.62	8.59	8.57	8.55	8.54
4	5.91	5.87	5.83	5.80	5.77	5.75	5.72	5.69	5.66	5.65
5	4.68	4.64	4.59	4.56	4.52	4.50	4.46	4.43	4.41	4.39
6	4.00	3.96	3.91	3.87	3.83	3.81	3.77	3.74	3.71	3.70
7	3.57	3.53	3.48	3.44	3.40	3.38	3.34	3.30	3.27	3.26
8	3.28	3.24	3.19	3.15	3.11	3.08	3.04	3.01	2.97	2.96
9	3.07	3.03	2.97	2.94	2.89	2.86	2.83	2.79	2.76	2.74
10	2.91	2.86	2.81	2.77	2.73	2.70	2.66	2.62	2.59	2.57
11	2.79	2.74	2.69	2.65	2.60	2.57	2.53	2.49	2.46	2.44
12	2.69	2.64	2.58	2.54	2.50	2.47	2.43	2.38	2.35	2.33
13	2.60	2.55	2.50	2.46	2.41	2.38	2.34	2.30	2.26	2.24
14	2.53	2.48	2.43	2.39	2.34	2.31	2.27	2.22	2.19	2.17
15	2.48	2.42	2.37	2.33	2.28	2.25	2.20	2.16	2.12	2.10
16	2.42	2.37	2.32	2.28	2.23	2.19	2.15	2.11	2.07	2.05
17	2.38	2.33	2.27	2.23	2.18	2.15	2.10	2.06	2.02	2.00
18	2.34	2.29	2.23	2.19	2.14	2.11	2.06	2.02	1.98	1.96
19	2.31	2.26	2.20	2.16	2.11	2.07	2.03	1.98	1.94	1.92
20	2.28	2.22	2.17	2.12	2.07	2.04	1.99	1.95	1.91	1.89
21	2.25	2.20	2.14	2.10	2.05	2.01	1.96	1.92	1.88	1.86
22	2.23	2.17	2.11	2.07	2.02	1.98	1.94	1.89	1.85	1.83
23	2.20	2.15	2.09	2.05	2.00	1.96	1.91	1.86	1.82	1.80
24	2.18	2.13	2.07	2.03	1.97	1.94	1.89	1.84	1.80	1.78
25	2.16	2.11	2.05	2.01	1.96	1.92	1.87	1.82	1.78	1.76
26	2.15	2.09	2.03	1.99	1.94	1.90	1.85	1.80	1.76	1.74
27	2.13	2.08	2.02	1.97	1.92	1.88	1.84	1.79	1.74	1.72
28	2.12	2.06	2.00	1.96	1.91	1.87	1.82	1.77	1.73	1.70
29	2.10	2.05	1.99	1.94	1.89	1.85	1.81	1.75	1.71	1.69
30	2.09	2.04	1.98	1.93	1.88	1.84	1.79	1.74	1.70	1.67
40	2.00	1.95	1.89	1.84	1.78	1.74	1.69	1.64	1.59	1.56
50	1.95	1.89	1.83	1.78	1.73	1.69	1.63	1.58	1.52	1.50
60	1.92	1.86	1.80	1.75	1.69	1.65	1.59	1.53	1.48	1.45
70	1.89	1.84	1.77	1.72	1.66	1.62	1.57	1.50	1.45	1.42
80	1.88	1.82	1.75	1.70	1.64	1.60	1.54	1.48	1.43	1.39
90	1.86	1.80	1.74	1.69	1.63	1.59	1.53	1.46	1.41	1.38
100	1.85	1.79	1.73	1.68	1.62	1.57	1.52	1.45	1.39	1.36
150	1.82	1.76	1.69	1.64	1.58	1.54	1.48	1.41	1.34	1.31
200	1.80	1.74	1.67	1.62	1.56	1.52	1.46	1.39	1.32	1.28

$$p = .025$$

ν_2	ν_1 :1	2	3	4	5	6	7	8	9	10
1	647.79	799.50	864.16	899.58	921.85	937.11	948.22	956.66	963.28	968.63
2	38.51	39.00	39.17	39.25	39.30	39.33	39.36	39.37	39.39	39.40
3	17.44	16.04	15.44	15.10	14.88	14.73	14.62	14.54	14.47	14.42
4	12.22	10.65	9.98	9.60	9.36	9.20	9.07	8.98	8.90	8.84
5	10.01	8.43	7.76	7.39	7.15	6.98	6.85	6.76	6.68	6.62
6	8.81	7.26	6.60	6.23	5.99	5.82	5.70	5.60	5.52	5.46
7	8.07	6.54	5.89	5.52	5.29	5.12	4.99	4.90	4.82	4.76
8	7.57	6.06	5.42	5.05	4.82	4.65	4.53	4.43	4.36	4.30
9	7.21	5.71	5.08	4.72	4.48	4.32	4.20	4.10	4.03	3.96
10	6.94	5.46	4.83	4.47	4.24	4.07	3.95	3.85	3.78	3.72
11	6.72	5.26	4.63	4.28	4.04	3.88	3.76	3.66	3.59	3.53
12	6.55	5.10	4.47	4.12	3.89	3.73	3.61	3.51	3.44	3.37
13	6.41	4.97	4.35	4.00	3.77	3.60	3.48	3.39	3.31	3.25
14	6.30	4.86	4.24	3.89	3.66	3.50	3.38	3.29	3.21	3.15
15	6.20	4.77	4.15	3.80	3.58	3.41	3.29	3.20	3.12	3.06
16	6.12	4.69	4.08	3.73	3.50	3.34	3.22	3.12	3.05	2.99
17	6.04	4.62	4.01	3.66	3.44	3.28	3.16	3.06	2.98	2.92
18	5.98	4.56	3.95	3.61	3.38	3.22	3.10	3.01	2.93	2.87
19	5.92	4.51	3.90	3.56	3.33	3.17	3.05	2.96	2.88	2.82
20	5.87	4.46	3.86	3.51	3.29	3.13	3.01	2.91	2.84	2.77
21	5.83	4.42	3.82	3.48	3.25	3.09	2.97	2.87	2.80	2.73
22	5.79	4.38	3.78	3.44	3.22	3.05	2.93	2.84	2.76	2.70
23	5.75	4.35	3.75	3.41	3.18	3.02	2.90	2.81	2.73	2.67
24	5.72	4.32	3.72	3.38	3.15	2.99	2.87	2.78	2.70	2.64
25	5.69	4.29	3.69	3.35	3.13	2.97	2.85	2.75	2.68	2.61
26	5.66	4.27	3.67	3.33	3.10	2.94	2.82	2.73	2.65	2.59
27	5.63	4.24	3.65	3.31	3.08	2.92	2.80	2.71	2.63	2.57
28	5.61	4.22	3.63	3.29	3.06	2.90	2.78	2.69	2.61	2.55
29	5.59	4.20	3.61	3.27	3.04	2.88	2.76	2.67	2.59	2.53
30	5.57	4.18	3.59	3.25	3.03	2.87	2.75	2.65	2.57	2.51
40	5.42	4.05	3.46	3.13	2.90	2.74	2.62	2.53	2.45	2.39
50	5.34	3.97	3.39	3.05	2.83	2.67	2.55	2.46	2.38	2.32
60	5.29	3.93	3.34	3.01	2.79	2.63	2.51	2.41	2.33	2.27
70	5.25	3.89	3.31	2.97	2.75	2.59	2.47	2.38	2.30	2.24
80	5.22	3.86	3.28	2.95	2.73	2.57	2.45	2.35	2.28	2.21
90	5.20	3.84	3.26	2.93	2.71	2.55	2.43	2.34	2.26	2.19
100	5.18	3.83	3.25	2.92	2.70	2.54	2.42	2.32	2.24	2.18
150	5.13	3.78	3.20	2.87	2.65	2.49	2.37	2.28	2.20	2.13
200	5.10	3.76	3.18	2.85	2.63	2.47	2.35	2.26	2.18	2.11

$p = .025$ (continued)

ν_2	ν_1 :12	14	17	20	25	30	40	60	100	150
1	976.71	982.53	988.73	993.10	998.08	1001.4	1005.6	1009.8	1013.2	1014.9
2	39.41	39.43	39.44	39.45	39.46	39.46	39.47	39.48	39.49	39.49
3	14.34	14.28	14.21	14.17	14.12	14.08	14.04	13.99	13.96	13.94
4	8.75	8.68	8.61	8.56	8.50	8.46	8.41	8.36	8.32	8.30
5	6.52	6.46	6.38	6.33	6.27	6.23	6.18	6.12	6.08	6.06
6	5.37	5.30	5.22	5.17	5.11	5.07	5.01	4.96	4.92	4.89
7	4.67	4.60	4.52	4.47	4.40	4.36	4.31	4.25	4.21	4.19
8	4.20	4.13	4.05	4.00	3.94	3.89	3.84	3.78	3.74	3.72
9	3.87	3.80	3.72	3.67	3.60	3.56	3.51	3.45	3.40	3.38
10	3.62	3.55	3.47	3.42	3.35	3.31	3.26	3.20	3.15	3.13
11	3.43	3.36	3.28	3.23	3.16	3.12	3.06	3.00	2.96	2.93
12	3.28	3.21	3.13	3.07	3.01	2.96	2.91	2.85	2.80	2.78
13	3.15	3.08	3.00	2.95	2.88	2.84	2.78	2.72	2.67	2.65
14	3.05	2.98	2.90	2.84	2.78	2.73	2.67	2.61	2.56	2.54
15	2.96	2.89	2.81	2.76	2.69	2.64	2.59	2.52	2.47	2.45
16	2.89	2.82	2.74	2.68	2.61	2.57	2.51	2.45	2.40	2.37
17	2.82	2.75	2.67	2.62	2.55	2.50	2.44	2.38	2.33	2.30
18	2.77	2.70	2.62	2.56	2.49	2.44	2.38	2.32	2.27	2.24
19	2.72	2.65	2.57	2.51	2.44	2.39	2.33	2.27	2.22	2.19
20	2.68	2.60	2.52	2.46	2.40	2.35	2.29	2.22	2.17	2.14
21	2.64	2.56	2.48	2.42	2.36	2.31	2.25	2.18	2.13	2.10
22	2.60	2.53	2.45	2.39	2.32	2.27	2.21	2.14	2.09	2.06
23	2.57	2.50	2.42	2.36	2.29	2.24	2.18	2.11	2.06	2.03
24	2.54	2.47	2.39	2.33	2.26	2.21	2.15	2.08	2.02	2.00
25	2.51	2.44	2.36	2.30	2.23	2.18	2.12	2.05	2.00	1.97
26	2.49	2.42	2.34	2.28	2.21	2.16	2.09	2.03	1.97	1.94
27	2.47	2.39	2.31	2.25	2.18	2.13	2.07	2.00	1.94	1.91
28	2.45	2.37	2.29	2.23	2.16	2.11	2.05	1.98	1.92	1.89
29	2.43	2.36	2.27	2.21	2.14	2.09	2.03	1.96	1.90	1.87
30	2.41	2.34	2.26	2.20	2.12	2.07	2.01	1.94	1.88	1.85
40	2.29	2.21	2.13	2.07	1.99	1.94	1.88	1.80	1.74	1.71
50	2.22	2.14	2.06	1.99	1.92	1.87	1.80	1.72	1.66	1.62
60	2.17	2.09	2.01	1.94	1.87	1.82	1.74	1.67	1.60	1.56
70	2.14	2.06	1.97	1.91	1.83	1.78	1.71	1.63	1.56	1.52
80	2.11	2.03	1.95	1.88	1.81	1.75	1.68	1.60	1.53	1.49
90	2.09	2.02	1.93	1.86	1.79	1.73	1.66	1.58	1.50	1.46
100	2.08	2.00	1.91	1.85	1.77	1.71	1.64	1.56	1.48	1.44
150	2.03	1.95	1.87	1.80	1.72	1.67	1.59	1.50	1.42	1.38
200	2.01	1.93	1.84	1.78	1.70	1.64	1.56	1.47	1.39	1.35

$$p = .01$$

ν_2	ν_1 :1	2	3	4	5	6	7	8	9	10
1	4052.2	4999.5	5403.4	5624.6	5763.6	5859	5928.4	5981.1	6022.5	6055.8
2	98.50	99.00	99.17	99.25	99.30	99.33	99.36	99.37	99.39	99.40
3	34.12	30.82	29.46	28.71	28.24	27.91	27.67	27.49	27.35	27.23
4	21.20	18.00	16.69	15.98	15.52	15.21	14.98	14.80	14.66	14.55
5	16.26	13.27	12.06	11.39	10.97	10.67	10.46	10.29	10.16	10.05
6	13.75	10.92	9.78	9.15	8.75	8.47	8.26	8.10	7.98	7.87
7	12.25	9.55	8.45	7.85	7.46	7.19	6.99	6.84	6.72	6.62
8	11.26	8.65	7.59	7.01	6.63	6.37	6.18	6.03	5.91	5.81
9	10.56	8.02	6.99	6.42	6.06	5.80	5.61	5.47	5.35	5.26
10	10.04	7.56	6.55	5.99	5.64	5.39	5.20	5.06	4.94	4.85
11	9.65	7.21	6.22	5.67	5.32	5.07	4.89	4.74	4.63	4.54
12	9.33	6.93	5.95	5.41	5.06	4.82	4.64	4.50	4.39	4.30
13	9.07	6.70	5.74	5.21	4.86	4.62	4.44	4.30	4.19	4.10
14	8.86	6.51	5.56	5.04	4.69	4.46	4.28	4.14	4.03	3.94
15	8.68	6.36	5.42	4.89	4.56	4.32	4.14	4.00	3.89	3.80
16	8.53	6.23	5.29	4.77	4.44	4.20	4.03	3.89	3.78	3.69
17	8.40	6.11	5.18	4.67	4.34	4.10	3.93	3.79	3.68	3.59
18	8.29	6.01	5.09	4.58	4.25	4.01	3.84	3.71	3.60	3.51
19	8.18	5.93	5.01	4.50	4.17	3.94	3.77	3.63	3.52	3.43
20	8.10	5.85	4.94	4.43	4.10	3.87	3.70	3.56	3.46	3.37
21	8.02	5.78	4.87	4.37	4.04	3.81	3.64	3.51	3.40	3.31
22	7.95	5.72	4.82	4.31	3.99	3.76	3.59	3.45	3.35	3.26
23	7.88	5.66	4.76	4.26	3.94	3.71	3.54	3.41	3.30	3.21
24	7.82	5.61	4.72	4.22	3.90	3.67	3.50	3.36	3.26	3.17
25	7.77	5.57	4.68	4.18	3.85	3.63	3.46	3.32	3.22	3.13
26	7.72	5.53	4.64	4.14	3.82	3.59	3.42	3.29	3.18	3.09
27	7.68	5.49	4.60	4.11	3.78	3.56	3.39	3.26	3.15	3.06
28	7.64	5.45	4.57	4.07	3.75	3.53	3.36	3.23	3.12	3.03
29	7.60	5.42	4.54	4.04	3.73	3.50	3.33	3.20	3.09	3.00
30	7.56	5.39	4.51	4.02	3.70	3.47	3.30	3.17	3.07	2.98
40	7.31	5.18	4.31	3.83	3.51	3.29	3.12	2.99	2.89	2.80
50	7.17	5.06	4.20	3.72	3.41	3.19	3.02	2.89	2.78	2.70
60	7.08	4.98	4.13	3.65	3.34	3.12	2.95	2.82	2.72	2.63
70	7.01	4.92	4.07	3.60	3.29	3.07	2.91	2.78	2.67	2.59
80	6.96	4.88	4.04	3.56	3.26	3.04	2.87	2.74	2.64	2.55
90	6.93	4.85	4.01	3.53	3.23	3.01	2.84	2.72	2.61	2.52
100	6.90	4.82	3.98	3.51	3.21	2.99	2.82	2.69	2.59	2.50
150	6.81	4.75	3.91	3.45	3.14	2.92	2.76	2.63	2.53	2.44
200	6.76	4.71	3.88	3.41	3.11	2.89	2.73	2.60	2.50	2.41

$p = .01$ (continued)

ν_2	ν_1 :12	14	17	20	25	30	40	60	100	150
1	6106.3	6142.7	6181.4	6208.7	6239.8	6260.6	6286.8	6313	6334.1	6344.7
2	99.42	99.43	99.44	99.45	99.46	99.47	99.47	99.48	99.49	99.49
3	27.05	26.92	26.79	26.69	26.58	26.50	26.41	26.32	26.24	26.20
4	14.37	14.25	14.11	14.02	13.91	13.84	13.75	13.65	13.58	13.54
5	9.89	9.77	9.64	9.55	9.45	9.38	9.29	9.20	9.13	9.09
6	7.72	7.60	7.48	7.40	7.30	7.23	7.14	7.06	6.99	6.95
7	6.47	6.36	6.24	6.16	6.06	5.99	5.91	5.82	5.75	5.72
8	5.67	5.56	5.44	5.36	5.26	5.20	5.12	5.03	4.96	4.93
9	5.11	5.01	4.89	4.81	4.71	4.65	4.57	4.48	4.41	4.38
10	4.71	4.60	4.49	4.41	4.31	4.25	4.17	4.08	4.01	3.98
11	4.40	4.29	4.18	4.10	4.01	3.94	3.86	3.78	3.71	3.67
12	4.16	4.05	3.94	3.86	3.76	3.70	3.62	3.54	3.47	3.43
13	3.96	3.86	3.75	3.66	3.57	3.51	3.43	3.34	3.27	3.24
14	3.80	3.70	3.59	3.51	3.41	3.35	3.27	3.18	3.11	3.08
15	3.67	3.56	3.45	3.37	3.28	3.21	3.13	3.05	2.98	2.94
16	3.55	3.45	3.34	3.26	3.16	3.10	3.02	2.93	2.86	2.83
17	3.46	3.35	3.24	3.16	3.07	3.00	2.92	2.83	2.76	2.73
18	3.37	3.27	3.16	3.08	2.98	2.92	2.84	2.75	2.68	2.64
19	3.30	3.19	3.08	3.00	2.91	2.84	2.76	2.67	2.60	2.57
20	3.23	3.13	3.02	2.94	2.84	2.78	2.69	2.61	2.54	2.50
21	3.17	3.07	2.96	2.88	2.79	2.72	2.64	2.55	2.48	2.44
22	3.12	3.02	2.91	2.83	2.73	2.67	2.58	2.50	2.42	2.38
23	3.07	2.97	2.86	2.78	2.69	2.62	2.54	2.45	2.37	2.34
24	3.03	2.93	2.82	2.74	2.64	2.58	2.49	2.40	2.33	2.29
25	2.99	2.89	2.78	2.70	2.60	2.54	2.45	2.36	2.29	2.25
26	2.96	2.86	2.75	2.66	2.57	2.50	2.42	2.33	2.25	2.21
27	2.93	2.82	2.71	2.63	2.54	2.47	2.38	2.29	2.22	2.18
28	2.90	2.79	2.68	2.60	2.51	2.44	2.35	2.26	2.19	2.15
29	2.87	2.77	2.66	2.57	2.48	2.41	2.33	2.23	2.16	2.12
30	2.84	2.74	2.63	2.55	2.45	2.39	2.30	2.21	2.13	2.09
40	2.66	2.56	2.45	2.37	2.27	2.20	2.11	2.02	1.94	1.90
50	2.56	2.46	2.35	2.27	2.17	2.10	2.01	1.91	1.82	1.78
60	2.50	2.39	2.28	2.20	2.10	2.03	1.94	1.84	1.75	1.70
70	2.45	2.35	2.23	2.15	2.05	1.98	1.89	1.78	1.70	1.65
80	2.42	2.31	2.20	2.12	2.01	1.94	1.85	1.75	1.65	1.61
90	2.39	2.29	2.17	2.09	1.99	1.92	1.82	1.72	1.62	1.57
100	2.37	2.27	2.15	2.07	1.97	1.89	1.80	1.69	1.60	1.55
150	2.31	2.20	2.09	2.00	1.90	1.83	1.73	1.62	1.52	1.46
200	2.27	2.17	2.06	1.97	1.87	1.79	1.69	1.58	1.48	1.42

Bibliography

Abramowitz, M. and I.A. Stegun (1972). *Handbook of Mathematical Functions.* Dover: New York.

Albert, J. (2009). *Bayesian Computation with R.* Springer: New York.

Anderson, T.W. and D.A. Darling (1952). Asymptotic theory of certain goodness-of-fit criteria based on stochastic processes, *Ann. Math. Statist.* **23**, 193-212.

Anderson, T.W. and D.A. Darling (1954). A test of goodness-of-fit, *J. Amer. Statist. Assoc.* **49**, 765-769.

Ash, R.B. (1972). *Real Analysis and Probability.* Academic Press: New York.

Baringhaus, L.; Danschke, R.; and N. Henze (1989). Recent and classical tests for normality, *Commun. in Statist., Simulation* **18**, 363-379.

Baringhaus, L. and N. Henze (1988). A consistent test for multivariate normality based on the empirical characteristic function, *Metrika* **55**, 363-379.

Barndorff-Nielsen, O. (1982). Exponential families, in v.2 of Kotz *et al.* (1982-1988).

Bartosyzynski, R. and M. Niewiadomska-Bugaj (1996). *Probability and Statistical Inference.* Wiley: New York.

Baxter, M. and A. Rennie (1996). *Financial Calculus.* Cambridge Press: Cambridge.

Berry, A.C. (1941). The accuracy of the Gaussian approximation to the sum of independent variates, *Trans. Amer. Math. Soc.* **48**, 122-136.

Bialik, C. (2007). Is a Carl doomed to be a C student? We don't think so, *Wall Street Journal* Dec. 7, B1.

Billingsley, P. (1968). *Weak Convergence of Measures.* Wiley: New York.

Billingsley, P. (1986, 1995). *Probability and Measure, 2d & 3d. eds.* Wiley-Interscience: New York.

Bisgaard, T.M. and Z. Sasvari (2000). *Characteristic functions and Moment Sequences: Positive Definiteness in Probability.* Nova Science: Huntington, NY.

Black, F. and M. Scholes (1973). The pricing of options and corporate liabilities, *J. Political Econ.* **81**, 637-654.

Blattberg, R.C. and N. Gonedes (1974). A comparison of the stable and Student distributions as statistical models for stock prices, *J. Business* **47**, 244-280.

Boos, D.D. and L.A. Stefanski (2011). *P*-value precision and reproducibility, *Amer. Statistician* **65**, 213-221.

Borch, K. (1968). *The Economics of Uncertainty.* Princeton: Princeton Press.

Bowman, K.O. and L.R. Shenton (1975). Omnibus test contours for departures from normality based on $\sqrt{b_1}$ and b_2, *Biometrika* **62**, 243-250.

Bowman, K.O. and L.R. Shenton (1986). Moment $(\sqrt{b_1}, b2)$ techniques, in D'Agostino and Stephens (1986).

Boyles, R.A. (2008). The role of likelihood in interval estimation, *Amer. Statistician* **62**, 22-26.

Breiman, L. (1961). Optimal gambling systems for favorable games, *Fourth Berkeley Symp. Math. Stat. Prob.* **1**, 65-78.

Brown, S. and W. Whitt (1996). Portfolio choice and the Bayesian Kelly criterion, *Advances in Appl. Prob.* **28**, 1145-1176.

Calvetti, D. and E. Somersalo (2007). *Introduction to Bayesian Scientific Computing.* Springer: New York.

Casella, G. and R.L. Berger (1990). *Statistical Inference.* Wadsworth: Pacific Grove, CA.

Chen, M.; Shao, Q; and J.G. Ibrahim (2000). *Monte Carlo Methods in Bayesian Computation.* Springer-Verlag: New York.

Chernoff, H; Gastwirth, J.; and M.V. Johns (1967). Asymptotic distribution of linear combinations of functions of order statistics with applications to estimation, *Ann. Math. Statist.* **38**, 52-73.

Chung, K.L. (1974). *A Course in Probability Theory, 2d. ed.* Academic Press: New York.

Cox, D.R. and D.V. Hinkley (1974). *Theoretical Statistics.* Chapman & Hall: London.

Cramér, H. (1946). *Mathematical Methods of Statistics.* Princeton Press: Princeton.

Csörgo, S. (1981). Limit behavior of the empirical characteristic function, *Ann. Probability* **9**, 130-144.

D'Agostino, R.B. and M.A. Stephens, eds. (1986). *Goodness-of-Fit Techniques.* Marcel Dekker: New York.

Datta, G.S. and M. Ghosh (2007). Characteristic functions without contour integration, *Amer. Statistician* **61**, 67-70.

Davidson, R. and J. MacKinnon (1993). *Estimation and Inference in Econometrics.* Oxford Press: New York.

Demirtas, H. and D. Hedeker (2011). A practical way for computing approximate lower and upper correlation limits, *Amer. Statistician* **65**, 104-109.

Durbin, J. and G.S. Watson (1950). Testing for serial correlation in least squares regression, I, *Biometrika* **37**, 409-428.

Durbin, J. and G.S. Watson (1951). Testing for serial correlation in least squares regression, II, *Biometrika* **38**, 159-178.

Durrett, R. (1984). *Brownian Motion and Martingales in Analysis.* Wadsworth: Belmont, CA.

Efron, B. and R.J. Tibshirani (1993). *An Introduction to the Bootstrap.* Chapman & Hall: New York.

Epps, T.W. (1976). The demand for brokers' services: The relation between security trading volume and transaction cost, *Bell J. of Econ.* **7**, 163-194.

Epps, T.W. (1993). Characteristic functions and their empirical counterparts: Geometrical interpretations and applications to statistical inference, *Amer. Statistician* **47**, 33-38.

Epps, T.W. (1999). Limiting behavior of the ICF test for normality under Gram-Charlier alternatives, *Statist. and Prob. Letters* **42**, 175-184.

Epps, T.W. (2007). *Pricing Derivative Securities, 2d ed.* World Scientific: Singapore.

Epps, T.W. (2009). *Quantitative Finance.* Wiley: New York.

Epps, T.W. and L.B. Pulley (1983). A test for normality based on the empirical characteristic function, *Biometrika* **70**, 723-726.

Epps, T.W. and L.B. Pulley (1986). A test of exponentiality *vs.* monotone-hazard alternatives derived from the empirical characteristic function, *J. Royal Statist. Soc. (Series B)* **48**, 206-213.

Esseen, C.G. (1945). Fourier analysis of distribution functions: A mathematical study of the Laplace–Gaussian law, *Acta Math.* **77**, 1-125.

Everitt, B.S. and D.J. Hand (1981). *Finite Mixture Distributions.* Chapman and Hall: London.

Feller, W. (1971). *An Introduction to Probability Theory and Its Applications, II.* Wiley: New York.

Feuerverger, A. and R.A. Mureika (1977). The empirical characteristic function and its applications, *Ann. Statist.* **5**, 88-97.

Feynman, R.P.; Leighton, R.B.; and M. Sands (1963). *The Feynman Lectures on Physics, Vol.1.* Addison-Wesley: Reading, MA.

Fishburn, P.C. (1969). A general theory of subjective probabilities and expected utilities, *Ann. Math. Statist.* **40**, 1419-1429.

Fisher, R.A. and E.A. Cornish (1960). The percentile points of distributions having known cumulants, *Technometrics* **2**, 209-225.

Fishman, G.S. (1996). *Monte Carlo: Concepts, Algorithms, and Applications.* Springer: New York.

Fulks, W. (1961). *Advanced Calculus: An Introduction to Analysis.* Wiley: New York.

Gelman, A.; Carlin, J.B.; Stern, H.S.; and D.B. Rubin (2004). *Bayesian Data Analysis.* Chapman-Hall/CRC: Boca Raton, FL.

Goldfeld, S.M. and R.E. Quandt (1965). Some tests for homoscedasticity, *J. Amer. Statist. Assn.* **60**, 539-547.

Good, P. (1994). *Permutation Tests.* Springer-Verlag: New York.

Gradshteyn, I.S.; Ryzhik, I.M.; Jeffrey, A; and D. Zwillinger (2000). *Table of Integrals, Series, and Products, 6th ed.* Academic Press: New York.

Greene, W.H. (1997). *Econometric Analysis, 3d ed.* Prentice-Hall: New York.

Griffiths, R.C. (1985). Orthogonal expansions, in v.6 of Kotz *et al.* (1982-1988).

Gürtler, N. and N. Henze (2000). Goodness-of-fit tests for the Cauchy distribution based on the empirical characteristic function, *Ann. Inst. Statist. Math.* **52**, 267-286.

Hansen, L. (1982). Large sample properties of generalized method of moments estimators, *Econometrica* **50**, 1029-1054.

Henze, N. (1990). An approximation to the limit distribution of the Epps-Pulley test statistic for normality, *Metrika* **37**, 7-18.

Henze, N. (1997). Extreme smoothing and testing for multivariate normality, *Statist. and Prob. Letters* **35**, 203-213.

Henze, N. and T. Wagner (1997). A new approach to the BHEP tests for multivariate normality, *J. Multivar. Anal.* **62**, 1-23.

Hoeffding, W. (1940). Scale-invariant correlation theory. In Fisher, N.I. and Sen, P.K., eds. (1994) *The Collected Works of Wassily Hoeffding*. Springer-Verlag: New York.

Hoeffding, W. (1948). A class of statistics with asymptotically normal distribution, *Ann. Math. Statist.* **19**, 293-325.

Hollander, M. and D.A. Wolfe (1973). *Nonparametric Statistical Methods*. Wiley: New York.

Hsu, J. (2010). Dark side of medical research: Widespread bias and omissions, *Live Science*, http://www.livescience.com/8365-dark-side-medical-research-widespread-bias-omissions.html.

Ince, D. (2011). The Duke University scandal–what can be done? *Significance* **8**, 113-115.

Ingersoll, J.E. (1987). *Theory of Financial Decision Making*. Rowman & Littlefield: Totowa, NJ.

Jarque, C.M. and A.K. Bera (1987). A test for normality of observations and regression residuals, *Internat. Statist. Rev.* **55**, 163-172.

Johnson, N. and S. Kotz (1970). *Distributions in Statistics: Continuous Univariate Distributions-1*. Wiley: New York.

Johnson, N. and S. Kotz (1970a). *Distributions in Statistics: Continuous Univariate Distributions-2*. Wiley: New York.

Johnson, N.; Kotz, S; and A. Kemp (1992). *Univariate Discrete Distributions, 2d ed.* Wiley-Interscience: New York.

Karatzas, I. and Shreve, S.E. (1991). *Brownian Motion and Stochastic Calculus, 2d ed.* Springer Verlag: New York.

Karlin, S. and Taylor, H.M. (1975). *A First Course in Stochastic Processes, 2d ed.* Academic Press: New York.

Kelly, J. (1956). A new interpretation of the information rate, *Bell Sys. Tech. J.* **35**, 917-926.

Kendall, M.G. and A. Stuart (1969, 1973). *The Advanced Theory of Statistics, Vols.1 & 2, 3d ed.* Griffin & Co.: London.

Kirkpatrick, S.; Gelatt, C.D.; and M.P. Vecchi (1983). Optimization by simulated annealing, *Science* **220**, 671-680.

Kolmogorov, A.N. (1933). *Grundbegriffe der Wahrscheinlichkeitsrechnung (Foundations of the Theory of Probability)*. Springer: Berlin.

Kotz, S.; Johnson, N; and C.R. Read, eds. (1982-1988). *Encyclopedia of Statistical Sciences*. Wiley-Interscience: New York.

Laha, R.G. and V.K. Rohatgi (1979). *Probability Theory*. Wiley: New York.

Latané, H. (1959). Criteria for choice among risky assets, *J. Politial Econ.* **35**, 144-155.

Lehmann, E.L. and G. Casella (1998). *Theory of Point Estimation, 2d ed.* Springer: New York.

Lehmann, E.L. and J.P. Romano (2005). *Testing Statistical Hypotheses, 3d. ed.* Springer: New York.

Lehmann, E.L. and H. Scheffé (1950). Completeness, similar regions and unbiased estimation, *Sankhyā* Series A **10**, 305-340.

Leslie, J.R.; Stephens, M.A.; and S. Fotopoulos (1986). Asymptotic distribution of the Shapiro-Wilk W for testing normality, *Ann. Statist.* **14**, 1497-1506.

Lindley, D.V. (1985). *Making Decisions.* Wiley: New York.

Luce, R.D. and H. Raiffa (1957). *Games and Decisions.* Wiley: New York.

Lukacs, E. (1942). A characterization of the normal distribution, *Ann. Math. Statist.* **13**, 91-93.

Lukacs, E. (1970). *Characteristic Functions, 2d. ed.* Griffen: London.

Mandelbrot, B. (1963). The variation of certain speculative prices, *J. Business* **36**, 394-419.

Mandelbrot, B. (1983). *The Fractal Geometry of Nature.* W.H. Freeman: New York.

Mann, H.B. and D.R. Whitney (1947). On a test of whether one of two random variables is stochastically larger than the other, *Ann. Math. Statist.* **18**, 50-60.

Mantel, N. (1987). Understanding Wald's test for exponential families, *Amer. Statistician* **41**, 147-148.

Maritz, J.S. (1995). *Distribution-free Statistical Methods.* Chapman & Hall: London.

McCullagh, P. (1989). Some statistical properties of a family of continuous univariate distributions, *J. Amer. Statist. Assn.* **84**, 125-129.

McCulloch, J.H. (1996). Financial applications of stable distributions, in *Statistical Methods in Finance* (v. 14 of *Handbook of Statistics*, G.S. Maddala and C.R. Rao, eds.). Elsevier: Amsterdam.

Moore, D.S. (1978). Chi-square tests, in R.H. Hogg, ed., *Studies in Statistics* (v.19 of *MAA Studies in Mathematics*). Math. Assn. of America.

Mossin, J. (1973). *Theory of Financial Markets.* Prentice-Hall: Englewood Cliffs, NJ.

Mukhopadhyay, N. (2010). When finiteness matters: Counterexamples to notions of covariance, correlation, and independence, *Amer. Statistician* **64**, 231-233.

Musiela, M. and M. Rutkowski (2005). *Martingale Methods in Financial Modeling, 2d ed.* Springer-Verlag: Berlin.

Neave, H.R. (1978). *Statistics Tables for Mathematicians, Engineers, Economists and the Behavioral and Management Sciences.* Allen & Unwin: London.

Neyman, J. and E.S. Pearson (1933). On the problem of the most efficient tests of statistical hypotheses, *Trans. Royal Society, Series A* **231**, 289-337.

Neyman, J. (1937). "Smooth" tests for goodness of fit, *Skand. Aktuarietidskr.* **20**, 150-199.

Noether, G.E. (1955). On a theorem of Pitman, *Ann. Math. Statist.* **26**, 64-68.

Ord, J.K. (1985). Pearson system of distributions, v.6 of Johnson, Kotz, and Read (1985), 655-659.

Patel, J.K. and C.B. Read (1982) *Handbook of the Normal Distribution*. Marcel Dekker: New York.

Pearson, E.S.; D'Agostino, R.B.; and K.O. Bowman (1977). Tests for departure from normality: Comparison of powers, *Biometrika* **64**, 231-246.

Pearson, E.S. and H.O. Hartley (1976). *Biometrika Tables for Statisticians, 3d ed.* Charles Griffin & Co.: London.

Pitman, E.J.G. (1948). *Non-Parametric Statistical Inference*. University of North Carolina: mimeo.

Pólya, G. (1949). Remarks on computing the probability integral in one and two dimensions, *Proc. First Berkeley Symp. on Math. Statist. and Prob.*, 63-8.

Praetz, P. (1972). The distribution of share price changes, *J. Business* **45**, 49-55.

Pratt, J.W.; Raiffa, H.; and R. Schlaifer (1964). The foundations of decision making under uncertainty: An elementary exposition, *J. Amer. Statist. Assn.* **59**, 353-375.

Press, W.H.; Teukolsky, S.A.; Vetterling, W.T.; and B.P. Flannery (2007). *Numerical Recipes: The Art of Scientific Computing, 3d ed.* Cambridge Press: Cambridge.

Randles, R.H. and D.A. Wolfe (1979). *Introduction to the Theory of Nonparametric Statistics*. Wiley: New York.

Rao, C.R. (1973). *Linear Statistical Inference and Applications, 2d ed.* Wiley: New York.

Rayner, J.W.C. and D.J. Best (1989). *Smooth Tests of Goodness of Fit*. Oxford Press: New York.

Rényi, A. (1970). *Probability Theory*. Elsevier: New York.

Royden, H.L. (1968). *Real Analysis, 2d ed.* Macmillan: New York.

Saha, S.; Chant, D.; Welham, J.; and J. McGrath (2005). A systematic review of the prevalence of schizophrenia. *PLoS Med* 2(5): e141. doi:10.1371/journal.pmed.0020141.

Sarkadi, K. (1975). The consistency of the Shapiro-Francia test, *Biometrika* **62**, 445-450.

Savage, L.J. (1954). *The Foundations of Statistics*. Wiley: New York.

Scholz, F.W. (1985). Maximum likelihood estimation, in v. 5 of Kotz *et al.* (1982-1988).

Schott, J.R. (1997). *Matrix Analysis for Statistics*. Wiley: New York.

Schrödinger, E. (1915). Zur theorie der fall- und steigerversuche an teilchen mit Brownscher bewegung (On a theory of the vibration of particles under Brownian motion), *Physikalische Zeitschrift* **16**, 289-295.

Sen, A. (2012). On the interrelation between the sample mean and the sample variance, *Amer. Statistician* **66**, 112-117.

Senn, S. (2011). Francis Galton and regression to the mean, *Significance* **8**, 124-126.

Senn, S. (2012). Tea for three: Of infusions and inferences and milk in first, *Significance* **9**, 30-33.

Serfling, R.J. (1980). *Approximation Theorems of Mathematical Statistics*. Wiley: New York.

Shapiro, S.S. and R.S. Francia (1972). An approximate analysis of variance test for normality, *J. Amer. Statist. Assn.* **67**, 215-216.

Shapiro, S.S. and M.B. Wilk (1965). An analysis of variance test for normality (complete samples), *Biometrika* **52**, 591-611.

Shapiro, S.S. and M.B. Wilk (1972). An analysis of variance test for the exponential distribution (complete samples), *Technometrics* **15**, 355-370.

Shea, G.A. (1983). Hoeffding's lemma, in v.3 of Kotz *et al.* (1982-1988).

Simmons, G.F. (2003). *Topology and Modern Analysis*. Krieger Pub. Co.: Malabar, FL.

Stephens, M.A. (1986). Tests based on regression and correlation, in D'Agostino, R.B. and M.A. Stephens, eds. *Goodness-of-Fit Techniques*.

Taylor, S.J. (1973). *Introduction to Measure and Integration*. Cambridge Press: Cambridge.

Theil, H. (1971). *Principles of Econometrics*. New York: Wiley.

Tweedie, M.C.K. (1957). Statistical properties of inverse Gaussian distributions-I, *Ann. Math. Statist.* **28**, 362-377.

Vaeth, M. (1985). On the use of Wald's test in exponential families, *Int. Statist. Rev.* **53**, 199-214.

von Neuman, J. and O. Morgenstern (1944). *Theory of Games and Economic Behavior*. Princeton Press: Princeton.

Wald, A. (1944). On cumulative sums of random variables, *Ann. Math. Statist.* **15**, 283-296.

Wald, A. (1949). Note on the consistency of the maximum likelihood estimate, *Ann. Math. Statist.* **20**, 595-601.

Wang, S.S. (2011). The risks of being left-handed, *Wall Street Journal* Dec. 6, D1.

Weisberg, S. and C. Bingham (1975). An approximate analysis of variance test for non-normality suitable for machine calculation, *Technometrics* **17**, 133-134.

Wiener, N. (1923). Differential space, *J. Math. Physics*, **2**, 131-174.

Wilcoxon, F. (1945). Individual comparisons by ranking methods, *Biometrics* **1**, 80-83.

Wilks, S. (1962). *Mathematical Statistics*. Wiley: New York.

Williams, D. (1991). *Probability with Martingales*. Cambridge Press: Cambridge.

Wolfowitz, J. (1949). On Wald's proof of the consistency of the maximum likelihood estimate, *Ann. Math. Statist.* **20**, 601-602.

Young, S.S. and A. Karr (2011). Deming, data and observational studies: A process out of control and needing fixing, *Significance* **8**, 116-120.

Zar, J.H. (1972). Signficance testing of the Spearman rank correlation coefficient, *J. Amer. Statist. Assn.* **67**, 578-80.

Ziliak, S.T. (2011). Matrixx *v.* Siracusano and Student *v.* Fisher: Statistical significance on trial, *Significance* **8**, 131-134.

Index